FreeCAD 1.0 Black Book

By
Gaurav Verma
Matt Weber
(CADCAMCAE Works)

Edited by
Kristen

Published by CADCAMCAE WORKS, USA. Copyright © 2025. All rights reserved. No part of this publication may be reproduced or distributed in any form or by any means, or stored in the database or retrieval system without the prior permission of CADCAMCAE WORKS. To get the permissions, contact at cadcamcaeworks@gmail.com

ISBN # 978-1-77459-160-4

NOTICE TO THE READER

Publisher does not warrant or guarantee any of the products described in the text or perform any independent analysis in connection with any of the product information contained in the text. Publisher does not assume, and expressly disclaims, any obligation to obtain and include information other than that provided to it by the manufacturer.

The reader is expressly warned to consider and adopt all safety precautions that might be indicated by the activities herein and to avoid all potential hazards. By following the instructions contained herein, the reader willingly assumes all risks in connection with such instructions.

The Publisher makes no representation or warranties of any kind, including but not limited to, the warranties of fitness for a particular purpose or merchantability, nor are any such representations implied with respect to the material set forth herein, and the publisher takes no responsibility with respect to such material. The publisher shall not be liable for any special, consequential, or exemplary damages resulting, in whole or part, from the reader's use of, or reliance upon, this material.

DEDICATION

To teachers, who make it possible to disseminate knowledge
to enlighten the young and curious minds
of our future generations

To students, who are the future of the world

THANKS

To my friends and colleagues

To my family for their love and support

Training and Consultant Services

At CADCAMCAE WORKS, we provide effective and affordable one to one online training on various software packages in Computer Aided Design(CAD), Computer Aided Manufacturing(CAM), Computer Aided Engineering (CAE), Computer programming languages(C/C++, Java, .NET, Android, Javascript, HTML and so on). The training is delivered through remote access to your system and voice chat via Internet at any time, any place, and at any pace to individuals, groups, students of colleges/universities, and CAD/CAM/CAE training centers. The main features of this program are:

Training as per your need

Highly experienced Engineers and Technician conduct the classes on the software applications used in the industries. The methodology adopted to teach the software is totally practical based, so that the learner can adapt to the design and development industries in almost no time. The efforts are to make the training process cost effective and time saving while you have the comfort of your time and place, thereby relieving you from the hassles of traveling to training centers or rearranging your time table.

Software Packages on which we provide basic and advanced training are:

CAD/CAM/CAE: CATIA, Creo Parametric, Creo Direct, SolidWorks, Autodesk Inventor, Solid Edge, UG NX, AutoCAD, AutoCAD LT, EdgeCAM, MasterCAM, SolidCAM, DelCAM, BOBCAM, UG NX Manufacturing, UG Mold Wizard, UG Progressive Die, UG Die Design, SolidWorks Mold, Creo Manufacturing, Creo Expert Machinist, NX Nastran, Hypermesh, SolidWorks Simulation, Autodesk Simulation Mechanical, Creo Simulate, Gambit, ANSYS and many others.

Computer Programming Languages: C++, VB.NET, HTML, Android, Javascript and so on.

Game Designing: Unity.

Civil Engineering: AutoCAD MEP, Revit Structure, Revit Architecture, AutoCAD Map 3D and so on.

We also provide consultant services for Design and development on the above mentioned software packages

For more information you can mail us at:
cadcamcaeworks@gmail.com

Table of Contents

Training and Consultant Services	iv
Preface	xviii
About Authors	xx

Chapter 1 : Starting with FreeCAD

Overview of FreeCAD	**1-2**
Installing FreeCAD	**1-3**
FreeCAD Interface	**1-4**
Starting FreeCAD	**1-6**
File Menu	**1-8**
New	1-8
Open	1-9
Open Recent	1-10
Close	1-10
Close All	1-11
Save	1-11
Save As	1-12
Save a Copy	1-12
Save All	1-13
Revert	1-13
Import	1-14
Export	1-15
Merge document	1-17
Document information	1-18
Print	1-19
Print Preview	1-19
Export PDF	1-20
View Menu	**1-21**
Create new view	1-21
Orthographic view	1-21
Perspective view	1-22
Fullscreen	1-22
Standard views	1-23
Freeze Display	1-24
Draw Style	1-27
Tools Menu	**1-28**
Edit parameters	1-28
Save Image	1-30
Load Image	1-31
Scene Inspector	1-32
Dependency Graph	1-33
Exporting Dependency Graph	1-35
Document Utility	1-36
Add text document	1-37
View turntable	1-37
Units converter	1-38

Customize	1-39
Addon Manager	1-42
Measure	1-43
Navigating in the 3d view	1-43
CAD Navigation	1-44
OpenInventor Navigation	1-45
Revit Navigation	1-45
Blender Navigation	1-45
MayaGesture Navigation	1-45
Touchpad Navigation	1-46
Gesture Navigation	1-46
OpenCascade Navigation	1-47

Chapter 2 : Sketching

Introduction	**2-2**
Starting Sketch	**2-2**
Sketch Creation Tools	**2-3**
Point	2-3
Polyline	2-3
Line	2-4
Arc	2-6
Circle/Ellipse	2-10
Rectangle	2-12
Polygon	2-15
Slot	2-17
B-spline	2-19
Construction Mode	2-21
Sketcher constraints	**2-22**
Dimensional constraints	2-22
Constrain coincident	2-28
Horizontal/vertical constraints	2-29
Constrain parallel	2-31
Constrain perpendicular	2-32
Constrain tangent or collinear	2-33
Constrain equal	2-34
Constrain symmetric	2-35
Constrain block	2-36
Constrain refraction (Snell's law)	2-37
Toggle constraints	2-39
Sketch Editing Tools	**2-41**
Create fillet/chamfer	2-41
Edit edge	2-43
Create external geometry	2-45
Create carbon copy	2-46
Move/Array transform	**2-47**
Rotate/Polar transform	2-49
Scale transform	2-50
Offset geometry	2-50
Symmetry	2-51
Remove axes alignment	2-52

Delete all geometry	2-53
Delete all constraints	2-54
Convert geometry to B-spline	2-56
Increase B-spline degree	2-56
Modify knot multiplicity	2-57
Insert knot	2-58
Join curves	2-58
Sketcher Visual tools	**2-59**
Select under-constrained elements	2-59
Select associated constraints	2-60
Select associated geometry	2-60
Select redundant constraints	2-61
Select conflicting constraints	2-61
Show/hide circular helper for arcs	2-61
Show/hide internal geometry	2-61
Switch virtual space	2-62
Show/hide B-spline information layer tools	2-62
Practical 1	2-65
Practical 2	2-69
Practical 3	2-74
Practice 1	**2-79**
Practice 2	**2-80**
Practice 3	**2-81**
Practice 4	**2-81**

Chapter 3 : Solid Modeling with Part Design

Introduction	**3-2**
Starting part design	**3-2**
Part Design Helper tools	3-2
Validate sketch	3-6
Datum tools	3-12
Part Design Modeling Tools	**3-18**
Part Design Modeling tools	3-18
Subtractive tools	3-36
Transformation Tools	3-56
Dress-up tools	3-64
Boolean operation	3-70
Creating Sprocket	3-72
Creating Involute Gear	3-74
Shaft Design Wizard	3-74
Practical	**3-75**

Chapter 4 : Solid Modeling with Part Workbench

Introduction	**4-2**
Starting Part Workbench	**4-2**
Primitive tools	4-3
Modifying Object tools	4-9
Measure	4-49
Other tools	4-52

Chapter 5 : Solid Modeling Practice

Practical 1	5-2
Practical 2	5-6
Practice 1 to 4	5-14
Practical 3	5-16
Practice 5	5-20
Practice 6	5-21
Practice 7	5-21
Practice 8	5-22
Practice 9	5-23
Practice 10	5-24
Practice 11	5-24

Chapter 6 : Drafting

Introduction	6-2
Starting draft Workbench	6-2
Drawing objects tools	6-3
Annotation objects tools	6-23
Modifying objects tools	6-29
Draft tray toolbar	6-61
Draft Snap toolbar	6-67
Draft utility tools toolbar	6-70
Practical 1	6-79

Chapter 7 : BIM Modeling

Introduction	7-2
Starting BIM Workbench	7-2
BIM Tools	7-3
Creating Project	7-3
Creating Site	7-4
Creating Building	7-5
Creating Level	7-6
Creating Space	7-8
Creating Wall	7-10
Creating Curtain Wall	7-13
Column	7-16
Creating Structural System	7-18
Creating Multiple Structures	7-19
Slab	7-20
Creating Window	7-21
Door	7-23
Creating Pipe	7-23
Creating Stairs	7-26
Creating Roof	7-27
Creating Panel	7-29
Creating Panel Cut	7-31
Creating Panel Sheet	7-33
Creating Nest	7-34
Creating Frame	7-36

Creating Fence	7-37
Creating Truss	7-39
Creating Equipment	7-41
Creating Custom Rebar	7-42
Creating Profile	7-44
Box	7-46
Shape builder	7-46
Facebinder	7-46
Objects library	7-46
Component	7-48
Adding an External Reference	7-48
Adding the Component	7-50
Removing the Component	7-51
Creating Axis	7-52
Creating Axis System	7-54
Creating Grid	7-55
Creating Section Plane	7-57
Creating Material	7-60
Creating Multi-Material	7-62
Creating Schedule	7-63
Cutting the object with Plane	7-65
Survey	7-67
Practical	**7-68**

Chapter 8 : CAM Workbench

Introduction	**8-2**
Starting CAM Workbench	**8-2**
Jobs, tools, and generating G-codes	**8-3**
Creating Job	8-3
ToolBit Library Editor	8-12
Exporting Template	8-16
Basic CAM operations	**8-18**
Creating Contour Operation	8-18
Creating Pocket Object	8-20
Performing Drill Operation	8-23
Creating Face Milling	8-25
Creating Helical Operation	8-27
Creating Adaptive Operation	8-28
Creating an Engraving Path	8-30
Creating Deburring Operation	8-32
Creating Vcarve	8-33
Creating 3D Pocket Operation	8-34
Path Modification Tools	**8-35**
Creating Copy	8-35
Creating Array	8-36
Creating Simple Copy	8-37
Path Utilities	**8-37**
Inspection of G-Code	8-37
CAM Simulator	8-38
New CAM Simulator	8-40

Finish selecting loop	8-41
Post-Processing	**8-42**
Check the CAM Job for common error	8-43
Path Dressup Tools	**8-44**
Axis Map	8-44
Modifying Boundary for Toolpath	8-45
Applying Dogbone Overcuts	8-46
Applying Dragknife Dressup	8-47
Modifying Lead In/Out	8-47
Defining Ramp Entry	8-48
Defining Holding Tags	8-49
Performing Z Depth Correction	8-49
Supplemental Commands	**8-50**
Adding Fixture in Setup	8-50
Adding Comments	8-50
Adding Stop Command	8-51
Adding Custom G-Codes	8-51
Probe Toolpath	8-51
Practical	**8-53**

Chapter 9 : FEM Workbench

Introduction to FEM	**9-2**
Starting FEM Workbench	**9-3**
Creating Analysis Container	**9-3**
Static Analysis	9-4
Frequency Analysis (Modal Analysis)	9-5
Thermal analysis	9-5
Defining Material For Solid	**9-6**
Editing Material	9-7
Applying Constraints	**9-8**
Mechanical Constraints	9-9
Geometrical Constraints	9-20
Thermal boundary condition and loads	9-23
Fluid Constraints	9-26
Applying Electrostatic Potential Boundary Condition	9-28
Current Density Boundary Condition	9-29
Magnetization boundary condition	9-30
Applying Constant Vacuum Permittivity	9-30
Creating Geometry Elements	**9-31**
Creating Beam Cross-section	9-31
Applying Rotation to Beam Elements	9-32
Defining Shell Plate Thickness	9-32
Creating Fluid Section for 1D Flow	9-33
Meshing	**9-33**
Creating FEM mesh using Netgen Mesher	9-34
Creating FEM mesh using Gmsh Mesher	9-36
Creating Boundary Mesh Layer	9-38
Create FEM mesh refinement	9-38
Creating FEM mesh group	9-39

Creating Erase Elements	9-40
FEM Mesh to Polygonal Mesh Model Conversion	9-41
Solvers	**9-41**
Calculix Standard Solver	9-41
Elmer Solver	9-44
Z88 Solver	9-50
Mystran Solver	9-51
Post Processing in FreeCAD FEM	**9-51**
Purge Results	9-52
Show Result	9-52
Creating Post Processing Pipeline for Results	9-52
Applying Warp Filter	9-53
Applying Scalar Clip Filter	9-54
Applying Function Cut Filter	9-54
Region Clip Filter	9-56
Contours filter	9-57
Line Clip Filter	9-57
Practical	**9-59**

Chapter 10 : Tech Drawing

Introduction to TechDraw Workbench	**10-2**
Starting A New Drawing Page	**10-2**
Inserting New Page of Selected Template	**10-5**
Update templates fields	**10-6**
Redraw page	**10-7**
Print All Pages	10-7
Inserting View in Page	**10-8**
Insert broken view	**10-10**
Inserting active view in page	**10-12**
Inserting Projection group	**10-13**
Insert Section Views	**10-15**
Insert Complex Section View	10-16
Inserting Detail View	**10-17**
Inserting Draft Workbench Object	**10-18**
Insert Arch Workbench Object	10-19
Insert Spreadsheet View	10-20
Moving the View	**10-22**
Sharing the view	**10-23**
Projecting the shape	**10-23**
Adjust Stacking Order of View	**10-25**
Stack Top View	10-25
Stack Bottom View	10-25
Stack UP View	10-25
Stack Down View	10-25
Inserting Annotations	**10-25**
Adding leader line to view	**10-26**
Inserting Rich Text Annotation	**10-28**
Adding cosmetic vertex	**10-29**
Adding midpoint vertices	**10-31**

Adding Quadrant vertices	10-32
Adding centerline to faces	10-32
Adding Centerline between 2 Lines	10-34
Adding Centerline between 2 Points	10-35
Adding Cosmetic Line through 2 points	10-36
Add Cosmetic Circle	10-38
Changing Appearance Of Line	10-39
Show/Hide Invisible Edges	10-40
Adding Welding Information to Leaderline	10-41
Create Surface Finish Symbol	10-43
Add Hole or Shaft Fit	10-43
Working with Clip Group	10-44
Inserting Dimensions	10-45
Inserting Length Dimension	10-46
Linking Dimension to 3D Geometry	10-47
Insert Balloon Annotation	10-48
Axonometric Length Dimension	10-49
Inserting Landmark Dimension	10-50
Repair Dimension Reference	10-51
Exporting Page to SVG file	10-52
Exporting Page as DXF file	10-53
Applying Hatch Pattern	10-53
Applying Geometric Hatch to a Face	10-54
Inserting Symbol from an SVG File	10-55
Inserts a bitmap from a file into a Page	10-56
Turning View Frames On/Off	10-57
Selecting Line Attributes, Cascade Spacing, and Delta Distance	10-57
Changing line attributes	10-58
Extending line	10-59
Shortening the line	10-60
Locking/unlocking the view	10-60
Positioning Section View	10-60
Positioning Horizontal Chain Dimensions	10-61
Positioning Vertical Chain Dimensions	10-63
Positioning Oblique Chain Dimensions	10-63
Cascading Horizontal Dimensions	10-64
Cascading Vertical Dimensions	10-65
Cascading Oblique Dimensions	10-65
Calculating The Area Of Selected Faces	10-66
Calculate the arc length of selected edges	10-66
Customizing Format Label	10-67
Adding circle centerlines	10-69
Adding bolt circle centerlines	10-70
Adding cosmetic thread hole side view	10-70
Adding cosmetic thread hole bottom view	10-71
Adding cosmetic thread bolt side view	10-72
Adding cosmetic thread bolt bottom view	10-72
Adding cosmetic intersection Vertex(es)	10-72
Adding an offset vertex	10-73

Adding Cosmetic Circle	10-74
Adding cosmetic arc	10-75
Adding Cosmetic Circle 3 Points	10-75
Adding Cosmetic Parallel Line	10-75
Adding Cosmetic Perpendicular Line	10-76
Creating horizontal chain dimensions	10-77
Creating vertical chain dimensions	10-78
Creating oblique chain dimensions	10-79
Creating horizontal coordinate dimensions	10-79
Creating vertical coordinate dimensions	10-80
Creating oblique coordinate dimensions	10-81
Creating horizontal chamfer dimension	10-81
Creating vertical chamfer dimension	10-82
Inserting Area Annotation	10-82
Creating arc length dimension	10-83
Inserting 'ø' PREFIX	10-84
Inserting '□' PREFIX	10-85
Inserting Nx prefix	10-85
Removing prefix	10-86
Increasing decimal places	10-87
Decreasing Decimal Places	10-88
Practical	10-88
Practice	10-91

Chapter 11 : Miscellaneous Workbench

Introduction to Surface Workbench	**11-2**
Creating Fill Surface using Curves and Vertices	11-2
Creating Surface using Boundary Curves	11-3
Creating Surface using Edges	11-4
Creating Surface by Extending a Face	11-5
Curve on mesh	11-6
Blend Curve	11-8
Introduction to Mesh Workbench	**11-8**
Importing Mesh Objects	11-9
Exporting Mesh Model	11-10
Creating Mesh from Shape	11-11
Refinement	11-12
Mesh Analysis Tools	11-13
Generating Curvature Plot	11-16
Harmonizing Normals	11-16
Flipping Normals of Mesh Faces	11-16
Filling Holes in Mesh	11-17
Closing Holes	11-17
Adding Triangles to Mesh	11-17
Removing Components of Mesh	11-18
Removing Components by Hand	11-19
Creating Mesh Segments	11-19
Smoothening Mesh Model	11-20
Decimating Mesh Model	11-21

Scaling Mesh Model	11-22
Creating Regular Mesh Solids	11-22
Performing Boolean Operations on Meshes	11-22
Mesh Cutting Operations	11-24
Creating Cross Sections	11-27
Merging Mesh	11-28
Splitting Mesh Object	11-28
Unwrap Mesh	11-28
Unwrap Face	11-29
Robot Workbench	**11-29**
Inserting Robot	11-29
Creating Robot Trajectory	11-30
Adding Edges to Trajectory	11-30
Setting Default Orientation of Robot	11-31
Setting Default Values for Robot	11-32
Adding Tool to Robot	11-33
Simulating Robot Trajectory	11-34
3D Printing Tools Workbench	**11-34**
Converting Imperial Mesh to Metric and vice-versa	11-34
Reducing and scaling mesh by 50%	11-35
Scaling mesh by variable scale factor	11-35
Reducing mesh by 95%	11-35
Creating mesh box	11-35
Converting mesh to refined solids	11-35
Changing objects transparency	11-35
Hiding or showing the selected objects	11-35
Modifying objects to random color	11-35
Changing color of objects	11-36
Changing the line width	11-36
Turning on or off the Bounding box size	11-36
Defining Printer bed size	11-36
Defining Die length size	11-36
Defining the Scale factor	11-36
Sheet Metal Workbench	**11-37**
Making Base Wall	11-37
Creating Bend Wall	11-37
Extending Sheet Metal Flange Face	11-39
Applying Sketch Bend	11-40
Unfolding Sheet Metal Bends	11-41
Adding Corner Relief	11-42
Applying Solid Corner Relief	11-43
Applying Rip Feature	11-43
Creating Bend	11-44
Making Extruded Cut	11-45
Applying Forming Tool	11-46
Extruded Cutout	11-47
Add Base Shape	11-48

Chapter 12 : Miscellaneous Tool

Introduction	12-2
Edit Menu	12-2
Undo Operation	12-2
Redo Operation	12-2
Cut, Copy, and Paste Operations	12-3
Duplicating Selected Objects	12-3
Refresh Operation	12-3
Box Selection	12-3
Select All	12-4
Deleting Objects	12-4
Sending to Python Console	12-4
Modifying Placement of Objects	12-4
Transform	12-5
Alignment Operation	12-5
Activating Edit Mode	12-6
Properties	12-7
Edit Mode options	12-7
Preferences	12-8
View Menu	12-22
Stereoscopic Views	12-22
Zoom Options	12-23
Document Window Options	12-23
TreeView actions Options	12-23
Toggle Axis Cross	12-24
Bounding Box	12-24
Clipping Plane	12-25
Persistent section cut	12-26
Texture Mapping	12-27
Visibility Options	12-27
Apply Random Colors to Model	12-28
Workbench Options	12-28
Showing/Hiding Toolbars	12-28
Showing/Hiding Panels	12-29
Status Bar	12-29
Macro Menu	12-29
Recording Macro	12-30
Executing and Managing Macros	12-30

Chapter 13 : Assembly Workbench

Introduction To Assembly Workbench	13-2
Create Assembly	13-2
Insert component	13-3
Solve Assembly	13-5
Toggle Grounded	13-5
Create a Fixed Joint	13-5
Create Revolute Joint	13-7
Create Cylindrical Joint	13-8
Create Slider Joint	13-8

Create Ball Joint	13-9
Create Distance Joint	13-10
Create Rack and Pinion Joint	13-11
Create Screw Joint	13-13
Create Gears Joint	13-15
Create Belt Joint	13-16
Practical	13-17

Chapter 14 : Introduction to Computational Fluid Dynamics

Introduction	14-2
Basic Properties of Fluids	14-2
Mass Density, Weight Density, and Specific Gravity	14-2
Viscosity	14-2
Types of Fluids	14-4
Thermodynamic Properties of Fluid	14-4
Universal Gas Constant	14-5
Compressibility of Gases	14-5
Vapour Pressure and Cavitation	14-5
Pascal's Law	14-5
Fluid Dynamics	14-5
Bernoulli's Incompressible Fluid Equation	14-5
Eulerian and Lagrangian Method of Analysis	14-6
Differential Approach of Fluid Flow Analysis	14-7
Acceleration	14-7
Introduction to CFD	14-7
Conservation of Mass	14-12
Conservation of Momentum	14-12
Conservation of Energy	14-14
Variations of Navier-Strokes Equation	14-15
Time Domain	14-15
Compressibility	14-15
Low and High Reynolds Numbers	14-16
Turbulence	14-17
Steps of Computational Fluid Dynamics	14-19
Creating Mathematical Model	14-19
Discretization of Model	14-19
Analyzing with Numerical Schemes	14-19
Solution	14-20
Visualization (Post-processing)	14-20
Finite Difference Method	14-20
Introduction to CfdOF	14-22
Activating CfdOF Workbench	14-23
Installing CfdOF Related Software	14-24
Preparing Model for CFD	14-26
Performing CFD on a Model	14-26
Applying Boundary Conditions	14-27
Creating CFD Mesh	14-28
Mesh refinement	14-29
Setting Physics Model	14-30

Adding Fluid Properties 14-31
Setting Initial Internal Flow Conditions 14-31
Porous Zone 14-32
Reporting function 14-33
Cfd scalar transport function 14-34
Solving the Analysis 14-34

Index I-1
Other Books by CADCAMCAE Works I-9

Preface

FreeCAD is a parametric (history based) open source CAD/CAM/CAE software which finds its uses in various fields of designing. The software has a modular structure which allows to plugin extensions (modules) to add functionality to the core application. An extension can be as complex as a whole new application programmed in C++ or as simple as a Python script or self-recorded macro. You have complete access to almost any part of FreeCAD from the built-in Python interpreter, macros or external scripts, be it geometry creation and transformation, the 2D or 3D representation of that geometry (scenegraph) or even the FreeCAD interface. The software allows import/export to standard formats such as STEP, IGES, OBJ, STL, DXF, SVG, STL, DAE, IFC or OFF, NASTRAN, VRML in addition to FreeCAD's native FCStd file format. FreeCAD's installer allows flexible installations on Windows systems. Packages for Ubuntu systems are also available and updated regularly. There is a huge library of external workbenches available for FreeCAD on web. You can download free copy of software from https://www.freecadweb.org/ web link.

The **FreeCAD 1.0 Black Book** is the 5th edition of our series on FreeCAD. This book is written to help beginners in creating some of the most complex solid models. The book follows a step by step methodology. In this book, we have tried to give real-world examples with real challenges in designing. We have tried to cover most of the topics utilized in industries for designing. The book covers almost all the information required by a learner to master the FreeCAD. The book starts with sketching and ends at advanced topics like CAM, FEM (Simulation), Sheet Metal, CFD, Assembly, and so on. Some of the salient features of this book are :

In-Depth explanation of concepts
Every new topic of this book starts with the explanation of the basic concepts. In this way, the user becomes capable of relating the things with real world.

Topics Covered
Every chapter starts with a list of topics being covered in that chapter. In this way, the user can easily find the topics of his/her interest easily.

Instruction through illustration
The instructions to perform any action are provided by maximum number of illustrations so that the user can perform the actions discussed in the book easily and effectively. There are about 2100 illustrations that make the learning process effective.

Tutorial point of view

At the end of concept's explanation, the tutorial make the understanding of users firm and long lasting. Almost each chapter of the book has tutorials that are real world projects. Moreover most of the tools in this book are discussed in the form of tutorials.

Project

Projects and exercises are provided to students for practicing.

For Faculty

If you are a faculty member then you can ask for video tutorials on any of the topic, exercise, tutorial, or concept. As a faculty, you can register on our website to get electronic desk copies of our latest books. Faculty resources are available in the **Faculty Member** page of our website (**www.cadcamcaeworks.com**) once you login. Note that faculty registration approval is manual and it may take upto two days for approval before you can access the faculty website.

Formatting Conventions Used in the Text

All the key terms like name of button, tool, drop-down etc. are kept bold.

Free Resources

Link to the resources used in this book are provided to the users via email. To get the resources, mail us at ***cadcamcaeworks@gmail.com*** with your contact information. With your contact record with us, you will be provided latest updates and informations regarding various technologies. The format to write us mail for resources is as follows:

Subject of E-mail as ***Application for resources of .... book***.
Also, given your information like
Name:
Course pursuing/Profession:
Contact Address:
E-mail ID:

Note: We respect your privacy and value it. If you do not want to give your personal informations then you can ask for resources without giving your information.

About Authors

The author of this book, Matt Weber, has written more than 16 books on CAD/CAM/CAE available in market. He has coauthored SolidWorks Simulation, SolidWorks Electrical, SolidWorks Flow Simulation, and SolidWorks CAM Black Books. The author has hands on experience on almost all the CAD/CAM/CAE packages. If you have any query/doubt in any CAD/CAM/CAE package, then you can contact the author by writing at cadcamcaeworks@gmail.com

The author of this book, Gaurav Verma, has written and assisted in more than 17 titles in CAD/CAM/CAE which are already available in market. He has authored Autodesk Fusion 360 Black Book, AutoCAD Electrical Black Book, Autodesk Revit Black Books, and so on. He has provided consultant services to many industries in US, Greece, Canada, and UK. He has assisted in preparing many Government aided skill development programs. He has been speaker for Autodesk University, Russia 2014. He has assisted in preparing AutoCAD Electrical course for Autodesk Design Academy. He has worked on Sheetmetal, Forging, Machining, and Casting designs in Design and Development departments of various manufacturing firms.

For Any query or suggestion

If you have any query or suggestion, please let us know by mailing us on ***cadcamcaeworks@gmail.com***. Your valuable constructive suggestions will be incorporated in our books and your name will be addressed in special thanks area of our books on your confirmation.

Chapter 1

Starting with FreeCAD

Topics Covered

The major topics covered in this chapter are:

- *Overview of FreeCAD*
- *Installing FreeCAD*
- *Starting FreeCAD*
- *File Menu*
- *View Menu*
- *Standard Views*
- *Freeze Display*
- *Draw Style*
- *Tools Menu*
- *Navigating in the 3D view*
- *FreeCAD Interface*

OVERVIEW OF FREECAD

FreeCAD is an open-source parametric 3D modeling application, made primarily to design real-life objects. Parametric modeling describes a certain type of modeling, where the shape of the 3D objects you design are controlled by parameters. For example, the shape of a brick might be controlled by three parameters: height, width, and length. In FreeCAD, as in other parametric modelers, these parameters are part of the object, and stay modifiable at any time, after the object has been created. It connects your entire product development process in a single program that works on all the three major platforms viz. Windows, Mac OS, and Linux; refer to Figure-1.

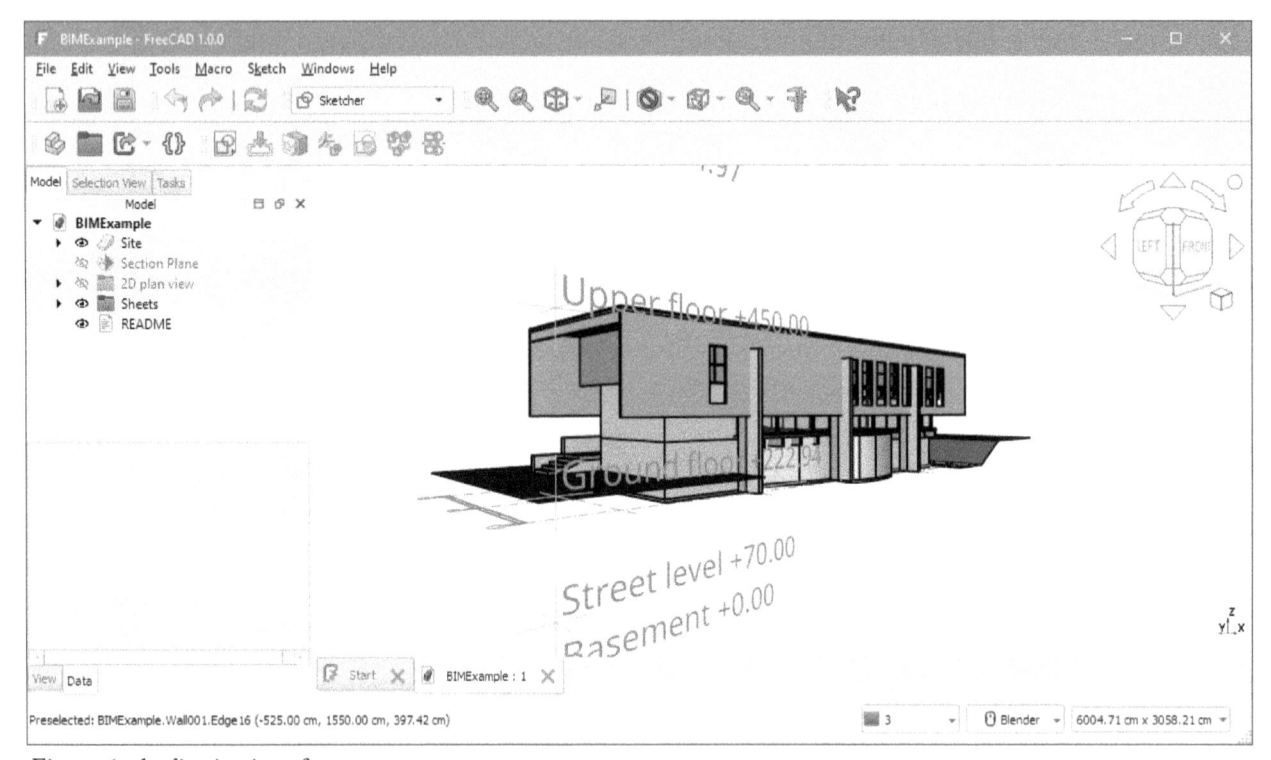

Figure-1. Application interface

FreeCAD is not designed for a particular kind of work or to make a certain kind of objects. Instead, it allows a wide range of uses and permits users to produce models of all sizes and purposes from small electronic components to 3D-printable pieces and all the way up to buildings. Each of these tasks have different dedicated sets of tools and workflow available. Being open-source, FreeCAD benefits from the contributions and efforts of a large community of programmers, enthusiasts, and users worldwide. FreeCAD is essentially an application built by the people who use it, instead of being made by a company trying to sell you a product. And of course, it also means that FreeCAD is free, not only to use but also to distribute, copy, modify, or even sell.

FreeCAD also benefits from the huge, accumulated experience of the open-source world. In its bowels, it includes several other open source components, as FreeCAD itself can be used as a component in other applications. It also possesses all kinds of features that have become a standard in the open-source world, such as supporting a wide range of file formats being hugely scriptable, customizing, and modifiable. All this is made possible through a dynamic and enthusiast community of users.

INSTALLING FREECAD

- Connect your PC with the internet connection and then go to `https://www.freecadweb.org/` as shown in Figure-2.

Figure-2. FreeCAD website

- Click on the **Download now** button, the **Downloads** tab of this website will be displayed; refer to Figure-3.

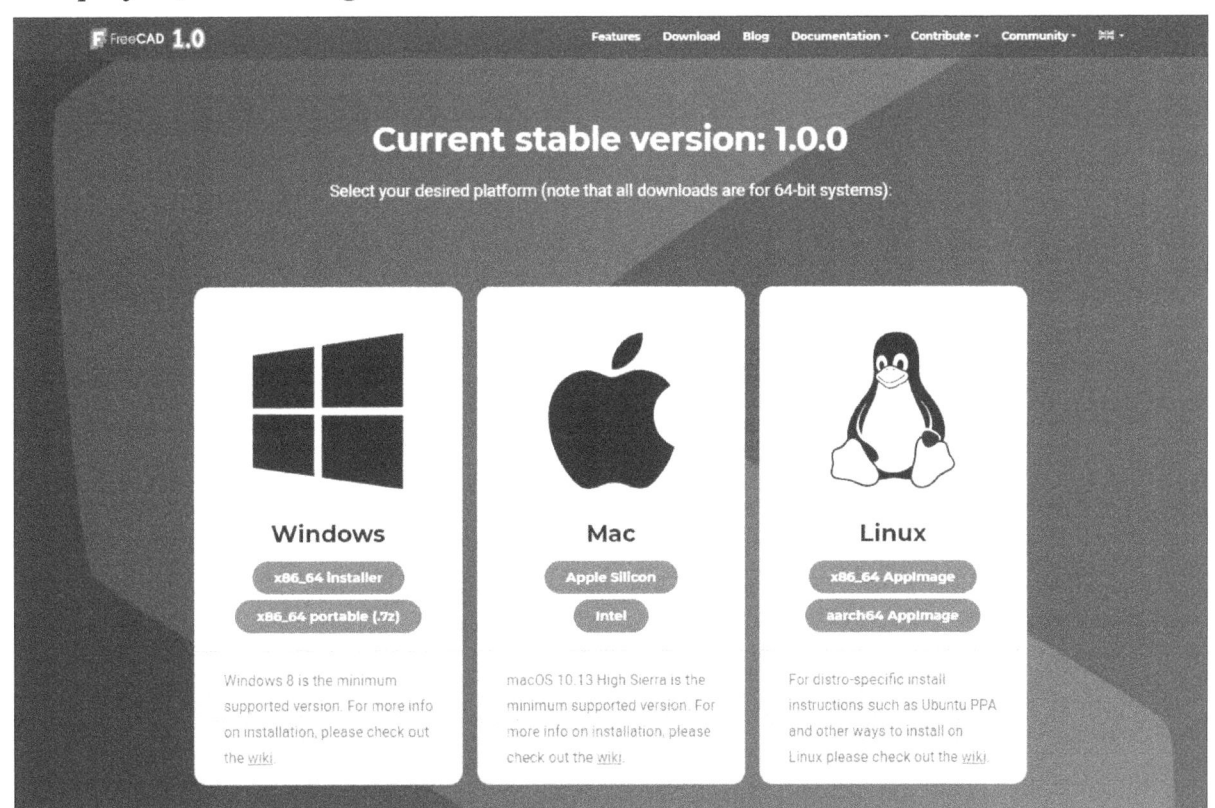

Figure-3. Downloads tab of FreeCAD website

- Select desired Operating System and the type of processor installed on your PC. The software will begin to download.
- Open the downloaded setup file and follow the instructions as per the setup window. The software will be installed in a couple of minutes.

FREECAD INTERFACE

The FreeCAD interface is based on the modern Qt, a well known graphical user interface **(GUI)** toolkit and has a state of the art organization. Some aspects of the interface can be customized. You can add custom toolbars with tools from several workbenches or tools defined in macros and you can create your own keyboard shortcuts. But the menus and default toolbars that come with FreeCAD and its workbenches cannot be changed. Different parts of the FreeCAD interface are discussed next; refer to Figure-4.

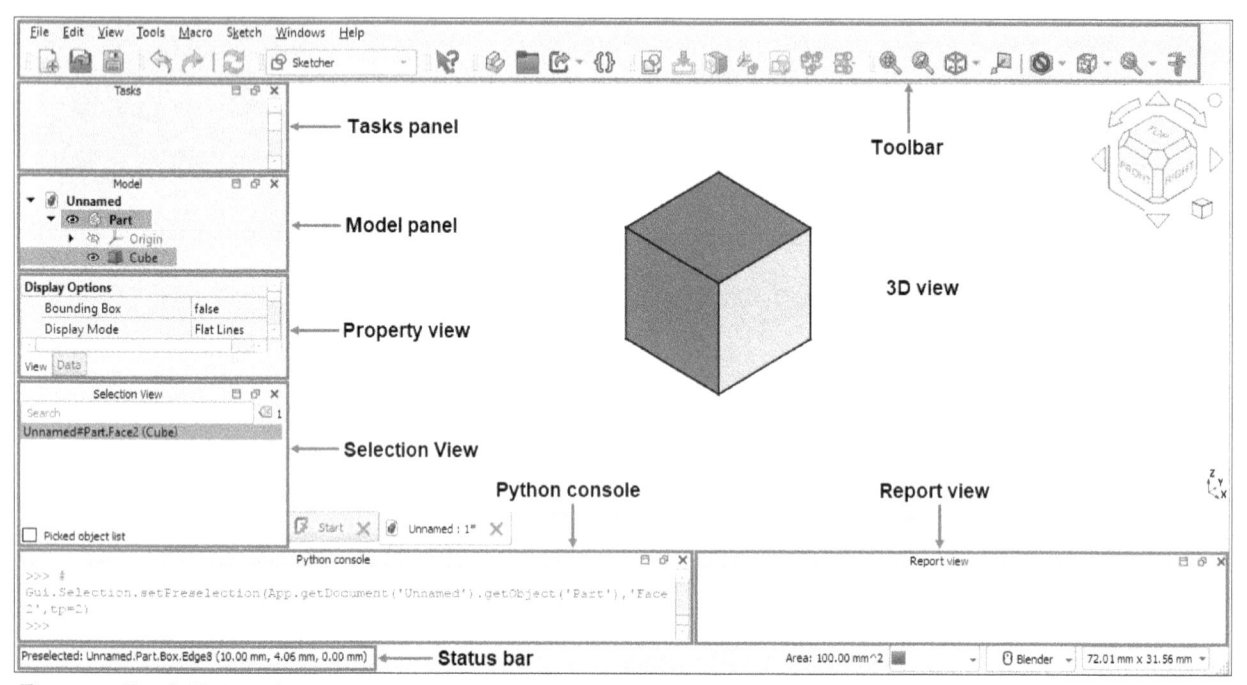

Figure-4. FreeCAD interface

Toolbars : The standard toolbars that appear in the interface are:

- The tools in the **File** toolbar are used to work with files, open documents, copy, paste, undo, and redo actions.
- The **Workbench** toolbar contains a single widget to select the active workbench.
- The tools in **Macro** toolbar are used to record, edit, and execute macros.
- In the **View** toolbar, the tools are used to control how objects appear in the 3D view.
- The tools in the **Structure** toolbar are used to organize objects in the document and create links to additional documents.

3D View : The **3D view** is the main component of the interface. It can be undocked out of the main window if you can have several views of the same document (or same objects) or several documents opened at the same time. You can select objects or parts of objects by clicking them and you can pan, zoom, and rotate the view with the mouse buttons.

Model panel : The **Model** panel shows you the tree view which is a representation of the document's content including 2D and 3D geometry with their parametric history but also supporting objects that contain data saved in the document.

Tasks panel : The **Tasks** panel is where FreeCAD will prompt you for values specific to the tool you're currently using at the time—for example, entering a 'length' value when the **Line** tool is being used. It will close automatically after pressing the **OK** (or **Cancel**) button. Also, by double-clicking the related object in the combo view, most tools will allow you to reopen the task panel in order to modify the settings.

Property view : The **Property view** appears when the **Model** panel is active in the interface. It allows managing the publicly exposed properties of the objects in the document. Generally, the property editor is intended to deal with just one object at one time. The values shown in the property editor belong to the selected object of the active document. Despite this, some properties like colors can be set for multiple selected objects. If there are no elements selected, the property editor will be empty.

Selection view : The **Selection view** is a panel in the interface by default located below the combo view. Just like the property editor, it shows more information about the currently selected objects. A selection can be made by picking an object in the 3D view or in the tree view. Multiple bodies can be selected by holding **Ctrl** key.

Report view : The **Report view** is a panel that shows text messages from FreeCAD processes and tools. The report view is normally hidden but it is a good idea to leave open as it will list any information, warnings, or errors to help you decipher (or debug) what you may have done wrong.

Python console : The **Python console** is a panel that runs an instance of the Python interpreter. It can be used to control FreeCAD processes. It creates and modifies objects and their properties. The Python console in FreeCAD has basic syntax highlighting able to differentiate with various styles and colors, comments, strings, numeric values, built in functions, printed text output, and delimiters like parentheses and commas.

Status bar : The **Status bar** is a simple information bar that appears at the bottom of the FreeCAD interface. When the mouse pointer is over a button or menu, the usage information of that command is displayed both in a textual pop-up and in the status bar. It shows the zoom level in the right corner next to the mouse navigation style. The status bar also shows the last pre-selected object or element of an object and the coordinates of the mouse pointer when the last pre-selection occurred; refer to Figure-5.

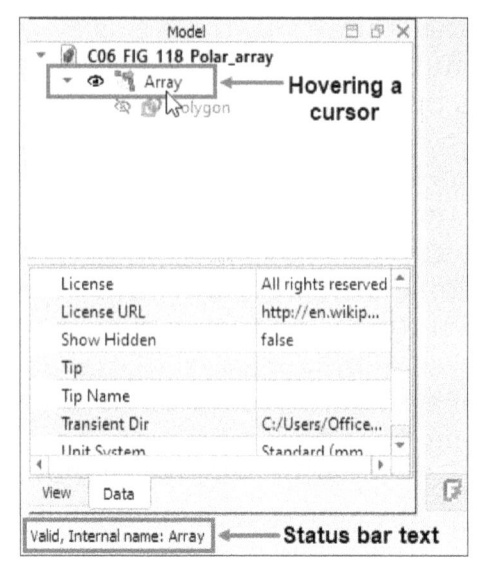

Figure-5. Valid Internal name Array status bar text

STARTING FREECAD

- To start **FreeCAD** from **Start** menu in Windows 8 or later, click on **FreeCAD** folder in the **Start** menu and select the **FreeCAD** icon; refer to Figure-6. Click on the program name to start the application.

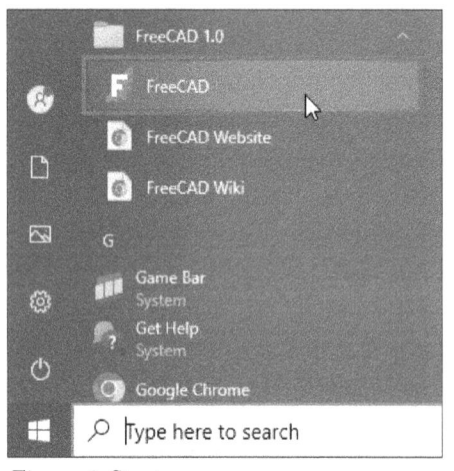

Figure-6. Start menu

- On starting the application, the initial screen of the application will be displayed as shown in Figure-7.

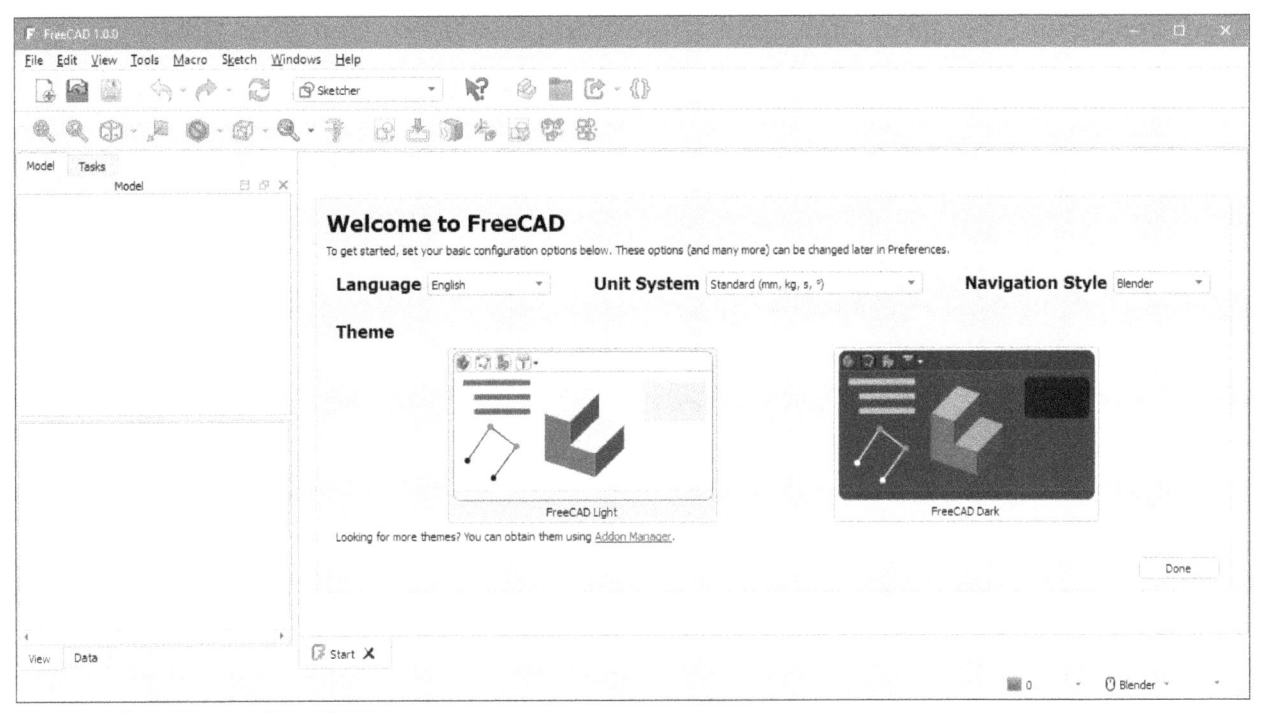

Figure-7. Initial screen of FreeCAD

- Select desired language of the software from the **Language** drop-down as per your preference.
- Select desired units system from the **Unit System** drop-down to define the measurement units of parts in the software.
- Select desired style for navigating the models from the **Navigation Style** drop-down.
- You can change the background theme for the application window by selecting desired theme from the **Theme** section.
- Click on the **Addon Manager** link from the **Theme** section. The **Addon Manager** dialog box will be displayed; refer to Figure-8. Select desired theme and click on the **Install** button from the top right corner, the theme will be installed.

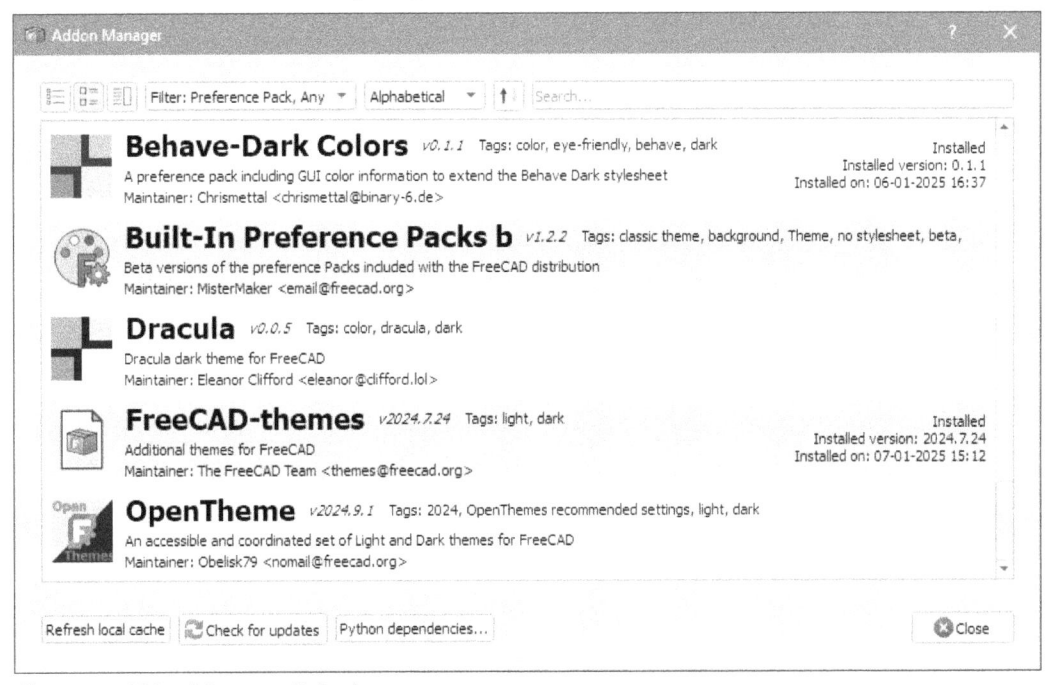

Figure-8. Addon Manager dialog box

- After specifying desired parameters, click on the **Done** button from the application window. The **New File** application window will be displayed; refer to Figure-9.

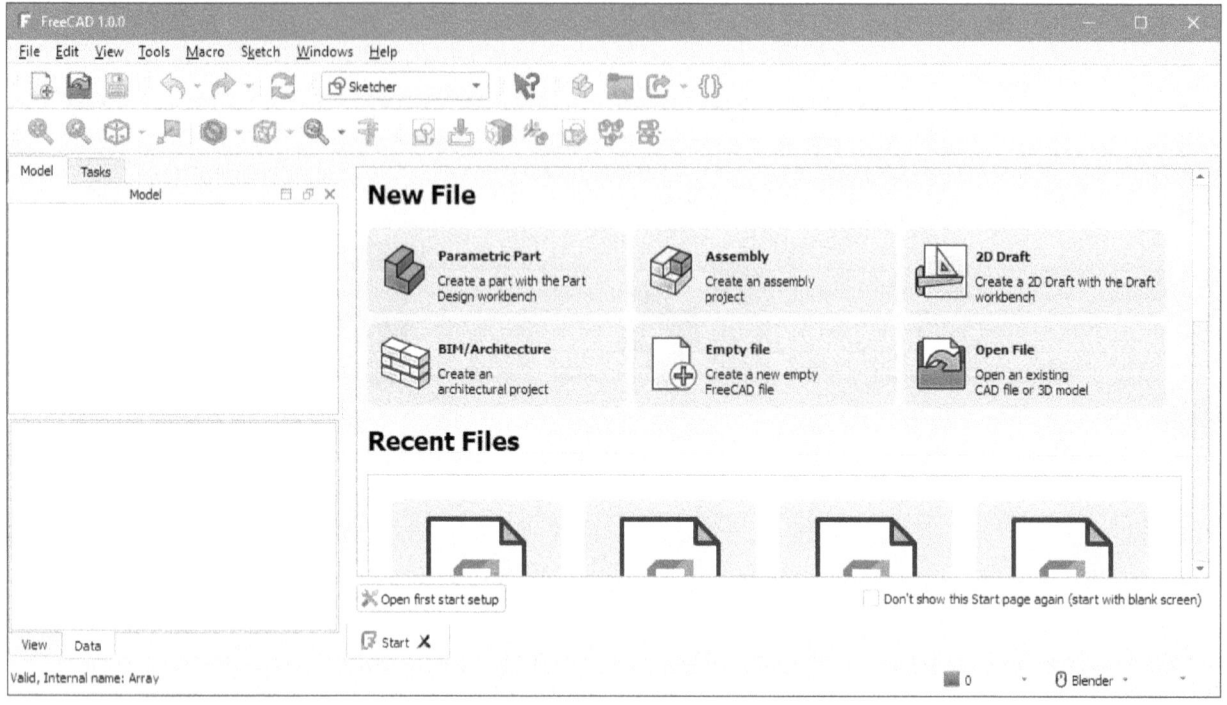

Figure-9. New File application window

- Select desired option from the **New File** section to open the new document.

FILE MENU

The options in the **File** menu are used to manage files and related parameters. Various tools of **File** menu are discussed next.

New

The **New** tool is used for initiating a new document file. The procedure to use this tool is discussed next.

- Click on the **New** tool from the **File** menu or **Ribbon** or press **Ctrl+N** key from keyboard; refer to Figure-10. The new document file will open; refer to Figure-11.

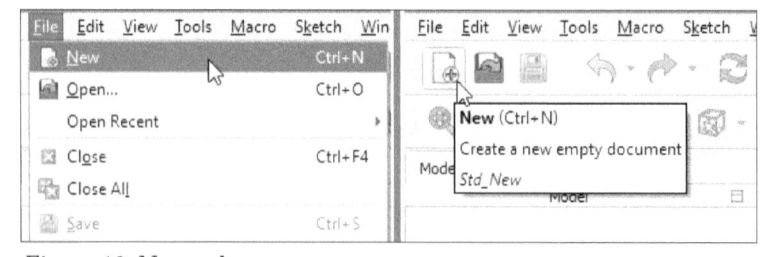

Figure-10. New tool

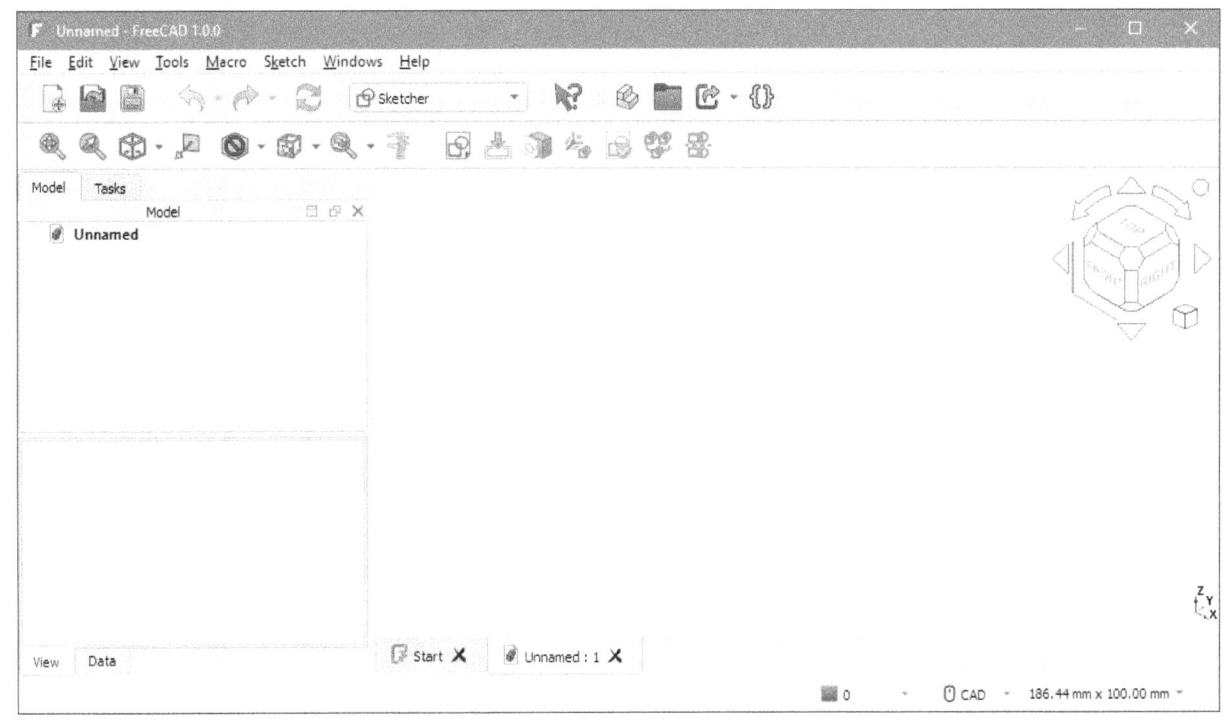

Figure-11. New document file

Open

The **Open** tool is used to open FreeCAD files and import files of other CAD applications. The procedure to use this tool is discussed next.

- Click on the **Open** tool from the **File** menu or **Ribbon** or press **CTRL+O** key from keyboard; refer to Figure-12. The **Open document** dialog box will be displayed; refer to Figure-13.

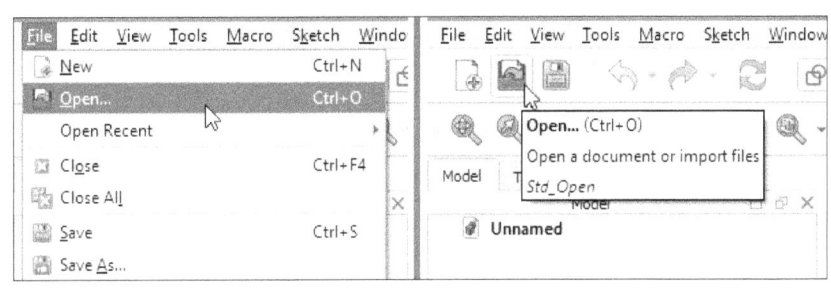

Figure-12. Open tool

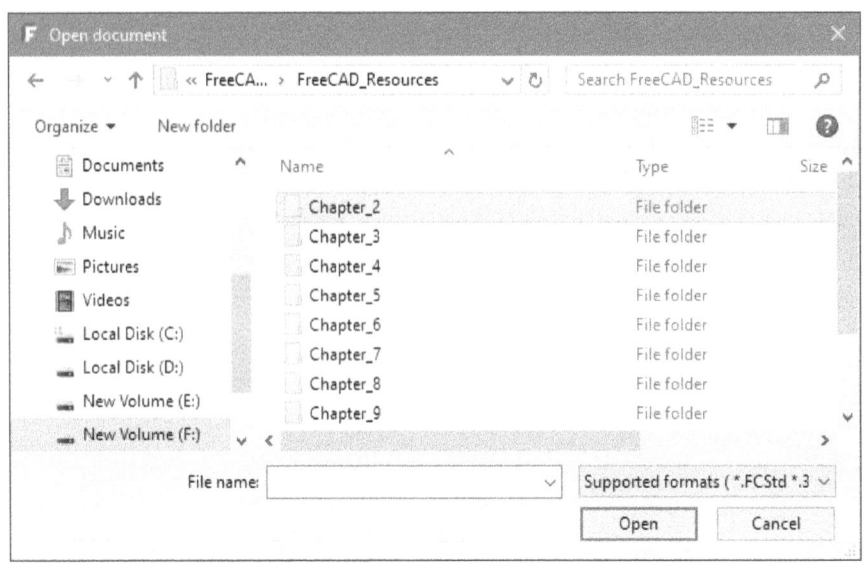

Figure-13. Open document dialog box

- Select desired format of file from the **File Type** drop-down to define format of file to be opened.
- Select desired file which you want to open and click on **Open** button from the dialog box. The file will open.

Open Recent

The **Open Recent** cascading menu displays a list of recent files which can be opened.

- Hover the cursor on the **Open Recent** cascading menu of the **File** menu to display the recent files list; refer to Figure-14. Select desired file name to open respective file.

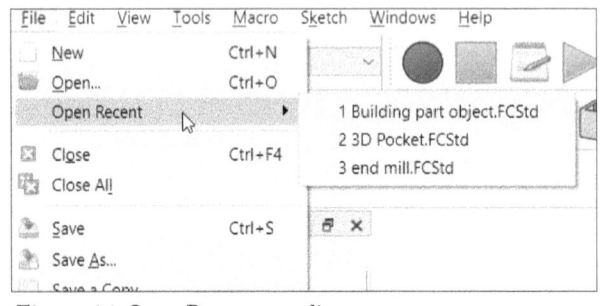

Figure-14. Open Recent cascading menu

Close

The **Close** tool is used to close the current file without closing the application. The procedure to use this tool is discussed next.

- Click on the **Close** tool from **File** menu or press **Ctrl+F4** key from keyboard; refer to Figure-15. The current file will be closed.

FreeCAD 1.0 Black Book 1-11

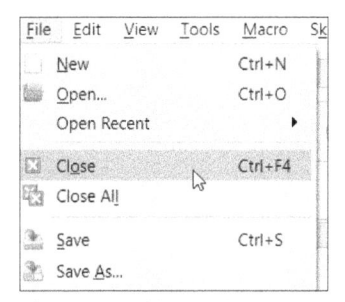

Figure-15. Close tool

Close All

The **Close All** tool is used to close all the files currently open without closing the application. The procedure to use this tool is discussed next.

- Click on the **Close All** tool from the **File** menu; refer to Figure-16. All the files which are currently open will be closed.

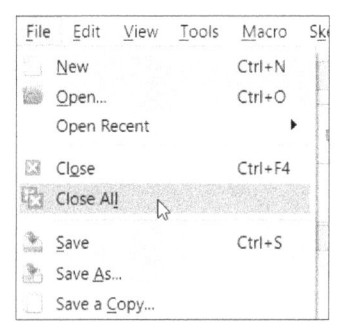

Figure-16. Close All tool

Save

The **Save** tool is used to save the current file opened in the application. The procedure to use this tool is discussed next.

- Click on the **Save** tool from **File** menu or **Ribbon** or press **Ctrl+S** key from keyboard; refer to Figure-17. The **Save FreeCAD Document** dialog box will be displayed; refer to Figure-18.

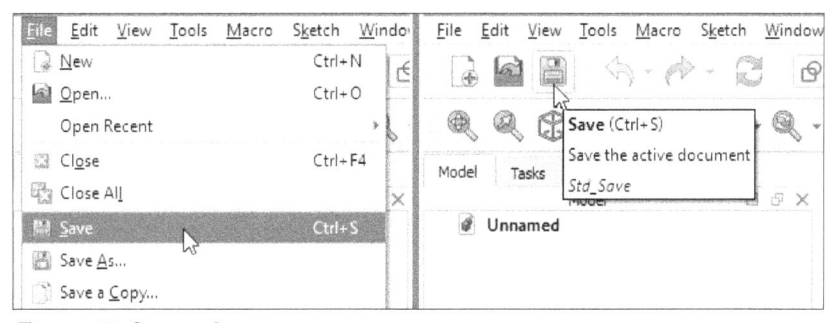

Figure-17. Save tool

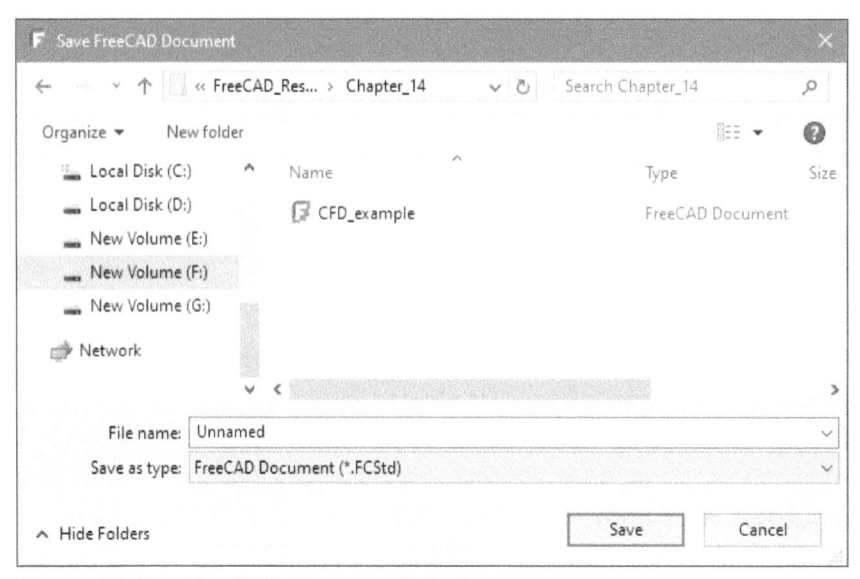

Figure-18. Save FreeCAD Document dialog box

- Specify desired location and name for the file, and then click on the **Save** button from the dialog box to save the file.

Save As

Using the **Save As** tool, you can save the file with different name. The procedure to use this tool is discussed next.

- Click on **Save As** tool from **File** menu; refer to Figure-19. The **Save FreeCAD Document** dialog box will be displayed as shown in Figure-18.

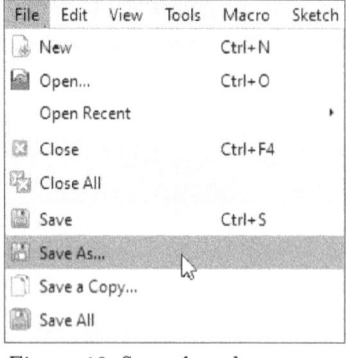

Figure-19. Save As tool

- Rest of the procedure is same as discussed for the **Save** tool.

Save a Copy

The **Save a Copy** tool is used to create a copy of the file with new name. The procedure to use this tool is discussed next.

- Click on the **Save a Copy** tool from the **File** menu; refer to Figure-20. The **Save FreeCAD Document** dialog box will be displayed as shown in Figure-18.

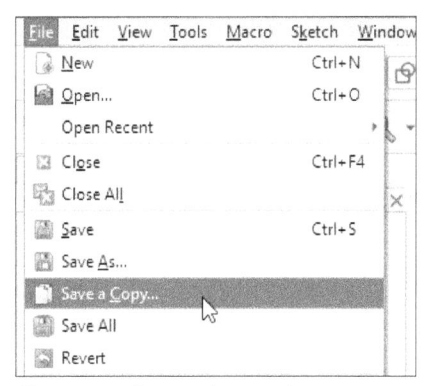

Figure-20. Save a Copy tool

- Rest of the procedure is same as we have discussed earlier.

Save All

The **Save All** tool is used to save all the documents currently open in the application. The procedure to use this tool is discussed next.

- Click on the **Save All** tool from the **File** menu; refer to Figure-21. The **Save FreeCAD Document** dialog box will be displayed as shown in Figure-18.

Figure-21. Save All tool

- Rest of the procedure is same as discussed earlier.

Revert

The **Revert** tool is used to close the active document and reopen the last saved version of this document. Using this tool, you will lose all the changes you have made since the last file save. The procedure to use this tool is discussed next.

- Click on the **Revert** tool from the **File** menu; refer to Figure-22. The **Revert document** dialog box will be displayed asking you for the confirmation to discard all the changes; refer to Figure-23.

Figure-22. Revert tool

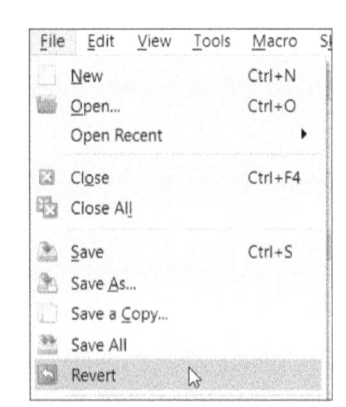

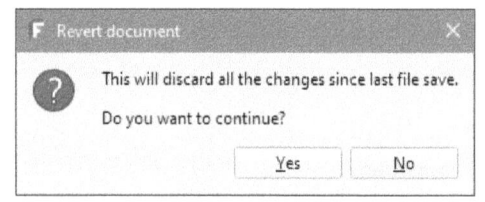

Figure-23. Revert document dialog box

- Click on the **Yes** button from the dialog box to discard all the changes.

Import

Many times, importing reduces lots of extra work of rebuilding the base sketch for models. In FreeCAD, we can directly use the CAD files of other software for creating or manipulating the model. Note that if you are a user of Image workbench then after FreeCAD 0.21, the **Image** workbench has been removed and its functionality is added in the **Import** tool so you can import image files as well by using this tool. The procedure to use this tool is discussed next.

- Click on the **Import** tool from the **File** menu; refer to Figure-24. The **Import file** dialog box will be displayed; refer to Figure-25.

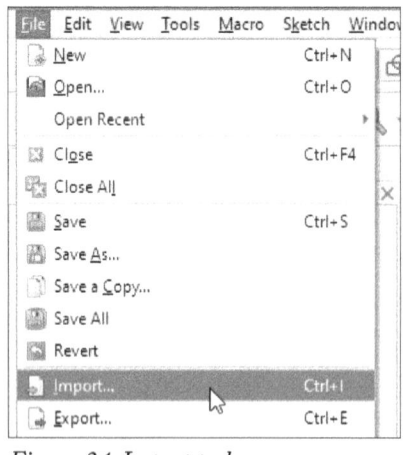

Figure-24. Import tool

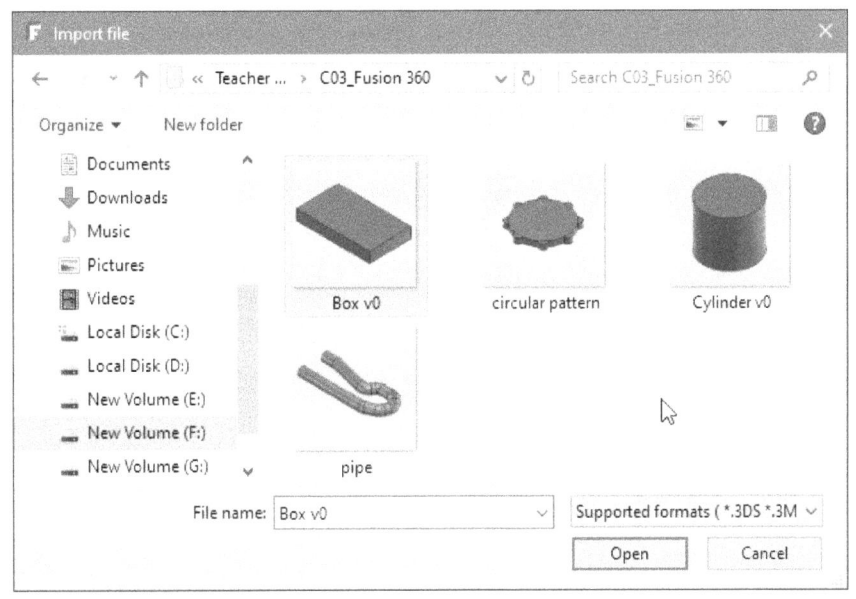

Figure-25. Import file dialog box

- Select desired format of non-native file which you want to open from the **File Type** drop-down; refer to Figure-26.

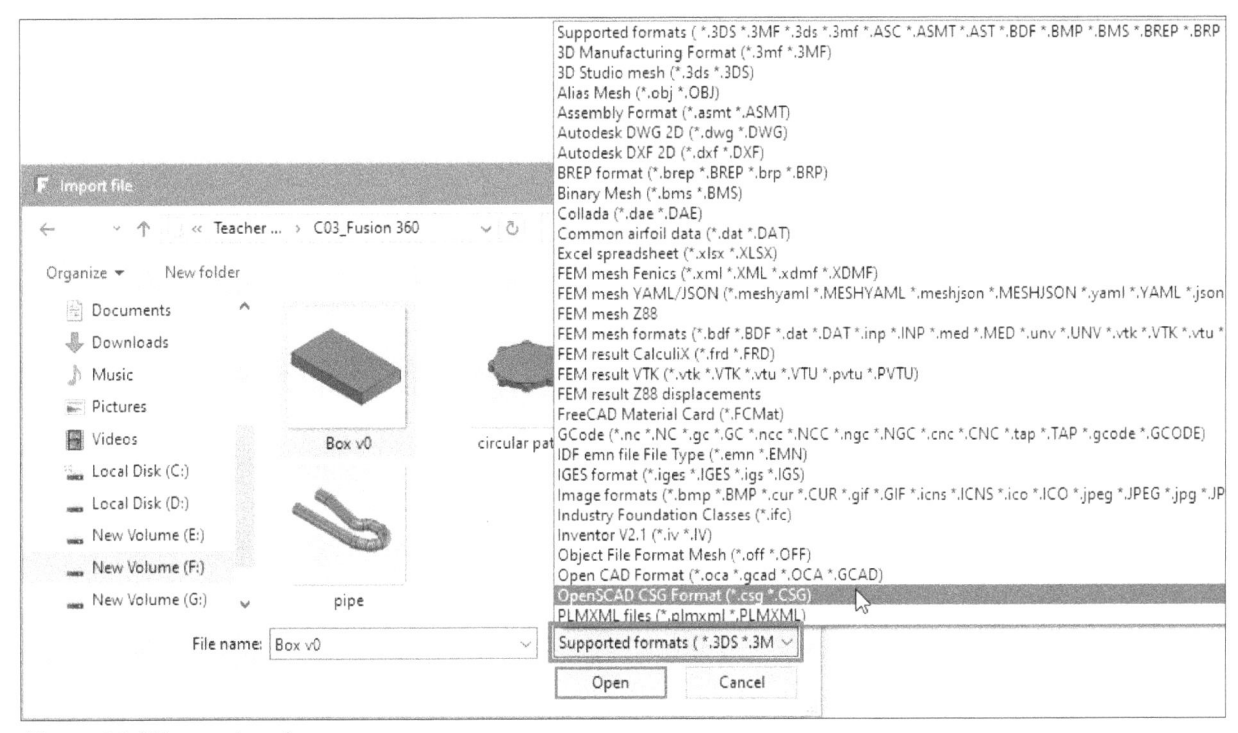

Figure-26. File type drop down

- Select desired file which you want to import and click on the **Open** button from the dialog box. The model will be updated in this graphic area.

Export

The **Export** tool is used to export existing file in various different formats. The procedure to use this tool is discussed next. IGES and STEP are the most common formats for import-export in CAD software.

- Select the solid part or sketch which you want to export from the **Model** panel as shown in Figure-27.

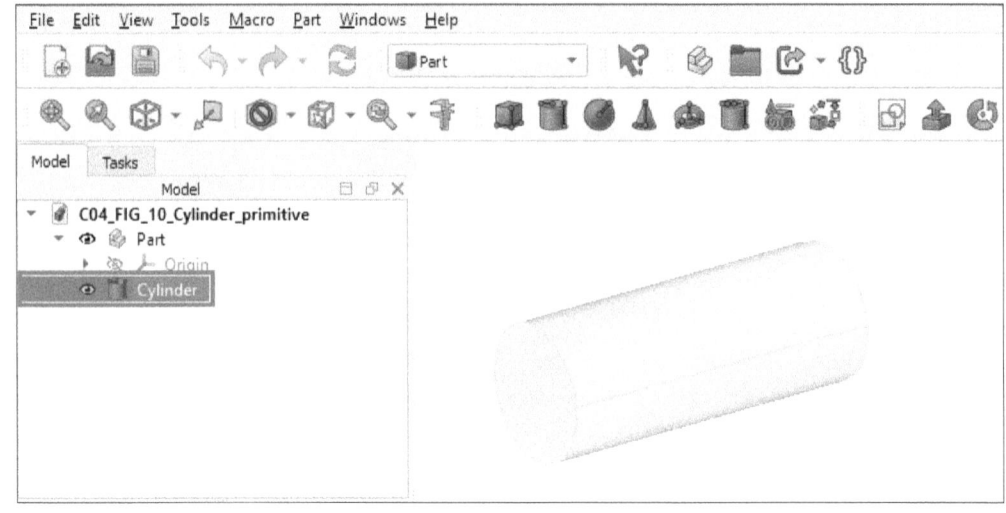

Figure-27. Solid model selected to export

- Click on the **Export** tool from **File** menu; refer to Figure-28. The **Export file** dialog box will be displayed; refer to Figure-29.

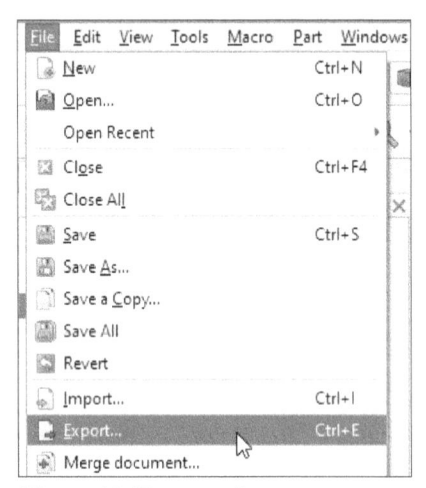

Figure-28. Export tool

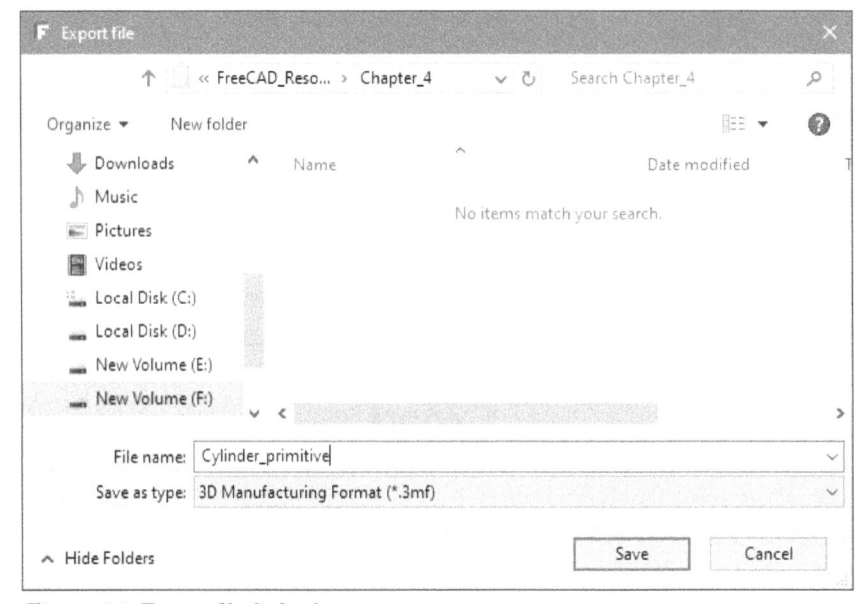

Figure-29. Export file dialog box

- Select desired format of file from the **Save as type** drop-down.
- Specify desired name for the file in the **File name** edit box and click on **Save** button from the dialog box to export the file.

Merge document

The **Merge document** tool is used to add the contents of a FreeCAD file into the active document. The procedure to use this tool is discussed next.

- Click on the **Merge document** tool from the **File** menu; refer to Figure-30. The **Merge project** dialog box will be displayed; refer to Figure-31.

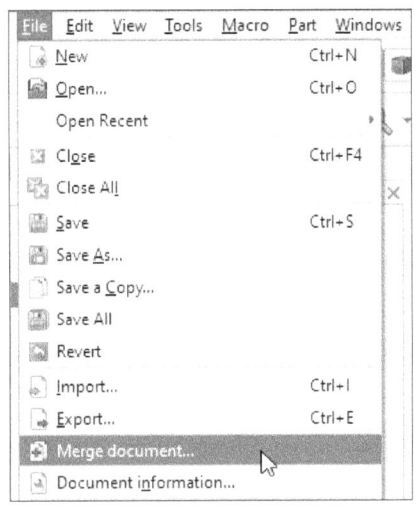

Figure-30. Merge document tool

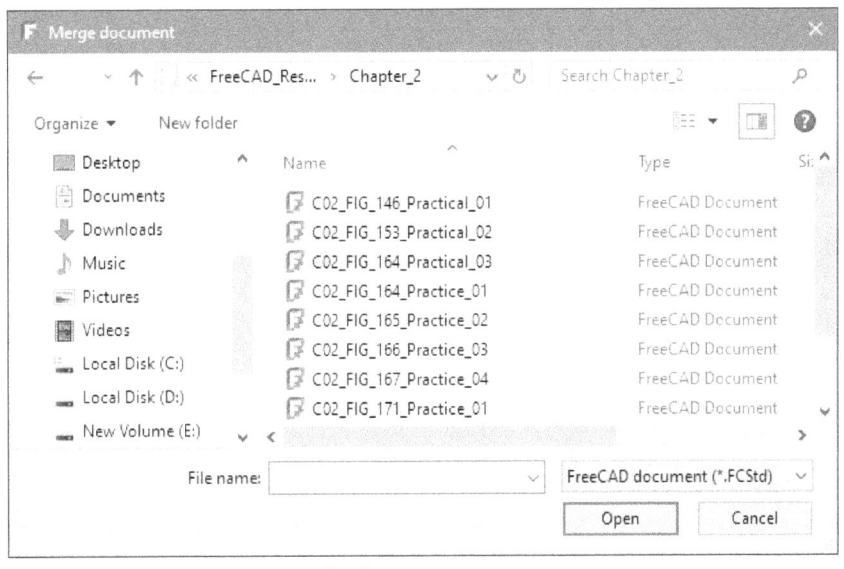

Figure-31. Merge document dialog box

- Select desired file which you want to add in the active document and click on **Open** button from the dialog box. The model of selected file will be added in the active document.

Document information

The **Document information** tool is used to display general information related to current project in **Document information** dialog box. Some of this information can be edited in the dialog box. The procedure to use this tool is discussed next.

- Click on the **Document information** tool from the **File** menu; refer to Figure-32. The **Document information** dialog box will be displayed; refer to Figure-33.

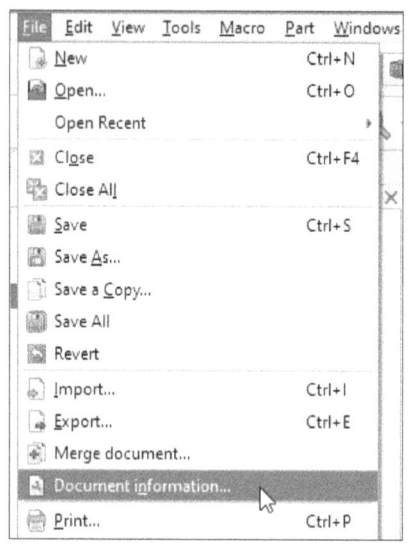

Figure-32. Document Information tool

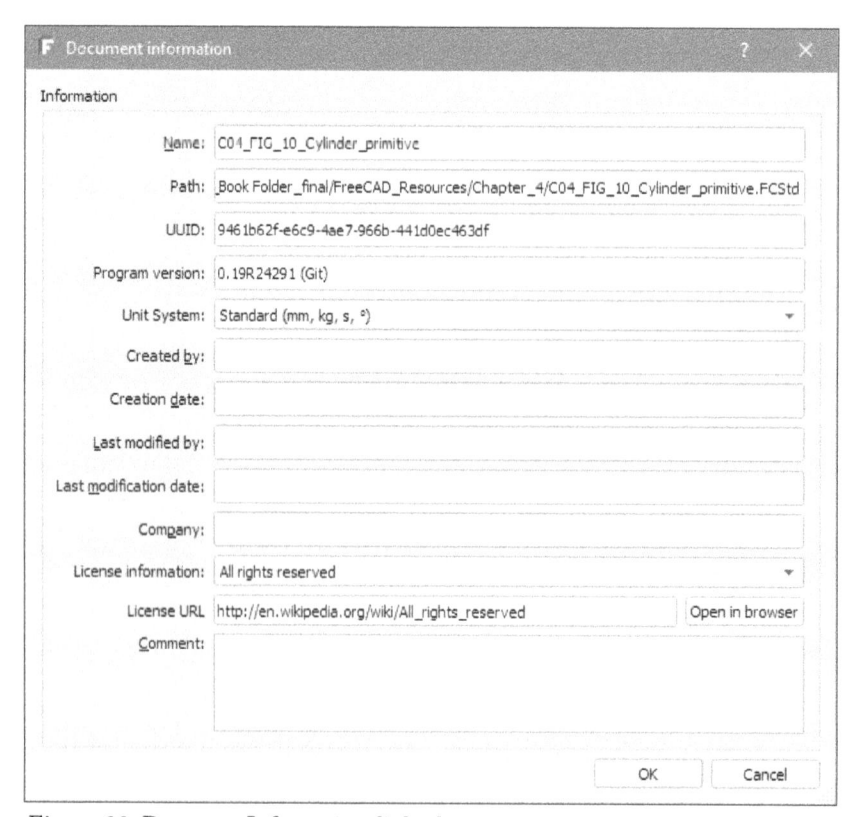

Figure-33. Document Information dialog box

- Specify and edit the information about the active document as desired and click on **OK** button from the dialog box.

Print

As the name suggests, the **Print** tool is used to get print out of the model/drawing on paper using printer connected to the system.

Print Preview

The **Print Preview** tool is used to check the print before sending command to printer. The procedure to use this tool is discussed next.

- Click on the **Print preview** tool from the **File** menu; refer to Figure-34. The **Print Preview** dialog box will be displayed along with the preview of the model; refer to Figure-35.

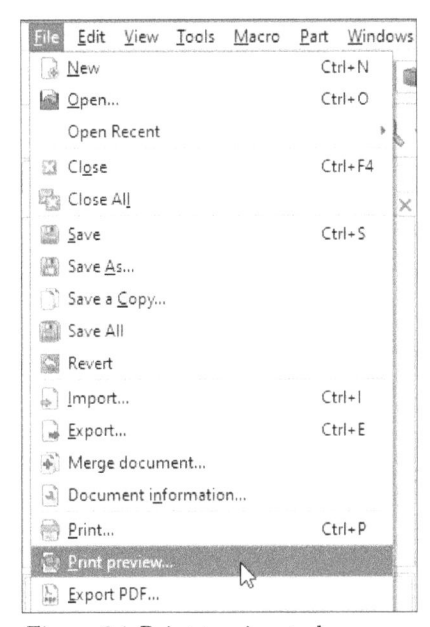

Figure-34. Print preview tool

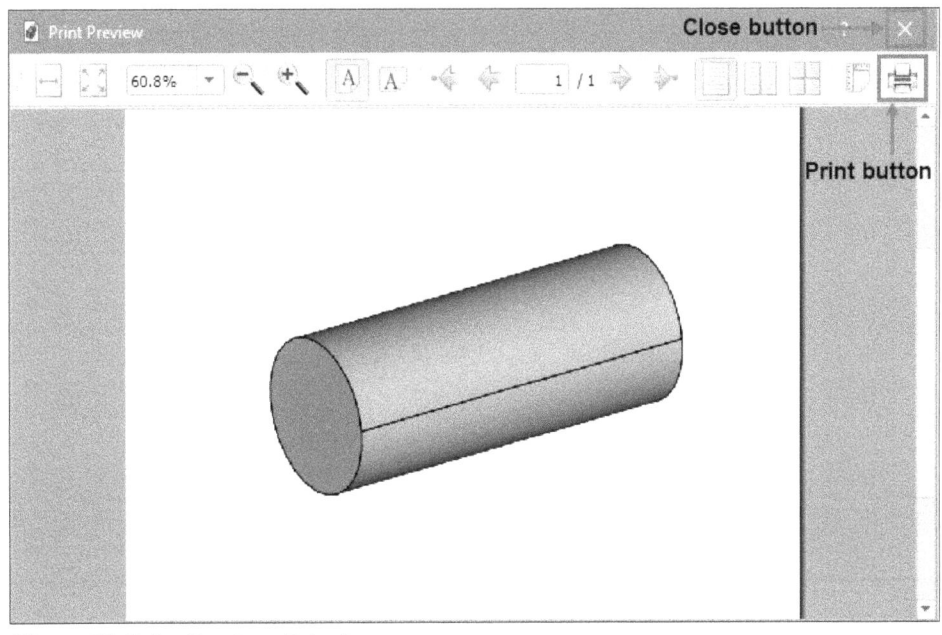

Figure-35. Print Preview dialog box

- If the preview is satisfactory then click on the **Print** button otherwise, click on the **Close** button from the toolbar displayed at the top of the modeling area and modify the model; refer to Figure-35.

Export PDF

The **Export PDF** tool is used to export or save the active document in PDF file format. The procedure to use this tool is discussed next.

- Click on the **Export PDF** tool from **File** menu; refer to Figure-36. The **Export PDF** dialog box will be displayed; refer to Figure-37.

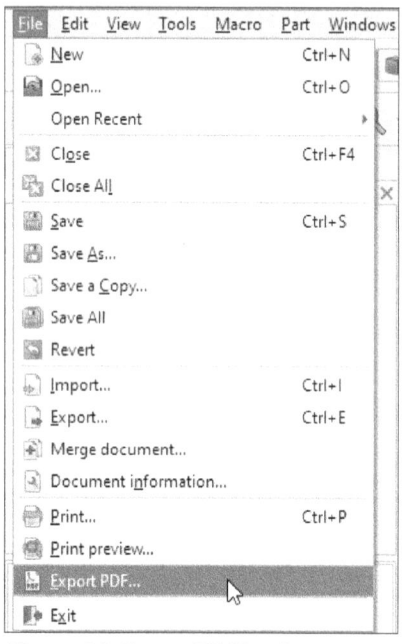

Figure-36. Export PDF tool

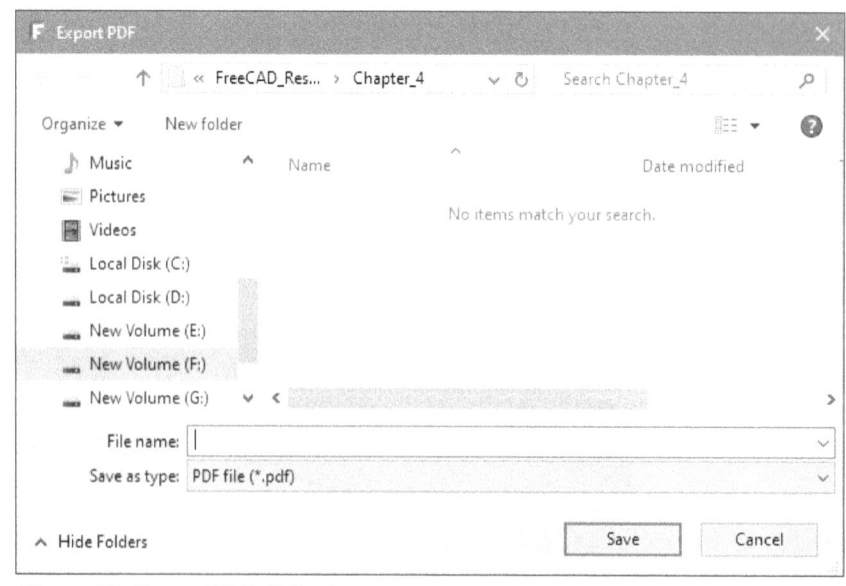

Figure-37. Export PDF dialog box

- Specify desired file name for saving the file as PDF and click on **Save** button from the dialog box.

VIEW MENU

The **View** menu provides tools to change the 3D view, the view properties of objects in the model, and tools related to the display of interface components. Various tools of **View** menu are discussed next.

Create new view

The **Create new view** tool is used to create a new 3D view for the active document. It can be useful if you want to inspect the model from multiple directions or at different zoom levels. The procedure to use this tool is discussed next.

- Click on the **Create new view** tool from the **View** menu; refer to Figure-38. A new window will open with the 3D view of an active model; refer to Figure-39.

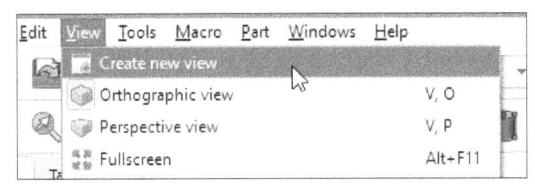

Figure-38. Create new view tool

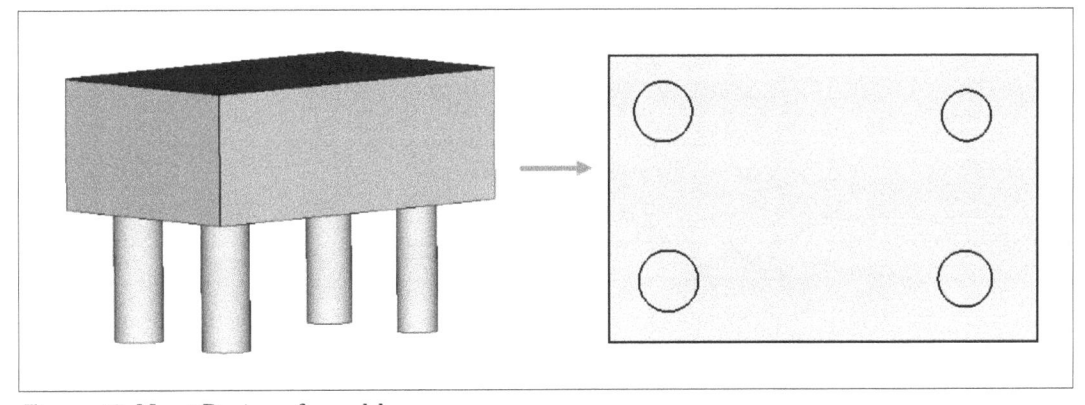

Figure-39. New 3D view of a model

Orthographic view

The **Orthographic view** tool switches the camera in the active 3D view to orthographic view mode. In this mode, objects that are far from the camera do not appear smaller than those that are closer. The procedure to use this tool is discussed next.

- Click on the **Orthographic view** tool from the **View** menu; refer to Figure-40. An orthographic view of the model will be displayed; refer to Figure-41.

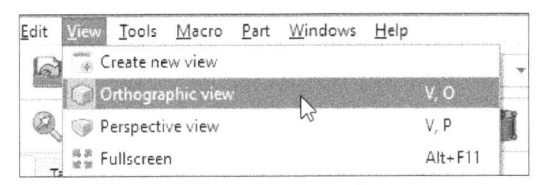

Figure-40. Orthographic view tool

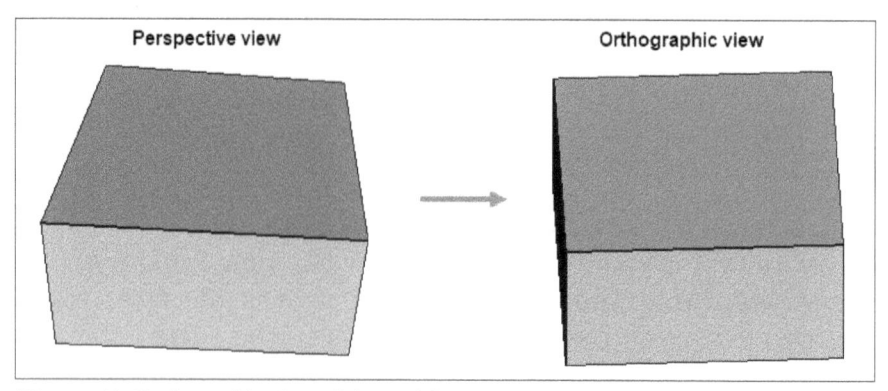

Figure-41. Orthographic view of the model

Perspective view

The **Perspective view** tool switches the camera in the active 3D view to perspective view mode. In this mode, objects that are far from the camera appear smaller than those that are closer. The procedure to use this tool is discussed next.

- Click on the **Perspective view** tool from the **View** menu; refer to Figure-42. A perspective view of the model will be displayed; refer to Figure-43.

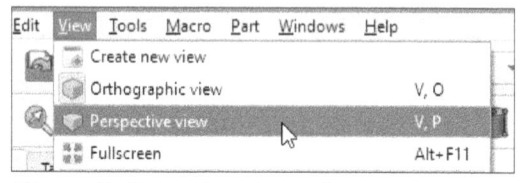

Figure-42. Perspective view tool

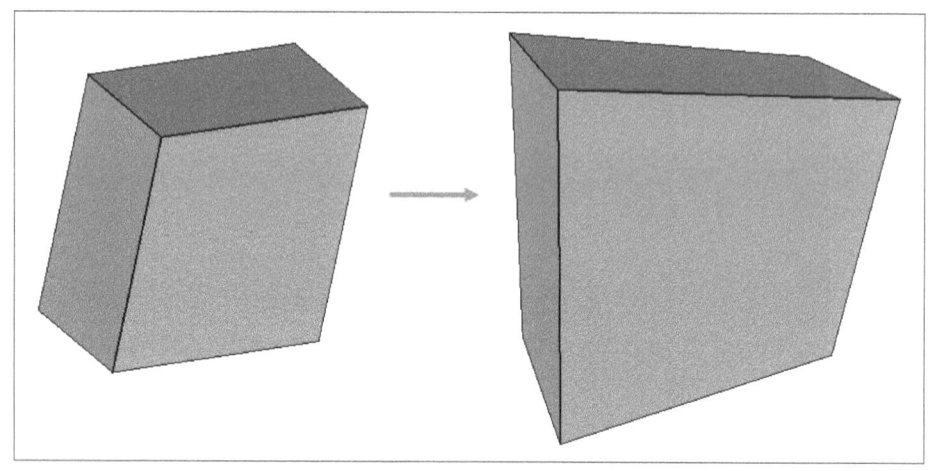

Figure-43. Perspective view of the model

Fullscreen

The **Fullscreen** tool switches the FreeCAD's main window in fullscreen mode. The procedure to use this tool is discussed next.

- Click on the **Fullscreen** tool from the **View** menu; refer to Figure-44. The view of the model will be displayed in Fullscreen window.

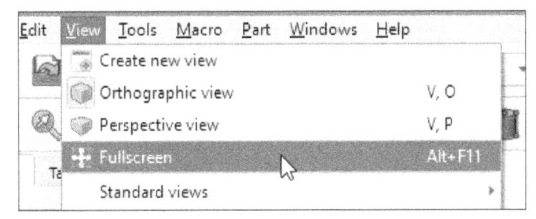

Figure-44. Fullscreen tool

Standard views

Standard views provide tools to orient the part, assembly, or sketch in one of the preset standard views. The standard views display the model or drawing through one, two, or four viewports.

Various tools available in the **Standard views** cascading menu (refer to Figure-45) in the **View** menu are discussed next.

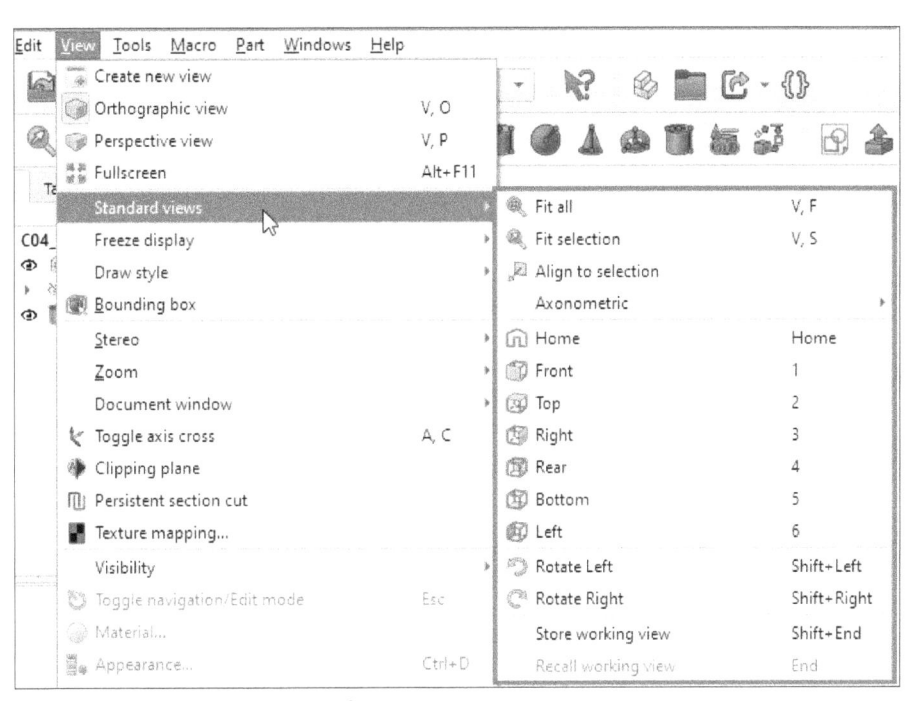

Figure-45. Standard views cascading menu

- **Fit all** : The **Fit all** tool zooms and pans the camera so that all visible objects fit inside the active 3D view.

- **Fit selection** : The **Fit selection** tool zooms and pans the camera so that all selected objects fit inside the active 3D view.

- **Align to selection** : The **Align to selection** tool points the camera in the opposite direction of the normal of a planar face, or the opposite direction of a straight edge in the active 3D view.

- **Axonometric** : The **Axonometric** cascading menu has three view tools:

    Isometric View : The **Isometric** tool realigns the camera in the active 3D view to obtain an isometric view. For a truly isometric view, the 3D view must be in orthographic mode but the tool also works, if the view is in perspective mode.

Dimetric View : The **Dimetric** tool realigns the camera in the active 3D view to obtain a dimetric view.

Trimetric View : The **Trimetric** tool realigns the camera in the active 3D view to obtain a trimetric view.

Note that in Isometric view all dimensions are the same scale and all projectors have equal angles, usually 30°. In Diametric view, two axes use the same scale, while the third (usually the vertical axis) is determined separately. In Trimetric view, all three projectors have different angles.

- **Front** : The **Front** tool points the camera in the active 3D view in the direction of the positive Y axis.

- **Top** : The **Top** tool points the camera in the active 3D view in the direction of the negative Z axis.

- **Right** : The **Right** tool points the camera in the active 3D view in the direction of the negative X axis.

- **Rear** : The **Rear** tool points the camera in the active 3D view in the direction of the negative Y axis.

- **Bottom** : The **Bottom** tool points the camera in the active 3D view in the direction of the positive Z axis.

- **Left** : The **Left** tool points the camera in the active 3D view in the direction of the positive X axis.

- **Rotate Left** : The **Rotate Left** tool rotates the camera in the active 3D view around the view direction in 90-degree increments towards the left (counterclockwise).

- **Rotate Right** : The **Rotate Right** tool rotates the camera in the active 3D view around the view direction in 90-degree increments towards the right (clockwise).

- **Store working view** : The **Store working view** tool stores the camera setting of the active 3D view as temporary view. You can also use **SHIFT+END** shortcut key to perform the same task.

- **Recall working view** : The **Recall working view** tool restores the earlier stored working view. You can also use **END** key to perform the same task.

Freeze Display

FreeCAD can store camera settings for up to 50 'frozen views'. Frozen view is the view with current specified orientation which you can restore after the view get disturbed. Frozen views are not stored in the document and if not saved with the **Save views** menu option, will be lost when the FreeCAD application closes. The tools that deal with frozen views are available in the **Freeze display** cascading menu of the **View** menu; refer to Figure-46.

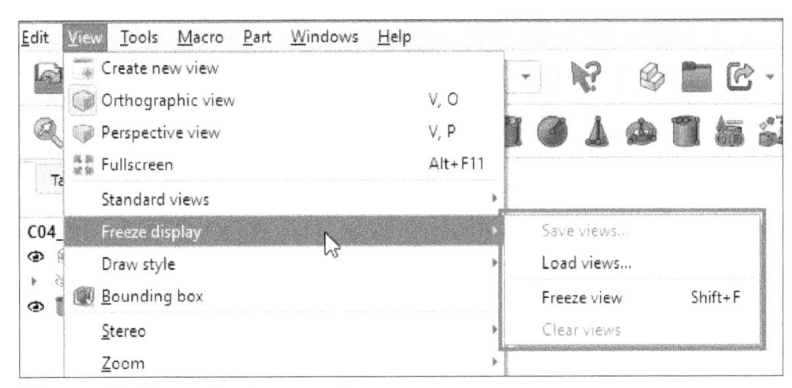

Figure-46. Freeze display cascading menu

Save Views

The **Save views** tool saves all existing frozen views in a file with the ***.cam** extension. To use this tool, one or more frozen views must exist. A frozen view can be created with the **Freeze view** menu option which will be discussed later. The procedure to use this tool is discussed next.

- Click on the **Save views** tool from the **Freeze display** cascading menu. The **Save frozen views** dialog box will be displayed; refer to Figure-47.

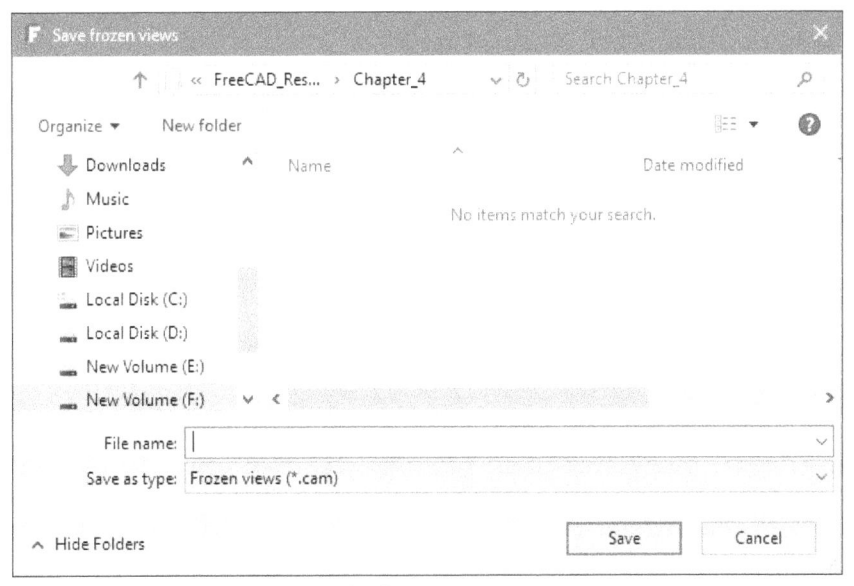

Figure-47. Save frozen views dialog box

- Specify desired name for the file in the **File name** edit box and click on **Save** button from the dialog box to save the frozen view.

Load views

The **Load views** tool loads the frozen views from a file with the ***.cam** extension. Note that when you load a frozen view then all existing frozen views will be deleted. The procedure to use this tool is discussed next.

- Click on the **Load views** tool from the **Freeze display** cascading menu.
- If the frozen views are saved earlier then the **Restore views** dialog box will be displayed; refer to Figure-48, asking you to confirm that you want to lose all existing frozen views.

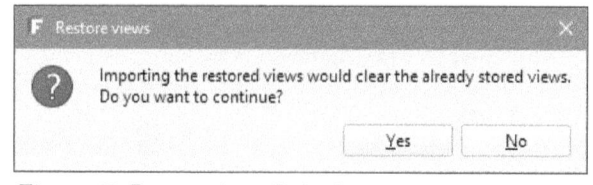

Figure-48. Restore views dialog box

- Click on the **Yes** button from the dialog box, the `Restore frozen views` dialog box will be displayed; refer to Figure-49.

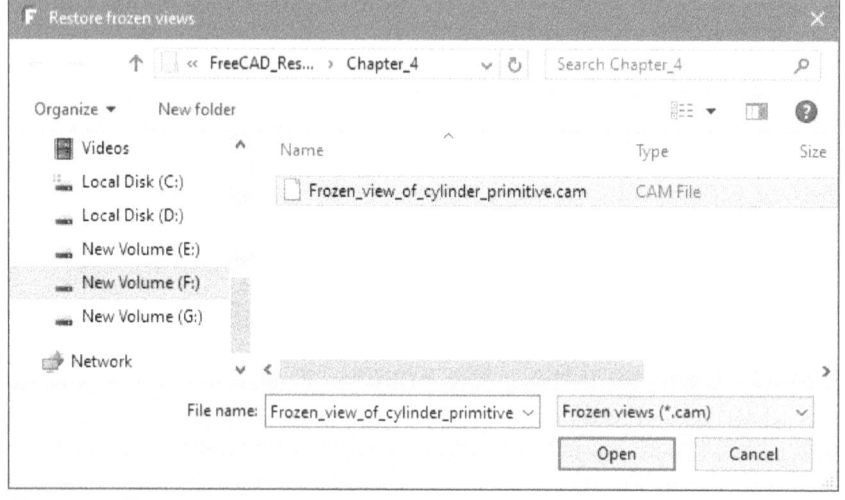

Figure-49. Restore frozen views dialog box

- If the frozen views are not saved earlier then also the `Restore frozen views` dialog box will be displayed.
- Select the file which you want to restore and click on **Open** button from the dialog box.

Freeze view

The **Freeze view** tool stores the current camera settings in a new frozen view. You can create maximum 50 frozen views. The procedure to use this tool is discussed next.

- Click on the `Freeze view` tool from `Freeze display` cascading menu. The frozen view of the model will be created; refer to Figure-50.

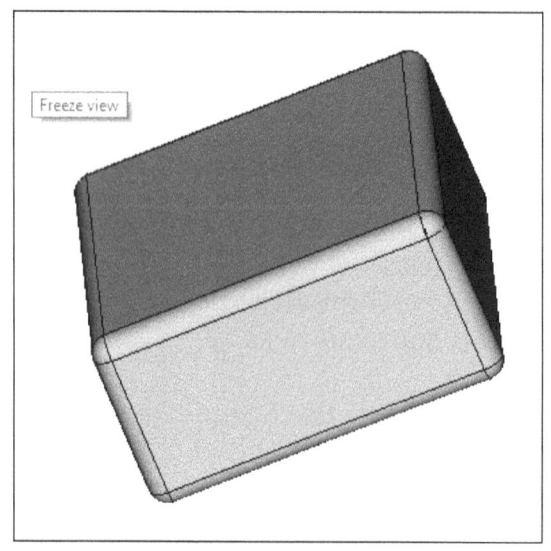

Figure-50. Frozen view of the model

Clear views

The **Clear views** tool deletes all existing frozen views. Click on the **Clear views** tool from **Freeze display** cascading menu, all the existing frozen views will be deleted.

Restore view

For each frozen view, a **Restore view** option is added with which the frozen view can be restored. The procedure to use this tool is discussed next.

- Click on the **Restore view** tool from **Freeze display** cascading menu, the frozen view of the model will be restored; refer to Figure-51.

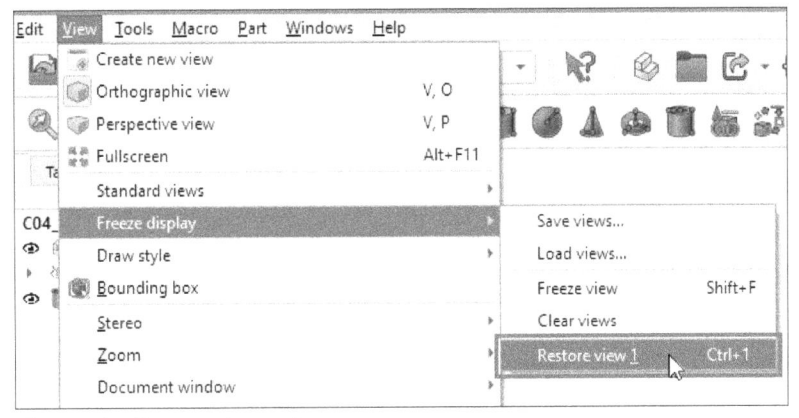

Figure-51. Restore view tool

Draw Style

The tools in **Draw style** cascading menu are used to display objects in graphics area of different view styles like wireframe, solid, hidden lines, and so on; refer to Figure-52 and Figure-53.

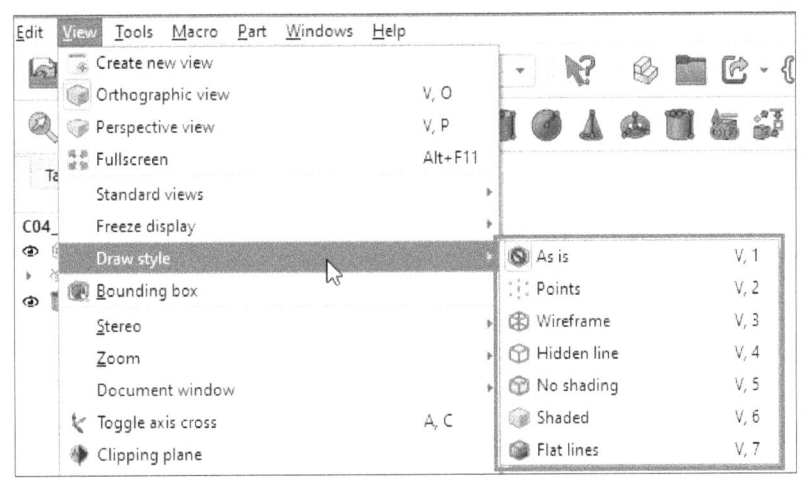

Figure-52. Draw style cascading menu

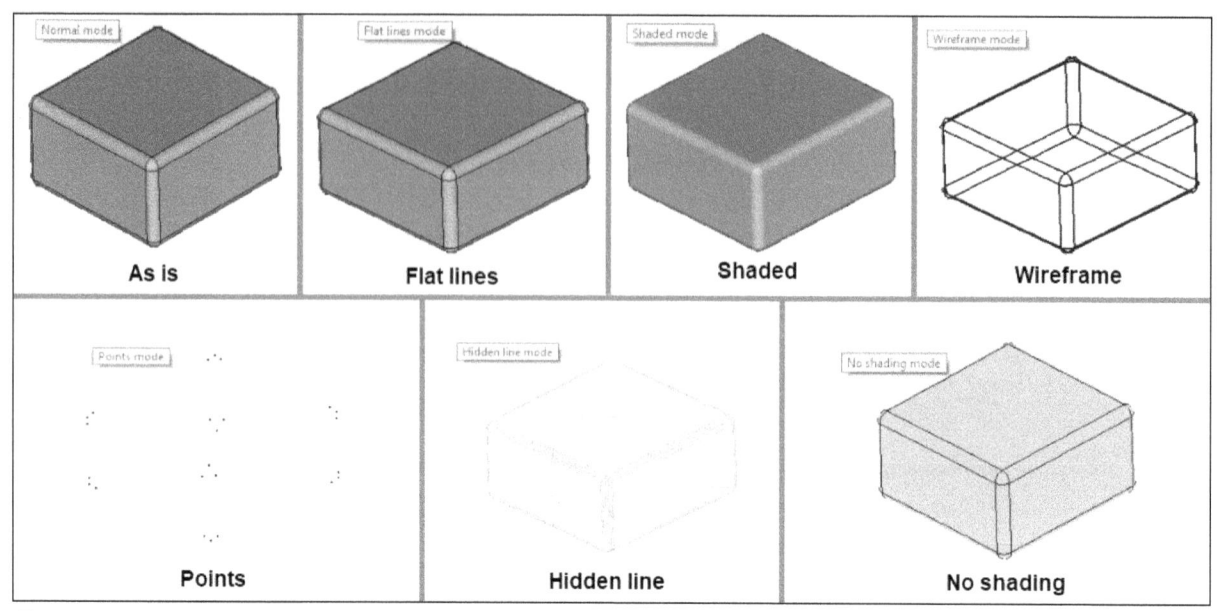

Figure-53. Various type of draw styles

You will learn about other tools of **View menu** later in this book.

TOOLS MENU

The **Tools** menu provides tools for debugging models, customizing FreeCAD's behavior as well as adding auxiliary tools; refer to Figure-54. Various tools available in the **Tools** menu are discussed next.

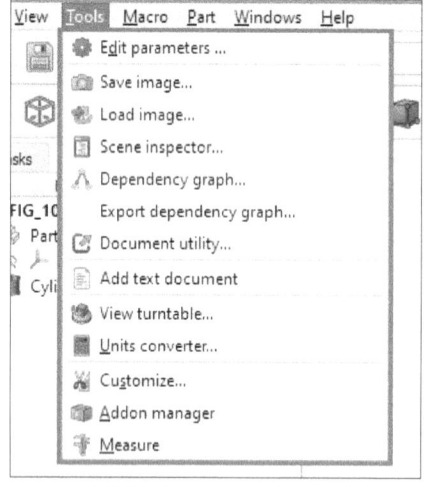

Figure-54. Tools menu

Edit parameters

The **Edit parameters** tool gives access to the parameters that control the program. By clicking on this tool, the **Parameter Editor** dialog box will be displayed as shown in Figure-55. The **Parameter Editor** controls the behavior of FreeCAD and its workbenches. The parameters are stored in a file called **user.cfg**.

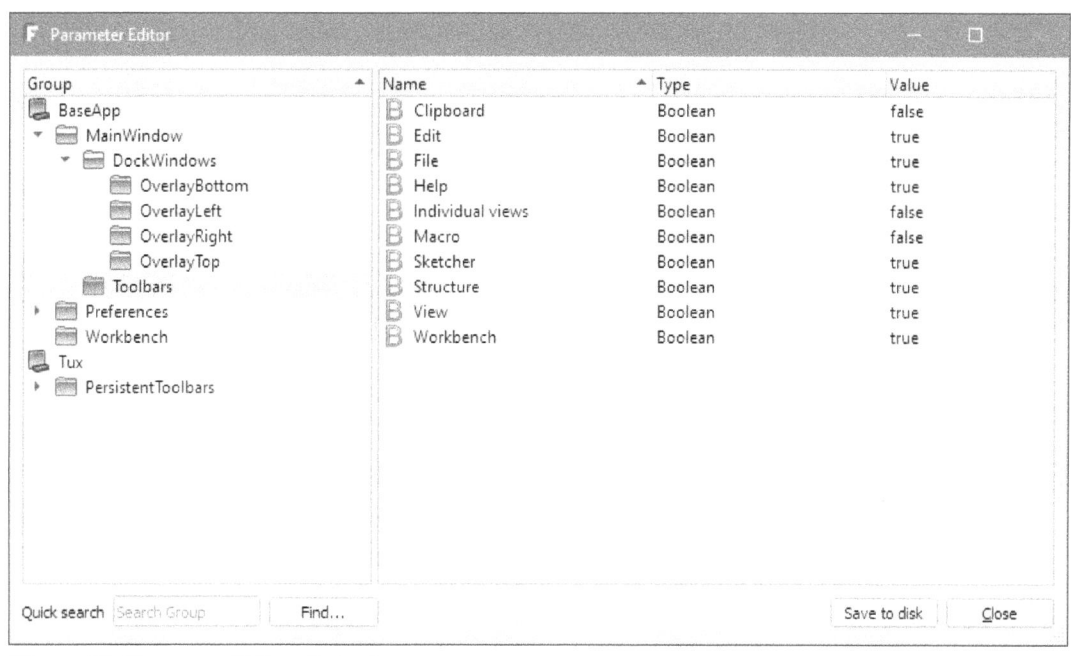

Figure-55. Parameter Editor dialog box

- Type desired name of the group in the **Quick search** edit box which you want to modify. The specified group name will be highlighted with red color in the left panel of the dialog box; refer to Figure-56.

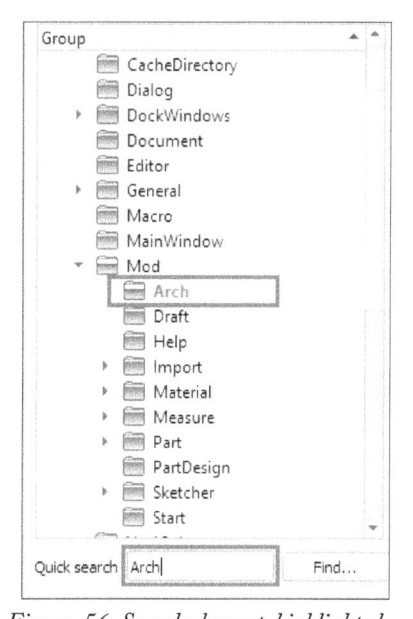

Figure-56. Searched group highlighted

- You can search object in the dialog box using advanced filters by using the **Find** dialog box. Click on the **Find** button next to the **Quick search** edit box in the dialog box. The **Find** dialog box will be displayed; refer to Figure-57.

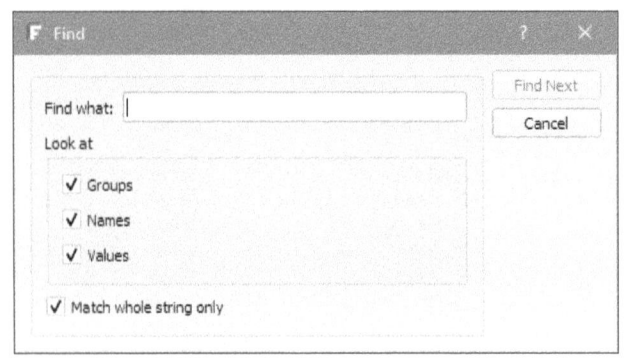

Figure-57. Find dialog box

- Type desired string name which you want to find in the **Find what** input box of the **Find** dialog box.
- Select the **Groups**, **Names**, and **Values** check boxes from **Look at** area of the **Find** dialog box to find the string values of groups, names, and values, respectively.
- Click on the **Find Next** button. The searched strings will be displayed in the **Parameter Editor** dialog box.
- Click on the **Cancel** button to close the **Find** dialog box.
- Specify desired parameters and click on **Save to disk** button to update the **user.cfg** file. For example, select the **General** group from the list in left area of the dialog box and modify the **FileOpenSavePath** parameter to change default directory where software will look to save/open the files.
- Click on the **Close** button to close the dialog box. You will learn more about the parameters in this dialog box later based on chapter requirement.

Save Image

The **Save image** tool is used to create an image file or a screenshot from the active 3D view. The procedure to use this tool is discussed next.

- Click on the **Save image** tool from the **Tools** menu. The **Save image** dialog box will be displayed; refer to Figure-58.

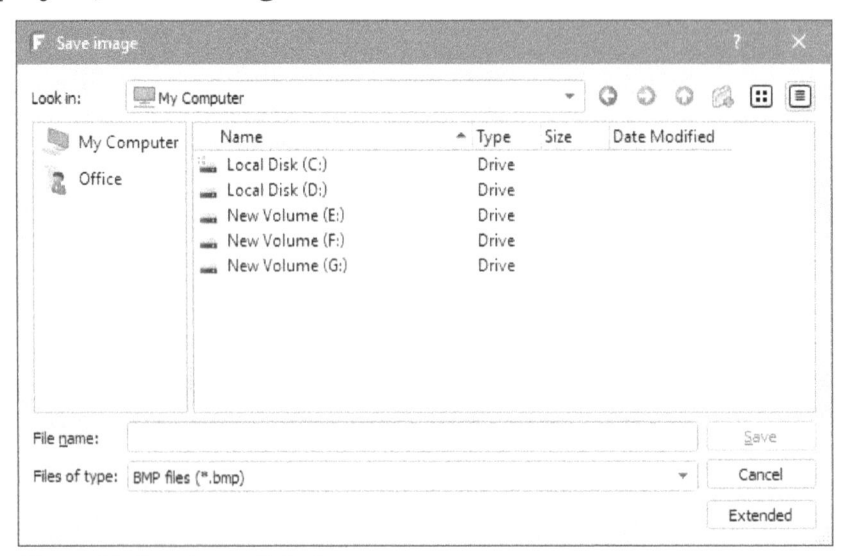

Figure-58. Save image dialog box

- Click on the **Extended** button from the dialog box. An additional panel will be added as shown in Figure-59.

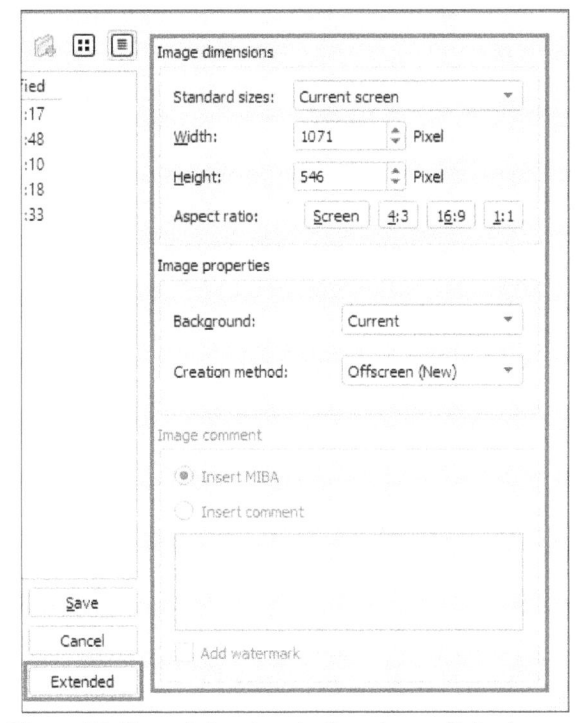

Figure-59. Extended options in Save image dialog box

- Select standard size for the image from **Standard sizes** drop-down list or specify width and height of the image in the **Width** and **Height** edit boxes, respectively in the **Image dimensions** area of the dialog box.
- Select desired **Aspect ratio** button to set the width to height ratio of the image.
- Select desired option from **Background** drop down list in the **Image properties** area of the dialog box to set background for the image.
- Select desired option from **Creation method** drop-down list in the **Image properties** area to set the creation method for the image.
- Select **Insert MIBA** radio button from **Image comment** area to add MIBA information to the file. The options in the **Image comment** area of the dialog box will be activated only on selecting **JPG**, **JPEG**, and **PNG** image file types.
- Select the **Insert comment** radio button to type a comment in the text field to embed a comment in the file.
- Select the **Add watermark** check box to add a watermark of the FreeCAD logo in the image file.
- Specify desired file name and click on **Save** button from the dialog box.

Load Image

The **Load image** tool is used to load earlier saved image files in the graphics area. The procedure to use this tool is given next.

- Click on the **Load image** tool from **Tools** menu. The **Choose an image file to open** dialog box will be displayed as shown in Figure-60.

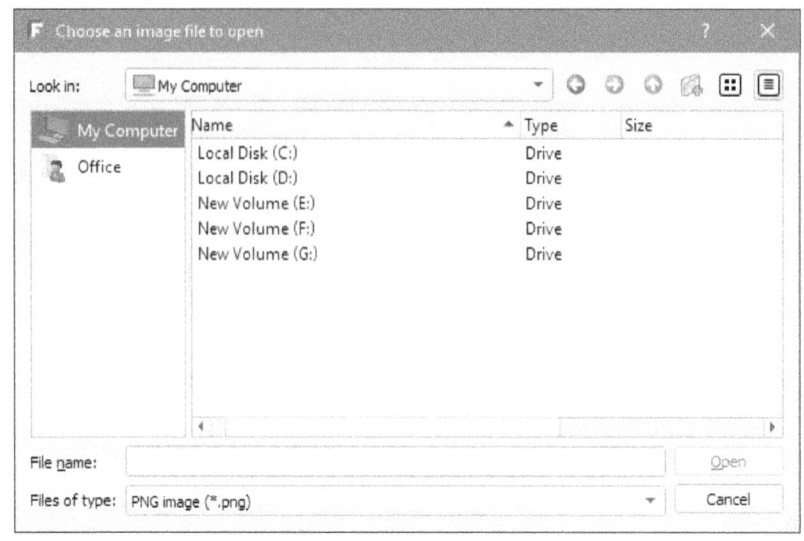

Figure-60. Choose an image file to open dialog box

- Select desired image file from the dialog box and click on the **Open** button. The image file will load in a new document tab.
- Right-click in the drawing area and select **Fit to window**, **zoom in**, and **zoom out** options to perform respective operations.
- To pan an image that is the larger than current view area, you can either use the scroll bars, or hold down the middle mouse button and move the cursor.

Scene Inspector

The **Scene inspector** tool displays an overview of all nodes in the scene graph of the active 3D view. It is more a utility for programmers than for average users. It can be used to find out why the rendering is slow or why something is not rendered properly. The procedure to use this tool is discussed next.

- Click on the **Scene inspector** tool from **Tools** menu. The **Scene Inspector** dialog box will be displayed as shown in Figure-61. This dialog box is modeless, it means it can stay open while you continue working in FreeCAD.

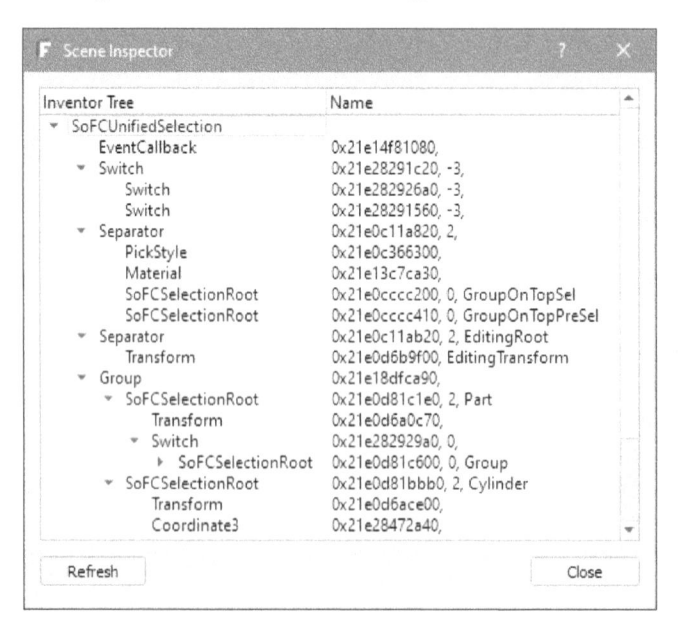

Figure-61. Scene Inspector dialog box

- Click on the **Refresh** button to update the overview or click on **Close** button to close the dialog box.

Dependency Graph

The **Dependency Graph** tool displays dependencies between objects of the active document in a dependency graph. In this graph, objects are listed in reverse chronological order, with the first created object at the bottom.

The dependency graph is purely a visualization tool therefore it cannot be edited. It automatically updates, if changes are made to the model.

Note that to use this tool, a third party software named **Graphviz** needs to be installed.

If the software is not pre-installed, install it after downloading from the **Graphviz** website (https://graphviz.org). If the software is installed in an unconventional location, FreeCAD will ask you for software installation path.

The procedure to use this tool is discussed next.

- Click on the **Dependency graph** tool from **Tools** menu. The **Graphviz not found** dialog box will be displayed, if **Graphviz** is not installed in standard location; refer to Figure-62.

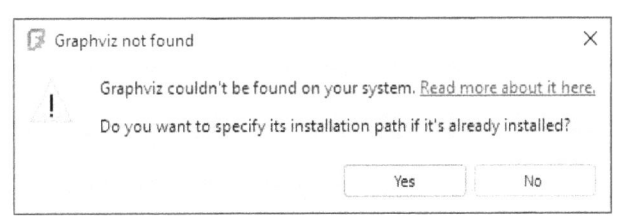

Figure-62. Graphviz not found dialog box

- Click on **Yes** button from the dialog box. The **Graphviz installation path** dialog box will be displayed asking for software installation folder; refer to Figure-63.

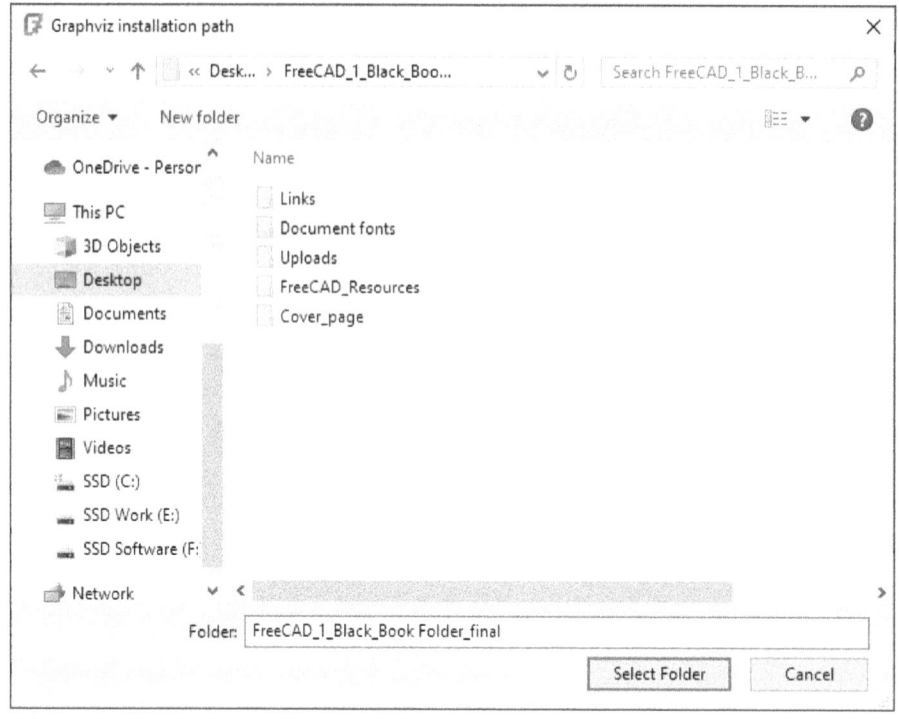

Figure-63. Graphviz installation path dialog box

- Browse to the folder where you have installed the software in your system (generally the path is: C:\Program Files\Graphviz\bin) and click on **Select Folder** button from the dialog box. A new window named **Dependency graph** will open displaying dependencies between objects; refer to Figure-64.

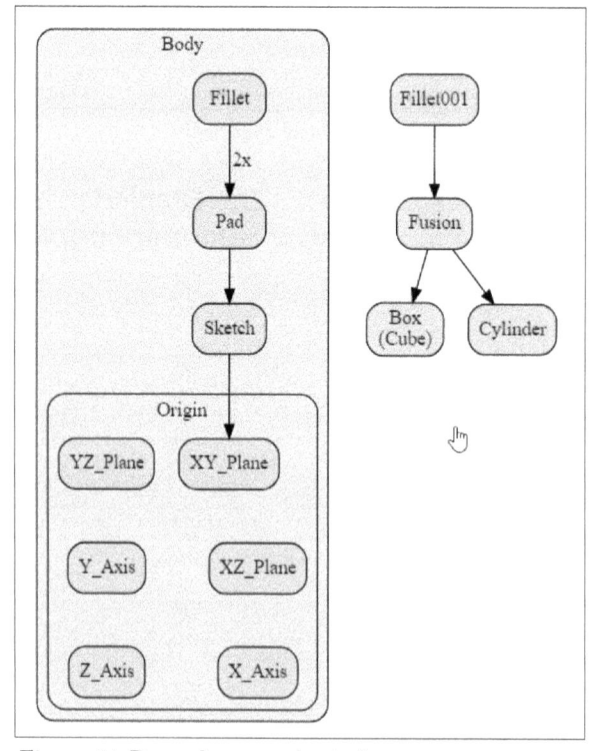

Figure-64. Dependency graph window

Exporting Dependency Graph

The **Export dependency graph** tool is used to export the dependency graph file in Graphviz format which can be directly opened in the Graphviz software. You can opt for this route if you are to getting a proper integration of Graphviz software with FreeCAD. The procedure is given next.

- After creating the model, click on the **Export dependency graph** tool from the **Tools** menu. The **Export graph** dialog box will be displayed; refer to Figure-65.

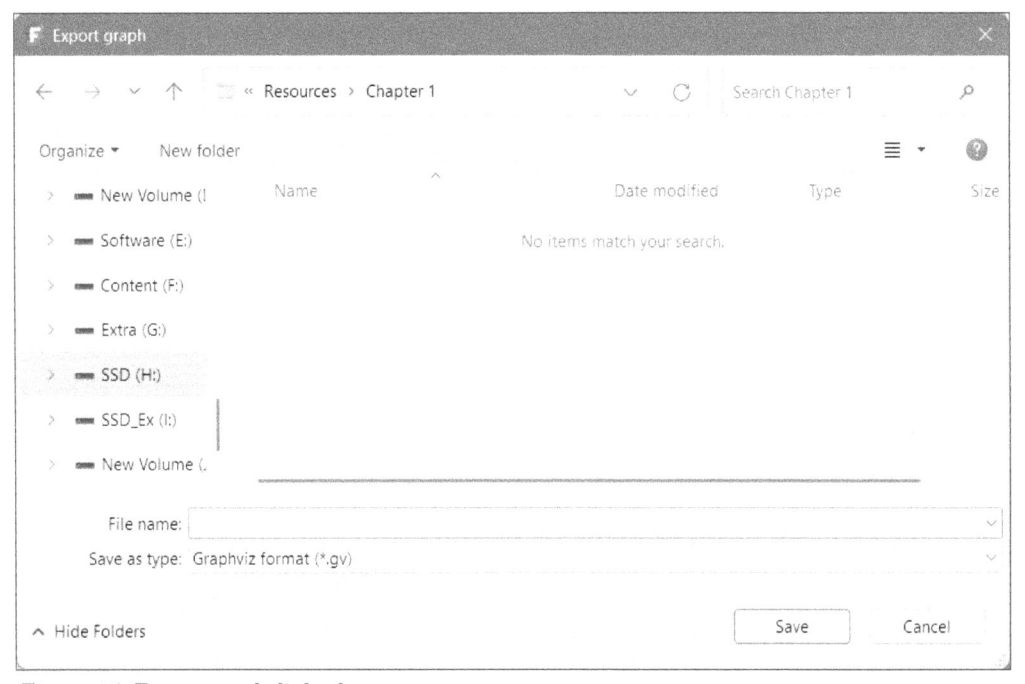

Figure-65. Export graph dialog box

- Specify desired name of the file in the edit box and save the file in desired location.

To view the dependency graph file, open the graphviz file in word processing software Notepad, Word, and so on. The content of file will be written in DOT language; refer to Figure-66. Copy all the content of this file using **CTRL+A** and **CTRL+C** shortcut keys. Visit the link **https://sketchviz.com/new** in your web browser and replace the content on left side in webpage by earlier copied content of file. The dependency graph will be displayed; refer to Figure-67.

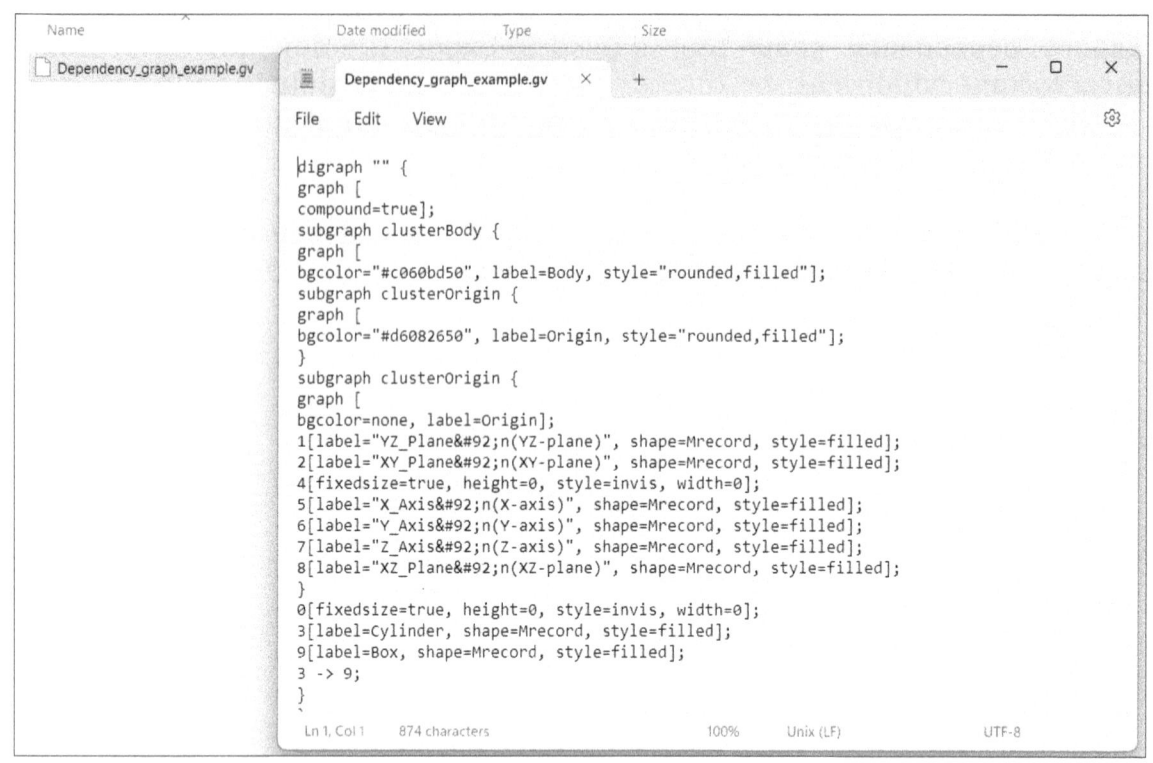

Figure-66. Dependency graph file content

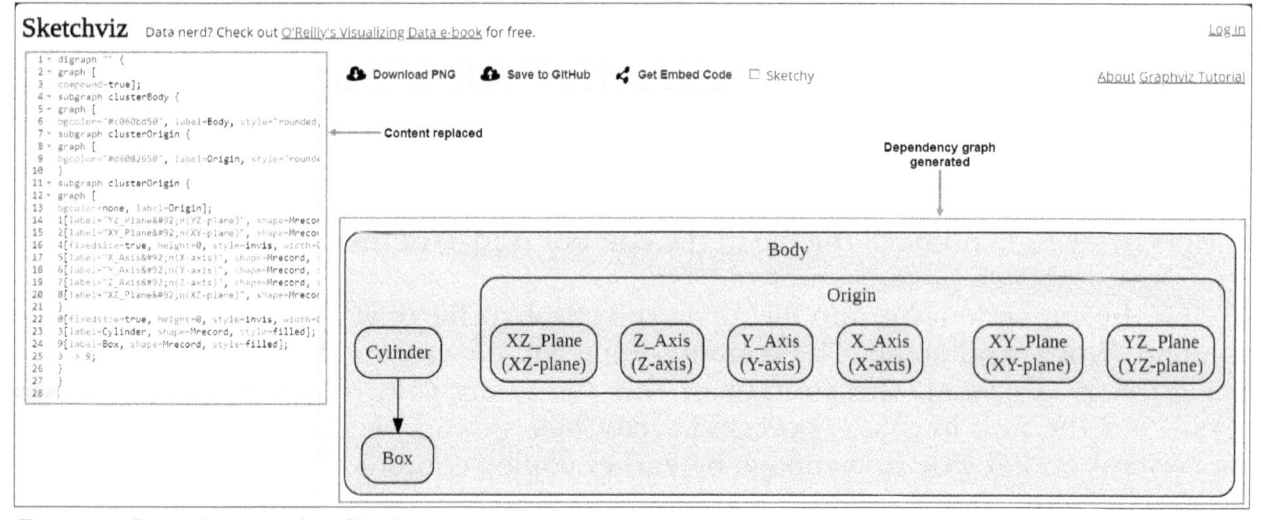

Figure-67. Dependency graph in Sketchviz

Document Utility

The **Document utility** tool assists in repairing damaged project files.

By using this tool, you can extract files from a FreeCAD project file **(*.FCStd)**, which is in fact a **ZIP** file and after manual edits, create a new project file from them.

The procedure to use this tool is discussed next.

- Click on the **Document utility** tool from the **Tools** menu. The **Document utility** dialog box will be displayed; refer to Figure-68.

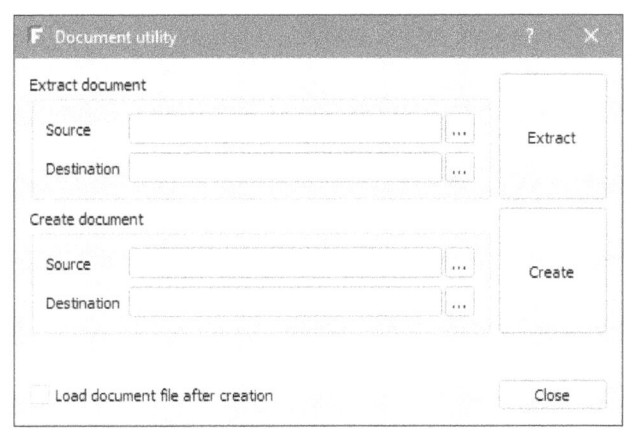

Figure-68. Document utility dialog box

- Click on the **Browse** ... button for **Source** edit box to define location of source file for extracting/creating the project in the **Extract project/Create project** area, respectively.
- Similarly, click on the **Browse** ... button for **Destination** edit box to specify the destination location for extracting/creating the project in the **Extract project/ Create project** area, respectively.
- Click on the **Extract** button to create zip file and click on the **Create** button to save project file.
- Select the **Load project file after creation** check box to open the created file.
- Click on the **Close** button to close the dialog box.

Add text document

The **Add text document** tool creates an object capable of holding arbitrary text. This element can be used to write general information or documentation about the model.

- Click on the **Add text document** tool from the **Tools** menu. A new document file named **Text document** will be opened; refer to Figure-69.

Figure-69. Text document file opened

- Specify desired text in the document file and close the tab after saving the file.

View turntable

The **View turntable** tool continuously rotates the camera in a desired angle and speed for desired time. The procedure to use this tool is discussed next.

- Click on the **View turntable** tool from **Tools** menu. The **View Turntable** dialog box will be displayed; refer to Figure-70.

Figure-70. View Turntable dialog box

- Move the **Angle** slider to specify the angle by which you want to rotate the model.
- Move the **Speed** slider to adjust rotation speed of the model.
- Select the **Fullscreen** check box to switch 3D view to fullscreen mode.
- Select the **Enable timer** check box to specify the duration of rotation.
- Click on **Play/Stop** button to start or stop the rotation of model.
- Click on **Close** button to close the dialog box.

Units converter

The **Units converter** tool is used to convert values from one unit system to another. The procedure to use this tool is discussed next.

- Click on **Units converter** tool from **Tools** menu. The **Units converter** dialog box will be displayed; refer to Figure-71. This dialog box is modeless, it means it can stay open while you continue working in FreeCAD.

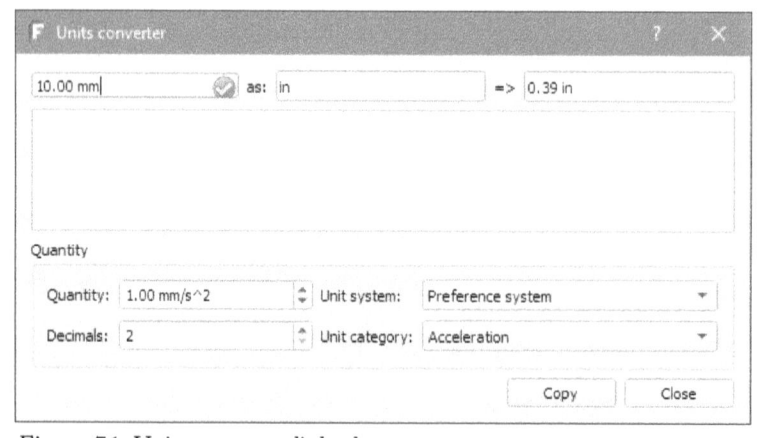

Figure-71. Units converter dialog box

- Specify desired values in the first two edit boxes to perform conversion.
- In first edit box, specify the value to be converted and in the **as:** edit box, specify the unit value to be converted.
- Result will be displayed in **=>** edit box. For example, if you want to convert **55** mm to feet then specify **55** in first edit box and **ft** in the **as:** edit box; refer to Figure-72. You can use in for inch, ft for feet, cm for centimeter, m for meter, and so on to define units of length. Similarly, you can use units of weight, density, speed, current, and so on.

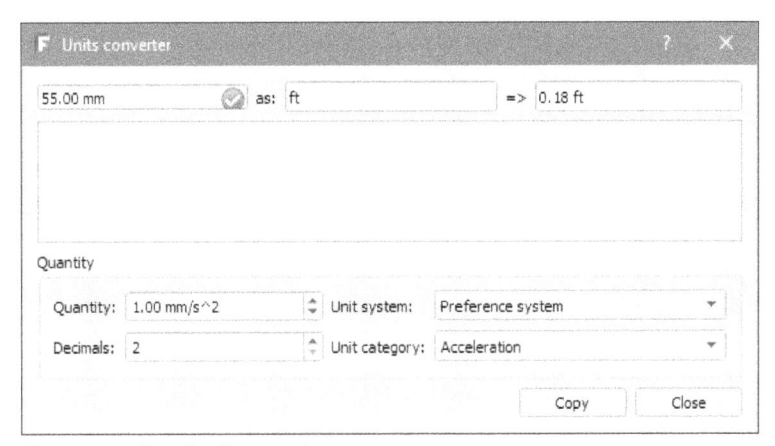

Figure-72. Result of conversion

Customize

The **Customize** tool gives access to several customization options. The customization depends on the workbenches that have been loaded in the current FreeCAD session. The procedure to use this tool is discussed next.

- Click on the **Customize** tool from **Tools** menu. The **Customize** dialog box will be displayed; refer to Figure-73.

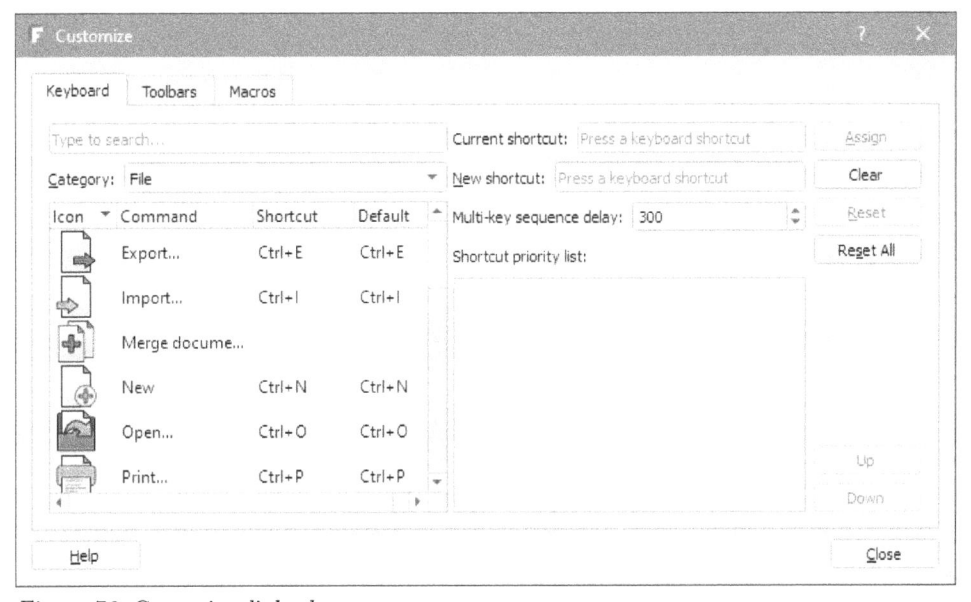

Figure-73. Customize dialog box

The options in various tabs of the **Customize** dialog box are discussed next.

Keyboard tab

Using the options in **Keyboard** tab, you can define custom keyboard shortcuts for using the tool. The **Keyboard** tab of **Customize** dialog box is shown in Figure-73.

- Select desired command category from **Category** drop-down list. Commands of selected category will be displayed in the **Commands** area.
- Select desired command from the **Commands** area of the dialog box.
- The **Current shortcut** input box displays the pre-assigned shortcut for the selected tool.

- To specify a new shortcut for the tool, click in the **New shortcut** input box and press new shortcut keys from the keyboard.
- Specify desired time in milliseconds in the **Multi-Key sequence delay** edit box to wait for the next keystroke of the current key sequence.
- Click on **Clear** button to remove specified shortcut for the selected command.
- Click on **Reset** button to remove a custom shortcut for the selected command.
- Click on **Reset All** button to remove all custom shortcuts specified for the commands.
- Click on the **Close** button to exit the dialog box.

Toolbars tab

In the **Toolbars** tab, you can create custom toolbars and perform modification in toolbars; refer to Figure-74.

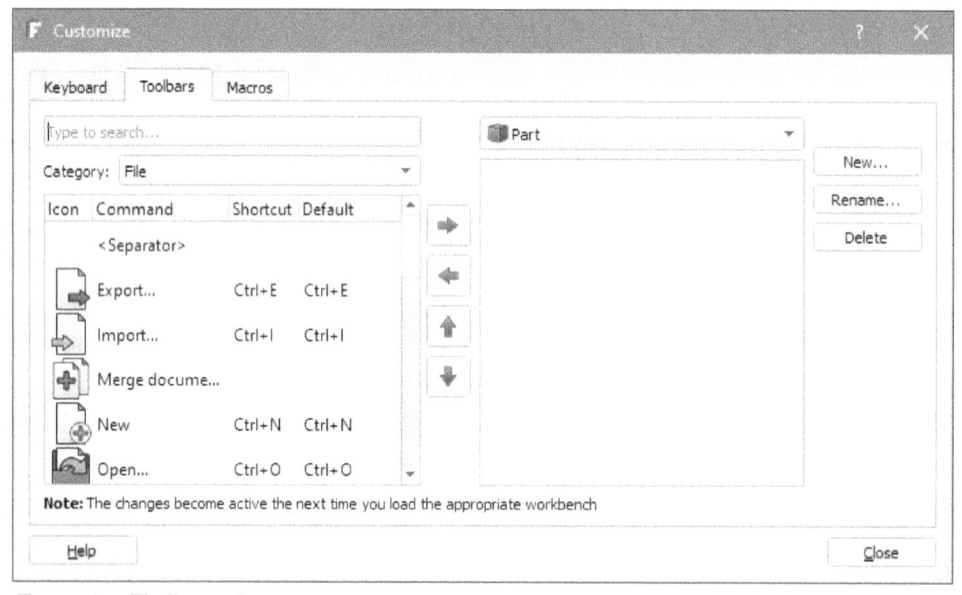

Figure-74. Toolbars tab

- In the drop-down list at right side in the dialog box, select a workbench whose custom toolbar is to be modified.
- Click on the **New** button from right side of the dialog box. The **New toolbar** dialog box will be displayed asking you to specify the toolbar name; refer to Figure-75.

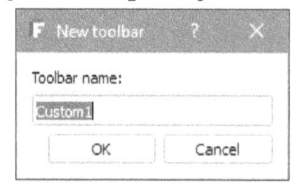

Figure-75. New toolbar dialog box

- Specify desired name in the **Toolbar name** edit box and click on the **OK** button. The new toolbar will be created in the right area of the dialog box.
- Select the newly created toolbar from the right area to change its name and click on **Rename** button, the **Rename toolbar** dialog box will be displayed; refer to Figure-76.

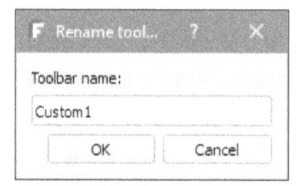

Figure-76. Rename toolbar dialog box

- Specify desired new name of the toolbar and click on **OK** button.
- If you want to delete the toolbar then select the toolbar from right area and click on **Delete** button.
- To disable a toolbar, deselect the check box in front of toolbar name in the right panel. The toolbar will be invisible in the FreeCAD interface.
- To add a command in the toolbar, select the created toolbar in the right area. If no toolbar is selected, the command will be added to the first toolbar available in the drop-down list in right side of the dialog box.
- Select desired command category from the drop-down list available in the left side of the dialog box. The list of related commands will be displayed.
- Select desired command from the left panel which you want to add to the toolbar and click on ▶ button.
- To remove a command, select it from the right panel and click on ◀ button.
- To change a command position, select it and click on ▲ button or ▼ button.
- Click on the **Close** button to exit the dialog box.

Macros tab

In the **Macros** tab, you can manage macros created earlier. FreeCAD uses a dedicated folder for user macros and only macros in that folder can be set up.

- Select desired macro from the **Macro** drop-down list. The parameters related to macro will be displayed; refer to Figure-77.

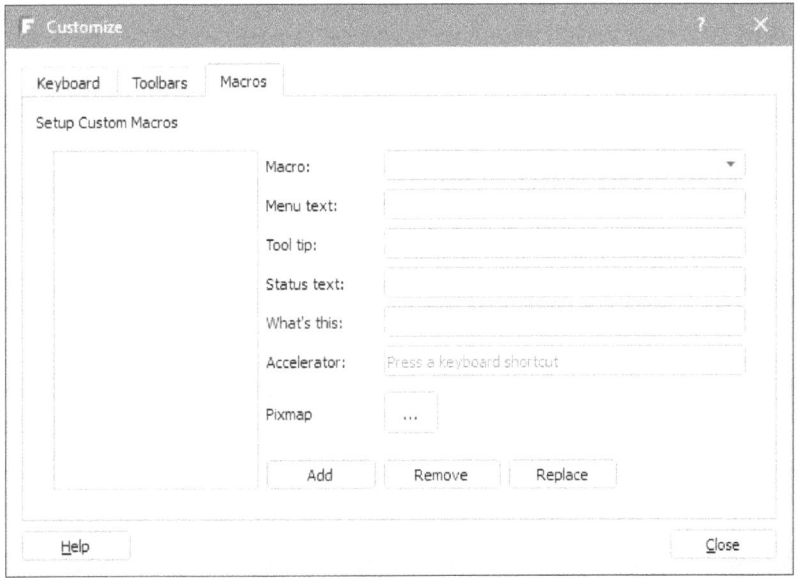

Figure-77. Macros tab

- Select desired macro from the **Macro** drop-down.
- Specify desired name to identify the macro command in the **Menu text** edit box.
- Specify desired tool tip text in the **Tool tip** edit box which will display near the cursor when you hover the cursor on the macro icon.
- Specify desired text in the **Status text** edit box which will display in the status bar when you hover the cursor on the macro icon.
- Specify the wiki page name for the macro if available on the internet in the **What's this** edit box.

- Specify the keyboard shortcut for a macro in the **Accelerator** edit box.
- To add an icon for the macro command, click on **Pixmap** button. The **Choose Icon** dialog box will be displayed; refer to Figure-78.

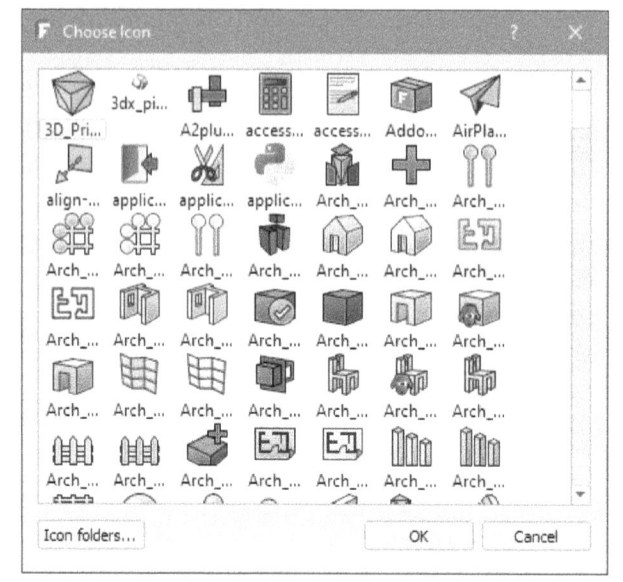

Figure-78. Choose Icon dialog box

- Select desired icon from the panel, the **Choose Icon** dialog box will be closed automatically. Click on **Icon folders** button from the **Choose Icon** dialog box if you want to choose an icon from the local drive.
- Click on the **Add** button from the **Customize** dialog box to add modified macro in the macro list available for document.
- To remove a macro, select it from the left area of the dialog box and click on **Remove** button.
- To change the macro command, double-click on the command from the left panel. Make desired changes and click on **Replace** button.
- Click on the **Close** button to exit the dialog box. You will learn more about macro creation and management later in this book.

Addon Manager

With the **Addon manager** tool, you can install and manage external workbenches and macros provided by the FreeCAD community. Note that you need an internet connection to download the addon resources.

- Click on the **Addon manager** tool from **Tools** menu. The **Addon Manager** dialog box will be displayed; refer to Figure-79.

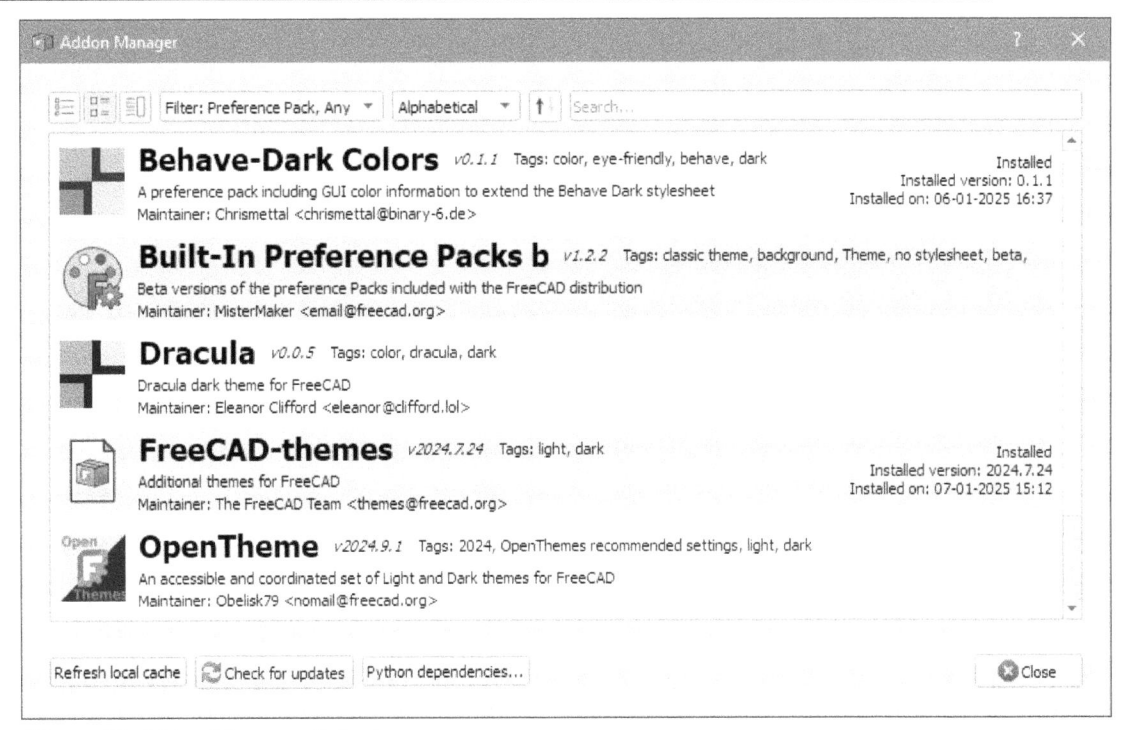

Figure-79. Addon Manager dialog box

- Select desired addon which you want to install from the list available in **Workbenches** tab.
- Click on **Install/update selected** button to install the addon.
- To uninstall the addon, select the addon from the list which you want to uninstall and click on **Uninstall selected** button from the dialog box.
- If you have installed or updated a workbench then by clicking on **Close** button, a new **Addon manager** dialog box will open informing you to restart FreeCAD for the changes to take effect.
- Click on **OK** button from the dialog box and restart FreeCAD.

Measure

The **Measure** tool is used to measure a feature. The procedure to use this tool will be discussed in **Chapter 4**.

Navigating in the 3d view

Navigating in the FreeCAD 3D view can be done with a mouse, a Space Navigator device, the keyboard, a touchpad, or a combination of these. FreeCAD can use several navigation modes which determine how the three basic view manipulation operations (pan, rotate, and zoom) are done as well as how to select objects on the screen. Navigation modes are accessed from the **Preferences screen** or directly by right-clicking anywhere in the graphic area; refer to Figure-80.

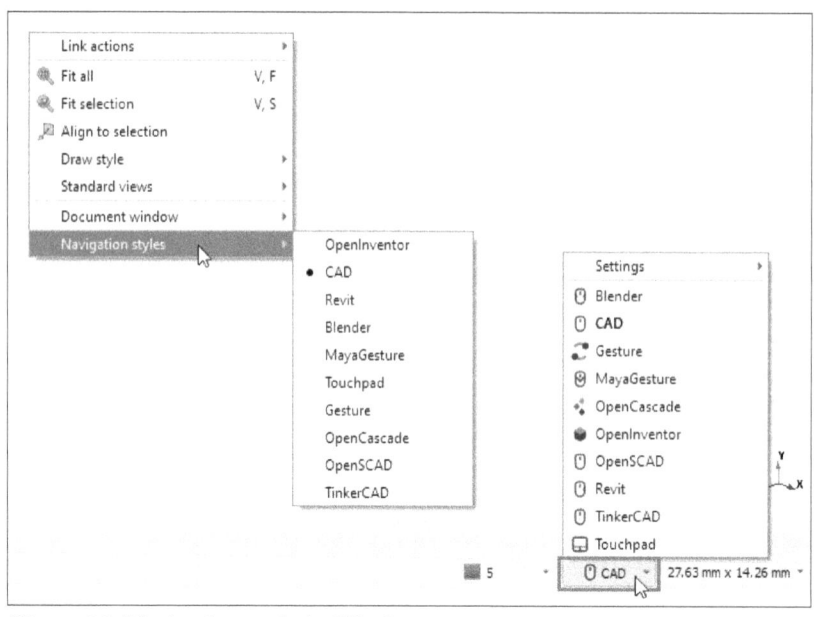

Figure-80. Navigation modes in 3D view

Each of these modes attribute different mouse buttons, mouse+keyboard combinations, or mouse gestures to these four operations. The procedure to use all these modes is discussed next.

CAD Navigation

This is the default navigation style. It allows the user a simple control of the view and does not require the use of keyboard keys except to make multi-selections. The mouse gestures used for object manipulation in **CAD** navigation style are shown in Figure-81.

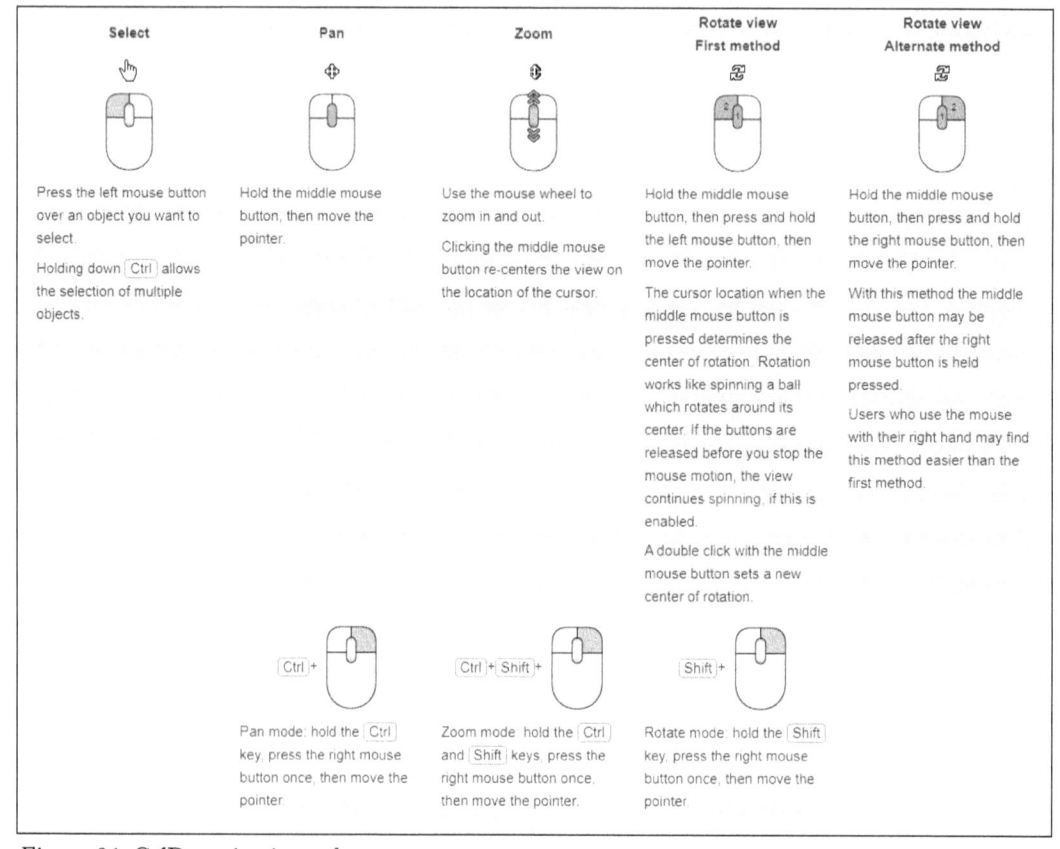

Figure-81. CAD navigation style

OpenInventor Navigation

In **OpenInventor** navigation, you must hold down the **Ctrl** key in order to select objects. The procedure to use this navigation style is shown in Figure-82.

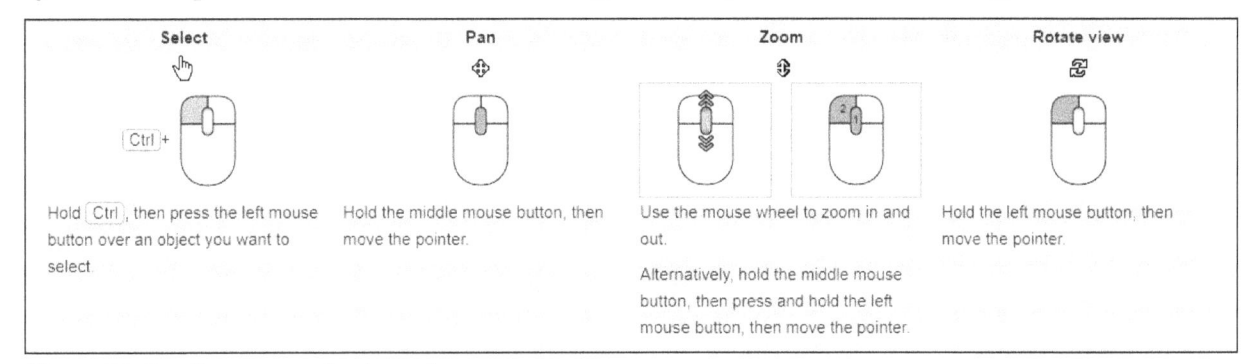

Figure-82. OpenInventor navigation style

Revit Navigation

The procedure to use **Revit** navigation style is shown in Figure-83.

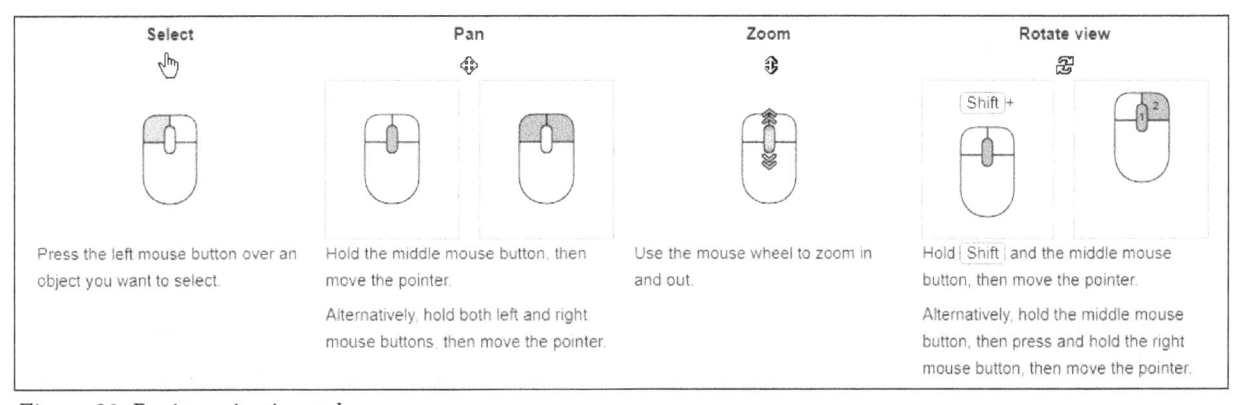

Figure-83. Revit navigation style

Blender Navigation

The procedure to use **Blender** navigation style is shown in Figure-84.

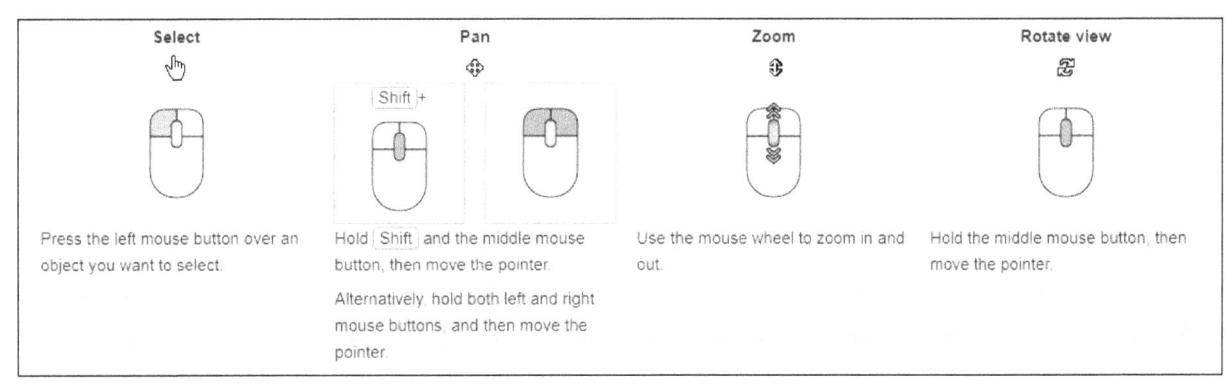

Figure-84. Blender navigation style

MayaGesture Navigation

In **Maya-Gesture** navigation, panning, zooming, and rotating the view require the **Alt** key together with a mouse button; therefore, a three-button mouse is required. The procedure to use this navigation is shown in Figure-85.

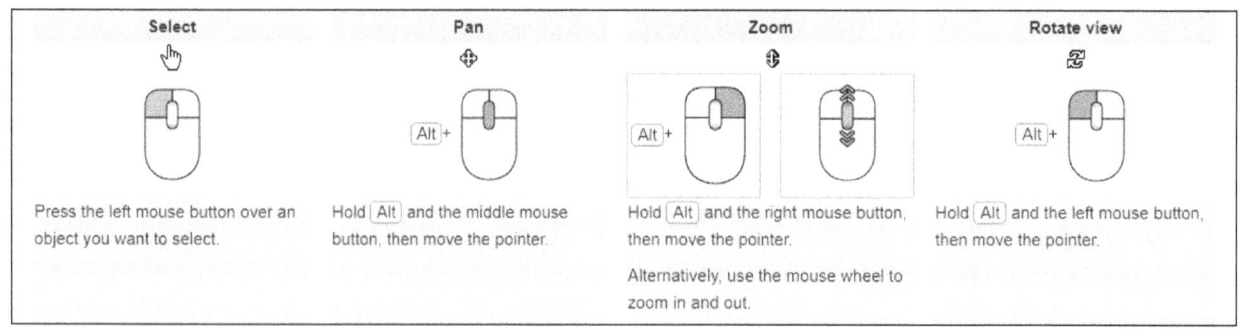

Figure-85. MayaGesture navigation style

Touchpad Navigation

In **Touchpad** navigation, panning, zooming, and rotating the view require a modifier key together with the touchpad. The procedure to use this navigation is shown in Figure-86.

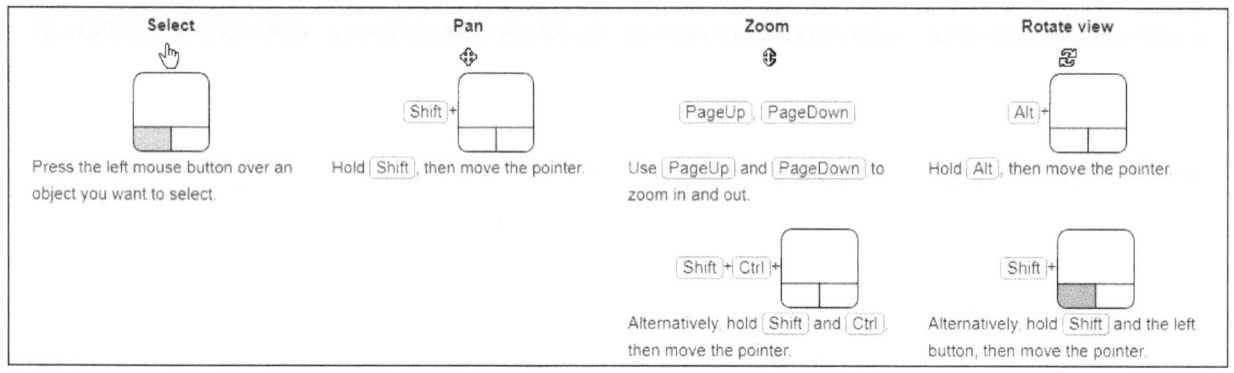

Figure-86. Touchpad navigation style

Gesture Navigation

The procedure to use **Gesture** navigation style is shown in Figure-87.

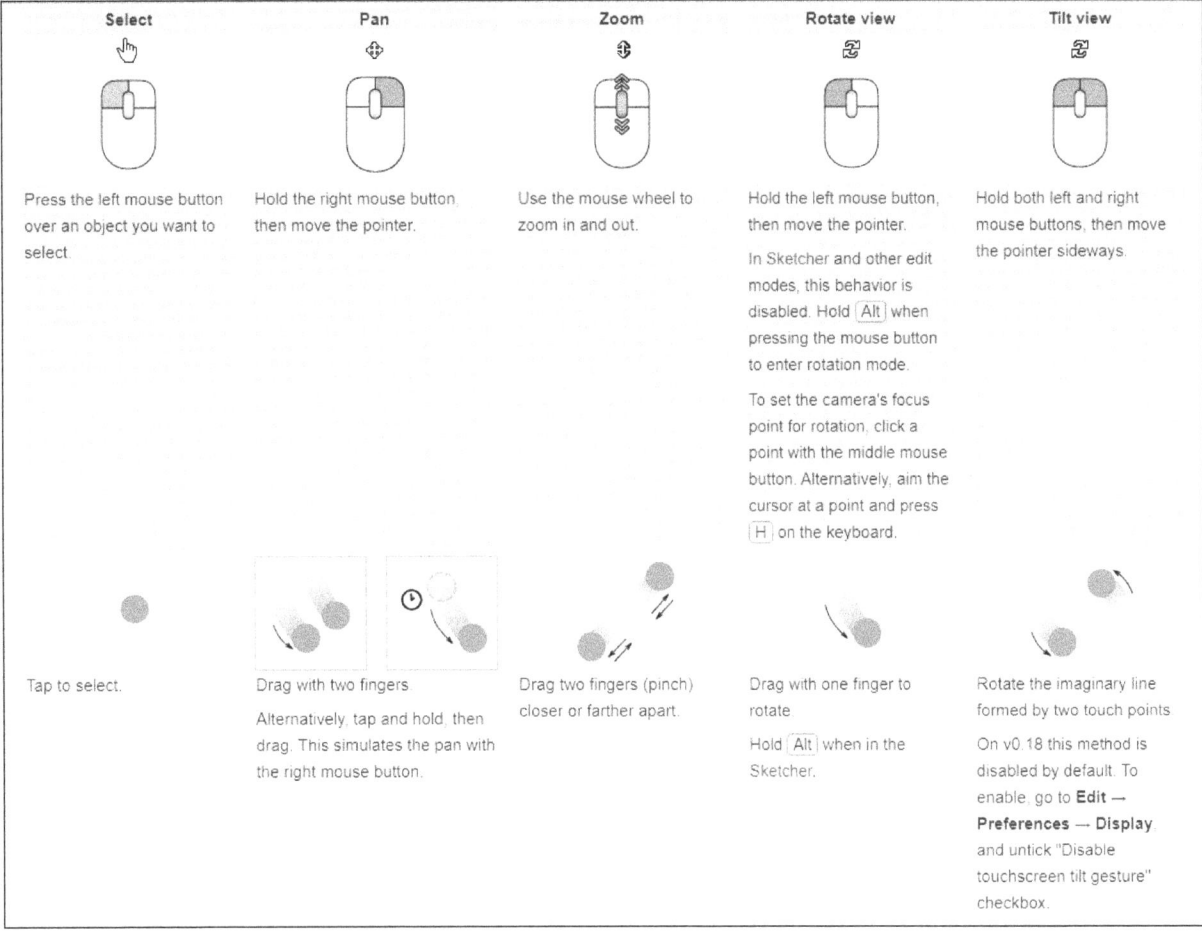

Select	Pan	Zoom	Rotate view	Tilt view
Press the left mouse button over an object you want to select.	Hold the right mouse button, then move the pointer.	Use the mouse wheel to zoom in and out.	Hold the left mouse button, then move the pointer. In Sketcher and other edit modes, this behavior is disabled. Hold [Alt] when pressing the mouse button to enter rotation mode. To set the camera's focus point for rotation, click a point with the middle mouse button. Alternatively, aim the cursor at a point and press [H] on the keyboard.	Hold both left and right mouse buttons, then move the pointer sideways.
Tap to select.	Drag with two fingers. Alternatively, tap and hold, then drag. This simulates the pan with the right mouse button.	Drag two fingers (pinch) closer or farther apart.	Drag with one finger to rotate. Hold [Alt] when in the Sketcher.	Rotate the imaginary line formed by two touch points. On v0.18 this method is disabled by default. To enable, go to **Edit → Preferences → Display** and untick "Disable touchscreen tilt gesture" checkbox.

Figure-87. Gesture navigation style

OpenCascade Navigation

The procedure to use **OpenCascade** navigation is shown in Figure-88.

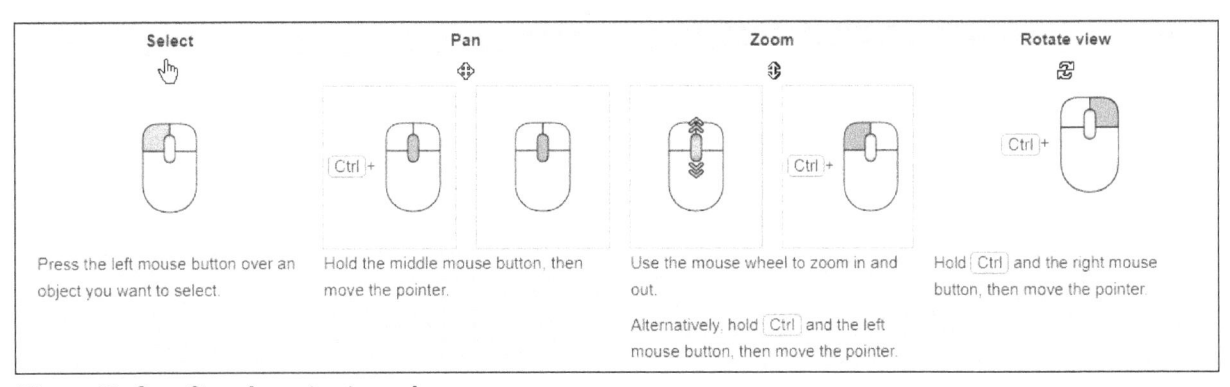

Select	Pan	Zoom	Rotate view
Press the left mouse button over an object you want to select.	Hold the middle mouse button, then move the pointer. [Ctrl]+	Use the mouse wheel to zoom in and out. Alternatively, hold [Ctrl] and the left mouse button, then move the pointer. [Ctrl]+	Hold [Ctrl] and the right mouse button, then move the pointer. [Ctrl]+

Figure-88. OpenCascade navigation style

SELF ASSESSMENT

Q1. What is the primary function of the New tool in FreeCAD?

A) Open an existing file
B) Initiate a new document file
C) Save the current file
D) Close the current file

Q2. Which keyboard shortcut is used to open a file in FreeCAD?

A) Ctrl+N
B) Ctrl+O
C) Ctrl+S
D) Ctrl+F4

Q3. What happens when you use the Close tool?

A) It closes the FreeCAD application.
B) It closes the current file but keeps the application open.
C) It saves the current file.
D) It opens a new document.

Q4. What does the Revert tool do in FreeCAD?

A) Saves the current document.
B) Closes the active document and reopens the last saved version.
C) Exports the document to a different format.
D) Imports a new file.

Q5. What function does the Print Preview tool serve?

A) It saves the model as a PDF.
B) It allows users to check the print layout before sending it to the printer.
C) It closes the print dialog.
D) It prints the model directly without previewing.

Q6. How can you restore a previously saved frozen view?

A) Use the Load views tool.
B) Use the Freeze view tool.
C) Use the Save views tool.
D) Click on the Clear views tool.

Q7. What is the purpose of the Save All tool?

A) Save the active document only.
B) Save all currently open documents in the application.
C) Save a copy of the current document.
D) Close all open documents.

Q8. Which keyboard shortcut is used to save the current document in FreeCAD?

A) Ctrl+N
B) Ctrl+O
C) Ctrl+S
D) Ctrl+F4

Q9. What does the Merge project tool do?

A) Closes the current document.
B) Imports a file into the active document.
C) Saves the current project.
D) Creates a new project.

Q10. Which of the following tools displays general information about the current project?

A) Document information
B) Print Preview
C) Export
D) Revert

Q11. To switch the camera to perspective view mode, which tool would you use?

A) Orthographic view
B) Perspective view
C) Create new view
D) Fit all

Q12. What is the effect of the Clear views tool in FreeCAD?

A) It saves the current view settings.
B) It deletes all existing frozen views.
C) It opens a new view.
D) It restores the last frozen view.

Q13. Which tool would you use to save the active document in PDF format?

A) Export PDF
B) Print
C) Save As
D) Export

Q14. Which feature allows you to specify parameters that control FreeCAD's behavior?

A) Edit parameters
B) Document information
C) Print Preview
D) Merge project

Q15. Which tool is primarily used to view the model before printing?

A) Print
B) Save
C) Print Preview
D) Export

Q16. How can you align the view to selected objects in FreeCAD?

A) Fit all
B) Align to selection
C) Rotate Left
D) Standard Views

Q17. What is the purpose of the Save Image tool in FreeCAD?

A) To save the current document.
B) To create an image file or screenshot from the active 3D view.
C) To export a model to a different format.
D) To load previously saved images.

Q18. What does the Load Image tool do?

A) Creates a new image file.
B) Loads earlier saved image files into the graphics area.
C) Saves the current view as an image.
D) Exports the current model as an image.

Q19. What does the Scene Inspector tool display in FreeCAD?

A) The dependencies between objects.
B) An overview of all nodes in the scene graph.
C) A list of saved images.
D) The dimensions of the model.

Q20. Which external software is required to use the Dependency Graph tool?

A) AutoCAD
B) Graphviz
C) Microsoft Excel
D) Adobe Photoshop

Q21. What is the function of the Document Utility tool in FreeCAD?

A) To create new project files from templates.
B) To assist in repairing damaged project files.
C) To export project files in different formats.
D) To merge multiple project files into one.

Q22. How do you create a distance object in FreeCAD?

A) By selecting the Measure distance tool and then choosing two points.
B) By typing in the desired distance in the text field.
C) By clicking on the Add text document tool.
D) By using the Units calculator tool.

Q23. Which tool allows you to rotate the camera at a specified angle and speed in FreeCAD?

A) View Turntable
B) Scene Inspector
C) Dependency Graph
D) Save Image

Q24. What does the Units Calculator tool in FreeCAD do?

A) It converts values from one unit system to another.
B) It measures distances between objects.
C) It saves documents in different formats.
D) It creates new project files.

Q25. In the Customize tool, which tab allows you to manage keyboard shortcuts?

A) Toolbars tab
B) Macros tab
C) Keyboard tab
D) Addon Manager tab

Q26. How can you add a new toolbar in the Customize tool?

A) By clicking on the Add button.
B) By clicking on the New button and specifying a name.
C) By selecting a workbench from the drop-down list.
D) By clicking on the Rename button.

Q27. What does the Addon Manager tool allow you to do?

A) Create new 3D models.
B) Install and manage external workbenches and macros.
C) Customize keyboard shortcuts.
D) Repair damaged project files.

Q28. What happens when you click the Refresh button in the Scene Inspector tool?

A) It saves the current scene.
B) It updates the overview of nodes in the scene graph.
C) It closes the Scene Inspector dialog box.
D) It loads a new image.

Q29. In the Save Image dialog box, which option allows you to add comments to the image file?

A) Insert MIBA
B) Add watermark
C) Insert comment
D) Creation method

Q30. Which statement is true regarding the Document Utility tool?

A) It can directly edit project files without extraction.
B) It assists in creating new project files only.
C) It can extract files from damaged project files for repair.
D) It is used to load images into the workspace.

Q31. What action is performed by the Measure Distance tool in FreeCAD?

A) It creates a graphical representation of the model.
B) It calculates and displays the distance between two selected points.
C) It adjusts the size of the model.
D) It saves the measurement to a file.

Q32. What feature does the View Turntable tool provide?

A) It displays 2D views of the model.
B) It allows continuous rotation of the camera at a specified speed and angle.
C) It saves the current view as an image file.
D) It measures distances within the model.

Q33. What do you need to do to convert units using the Units Calculator tool?

A) Select a pre-defined unit type from a list.
B) Input the value to convert and the target unit type.
C) Draw a shape first before using the calculator.
D) Specify the color of the units.

Q34. In the Customize dialog box, which tab is used for managing macros?

A) Toolbars tab
B) Keyboard tab
C) Macros tab
D) Addon Manager tab

Q35. How do you delete a custom toolbar in FreeCAD?

A) Click on the Delete button after selecting the toolbar.
B) Right-click the toolbar and select Delete.
C) Drag the toolbar out of the interface.
D) Use the Reset All button in the Customize tool.

Q36. What is the primary function of the Addon Manager tool in FreeCAD?

A) To create new 3D objects.
B) To install and manage external add-ons for additional functionality.
C) To repair corrupted files.
D) To customize keyboard shortcuts.

Q37. What does the Enable timer checkbox do in the View Turntable tool?

A) It allows you to specify how long the rotation will occur.
B) It sets the angle of rotation.
C) It changes the rotation speed.
D) It pauses the rotation.

Q38. Which navigation method can be used in FreeCAD's 3D view?

A) Only mouse navigation
B) Mouse, Space Navigator, keyboard, and touchpad
C) Touchpad only
D) Keyboard shortcuts only

Q39. What is the first step to use the Dependency Graph tool?

A) Install Graphviz.
B) Click on the Dependency Graph tool from the Tools menu.
C) Save the current project.
D) Open the Scene Inspector.

Q40. What is the function of the Load Image tool in FreeCAD?

A) To save the current view as an image
B) To load previously saved image files into the graphics area
C) To create a new image file from the current view
D) To convert image formats

For Students Notes

Chapter 2

Sketching

Topics Covered

The major topics covered in this chapter are:

- *Introduction*
- *Starting Sketch*
- *Sketch Creation Tools*
- *Sketch Editing Tools*
- *Sketcher Constraints*
- *Sketcher Tools*
- *Sketcher B-Spline Tools*
- *Sketching Practical and Practice*

INTRODUCTION

In Engineering, sketches are based on real dimensions of real world objects. These sketches work as building blocks for various 3D operations. In this chapter, we will be working with sketch entities like; Line, Circle, Arc, Polygon, Ellipse, and so on to form base feature for various 3D operations. Note that the sketching environment is the base of 3D models so you should be proficient in sketching.

In this chapter, we will be working in **Sketcher Workbench**. We will learn about various tools of toolbar used in 2D sketching.

STARTING SKETCH

- To start a new sketch file, click on the **New** button from **File** menu and select **Sketcher** workbench from **Switch between workbenches** drop-down in the **Toolbar**; refer to Figure-1. The **Sketcher** toolbar will be displayed; refer to Figure-2.

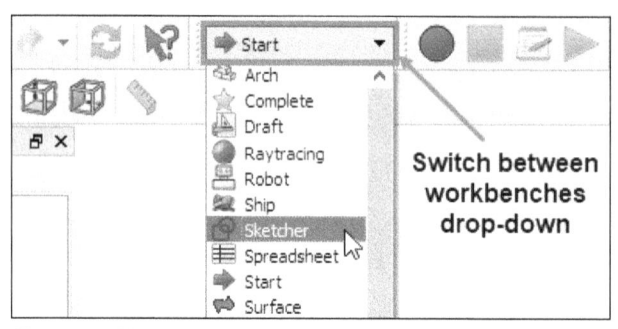

Figure-1. Sketcher workbench

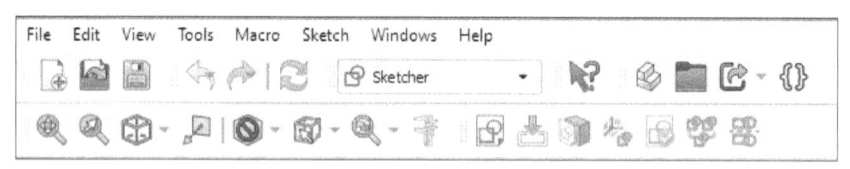

Figure-2. Sketcher toolbar

- Click on the **Create sketch** tool from the toolbar. The **Choose orientation** dialog box will be displayed asking you to select the plane; refer to Figure-3.

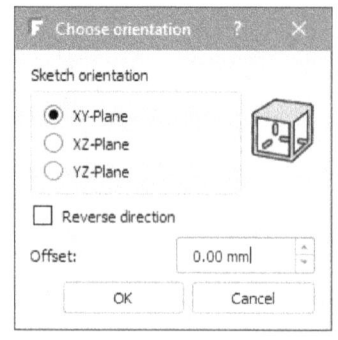

Figure-3. Choose orientation dialog box

- Select desired radio button from **Sketch orientation** area to define which plane will be used as sketching plane.
- Select the **Reverse direction** check box to reverse the plane direction.
- Specify the offset distance for plane in **Offset** edit box. The plane for sketching will be at specified offset distance normal to selected base plane.

- Click on the **OK** button to close the dialog box. The selected plane will become parallel to the screen and act as current sketching plane. Also, the tools to create 2D sketching will be displayed in the toolbar; refer to Figure-4. Now, we are ready to draw sketch on the selected plane.

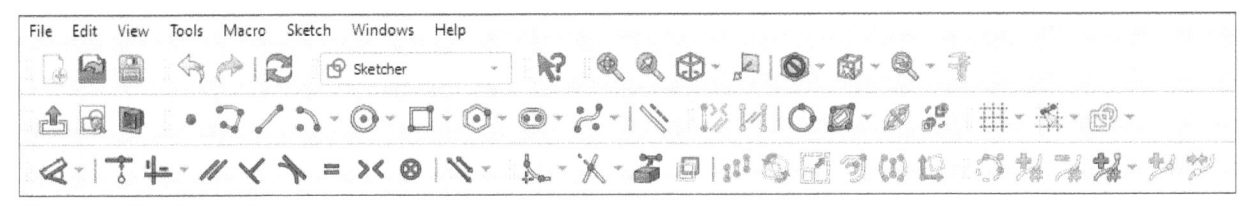

Figure-4. Sketching tools

First, we will start with the sketch creation tools and later we will discuss the other tools.

SKETCH CREATION TOOLS

After selecting the plane, all the sketch creation tools will become active. These tools are discussed next.

Point

The **Point** tool creates a point in the current sketch which can be used for constructing geometry elements. The procedure to use this tool is discussed next.

- Click on **Create point** tool from **Toolbar** in the **Sketcher** workbench; refer to Figure-5. You will be asked to specify location to the point.

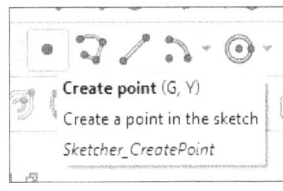

Figure-5. Point tool

- Click at desired locations in the 3D view area. The points will be created; refer to Figure-6.

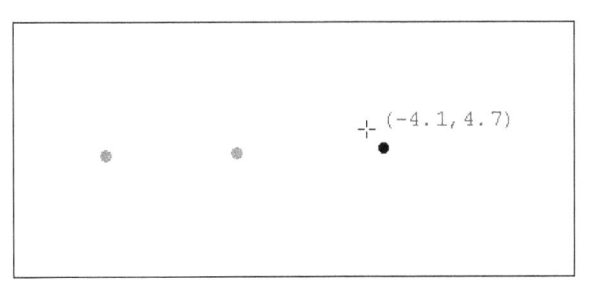

Figure-6. Creation of point

- Press **ESC** key from keyboard or press **RMB** to exit the tool.

Polyline

The **Create polyline** tool is used to create continuous line and arc segments connected by their vertices. The polyline tool has multiple modes that can be toggled with the **M** key.

For example, you can draw tangent or perpendicular arcs following a line or arc segment. The procedure to use this tool is discussed next.

- Click on the **Create polyline** tool from **Toolbar** in the **Sketcher** workbench; refer to Figure-7. You will be asked to create a line segment.

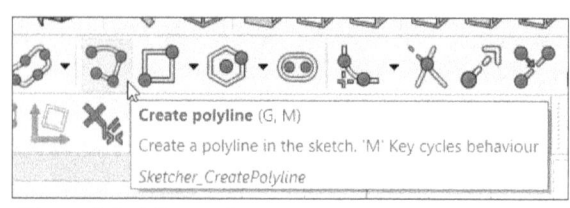

Figure-7. Polyline tool

- Click in the 3D view area and create a line segment. Press **M** key from the keyboard to toggle different modes of creating polyline; refer to Figure-8.
- After creating a line segment, press **M** key for one time to create a line perpendicular to the previous entity.
- Press **M** key for two times to create a line tangential to the previous entity.
- Press **M** key for three times to create an arc tangential to the previous entity.
- Press **M** key for four times to create an arc perpendicular to the left of the previous entity.
- Press **M** key for five times to create an arc perpendicular to the right of the previous entity.
- Press **M** key for six times to create a line coincident to the previous entity as in the normal state.
- Press **ESC** key or press **RMB** to exit the creation and continue to create a new line.
- Press **ESC** key or press **RMB** again to exit the tool.

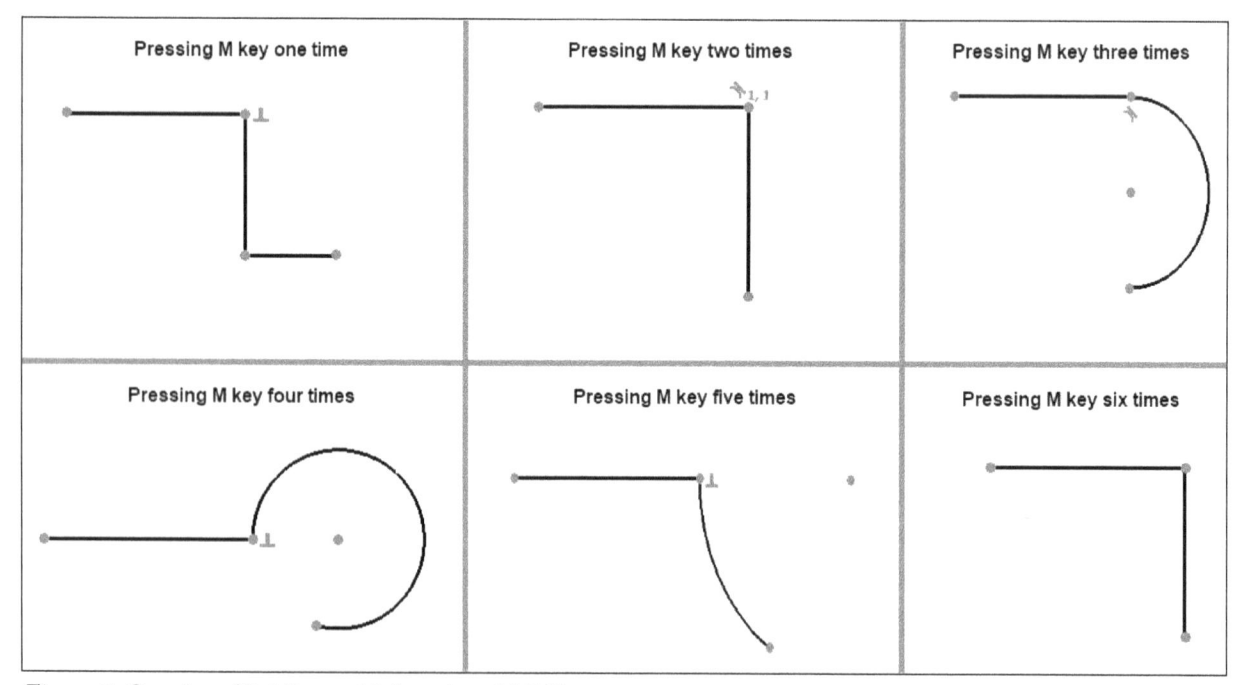

Figure-8. Creation of Polyline multiple modes with M key

Line

The **Line** tool is used to create a line by picking two points in the 3D view. The procedure to use this tool is discussed next.

- Click on **Create line** tool from **Toolbar** in **Sketcher** workbench; refer to Figure-9. The line dialog box will be displayed; refer to Figure-10. You can also press **GL** key from keyboard to select the **Line** tool. You will be asked to specify starting point of the line.

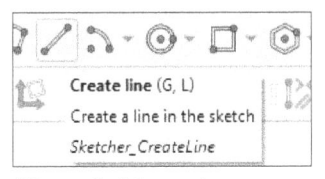

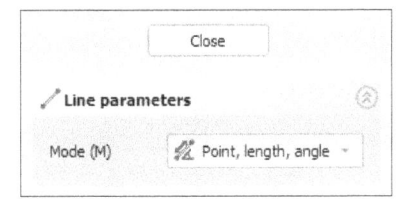

Figure-9. Line tool

Figure-10. Line parameters dialog box

- Select **Point, length, angle** option from the **Mode** drop-down to create a line. You will be asked to specify starting point of the line.
- Click at desired location to specify the start point and move the cursor away in desired direction in which you want to create the line. You will be asked to specify the end point or length and angle of the line.
- Click at desired location to specify the end point of the line or enter desired length and angle of line in their respective edit boxes in the 3D view area. Note that you can switch between the edit boxes by using **Tab** key from the keyboard.
- After specifying desired parameters, press **Enter** key from the keyboard. The line will be created; refer to Figure-11.

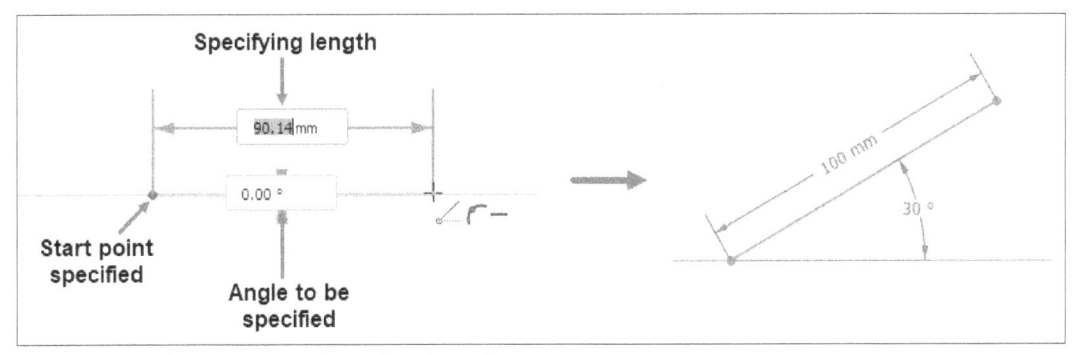

Figure-11. Line created using Point length angle option

- Similarly, you can create line by using the **Point, width, height** and the **2 points** options.
- If you want to create a line for construction purpose then select the line you have created and click **RMB** on an empty area of the 3D view, a shortcut menu will be displayed.
- Click on the **Toggle construction geometry** option from the shortcut menu, the normal line will become construction line and vice-versa; refer to Figure-12. Note that construction geometry is displayed in blue color.

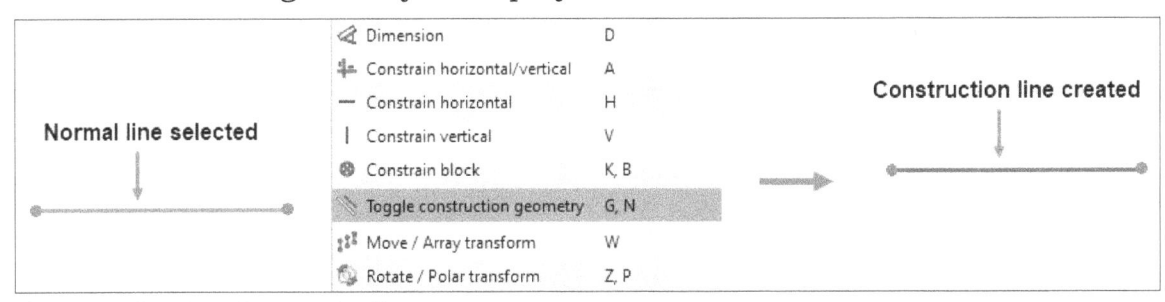

Figure-12. Creation of construction line

- Press **ESC** key or press **RMB** to exit the tool.

Arc

The tools available in **Create arc** drop-down of **Toolbar** in the **Sketcher** workbench are used to create an arc using three points. There are five tools in **Create arc** drop-down **Center arc by center, Create arc by 3 points, Create arc of ellipse, Create arc of hyperbola,** and **Create arc of parabola**; refer to Figure-13. The procedures to use these tools are discussed next.

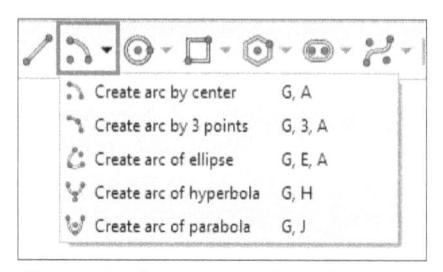

Figure-13. Create an arc drop-down

Create arc by center

The **Create arc by center** tool is used to create an arc by its center and its endpoints. The procedure to use this tool is discussed next.

- Click on the **Create arc by center** tool from the **Create arc** drop-down or press **GA** from keyboard. The **Arc parameters** dialog box will be displayed with the **Center** option selected in the **Mode** drop-down; refer to Figure-14 and you will be asked to specify the center point of arc.

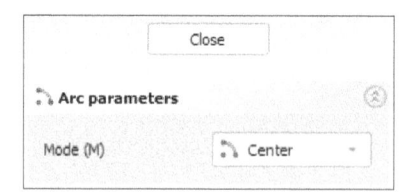

Figure-14. Arc parameters dialog box

- Click at desired location to specify the center point of arc.
- Move the cursor away at desired direction and click to specify the starting point of arc or you can also specify desired radius value in the edit box displayed and then click to define the starting point of arc. You will be asked to specify the end point or aperture angle of the arc.
- Move the cursor away and click at desired location to specify the end point of arc or you can also specify desired aperture angle in the edit box displayed and then click to define the end point of arc. The arc will be created; refer to Figure-15.

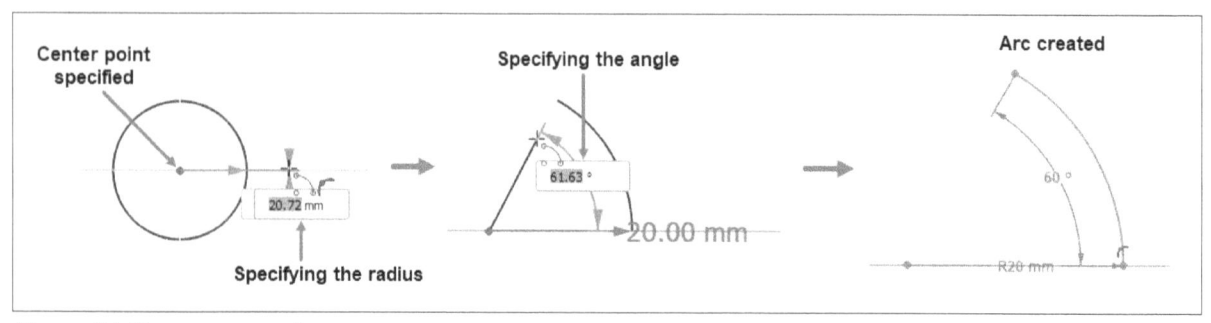

Figure-15. Center arc created

- Similarly, you can create an arc using **3 rim points** option from the **Mode** drop-down; refer to Figure-16.

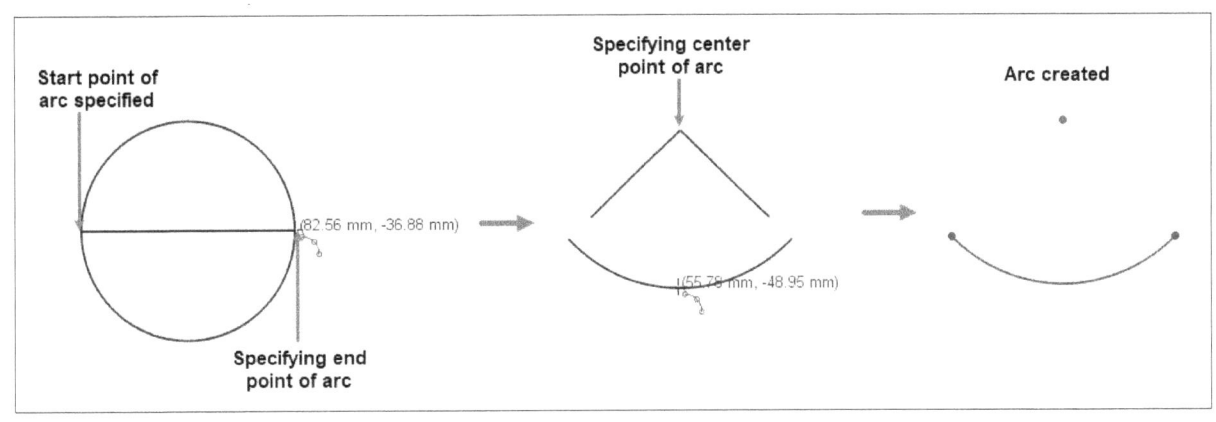

Figure-16. 3 rim points arc created

- Press **ESC** key or press **RMB** to exit the tool.

Create arc by 3 points

The **Create arc by 3 points** tool is used to create an arc by its endpoints and a point along the arc. The procedure to use this tool is same as discussed in previous tool; refer to Figure-17.

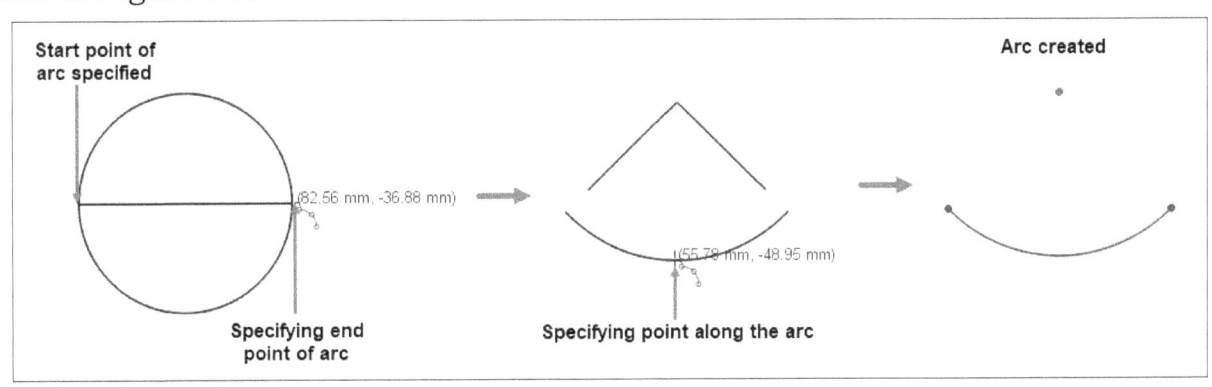

Figure-17. Arc by 3 points created

Create arc of ellipse

The **Create arc of ellipse** tool is used to create an arc of ellipse using four points (the center, the end of major radius, the start point, and the end point). The procedure to use this tool is discussed next.

- Click on the **Create arc of ellipse** tool from the **Create arc** drop-down or press **GE** from keyboard. You will be asked to specify center point of the arc.
- Click in the 3D view area to specify center point of the arc; refer to Figure-18. You will be asked to specify major radius of ellipse.

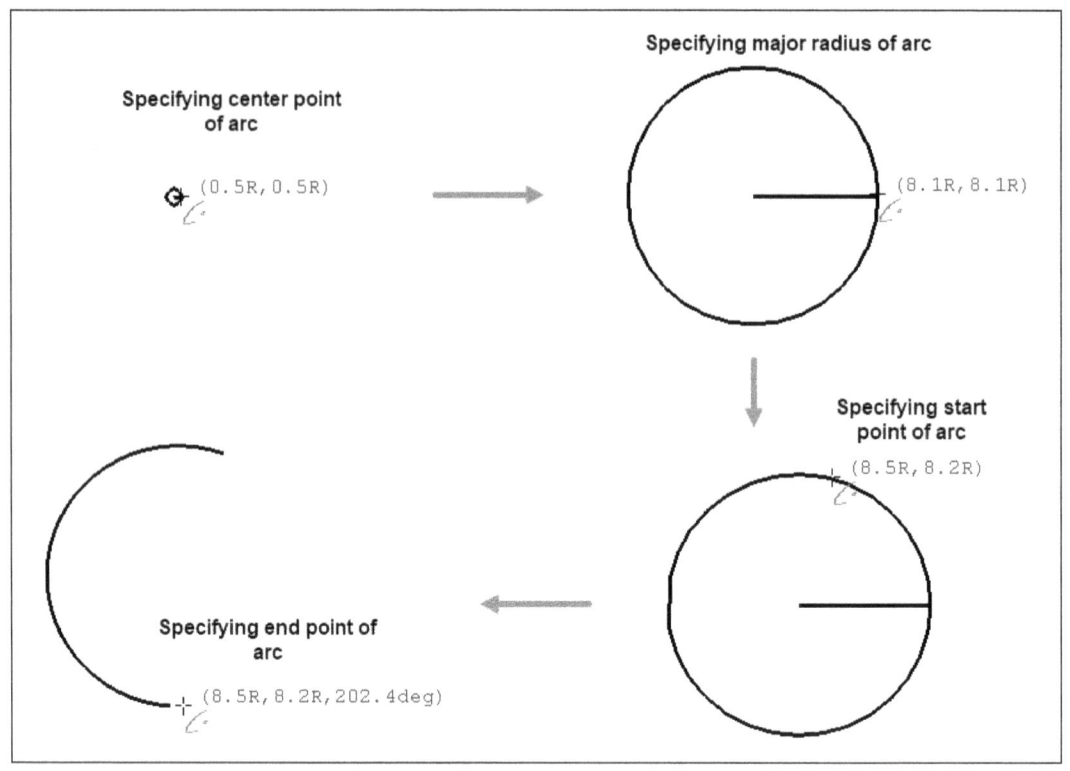

Figure-18. Creation of arc of ellipse

- Move the cursor away from the center point and click to specify major radius of the arc. You will be asked to specify start point of arc.
- Click on desired location to specify starting point and then end point of the arc. The arc of ellipse will be created.
- Press **ESC** key or press **RMB** to exit the tool.

Create arc of hyperbola

The **Create arc of hyperbola** tool is used to create an arc of hyperbola using four points (the center point, the major radius of arc, the start point, and the end point of arc). The procedure to use this tool is discussed next.

- Click on the **Create arc of hyperbola** tool from the **Create arc** drop-down or press **GH** from keyboard. You will be asked to specify center point of arc.
- Click in the 3D view area to specify center point of arc; refer to Figure-19. You will be asked to specify major radius of arc.

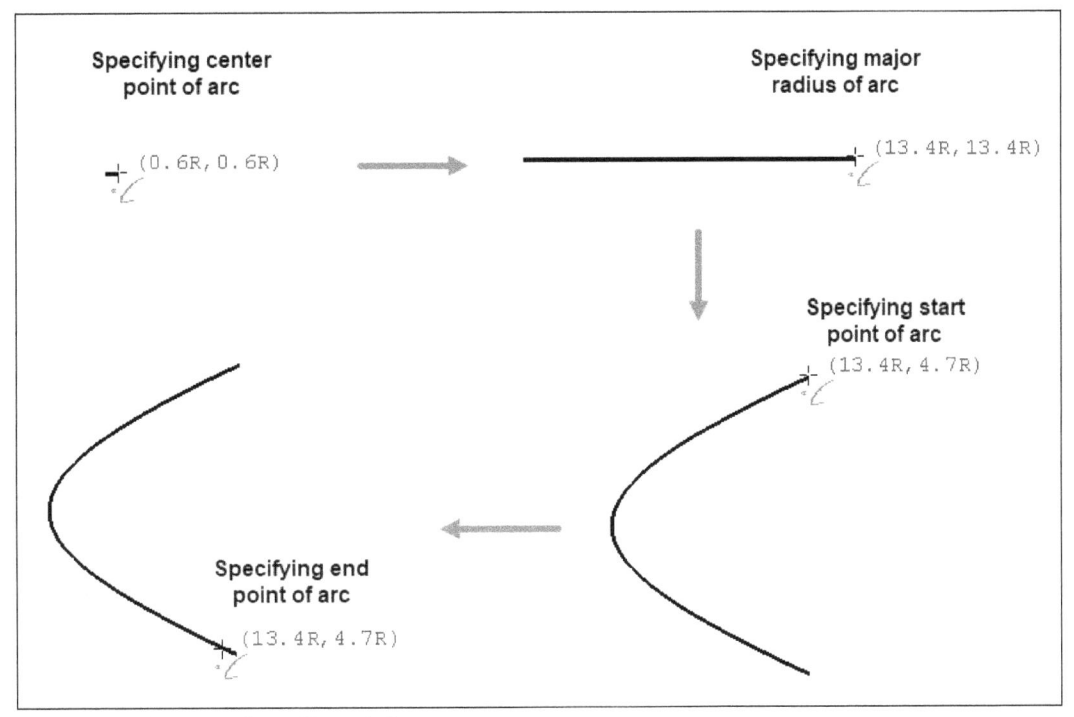

Figure-19. Creation of arc of hyperbola

- Move the cursor away from the center point and click to specify major radius of the arc. You will be asked to specify starting point of arc.
- Click on desired location to specify starting point of arc and then the end point of arc. The arc of hyperbola will be created.
- Press **ESC** key or press **RMB** to exit the tool.

Create arc of parabola

The **Create arc of parabola** tool is used to create an arc of parabola using four points (the focus of arc, the vertex of arc, the start point, and the end point of arc). The procedure to use this tool is discussed next.

- Click on the **Create arc of parabola** tool from the **Create arc** drop-down or press **GP** from keyboard. You will be asked to specify focus of the arc.
- Click in the 3D view area to specify focus point of arc; refer to Figure-20.

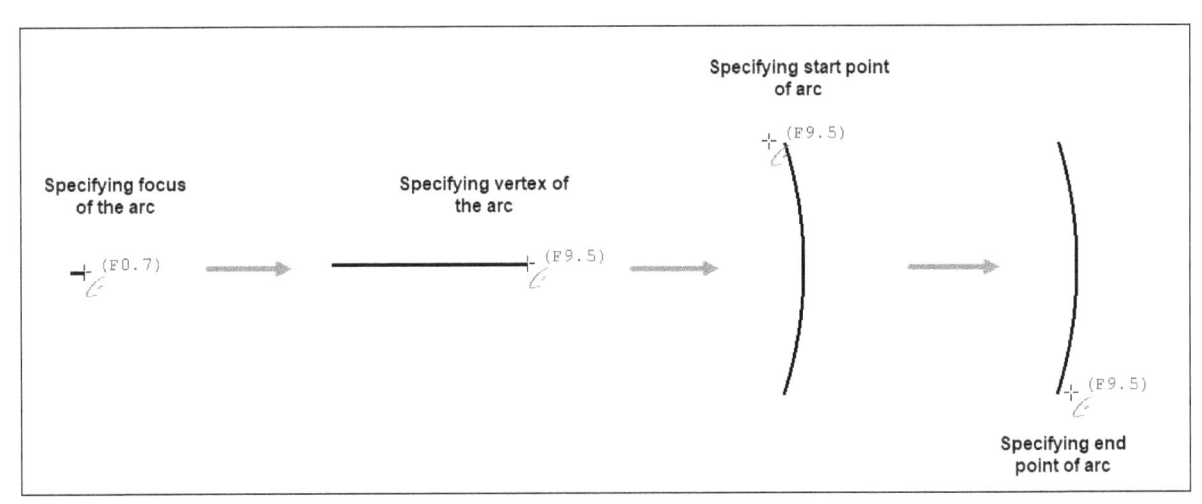

Figure-20. Creation of arc of parabola

- Move the cursor away and click to specify vertex of the arc.
- Click at desired location to specify starting point of arc and then the end point of arc. The arc of parabola will be created.
- Press **ESC** key or press **RMB** to exit the tool.

Circle/Ellipse

The tools available in **Create circle/ellipse** drop-down from **Toolbar** in the **Sketcher** workbench are used to create circle and ellipse using two and three points. There are four tools in **Create circle/ellipse** drop-down; **Create circle by center**, **Create circle by 3 points**, **Create ellipse by center**, and **Create ellipse by 3 points**; refer to Figure-21. The procedures to use these tools are discussed next.

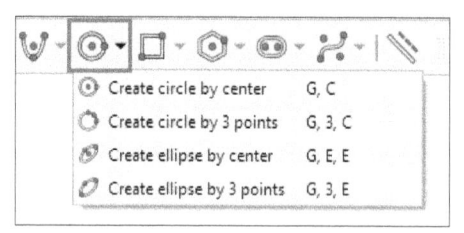

Figure-21. Create a circle drop-down

Create circle by center

The **Create circle by center** tool is used to create a circle by its center and a point along the circle. The procedure to use this tool is discussed next.

- Click on the **Create circle by center** tool from **Create circle/ellipse** drop-down or press **GC** from the keyboard. The **Circle parameters** dialog box will be displayed with the **Center** option selected in the **Mode** drop-down; refer to Figure-22. Also, you will be asked to specify the center point of the circle.

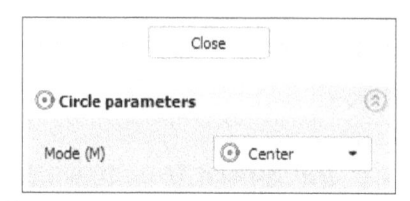

Figure-22. Circle parameters dialog box

- Click in the 3D view area at desired location to specify center point of the circle. You will be asked to specify diameter of the circle.
- Move the cursor away and click at desired location to specify the diameter of circle or you can specify desired diameter value in the edit box displayed and then press **ENTER** to define the diameter of circle. The circle will be created; refer to Figure-23.

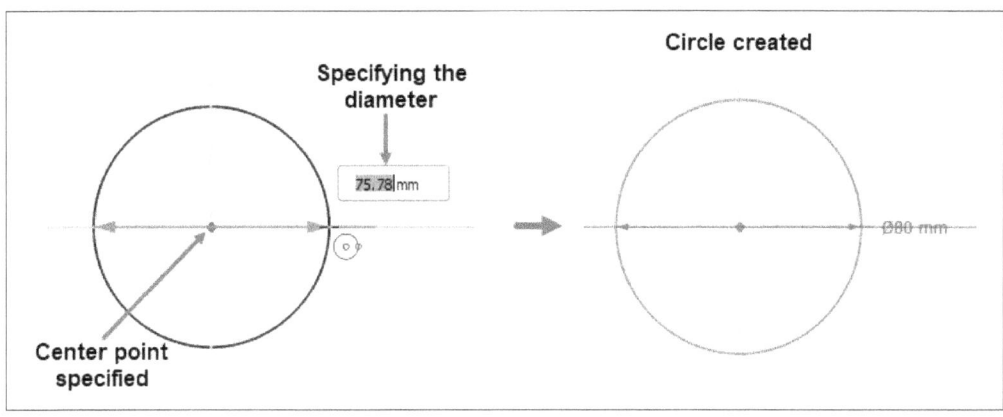

Figure-23. Circle by center created

- Press **ESC** key or press **RMB** to exit the tool.

Create circle by 3 points

The **Create circle by 3 points** tool is used to create a circle by three points along the circle; refer to Figure-24.

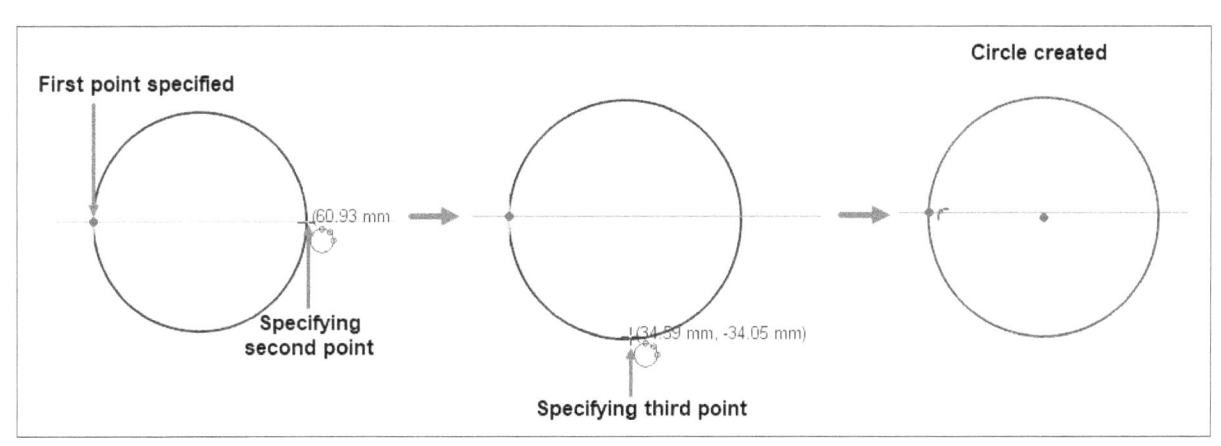

Figure-24. Circle by 3 points created

Create ellipse by center

The **Create ellipse by center** tool is used to create an ellipse by its center, the end of major radius, and the minor radius. The procedure to use this tool is discussed next.

- Click on the **Create ellipse by center** tool from **Create circle/ellipse** drop-down. The **Ellipse parameters** dialog box will be displayed with the **Center** option selected in the **Mode** drop-down; refer to Figure-25. And you will be asked to specify the center point of the ellipse.

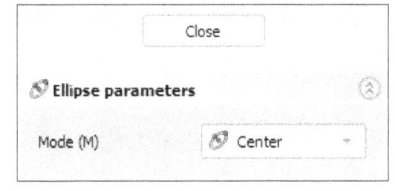

Figure-25. Ellipse parameters dialog box

- Click in the 3D view area at desired location to specify center point of the ellipse. You will be asked to specify the major radius of ellipse.

- Move the cursor away and click at desired location to specify the major radius of ellipse or you can also specify desired radius value in the edit box displayed and then click to define the major radius of ellipse. You will be asked to specify minor radius of ellipse.
- Move the cursor away and click at desired location to specify the minor radius of ellipse or enter desired value in the edit box displayed to specify the minor radius. The ellipse will be created; refer to Figure-26.

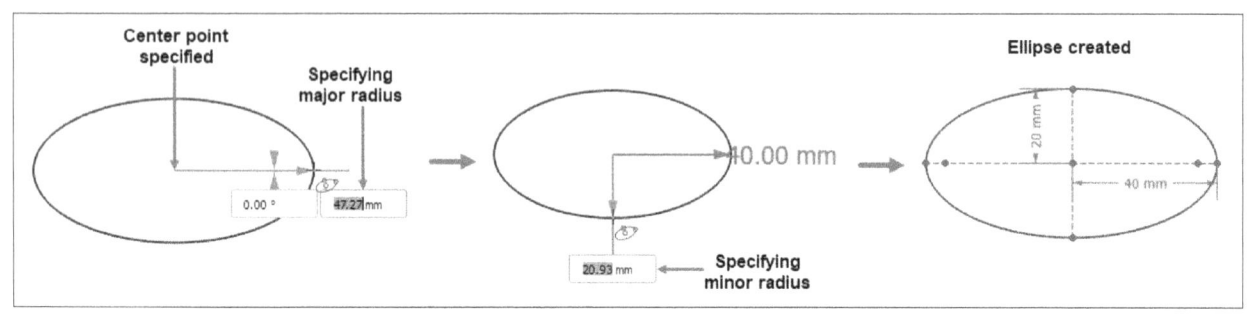

Figure-26. Ellipse by center created

- Press **ESC** key or press **RMB** to exit the tool.

Create ellipse by 3 points

The **Create ellipse by 3 points** tool is used to create an ellipse by the endpoints of one of its axes and a point along the ellipse. The procedure to use this tool is same as discussed in **Create circle by 3 points** tool.

Rectangle

There are three tools available in **Create rectangle** drop-down of **Toolbar** in the **Sketcher** workbench to create rectangles; **Rectangle**, **Centered rectangle**, and **Rounded rectangle**; refer to Figure-27. The procedures to use these tools are discussed next.

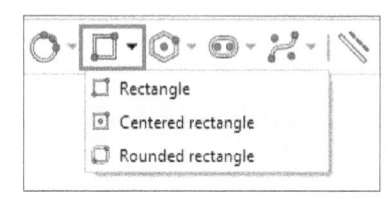

Figure-27. Create a rectangle drop down

Rectangle

The **Rectangle** tool is used to create a rectangle by using two opposite points. The procedure to use this tool is given next.

- Click on **Rectangle** tool from **Create rectangle** drop-down. The **Rectangle parameters** dialog box will be displayed with **Corner, width,** and **height** option selected in the **Mode** drop-down to create a rectangle; refer to Figure-28. Also, you will be asked to specify first corner point of the rectangle.

FreeCAD 1.0 Black Book 2-13

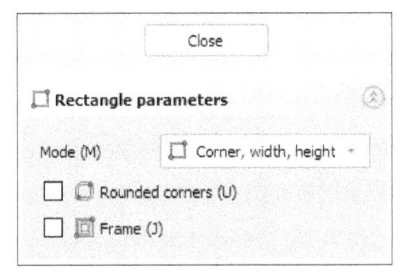

Figure-28. Rectangle parameters dialog box

- Click in the 3D view area at desired location to specify first corner point of the rectangle. You will be asked to specify second corner point of the rectangle.
- Move the cursor away and click at desired location to specify second corner point of the rectangle or you can also enter desired values in the edit boxes displayed to define the width and height of rectangle. The rectangle will be created; refer to Figure-29.

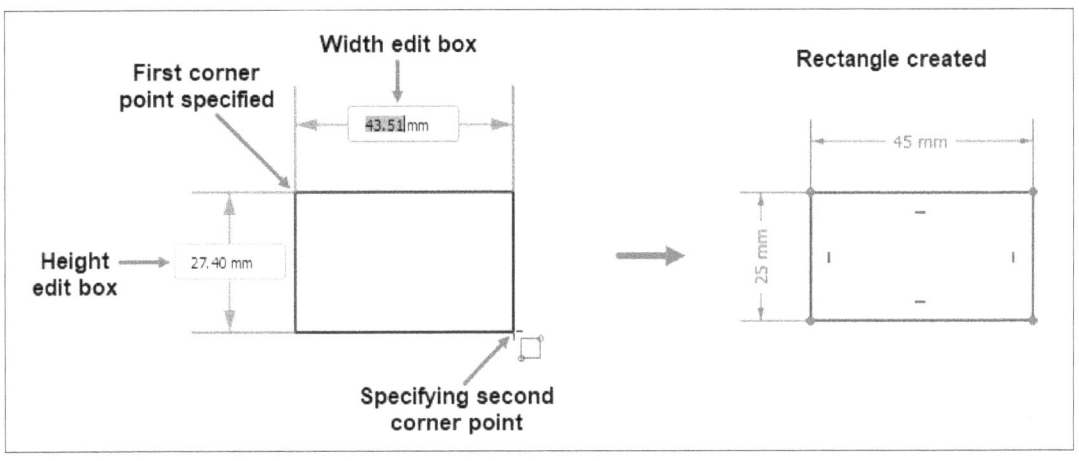

Figure-29. Rectangle created

- Press **ESC** key or press **RMB** to exit the tool.

Centered rectangle

The **Centered rectangle** tool is used to create a rectangle by using center point and a corner point. Click on **Centered rectangle** tool from **Create rectangle** drop-down. The **Rectangle parameters** dialog box will be displayed as discussed earlier with the **Center, width, height** option selected in the **Mode** drop-down. The procedure to use this tool is same as discussed in previous tool; refer to Figure-30.

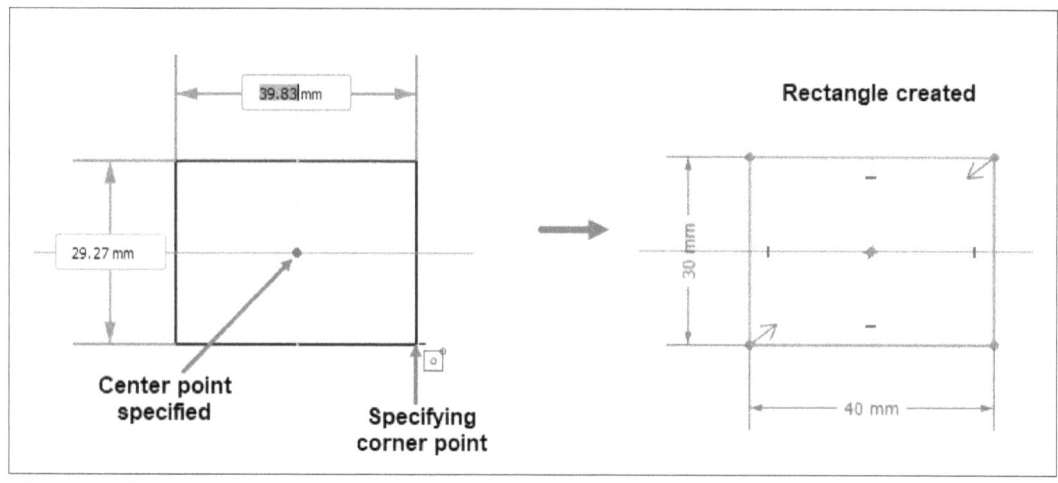

Figure-30. Centered rectangle created

- Press **ESC** key or press **RMB** to exit the tool.

Rounded rectangle

The **Rounded rectangle** tool is used to create a rounded rectangle by using two opposite points. The procedure to use this tool is discussed next.

- Click on the **Rounded rectangle** tool from **Create rectangle** drop-down. The **Rectangle parameters** dialog box will be displayed with the **Rounded corners** check box selected; refer to Figure-31. And you will be asked to specify first corner point of the rectangle.

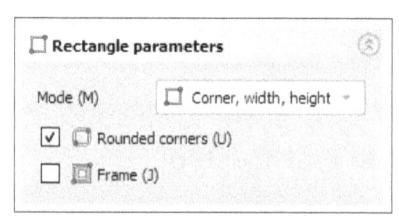

Figure-31. Rectangle parameters dialog box for rounded rectangle

- Click in the 3D view area at desired location to specify first corner point of the rectangle. You will be asked to specify second corner point of the rectangle.
- Move the cursor away and click at desired location to specify second corner point of the rectangle or you can also enter desired values in the edit boxes displayed to define the width and height of rectangle.
- After specifying width and height of rectangle, you will be asked to specify the fillet radius of rectangle.
- Specify desired fillet radius in the edit box displayed and press **ENTER**. The rounded rectangle will be created; refer to Figure-32.

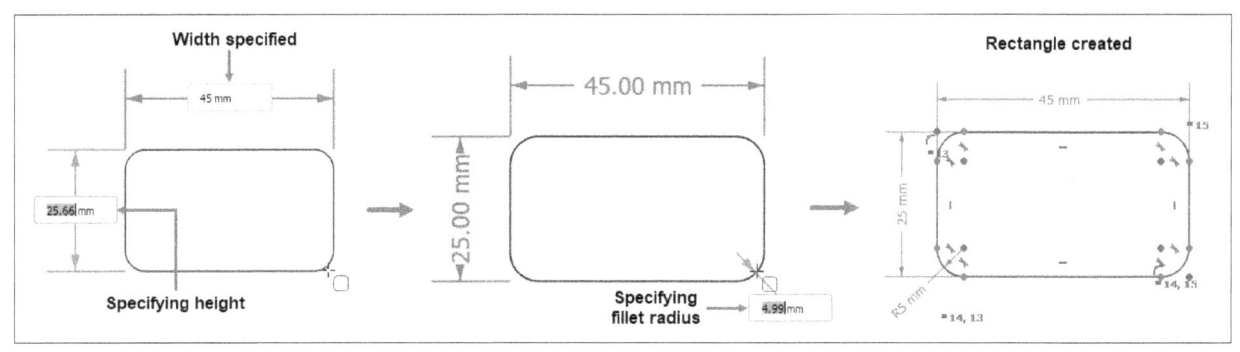

Figure-32. Rounded rectangle created

- Select **Frame** check box from **Rectangle parameters** dialog box to create two rectangles with a constant offset. After specifying width and height of rectangle, you will be asked to specify offset distance.
- Specify desired offset distance in the edit box displayed and press **Enter**. The offset rectangle created; refer to Figure-33.

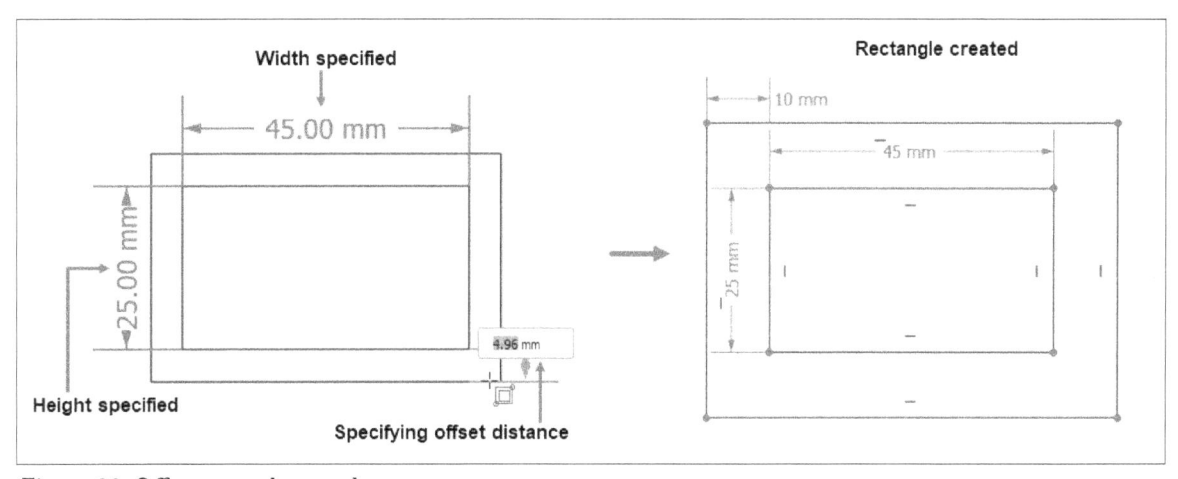

Figure-33. Offset rectangle created

- Press **ESC** key or press **RMB** to exit the tool.

Polygon

There are seven tools available in **Create regular polygon** drop-down from **Toolbar** in the **Sketcher** workbench to create a polygon; **Triangle**, **Square**, **Pentagon**, **Hexagon**, **Heptagon**, **Octagon**, and **Regular Polygon**; refer to Figure-34. The procedures to use these tools are discussed next.

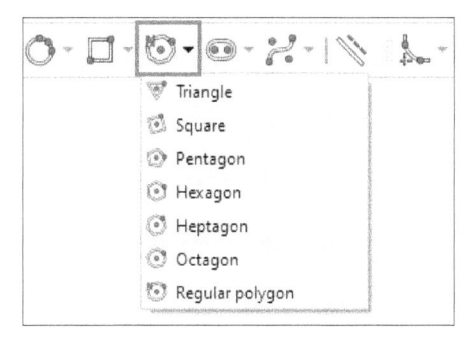

Figure-34. Create regular polygon drop-down

Triangle

The **Triangle** tool is used to create an equilateral triangle inscribed in a construction geometry circle. The procedure to use this tool is discussed next.

- Click on the **Triangle** tool from the **Create regular polygon** drop-down. The **Polygon parameters** dialog box will be displayed; refer to Figure-35. You will be asked to specify the center point.

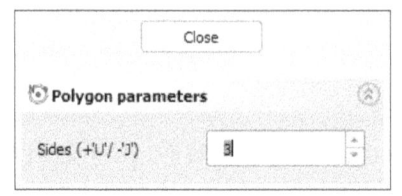

Figure-35. Polygon parameters dialog box

- Click in the 3D view area at desired location to specify the center point of triangle. You will be asked to specify one of the corner point.
- Move the cursor away and click at desired location to specify one of the corner point of triangle or you can also specify desired values in the edit boxes displayed to define the distance and angle of triangle with respect to the center point of triangle.
- After specifying desired values in the edit boxes displayed, press **Enter**. The triangle will be created; refer to Figure-36.

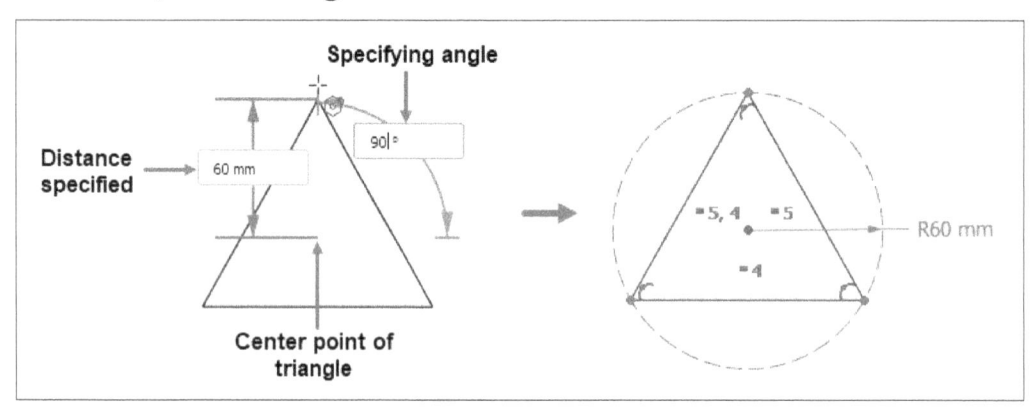

Figure-36. Triangle created

- Or, you can also specify desired number of sides in the **Sides** edit box of dialog box to create polygon.
- Press **ESC** key or press **RMB** to exit the tool.

Similarly, you can use **Triangle**, **Square**, **Pentagon**, **Hexagon**, **Heptagon**, and **Octagon** tools.

Regular polygon

A **Regular polygon** tool is used to create a polygon inscribed in a construction geometry circle with specified number of sides. The procedure to use this tool is discussed next.

- Click on the **Regular polygon** tool from **Create regular polygon** drop-down. The **Create regular polygon** dialog box will be displayed asking you to specify the number of sides; refer to Figure-37.

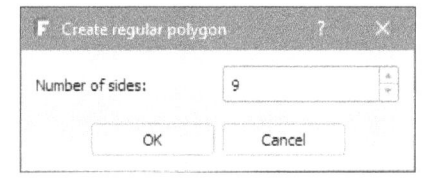

Figure-37. Create regular polygon dialog box

- Specify desired number of sides you want to have in polygon in the **Number of Sides** edit box. Click on the **OK** button to close the dialog box. The **Polygon parameters** dialog box will be displayed as discussed earlier and you will be asked to specify center point of polygon.
- Click in the 3D view area to specify the center point for polygon.
- Move the cursor away from center point and click to specify the vertex of polygon. The regular polygon will be created; refer to Figure-38.

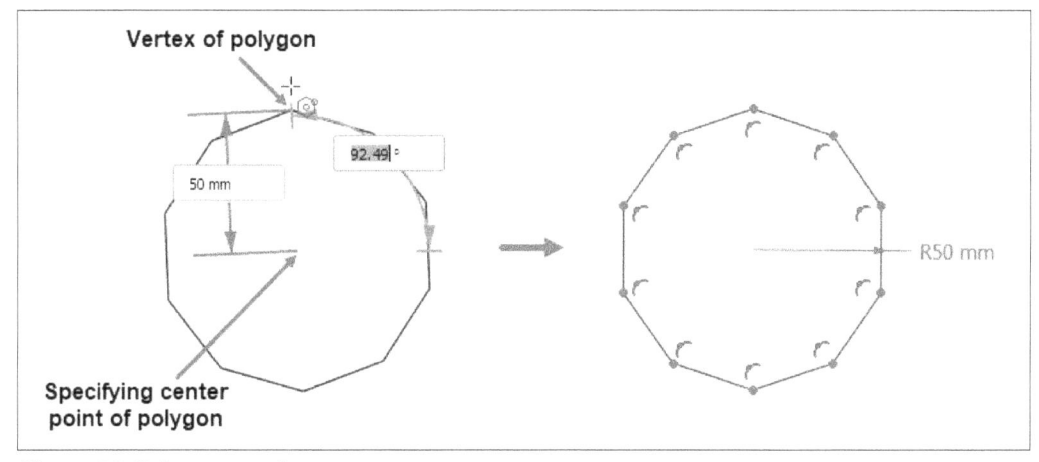

Figure-38. Polygon created

- Press **ESC** key or press **RMB** to exit the tool.

Slot

There are two tools available in **Create slot** drop-down from **Toolbar** in the **Sketcher** workbench to create a polygon; **Slot** and **Arc slot**; refer to Figure-39. The procedures to use these tools are discussed next.

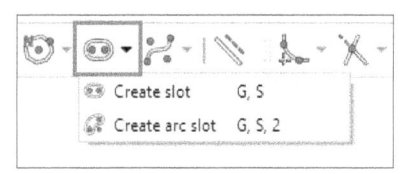

Figure-39. Create slot drop-down

Create slot

The **Create slot** tool is used to create, a closed polyline consisting of two semicircles connected by two parallel straight lines. The procedure to use this tool is discussed next.

- Click on the **Create slot** tool from **Create slot** drop-down in the **Toolbar**. You will be asked to specify center point of one semicircle.
- Click in the 3D view area at desired location to specify center point of one semicircle.
- Move the cursor away and click at desired location to specify the center point of second semicircle or enter desired distance between the centers of the semicircle

and angle of slot in the edit boxes displayed. You will be asked to specify the radius of slot.
- Move the cursor away and click at desired location to specify the radius of slot or enter desired value in the edit box displayed to define the radius of slot. The slot will be created; refer to Figure-40.

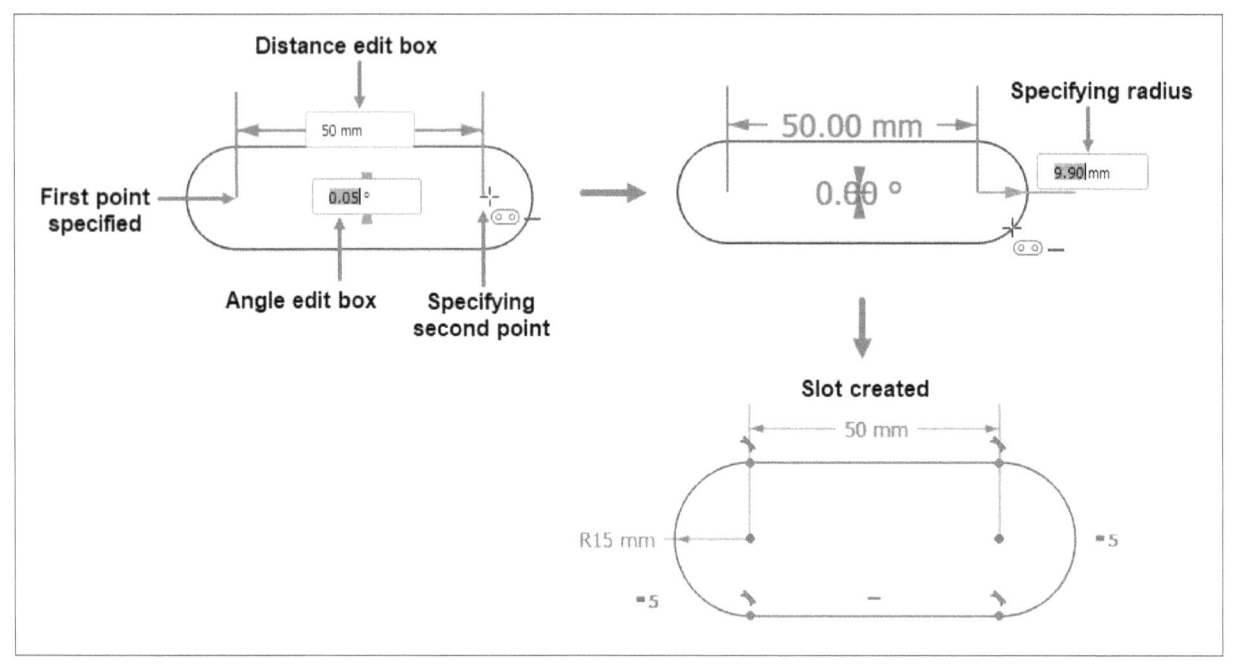

Figure-40. Slot created

- Press **ESC** key or press **RMB** to exit the tool.

Create arc slot

The **Create arc slot** tool is used to create an arc slot, a closed polyline consisting of two parallel concentric arcs closed by two semicircles or two radial straight lines. The procedure to use this tool is discussed next.

- Click on the **Create arc slot** tool from **Create slot** drop-down in the **Toolbar**. The **Arc Slot parameters** dialog box will be displayed with the **Arc ends** option selected in the **Mode** drop-down; refer to Figure-41. And you will be asked to specify the center point of the arc slot

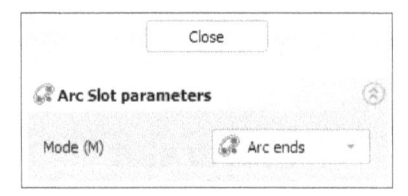

Figure-41. Arc Slot parameters dialog box

- Click in the drawing area at desired location to specify the center point of arc slot. You will be asked to specify the center of first semicircle.

- Move the cursor away and click at desired location to specify the center of first semicircle or enter desired radius of the centerline of the slot and starting angle of the slot in the edit boxes displayed. You will be asked to specify the center of second semicircle.
- Move the cursor away and click at desired location to specify the center of second semicircle or enter desired value in the edit box displayed to define the aperture angle of the centerline arc. You will be asked to specify the radius of semicircles.
- Move the cursor away and click at desired location to specify the radius of semicircles or enter desired radius value in the edit box displayed. The arc slot will be created; refer to Figure-42.

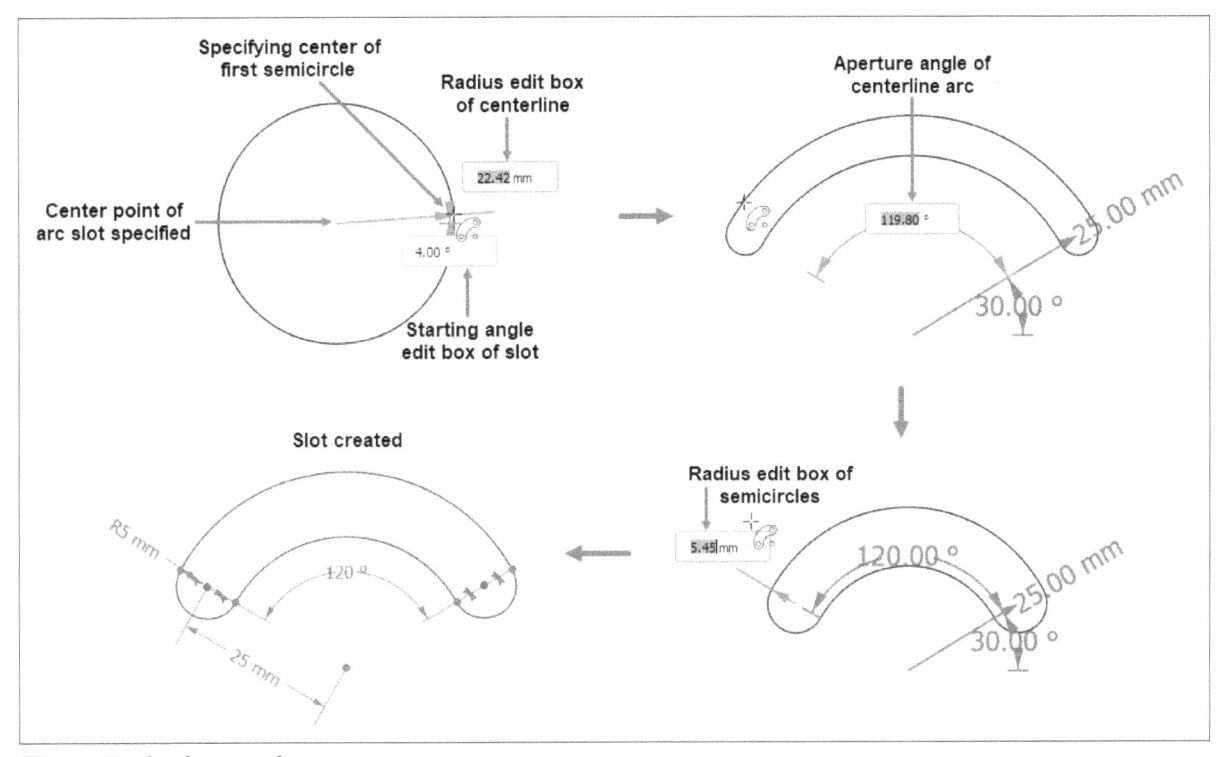

Figure-42. Arc slot created

- Press **ESC** key or press **RMB** to exit the tool.

Similarly, you can create slot using **Flat ends** option in the **Mode** drop-down of **Arc Slot parameters** dialog box.

B-spline

There are four tools available in **Create B-spline** drop-down from **Toolbar** in the **Sketcher** workbench to create B-spline curves; **B-spline by control points**, **Periodic B-spline by control points**, **B-spline by knots**, and **Periodic B-spline by knots**; refer to Figure-43. The procedures to use these tools are discussed next.

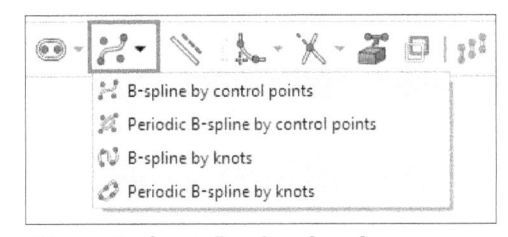

Figure-43. Create B-spline drop down

B-spline by control points

The **B-spline by control points** tool is used to create open B-spline curves by tracing its control points. The procedure to use this tool is discussed next.

- Click on the **B-spline by control points** tool from **Create B-spline** drop-down in the **Toolbar**. The **B-spline parameters** dialog box will be displayed with **By control points** option selected in the **Mode** drop-down; refer to Figure-44. And you will be asked to create the series of points.

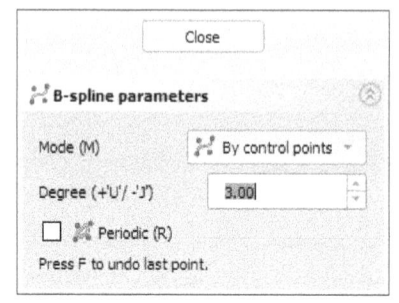

Figure-44. B spline parameters dialog box

- Specify desired value between **1** and **25** in the **Degree** edit box of the dialog box.
- Click in the 3D view area to specify a series of points to create spline or specify desired distance value between the control points and specify desired angle value with respect to origin in the edit boxes displayed. The B-spline will be created; refer to Figure-45.

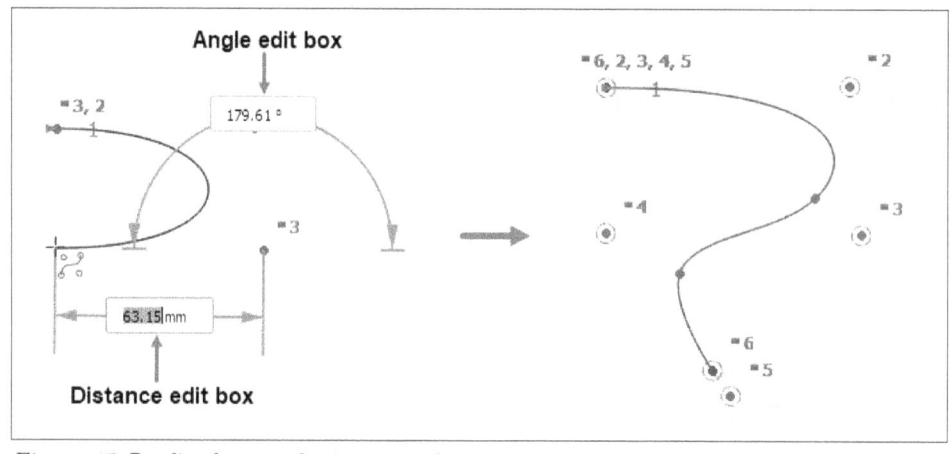

Figure-45. B spline by control points created

- Press **ESC** key or press **RMB** to exit the tool.

Periodic B-spline by control points

The **Periodic B-spline by control points** tool is used to create closed B-spline curves by tracing its control points. Click on the **Periodic B-spline by control points** tool from **Create B-spline** drop-down in the **Toolbar**. The **B-spline parameters** dialog box will be displayed as discussed earlier with **Periodic** check box selected. The procedure to use this tool is same as discussed in previous tool; refer to Figure-46.

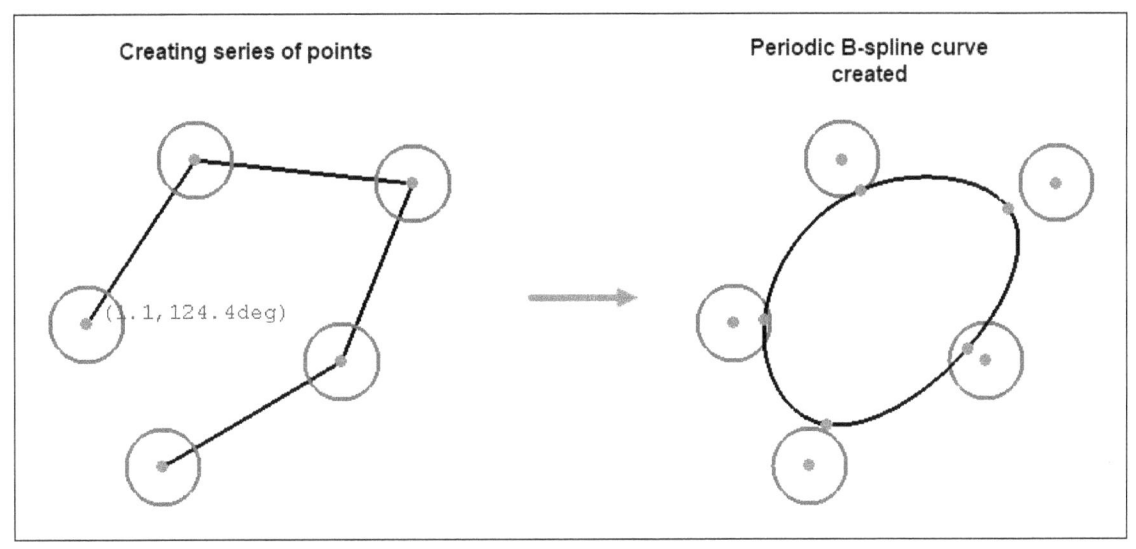

Figure-46. Creation of Periodic B spline

Similarly, you can create B-spline using **B-spline by knots** and **Periodic B-spline by knots** tool.

Construction Mode

The **Toggle construction geometry** tool is used to toggle the sketch geometry between construction mode and sketch mode. It can be used on any type of geometry like line, arc, and circle. The procedure to use this tool is discussed next.

- Click on the **Toggle construction geometry** tool from **Toolbar** in the **Sketcher** workbench; refer to Figure-47. You will be asked to select the sketch.

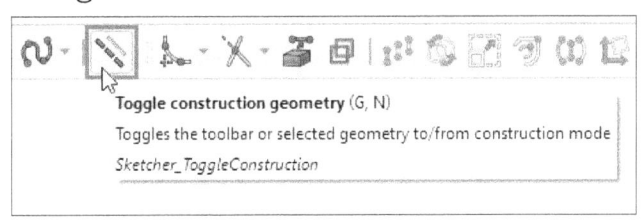

Figure-47. Toggle construction geometry tool

- Select one or more sketch geometry in the 3D view to toggle the construction mode; refer to Figure-48. The sketch will be in construction mode.

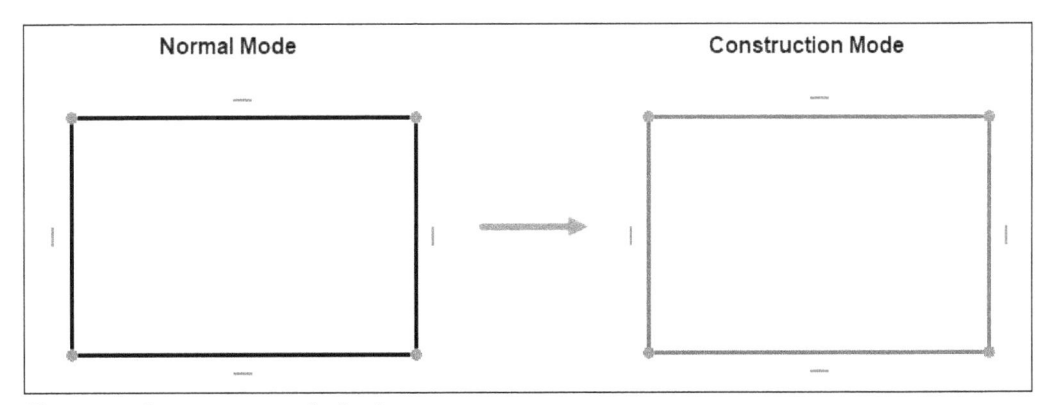

Figure-48. Construction mode sketch

Note that if you have not selected any sketch entity before using this tool and you select the **Toggle construction geometry** tool then all the sketches will be created in construction mode until you click the button again.

SKETCHER CONSTRAINTS

Constraints are used to restrict the shape and size of geometric entities.

Dimensional constraints

Dimensional constraints are associated with numeric data for which you can use the expressions. The tools to apply dimensional constraints are available in the **Dimensional constraints** drop-down of **Toolbar** in the **Sketcher** workbench; refer to Figure-49. The procedures to use these tools are discussed next.

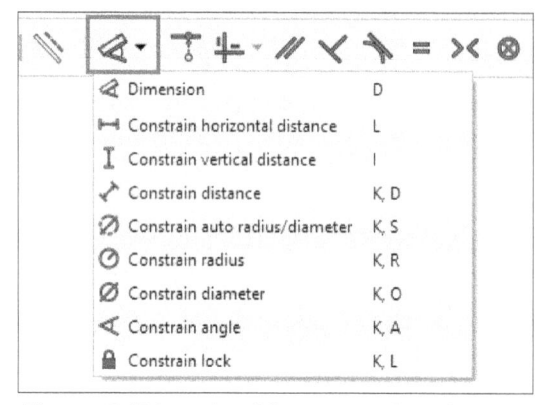

Figure-49. Dimensional Constraints drop-down

Dimension

The **Dimension** tool works on the current selection. This tool is used for dimensional constraints as well as for geometric constraints. The procedure to use this tool is discussed next.

- Click on the **Dimension** tool from **Dimensional constraints** drop-down in the **Toolbar** of **Sketcher** workbench. You will be asked to select the geometry(ies).
- Select desired geometry(ies) in the 3D view area to be dimensioned. The dialog box will be displayed related to the geometry selection.
- Specify desired parameters in the dialog box and click on the **OK** button. The dimension(s) will be applied; refer to Figure-50.

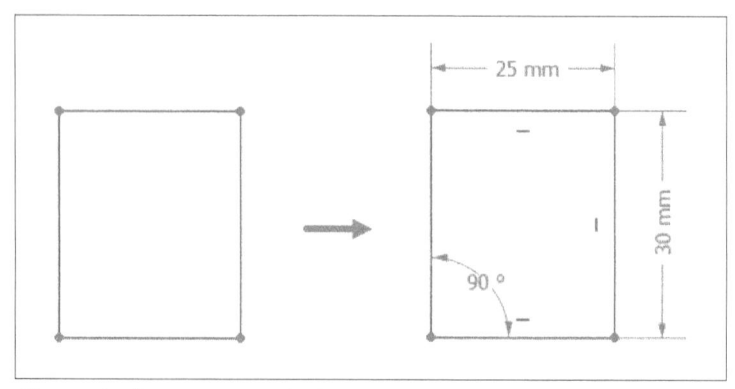

Figure-50. Dimensions applied

Constrain horizontal distance

The **Constrain horizontal distance** fixes the horizontal distance between two points or line endpoints. You can also select a line object to apply horizontal length dimension. The procedure to use this tool is discussed next.

- Click on the **Constrain horizontal distance** tool from **Dimensional constraints** drop-down in the **Toolbar** of **Sketcher** workbench. You will be asked to select endpoints or a horizontal line.
- Select endpoints of a sketch or select a line to fix the horizontal distance or length. The **Insert length** dialog box will be displayed; refer to Figure-51, asking you to specify the horizontal length of entity.

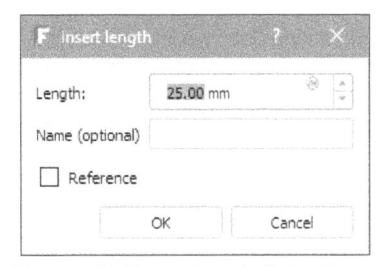

Figure-51. Insert length dialog box

- Specify desired value in the **Length** edit box of the dialog box.
- You can also specify name of the dimension parameter in the **Name** edit box of the **Insert Length** dialog box.
- Select the **Reference** check box to create the reference or constraint dimension.
- Click on **OK** button from the dialog box. The horizontal length or distance of the line/point will be modified; refer to Figure-52.

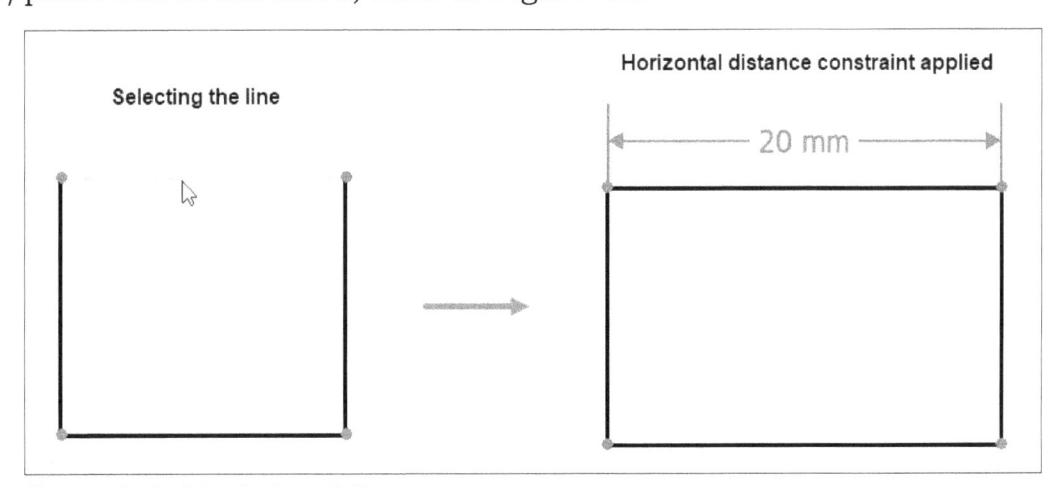

Figure-52. Applying horizontal distance constraint

- Press **ESC** key or press **RMB** to exit the tool.

Constrain vertical distance

The **Constrain vertical distance** tool fixes the vertical distance between two points or line endpoints. The procedure to use this tool is discussed next.

- Click on the **Constrain vertical distance** tool from **Dimensional constraints** drop-down in the **Toolbar** of **Sketcher** workbench. You will be asked to select endpoints or a line.
- Select endpoints of a sketch or select a line to fix the vertical distance. The **Insert length** dialog box will be displayed as discussed earlier, asking you to specify the vertical length of entity.
- Specify desired value in the **Length** edit box and click on **OK** button from the dialog box.
- The vertical length or distance of the sketch will be modified; refer to Figure-53.

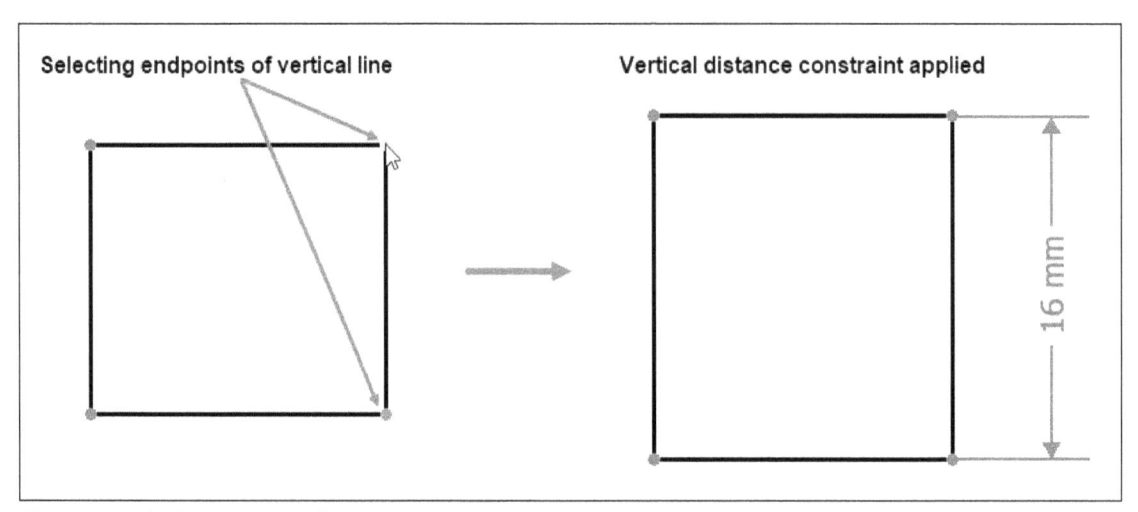

Figure-53. Applying vertical distance constraint

- Press **ESC** key or press **RMB** to exit the tool.

Constrain distance

The **Constrain distance** tool defines the distance of a selected line by constraining its length or defines the distance between two points by constraining the distance between them. Using this tool, you can specify aligned distance of selected line. The procedure to use this tool is discussed next.

- Click on the **Constrain distance** tool from **Dimensional constraints** drop-down in the **Toolbar** of **Sketcher** workbench. You will be asked to select the entities.
- Select two points/a line/a point and a line to apply the distance constraint in the sketch. The **Insert length** dialog box will be displayed as discussed earlier, asking you to specify the distance of selected entity.
- Specify desired distance value in the **Length** edit box and click on **OK** button from the dialog box. The length or distance of the selected entity will be modified; refer to Figure-54.

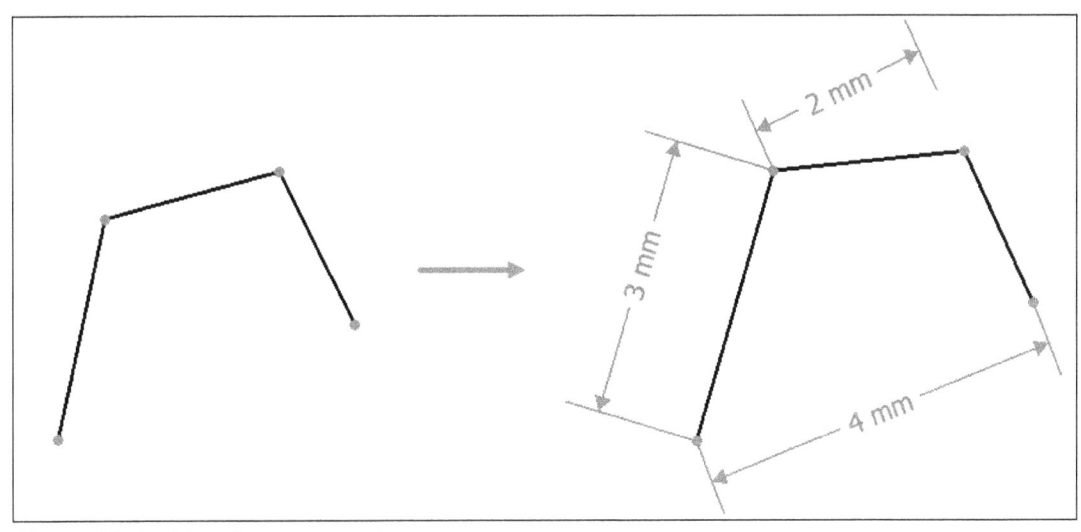

Figure-54. Applying distance constraint

- Press **ESC** key or press **RMB** to exit the tool.

Constrain auto radius/diameter

The **Constrain auto radius/diameter** tool automatically defines the radius/diameter of a selected arc or circle (weight for a B-spline pole, diameter for a complete circle, and radius for a arc). The procedure to use this tool is same as discussed for previous tools.

Constrain radius

The **Constrain radius** tool defines the radius of a selected arc or circle by constraining the radius. The procedure to use this tool is discussed next.

- Click on the **Constrain radius** tool from **Dimensional constraints** drop-down in the **Toolbar** of **Sketcher** workbench. You will be asked to select a circle or arc.
- Select one or more circles and arcs to define the radius. The **Insert radius** dialog box will be displayed; refer to Figure-55.

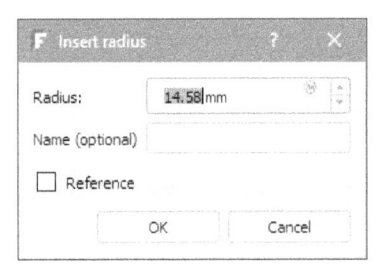

Figure-55. Insert radius dialog box

- Specify desired value in **Radius** edit box and click on **OK** button from the dialog box. The radius of selected entity will be modified; refer to Figure-56.

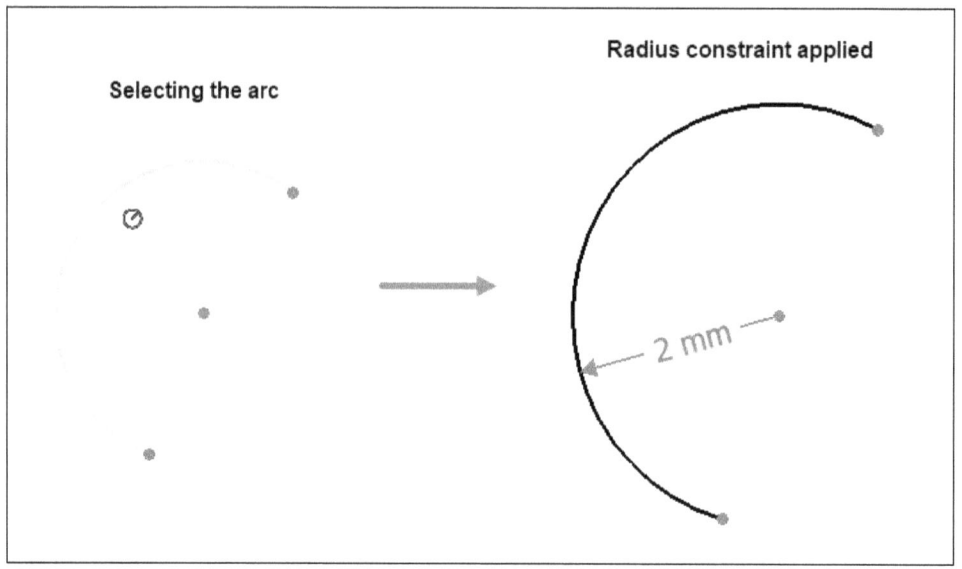

Figure-56. Applying radius constraint

- Press **ESC** key or press **RMB** to exit the tool.

Constrain diameter

The **Constrain diameter** tool defines the diameter of selected arc or circle. The procedure to use this tool is discussed next.

- Click on the **Constrain diameter** tool from **Dimensional constraints** drop-down in the **Toolbar** of **Sketcher** workbench. You will be asked to select an arc or circle.
- Select one or more circles or arcs to define the diameter. The **Insert diameter** dialog box will be displayed; refer to Figure-57.

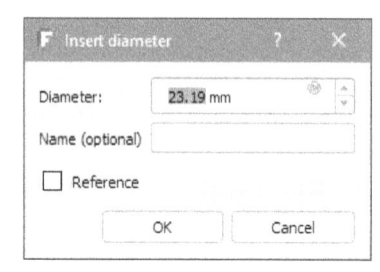

Figure-57. Insert diameter dialog box

- Specify desired value in **Diameter** edit box and click on **OK** button from the dialog box. The diameter of selected entity will be modified; refer to Figure-58.

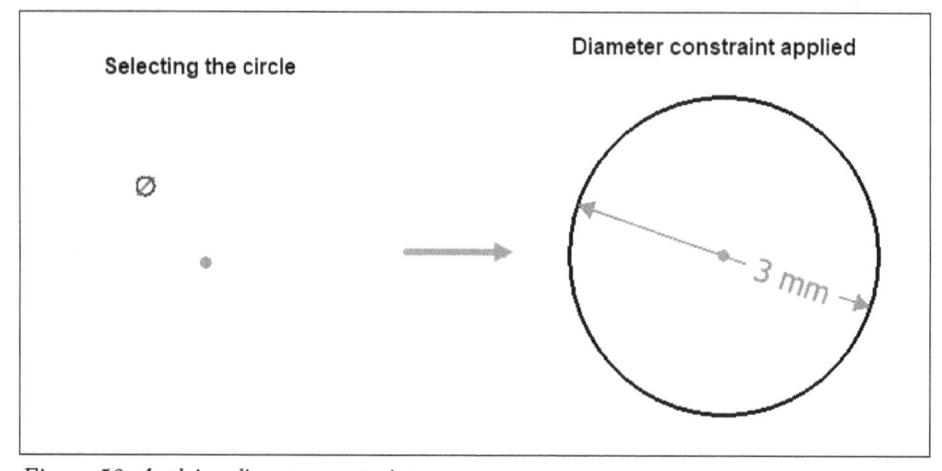

Figure-58. Applying diameter constraint

Constrain angle

The **Constrain angle** tool defines the internal angle between two selected lines. It is capable of setting slopes of individual lines, angles between lines, angles of intersections of curves, and angle spans of circular arcs. The procedure to use this tool is discussed next.

- Click on the **Constrain angle** tool from **Dimensional constraints** drop-down in the **Toolbar** of **Sketcher** workbench. You will be asked to select one or multiple entities.
- Select the entities from the sketch, the **Insert angle** dialog box will be displayed; refer to Figure-59.

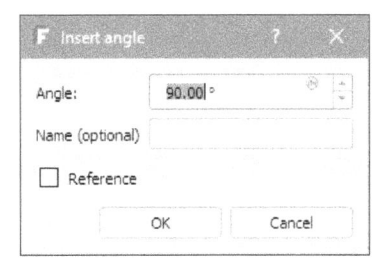

Figure-59. Insert angle dialog box

- Specify desired value in **Angle** edit box and click on **OK** button from the dialog box. The angle between the entities will be modified; refer to Figure-60.

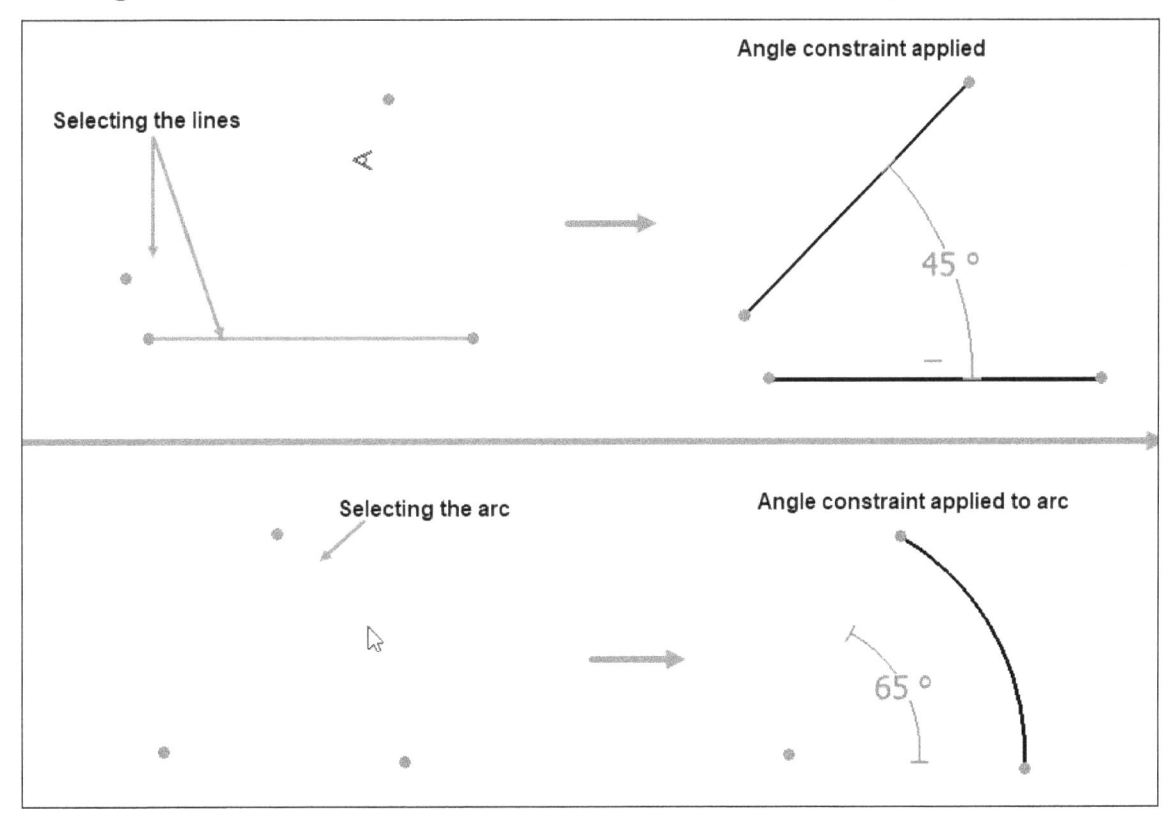

Figure-60. Applying angle constraint

- Press **ESC** key or press **RMB** to exit the tool.

Lock constraint

Lock constraint tool applies **Horizontal distance** and **Vertical distance** dimensions to selected vertices (points) in the sketch.

If a single vertex is selected, the horizontal and vertical distance constraints will refer to the sketch origin point. If two or more points are selected before activating this tool, horizontal and vertical distance constraints will be added for each pair of points.

The procedure to use this tool is discussed next.

- Click on the **Lock constraint** tool from **Dimensional constraints** drop-down in the **Toolbar** of **Sketcher** workbench. You will be asked to select the vertices.
- Select one or more vertices from the sketch on which you want to apply lock constraint. The lock constraint will be applied to the sketch entity; refer to Figure-61.

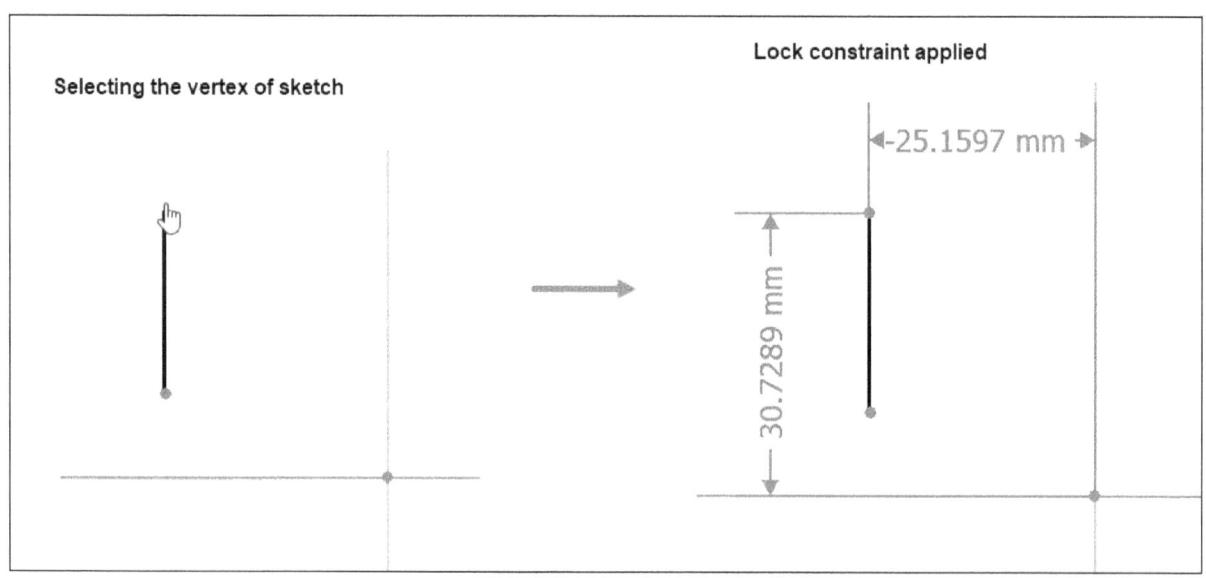

Figure-61. Applying lock constraint

- Double-click on the dimension value in the 3D view and modify the value as desired in the edit box displayed.
- Press **ESC** key or press **RMB** to exit the tool.

Constrain coincident

The **Constrain coincident** tool is useful when there is a break in the profile, for example: where two lines end near each other and need to be joined. A coincident constraint on their endpoints will close the gap and make two selected points coincident. The procedure to use this tool is discussed next.

- Click on the **Constrain coincident** tool from **Toolbar** in the **Sketcher** workbench; refer to Figure-62. You will be asked to select the endpoints.

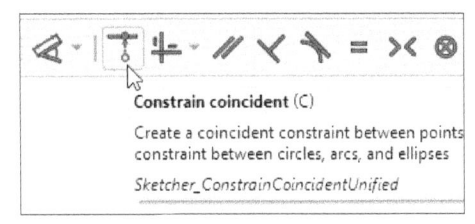

Figure-62. Constrain coincident tool

- Select the endpoints of the sketch which you want to join or make coincident to each other. The endpoints of the sketch will become coincident to each other; refer to Figure-63.
- Press **ESC** key or press **RMB** to exit the tool.

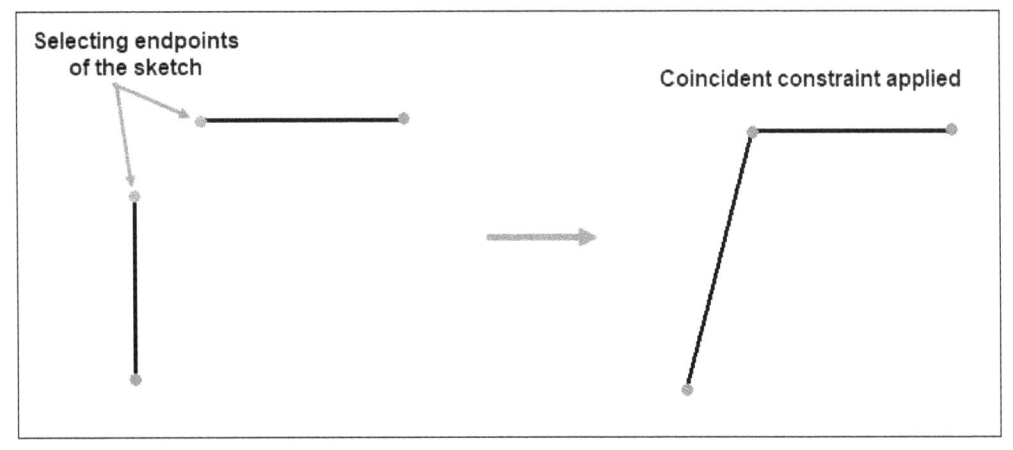

Figure-63. Applying coincident constraint

Horizontal/vertical constraints

There are three tools available in **Horizontal/vertical constraints** drop-down of **Toolbar** in the **Sketcher** workbench to apply horizontal and vertical constraints; **Constrain horizontal/vertical**, **Constrain horizontal**, and **Constrain vertical**; refer to Figure-64. The procedures to use these tools are discussed next. Note that only lines and vertices will be selectable for these tools.

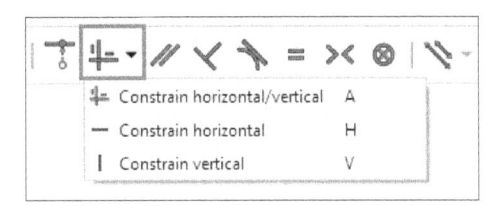

Figure-64. Horizontal vertical constraints drop down

Constrain horizontal/vertical

The **Constrain horizontal/vertical** tool is used to constrain lines or pairs of points to be horizontal or vertical, whichever is closest to the current alignment. The procedure to use this tool is discussed next.

- Click on the **Constrain horizontal/vertical** tool from **Horizontal/vertical constraints** drop-down in the **Toolbar** of **Sketcher** workbench. You will be asked to select the lines or vertices.

- Select the lines to be constrained. The selected lines will become vertically or horizontally constrained, whichever is closest to the current alignment; refer to Figure-65. You can also select two vertices to make them aligned.
- Press **ESC** key or press **RMB** to exit the tool.

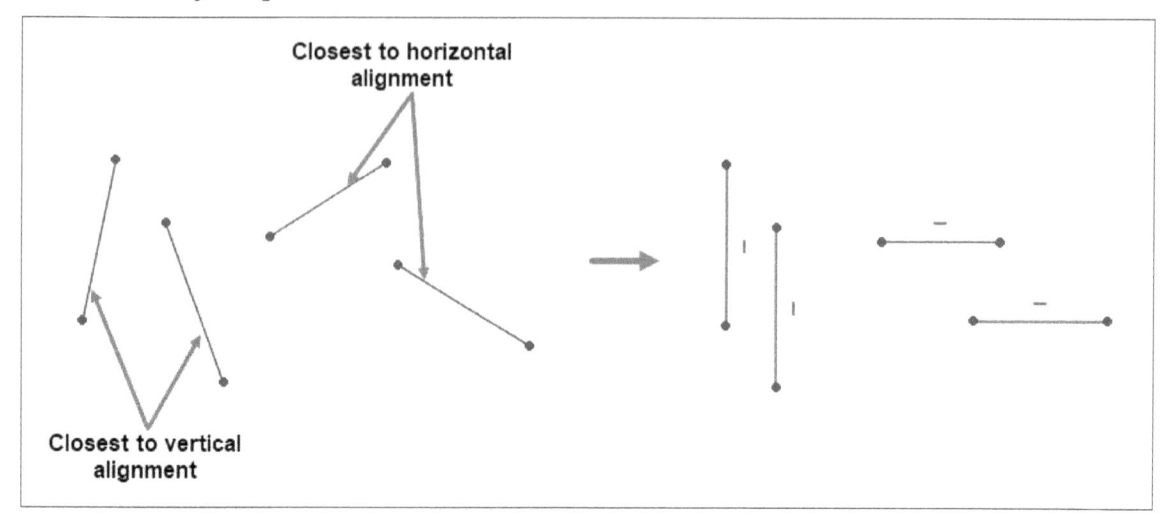

Figure-65. Lines constrained

Constrain horizontal

The **Constrain horizontal** tool forces a selected line or lines in the sketch to be aligned to the horizontal axis of the sketch. The procedure to use this tool is discussed next.

- Click on the **Constrain horizontal** tool from **Horizontal/vertical constraints** drop-down in the **Toolbar** of **Sketcher** workbench. You will be asked to select the lines.
- Select the lines which you want to constrain horizontally. The selected lines will become horizontally constrained; refer to Figure-66. You can also align two vertices in horizontal line.

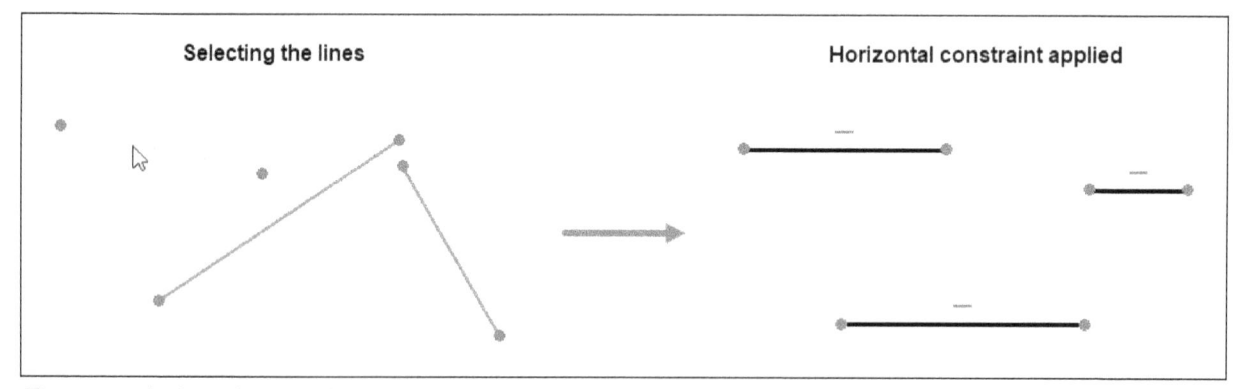

Figure-66. Applying horizontal constraint

- Press **ESC** key or press **RMB** to exit the tool.

Constrain vertical

The **Constrain vertical** tool is used to constrain the selected lines or polyline elements to a true vertical orientation. The procedure to use this tool is discussed next.

- Click on the **Constrain vertical** tool from **Horizontal/vertical constraints** drop-down in the **Toolbar** of **Sketcher** workbench. You will be asked to select the lines or vertices.
- Select the lines to be constrained vertically. The selected lines will become vertically constrained; refer to Figure-67. You can also select two vertices to make them vertically aligned; refer to Figure-68.

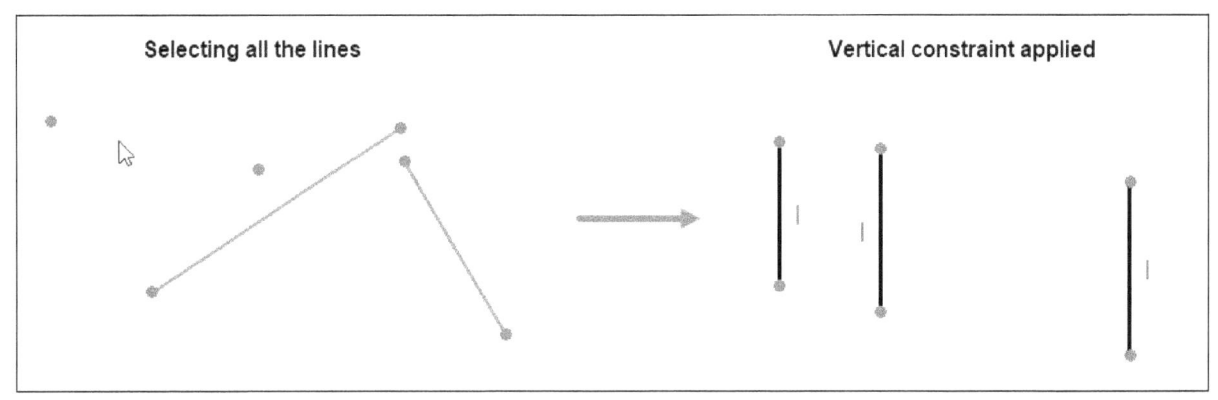

Figure-67. Applying vertical constraint

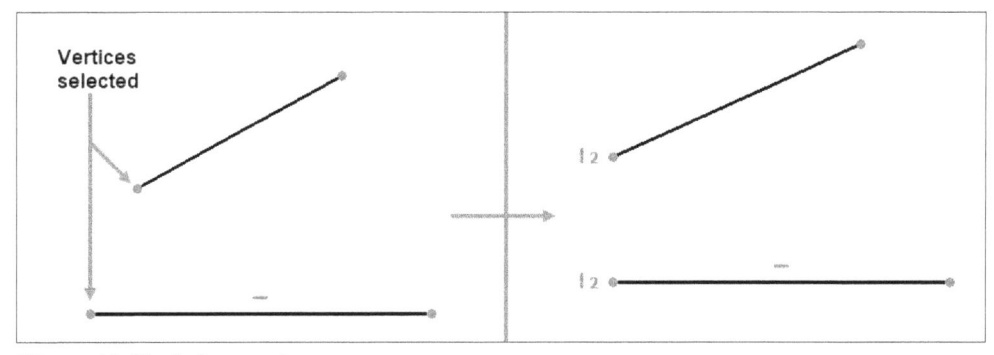

Figure-68. Vertical constrain

- Press **ESC** key or press **RMB** to exit the tool.

Constrain parallel

The **Constrain parallel** tool forces two selected lines to be parallel to each other. The procedure to use this tool is discussed next.

- Click on the **Constrain parallel** tool from **Toolbar** in the **Sketcher** workbench; refer to Figure-69. You will be asked to select the lines.

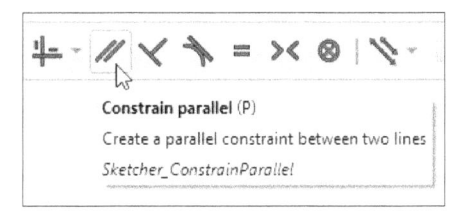

Figure-69. Constrain parallel tool

- Select the lines or line and axis which you want to make parallel. The selected lines will become parallel to each other; refer to Figure-70.

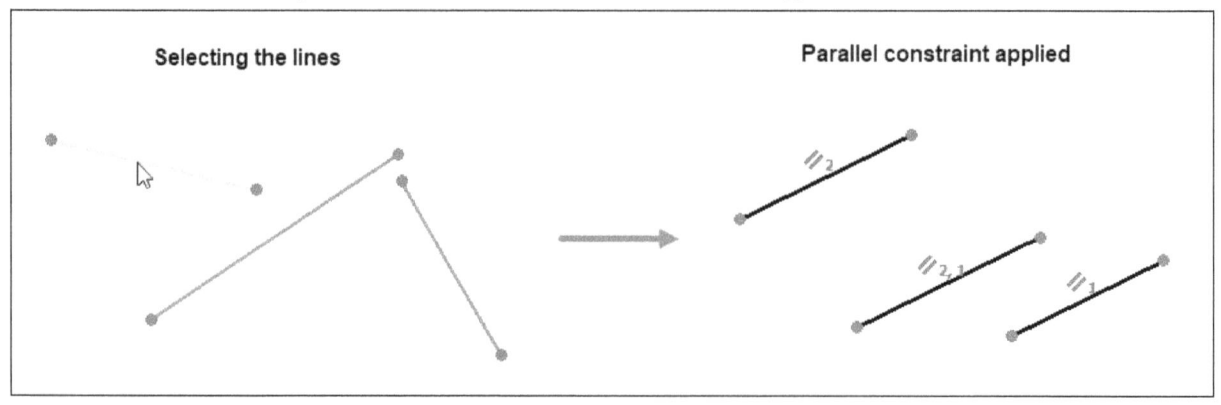

Figure-70. Applying parallel constraint

- Press **ESC** key or **RMB** to exit the tool.

Constrain perpendicular

The **Constrain perpendicular** tool makes two lines perpendicular to each other or two curves to be perpendicular at their intersection point. The procedure to use this tool is discussed next.

- Click on the **Constrain perpendicular** tool from **Toolbar** in the **Sketcher** workbench; refer to Figure-71. You will be asked to select desired entities.

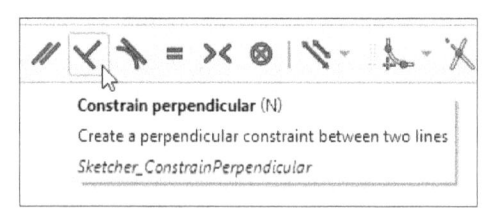

Figure-71. Constrain perpendicular tool

- Select the lines, curves, arcs, circles, ellipses, etc. which you want to make perpendicular. The entities selected will become perpendicular to each other; refer to Figure-72.
- Press **ESC** key or **RMB** to exit the tool.

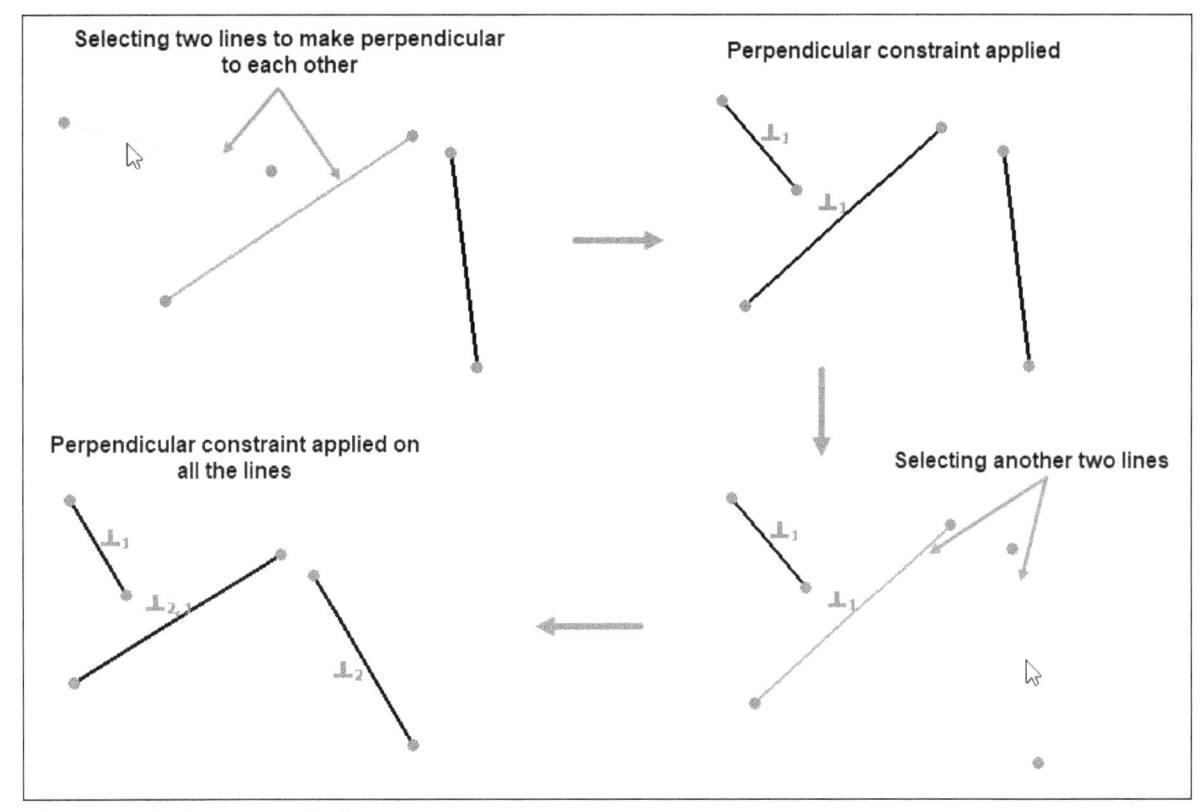

Figure-72. Applying perpendicular constraint

Constrain tangent or collinear

The **Constrain tangent or collinear** tool applies a tangent constraint between arcs, circles, curves, and lines. If you are applying tangent constraint between two lines then they will become collinear. A line segment does not have to lie directly on an arc or circle to be constrained tangent to that arc or circle. The procedure to use this tool is discussed next.

- Click on the **Constrain tangent or collinear** tool from **Toolbar** in the **Sketcher** workbench; refer to Figure-73. You will be asked to select the entities.

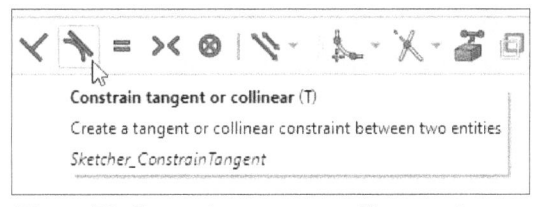

Figure-73. Constrain tangent or collinear tool

- Select the entities on which you want to apply tangent constraint. The entities will become tangentially constraint to each other; refer to Figure-74.

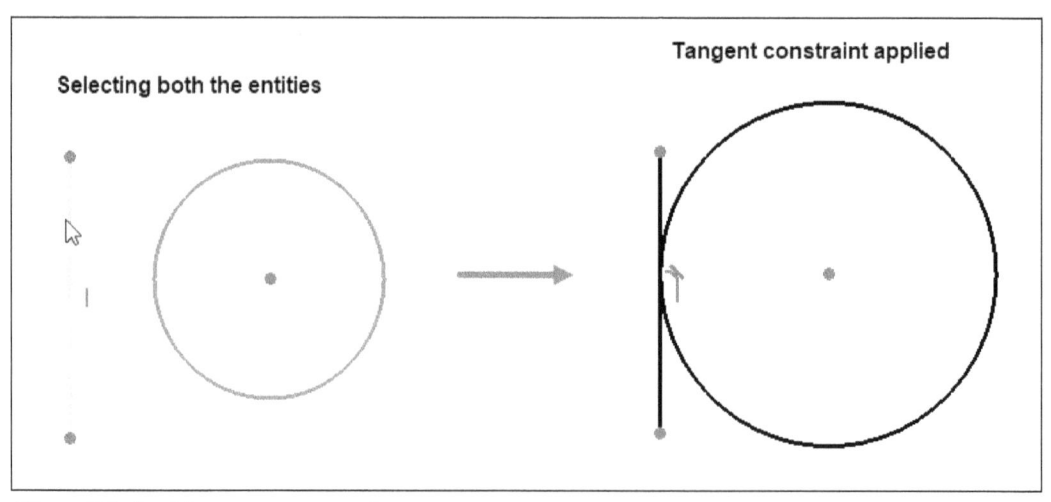

Figure-74. Applying tangent constraint

- Press **ESC** key or **RMB** to exit the tool.

Constrain equal

The **Constrain equal** tool forces two or more sketch segments to have equal length. If applied to arcs or circles, the radii are constrained to be equal. The procedure to use this tool is discussed next. Note that **Constrain equal** cannot be applied to geometric entities which are not of the same type.

- Click on the **Constrain equal** tool from **Toolbar** in the **Sketcher** workbench; refer to Figure-75. You will be asked to select the entities of same geometry type.

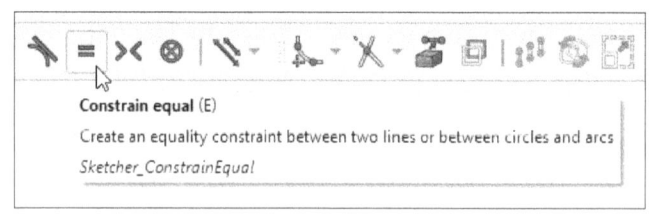

Figure-75. Constrain equal tool

- Select the entities of same geometry type on which you want to apply equal constraint. The selected entities will become equally constrained; refer to Figure-76.
- Press **ESC** key or **RMB** to exit the tool.

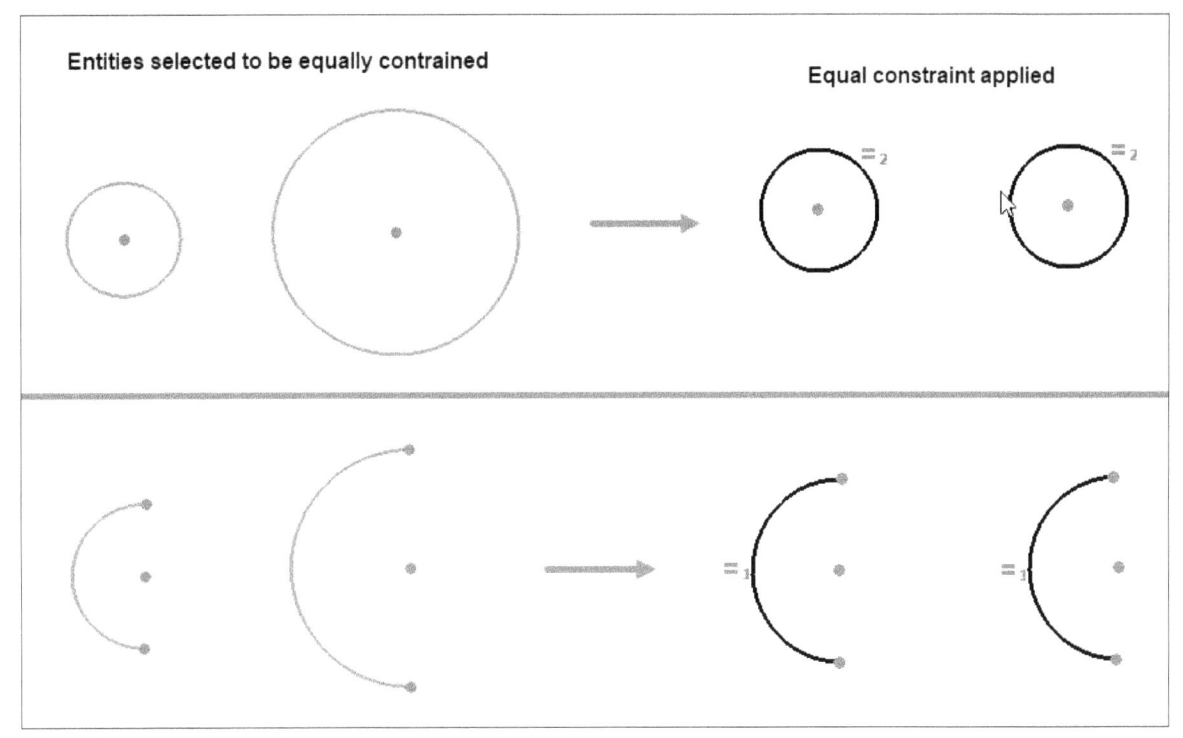

Figure-76. Applying equal constraint

Constrain symmetric

The **Constrain symmetric** tool is used to constrain the two selected points to be symmetrical about a given line, i.e., both selected points are constrained to lie on a normal to the line through both points and are constrained to be equidistant from the line. The procedure to use this tool is discussed next.

- Click on the **Constrain symmetric** tool from **Toolbar** in the **Sketcher** workbench; refer to Figure-77. You will be asked to select the points from the sketch.

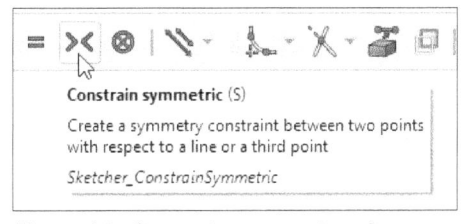

Figure-77. Constrain symmetric tool

- Select two points and a line from the sketch. The selected entities will be constrained symmetrically; refer to Figure-78.

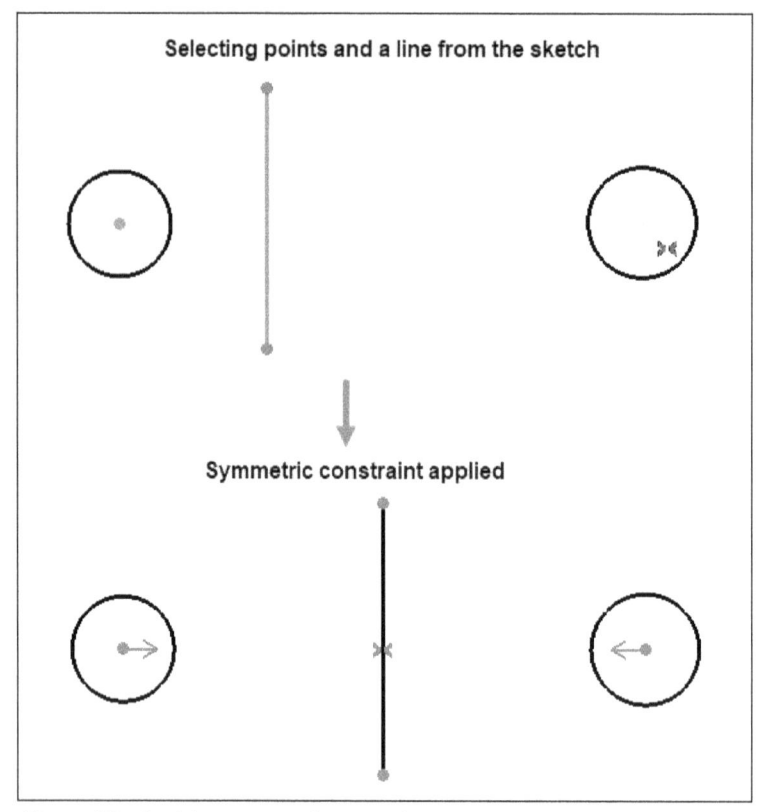

Figure-78. Applying symmetric constraint

- Press **ESC** key or **RMB** to exit the tool.

Constrain block

The **Constrain block** tool is used to block a sketch entity from moving, that is, it prevents its vertices from changing their current positions. It is mainly useful to fix the position of B-splines, otherwise it is difficult to fully constrain it. The procedure to use this tool is discussed next.

- Click on the **Constrain block** tool from **Toolbar** in the **Sketcher** workbench; refer to Figure-79. You will be asked to select the entity.

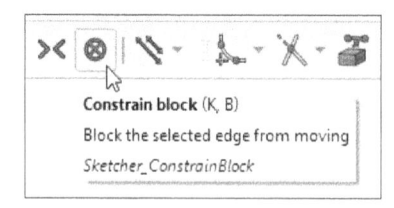

Figure-79. Constrain block tool

- Select the entity or entities whose position(s) are to be blocked. Now, the position of entity will be blocked; refer to Figure-80.
- Press **ESC** key or **RMB** to exit the tool.

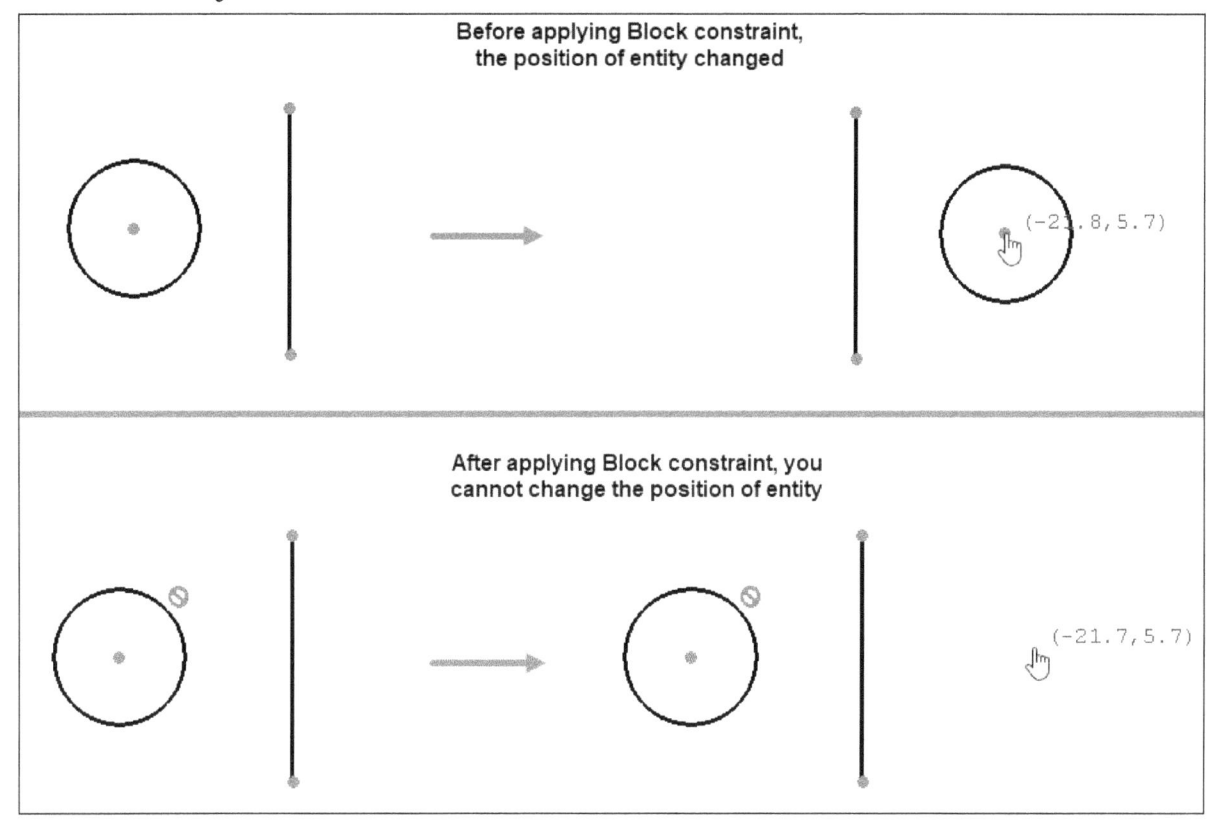

Figure-80. Applying block constraint

Constrain refraction (Snell's law)

The **Constrain refraction (Snell's law)** tool constrains two lines to follow the law of refraction of light as it penetrates through an interface, where two materials of different refraction indices meet. The procedure to use this tool is discussed next.

- You need two lines to follow a beam of light and a curve to act as an interface. The lines should be on different sides of the interface. The interface can be a line, circle/arc, or ellipse/arc of ellipse.
- Select the endpoint of lines and then the interface geometry.
- Click on the **Constrain refraction (Snell's law)** tool from the **Sketcher constraints** cascading menu in the **Sketch** menu; refer to Figure-81. The **Refractive index ratio** dialog box will be displayed; refer to Figure-82.

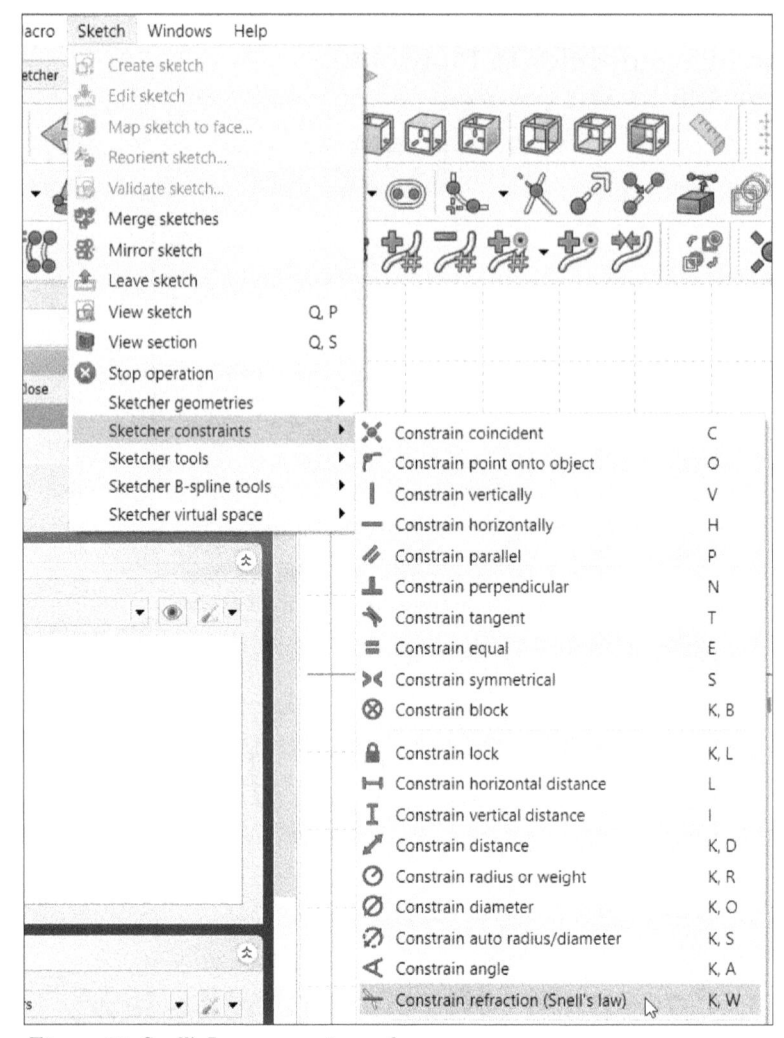

Figure-81. Snell's Law constraint tool

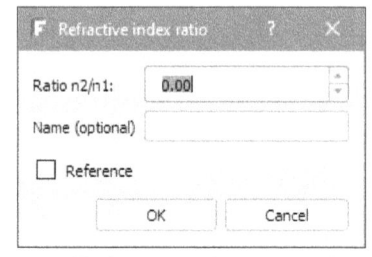

Figure-82. Refractive index ratio dialog box

- Specify desired value in **Ratio n2/n1** edit box and click on **OK** button from the dialog box. The endpoints will be constrained on the interface as per the Snell's law; refer to Figure-83.

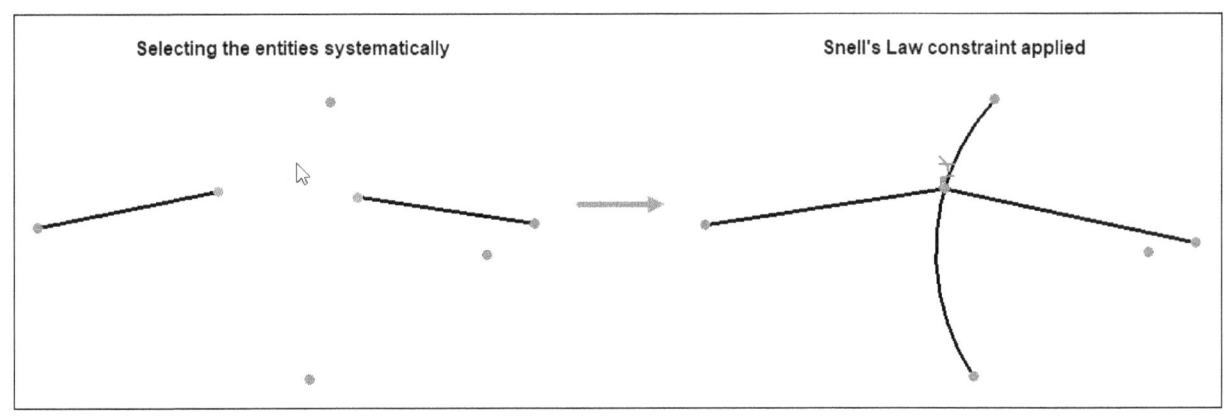

Figure-83. Applying snell's law constraint

- Press **ESC** key or press **RMB** to exit the tool.

Toggle constraints

There are two tools available in `Toggle constraints` drop-down of `Toolbar` in the `Sketcher` workbench; `Toggle driving/reference constraint` and `Activate/deactivate constraint`; refer to Figure-84. The procedures to use these tools are discussed next.

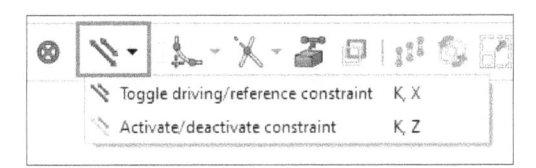

Figure-84. Toggle constraints drop-down

Toggle driving/reference constraint

The `Toggle driving/reference constraint` tool is used to switch between driving and driven dimension constraints (except `Snell's law constraint`). The icons in the `Dimensional constraints` drop-down turns blue and in place of dimensional constraints, reference dimensions are created; refer to Figure-85.

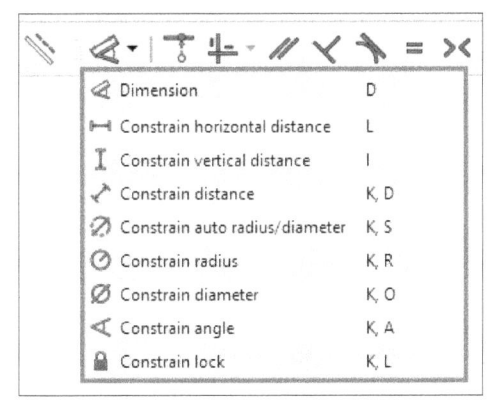

Figure-85. Dimensional constraints icons turns blue

Reference dimensions do not constrain the sketch. The procedure to use this tool is discussed next.

- Click on the **Toggle driving/reference constraint** tool from **Toggle constraints** drop-down in the **Toolbar** of **Sketcher** workbench. The dimensional constraint icons turn from red to blue.
- Click on the **Constrain vertical distance** tool from **Dimensional constraints** drop-down in the **Toolbar** of **Sketcher** workbench. You will be asked to select the entity from sketch to create reference dimension.
- Select the entity from the sketch to apply reference dimension. The reference dimension will be applied to the selected entity; refer to Figure-86.

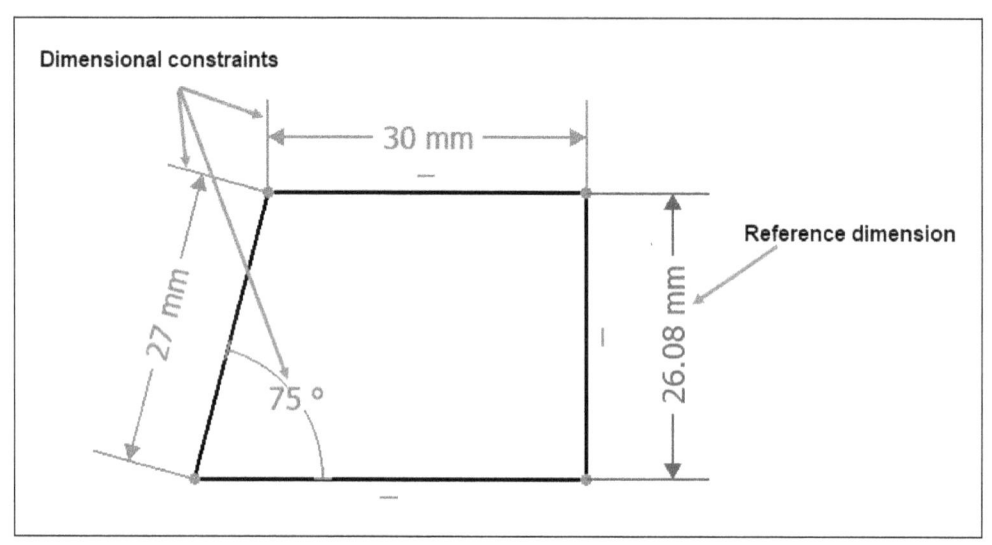

Figure-86. Applying reference dimension

- Press **ESC** key or press **RMB** to exit the tool.

Activate/deactivate constraint

The **Activate/deactivate constraint** tool allows you to activate and deactivate an already placed constraint. The procedure to use this tool is discussed next.

- Select an already placed constraint of an entity which you want to deactivate and click on the **Activate/deactivate constraint** tool from **Toggle constraints** drop-down in the **Toolbar** of **Sketcher** workbench. The placed constraint will be deactivated; refer to Figure-87.

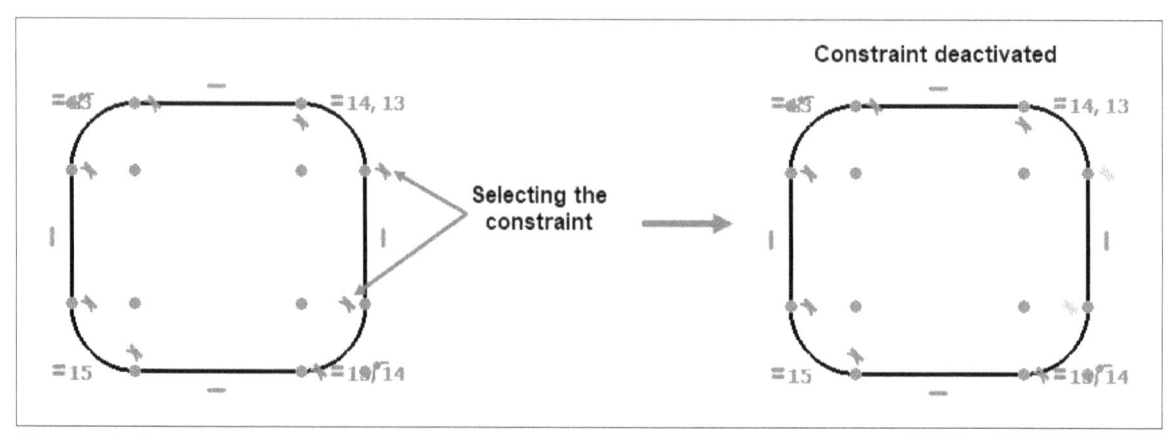

Figure-87. Constraint deactivated

- If you want to reactivate the constraint, then select the deactivated constraint and click on the tool again.

SKETCH EDITING TOOLS

In **FreeCAD** software, the standard tools to edit sketch entities are available in the **Toolbar** of **Sketcher** workbench; refer to Figure-88.

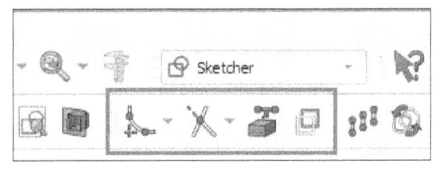

Figure-88. Sketch editing tools

Create fillet/chamfer

There are two tools available in **Create fillet/chamfer** drop-down of **Toolbar** in the **Sketcher** workbench to create fillets; **Create fillet** and **Create chamfer**; refer to Figure-89. The procedures to use these tools are discussed next.

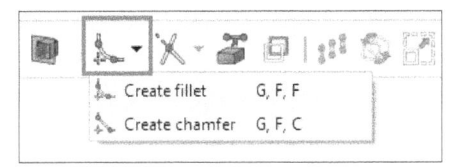

Figure-89. Create fillet chamfer drop down

Create fillet

The **Create fillet** tool is used to create fillet between two geometries like lines, arcs etc. joined at one point. The procedure to use this tool is discussed next.

- Click on the **Create fillet** tool from **Create fillet/chamfer** drop-down in the **Toolbar** of **Sketcher** workbench. The **Fillet/Chamfer parameters** dialog box will be displayed with the **Fillet** option selected in the **Mode** drop-down; refer to Figure-90. And you will be asked to select the intersecting geometries like lines, arcs, etc.

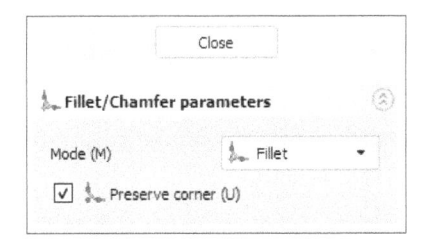

Figure-90. Fillet Chamfer parameters dialog box

- Select the two connecting lines/arcs from the 3D view area or vertex at intersection of two lines. The fillet will be created; refer to Figure-91.

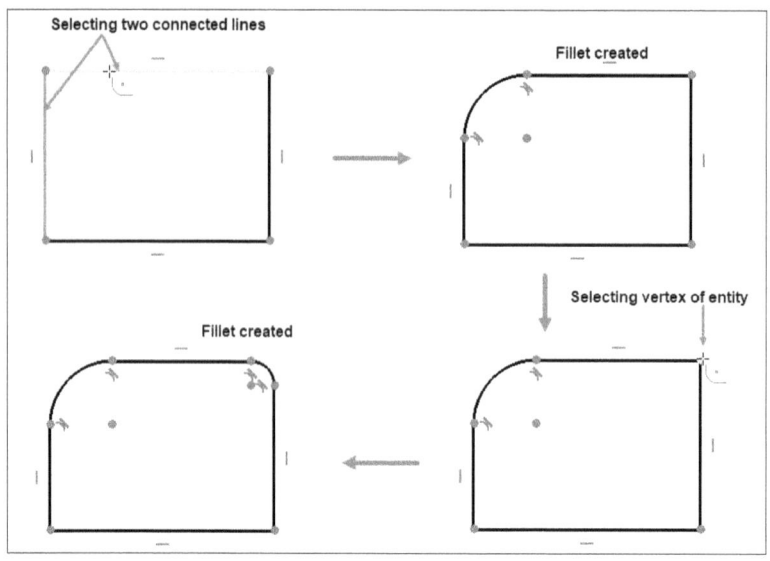

Figure-91. Fillet creation

- Select **Preserve corner** check box from the dialog box to preserve intersection point and most constraints.

Note that the distance at which you click from the vertex will specify the fillet radius. You can change the value of fillet radius later.

- Press **ESC** key or press **RMB** to exit the tool.

Create chamfer

The **Create chamfer** tool is used to create chamfer between two geometries like lines, arcs etc. joined at one point. Click on the **Create chamfer** tool from **Create fillet/chamfer** drop-down in the **Toolbar** of **Sketcher** workbench. The **Fillet/Chamfer parameters** dialog box will be displayed as discussed earlier with the **Chamfer** option selected in the **Mode** drop-down The procedure to use this tool is same as discussed for **Fillet** tool; refer to Figure-92.

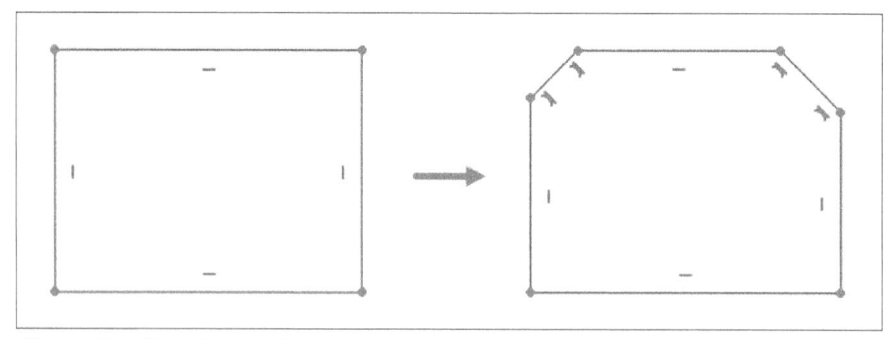

Figure-92. Chamfer created

Edit edge

There are three tools available in **Edit edge** drop-down of **Toolbar** in the **Sketcher** workbench for editing the edges; **Trim edge**, **Split edge**, and **Extend edge**; refer to Figure-93. The procedures to use these tools are discussed next.

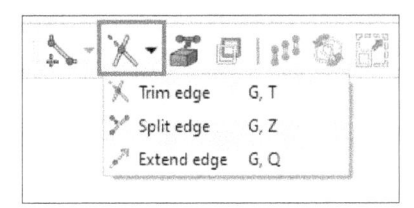

Figure-93. Edit edge drop down

Trim edge

The **Trim edge** tool is used to trim a curve to the nearest intersection point of other entity. The procedure to use this tool is discussed next.

- Click on the **Trim edge** tool from **Edit edge** drop-down in the **Toolbar** of **Sketcher** workbench. You will be asked to select the curve to be trimmed.
- Select an intersecting curve that you want to be trimmed. The curve will be trimmed; refer to Figure-94.

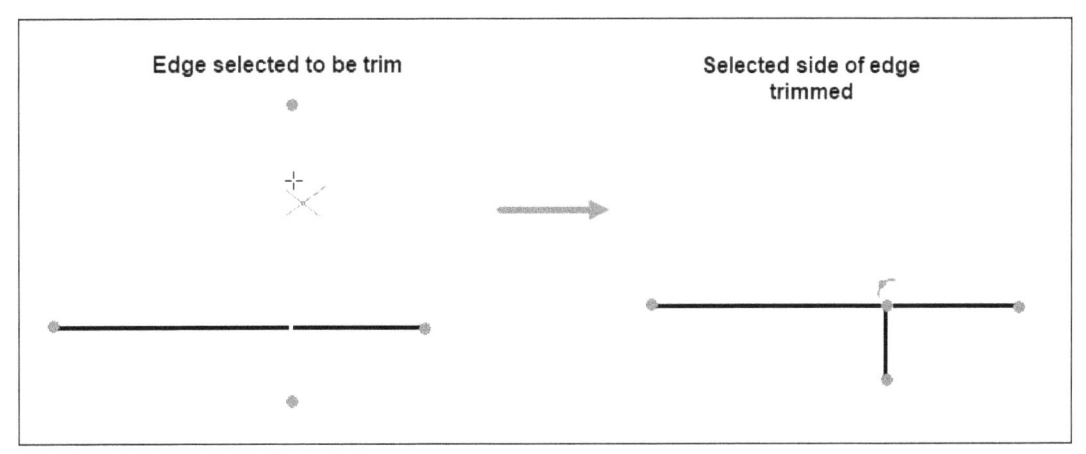

Figure-94. Edge trimmed

- Press **ESC** key or press **RMB** to exit the tool.

Split edge

The **Split edge** tool is used to divide an edge at selected location. The procedure to use this tool is discussed next.

- Click on the **Split edge** tool from **Edit edge** drop-down in the **Toolbar** of **Sketcher** workbench. You will be asked to select the edge to be split.
- Select a location on the edge where you want to split the edge. The edge will be split; refer to Figure-95.

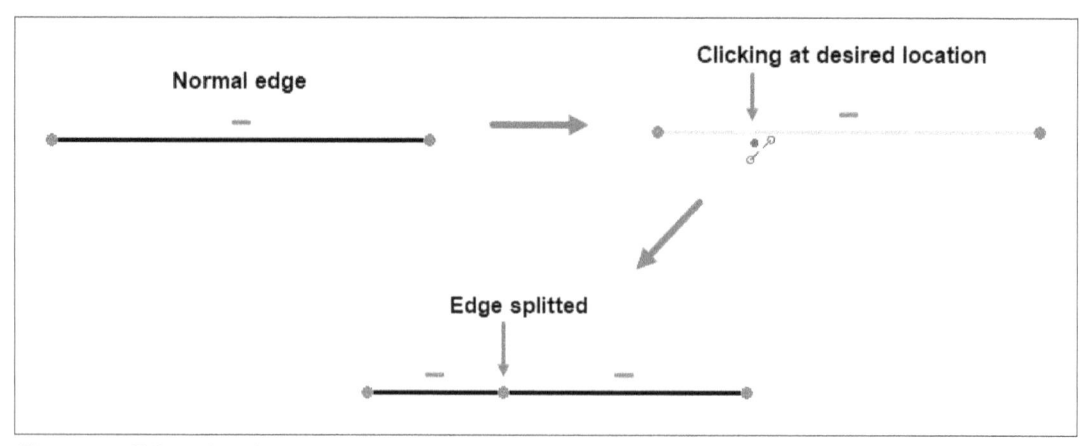

Figure-95. Edge splitted

Extend edge

The **Extend edge** tool is used to extend a curve to specified location in the sketch or to another curve. The procedure to use this tool is discussed next.

- Click on the **Extend edge** tool from **Edit edge** drop-down in the **Toolbar** of **Sketcher** workbench. You will be asked to select a line or an arc.
- Select a line or an arc you want to extend in a specified location or up to selected entity; refer to Figure-96.

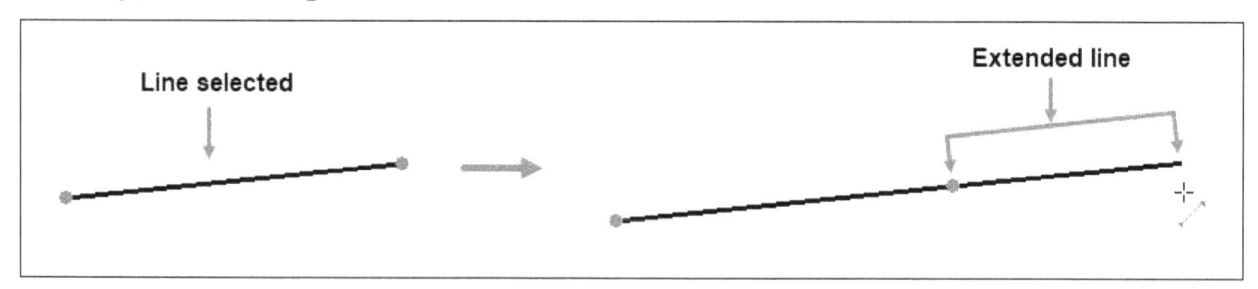

Figure-96. Line extended to an arbitrary location

- To extend a line/arc to another curve, hover the cursor on the curve. When the curve is highlighted and the **Point on object** [r] constraint icon appears then press **LMB** to extend the line. The line will be extended; refer to Figure-97.

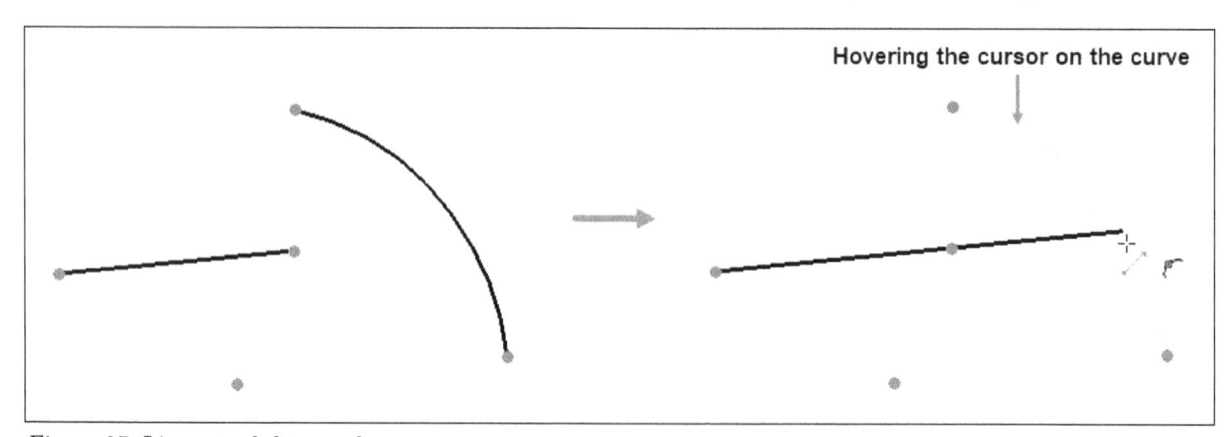

Figure-97. Line extended to an edge

- To extend an edge to a point in the sketch, hover the cursor over the point. When the point is highlighted and the **Coincident** constraint icon appears then press **LMB** to extend the line. The line will be extended; refer to Figure-98.

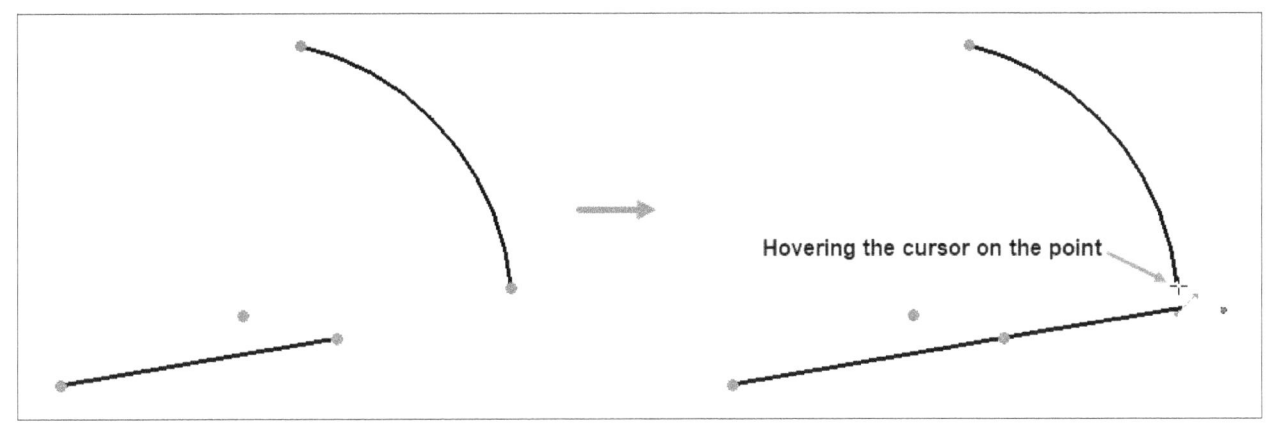

Figure-98. Line extended to a point

Note that only arcs and lines can be extended from this tool.

- Press **ESC** key or **RMB** to exit the tool.

Create external geometry

The **Create external geometry** tool is used to apply a constraint between sketch geometry and geometry external to current sketch. It works by inserting a linked construction geometry in the sketch. The procedure to use this tool is discussed next.

- Click on the **Create external geometry** tool from **Toolbar** in the **Sketcher** workbench; refer to Figure-99. You will be asked to select a line, arc, or curve.

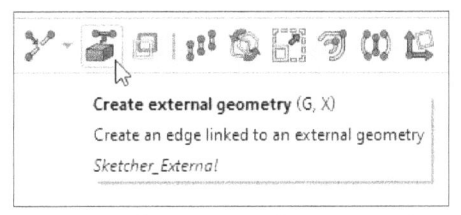

Figure-99. Create external geometry tool

- Select a curve that you want to link to the sketch; refer to Figure-100. The selected curve will become part of current sketch. Now, you can apply constraints between new sketch curves and previous selected sketch curves.

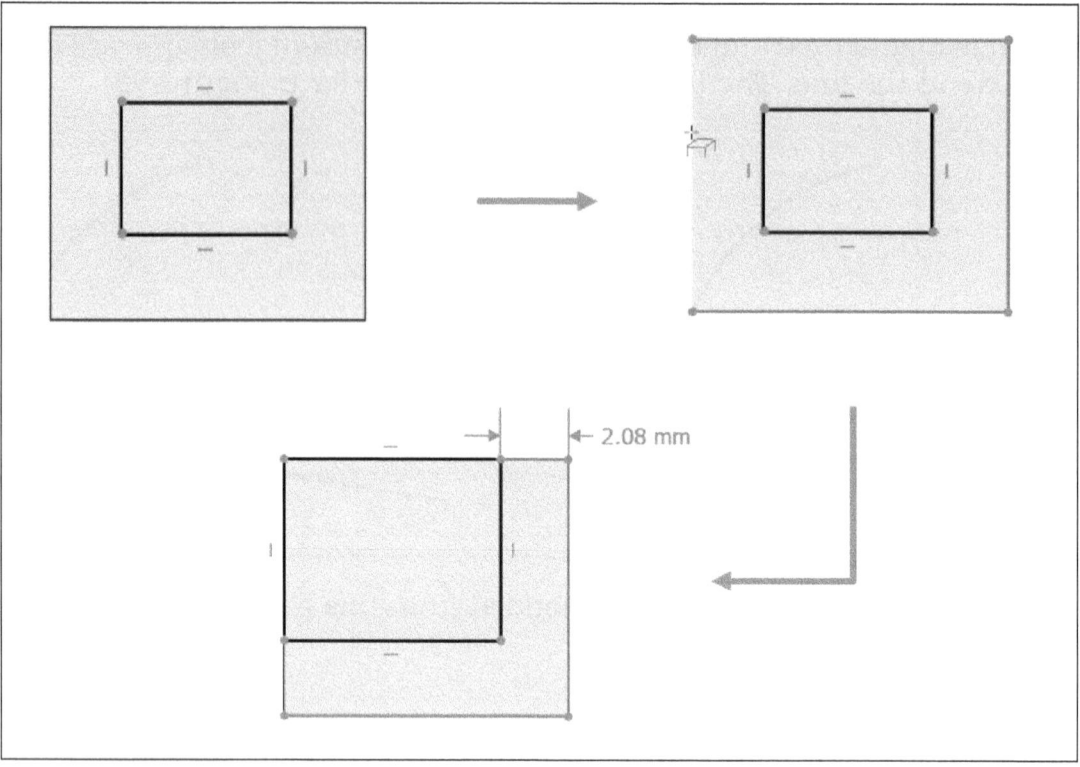

Figure-100. External geometry creation

- Press **ESC** key or press **RMB** to exit the tool.

Create carbon copy

The **Create carbon copy** tool copies all the geometry and constraints from another sketch into the active sketch. The procedure to use this tool is discussed next.

- Click on the **Create carbon copy** tool from **Toolbar** in the **Sketcher** workbench; refer to Figure-101. You will be asked to select curves from another sketch.

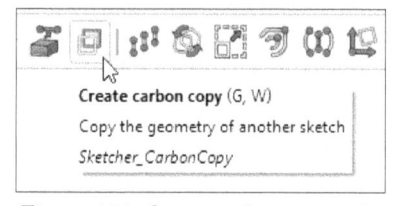

Figure-101. Create carbon copy tool

- Click on curves from the existing sketch, the geometry elements as well as constraints are copied into the active sketch; refer to Figure-102.
- Press **ESC** key or press **RMB** to exit the tool.

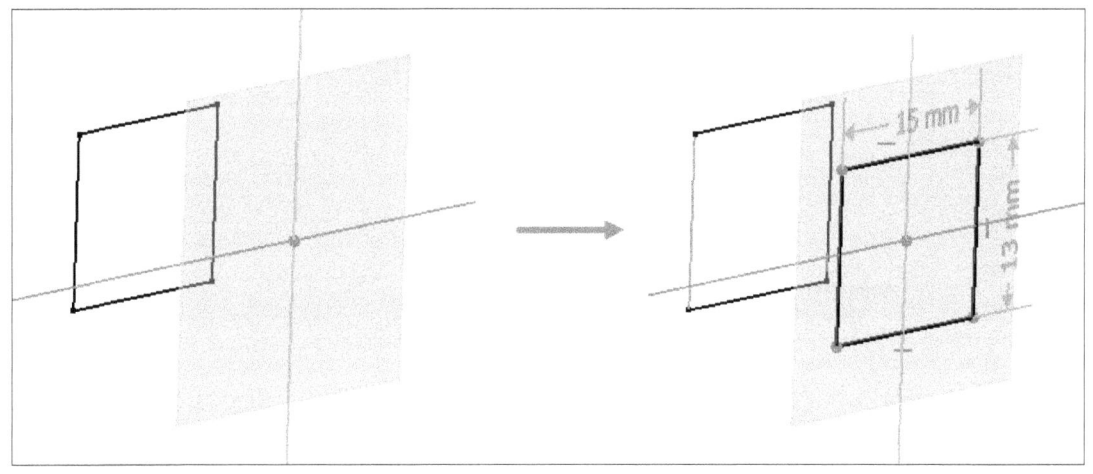

Figure-102. Carbon copy of sketch geometry

Move/Array transform

The **Move/Array transform** tool is used to move or create copies of selected elements. The procedure to use this tool is discussed next.

- Select desired entity from the 3D view area which you want to move or create an array and click on the **Move/Array transform** tool from **Toolbar** in the **Sketcher** workbench; refer to Figure-103. The **Translate parameters** dialog box will be displayed; refer to Figure-104, asking you to specify the parameters.

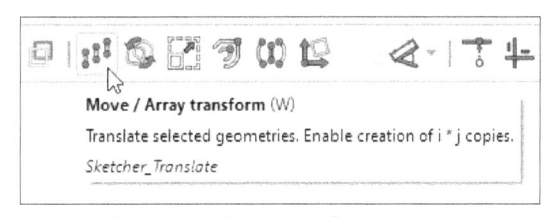

Figure-103. Move Array transform tool

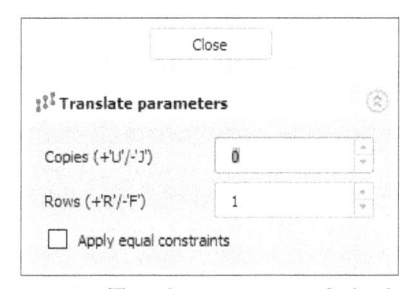

Figure-104. Translate parameters dialog box

- Specify desired copies of the entity to be created in the **Copies** edit box and specify desired number of rows upto which copies to be created in the **Rows** edit box.
- Select **Apply equal constraints** check box to apply equal constraints between original objects and their copies.
- After specifying desired parameters in the dialog box, click at desired location to specify the start point of the array. You will be asked to specify the end point.
- Move the cursor away and click at desired location to specify the end point of the array or enter desired distance between the copies and desired angle from the origin point in the edit boxes displayed. The array for the single row will be created; refer to Figure-105. In case of multiple rows, you will be asked to specify the location of subsequent rows.

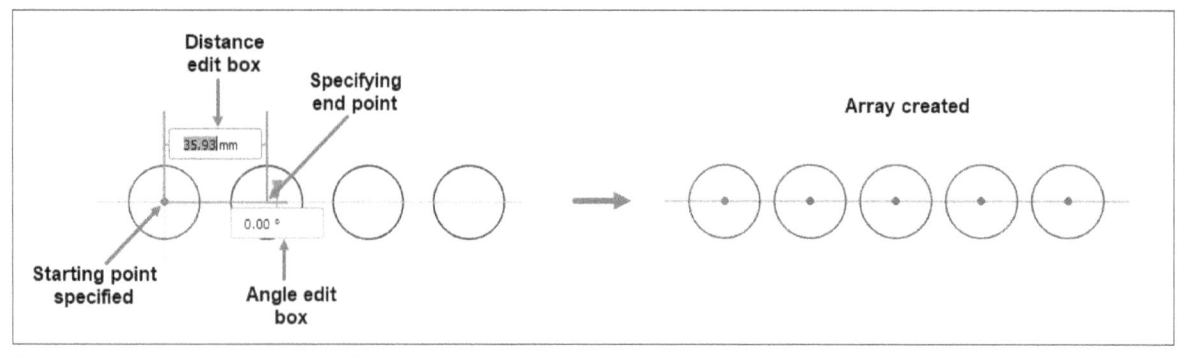

Figure-105. Array created for single row

- Move the cursor away and click at desired location to specify the location for subsequent rows or enter desired distance between the rows and desired angle of rows from the origin in the edit boxes displayed. The array in multiple rows will be created; refer to Figure-106.

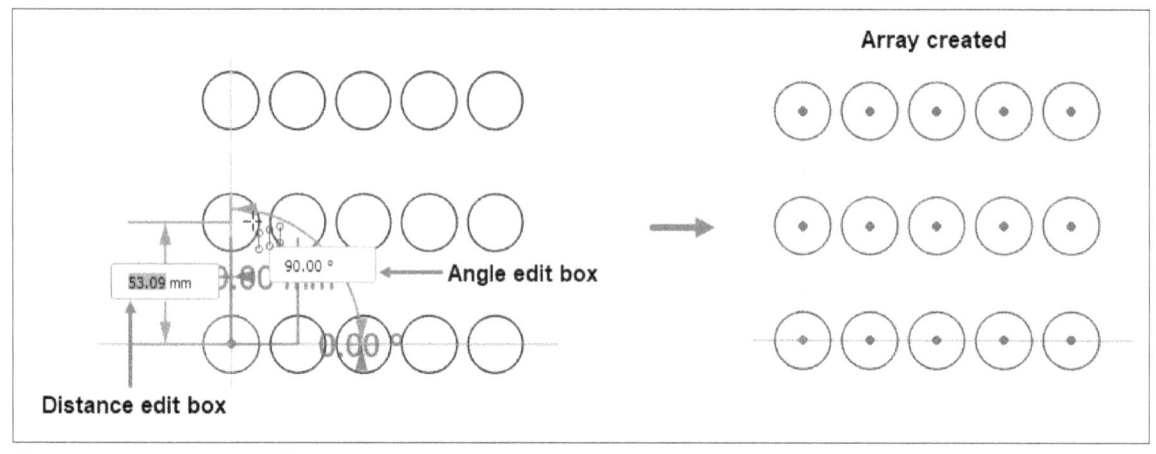

Figure-106. Array created for multiple rows

- If you want to move the entity, click on the **Move/Array transform** tool. You will be asked to specify start point.
- Click in the 3D view area at desired location to specify the starting point of movement. You will be asked to specify the end point.
- Move the cursor away and click at desired location to specify the target point at which the entity to be moved or enter desired distance value upto which the entity to be moved and specify desired angle value at which the entity to be moved in the edit boxes displayed. The entity will be moved; refer to Figure-107.

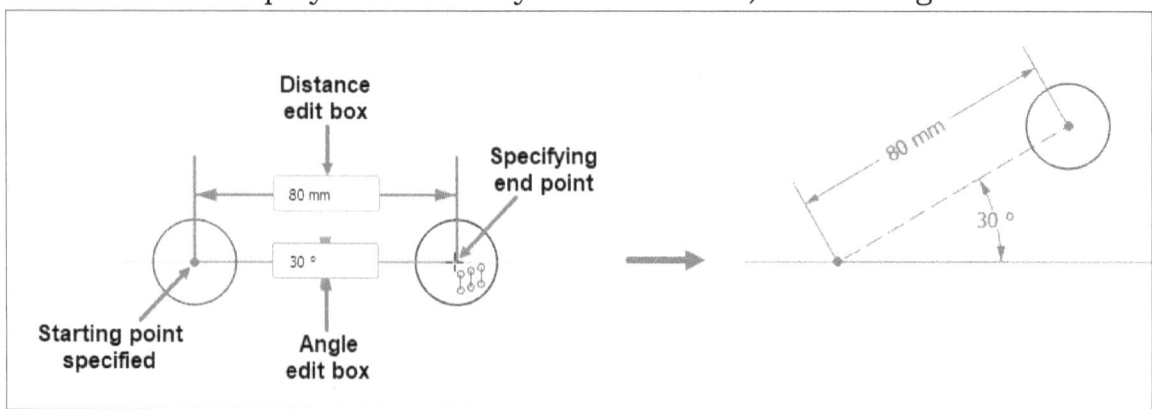

Figure-107. Entity moved

- Press **ESC** key or press **RMB** to exit the tool.

Rotate/Polar transform

The **Rotate/Polar transform** tool is used to rotate or create rotated copies of selected elements. The procedure to use this tool is discussed next.

- Select desired entity from the 3D view area which you want to rotate or create polar array and click on the **Rotate/Polar transform** tool from **Toolbar** in the **Sketcher** workbench; refer to Figure-108. The **Rotate parameters** dialog box will be displayed; refer to Figure-109, asking you to specify the parameters.

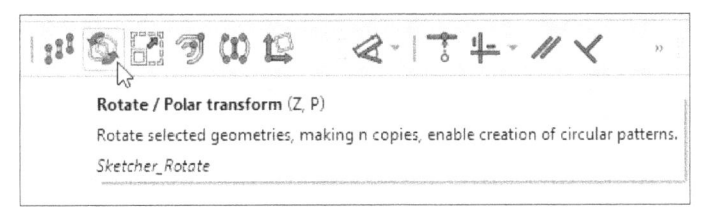

Figure-108. Rotate Polar transform tool

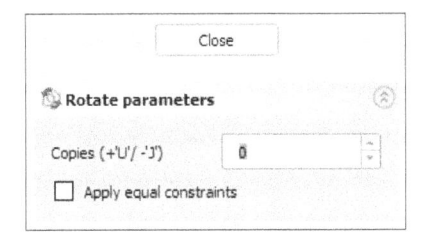

Figure-109. Rotate parameters dialog box

- Specify desired copies of the entity to be created in the **Copies** edit box. You will be asked to specify the rotation center.
- Click in the 3D view area at desired location to specify the center for rotation of entity(ies). You will be asked to specify the start angle.
- Rotate the cursor and click at desired location to specify the starting angle of rotation. You will be asked to specify the rotation angle.
- Rotate the cursor away and click at desired location to specify the rotation angle or enter desired value in the edit box displayed. The polar transform will be created; refer to Figure-110.

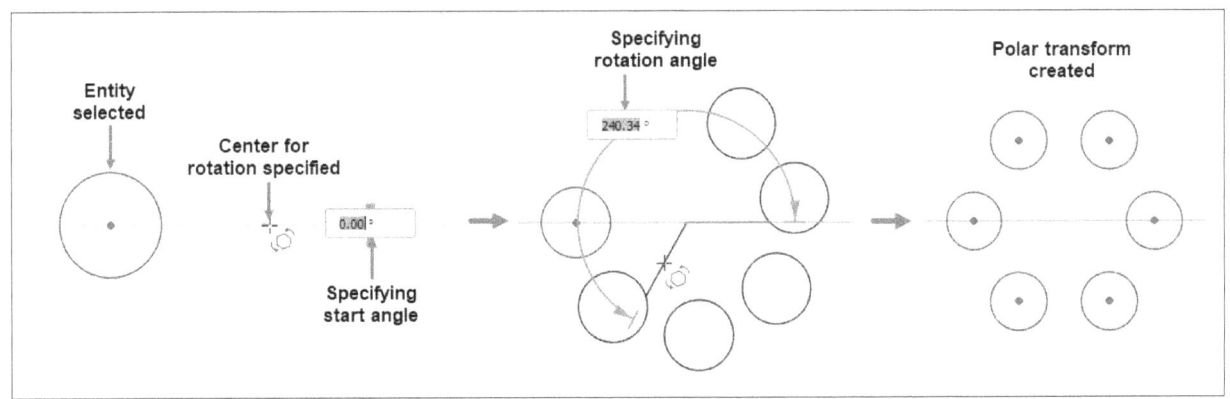

Figure-110. Polar transform created

- Or, after specifying the center for rotation of entity(ies), enter desired rotation angle in the edit box displayed. The polar transform will be created.
- Other parameters in the dialog box is same as discussed earlier.
- Press **ESC** key or press **RMB** to exit the tool.

Scale transform

The `Scale transform` tool is used to scale or create scaled copies of selected elements. The procedure to use this tool is discussed next.

- Select desired entity from the 3D view area to be scaled and click on the `Scale transform` tool from **Toolbar** in the **Sketcher** workbench; refer to Figure-111. The `Rotate parameters` dialog box will be displayed; refer to Figure-112, asking you to specify the parameters.

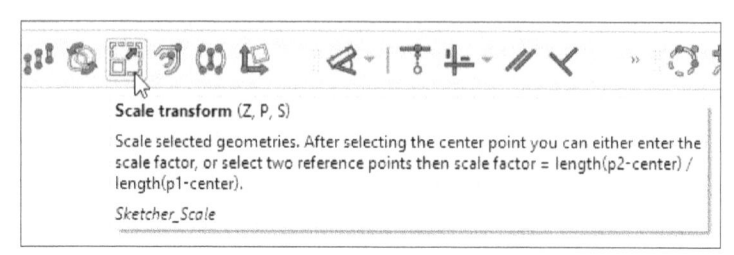

Figure-111. Scale transform tool

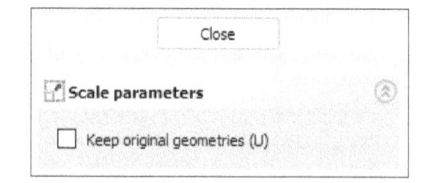

Figure-112. Scale parameters dialog box

- Select `Keep original geometries` check box if you do not want to delete original geometries after scaling. You will be asked to specify the base point.
- Click in the 3D view area at desired location to specify the base point for scaling. You will be asked to specify the first end point.
- Move the cursor away and click at desired location to specify the end point of first auxiliary line. You will be asked to specify the second endpoint.
- Move the cursor away and click at desired location to specify the end point of second auxiliary line or enter desired scale factor in the edit box displayed. The object will be scaled; refer to Figure-113.

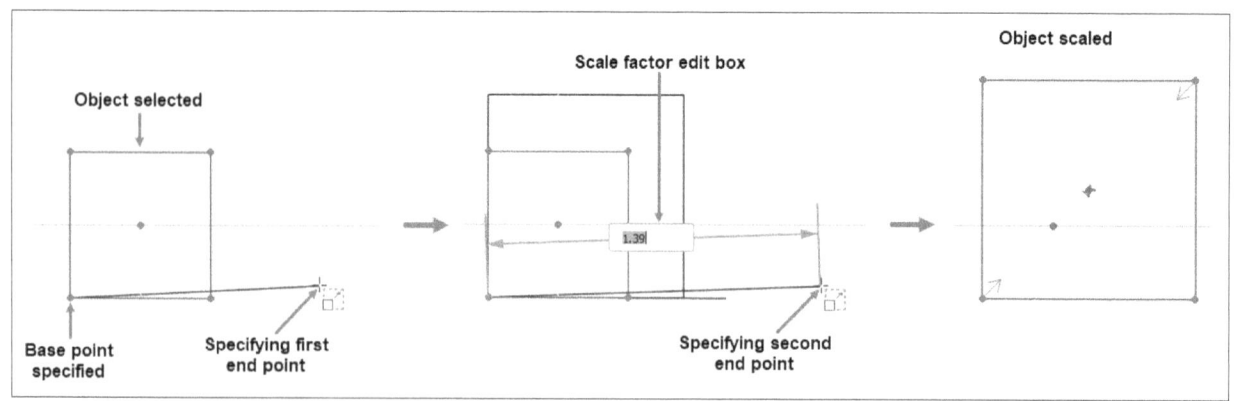

Figure-113. Object scaled

- Press **ESC** key or press **RMB** to exit the tool.

Offset geometry

The `Offset geometry` tool is used to create equidistant edges around selected edges. The procedure to use this tool is discussed next.

- Select desired entity from the 3D view area to be offset and click on the **Offset geometry** tool from **Toolbar** in the **Sketcher** workbench; refer to Figure-114. The **Offset parameters** dialog box will be displayed with **Arc** option selected in the **Mode** drop-down; refer to Figure-115, asking you to specify the parameters.

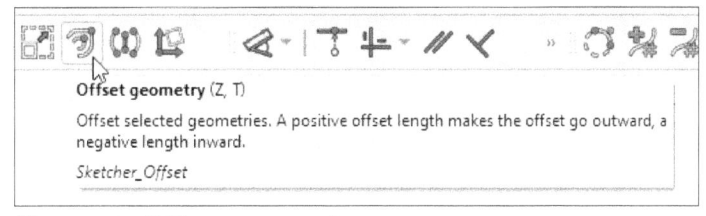

Figure-114. Offset geometry tool

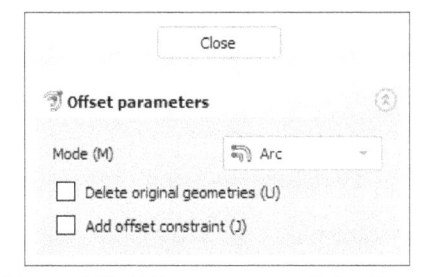

Figure-115. Offset parameters dialog box

- Select **Delete original geometries** check box to delete the original geometry after creating the offset geometry.
- Select **Add offset constraints** check box to add a dimensional constraint between the offset geometry and the original geometry. You will be asked to specify the offset distance.
- Move the cursor away and click at desired location to specify the offset distance or enter desired offset distance value in the edit box displayed. The offset object will be created; refer to Figure-116.

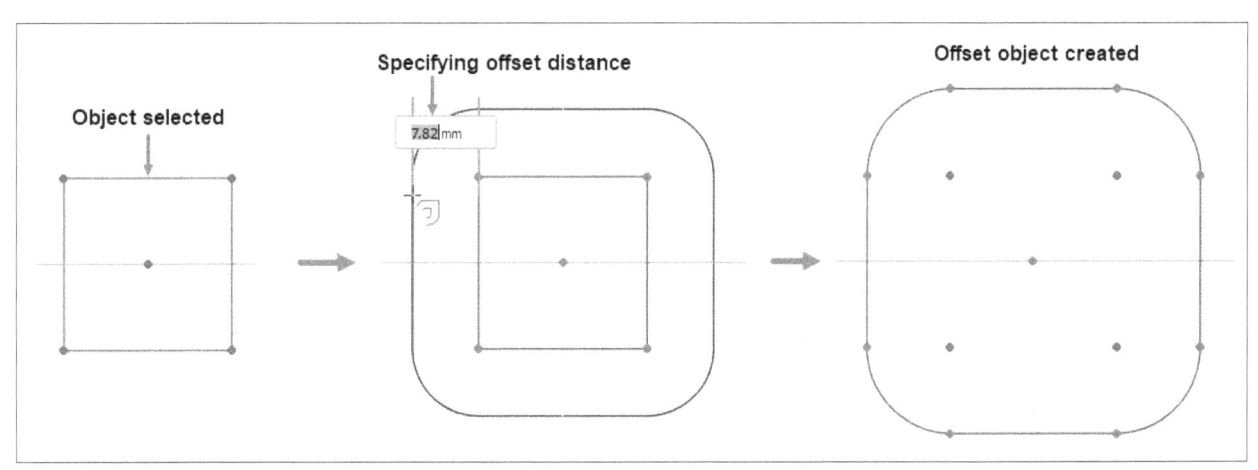

Figure-116. Offset geometry created

- Similarly, you can create offset geometry using **Intersection** option from **Mode** drop-down.
- Press **ESC** key or press **RMB** to exit the tool.

Symmetry

The **Symmetry** tool is used to mirror sketcher geometry in reference to a chosen line or sketch axis. Note that last selected entity will be used as symmetry axis. The procedure to use this tool is discussed next.

- Select the geometry that you want to copy and click on the **Symmetry** tool from **Toolbar** in the **Sketcher** workbench; refer to Figure-117. The **Symmetry parameters** dialog box will be displayed; refer to Figure-118, asking you to specify the parameters.

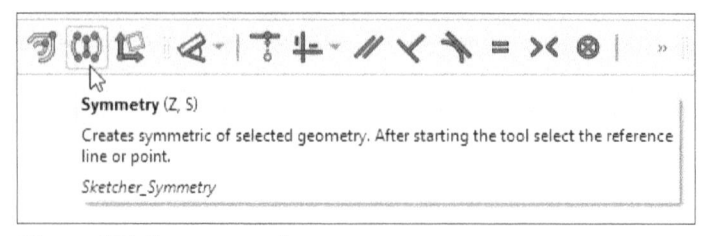

Figure-117. Symmetry tool

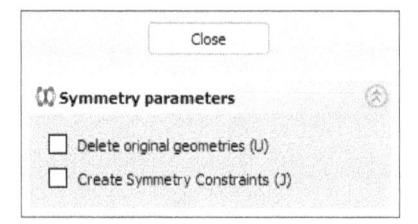

Figure-118. Symmetry parameters dialog box

- Select **Create Symmetry Constraints** check box to add symmetric constraint between original geometry and symmetric geometry.
- Other parameters in the dialog box is same as discussed earlier. You will be asked to select a line or sketch axis.
- Select desired line or sketch axis in the 3D view area to be used as symmetry axis. The selected geometry will be copied symmetrically against the selected line or sketch axis; refer to Figure-119.

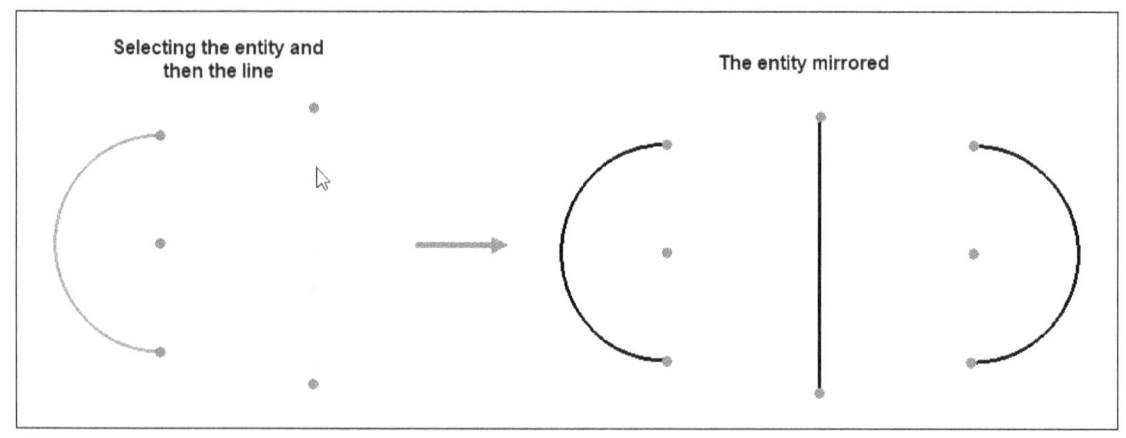

Figure-119. Symmetry entity creation

- Press **ESC** key or press **RMB** to exit the tool.

Remove axes alignment

The **Remove axes alignment** tool modifies constraints to remove axes alignment while trying to preserve the shape of section. The procedure to use this tool is discussed next.

- Select desired geometry from which you want to remove the axes alignment; refer to Figure-120.

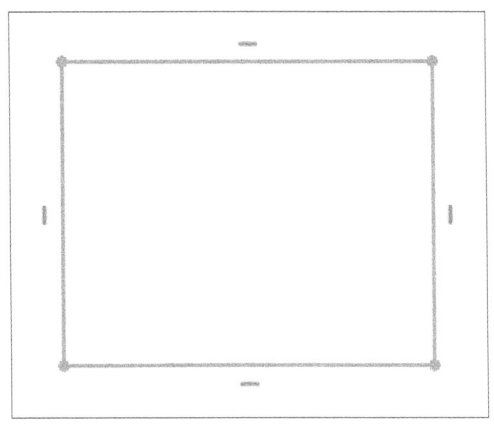

Figure-120. Rectangle selected

- Click on the **Remove axes alignment** tool from **Toolbar** in the **Sketcher** workbench; refer to Figure-121. The axes alignment from the geometry will be removed; refer to Figure-122.

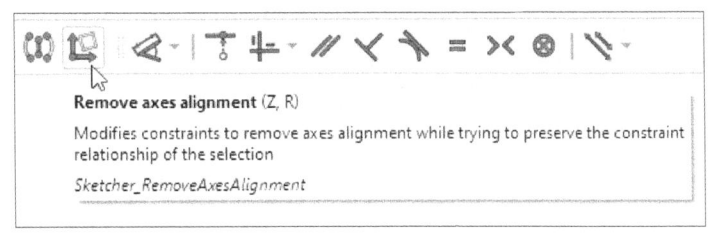

Figure-121. Remove axes alignment tool

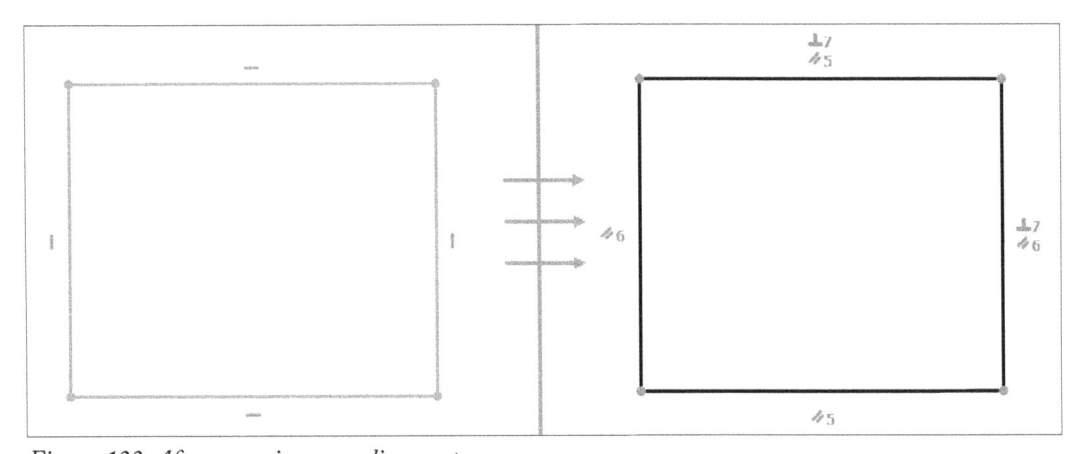

Figure-122. After removing axes alignment

Delete all geometry

The **Delete all geometry** tool is used to delete all the geometries of current sketch. The procedure to use this tool is discussed next.

- Click on the **Delete all geometry** tool from the **Sketcher tools** cascading menu in the **Sketch** menu; refer to Figure-123. The **Delete All Geometry** information box will be displayed asking you whether to delete all the geometries or not; refer to Figure-124.

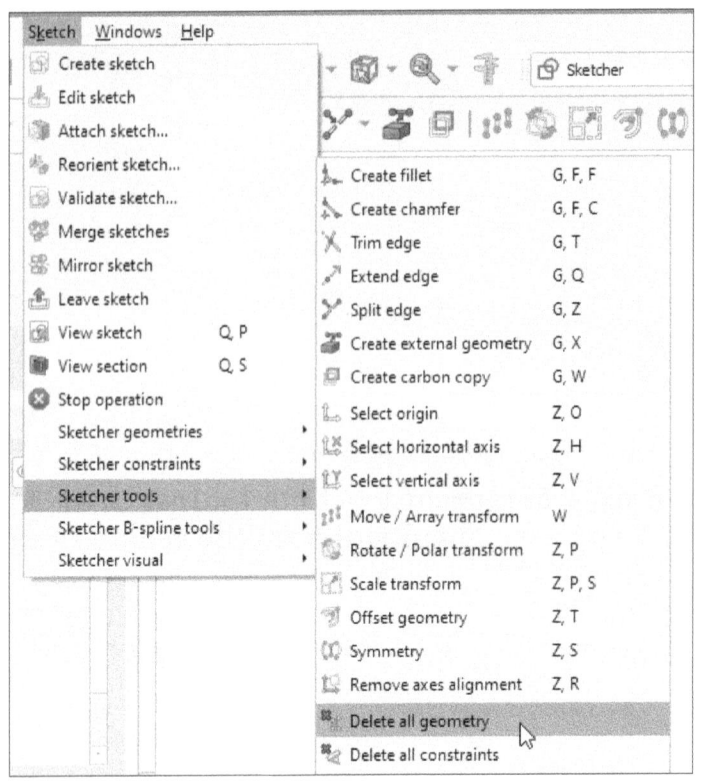

Figure-123. Delete all geometry tool

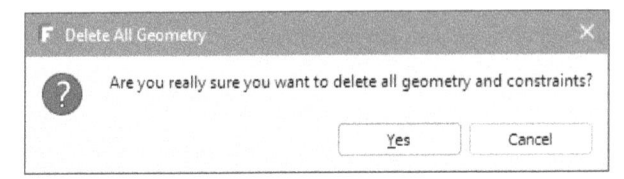

Figure-124. Delete All Geometry information box

- Click on the **Yes** button from the dialog box to delete all the geometry from the sketch. The geometries will be deleted.

Delete all constraints

The `Delete all constraints` tool is used to delete all the constraints from the sketch being edited and leaves the geometry intact. The procedure to use this tool is discussed next.

- Click on the **Delete all constraints** tool from the **Sketcher tools** cascading menu in the **Sketch** menu; refer to Figure-125. The **Delete All Constraints** information box will be displayed asking you to whether to delete all the constraints or not; refer to Figure-126.

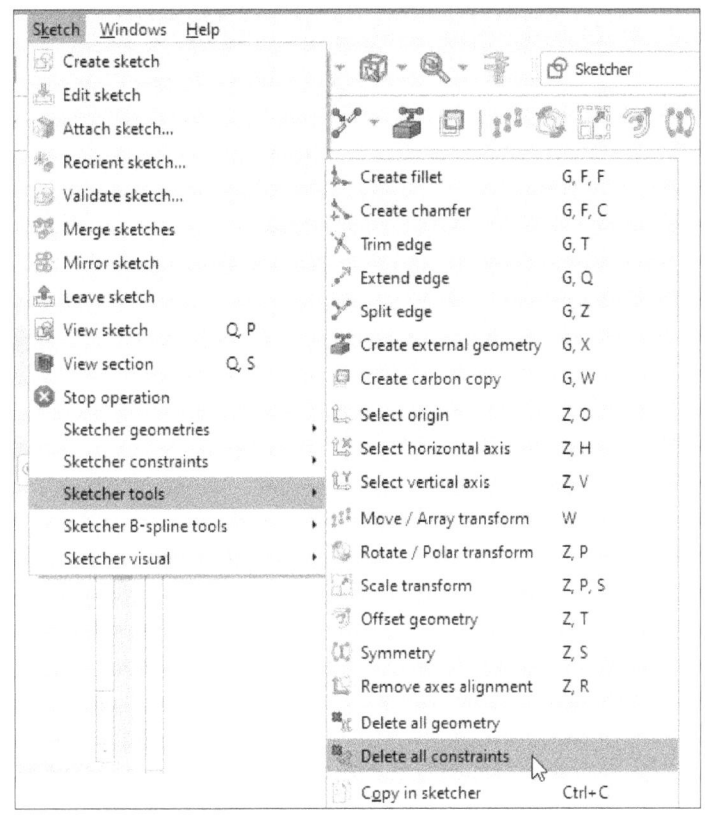

Figure-125. Delete all constraints tool

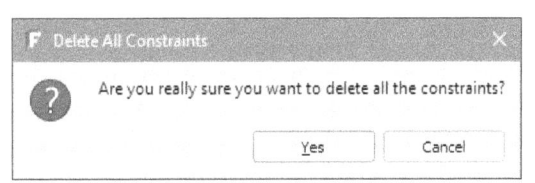

Figure-126. Delete All Constraints information box

- Click on the **Yes** button from the dialog box to delete all the constraints from the sketch. The constraints will be deleted; refer to Figure-127.

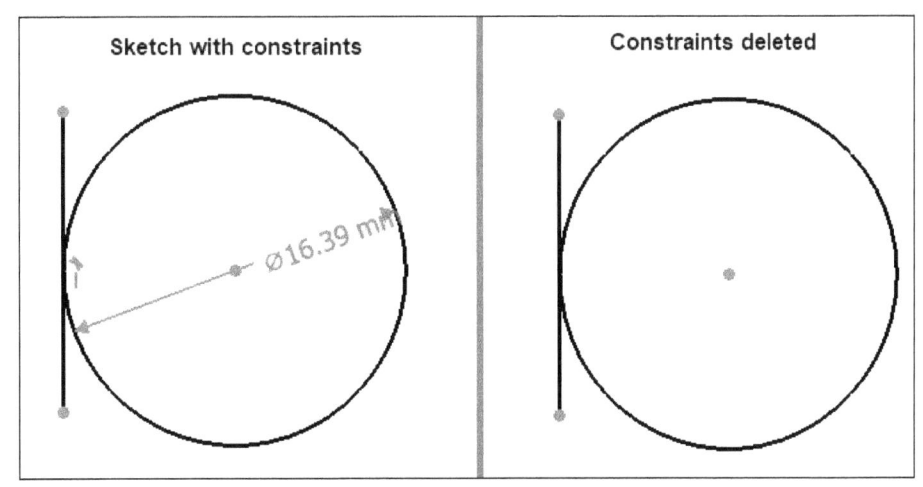

Figure-127. Constraints deleted from the sketch

Convert geometry to B-spline

The **Convert geometry to B-spline** tool converts compatible geometry, edges, and curves into a B-spline. The procedure to use this tool is discussed next.

- Select the geometry which you want to convert into a B-spline.
- Click on the **Convert geometry to B-spline** tool from **Toolbar** in the **Sketcher** workbench; refer to Figure-128. The selected geometry will be converted into a B-spline as shown in Figure-129.

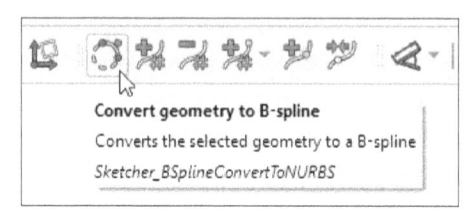

Figure-128. Convert geometry to B spline tool

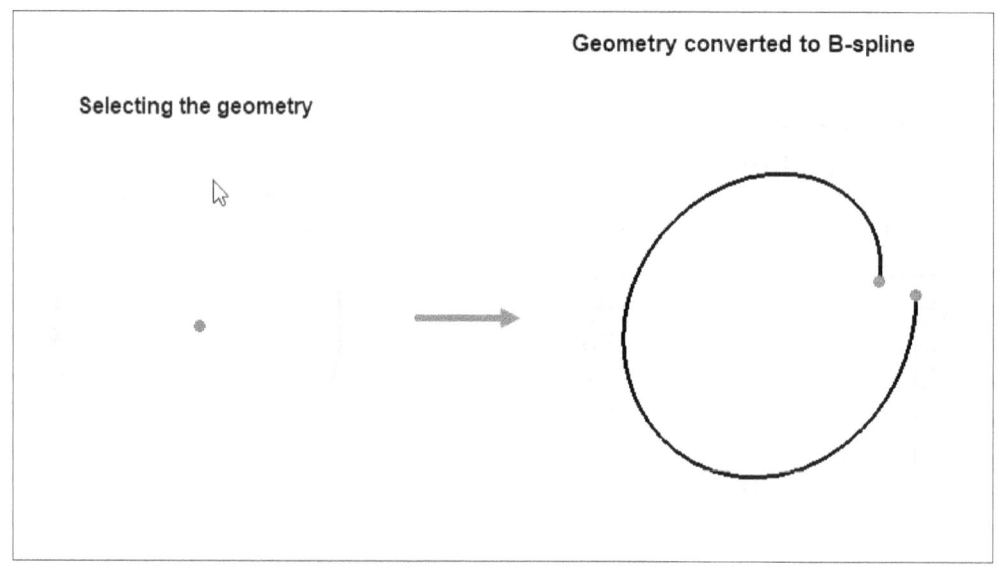

Figure-129. Geometry converted to B-spline

Increase B-spline degree

The **Increase B-spline degree** tool increases the degree of the B-spline. Degrees define the number of vertices of spline that can be modified. The procedure to use this tool is discussed next.

- Select the B-spline entity whose degree is to be increased.
- Click on the **Increase B-spline degree** tool from **Toolbar** in the **Sketcher** workbench; refer to Figure-130. The degree of B-spline entity will be increased; refer to Figure-131.

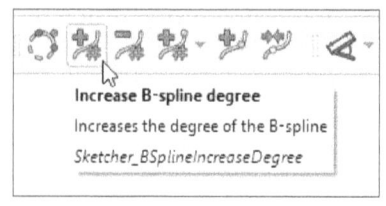

Figure-130. Increase B spline degree tool

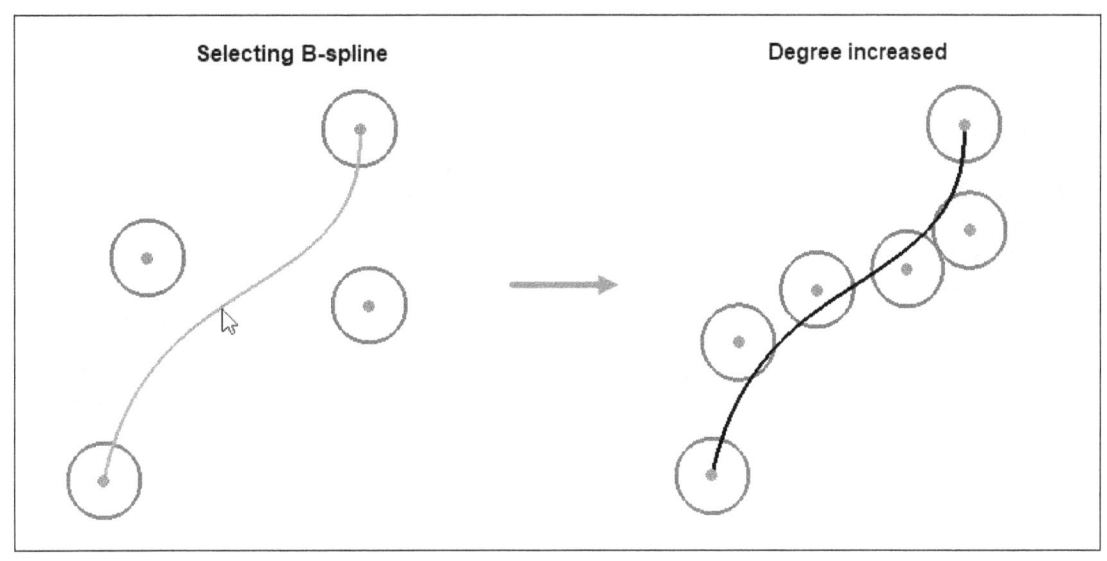

Figure-131. Degree increased of a B spline

Similarly, you can use **Decrease B-spline degree** tool to decrease the degree of the B-spline.

Modify knot multiplicity

There are two tools available in **Modify knot multiplicity** drop-down from **Toolbar** in the **Sketcher** workbench to modify knot multiplicity; **Increase knot multiplicity** and **Decrease knot multiplicity**; refer to Figure-132. The procedures to use these tools are discussed next.

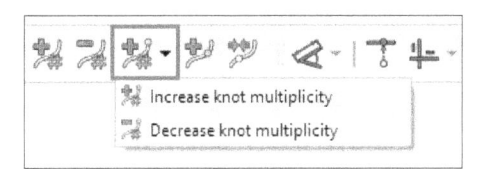

Figure-132. Modify knot multiplicity drop down

Increase knot multiplicity

The **Increase knot multiplicity** tool increases the knot multiplicity of a B-spline curve knot. The number of times a knot value is duplicated is called the knot multiplicity. Knot multiplicity defines the number of times knot appears in the knot sequence of spline. Knot multiplicity higher than degree of spline will cause spline to split in disjoint parts. Select the knot from spline and click on the **Increase knot multiplicity** tool from the drop-down.

Decrease knot multiplicity

The **Decrease knot multiplicity** tool decreases the knot multiplicity of a B-spline curve knot. Select the knot from spline and click on the **Decrease knot multiplicity** tool from the drop-down.

Insert knot

The **Insert knot** tool is used to insert a knot into an existing B-spline. If a knot already exists at the specified parameter, it increases the multiplicity of that knot by one. The procedure to use this tool is discussed next.

- Select the B-spline into which you want to insert the knot and click on the **Insert knot** tool from **Toolbar** in the **Sketcher** workbench; refer to Figure-133. A knot symbol will be attached to the cursor.

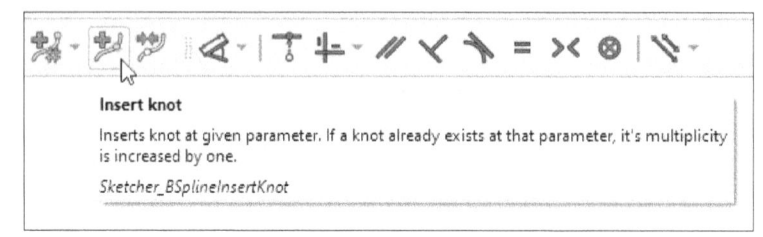

Figure-133. Insert knot tool

- Move the cursor over the spline and click at desired location where you want to insert the knot. The knot will be inserted; refer to Figure-134.

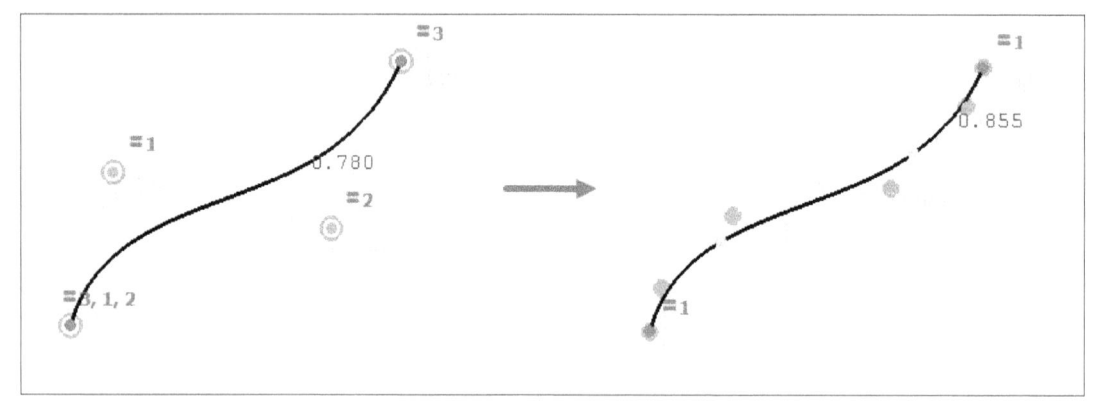

Figure-134. Knot inserted

Join curves

The **Join curves** tool is used to join two curves at selected end points. The procedure to use this tool is discussed next.

- Select the end points of two curves to be joined and click on the **Join curves** tool from **Toolbar** in the **Sketcher** workbench; refer to Figure-135. The end point of curves will be joined.

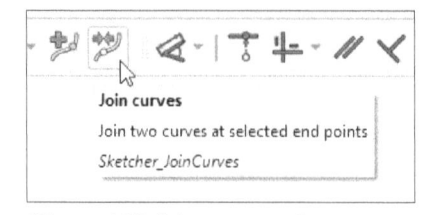

Figure-135. Join curves tool

SKETCHER VISUAL TOOLS

The **Sketcher visual** tools in **FreeCAD** are available in the **Sketcher** workbench of **toolbar**. You can also access the sketcher visual tools from the **Sketcher visual** cascading menu in the **Sketch** menu; refer to Figure-136. The procedures to use these tools are discussed next.

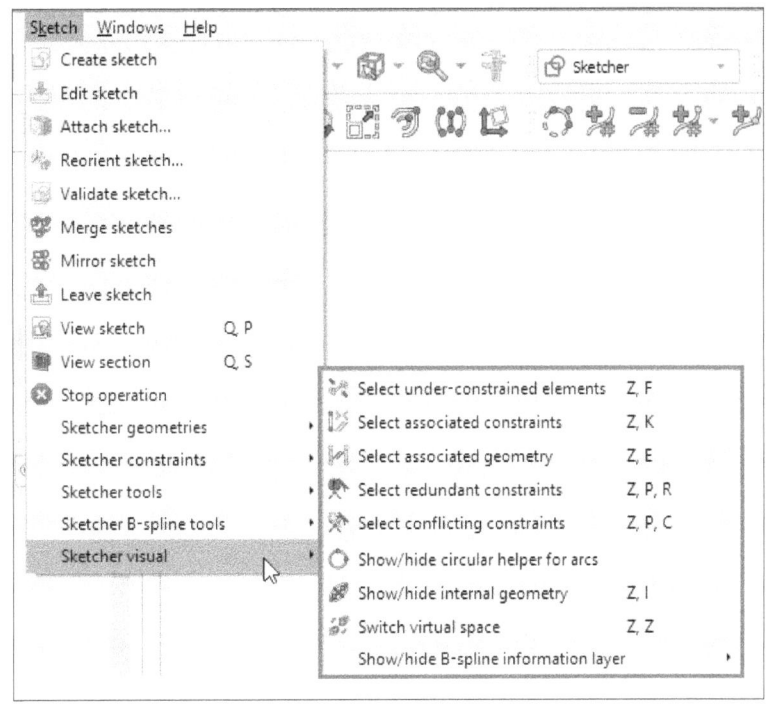

Figure-136. Sketcher visual cascading menu

Select under-constrained elements

The **Select under-constrained elements** tool is used to aid in fully constraining a sketch by highlighting the sketch elements with remaining degrees of freedom (DoF) in green color. The procedure to use this tool is discussed next.

- Click on the **Select under-constrained elements** tool from the **Sketcher visual** cascading menu in the **Sketch** menu. The **Solver messages** dialog box will be displayed in the **Tasks** panel of **Combo View** displaying the messages related to constraints and the geometry will be highlighted in green color; refer to Figure-137. Note that in case of lines, only points will be highlighted in green.

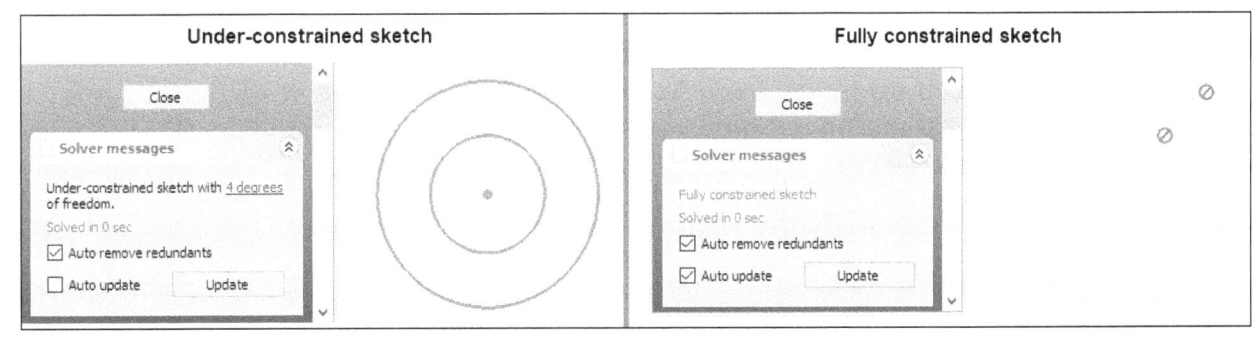

Figure-137. Solver messages dialog box displaying constraint messages

- Press **LMB** to exit the tool.

Select associated constraints

The **Select associated constraints** tool is used to select the constraints associated to the selected sketcher elements. The procedure to use this tool is discussed next.

- Select the entities of the sketch whose constraints are to be selected.
- Click on the **Select associated constraints** tool from **Sketcher visual** cascading menu in the **Sketch** menu or from **Toolbar** in the **Sketcher** workbench; refer to Figure-138. The constraints associated with selected entities will be selected; refer to Figure-139.
- Press **LMB** to exit the tool.

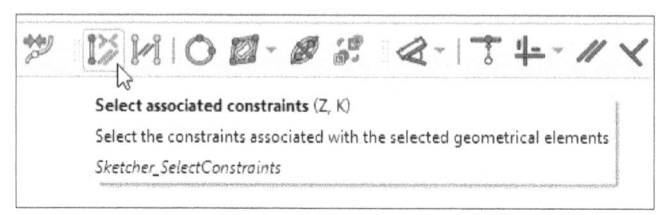

Figure-138. Select associated constraints tool

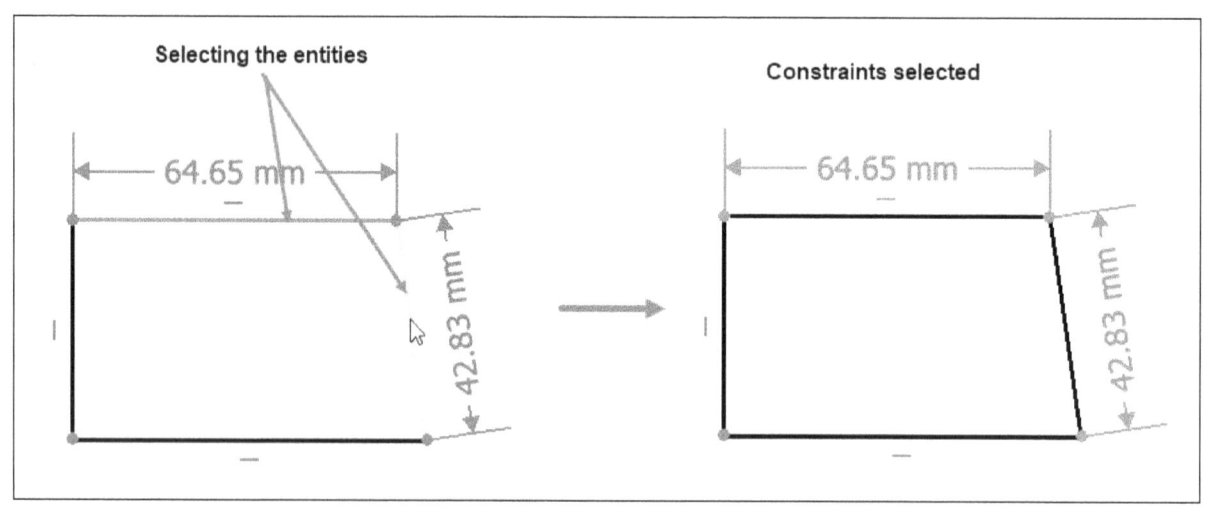

Figure-139. Selection of constraints

Select associated geometry

The **Select associated geometry** tool is used to select the sketcher element associated with constraints. The procedure is same as previous topic discussed.

- Click on the **Select associated geometry** tool from **Sketcher visual** cascading menu in the **Sketch** menu or from **Toolbar** in the **Sketcher** workbench; refer to Figure-140. The associated geometry of sketch will be selected in the 3D view area.
- Press **LMB** to exit the tool.

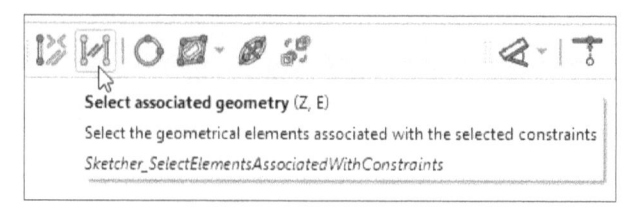

Figure-140. Select associated geometry tool

Select redundant constraints

The **Select redundant constraints** tool is used to select unnecessary constraints of the sketch. The procedure to use this tool is discussed next.

- Click on the **Select redundant constraints** tool from **Sketcher visual** cascading menu in the **Sketch** menu. The redundant constraints of the sketch will be selected in the 3D view area; refer to Figure-141.

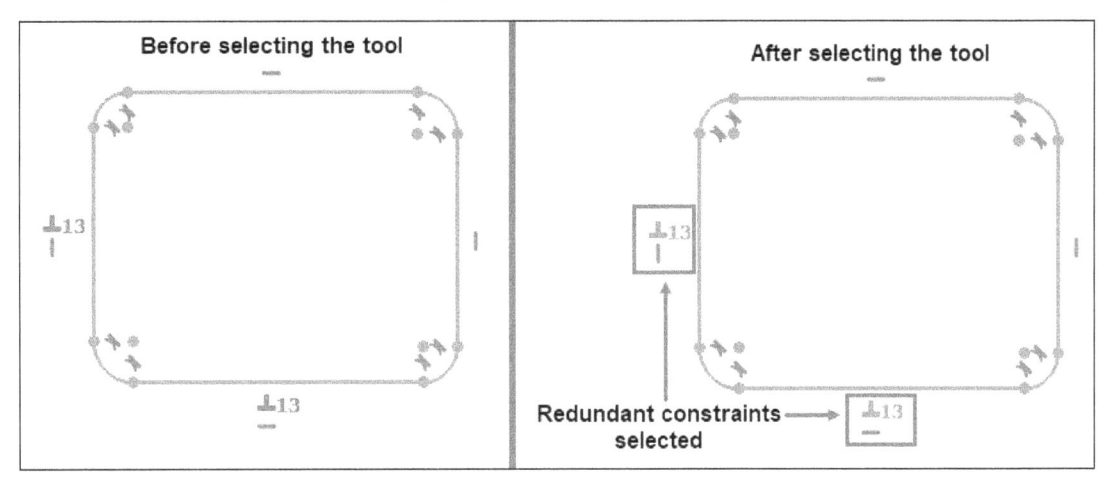

Figure-141. Redundant constraints of sketch selected

- Press **LMB** to exit the tool.

Select conflicting constraints

The **Select conflicting constraints** tool is used to select the conflicting constraints of a sketch. The procedure to use this tool is same as discussed for **Select redundant constraints** tool.

Show/hide circular helper for arcs

The **Show/hide circular helper for arcs** tool is used to show or hide the circular helpers for arcs in all sketches. The procedure to use this tool is discussed next.

- Click on the **Show/hide circular helper for arcs** tool from **Sketcher visual** cascading menu in the **Sketch** menu or from **Toolbar** in the **Sketcher** workbench. The circular helper will be shown; refer to Figure-142.

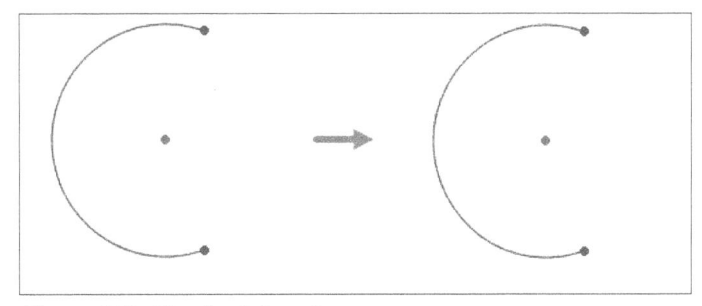

Figure-142. Circular helper shown

Show/hide internal geometry

The **Show/hide internal geometry** tool is used to show or hide unnecessary internal geometry or recreates missing geometry of a selected ellipse, arc of ellipse/hyperbola/parabola, or B-spline. The procedure to use this tool is discussed next.

- Select desired entity of sketch from which you want to show or hide the unnecessary geometry.
- Click on the **Show/hide internal geometry** tool from **Sketcher visual** cascading menu in the **Sketch** menu or from **Toolbar** in the **Sketcher** workbench; refer to Figure-143. The internal geometry of selected entity will be hidden from the sketch like control points of spline; refer to Figure-144.

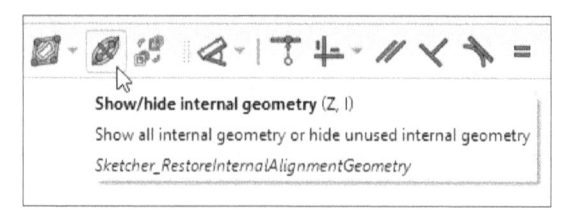

Figure-143. Show hide internal geometry tool

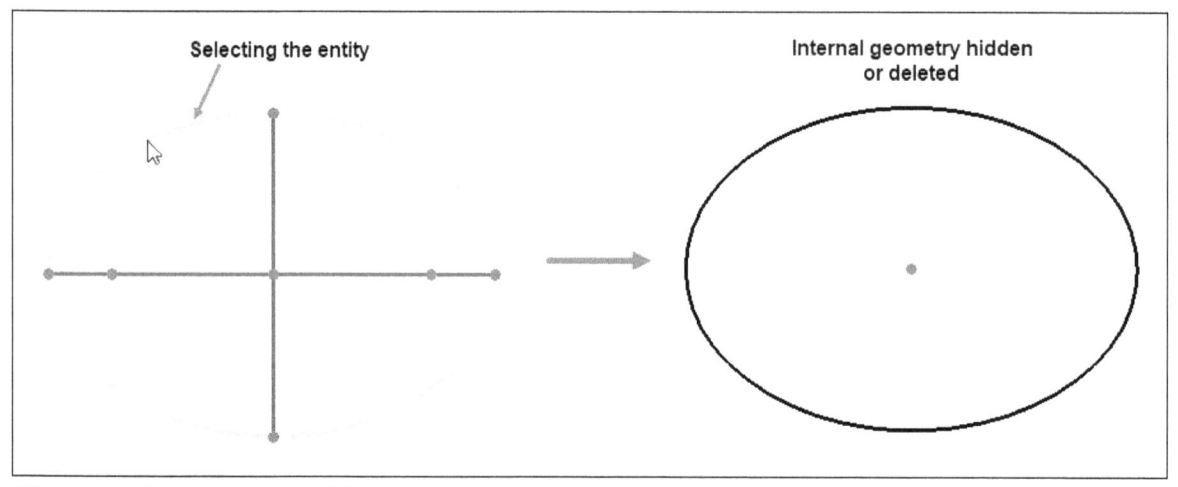

Figure-144. Internal geometry deleted

Switch virtual space

The **Switch virtual space** tool is used to switch or hide the selected constraints to the other virtual space.

Show/hide B-spline information layer tools

The **Show/hide B-spline information layer** tools in **FreeCAD** are available in the **Show/hide B-spline information layer** cascading menu of **Sketcher visual** cascading menu in the **Sketch** menu or in the **Show/hide B-spline information layer** drop-down of **Toolbar** in the **Sketcher** workbench; refer to Figure-145. The procedures to use these tools are discussed next.

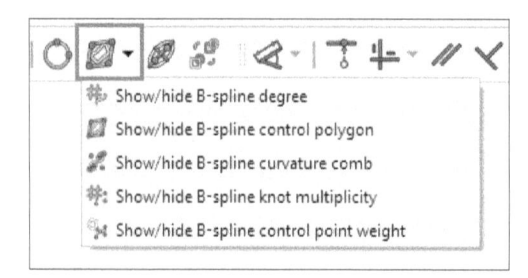

Figure-145. Show hide B spline information layer drop down

Show/hide B-spline degree

The **Show/hide B-spline degree** tool shows or hides the degree of B-spline curve. The procedure to use this tool is discussed next.

- Select the B-spline entity whose degree is to be shown or hidden.
- Click on the **Show/hide B-spline degree** tool from **Show/hide B-spline information layer** drop-down in the **Toolbar** of **Sketcher** workbench. The degree of B-spline will be displayed as shown in Figure-146.

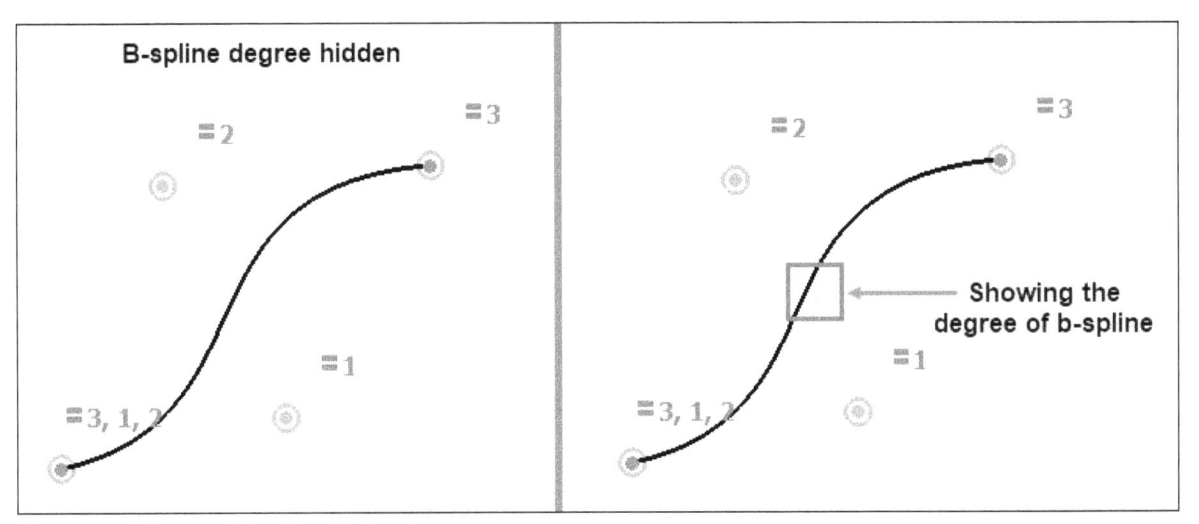

Figure-146. Degree of B spline displayed

Show/hide B-spline control polygon

The **Show/hide B-spline control polygon** tool shows or hides the defining polygon of a B-spline curve. The procedure to use this tool is discussed next.

- Select a B-spline entity whose polygon is to be shown or hidden.
- Click on the **Show/hide B-spline control polygon** tool from **Show/hide B-spline information layer** drop-down in the **Toolbar** of **Sketcher** workbench. The polygon of selected B-spline entity will be shown or hidden; refer to Figure-147.

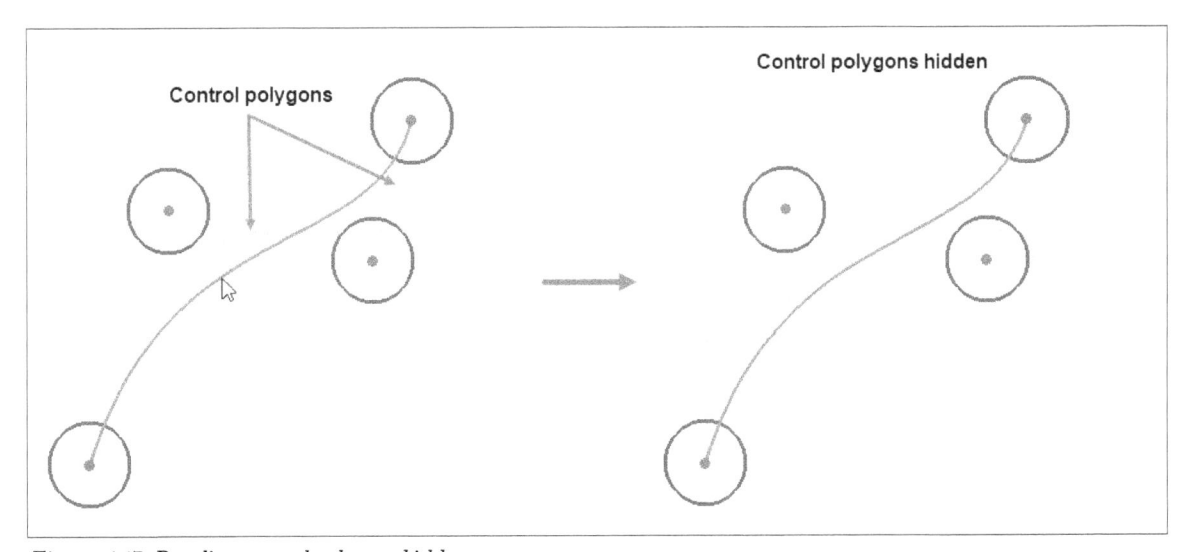

Figure-147. B-spline control polygons hidden

Show/hide B-spline curvature comb

The **Show/hide B-spline curvature comb** tool shows or hides the curvature comb of B-spline curve. The procedure to use this tool is discussed next.

- Select the B-spline entity whose curvature comb is to be shown or hidden.
- Click on the **Show/hide B-spline curvature comb** tool from **Show/hide B-spline information layer** drop-down in the **Toolbar** of **Sketcher** workbench. The curvature comb of selected B-spline entity will be shown or hidden as shown in Figure-148.

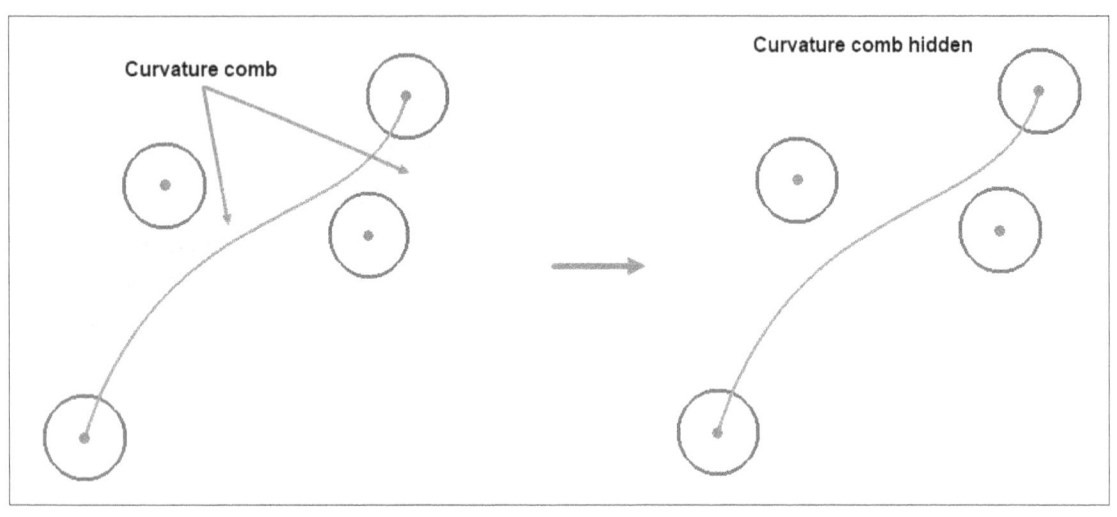

Figure-148. B-spline curvature comb hidden

Show/hide B-spline knot multiplicity

The **Show/hide B-spline knot multiplicity** tool shows or hides the knot multiplicity of a B-spline curve. The procedure to use this tool is discussed next.

- Select the B-spline entity whose knot multiplicity is to be shown or hidden.
- Click on the **Show/hide B-spline knot multiplicity** tool from **Show/hide B-spline information layer** drop-down in the **Toolbar** of **Sketcher** workbench. The knot numbers of spline vertices will be displayed; refer to Figure-149.

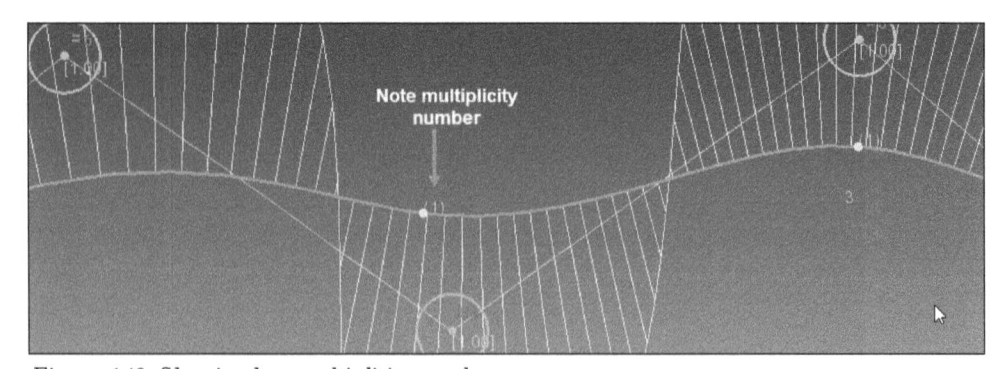

Figure-149. Showing knot multiplicity number

Show/hide B-spline control point weight

The **Show/hide B-spline control point weight** tool shows or hides of the weights for the control points of a B-spline curve.

B-splines are basically a combination of Bezier curves'. The Bezier curve is calculated using this formula:

$$\text{Bezier}(n, t) = \sum_{i=0}^{n} \underbrace{\binom{n}{i}}_{\text{polynomial term}} \underbrace{(1-t)^{n-i} t^i}_{\text{polynomial term}} \underbrace{P_i}_{\text{point coordinate}}$$

n is hereby the degree of the curve. So, a bezier curve of degree **n** is a polygon with order **n**. The factors P_i are hereby infact the coordinates of the Bezier curves' control points.

The procedure to use this tool is discussed next.

- Select the B-spline entity whose control point weight is to be shown or hidden.
- Click on the **Show/hide B-spline control point weight** tool from **Show/hide B-spline information layer** drop-down in the **Toolbar** of **Sketcher** workbench. The control point weight of selected B-spline entity will be shown or hidden as shown in Figure-150.

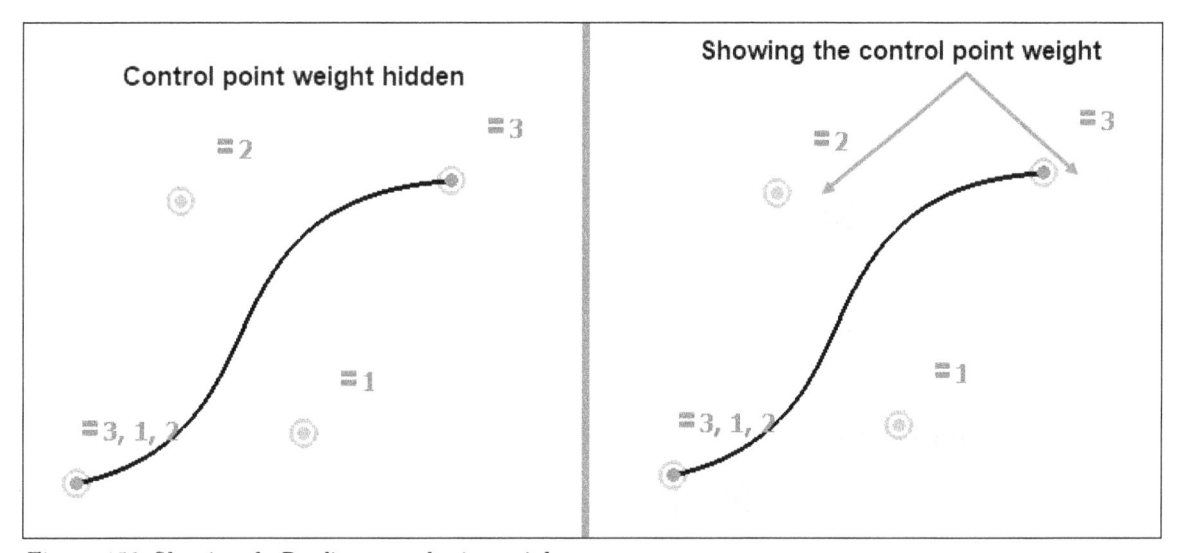

Figure-150. Showing the B spline control point weight

Practical 1

Create the sketch as shown in Figure-151. Also, dimension the sketch as per the figure.

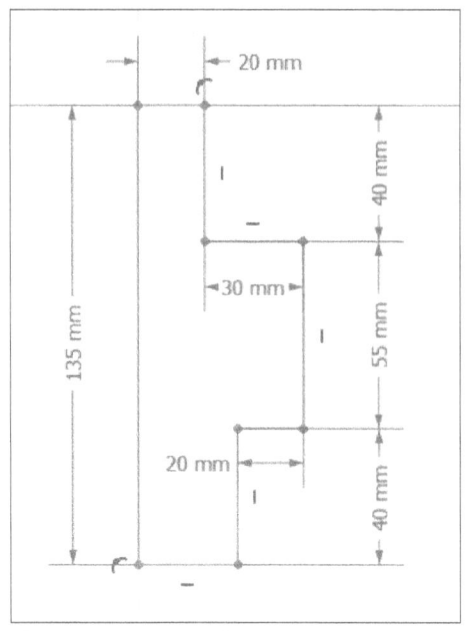

Figure-151. Practical 1

Steps to be performed:
- Start a new part file.
- Select a sketching plane and activate sketching mode.
- Create the sketch using **Line** tool.
- Apply the dimensions using dimension tools.
- Save the file.

Starting a new sketch

- Start FreeCAD if not started yet.
- Select the **Sketcher** option from the **Switch between workbenches** drop-down in the **Toolbar**. The sketcher environment will be displayed.
- Click on the **Create sketch** tool from the **Toolbar** or **Sketch** menu. The **Choose orientation** dialog box will be displayed; refer to Figure-152.

Figure-152. Choose orientation dialog box

- Select the **XY-Plane** radio button from the **Sketch orientation** area of dialog box to create sketch on XY plane.
- Click on **OK** button from the dialog box. The tools to create sketch will become active.

Creating line sketch

- Click on the **Create polyline** tool from the **Toolbar** or **Sketcher geometries** cascading menu of **Sketch** menu to activate polyline tool. You will be asked to specify start point of line.
- Click at the origin to specify start point of line and create the sketch as shown in Figure-153. Make sure to select the **Snap to grid** check box from **Toggle snap** drop-down in the toolbar of **Sketcher** workbench. The process to set unit to **mm** has been discussed earlier.

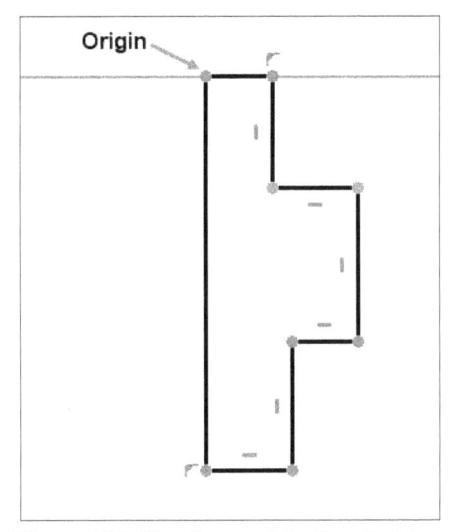

Figure-153. Sketch created for Practical 1

Applying dimensions

- Click on the **Constrain vertical distance** tool from the **Toolbar** or **Sketcher constraints** cascading menu of **Sketch** menu. You can also use **SHIFT+V** shortcut key to activate this tool. On activating this tool, you will be asked to select a vertical line.
- Select the vertical base line of sketch. The **Insert length** dialog box will be displayed with dimension of line; refer to Figure-154.

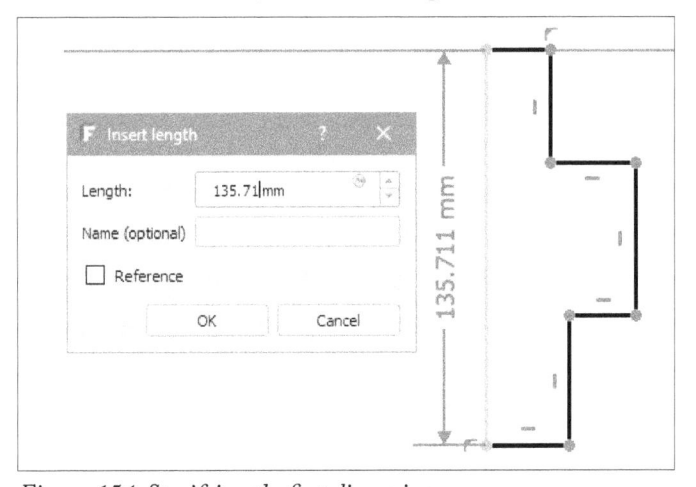

Figure-154. Specifying the first dimension

- Specify the length of vertical line as **135** mm in the **Length** edit box and click on the **OK** button to create the dimension.
- Repeat the steps to create all the vertical dimensions of sketch as shown in Figure-155.

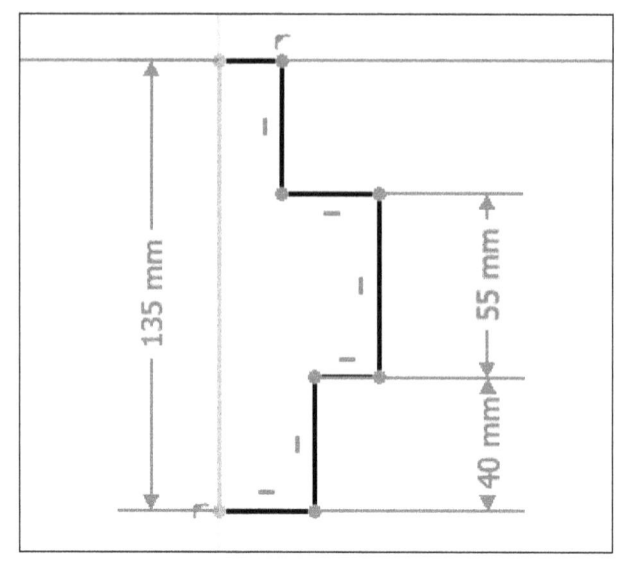

Figure-155. Vertical dimensions of sketch

- Click on the **Constrain horizontal distance** tool from the **Toolbar** or **Sketcher constraints** cascading menu of **Sketch** menu. You can also use **SHIFT+H** shortcut key to activate this tool. On activating this tool, you will be asked to select a horizontal line.
- Select the horizontal line of sketch. The **Insert length** dialog box will be displayed with dimension of line; refer to Figure-156.

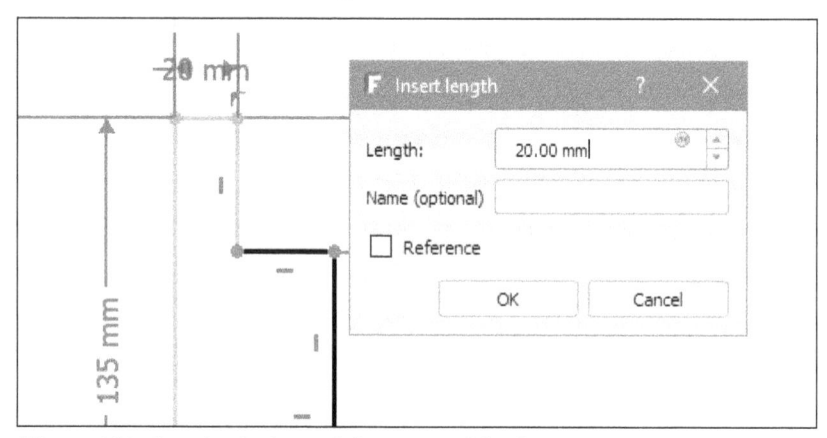

Figure-156. Creating horizontal dimension of sketch

- Specify the length of horizontal line as **20** mm in the **Length** edit box and click on the **OK** button to create the dimension.
- Repeat the steps to create all the horizontal dimensions of sketch.
- After specifying all the constraints and dimensions, the sketch should be displayed as fully constrained; refer to Figure-157.

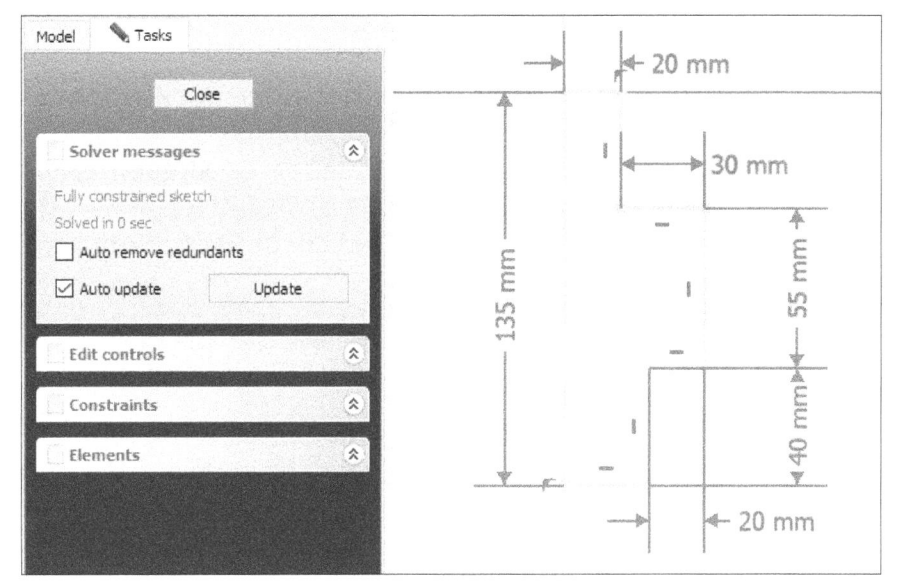

Figure-157. Fully constrained sketch

- Click on the **Close** button from **Tasks** panel of **Combo View** to exit sketch mode.
- Click on the **Save** button from the **Toolbar** and save the file at desired location.

Practical 2

In this practical, we will create the sketch as shown in Figure-158.

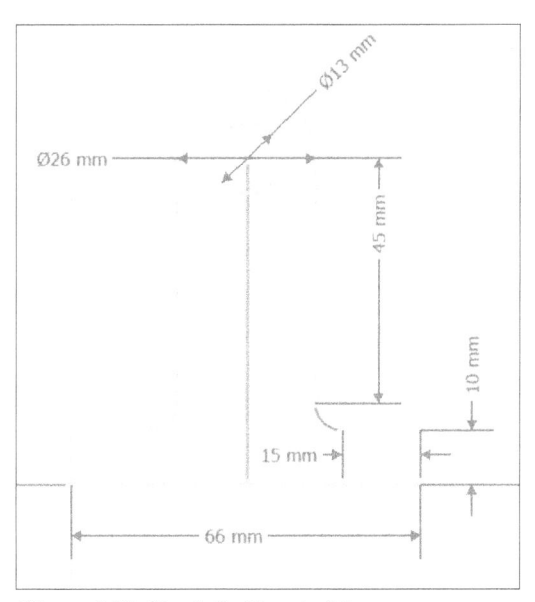

Figure-158. Sketch for Practical 2

Steps to be performed:
- Start a new part file.
- Select a sketching plane and activate sketching mode.
- Create the sketch using tools like Line, Circle, Arc, Fillet, and so on.
- Apply the dimensions using dimension tools.
- Save the file.

Starting a new sketch

- Start FreeCAD if not started yet.
- Select the **Sketcher** option from the **Switch between workbenches** drop-down in the **Toolbar**. The sketcher environment will be displayed.
- Click on the **Create sketch** tool from the **Toolbar** or **Sketch** menu. The **Choose orientation** dialog box will be displayed.
- Select the **XY-Plane** radio button from the **Sketch orientation** area of dialog box to create sketch on XY plane.
- Click on the **OK** button from the dialog box. The tools to create sketch will become active.

Creating sketch

- Click on the **Create line** tool from the **Toolbar** or **Sketcher geometries** cascading menu of **Sketch** menu to activate line tool. You will be asked to specify start point of line.
- Create a horizontal line of length approximately **66** mm starting from origin.
- Select the **Toggle construction geometry** tool from the **Toolbar** or **Sketcher geometries** cascading menu of **Sketch** menu to activate construction mode.
- Create a vertical line of approximately **60** mm length passing through mid-point of horizontal line; refer to Figure-159.

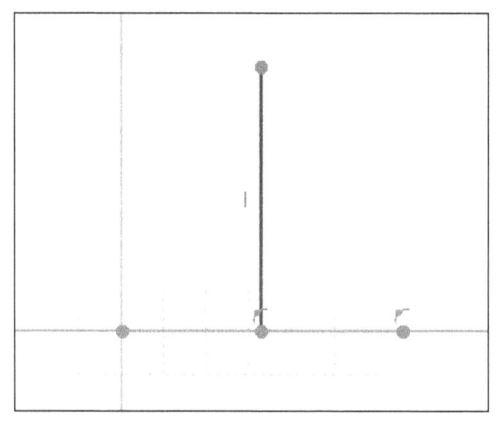

Figure-159. Lines created

- Exit the construction mode by selecting the **Toggle construction geometry** tool again from **Toolbar**.
- Click on the **Create circle by center** tool from the **Sketcher geometries** cascading menu of the **Sketch** menu or **Toolbar**. You will be asked to specify location for the center of circle.
- Click on the open end point of construction line and create a circle of diameter approximately **13**; refer to Figure-160.

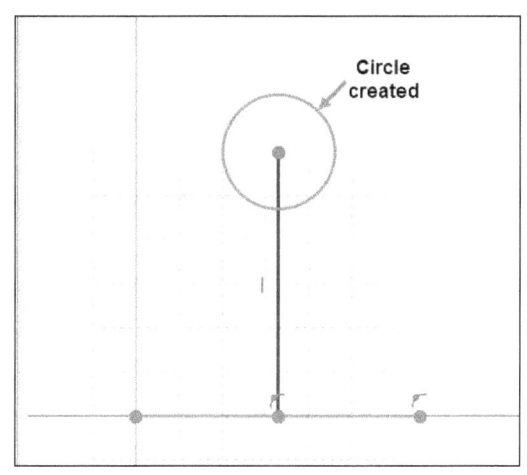

Figure-160. Circle created

- Create a circle of diameter approximately **26** using the same center as previous circle; refer to Figure-161.

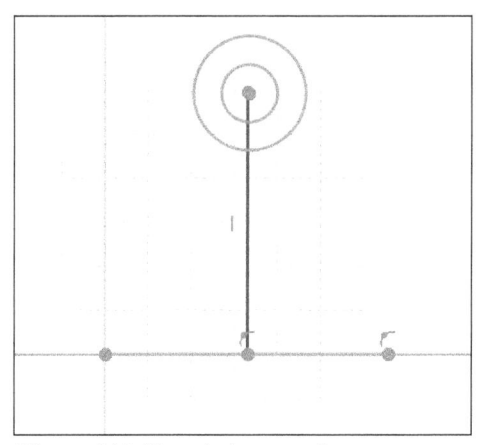

Figure-161. Two circles created

- Press **GL** shortcut key to activate line tool and create line sketch using approximate dimension as per the drawing; refer to Figure-162.

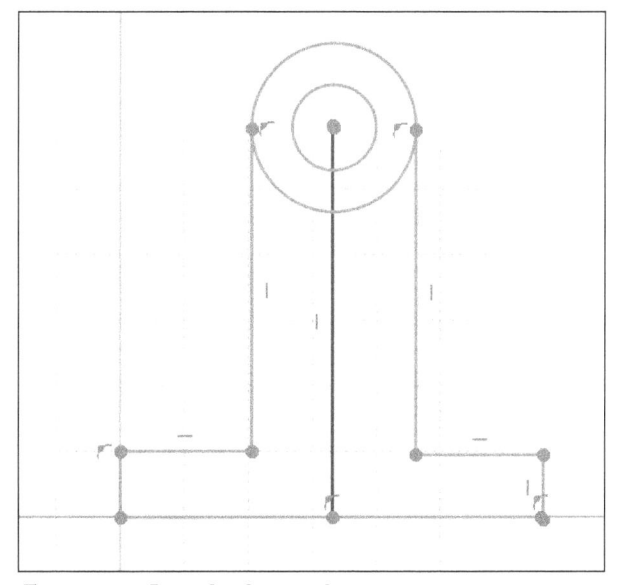

Figure-162. Line sketch created

Trimming and applying fillets

- Click on the **Trim edge** tool from the **Sketcher tools** cascading menu of the **Sketch** menu or from the **Edit edge** drop-down in the **Toolbar** and click on the inner section of larger circle to trim extra portion; refer to Figure-163.

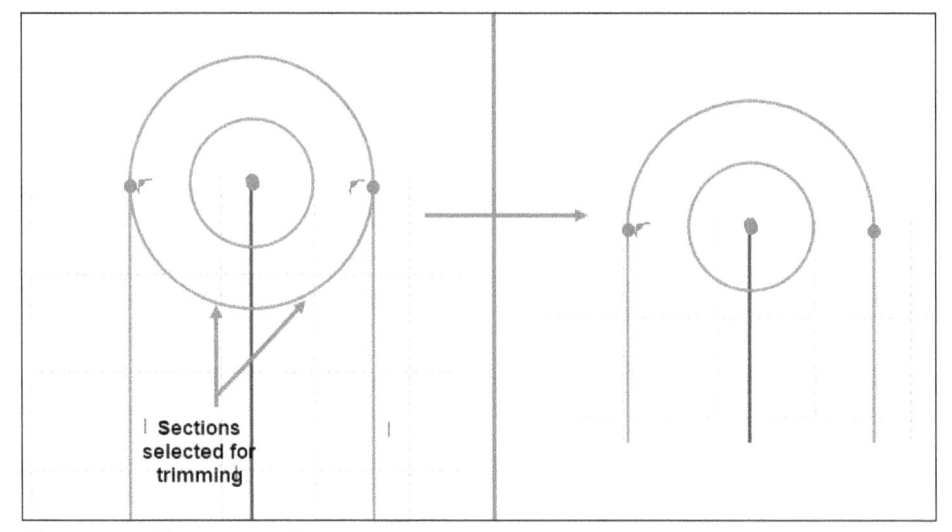

Figure-163. Trimming circle

- Press **ESC** once to exit the **Trim edge** tool and click on the **Create fillet** tool from the **Sketcher tools** cascading menu of **Sketch** menu. You will be asked to select two entities between which the fillet will be applied.
- Select the lines in pair near their connection points as shown in Figure-164.

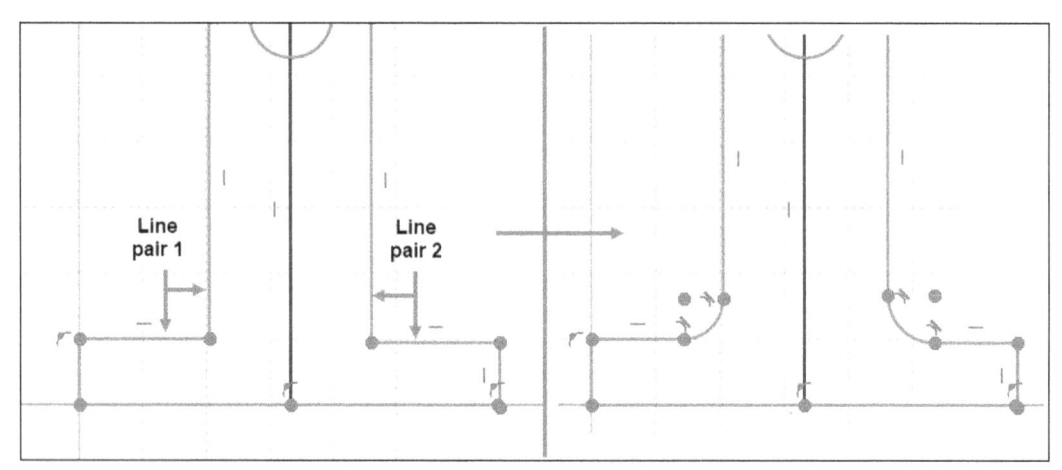

Figure-164. Applying fillets

Applying constraints and dimensions

- Select two bottom corner points and centerline passing through the sketch and then select the **Constrain symmetric** tool from **Sketcher constraints** cascading menu of the **Sketch** menu to keep centerline at the middle of sketch; refer to Figure-165.

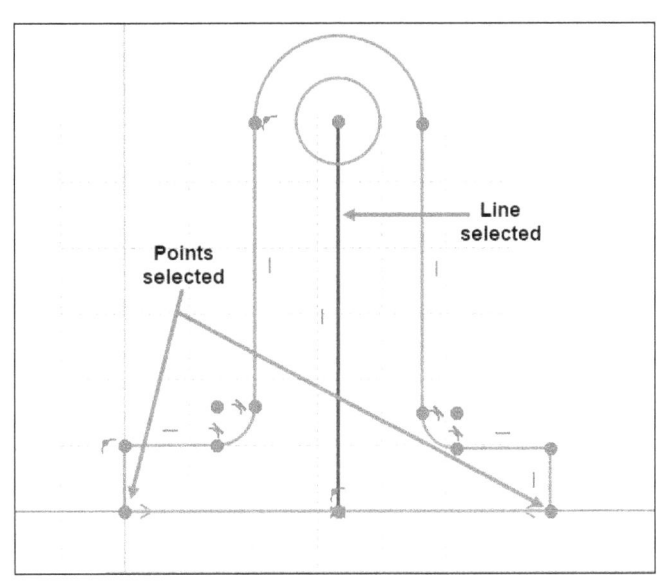

Figure-165. Selection for symmetrical constraint

- If there are any disconnection between points that should be coincident then apply the coincident constraint on those points. Note that you may need to zoom on the intersection points to check whether they are actually coincident or not.
- Click on the **Constrain distance** tool from **Toolbar** or **Sketcher constraints** cascading menu in the **Sketch** menu and apply the dimensions as shown in Figure-166.

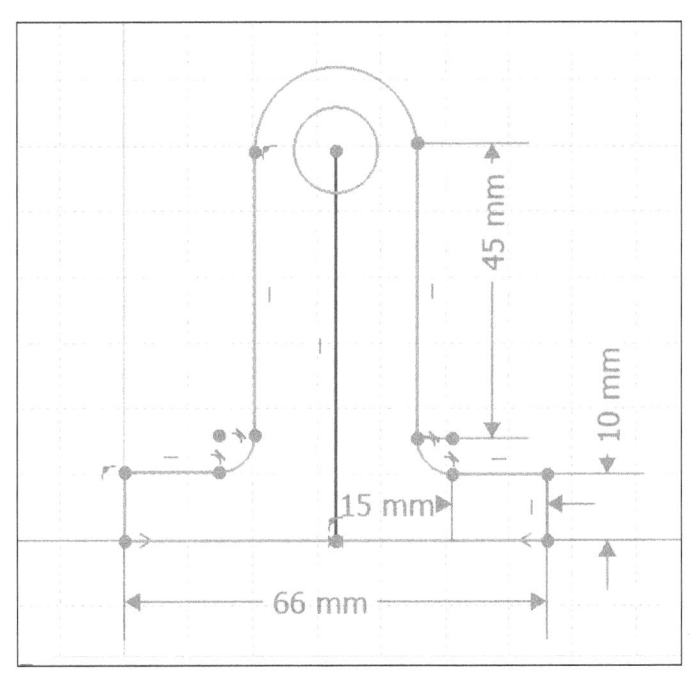

Figure-166. Distance dimensions applied

- Click on the **Constrain equal** tool from **Sketcher constraints** cascading menu of **Sketch** menu and one by one make the entities on both sides of centerline equal; refer to Figure-167.

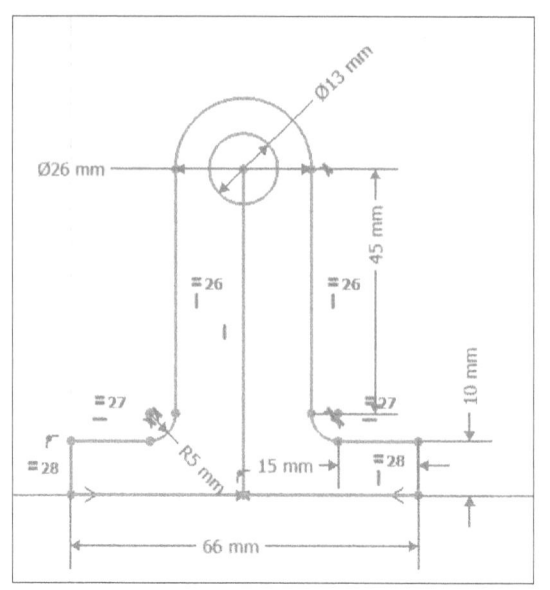

Figure-167. After applying equal constraint

- Select the **Constrain diameter** tool from the **Sketcher constraints** cascading menu of the **Sketch** menu and apply diameter dimensions to arc and circles. The fully defined sketch should display as shown in Figure-168.

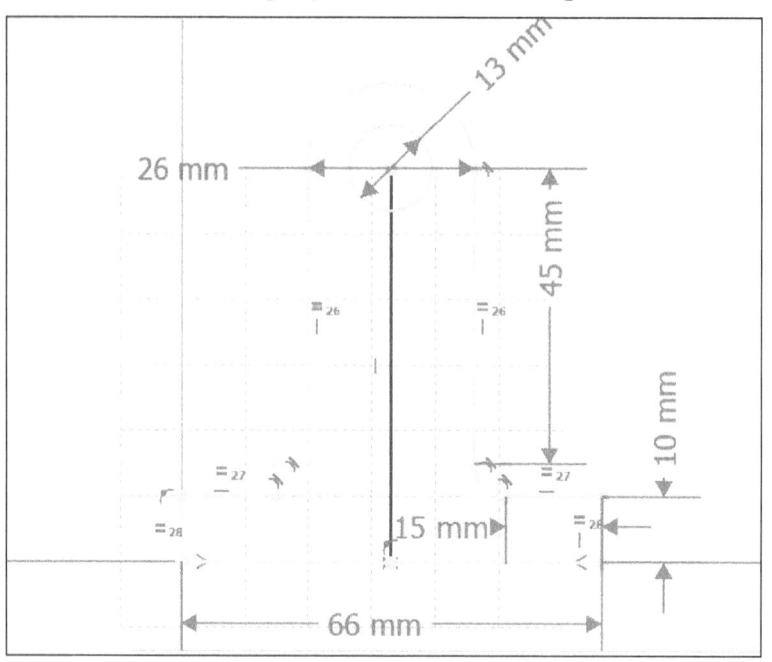

Figure-168. Fully defined sketch created

Close the sketcher and save the file as discussed earlier.

Practical 3

In this practical, we will create the sketch as shown in Figure-169.

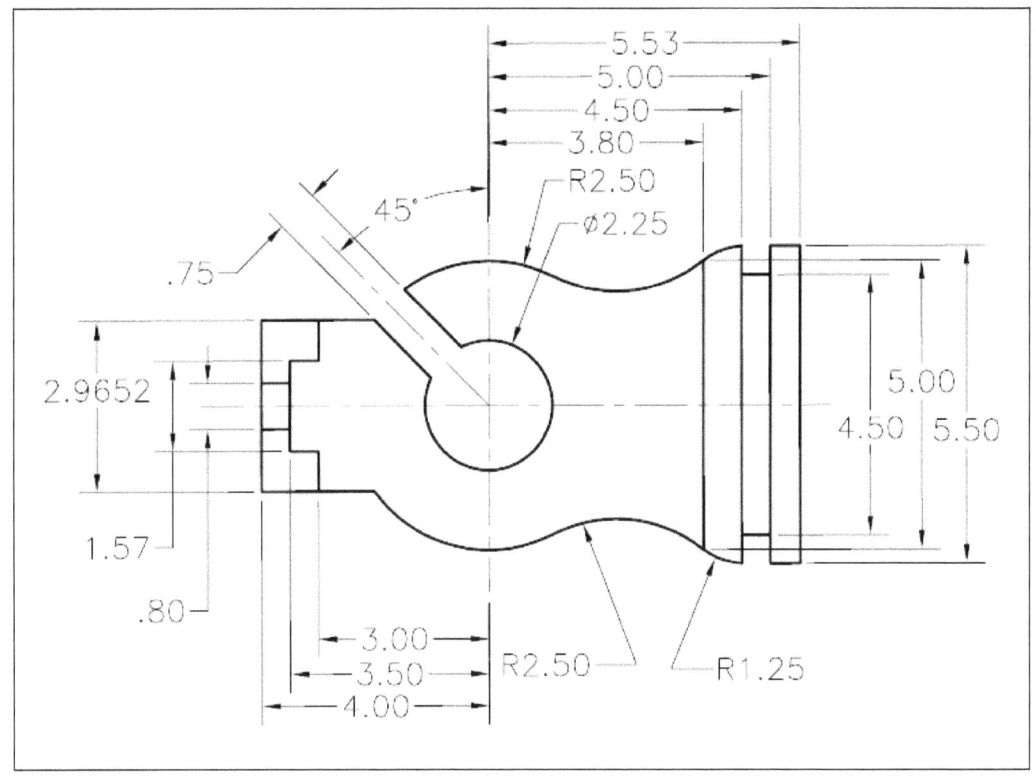

Figure-169. Practical 03

Steps to be performed:
- Start a new part file.
- Select a sketching plane and activate sketching mode.
- Create the sketch using various sketching tools.
- Apply the dimensions using dimension tools.
- Save the file.

Starting a new sketch

- Start FreeCAD if not started yet.
- To change the current unit to inch, click on **Preferences** option from the **Edit** menu. The **Preferences** dialog box will be displayed. Select the **General** option in the **General** node of the dialog box and select **Imperial decimal** option from the **Default unit system** drop-down in this page; refer to Figure-170. After setting parameters, click on the **OK** button from the dialog box. The unit system will be changed to inches.

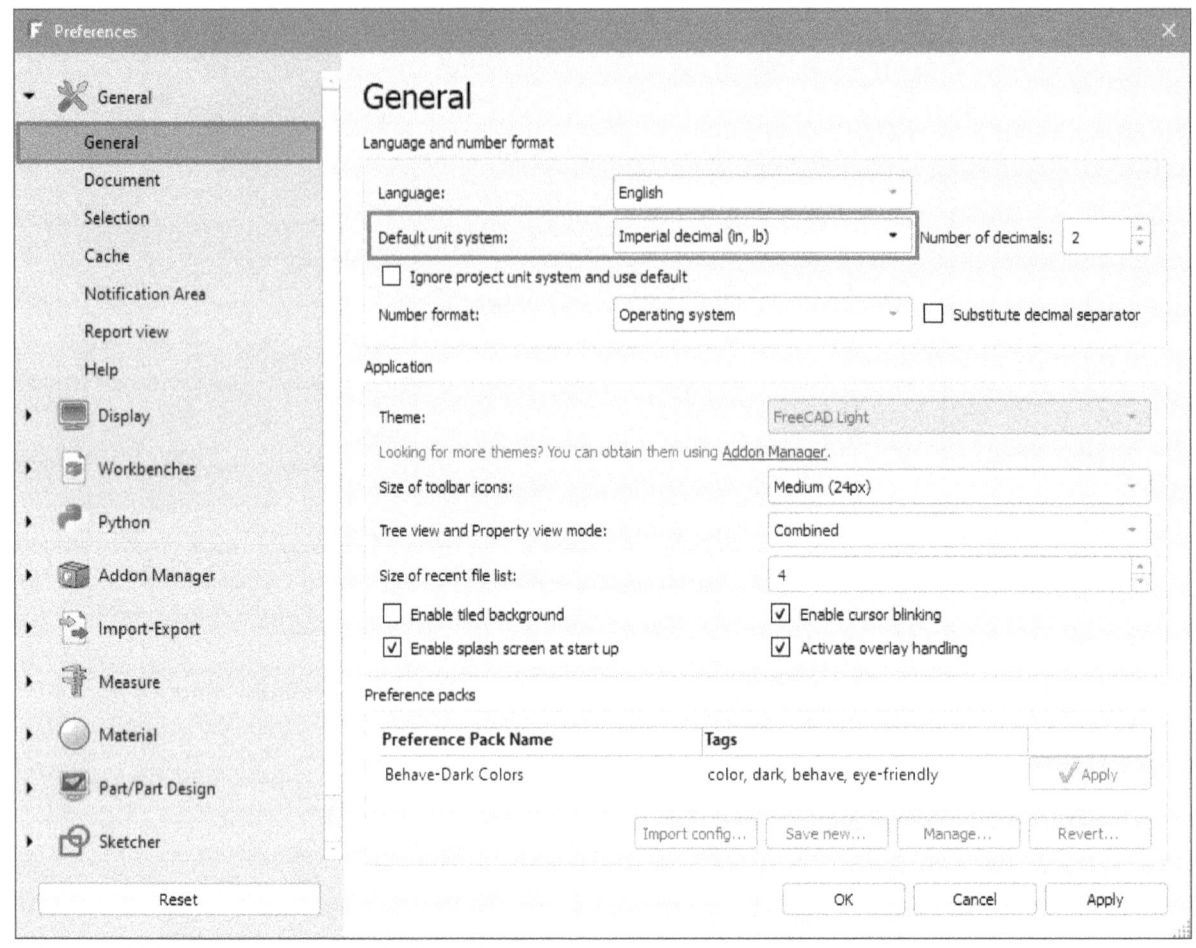

Figure-170. Unit system set to imperial decimal

- Select the **Sketcher** option from the **Switch between workbenches** drop-down in the **Toolbar**. The sketcher environment will be displayed.
- Click on the **Create sketch** tool from the **Toolbar** or **Sketch** menu. The **Choose orientation** dialog box will be displayed.
- Select the **XY-Plane** radio button from the **Sketch orientation** area of dialog box to create sketch on XY plane.
- Click on **OK** button from the dialog box. The tools to create sketch will become active.

Creating sketch

- Click on the **Create circle by center** tool from the **Toolbar** or **Sketcher geometries** cascading menu of **Sketch** menu to activate circle tool. You will be asked to specify location for the center of circle.
- Click at origin in the 3D view area and create concentric circles of diameter approximately **5** and **2.25**; refer to Figure-171.

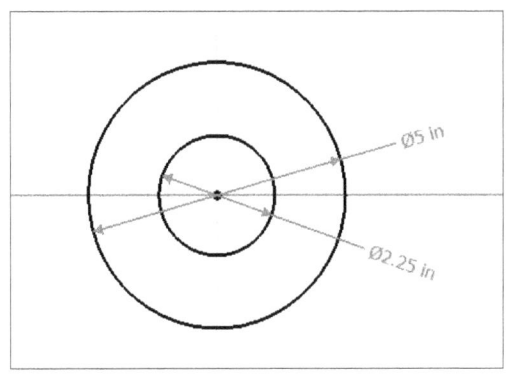

Figure-171. Creating circles

- After creating the circles, apply the dimensions to the circles as per the drawing using **Constrain diameter** tool from **Sketcher constraints** cascading menu of the **Sketch** menu.
- Click on the **Create line** tool from the **Toolbar** and create line sketch using approximate dimensions as per the drawing; refer to Figure-172.

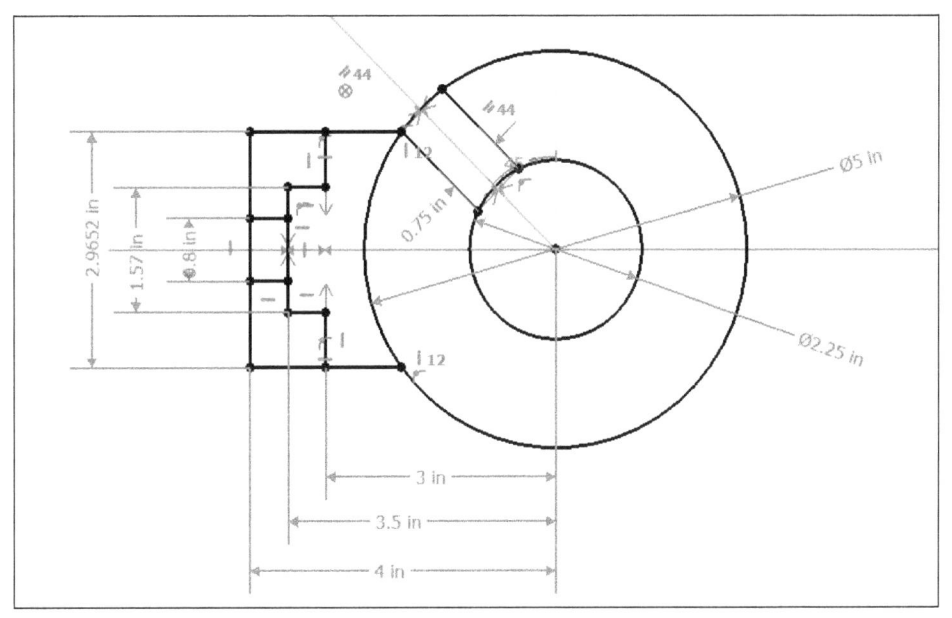

Figure-172. Creating the sketch using Line tool

- After creating the lines, apply the dimensions to the lines using desired dimension tools from **Sketcher constraints** cascading menu of the **Sketch** menu.
- Create rectangle and line sketches using approximate dimensions as per the drawing using **Create rectangle** and **Create line** tool from the **Toolbar**; refer to Figure-173.

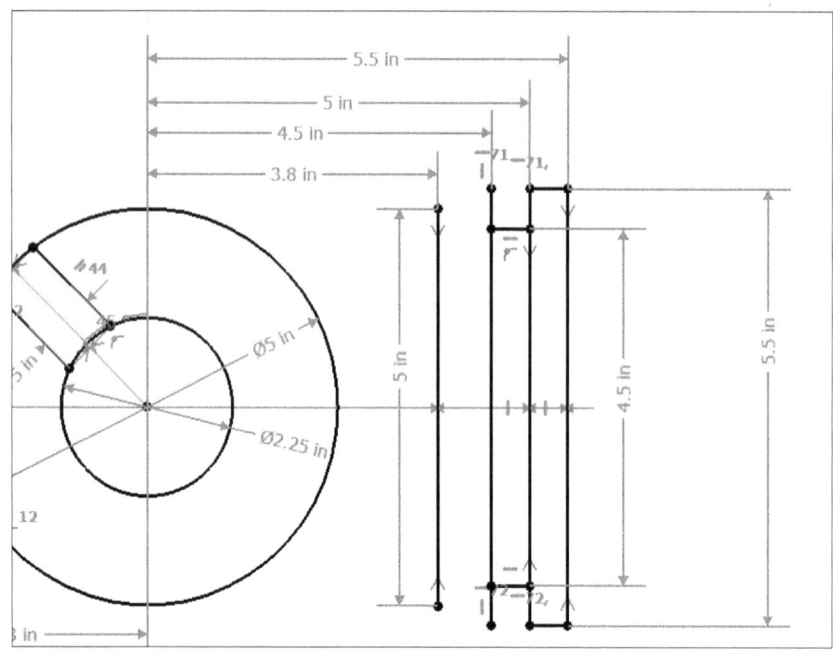

Figure-173. Creating rectangle and line sketches

Creating arcs

- Click on the **Create arc by 3 points** tool from **Sketch geometries** cascading menu or from the **Create arc** drop-down in the **Toolbar** and create the arcs as per the drawing using approximate dimensions.
- After creating the arcs, apply the constraints and dimensions to the arcs using tools from **Sketcher constraints** cascading menu of **Sketch** menu; refer to Figure-174.

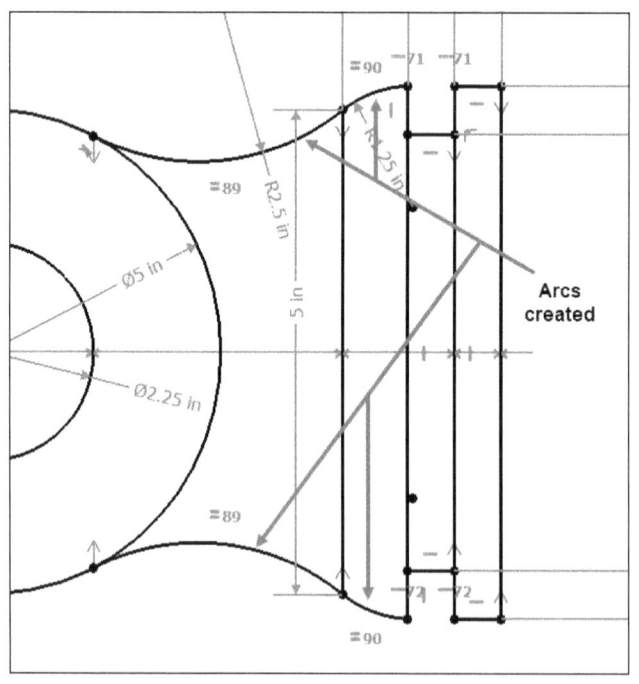

Figure-174. Creating arcs

Trimming the entity

- Click on the **Trim edge** tool from the **Sketcher tools** cascading menu of the **Sketch** menu and click on the edge of both the circles to trim extra portion; refer to Figure-175.

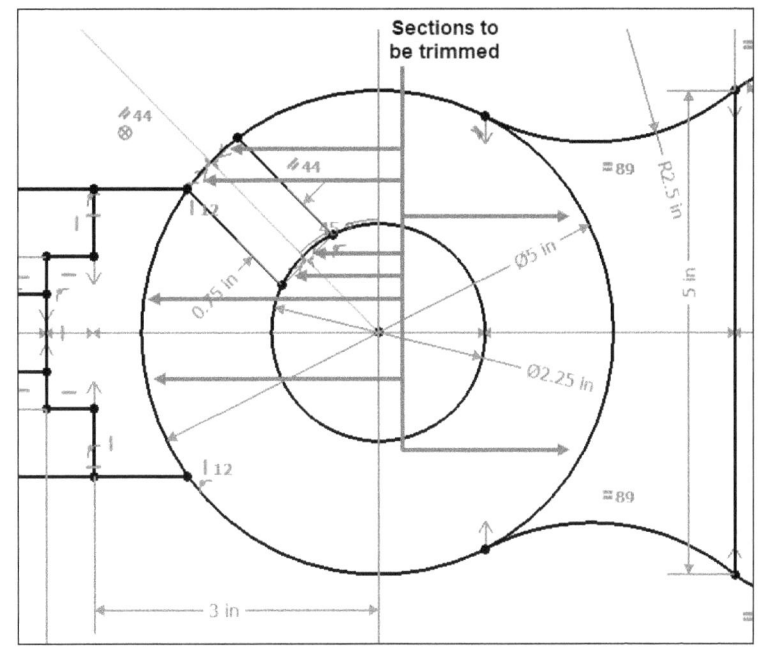

Figure-175. Trimming the extra portions

Close the sketcher and save the file as discussed earlier; refer to Figure-176.

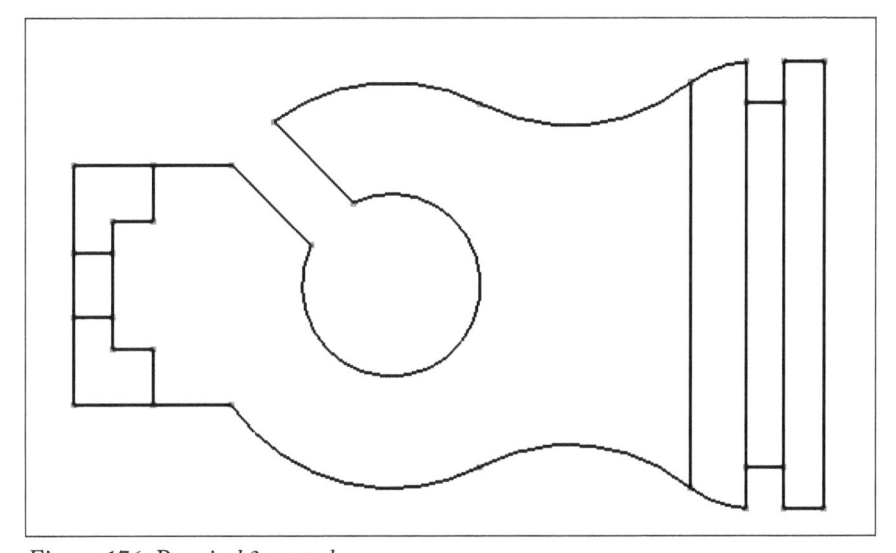

Figure-176. Practical 3 created

PRACTICE 1

In this practice session, you will create a sketch for the drawing shown in Figure-177.

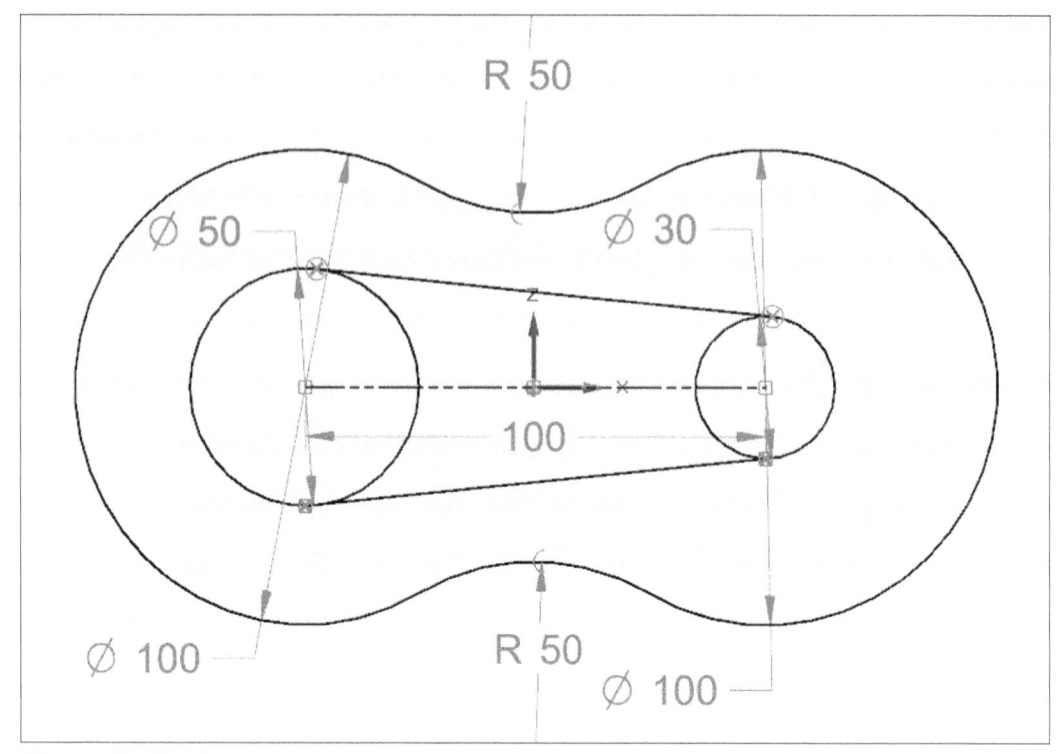

Figure-177. Practice 1

PRACTICE 2

In this practice session, you will create a sketch for drawing shown in Figure-178.

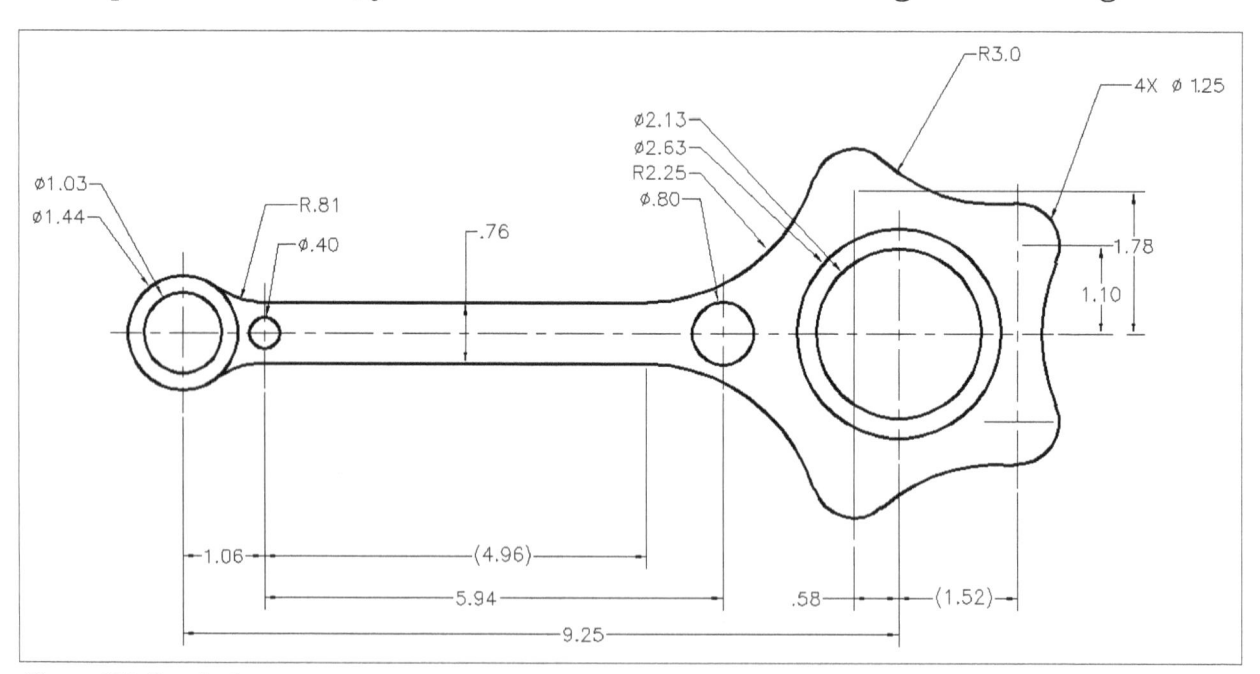

Figure-178. Practice 2

PRACTICE 3

In this practice session, you will create a sketch for the drawing given in Figure-179.

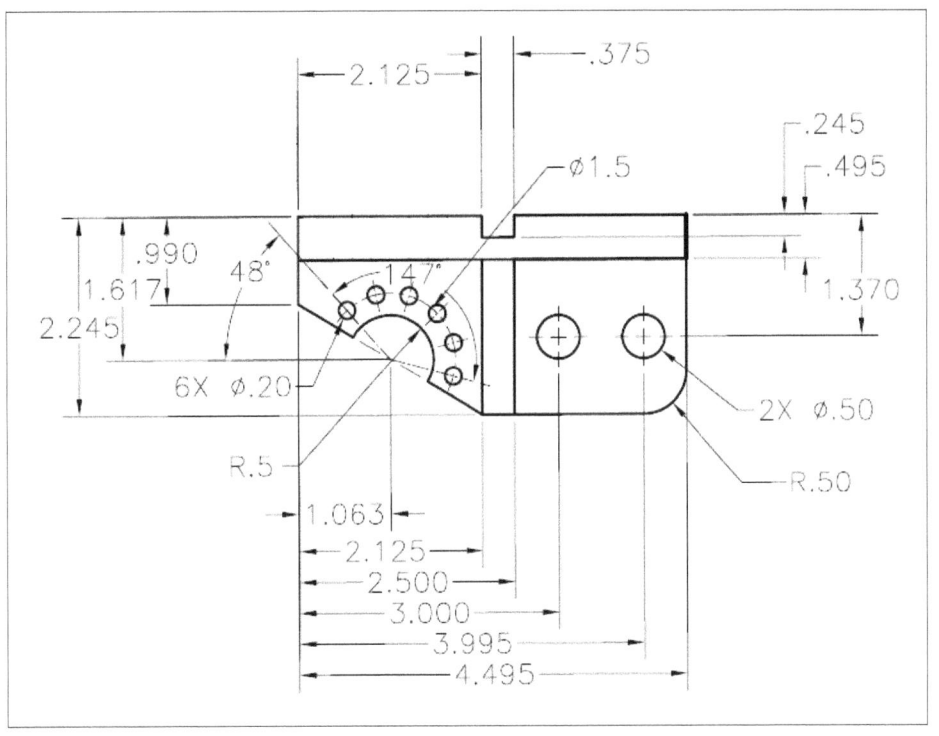

Figure-179. Practice 3

PRACTICE 4

In this practice session, you will create sketches of different views for the drawing given in Figure-180.

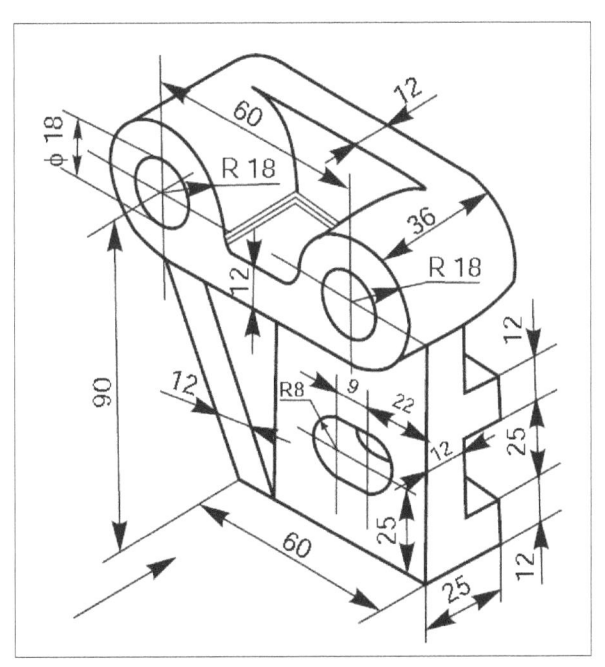

Figure-180. Practice 4

SELF ASSESSMENT

Q1. Which of the following sketch entities are used as building blocks for 3D operations in engineering sketches?

A) Line, Circle, Arc, Polygon, Ellipse
B) Polygon, Circle, Square
C) Point, Arc, Ellipse
D) Circle, Triangle, Rectangle

Q2. In the Sketcher workbench, what is the first step to start a new sketch?

A) Click on Create Sketch button
B) Click on New button from File menu and select Sketcher workbench
C) Select a plane for sketching
D) Choose the offset distance for the plane

Q3. What does the Reverse Direction checkbox do in the Sketcher workbench?

A) Changes the plane orientation
B) Reverses the direction of the selected plane
C) Rotates the sketch in a different direction
D) Reverses the offset distance

Q4. In the Sketcher workbench, which tool is used to create a point in the current sketch?

A) Create Line tool
B) Create Circle tool
C) Create Point tool
D) Create Polygon tool

Q5. How can you exit the Create Polyline tool in the Sketcher workbench?

A) Press ESC or right-click
B) Press M key
C) Click on the "Exit" button
D) Press the spacebar

Q6. In the Sketcher workbench, which of the following is true when using the Create Line tool?

A) The starting and ending points must be manually typed
B) You must first create a circle before drawing a line
C) You can select from options like Point, length, and angle or 2 points to create a line
D) You can only create lines with fixed length

Q7. What does the "Arc of Ellipse" tool do in the Sketcher workbench?

A) Creates a circle with three points
B) Creates an ellipse with two points
C) Creates an arc of an ellipse using four points
D) Creates a polygon using four points

Q8. Which of the following tools allows you to create a circle by specifying its center and a point along the circle?

A) Create Arc by Center
B) Create Circle by Center
C) Create Circle by 3 Points
D) Create Ellipse by 3 Points

Q9. In the Sketcher workbench, what does the "Ellipse by Center" tool do?

A) Creates an ellipse by using three points: the center, major radius, and minor radius
B) Creates an ellipse by using four points
C) Creates a square inscribed in a circle
D) Creates a line tangent to the ellipse

Q10. Which tool is used to create a rectangle by specifying two opposite corner points?

A) Centered Rectangle Tool
B) Rectangle Tool
C) Rounded Rectangle Tool
D) Ellipse Tool

Q11. What is the Hexagon tool used for?

A) To create a regular hexagon with a fixed size
B) To create a hexagon inscribed in a construction geometry circle
C) To create a hexagon by specifying vertices manually
D) To draw a hexagonal shape with customizable angles

Q12. Which of the following steps is involved in creating a hexagon using the Hexagon tool?

A) Click on the Hexagon tool and specify the center point of the hexagon
B) Draw each side manually by clicking the vertices
C) Use the ESC key to complete the shape
D) None of the above

Q13. Which tool is used to create a heptagon inscribed in a construction geometry circle?

A) Polygon tool
B) Heptagon tool
C) Octagon tool
D) Triangle tool

Q14. In the Heptagon tool, what is the first step when creating a heptagon?

A) Click on the center point for the heptagon
B) Move the cursor away from the center point
C) Select the number of sides in the dialog box
D) Press the ESC key

Q15. Which tool is used to create an octagon inscribed in a construction geometry circle?

A) Octagon tool
B) Polygon tool
C) Triangle tool
D) Regular Polygon tool

Q16. What is the procedure for creating a regular polygon using the Regular Polygon tool?

A) Select the number of sides in the dialog box
B) Define the center of the polygon in the 3D view
C) Specify the vertex of the polygon
D) All of the above

Q17. What is the Slot tool used for?

A) To create a simple rectangle
B) To draw a slot by specifying center points of two semicircles
C) To draw an arc with a specified radius
D) To create a circle inscribed in a polygon

Q18. When using the Create arc slot tool, what option allows you to create using the arc ends of the slot?

A) Arc ends
B) Flat ends
C) Radius ends
D) Circle ends

Q19. Which tool allows you to create B-spline curves by tracing control points?

A) B-spline by knots
B) Periodic B-spline
C) B-spline by control points
D) Arc slot

Q20. How do you create a periodic B-spline by control points?

A) Select the "By knots" option in the drop-down
B) Create a series of points and press the ESC key
C) Click on the Periodic B-spline by control points tool and create points
D) Use the "By control points" option for B-spline

Q21. What does the Toggle construction geometry tool do?

A) Converts selected geometry to a sketch mode
B) Toggles sketch geometry between construction mode and sketch mode
C) Creates a regular polygon
D) Applies a dimensional constraint

Q22. What does the Lock constraint tool apply to selected vertices?

A) Horizontal and vertical distance dimensions
B) Radius and diameter constraints
C) Distance between two points
D) Angle constraints

Q23. Which of the following is true about the Horizontal distance constraint?

A) It sets the vertical distance between two points
B) It defines the horizontal distance between two points or line endpoints
C) It is used to apply a radius constraint
D) It defines the distance between a point and a line

Q24. What does the Vertical distance constraint tool do?

A) Fixes the horizontal distance between two points
B) Fixes the vertical distance between two points or line endpoints
C) Defines the radius of an arc
D) Sets the angle between two lines

Q25. What is the Distance constraint tool used for?

A) Defining the distance between two points or constraining the length of a line
B) Creating a polygon with a specified number of sides
C) Applying angle constraints
D) None of the above

Q26. What is the function of the Constrain radius tool?

A) It defines the angle between two lines
B) It automatically adjusts the radius of an arc or circle
C) It defines the radius of a selected arc or circle
D) It modifies the vertical distance

Q27. How does the Constrain diameter tool work?

A) It defines the radius of an arc or circle
B) It defines the diameter of a selected arc or circle
C) It defines the angle between two arcs
D) It defines the distance between two points

Q28. What does the Angle constraint tool define?

A) The distance between two points
B) The angle between two selected lines
C) The diameter of an arc
D) The radius of a circle

Q29. What is the primary purpose of constraints in Sketcher?

A) To apply colors to the sketch
B) To restrict the shape and size of geometric entities
C) To add text labels to the sketch
D) To make the sketch look aesthetic

Q30. What type of data is associated with geometric constraints in Sketcher?

A) Numeric data
B) Text data
C) No numeric data
D) Image data

Q31. Which constraint tool is used to close a gap between two lines by making their endpoints coincident?

A) Point on object
B) Constrain coincident
C) Perpendicular constraint
D) Parallel constraint

Q32. How do you exit the Constrain coincident tool?

A) Press the ESC key or right-click
B) Click the "Exit" button
C) Click the "Done" button
D) Press the Enter key

Q33. What does the "Constrain point onto object" tool do in Sketcher?

A) Moves a point to another location
B) Affixes a point onto a line, arc, or curve
C) Makes two points coincide
D) Draws a point at a specific location

Q34. What can be constrained vertically using the Vertical constraint tool?

A) Curves and circles
B) Only lines and vertices
C) Points and arcs
D) Only text labels

Q35. What happens when you apply the Horizontal constraint tool to a line?

A) It becomes horizontal in orientation
B) It becomes vertical in orientation
C) It becomes tangent to another line
D) It becomes parallel to another line

Q36. Which tool is used to make two lines parallel to each other?

A) Constrain coincident
B) Constrain parallel
C) Constrain perpendicular
D) Constrain tangent

Q37. What type of geometric entities can be made perpendicular using the Constrain Perpendicular tool?

A) Only lines
B) Only points
C) Lines, curves, arcs, circles, ellipses, etc.
D) Only circles

Q38. When the Tangent constraint tool is used, what happens between arcs, circles, curves, and lines?

A) They become parallel
B) They become perpendicular
C) They become collinear
D) They become tangent to each other

Q39. Which constraint forces two or more sketch segments to have equal length?
A) Parallel constraint
B) Equal constraint
C) Symmetric constraint
D) Tangent constraint

Q40. The Symmetric constraint in Sketcher is used to constrain points in relation to:

A) A circle
B) A specific angle
C) A given line
D) A reference point

Q41. What is the function of the Constrain block tool?

A) It blocks the view of the sketch
B) It blocks the position of an entity, preventing movement
C) It blocks the tool from being used again
D) It blocks entities from being deleted

Q42. What does the Snell's Law constraint tool do in Sketcher?

A) Constrains two lines to follow the law of refraction of light
B) Constrains two lines to be parallel
C) Constrains two points to be symmetrical
D) Constrains two entities to be tangent

Q43. What is the effect of using the Toggle Driving constraint tool in Sketcher?

A) It adds a reference dimension to the sketch
B) It changes the entity's shape
C) It switches between driving and driven dimension constraints
D) It blocks the sketch entity from moving

Q44. What does the Toggle Active constraint tool do?

A) Activates a constraint on an entity
B) Deactivates a constraint on an entity
C) Removes the constraint entirely
D) Makes the constraint optional

Q45. Which Sketcher tool is used to create a fillet between two geometries joined at a point?

A) Sketch fillet
B) Create chamfer
C) Trim edge
D) Extend edge

Q46. What is the primary purpose of the Trim edge tool in Sketcher?

A) To extend an edge to a point
B) To trim a curve to the nearest intersection point
C) To split an edge into smaller sections
D) To create a fillet between two entities

Q47. Which tool in Sketcher splits an edge while transferring most constraints?

A) Split edge
B) Extend edge
C) Trim edge
D) Create chamfer

Q48. How does the Extend edge tool work in Sketcher?

A) It shortens a curve to a point
B) It extends a curve to a specified location or another curve
C) It joins two lines
D) It trims an edge to the nearest intersection point

For Student Notes

Chapter 3
Solid Modeling with Part Design

Topics Covered

The major topics covered in this chapter are:

- **Introduction to Solid Modeling**
- **Starting Part Design**
- **Part Design Helper Tools**
- **Part Design Modeling Tools**
- **Datum Tools**
- **Additive Tools**
- **Subtractive Tools**
- **Transformation Tools**
- **Dress-up Tools**
- **Boolean Operation**
- **Involute Gear Tool**
- **Shaft Design Wizard**

INTRODUCTION

Solid Modeling is the most advanced method of geometric modeling in 3D. Solid modeling is the representation of solid objects on your computer. Now, we will learn about various tools and commands used to create solid models.

In FreeCAD, the **Part Design Workbench** provides advanced tools for modeling complex solid parts. It is mostly focused on creating mechanical parts that can be manufactured and assembled into a finished product. Nevertheless, the created solids can be used in general for any other purpose such as architectural design, finite element analysis, or machining and 3D printing.

STARTING PART DESIGN

The **Part Design** workbench is intrinsically related to the **Sketcher** workbench. The user normally creates a sketch then uses the **Part Design Pad** tool to extrude it, create a basic solid, and then this solid is further modified.

- To start a new part design file, click on the **New** button from **File** menu and select **Part Design** workbench from **Switch between workbenches** drop-down in the **Toolbar**; refer to Figure-1. The tools to create 3D model will be displayed in the **Toolbar**; refer to Figure-2.

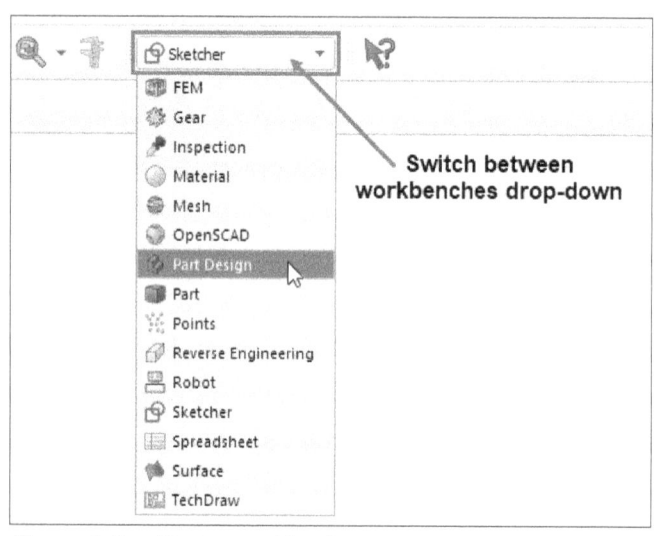

Figure-1. Part Design workbench

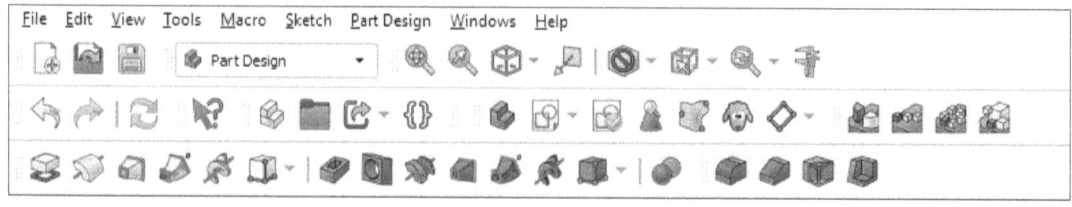

Figure-2. Part Design tools

Part Design Helper tools

The **Part Design Helper** tools help in creating a body and make it active for creating and editing the sketch. You can also map a sketch to desired face using tools of **Part Design Helper** toolbar. These tools are available in the **Toolbar** of **Part Design** workbench; refer to Figure-3. The procedures to use these tools are discussed next.

FreeCAD 1.0 Black Book 3-3

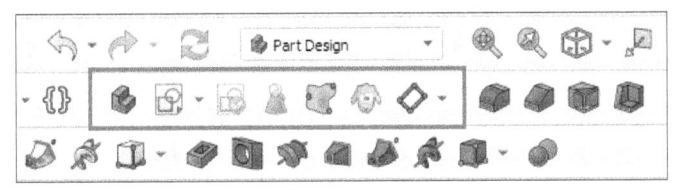

Figure-3. Part Design Helper tools

Creating a body

A **Part Design** body is the base element to create solid shapes with the **Part Design** workbench. It can contain sketches, datum objects, and part design features that help in building a single contiguous solid. The procedure to use this tool is discussed next.

- Click on the **Create body** tool from **Toolbar** in the **Part Design** workbench; refer to Figure-4. An empty body is created and it automatically becomes active; refer to Figure-5.

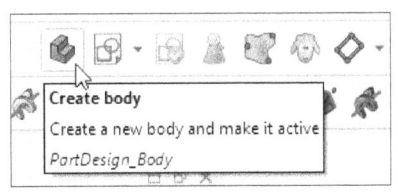

Figure-4. Create body tool

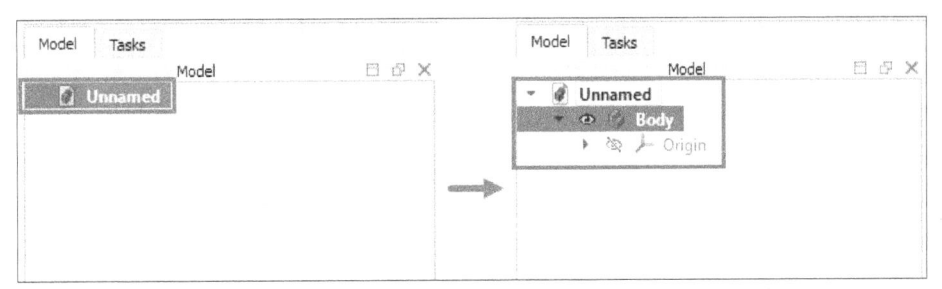

Figure-5. A new empty body created

- Now, you can click **Create sketch** tool to create a sketch in the body that can be used with **Pad** tool or other 3D protrusion tools.

Creating a sketch

The **Create sketch** tool creates a new sketch on selected face or plane. If no face is selected while this tool is executed, the user is prompted to select a plane from the **Tasks** panel or from the 3D view area. The procedure to use this tool is discussed next.

- Click on the **Create sketch** tool from **Toolbar** in the **Part Design** workbench; refer to Figure-6. The **Select attachment** dialog will be displayed in the **Tasks** panel of **Combo View** to select desired plane; refer to Figure-7. You can also select the plane from 3D view area.

Figure-6. Create sketch tool

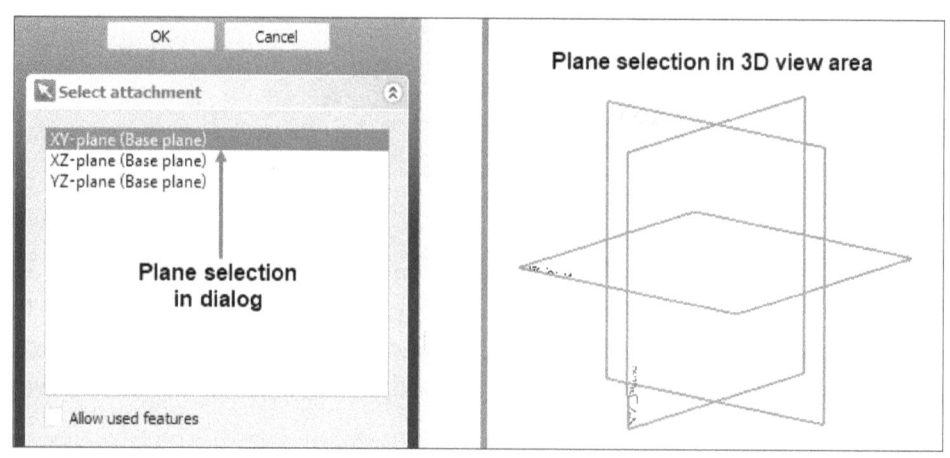

Figure-7. Selecting the plane to create sketch

- Select **Allow used features** check box to reuse earlier used features.
- Select the **From other bodies of the same part** check box from the **Allow external features** section of the dialog to use any elements used in the same body as dimensioning reference for creating sketch.
- Select the **From different parts or free features** check box to use parts and features of other objects in the model as reference for sketch dimensioning.
- Select the **Make independent copy** radio button to create independent copies of selected sketching references so that they do not get changed when original references have been changed.
- Select the **Make dependent copy** radio button to create reference elements dependent on the original features. This is basically use of a Shape binder.
- Select the **Create cross-reference** radio button to linked the original elements as reference without creating copies. Any changes in original elements are reflected to this sketch.
- After selecting desired plane, click on **OK** button from the dialog. The interface switches to **Sketcher** workbench to create or edit the sketch. You have learned about sketches in previous chapter.

Attach sketch

The **Attach sketch** tool maps an existing sketch on the face of a selected body or a selected plane. Note that this tool is not used to create new sketches. It only maps or remaps existing sketch to the face of a solid/part design feature or a plane. The procedure to use this tool is discussed next.

- Select a face or plane on which you want to map the sketch.
- Click on the **Attach sketch** tool from **Create Sketch** drop-down in the **Toolbar** of **Part Design** workbench; refer to Figure-8. The **Select sketch** dialog box will be displayed; refer to Figure-9, asking you to select the sketch entity to be mapped from the list.

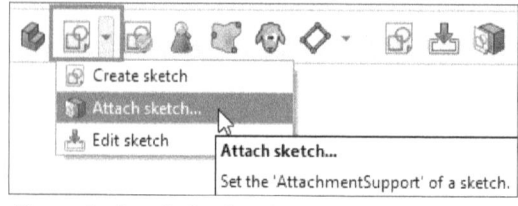

Figure-8. Attach sketch tool

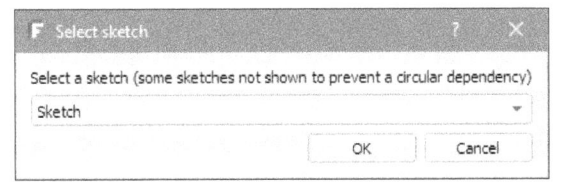

Figure-9. Select sketch dialog box

- Select desired sketch entity from the list which you want to map and click on **OK** button from the dialog box. The `Sketch attachment` dialog box will be displayed; refer to Figure-10, asking you to select method from the list to be used to attach sketch with selected face. Note that depending on the attachment face/plane selected earlier, you may get different options in the `Sketch attachment` dialog box.

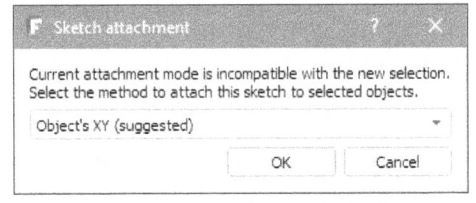

Figure-10. Sketch attachment dialog box

- Select desired method from the list to map the sketch on selected object and click on **OK** button from the dialog box. Generally, it is fine to select the suggested method. The selected sketch entity will be mapped to the part design feature; refer to Figure-11.

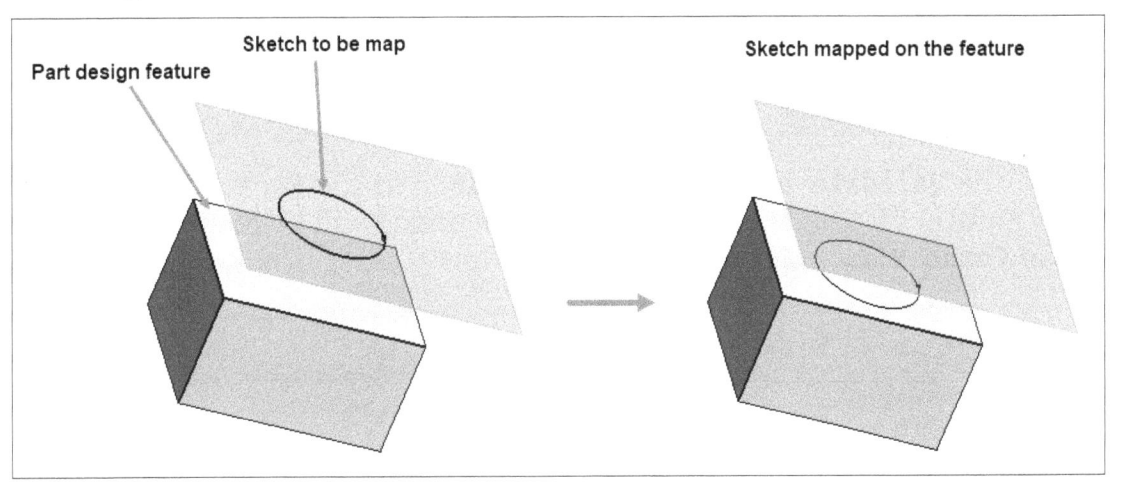

Figure-11. Sketch mapped on the solid feature

Edit sketch

The `Edit sketch` tool is used to edit an existing sketch. The procedure to use this tool is discussed next.

- Select the sketch that you want to edit from the **Model Tree** or from the 3D view area; refer to Figure-12.

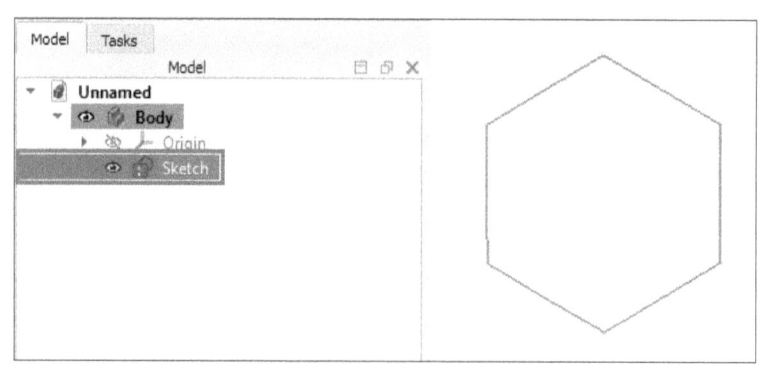

Figure-12. Selecting the sketch to edit

- Click on the **Edit sketch** tool from **Create Sketch** drop-down in the **Part Design** workbench; refer to Figure-13. The **Sketcher** workbench will be displayed with options to edit the existing sketch.

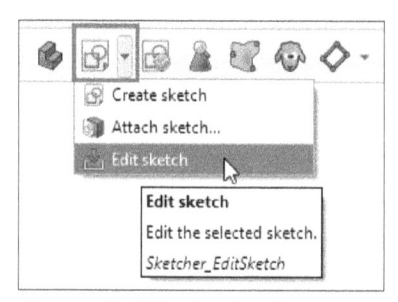

Figure-13. Edit sketch tool

Validate sketch

The **Validate sketch** tool is used to verify tolerance of different points and adjusts them. The procedure to use this tool is discussed next.

- Click on the **Validate sketch** tool from **Toolbar** in the **Part Design** workbench; refer to Figure-14. The **Sketcher validation** dialog will be displayed in the **Tasks** panel of **Combo View**; refer to Figure-15.

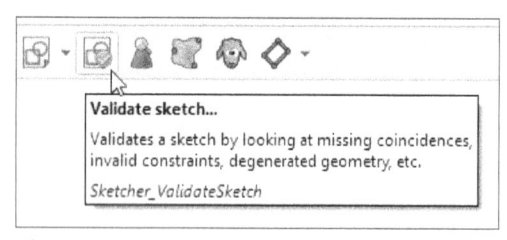

Figure-14. Validate sketch tool

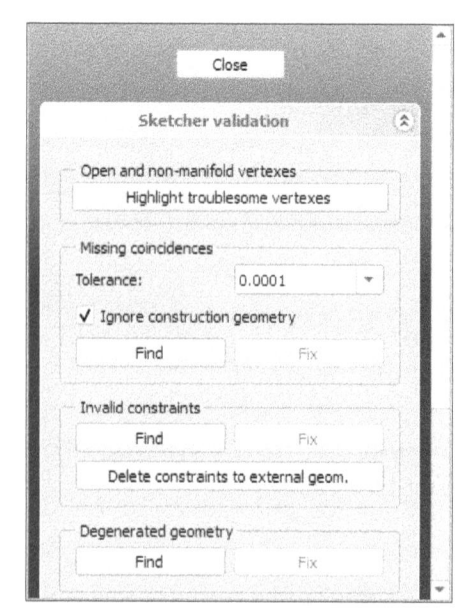

Figure-15. Sketcher validation dialog box

- Click on the **Highlight troublesome vertexes** button from the dialog to highlight vertexes that are overlapping or not connected within tolerance specified in the dialog; refer to Figure-16.

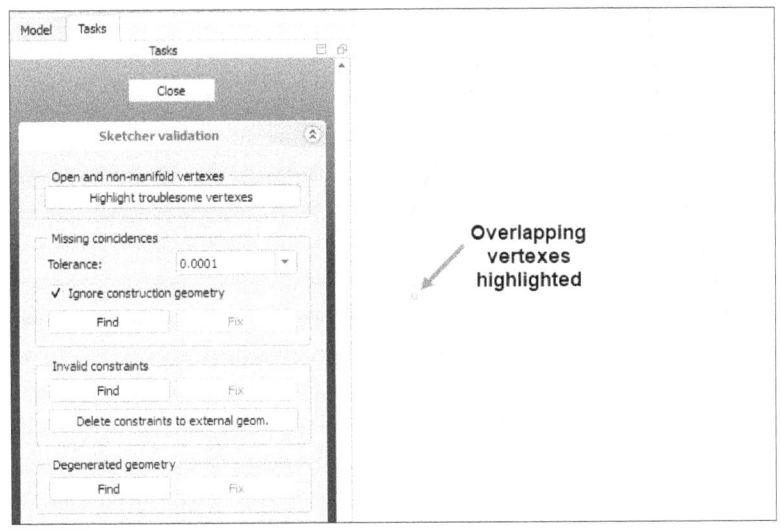

Figure-16. Highlighting troublesome vertexes

- Clear the **Ignore construction geometry** check box if you do not want to perform validation checks on construction geometries in the sketch.
- Select the **Find** button from the **Missing coincidences** area of the dialog to check end points of curves in sketch which are not connected properly by coincidence constraint.
- Similarly, you can use the other options of the dialog to perform validation checks.
- Click on the **Close** button from the dialog to exit the tool.

Check geometry

The **Check geometry** tool runs a verification and reports if geometry is a valid solid. The procedure to use this tool will be discussed in **Chapter 4**.

Create a shape binder

The **Create a shape binder** tool creates a datum shape binder from selected body inside the active body. A shape binder is a reference object that links edges or faces of another body to active body. It can also be used to link a sketch from one body to another body. The datum shape binder object displays as translucent yellow in the 3D view. The procedure to use this tool is discussed next.

- Click on the **Create a shape binder** tool from **Part Design** menu in the **Part Design** workbench; refer to Figure-17. The **Datum shape parameters** dialog will be displayed in the **Tasks** panel of **Combo View**; refer to Figure-18.

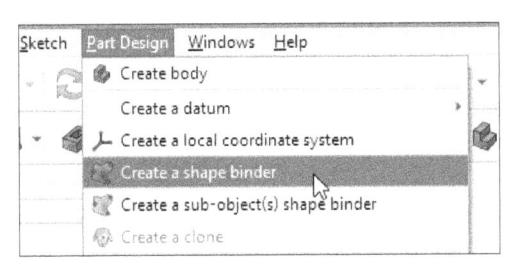

Figure-17. Create a shape binder tool

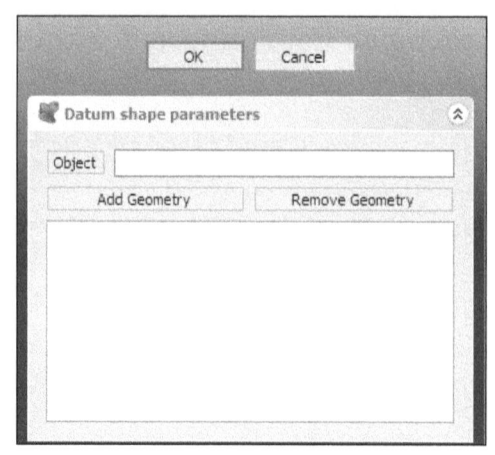

Figure-18. Datum shape parameters dialog

- The **Object** button in the dialog will select the whole body and the **Add Geometry** button will select elements of the object (vertex, edge, face). Click on **Object** button and select desired object in the 3D view area or click on **Add Geometry** button and select elements of the object (vertex, edge, face) in the 3D view area. The datum shape binder object will be created; refer to Figure-19.

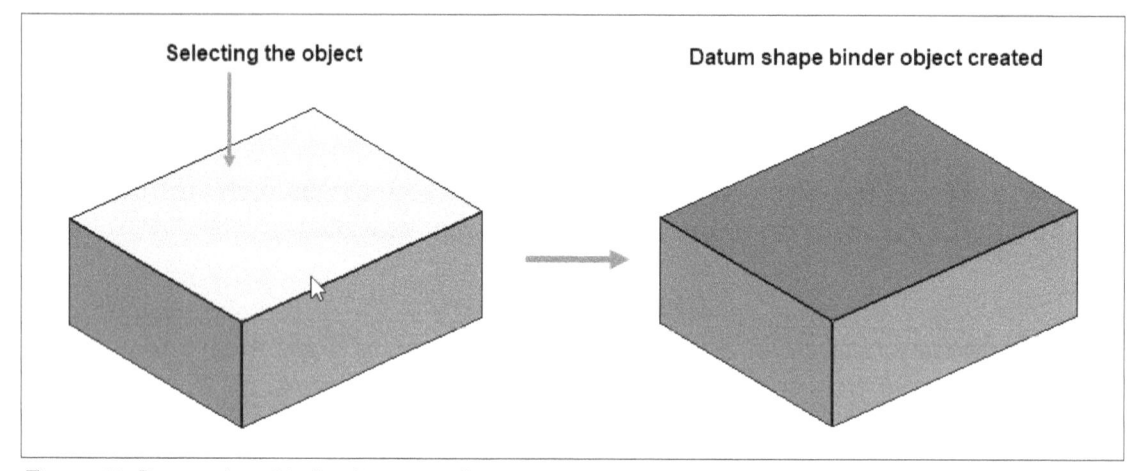

Figure-19. Datum shape binder object created

- To remove selected geometry, click on **Remove Geometry** button from the dialog and select the geometry in the 3D view area.
- To display the datum shape binder object, toggle the visibility of **Pad** feature from **True** to **False** in the **Property editor** of **Combo View** or select the **Pad** feature from **Model** panel in **Combo View** and press **SPACEBAR** key. The datum shape binder object will be displayed; refer to Figure-20.

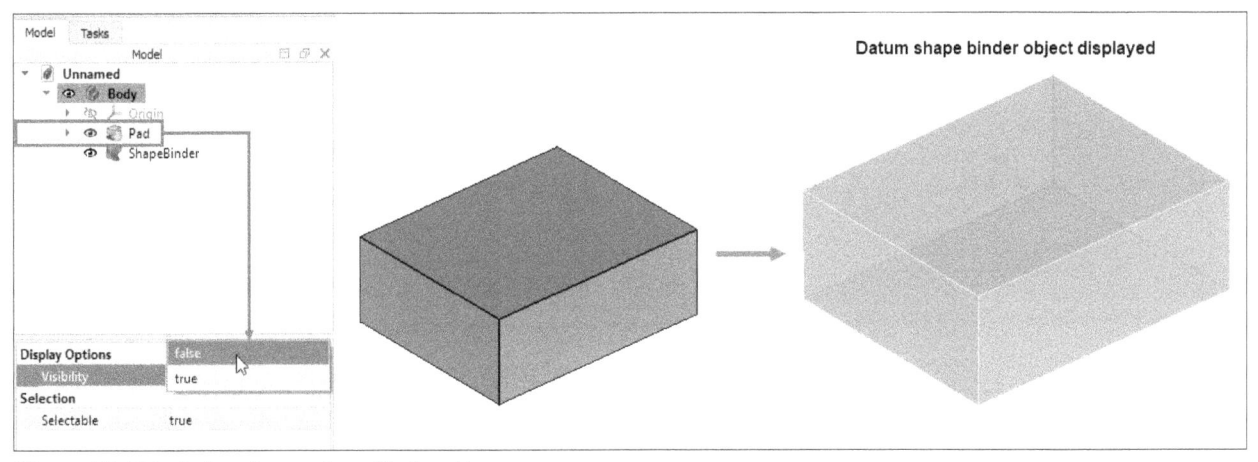

Figure-20. Displaying datum shape binder object

Create a sub-object(s) shape binder

The **Create a sub-object(s) shape binder** tool is used to create a shape binder to a sub-element, like edge or face from another body, while retaining the relative position of that element. Note that although shape binder and sub-object shape binder are created using similar procedure, the shape binder can either reference a single whole object, or sub-elements belonging to a single parent object whereas a sub-object shape binder does not have these restrictions. The procedure to use this tool is discussed next.

- Select the body from the **Model Tree** or from the 3D view area which you want to create shape binder; refer to Figure-21.

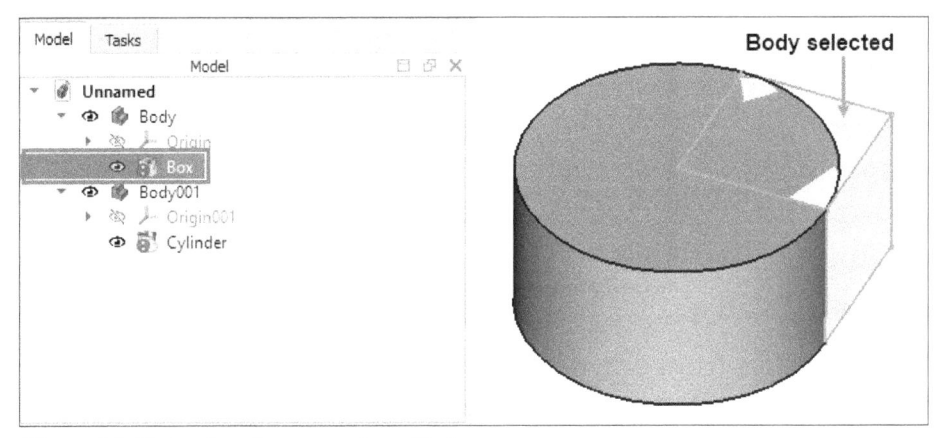

Figure-21. Body selected to create shape binder

- Click on the **Create a sub-object(s) shape binder** tool from the **Part Design** menu or from the **Toolbar** in the **Part Design** workbench; refer to Figure-22. The selected body will be created as a shape binder object; refer to Figure-23.

Figure-22. Create a sub object(s) shape binder tool

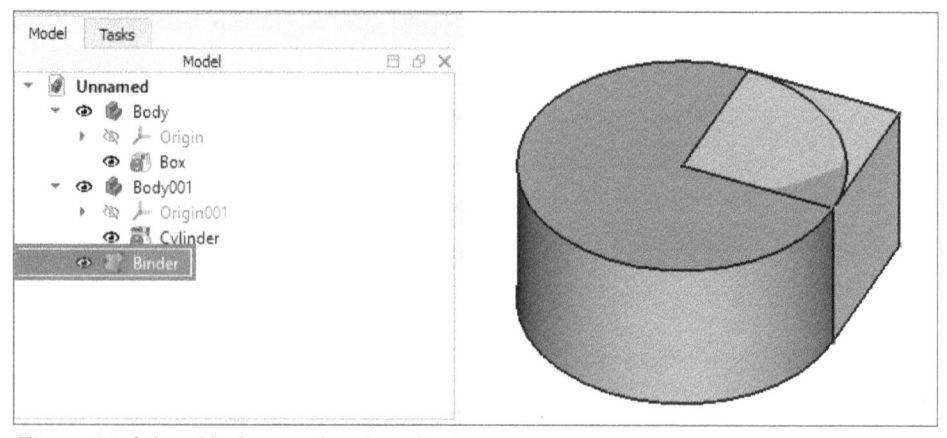

Figure-23. Selected body created as shape binder

An Example of using Shape Binder

- Now, select the shape binder object from the **Model Tree** and click on the **Boolean operation** tool from the **Part Design** menu or from **Toolbar** in the **Part Design** workbench (You will learn about this tool later in this chapter). The **Boolean parameters** dialog will be displayed in the **Tasks** panel of **Combo View** and the selected shape binder object will be displayed in the selection area of the dialog; refer to Figure-24.

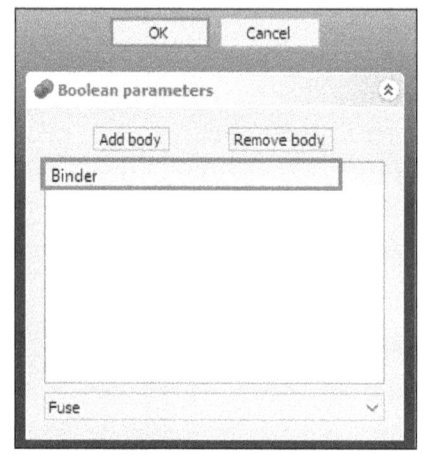

Figure-24. Boolean parameters dialog with shape binder object selected

- Select desired option from the drop-down in the dialog (**Cut** option in this case) and click on the **OK** button from the dialog. The boolean feature will be created; refer to Figure-25.

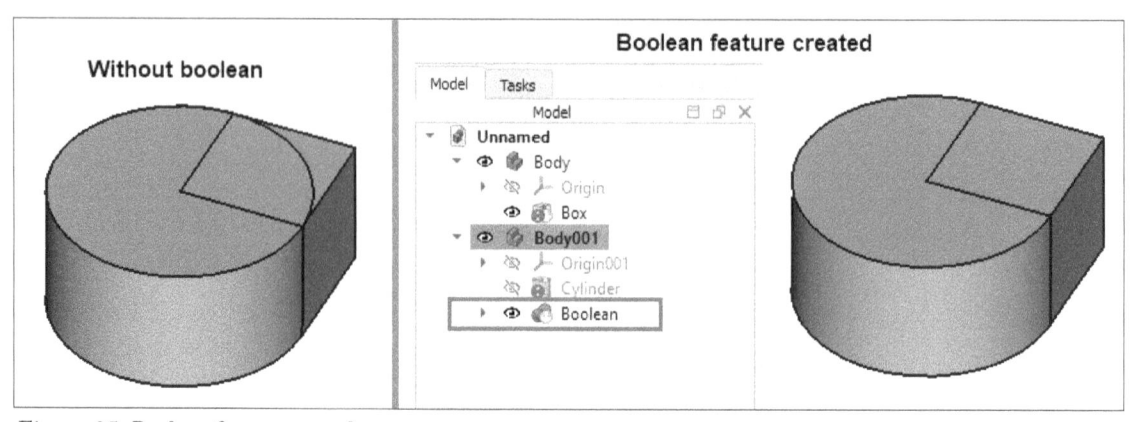

Figure-25. Boolean feature created

- To display the boolean cut feature, select the first body from the **Model Tree** and select **False** option from **Visibility** drop-down in the **Property editor** dialog or press **Space** button from the keyboard. The boolean cut feature will be displayed; refer to Figure-26.

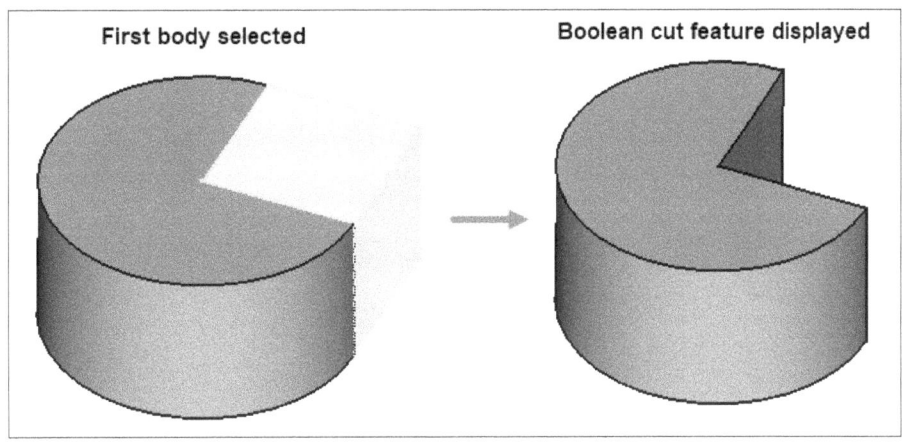

Figure-26. Boolean cut feature displayed

Create a clone

The **Create a clone** tool creates a linked copy of selected object which will follow any future edits to the original object. The procedure to use this tool is discussed next.

- Select the object to be cloned from **Model Tree** or from the 3D view area.
- Click on the **Create a clone** tool from the **Part Design** menu or from **Toolbar** in the **Part Design** workbench; refer to Figure-27. The selected object will be cloned.

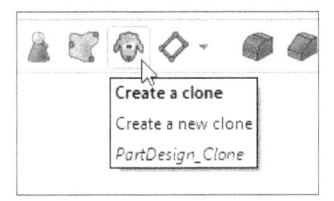

Figure-27. Create a clone tool

- Double-click on cloned object from **Model Tree**. The **Transform** dialog will be displayed in the **Tasks** panel of **Combo View** and a **Placement triad** will be displayed on the object in the 3D view area; refer to Figure-28.

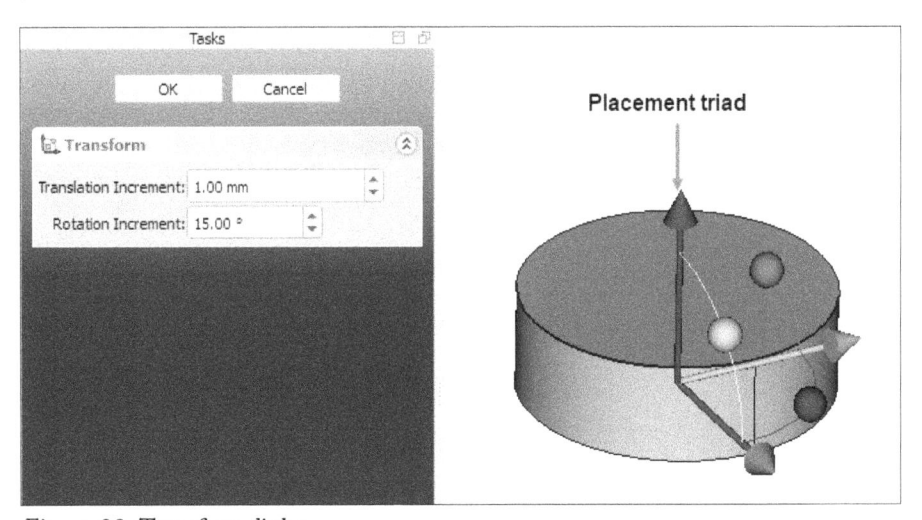

Figure-28. Transform dialog

- Drag the **Placement triad** to place the cloned object in desired location. The object will be placed; refer to Figure-29.

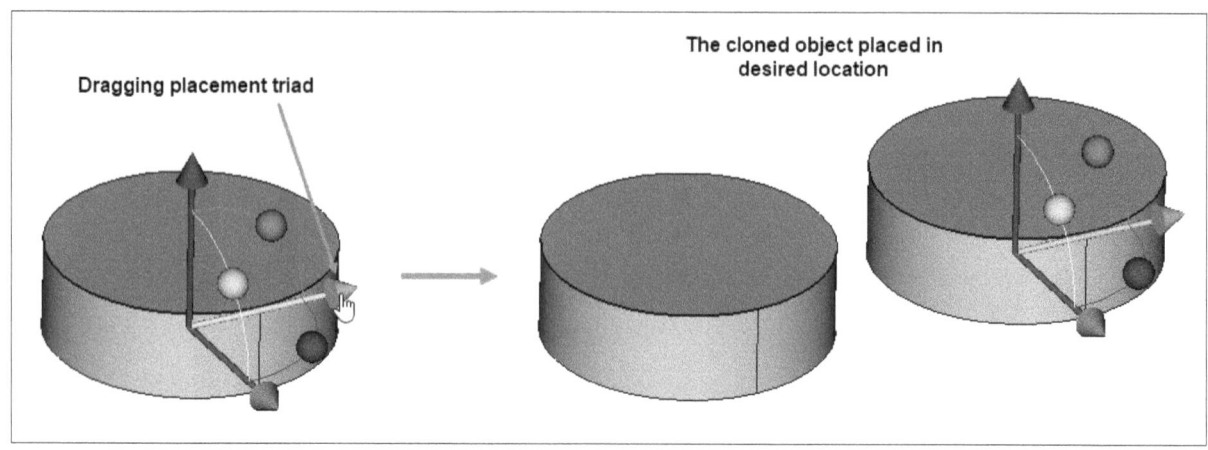

Figure-29. Placement of cloned object

- Click on **OK** button from the dialog to complete the procedure.

Datum tools

The **Datum tools** in FreeCAD are referred to as auxiliary geometrical elements which will not form part of the final shape of the body but can be used as references and supports for sketches and other types of features. The **Datum tools** are available in the **Create a datum** drop-down in the **Part Design Helper Toolbar** of **Part Design** workbench or **Create a datum** cascading menu in the **Part Design** menu; refer to Figure-30. The procedures to use these tools are discussed next.

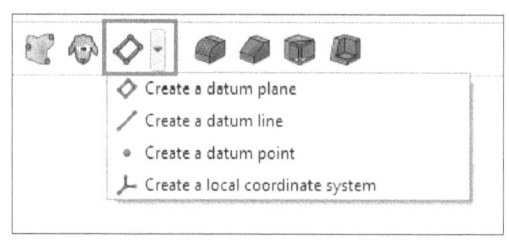

Figure-30. Datum tools

Create a datum plane

The **Create a datum plane** tool creates a datum plane which can be used as reference for sketches or other datum geometries. Sketches can also be attached to datum planes. The procedure to use this tool is discussed next.

- Click on the **Create a datum plane** tool from **Create a datum** drop-down in the **Toolbar** of **Part Design** workbench. The **Datum Plane parameters** dialog will be displayed in the **Tasks** panel of **Combo View**; refer to Figure-31.

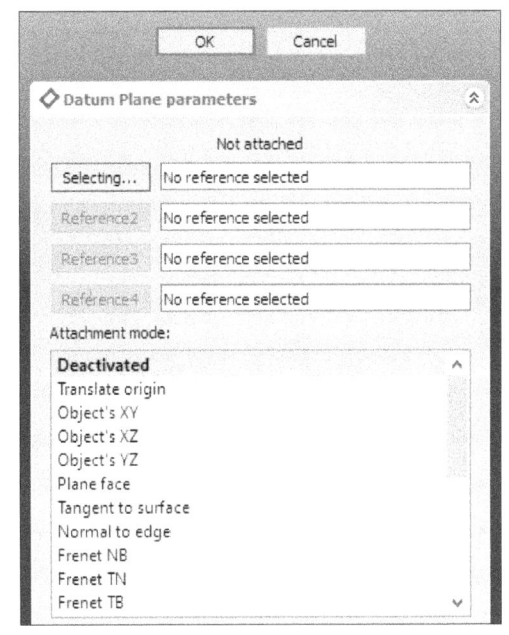

Figure-31. Datum Plane parameters dialog

- Select first reference for attachment mode from the 3D view area or from the attachment modes list available in **Attachment mode** area of the dialog. You can select a point, line, plane, or face as reference. Note that when you hover cursor on an option in the **Attachment mode** area of the dialog box then selection options related to that attachment method.
- Click on the **Deactivated** mode in the attachment mode then the attachment is disabled. The object can be moved by editing its Placement property.
- Click on the **Translate origin** mode in the attachment mode then the origin is aligned to match a vertex. The orientation is still controlled by the Placement property of the attached object.
- Click on the **Object's XY** mode in the attachment mode then the Placement is made equal to the Placement of a linked object.
- Click on the **Object's YZ** mode in the attachment mode then the the X, Y and Z axes are matched with a linked object's local X, Z and -Y axes, respectively.
- Click on the **Object ZX** mode in the attachment mode then the X, Y and Z axes are matched with a linked object's local Y, Z and X axes, respectively.
- Click on the **Plane face** mode in the attachment mode then the plane is aligned to coincide with a planar face.
- Click on the **Tangent to surface** mode in the attachment mode then the plane is made tangent to a face at a vertex.
- Click on the **Normal to edge** mode in the attachment mode then the aligned to be normal to an edge.
- To add an additional reference, click on the **Reference2** button.
- Specify desired offset values in the **Attachment Offset** area of the dialog to move plane by specified value in respective directions.
- Click on **OK** button from the dialog. The datum plane will be created at specified offset distance; refer to Figure-32.

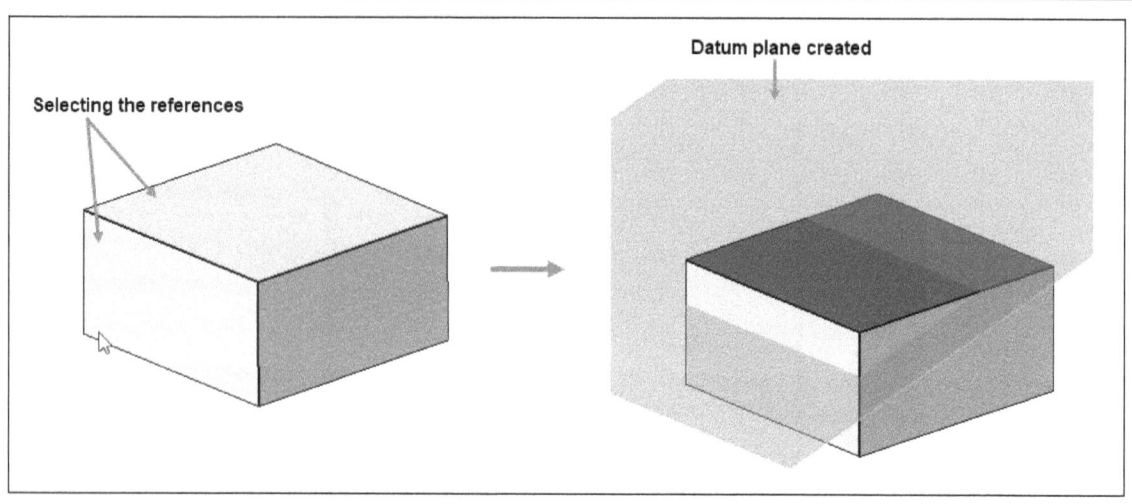

Figure-32. Creation of datum plane

Creating a plane tangent to curved face of body

To create a tangent plane, you will need a curved face and a vertex from which plane will pass. The procedure to create tangent plane is given next.

- Click on the **Create a datum plane** tool from **Create a datum** drop-down in the **Toolbar**. The **Datum Plane parameters** dialog will be displayed in the **Tasks** panel of **Combo View**.
- Select the **Tangent to surface** option from the **Attachment mode** section of the **Datum Plane parameters** dialog. You will be asked to select references for creating the plane.
- Select the vertex (datum point) and curved face. Preview of tangent plane will be displayed; refer to Figure-33.

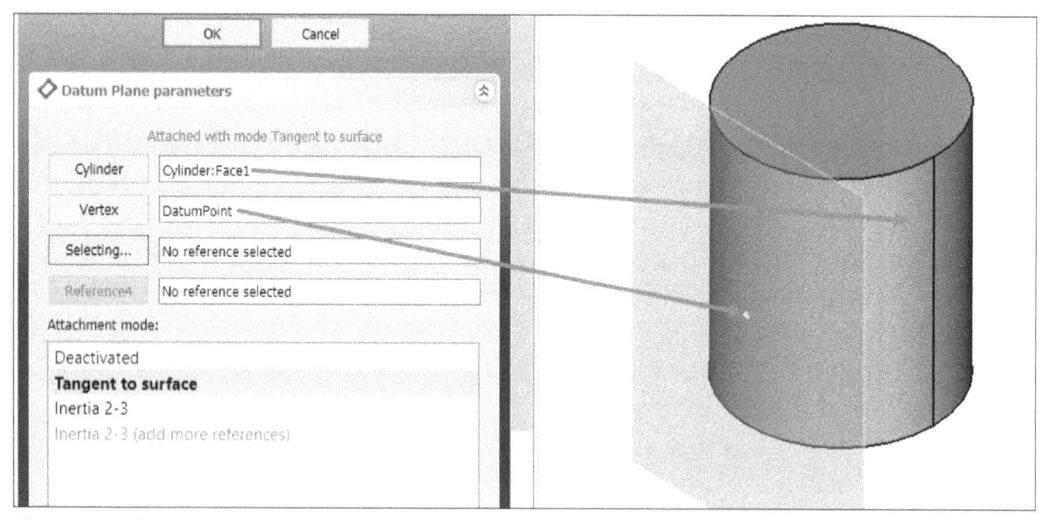

Figure-33. Preview of tangent plane

- Click on the **OK** button from the dialog to create the plane.

Create a datum line

The **Create a datum line** tool creates a datum line which can be used as reference for sketches, other datum geometries, or features. For example, it can be used as revolution axis for **Revolution** and **Groove** features. The procedure to use this tool is discussed next.

- Click on the **Create a datum line** tool from **Create a datum** drop-down in the **Toolbar** of **Part Design** workbench. The **Datum Line parameters** dialog will be displayed in the **Tasks** panel of **Combo View**; refer to Figure-34.

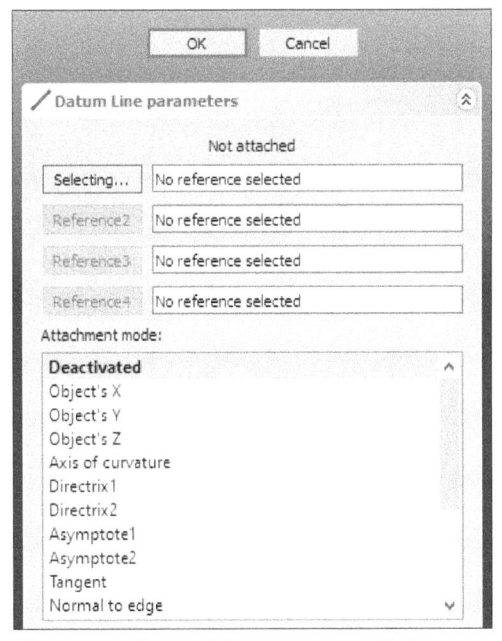

Figure-34. Datum Line parameters dialog

- Select first reference for attachment mode from the 3D view area or from the attachment modes list available in **Attachment mode** area of the dialog.
- To add an additional reference, click on the **Reference2** button and select desired reference.
- Specify desired offset values in the **Attachment Offset** area of the dialog.
- Click on **OK** button from the dialog. The datum line will be created at specified offset distance; refer to Figure-35.

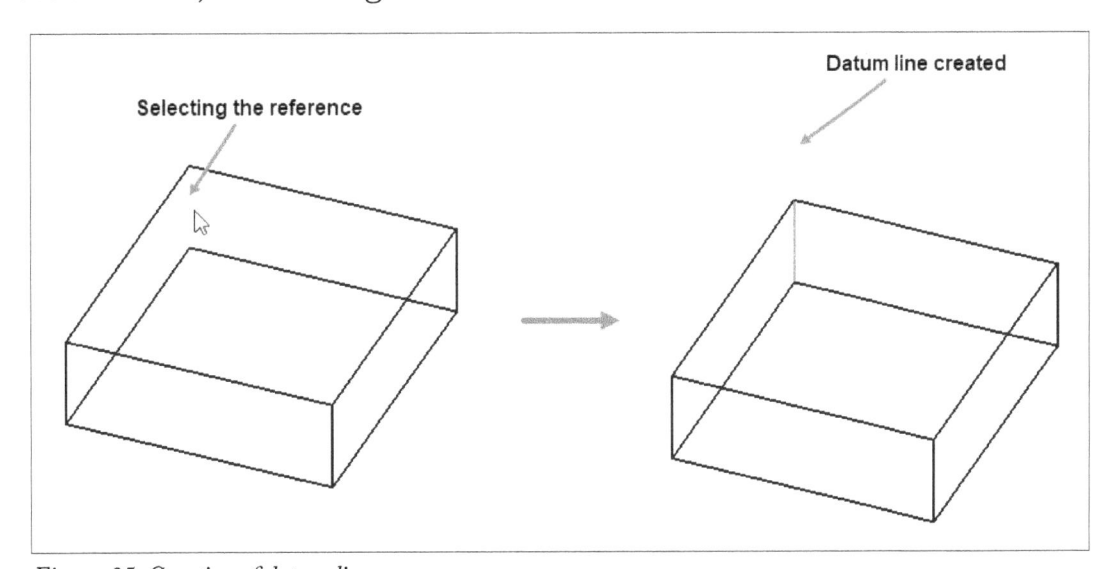

Figure-35. Creation of datum line

Create a datum point

The **Create a datum point** tool creates a datum point in the active body which can be used as reference for sketches or other datum geometry. The procedure to use this tool is discussed next.

- Click on the **Create a datum point** tool from **Create a datum** drop-down in the **Toolbar** of **Part Design** workbench. The **Datum Point parameters** dialog will be displayed in the **Tasks** panel of **Combo View**; refer to Figure-36.

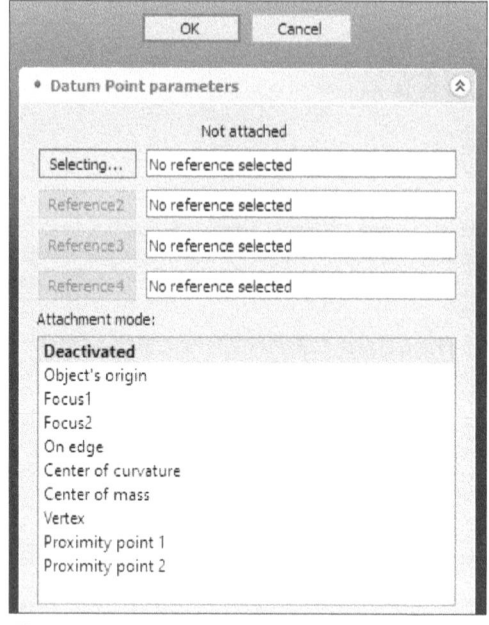

Figure-36. Datum Point parameters dialog

- Select the first reference for attachment mode from the 3D view area or from the attachment modes list available in **Attachment mode** area of the dialog.
- Select the **Deactivated** option from the **Attachment mode** area if you want to select desired references for defining datum point without using predefined attachment modes. Select the **Object's origin** option from the area to use origin of selected object as datum point. Select the **Focus1** option from the area to use focal point 1 of selected curve. Select **Focus2** option from the area to use focal point 2 of selected curve. Similarly, use other options in the area to use respective references for creating datum point.
- Specify desired offset values in **Attachment Offset** area of the dialog to move point by specified value; refer to Figure-37.

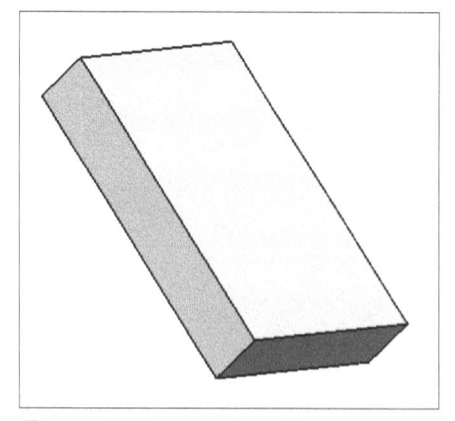

Figure-37. Datum point offset

- Click on the **Flip sides** check box then the datum point is opposite to the Sketch.
- Click on **OK** button from the dialog. The datum point will be attached to the feature at a specified offset distance; refer to Figure-38.

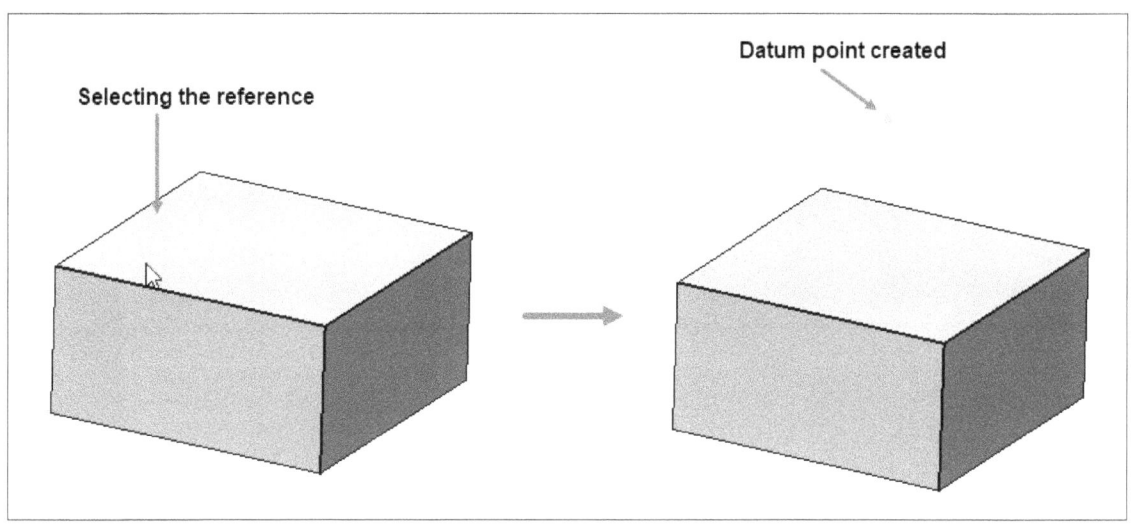

Figure-38. Creation of datum point

Create a local coordinate system

The **Create a local coordinate system** tool creates a local coordinate system which can be used as reference for other datum geometry. It also helps identify the orientation of the referenced datum geometry in 3D space. The procedure to use this tool is discussed next.

- Click on **Create a local coordinate system** tool from **Create a datum** drop-down in the **Toolbar** of **Part Design** workbench. The **Local Coordinate System parameters** dialog will be displayed in the **Tasks** panel of **Combo View**; refer to Figure-39.

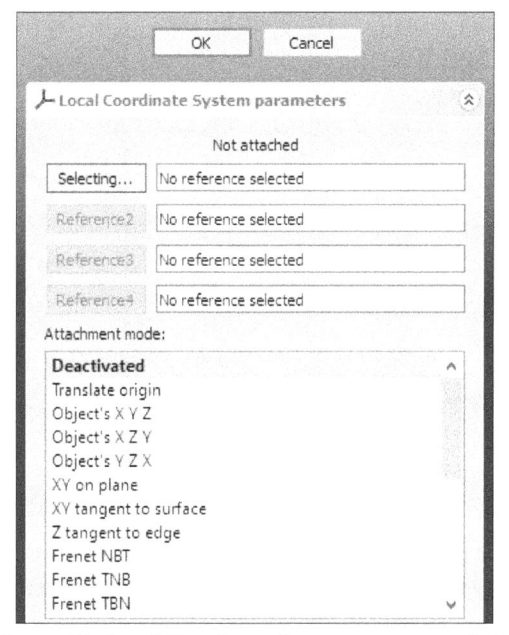

Figure-39. Local Coordinate System parameters dialog

- Select a point, line, face, or plane as first reference for attachment mode from the 3D view area or from the attachment modes list available in **Attachment mode** area of the dialog.
- To add an additional freeference, click on the **Reference2** button and select desired entity.

- Specify desired offset values in the **Attachment Offset** area of the dialog.
- Click on **OK** button from the dialog. The local coordinate system will be created at specified offset distance; refer to Figure-40.

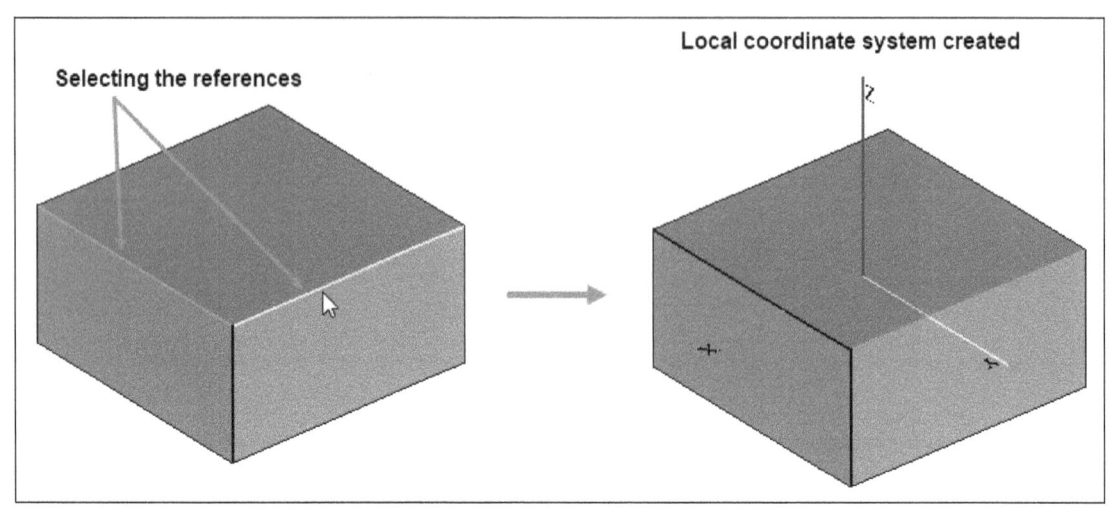

Figure-40. Creation of local coordinate system

PART DESIGN MODELING TOOLS

With the help of Part Design Modeling tools, you can create a part from a conceptual sketch through solid feature-based modeling, as well as build and modify parts through direct intuitive graphical manipulation.

Part Design Modeling tools

The **Part Design Modeling tools** are used for creating base features or adding material to an existing solid body. The **Part Design Modeling tools** are available in the **Toolbar** of **Part Design** workbench; refer to Figure-41. The procedures to use these tools are discussed next.

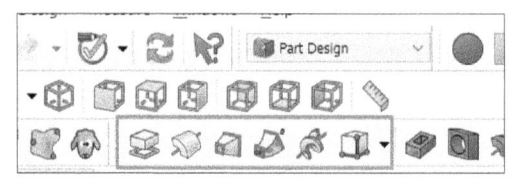

Figure-41. Part Design Modeling tools

Pad

The **Pad** tool extrudes a sketch to form solid in direction normal to the sketch plane. The procedure to use this tool is discussed next.

- Select desired sketch from the **Model Tree** to create the pad feature; refer to Figure-42. Note that sketch should be a closed loop curve.

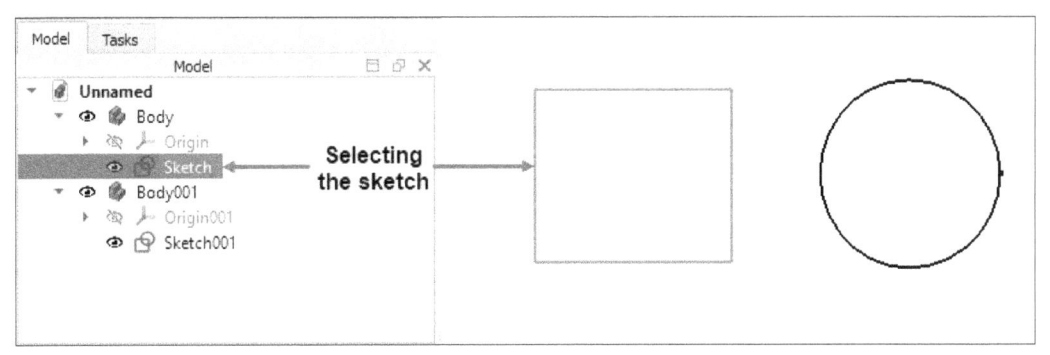

Figure-42. Selecting the sketch to create pad feature

- Click on the **Pad** tool from **Toolbar** in the **Part Design** workbench; refer to Figure-43. The **Pad parameters** dialog will be displayed in the **Tasks** panel of **Combo View** along with the preview of extrude feature; refer to Figure-44.

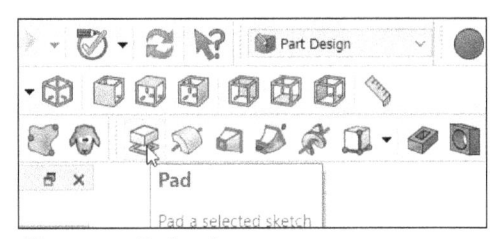

Figure-43. Pad tool

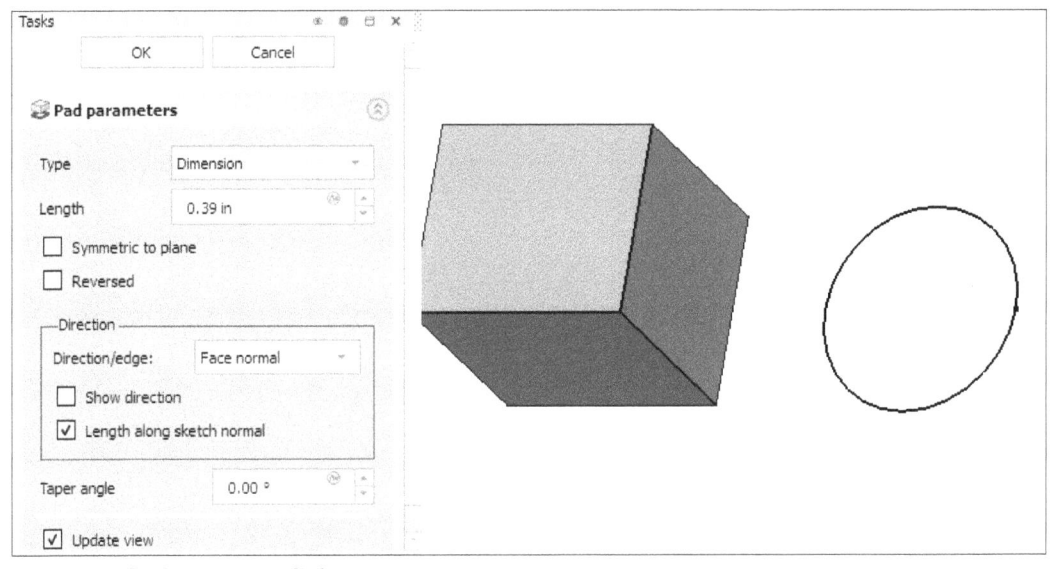

Figure-44. Pad parameters dialog

- Select desired type of extrusion from **Type** drop-down. Select **Dimension** option from **Type** drop-down to extrude selected sketch on either side of the sketch plane at specified length. Select **To last** option to extrude the sketch up to the last face of the base feature. Select **To first** option to extrude the sketch up to the first face of the base feature. Select **Up to face** option to extrude the sketch up to selected face of base feature. Select **Two dimensions** option to extrude the sketch by specifying length for both the sides. Select **Up to shape** option to extrude the sketch up to the selected shape of base feature.
- Specify desired length of extrusion in the **Length** edit box.

- Select **Symmetric to plane** check box to extrude the sketch on both the sides by same specified length. This option is available only if the **Dimension** option is selected in the **Length** drop-down.
- Select **Reversed** check box to reverse the direction of extrusion. This option works only if the **Dimension** option is selected in the **Length** drop-down.
- Select desired direction of the extrusion from **Direction/edge** drop-down. Select **Sketch normal** option from the **Direction/Edge** drop-down to extrude the sketch or face along its normal. Select the **Select reference** option to extrude the sketch along an edge of the 3D model. Select **Custom direction** option to set the pad length along the custom direction by specifying desired values in **x**, **y**, and **z** edit boxes.
- Select the **Show direction** check box to display **x**, **y**, and **z** edit boxes.
- Select **Length along sketch normal** check box to measure the pad length along the sketch normal.
- Specify desired value in the **Taper angle** edit box to taper the pad in the extrusion direction by the given angle. This option is available only if the **Dimension** or **Two dimensions** option is selected in the **Length** drop-down.
- Click on **OK** button from the dialog. The selected sketch will be extruded.

Revolution

The **Revolution** tool is used to create a solid by revolving selected sketch or 2D object about given axis. Note that the sketch should be closed loop and revolution axis should not intersect with the sketch. The procedure to use this tool is discussed next.

- Select the sketch to be revolved from **Model** panel in the **Combo View**; refer to Figure-45.

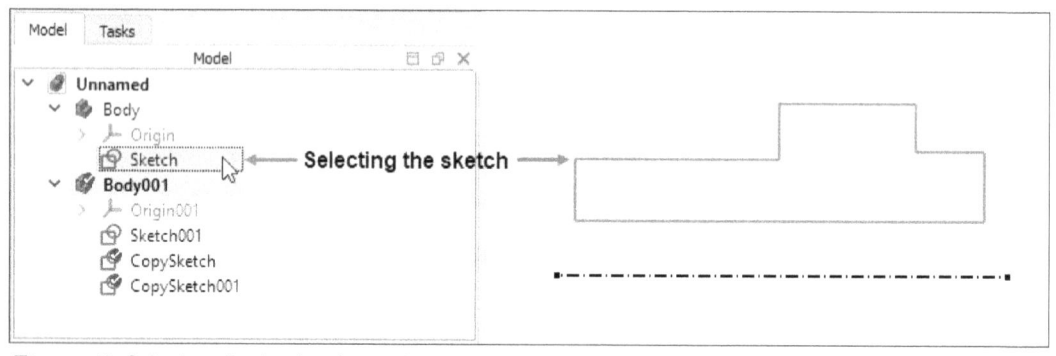

Figure-45. Selecting the sketch to be revolve

- Click on the **Revolution** tool from **Toolbar** in the **Part Design** workbench; refer to Figure-46. The **Revolution parameters** dialog will be displayed in the **Tasks** panel of **Combo View** along with the preview of revolution feature; refer to Figure-47.

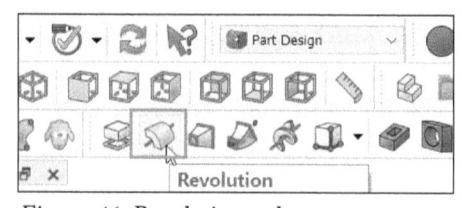

Figure-46. Revolution tool

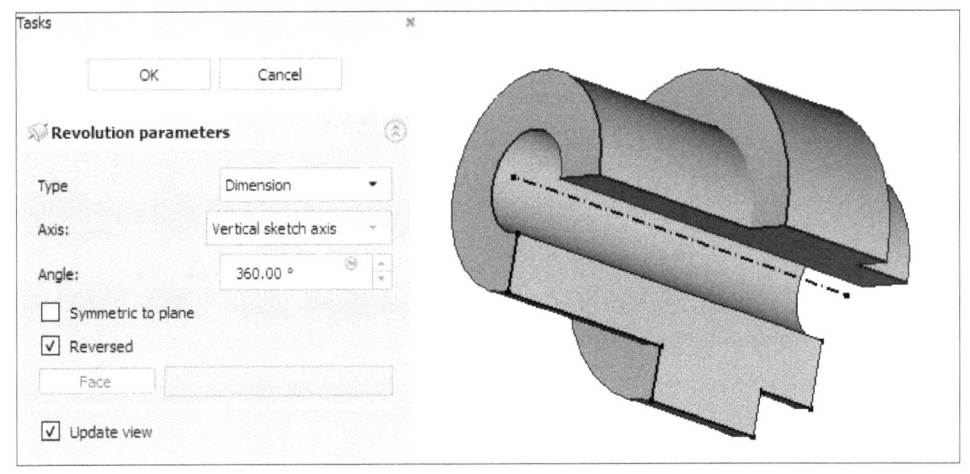

Figure-47. Revolution parameters dialog

- Select desired option from **Type** drop-down to specify the angle of revolution. The options in this drop-down are same as discussed for **Pad** tool.
- Select desired option from **Axis** drop-down in the dialog about which the sketch is to be revolved. Select **Vertical sketch axis** option from the **Axis** drop-down to create revolve feature about a vertical sketch axis. Select **Horizontal sketch axis** option to create revolve feature about a horizontal sketch axis. Select **Base X axis**, **Base Y axis**, or **Base Z axis** option to create the revolve feature about **X**, **Y**, or **Z** axis of body's origin, respectively. Select the **Select reference** option to create revolve feature by selecting an edge on the body or a datum line as an axis of revolution.
- Specify the angle of revolution by which the revolution is to be formed in the **Angle** edit box or **Angle** and **2nd angle** edit boxes in case of **Two dimensions** revolution type.
- Specify the other parameters as desired and click on **OK** button from the dialog. The revolution feature will be created.

Additive loft

The **Additive loft** tool creates a solid active body by making a transition between two or more sketches. The procedure to use this tool is discussed next.

- Click on **Additive loft** tool from **Toolbar** in the **Part Design** workbench; refer to Figure-48. The **Select attachment** dialog will be displayed in the **Tasks** panel of **Combo View** asking you to select the base profile; refer to Figure-49.

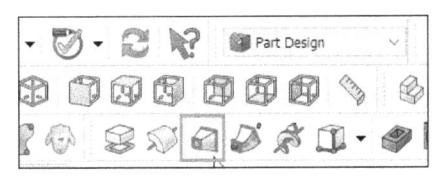

Figure-48. Additive loft tool

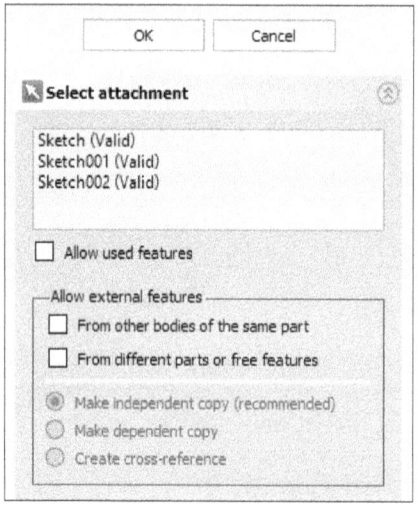

Figure-49. Select attachment dialog

- Select a sketch from the dialog to be used as base profile for additive loft feature and click on **OK** button; refer to Figure-50. The **Loft parameters** dialog will be displayed; refer to Figure-51.

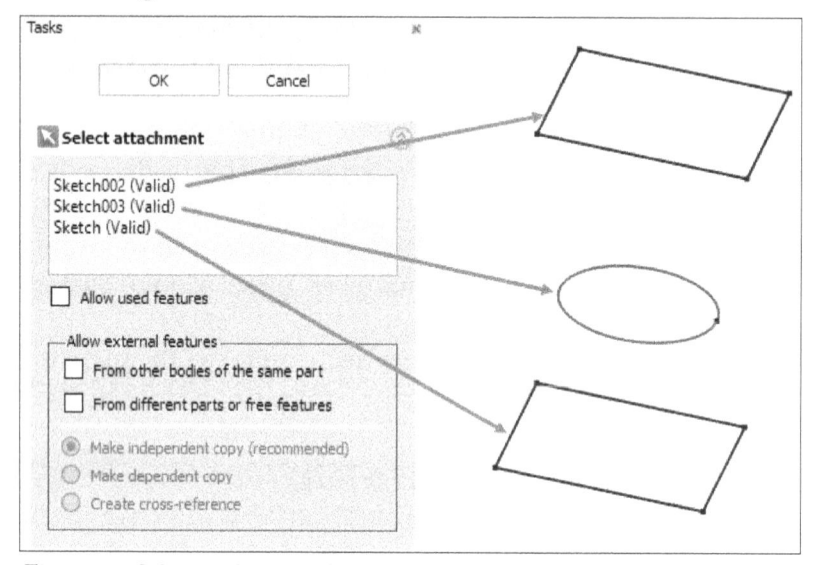

Figure-50. Selecting the sketch for base profile

Figure-51. Loft parameters dialog

- Select **Ruled surface** check box to make straight transitions between cross-sections.
- Select **Closed** check box to make a transition from the last cross-section to the first to create a loop.
- The base profile is preselected in the **Object** selection box of **Profile** area. Click on **Add Section** button and select the second sketch in the 3D view area for the first loft section and select the third sketch for the second loft section. The preview of additive loft feature will be displayed; refer to Figure-52.

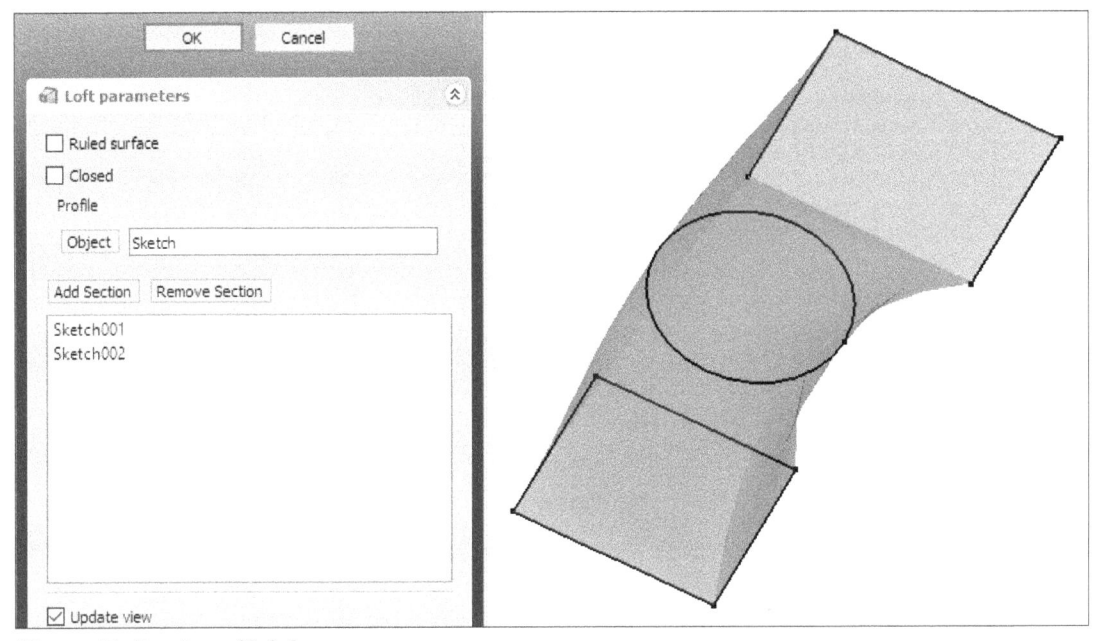

Figure-52. Preview of loft feature

- Specify the parameters as desired and click on **OK** button from the dialog to create the feature.

Additive pipe

The **Additive pipe** tool creates a solid as active body by sweeping one or more sketches along an open or closed path. The procedure to use this tool is discussed next.

- Click on **Additive pipe** tool from **Toolbar** in the **Part Design** workbench; refer to Figure-53. The **Select attachment** dialog will be displayed in the **Tasks** panel of **Combo View** as discussed earlier asking you to select the base profile.

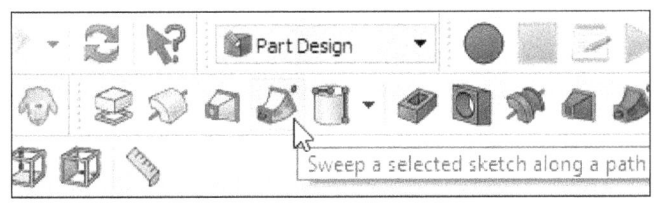

Figure-53. Additive pipe tool

- Select desired sketch for base profile of sweep feature from the dialog; refer to Figure-54 and click on **OK** button. The **Pipe parameters** dialog will be displayed in the **Tasks** panel of **Combo View**; refer to Figure-55.

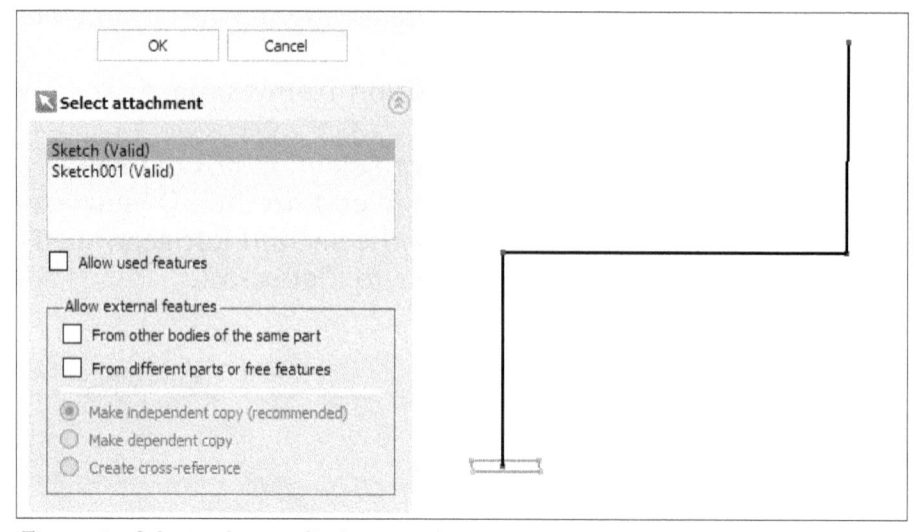

Figure-54. Selecting base profile for sweep feature

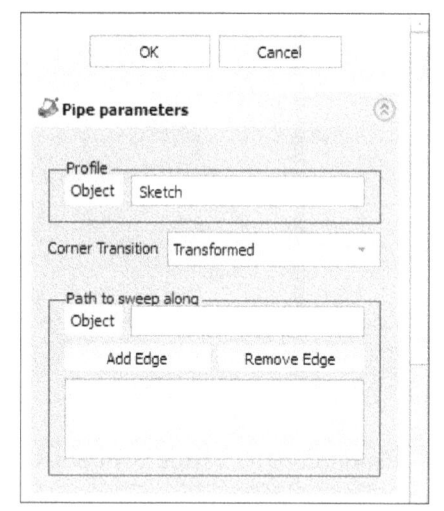

Figure-55. Pipe parameters dialog

- The base profile is preselected in the **Object** selection box of the **Profile** area.
- Click on the **Object** button from `Path to sweep along` area and select an edge of the sketch in 3D view area as a path for sweep feature. The whole sketch will be selected as sweep feature path and the preview of sweep feature will be displayed; refer to Figure-56.

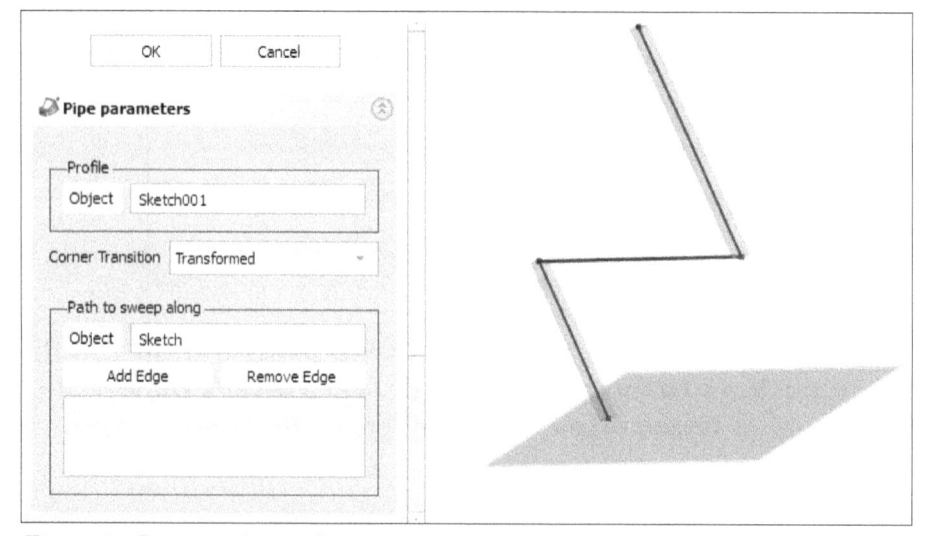

Figure-56. Preview of sweep feature

- Click on **Add Edge** button to select single edge of the sketch from the 3D view area as a path for sweep feature.
- To remove an edge of the sketch as a sweep feature path, select an edge from 3D view area and click on **Remove Edge** button from the dialog.
- Select desired option from **Corner Transition** drop-down. The sweep feature will be created as the selected option; refer to Figure-57.

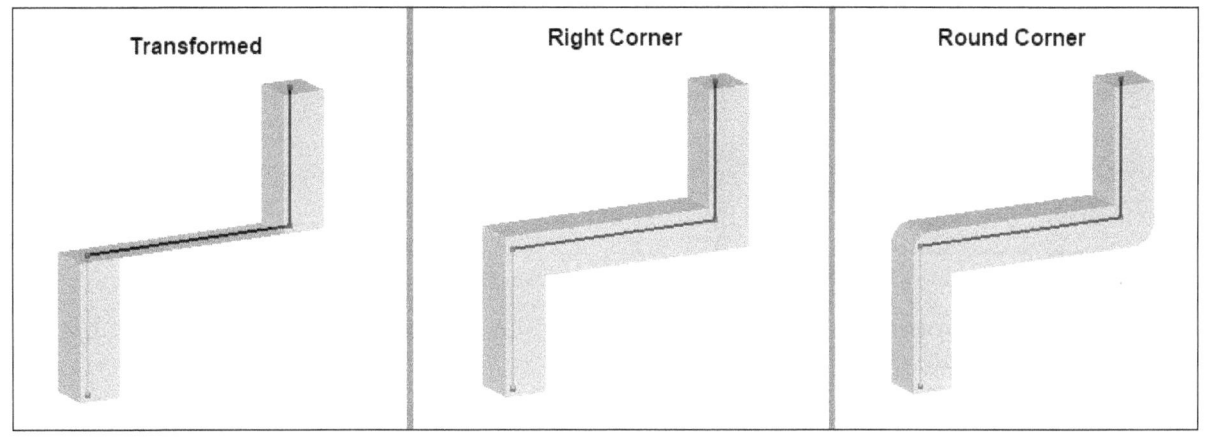

Figure-57. Corner Transition options

- Specify the parameters as desired and click on **OK** button from the dialog to create the sweep feature.

Additive Helix

The **Additive Helix** tool creates a solid by sweeping selected sketch or 2D object along a helix path. The procedure to use this tool is discussed next.

- Create desired sketch using which you want to create a solid sweep along a helical path; refer to Figure-58.

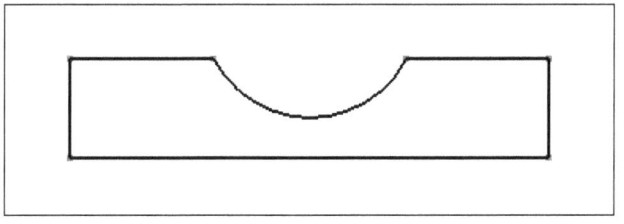

Figure-58. Sketch created for additive helix

- Select the sketch created from the **Model Tree** and click on the **Additive Helix** tool from **Toolbar** in the **Part Design** workbench; refer to Figure-59. The **Helix parameters** dialog will be displayed in the **Tasks** panel of **Combo View** along with the preview of additive helix feature; refer to Figure-60.

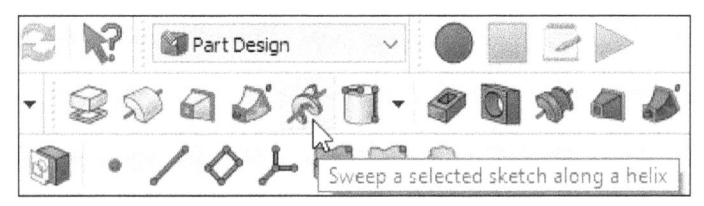

Figure-59. Additive Helix tool

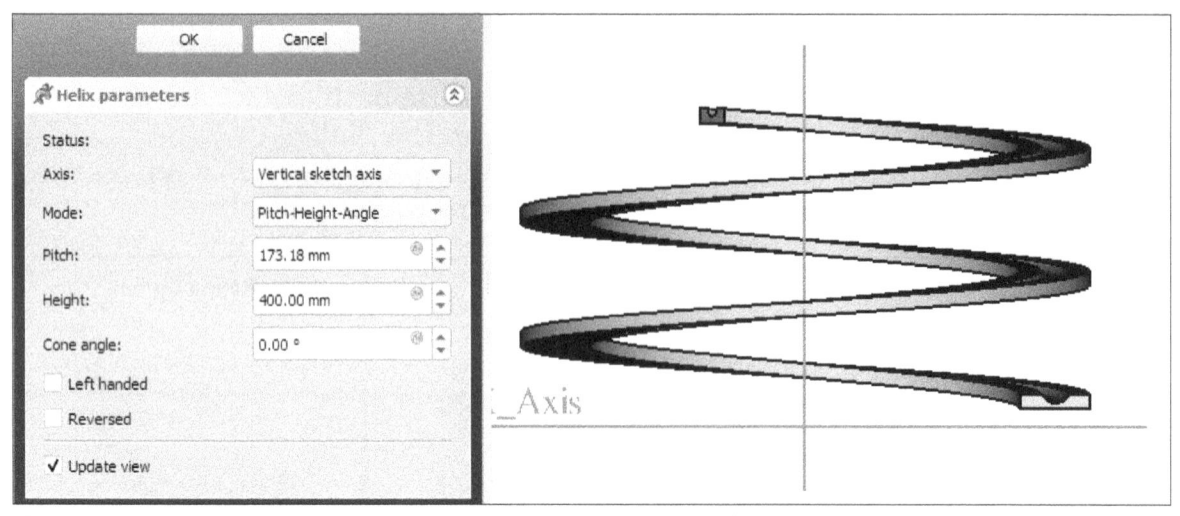

Figure-60. Helix parameters dialog along with preview of additive helix feature

- Select desired option from **Axis** drop-down to specify the axis about which the sketch is to be swept.
- Select desired option from **Mode** drop-down to specify what parameters will be used to define the helix. Select **Pitch-Height-Angle** option from the **Mode** drop-down to define the helix via the height per turn and the overall height. Select the **Pitch-Turns-Angle** option to define the helix via the height per turn and the number of turns. Select the **Height-Turns-Angle** option to define the helix via the overall height and the number of turns. Select the **Height-Turns-Growth** option to define the helix via the overall height, the number of turns, and the growth of the helical radius.
- Specify desired value in the **Pitch** edit box to define the distance between turns in the helix.
- Specify desired value in the **Height** edit box to define the height of the helix.
- Specify desired value in the **Turns** edit box to define the number of turns in the helix.
- Specify desired value in the **Cone angle** edit box to define the rate at which the radius of the helix increase along the axis.
- Specify desired value in the **Radial growth** edit box to define the increase/decrease of the radius of helix per turn.
- Select **Left handed** check box to reverse the turning direction of helix from clockwise to counter clockwise.
- Select the **Reversed** check box to reverse the axis direction of helix from default.
- After specifying desired parameters in the dialog, click on the **OK** button to create the additive helix feature.

Additive Box

The **Additive Box** tool inserts a primitive box in the active body as the first feature or fuses it to the existing feature. The procedure to use this tool is discussed next.

- Click on the **Additive Box** tool from **Create an additive primitive** drop-down in the **Toolbar** of **Part Design** workbench; refer to Figure-61. The **Primitive parameters** dialog will be displayed in the **Tasks** panel of **Combo View** along with the preview of additive box feature; refer to Figure-62.

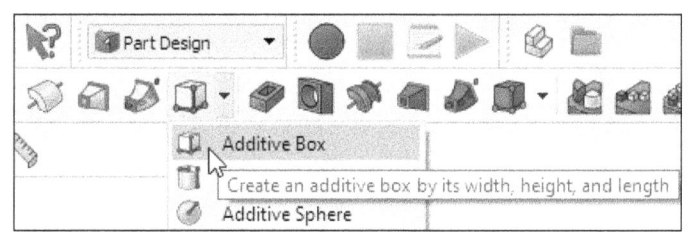

Figure-61. Additive Box tool

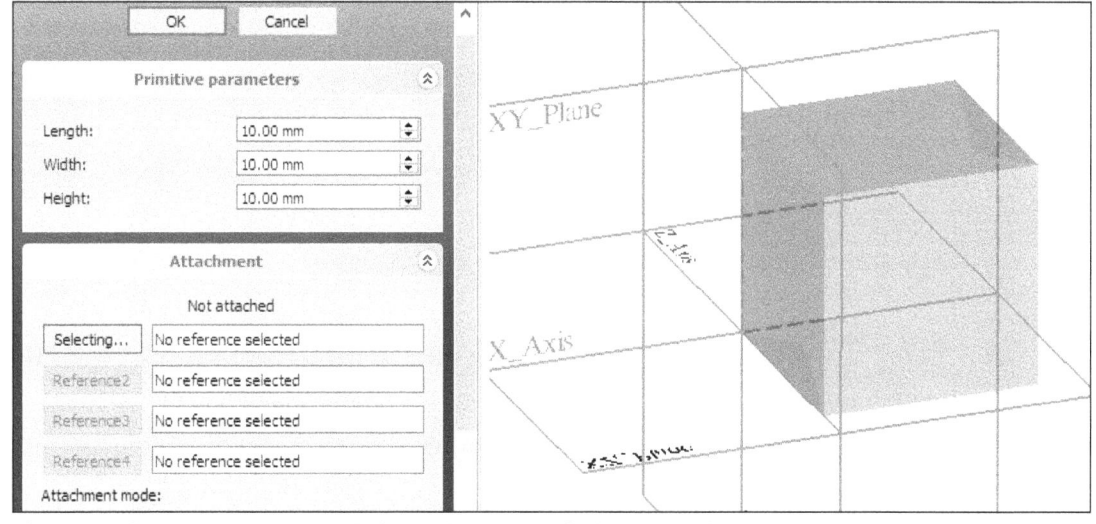

Figure-62. Primitive parameters dialog with preview of additive box feature

- Specify the length, width, and height of the feature in the **Length**, **Width**, and **Height** edit boxes of the **Primitive parameters** dialog, respectively.
- Specify the other parameters in the dialog as discussed earlier.
- Click on **OK** button from the dialog. The additive box feature will be created; refer to Figure-63.

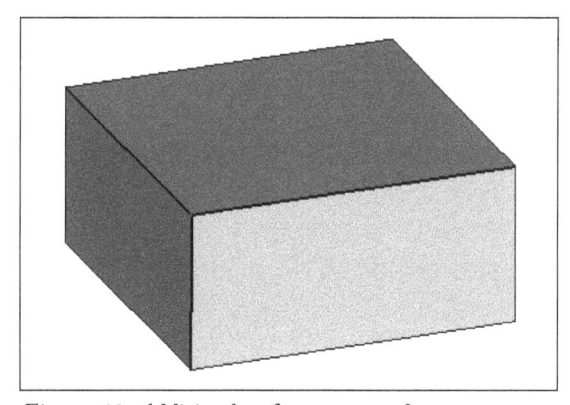

Figure-63. Additive box feature created

Additive Cylinder

The **Additive Cylinder** tool inserts a primitive cylinder in the active body as the first feature or fuses it to the existing feature. The procedure to use this tool is discussed next.

- Click on the **Additive Cylinder** tool from **Create an additive primitive** drop-down in the **Toolbar** of **Part Design** workbench; refer to Figure-64. The **Primitive parameters** dialog will be displayed in the **Tasks** panel of **Combo View** along with the preview of additive cylinder feature; refer to Figure-65.

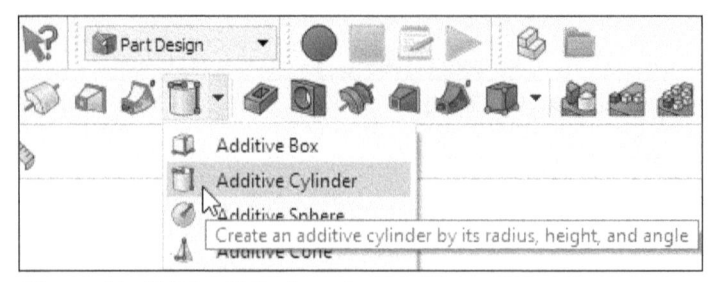

Figure-64. Additive Cylinder tool

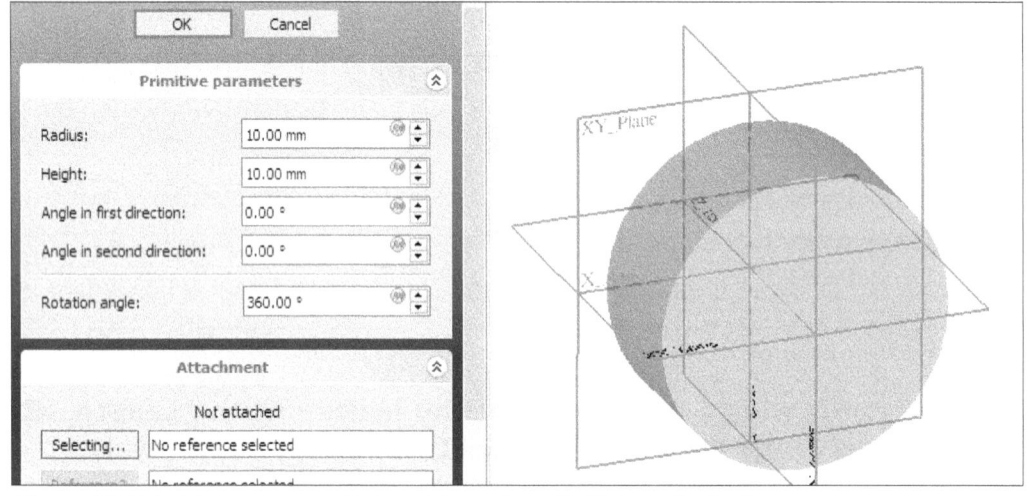

Figure-65. Primitive parameters dialog with preview of additive cylinder feature

- Specify the radius and height of the feature in the **Radius** and **Height** edit boxes, respectively.
- Specify desired angle of first and second direction in the **Angle in first direction** and **Angle in second direction** edit boxes to define revolution in clockwise and counter clockwise directions, respectively.
- Specify the angle of revolution of the cross-section in the **Rotation angle** edit box of the dialog.
- Specify the other parameters in the dialog as discussed earlier.
- Click on **OK** button from the dialog. The additive cylinder feature will be created; refer to Figure-66.

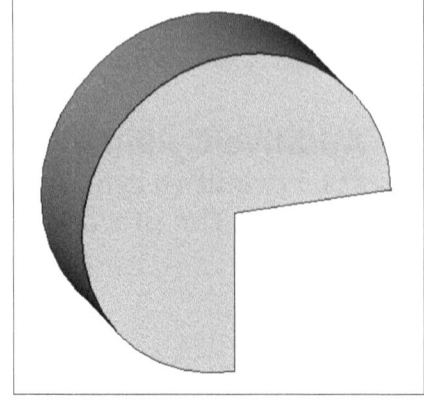

Figure-66. Additive cylinder feature created

Additive Sphere

The **Additive Sphere** tool inserts a primitive sphere in the active body as the first feature or fuses it to the existing feature. The procedure to use this tool is discussed next.

- Click on the **Additive Sphere** tool from **Create an additive primitive** drop-down in the **Toolbar** of **Part Design** workbench; refer to Figure-67. The **Primitive parameters** dialog will be displayed in the **Tasks** panel of **Combo View** along with the preview of additive sphere feature; refer to Figure-68.

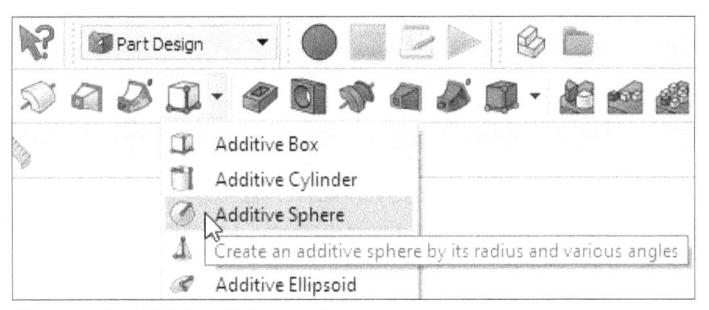

Figure-67. Additive Sphere tool

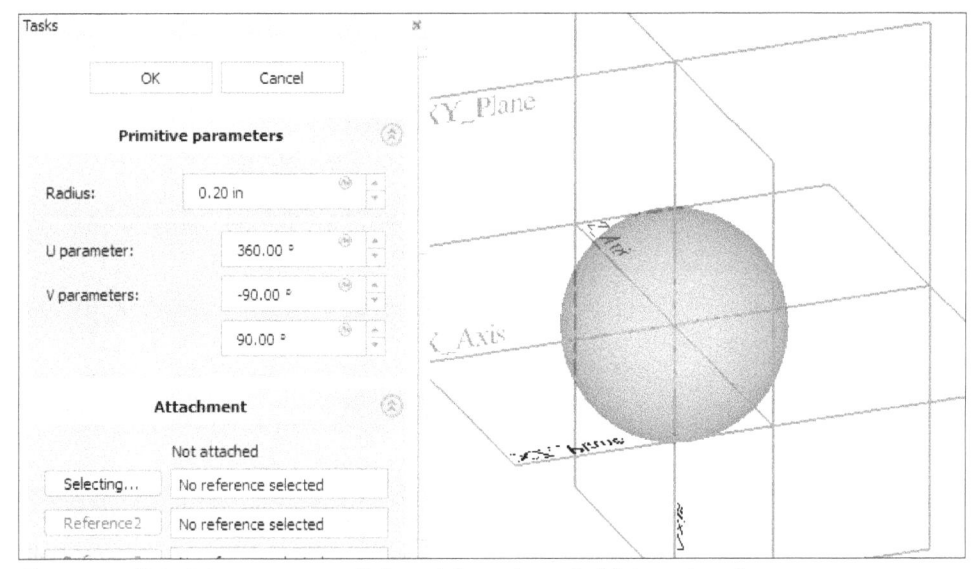

Figure-68. Primitive parameters dialog with preview of additive sphere feature

- Specify desired radius of the sphere in the **Radius** edit box of the dialog.
- Specify desired angle of revolution of the cross section in the **U parameter** edit box of the dialog.
- Specify lower truncation of the sphere parallel to the circular cross section in the **V parameters** edit box and specify upper truncation of the ellipsoid parallel to the circular cross section in the edit box just below to the **V parameters** edit box of the dialog.
- Specify the other parameters in the dialog as discussed earlier.
- Click on **OK** button from the dialog. The additive sphere feature will be created; refer to Figure-69.

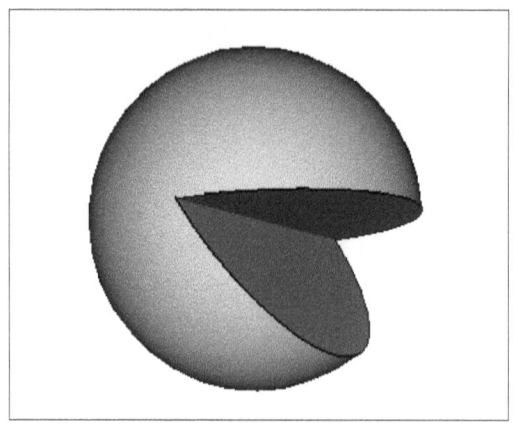

Figure-69. Additive sphere feature created

Additive Cone

The **Additive Cone** tool inserts a primitive cone in the active body as the first feature or fuses it to the existing feature. The procedure to use this tool is discussed next.

- Click on the **Additive Cone** tool from **Create an additive primitive** drop-down in the **Toolbar** of **Part Design** workbench; refer to Figure-70. The **Primitive parameters** dialog will be displayed in the **Tasks** panel of **Combo View** along with the preview of additive cone feature; refer to Figure-71.

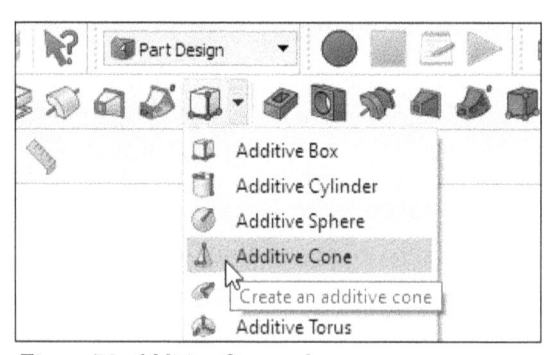

Figure-70. Additive Cone tool

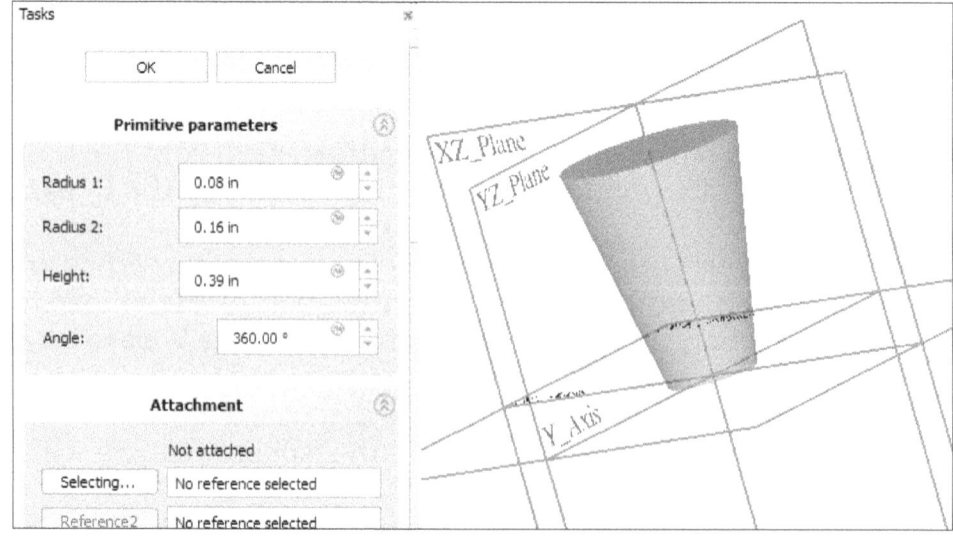

Figure-71. Primitive parameters dialog with preview of additive cone feature

- Specify desired radius value for base of the cone in the **Radius 1** edit box and specify desired radius value for top of the cone in the **Radius 2** edit box of the dialog.
- Specify the height of the cone along its axis in the **Height** edit box of the dialog.
- Specify the angle of revolution of the cross section in the **Angle** edit box of the dialog.
- Specify the other parameters in the dialog as discussed earlier.
- Click on **OK** button from the dialog. The additive cone feature will be created; refer to Figure-72.

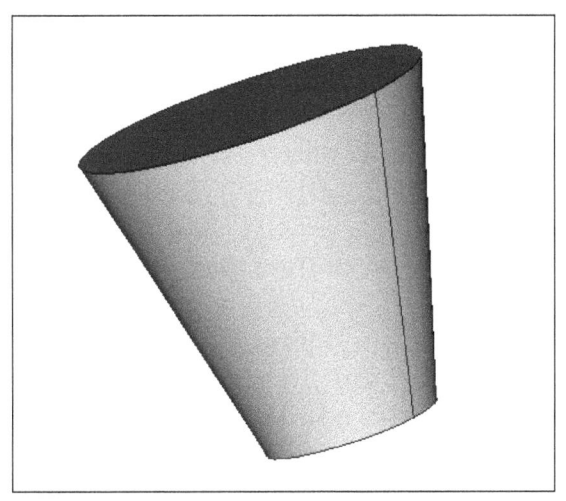

Figure-72. Additive cone feature created

Additive Ellipsoid

The **Additive Ellipsoid** tool inserts a primitive ellipsoid in the active body as the first feature or fuses it to the existing feature. The procedure to use this tool is discussed next.

- Click on the **Additive Ellipsoid** tool from **Create an additive primitive** drop-down in the **Toolbar** of **Part Design** workbench; refer to Figure-73. The **Primitive parameters** dialog will be displayed in the **Tasks** panel of **Combo View** along with the preview of additive ellipsoid feature; refer to Figure-74.

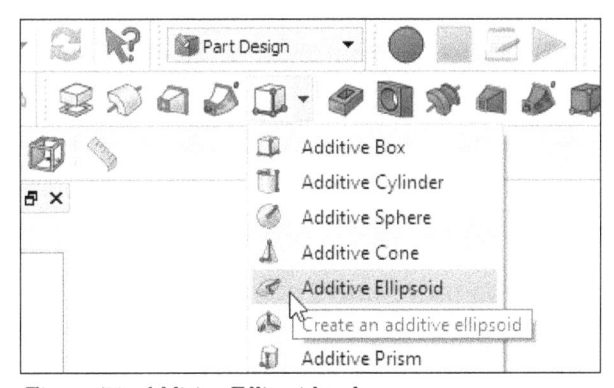

Figure-73. Additive Ellipsoid tool

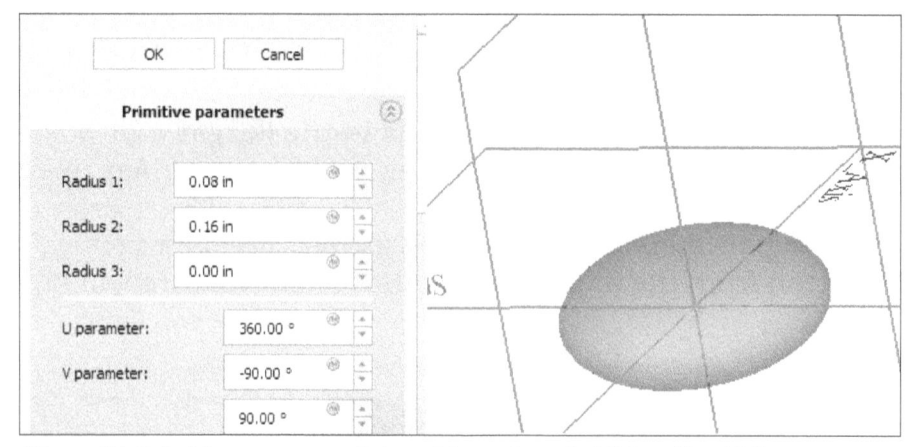

Figure-74. Primitive parameters dialog with preview of additive ellipsoid feature

- Specify desired radius value along the ellipsoid's vertical axis, along the ellipsoid's length, and along the ellipsoid's width in the **Radius 1**, **Radius 2**, and **Radius 3** edit boxes, respectively.
- Specify desired angle of revolution of the elliptical cross section in the **U parameter** edit box of the dialog.
- Specify the lower truncation of the ellipsoid parallel to the circular cross section in the **V parameter** edit box and specify upper truncation of the ellipsoid parallel to the circular cross section in the edit box just below to the **V parameter** edit box.
- Specify other parameters in the dialog as discussed earlier.
- Click on **OK** button from the dialog. The additive ellipsoid feature will be created; refer to Figure-75.

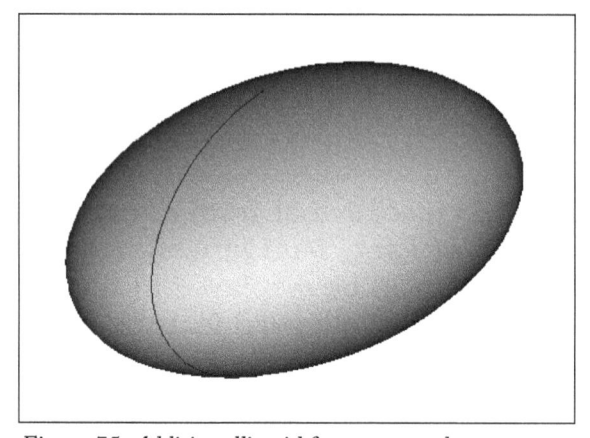

Figure-75. Additive ellipsoid feature created

Additive Torus

The **Additive Torus** tool inserts a primitive torus in the active body as the first feature or fuses it to the existing feature. The procedure to use this tool is discussed next.

- Click on the **Additive Torus** tool from **Create an additive primitive** drop-down in the **Toolbar** of **Part Design** workbench; refer to Figure-76. The **Primitive parameters** dialog will be displayed in the **Tasks** panel of **Combo View** along with the preview of additive ellipsoid feature; refer to Figure-77.

FreeCAD 1.0 Black Book   3-33

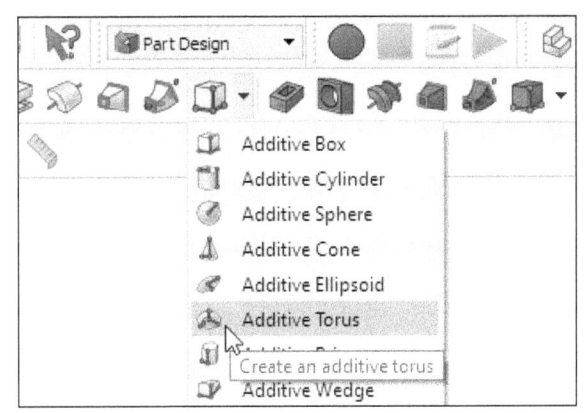

Figure-76. Additive Torus tool

- Specify desired radius of the imaginary orbit around which the circular cross section revolves in the **Radius 1** edit box and specify desired radius of circular cross section defining the form of the torus in **Radius 2** edit box of the dialog.

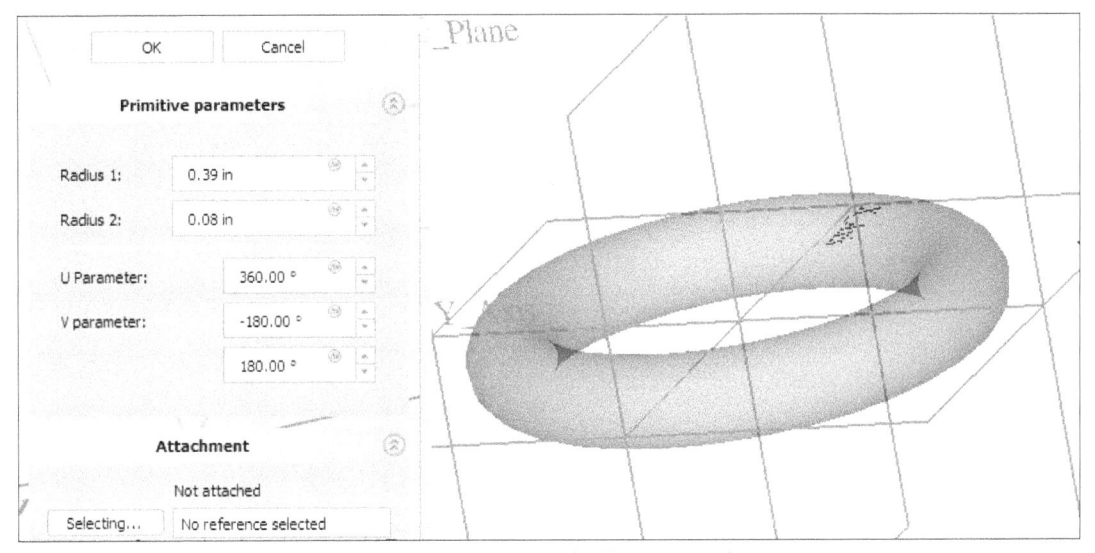

Figure-77. Primitive parameters dialog with preview of additive torus feature

- Specify other parameters in the dialog as discussed earlier.
- Click on **OK** button from the dialog. The additive torus feature will be created; refer to Figure-78.

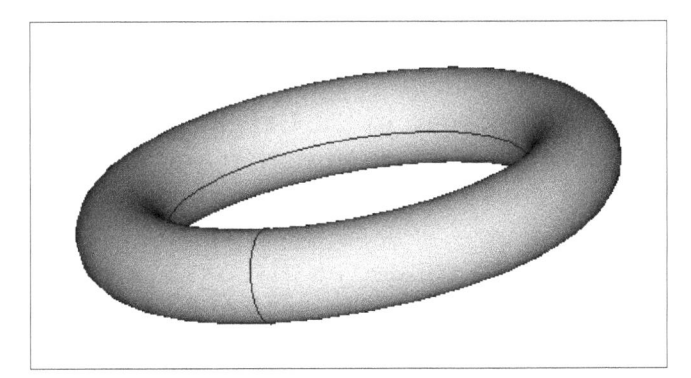

Figure-78. Additive torus feature created

Additive Prism

The **Additive Prism** tool inserts a primitive prism in the active body as the first feature or fuses it to the existing feature. The procedure to use this tool is discussed next.

- Click on the **Additive Prism** tool from **Create an additive primitive** drop-down in the **Toolbar** of **Part Design** workbench; refer to Figure-79. The **Primitive parameters** dialog will be displayed in the **Tasks** panel of **Combo View** along with the preview of additive prism feature; refer to Figure-80.

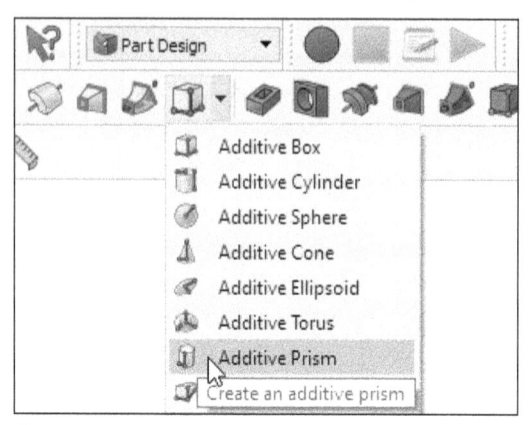

Figure-79. Additive Prism tool

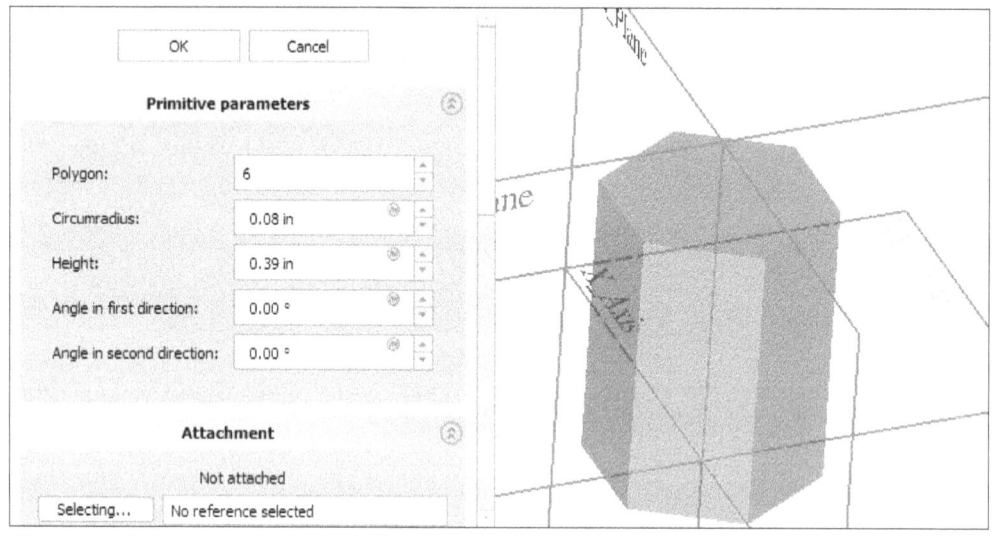

Figure-80. Primitive parameters dialog with preview of additive prism feature

- Specify desired number of sides in the polygon cross section of the prism in the **Polygon** edit box of the dialog.
- Specify the circumscribed radius of the polygon cross section of the prism in the **Circumradius** edit box of the dialog.
- Specify desired height of the prism in the **Height** edit box of the dialog.
- Specify desired angle value for the first direction and second direction in the **Angle in first direction** and **Angle in second direction** edit boxes of the dialog, respectively.
- Specify other parameters in the dialog as discussed earlier.
- Click on **OK** button from the dialog. The additive prism feature will be created; refer to Figure-81.

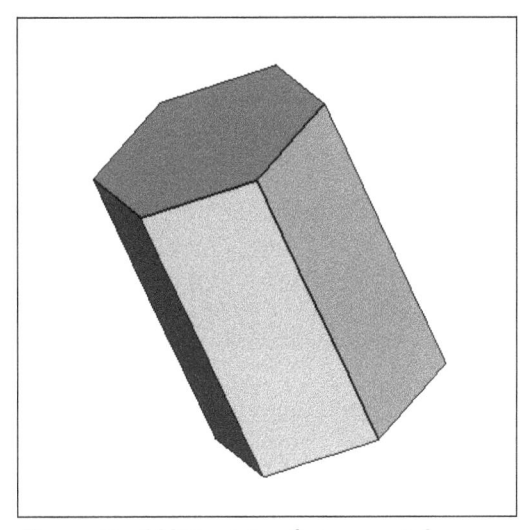

Figure-81. Additive prism feature created

Additive Wedge

The **Additive Wedge** tool inserts a primitive wedge in the active body as the first feature or fuses it to the existing feature. The procedure to use this tool is discussed next.

- Click on the **Additive Wedge** tool from **Create an additive primitive** drop-down in the **Toolbar** of **Part Design** workbench; refer to Figure-82. The **Primitive parameters** dialog will be displayed in the **Tasks** panel of **Combo View** along with the preview of additive wedge feature; refer to Figure-83.

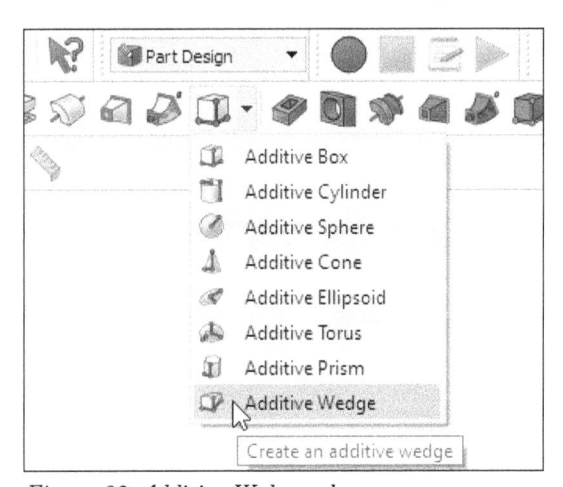

Figure-82. Additive Wedge tool

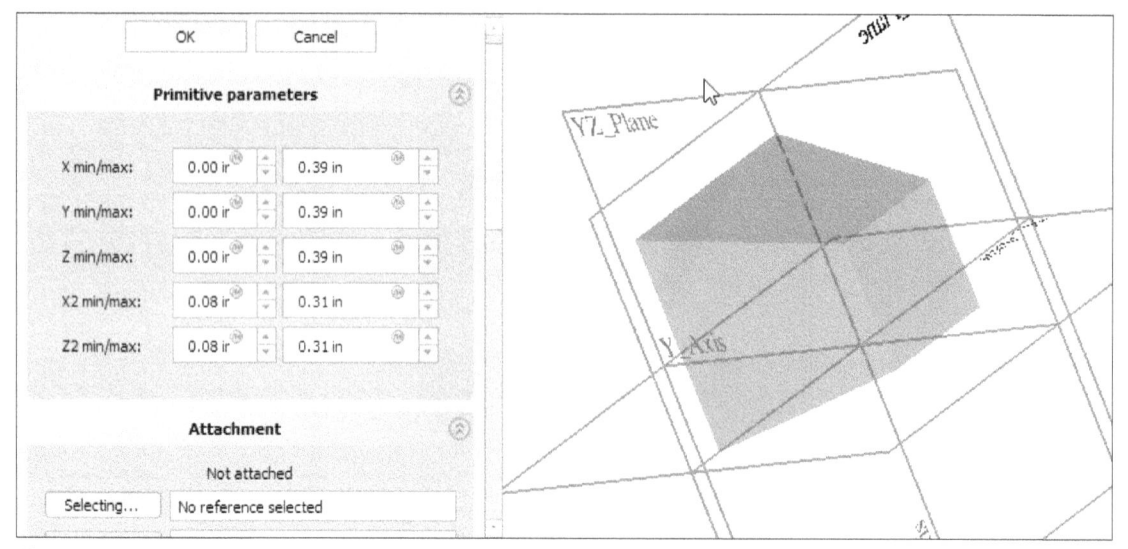

Figure-83. Primitive parameters dialog with preview of additive wedge feature

- Specify the minimum/maximum value for base face of the feature along X axis in the **X min/max** edit boxes, respectively.
- Specify desired height of the wedge in the **Y min/max** edit boxes, respectively.
- Specify the minimum/maximum value for base face of the feature along Z axis in the **Z min/max** edit boxes, respectively.
- Specify the minimum/maximum value for top face of the feature along X axis in the **X2 min/max** edit boxes, respectively.
- Specify the minimum/maximum value for top face of the feature along Z axis in the **Z2 min/max** edit boxes, respectively.
- Specify the other parameters in the dialog as discussed earlier.
- Click on **OK** button from the dialog. The additive wedge feature will be created; refer to Figure-84.

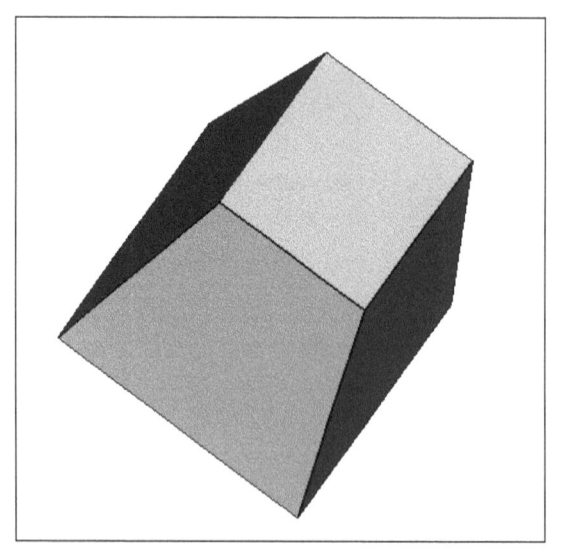

Figure-84. Additive wedge feature created

Subtractive tools

The **Subtractive tools** are used for subtracting materials from an existing body. The **Subtractive tools** are available in the **Toolbar** of **Part Design** workbench; refer to Figure-85. The procedures to use these tools are discussed next.

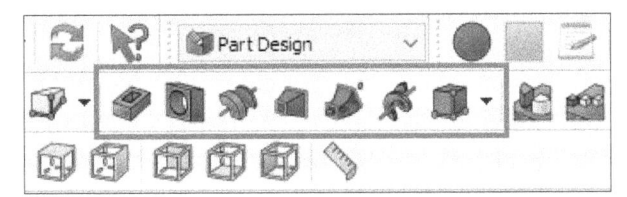

Figure-85. Subtractive tools in Part Design workbench

Pocket

The **Pocket** tool cuts out a solid by extruding a sketch (or a face of the solid) in a straight path and subtracting it from the solid. The procedure to use this tool is discussed next.

- Click on the **Pocket** tool from **Toolbar** in the **Part Design** workbench; refer to Figure-86. The **Pocket parameters** dialog will be displayed in the **Tasks** panel of **Combo View** along with the preview of pocket feature; refer to Figure-87.

Figure-86. Pocket tool

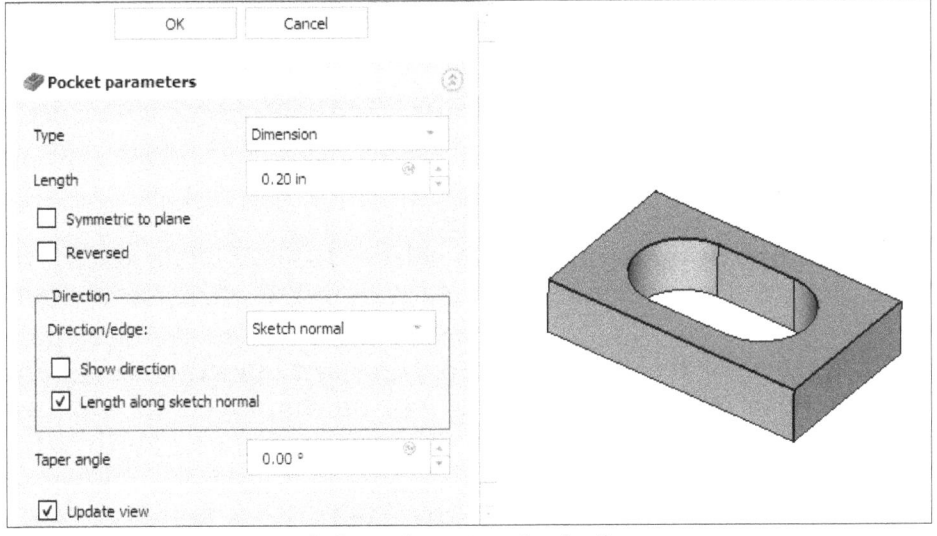

Figure-87. Pocket parameters dialog with preview of pocket feature

- Select desired option from **Type** drop-down in the dialog. Select the **Dimension** option from the **Type** drop-down to extrude cut the feature to the defined sketch plane at a specified length. Select the **Through all** option to extrude cut the feature through all material in the extrusion direction. Select the **To first** option to extrude cut the feature up to the first face of the solid entity in the extrusion direction. Select the **Up to face** option to extrude cut the feature up to the selected face. Select the **Two dimensions** option to extrude cut the feature in opposite direction by specifying second length of the extrusion. Select **Up to shape** option to pocket the sketch up to the selected shape of base feature.
- Specify desired value of length pocket in the **Length** edit box.

- Select **Symmetric to plane** check box to extrude cut the feature in both directions with respect to sketch plane.
- Select **Reversed** check box to reverse the direction of extrusion.
- Select desired direction of the extrusion from **Direction/edge** drop-down. Select **Sketch normal** option to extrude cut the sketch or face along its normal. Select the **Select reference** option to extrude cut the sketch along an edge of the 3D model. Select **Custom direction** option to set the pocket length along the custom direction by specifying desired values in **x**, **y**, and **z** edit boxes.
- Select the **Show direction** check box to display **x**, **y**, and **z** edit boxes.
- Select **Length along sketch normal** check box to measure the pocket length along the sketch normal.
- Specify desired value in the **Taper angle** edit box to taper the pocket in the extrusion direction by the given angle. This option is available only if the **Dimension** or **Two dimensions** option is selected in the **Length** drop-down.
- Click on **OK** button from the dialog. The pocket feature will be created; refer to Figure-88.

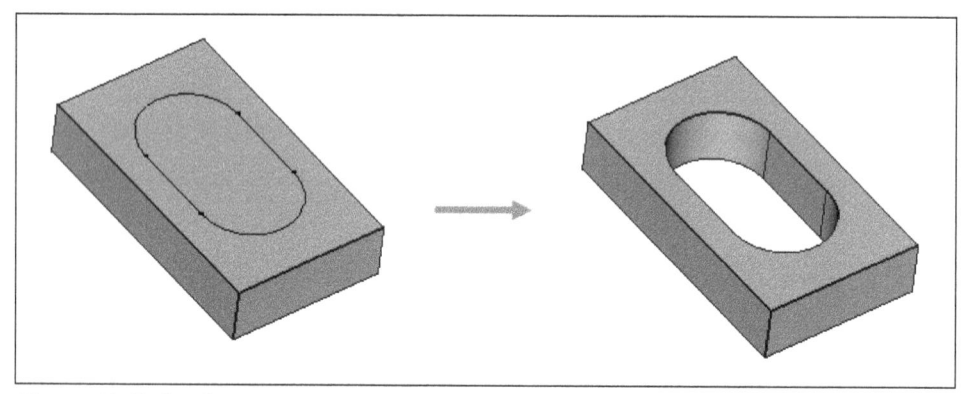

Figure-88. Pocket feature created

Hole

The **Hole** tool creates one or more holes from selected sketch. The sketch must contain one or multiple circles. The procedure to use this tool is discussed next.

- Click on the **Hole** tool from **Toolbar** in the **Part Design** workbench; refer to Figure-89. The **Hole parameters** dialog will be displayed in the **Tasks** panel of **Combo View** along with the preview of hole feature; refer to Figure-90.

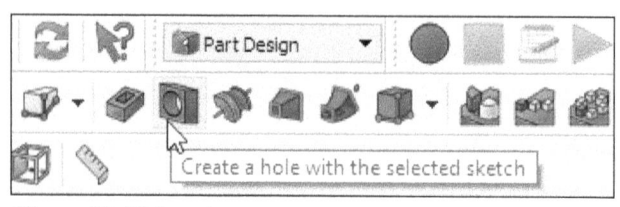

Figure-89. Hole tool

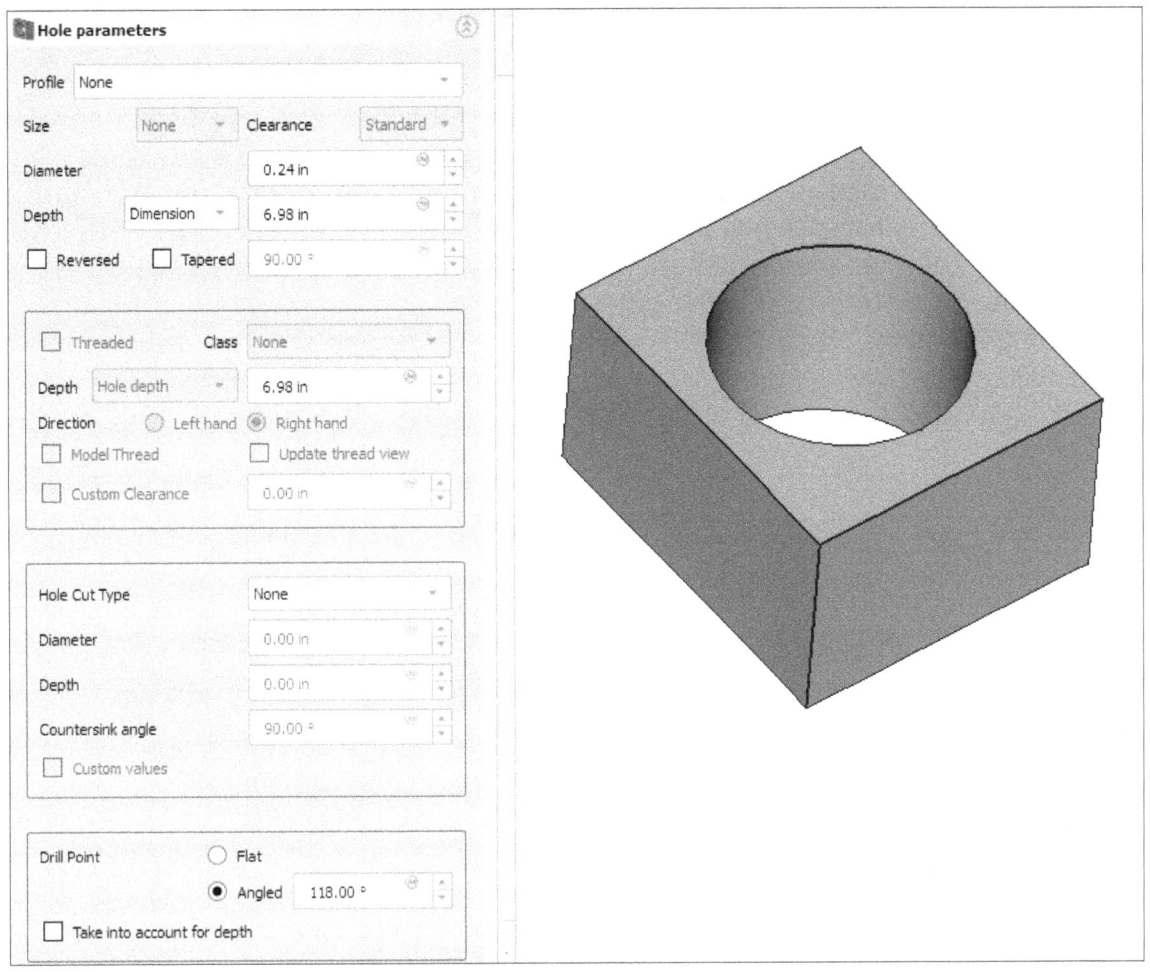

Figure-90. Hole parameters dialog with preview of hole feature

- Select desired option from **Profile** drop-down to specify the type of thread.
- Select desired size of thread from **Size** drop down.
- Select desired option from **Clearance** drop-down to define standard, close, or wide fit for threaded profiles.
- Specify desired hole diameter in the **Diameter** edit box.
- Select **Dimension** option from **Depth** drop-down to specify the depth value in the edit box and select **Through all** option from the drop-down to cut the hole through the whole body.
- Select the **Reversed** check box to reverse the direction of hole extrusion.
- Select the **Tapered** check box to specify the taper angle of hole in the edit box.
- Select the **Threaded** check box to add threading data to the hole feature and the hole minor diameter is used.
- Specify desired tolerance class for hole in the **Class** drop-down.
- Select desired option from the **Depth** drop-down and specify the value of depth in the edit box.
- The radio buttons in **Direction** area will be activate when the **Threaded** check box will be selected. Select desired radio button to specify the thread direction.
- Select the **Model Thread** check box to create a modelled thread on the component.
- Select the **Update thread view** check box to live update the changes made to the thread.
- Select the **Custom Clearance** check box to customize the thread clearance by specifying desired value in the **Clearance** edit box.

- Specify desired option from the **Hole Cut Type** drop-down to define the cut type of screw heads.
- Specify the major diameter for all hole cut types in the **Diameter** edit box.
- Specify the depth of hole cut type from the sketch plane in the **Depth** edit box.
- Specify the angle of conical hole cut in the **Countersink angle** edit box.
- Select **Flat** radio button from the **Drill point** area to create the flat bottom of the hole and select **Angled** radio button to create the conical point of the hole.
- Select **Take into account for depth** check box to subtract the conical height from the dimension.
- Click on **OK** button from the dialog. The hole feature will be created; refer to Figure-91.

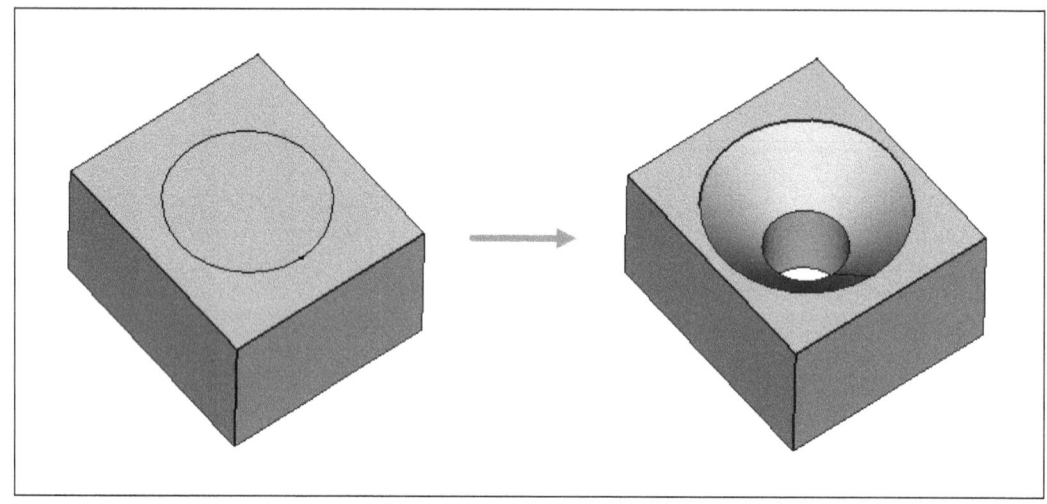

Figure-91. Hole feature created

Groove

The **Groove** tool revolves selected sketch or profile about a given axis, cutting out material from the solid model. The procedure to use this tool is discussed next.

- Select the closed loop sketch to be revolved and click on the **Groove** tool from **Toolbar** in the **Part Design** workbench; refer to Figure-92. The **Revolution parameters** dialog will be displayed in the **Tasks** panel of **Combo View**; refer to Figure-93.

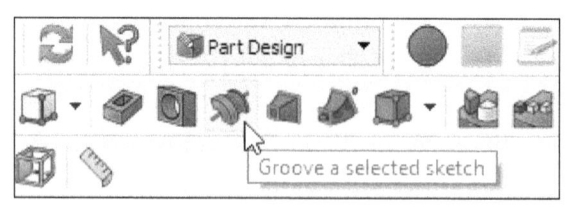

Figure-92. Groove tool

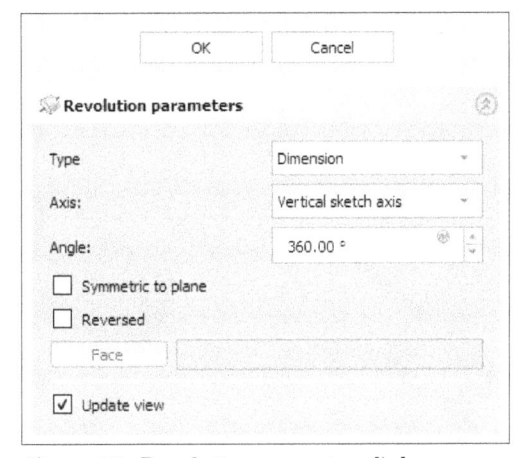

Figure-93. Revolution parameters dialog

- Select desired option from **Type** drop-down to specify the angle of revolution. The options in this drop-down is same as discussed in **Pad** tool.
- Select the sketch of axis from **Model** panel in the **Combo View** and press **Space** key to toggle the visibility of axis of revolution; refer to Figure-94.

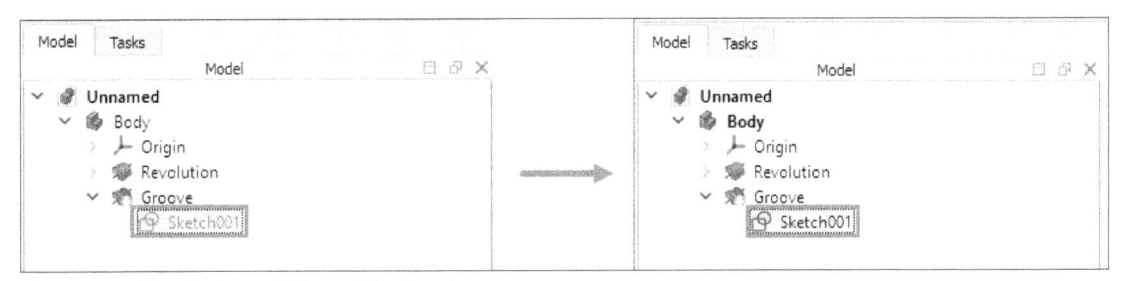

Figure-94. Toggle the visibility of sketch axis

- Select the **Select reference** option from **Axis** drop-down in the dialog and select the sketch of axis in the 3D view area. The preview of groove feature will be displayed; refer to Figure-95.

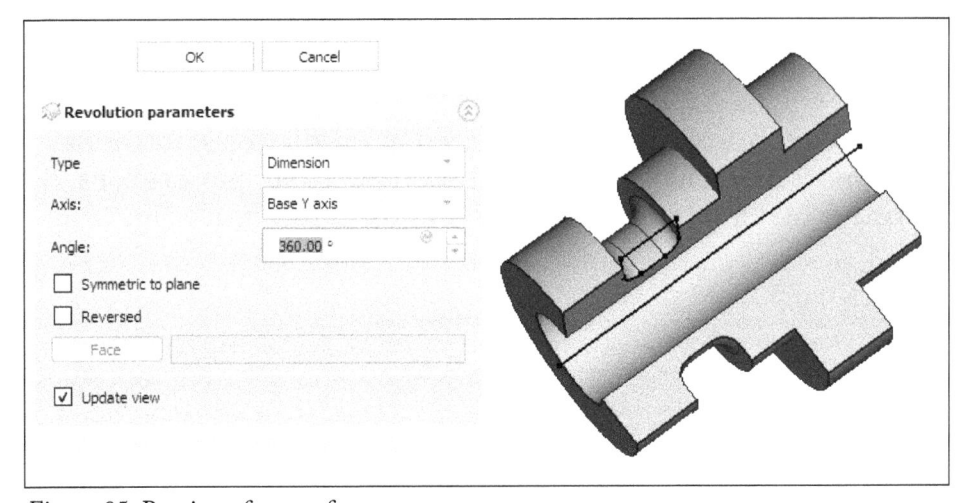

Figure-95. Preview of groove feature

- Specify the angle of revolution through which the groove is to be formed in the **Angle** edit box or **Angle** and **2nd angle** edit boxes in case of **Two dimensions** revolution type.
- Specify the other parameters in the dialog as discussed earlier.
- Click on **OK** button from the dialog. The groove feature will be created; refer to Figure-96.

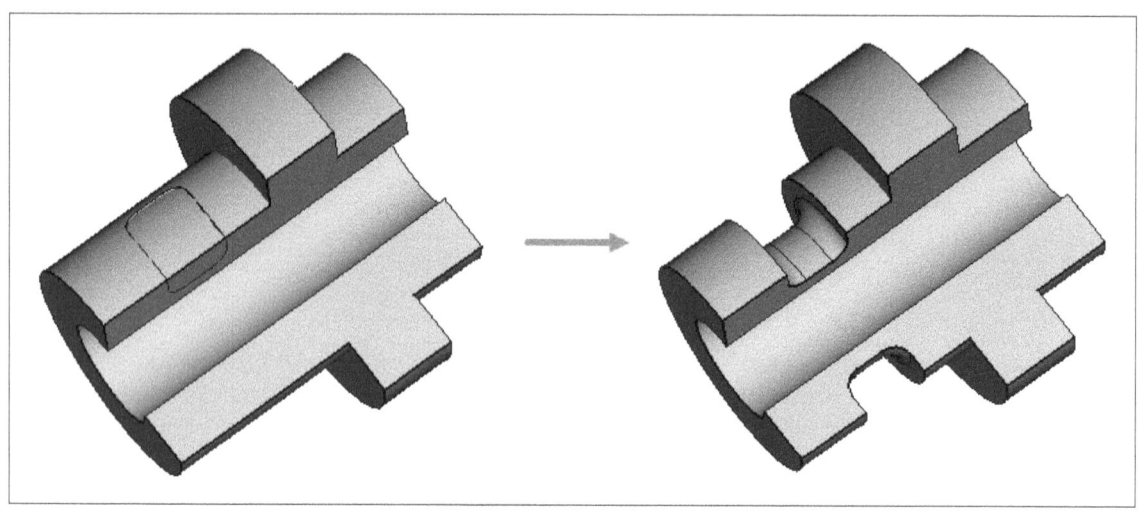

Figure-96. Groove feature created

Subtractive Loft

The **Subtractive Loft** tool removes material from the active body by making a transition between two or more sketches. The procedure to use this tool is discussed next.

- Click on the **Subtractive Loft** tool from **Toolbar** in the **Part Design** workbench; refer to Figure-97. If no object is selected before selecting the tool then the **Select feature** dialog will be displayed in the **Tasks** panel of **Combo View** as discussed earlier asking you to select the base profile.

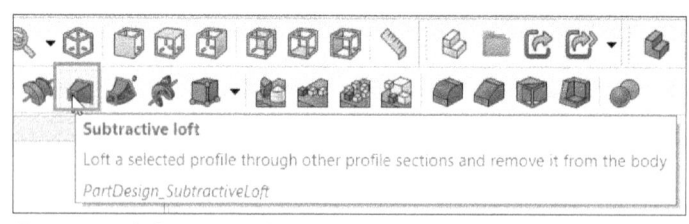

Figure-97. Subtractive loft tool

- Select desired sketch from the dialog as base profile for subtractive loft feature; refer to Figure-99 and click on **OK** button. The **Loft parameters** dialog will be displayed in the **Tasks** panel of **Combo View**; refer to Figure-99.

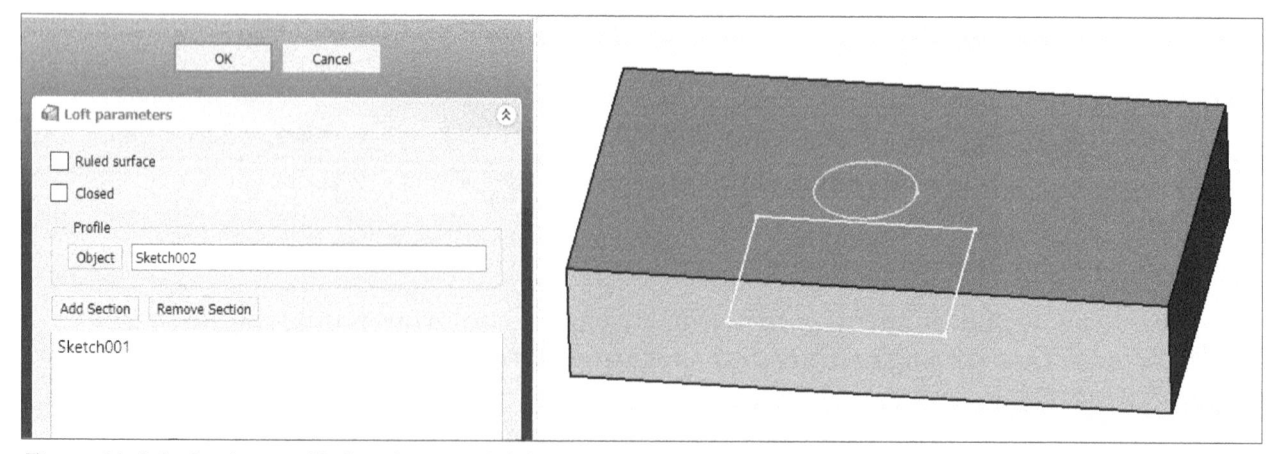

Figure-98. Selecting base profile for subtractive loft feature

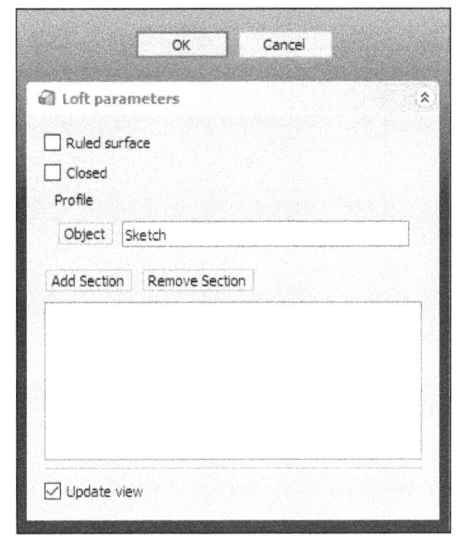

Figure-99. Loft parameters dialog

- The base profile is preselected in the **Object** selection box of **Profile** area. Click on **Add Section** button and select the first, second, and third sketches for loft section.
- Select the **Ruled surface** check box to make straight transitions between cross-sections.
- Select the **Closed** check box to make a transition from the last cross-section to the first by creating a loop.
- Click on **OK** button from the dialog. The subtractive loft feature will be created; refer to Figure-100

Figure-100. Subtractive loft feature created

Subtractive Pipe

The **Subtractive Pipe** tool creates a subtractive solid in the active body by sweeping one or more sketches along an open or closed path and subtracts it from the active body. The procedure to use this tool is discussed next.

- Click on the **Subtractive Pipe** tool from **Toolbar** in the **Part Design** workbench; refer to Figure-101. The **Pipe parameters**, **Section orientation**, and **Section transformation** dialogs will be displayed in the **Tasks** panel of **Combo View**; refer to Figure-102.

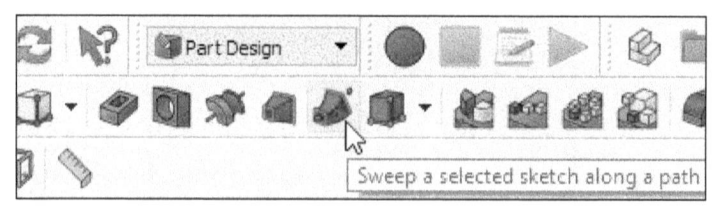

Figure-101. Subtractive pipe tool

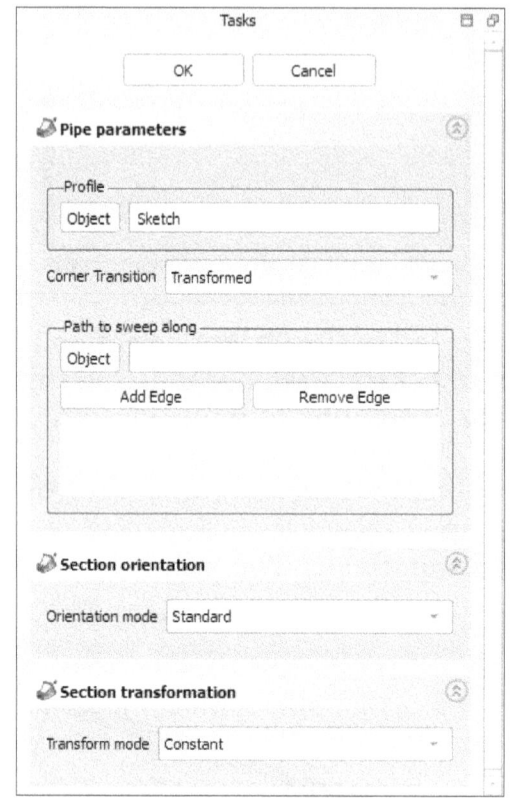

Figure-102. Pipe parameters dialog

- The first cross-section is preselected in the **Object** selection box in the **Profile** area of dialog.
- Select desired transition mode from **Corner Transition** drop-down in the dialog.
- Click on the **Object** button from **Path to sweep along** area and select an edge of the sketch to be used as path for sweep feature in the 3D view area. The preview of subtractive pipe feature will be displayed; refer to Figure-103.

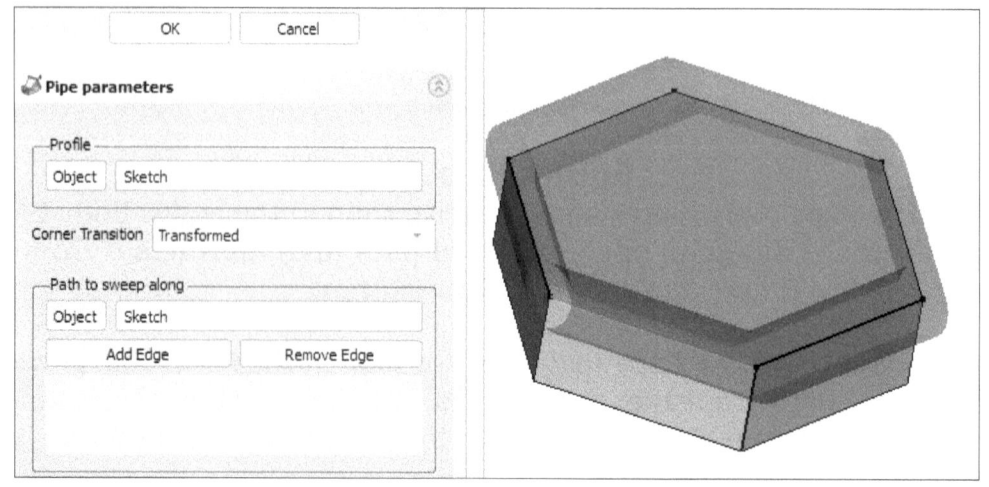

Figure-103. Preview of subtractive pipe feature

- Select desired option from **Orientation mode** drop-down in the **Section orientation** dialog.
- Select desired option from **Transform mode** drop-down in the **Section transformation** dialog.
- Specify the other parameters in the **Pipe parameters** dialog as desired and click on **OK** button. The subtractive pipe feature will be created; refer to Figure-104.

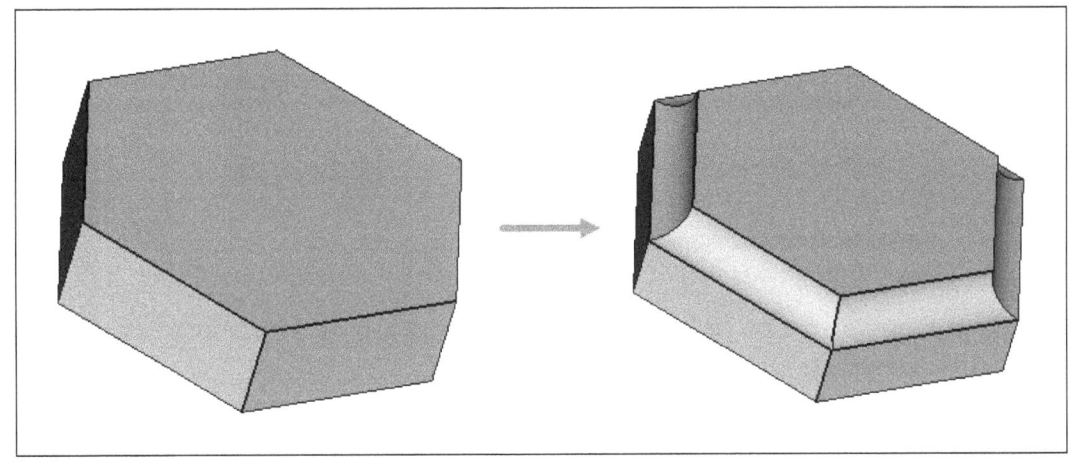

Figure-104. Subtractive pipe feature created

Subtractive Helix

The **Subtractive Helix** tool creates a solid shape by sweeping a sketch along a helix and subtracts it from the active body. The procedure to use this tool is discussed next.

- Create and select the sketch to be swept into a helix; refer to Figure-105.
- Click on the **Subtractive Helix** tool from **Toolbar** in the **Part Design** workbench; refer to Figure-106. The **Helix parameters** dialog will be displayed in the **Tasks** panel of **Combo View** along with the preview of subtractive helix feature; refer to Figure-107.

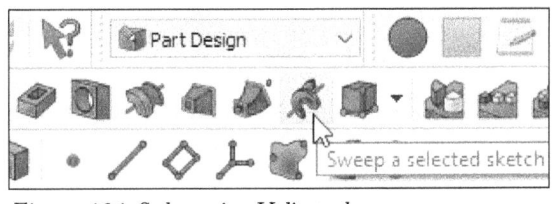

Figure-106. Subtractive Helix tool

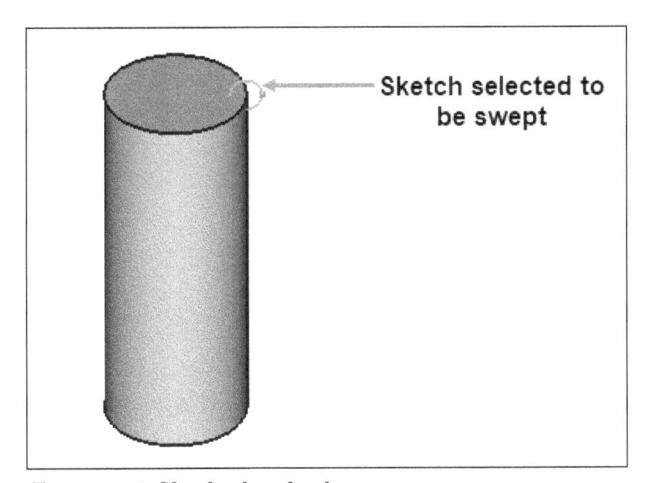

Figure-105. Sketch selected to be swept

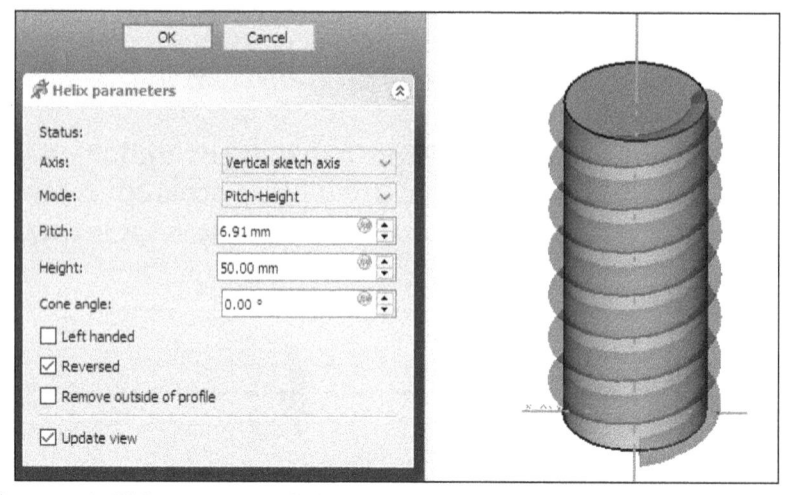

Figure-107. Helix parameters dialog along with preview of subtractive helix feature

- All the parameters in the **Helix parameters** dialog of **Subtractive Helix** tool is same as discussed in the **Additive Helix** tool.
- Select the **Remove outside of profile** check box to create the feature by the intersection of the swept profile and the pre-existing body.
- Click on the **OK** button from the dialog. The subtractive helix feature will be created; refer to Figure-108.

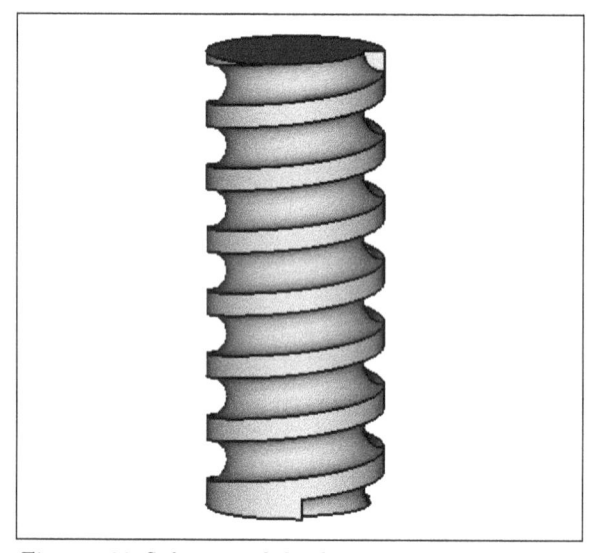

Figure-108. Subtractive helix feature created

Subtractive Box

The **Subtractive Box** tool inserts a subtractive box in the active body. Its shape is subtracted from the existing solid. The procedure to use this tool is discussed next. Note that an existing solid body will be required for creating a subtractive box feature.

- Click on the **Subtractive Box** tool from **Create a subtractive primitive** drop-down in the **Toolbar** of **Part Design** workbench; refer to Figure-109. The **Primitive parameters** dialog will be displayed in the **Tasks** panel of **Combo View** along with the preview of subtractive box feature; refer to Figure-110.

FreeCAD 1.0 Black Book 3-47

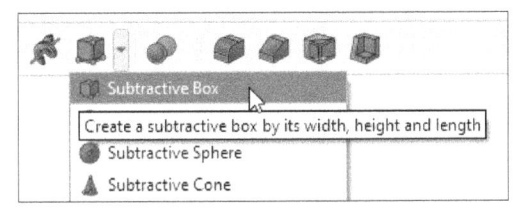

Figure-109. Subtractive Box tool

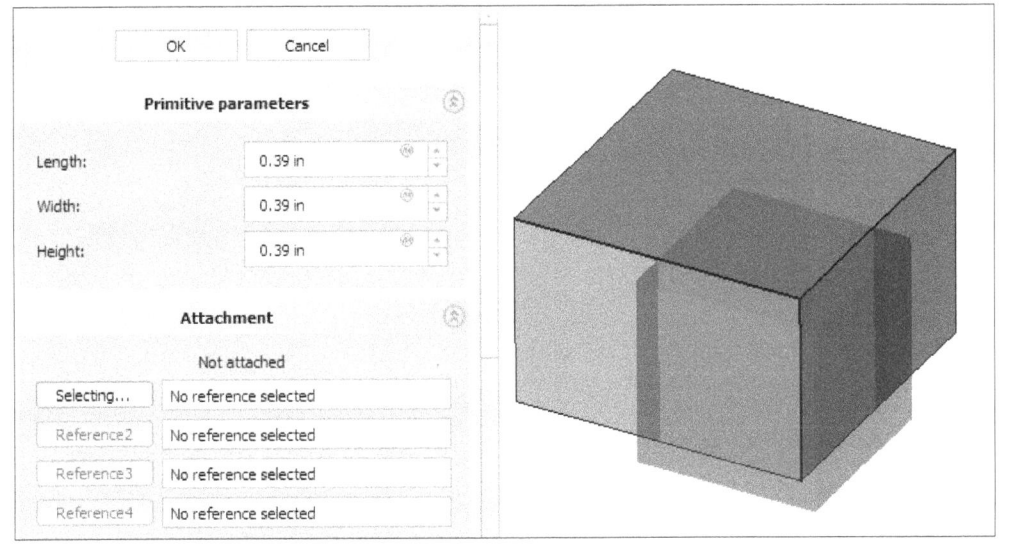

Figure-110. Primitive parameters dialog with preview of subtractive box feature

- Specify the length, width, and height of the feature in the **Length**, **Width**, and **Height** edit boxes, respectively.
- Specify the other parameters in the dialog as discussed earlier.
- Click on **OK** button from the dialog. The subtractive box feature will be created; refer to Figure-111.

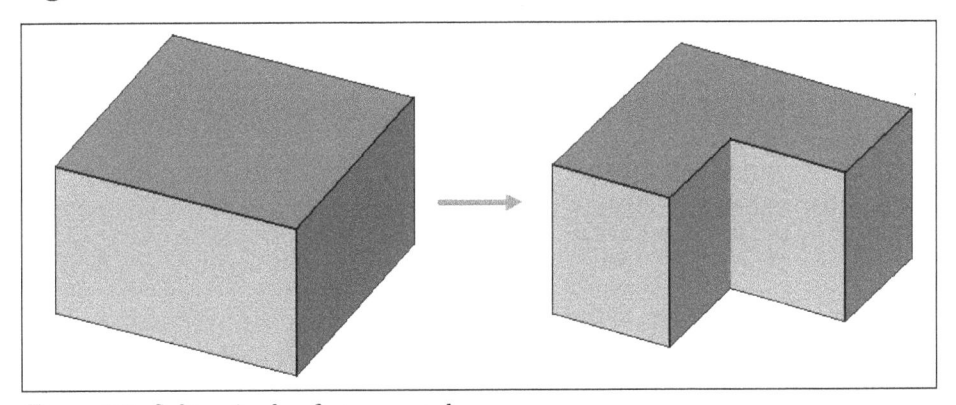

Figure-111. Subtractive box feature created

Subtractive Cylinder

The **Subtractive Cylinder** tool inserts a subtractive cylinder in the active body. Its shape is subtracted from the existing solid. The procedure to use this tool is discussed next.

- Click on the **Subtractive Cylinder** tool from **Create a subtractive primitive** drop-down in the **Toolbar** of **Part Design** workbench; refer to Figure-112. The **Primitive parameters** dialog will be displayed in the **Tasks** panel of **Combo View** along with the preview of subtractive cylinder feature; refer to Figure-113.

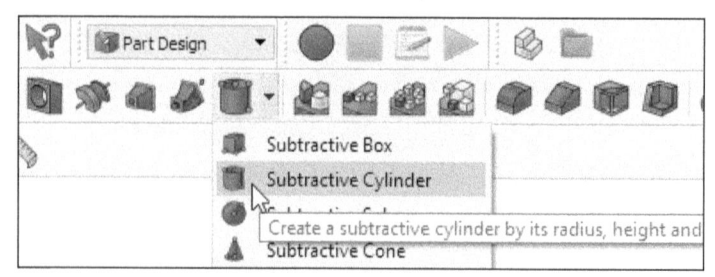

Figure-112. Subtractive cylinder tool

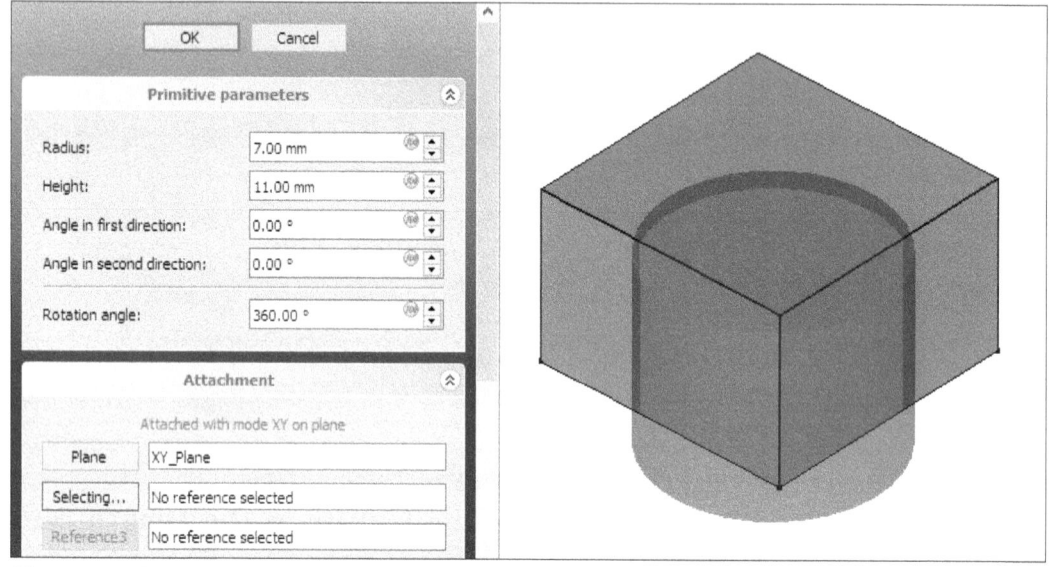

Figure-113. Primitive parameters dialog with preview of subtractive cylinder feature

- Specify desired radius and height of the feature in the **Radius** and **Height** edit boxes, respectively.
- Specify desired angle of first and second direction in the **Angle in first direction** and **Angle in second direction** edit boxes, respectively.
- Specify desired value of angle of revolution of the cross-section in the **Rotation Angle** edit box.
- Specify the other parameters in the dialog as discussed earlier.
- Click on **OK** button from the dialog. The subtractive cylinder feature will be created; refer to Figure-114.

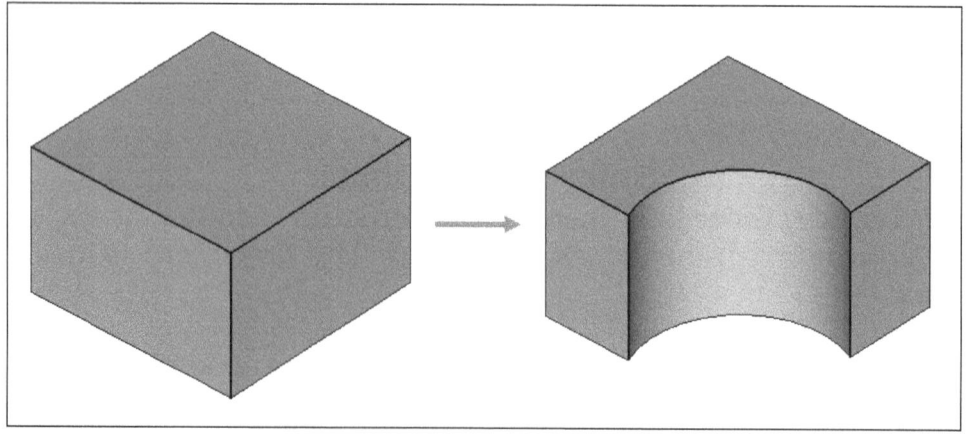

Figure-114. Subtractive cylinder feature created

Subtractive Sphere

The **Subtractive Sphere** tool inserts a subtractive sphere in the active body. Its shape is subtracted from the existing solid. The procedure to use this tool is discussed next.

- Click on the **Subtractive Sphere** tool from **Create a subtractive primitive** drop-down in the **Toolbar** of **Part Design** workbench; refer to Figure-115. The **Primitive parameters** dialog will be displayed in the **Tasks** panel of **Combo View** along with the preview of subtractive sphere feature; refer to Figure-116.

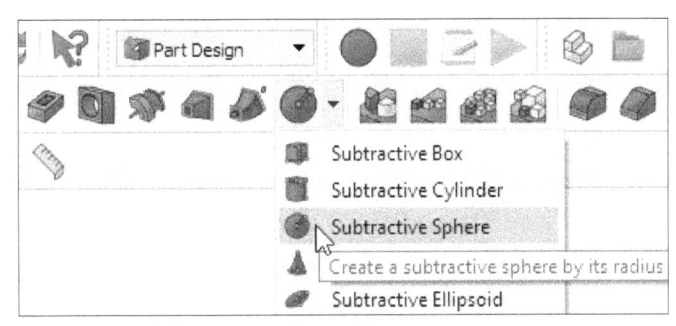

Figure-115. Subtractive sphere tool

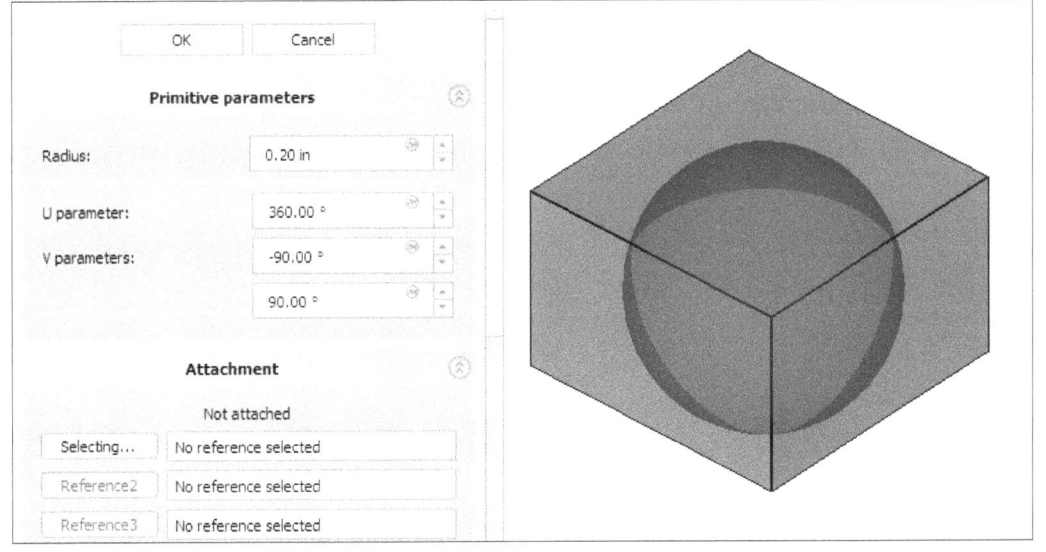

Figure-116. Primitive parameters dialog with preview of subtractive sphere feature

- Specify desired radius of the sphere in the **Radius** edit box.
- Specify desired angle of revolution of the cross section in the **U parameter** edit box.
- Specify lower truncation of the sphere parallel to the circular cross-section in the **V parameters** edit box and specify upper truncation of the ellipsoid parallel to the circular cross-section in the edit box just below to the **V parameters** edit box of the dialog.
- Specify other parameters in the dialog as discussed earlier.
- Click on **OK** button from the dialog. The subtractive sphere feature will be created; refer to Figure-117.

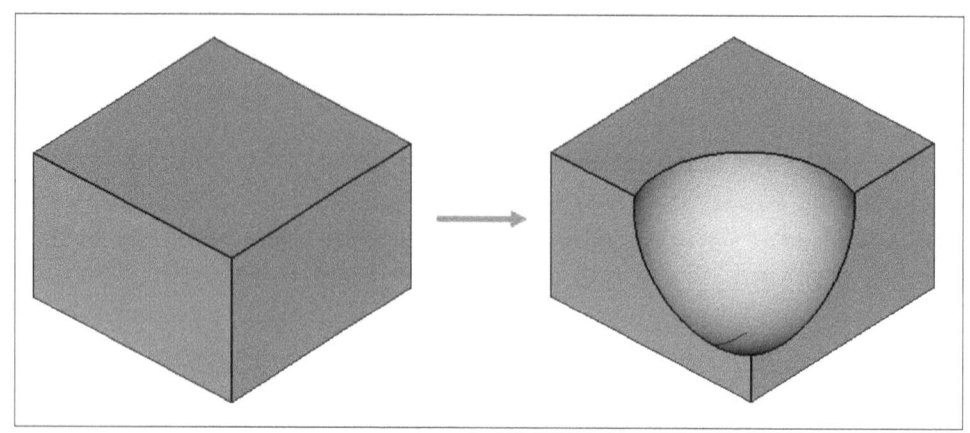

Figure-117. Subtractive sphere feature created

Subtractive Cone

The **Subtractive Cone** tool inserts a subtractive cone in the active body. Its shape is subtracted from the existing solid. The procedure to use this tool is discussed next.

- Click on the **Subtractive Cone** tool from **Create a subtractive primitive** drop-down in the **Toolbar** of **Part Design** workbench; refer to Figure-118. The **Primitive parameters** dialog will be displayed in the **Tasks** panel of **Combo View** along with the preview of subtractive cone feature; refer to Figure-119.

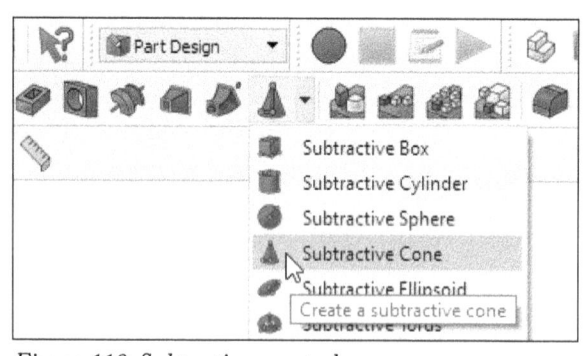

Figure-118. Subtractive cone tool

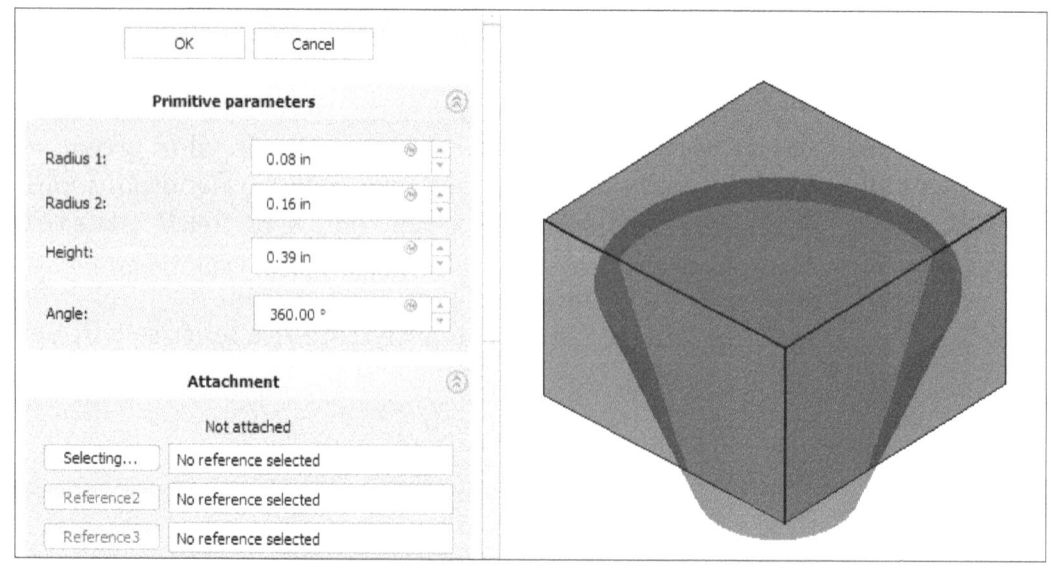

Figure-119. Primitive parameters dialog with preview of subtractive cone feature

- Specify desired radius value for the base of the cone in **Radius 1** edit box and specify desired radius value for the top of the cone in **Radius 2** edit box of the dialog.
- Specify desired height of the cone in the **Height** edit box.
- Specify desired angle of revolution of the cross section in the **Angle** edit box.
- Specify other parameters in the dialog as discussed earlier.
- Click on **OK** button from the dialog. The subtractive cone feature will be created; refer to Figure-120.

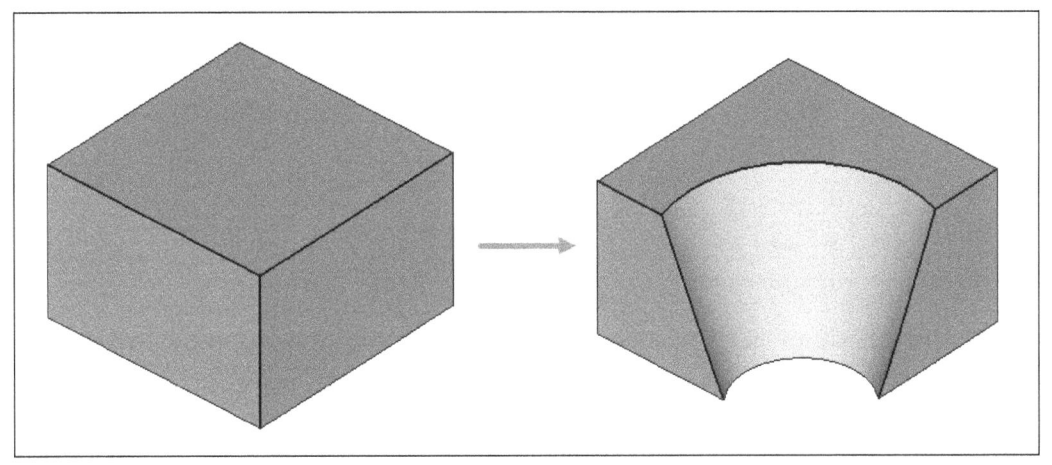

Figure-120. Subtractive cone feature created

Subtractive Ellipsoid

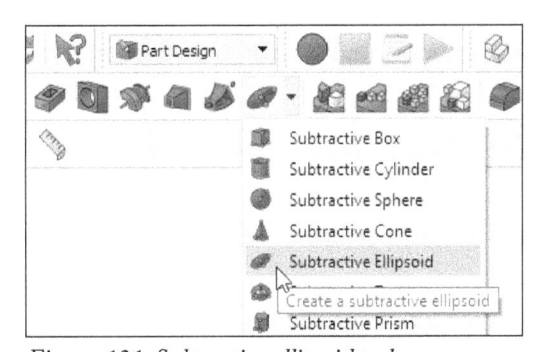

The **Subtractive Ellipsoid** tool inserts a subtractive ellipsoid in the active body. Its shape is subtracted from the existing solid. The procedure to use this tool is discussed next.

- Click on the **Subtractive Ellipsoid** tool from **Create a subtractive primitive** drop-down in the **Toolbar** of **Part Design** workbench; refer to Figure-121. The **Primitive parameters** dialog will be displayed in the **Tasks** panel of **Combo View** along with the preview of subtractive ellipsoid feature; refer to Figure-122.

Figure-121. Subtractive ellipsoid tool

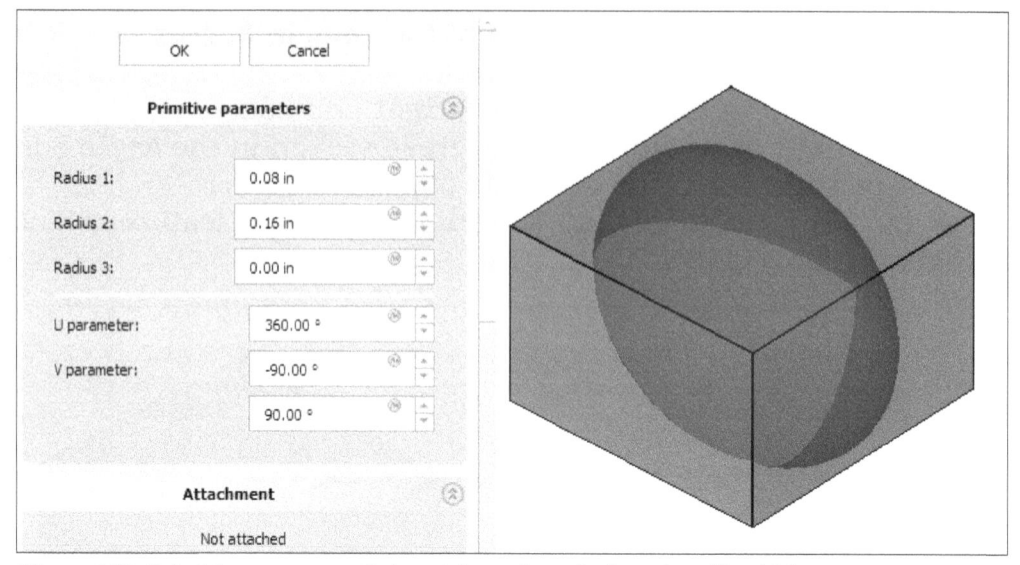

Figure-122. Primitive parameters dialog with preview of subtractive ellipsoid feature

- Specify desired radius value along the ellipsoid's vertical axis in **Radius 1** edit box, along the ellipsoid's length in **Radius 2** edit box, and along the ellipsoid's width in **Radius 3** edit box of the dialog.
- Specify the angle of revolution of the elliptical cross section in the **U parameter** edit box.
- Specify lower truncation of the ellipsoid parallel to the circular cross-section in the **V parameter** edit box and specify upper truncation of the ellipsoid parallel to the circular cross section in the edit box just below to the **V parameter** edit box of the dialog.
- Specify other parameters in the dialog as discussed earlier.
- Click on **OK** button from the dialog. The subtractive ellipsoid feature will be created; refer to Figure-123.

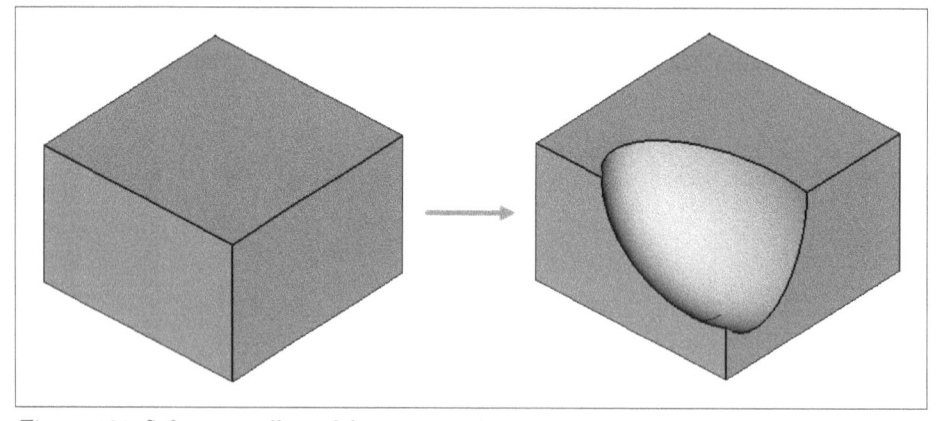

Figure-123. Subtractive ellipsoid feature created

Subtractive Torus

The **Subtractive Torus** tool inserts a subtractive torus in the active body. Its shape is subtracted from the existing solid. The procedure to use this tool is discussed next.

- Click on the **Subtractive Torus** tool from **Create a subtractive primitive** drop-down in the **Toolbar** of **Part Design** workbench; refer to Figure-124. The **Primitive parameters** dialog will be displayed in the **Tasks** panel of **Combo View** along with the preview of subtractive torus feature; refer to Figure-125.

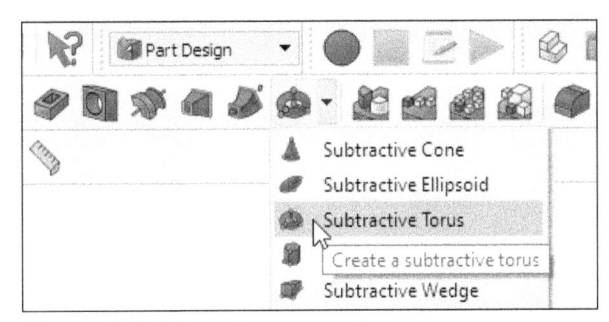

Figure-124. Subtractive torus tool

- Specify desired radius of the imaginary orbit around which the circular cross-section revolves in **Radius 1** edit box and specify the radius of circular cross-section defining the form of the torus in the **Radius 2** edit box of the dialog.
- Specify the angle of revolution of the circular cross section in the `U parameter` edit box.
- Specify lower truncation of the torus parallel to the circular cross-section in the `V parameter` edit box and specify upper truncation of the ellipsoid parallel to the circular cross section in the edit box just below to the `V parameter` edit box of the dialog.

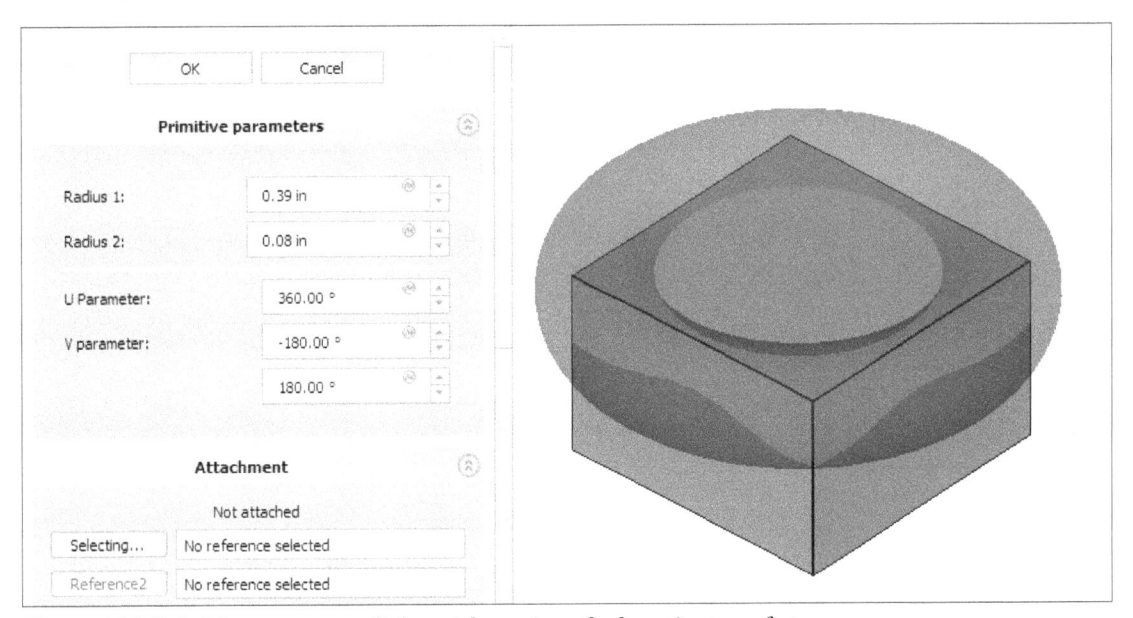

Figure-125. Primitive parameters dialog with preview of subtractive torus feature

- Specify other parameters in the dialog as discussed earlier.
- Click on **OK** button from the dialog. The subtractive torus feature will be created; refer to Figure-126.

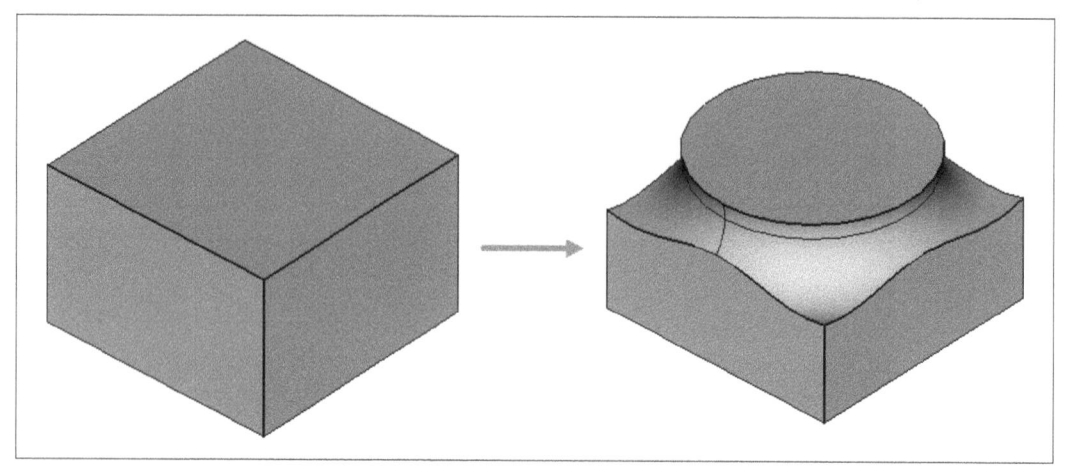

Figure-126. Subtractive torus feature created

Subtractive Prism

The **Subtractive Prism** tool inserts a subtractive prism in the active body. Its shape is subtracted from the existing solid. The procedure to use this tool is discussed next.

- Click on the **Subtractive Prism** tool from **Create a subtractive primitive** drop-down in the **Toolbar** of **Part Design** workbench; refer to Figure-127. The **Primitive parameters** dialog will be displayed in the **Tasks** panel of **Combo View** along with the preview of subtractive prism feature; refer to Figure-128.

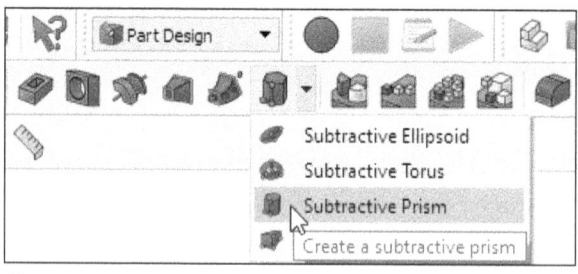

Figure-127. Subtractive prism tool

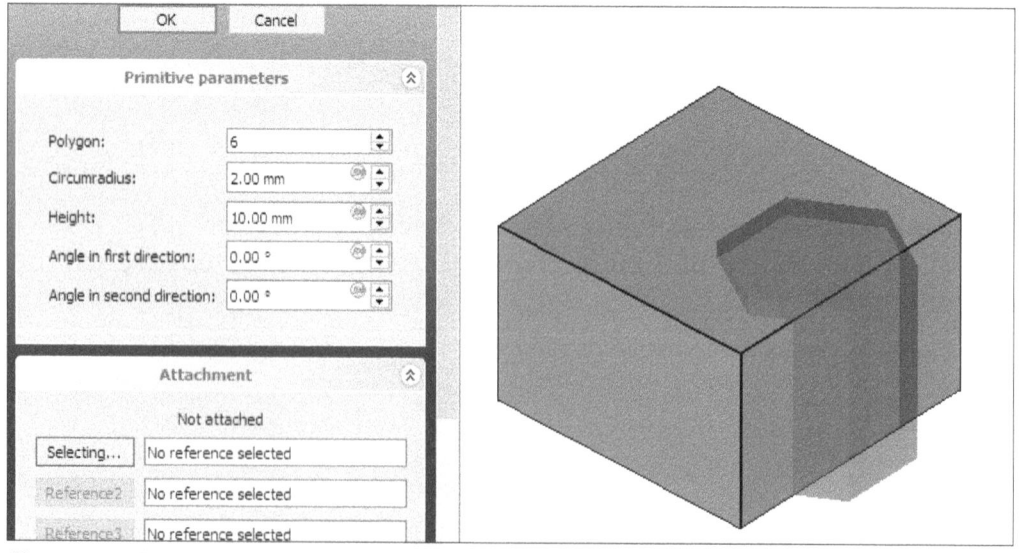

Figure-128. Primitive parameters dialog with preview of subtractive prism feature

- Specify desired number of sides for polygon cross-section of the prism in the **Polygon** edit box.

- Specify circumscribed radius of the polygon cross-section of the prism in the **Circumradius** edit box.
- Specify desired height of the prism in the **Height** edit box of the dialog.
- Specify desired angle value for the first and second directions in the **Angle in first direction** and **Angle in second direction** edit boxes of the dialog, respectively.
- Specify other parameters in the dialog as discussed earlier.
- Click on **OK** button from the dialog. The subtractive prism feature will be created; refer to Figure-129.

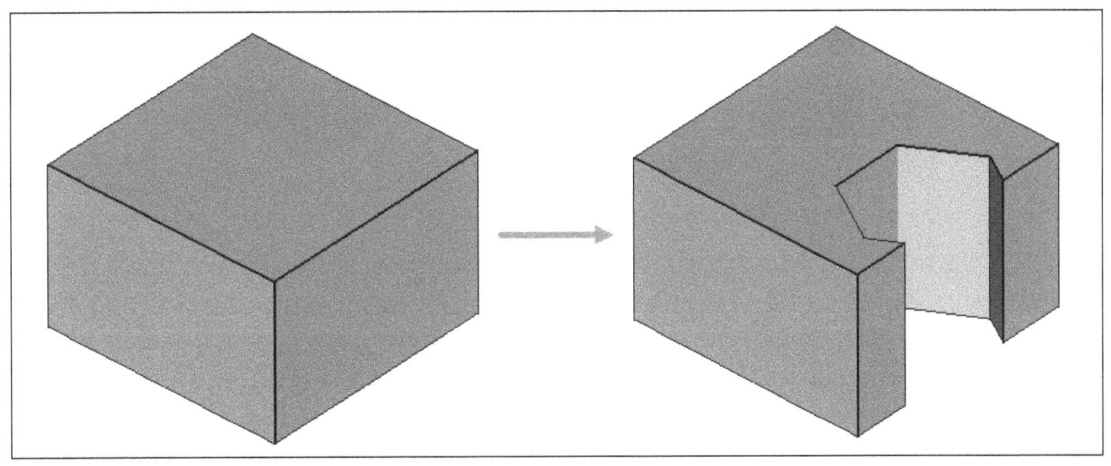

Figure-129. Subtractive prism feature created

Subtractive Wedge

The **Subtractive Wedge** tool inserts a subtractive wedge in the active body. Its shape is subtracted from the existing solid. The procedure to use this tool is discussed next.

- Click on the **Subtractive Wedge** tool from **Create a subtractive primitive** drop-down in the **Toolbar** of **Part Design** workbench; refer to Figure-130. The **Primitive parameters** dialog will be displayed in the **Tasks** panel of **Combo View** along with the preview of subtractive wedge feature; refer to Figure-131.

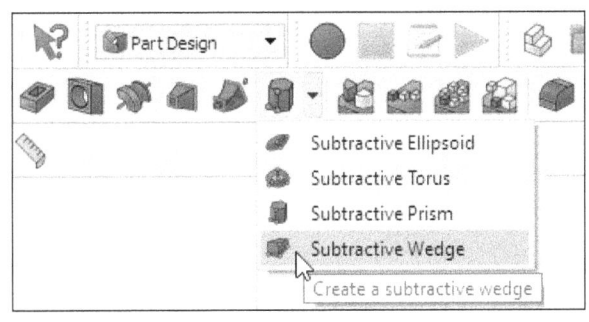

Figure-130. Subtractive wedge tool

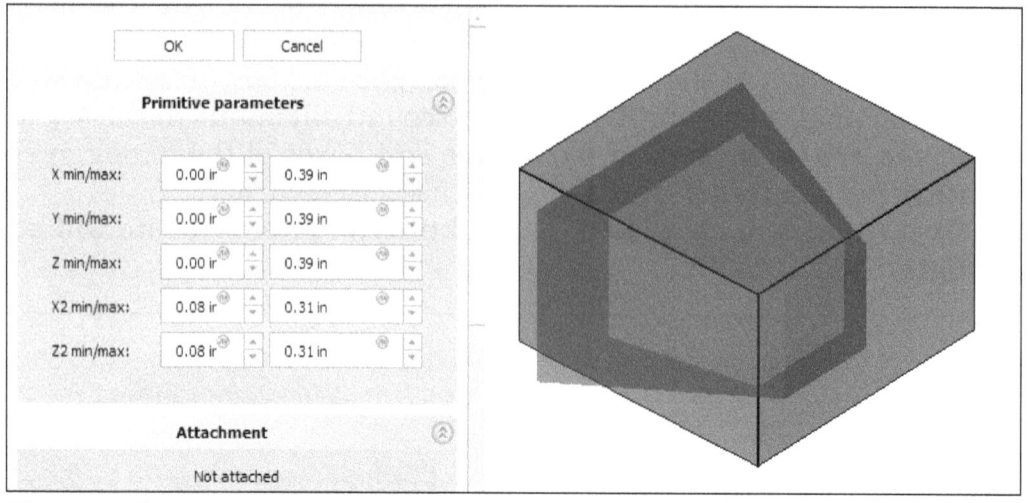

Figure-131. Primitive parameters dialog with preview of subtractive wedge feature

- Specify the minimum/maximum value for base face of the feature along X axis in the **X min/max** edit boxes, respectively. In other words, you can specify the length of base feature in both positive and negative directions along the X axis.
- Specify desired height of the wedge in the **Y min/max** edit boxes, respectively.
- Specify the minimum/maximum value for base face of the feature along Z axis in the **Z min/max** edit boxes, respectively.
- Specify the minimum/maximum value for top face of the feature along X axis in the **X2 min/max** edit boxes, respectively.
- Specify the minimum/maximum value for top face of the feature along Z axis in the **Z2 min/max** edit boxes, respectively.
- Specify other parameters in the dialog as discussed earlier.
- Click on **OK** button from the dialog. The subtractive wedge feature will be created; refer to Figure-132.

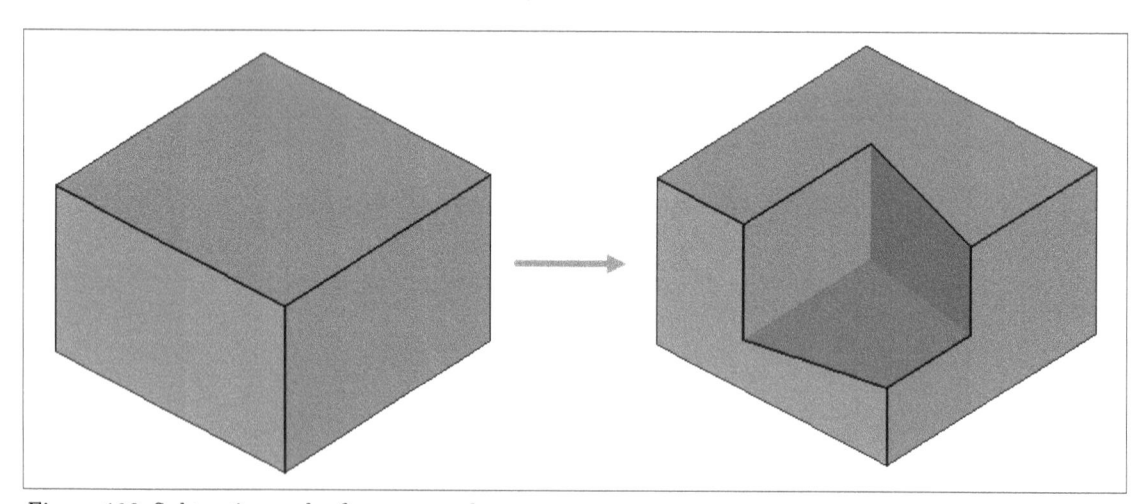

Figure-132. Subtractive wedge feature created

Transformation Tools

The **Transformation** tools are used for transforming existing features using mirror and pattern operations. They will allow you to choose which features to transform. The **Transformation** tools are available in the **Toolbar** of **Part Design** workbench; refer to Figure-133. The procedures to use these tools are discussed next.

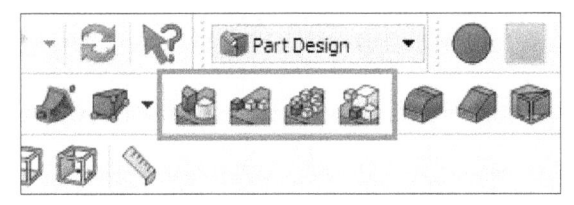

Figure-133. Transformation tools in part design workbench

Mirrored

The **Mirrored** tool creates mirror copy of one or more features about a plane or face. The procedure to use this tool is discussed next.

- Click on the **Mirrored** tool from **Toolbar** in the **Part Design** workbench; refer to Figure-134. The **Select attachment** dialog will be displayed in the **Tasks** panel of **Combo View**; refer to Figure-135, asking you to select the features to be mirrored.

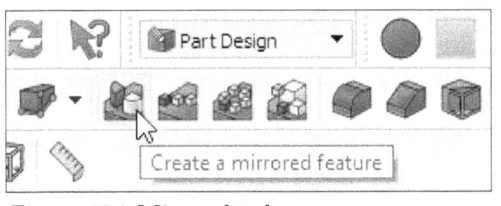

Figure-134. Mirrored tool

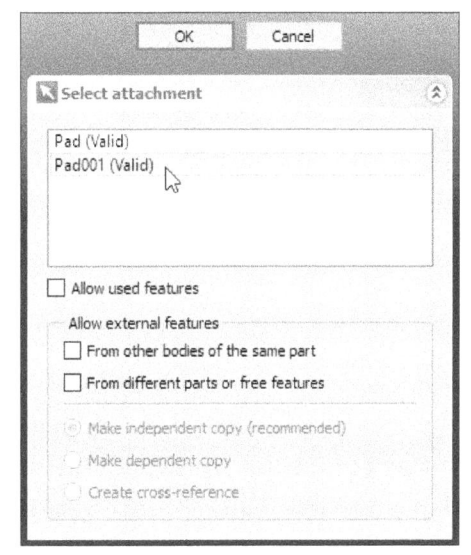

Figure-135. Select attachment dialog

- Select the feature from the dialog which you want to mirror and click on **OK** button. The **Mirrored parameters** dialog will be displayed in the **Tasks** panel of **Combo View** along with the preview of mirrored feature; refer to Figure-136.

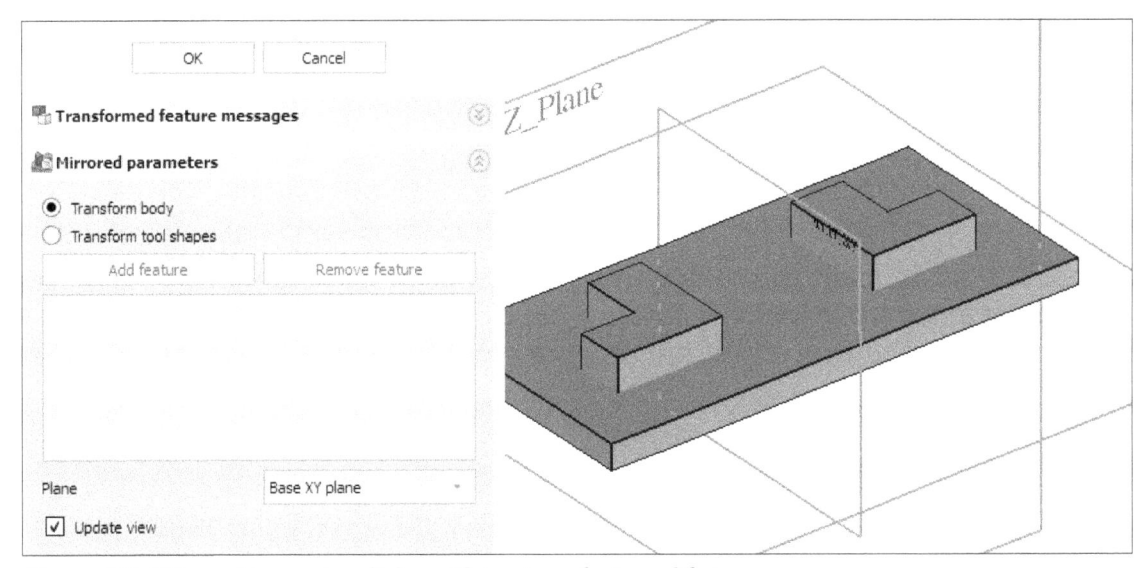

Figure-136. Mirrored parameters dialog with preview of mirrored feature

- Select **Transform body** radio button to transform the shape of whole base feature.
- Select **Transform tool shapes** radio button to transform the individual tool shapes of selected features.
- To add more features to be mirrored, click on the **Add feature** button from the dialog and select the feature in the 3D view.
- To remove the feature to be mirrored, select the feature from the list and click on **Remove feature** button.
- Select desired plane or face to be used as mirror axis from **Plane** drop-down in the dialog.
- Click on **OK** button from the dialog. The mirrored feature will be created; refer to Figure-137.

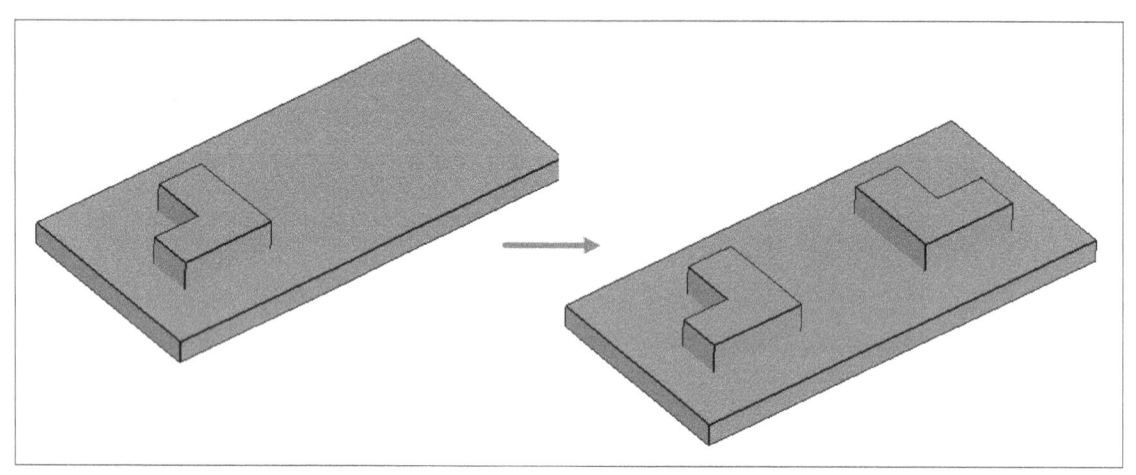

Figure-137. Mirrored feature created

Linear Pattern

The **Linear Pattern** tool creates evenly spaced copies of a feature in linear direction. The procedure to use this tool is discussed next.

- Click on the **Linear Pattern** tool from **Toolbar** in the **Part Design** workbench; refer to Figure-138. The **Select attachment** dialog will be displayed in the **Tasks** panel of **Combo View** as discussed earlier asking you to select the feature to be linear patterned.

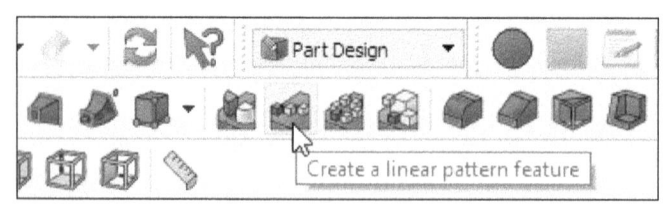

Figure-138. Linear pattern tool

- Select the feature from the dialog to be linear patterned and click on **OK** button. The **LinearPattern parameters** dialog will be displayed in the **Tasks** panel of **Combo View** along with the preview of linear pattern of the feature; refer to Figure-139.

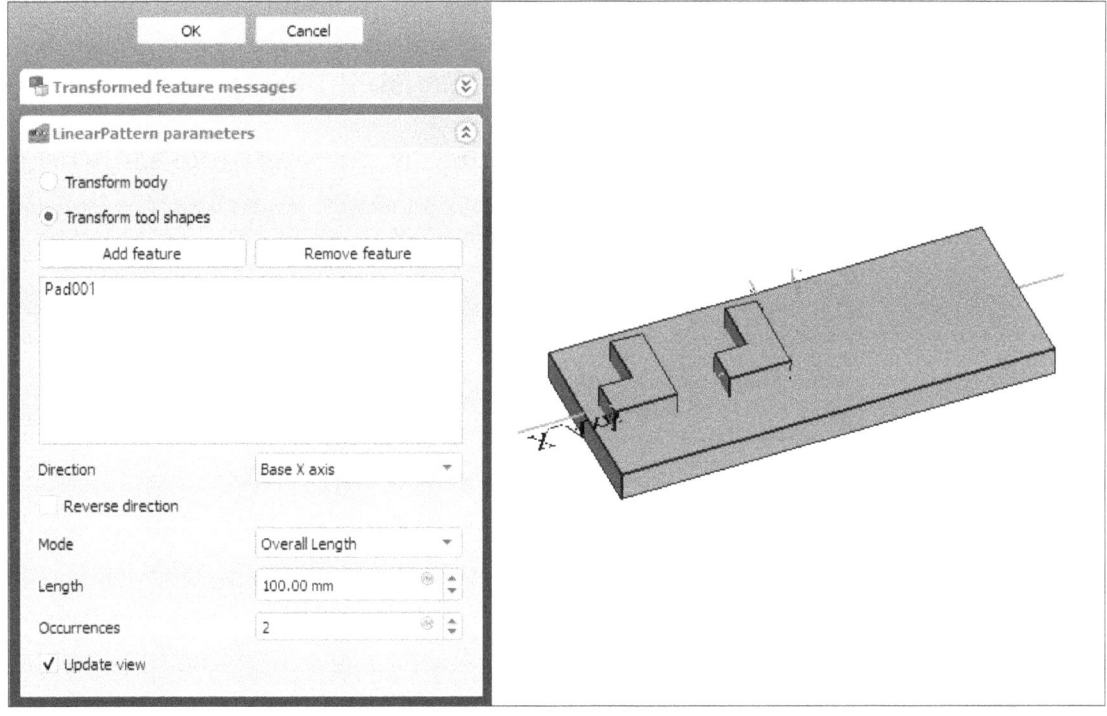

Figure-139. Linear Pattern parameters dialog with preview of linear patterned feature

- To add more features to be patterned, click on **Add feature** button from the dialog and select the feature from **Model Tree** or from the 3D view.
- To remove a feature from the list of features being patterned, select the feature from the list and click on **Remove feature** button.
- Specify desired direction of pattern from the **Direction** drop-down.
- Select the **Reverse direction** check box to reverse the direction of linear pattern.
- Select **Overall Length** option from **Mode** drop-down and specify the distance between patterned feature and the original feature in the **Length** edit box.
- Select **Offset** option from **Mode** drop-down and specify desired offset distance between original feature and the next patterned feature in the **Offset** edit box.
- Specify desired number of occurrences of linear pattern in the **Occurrences** edit box.
- Other parameters in the dialog have been discussed earlier.
- Click on **OK** button from the dialog. The linear pattern of the feature will be created; refer to Figure-140.

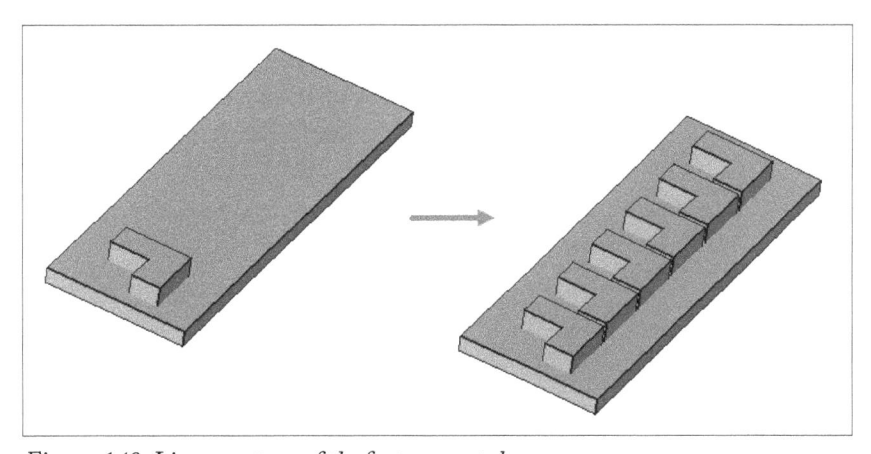

Figure-140. Linear pattern of the feature created

Polar Pattern

The **Polar Pattern** tool creates a set of copies of selected features rotated around a chosen axis. The procedure to use this tool is discussed next.

- Click on the **Polar Pattern** tool from **Toolbar** in the **Part Design** workbench; refer to Figure-141. The **Select attachment** dialog will be displayed in the **Tasks** panel of **Combo View** as discussed earlier asking you to select the feature.

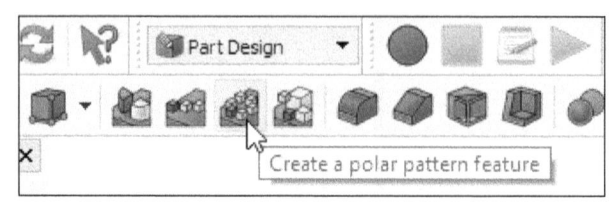

Figure-141. Polar pattern tool

- Select the feature to be polar patterned from the dialog and click on **OK** button. The **PolarPattern parameters** dialog will be displayed in the **Tasks** panel of **Combo View** along with the preview of polar pattern of the feature; refer to Figure-142.

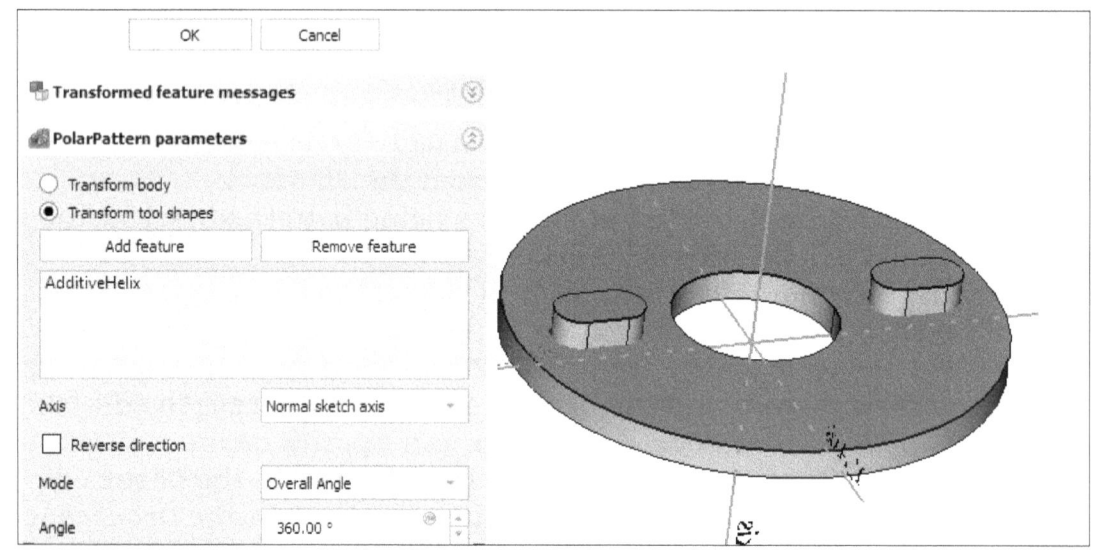

Figure-142. Polar Pattern parameters dialog with preview of polar patterned feature

- Specify desired axis of rotation for polar pattern from **Axis** drop-down in the dialog.
- Select the **Reverse direction** check box to reverse the direction of polar pattern.
- Select **Overall Angle** option from **Mode** drop-down and specify desired angle between patterned feature and the original feature in the **Angle** edit box.
- Select **Offset Angle** option from **Mode** drop-down and specify desired offset distance between the consecutive patterned feature and the original feature in the **Offset** edit box.
- Specify desired number of occurrences of polar pattern in the **Occurrences** edit box.
- Other parameters in the dialog have been discussed earlier.
- Click on **OK** button from the dialog. The polar pattern of the feature will be created; refer to Figure-143.

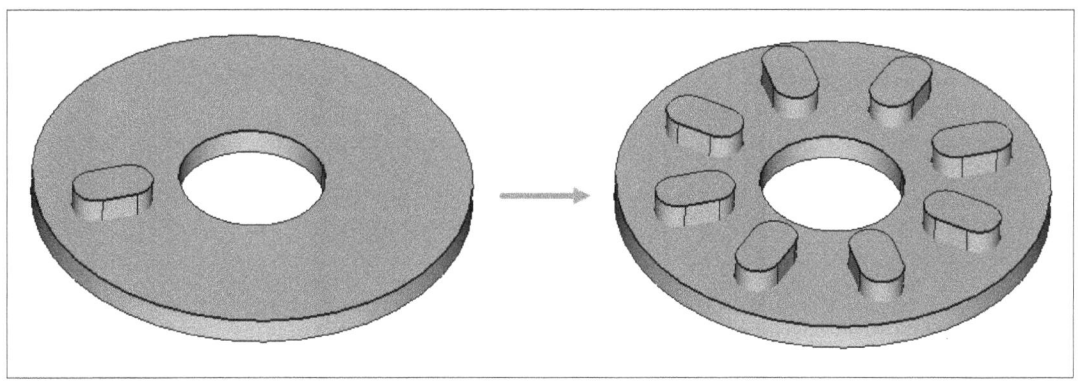

Figure-143. Polar pattern of the feature created

Create MultiTransform

The **Create MultiTransform** tool takes one or a set of part features as its input and allows the user to apply multiple transformations to that feature or a set of features progressively in sequence - creating a combined or compound transformation. The procedure to use this tool is discussed next.

- Click on the **Create MultiTransform** tool from **Toolbar** in the **Part Design** workbench; refer to Figure-144. The **Select attachment** dialog will be displayed in the **Tasks** panel of **Combo View** as discussed earlier asking you to select the feature.
- Select the feature you want to be multi transformed and click on **OK** button from the dialog. The **MultiTransform parameters** dialog will be displayed in the **Tasks** panel of **Combo View**; refer to Figure-145.

Figure-144. Multi transform tool

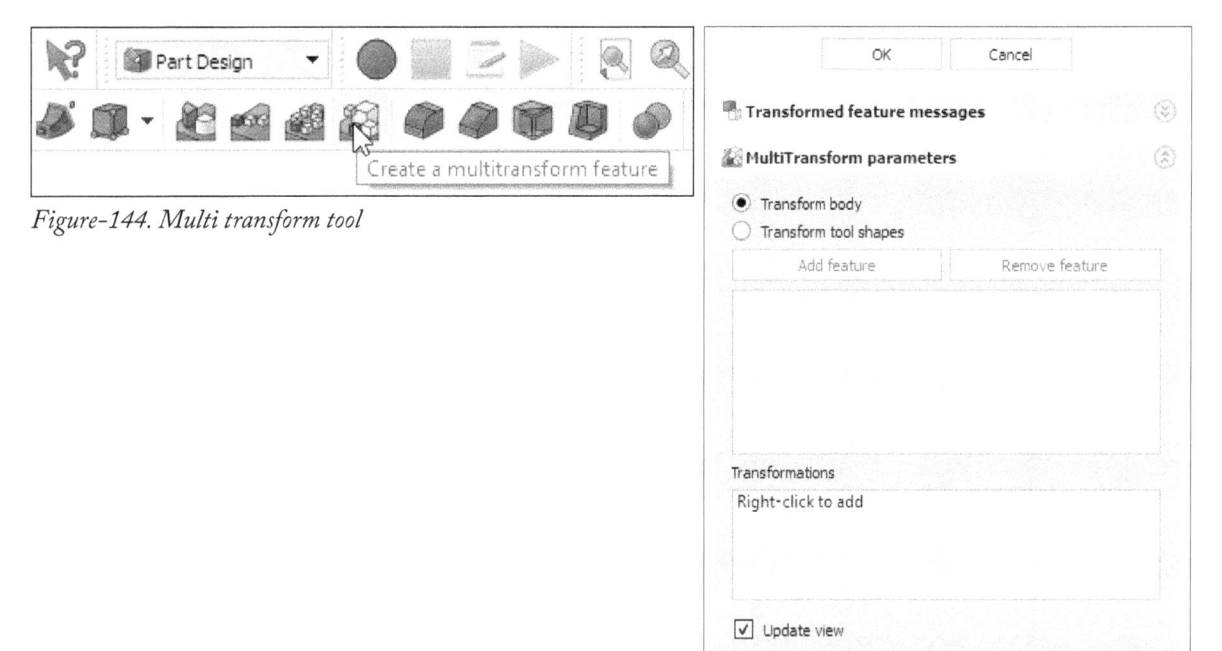

Figure-145. MultiTransform parameters dialog

- If you want to include additional features for the transformations, click on the **Add feature** button from the dialog and select the feature you want to add from the **Model Tree** or from the 3D view.
- Click **RMB** in the **Transformations** list view in the dialog. A shortcut menu will be displayed; refer to Figure-146.

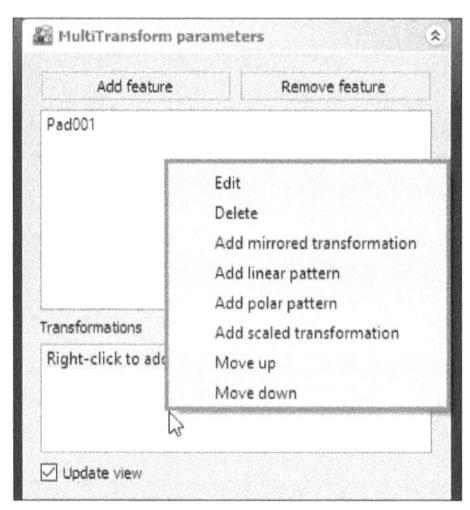

Figure-146. Shortcut menu

- Select desired transformation option from the shortcut menu. The parameters related to the selected transformation will be displayed in the dialog along with the preview of linear pattern of the feature; refer to Figure-147.

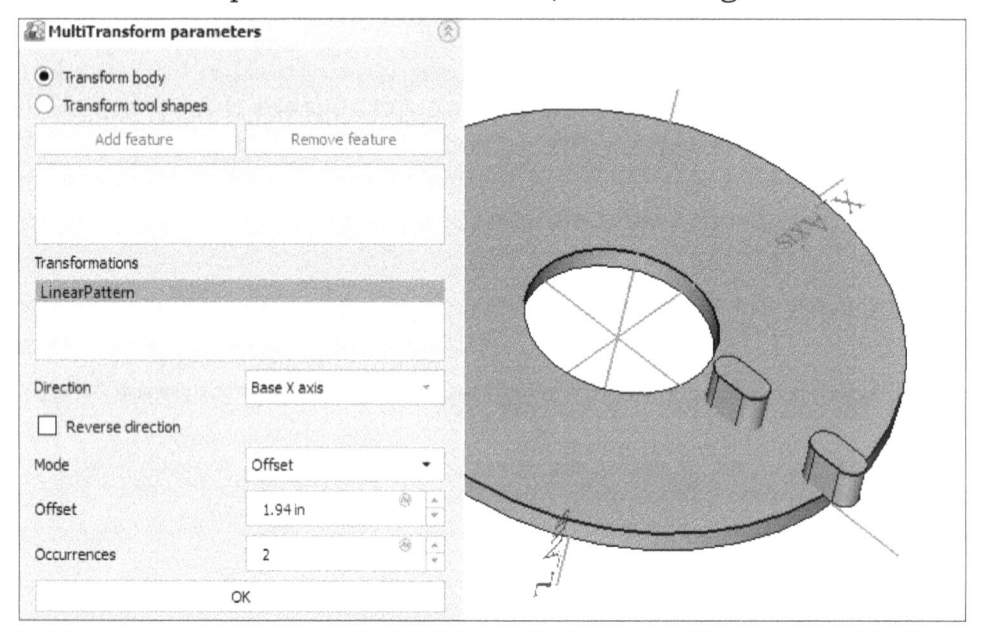

Figure-147. Linear pattern type parameters displayed along with the preview of linear pattern of the feature

- Specify all the parameters related to transformation as discussed earlier.
- Click on **OK** button below **Transformations** list view area. The linear pattern of the feature will be created; refer to Figure-148.

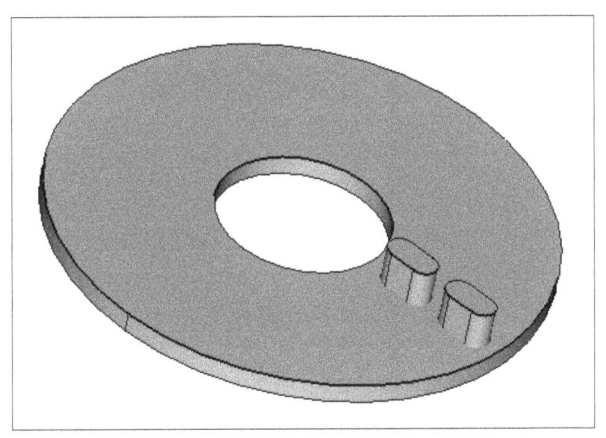

Figure-148. Linear pattern of the feature created

- If you want to add more transformation to the feature then click **RMB** in the **Transformations** list view in the dialog and select desired pattern type from the shortcut menu displayed.
- If you have selected **Add polar pattern** option then the parameters related to selected polar pattern type will be displayed in the dialog along with the preview of polar pattern of the feature; refer to Figure-149.

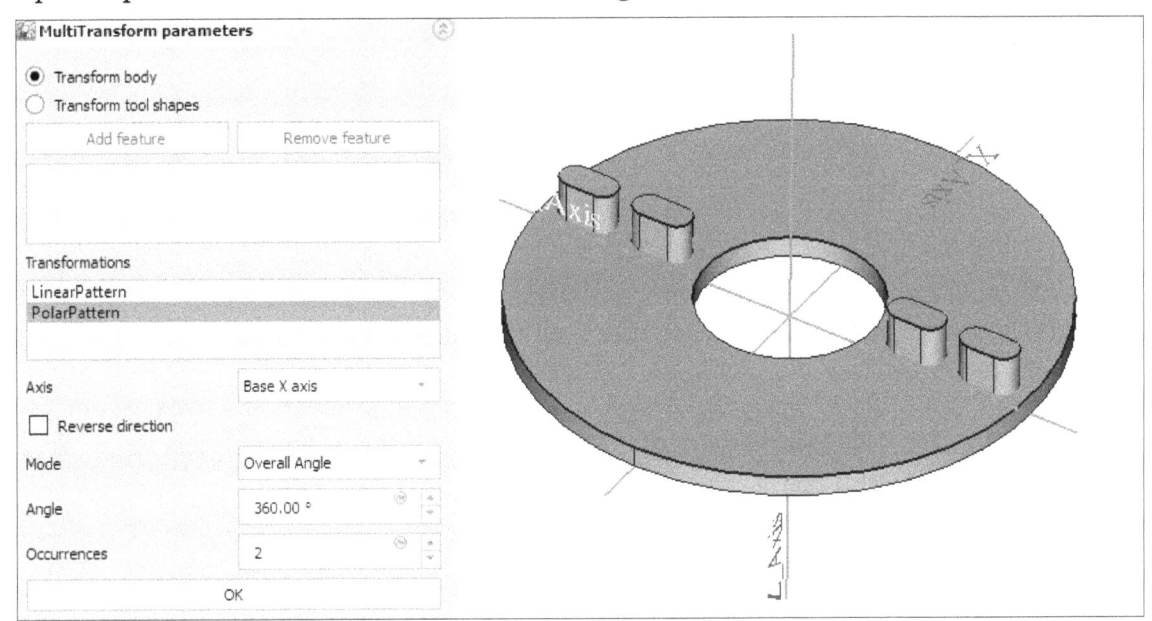

Figure-149. Polar pattern type parameters displayed along with the preview of polar pattern of the feature

- Specify all the parameters related to polar pattern type as discussed earlier.
- Click on **OK** button below **Transformations** list box. The polar pattern of the feature will be created; refer to Figure-150.
- Click on **OK** button to close the dialog.

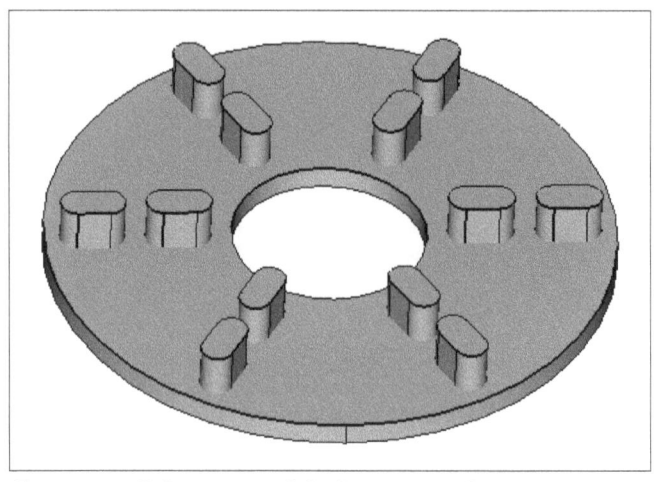

Figure-150. Polar pattern of the feature created

Dress-up tools

The **Dress-up** tools are used to apply treatment to the selected edges or faces. These tools are available in the **Toolbar** of **Part Design** workbench; refer to Figure-151. The procedures to use these tools are discussed next.

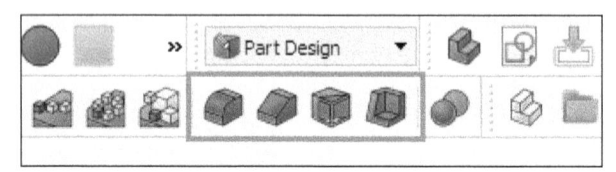

Figure-151. Dress up tools in part design workbench

Fillet

The **Fillet** tool creates fillets (rounds) on the selected edges of an object. The procedure to use this tool is discussed next.

- Select an edge or multiple edges or face of an object on which you want to apply fillet; refer to Figure-152.

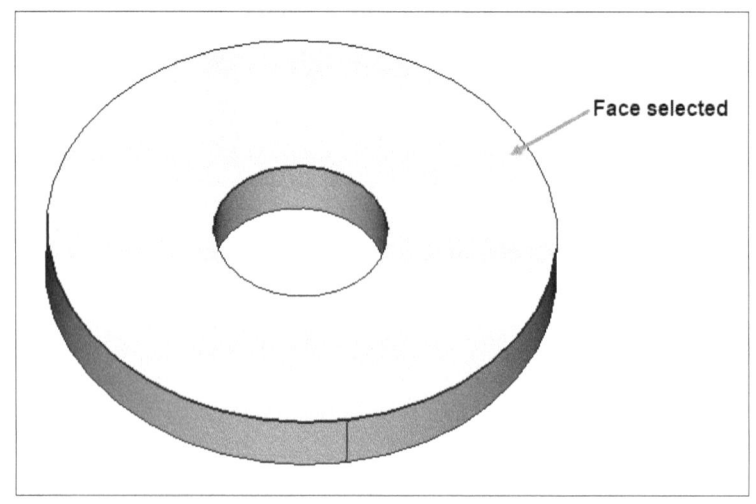

Figure-152. Face selected to create fillet

FreeCAD 1.0 Black Book   3-65

- Click on the **Fillet** tool from **Toolbar** in the **Part Design** workbench; refer to Figure-153. The **Fillet parameters** dialog will be displayed in the **Tasks** panel of **Combo View** along with the preview of fillets created on the object; refer to Figure-154.

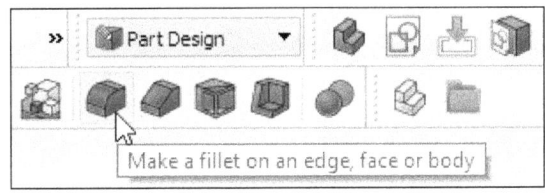

Figure-153. Fillet tool

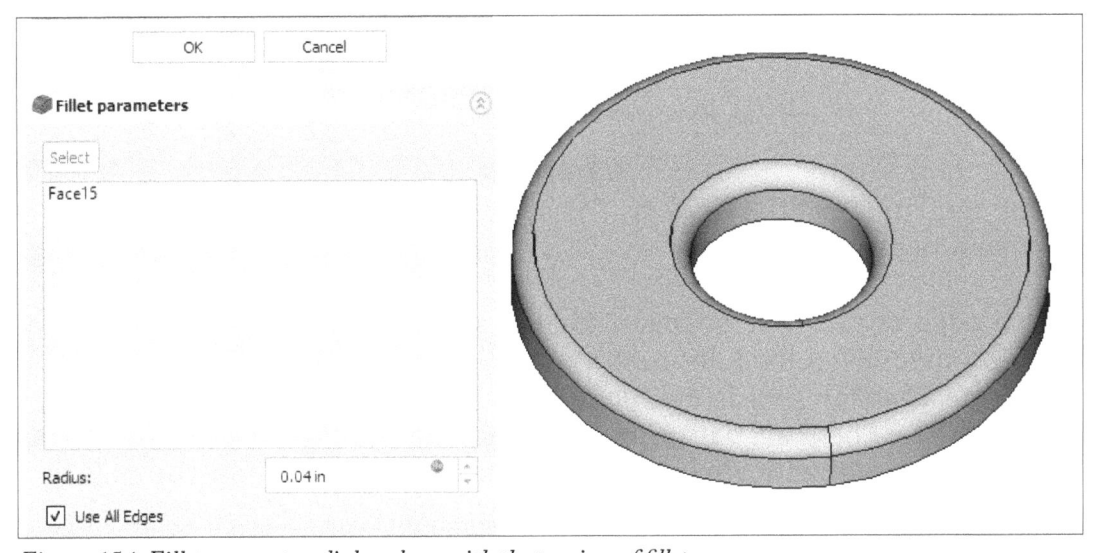

Figure-154. Fillet parameters dialog along with the preview of fillets

- If you want to add more edges or faces to be fillet then click on the **Select** button from the dialog and select the edges or faces from the 3D view area. If you want to add all the edges of the model to apply fillet then press **RMB** in the list view area of the dialog and select **Add all edges** option from the options displayed.
- Click on the **Preview** button to display the preview of the fillet creation.
- If you want to remove the fillet then select the face or an edge from the list view area in the dialog and press **RMB** in the list view area and select **Remove** option.
- Specify desired radius of the fillet in the **Radius** edit box.
- Select **Use All Edges** check box to select all edges of the feature to apply fillet.
- Click on **OK** button from the dialog. The fillets will be created; refer to Figure-155.

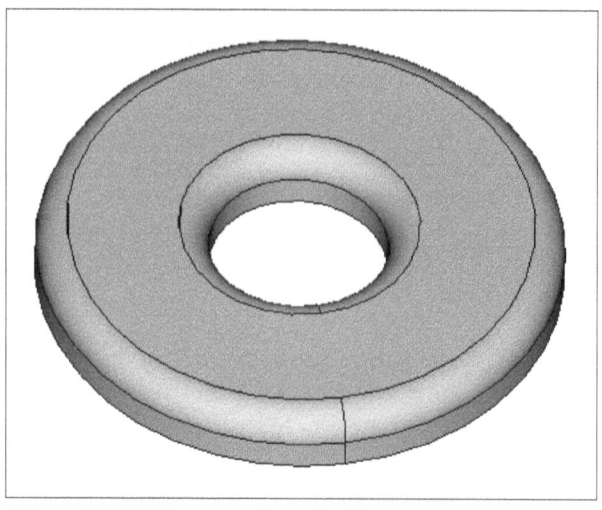

Figure-155. Fillets created on the object

Chamfer

The **Chamfer** tool creates chamfers on the selected edges of an object. The procedure to use this tool is discussed next.

- Select an edge or multiple edges or face of an object on which you want to apply chamfer as discussed for fillet tool.
- Click on the **Chamfer** tool from **Toolbar** in the **Part Design** workbench; refer to Figure-156. The **Chamfer parameters** dialog will be displayed in the **Tasks** panel of **Combo View** along with the preview of chamfers created on the object; refer to Figure-157.

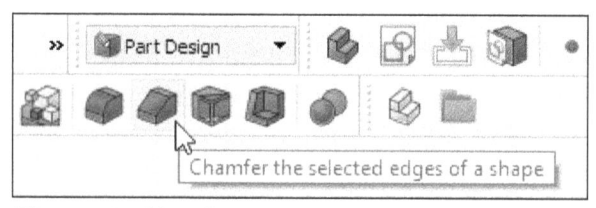

Figure-156. Chamfer tool

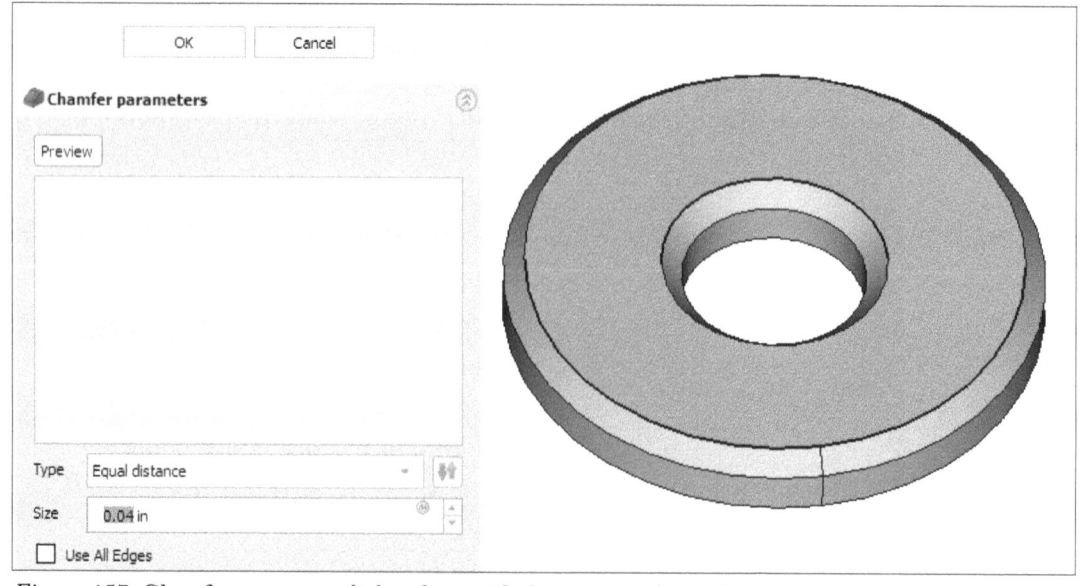

Figure-157. Chamfer parameters dialog along with the preview of chamfers

- Select desired chamfer type from **Type** drop-down in the dialog. Select the **Equal distance** option to equally distance the chamfer edges from the body edge. Select the **Two distances** option to specify the distances of the chamfer edge to the body edge. Select the **Distance and angle** option to specify one distance of the chamfer edge to the body edge. The second chamfer edge is defined by the angle of the chamfer.
- Specify desired size of the chamfer in the **Size** edit box.
- The other parameters of the dialog is same as discussed for **Fillet** tool.
- Click on **OK** button from the dialog. The chamfers will be created; refer to Figure-158.

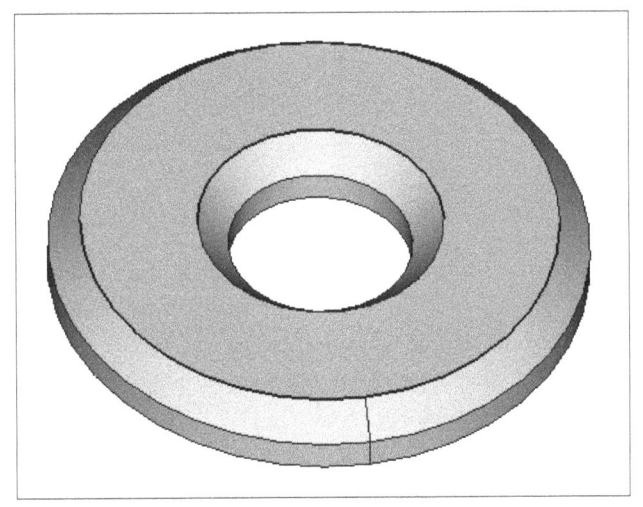

Figure-158. Chamfers created on the object

Draft

The **Draft** tool creates angular draft on the selected faces of an object. The procedure to use this tool is discussed next.

- Select one or multiple faces of an object on which you want to apply an angular draft; refer to Figure-159.
- Click on the **Draft** tool from **Toolbar** in the **Part Design** workbench; refer to Figure-160. The **Draft parameters** dialog will be displayed in the **Tasks** panel of **Combo View** along with the preview of draft applied; refer to Figure-161.

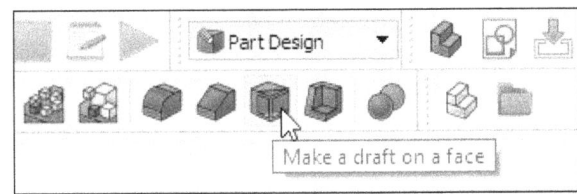

Figure-160. Draft tool

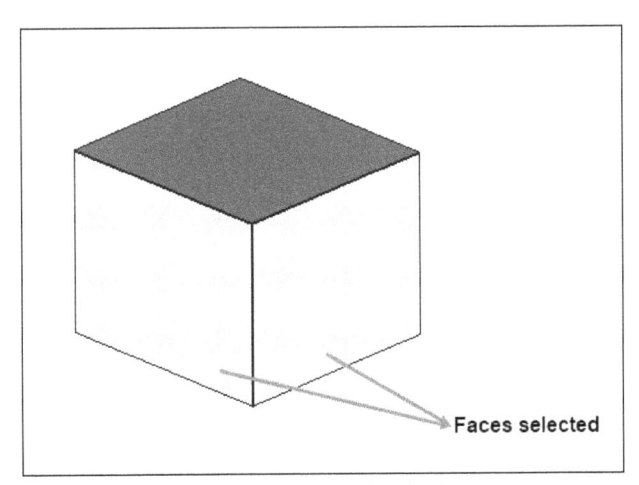

Figure-159. Faces selected for angular draft

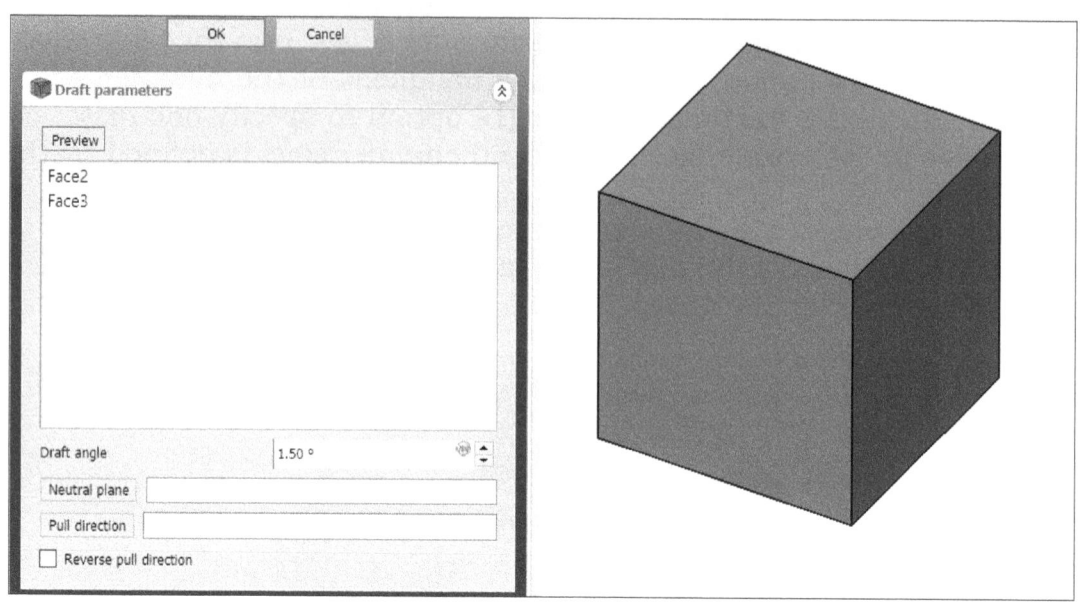

Figure-161. Draft parameters dialog along with the preview of draft applied

- If you want to add more faces to apply the draft then click on the **Select** button from the dialog and select the faces of an object in the 3D view area.
- If you want to remove the faces then select the faces from the list view area in the dialog and press **RMB** in the list view area and select **Remove** option.
- Specify desired angle of draft in the **Draft angle** edit box.
- Click on the **Neutral plane** button and select the face to be used as reference for defining draft angle.
- Click on **Pull direction** button and select desired edge of an object to define direction in which draft will be applied.
- Select the **Reverse pull direction** check box to reverse the direction of draft pulling.
- Click on **OK** button from the dialog. The draft will be created on the object; refer to Figure-162.

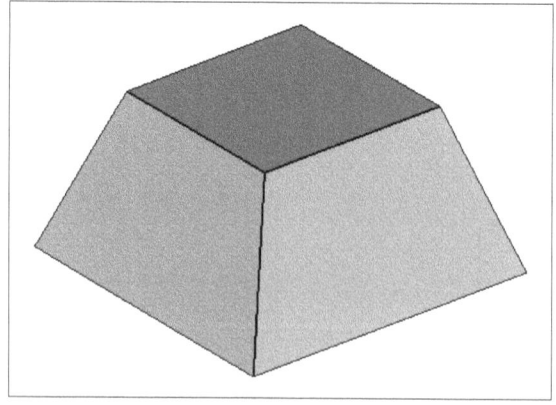

Figure-162. Draft created on the object

Thickness

The **Thickness** tool works on a solid body and transforms it into a thick-walled hollow object with at least one open face. The procedure to use this tool is discussed next.

- Select the face or multiple faces of an object to be hollowed; refer to Figure-163.

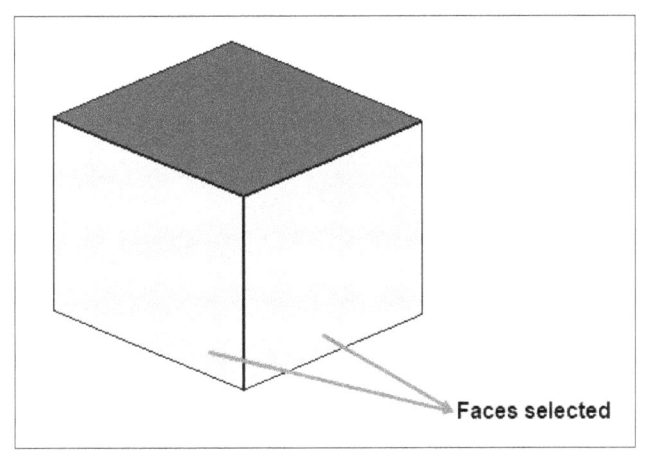

Figure-163. Faces selected to be hollowed

- Click on the **Thickness** tool from **Toolbar** in the **Part Design** workbench; refer to Figure-164. The **Thickness parameters** dialog will be displayed in the **Tasks** panel of **Combo View** along with the preview of hollowed faces; refer to Figure-165.

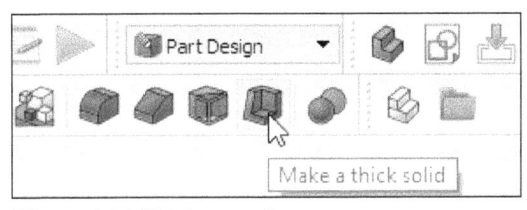

Figure-164. Thickness tool

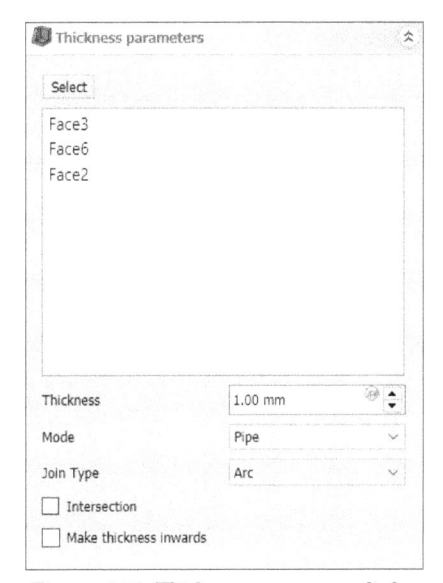

Figure-165. Thickness parameters dialog along with the preview of opened faces

- If you want to add more faces to be hollowed then select the faces of the object from the 3D view area.
- If you want to remove the faces then select the faces from the list of the dialog, press **RMB** in the list view area, and select **Remove** option.
- Specify desired wall thickness value in the **Thickness** edit box.
- Specify desired mode of thickness from the **Mode** drop-down. Select the **Skin** option if you want to get an item like a vase, headless but with the bottom. Select the **Pipe** option if you want to get an object like a pipe, headless, and bottomless. Select the **Recto Verso** option to flip the thickness side.
- Select desired option from the **Join Type** drop-down. Select the **Arc** option if you want to remove the outer edges and create a fillet with radius equal to the defined thickness. Select the **Intersection** option when faces are offset outward, sharp edges are kept between faces.
- Select **Intersection** check box to avoid self-intersection in the model.

- Select the **Make thickness inwards** check box if you want to offset the faces inwards.
- Click on **OK** button from the dialog. The thick-walled hollow object with opened faces will be created; refer to Figure-166.

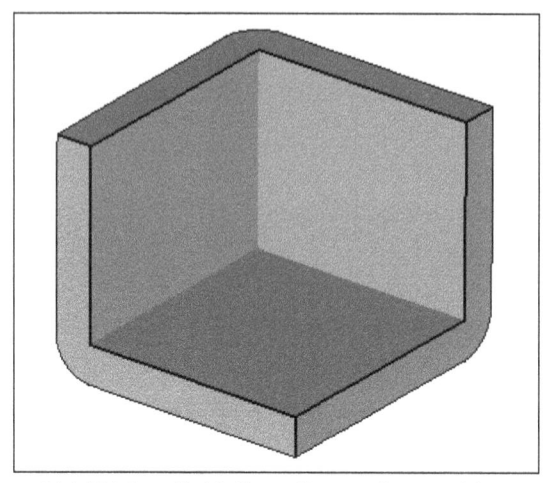

Figure-166. Thick walled hollow object with opened faces created

Boolean operation

The **Boolean** tool applies a boolean operation (fuse, cut, or common) to the Part Design bodies. The procedure to use this tool is discussed next.

- Activate the body on which you want to apply the boolean feature. To activate the body, right-click on desired body from the **Model** panel, and select the **Toggle Active body** option from the shortcut menu. The body will be activated.
- Click on the **Boolean operation** tool from **Toolbar** in the **Part Design** workbench; refer to Figure-167. The **Boolean parameters** dialog will be displayed in the **Tasks** panel of **Combo View**; refer to Figure-168.

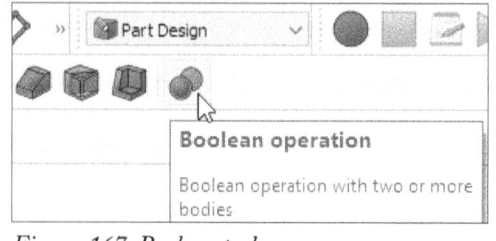

Figure-167. Boolean tool

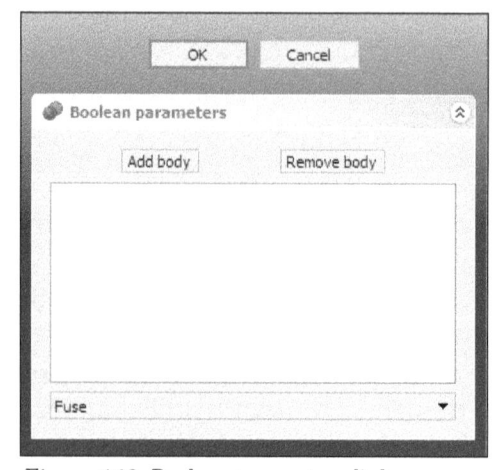

Figure-168. Boolean parameters dialog

- Click on the **Add body** button from the dialog, the active body temporarily disappears from the 3D view area; refer to Figure-169.

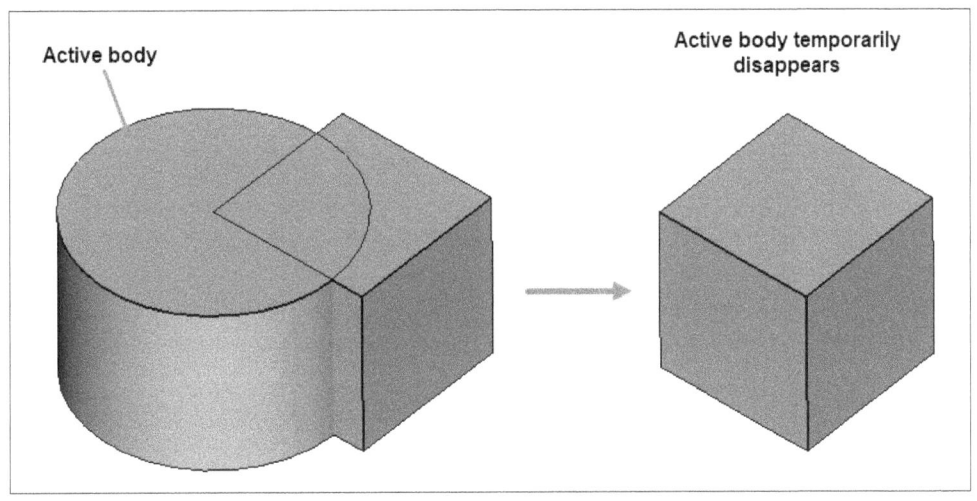

Figure-169. Active body temporarily disappears on clicking Add body button

- Select the body or multiple bodies which you want to use for creating the boolean feature from the 3D view area.
- Select desired type of boolean operation from the drop-down in the dialog. The preview of cut boolean feature will be displayed; refer to Figure-170.

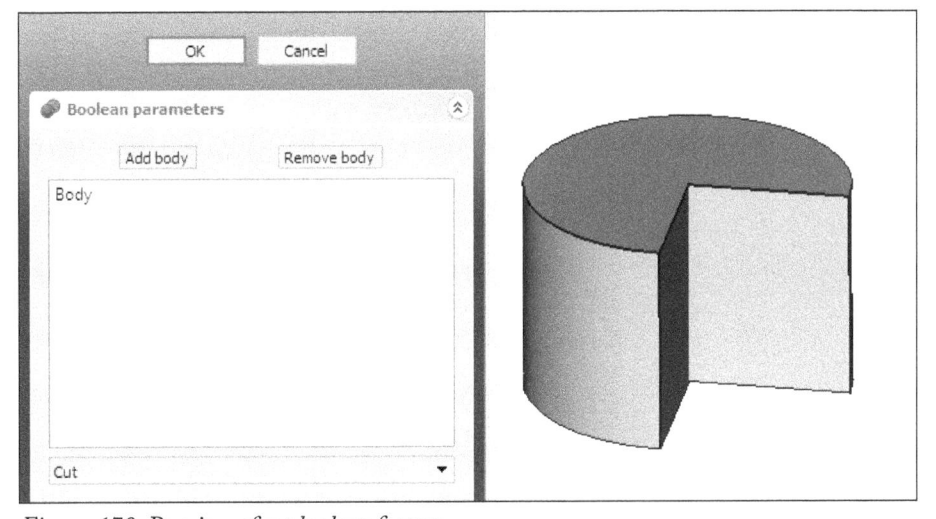

Figure-170. Preview of cut boolean feature

- Select the **Fuse** option from drop-down to merge the tool body or bodies to the active body. Select the **Cut** option to subtract the tool body or bodies from the active body. Select the **Common** option to extract the intersection from the selected body or bodies with the active body.
- If you want to remove the body to be boolean then select the body from the 3D view area and click on **Remove body** button from the dialog.
- Click on **OK** button from the dialog. The cut boolean feature will be created; refer to Figure-171.

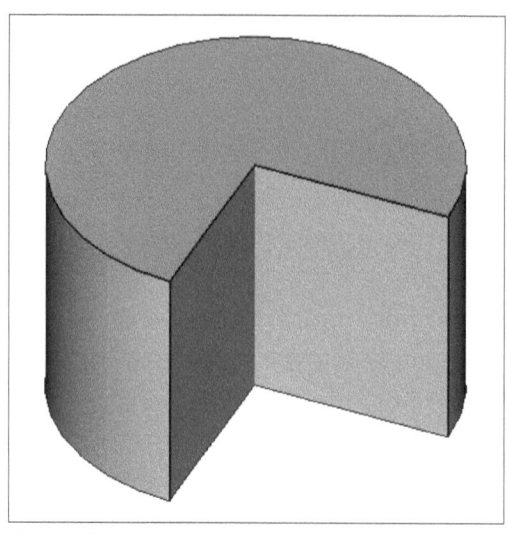

Figure-171. Cut boolean feature created

Creating Sprocket

The **Sprocket** tool allows you to create a 2D profile of a sprocket. It can be padded with the **Part Design Pad** feature. The procedure to use this tool is discussed next.

- Click on the **Sprocket** tool from the **Part Design** menu; refer to Figure-172. The profile of sprocket will be displayed with **Sprocket parameters** dialog in the **Tasks** panel of **Combo View**; refer to Figure-173.

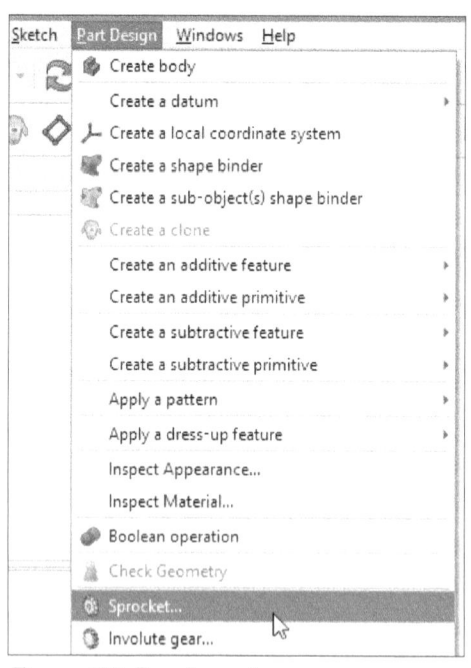

Figure-172. Sprocket tool

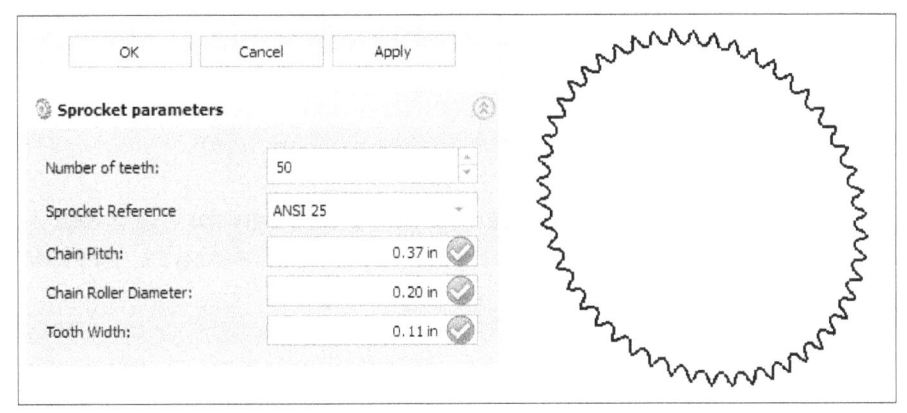

Figure-173. Sprocket parameter dialog with sprocket profile

- Specify desired number of teeth for the sprocket in the **Number of teeth** edit box.
- Select desired sprocket type from **Sprocket Reference** drop-down.
- Specify desired distance between the two teeth in the **Chain Pitch** edit box of the dialog.
- Specify desired diameter of the chain rollers in the **Chain Roller Diameter** edit box.
- Specify desired thickness of sprocket in the **Tooth Width** edit box.
- Click on the **OK** button from the dialog. The sprocket will be created.
- Now, if you want to pad the sprocket feature then select the sprocket created from the **Model Tree**; refer to Figure-174, and click on the **Pad** tool from the **Toolbar** as discussed earlier in this chapter. The pad feature of sprocket profile will be created; refer to Figure-175.

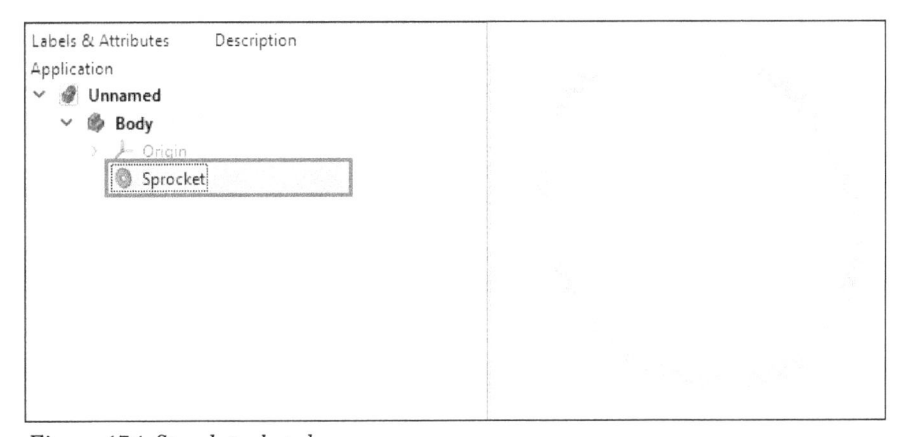

Figure-174. Sprocket selected

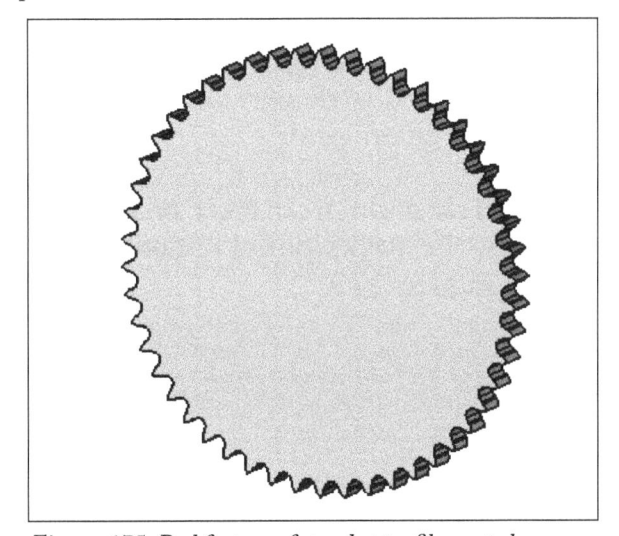

Figure-175. Pad feature of sprocket profile created

Creating Involute Gear

The **Involute gear** tool is used to generate profile for creating involute gear. You can later use this profile for **Pad** tool to create solid gear feature. The procedure to use this tool is given next.

- Click on the **Involute gear** tool from the **Part Design** menu; refer to Figure-176. The profile of involute gear will be displayed with **Involute parameter** dialog in the **Tasks** panel of **Combo View**; refer to Figure-177.

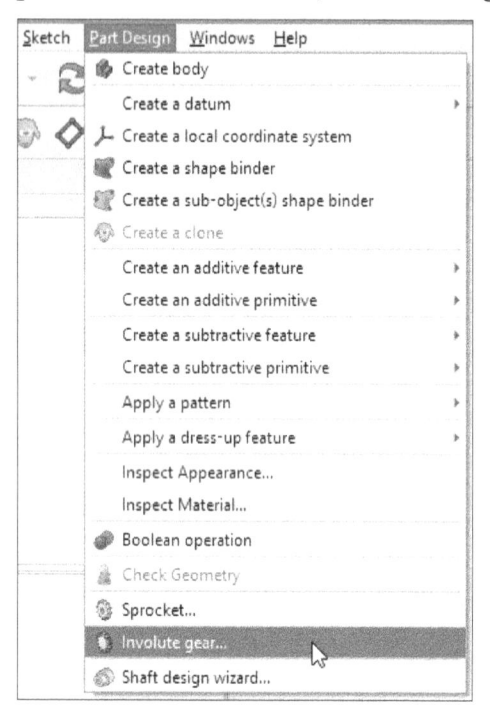

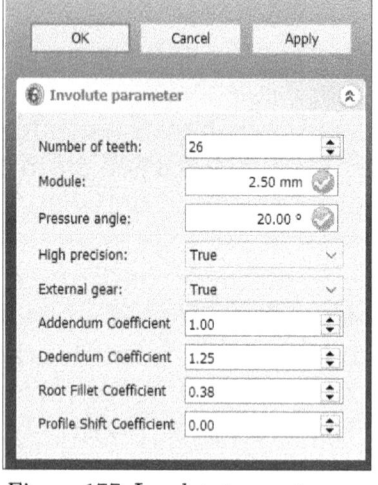

Figure-177. Involute parameter dialog with involute gear profile

Figure-176. Involute gear tool

- Specify the number of teethes to be created in profile in the **Number of teeth** edit box of dialog.
- Specify the other parameters like module of gear, pressure angle, and so on in respective edit boxes of the dialog and then click on the **OK** button to create the profile.

Select the profile created and click on the **Pad** tool from **Toolbar** to create solid gear.

Shaft Design Wizard

The **Shaft design wizard** tool is used to design shaft based on specified parameters. The procedure to use this tool is given next.

- Click on the **Shaft design wizard** tool from **Part Design** menu; refer to Figure-178. A dialog will be displayed in the **Tasks** panel of **Combo View** to specify parameters for creating shaft; refer to Figure-179.

FreeCAD 1.0 Black Book　　　　　　　　　　　　　　　　　　　　　　　　　　　　　　　3-75

Figure-178. Shaft design wizard tool

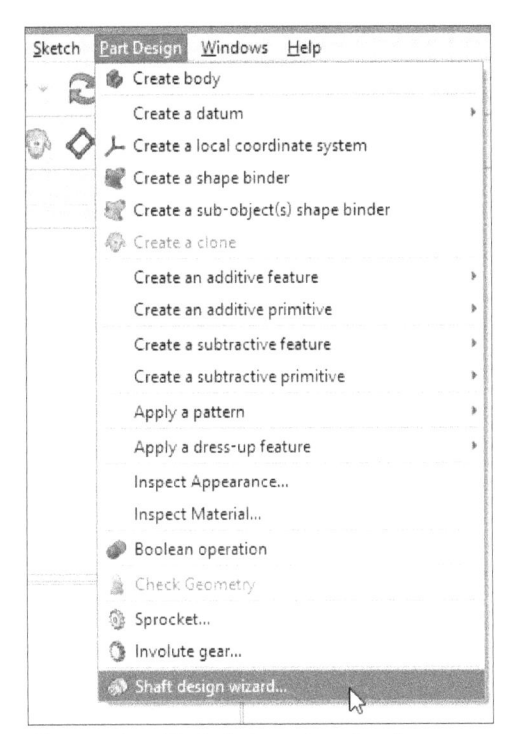

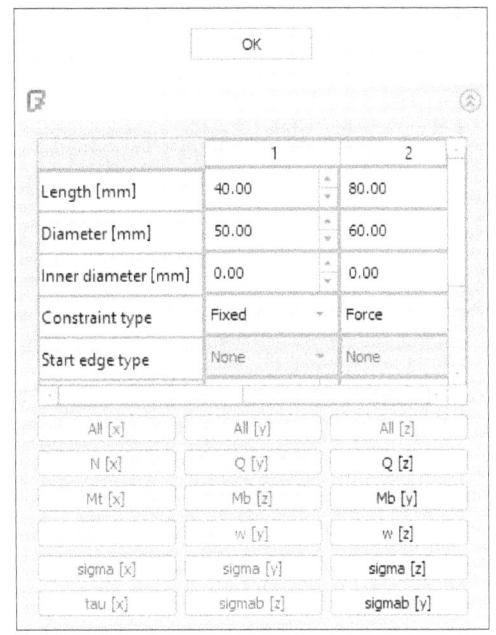

Figure-179. Dialog for shaft design

- Specify length, diameter, inner diameter, and constraint type for shaft sections in respective fields of dialog.
- If you want to add more sections to the shaft then right-click in the dialog and select **Add column** option; refer to Figure-180.

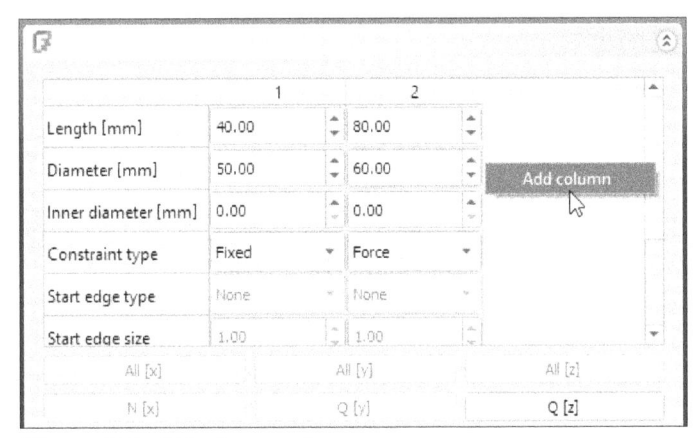

Figure-180. Add column option

- You can check various parameters of shaft using the buttons at the bottom in the dialog.

PRACTICAL

Create the sketch as shown in Figure-181.

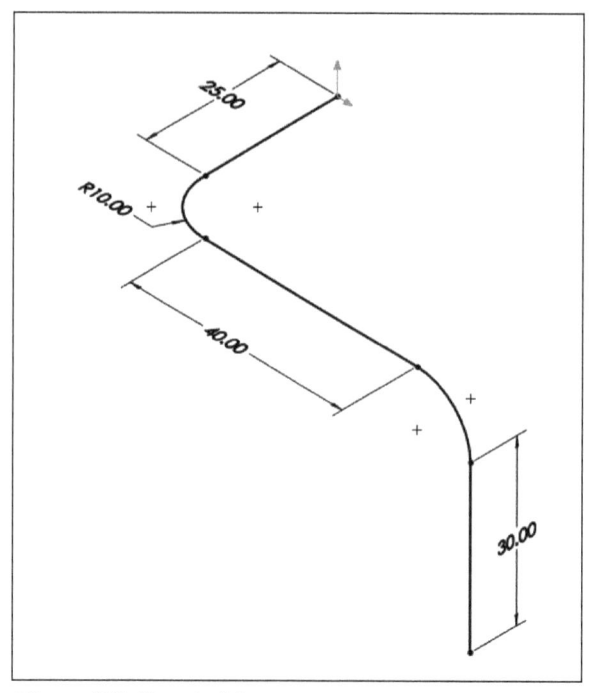

Figure-181. Practical 1

Steps to be performed:
- Start a new part file.
- Select a sketching plane and activate sketching mode.
- Create the sketch using **Line** tool.
- Apply the dimensions using dimension tools.
- Save the file.

Starting a new sketch

- Start FreeCAD if not started yet.
- Select the **Sketcher** option from the **Switch between workbenches** drop-down in the **Toolbar**. The sketcher environment will be displayed.
- Click on the **Create sketch** tool from the **Toolbar** or **Sketch** menu. The **Choose orientation** dialog box will be displayed.
- Select the **XY-Plane** tool from the **Sketch orientation** area of dialog box to create sketch on XY plane.
- Click on the **OK** button from the dialog box. The tools to create sketch will become active.

Creating sketch

- Click on the **Create line** tool from the **Toolbar** or **Sketcher geometries** cascading menu of **Sketch** menu to activate line tool. You will be asked to specify start point of line.
- Create a vertical line of length approximately **40** mm starting from origin and create a horizontal line of approximately **25** mm length; refer to Figure-182.

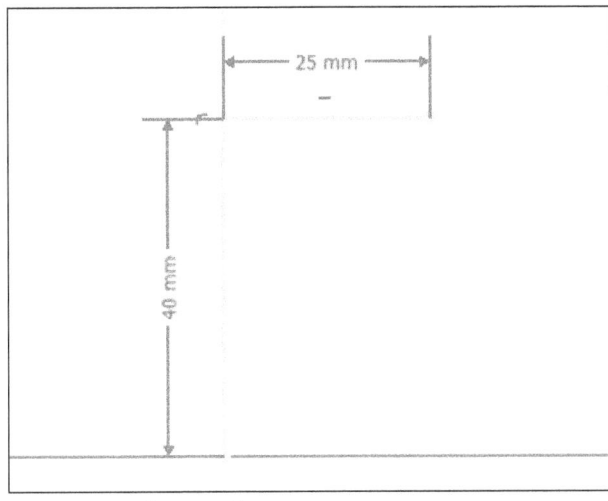

Figure-182. Creating lines

- Click on the **Sketch fillet** tool from the **Toolbar** and apply the fillet of radius **10** mm between the lines; refer to Figure-183.

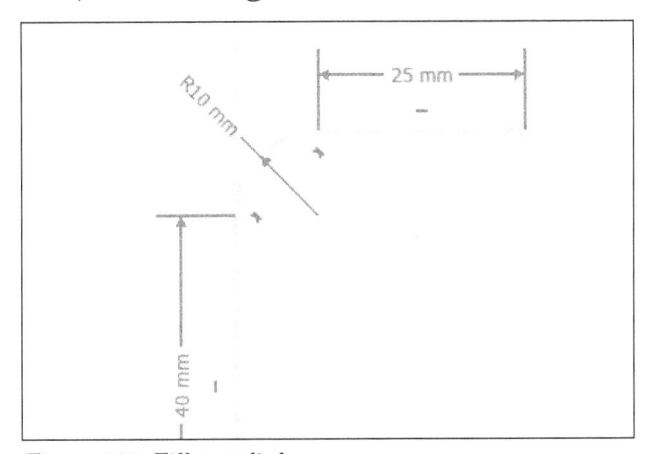

Figure-183. Fillet applied

- Apply the dimensions as per the drawing. Click on the **Close** button from the **Tasks** panel or press **ESC** key to exit the sketching mode.

Creating second sketch

- Click on the **Create a new datum plane** tool from the **Part Design** workbench and create a datum plane on the **XZ** plane as shown in Figure-184.

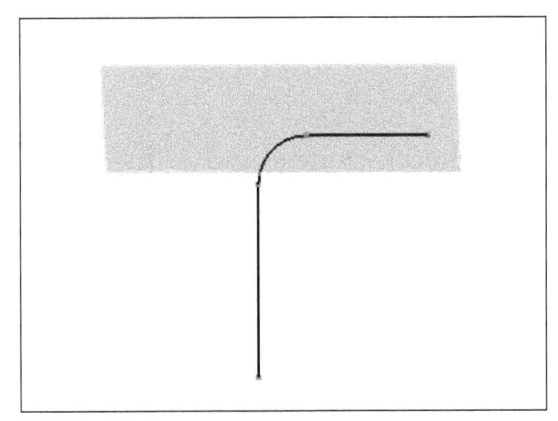

Figure-184. Datum plane created

- Select the datum plane recently created and click on the **Create sketch** tool from the **Toolbar** and create the sketch as shown in Figure-185.

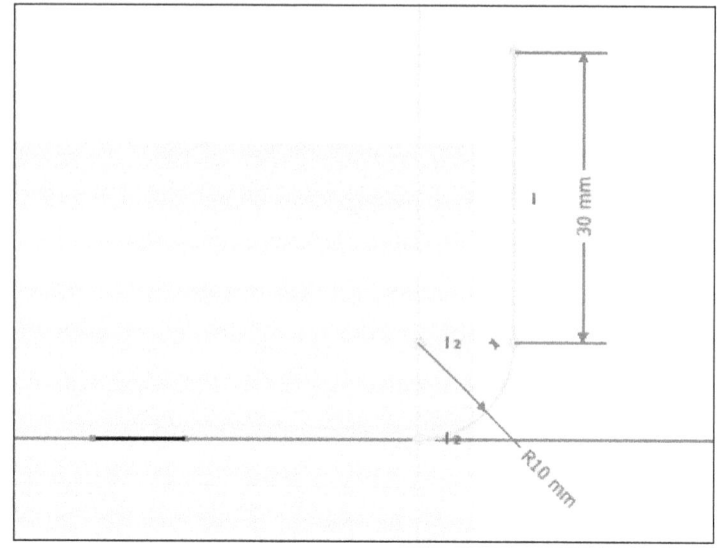

Figure-185. Creating second sketch

- Click on the **Close** button from the **Tasks** panel or press **Esc** key to exit the sketching mode. The 3D sketch will be created.

SELF-ASSESSMENT

Q1. What is the purpose of the "Create body" tool in the Part Design workbench?

A) To create a new part from scratch
B) To create a solid shape with sketches and features
C) To edit existing bodies
D) To delete bodies

Q2. Which tool is used to create a sketch in the active body?

A) Create body tool
B) Attach Sketch tool
C) Create sketch tool
D) Edit sketch tool

Q3. What does the "Validate Sketch" tool do?

A) It deletes the sketch
B) It creates a new sketch
C) It verifies the tolerance of sketch points and adjusts them
D) It copies the sketch to another body

Q4. What is the function of the "Attach Sketch" tool?

A) To create a new sketch
B) To delete an existing sketch
C) To map an existing sketch onto a selected face or plane
D) To modify the original sketch

Q5. What is a datum shape binder used for in the Part Design workbench?

A) To create a new solid shape
B) To link edges or faces of another body to the active body
C) To edit an existing sketch
D) To delete a body

Q6. What does the "Create a clone" tool do?

A) It creates a completely independent copy of an object
B) It creates a linked copy of a selected object
C) It merges two objects together
D) It exports the object to a file

Q7. In the Pad tool parameters, which option allows the user to extrude a sketch on both sides of the sketch plane?

A) Symmetric to plane
B) Reversed
C) To first
D) Up to face

Q8. What is the first step in creating a new sketch in the Part Design workbench?

A) Select a body
B) Click on the Create sketch tool
C) Click on the Edit sketch tool
D) Click on the Validate Sketch tool

Q9. What is the primary use of datum tools in FreeCAD?

A) To create the final shape of a body
B) To create auxiliary geometrical elements for reference
C) To delete unwanted shapes
D) To export models to other software

Q10. How can you create a datum point in FreeCAD?

A) By selecting the Create a datum point tool from the Part Design workbench
B) By using the Pad tool
C) By editing an existing sketch
D) By selecting the Attach Sketch tool

Q11. Which attachment mode option allows you to align the placement of an object with a linked object's placement?

A) Deactivated mode
B) Object's XY mode
C) Object's YZ mode
D) Tangent to surface mode

Q12. What does the "Create local coordinate system" tool define?

A) The orientation of the final shape
B) A coordinate system for referencing other geometries
C) A new sketch
D) A new body

Q13. What is the primary purpose of the Revolution tool in FreeCAD?

A) To create a solid by extruding a sketch
B) To create a solid by revolving a sketch around an axis
C) To create a hollow shape
D) To cut a solid shape from another solid

Q14. Which of the following is a requirement for using the Revolution tool?

A) The sketch must be an open loop.
B) The revolution axis must intersect with the sketch.
C) The sketch must be a closed loop.
D) The sketch must be placed on a datum plane.

Q15. What feature does the Additive Loft tool create?

A) A solid by revolving a sketch
B) A solid by transitioning between multiple sketches
C) A hollow tube by sweeping a sketch
D) A primitive shape

Q16. What option in the Loft parameters dialog allows for straight transitions between cross-sections?

A) Closed
B) Ruled surface
C) Add Section
D) Profile area

Q17. The Additive Pipe tool in FreeCAD is used for what purpose?

A) To create a solid by revolving a sketch
B) To sweep a sketch along an open or closed path
C) To create a solid by making a transition between sketches
D) To insert a primitive shape

Q18. What does the Additive Box tool create?

A) A primitive box shape
B) A complex 3D model
C) A hollow box shape
D) A sketch for lofting

Q19. Which tool is used to create one or more holes in a solid in FreeCAD?

A) Pocket tool
B) Revolve tool
C) Hole tool
D) Wedge tool

Q20. What does the 'Select reference' option do in the Revolution tool's parameters?

A) Allows selection of a predefined shape
B) Allows selection of an edge or datum line as the axis of revolution
C) Defines the angle of revolution
D) Selects the type of sketch to be revolved

Q21. Which option in the Additive Loft tool dialog allows for creating a loop from the last cross-section to the first?

A) Add Section
B) Closed
C) Ruled surface
D) Base profile

Q22. The Pocket tool is primarily used for which of the following actions?

A) To add material to a body
B) To cut out a solid by extruding a sketch
C) To create a solid by revolving a sketch
D) To loft between multiple profiles

Q23. What does the 'Dimension' option in the Pocket parameters dialog allow you to do?

A) Cut the hole to a specified length
B) Extrude cut the feature to the defined sketch plane at a specified length
C) Cut the hole through the entire body
D) Specify the direction of the pocket

Q24. Which tool is used to create a primitive torus in FreeCAD?

A) Additive Box
B) Additive Torus
C) Additive Cylinder
D) Additive Sphere

Q25. What does the Additive Prism tool require for its creation?

A) Only one sketch
B) The number of sides in the polygon cross-section
C) The total volume
D) The angle of rotation

Q26. What is the primary function of the Groove tool in the Part Design workbench?

A) To create a solid shape
B) To revolve a sketch about an axis
C) To add material to a model
D) To mirror features

Q27. Which parameter is specified to determine the extent of the groove created using the Groove tool?

A) Sketch type
B) Angle of revolution
C) Length of the groove
D) Width of the sketch

Q28. The Subtractive Loft tool is used to:

A) Create a solid shape by adding material
B) Remove material by transitioning between two or more sketches
C) Mirror existing features
D) Rotate a sketch around an axis

Q29. What must be selected first when using the Subtractive Pipe tool?

A) The angle of rotation
B) The path for the sweep feature
C) The cross-section of the pipe
D) The existing solid body

Q30. Which feature does the Subtractive Helix tool create?

A) A box shape
B) A solid shape by sweeping a sketch along a helix
C) A cylindrical shape
D) A cone shape

Q31. What is the first step to create a subtractive cylinder using the Subtractive Cylinder tool?

A) Specify the height and angle
B) Click on the Subtractive Cylinder tool
C) Define the radius
D) Select the existing solid body

Q32. In the context of the Subtractive Wedge tool, what do the X min/max edit boxes define?

A) The radius of the wedge
B) The length of the base feature along the X axis
C) The height of the wedge
D) The angle of the top face

Q33. Which parameter in the Subtractive Prism tool determines the number of sides for the polygon cross-section?

A) Circumradius
B) Height
C) Polygon
D) Angle

Q34. What does the Polar Pattern tool do?

A) It creates a 3D model from 2D sketches.
B) It creates a set of copies of selected features rotated around a chosen axis.
C) It applies transformations to features sequentially.
D) It applies treatments to edges or faces.

Q35. Which dialog box appears first when using the Polar Pattern tool?

A) Fillet parameters dialog
B) MultiTransform parameters dialog
C) Select feature dialog
D) Boolean parameters dialog

Q36. What is the primary function of the Fillet tool?

A) To create sharp edges on selected faces.
B) To apply fillets (rounds) on selected edges.
C) To hollow out the body of a solid.
D) To create a draft on faces.

Q37. What is the function of the Draft tool in Part Design?

A) It creates chamfers on edges.
B) It creates angular drafts on selected faces.
C) It pads a profile to create a solid.
D) It merges multiple bodies into one.

Q38. What does the Boolean tool allow you to do?

A) It migrates files from older versions of FreeCAD.
B) It applies boolean operations such as fuse, cut, or common.
C) It creates sprocket profiles.
D) It pads 2D profiles into 3D objects.

Q39. What is the first step in creating a sprocket using the Sprocket tool?

A) Specify the thickness of the sprocket.
B) Click on the Sprocket tool from the Part Design menu.
C) Select the sprocket type from the drop-down menu.
D) Specify the number of teeth.

Q40. In the Polar Pattern tool, how can the direction of the pattern be reversed?

A) By clicking the Reverse direction check box.
B) By specifying the angle in the dialog.
C) By selecting the Reverse axis option.
D) By changing the number of occurrences.

FOR STUDENT NOTES

Chapter 4

Solid Modeling with Part Workbench

Topics Covered

The major topics covered in this chapter are:

- *Introduction to Solid Modeling*
- *Starting Part Workbench*
- *Primitive Tools*
- *Modifying Object Tools*
- *Offset Tools*
- *Compound Tools*
- *Join Features Tools*
- *Splitting Tools*
- *Measure Tools*

INTRODUCTION

The solid modeling capabilities of FreeCAD are based on the **OpenCASCADE Technology** (OCCT) kernel, a professional-grade CAD system that features advanced 3D geometry creation and manipulation. The Part Workbench is a layer sitting on top of the OCCT libraries that gives the user access to OCCT geometric primitives and functions. Essentially, all 2D and 3D drawing functions in every workbench (Draft, Sketcher, PartDesign, etc.) are based on these functions exposed by the Part Workbench. Therefore, the **Part Workbench** is considered the core component of the modeling capabilities of FreeCAD.

Part objects are more complex than mesh objects created with the **Mesh Workbench** as they permit more advanced operations like coherent boolean operations, modifications history, and parametric behavior.

The **Part Workbench** is the basic layer that exposes the OCCT drawing functions to all workbenches in FreeCAD; refer to Figure-1.

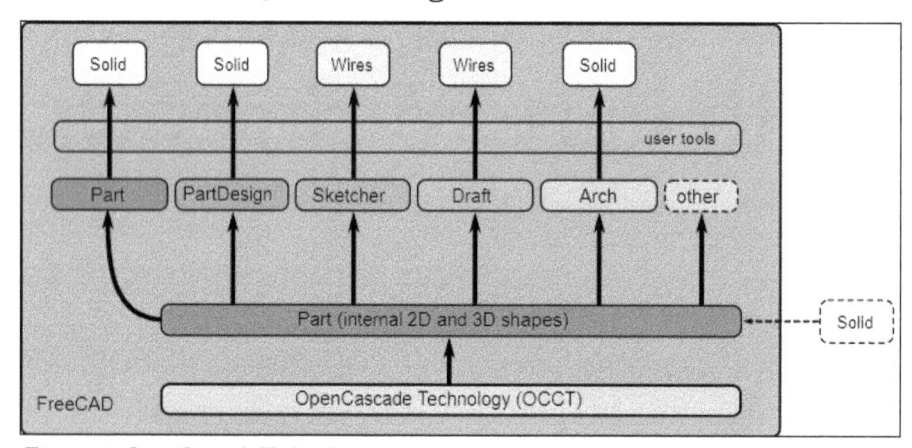

Figure-1. OpenCascade Technology

STARTING PART WORKBENCH

The objects created with the **Part Workbench** are relatively simple; they are intended to be used with boolean operations (unions and cuts) in order to build more complex shapes. This modeling paradigm is known as the constructive solid geometry (CSG) workflow and it was the traditional methodology used in early CAD systems.

- To start a new part file, click on the **New** button from **File** menu and select **Part** workbench from **Switch between workbenches** drop-down in the **Toolbar**; refer to Figure-2. The tools to create 3D models will be displayed in the **Toolbar**; refer to Figure-3.

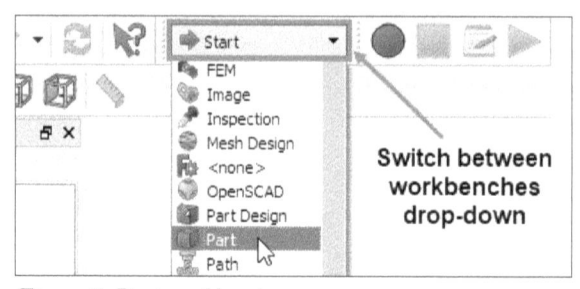

Figure-2. Part workbench

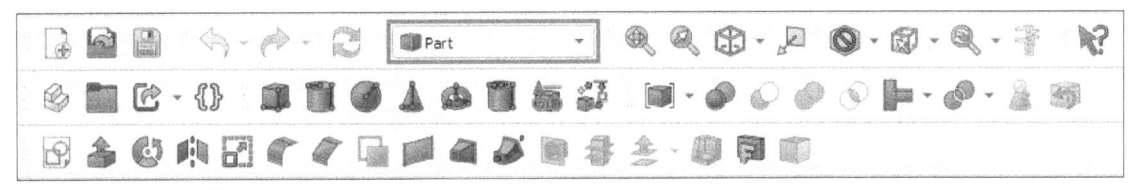

Figure-3. Part workbench tools

Primitive tools

Primitives are the tools for creating primitive objects. These tools are available in the **Toolbar** of **Part** workbench; refer to Figure-4. The procedures to use these tools are discussed next.

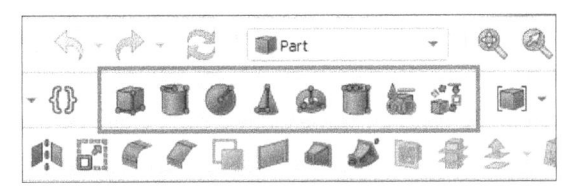

Figure-4. Primitive tools

Cube

The **Cube** tool inserts a parametric, rectangular cuboid, using geometric primitive into the active document. The procedure to use this tool is discussed next.

- Click on the **Cube** tool from **Toolbar** in the **Part** workbench; refer to Figure-5. A rectangular cuboid primitive will be inserted in the 3D view area; refer to Figure-6.

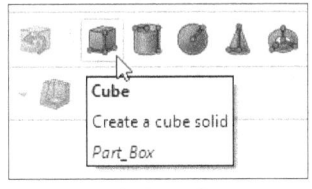

Figure-5. Cube tool

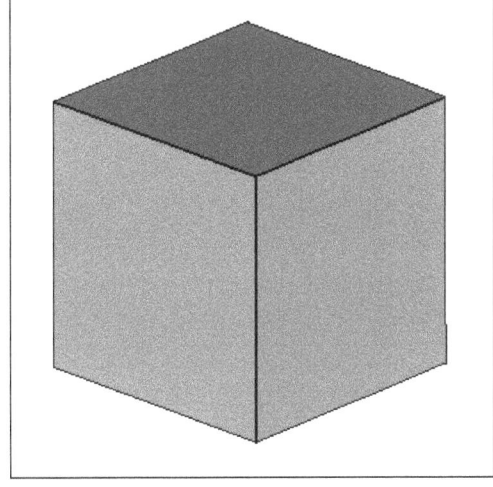

Figure-6. Rectangular cuboid primitive inserted

- If you want to edit the properties of inserted cuboid then select the cube created from the **Model Tree**; refer to Figure-7. The **Property editor** dialog will be displayed in the **Model** panel of **Combo View**; refer to Figure-8.

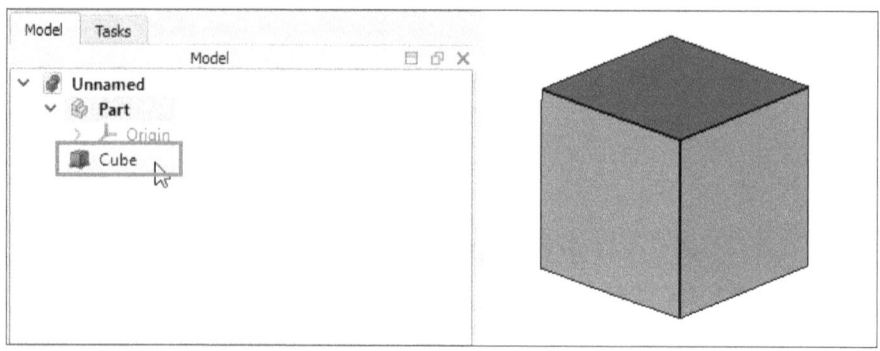

Figure-7. Selecting the object

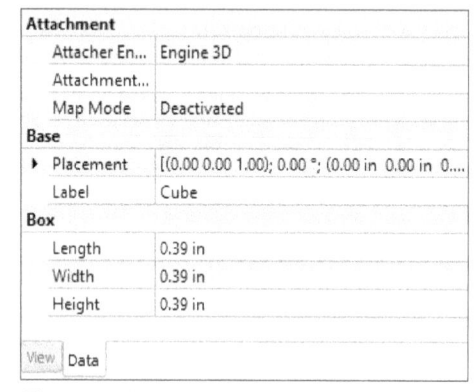

Figure-8. Property editor dialog

- Click on the **Placement** node from **Base** section of the dialog. The parameters related to placement will be displayed to specify the orientation and position of the cube in the 3D view area.
- Specify desired label for the cube in the **Label** edit box.
- Specify desired length, width, and height of the cube in the **Length**, **Width**, and **Height** edit boxes, respectively from **Box** section of the dialog. The properties of the rectangular cuboid will be edited.

Cylinder

The **Cylinder** tool creates a simple parametric cylinder with position, angle, radius, and height parameters. The procedure to use this tool is discussed next.

- Click on the **Cylinder** tool from **Toolbar** in the **Part** workbench; refer to Figure-9. A cylinder primitive will be inserted in the 3D view area; refer to Figure-10.

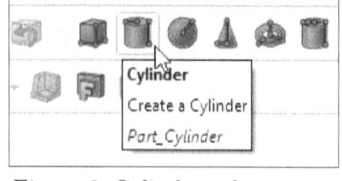

Figure-9. Cylinder tool

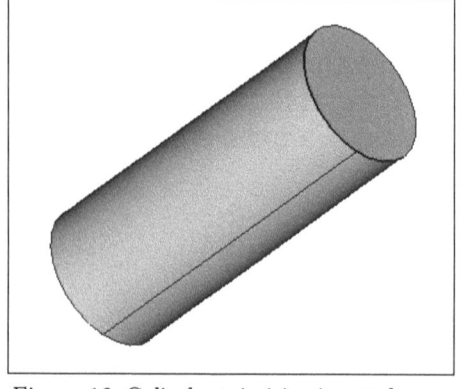

Figure-10. Cylinder primitive inserted

- The procedure to edit the properties of cylinder is same as discussed for **Box** tool.

Sphere

The **Sphere** tool creates a simple parametric sphere with position, angle 1, angle 2, angle 3, and radius parameters. The procedure to use this tool is discussed next.

- Click on the **Sphere** tool from **Toolbar** in the **Part** workbench; refer to Figure-11. The sphere primitive will be inserted in the 3D view area; refer to Figure-12.

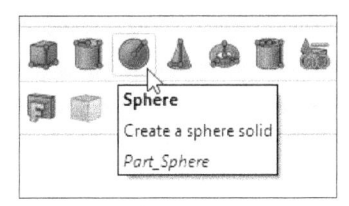

Figure-11. Sphere tool

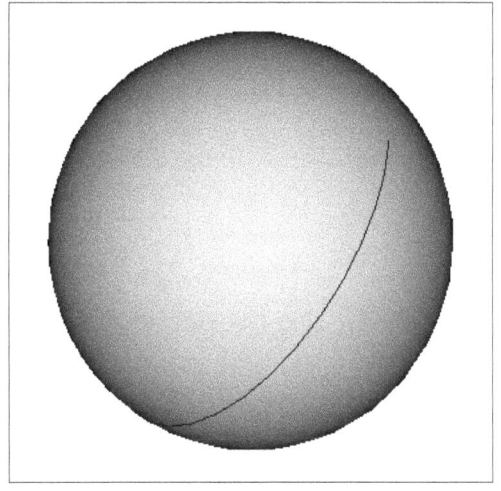

Figure-12. Sphere primitive inserted

- The procedure to edit the properties of sphere is same as discussed earlier.

Cone

The **Cone** tool creates a simple parametric truncated cone. The procedure to use this tool is discussed next.

- Click on the **Cone** tool from **Toolbar** in the **Part** workbench; refer to Figure-13. The cone primitive will be inserted in the 3D view area; refer to Figure-14.

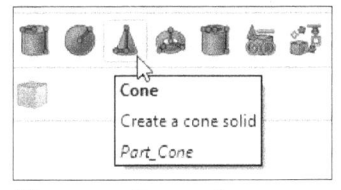

Figure-13. Cone tool

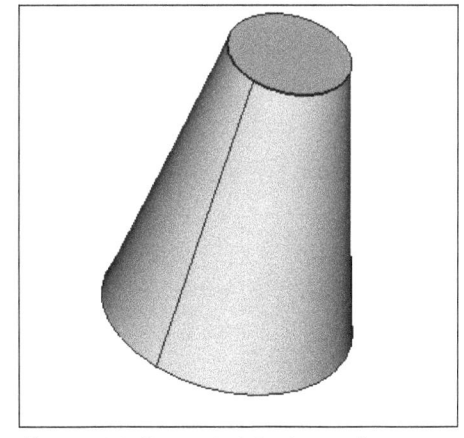

Figure-14. Cone primitive inserted

- The procedure to edit the properties of cone is same as discussed earlier.

Torus

The **Torus** tool creates a simple parametric torus with position, angle1, angle2, angle3, radius1, and radius2 as parameters. The procedure to use this tool is discussed next.

- Click on the **Torus** tool from **Toolbar** in the **Part workbench**; refer to Figure-15. The torus primitive will be inserted in the 3D view area; refer to Figure-16.

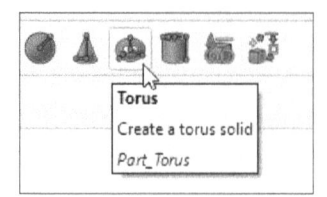

Figure-15. Torus tool

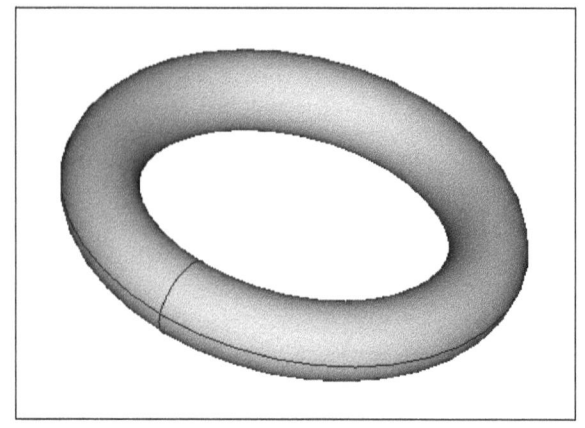

Figure-16. Torus primitive inserted

- The procedure to edit the properties of torus is same as discussed earlier.

Create tube

The **Create tube** tool inserts a tube into the active document. The tube is geometrically treated as a cut of a smaller cylinder into a larger one. The procedure to use this tool is discussed next.

- Click on the **Create tube** tool from **Toolbar** in the **Part** workbench; refer to Figure-17. The **Tube** dialog will be displayed in the **Tasks** panel of **Combo View** along with the tube primitive inserted in the 3D view area; refer to Figure-18.

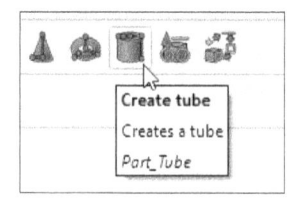

Figure-17. Create tube tool

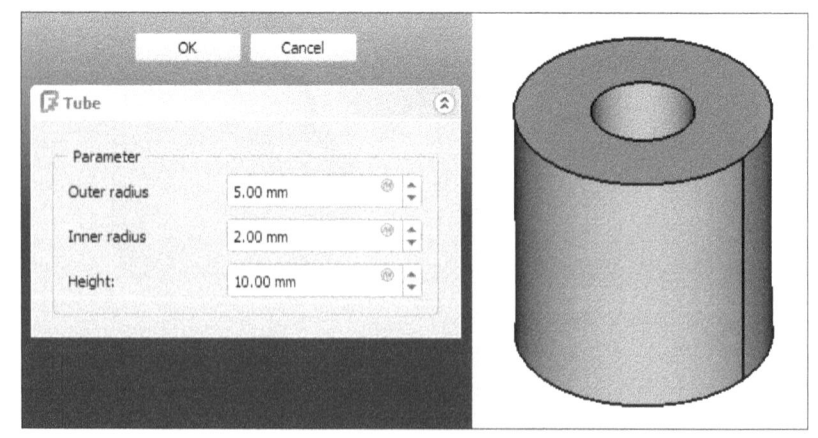

Figure-18. Tube dialog displayed along with the tube primitive

- Specify desired outer and inner radius of the tube in the **Outer radius** and **Inner radius** edit boxes of the dialog, respectively.
- Specify desired height of the tube in the **Height** edit box and click on the **OK** button from the dialog to create the feature.
- The procedure to edit the other properties of tube is same as discussed earlier.

Create primitives

The **Create primitives** tool is used to create any of the parametric geometric primitives defined in the part workbench. The procedure to use this tool is discussed next.

- Click on the **Create primitives** tool from **Toolbar** in the **Part** workbench; refer to Figure-19. The **Geometric Primitives** dialog will be displayed in the **Tasks** panel of **Combo View**; refer to Figure-20.

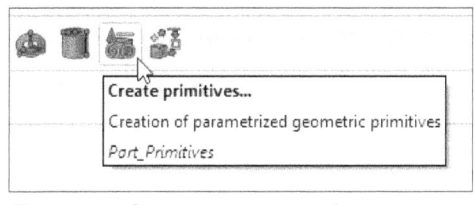

Figure-19. Create primitives tool

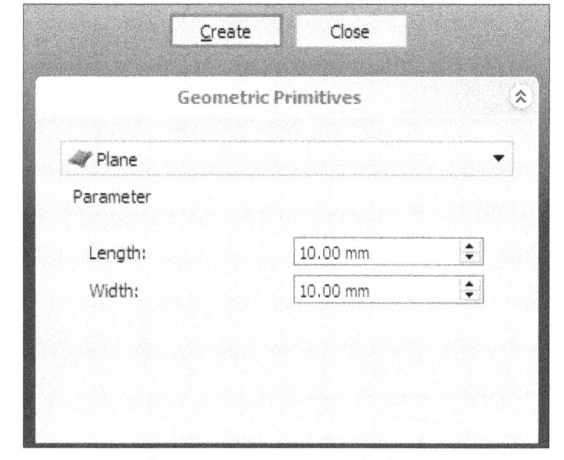

Figure-20. Geometric Primitives dialog

- Click on the **Primitive type** drop-down in the dialog. The list of available parametric geometric primitives will be displayed; refer to Figure-21.

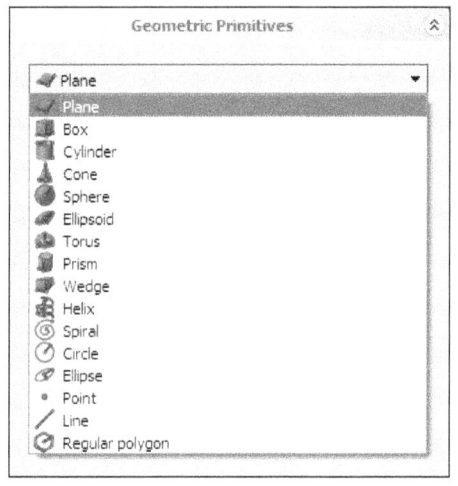

Figure-21. List of available geometric primitives

- Select desired primitive type from the drop-down to be created. The parameters related to the selected primitive type will be displayed in the dialog.
- Specify the parameters as desired and click on the **Create** button from the dialog. The selected geometric primitive will be created; refer to Figure-22.

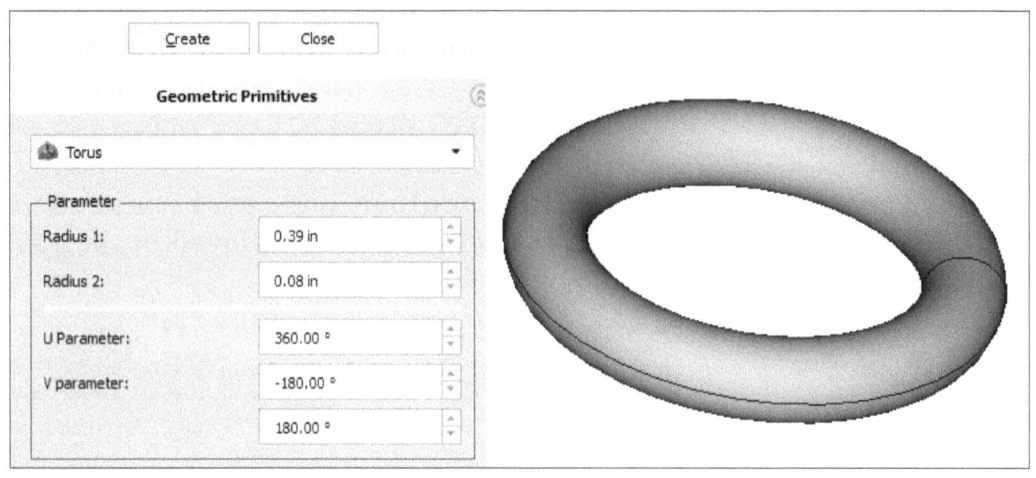

Figure-22. Geometric primitive created

- Specify all the parameters as desired in the dialog and click on **Close** button to close the dialog.

Shape builder

The **Shape builder** tool is used to create more complex shapes from various parametric geometric primitives. The procedure to use this tool is discussed next.

- Click on the **Shape builder** tool from **Toolbar** in the **Part** workbench; refer to Figure-23. The **Create shape** dialog will be displayed in the **Tasks** panel of **Combo View**; refer to Figure-24.

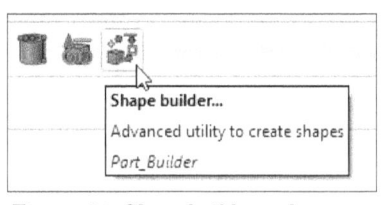

Figure-23. Shape builder tool

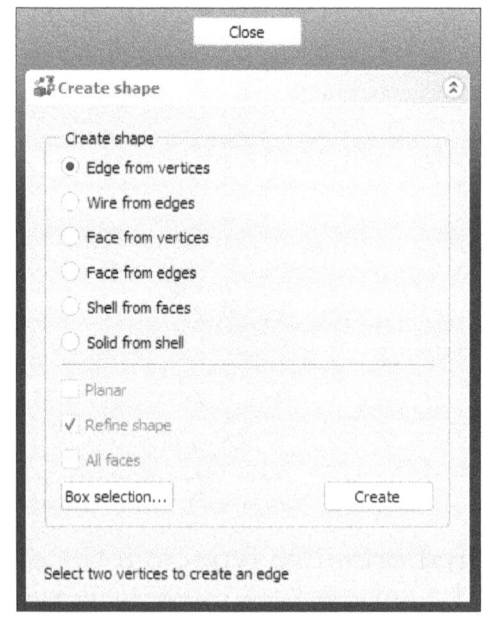

Figure-24. Create shape dialog

- Select desired radio button from **Create shape** area of the dialog for the shape to be created. The conditions to create the selected shape will be displayed at the bottom of the dialog; refer to Figure-24.
- Select the objects as required for the shape and click on the **Create** button from the dialog. The shape will be created; refer to Figure-25.

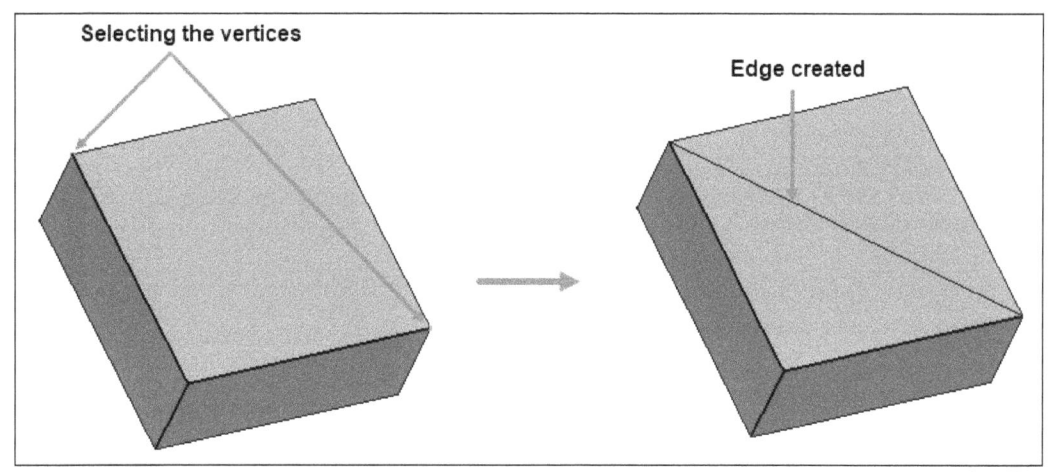

Figure-25. Edge shape created

- Click on **Close** button to close the dialog.

Modifying Object tools

The **Modifying object** tools are used for modifying existing objects. They will allow you to choose which object to modify. These tools are available in the **Toolbar** of **Part** workbench; refer to Figure-26. The procedures to use these tools are discussed next.

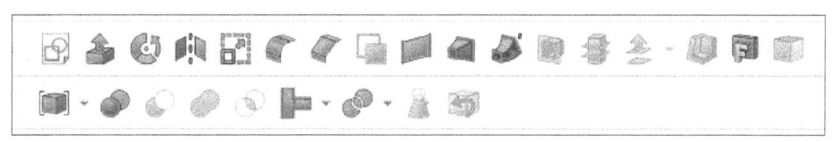

Figure-26. Modifying object tools

Extrude

The **Extrude** tool extends a shape by a specified distance in a specified direction. The output shape type will vary depending on the input shape type and the options selected. The procedure to use this tool is discussed next.

- Select desired shape from the 3D view area or from the **Model Tree** which you want to extrude; refer to Figure-27.

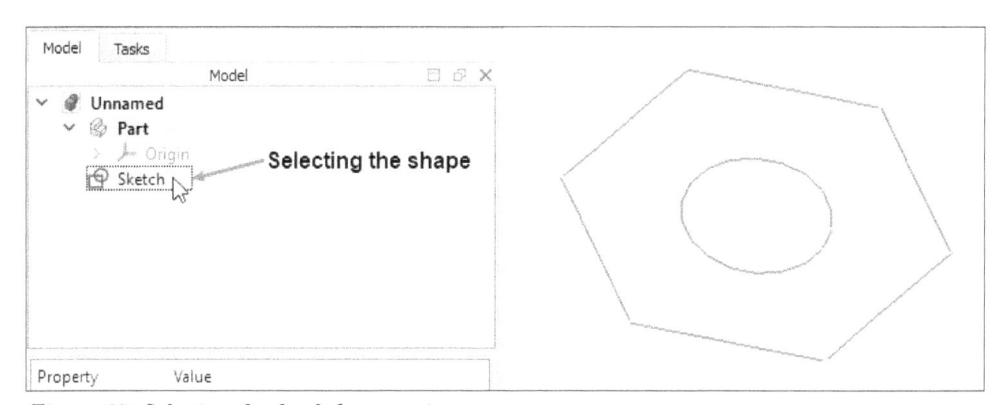

Figure-27. Selecting the sketch for extrusion

- Click on the **Extrude** tool from **Toolbar** in the **Part** workbench; refer to Figure-28. The **Extrude** dialog will be displayed in the **Tasks** panel of **Combo View**; refer to Figure-29.

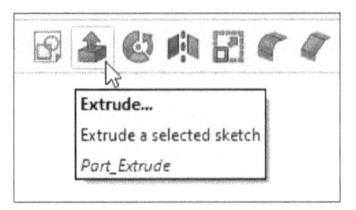

Figure-28. Extrude tool

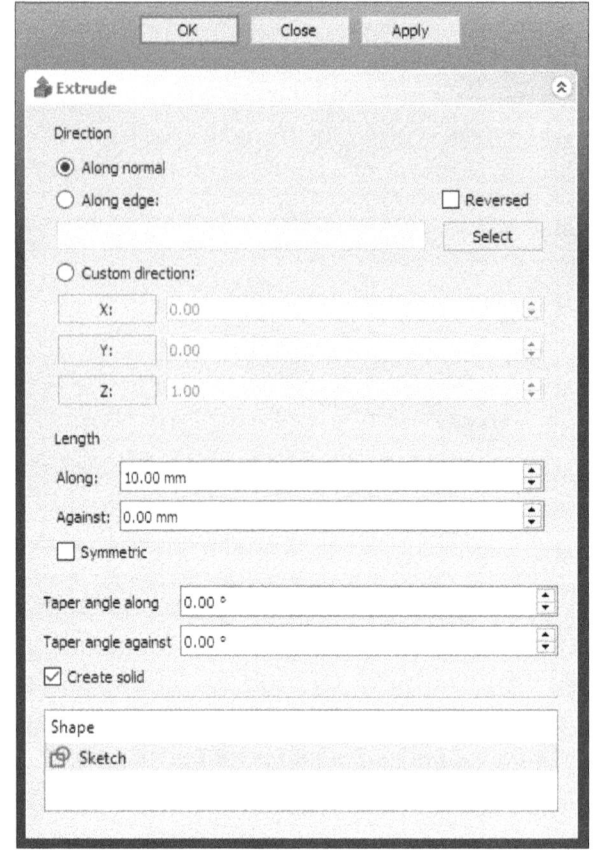

Figure-29. Extrude dialog

- Select the **Along normal** radio button from **Direction** area of the dialog to create the extrusion in a normal direction.
- Select the **Along edge** radio button and click on **Select** button below the radio button to select the edge along the direction of which the extrusion will be created.
- Select the **Reversed** check box next to the **Along edge** radio button to reverse the direction of extrusion along an edge.
- Select the **Custom direction** radio button and specify desired value in **X**, **Y**, and **Z** edit boxes to specify the extrusion direction along x, y, and z axis, respectively.
- Specify desired length of extrusion along normal direction in the **Along** edit box from **Length** area of the dialog.
- Specify the length of extrusion along reverse direction in the **Against** edit box.
- Select the **Symmetric** check box to create the extrusion in both the directions.
- Specify desired value in **Taper angle along** edit box to apply taper angle along the extrusion direction and specify desired value in **Taper angle against** edit box to apply taper angle against the extrusion direction.
- Select the **Create solid** check box to extrude the closed wire or edge as a solid model. The selected shape to be extruded will display in the **Shape list** area of the dialog. If you want to select more shapes to extrude then select desired shape from the **Shape list** area.
- Click on **Apply** button from the dialog. The preview of extrusion will be created; refer to Figure-30.

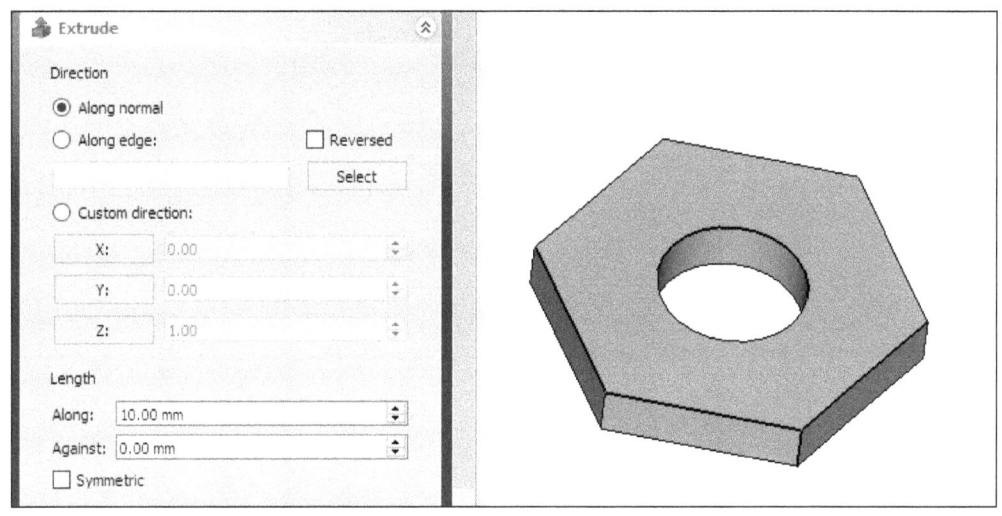

Figure-30. Preview of extrusion

- Click on **OK** button to close the dialog. The extrusion will be created.

Revolve

The **Revolve** tool revolves selected object about an axis. The procedure to use this tool is discussed next.

- Select desired shape from 3D view area or from the **Model Tree** which you want to revolve; refer to Figure-31.

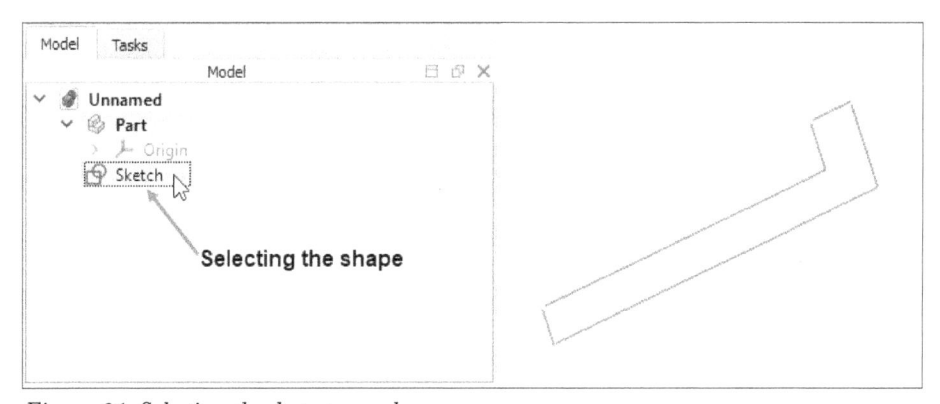

Figure-31. Selecting the shape to revolve

- Click on the **Revolve** tool from **Toolbar** in the **Part** workbench; refer to Figure-32. The **Revolve** dialog will be displayed in the **Tasks** panel of **Combo View**; refer to Figure-33.

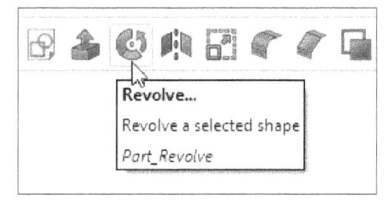

Figure-32. Revolve tool

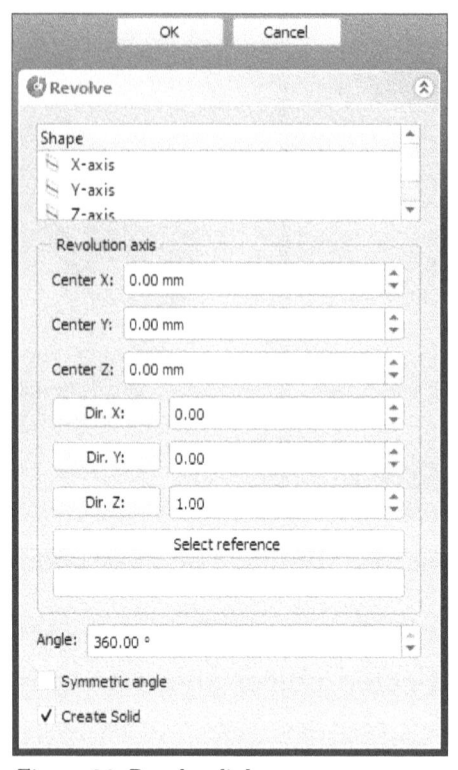

Figure-33. Revolve dialog

- The selected shape will display in the **Shape list** area of the dialog. If you want to select more shapes then select the shape from shape list area.
- Specify the parameters as desired from **Revolution axis** area of the dialog.
- Click on the **Select reference** button from **Revolution axis** area. You are asked to select the line or an arc.
- Select the line or an arc from 3D view area which you want to make reference for the revolution of shape; refer to Figure-34.

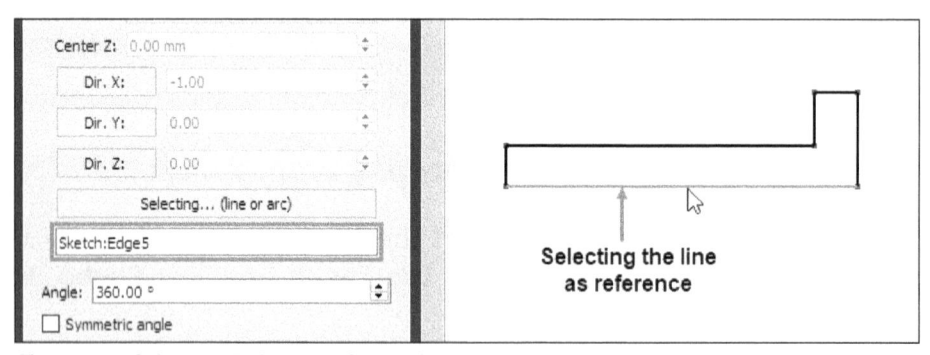

Figure-34. Selecting the line as reference for revolution

- Specify desired angle of revolution in the **Angle** edit box.
- Select the **Symmetric angle** check box to revolve the shape in both the direction at the given angle.
- Select the **Create solid** check box to revolve the shape as a solid model.
- Click on **OK** button from the dialog. The revolved model of the shape will be created; refer to Figure-35.

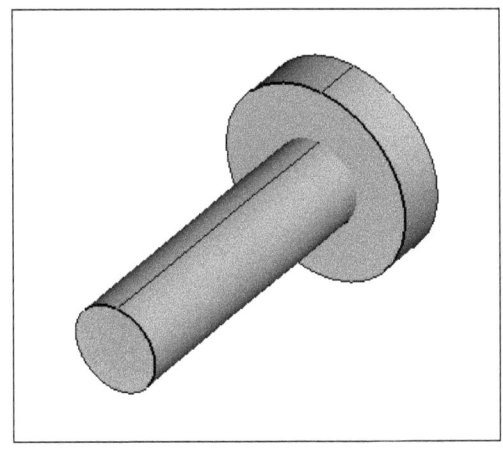

Figure-35. Revolved model of the shape created

Mirroring

The **Mirroring** tool creates a new object (image) which is a reflection of the original object (source). The image object is created on the other side of mirror plane. The procedure to use this tool is discussed next.

- Click on the **Mirroring** tool from **Toolbar** in the **Part** workbench; refer to Figure-36. The **Mirroring** dialog will be displayed in the **Tasks** panel of **Combo View**; refer to Figure-37.

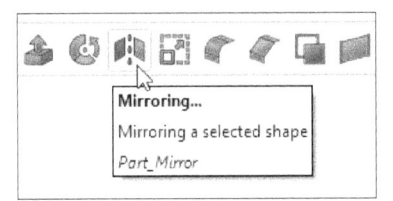

Figure-36. Mirroring tool

Figure-37. Mirroring dialog

- Select desired shape which you want to mirror from **Shapes list** area of the dialog.
- Select desired mirror plane from the **Mirror plane** drop-down.
- Specify desired value in **x**, **y**, and **z** edit boxes from the **Base point** area of the dialog to move the mirror plane parallel to the selected standard mirror plane.
- Click on **OK** button from the dialog. The mirrored entity of the shape will be created; refer to Figure-38.

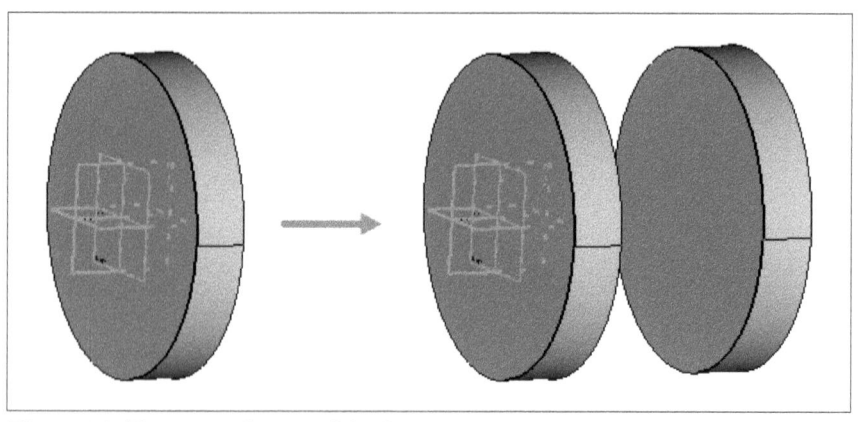

Figure-38. The mirrored entity of the shape

Scale

The **Scale** tool is used to increase/decrease the size one or more shapes by specified factor. The procedure to use this tool is discussed next.

- Select desired object from the **Model Tree** or from the 3D view area to be scaled; refer to Figure-39.

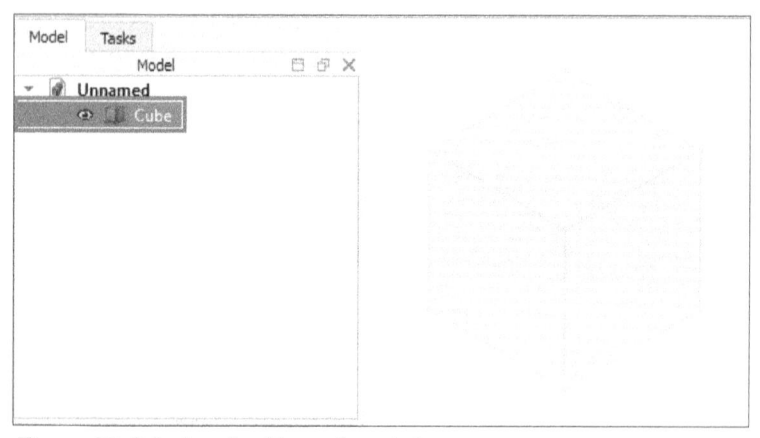

Figure-39. Selecting the object to be scaled

- Click on the **Scale** tool from **Toolbar** in the **Part** workbench; refer to Figure-40. The **Scale** dialog will be displayed in the **Tasks** panel of **Combo View**; refer to Figure-41.

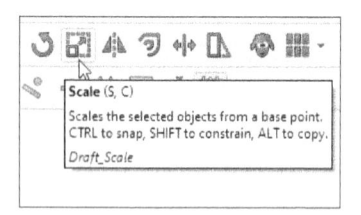

Figure-40. Scale tool

Figure-41. Scale dialog

- Select **Uniform Scaling** radio button to scale the object by single factor in all the directions and specify desired value of factor in the **Factor** edit box.
- Select **Non-Uniform Scaling** radio button to scale the object by different scale factor in each direction and specify desired scale values in the **X Factor**, **Y Factor**, and **Z Factor** edit boxes.
- Select desired shape from **Shape** list box in the dialog and click on the **Apply** or **OK** button. The shape will be scaled; refer to Figure-42.

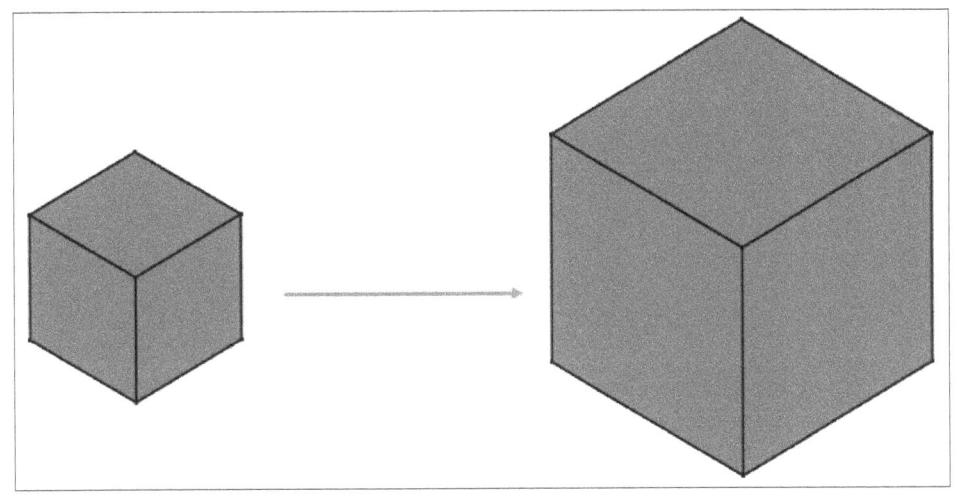

Figure-42. Shape scaled

Fillet

The **Fillet** tool creates a fillet (round) on the selected edges of an object. The procedure to use this tool is discussed next.

- Click on the **Fillet** tool from **Toolbar** in the **Part** workbench; refer to Figure-43. The **Fillet Edges** dialog will be displayed in the **Tasks** panel of **Combo View**; refer to Figure-44.

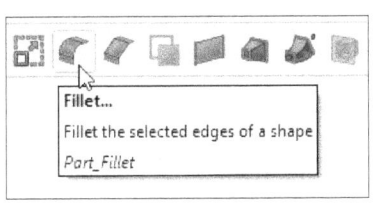

Figure-43. Fillet tool

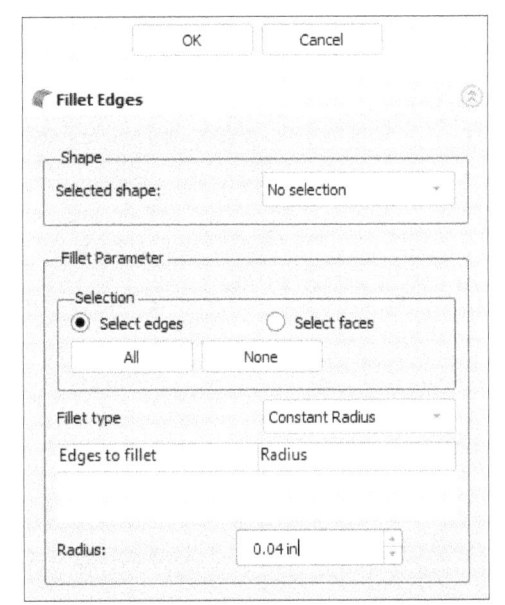

Figure-44. Fillet Edges dialog

- Click on the **Selected shape** drop-down from **Shape** area of the dialog and select the shape from the drop-down list you want to be fillet. All the edges of the selected shape will be displayed in the **Shapes list** area of the dialog.
- If you want to select the edges of the shape then select the **Select edges** radio button from **Selection** area in the **Fillet Parameter** section of the dialog and if you want to select the faces of the shape then select the **Select faces** radio button from the **Selection** area.
- Click on the **All** button from the **Selection** area to select all the edges available in the **Shapes list** area of the dialog and click on **None** button to deselect all the edges from the **Shapes** list area; refer to Figure-45.

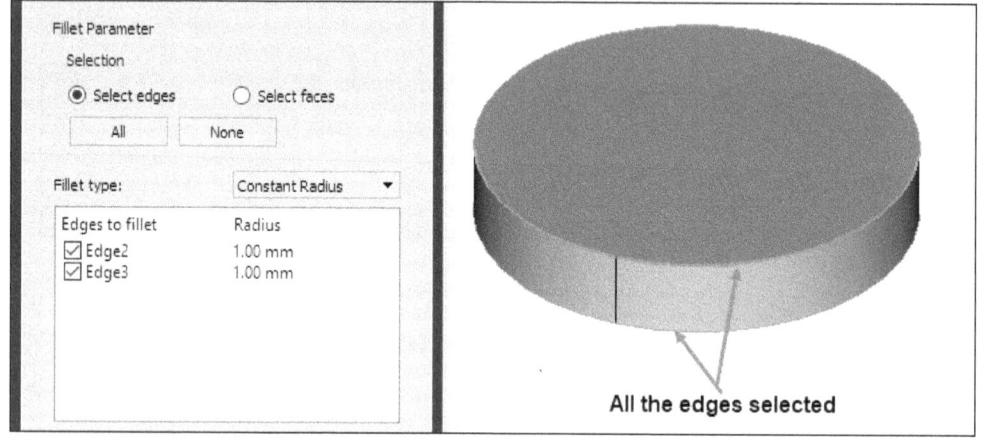

Figure-45. All the edges selected to be fillet

- Select desired type of fillet from **Fillet type** drop-down in the **Fillet Parameter** section. Select **Constant Radius** option from **Fillet type** drop-down to specify same fillet radius on both ends of the edges. Select **Variable Radius** option from the drop-down to specify different radius values at start and end points of the edges.
- Click on **OK** button from the dialog. The fillets will be created on the edges; refer to Figure-46.

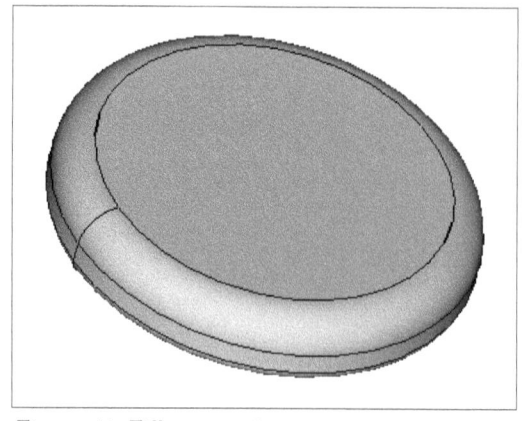

Figure-46. Fillets created

Chamfer

The **Chamfer** tool chamfers selected edges of an object. The procedure to use this tool is discussed next.

- Click on the **Chamfer** tool from **Toolbar** in the **Part** workbench; refer to Figure-47. The **Chamfer Edges** dialog will be displayed in the **Tasks** panel of **Combo View**; refer to Figure-48.

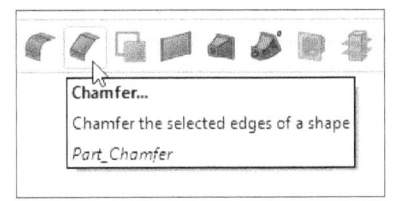

Figure-47. Chamfer tool

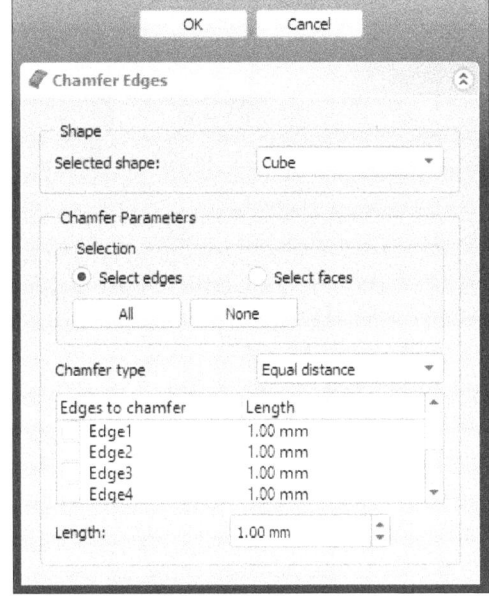

Figure-48. Chamfer Edges dialog

- The procedure to create the chamfer is same as discussed for **Fillet** tool.
- Specify the parameters and click on **OK** button from the dialog. The chamfers will be created on the edges; refer to Figure-49.

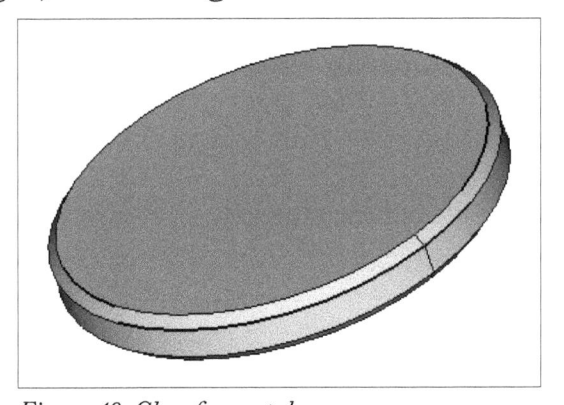

Figure-49. Chamfer created

Make face from wires

The **Make face from wires** tool is used to create face from closed wireframe (sketching) curves in custom shape. The procedure to use this tool is discussed next.

- Select desired closed sketch from the **Model Tree** or from the 3D view area that you want to make a face; refer to Figure-50.

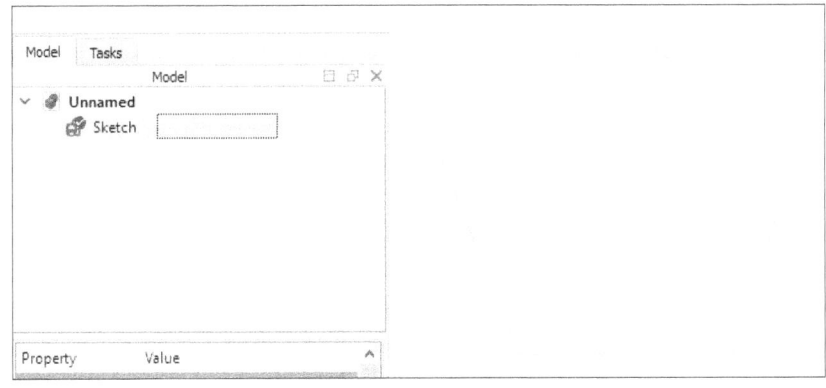

Figure-50. Sketch selected to make face

- Click on the **Make face from wires** tool from **Toolbar** in the **Part** workbench; refer to Figure-51. The face will be created; refer to Figure-52.

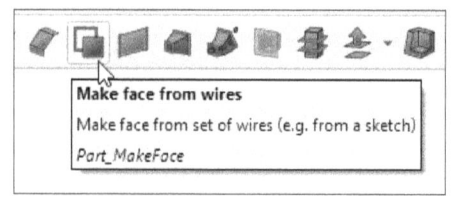

Figure-51. Make face from wires tool

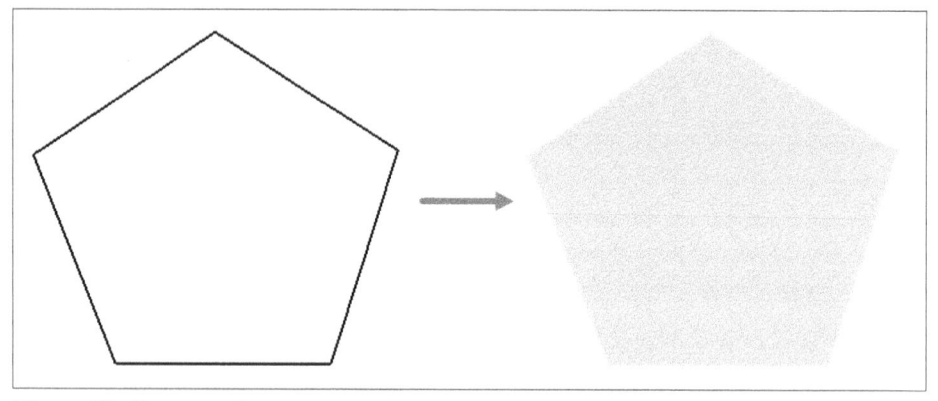

Figure-52. Face created

Create ruled surface

The **Create ruled surface** tool is used to create a ruled surface using two edges or sketch curves. The procedure to use this tool is discussed next.

- Select the two edges or two wires from the 3D view area; refer to Figure-53.

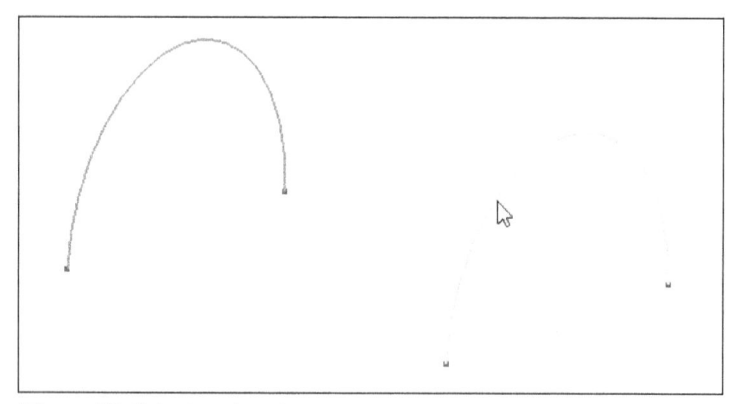

Figure-53. Selecting the wires

- Click on the **Ruled Surface** tool from **Toolbar** in the **Part** workbench; refer to Figure-54. The ruled surface will be created; refer to Figure-55.

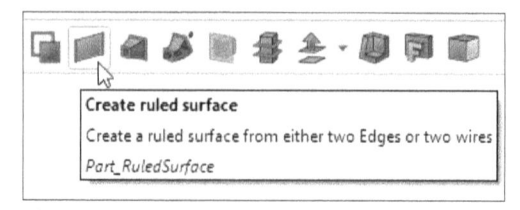

Figure-54. Create ruled surface tool

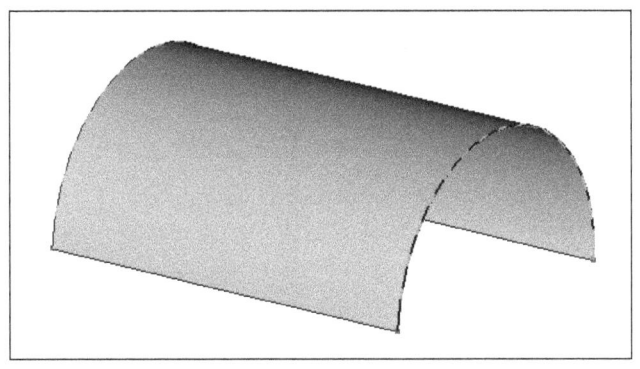

Figure-55. Ruled surface created

Loft

The **Loft** tool is used to create a face, shell, or a solid shape from two or more profiles. The profiles can be a point (vertex), line (Edge), wire, or face. Edges and wires may be either open or closed. The procedure to use this tool is discussed next.

- Click on the **Loft** tool from **Toolbar** in the **Part** workbench; refer to Figure-56. The **Loft** dialog will be displayed in the **Tasks** panel of **Combo View**; refer to Figure-57.

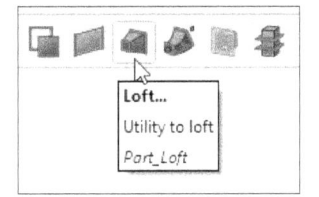

Figure-56. Loft tool

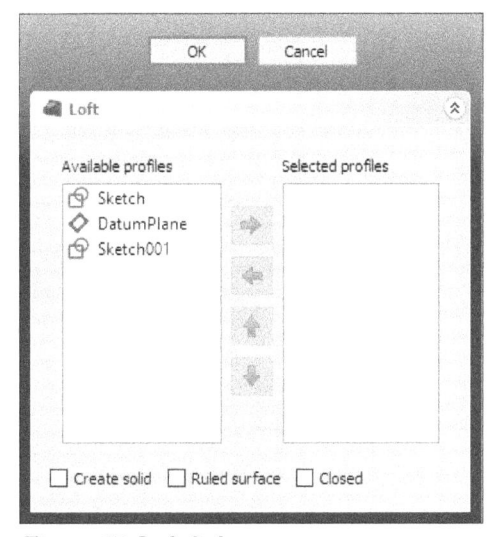

Figure-57. Loft dialog

- The list of profiles which are available to be loft will display in the **Available profiles** list area in the dialog.
- To create the loft feature, add the two or more available profiles one by one in the **Selected profiles** list area by clicking on **Add** button.
- Select desired profile from the **Selected profiles** list area as a base profile for loft; refer to Figure-58.

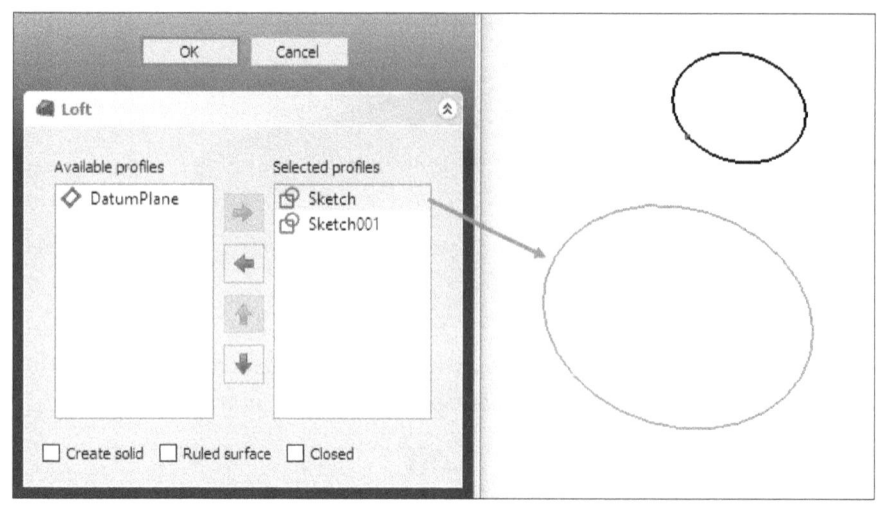

Figure-58. Selecting the base profile for loft

- Select the **Create solid** check box to create solid model of the profiles if the profiles are of closed geometry. If the profiles are open geometry then it creates a face or shell of the profiles.
- Select the **Ruled surface** check box to create a face or ruled surface using the profiles.
- Select the **Closed** check box to loft the last profile to the first profile creating a closed body.
- Click on **OK** button from the dialog. The loft feature will be created; refer to Figure-59.

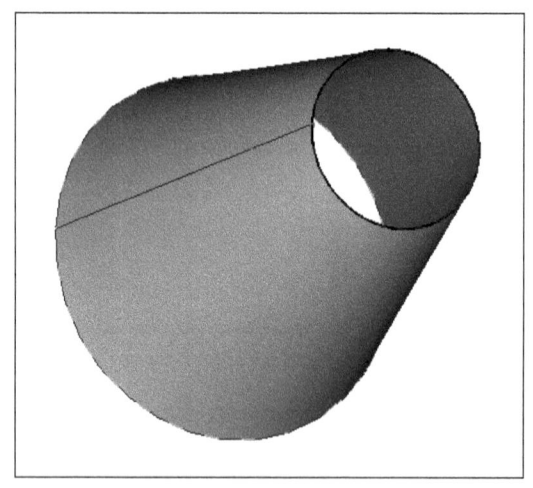

Figure-59. Loft feature created

Sweep

The **Sweep** tool is used to create a face, a shell, or a solid shape from one or more profiles (cross-sections) projected along a path. The procedure to use this tool is discussed next.

- Click on the **Sweep** tool from **Toolbar** in the **Part** workbench; refer to Figure-60. The **Sweep** dialog will be displayed in the **Tasks** panel of **Combo View**; refer to Figure-61.

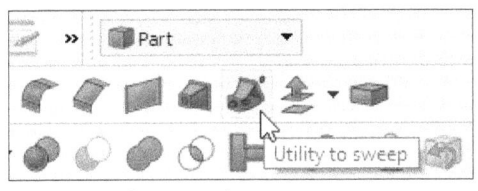

Figure-60. Sweep tool

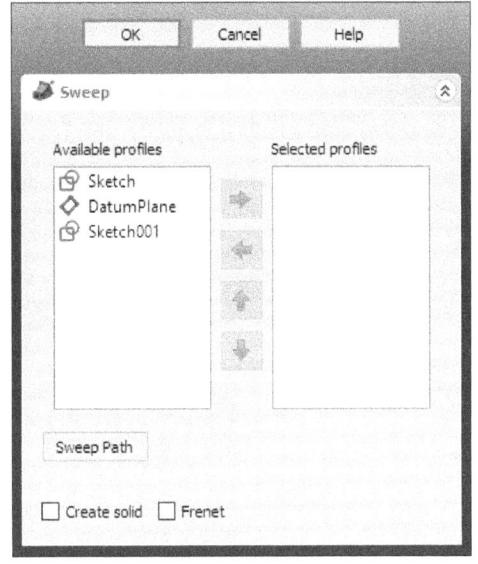

Figure-61. Sweep dialog

- Select the profile to be used as a base profile for sweep feature from the **Available profiles** list area of the dialog and add it in the **Selected profiles** list area by clicking on **Add** button.
- Select the profile from the **Selected profiles** list area and click on **Sweep Path** button. You are asked to select the profile for sweep path from the 3D view area.
- Select the profile from the 3D view area to be used as path for creating a sweep feature; refer to Figure-62.

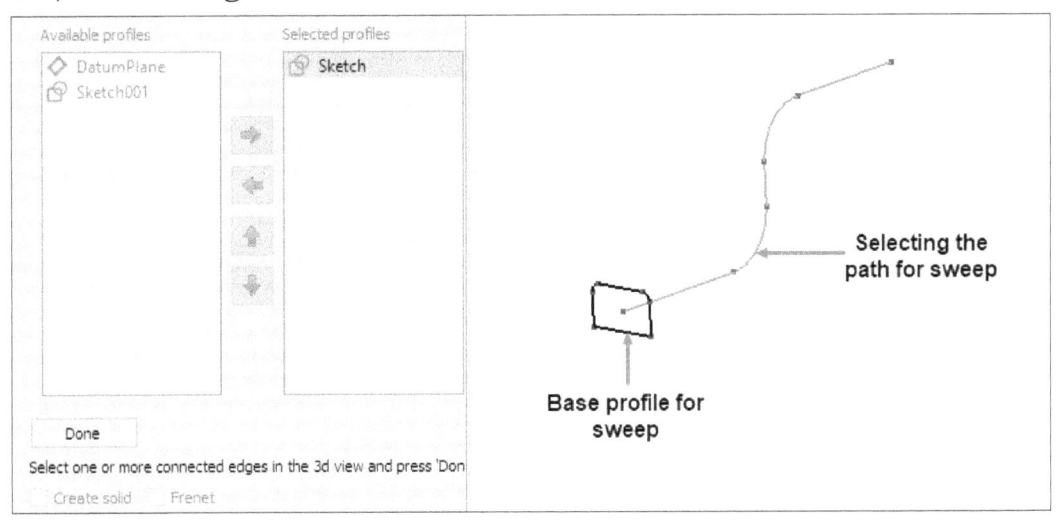

Figure-62. Selecting the path for sweep feature

- Click on **Done** button from the dialog.
- Select the **Create solid** check box to create the solid model of the profiles, if the profiles are of closed geometry. If the profiles are of open geometry, it creates a face or shell of the profiles.
- Select the **Frenet** check box to compute the profile orientation basing on local curvature and tangency vectors of the path.
- Click on **OK** button from the dialog. The sweep feature will be created; refer to Figure-63.

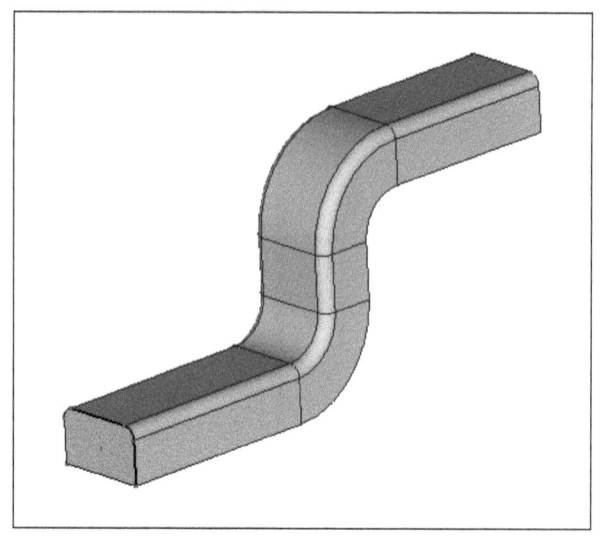

Figure-63. Sweep feature created

Section

The **Section** tool is used to create a section by intersecting an object with a section plane. The procedure to use this tool is discussed next.

- Select the model and section plane from the **Combo View** and then click on the **Section** tool from the **Toolbar** in the **Part** workbench; refer to Figure-64. The section view will be created; refer to Figure-65.

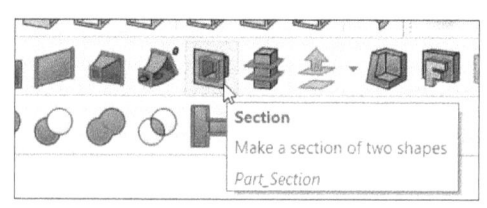

Figure-64. Section tool

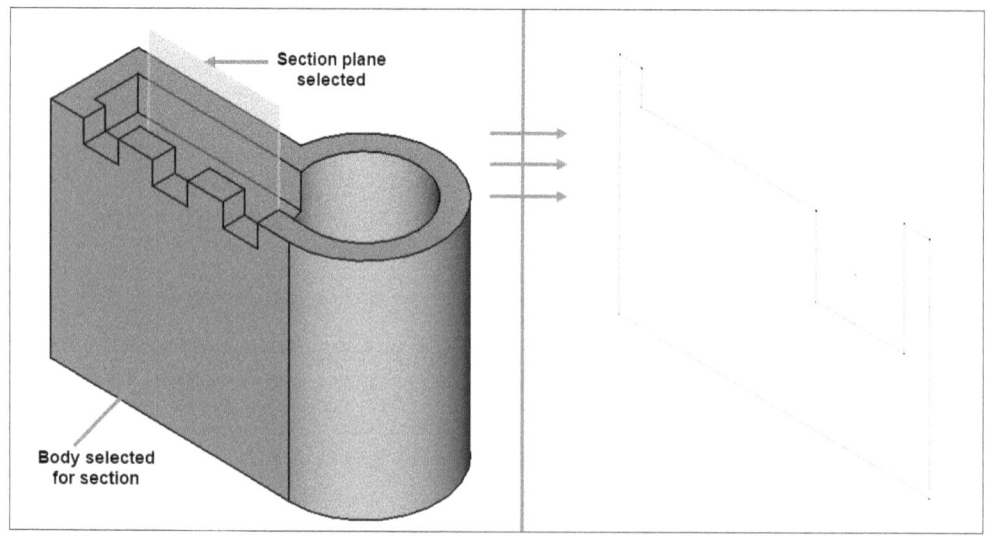

Figure-65. Section created

Cross-Sections

The **Cross-sections** tool creates one or more cross-sections of selected body. The procedure to use this tool is discussed next.

- Select the shape from which you want to create cross-sections from the **Model Tree** or from the 3D view area; refer to Figure-66.

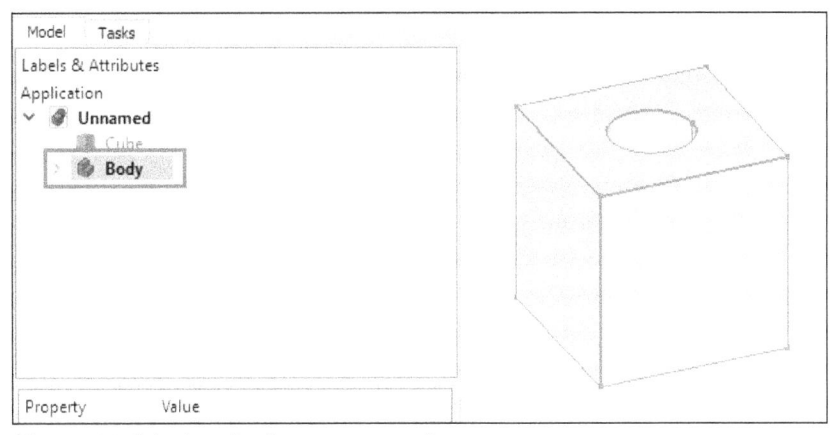

Figure-66. Selecting the shape to cross-section

- Click on the **Cross-sections** tool from **Toolbar** in the **Part** workbench; refer to Figure-67. The **Cross sections** dialog will be displayed in the **Tasks** panel of **Combo View** along with the preview of cross section; refer to Figure-68.

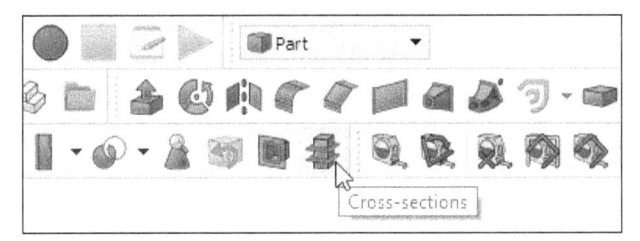

Figure-67. Cross sections tool

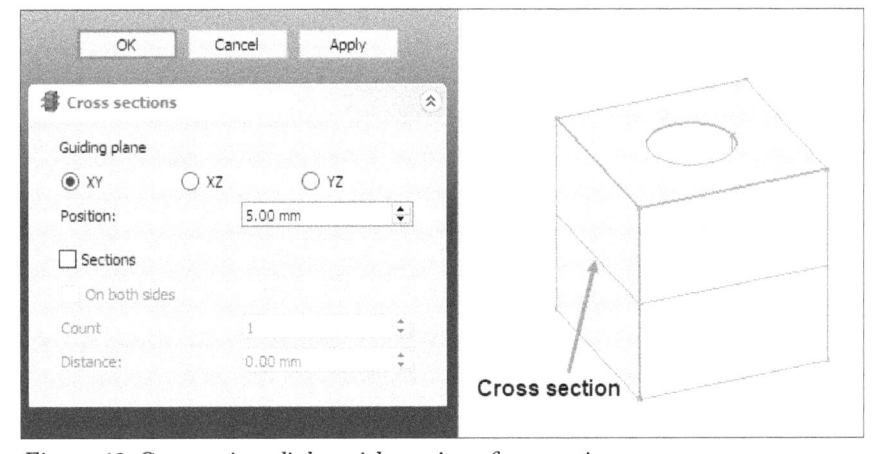

Figure-68. Cross sections dialog with preview of cross section

- Select desired guiding plane of creating the cross-section from the **Guiding plane** section of the dialog. Select the **XY**, **XZ**, or **YZ** radio button to create the cross-section along xy, xz, or yz plane, respectively.
- Specify desired position of cross section in the **Position** edit box.
- Select the **Sections** check box to edit the parameters of cross-section.

- Select **On both sides** check box to create the cross-section on both sides of the selected guiding plane.
- Specify the number of cross sections to be created in the **Count** edit box.
- Specify distance between the cross sections in the **Distance** edit box of the dialog.
- Click on the **Apply** button. The cross sections of the shape will be created; refer to Figure-69.

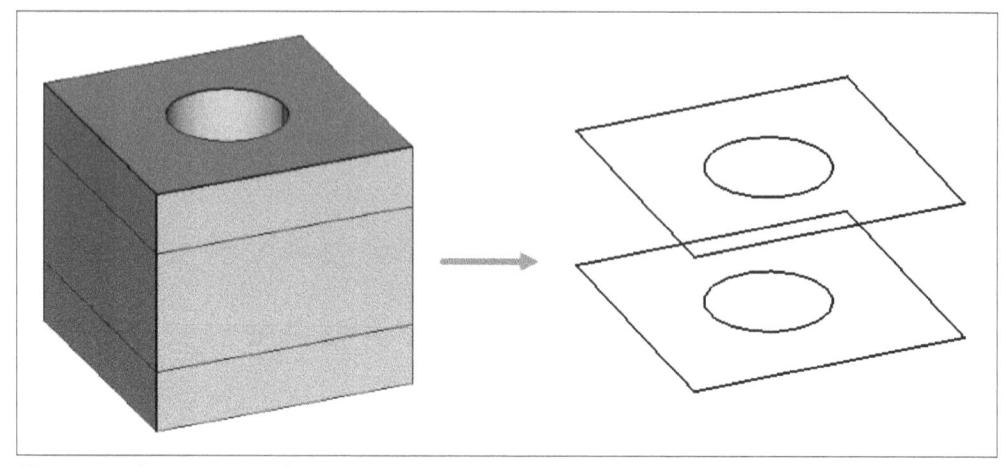

Figure-69. Cross sections of the shape created

- Click on **OK** button to close the dialog.

Offset tools

The **Offset** tools are used to create 2D as well as 3D copies of selected shapes. The tools available to create 2D and 3D offsets are discussed next.

3D Offset

The **3D Offset** tool creates parallel copies of selected shape at a certain distance from the base shape, giving a new object. The procedure to use this tool is discussed next.

- Select the object from the **Model Tree** or from the 3D view area of which you want to create the offset; refer to Figure-70.

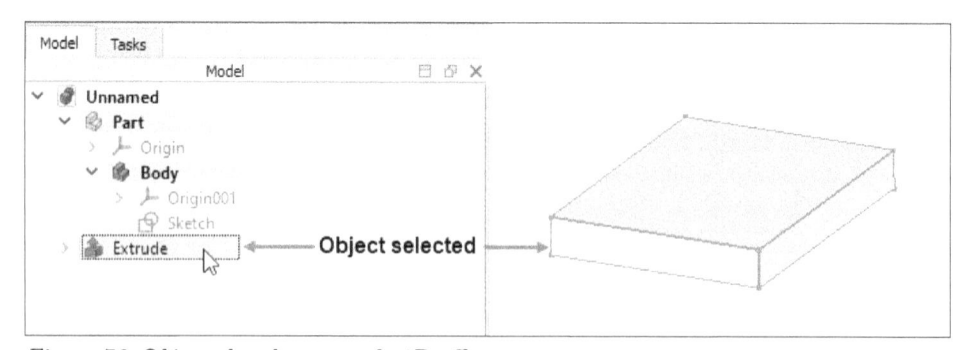

Figure-70. Object selected to create the 3D offset

- Click on the **3D Offset** tool from **Offset** drop-down in the **Toolbar** of **Part** workbench; refer to Figure-71. The **Offset** dialog will be displayed in the **Tasks** panel of **Combo View** along with the preview of offset creation; refer to Figure-72.

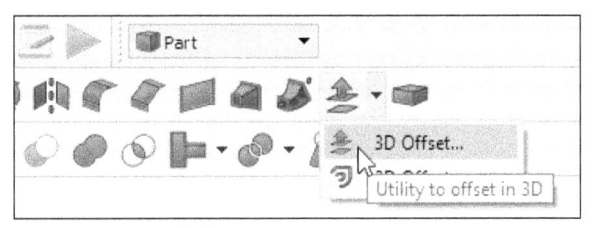

Figure-71. 3D Offset tool

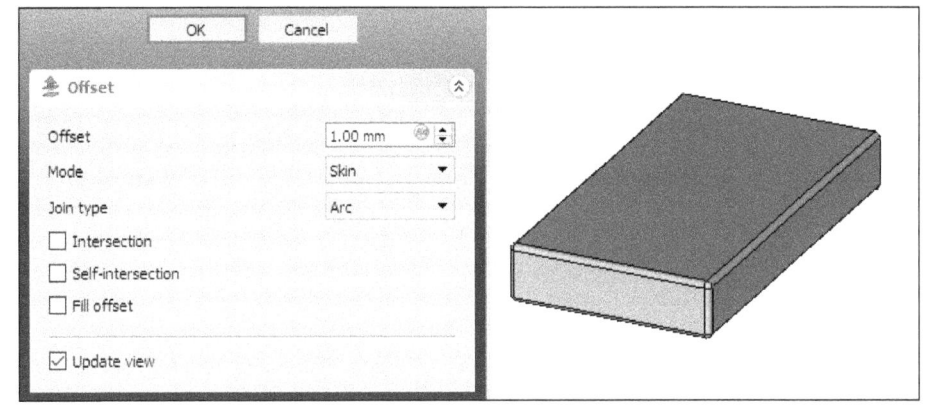

Figure-72. 3D Offset dialog with the preview of offset creation

- Specify desired value in the **Offset** edit box to define distance between original and offsetted faces.
- Select desired mode of offset creation from the **Mode** drop-down. Select the **Skin** option to create a new shape around the source shape. Select the **Pipe** option to create a new shape like a pipe, headless, and bottomless. Select the **Recto Verso** option to flip the side of offset.
- Select desired option from the **Join type** drop-down. Select the **Arc** and **Tangent** option to create the offset of rounded corners. Select the **Intersection** option to create the offset of sharp corners by linear extension of the edges.
- Select the **Intersection** check box to allow the offsets pointing inwards to over flood the gap by intersecting the resulting shape until opposite faces are reached.
- Select the **Self-intersection** check box to allow creation of self-intersecting body after apply offset operation.
- Select the **Fill offset** check box while using the 2 dimensional shape in which the gap in between the two shapes gets filled.
- Click on **OK** button from the dialog. The 3D offset of the original object will be created; refer to Figure-73.

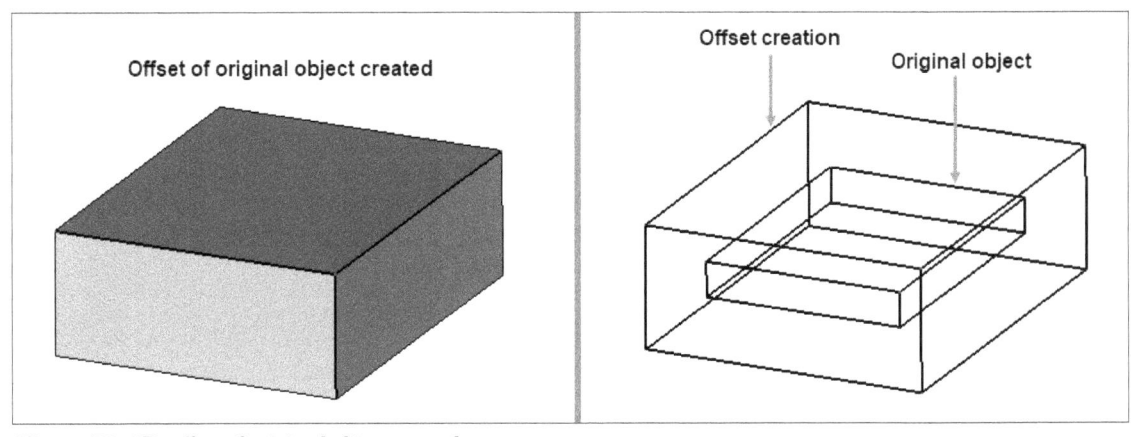

Figure-73. 3D offset of original object created

2D Offset

The **2D Offset** tool constructs a wire parallel to the original wire at a certain distance from it or enlarges/shrinks a planar face. The procedure to use this tool is discussed next.

- Select the object from the **Model Tree** or from the 3D view area which you want to offset; refer to Figure-74.

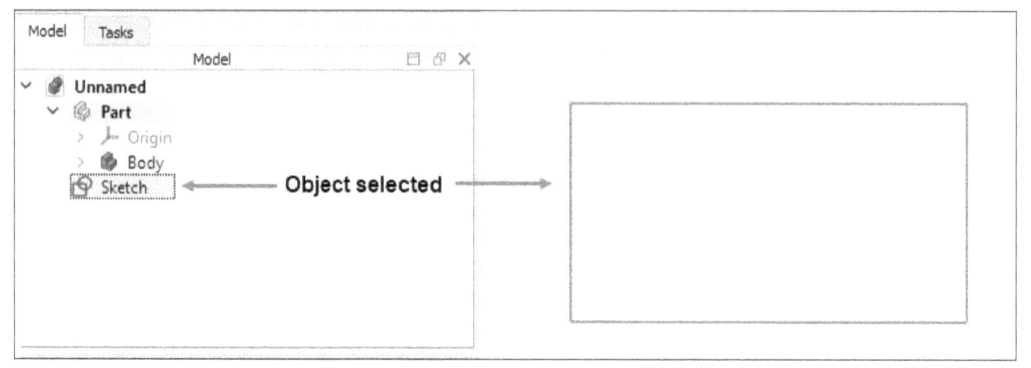

Figure-74. Object selected to create the 2D offset

- Click on the **2D Offset** tool from the **Offset** drop-down in the **Toolbar** of **Part** workbench; refer to Figure-75. The **Offset** dialog will be displayed in the **Tasks** panel of **Combo View** along with the preview of offset creation; refer to Figure-76.

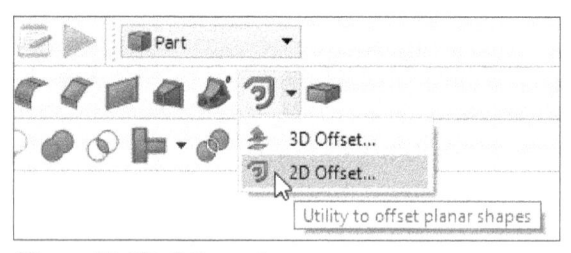

Figure-75. 2D Offset tool

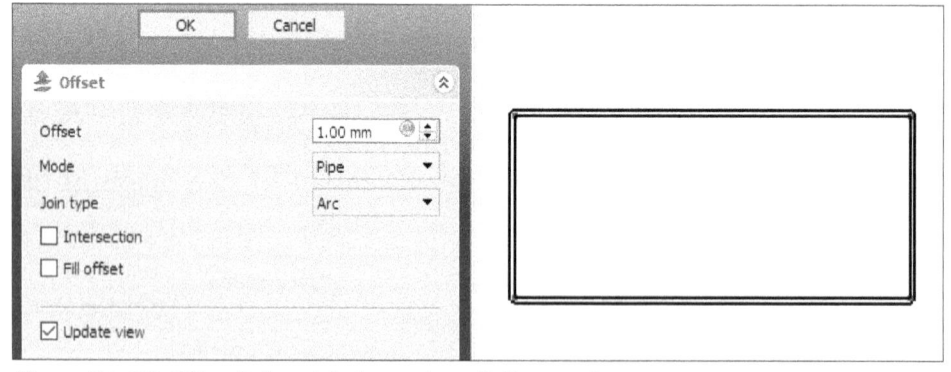

Figure-76. 2D Offset dialog with the preview of offset creation

- The parameters in the **2D Offset** dialog are same as discussed for **3D Offset** tool.
- Specify the parameters and click on **OK** button from the dialog. The 2D offset of the original object will be created; refer to Figure-77.

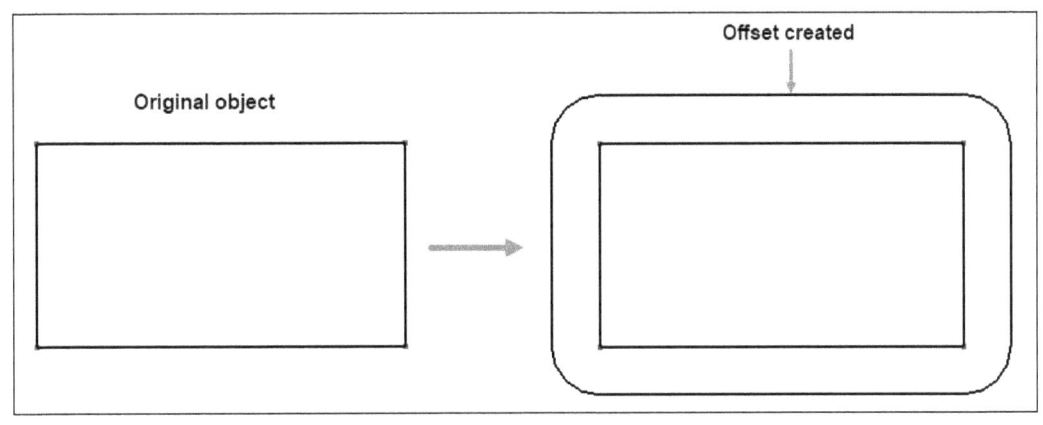

Figure-77. 2D offset of original object created

Thickness

The **Thickness** tool works on a solid shape and transforms it into a hollow object giving each of its faces a defined thickness. The procedure to use this tool is discussed next.

- Select one or more faces of the solid object from the 3D view area which you want to remove after applying thickness; refer to Figure-78.

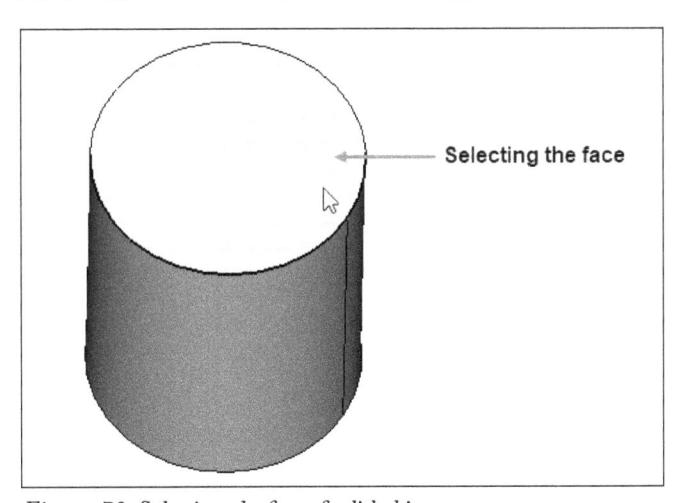

Figure-78. Selecting the face of solid object

- Click on the **Thickness** tool from **Toolbar** in the **Part** workbench; refer to Figure-79. The **Thickness** dialog will be displayed in the **Tasks** panel of **Combo View** along with the preview of thickness defined to the object; refer to Figure-80.

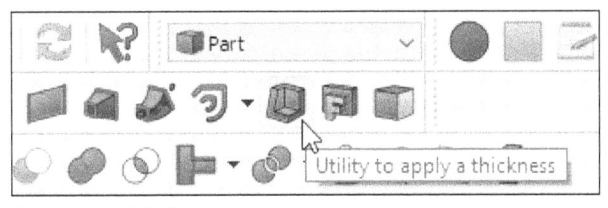

Figure-79. Thickness tool

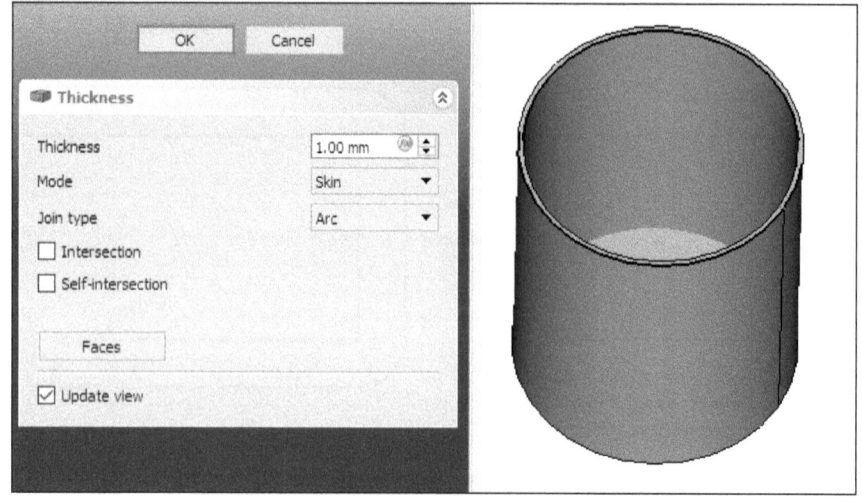

Figure-80. Thickness dialog with the preview of thickness defined

- The parameters of **Thickness** dialog are same as discussed for **Offset** tools.
- Specify the parameters and click on **OK** button from the dialog. The thickness will be defined to the object; refer to Figure-81.

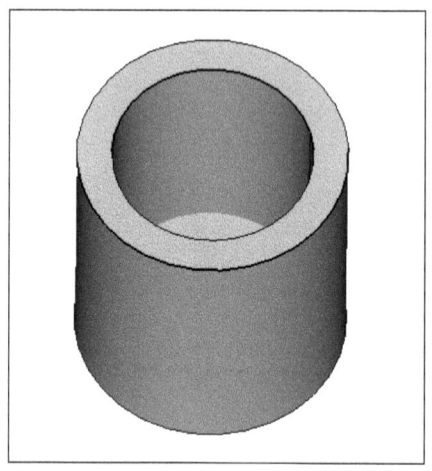

Figure-81. Thickness defined to the object

Create projection on surface

The **Create projection on surface** tool is used to project a shape on top of a face from another object. This can be used to project a logo or textual object onto different surfaces to create interesting effects. The procedure to use this tool is discussed next.

- Click on the **Create projection on surface** tool from **Toolbar** in the **Part** workbench; refer to Figure-82. The **Projection on surface** dialog will be displayed in the **Tasks** panel of **Combo View**; refer to Figure-83, asking you to select the projection surface.

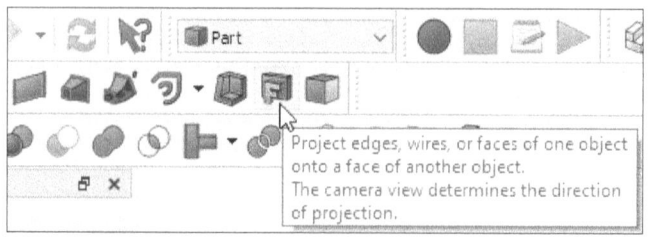

Figure-82. Create projection on surface tool

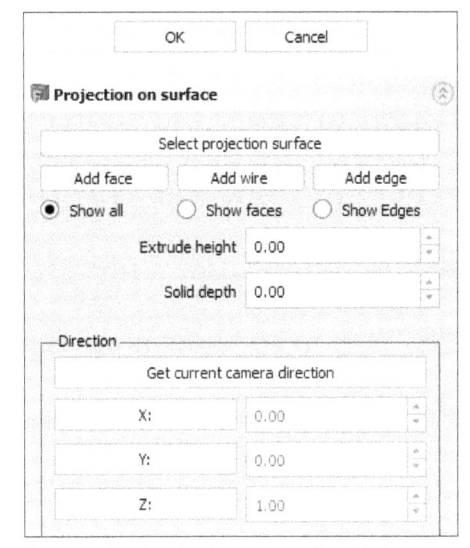

Figure-83. Projection on surface dialog

- Click on the **Select projection surface** button from the dialog and select the surface of the model on which you want to project the shape; refer to Figure-84. The other options in the dialog will be activated.

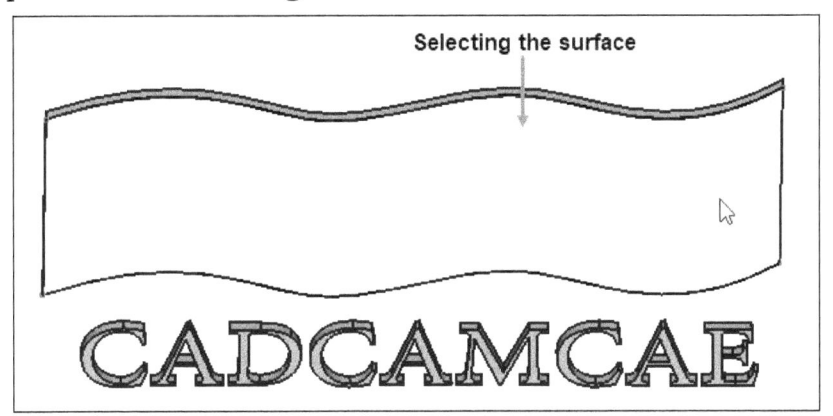

Figure-84. Selecting the surface to project the shape

- Click on the **Add face** button from the dialog and keep the surface parallel to the screen.
- Select the faces of the shape to project on the surface. The preview of projection will be displayed on the surface in real time; refer to Figure-85.

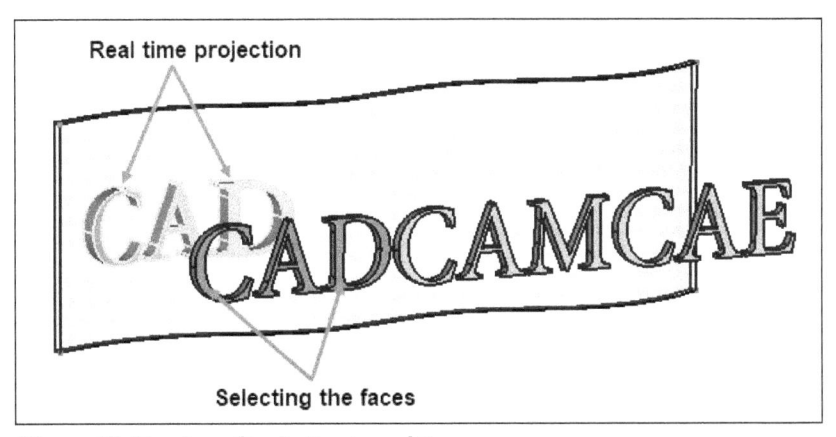

Figure-85. Preview of projection in real time

- Click on the **Add wire** button to select the source edge. This button will select the entire wire to which the edges belongs. For example, by choosing a single edge of a polygon, it will project the entire polygon.
- Click on the **Add edge** button to select the source edge. This button will only project the selected edge.
- Select the **Show all** radio button to show all types of closed wires and edges on the target surface.
- Select the **Show faces** radio button to show a preview of a filled face on the target surface.
- Select the **Show Edges** radio button to show a preview of the edges on the target surface.
- Specify desired height of the projected shape in the **Extrude height** edit box of the dialog.
- Specify desired depth of the projected face along the projection direction in the **Solid depth** edit box of the dialog.
- Click on the **Get current camera direction** button from **Direction** area of the dialog to change the projection direction. Alternatively, click on the **X:**, **Y:**, or **Z:** buttons to set the projection direction to the main global axes, **+X**, **-X**, **+Y**, **-Y**, **+Z**, or **-Z**.
- After specifying desired parameters, click on the **OK** button from the dialog to create the projection; refer to Figure-86.

Figure-86. Projection created

Color per face

The **Color per face** tool allows you to define a color for each face or surface of an object and set the properties of selected faces. The procedure to use this tool is discussed next.

- Select the object from the **Model Tree** on which you want to define the color; refer to Figure-87.

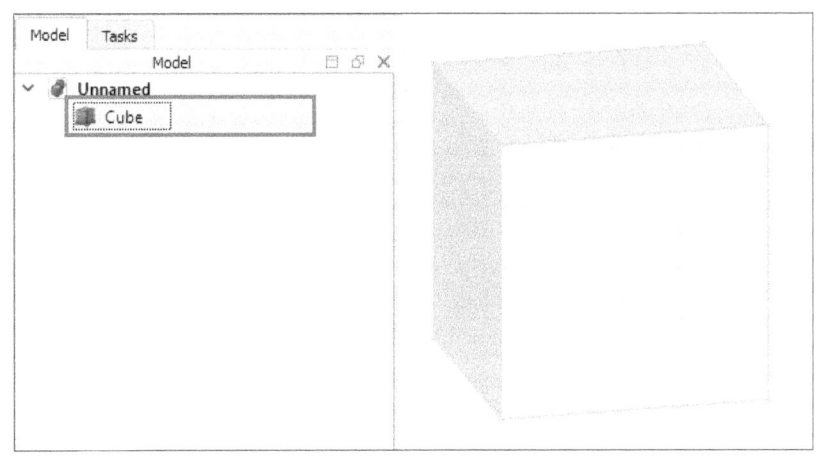

Figure-87. Selecting the object to define the color

- Click on the **Color per face** tool from **Toolbar** in the **Part** workbench; refer to Figure-88. The **Set appearance per face** dialog will be displayed in the **Tasks** panel of **Combo View**; refer to Figure-89.

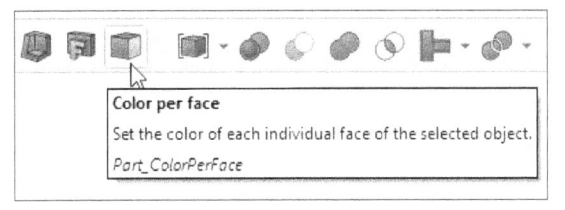

Figure-88. Color per face tool

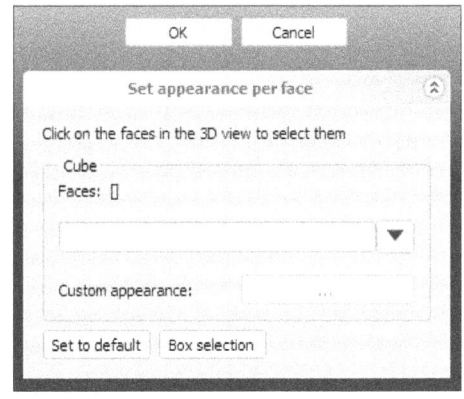

Figure-89. Set appearance per face dialog

- Select desired face(s) of the object in the 3D view area to define the color and material properties; refer to Figure-90.

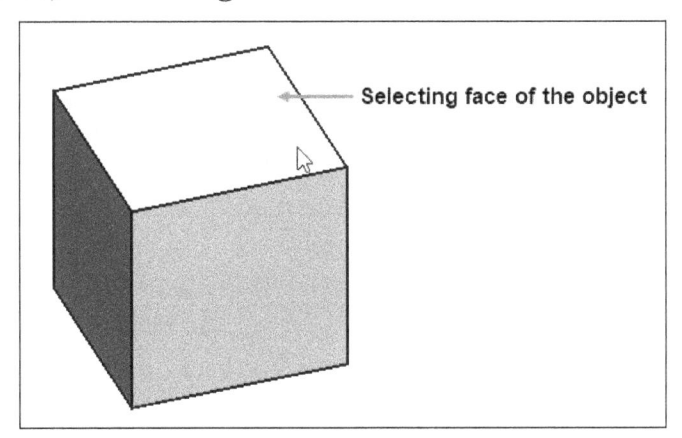

Figure-90. Selecting face of the object to define the color

- Click on the down arrow ▼ button from the dialog and select desired material from the list displayed to be applied to the face(s); refer to Figure-91.

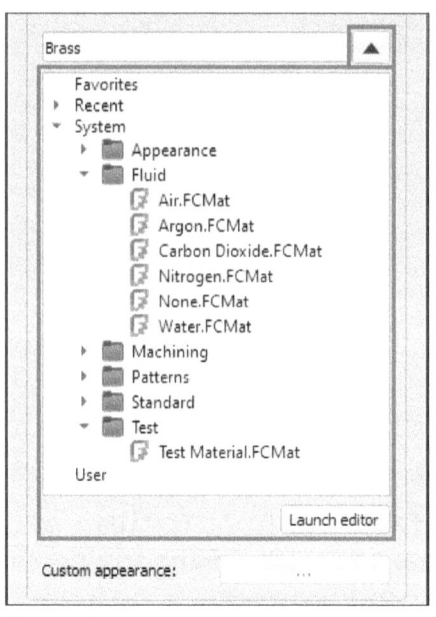

Figure-91. Selecting the material

- Click on the **Launch editor** button, the **Materials** dialog box will be displayed to define the material and appearance properties of the object; refer to Figure-92. The options in this dialog box will be discussed later in this book.

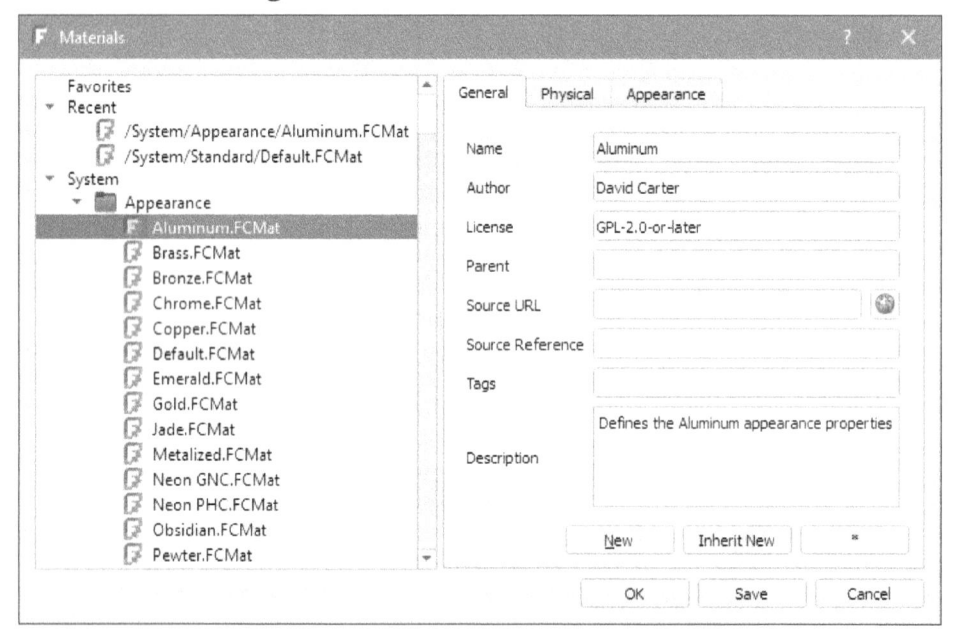

Figure-92. Materials dialog box

- After specifying desired parameters, click on **OK** button to close the dialog box.
- Click on the **Custom appearance** button. The **Material properties** dialog box will be displayed; refer to Figure-93.

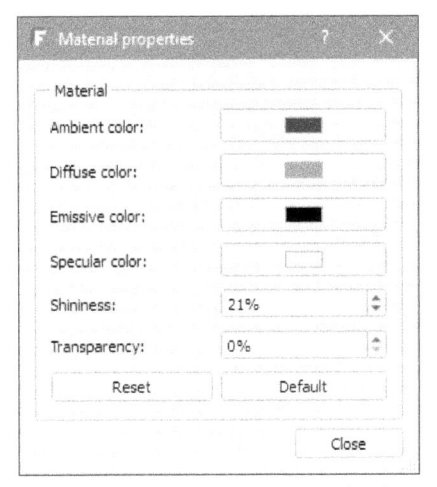

Figure-93. Material properties dialog box

- Click on the **Ambient color** button. The **Select Color** dialog box will be displayed to define the color of the shadows of the object; refer to Figure-94.

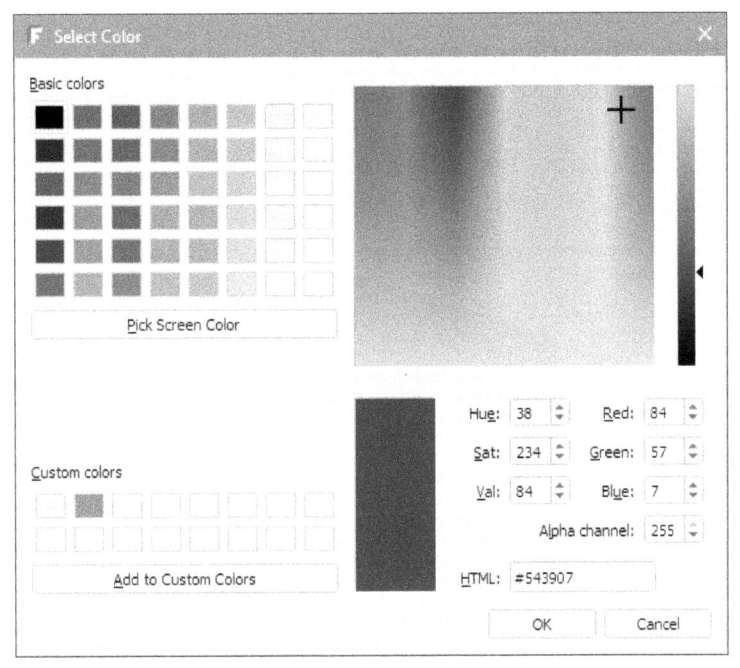

Figure-94. Select Color dialog box

- After specifying desired parameters, click on **OK** button from the dialog box.
- Similarly, click on the **Diffuse color** button to define the actual color of the object, click on the **Emissive color** button to define the color of the light radiating from the object, and click on the **Specular color** button to define the color of the reflection on the surface of the object.
- Specify desired amount of shininess and transparency of the color in the **Shininess** and **Transparency** edit boxes, respectively.
- After specifying desired parameters, click on **Close** button from the dialog box, the selected color(s) will be defined to the selected face(s); refer to Figure-95.

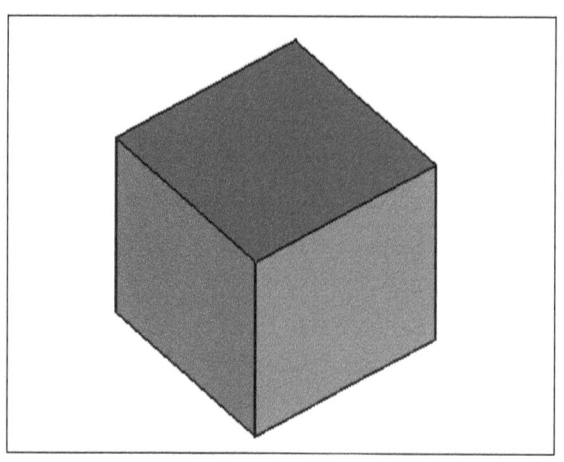

Figure-95. Color defined to the face

- Repeat the procedure to define colors to the other faces of the object.
- Click on the **Set to default** button in the dialog to set the color of all the faces of the object to the default color.
- If you want to select multiple faces of an object, then click on the **Box selection** button and press & hold **LMB** and drag with the mouse a selection rectangle in the object.
- Click on the **OK** button from the dialog to complete the procedure. The colors and material properties will be defined to the faces of the object; refer to Figure-96.

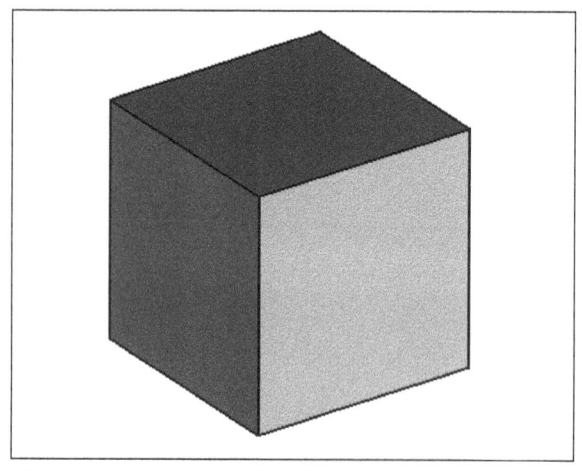

Figure-96. Colors defined to the object

Compound tools

The **Compound** tools are used to create a set of shapes, to split up the shapes, or to extract individual pieces from split shapes. These tools are discussed next.

Make Compound

The **Make Compound** tool creates a compound of any kind of topological shapes. These can be solids, meshes, or any other kind of topological shapes. A compound is a set of shapes grouped into one object. The procedure to use this tool is discussed next.

- Select the topological shapes from the **Model Tree** to be added to the compound; refer to Figure-97.

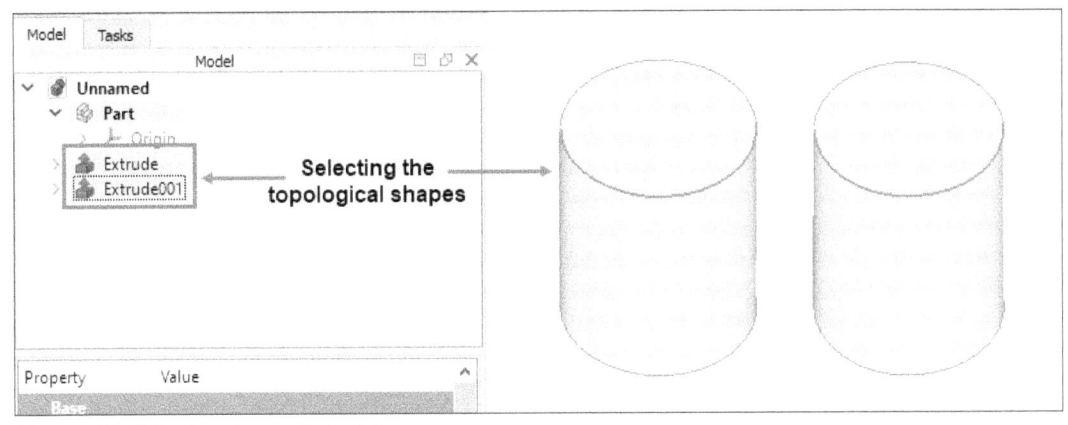

Figure-97. Selecting the topological shapes

- Click on the **Make compound** tool from **Compound** drop-down in the **Toolbar** of **Part** workbench; refer to Figure-98. The compound of the selected topological shapes will be created; refer to Figure-99.

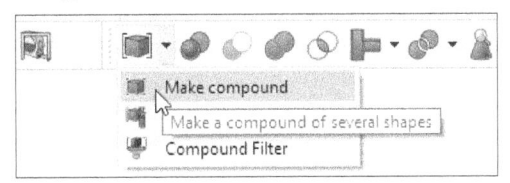

Figure-98. Make compound tool

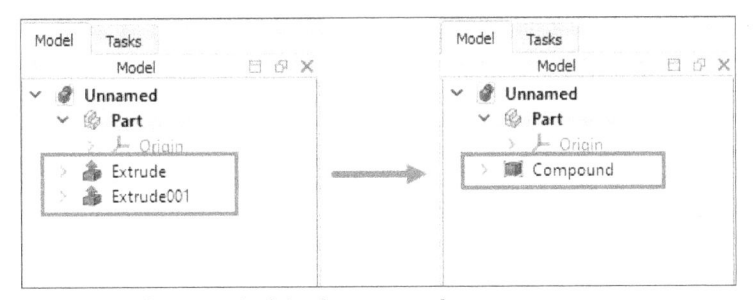

Figure-99. Compound of the shapes created

Explode Compound

The **Explode compound** tool is used to split up compounds of shapes to make each contained shape available as a separate object in Model tree view. The procedure to use this tool is discussed next.

- Select the compound body to be split into individual objects from the **Model Tree**; refer to Figure-100.

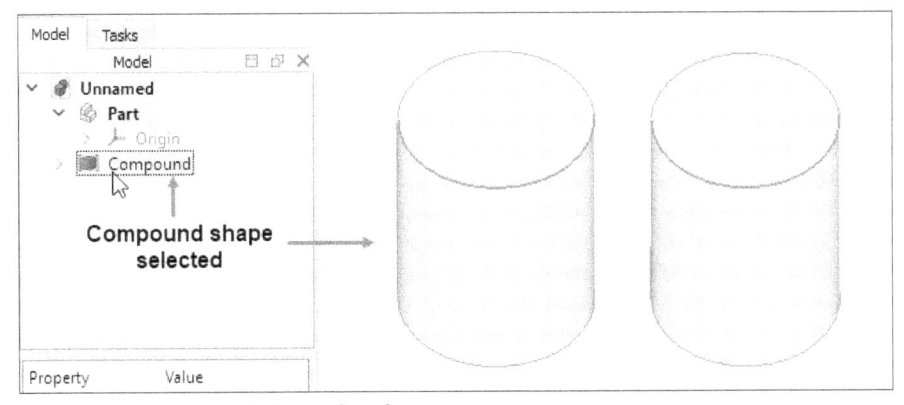

Figure-100. Compound shape selected

- Click on the **Explode compound** tool from **Compound** drop-down in the **Toolbar** of **Part** workbench; refer to Figure-101. The Compound shape will explode to generate separate objects; refer to Figure-102.

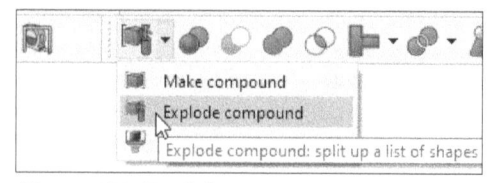

Figure-101. Explode compound tool

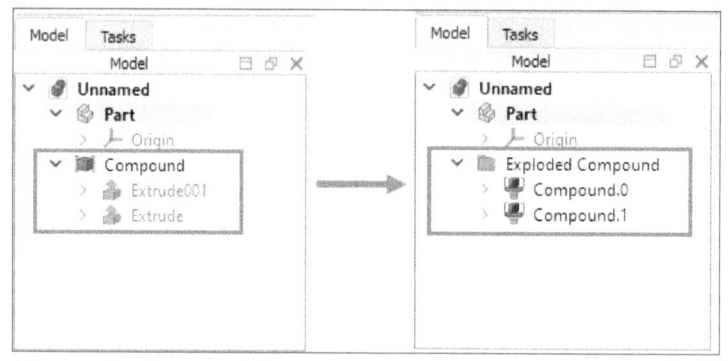

Figure-102. Compound shape exploded

Compound Filter

The **Compound Filter** tool can be used to extract individual pieces of the result of a **Part Slice** operation with which you have split an object. The procedure to use this tool is discussed next.

- Select the sliced object from **Model Tree** or from the 3D view area to extract the individual pieces; refer to Figure-103.

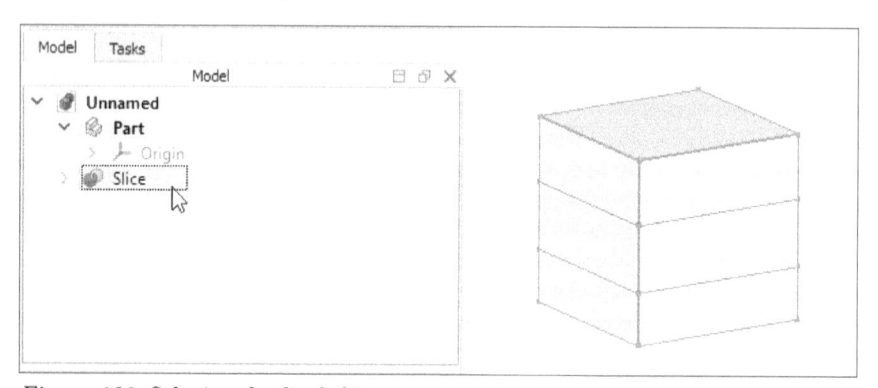

Figure-103. Selecting the sliced object

- Click on the **Compound Filter** tool from **Compound** drop-down in the **Toolbar** of **Part** workbench; refer to Figure-104. The compound filter object will be created; refer to Figure-105.

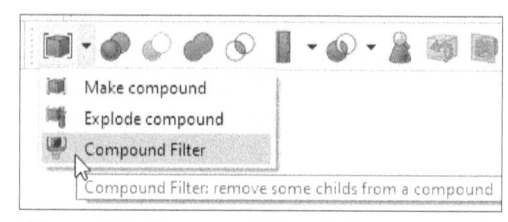

Figure-104. Compound Filter tool

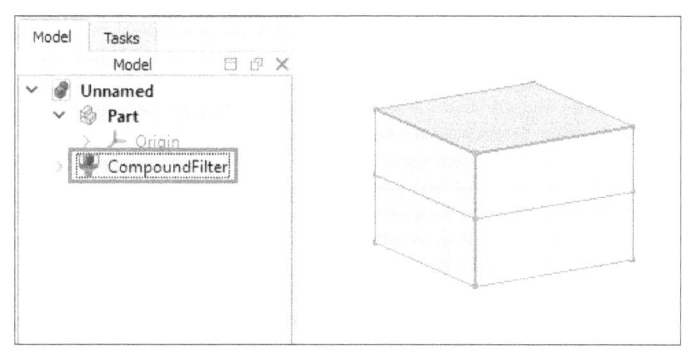

Figure-105. Compound filter object created

- Select the compound filter object from the **Model Tree**. The **Property editor** dialog will be displayed in the **Model** tab of **Combo View** as discussed earlier.
- Select the **specific items** option from **Filter Type** drop-down in the **Compound Filter** section of the dialog.
- Specify the number of elements you want to extract in the **items** edit box of the dialog. For a single piece, this is a number starting with 0, i.e. if you want to extract the first element, enter **0** in this edit box, **1** for the next element.
- After specifying the parameters, click **LMB** in the 3D view area. The element will be extracted; refer to Figure-106.

You can separate desired sliced object from the other sliced objects using **Property editor**.

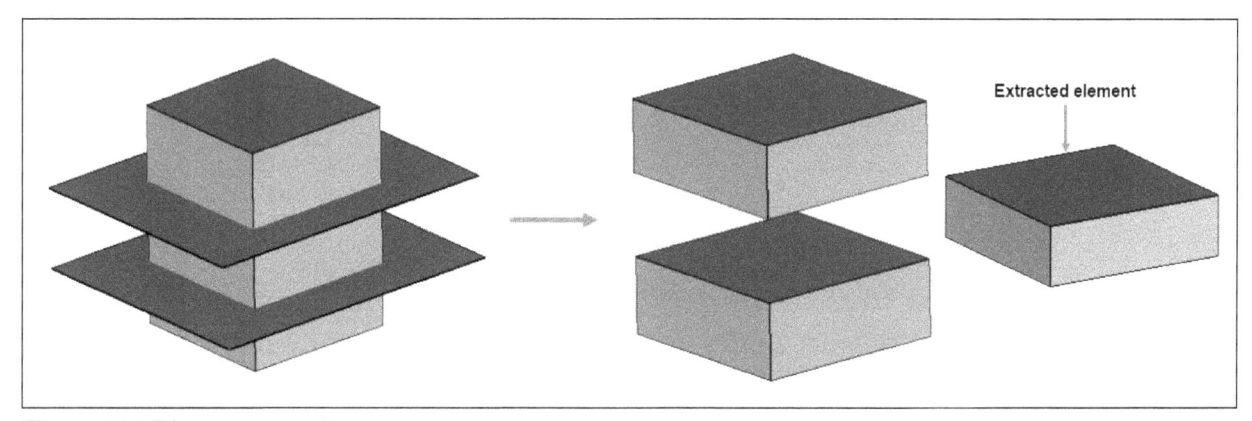

Figure-106. Element extracted

Boolean

The **Boolean** tool is a generic all-in-one boolean tool. It allows you to specify the objects and operation to perform via a single dialog. The procedure to use this tool is discussed next.

- Click on the **Boolean** tool from **Toolbar** in the **Part** workbench; refer to Figure-107. The **Boolean Operation** dialog will be displayed in the **Tasks** panel of **Combo View**; refer to Figure-108.

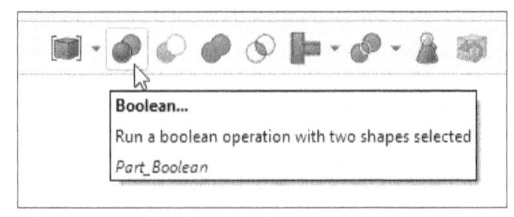

Figure-107. Boolean tool

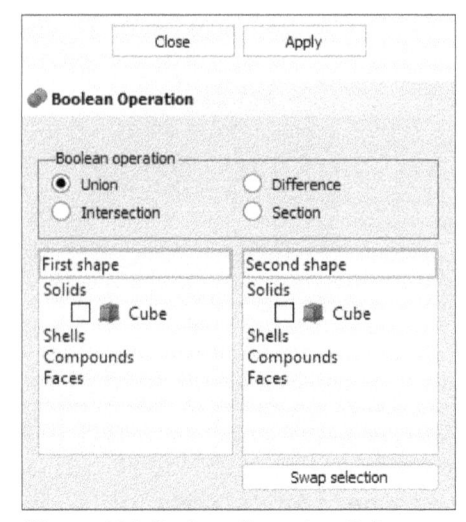

Figure-108. Boolean Operation dialog

- The objects available to perform boolean operation will display in the both **First shape** list and **Second shape** list area of the dialog; refer to Figure-109.

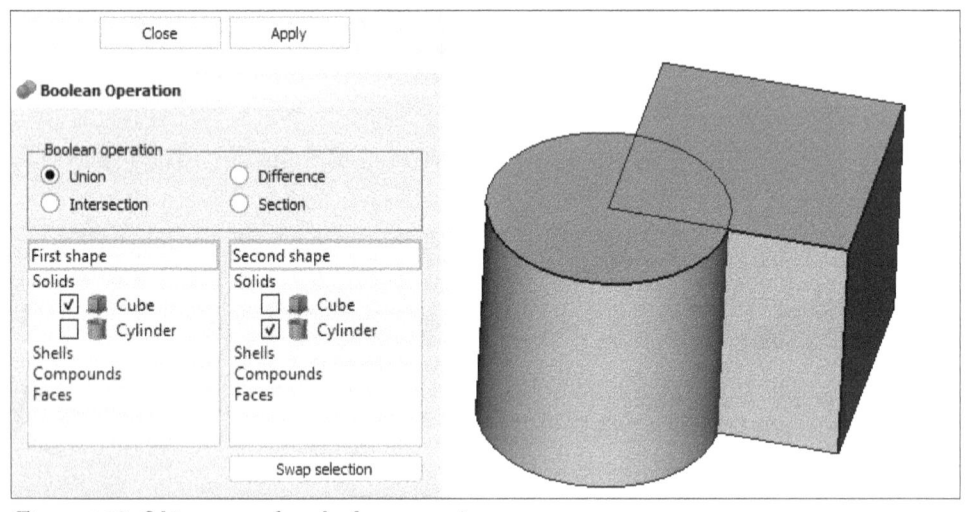

Figure-109. Objects to perform boolean operation

- Select the type of boolean operation you want to perform from the **Boolean operation** area of the dialog. Select the **Union** radio button if you want to unite the selected part objects into one. Select the **Difference** radio button if you want to subtract second object from first object. Select the **Intersection** radio button if you want to extract the common part between selected part objects. Select the **Section** radio button if you want to extract a section from the intersection of two selected shapes.
- Select the first object to be used in boolean operation from the **First shape** list area and select the second object from the **Second shape** list area.
- If you want to swap the selection of objects in the shape lists area then click on the **Swap selection** button at the bottom of the dialog.
- Click on **Apply** button from the dialog. The boolean operation will be performed; refer to Figure-110.

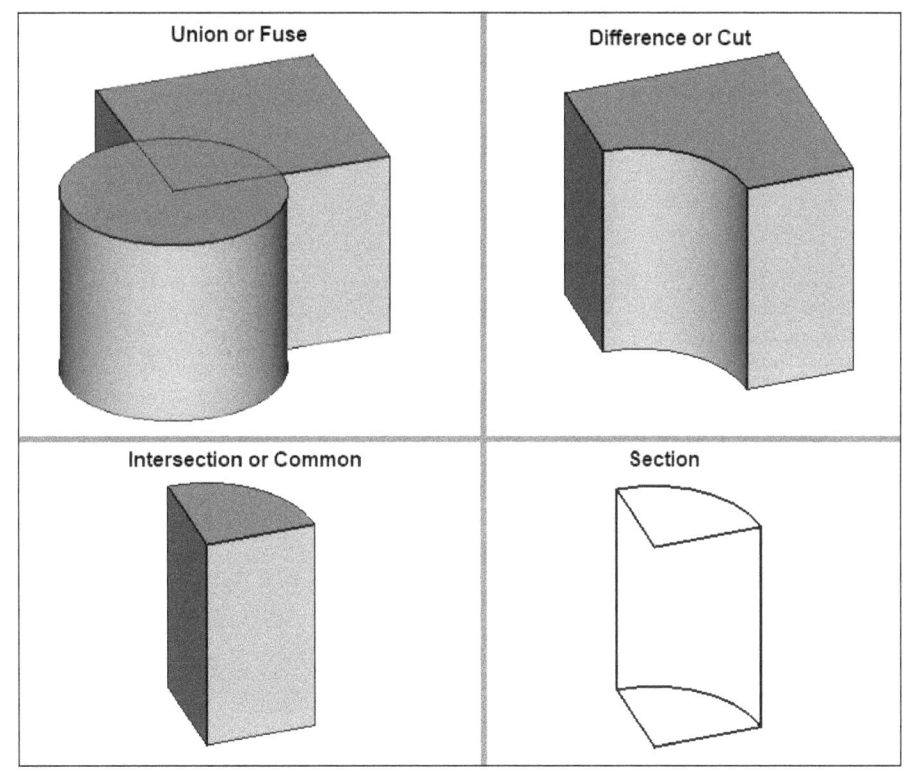

Figure-110. Types of boolean operation

- Click on **Close** button to close the dialog.

If you want the quicker access to these operations, use the **Fuse** tool, **Cut** tool, **Common** tool, and **Section** tool from **Toolbar** in the **Part** workbench.

Join objects tools

The **Join objects** tools are used to connect, embed, and cutout the walled objects. These tools are discussed next.

Connect objects

The **Connect objects** tool connects the interiors of two walled objects (e.g., pipes). It can also join shells and wires. The procedure to use this tool is discussed next.

- Select two or more objects from the **Model Tree** or from the 3D view area which you want to connect; refer to Figure-111.

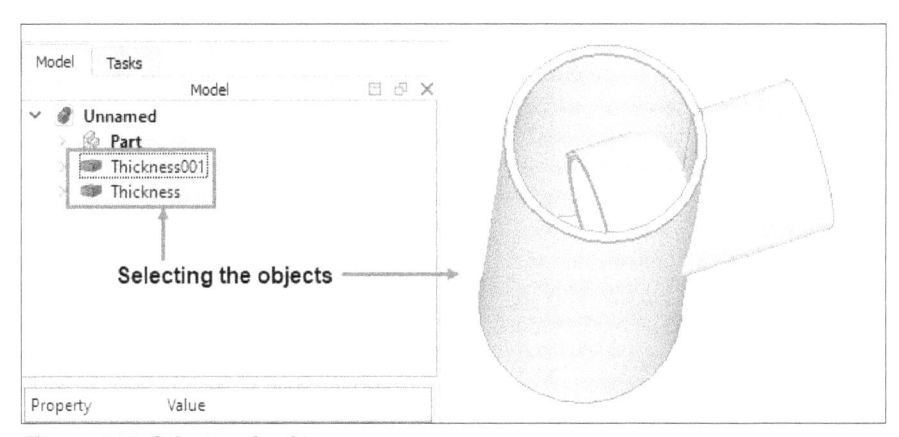

Figure-111. Selecting the objects to connect

- Click on the **Connect objects** tool from **Join objects** drop-down in the **Toolbar** of **Part** workbench; refer to Figure-112. A connected parametric object will be created; refer to Figure-113.

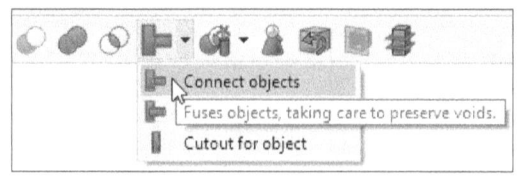

Figure-112. Connect objects tool

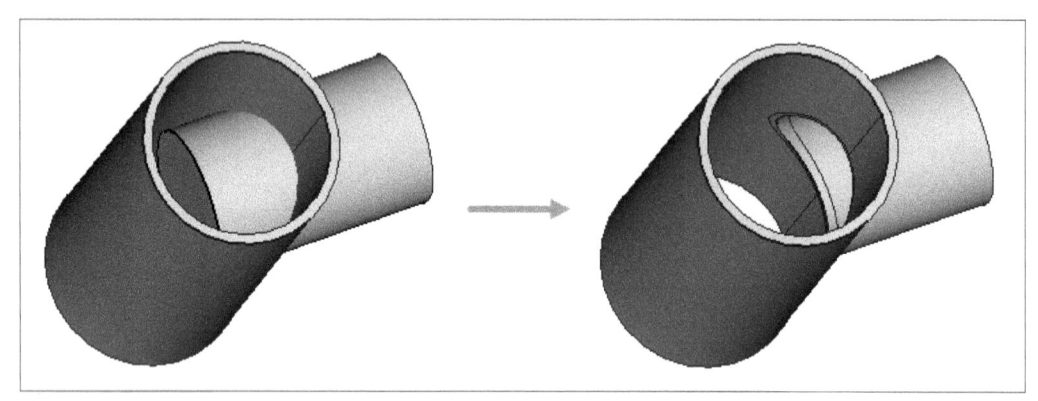

Figure-113. Connect objects created

Embed object

The **Embed object** tool embeds a walled object (e.g., a pipe) into another walled object. The procedure to use this tool is discussed next.

- Select the base object first and then select the object to be embedded from the **Model Tree** or from the 3D view area; refer to Figure-114.

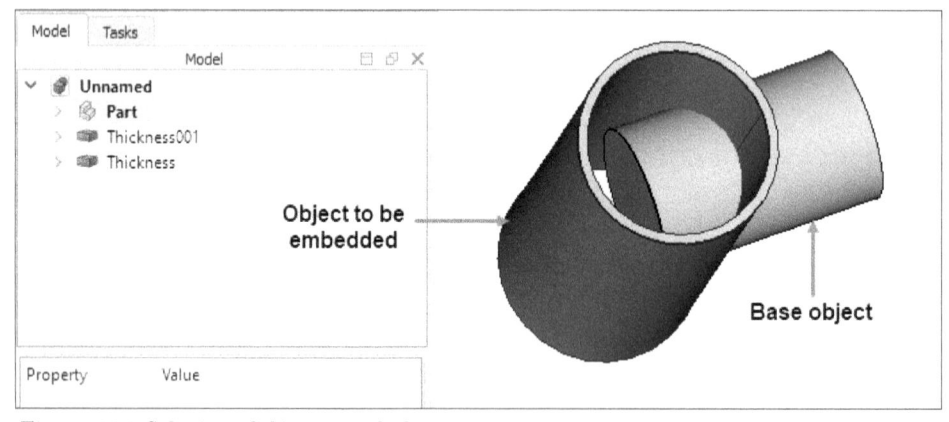

Figure-114. Selection of objects to embed

- Click on the **Embed object** tool from **Join Objects** drop-down in the **Toolbar** of **Part** workbench; refer to Figure-115. The embeded object will be created; refer to Figure-116.

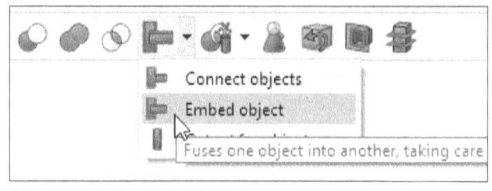

Figure-115. Embed object tool

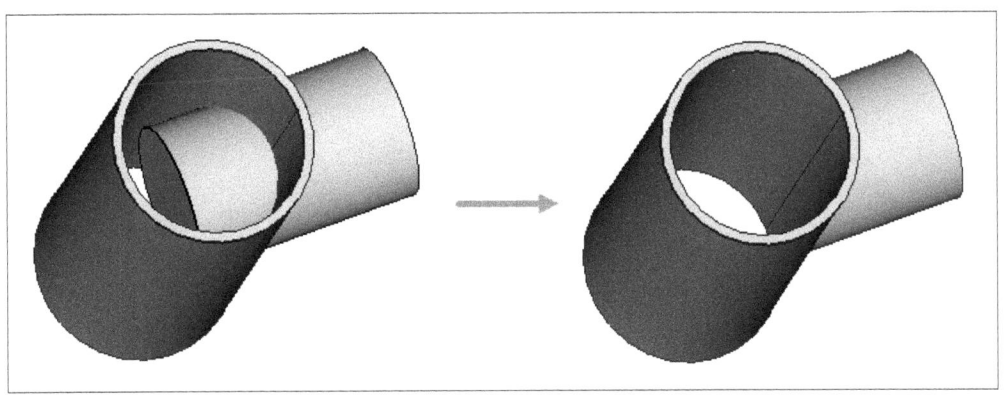

Figure-116. Embed object created

Cutout for object

The **Cutout for object** tool creates a cutout in a walled object (e.g. a pipe) to fit another walled object. The procedure to use this tool is discussed next.

- Select the base object first and then select the object to define the cutout from the **Model Tree** or from the 3D view area; refer to Figure-117.

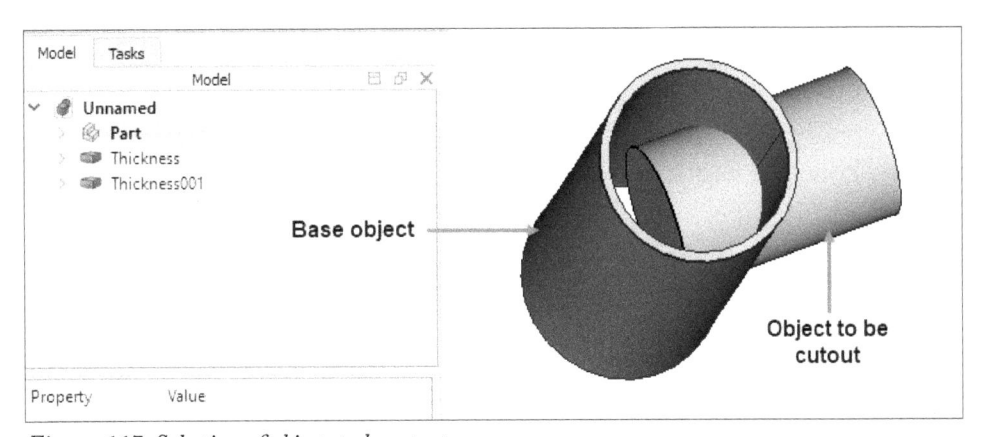

Figure-117. Selection of objects to be cutout

- Click on the **Cutout for object** tool from **Join Objects** drop-down in the **Toolbar** of **Part** workbench; refer to Figure-118. The cutout object will be created; refer to Figure-119.

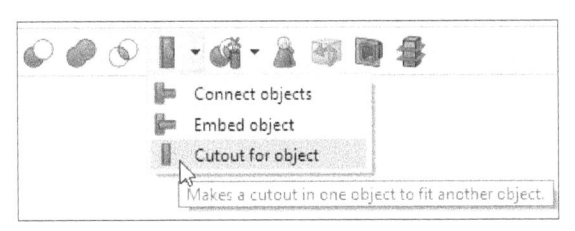

Figure-118. Cutout for object tool

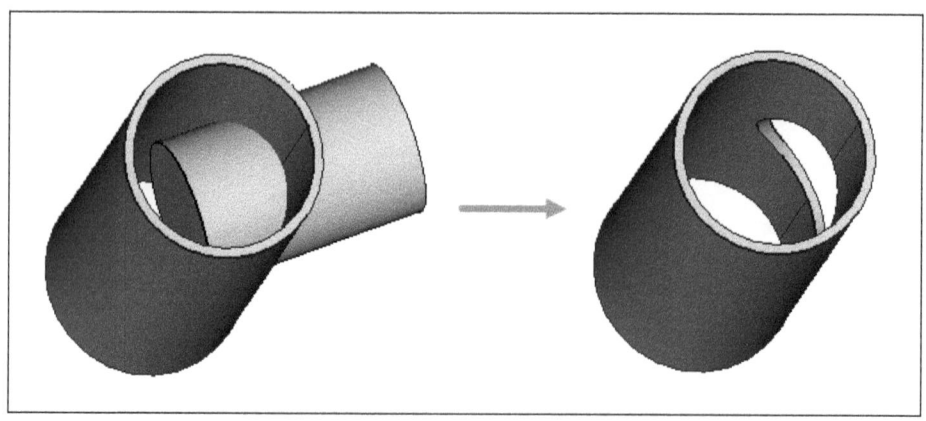

Figure-119. Cutout object created

Split objects tools

The **Split objects** tools are used to compute boolean fragments and split features. These tools are discussed next.

Boolean Fragments

The **Boolean fragments** tool is used to compute all fragments that can result from applying boolean operations between input shapes. The procedure to use this tool is discussed next.

- Select the objects to create boolean fragments from the **Model Tree** or from the 3D view area; refer to Figure-120.

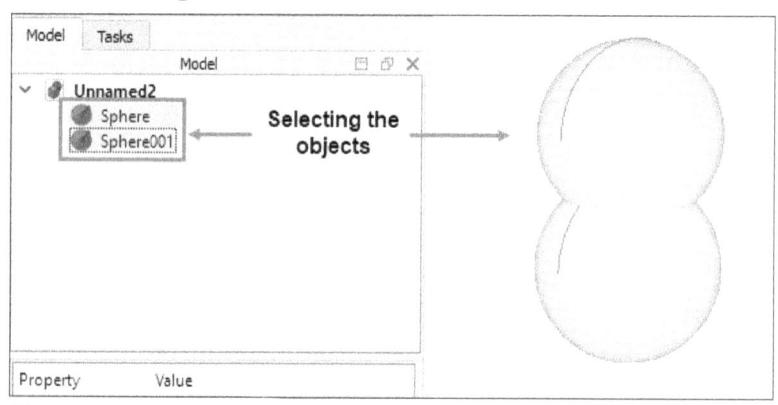

Figure-120. Selecting the objects to create boolean fragments

- Click on the **Boolean Fragments** tool from **Split objects** drop-down in the **Toolbar** of **Part** workbench; refer to Figure-121. The boolean fragments of the objects will be created; refer to Figure-122.

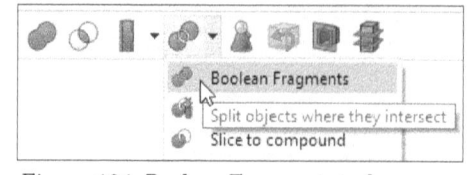

Figure-121. Boolean Fragments tool

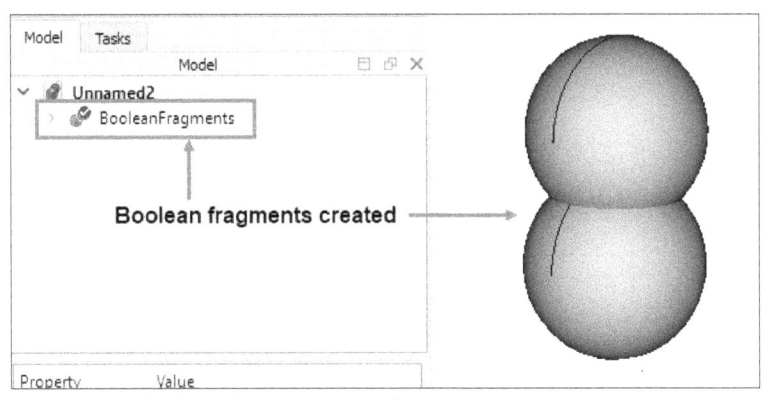

Figure-122. Boolean fragments created

- The individual pieces of the boolean fragments will not be displayed immediately as the pieces remain grouped together.
- If you want to extract the individual pieces of the object then explode the object using **Downgrade** tool.
- Select both the objects from 3D view area and click on the **Downgrade** tool from **Draft modifications tools** in the **Draft** workbench; refer to Figure-123. The exploded pieces of the boolean fragments will be displayed in the **Model Tree**; refer to Figure-124.

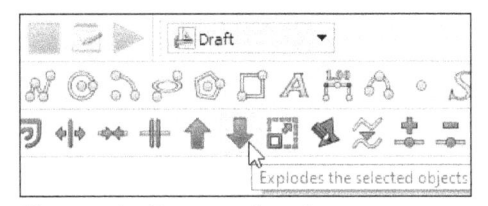

Figure-123. Downgrade tool

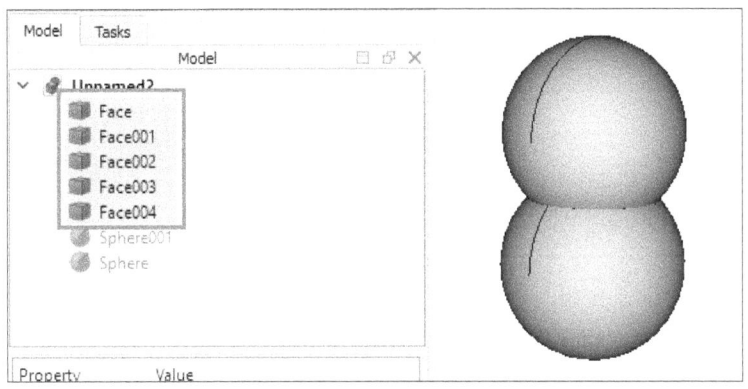

Figure-124. Exploded pieces of the boolean fragments

- Select the pieces of the object one by one from 3D view area and move apart manually using the **Property editor**. The exploded view of the boolean fragments will be displayed; refer to Figure-125.

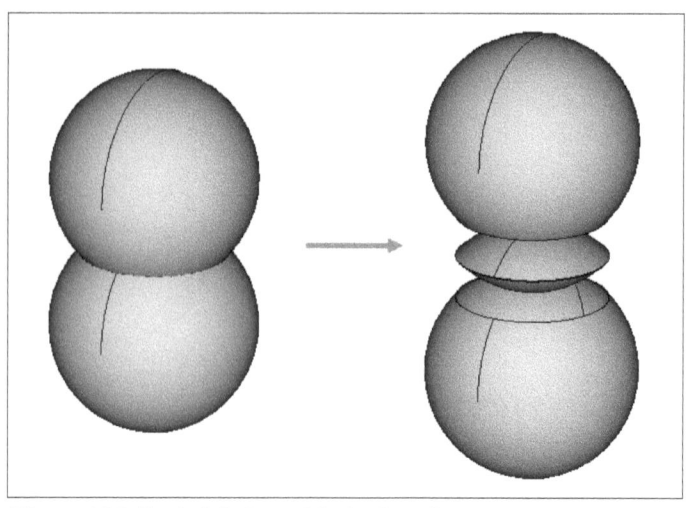

Figure-125. Exploded view of the boolean fragments

Slice Apart

The **Slice apart** tool is used to split shapes by intersection with other shapes. The procedure to use this tool is discussed next.

- Select the base object first which is to be sliced and then select the other object to slice with from the **Model Tree** or from the 3D view area; refer to Figure-126.

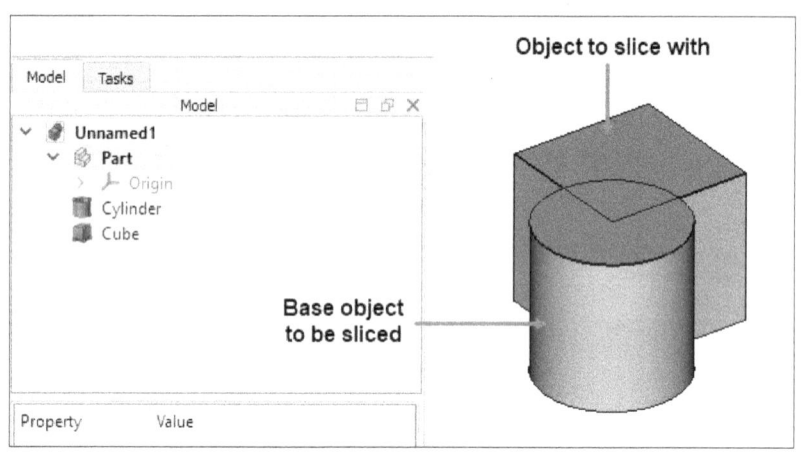

Figure-126. Selection of objects to slice apart

- Click on the **Slice apart** tool from **Split objects** drop-down in the **Toolbar** of **Part** workbench; refer to Figure-127. The sliced part of base object will be created; refer to Figure-128.

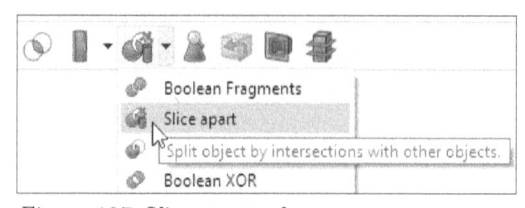

Figure-127. Slice apart tool

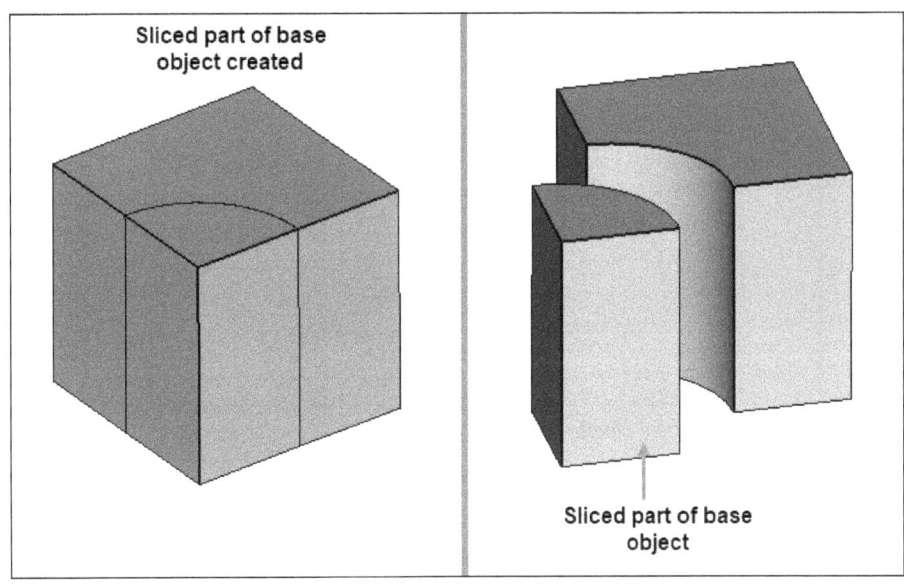

Figure-128. Sliced part of base object created

Slice to compound

The **Slice** tool also known as **Slice to compound** tool is used to split shapes by intersection with other shapes. The procedure to use this tool is discussed next.

- First select the base object to be sliced and then select the other object to slice with from the **Model Tree** or from the 3D view area; refer to Figure-129.

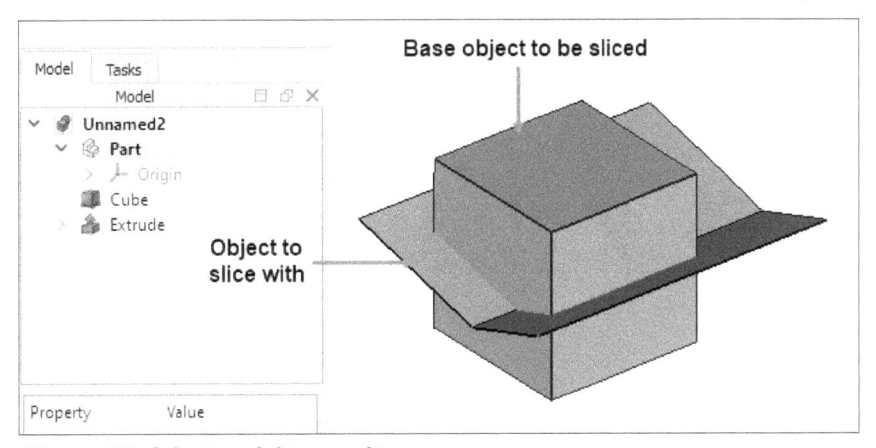

Figure-129. Selection of objects to slice

- Click on the **Slice to compound** tool from **Split objects** drop-down in the **Toolbar** of **Part** workbench; refer to Figure-130. The sliced object will be created; refer to Figure-131.

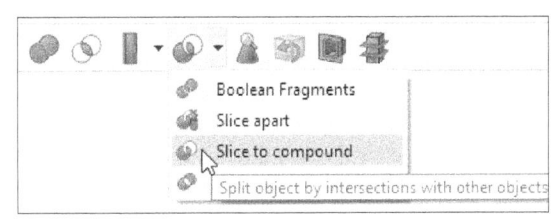

Figure-130. Slice to compound tool

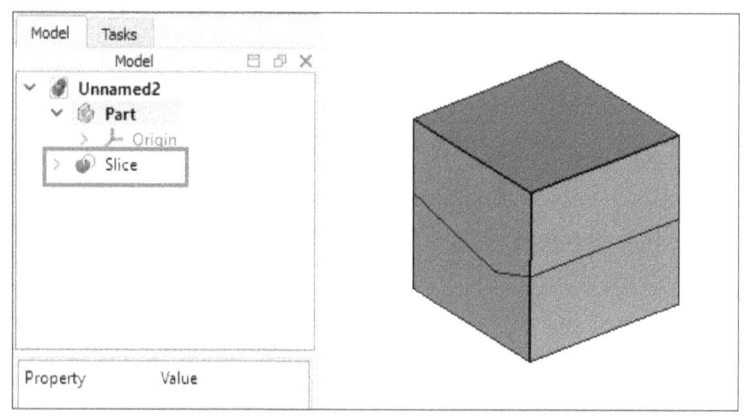

Figure-131. Sliced object created

- If you want to break the sliced object into individual pieces then select the object from the **Model Tree** and explode the object using **Explode compound** tool which we have discussed earlier. The sliced object will be exploded.
- If you want to view the exploded slice object individually then select the object from the **Model Tree** or from the 3D view area and move apart manually using the **Property editor**. The exploded view of sliced object will be displayed; refer to Figure-132.

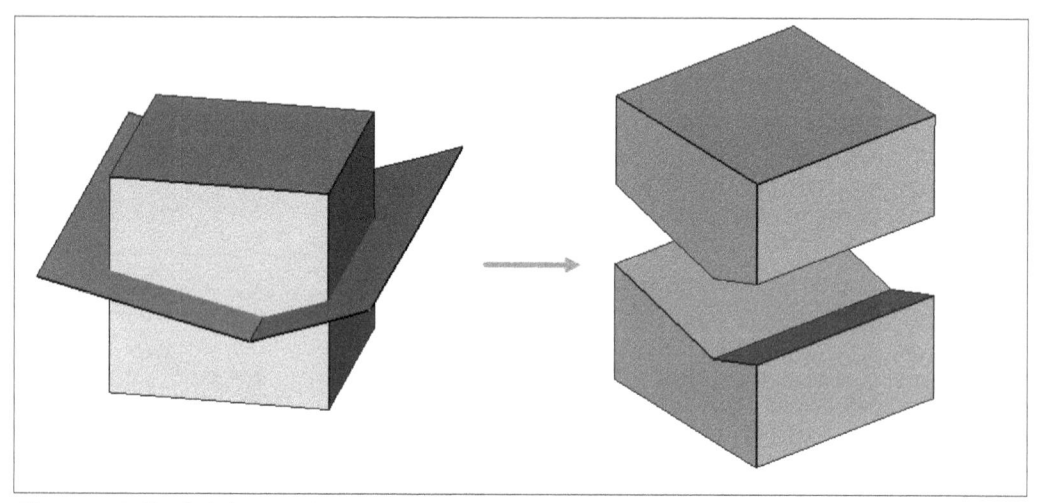

Figure-132. Exploded view of sliced object

Boolean XOR

The **Boolean XOR** tool is used to remove geometry shared by objects and leaves a void space between the involved objects. The procedure to use this tool is discussed next.

- Select two or more objects from the **Model Tree** or from the 3D view area; refer to Figure-133.

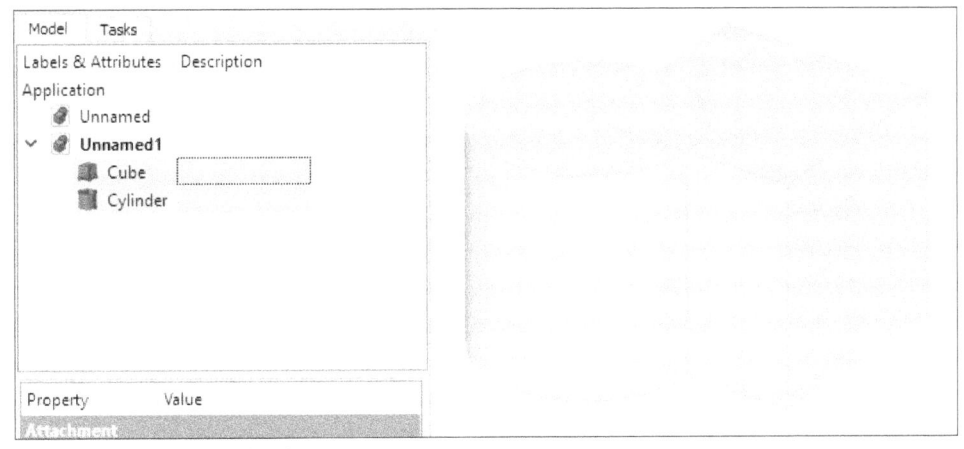

Figure-133. Selecting the objects

- Click on the **Boolean XOR** tool from **Split objects** drop-down in the **Toolbar** of **Part** workbench; refer to Figure-134. The overlapping geometry will be removed from the selected objects; refer to Figure-135.

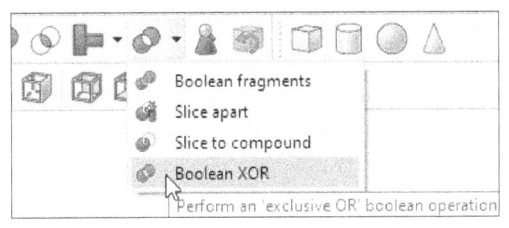

Figure-134. Boolean XOR tool

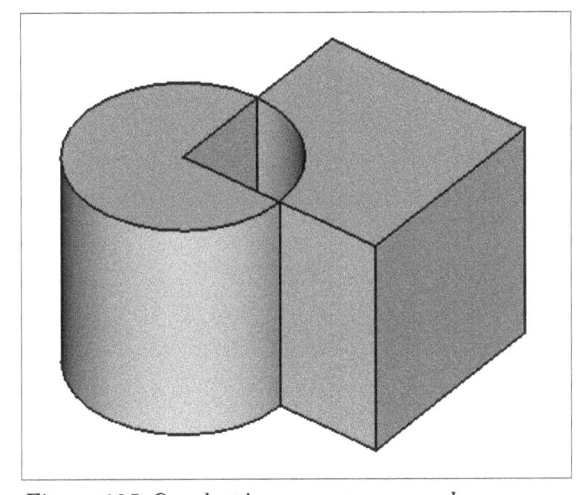

Figure-135. Overlapping geometry removed

Check Geometry

The **Check Geometry** tool runs a verification and reports if geometry is a valid solid. The procedure to use this tool is discussed next.

- Select the geometry from the **Model Tree** which you want to check for valid solid; refer to Figure-136.

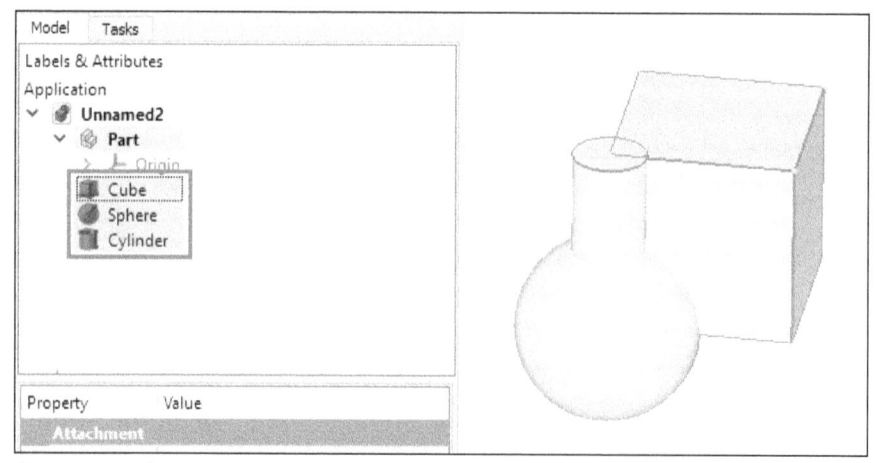

Figure-136. Selecting the geometry to check

- Click on the **Check Geometry** tool from **Toolbar** in the **Part** workbench; refer to Figure-137. The **Check Geometry Results** and **Shape Content** dialogs will be displayed in the **Tasks** panel of **Combo View** showing the result of the scan; refer to Figure-138.

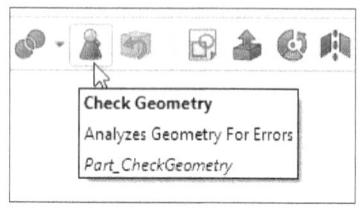

Figure-137. Check Geometry tool

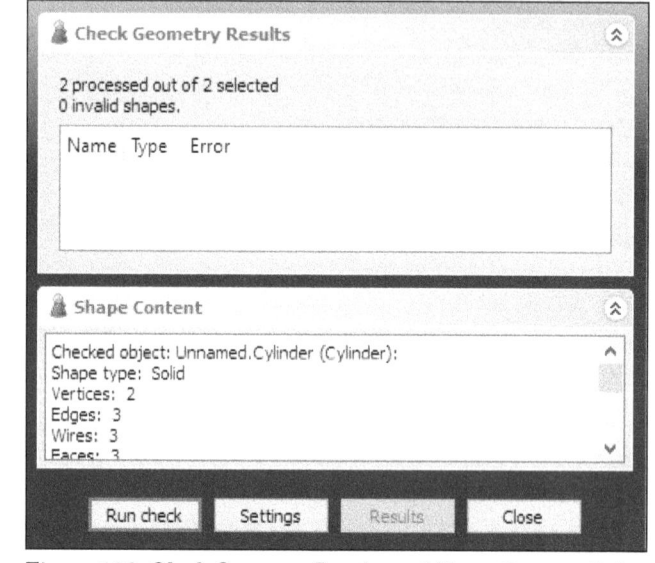

Figure-138. Check Geometry Results and Shape Content dialogs

- If the result of the scan showing errors, click on a specific error message in the dialog. The corresponding geometric object will be highlighted in the 3D view area.

Note that till the FreeCAD version 1.0, it has no automatic repair methods for solids, so you need to look at the steps used in FreeCAD and try to fix the error yourself.

- Click on **Close** button to close the dialog.

Defeaturing

The **Defeaturing** tool is intended for removal of selected features from the model. In this context, features are meant as holes, protrusions, gaps, chamfers, fillets, etc. found on the model. The procedure to use this tool is discussed next.

- Select the faces on the model from the 3D view area which you want to defeature; refer to Figure-139.

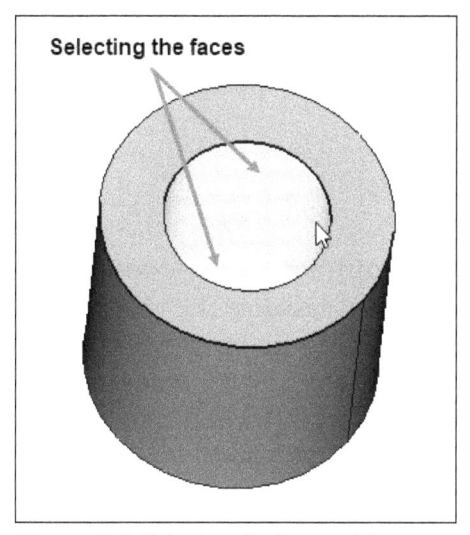

Figure-139. Selecting the faces to defeature

- Click on the **Defeaturing** tool from **Toolbar** in the **Part** workbench; refer to Figure-140. The selected feature will be removed from the model; refer to Figure-141.

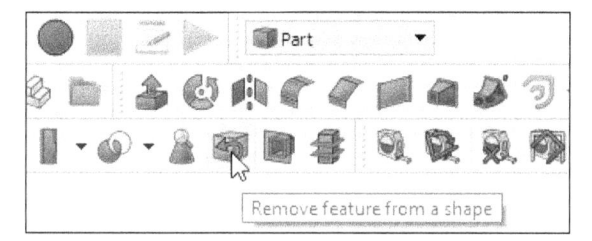

Figure-140. Defeaturing tool

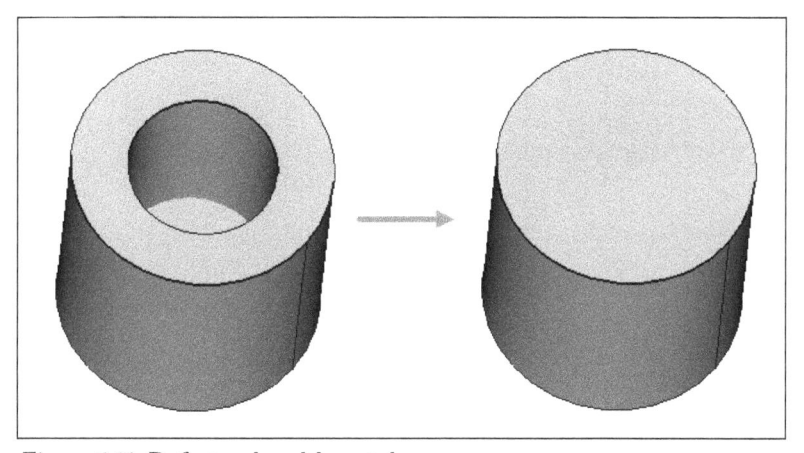

Figure-141. Defeatured model created

Measure

The **Measure** tool allows linear, radial, and angular measurement between points, edges, and faces. This tool is available in the **Toolbar** of **Part** workbench; refer to Figure-142. On clicking this tool, the **Measurement** dialog will be displayed; refer to Figure-143.

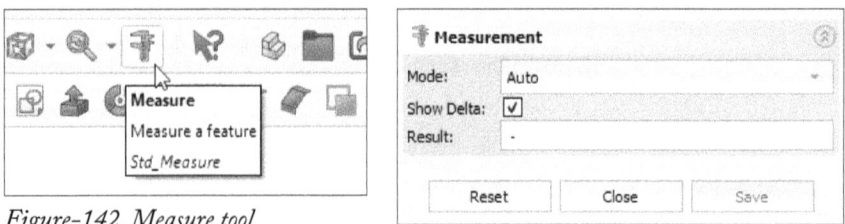

Figure-142. Measure tool

Figure-143. Measurement dialog box

- Select desired mode of measurement from **Mode** drop-down to measure the objects.
- Select **Show Delta** check box to display the delta property of the measurement.
- The **Result** area displays the measurement results of the selected object.
- Click on the **Reset** button to reset all the selections in the 3D view area.
- Click on the **Save** button to save and keep the measurement in the 3D view.
- Click on the **Close** button to exit the tool.

The procedures to use various measurement modes are discussed next.

Auto

The **Auto** option is used to measure the parameters which are automatically determined by the tool based on the selection of entities.

Distance

The **Distance** option is used to measure the shortest distance between the selected entities in the 3D view; refer to Figure-144.

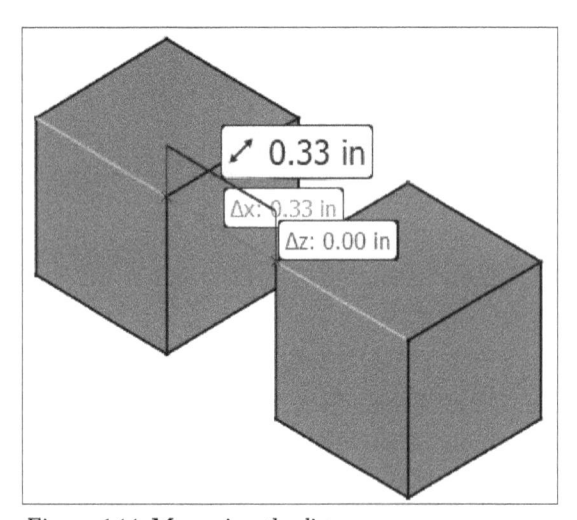

Figure-144. Measuring the distance

Distance Free

The **Distance Free** option is used to measure the distance between two freely selected points/edges/faces in the 3D view; refer to Figure-145.

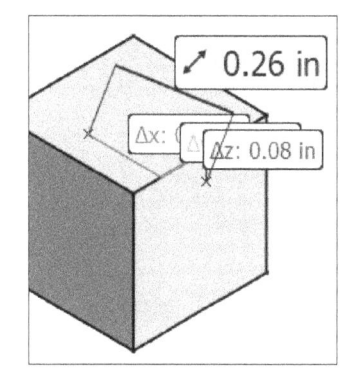

Figure-145. Measuring the distance free

Angle

The **Angle** option is used to measure the angle between the selected entities in the 3D view; refer to Figure-146.

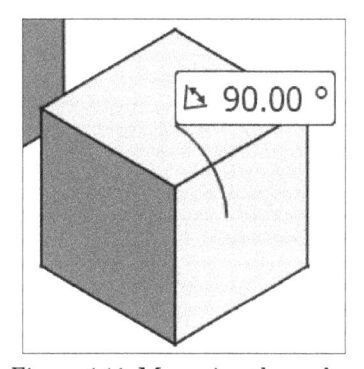

Figure-146. Measuring the angle

Length

The **Length** option is used to measure the length of the selected edge in 3D view; refer to Figure-147.

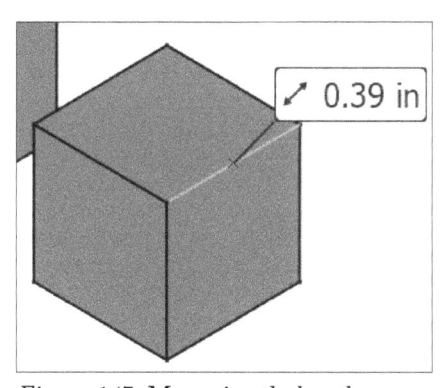

Figure-147. Measuring the length

Position

The **Position** option is used to measure the coordinates of the selected vertex in 3D view; refer to Figure-148.

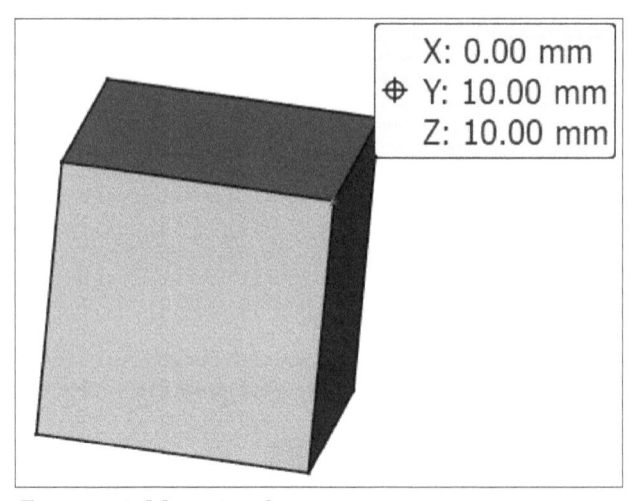

Figure-148. Measuring the position

Area

The **Area** option is used to measure the area of the selected face in the 3D view; refer to Figure-149.

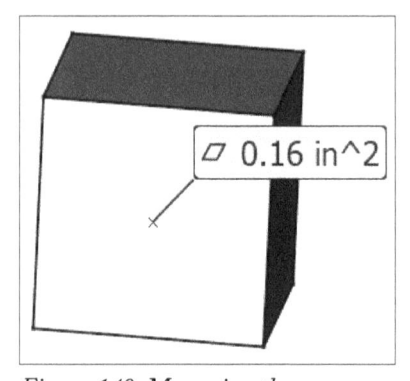

Figure-149. Measuring the area

Center of Mass

The **Center of Mass** option is used to find center of mass of body with respect to selected edge, face, and vertex in the 3D view; refer to Figure-150.

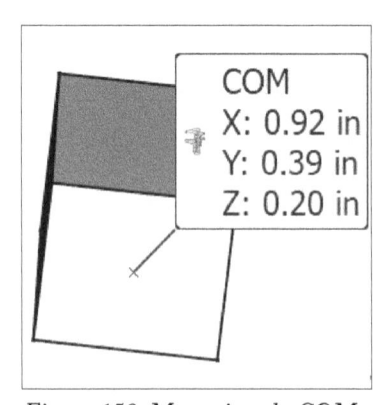

Figure-150. Measuring the COM

Other tools

There are some other tools available in the **Part** menu of the **Ribbon**; refer to Figure-151. The procedures to use these tools are discussed next.

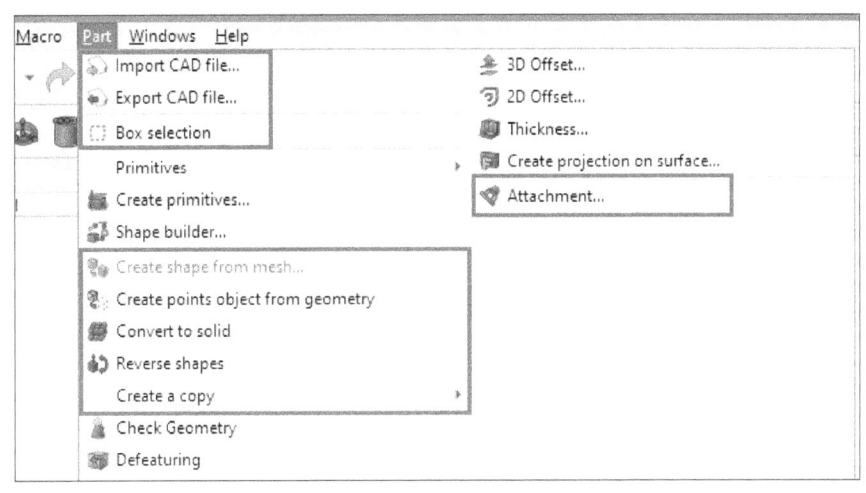

Figure-151. Other tools in Part menu

Import CAD file

The **Import CAD file** tool opens a dialog that allows you to import the following CAD file formats: **IGES**, **STEP**, and **BREP** into the current document. The procedure to use this tool is discussed next.

- Click on the **Import CAD file** tool from the **Part** menu in the **Ribbon**. The **Open** dialog box will be displayed; refer to Figure-152.

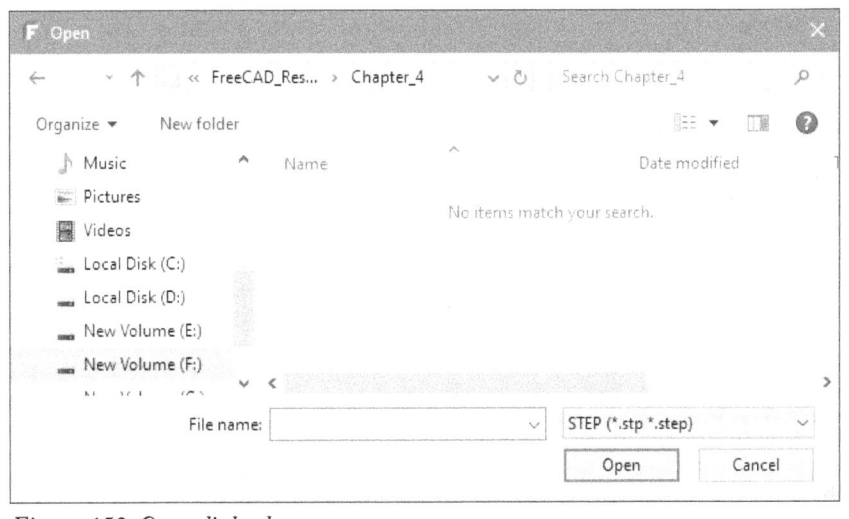

Figure-152. Open dialog box

- Select desired file of the supported format which you want to import from the directory and click on **Open** button from the dialog box. The file will be imported; refer to Figure-153.

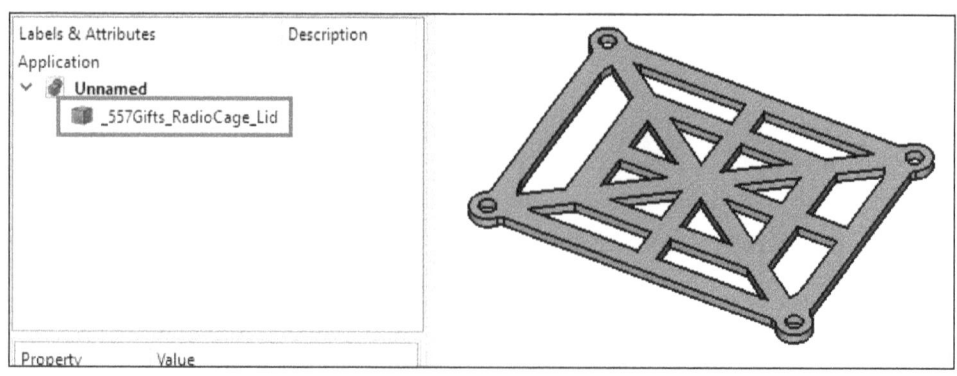

Figure-153. File imported

Export CAD file

The **Export CAD file** tool opens a dialog that allows you to export objects in the document to the following CAD file formats: **IGES**, **STEP**, and **BREP**. The procedure to use this tool is discussed next.

- Click on the **Export CAD file** tool from the **Part** menu in the **Ribbon**. The **Save as** dialog box will be displayed; refer to Figure-154.

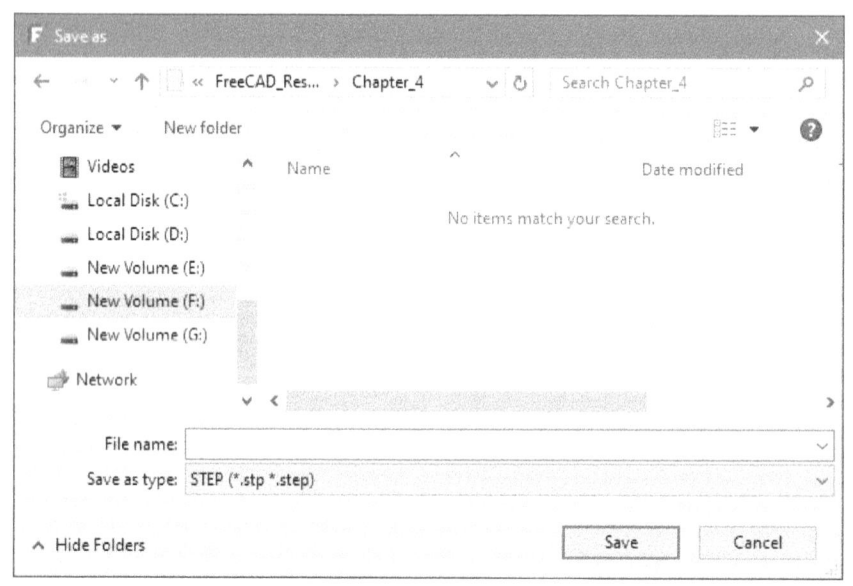

Figure-154. Save as dialog box

- Specify desired directory in the dialog box in which you want to save the file.
- Specify desired name of the file to be save in the **File name** edit box.
- Specify the supported format to save the file from **Save as type** drop-down.
- Click on the **Save** button from the dialog box to save the file.

Box selection

The **Box selection** tool allows you to select the faces of the objects on the screen that are touched by the rectangular selection. The procedure to use this tool is discussed next.

- Click on the **Box selection** tool from the **Part** menu in the **Ribbon**. You are asked to select the objects.

- Click **RMB** and drag the pointer in the 3D view area creating a rectangle that touches the objects which you want to select; refer to Figure-155. The objects will be selected.

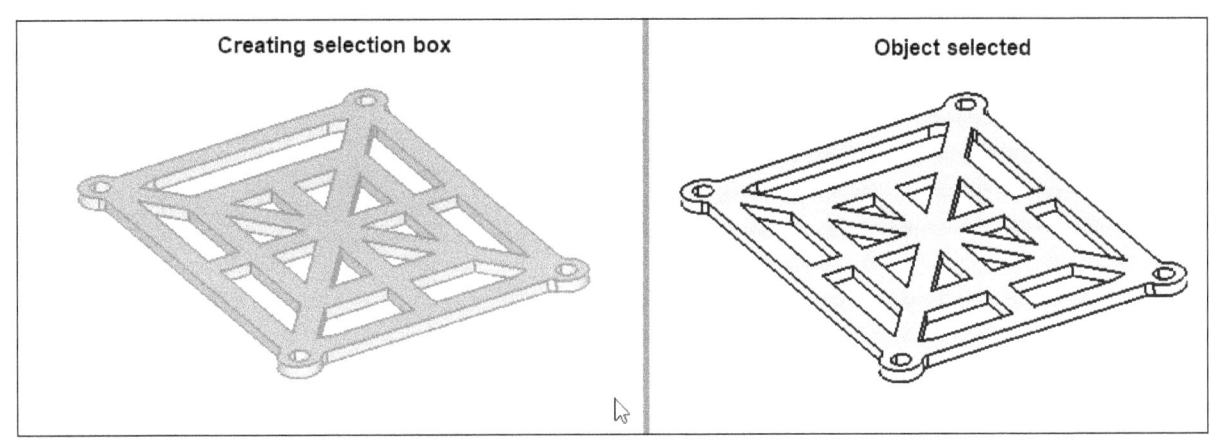

Figure-155. Object selected

Create shape from mesh

The **Create shape from mesh** tool creates a shape from a mesh object. Mesh objects have limited editing capabilities in FreeCAD, converting them to shapes will allow their use with many more boolean and modification tools. The procedure to use this tool is discussed next.

- Select desired meshed object from the **Model Tree** or from the 3D view area which you want to convert to shape; refer to Figure-156.

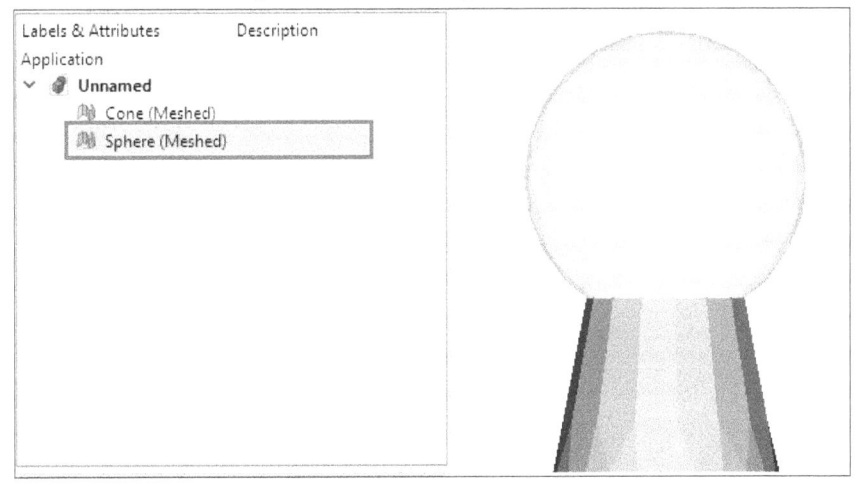

Figure-156. Meshed object selected

- Click on the **Create shape from mesh** tool from the **Part** menu in the **Ribbon**. The **Shape from mesh** dialog box will be displayed; refer to Figure-157.

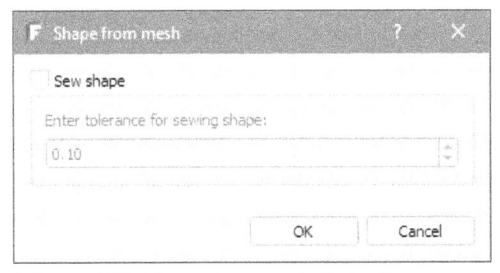

Figure-157. Shape from mesh dialog box

- Select **Sew shape** check box and specify desired tolerance value for the sewing shape in the **Enter tolerance for sewing shape** edit box and click on **OK** button from the dialog box. The selected meshed object will be created as a shape; refer to Figure-158.

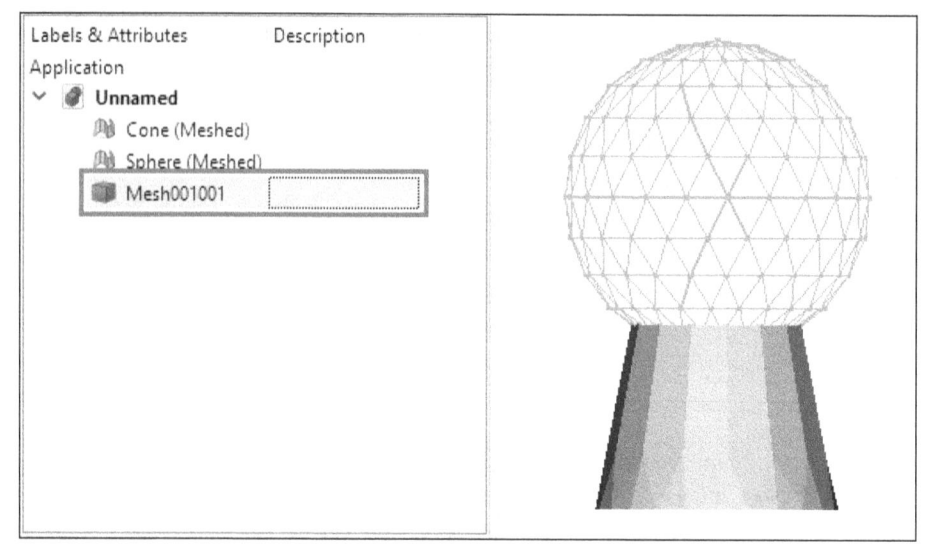

Figure-158. Shape created from meshed object

Create points object from geometry

The **Create points object from geometry** tool creates a shape object made of points from a mesh object imported or produced from the **Mesh** workbench. The resulting shape is a collection of vertices or points, which can be used as reference to further create lines, sketches, and faces with other tools. The procedure to use this tool is discussed next.

- Select desired meshed object from the **Model Tree** or from the 3D view area which you want to create points shape object; refer to Figure-159.

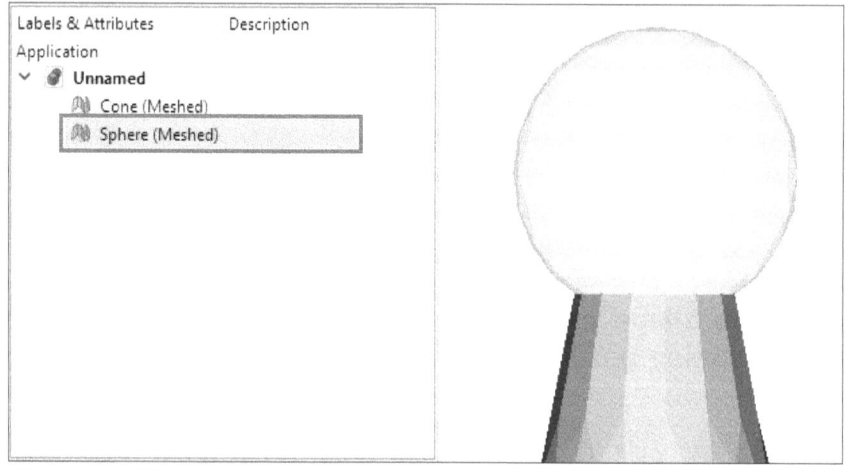

Figure-159. Meshed object selected

- Click on the **Create points object from geometry** tool from the **Part** menu in the **Ribbon**. The selected meshed object will be created as a shape object made of points; refer to Figure-160.

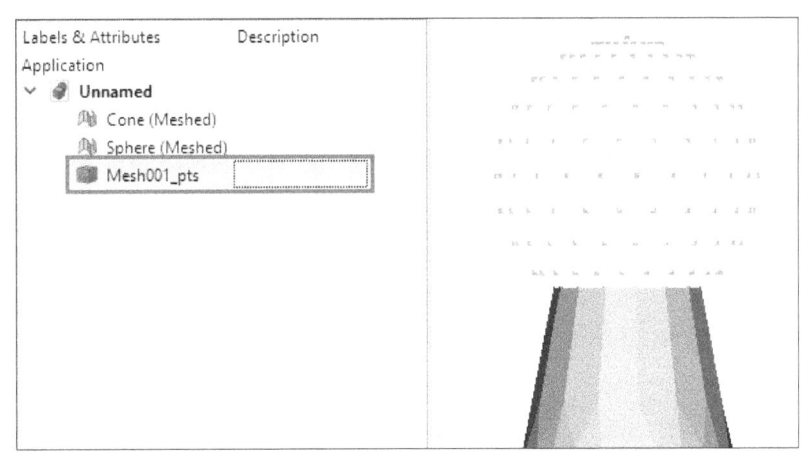

Figure-160. Points shape object created

Convert to solid

The **Convert to solid** tool converts a shape object to a solid object. The procedure to use this tool is discussed next.

- Select the shape object from the **Model Tree** which you have created from the mesh object; refer to Figure-161.

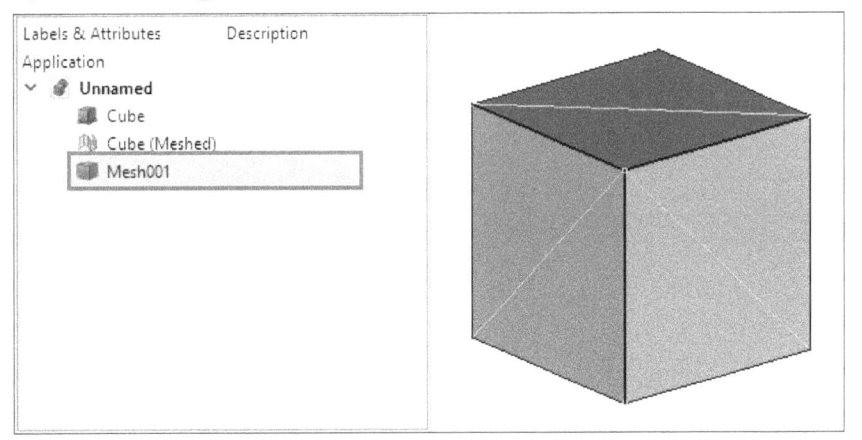

Figure-161. Selecting the shape object created from mesh object

- Click on the **Convert to solid** tool from the **Part** menu in the **Ribbon**. The selected shape object will be converted into a solid object; refer to Figure-162.

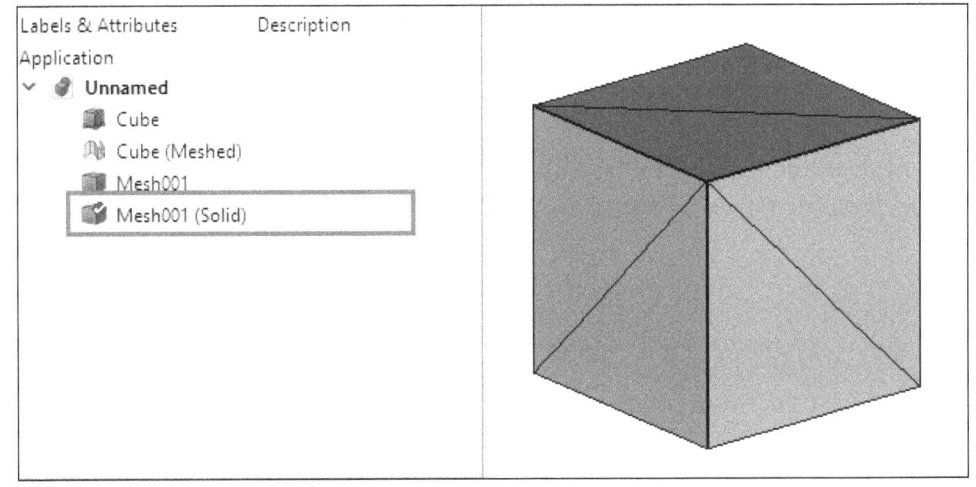

Figure-162. Shape object converted to solid object

Reverse shapes

The **Reverse shapes** tool flips the normals of all faces of the selected object. The procedure to use this tool is discussed next.

- Select desired shape from the **Model Tree** or from the 3D view area which you want to reverse; refer to Figure-163.

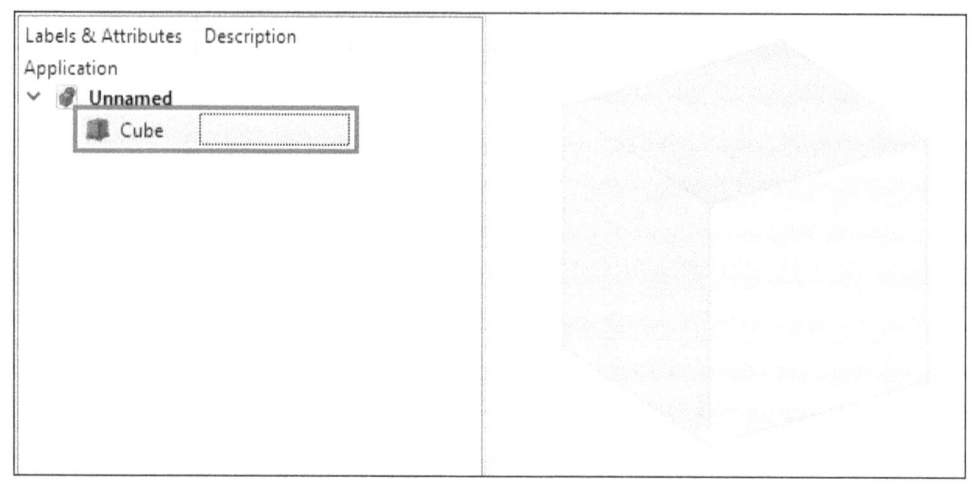

Figure-163. Selecting the shape to reverse

- Click on the **Reverse shapes** tool from the **Part** menu in the **Ribbon**. The selected shape will be reversed; refer to Figure-164.

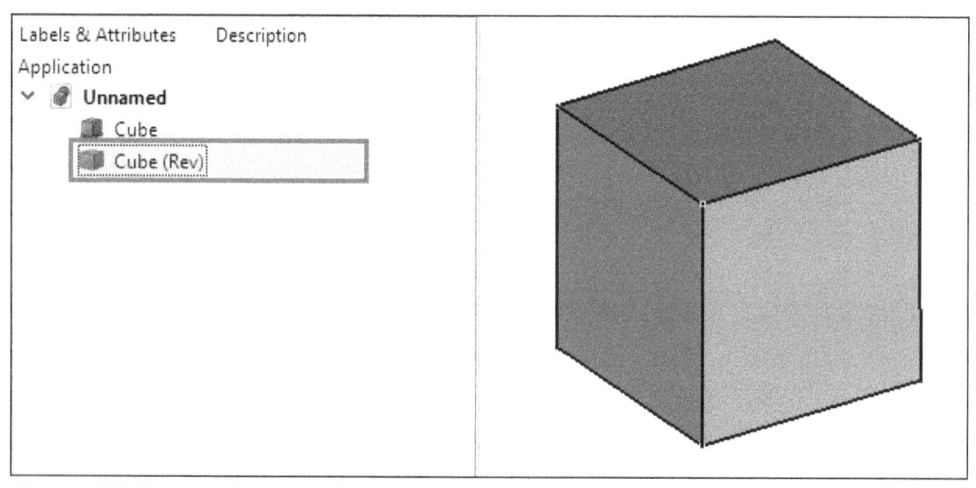

Figure-164. Reverse shape created

Now, if you want to verify that the shape is reversed or not then follow the steps given below:

- Hide the shape selected to reverse from the **Model Tree** by pressing the **Space Bar** from the keyboard; refer to Figure-165.

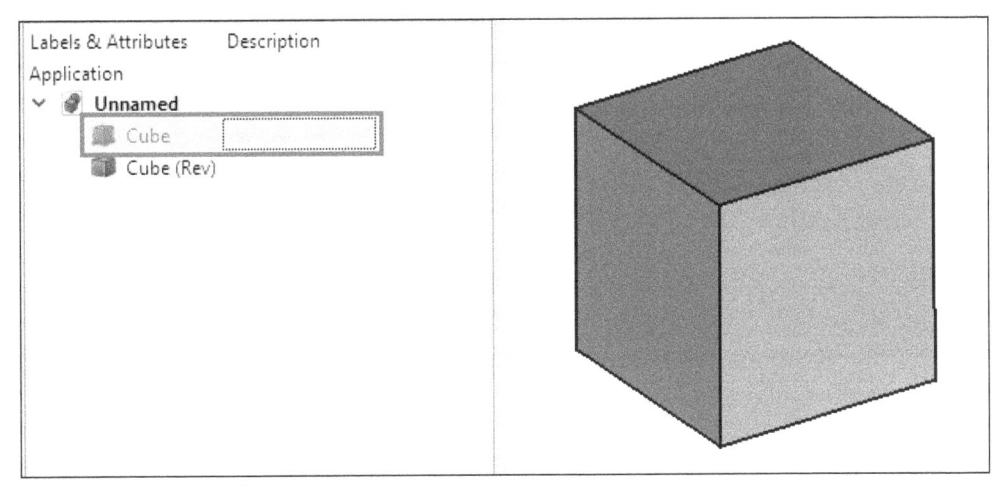

Figure-165. Hiding the shape selected to reverse

- Select the reversed shape from the **Model Tree** as shown in Figure-166.

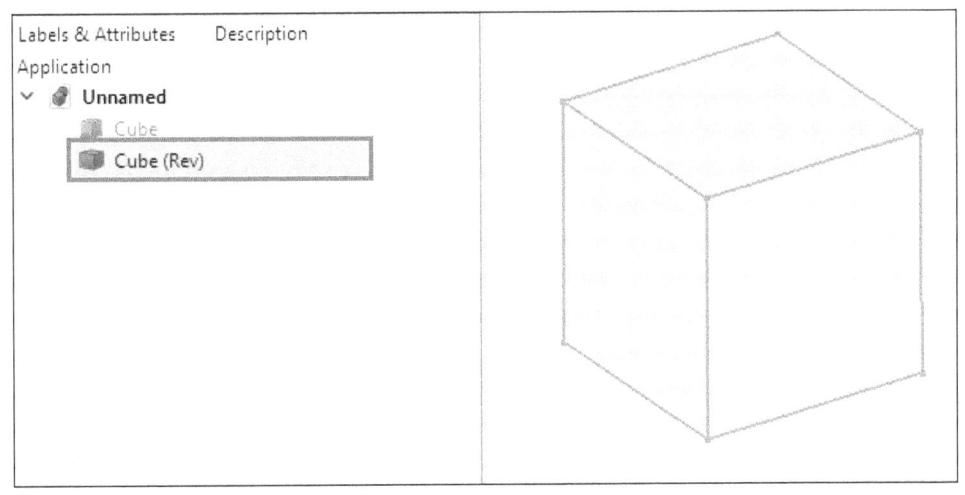

Figure-166. Reversed shape selected

- Select **One side** option from **Lighting** drop-down in the **Object Style** section of **View** tab in the **Property editor** dialog; refer to Figure-167.
- Click **RMB** in the 3D view area, the reversed shape will be turned black, it means that you are now looking at the back of the faces; refer to Figure-168.

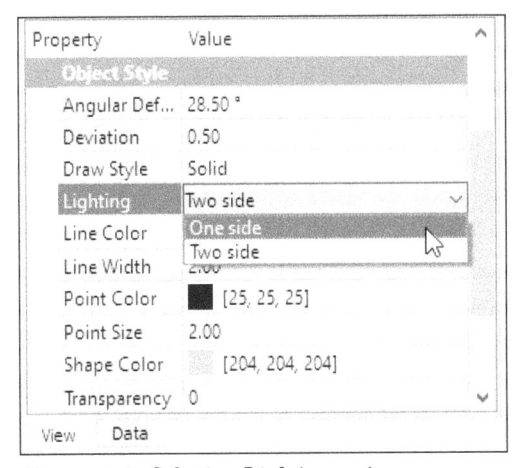

Figure-167. Selecting Lighting option

Figure-168. Reversed shape turned black

Create simple copy

The **Create simple copy** tool creates a simple copy of a selected object. The procedure to use this tool is discussed next.

- Select the object from the **Model Tree** or from the 3D view area of which you want to create a copy; refer to Figure-169.

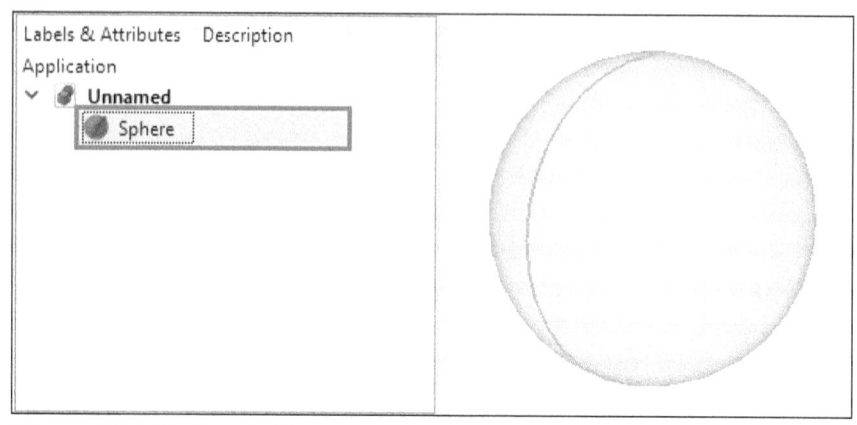

Figure-169. Selecting the object to copy

- Click on the **Create simple copy** tool from **Create a copy** cascading menu in the **Part** menu of the **Ribbon**. A copy of the selected object will be created; refer to Figure-170.

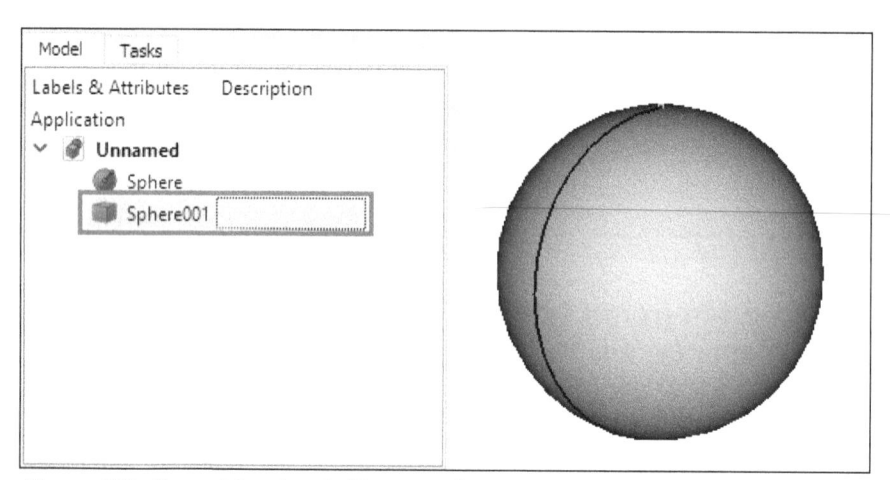

Figure-170. Copy of the selected object created

- You can reposition the copied object by selecting the object from **Model Tree** and specifying desired values in the **x**, **y**, or **z** coordinates from **Data** tab in the **Property editor** dialog; refer to Figure-171. The copied object will be repositioned; refer to Figure-172.

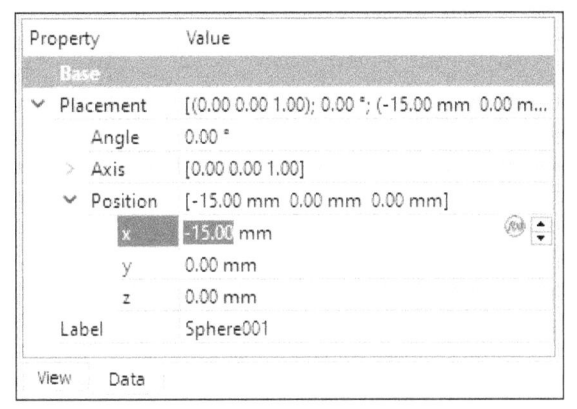

Figure-171. Specifying values to reposition the object

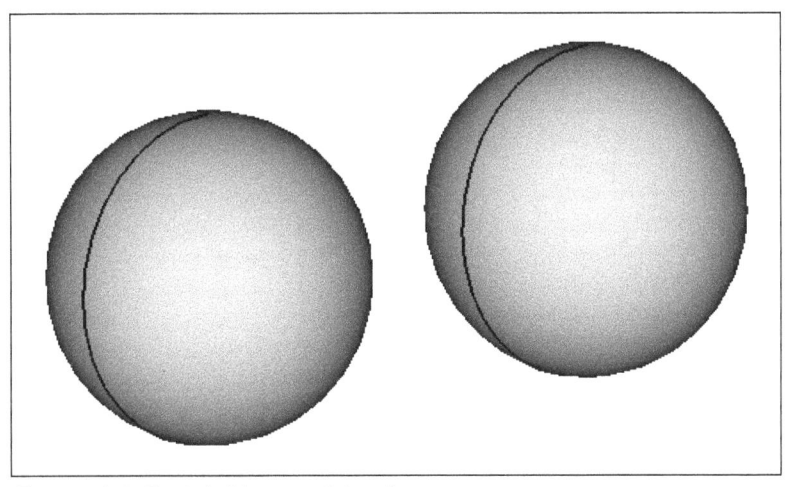

Figure-172. Copied object repositioned

Create transformed copy

The **Create transformed copy** tool creates transformed copy of a selected object that has been displaced from its original position. The procedure to use this tool is same as discussed in previous tool.

Create shape element copy

The **Create shape element copy** tool creates a copy from a sub-element (vertex, edge, face) of a selected object. The procedure to use this tool is same as discussed earlier.

Refine shape

The **Refine shape** tool creates a copy of selected object with a refined shape, that is, with certain edges and faces cleaned up. The procedure to use this tool is discussed next.

- Select the objects and perform boolean operations like, **Part Fuse**, as discussed earlier; refer to Figure-173. The objects will be fused but still some lines from the previous objects will remain visible; refer to Figure-174.

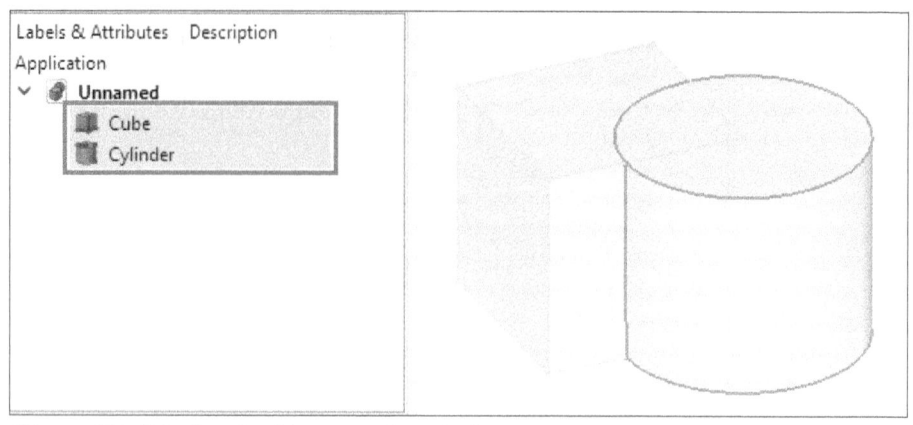

Figure-173. Selecting the objects to perform boolean operation

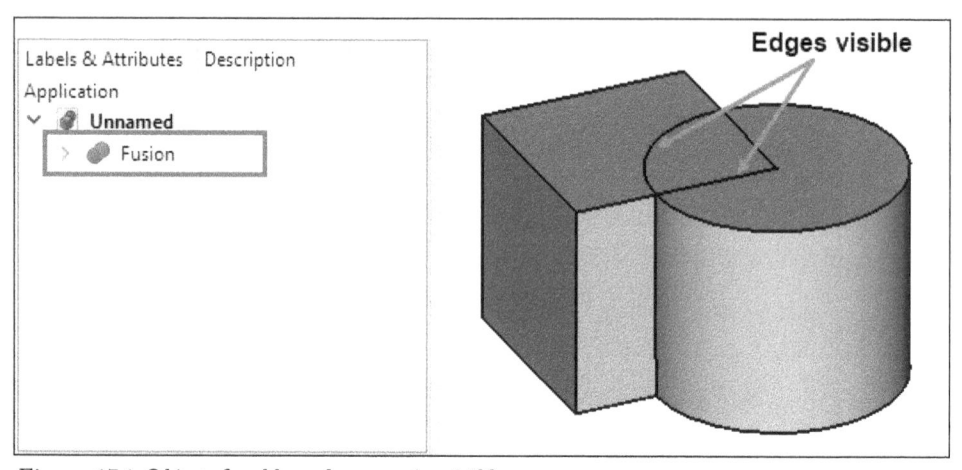

Figure-174. Objects fused but edges remain visible

- Select the fused feature from the **Model Tree** and click on the **Refine shape** tool from **Create a copy** cascading menu in the **Part** menu of the **Ribbon**. The objects will be refined; refer to Figure-175.

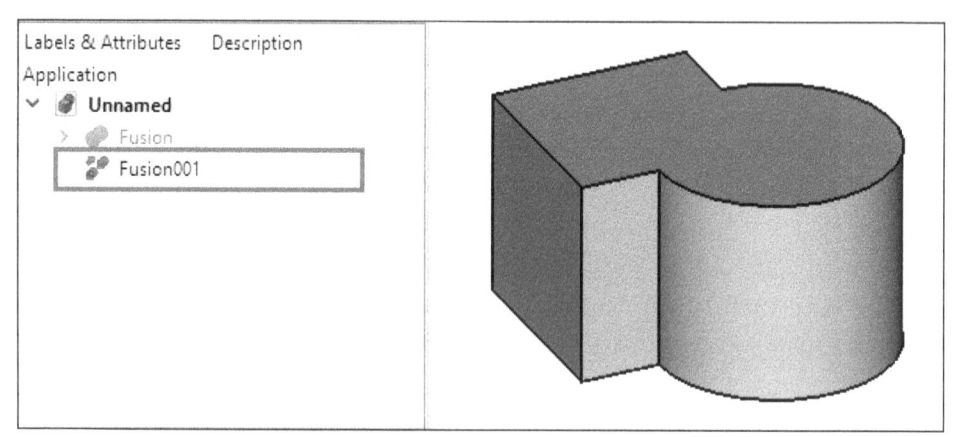

Figure-175. Refined object

- You can reposition the fused features with the **Property editor** dialog as discussed earlier.

Attachment

The **Attachment** tool is a utility to attach an object to another one. The attached object is linked to the other object which means that if the latter's placement is changed afterwards, the attached object will update to its new position. The procedure to use this tool is discussed next.

- Select the object which you want to attach from the **Model Tree** or from the 3D view area; refer to Figure-176.

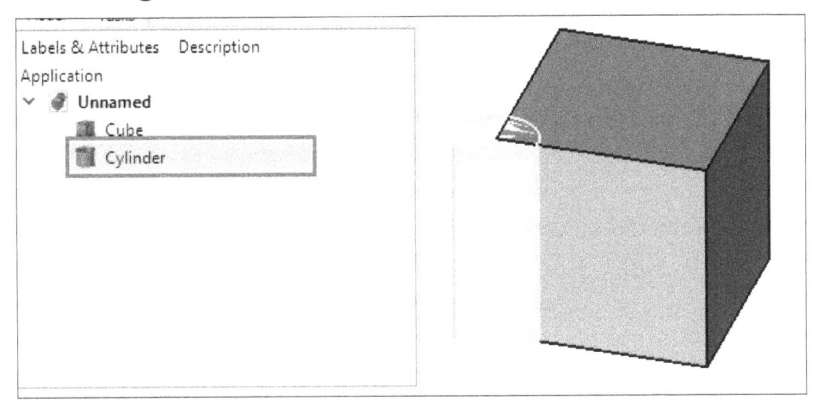

Figure-176. Object selected to attach

- Click on the **Attachment** tool from **Part** menu in the **Ribbon**. The **Attachment** dialog will be displayed in the **Tasks** panel of **Combo View**; refer to Figure-177, asking you to select the vertex, edge, face/plane as a reference.

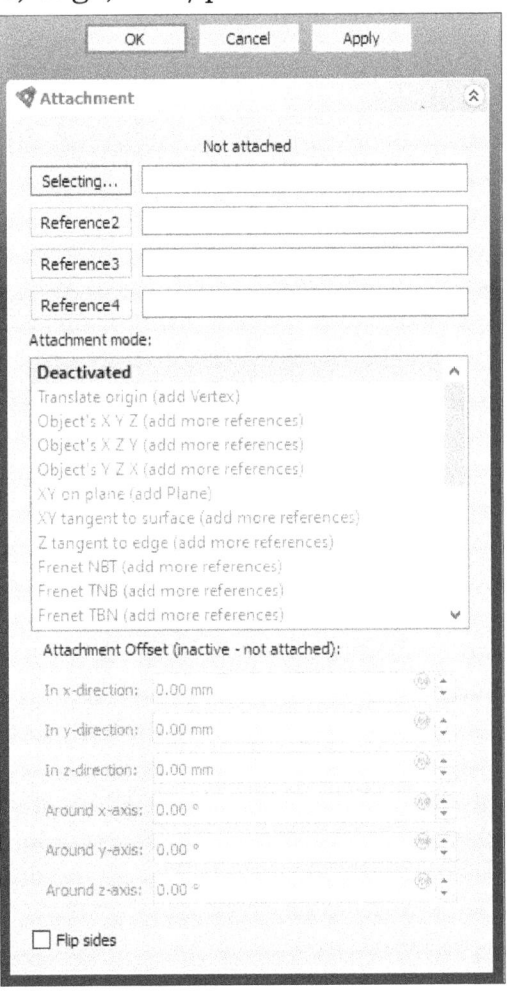

Figure-177. Attachment dialog

- Select desired reference on the object on which you want to attach the selected object; refer to Figure-178.

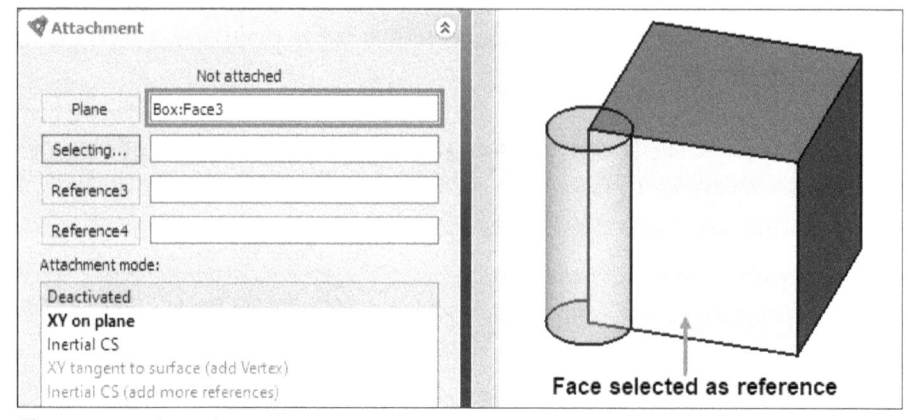

Figure-178. Face selected as reference

- Select desired attachment mode from **Attachment mode** area of the dialog; refer to Figure-179.

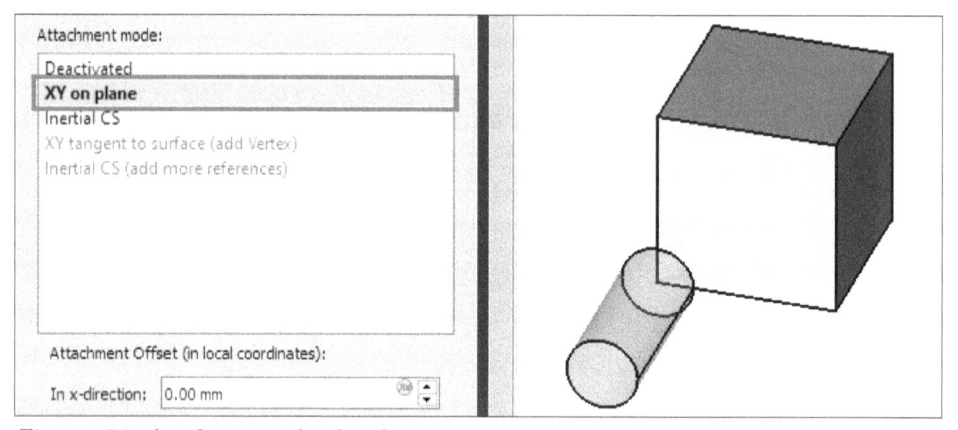

Figure-179. Attachment mode selected

- Specify desired offset values relative to the local coordinate system in the edit boxes from **Attachment Offset (in local coordinates)** area of the dialog.
- Select **Flip sides** check box to reverse the attached object from its **XY** plane.
- After specifying desired parameters, click on the **OK** button from the dialog. The object will be attached; refer to Figure-180.

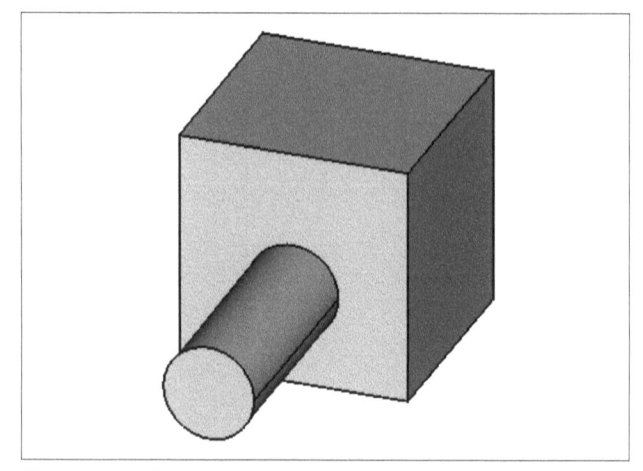

Figure-180. Object attached

SELF-ASSESSMENT

Q1. What is the purpose of the Cube tool in the Part workbench?

A) To create a sphere
B) To insert a parametric, rectangular cuboid
C) To modify existing shapes
D) To create a torus

Q2. What does the Cylinder tool create in the Part workbench?

A) A truncated cone
B) A rectangular cuboid
C) A parametric cylinder
D) A fillet shape

Q3. Which of the following tools allows the creation of more complex shapes from parametric geometric primitives?

A) Cube Tool
B) Create Primitives Tool
C) Shape Builder Tool
D) Mirror Tool

Q4. What is the function of the Extrude tool?

A) To rotate a shape around an axis
B) To extend a shape by a specified distance in a specified direction
C) To create a mirrored object
D) To scale an object

Q5. When using the Revolve tool, what do you need to specify?

A) The radius of the shape
B) The angle of revolution and the axis about which to revolve
C) The color and texture of the object
D) The height and width of the shape

Q6. Which tool would you use to create a new object that is a reflection of an original object?

A) Scale Tool
B) Mirror Tool
C) Chamfer Tool
D) Fillet Tool

Q7. What is the first step to use the Create Primitives tool?

A) Click on the Shape Builder tool
B) Select a primitive type from the drop-down menu
C) Click on the Cylinder tool
D) Specify the parameters for extrusion

Q8. Which of the following tools is used to modify existing shapes?

A) Create Primitives Tool
B) Shape Builder Tool
C) Modifying Object Tools
D) Cube Tool

Q9. What does the Tube tool geometrically represent?

A) A solid sphere
B) A cut of a smaller cylinder into a larger one
C) A single geometric edge
D) A rectangular cuboid

Q10. To create a ruled surface using the Ruled Surface tool, what is required?

A) Two edges or sketch curves
B) A single wire
C) A closed sketch
D) A geometric primitive

Q11. Which feature allows the creation of more intricate shapes by combining simpler geometric forms?

A) Fillet Tool
B) Shape Builder Tool
C) Revolve Tool
D) Chamfer Tool

Q12. What is the primary function of the Chamfer tool?

A) To create a rounded edge on an object
B) To create a new shape from existing shapes
C) To create a beveled edge on selected edges
D) To mirror an object

Q13. What is the primary function of the Sweep tool?

A) To create a hollow object
B) To create a face, shell, or solid shape from profiles along a path
C) To create a section view of a model
D) To attach one object to another

Q14. What should you do first when using the Sweep tool?

A) Select the Sweep Path
B) Click on the Done button
C) Select a profile to use as a base from the Available profiles
D) Check the Create solid checkbox

Q15. What is the purpose of the Section tool?

A) To create a solid model from profiles
B) To create a section by intersecting an object with a section plane
C) To define colors for faces
D) To create a compound of shapes

Q16. What does the 3D Offset tool create?

A) A single wire parallel to the original
B) A shape by intersecting two solids
C) Parallel copies of a selected shape at a specified distance
D) A section view of a model

Q17. When using the 2D Offset tool, what happens if you specify a negative offset?

A) The shape is deleted
B) A shape is created in the opposite direction of the original
C) The shape becomes invisible
D) An error message is displayed

Q18. Which tool is used to transform a solid shape into a hollow object?

A) Sweep Tool
B) Projection On Surface Tool
C) Thickness Tool
D) Color Per Face Tool

Q19. What does the Projection On Surface tool do?

A) Creates a new shape based on profiles
B) Projects a shape onto a face of another object
C) Defines colors for individual faces
D) Attaches one object to another

Q20. In the Compound tools, what does the Make Compound tool do?

A) Splits a compound into individual shapes
B) Creates a compound of selected topological shapes
C) Filters specific elements from a compound
D) Performs boolean operations on shapes

Q21. What is the purpose of the Boolean tool in the Part workbench?

A) To connect objects
B) To specify operations between selected objects (Union, Difference, Intersection)
C) To create offset shapes
D) To change the color of an object

Q22. What does the Color per face tool allow you to do?

A) Define a color for each face of an object
B) Create a new shape based on colors
C) Apply a single color to the entire object
D) Remove color from the object

Q23. Which checkbox should be selected to allow offsets to overlap during the 3D Offset process?

A) Self-intersection
B) Fill offset
C) Intersection
D) Skin

Q24. What is a characteristic feature of the Explode compound tool?

A) It connects two objects
B) It creates a new compound
C) It splits a compound into separate objects
D) It changes the color of a compound

Q25. Which tool is used to create a hollow shape with defined thickness from a solid object?

A) Thickness Tool
B) Join Objects Tool
C) Projection On Surface Tool
D) Sweep Tool

Q26. What is the primary function of the Boolean fragments tool?

A) To split shapes by intersection with other shapes
B) To create all fragments from boolean operations between input shapes
C) To measure distance between selected entities
D) To check the validity of a solid's geometry

Q27. How can you view individual pieces of boolean fragments after using the Boolean fragments tool?

A) By clicking on the Check Geometry tool
B) By using the Downgrade tool
C) By exporting the fragments
D) By using the Slice tool

Q28. Which tool is used to split shapes by intersection with another shape?

A) Boolean XOR
B) Slice Apart
C) Measure Tool
D) Defeaturing Tool

Q29. What does the Boolean XOR tool do?

A) It measures angles between two entities
B) It removes overlapping geometry shared by selected objects
C) It creates a simple copy of a selected object
D) It converts a shape object to a solid object

Q30. Which tool is used to measure the shortest distance between selected entities in the 3D view?

A) Length Tool
B) Angle Tool
C) Distance Tool
D) Position Tool

Q31. How do you measure the area of a selected face in the 3D view?

A) Using the Center of Mass tool
B) Using the Area tool
C) Using the Length tool
D) Using the Position tool

Q32. What happens when you click the Reset button in the Measure tool dialog box?

A) All measurements are saved
B) The measurement is deleted from the 3D view
C) The tool is closed
D) The measurement is recalibrated

Q33. Which tool allows you to import CAD file formats such as IGES and STEP?

A) Export CAD
B) Import CAD
C) Create shape from mesh
D) Box Selection

Q34. What is the purpose of the Defeaturing tool in the Part workbench?

A) To measure angles in the model
B) To remove selected features such as holes and protrusions
C) To create a simple copy of an object
D) To verify the geometry of a solid

Q35. Which tool can be used to split a shape into a compound shape?

A) Slice tool
B) Boolean fragments tool
C) Check Geometry
D) Reverse shapes

Q36. What does the Create points object from mesh tool produce?

A) A solid object from a shape
B) A shape object made of points
C) A refined shape with cleaned edges
D) A transformed copy of the original object

Q37. Which of the following statements is true regarding the Check Geometry tool?

A) It automatically repairs any detected errors.
B) It only checks the size of the geometry.
C) It highlights errors in the 3D view for correction.
D) It is used to create a copy of the selected object.

Q38. What should you do if you want to ensure the shape is reversed using the Reverse shapes tool?

A) Use the Defeaturing tool after reversing
B) Select the shape and change the lighting option to One side
C) Click the Reset button in the Measure tool
D) Hide the shape selected to reverse from the Model Tree

Q39. What is the result of applying the Create simple copy tool?

A) A refined version of the selected object
B) A transformed copy of the selected object
C) A direct copy of the selected object
D) A shape created from a mesh

Q40. What action should be taken to create a transformed copy of an object?

A) Use the Measure tool to check the original dimensions
B) Select the object and apply the Create transformed copy tool
C) Import the object into a new document
D) Use the Downgrade tool on the original object

Chapter 5

Solid Modeling Practice

Topics Covered

The major topics covered in this chapter are:

- *Practical 1*
- *Practical 2*
- *Practical 3*
- *Practices*

PRACTICAL 1

Create the model (isometric view) as shown in Figure-1. The views of the model with dimensions are given in Figure-2.

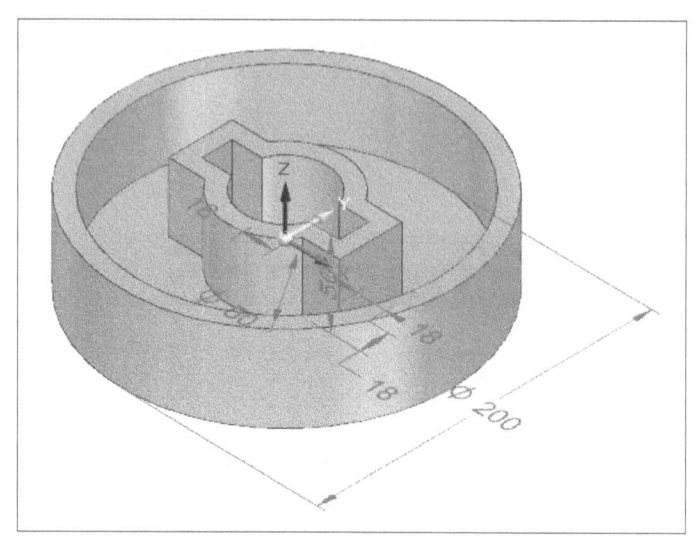

Figure-1. Practical Model 1

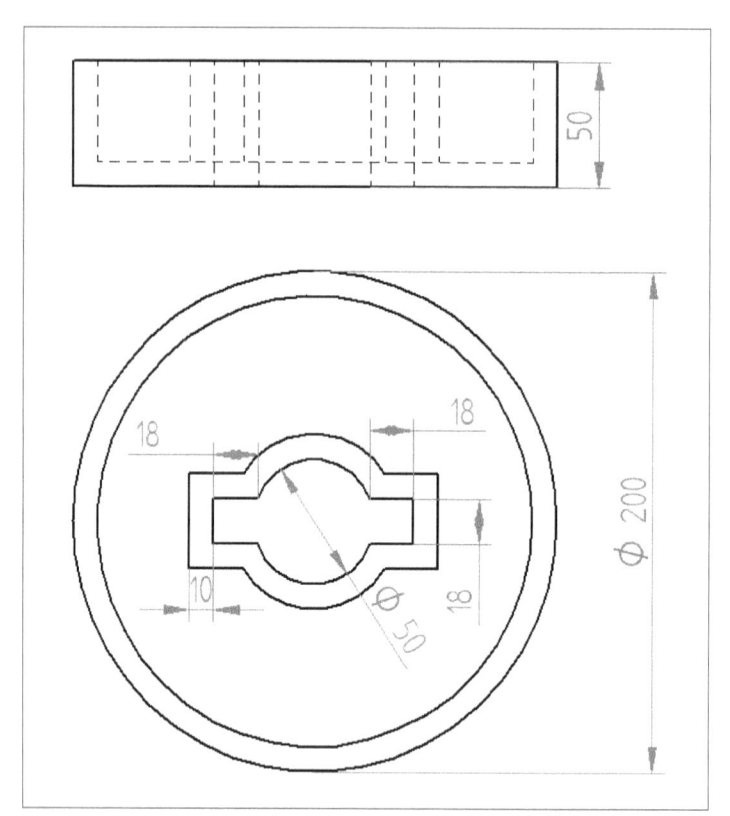

Figure-2. Views for practical 1

Before we start working on the practical, it is important to understand two terms; First angle projection and Third angle projection. These are the standards of placing views in the engineering drawing. The views placed in the above figure are using third angle projection. In first angle projection, the top view of model is placed below the front view and right side view is placed at left of the front view. You will learn more about projection in chapter related to drafting.

Starting Part Design Environment and Creating Sketch

- Double-click on the FreeCAD icon from desktop if you have not started FreeCAD yet.
- Start a new document by clicking on the **New** button from **Toolbar** or press **CTRL+N** shortcut key.
- Select the **Part Design** option from **Switch between workbenches** drop-down in the **Toolbar**. The tools to create solid model will be displayed.
- Click on the **Create sketch** tool from the **Toolbar** or **Part Design** menu. The **Select attachment** dialog will be displayed in the **Combo View** and you will be asked to select a plane for creating sketch.
- Select the **XY** plane from the list in the dialog and click on the **OK** button. The sketching environment will be displayed.
- Create a circle of diameter **200** with its center at origin; refer to Figure-3. Note that as we create a circle, we have also applied the diameter dimension to it. In FreeCAD, it is better to apply dimension as soon as you create the entity if the entity is independent.

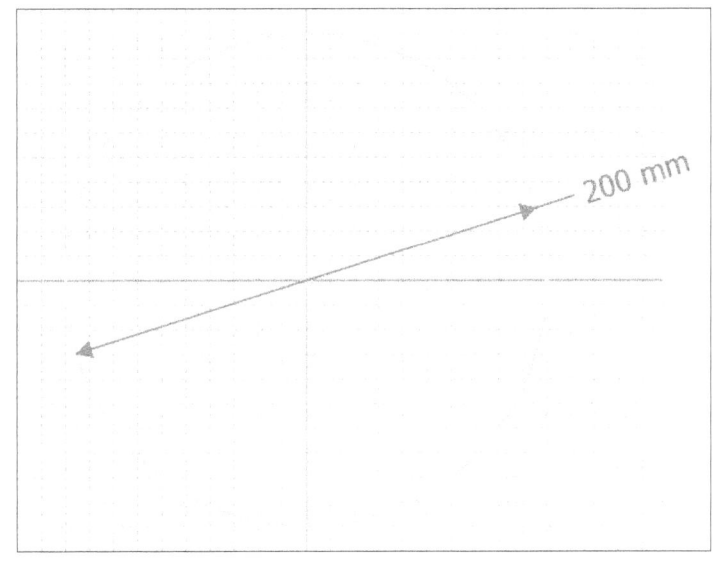

Figure-3. Circle to be created

- Create a circle of diameter **50** at the center of earlier created circle and then draw a rectangle as shown in Figure-4.

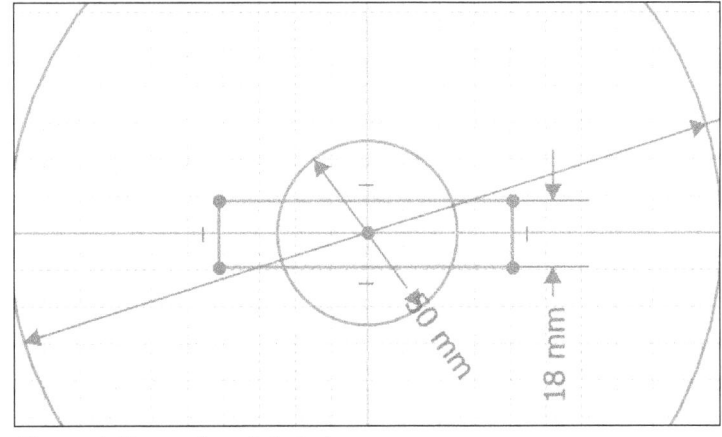

Figure-4. Rectangle and circle drawn

- Trim the inner portion of circle and rectangle, and then apply the dimensions as shown in Figure-5.

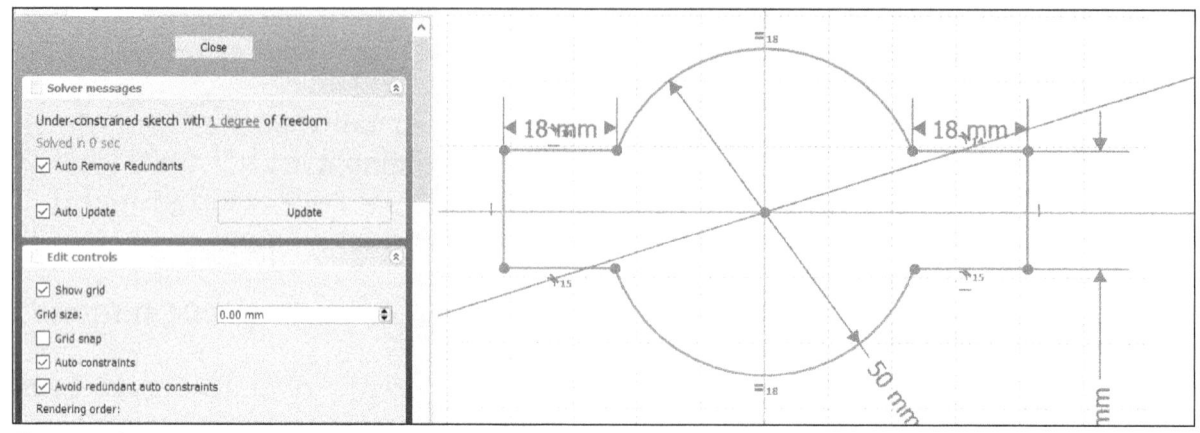

Figure-5. Sketch after applying dimensions

- Note that there is 1 degree of freedom left in the sketch. If you drag the vertical line of rectangle then you will find that it is free to move up and down. To constraint this movement, select the points of rectangle as shown in Figure-6 and click on the **Constrain vertical** tool from the **Toolbar**. The sketch will become fully constraint; refer to Figure-7.

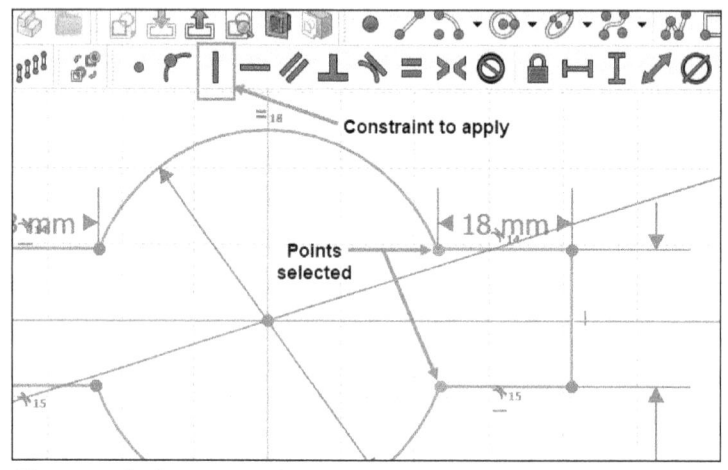

Figure-6. Applying vertical constrain

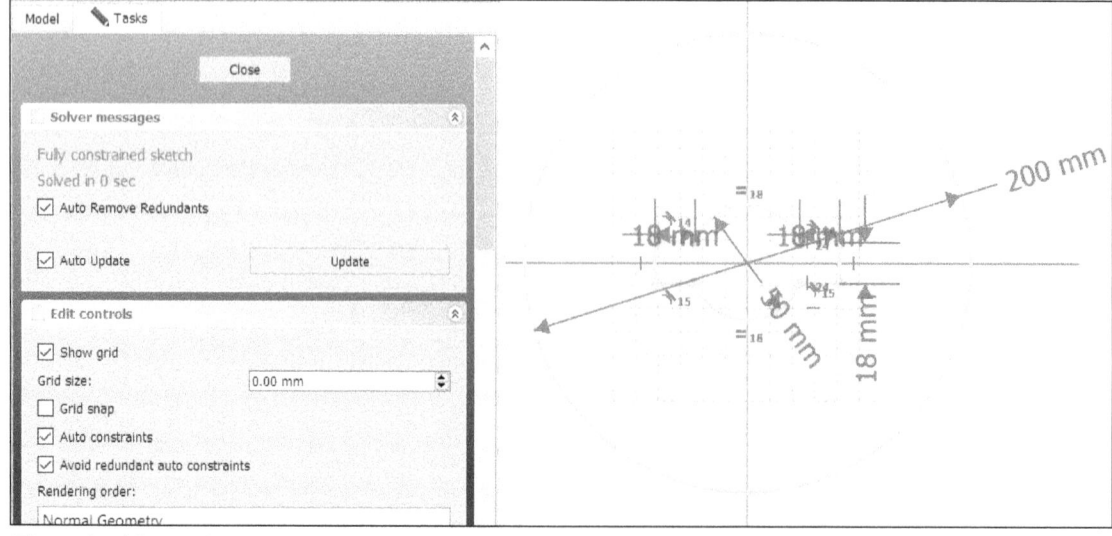

Figure-7. After applying constraints

- Click on the **Close** button to exit the sketching environment.

Creating Pad and Thick Solid Features

- Select the newly created sketch from the **Model Tree** if not selected by default and then click on the **Pad** tool from **Toolbar** or **Part Design** menu. Preview of pad feature will be displayed.
- Specify the length of pad feature as **50** in the **Length** edit box of dialog; refer to Figure-8 and click on the **OK** button. The feature will be created.

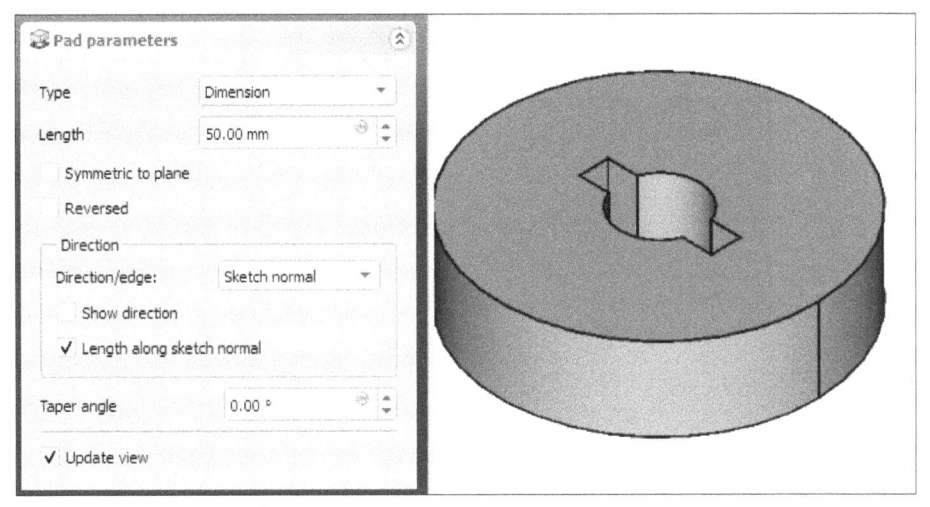

Figure-8. Creating pad feature

- Select the top flat face of the model from drawing area and click on the **Thickness** tool from the **Toolbar** or **Part Design** menu; refer to Figure-9. Preview of thickness feature will be displayed. (Do not worry if everything goes blank!!)

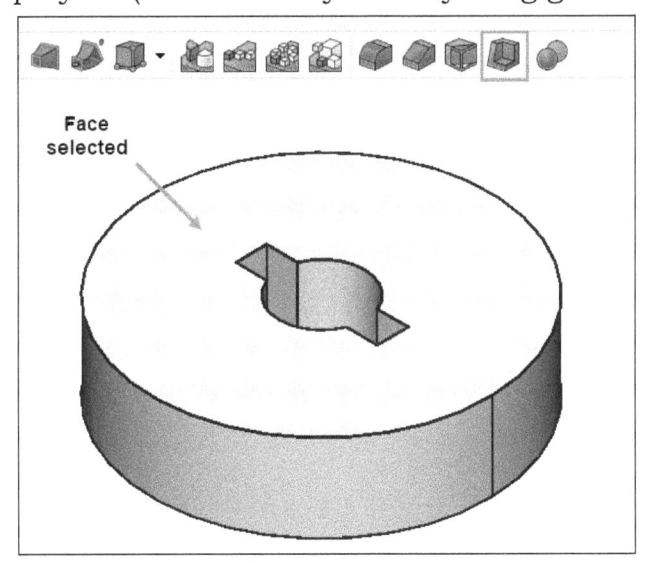

Figure-9. Face selected for thickness feature

- Specify the parameters shown in Figure-10 and click on the **OK** button to create the final model.

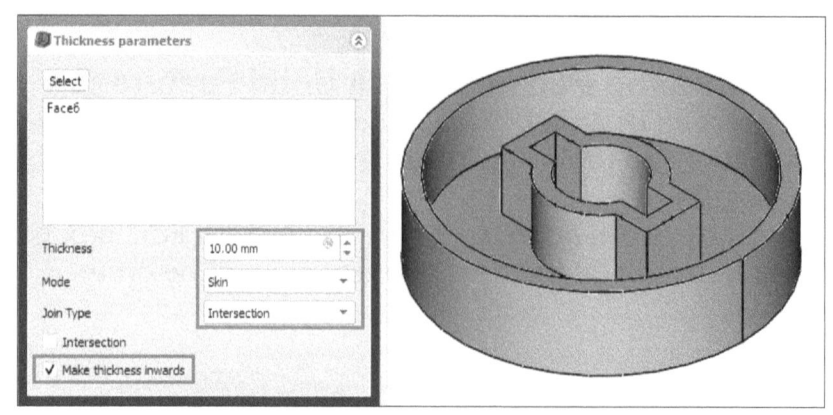

Figure-10. Preview of thickness feature

PRACTICAL 2

Create the model (isometric view) as shown in Figure-11. The dimensions of the model are given in Figure-12.

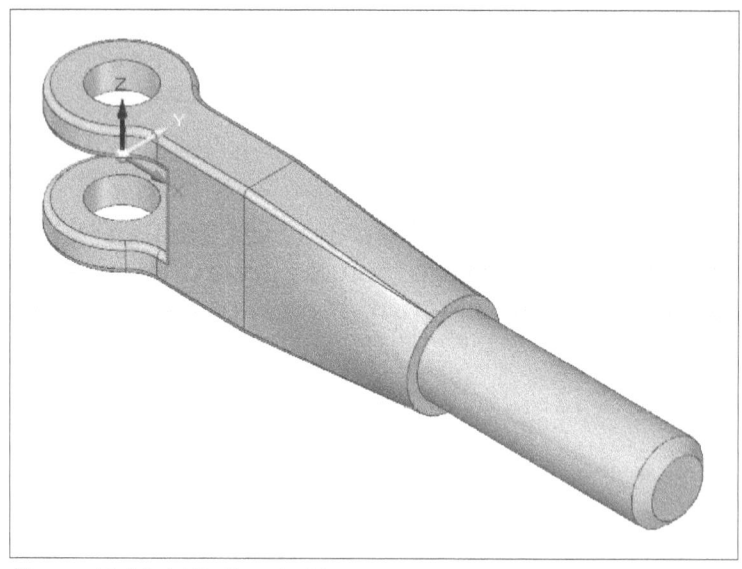

Figure-11. Model for Practical 2

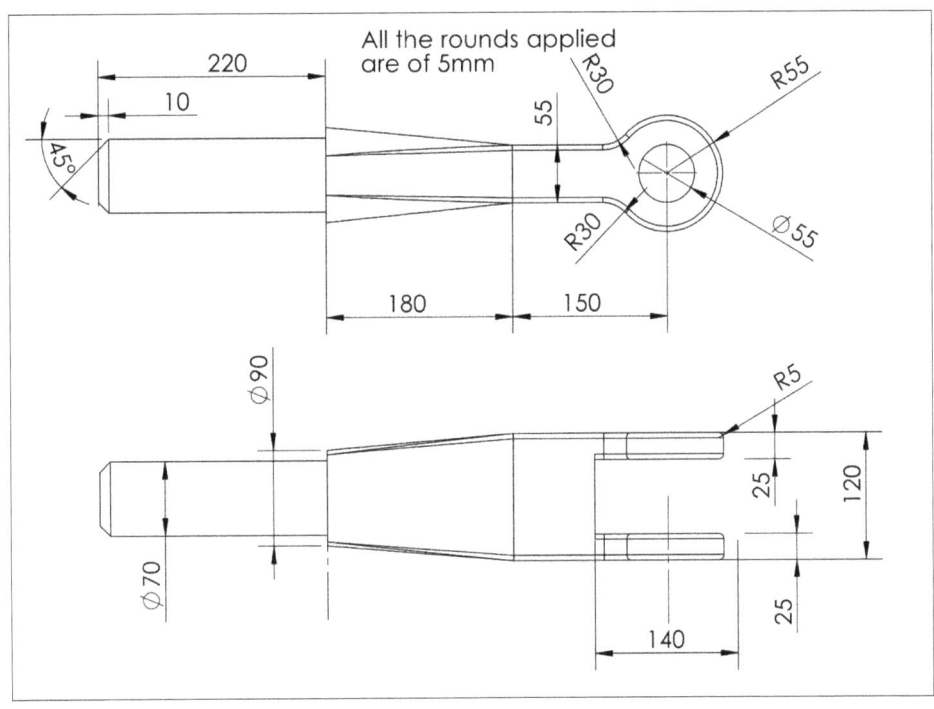

Figure-12. Practical 2 drawing views

Creating first pad feature

This model is a combination of pad and loft features. We will create the model in multiple steps with first feature acting as base for all the other features.

- Start FreeCAD if not started yet and create a new document.
- Switch to the **Part Design** workbench using **Part Design** option from **Switch between workbenches** drop-down in the **Toolbar**.
- Click on the **Create sketch** tool from the **Toolbar** or **Part Design** menu. The **Select attachment** dialog box will be displayed and you will be asked to specify a plane for creating sketch.
- Select the **XY** plane from dialog or drawing area and click on the **OK** button. The sketching environment will be displayed.
- Create the sketch as shown in Figure-13.

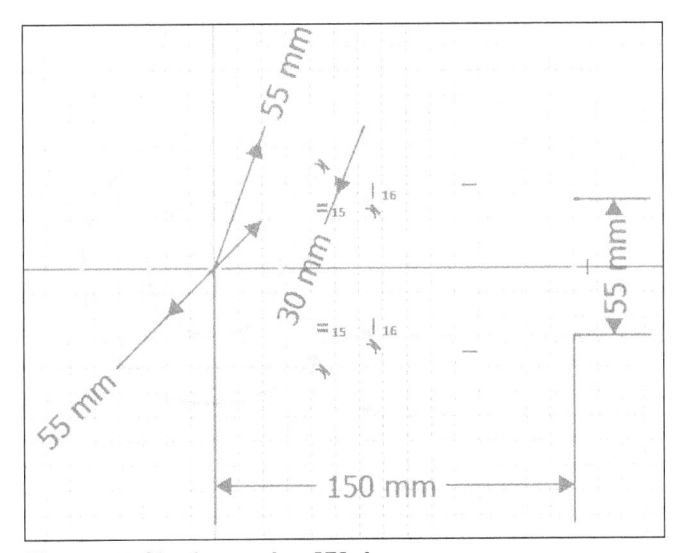

Figure-13. Sketch created on XY plane

- Click on the **Close** button from dialog after creating sketch to exit sketching mode.
- Click on the **Pad** tool from **Toolbar** after selecting newly created sketch. The **Pad parameters** dialog will be displayed with preview of feature.
- Select the **Symmetric to plane** check box from dialog and specify length of extrude feature as **120** mm; refer to Figure-14.

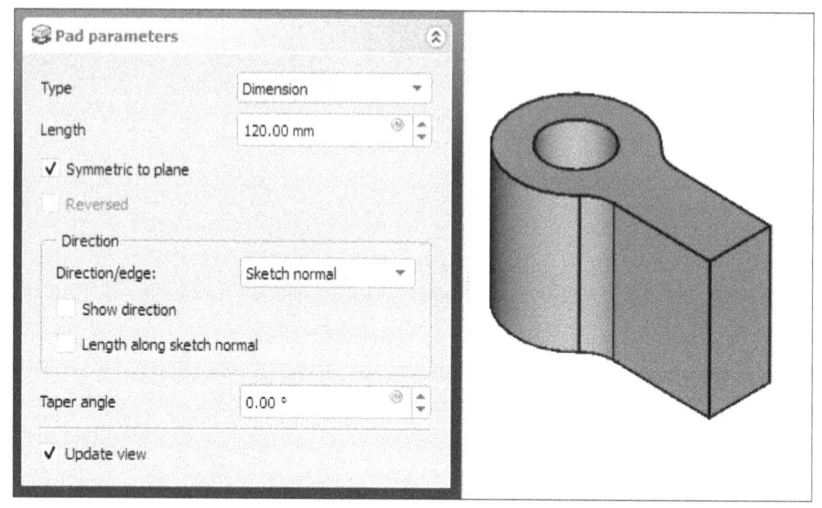

Figure-14. Pad parameters specified

- Click on the **OK** button from the dialog to create the feature.

Creating Loft feature

- Select the flat side face of model as shown in Figure-15 and click on the **Create sketch** tool from **Toolbar**. The sketching environment will be displayed.

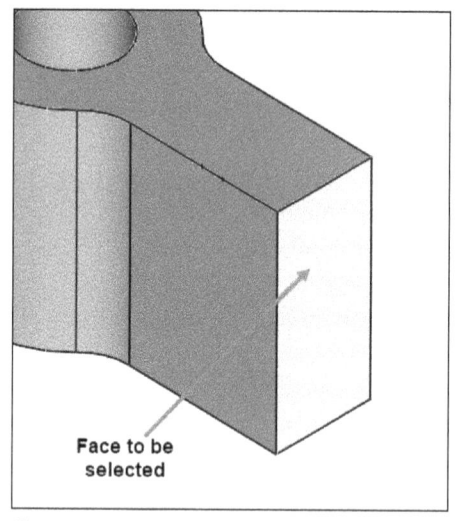

Figure-15. Face selected for sketching

- Click on the **Create external geometry** tool from the **Sketcher tools** cascading menu or **Toolbar**. You will be asked to select edges to be projected in the sketch.
- Select the boundary edges of flat face selected earlier; refer to Figure-16. Note that the projected edges are not part of sketch, they are for reference only.
- Click on the **Create rectangle** tool from **Toolbar** or **Sketcher geometries** cascading menu from the **Sketch** menu and create a rectangle overlapping the projected edges; refer to Figure-17.

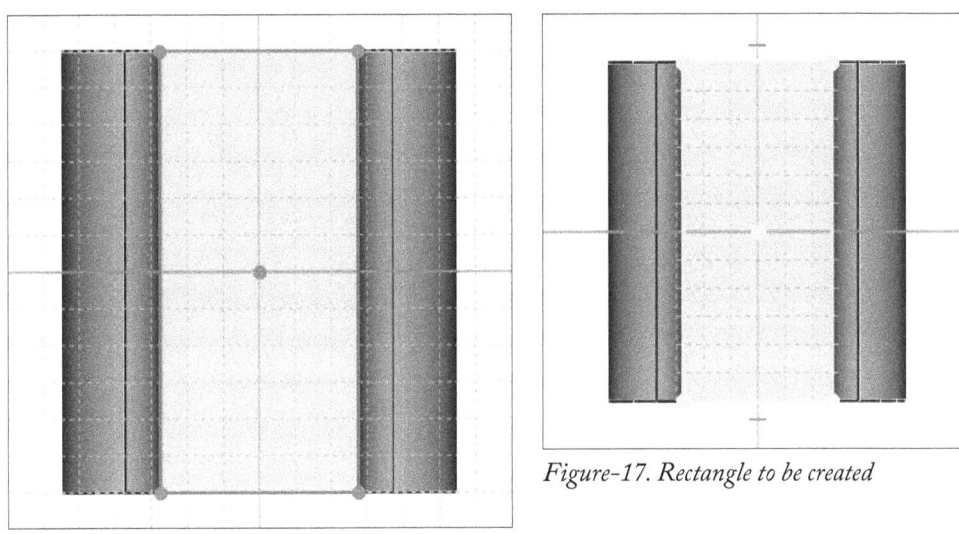

Figure-16. Edges of face projected in sketch

Figure-17. Rectangle to be created

- Click on the **Close** button to exit sketching environment. Click in the empty area to exit any entity selected.
- Click on the **Create a datum plane** tool from the **Part Design** menu or **Toolbar**. The **Datum Plane parameters** dialog will be displayed in **Combo View**.
- Select flat side face of model as reference and specify **Z** offset distance as **180**; refer to Figure-18.

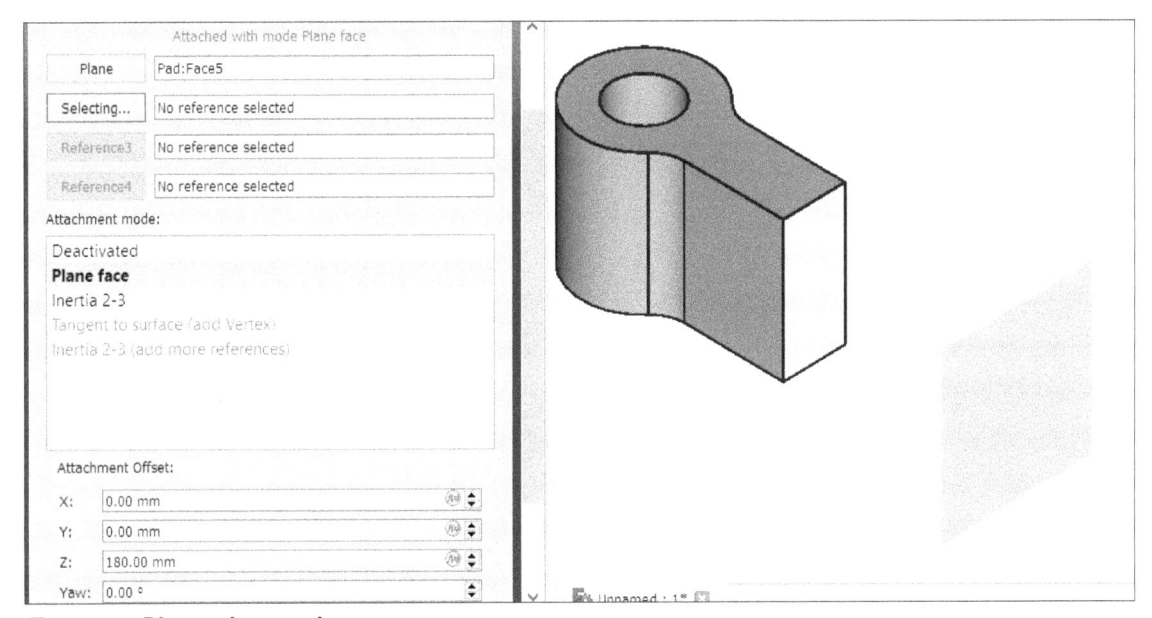

Figure-18. Plane to be created

- Click on the **OK** button from the dialog to create plane.
- Select the newly created plane and click on the **Create sketch** tool from **Part Design** menu or **Toolbar**. The sketching environment will be displayed.
- Create a circle of diameter **90** mm at the origin; refer to Figure-19.

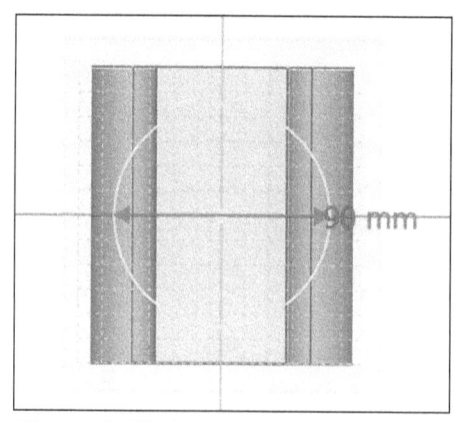

Figure-19. Circle to create

- Click on the **Close** button from the dialog to exit sketching mode.
- Click on the **Additive loft** tool from **Part Design** menu or **Toolbar**. The **Select attachment** dialog will be displayed in **Combo View**.
- Select the sketch earlier created on flat face of model from the list; refer to Figure-20 and click on the **OK** button from the dialog. The **Loft parameters** dialog will be displayed and you will be asked to add section sketches for loft feature.

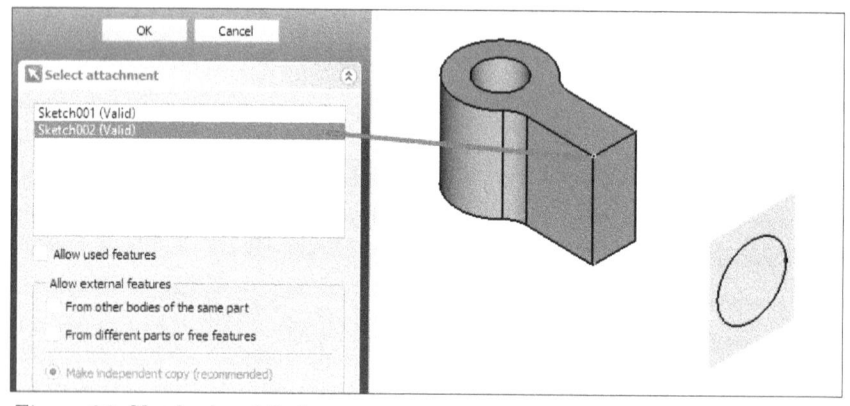

Figure-20. Sketch selected for base

- Click on the **Add Section** button from the dialog and select the circle created earlier. Preview of loft feature will be displayed; refer to Figure-21.

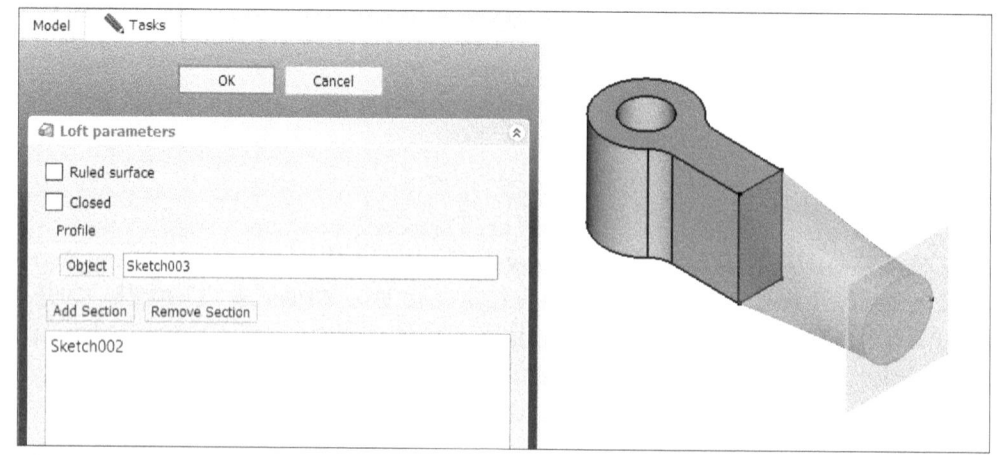

Figure-21. Preview of loft

- Click on the **OK** button from the dialog to create the feature.

Creating Second Pad Feature

- Select the datum plane earlier created and press **SPACEBAR** to hide it.
- Select the flat face of recently created loft feature and click on the **Create sketch** tool from **Part Design** menu. The sketching environment will be activated.
- Create a circle of diameter **70** at the origin; refer to Figure-22.

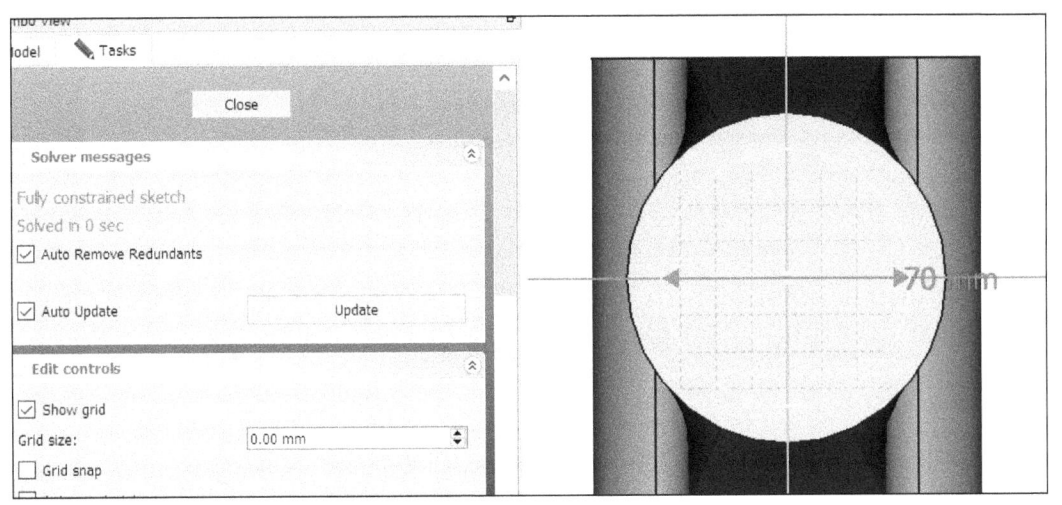

Figure-22. Circle created on loft face

- Click on the **Close** button from the dialog to exit sketching environment.
- Click on the **Pad** tool from the **Toolbar** or **Part Design** menu. Preview of pad feature will be displayed using the recently created sketch.
- Specify the length of feature as **220** in the **Length** edit box; refer to Figure-23 and click on the **OK** button to create the feature.

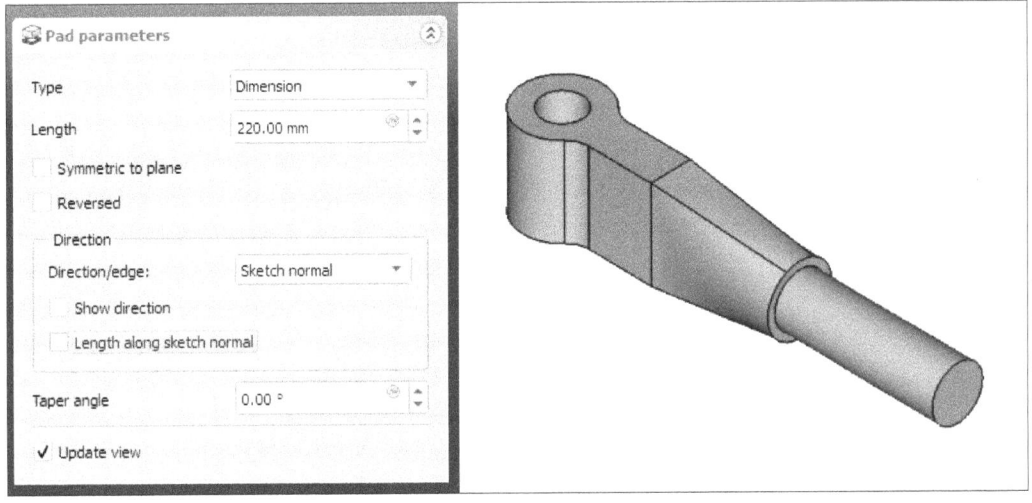

Figure-23. Pad feature after applying length value

Creating Pocket Feature

- Make sure nothing is selected from the model and then click on the **Create sketch** tool from **Part Design** menu. The **Select attachment** dialog will be displayed with list of planes available.
- Select the **XZ Plane** from the dialog and click on the **OK** button. The sketching environment will be displayed; refer to Figure-24.

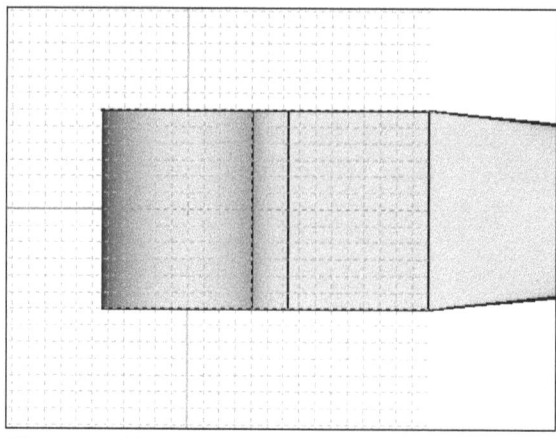

Figure-24. Sketching environment

- Create a rectangle as shown in Figure-25. You may need to switch draw style to **Wireframe** for creating rectangle.

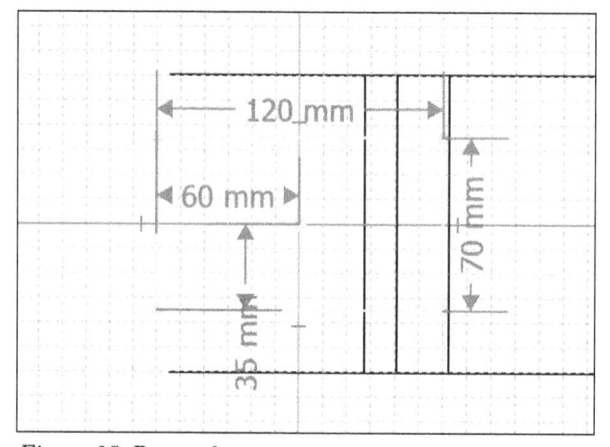

Figure-25. Rectangle to create

- After creating sketch, click on the **Close** button.
- Make sure the sketch is selected and click on the **Pocket** tool from **Part Design** menu. Preview of the pocket feature will be displayed; refer to Figure-26.

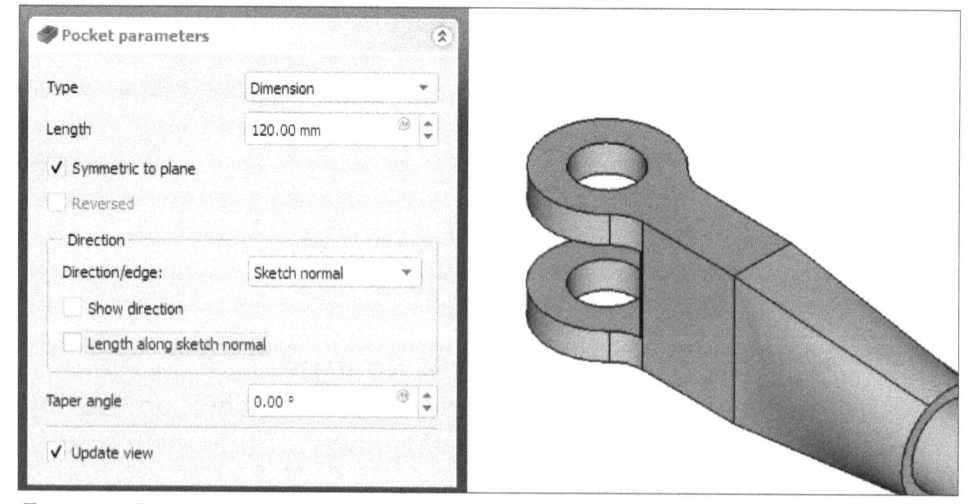

Figure-26. Preview of pocket feature

- Click on the **OK** button from the dialog to create the feature.

Applying Fillets and Chamfers

- Select the edges of model and click on the **Fillet** tool from the **Part Design** menu. Preview of the fillet will be displayed; refer to Figure-27.

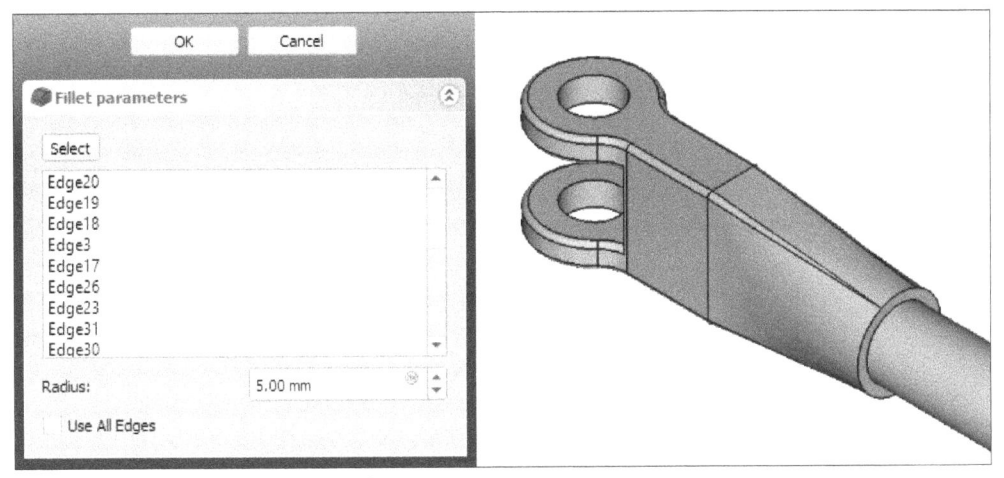

Figure-27. Preview of model after fillet

- Click on the **OK** button from the dialog to apply fillets.
- Select the round edge of pad feature created earlier and click on the **Chamfer** tool from **Part Design** menu.
- Specify the parameters as shown in Figure-28 and click on the **OK** button to apply chamfer.

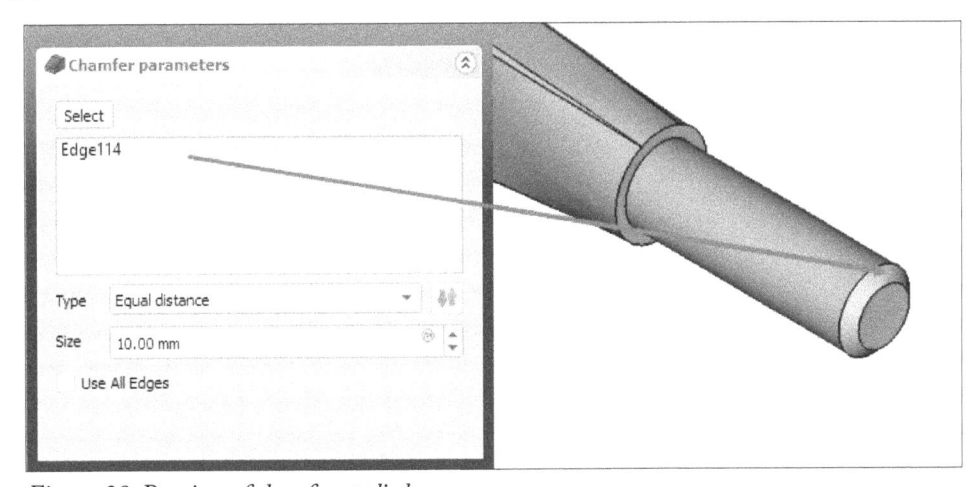

Figure-28. Preview of chamfer applied

- Save the file at desired location.

PRACTICE 1 TO 4

Create 3D models for drawings shown in Figure-29, Figure-30, Figure-31, and Figure-32.

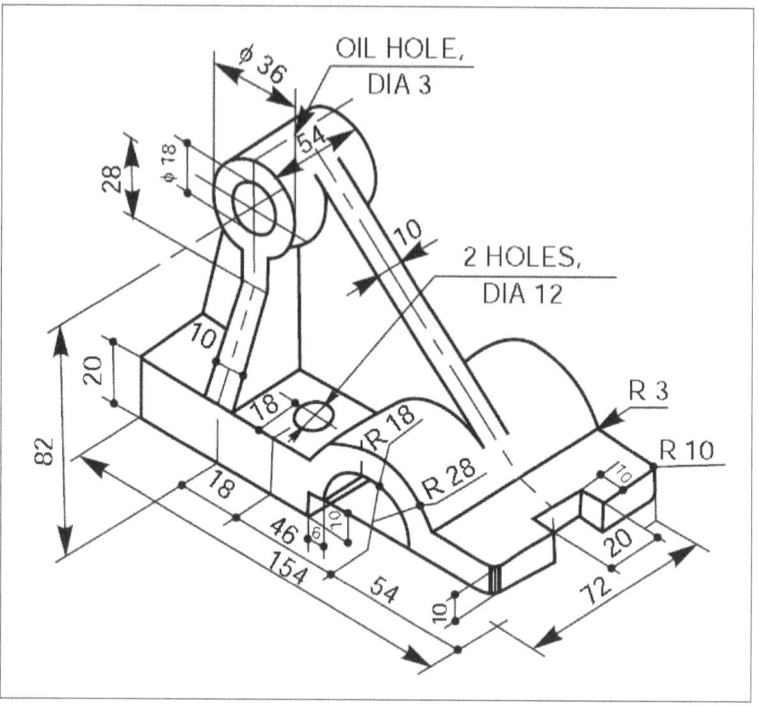

Figure-29. Practice 1

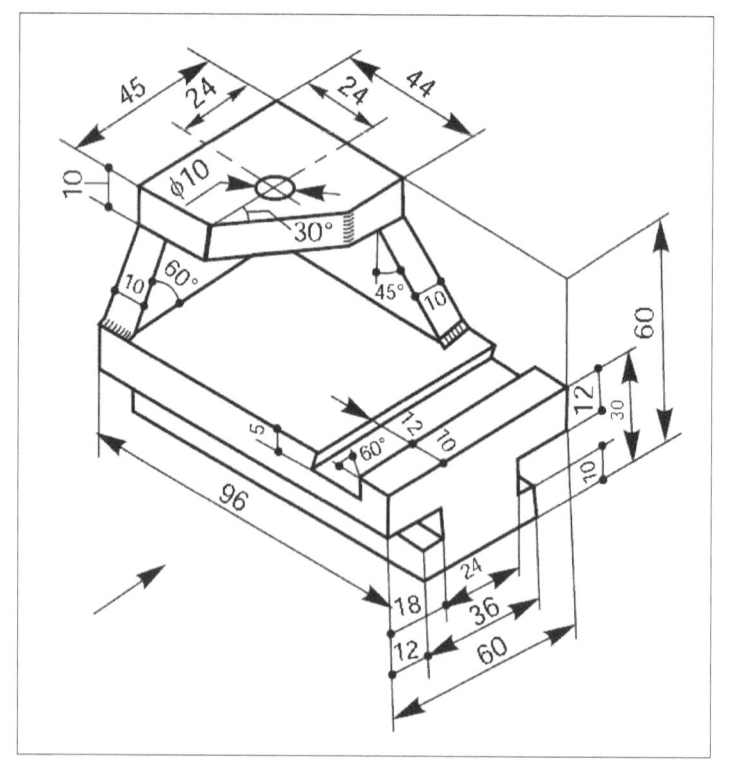

Figure-30. Practice 2

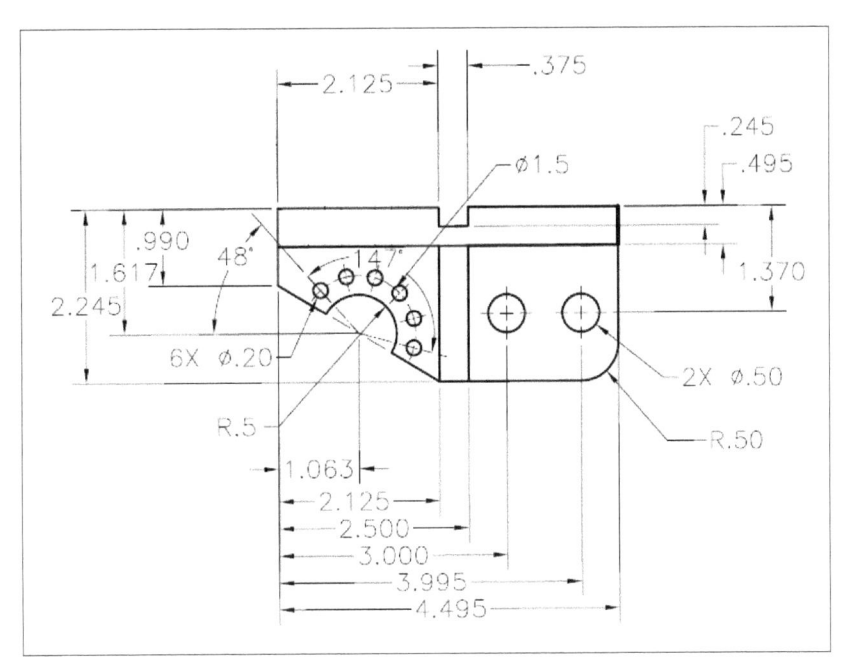

Figure-31. Practice 3

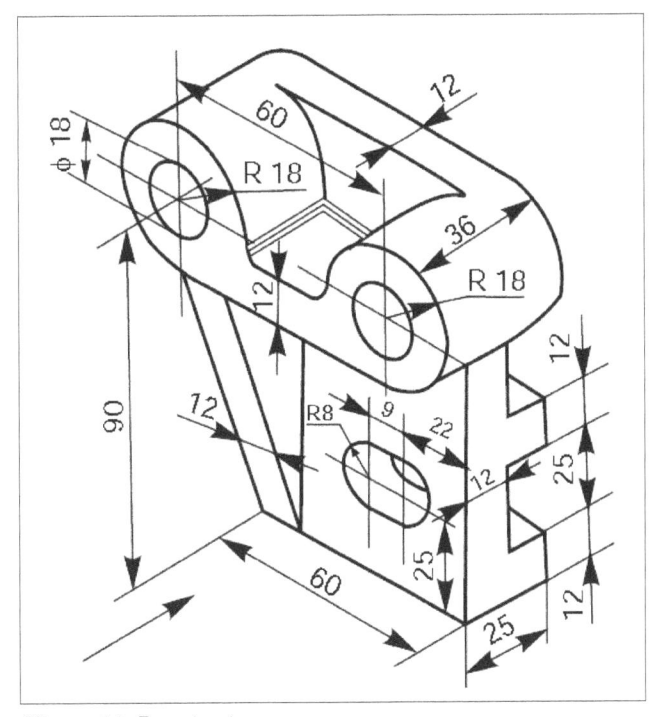

Figure-32. Practice 4

PRACTICAL 3

Create the model (isometric view) as shown in Figure-33. The dimensions and view are given in Figure-34.

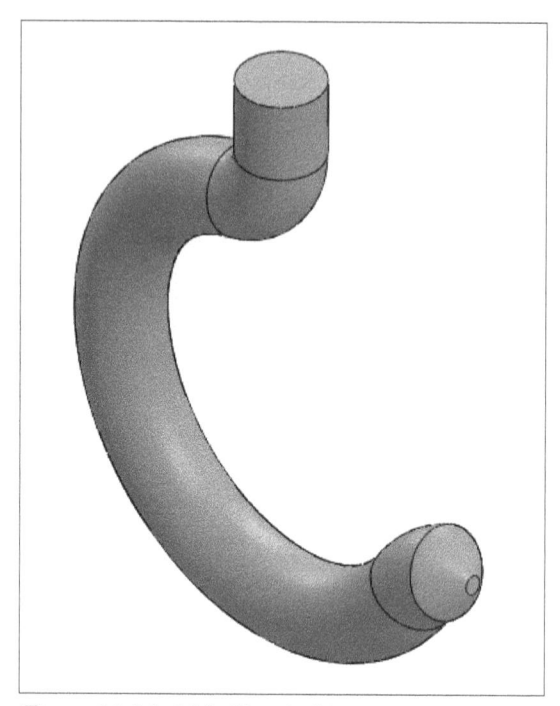

Figure-33. Model for Practical 3

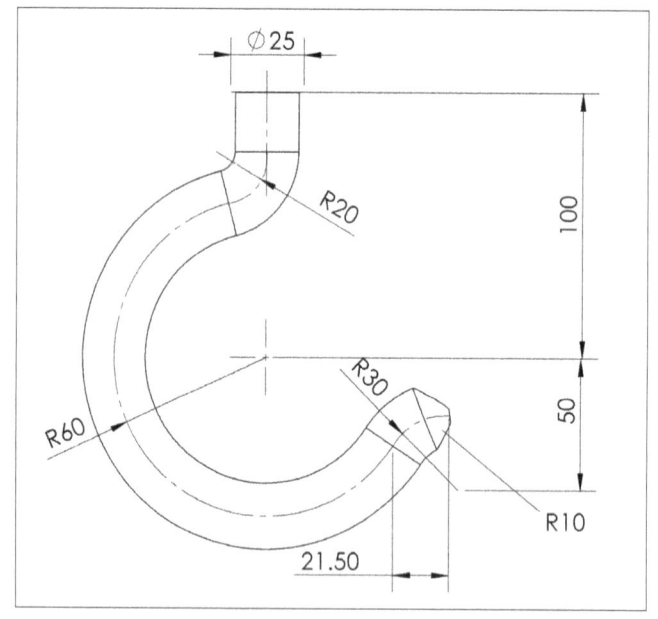

Figure-34. Practical 3 drawing view

Creating Sketches for Hook

- Start FreeCAD application using desktop icon if not started yet.
- Create a new document and switch to **Part Design** workbench.
- Click on the **Create sketch** tool from **Toolbar**. The **Select attachment** dialog will be displayed in **Combo View** and you will be asked to select sketching plane.
- Select the **XY Plane** from the list and click on the **OK** button. The sketching environment will be displayed.
- Create the sketch as shown in Figure-35.

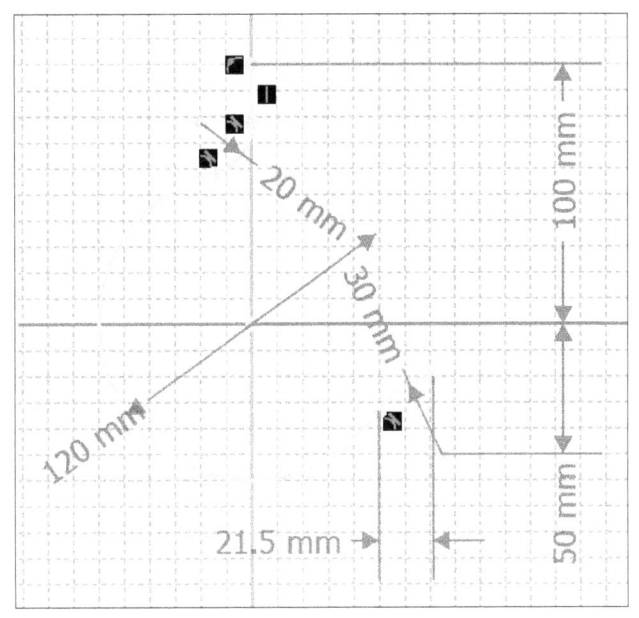

Figure-35. Sketch to be created

- Click on **Close** button from the dialog to exit sketching environment.
- Click in the empty area to make sure nothing is selected.
- Select straight line of sketch as shown in Figure-36 and click on the **Create a datum plane** tool from **Toolbar** or **Part Design** menu. Preview of plane will be displayed normal to selected line segment; refer to Figure-37.

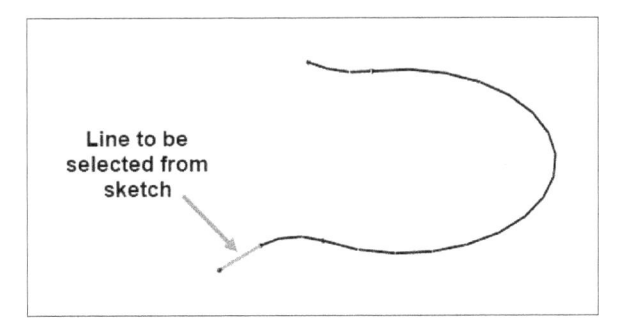

Figure-36. Line selected for plane creation

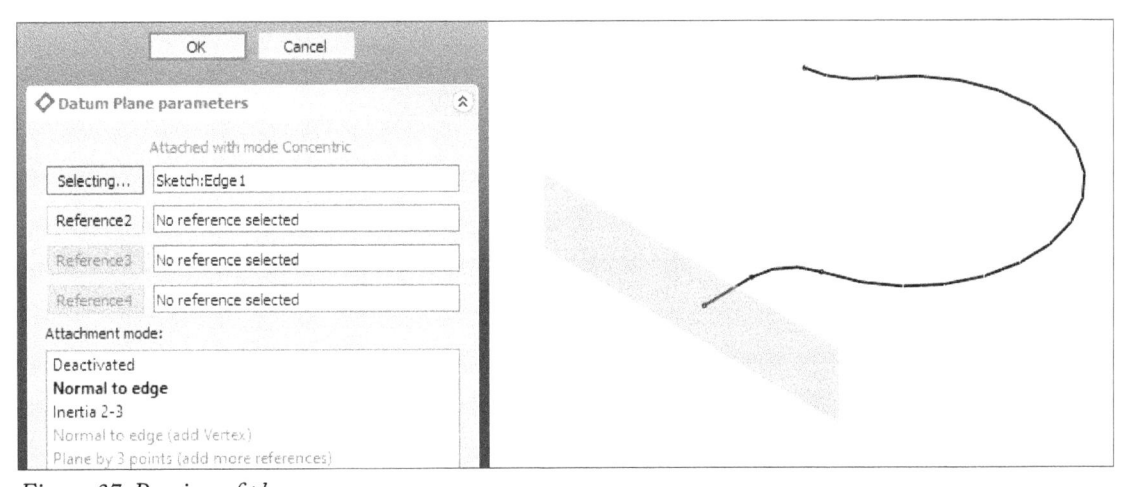

Figure-37. Preview of plane

- Click on the **OK** button from the dialog to create plane.

- Select the newly created plane from drawing area and click on the **Create sketch** tool from **Part Design** menu or **Toolbar**. The sketching environment will become active.
- Create a circle of diameter **25** mm at the origin; refer to Figure-38.

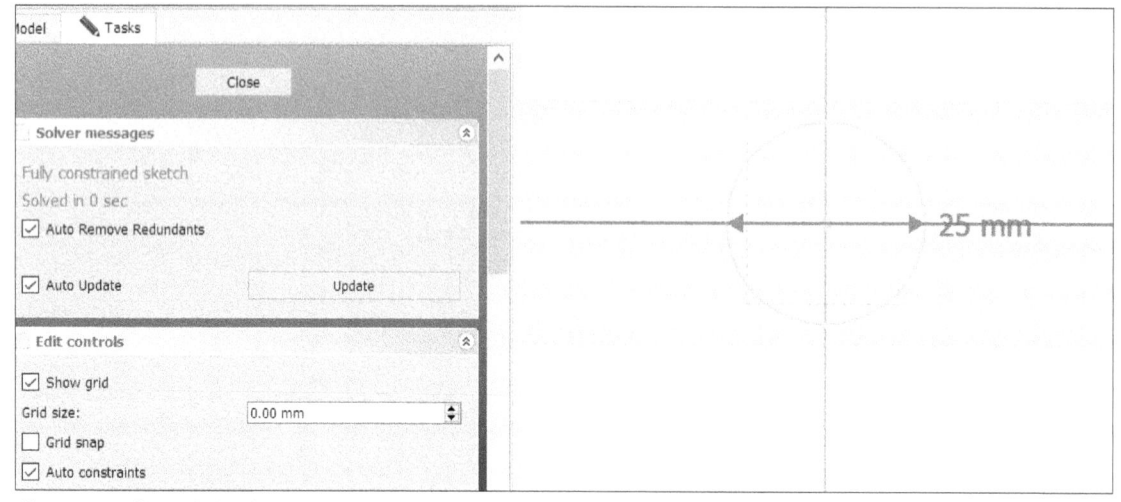

Figure-38. Second sketch to be created

- Click on the **Close** button from the dialog to exit sketching environment. Click in the empty area of drawing to make sure nothing is selected.

Creating Sweep Feature

- Select the newly created circle sketch and click on the **Additive pipe** tool from the **Part Design** menu. The **Pipe parameters** dialog will be displayed with earlier selected sketch defined as section profile; refer to Figure-39.

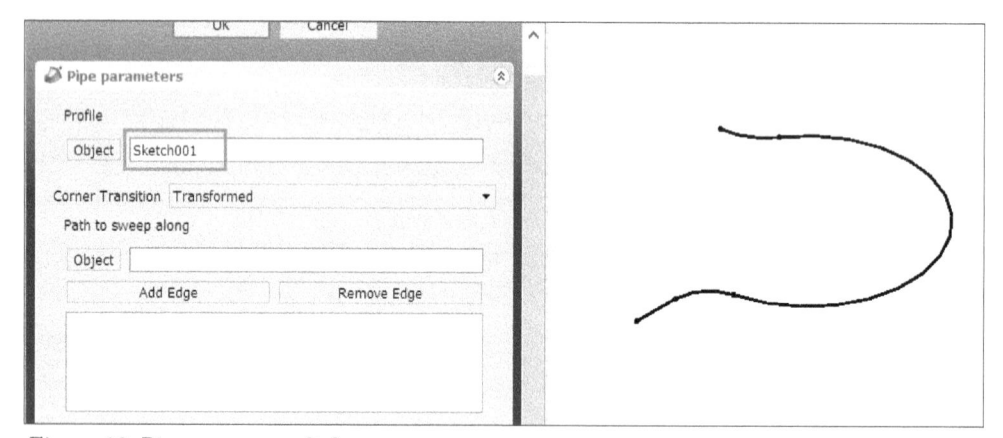

Figure-39. Pipe parameters dialog

- Click on the **Add Edge** button from dialog and select the line in profile sketch. Preview of pipe feature will be displayed; refer to Figure-40.

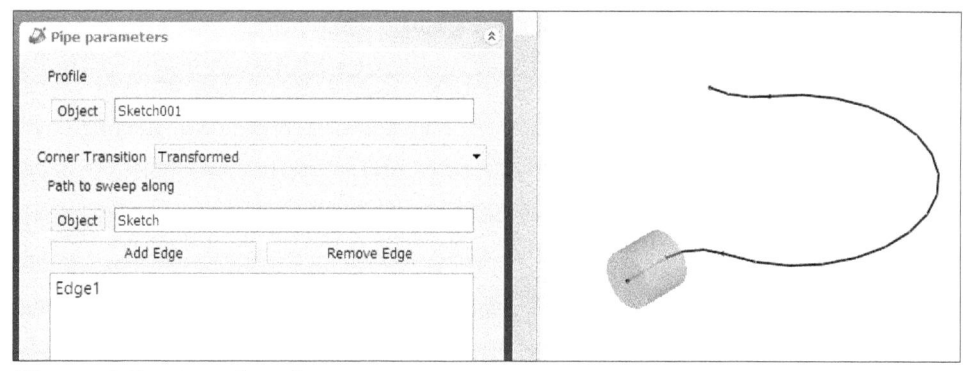

Figure-40. Preview of pipe feature

- Click on the **Add Edge** button again and select next curve in the sketch. Repeat the steps until all the curves of sketch are added as path; refer to Figure-41.
- Click on the **OK** button from dialog to create the feature.

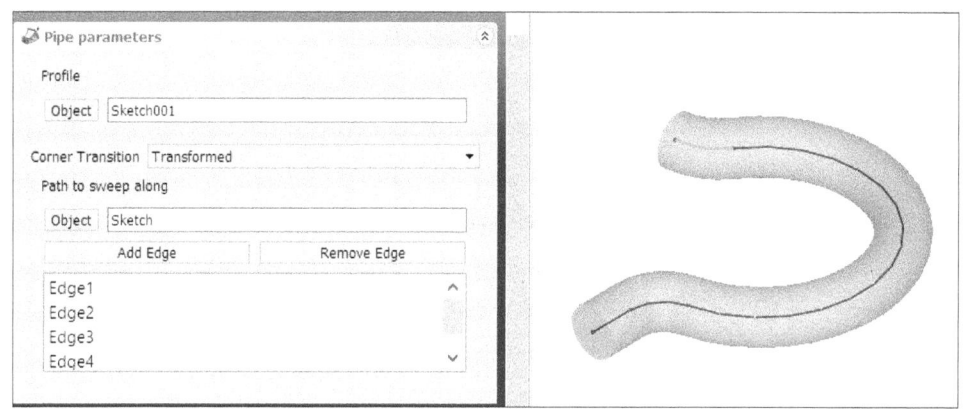

Figure-41. Preview of feature

- Apply a fillet of radius **10** mm at the end edge of model; refer to Figure-42.

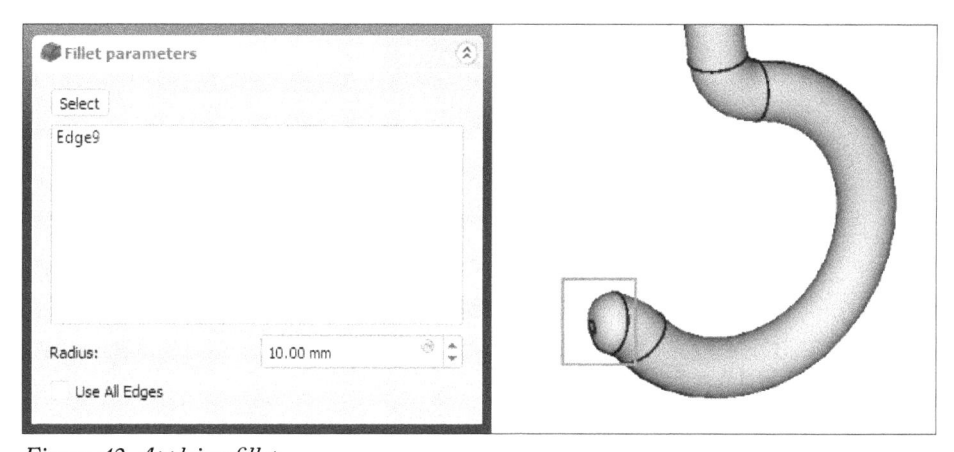

Figure-42. Applying fillet

PRACTICE 5

Create the model as shown in Figure-43. The dimensions are given in Figure-44.

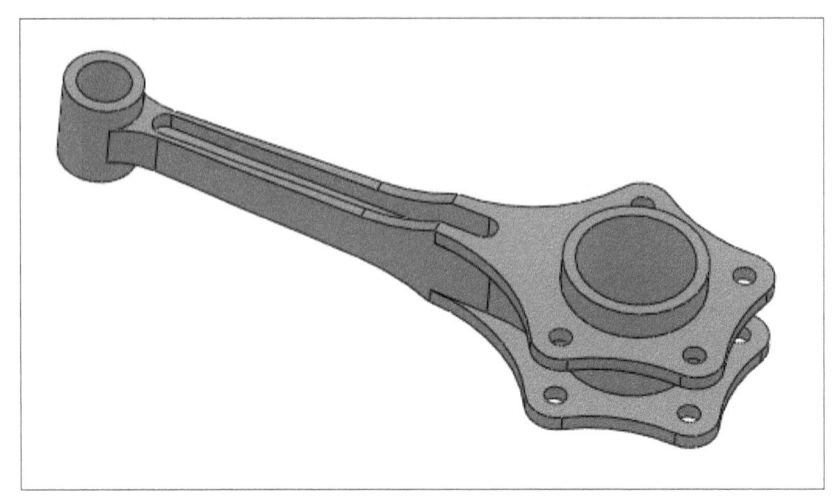

Figure-43. Practice 5

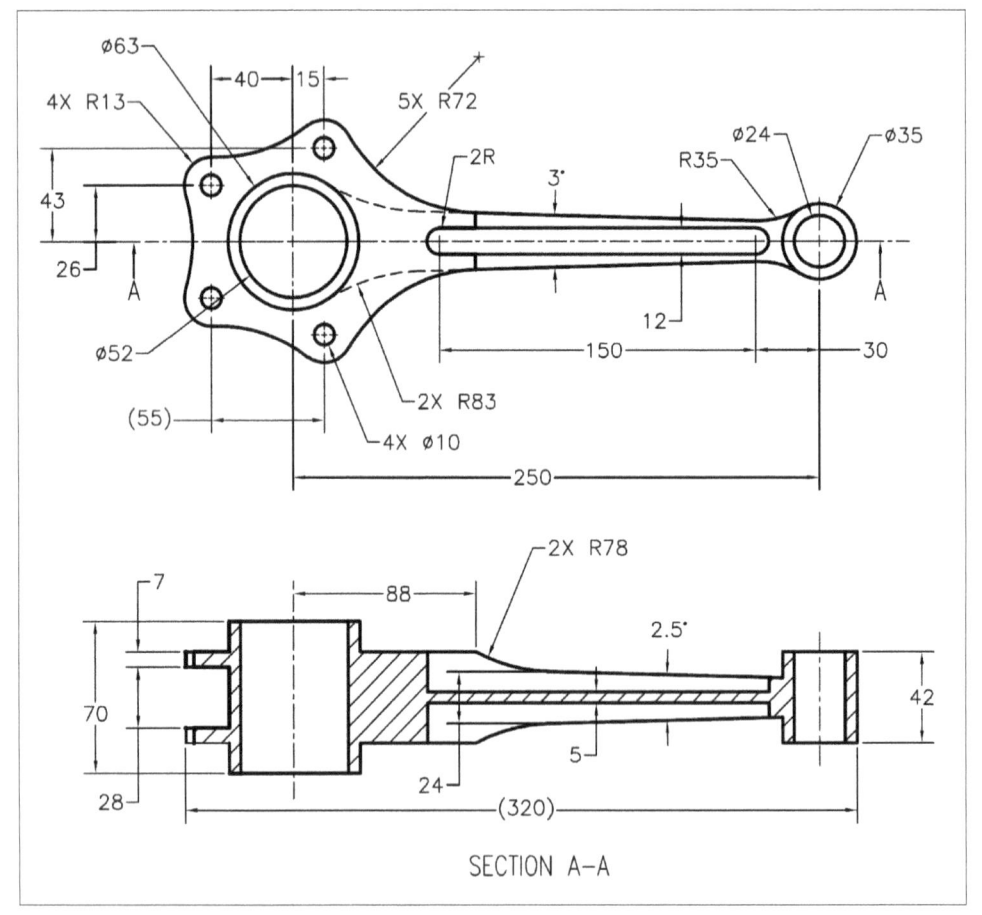

Figure-44. Dimensions of the Practice 5 model

PRACTICE 6

Create the model using the drawings shown in Figure-45.

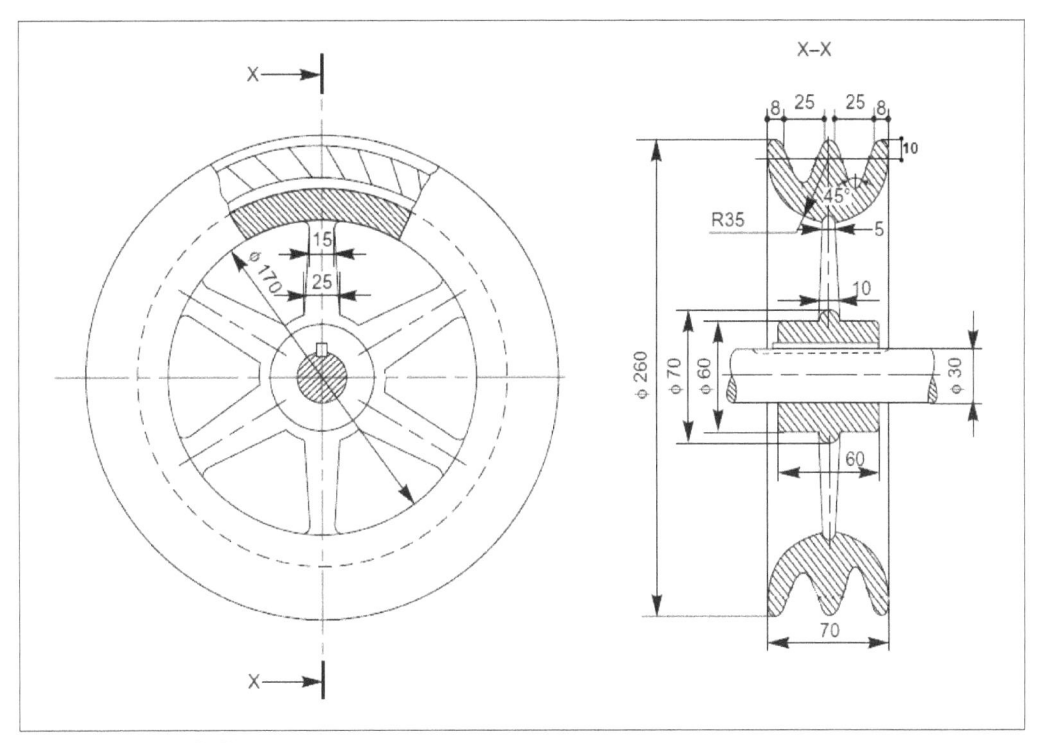

Figure-45. Rope Pulley

PRACTICE 7

Create the model as shown in Figure-46. Dimensions are given in Figure-47. Assume the missing dimensions.

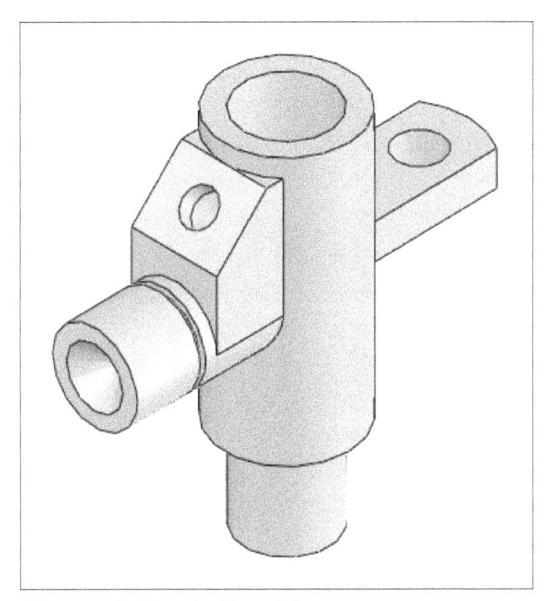

Figure-46. Practice 7 model

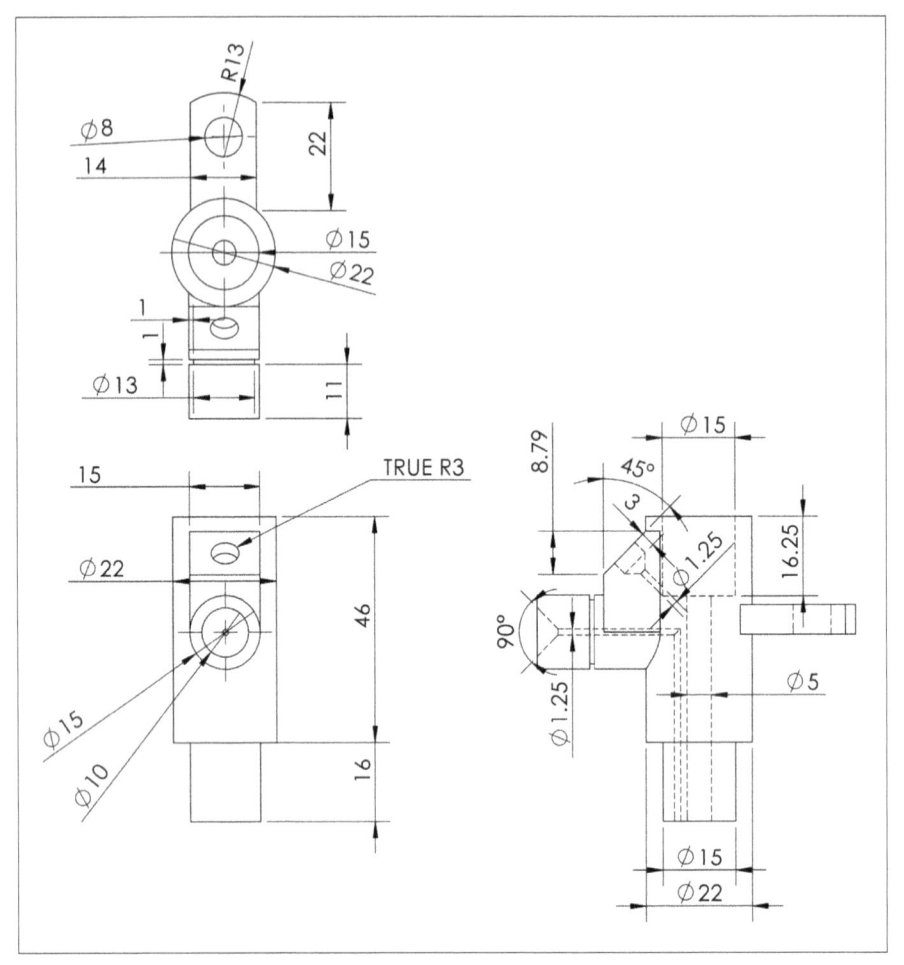

Figure-47. Dimensions of the Practice 7 model

PRACTICE 8

Create the model by using the dimensions given in Figure-48.

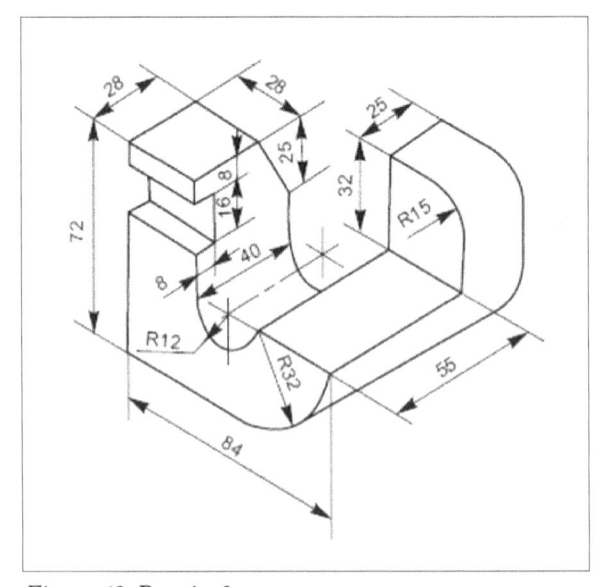

Figure-48. Practice 8

PRACTICE 9

Create the model by using the dimensions given in Figure-49.

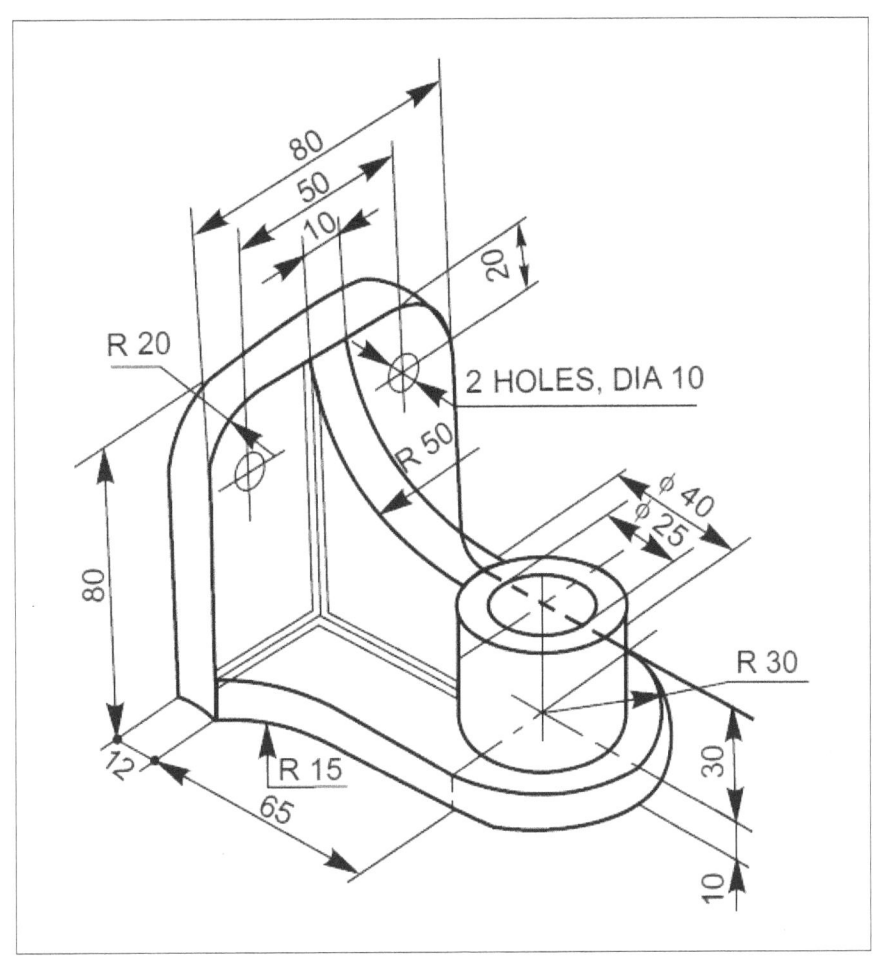

Figure-49. Practice 9

PRACTICE 10

Create the model by using the dimensions given in Figure-50.

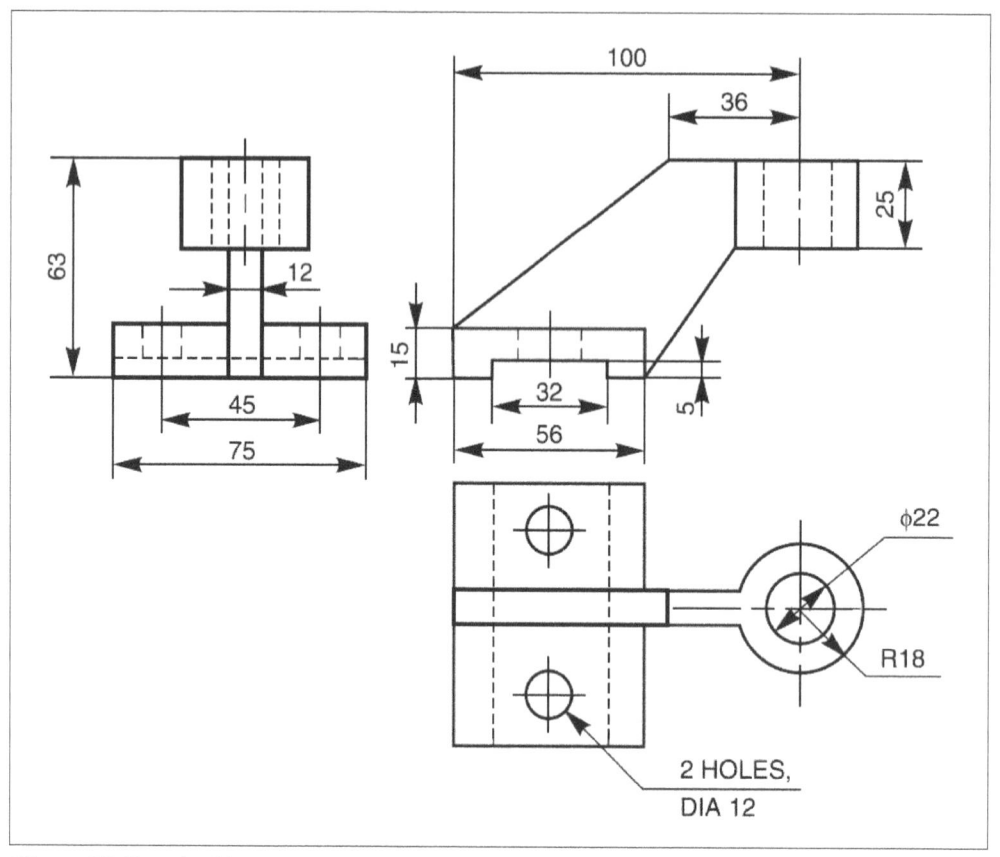

Figure-50. Practice 10

PRACTICE 11

Create a ring nut with value of **D** as **5, 6, 8,** and **10** using equation and design table. Dimensions are given in Figure-51.

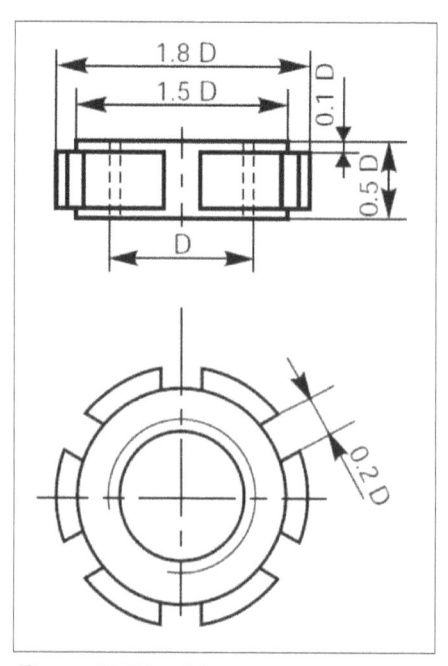

Figure-51. Ring Nut

Chapter 6

Drafting

Topics Covered

The major topics covered in this chapter are:

- **Introduction to Drafting**
- **Drawing Objects Tools**
- **Annotation Objects Tools**
- **Modifying Objects Tools**
- **Draft Tray Toolbar**
- **Draft Snap Toolbar**

INTRODUCTION

The tools in **Draft Workbench** are used to draw simple 2D objects and modify them afterwards. It also provides tools to define a working plane, a grid, and a snapping system to precisely control the position of your geometry. Example of the functionality of the **Draft Workbench** includes many elements like lines, curves, arcs, and polygons; refer to Figure-1.

Figure-1. Draft workbench example

The created 2D objects can be used for general drafting in a way similar to Inkscape or AutoCAD. These 2D shapes can also be used as base features of 3D objects created with other workbenches, for example, the **Part** and **Arch** workbenches. Conversion of draft objects to sketches is also possible which means that the shapes can also be used with the **PartDesign** workbench for the creation of solid bodies.

STARTING DRAFT WORKBENCH

FreeCAD is primarily a 3D modeling application and thus its 2D tools aren't as advanced as in other drawing programs. If your primary goal is the production of complex 2D drawings and DXF files and you don't need 3D modeling, you may wish to consider a dedicated software program for technical drafting such as LibreCAD, QCad, or others.

- To start a new draft file, click on the **New** button from **File** menu and select **Draft** workbench from **Switch between workbenches** drop-down in the **Toolbar**; refer to Figure-2. The tools to perform drafting will be displayed in the **Toolbar**; refer to Figure-3.

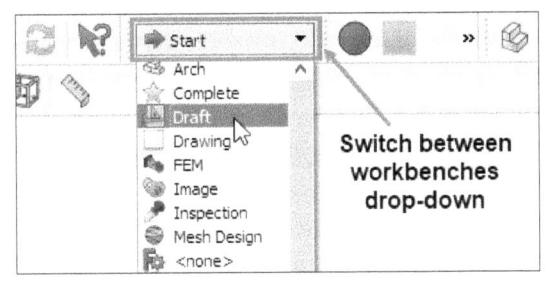

Figure-2. Draft workbench

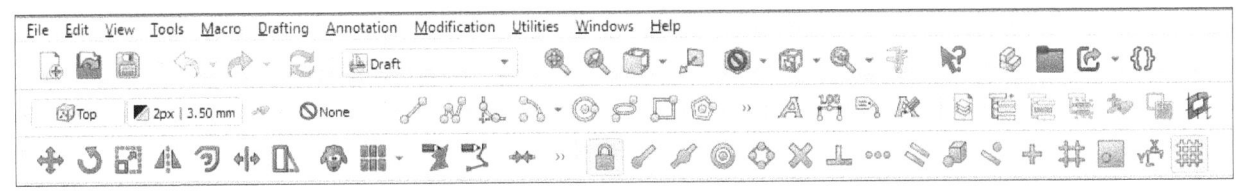

Figure-3. Draft workbench tools

Drawing objects tools

The **Drawing objects** tools are used for creating 2D objects. These tools are available in the **Toolbar** of **Draft** workbench; refer to Figure-4. The procedures to use these tools are discussed next.

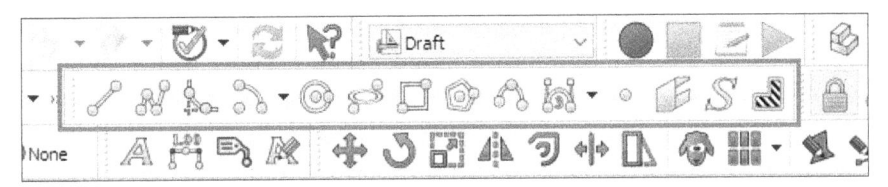

Figure-4. Drawing objects tools

Creating Line

The **Line** tool creates a straight line defined by two points. The procedure to use this tool is discussed next.

- Click on the **Line** tool from **Toolbar** in the **Draft** workbench; refer to Figure-5. The **Line** dialog will be displayed in the **Tasks** panel of **Combo View** along with the plus sign in place of original cursor; refer to Figure-6. You will be asked to specify the first point.

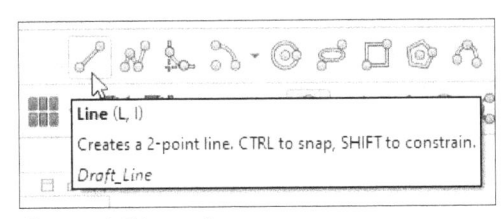

Figure-5. Line tool

Figure-6. Line dialog

- Click in the 3D view area to specify the first point or enter desired values for x, y, and z coordinates in the **Local ΔX**, **Local ΔY**, and **Local ΔZ** edit boxes of the dialog, respectively.
- After specifying coordinates for the first point in dialog, click on the **Enter point** button from the dialog. You will be asked to specify the second point.
- Move the cursor away and click at desired location to specify the second point or enter desired values for the coordinates in their respective edit boxes.
- After specifying coordinates for the second point in dialog, click on the **Enter point** button again. The line will be created; refer to Figure-7.

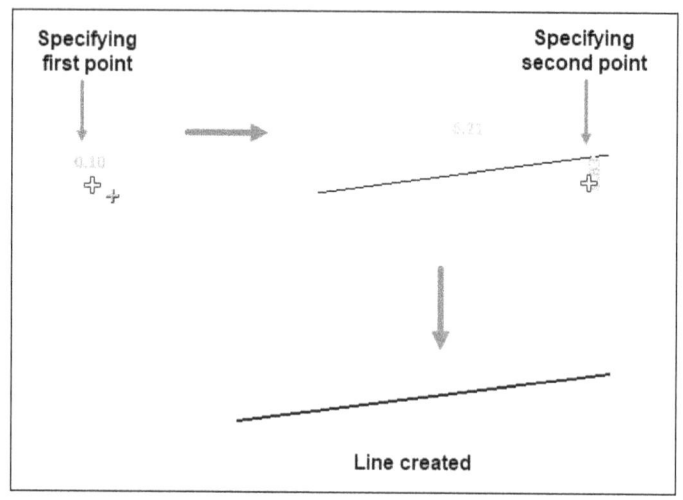

Figure-7. Line created

- Specify desired length of the line in the **Length** edit box.
- Specify desired value for the angle of line in the **Angle** edit box. Select the check box next to the **Angle** to constrain the pointer to the specified angle.
- On selecting the **Relative (R)** check box, the coordinates of second point are relative to the first one. If not selected, they are absolute taken from the origin.
- Select **Global (G)** check box to specify coordinates relative to global coordinate system.
- On selecting the **Continue (T)** check box, the **Line** tool will restart after you give the second point allowing you to draw another line segment without clicking on the tool button again.
- Click on **Undo (CTRL+Z)** button to undo the last point.

Creating Polyline

The **Wire** or **Polyline** tool creates a sequence of several line segments. This tool allows you to enter more than two points. The procedure to use this tool is discussed next.

- Click on the **Polyline** tool from **Toolbar** in the **Draft** workbench; refer to Figure-8. The **Polyline** dialog will be displayed in the **Tasks** panel of **Combo View** along with the plus sign in place of original cursor; refer to Figure-9. You will be asked to specify first point of the line.

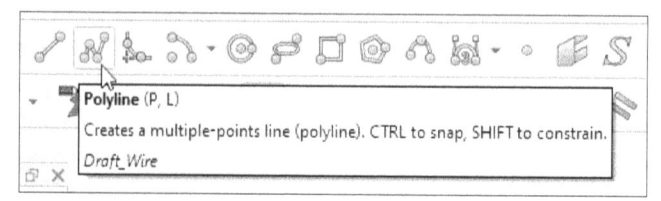

Figure-8. Polyline tool

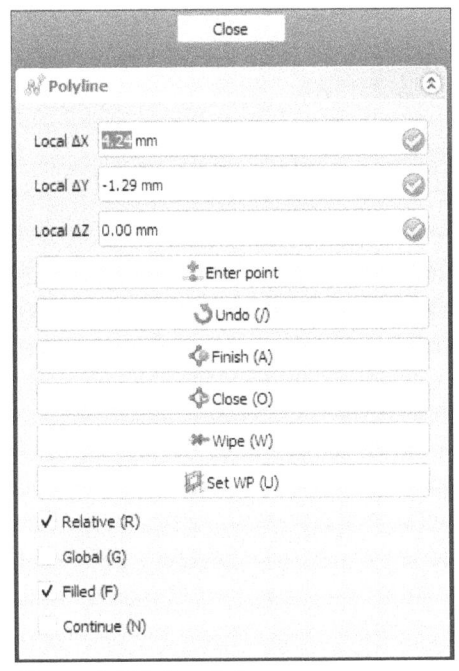

Figure-9. Polyline dialog

- Click in the 3D view area to specify the first point or enter desired values for x, y, and z coordinates in the **Local ΔX**, **Local ΔY**, and **Local ΔZ** edit boxes of the dialog, respectively.
- After specifying coordinates for the first point in dialog, click on the **Enter point** button from the dialog. You will be asked to specify the additional points.
- Move the cursor away and click at desired locations to specify the additional points for the polyline or enter desired values for the coordinates in their respective edit boxes.
- After specifying coordinates for additional points, click on the **Enter point** button again. The polyline will be created; refer to Figure-10.

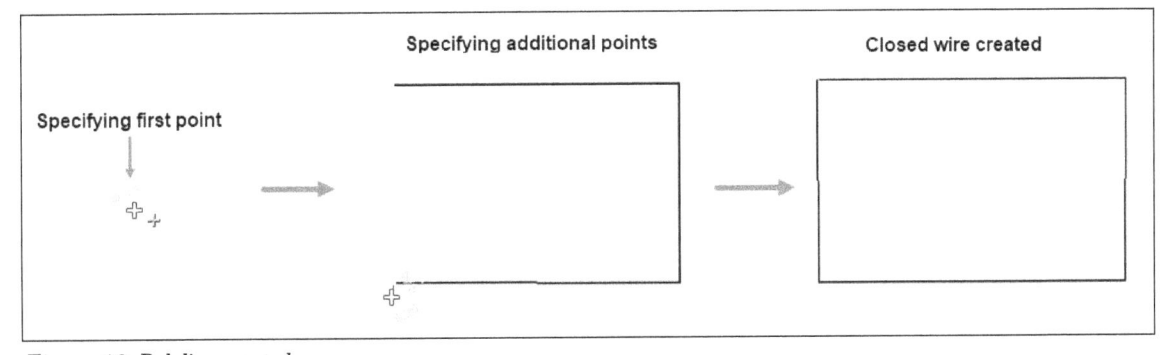

Figure-10. Polyline created

- On selecting the **Filled** check box, a closed wire or a polyline will create a face.
- Click on the **Finish (A)** button to finish the polyline, leaving it open.
- Click on the **Close (O)** button to close the wire, that is, a line segment will be added from the last point to the first one to form a face.
- Click on the **Wipe (W)** button to remove the line segments already placed but keep editing the wire from the last point.
- Click on the **Set WP (U)** button to adjust the current working plane in the orientation of the last point.
- Other parameters of the **Polyline** tool have been discussed in the **Line** tool.
- Click on **Close** button to close the dialog.

Creating Fillet

The **Fillet** tool creates a fillet, a rounded corner, or a chamfer, a straight edge, between two draft lines. The procedure to use this tool is discussed next.

- Select two lines that meet in a single point from the **Model Tree** or from the 3D view area between which you want to create a fillet or a chamfer; refer to Figure-11.

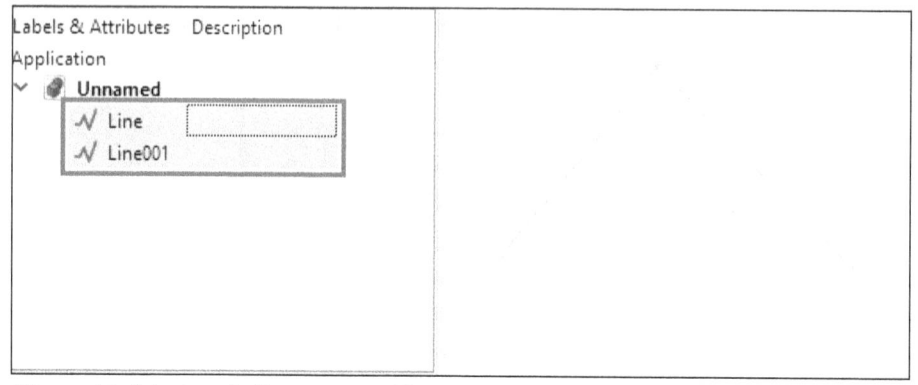

Figure-11. Selecting the lines to create fillet or chamfer

- Click on the **Fillet** tool from **Toolbar** in the **Draft** workbench; refer to Figure-12. The **Fillet** dialog will be displayed in the **Tasks** panel of **Combo View**; refer to Figure-13.

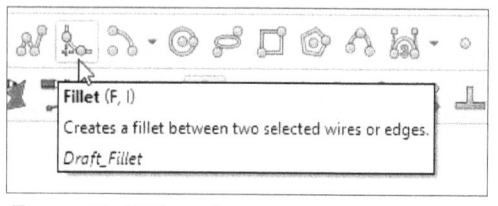

Figure-12. Fillet tool

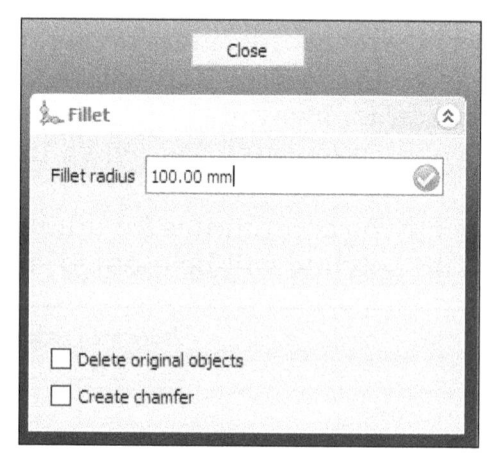

Figure-13. Fillet dialog

- Specify desired radius for the fillet in the **Fillet radius** edit box of the dialog.
- Select **Delete original objects** check box to delete the original object after creating fillet.
- Select **Create chamfer** check box to create the chamfer feature between the selected lines.
- After specifying desired parameters, press **Enter** key from the keyboard. The fillet feature will be created; refer to Figure-14.

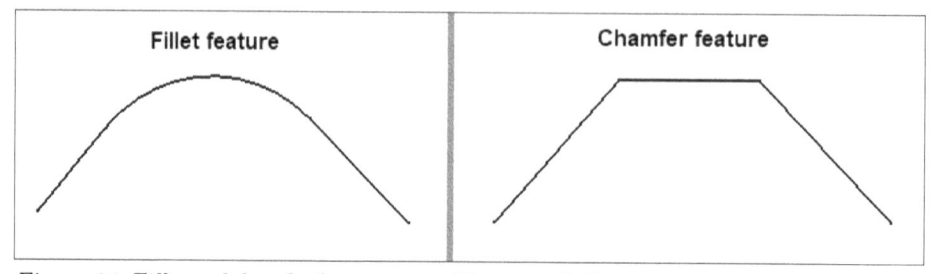

Figure-14. Fillet and chamfer features created between the lines

Creating Arc

The **Arc** tool creates a circular arc in the current work plane by entering four points, the center, a point that defines the radius, the first end point and the second end point, or by picking tangents, or any combination of these. The procedure to use this tool is discussed next.

- Click on the **Arc** tool from **Arc tools** drop-down in the **Toolbar** of **Draft** workbench; refer to Figure-15. The **Arc** dialog will be displayed in the **Tasks** panel of **Combo View** along with the plus ⊕ sign in place of original cursor; refer to Figure-16. You will be asked to specify the first point.

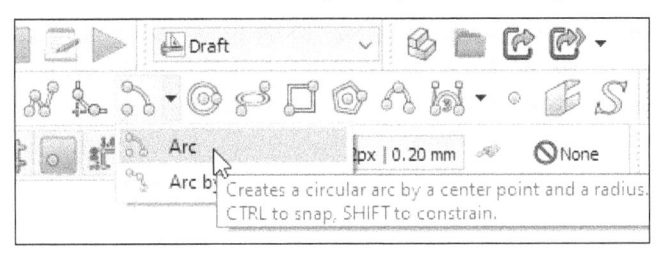

Figure-15. Arc tool

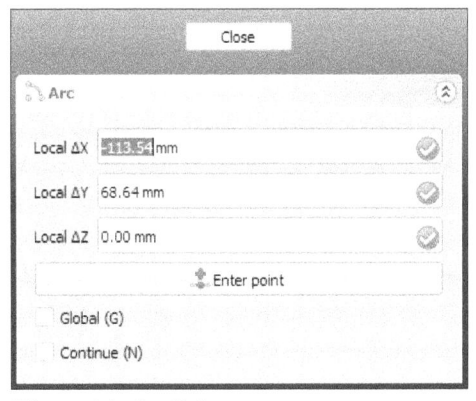

Figure-16. Arc dialog

- Click in the 3D view area to specify the first point or enter desired values for x, y, and z coordinates in the **Local ΔX**, **Local ΔY**, and **Local ΔZ** edit boxes of the dialog, respectively.
- After specifying coordinates in the dialog, click on the **Enter point** button. You will be asked to specify the radius of arc.
- Move the cursor away and click at desired location to specify the second point or enter desired radius value for the arc in the **Radius** edit box. You will be asked to specify the start angle.
- Click in the 3D view area to specify the start angle for the arc or enter desired angle value in the **Start angle** edit box of the dialog. You will be asked to specify the aperture angle.
- Click in the 3D view area to specify the aperture angle of an arc or enter desired angle value in the **Aperture angle** edit box of the dialog. The arc will be created; refer to Figure-17.

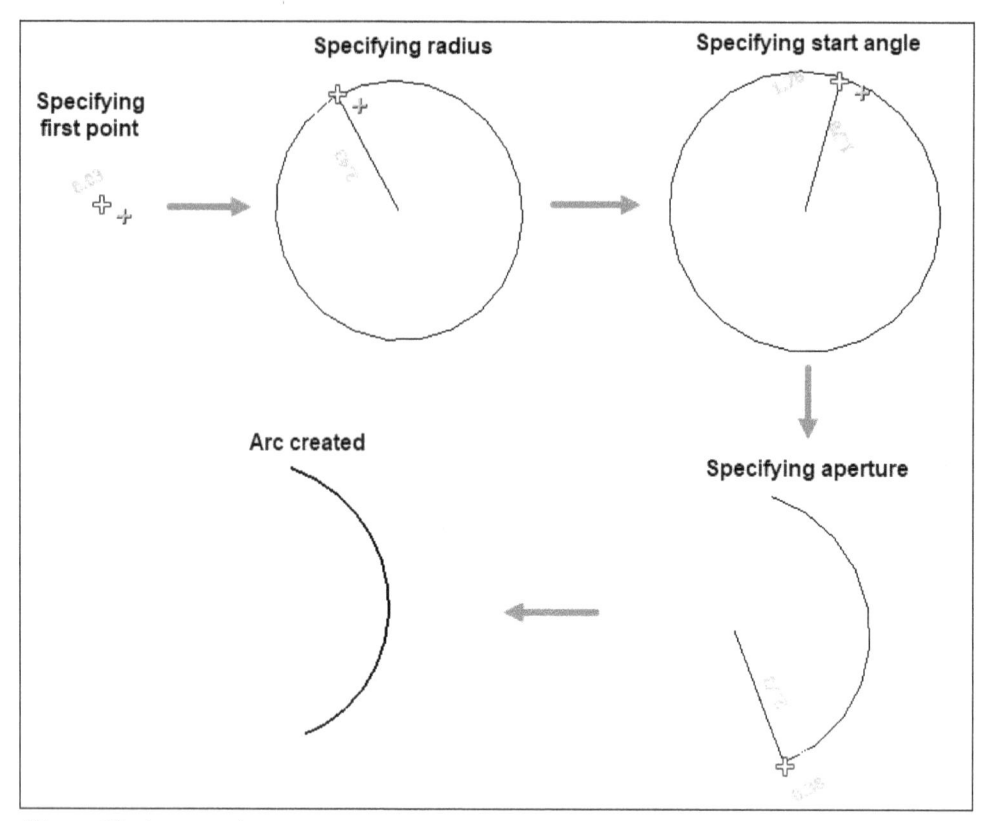

Figure-17. Arc created

- Other parameters in the **Arc** dialog have been discussed earlier.

Creating Arc by 3 Points

The **Arc by 3 points** tool creates a circular arc in the current working plane using three points that define its circumference. The center and radius are calculated from these points. The procedure to use this tool is discussed next.

- Click on the **Arc by 3 points** tool from **Arc tools** drop-down in the **Toolbar** of **Draft** workbench; refer to Figure-18. The **Arc by 3 points** dialog will be displayed in the **Tasks** panel of **Combo View** along with the plus ⊕ sign in place of original cursor; refer to Figure-19. You will be asked to specify the first point.

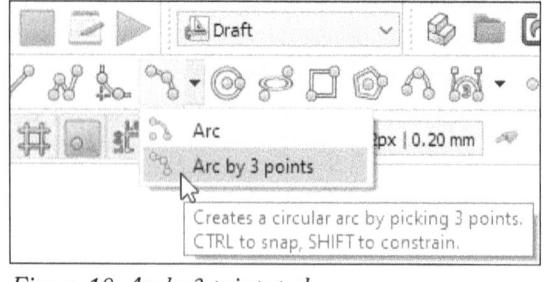

Figure-18. Arc by 3 points tool

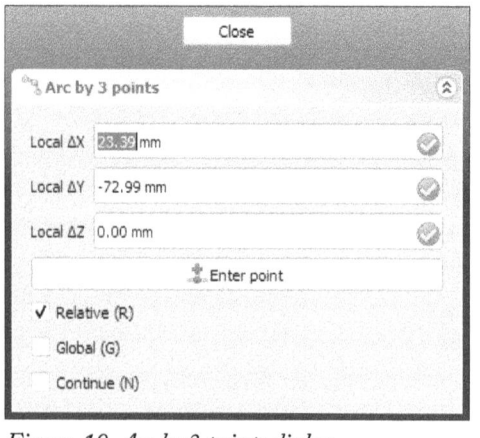

Figure-19. Arc by 3 points dialog

- Click in the 3D view area to specify the first point or enter desired values for x, y, and z coordinates in the **Local ΔX**, **Local ΔY**, and **Local ΔZ** edit boxes of the dialog, respectively.

FreeCAD 1.0 Black Book 6-9

- After specifying coordinates for the first point, click on the **Enter point** button. You will be asked to specify the additional two points.
- Move the cursor away and click at desired locations to specify the second and third points or enter desired values for the coordinates in their respective edit boxes.
- On specifying the third point, the arc by three points will be created; refer to Figure-20.

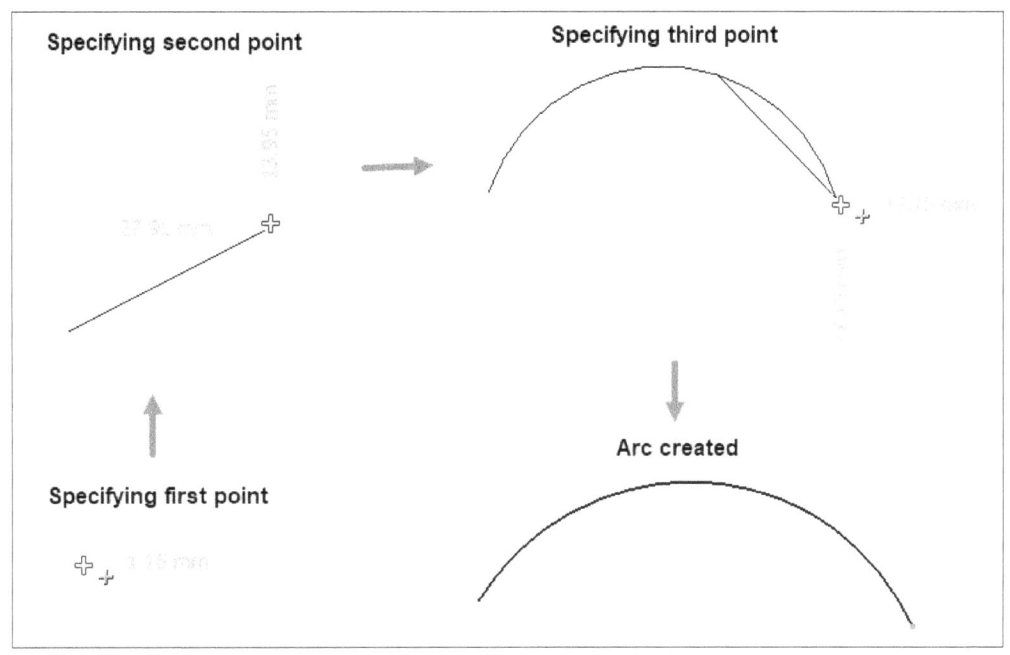

Figure-20. Three points arc created

- Other parameters in the dialog have been discussed earlier.

Creating Circle

The **Circle** tool creates a circle in the current work plane by entering two points, the center and the radius, or by picking tangents, or any combination of these. The procedure to use this tool is discussed next.

- Click on the **Circle** tool from **Toolbar** in the **Draft** workbench; refer to Figure-21. The **Circle** dialog will be displayed in the **Tasks** panel of **Combo View** along with the plus sign in place of original cursor; refer to Figure-22. You will be asked to specify the first point.

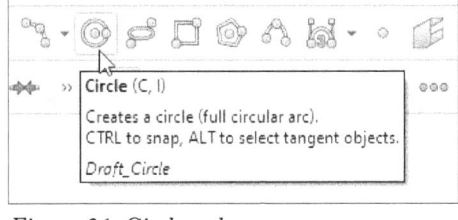

Figure-21. Circle tool

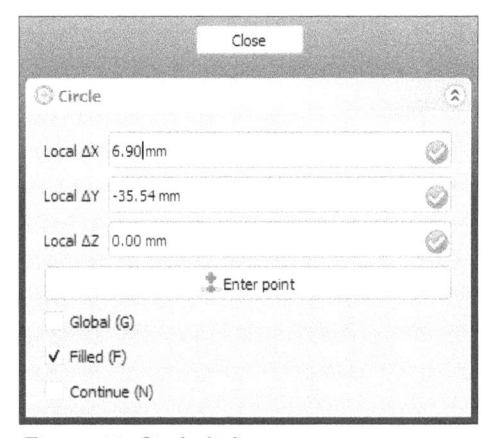

Figure-22. Circle dialog

- Click in the 3D view area to specify the first point or enter desired values for x, y, and z coordinates in the **Local ΔX**, **Local ΔY**, and **Local ΔZ** edit boxes of the dialog, respectively.
- After specifying coordinates for the first point in dialog, click on the **Enter point** button. You will be asked to specify the radius of the circle.
- Move the cursor away and click at desired location to specify second point for the circle or enter desired radius value in the **Radius** edit box of the dialog. The circle will be created; refer to Figure-23.

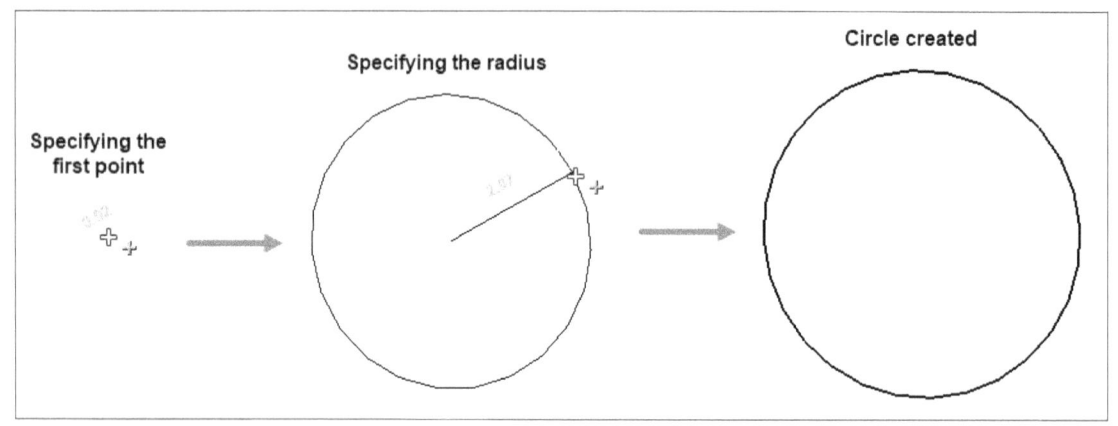

Figure-23. Circle created

- Other parameters in the **Circle** dialog have been discussed earlier.

Creating Ellipse

The **Ellipse** tool creates an ellipse in the current work plane by entering two points, defining the corners of a rectangular box in which the ellipse will fit. This tool can also be used to create elliptical arcs by specifying the start and end angles. The procedure to use this tool is discussed next.

- Click on the **Ellipse** tool from **Toolbar** in the **Draft** workbench; refer to Figure-24. The **Ellipse** dialog will be displayed in the **Tasks** panel of **Combo View** along with the plus ⊕ sign in place of original cursor; refer to Figure-25. You will be asked to specify the first point.

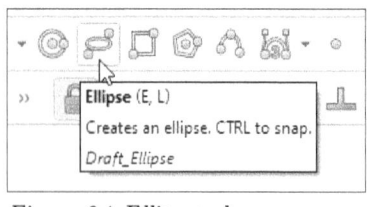

Figure-24. Ellipse tool

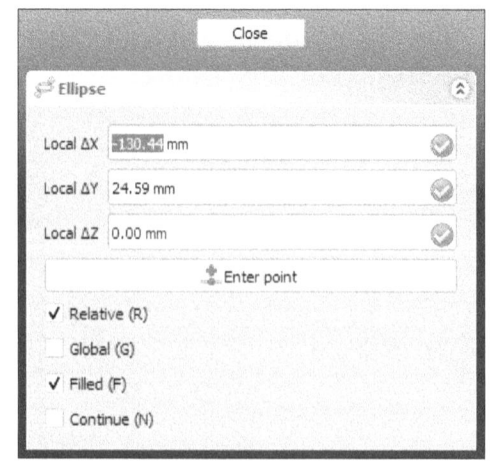

Figure-25. Ellipse dialog

- Click in the 3D view area to specify the first point or enter desired values for the x, y, and z coordinates in the **Local ΔX** , **Local ΔY** , **Local ΔZ** and edit boxes of the dialog, respectively.

- After specifying coordinates for the first point in dialog, click on the **Enter point** button from the dialog. You will be asked to specify the second point.
- Move the cursor away and click at desired location to specify the second point or enter desired values for the coordinates in their respective edit boxes.
- After specifying coordinates for the second point in dialog, click on the **Enter point** button again. The ellipse will be created; refer to Figure-26.

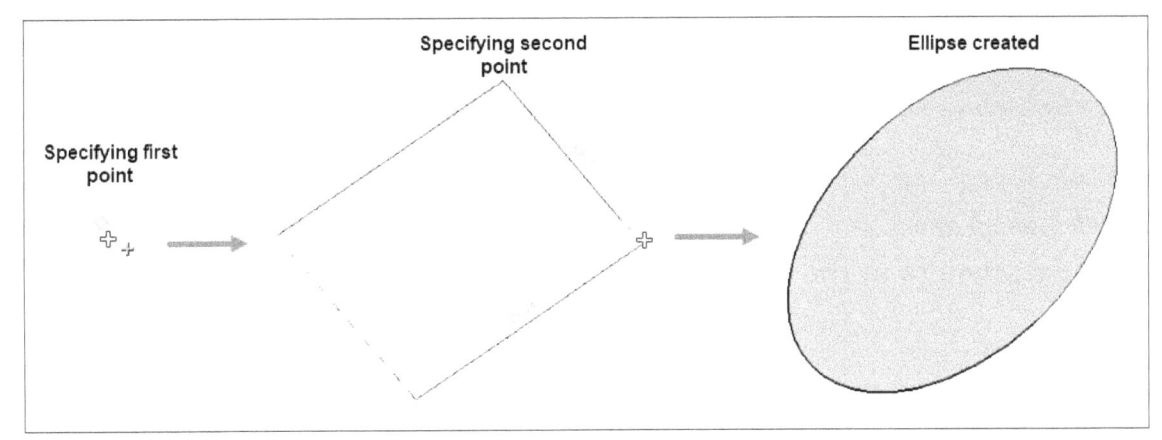

Figure-26. Ellipse created

- Other parameters in the **Ellipse** dialog have been discussed earlier.

Creating Rectangle

The **Rectangle** tool creates a rectangle by picking two points. The procedure to use this tool is discussed next.

- Click on the **Rectangle** tool from **Toolbar** in the **Draft** workbench; refer to Figure-27. The **Rectangle** dialog will be displayed in the **Tasks** panel of **Combo View** along with the plus sign in place of original cursor; refer to Figure-28. You will be asked to specify the first point.

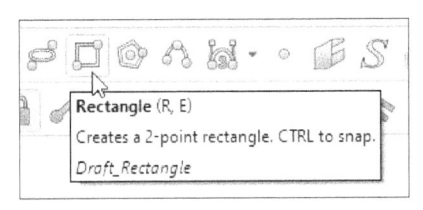

Figure-27. Rectangle tool

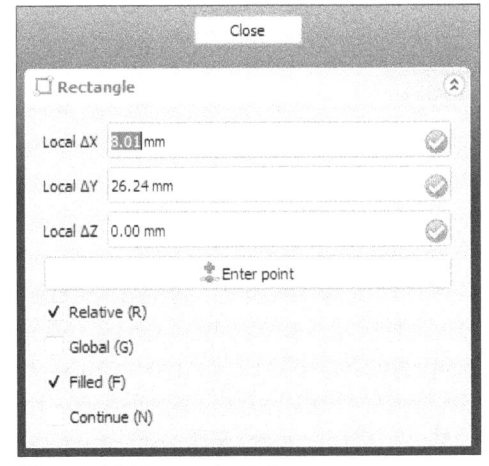

Figure-28. Rectangle dialog

- Click in the 3D view area to specify the first point or enter desired values for the x, y, and z coordinates in the **Local ΔX**, **Local ΔY**, and **Local ΔZ** edit boxes of the dialog, respectively.
- After specifying coordinates for the first point in dialog, click on the **Enter point** button from the dialog. You will be asked to specify the second point.
- Move the cursor away and click at desired location to specify the second point or enter desired values for the coordinates in their respective edit boxes of the dialog.

- After specifying coordinates for the second point in dialog, click on the **Enter point** button again. The rectangle will be created; refer to Figure-29.

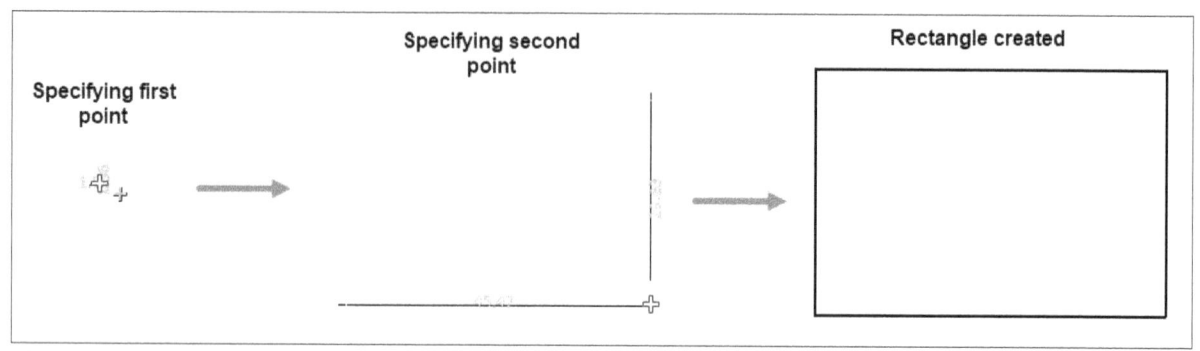

Figure-29. Rectangle created

- Other parameters in the **Rectangle** dialog have been discussed earlier.

Creating Polygon

The **Polygon** tool creates a regular polygon inscribed in a circumference by picking two points, the center, and the radius. The procedure to use this tool is discussed next.

- Click on the **Polygon** tool from **Toolbar** in the **Draft** workbench; refer to Figure-30. The **Polygon** dialog will be displayed in the **Tasks** panel of **Combo View** along with the plus sign in place of original cursor; refer to Figure-31. You will be asked to specify the first point.

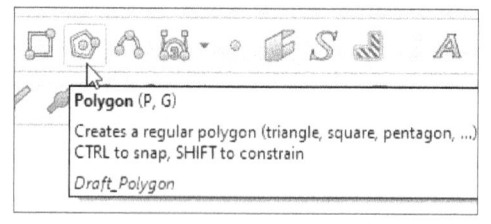

Figure-30. Polygon tool

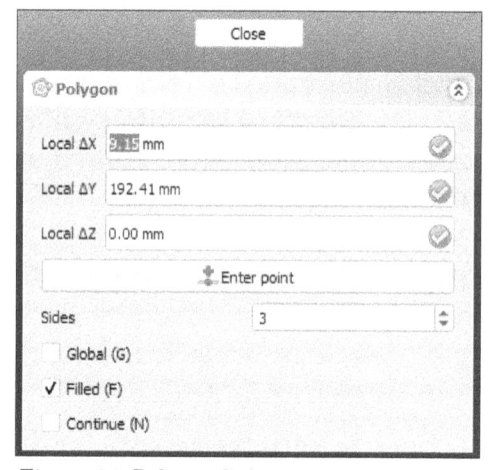

Figure-31. Polygon dialog

- Click in the 3D view area to specify the first point or enter desired values for the x, y, and z coordinates in the **Local ΔX**, **Local ΔY**, and **Local ΔZ** edit boxes of the dialog, respectively.
- After specifying coordinates in the dialog, click on the **Enter point** button from the dialog. You will be asked to specify the sides and radius.
- Specify desired number of sides for the polygon in the **Sides** edit box of the dialog.
- Move the cursor away and click at desired location to specify the radius of the polygon or enter desired value of radius in the **Radius** edit box of the dialog. The regular polygon will be created; refer to Figure-32.

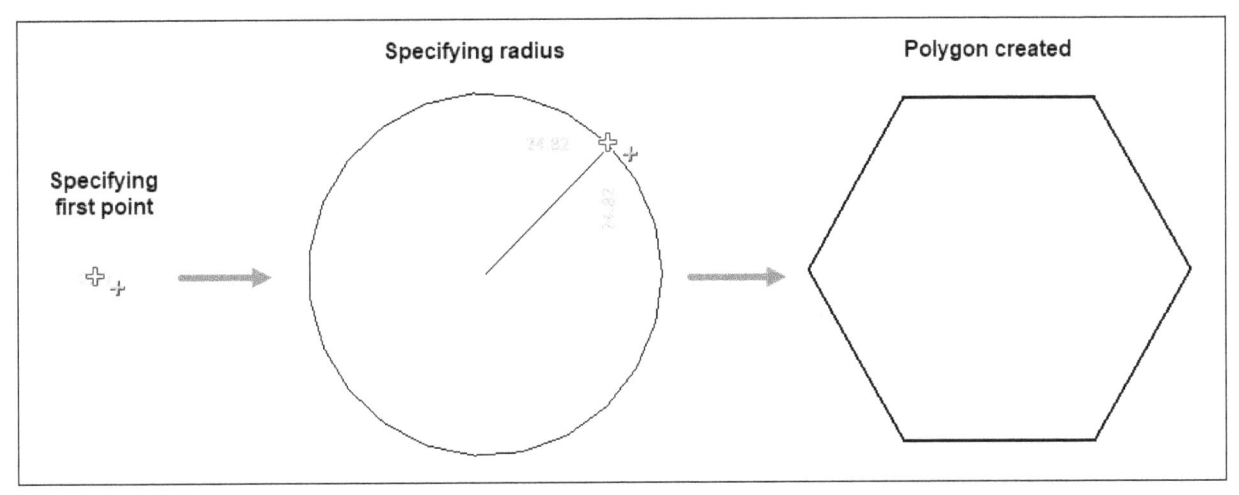

Figure-32. Regular polygon created

- Other parameters in the **Polygon** dialog have been discussed earlier.

Creating B-Spline

The **B-Spline** tool creates a B-spline curve from several points. The procedure to use this tool is discussed next.

- Click on the **B-spline** tool from **Toolbar** in the **Draft** workbench; refer to Figure-33. The **Bspline** dialog will be displayed in the **Tasks** panel of **Combo View** along with the plus ⊕ sign in place of original cursor; refer to Figure-34. You will be asked to specify the first point.

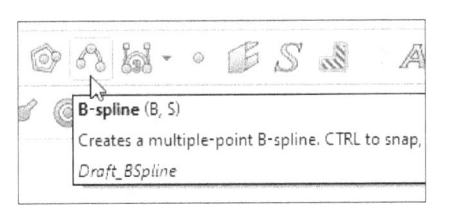

Figure-33. B-Spline tool

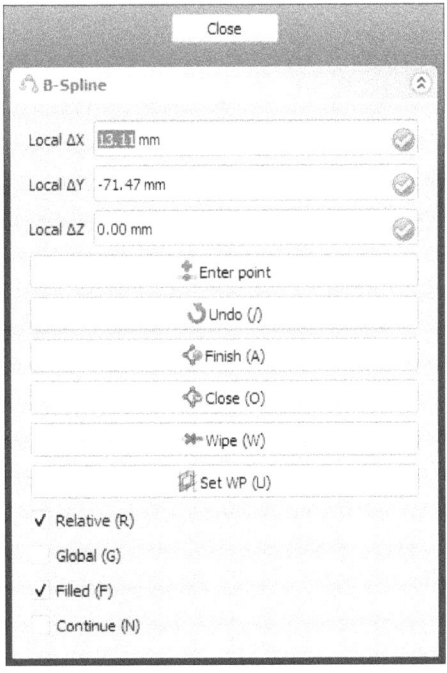

Figure-34. B Spline dialog

- Click in the 3D view area to specify the first point or enter desired values for the x, y, and z coordinates in the **Local X**, **Local Y**, and **Local Z** edit boxes of the dialog, respectively.
- After specifying coordinates for the first point in the dialog, click on the **Enter point** button from the dialog. You will be asked to specify the additional points.

- Move the cursor away and click at desired locations to specify the additional points or enter desired values for the coordinates in their respective edit boxes.
- After specifying coordinates for the additional points in the dialog, click on the **Enter point** button again and press **Esc** key. The B-spline will be created; refer to Figure-35.

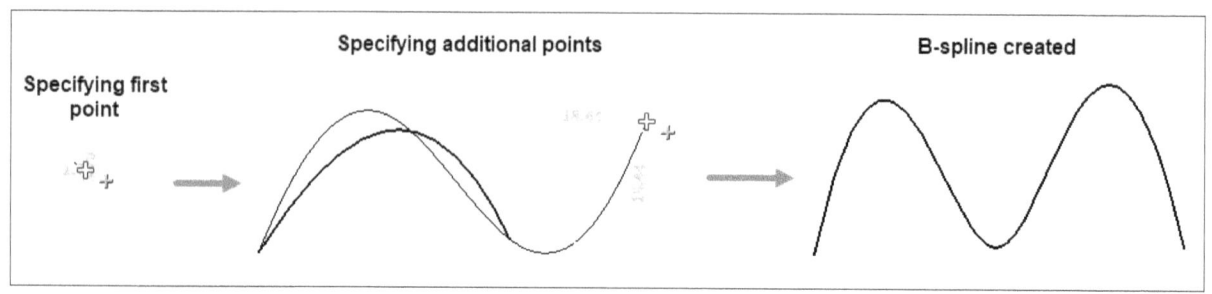

Figure-35. B-spline created

- Other parameters in the **B-spline** dialog have been discussed earlier.

Creating Cubic bezier curve

The **Cubic bezier curve** tool creates a bezier curve of third degree by dragging two points. The procedure to use this tool is discussed next.

- Click on the **Cubic bezier curve** tool from **Bezier tools** drop-down in the **Toolbar** of **Draft** workbench; refer to Figure-36. The **Cubic Bezier curve** dialog will be displayed in the **Tasks** panel of **Combo View** along with the plus sign in place of original cursor; refer to Figure-37. You will be asked to specify the first point.

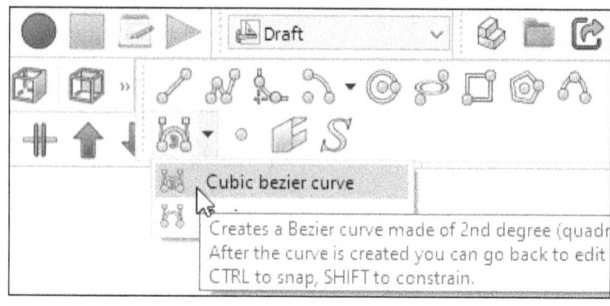

Figure-36. Cubic bezier curve tool

Figure-37. Cubic Bezier curve dialog

Note that it is not possible to create the curve by entering coordinates value in the edit boxes of dialog.

- To specify the starting point of curve, click & hold the **LMB**, move the cursor in desired direction to define slope of curve. Release the **LMB**, click & hold again to specify the next point and slope. Repeat the procedure to specify consecutive points. Press **ESC** button to exit the tool. The cubic bezier curve will be created; refer to Figure-38.

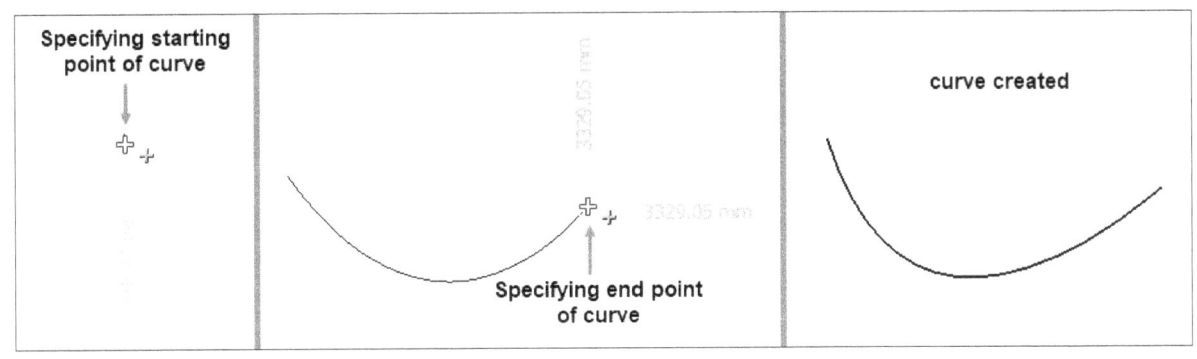

Figure-38. Cubic bezier curve created

- Other parameters in the dialog have been discussed earlier. Press **ESC** key to exit the tool.

Creating Bezier curve

The **Bezier curve** tool creates a bezier curve or a piecewise bezier curve from several points. This tool uses control points to define the direction of the curve. The procedure to use this tool is discussed next.

- Click on the **Bezier curve** tool from **Bezier tools** drop-down in the **Toolbar** of **Draft** workbench; refer to Figure-39. The **Bezier curve** dialog will be displayed in the **Tasks** panel of **Combo View** along with the plus ⊕ sign in place of original cursor; refer to Figure-40. You will be asked to specify the first point.

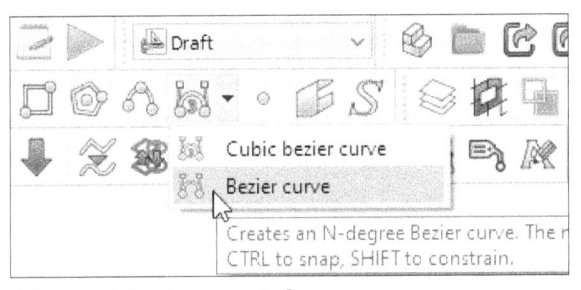

Figure-39. Bezier curve tool

Figure-40. Bezier curve dialog

- Click in the 3D view area to specify the first point or enter desired values for the x, y, and z coordinates in the **Local ΔX**, **Local ΔY**, and **Local ΔZ** edit boxes of the dialog, respectively.
- After specifying coordinates for the first point in the dialog, click on the **Enter point** button from the dialog.
- Move the cursor away and click at desired locations to specify the additional points to create the bezier curve or enter desired values for the coordinates in their respective edit boxes of the dialog.
- After specifying coordinates for additional points in the dialog, click on the **Enter point** button again. The bezier curve will be created; refer to Figure-41.

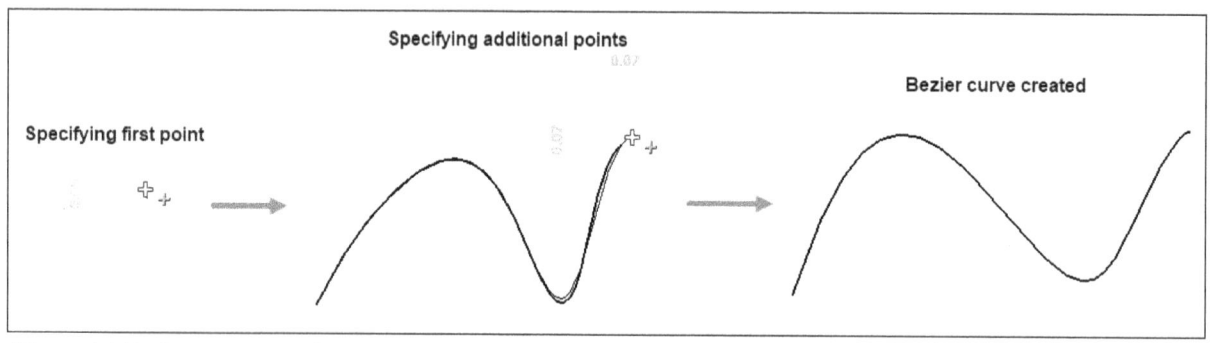

Figure-41. Bezier curve created

- Other parameters in the **BezCurve** dialog have been discussed earlier. Press **Esc** key from keyboard to exit the tool.

Creating Point

The **Point** tool creates a simple point in the current work plane, handy to serve as reference for placing lines, wires, or other objects later. The procedure to use this tool is discussed next.

- Click on the **Point** tool from **Toolbar** in the **Draft** workbench; refer to Figure-42. The **Point** dialog will be displayed in the **Tasks** panel of **Combo View** along with the plus sign in place of original cursor; refer to Figure-43. You will be asked to specify the location of point.

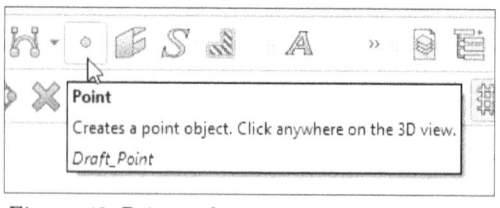

Figure-42. Point tool

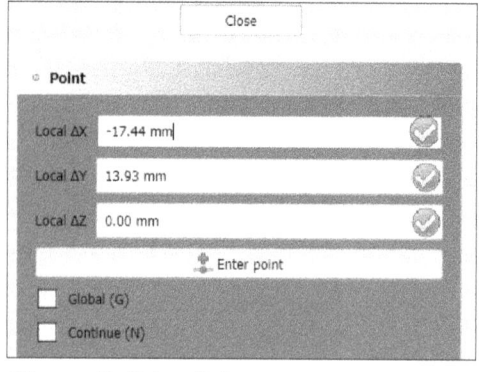

Figure-43. Point dialog

- Click in the 3D view area to specify the location of point or enter desired values for x, y, and z coordinates in the **Local ΔX**, **Local ΔY**, and **Local ΔZ** edit boxes of the dialog, respectively.
- After specifying coordinates in the dialog, click on the **Enter point** button from the dialog. The point will be created; refer to Figure-44.

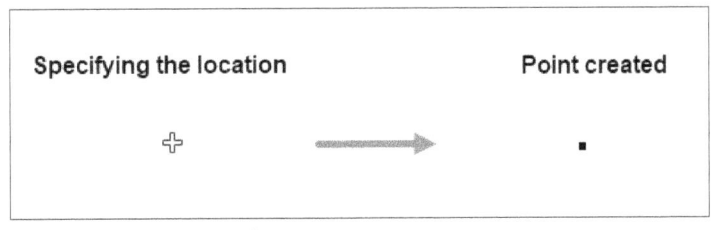

Figure-44. Point created

- Other parameters in the **Point** dialog have been discussed earlier. Click on **Close** button to close the dialog.

Creating Facebinder

The **Facebinder** tool creates a surface object from the selected faces of a solid object. It is parametric tool, meaning that if you modify the original object, the facebinder updates accordingly. It can be used to create an extrusion from a collection of faces to other objects. The procedure to use this tool is discussed next.

- Select desired faces from the solid object which you want to create the facebinder; refer to Figure-45.

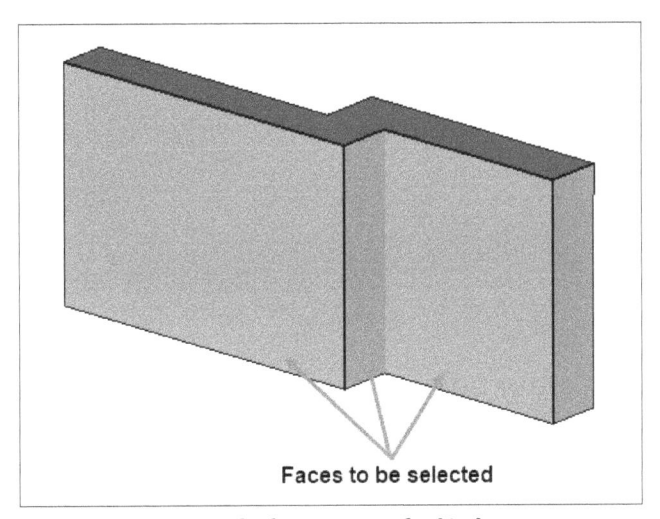

Figure-45. Selecting the faces to create facebinder

- Click on the **Facebinder** tool from **Toolbar** in the **Draft** workbench; refer to Figure-46. The facebinder object will be created; refer to Figure-47.

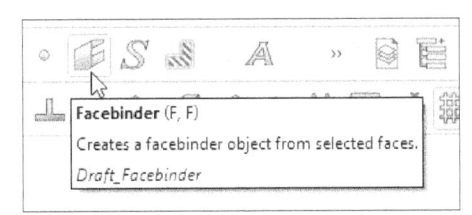

Figure-46. Facebinder tool

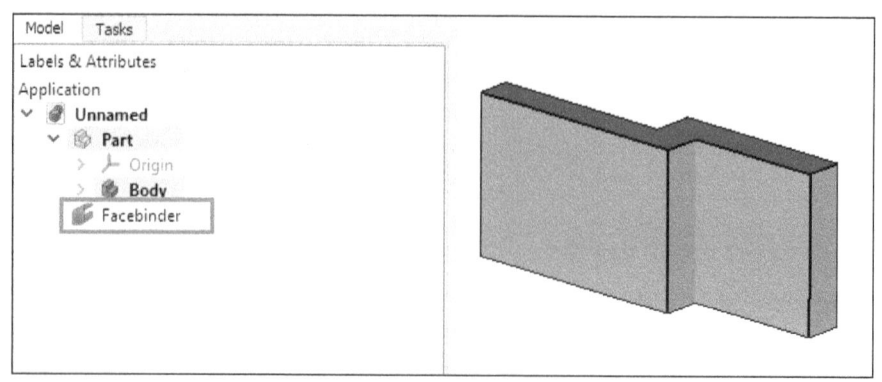

Figure-47. Facebinder object created

- If you want to see the facebinder object, select the original solid object from the **Model Tree** and hide it using **Space** key from the keyboard. The facebinder object will be displayed in the 3D view area; refer to Figure-48.

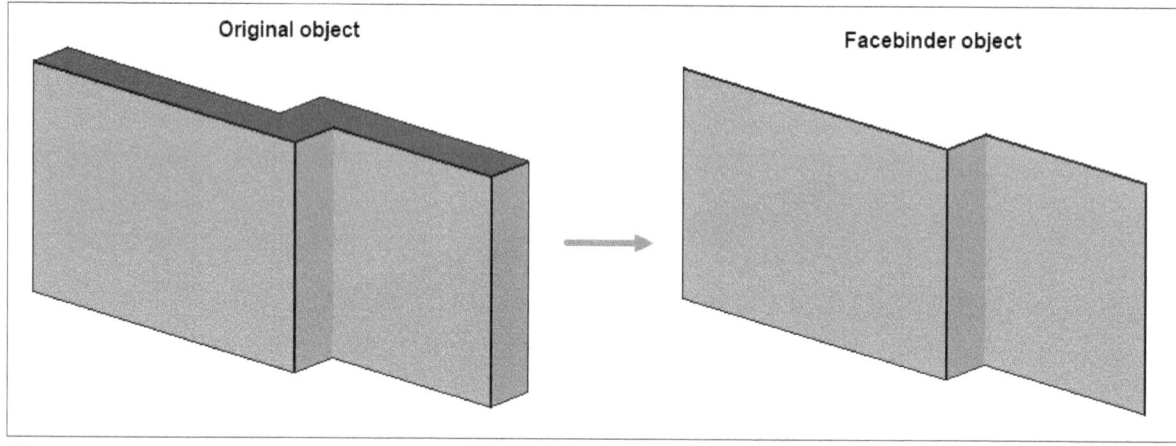

Figure-48. Facebinder object displayed

- You can edit the facebinder object using **Faces** dialog.
- To add the face to the facebinder object, press and hold the **CTRL** key from the keyboard and select the face or multiple faces from the solid object in the 3D view area; refer to Figure-49.

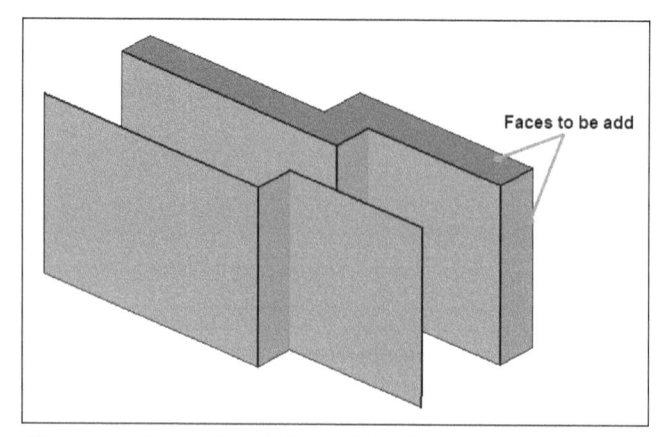

Figure-49. Faces to be add from solid object

- Holding the **CTRL** key, double-click on the facebinder object from the **Model Tree**. The **Faces** dialog will be displayed in the **Tasks** panel of **Combo View**; refer to Figure-50.

Figure-50. Faces dialog

- Click on **Add** button from the dialog, the faces added in the dialog will be added in the facebinder object; refer to Figure-51.

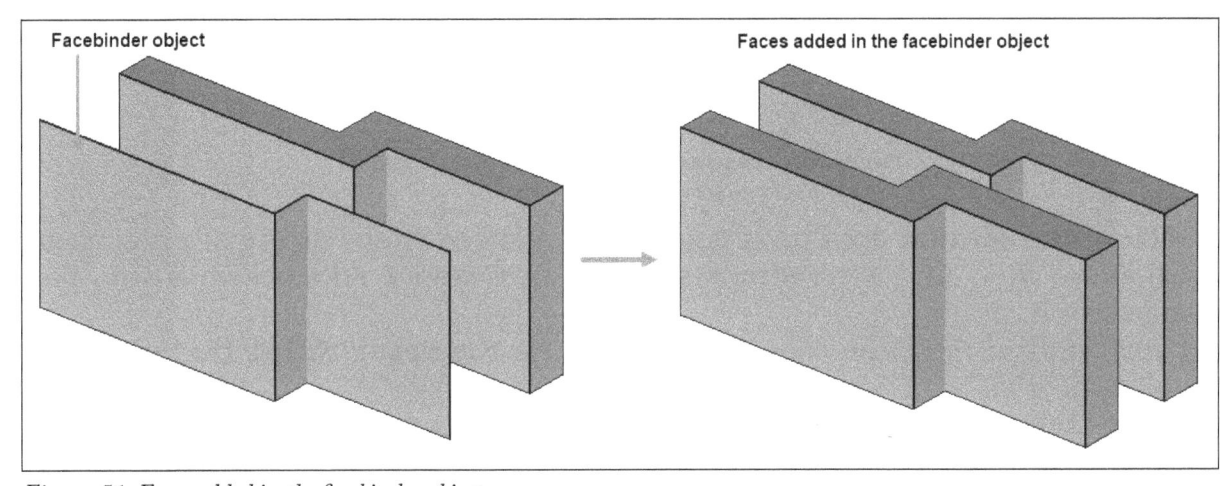

Figure-51. Faces added in the facebinder object

- If you want to remove the face from the facebinder object then select the face from the `Facebinder elements` list view area of the dialog and click on **Remove** button. The face will be removed.
- Click on **OK** button from the dialog to complete the procedure.

Creating Shape from text

The `Shape from text` tool inserts a compound shape that represents a text string. Text height, tracking, and font can be specified. The resulting shape can be used with the **Extrude** tool to create 3D letters. The procedure to use this tool is discussed next.

- Click on the `Shape from text` tool from **Toolbar** in the **Draft** workbench; refer to Figure-52. The **ShapeString** dialog will be displayed in the **Tasks** panel of **Combo View** along with the plus sign in place of original cursor; refer to Figure-53. You will be asked to specify the location of shape string.

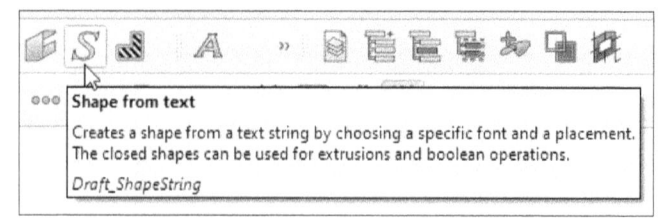

Figure-52. Shape from text tool

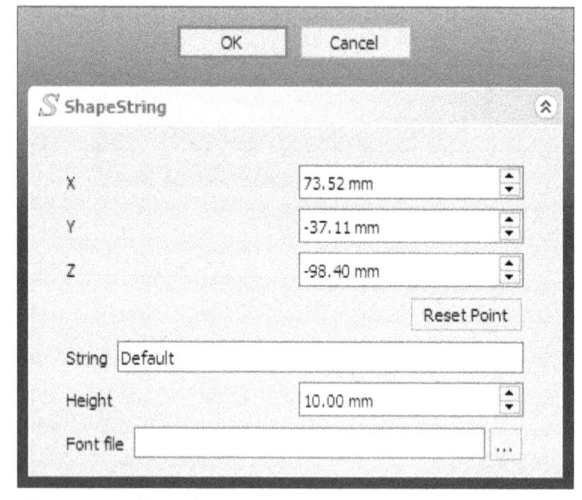

Figure-53. ShapeString dialog

- Click in the 3D view area to specify the location of shape string or enter desired values for x, y, and z coordinates in the **X**, **Y**, and **Z** edit boxes of the dialog, respectively.
- If you want to reset the coordinates, click on the **Reset Point** button from the dialog.
- Specify desired text you want to create in the **String** edit box.
- Specify the height of the text in the **Height** edit box.
- Click on the button [...] available next to the **Font file** edit box. The **Select a file** dialog box will be displayed; refer to Figure-54.

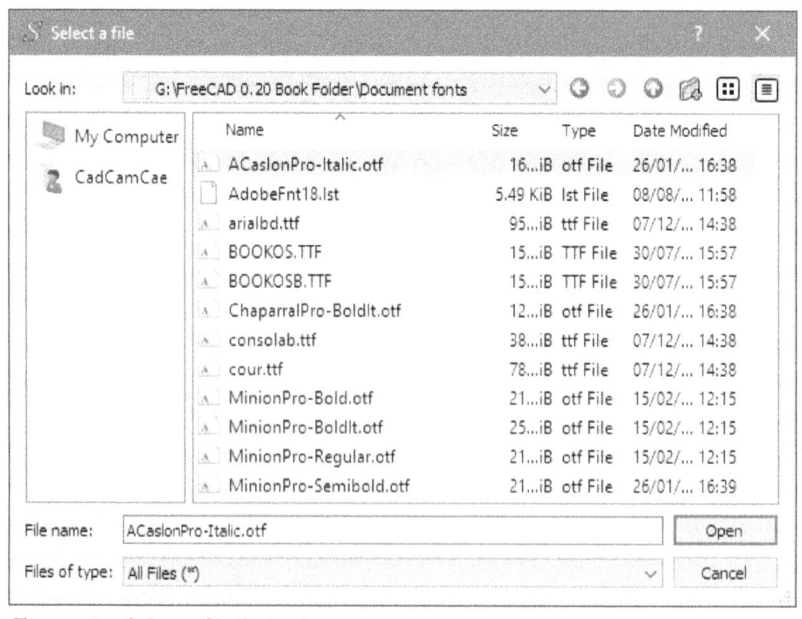

Figure-54. Select a file dialog box

- Select desired font file from the dialog box and click on **Open** button. The selected font file will be displayed in the **Font file** edit box of the dialog; refer to Figure-55. The font file is available in the installation folder of FreeCAD software.

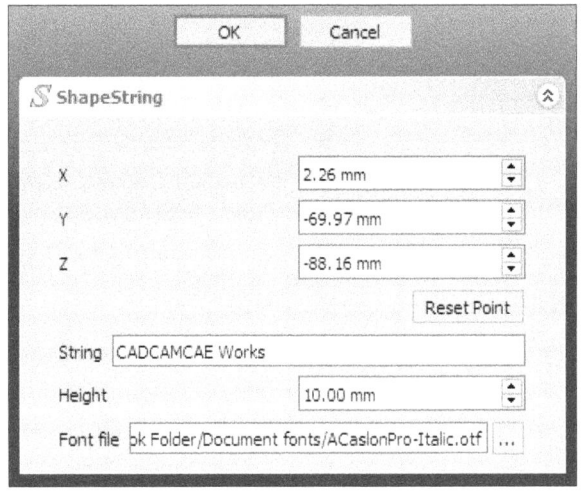

Figure-55. Font file selected

- Click on **OK** button from the dialog. The shape string will be created; refer to Figure-56.

Figure-56. Shape string created

Creating Hatch

The **Hatch** tool is used to create hatches on the planar faces of a selected object. The procedure to use this tool is discussed next.

- Select desired object with planar faces from the **Model Tree** or from the 3D view area to create hatches on it; refer to Figure-57.

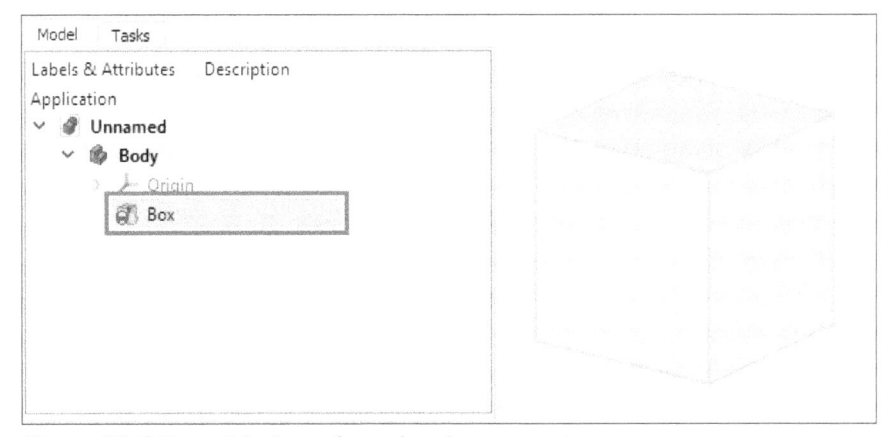

Figure-57. Object with planar faces selected

- Click on the **Hatch** tool from **Toolbar** in the **Draft** workbench; refer to Figure-58. The **Hatch** dialog will be displayed in the **Tasks** panel of **Combo View**; refer to Figure-59.

6-22   **FreeCAD 1.0 Black Book**

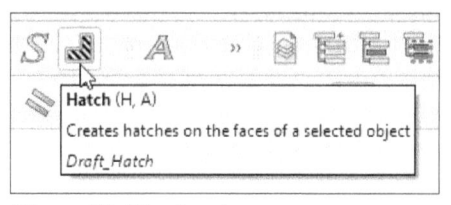

Figure-58. Hatch tool

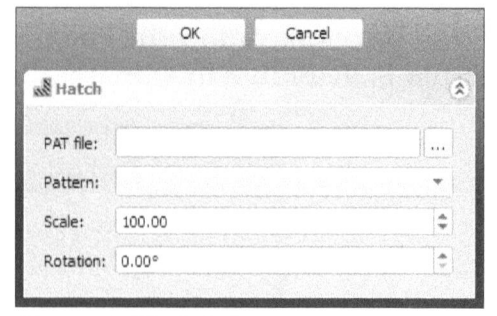

Figure-59. Hatch dialog

- Click on the ... button to specify a PAT file for **PAT file** edit box. The **Select a file** dialog box will be displayed; refer to Figure-60.

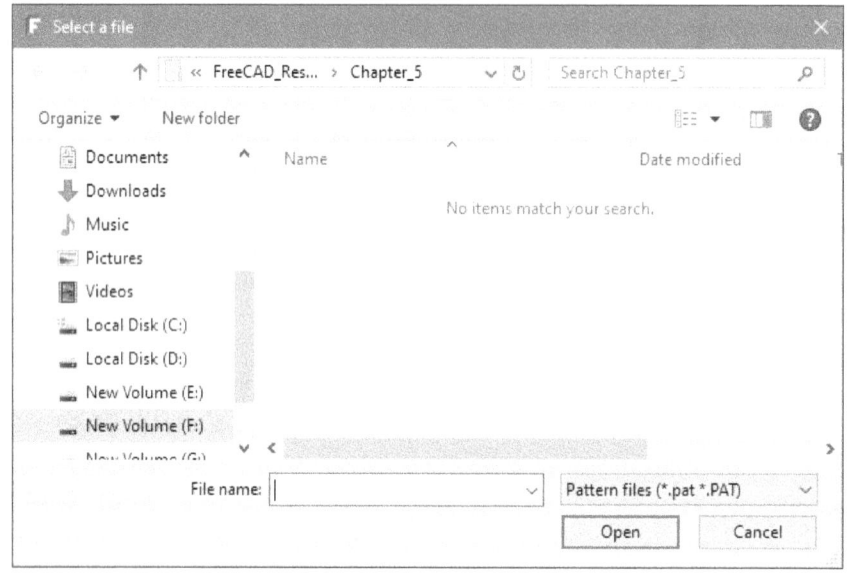

Figure-60. Select a file dialog box

- Select a PAT file which is installed with FreeCAD and click on the **Open** button from the dialog box. (On windows, PAT file is usually located in: **C:\Program Files\FreeCAD 1.0\data\Mod\TechDraw\PAT**).
- Select desired type of pattern from **Pattern** drop-down.
- Specify desired scale value for the pattern in the **Scale** edit box.
- Specify desired rotation value for the pattern in the **Rotation** edit box.
- After specifying desired parameters, click on **OK** button from the **Hatch** dialog. The hatching will be created on the object; refer to Figure-61.

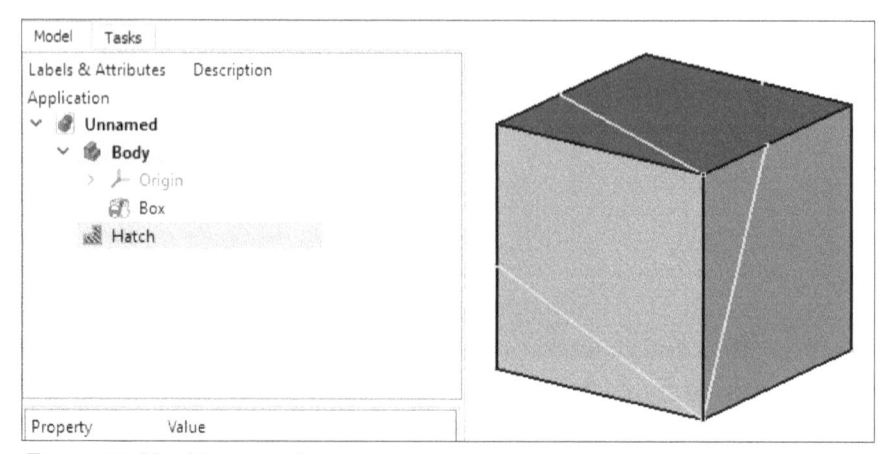

Figure-61. Hatching created

Annotation objects tools

The **Annotation objects** tools are used for annotating the objects. These tools are available in the **Toolbar** of **Draft** workbench; refer to Figure-62. The procedures to use these tools are discussed next.

Figure-62. Annotation objects tools

Creating Text

The **Text** tool inserts a multi-line text box at a given point. The procedure to use this tool is discussed next.

- Click on the **Text** tool from **Toolbar** in the **Draft** workbench; refer to Figure-63. The **Text** dialog will be displayed in the **Tasks** panel of **Combo View** along with the plus sign in place of original cursor; refer to Figure-64. You will be asked to specify the location of text.

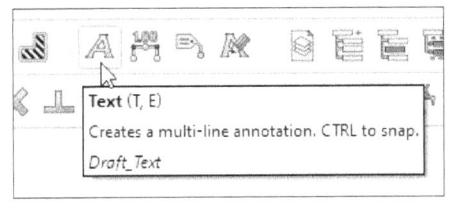

Figure-63. Text tool

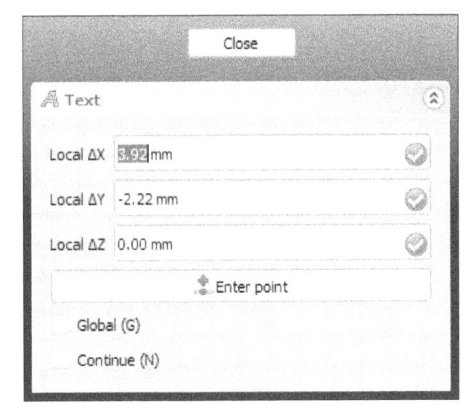

Figure-64. Text dialog

- Click in the 3D view area to specify desired location to place text or enter desired values for the x, y, and z coordinates in the **Local ΔX**, **Local ΔY**, and **Local ΔZ** edit boxes of the dialog, respectively.
- After specifying coordinates in the dialog, click on the **Enter point** button from the dialog. You will be asked to specify text in the **Text** dialog; refer to Figure-65.

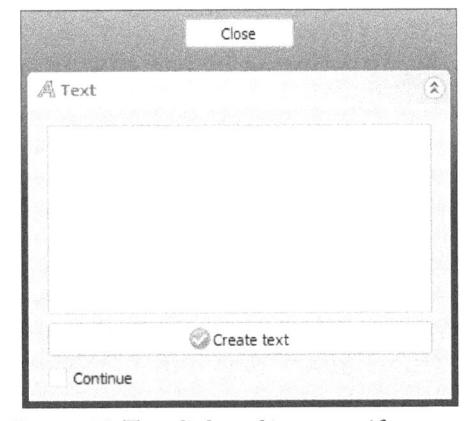

Figure-65. Text dialog asking to specify text

- Specify desired text in the edit box of the dialog and click on **Create text** button or press **ENTER** key twice from the keyboard. The text will be created; refer to Figure-66.

CADCAMCAE Works

Figure-66. Text created

- Other parameters in the dialog have been discussed earlier.

Creating Dimension

The **Dimension** tool creates an object that measures and displays the distance between two points; a third point specifies the position of the dimension line. The procedure to use this tool is discussed next.

- Click on the **Dimension** tool from **Toolbar** in the **Draft** workbench; refer to Figure-67. The **Dimension** dialog will be displayed in the **Tasks** panel of **Combo View** along with the plus sign in place of original cursor; refer to Figure-68. You will be asked to specify the first point or select the edge.

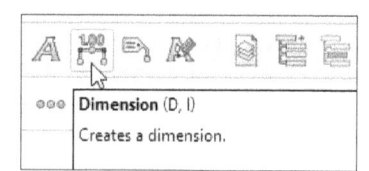

Figure-67. Dimension tool

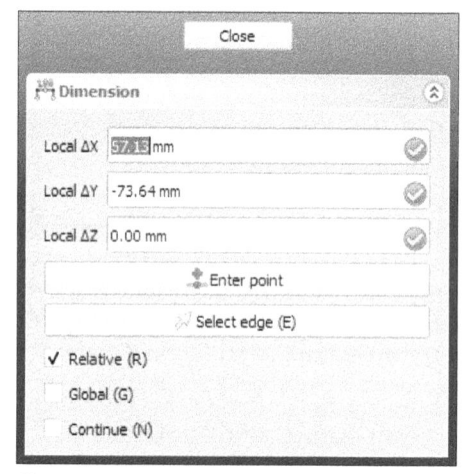

Figure-68. Dimension dialog

- Click in the 3D view area to specify first point from where you want to measure the distance or enter desired values for the x, y, and z coordinates in the **Local ΔX**, **Local ΔY**, and **Local ΔZ** edit boxes of the dialog, respectively.
- After specifying coordinates for the first point in the dialog, click on the **Enter point** button from the dialog. You will be asked to specify the second point.
- Move the cursor away and click at desired location to specify second point up to which you want to measure the distance or enter desired values for the coordinates in their respective edit boxes of the dialog. The first two points define the measured distance.
- After specifying coordinates for the second point in the dialog, click on the **Enter point** button again from the dialog. You will be asked to specify the third point.
- Move the cursor away and click at desired location to specify the third point or enter desired values for the coordinates in their respective edit boxes of the dialog. The third point defines the position of the measurement line.

- After specifying coordinates for the third point in the dialog, click on the **Enter point** button again from the dialog. The dimension will be created; refer to Figure-69.

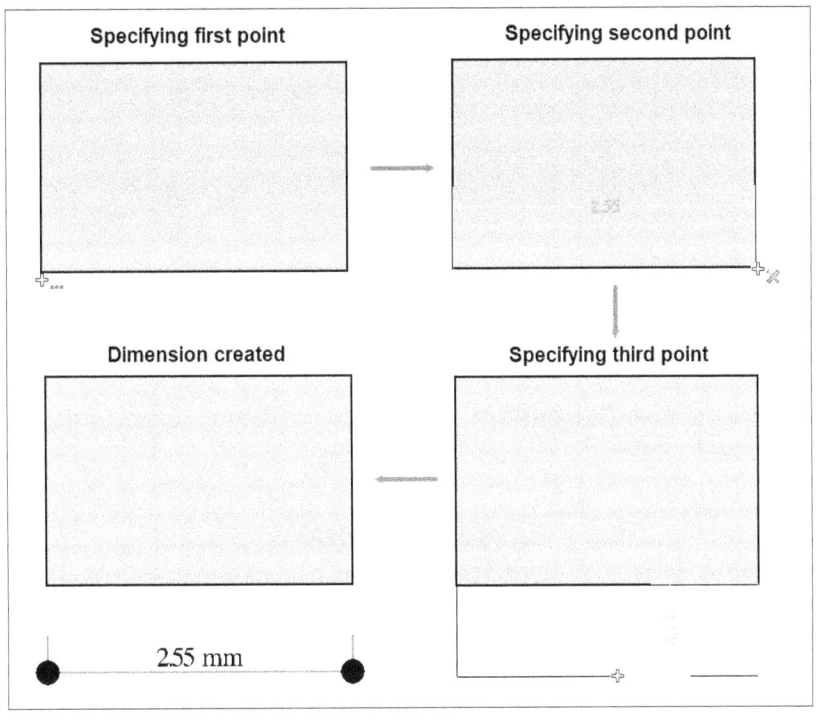

Figure-69. Dimension created

- If you want to create the dimension by selecting the edge then on selecting the tool, the **Dimension** dialog will be displayed as discussed earlier.
- Click on the **Select edge** button from the **Dimension** dialog. You will be asked to select the edge to be measured.
- Select the edge of the model whose distance is to be measured by this dimension; refer to Figure-70. You will be asked to specify the point to place the measurement line.

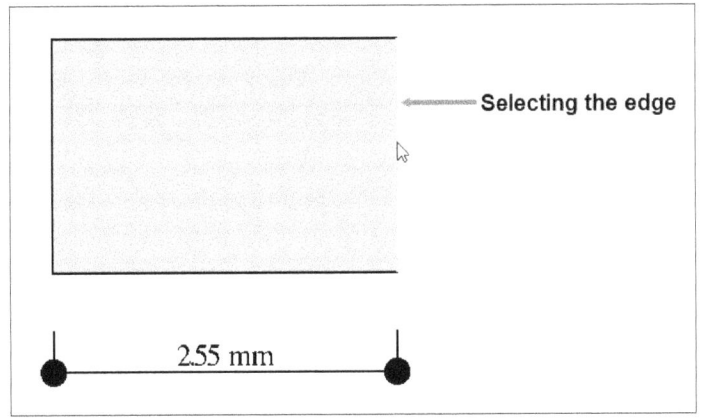

Figure-70. Selecting the edge to be measured

- Move the cursor away and click at desired location to specify the point to position the measurement line or enter desired values for the x, y, and z coordinates in the **Local X**, **Local Y**, and **Local Z** edit boxes of the dialog, respectively.
- After specifying coordinates for the point in the dialog, click on the **Enter point** button from the dialog. The dimension will be created; refer to Figure-71.

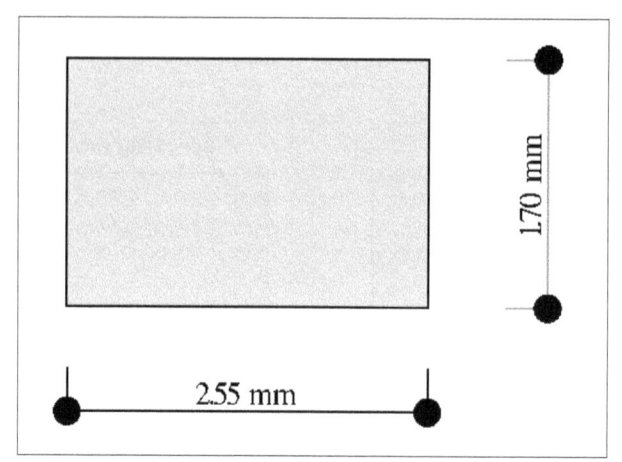

Figure-71. Dimension created

- Other parameters in the **Dimension** dialog have been discussed earlier.
- Click on the **Close** button to close the dialog.

Creating Label

The **Label** tool inserts a multi-line text box with a 2-segment leader line and an arrow. If an object or a sub-element (face, edge, or vertex) is selected when starting the command, the Label can be made to display a certain attribute of the selected element including position, length, area, volume, or material. The procedure to use this tool is discussed next.

- Click on the **Label** tool from **Toolbar** in the **Draft** workbench; refer to Figure-72. The **Label** and **Label type** dialogs will be displayed in the **Tasks** panel of **Combo View** along with the plus ⊕ sign in place of original cursor; refer to Figure-73. You will be asked to specify the first point.

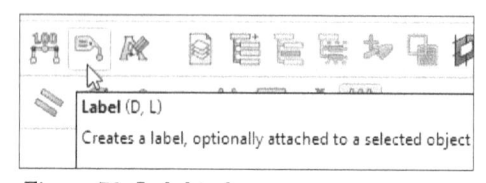

Figure-72. Label tool

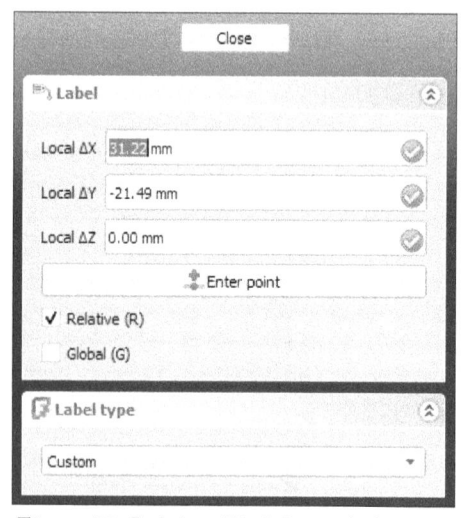

Figure-73. Label and Label type dialogs

- Select desired label type from the drop-down available in the **Label type** dialog. Select **Custom** option to display the contents of custom text. Select **Name** option to display the internal name of target object. Select **Label** option to display the label of target object. Select **Position** option to display the coordinates of the base point of the target object. Select **Length** option to display the length of the target sub-element. Select **Area** option to display the area of the target sub-element. Select **Volume** option to display the volume of target object. Select **Tag** option to display the tag attribute of the target object. Select **Material** option to display the label of the material of the target object.
- Click in the 3D view area to specify the first point of label or enter desired values for the x, y, and z coordinates in the **Local ΔX**, **Local ΔY**, and **Local ΔZ** edit boxes of the dialog, respectively.
- After specifying coordinates for the first point in the dialog, click on the **Enter point** button from the dialog. You will be asked to specify the second point of label or the start point of a horizontal or vertical leader.
- Move the cursor away and click at desired location to specify the second point or enter desired value for the coordinates in their respective edit boxes.
- After specifying coordinates for the second point, click on the **Enter point** button from the dialog. You will be asked to specify the third point of label or the base point of the text.
- Move the cursor away and click at desired location to specify the third point or enter desired values for the coordinates in their respective edit boxes.
- After specifying coordinates for the third point in the dialog, click on the **Enter point** button again. The label will be created; refer to Figure-74.

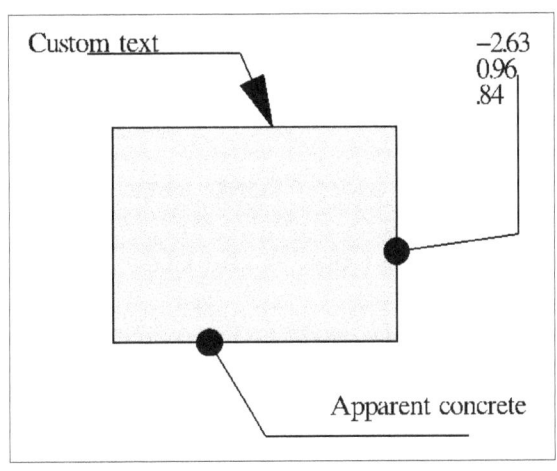

Figure-74. Labels created

Annotation styles

The **Annotation styles** tool allows you to define styles that affect the visual properties of annotation objects like dimension lines, text, and so on. The procedure to use this tool is discussed next.

- Click on the **Annotation styles** tool from **Toolbar** in the **Draft** workbench; refer to Figure-75. The **Annotation Styles Editor** dialog box will be displayed; refer to Figure-76.

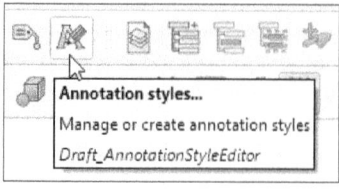

Figure-75. Annotation styles tool

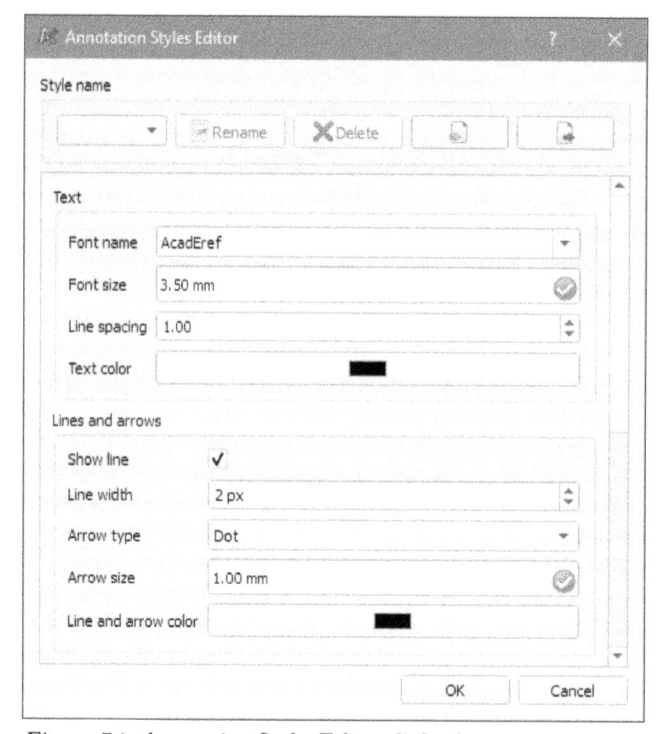

Figure-76. Annotation Styles Editor dialog box

- Select desired annotation style from the drop-down in the **Style name** area of the dialog box or select **Add new** option from the drop-down to define the new style.
- Click on the **Rename** button from **Style name** area to rename the selected style.
- Click on the **Delete** button from **Style name** area to delete the selected style.
- Click on the [   ] button from **Style name** area to import all styles from **.json** file.
- Click on the [   ] button from **Style name** area to export all styles to **.json** file.
- Specify desired font name to use for texts and dimensions in the **Font name** edit box from **Text** area.
- Specify desired size of the font in the **Font size** edit box from **Text** area in the dialog box.
- Specify desired spacing between lines in the **Line spacing** edit box from the **Text** area.
- Specify desired value in the **Scale multiplier** edit box from **Units** area to change the size of texts and markers.
- Select the **Show unit** check box from **Units** area to show the unit next to the dimension value.
- Specify a valid length unit like mm, m, in, ft in the **Unit override** edit box from **Units** area to force displaying the dimension value in this unit.
- Specify desired number of decimals to show for dimension values in the **Decimals** edit box from **Units** area.
- Select **Show lines** check box from **Line and arrows** area to display the dimension line.
- Specify desired width of the dimension lines in the **Line width** edit box from **Line and arrows** area.
- Click on the **Line/text color** button from **Line and arrows** area to specify the color of dimension lines, arrows, and texts.
- Select desired type of arrows or markers to use at the end of dimension lines from **Arrow type** drop-down in the **Line and arrows** area.

- Specify desired size of the dimension arrows or markers in the **Arrow size** edit box from **Line and arrows** area.
- Specify desired distance value that the dimension line is additionally extended in the **Dimension overshoot** edit box from **Line and arrows** area.
- Specify desired length of the extension lines in the **Extension lines** edit box from **Line and arrows** area.
- Specify desired distance value that the extension lines are additionally extended beyond the dimension line in the **Extension overshoot** edit box from **Line and arrows** area.
- After specifying desired parameters, click on the **OK** button to close the dialog box.

Modifying objects tools

The **Modifying objects** tools are used for modifying existing objects. They work on selected objects but if no object is selected, you will be asked to select one. These tools are available in the **Toolbar** of **Draft** workbench; refer to Figure-77. These tools are discussed next.

Figure-77. Modifying object tools

Moving object

The **Move** tool moves or copies the selected objects from one point to another. The procedure to use this tool is discussed next.

- Select the object from the **Model Tree** which you want to move; refer to Figure-78.

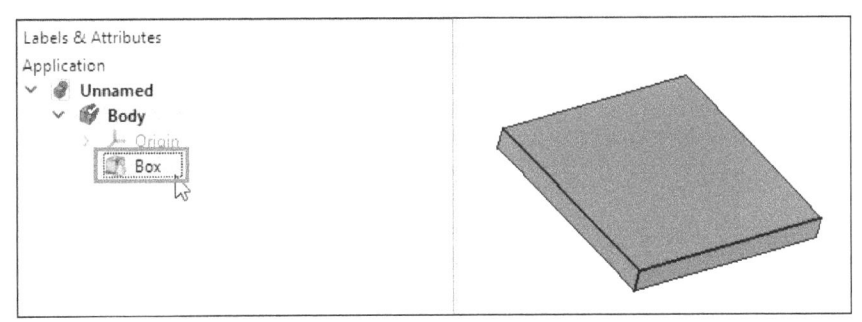

Figure-78. Selecting the object to move

- Click on the **Move** tool from **Toolbar** in the **Draft** workbench; refer to Figure-79. The **Move** dialog will be displayed in the **Tasks** panel of **Combo View** along with the plus ⊕ sign in place of original cursor; refer to Figure-80. You will be asked to specify the base point to move or copy an object.

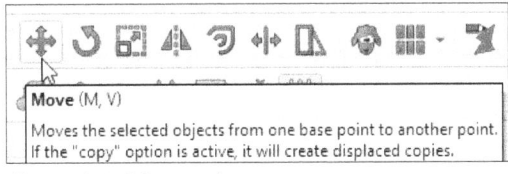

Figure-79. Move tool

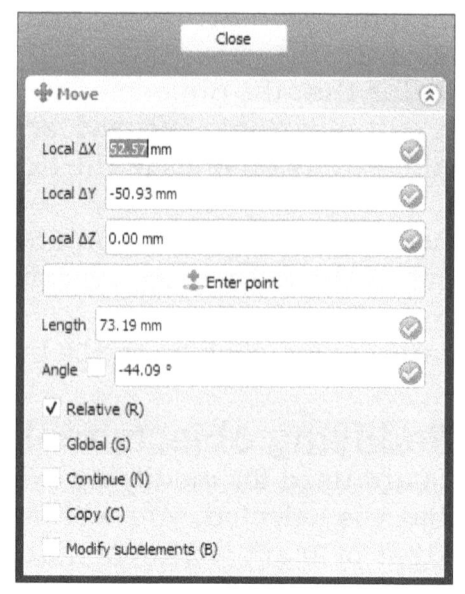

Figure-80. Move dialog

- Click in the 3D view area to specify the base point of moving the object or enter desired values for x, y, and z coordinates in the **Local ΔX**, **Local ΔY**, and **Local ΔZ** edit boxes of the dialog, respectively.
- After specifying coordinates in the dialog, click on the **Enter point** button from the dialog. You will be asked to specify the new position for the base point.
- Move the cursor away and click at desired location to specify the new position of the base point or enter desired values for the coordinates in their respective edit boxes of the dialog.
- After specifying coordinates in the dialog, click on the **Enter point** button again. The object will be moved to the new position; refer to Figure-81.

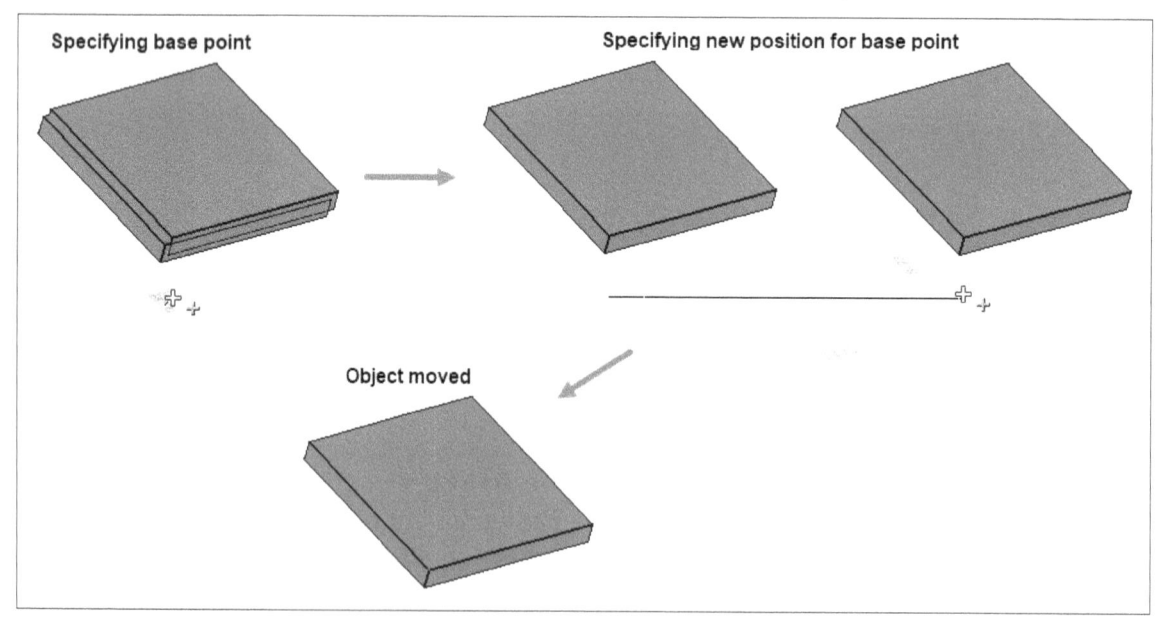

Figure-81. Object moved

- Specify desired length of the current segment in the **Length** edit box.
- Specify desired angle of the current segment in the **Angle** edit box and select the **Angle** check box to lock the current angle.
- Select **Copy** check box from the dialog to keep the original object in its place and make a copy at the second point.

- Select the **Modify subelements** check box to modify the sub-elements instead of entire objects.
- Other parameters in the **Move** dialog have been discussed earlier.

Rotating object

The **Rotate** tool rotates or copies the selected objects by a given angle around a reference point. The procedure to use this tool is discussed next.

- Select the object from the **Model Tree** which you want to rotate; refer to Figure-82.

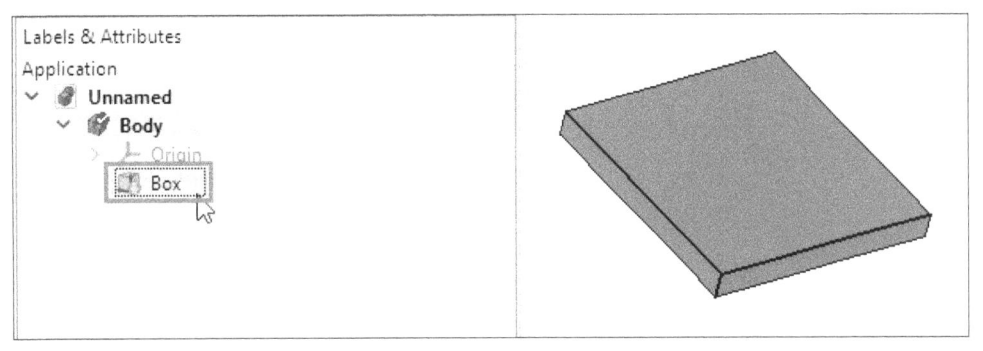

Figure-82. Selecting the object to rotate

- Click on the **Rotate** tool from **Toolbar** in the **Draft** workbench; refer to Figure-83. The **Rotate** dialog will be displayed in the **Tasks** panel of **Combo View** along with the plus sign in place of original cursor; refer to Figure-84. You will be asked to specify the base point through which the axis of rotation will pass.
- Click in the 3D view area to specify the base point of rotating the object or enter desired values for x, y, and z coordinates in the **Local ΔX**, **Local ΔY**, and **Local ΔZ** edit boxes of the dialog, respectively.

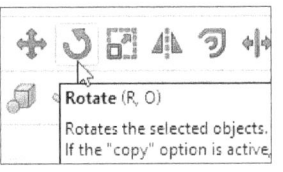

Figure-83. Rotate tool

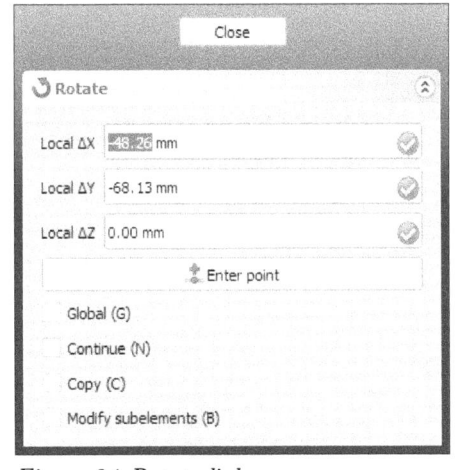

Figure-84. Rotate dialog

- After specifying coordinates in the dialog, click on the **Enter point** button from the dialog. You will be asked to specify the base angle.
- Move the cursor away and click at desired location to specify the base angle for the object that will rotate around the base point or enter desired value in the **Base angle** edit box of the dialog. You will be asked to specify the rotation angle.
- Rotate the cursor away and click at desired location to specify the rotation angle for the object or specify desired value in the **Rotation** edit box of the dialog. The object will be rotated; refer to Figure-85.

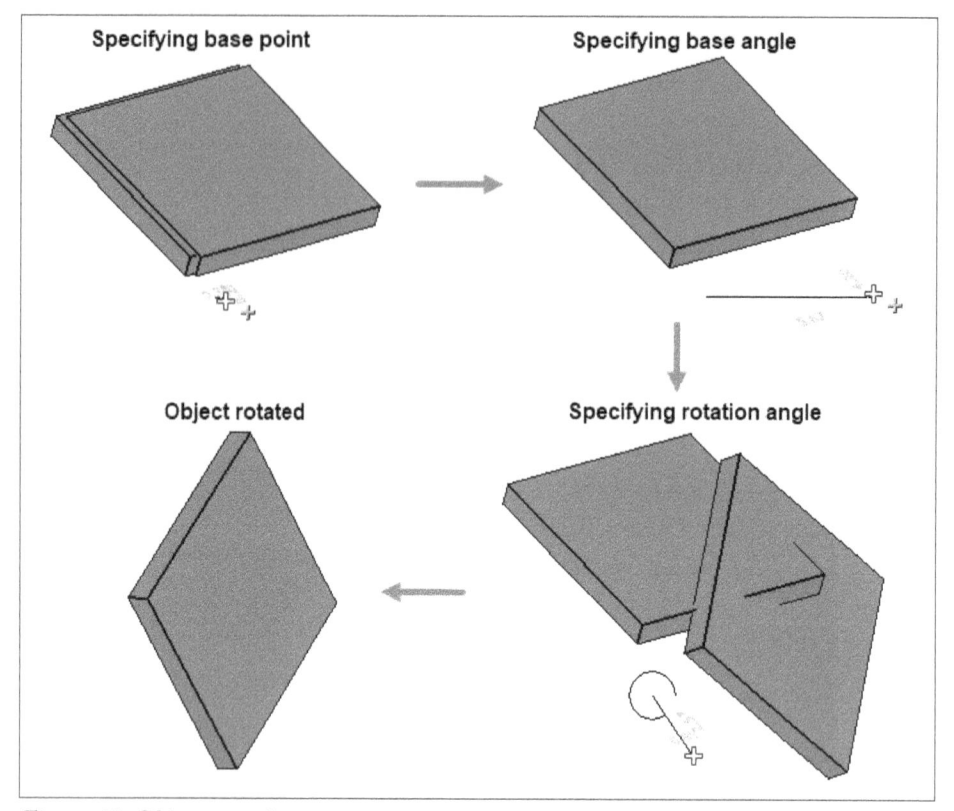

Figure-85. Object rotated

- Other parameters in the **Rotate** dialog have been discussed earlier.

Scaling object

The **Scale** tool scales or copies selected objects around a base point. The procedure to use this tool is discussed next.

- Select the object from the **Model Tree** or from the 3D view area which you want to scale; refer to Figure-86.

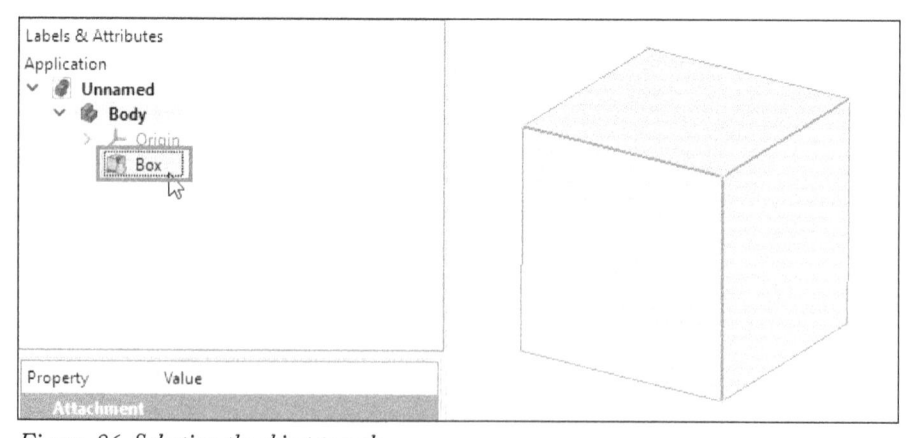

Figure-86. Selecting the object to scale

- Click on the **Scale** tool from **Toolbar** in the **Draft** workbench; refer to Figure-87. The **Scale** dialog will be displayed in the **Tasks** panel of **Combo View** along with the plus sign in place of original cursor; refer to Figure-88. You will be asked to specify the base point.

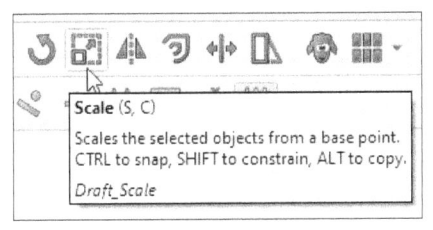

Figure-87. Scale tool

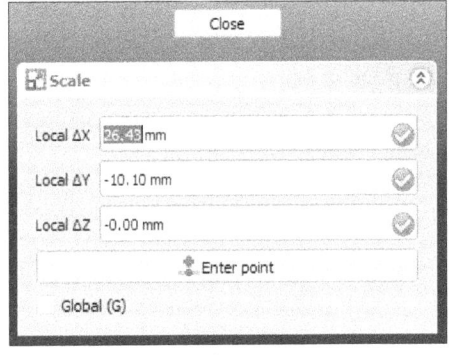

Figure-88. Scale dialog

- Click in the 3D view area to specify the base point for scaling the object or enter desired values for the x, y, and z coordinates in the **Local ΔX**, **Local ΔY**, and **Local ΔZ** edit boxes of the dialog, respectively.
- Other parameters in the **Scale** dialog have been discussed earlier.
- After specifying coordinates in the dialog, click on **Enter point** button from the dialog. The **Scale** dialog will be modified; refer to Figure-89.

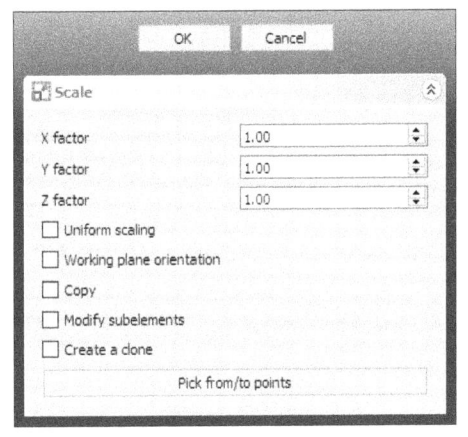

Figure-89. Modified Scale dialog

- Specify desired values for scaling the object along x, y, and z directions in the **X factor**, **Y factor**, and **Z factor** edit boxes of the dialog, respectively. The preview of the scaled object will be displayed; refer to Figure-90.

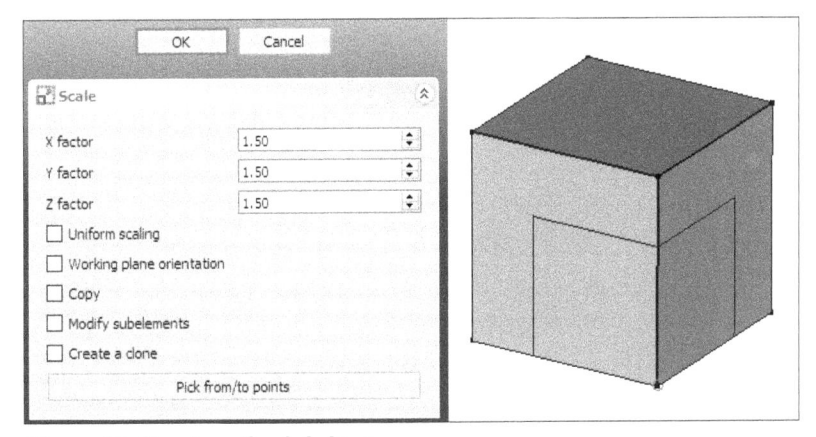

Figure-90. Preview of scaled object

- Select **Uniform scaling** check box to lock the same value for X, Y, and Z factors.
- Select **Working plane orientation** check box to lock the scaling of an object along the current working plane.
- Select **Copy** check box to create the scaled copy of the original object.

- Select **Modify subelements** check box to modify the selected subelements instead of the whole object.
- Select **Create a clone** check box to create the clone of original object.
- After specifying desired parameters, click on **OK** button from the dialog. The object will be scaled; refer to Figure-91.

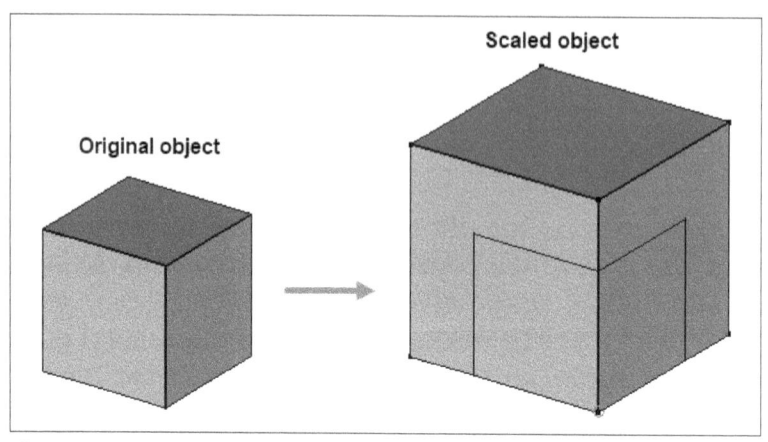

Figure-91. Original object scaled

Mirroring object

The **Mirror** tool produces a mirrored copy of selected object using the mirror operation. The copy just like a **Draft Clone**, is linked to the original object. This means that if the original object changes its shape and properties, the mirrored shape changes as well. The procedure to use this tool is discussed next.

- Select the object from the **Model Tree** or from the 3D view area which you want to be mirrored; refer to Figure-92.

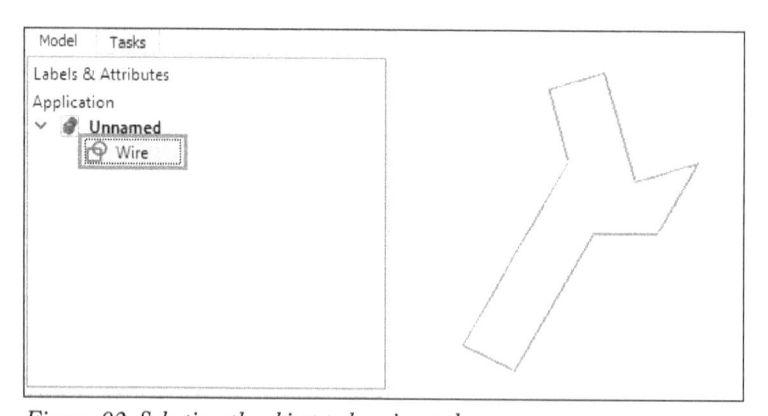

Figure-92. Selecting the object to be mirrored

- Click on the **Mirror** tool from **Toolbar** in the **Draft** workbench; refer to Figure-93. The **Mirror** dialog will be displayed in the **Tasks** panel of **Combo View** along with the plus sign in place of original cursor; refer to Figure-94. You will be asked to specify the first point.

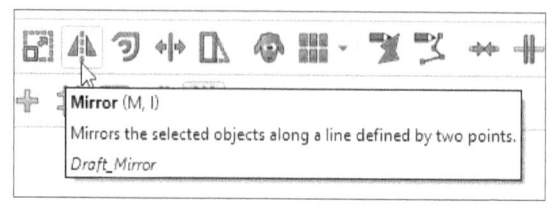

Figure-93. Mirror tool

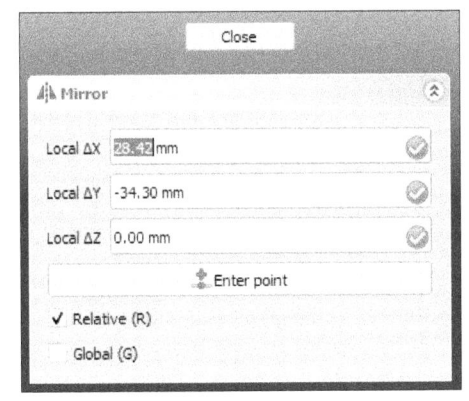

Figure-94. Mirror dialog

- Click in the 3D view area to specify the first point or enter desired values for the x, y, and z coordinates in the **Local ΔX**, **Local ΔY**, and **Local ΔZ** edit boxes of the dialog, respectively.
- After specifying coordinates for the first point in dialog, click on the **Enter point** button from the dialog. You will be asked to specify the second point.
- Move the cursor away and click at desired location to specify the second point or enter desired values for the coordinates in their respective edit boxes.
- After specifying coordinates for the second point in dialog, click on the **Enter point** button again. The object will be mirrored; refer to Figure-95.

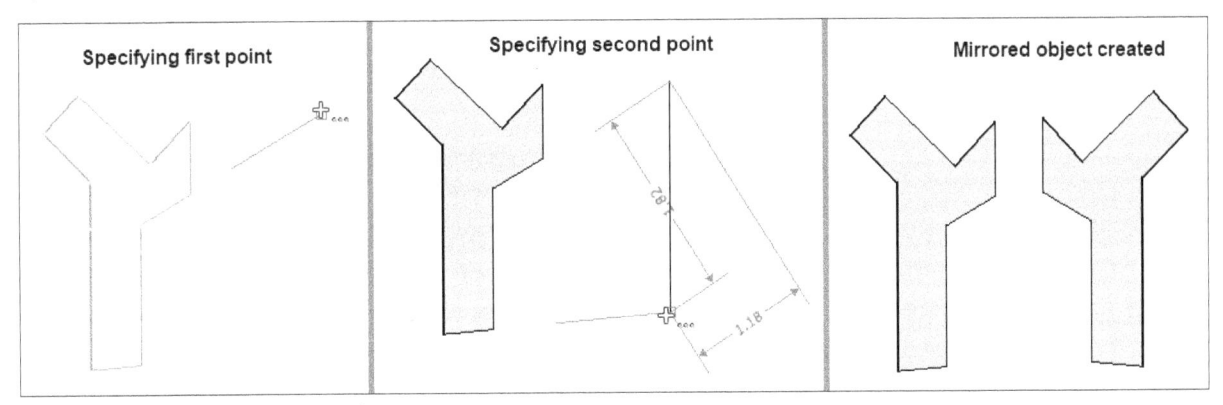

Figure-95. Mirrored object created

- Other parameters in the **Mirror** dialog have been discussed earlier.

Offsetting object

The **Offset** tool moves the selected object by a given distance (offset) perpendicular to itself. The procedure to use this tool is discussed next.

- Select the object from the **Model Tree** which you want to offset; refer to Figure-96.

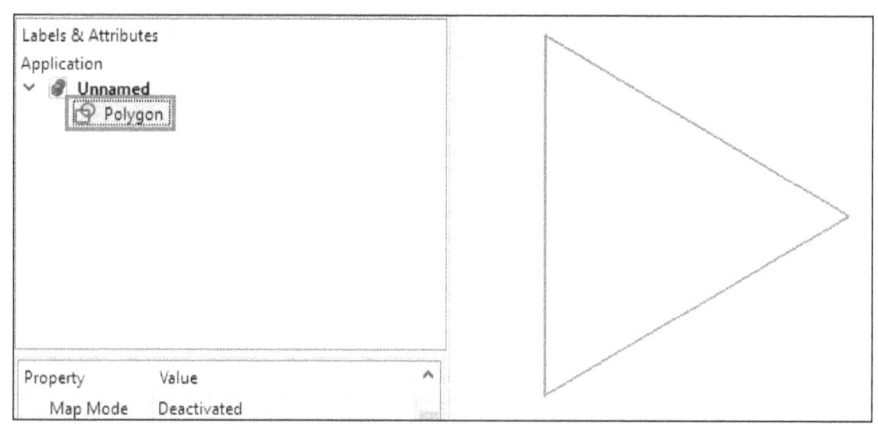

Figure-96. Selecting the object to offset

- Click on the **Offset** tool from **Toolbar** in the **Draft** workbench; refer to Figure-97. The **Offset** dialog will be displayed in the **Tasks** panel of **Combo View** along with the plus sign in place of original cursor; refer to Figure-98. You will be asked to specify the offset distance.

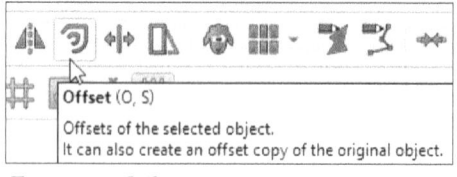

Figure-97. Offset tool

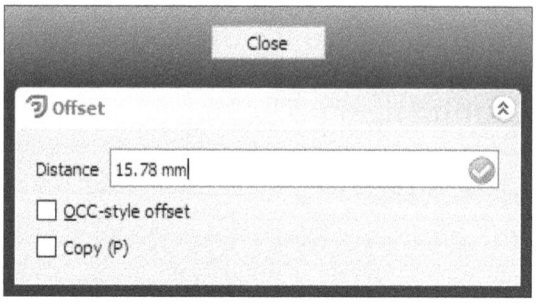

Figure-98. Offset dialog

- Move the cursor away and click in the 3D view area to specify the offset distance or enter desired value in the **Distance** edit box of the dialog and press **ENTER** key from keyboard. The offset of the object will be created; refer to Figure-99.

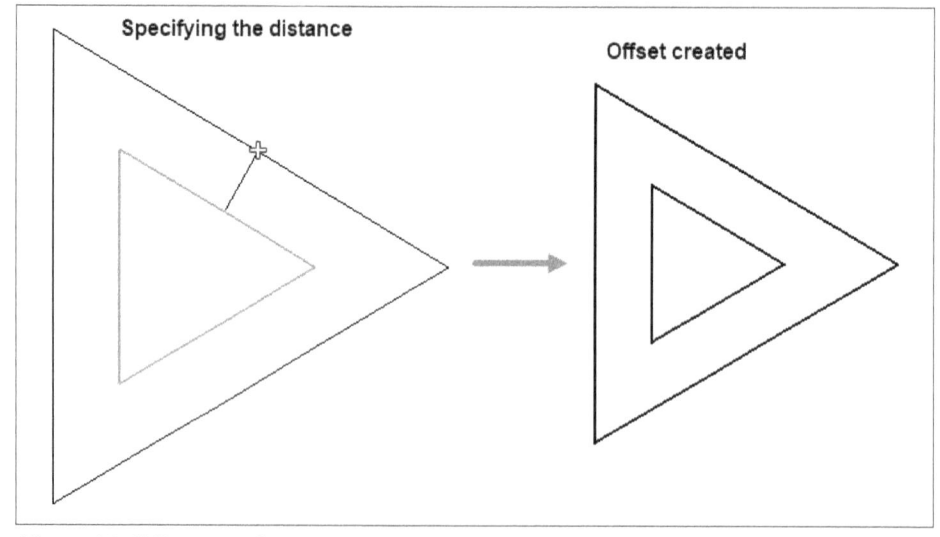

Figure-99. Offset created

Trimming/Extending object

The **Trimex** tool trims or extends lines and wires so that they end at an intersection with another curve or edge. This tool also extrudes faces created from closed wires. The procedure to use this tool is discussed next.

- Select the line from the 3D view area which you want to extend; refer to Figure-100.

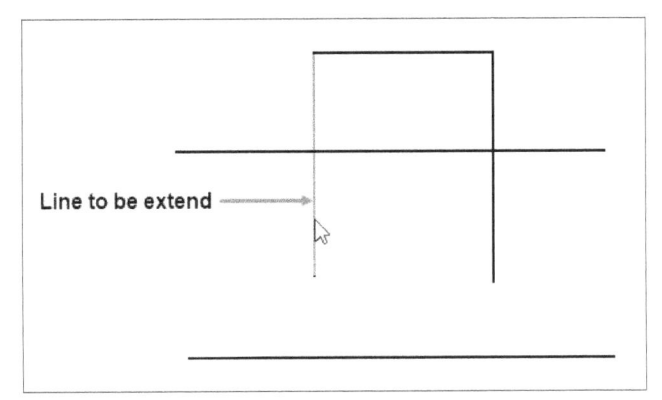

Figure-100. Selecting the line to be extend

- Click on the **Trimex** tool from **Toolbar** in the **Draft** workbench; refer to Figure-101. The **Trimex** dialog will be displayed in the **Tasks** panel of **Combo View** and the cursor will be attached to the selected line; refer to Figure-102. You will be asked to specify the distance.

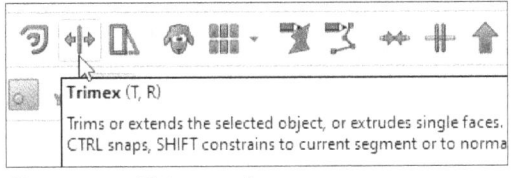

Figure-101. Trimex tool

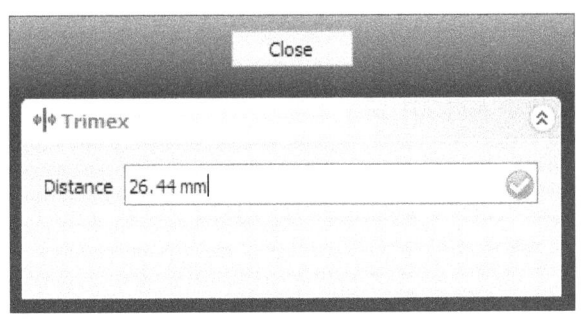

Figure-102. Trimex dialog

- Move the cursor away to the distance up to which you want to extend the line and click **LMB** or enter desired value in the **Distance** edit box of the dialog and press **Enter** key from keyboard. The line will be extended; refer to Figure-103.

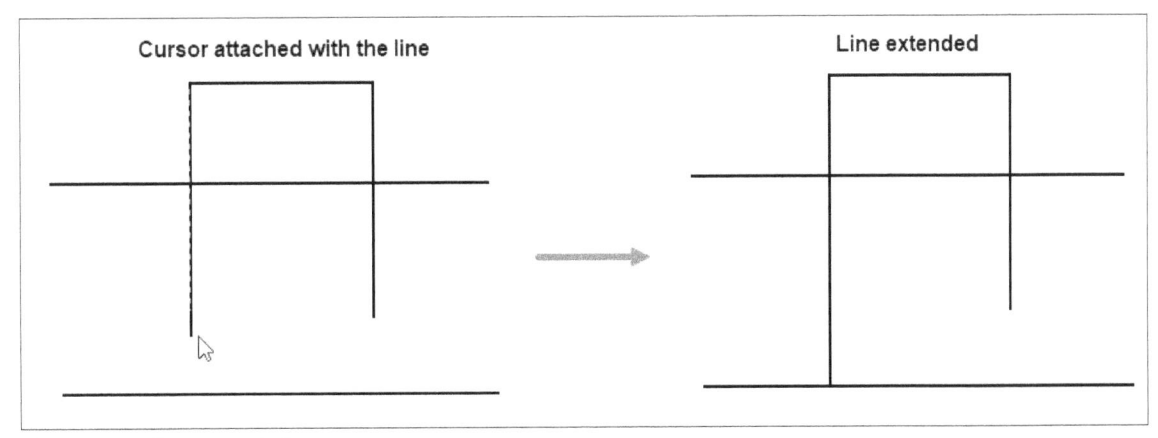

Figure-103. Line extended

- If you want to trim the line then select the line from the 3D view area which you want to trim; refer to Figure-104.

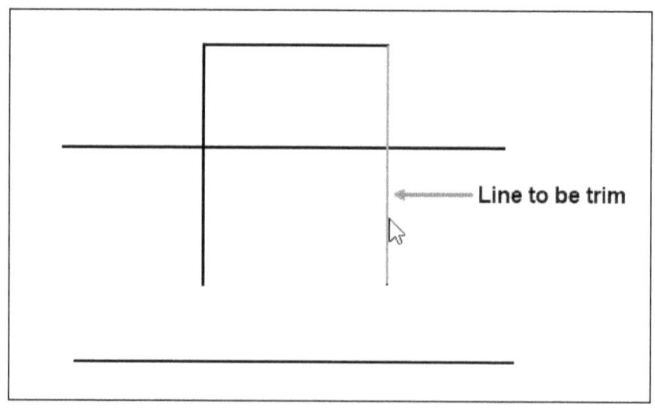

Figure-104. Selecting the line to be trim

- After selecting the line, click on the **Trimex** tool. The **Trimex** dialog will be displayed and the cursor will be attached to the selected line as discussed earlier. You will be asked to specify the distance.
- Move the cursor away to the distance upto which you want to trim the line and click **LMB** or enter desired value in the **Distance** edit box of the dialog and press **Enter** key from keyboard. The line will be trimmed; refer to Figure-105.

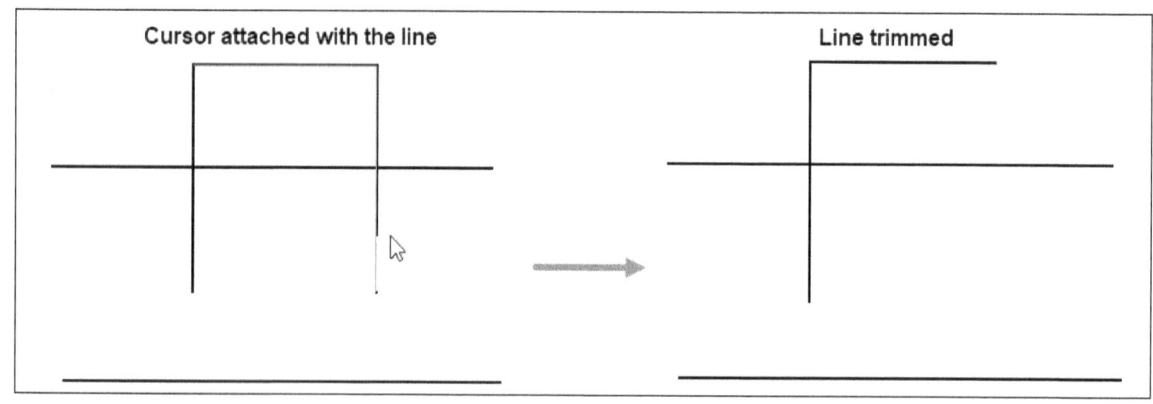

Figure-105. Line trimmed

- If you want to extrude the face created from closed wire then select the face from the **Model Tree** or from the 3D view area; refer to Figure-106.

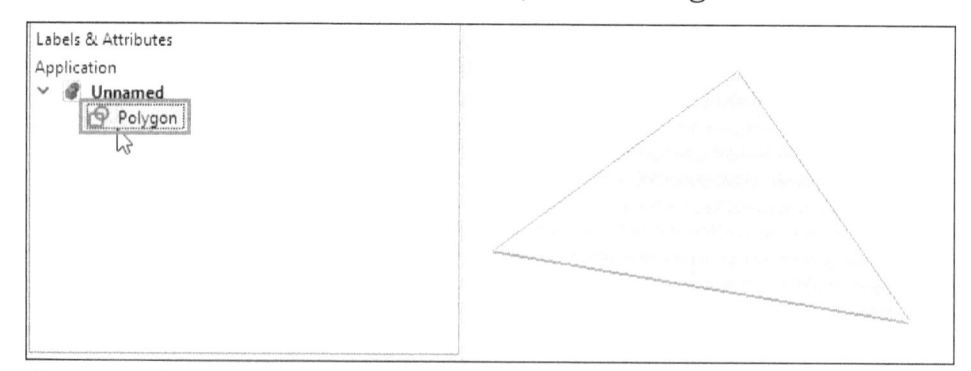

Figure-106. Selecting the face to extrude

- After selecting the face, click on the **Trimex** tool. The Trim dialog will be displayed and the cursor will be attached to the extrusion of face. You will be asked to specify the distance.
- Move the cursor away to the distance up to which you want to extrude the face and click **LMB** or enter desired value in the **Distance** edit box of the dialog and press **Enter** key from keyboard. The face will be extruded; refer to Figure-107.

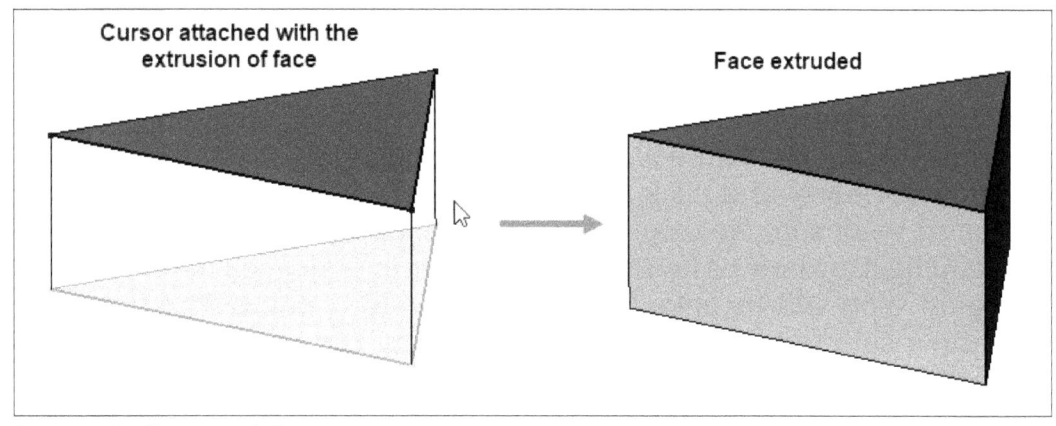

Figure-107. Face extruded

Stretching object

The **Stretch** tool stretches an object by moving some of its selected vertices. The equivalent action is editing the object and moving the points manually to a new position. The procedure to use this tool is discussed next.

- Select the object from the **Model Tree** or from the 3D view area which you want to stretch; refer to Figure-108.

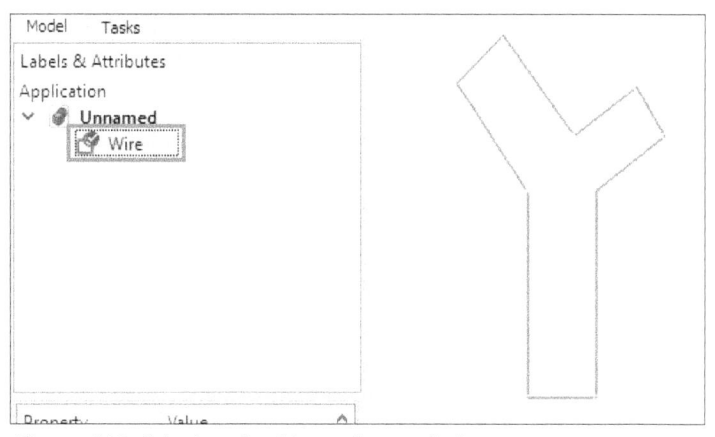

Figure-108. Selecting the object to be stretched

- Click on the **Stretch** tool from **Toolbar** in the **Draft** workbench; refer to Figure-109. The **Stretch** dialog will be displayed in the **Tasks** panel of **Combo View** along with the plus sign in place of original cursor; refer to Figure-110. You will be asked to specify the first point.

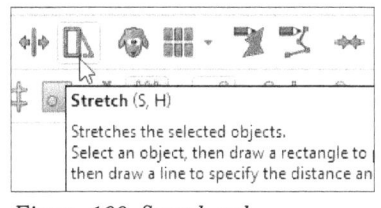

Figure-109. Stretch tool

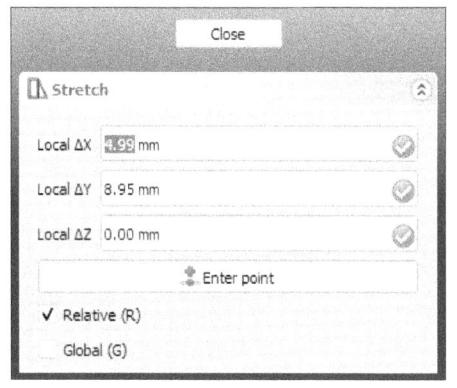

Figure-110. Stretch dialog

- Click in the 3D view area to specify the first point or enter desired values for the x, y, and z coordinates in the **Local ΔX**, **Local ΔY**, and **Local ΔZ** edit boxes of the dialog, respectively.
- After specifying coordinates for the first point in dialog, click on the **Enter point** button from the dialog. You will be asked to specify the second point.
- Move the cursor away and click at desired location to specify the second point or enter desired values for the coordinates in their respective edit boxes.
- After specifying coordinates for the second point in dialog, click on the **Enter point** button again. You will be asked to specify the third point. The first and second point define a selection rectangle. The vertices of the original object enclosed by this rectangle will become highlighted; refer to Figure-111.

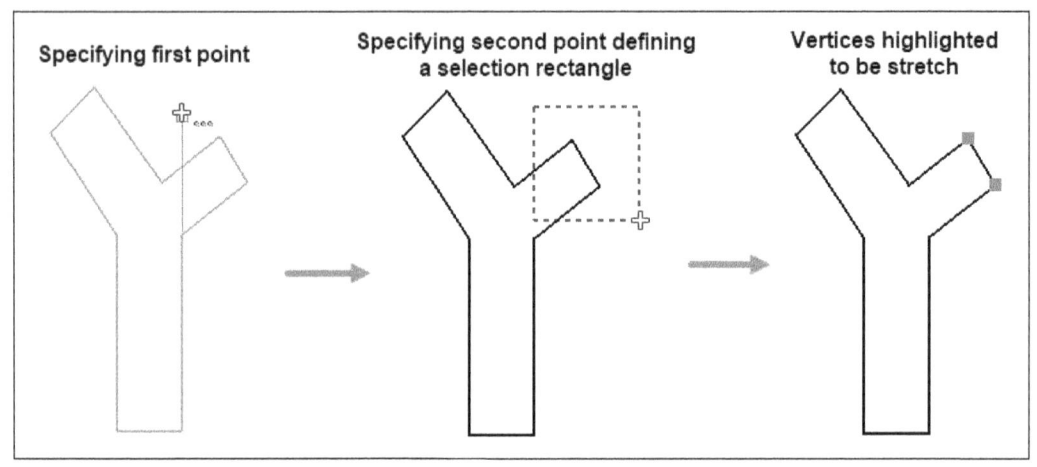

Figure-111. Selecting the vertices to be stretch

- Click in the 3D view area to specify the third point or enter desired values for the coordinates in their respective edit boxes.
- After specifying coordinates for the third point in dialog, click on the **Enter point** button again. You will be asked to specify the fourth point.
- Move the cursor away and click at desired location to specify the fourth point or enter desired values for the coordinates in their respective edit boxes. The third and fourth point define a line whose distance and direction will be used to stretch the figure attached to the highlighted points.
- After specifying coordinates for the fourth point in dialog. Click on the **Enter point** button again. The object will be stretched; refer to Figure-112.

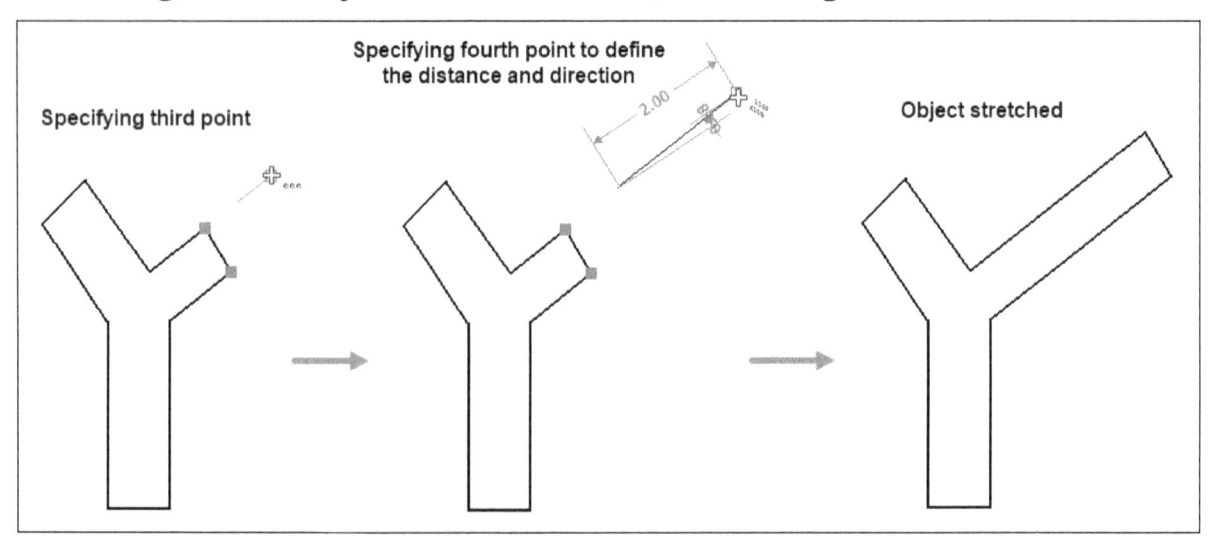

Figure-112. Object stretched

- Other parameters in the **Stretch** dialog have been discussed earlier.

Creating Clone

The **Clone** tool produces linked copies of selected shape. This means that if the original object changes its shape and properties, all clones change as well. The procedure to use this tool is discussed next.

- Select the object from the **Model Tree** or from the 3D view area for which you want to create clone; refer to Figure-113.

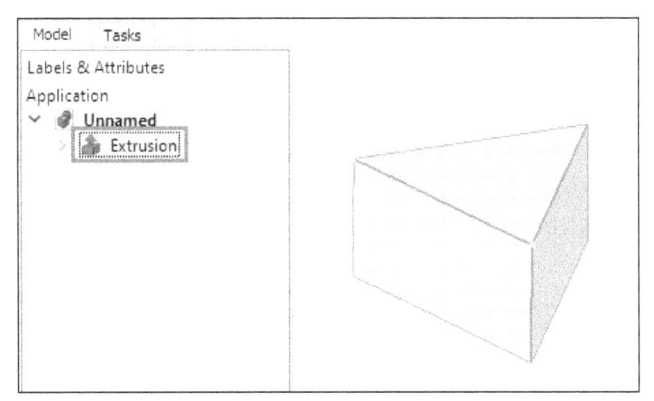

Figure-113. Selecting the object to be cloned

- Click on the **Clone** tool from **Toolbar** in the **Draft** workbench; refer to Figure-114. The cloned object will be created; refer to Figure-115.

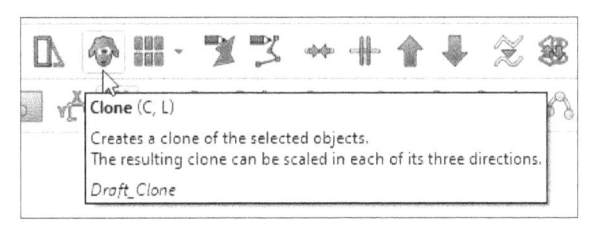

Figure-114. Clone tool

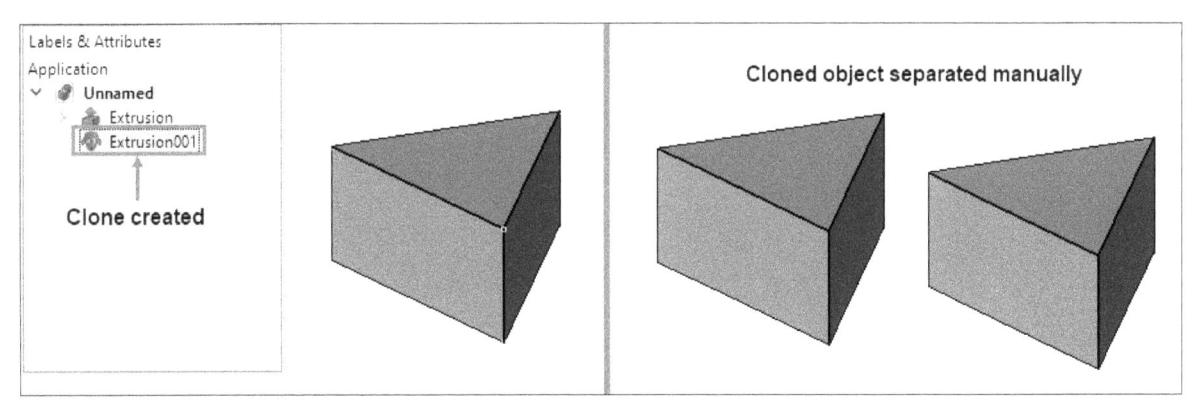

Figure-115. Cloned object created

- You can change the properties of clone created from the **Property editor** dialog as discussed earlier.

Creating Array

The **Array** tool creates an orthogonal (3-axes), polar, or circular array of selected object. The procedure to use this tool is discussed next.

- Select the object from **Model Tree** or from the 3D view area of which you want to create an array; refer to Figure-116.

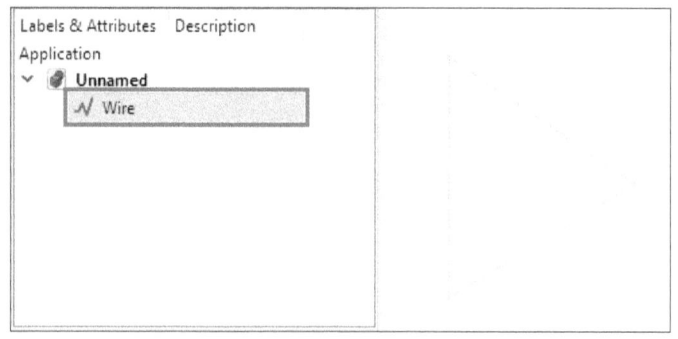

Figure-116. Selecting the object to create array

- Click on the **Array** tool from **Array tools** drop-down in the **Toolbar** of **Draft** workbench; refer to Figure-117. The **Orthogonal array** dialog will be displayed in the **Tasks** panel of **Combo View**; refer to Figure-118.

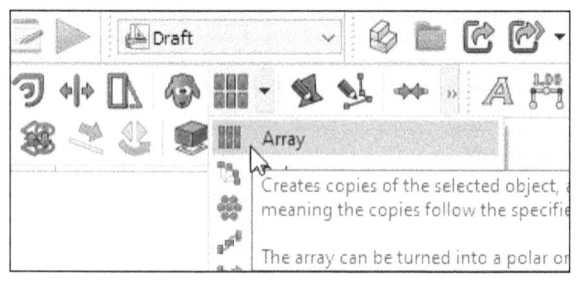

Figure-117. Array tool

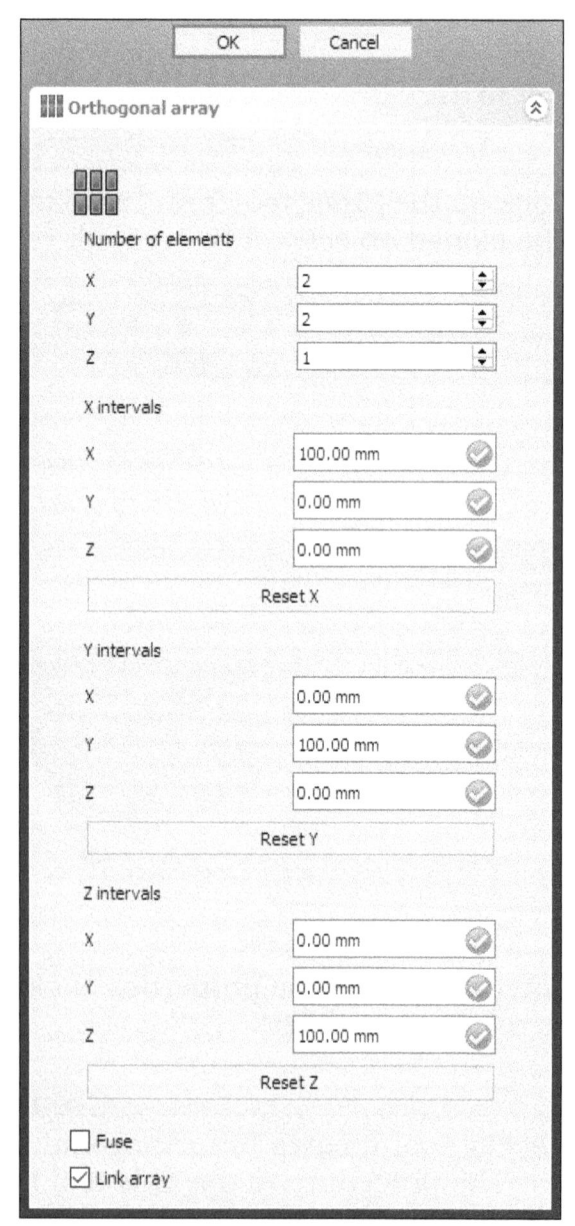

Figure-118. Orthogonal array dialog

- Specify desired number of elements for the x, y, and z directions in the **X**, **Y**, and **Z** edit boxes, respectively from **Number of elements** area of the dialog.
- Enter desired values in **X**, **Y**, and **Z** edit boxes of **X intervals**, **Y intervals**, and **Z intervals** area to specify the displacement for the elements in the X direction, Y direction, and Z direction, respectively.
- For a rectangular array along X direction, the **Y** and **Z** values must be **0**; along Y direction, the **X** and **Z** values must be **0**; along Z direction, the **X** and **Y** values must be **0**.
- Click on the **Reset X**, **Reset Y**, or **Reset Z** button to reset the displacement in the given direction to the default values.
- Select the **Fuse** check box to fuse the overlapping elements in the array.
- Select the **Link array** check box to create the link array instead of a regular array.
- After specifying desired parameters in the dialog, click on the **OK** button. The orthogonal array will be created; refer to Figure-119.

Figure-119. Orthogonal array created

Creating Polar array

The **Polar array** tool creates an array of selected object by placing copies along a circumference. The procedure to use this tool is discussed next.

- Select the object from the **Model Tree** or from the 3D view area of which you want to create a polar array; refer to Figure-120.

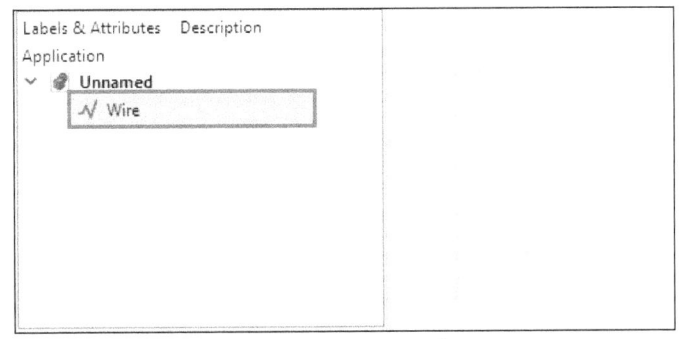

Figure-120. Selecting the object to create polar array

- Click on the **Polar array** tool from **Array tools** drop-down in the **Toolbar** of **Draft** workbench; refer to Figure-121. The **Polar array** dialog will be displayed in the **Tasks** panel of **Combo View** along with the plus ⊕ sign in place of original cursor; refer to Figure-122.

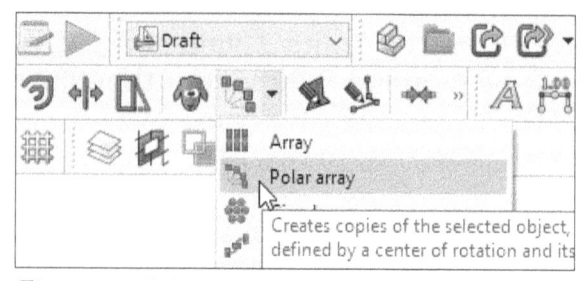

Figure-121. Polar array tool

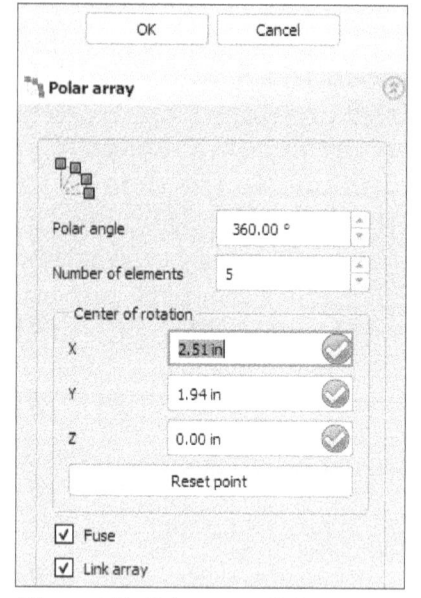

Figure-122. Polar array dialog

- Enter desired angle value in the **Polar angle** edit box to specify the total angle of the array. The angle is positive in the counter-clockwise direction.
- Enter desired number of elements in the array, including a copy of the original object in the **Number of elements** edit box of the dialog.
- Click in the 3D view area to specify the point through which the rotation axis of the array will pass or enter desired coordinates for the center of rotation in the **X**, **Y**, and **Z** edit boxes from **Center of rotation** area of the dialog. The polar array will be created; refer to Figure-123.

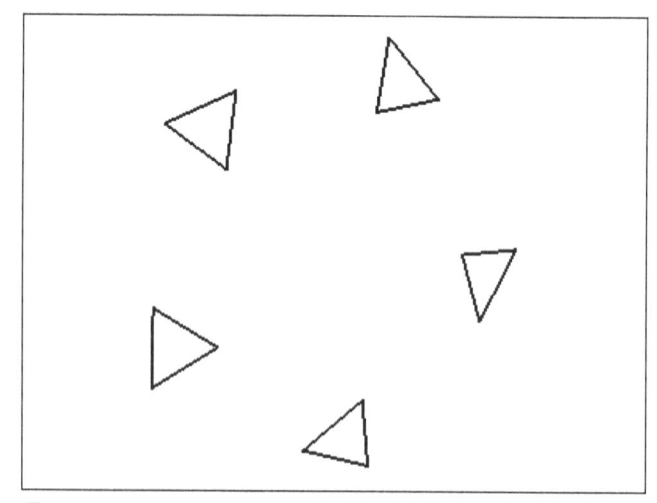

Figure-123. Polar array created

- Click on the **Reset point** button to reset the coordinates of the center of rotation to the origin.
- Other parameters in the dialog have been discussed earlier.

Creating Circular array

The **Circular array** tool creates an array of selected object by placing copies along concentric circumferences. The procedure to use this tool is discussed next.

- Select the object from the **Model Tree** or from the 3D view area of which you want to create a circular pattern; refer to Figure-124.

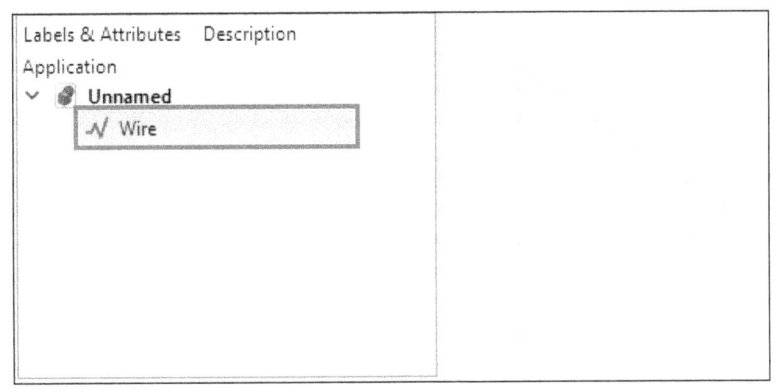

Figure-124. Selecting the object to create circular array

- Click on the **Circular array** tool from **Array tools** drop-down in the **Toolbar** of **Draft** workbench; refer to Figure-125. The **Circular array** dialog will be displayed in the **Tasks** panel of **Combo View** along with the plus ⊕ sign in place of original cursor; refer to Figure-126.

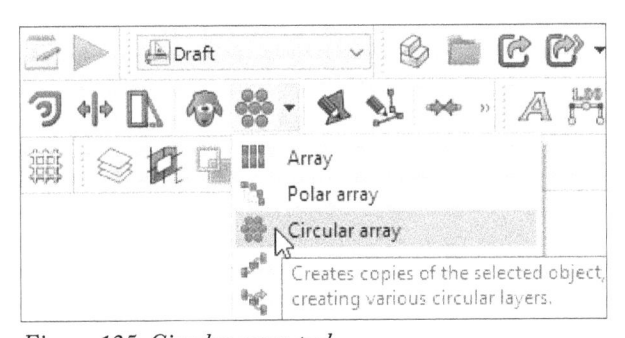

Figure-125. Circular array tool

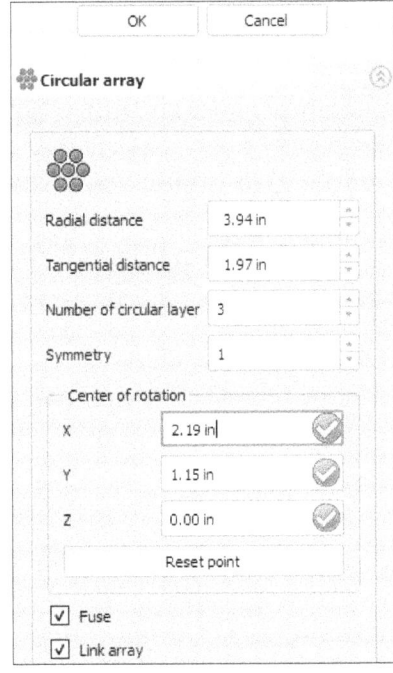

Figure-126. Circular array dialog

- Enter desired value in **Radial distance** edit box to specify the distance from one layer of objects to the next layer of objects.
- Enter desired value in **Tangential distance** edit box to specify distance between the elements on the same circular layer. It must be larger than **0**.
- Specify desired number of circular layers or rings to create, including a copy of the original object in the **Number of circular layers** edit box. It must be at least **2**.
- Specify desired number of symmetry lines in the circular array in the **Symmetry** edit box.

- Click in the 3D view area to specify the point through which the rotation axis of the array will pass or enter desired coordinates for the center of rotation in the **X**, **Y**, and **Z** edit boxes from `Center of rotation` area of the dialog. The circular array will be created; refer to Figure-127.

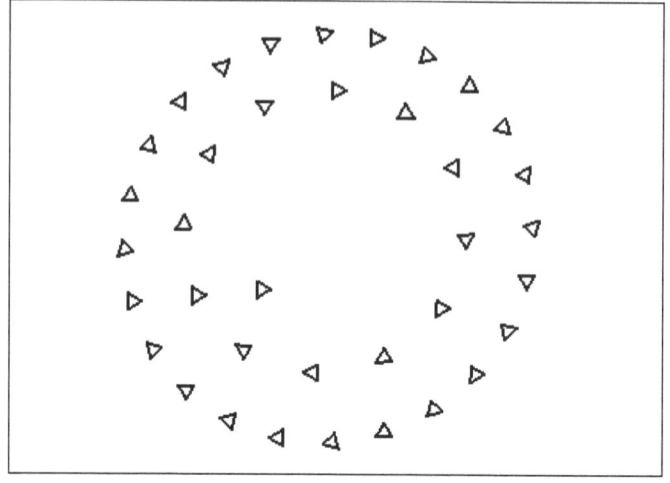

Figure-127. Circular array created

- Click on the `Reset point` button to reset the coordinates of the center of rotation to the origin.
- Other parameters in the dialog have been discussed earlier.

Creating Path array

The `Path array` tool places copies of selected shape along selected path which can be a wire, a B-Spline, and similar edges. The procedure to use this tool is discussed next.

- First, select the object from the `Model Tree` or from the 3D view area that you want to array and then select the object to be used as path along which the object will be distributed while holding the **CTRL** key; refer to Figure-128.

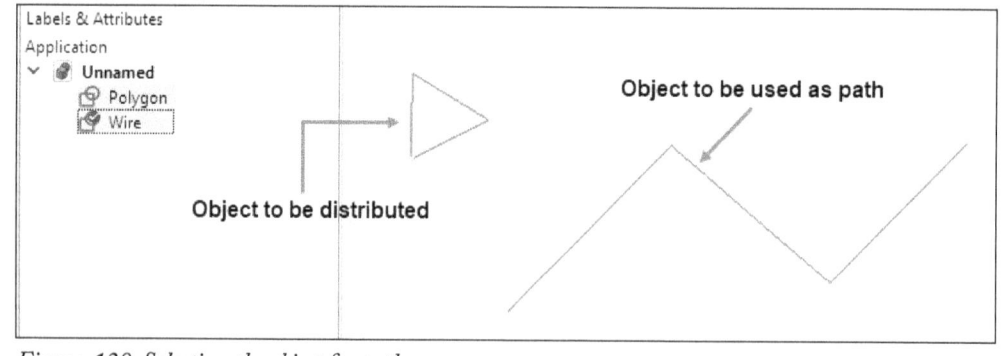

Figure-128. Selecting the object for path array

- Click on the `Path array` tool from `Array tools` drop-down in the `Toolbar` of `Draft` workbench; refer to Figure-129. The array objects will be created along path; refer to Figure-130.

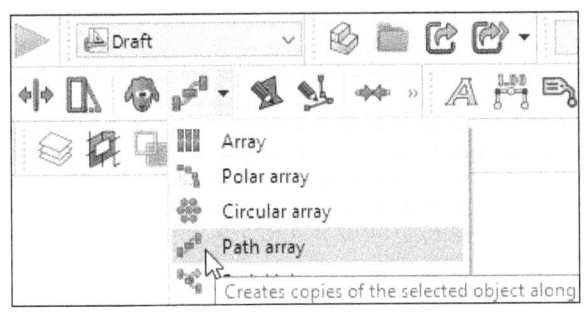

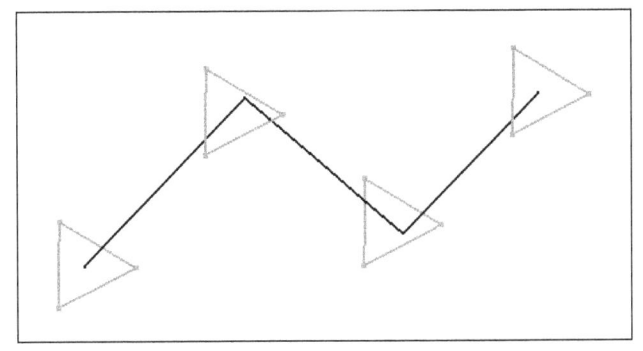

Figure-129. Path array tool

Figure-130. Array object created along path

- You can change the properties of array created from the **Property editor** dialog as discussed earlier.

Creating Path Link array

The **Path Link array** tool creates a link array from selected object by placing copies along a path. The procedure to use this tool is same as discussed for **Path array** tool.

Creating Point array

The **Point array** tool places copies of selected object on selected points. The procedure to use this tool is discussed next.

- First, select the object from the **Model Tree** or from the 3D view area that you want to distribute and then select the point along which the object will be distributed; refer to Figure-131.

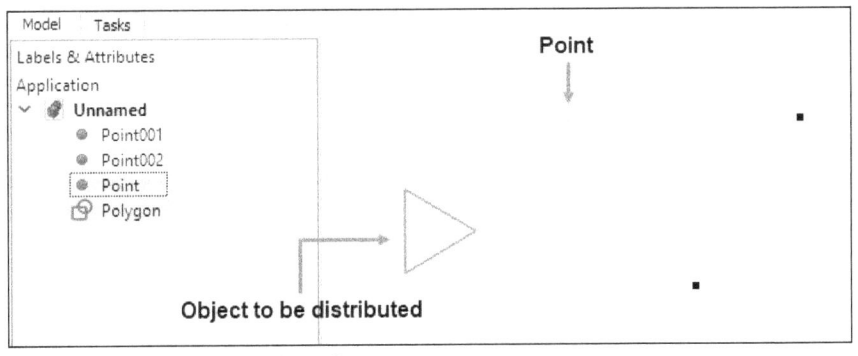

Figure-131. Selecting the object for point array

- Click on the **Point array** tool from **Array tools** drop-down in the **Toolbar** of **Draft** workbench; refer to Figure-132. The array object will be created along point; refer to Figure-133.

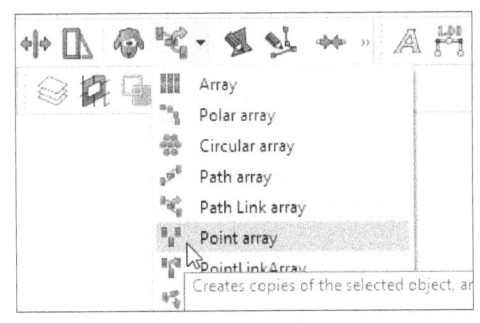

Figure-132. Point array tool

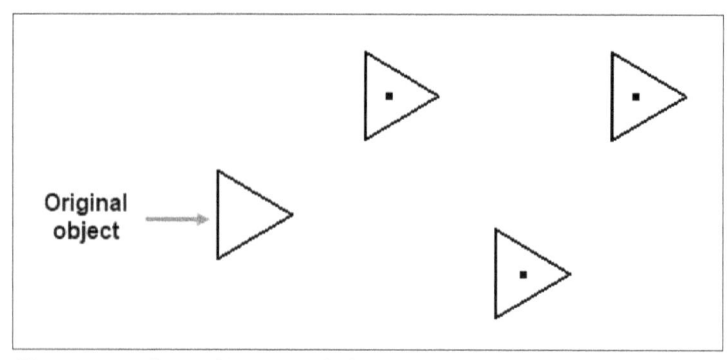

Figure-133. Array object created along point

- You can change the properties of array created from the **Property editor** dialog as discussed earlier.

Creating Point Link array

The **Point Link array** tool creates a link array from a selected object by placing copies at the points from a point compound. The procedure to use this tool is same as discussed for **Point array** tool.

Path twisted array

The **Path twisted array** tool is used to create copies of selected object along selected path with defined twist angle. The procedure to use this tool is discussed next.

- Create a sketch and path in the 3D view area refer to Figure-134. The path can be a polyline, B-spline, Bezier curve or even edges of other object.

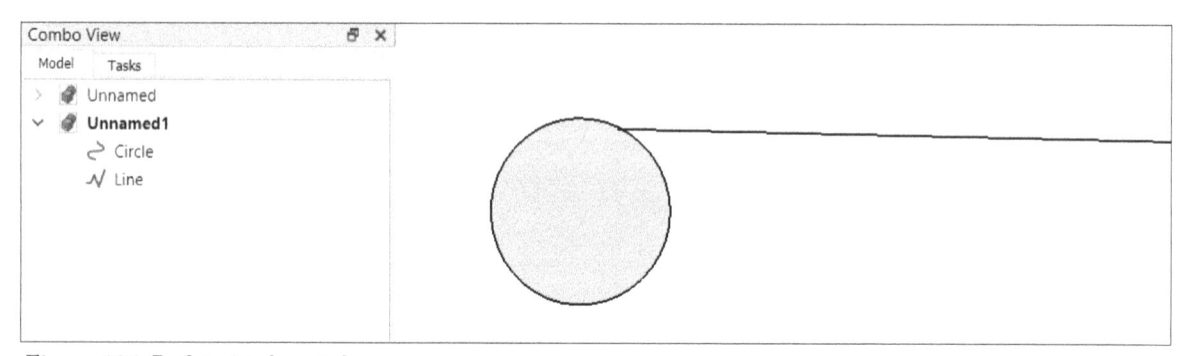

Figure-134. Path twisted created

- Select the object and then select the path from drawing area or **Model** tab of **Combo View**; refer to Figure-135.

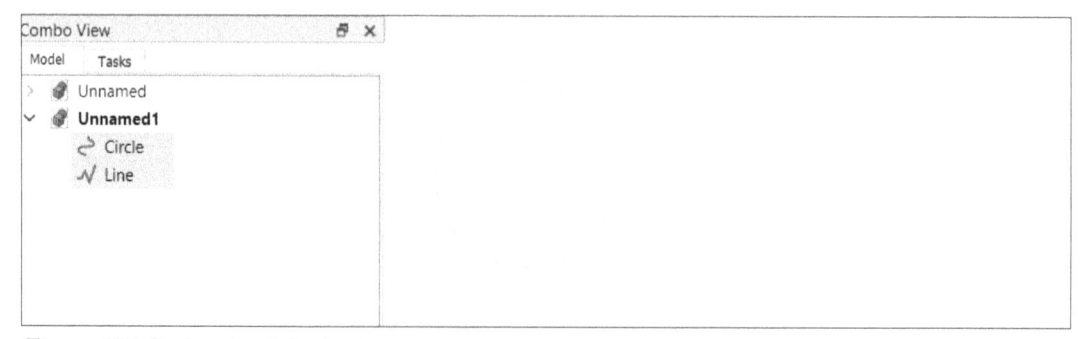

Figure-135. Path twisted sketch select

- Click on the **Path twisted array** from the **Array** tools drop-down in the **Toolbar** of **Draft** workbench; refer to Figure-136. The array object will be created; refer to Figure-137.

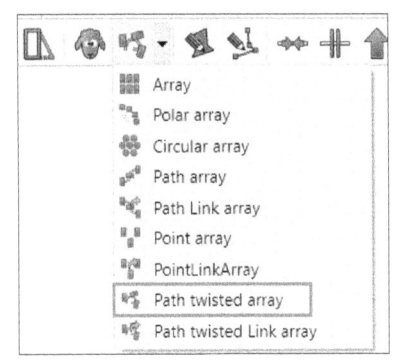

Figure-136. Path array tool

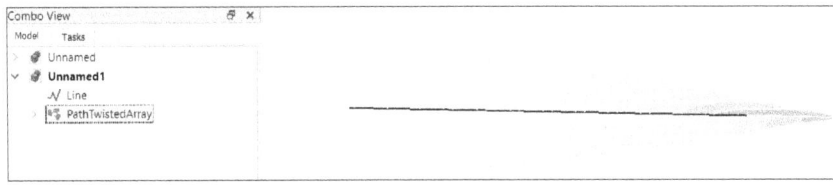

Figure-137. Path twisted sketch created

- You can change the properties of array in the **Data** tab of **Property** dialog as discussed earlier.

Path twisted link array

The **Path twisted link array** tool works similar to the **Path twisted array** tool but creates a link array instead of copy array. The procedure to use this tool is discussed next.

- Create a sketch and path in the 3D view area; refer to Figure-138.

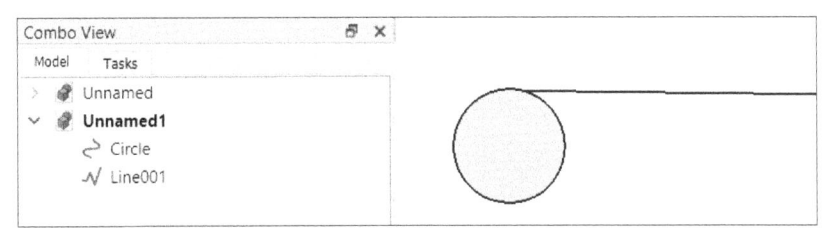

Figure-138. Path twisted link sketch created

- Select the object and then select the path; refer to Figure-139.

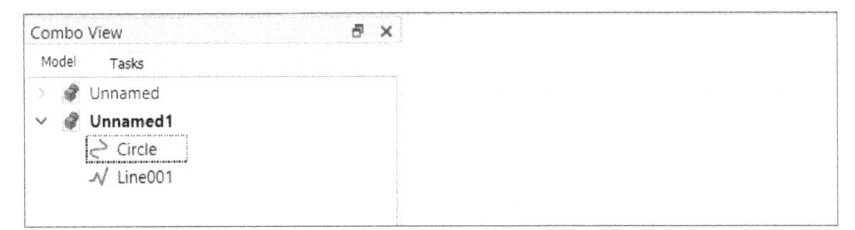

Figure-139. Path twisted link sketch selected

- Click on the **Path twisted link array** tool from the **Array** drop-down in the **Toolbar** of **Draft** workbench; refer to Figure-140. The array will be created; refer to Figure-141.

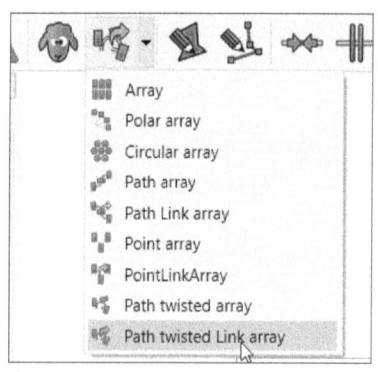

Figure-140. Path twisted link tool

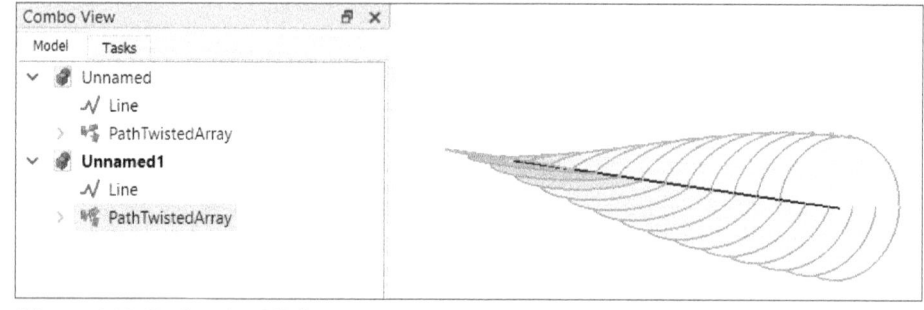

Figure-141. Path twisted link array

Editing object

The **Edit** tool allows you to graphically edit certain properties of the selected object such as, the vertices of a wire, the length and width of a rectangle, the radius of the circle, and so on. The procedure to use this tool is discussed next.

- Select the object you want to edit from the **Model Tree** or from the 3D view area; refer to Figure-142.

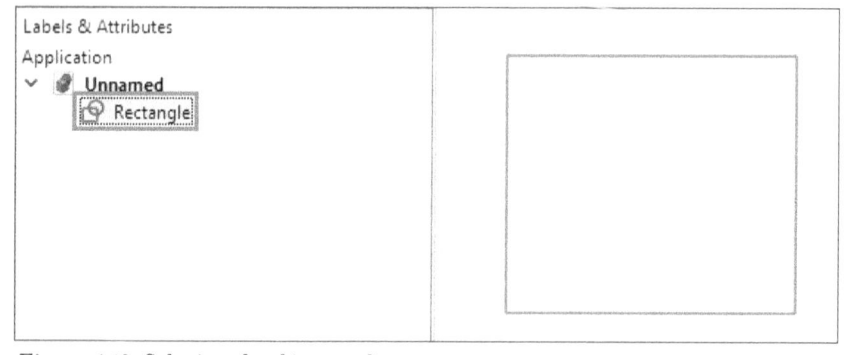

Figure-142. Selecting the object to edit

- Click on the **Edit** tool from **Toolbar** in the **Draft** workbench; refer to Figure-143. The snapping points will be displayed on the object; refer to Figure-144.

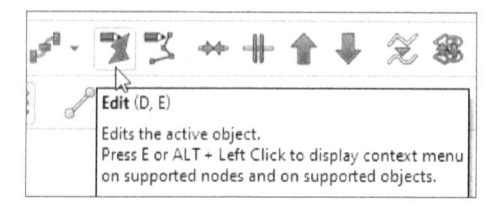

Figure-143. Edit tool

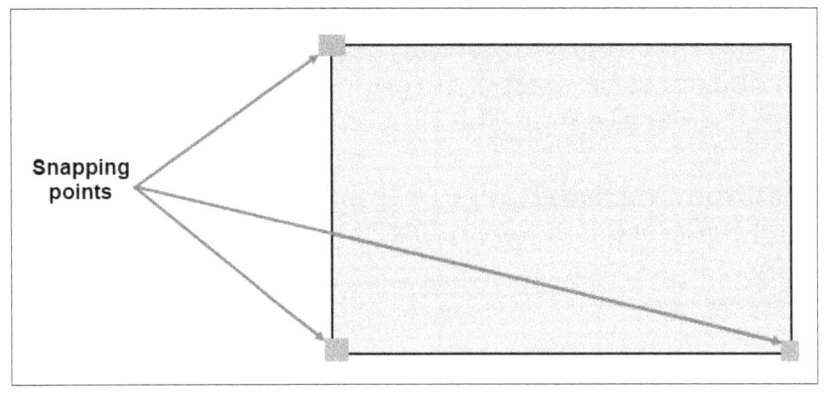

Figure-144. Snapping points displayed on object

- Click on the snapping point from where you want to edit the object. The **Edit node** dialog will be displayed in the **Tasks** panel of **Combo View** along with the plus sign in place of original cursor asking you to specify another point; refer to Figure-145.

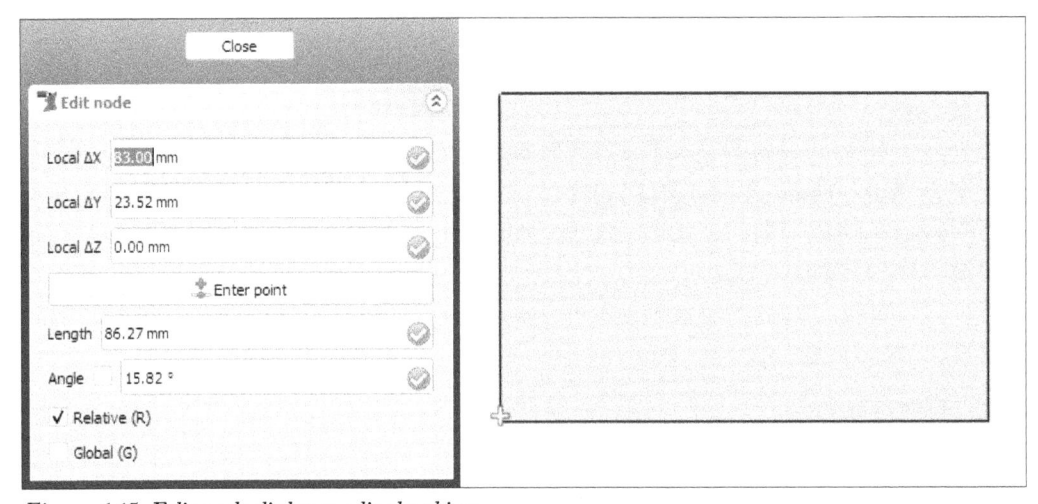

Figure-145. Edit node dialog to edit the object

- Move the cursor away and click **LMB** at desired location to specify the another point of object or enter desired values for the x, y, and z coordinates in the **Local ΔX**, **Local ΔY**, and **Local ΔZ** edit boxes of the dialog, respectively.
- After specifying coordinates for another point in the dialog, click on **Enter point** button from the dialog. The object will be edited and the **Close** button will be displayed in the **Tasks** panel; refer to Figure-146.

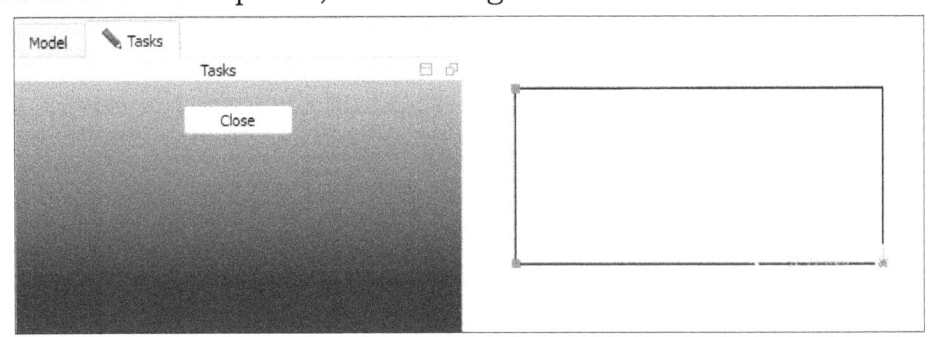

Figure-146. Object edited

- Click on **Close** button to finish the editing operation.

Highlighting the Subelement

The **Subelement highlight** tool temporarily highlights selected objects, or the base objects of selected objects to be used in conjunction with the subelement mode of the **Move** tool, **Rotate** tool, or **Scale** tool. The procedure to use this tool is discussed next.

- Select the object from the **Model Tree** or from the 3D view area whose subelement you want to highlight; refer to Figure-147.

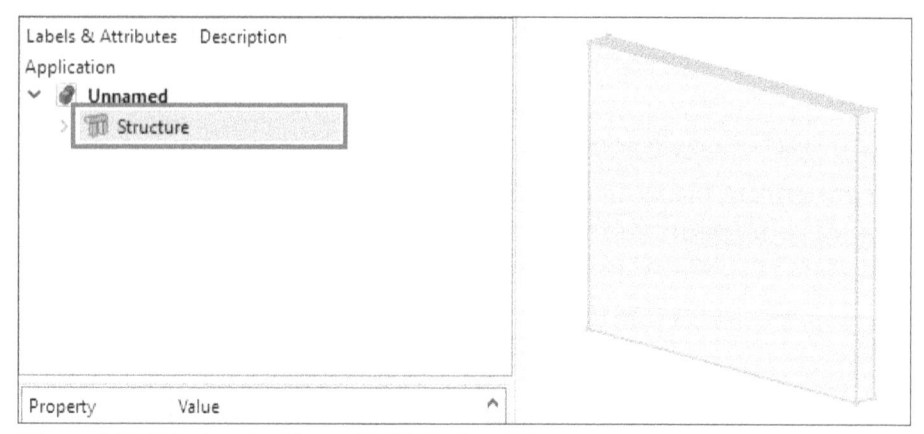

Figure-147. Selecting the object to highlight the subelement

- Click on the **Subelement highlight** tool from **Toolbar** in the **Draft** workbench; refer to Figure-148. The subelement of the selected object will be highlighted; refer to Figure-149.

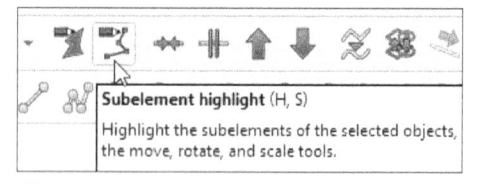

Figure-148. Subelement highlight tool

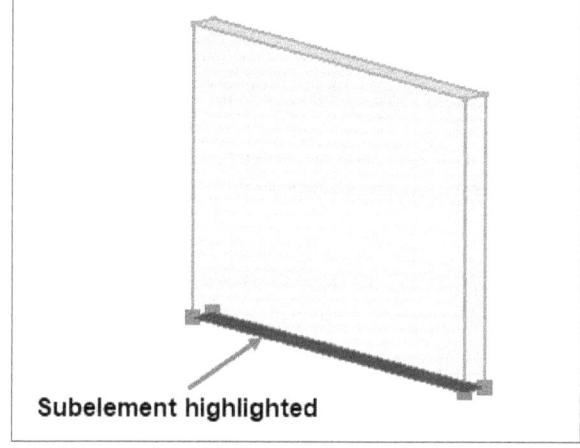

Figure-149. Subelement highlighted

- Now, if you want to move, rotate, or scale the subelement then select the point of highlighted subelement; refer to Figure-150 and click on the **Move** tool.

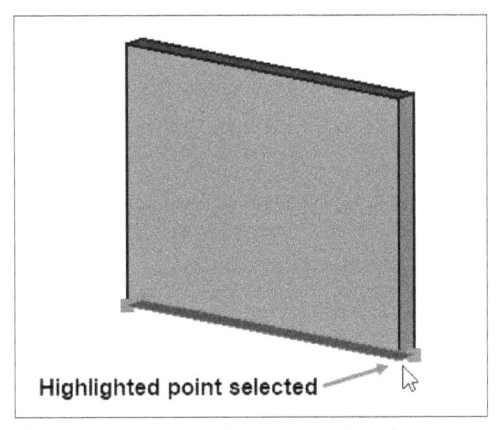

Figure-150. Highlighted point selected

- Move the cursor away and click at desired location to specify the new position of the subelement or enter desired values for the coordinates in their respective edit boxes of the dialog and click **Enter point** button. The subelement will be moved and copied to the new position; refer to Figure-151.

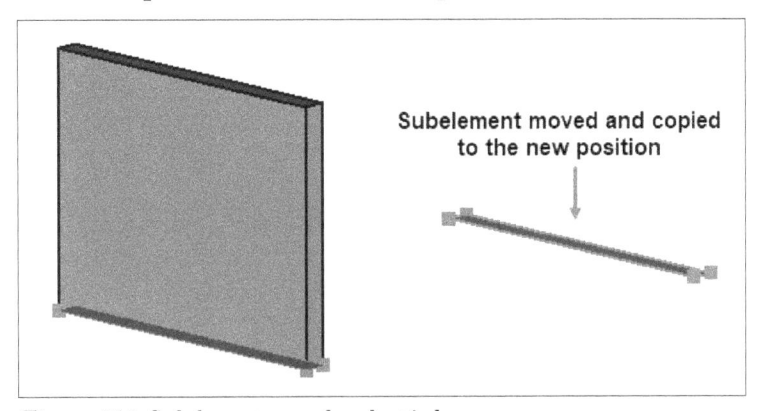

Figure-151. Subelement moved and copied

- Click on the **Subelement highlight** tool again to exit the tool.

Joining objects

The **Join** tool attempts to join all wires currently in the selection into a single wire. The procedure to use this tool is discussed next.

- Select two or more lines or curves from the **Model Tree** or from the 3D view area to make a single wire; refer to Figure-152.

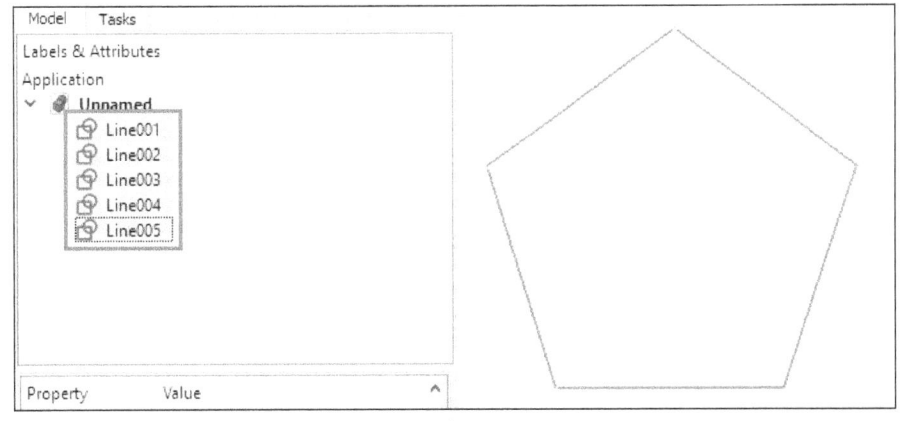

Figure-152. Selecting the lines to create single wire

- Click on the **Join** tool from **Toolbar** in the **Draft** workbench; refer to Figure-153. The single wire will be created; refer to Figure-154.

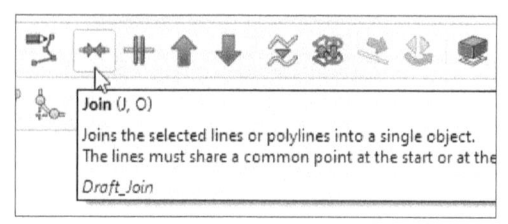

Figure-153. Join tool

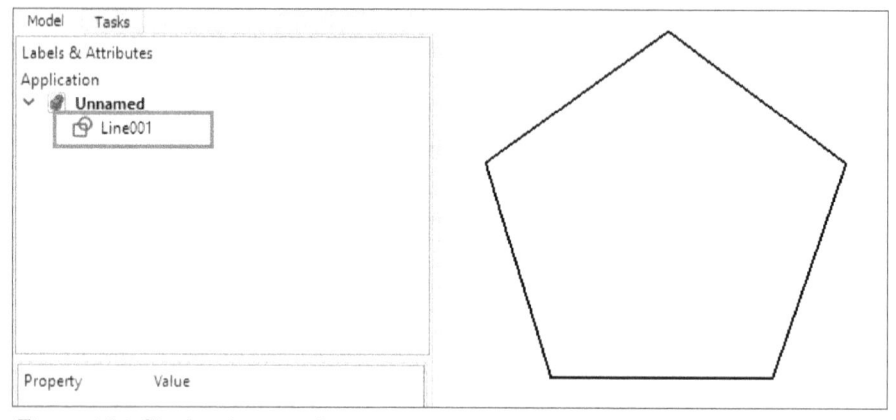

Figure-154. Single wire created

Splitting object

The **Split** tool attempts to split an existing wire at a specified edge or point. The procedure to use this tool is discussed next.

- Click on the **Split** tool from **Toolbar** in the **Draft** workbench; refer to Figure-155.

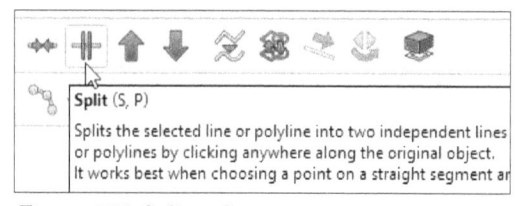

Figure-155. Split tool

- Click on the wire at the location from where you want to split the wire. The wire will be split from that location; refer to Figure-156.

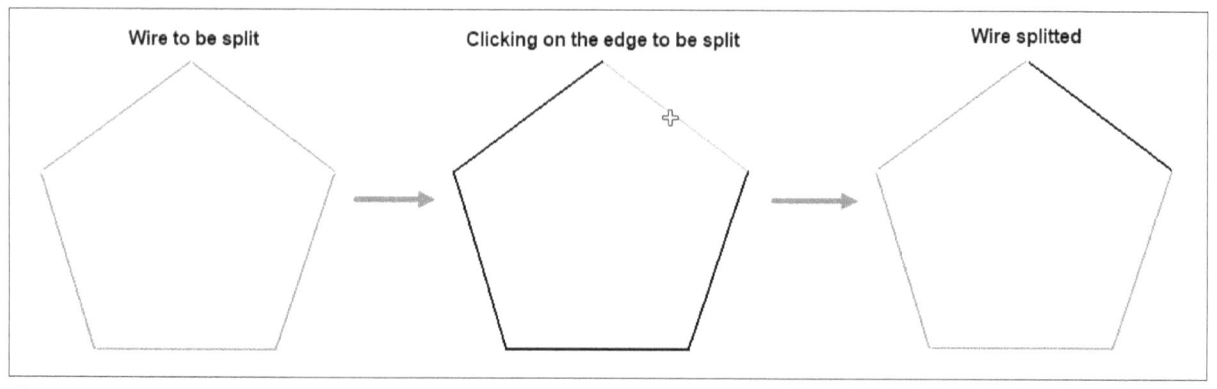

Figure-156. Wire split

Upgrading object

The **Upgrade** tool performs things such as creating faces and fusing different elements. The procedure to use this tool is discussed next.

- Select the object from the `Model Tree` or from the 3D view area which you want to upgrade; refer to Figure-157.

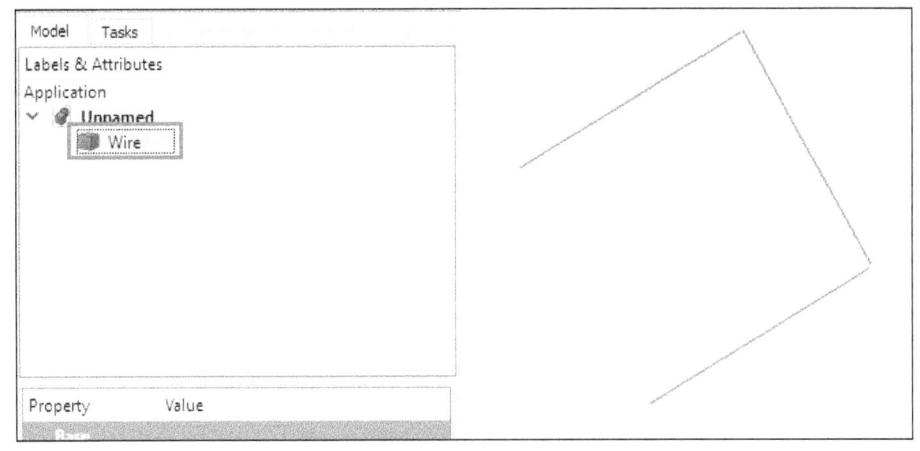

Figure-157. Selecting the object to upgrade

- Click on the **Upgrade** tool from **Toolbar** in the **Draft** workbench; refer to Figure-158. The opened wire will be upgraded to a closed wire; refer to Figure-159.

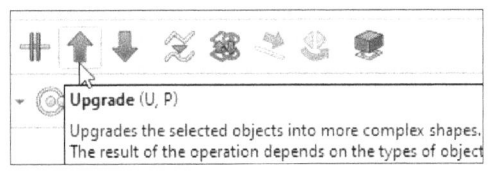

Figure-158. Upgrade tool

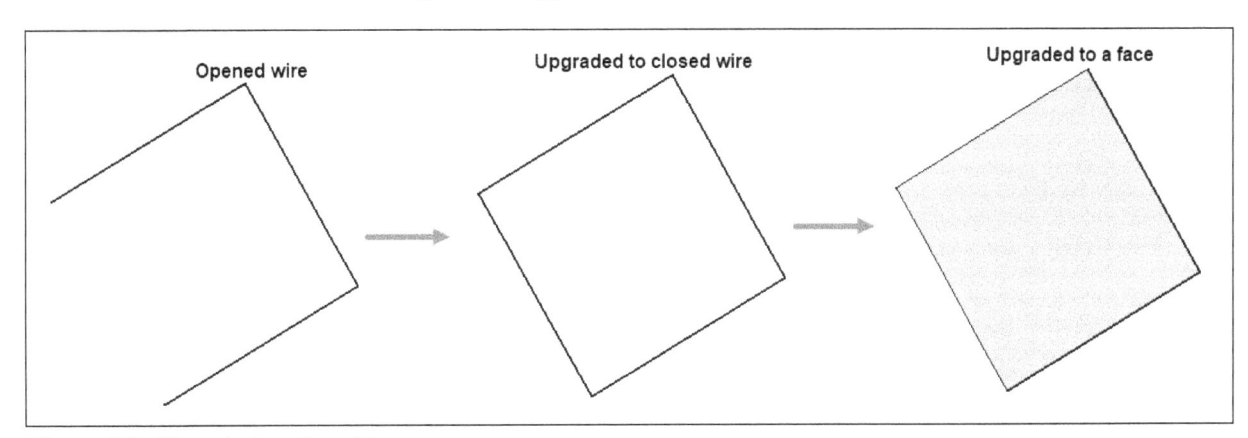

Figure-159. Upgradation of an object

- If you want to upgrade the closed wire then click on the **Upgrade** tool again. The closed wire will be upgraded to a face; refer to Figure-159.

Downgrading object

The **Downgrade** tool performs things such as breaking faces and de-constructing wires into their individual objects. The procedure to use this tool is discussed next.

- Select the object from the `Model Tree` or from the 3D view area which you want to downgrade; refer to Figure-160.

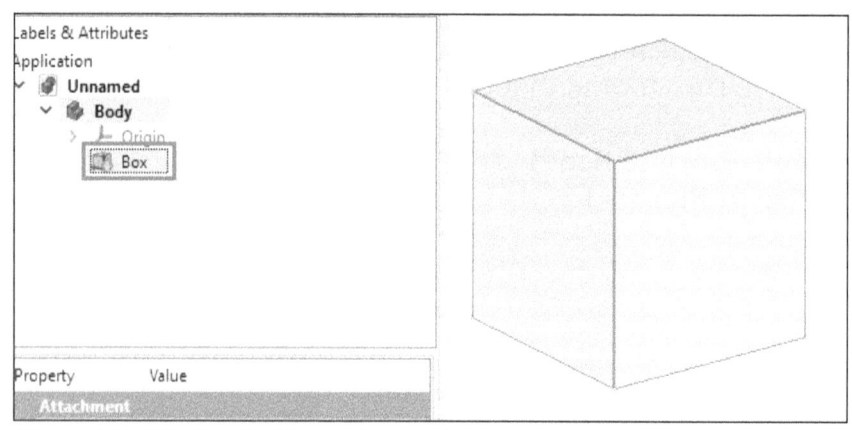

Figure-160. Selecting the object to downgrade

- Click on the **Downgrade** tool from **Toolbar** in the **Draft** workbench; refer to Figure-161. The object will be downgraded into the split faces; refer to Figure-162.

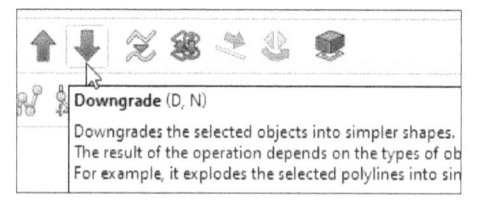

Figure-161. Downgrade tool

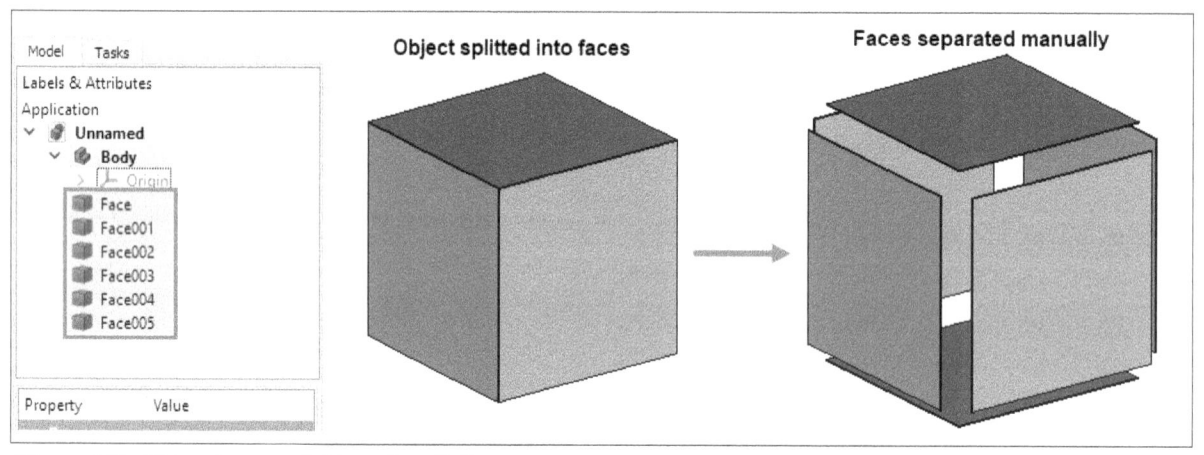

Figure-162. Object downgraded into the split faces

Converting Wire to B-spline

The **Wire to B-spline** tool converts wire to B-Spline and vice-versa. The procedure to use this tool is discussed next.

- Select the wire from the **Model Tree** or from the 3D view area which you want to convert into B-Spline; refer to Figure-163.

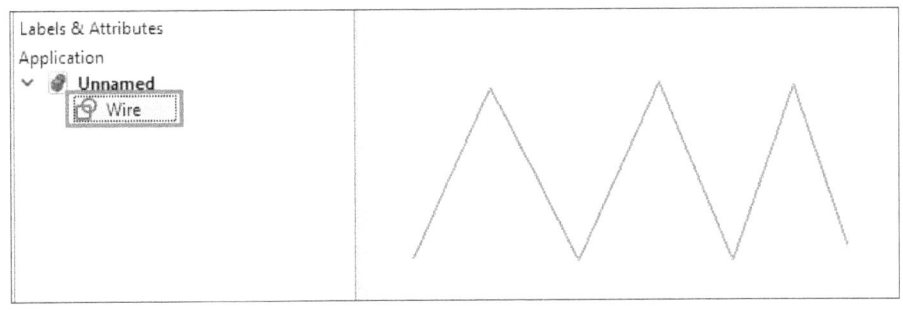

Figure-163. Selecting the wire to create B Spline

- Click on the **Wire to B-spline** tool from **Toolbar** in the **Draft** workbench; refer to Figure-164. The wire will be converted into B-Spline; refer to Figure-165.

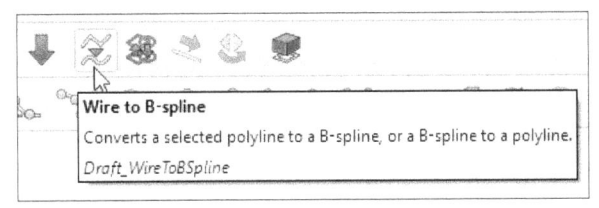

Figure-164. Wire to BSpline tool

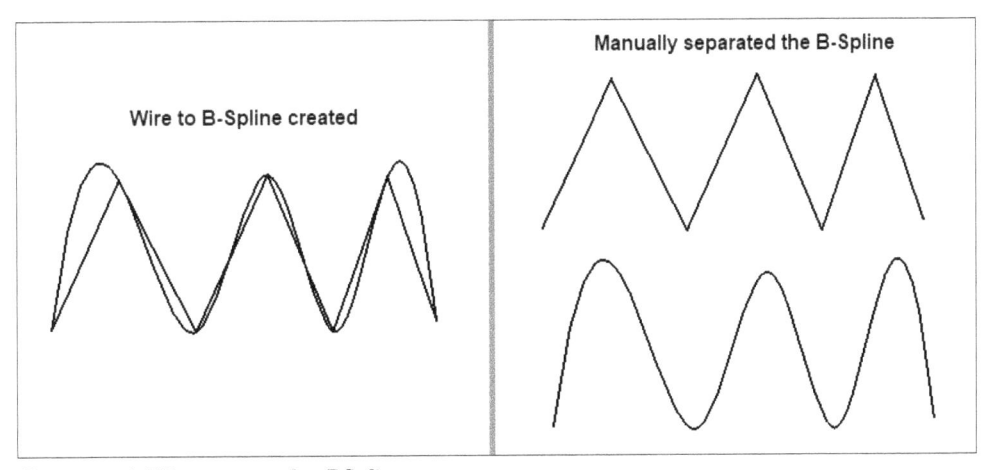

Figure-165. Wire converted to BSpline

Converting Draft to sketch

The **Draft to sketch** tool converts objects of **Draft** workbench into the objects of **Sketcher** workbench with constraints and vice-versa. The procedure to use this tool is discussed next.

- Select the object you want to convert from the **Model Tree** or from the 3D view area; refer to Figure-166.

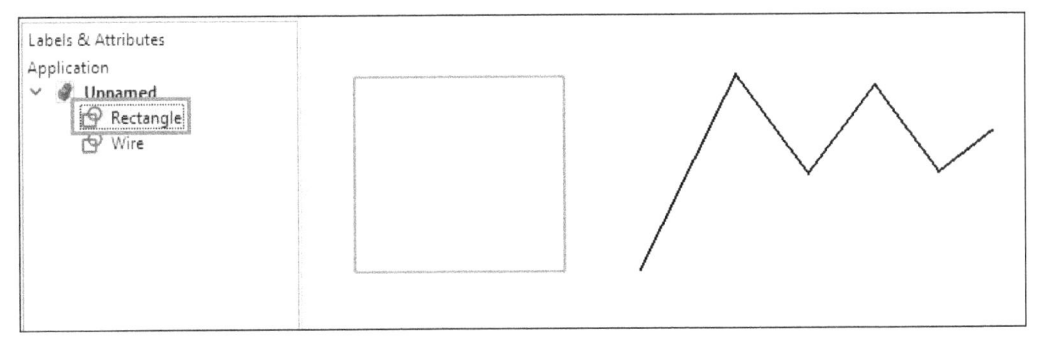

Figure-166. Selecting the object to convert

- Click on the **Draft to sketch** tool from **Toolbar** in the **Draft** workbench; refer to Figure-167. The objects of **Draft** workbench will be converted into sketches of **Sketcher** workbench with constraints; refer to Figure-168.

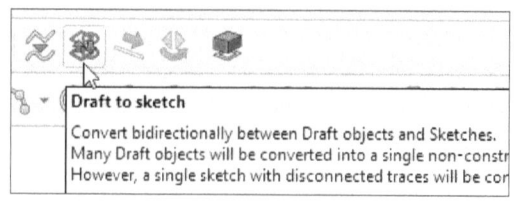

Figure-167. Draft to sketch tool

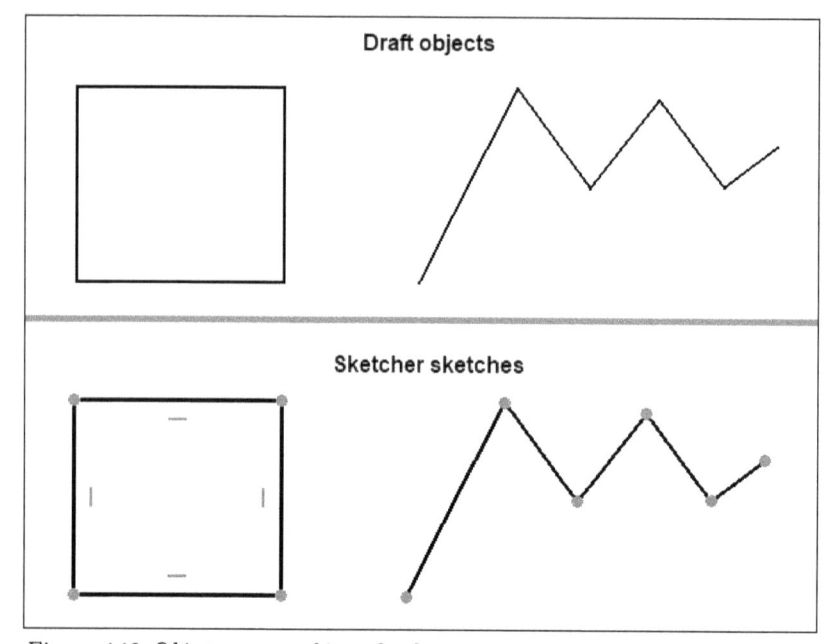

Figure-168. Objects converted into sketches

Set slope

The **Set slope** tool slopes/inclines selected lines or wires by increasing, or decreasing, the Z coordinate of all points after the first one. The procedure to use this tool is discussed next.

- Create and select one or more lines or wires from the **Model Tree** or from the 3D view area; refer to Figure-169.

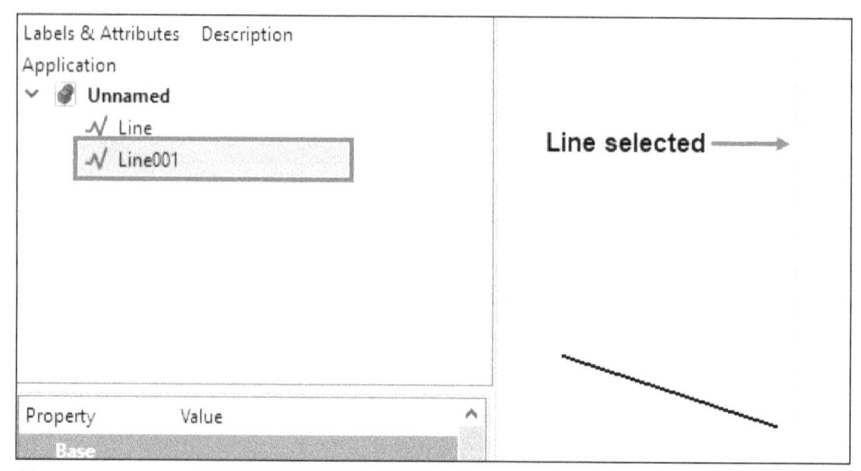

Figure-169. Selecting the line to create slope

- Click on the **Set slope** tool from **Toolbar** in the **Draft** workbench; refer to Figure-170. The **Slope** dialog will be displayed; refer to Figure-171.

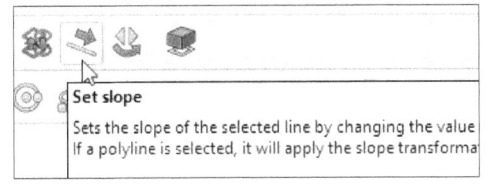

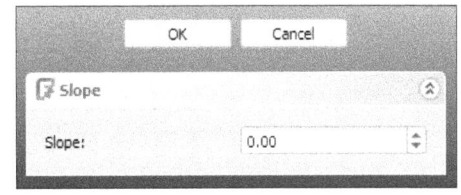

Figure-170. Set slope tool

Figure-171. Slope dialog

- Specify desired slope factor value for the selected lines in the **Slope** edit box of the dialog. Note that entering **0** value means each segment is horizontal, entering **1** value means each segment is **45** degree up, and entering **-1** value means each segment is **45** degree down.
- After entering desired value, click on the **OK** button from the dialog. The slope of the line will be created; refer to Figure-172.

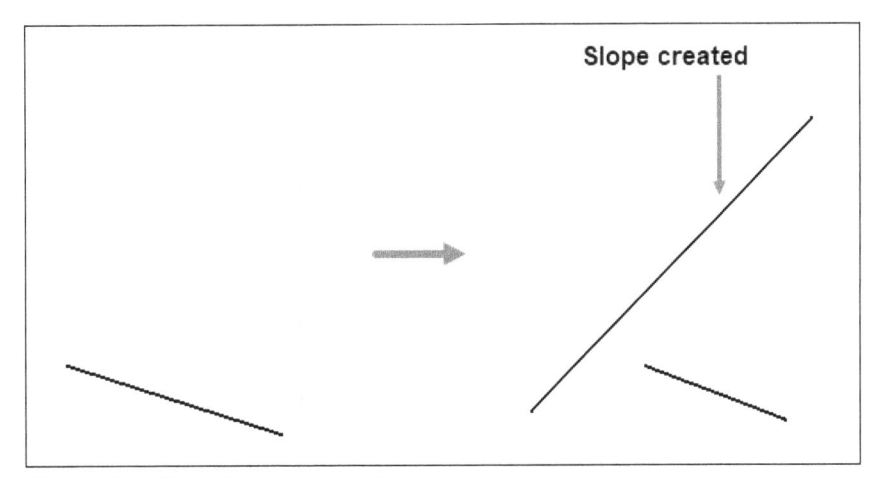

Figure-172. Slope of the line created

Creating Flip dimension

The **Flip dimension** tool rotates the dimension text of selected dimensions 180 degree around the dimension line. It can be used to correct dimensions whose text appears mirrored. The procedure to use this tool is discussed next.

- Create and select one or more dimensions of the object from the **Model Tree** or from the 3D view area; refer to Figure-173.

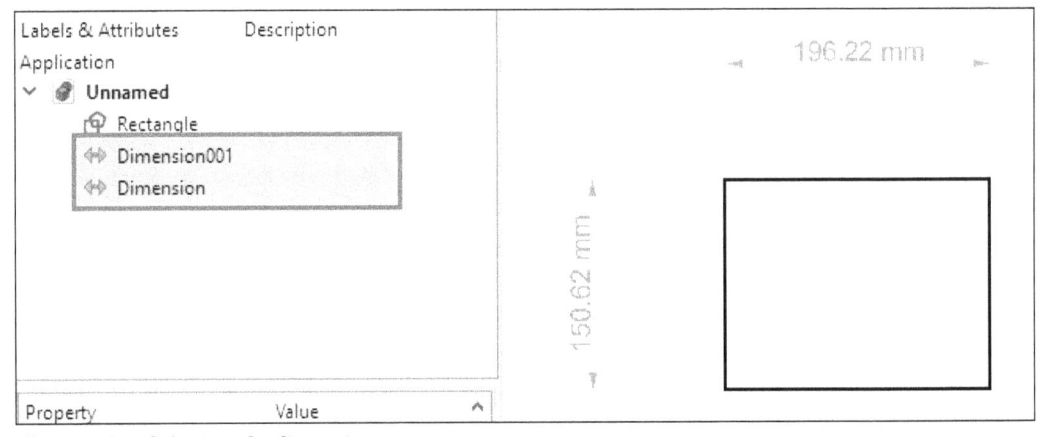

Figure-173. Selecting the dimensions

- Click on the **Flip dimension** tool from **Toolbar** in the **Draft** workbench; refer to Figure-174. The dimension will be flipped; refer to Figure-175.

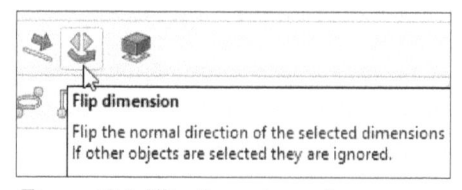

Figure-174. Flip dimension tool

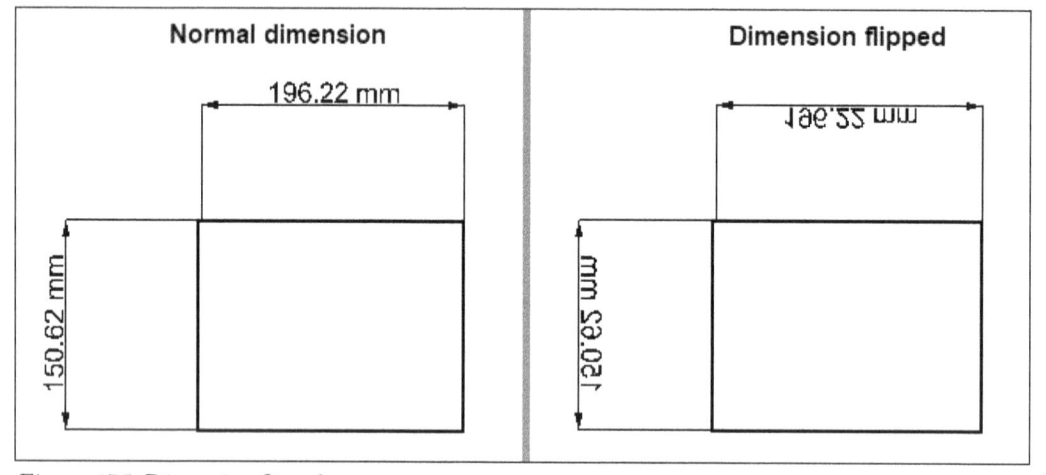

Figure-175. Dimension flipped

Creating Shape 2D view

The **Shape 2D view** tool produces a 2D projection from selected 3D solid object such as those created with the **Part**, **Part Design**, and **Arch** workbenches. The procedure to use this tool is discussed next.

- Rotate the view of 3D object to be projected from the 3D view area, so it reflects the direction of the desired projection. For example, a top view will project the object on the XY plane.
- Select the 3D object from the **Model Tree** or from the 3D view area; refer to Figure-176.

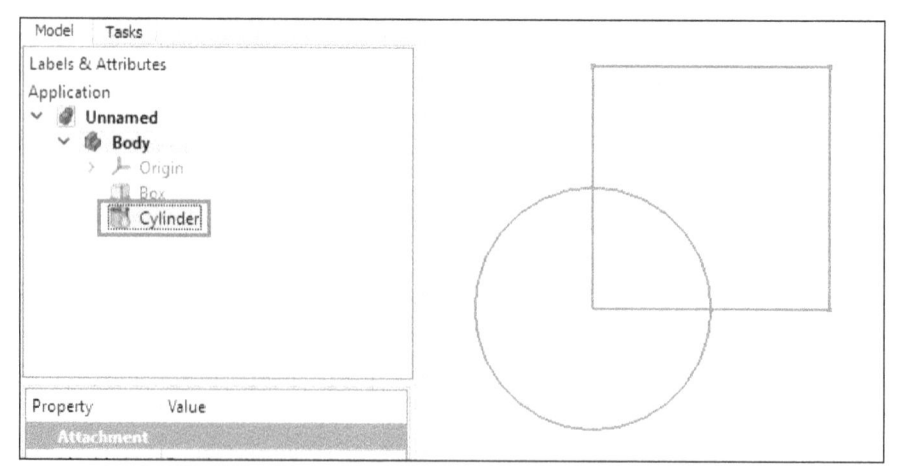

Figure-176. Selecting the 3D object to be project

- Click on the **Shape 2D view** tool from **Toolbar** in the **Draft** workbench; refer to Figure-177. The 2D projection will be created below the selected 3D object, lying on the XY plane; refer to Figure-178.

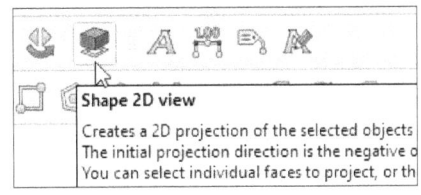

Figure-177. Shape 2D view tool

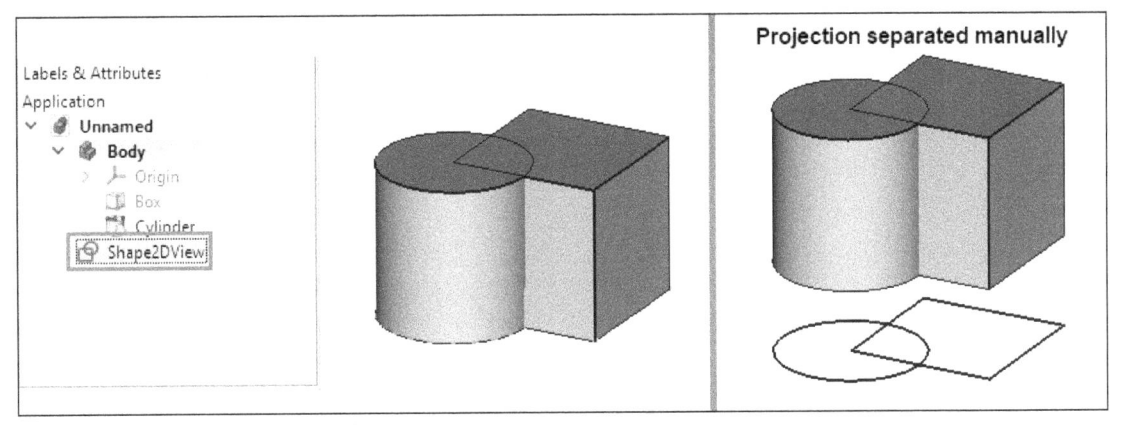

Figure-178. 2D projection created

- You can manually change the position of projection created from the **Property editor** dialog.

Draft tray toolbar

The **Draft tray toolbar** allows selecting the working plane together with some visual properties like the line color, shape color, text size, line width, and automatic group. This toolbar is available in the **Draft** workbench; refer to Figure-179. The procedures to use the tools of this toolbar are discussed next.

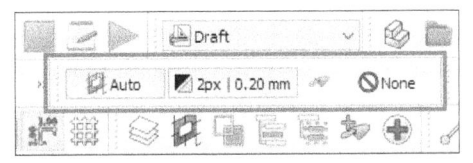

Figure-179. Draft tray toolbar

Creating Working Plane

The **Working Plane** tool sets a working plane on a standard view or selected face. The working plane in the 3D view area indicates where a draft shape will be built. The procedure to use this tool is discussed next.

- Click on the **Working Plane** tool from **Toolbar** in the **Draft** workbench; refer to Figure-180. The **Working plane setup** dialog will be displayed in the **Tasks** panel of **Combo View**; refer to Figure-181.

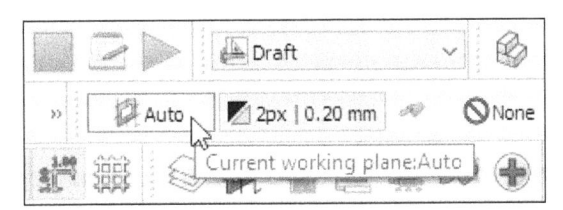

Figure-180. Select Plane tool

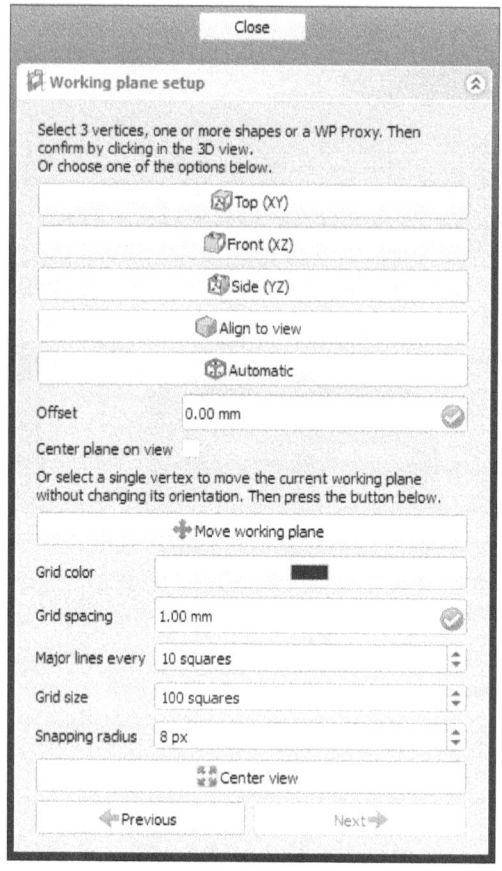

Figure-181. Working plane setup dialog

- Click on the **Top (XY)**, **Front (XZ)**, or **Slide (YZ)** button from **Select a face or working plane proxy or 3 vertices** area to set the working plane on XY, XZ, or YZ plane, respectively.
- Click on the **Align to view** button to set the working plane to the current 3D view, perpendicular to the camera axis and passing through the origin.
- Click on the **Automatic** button to unset any current working plane and automatically set a working plane when a tool is used.
- Specify desired offset value in the **Offset** edit box to set the working plane at specified perpendicular distance from the plane you selected.
- Select **Center plane on view** check box to put the origin of the working plane in the center of the current 3D view.

Or,

- Select desired vertex in the 3D view area to move the current working plane so that its origin matches the position of the selected vertex and click on the **Move working plane** button from **Select a single vertex to move the current working plane without changing its orientation** area of the dialog.
- Select desired color of the grid from **Grid color** area of the dialog.
- Specify desired value in **Grid spacing** edit box to define the space between two consecutive lines in the grid.
- Specify desired value in **Main lines every** edit box to draw a slightly thicker line in the grid at the set value.
- Specify desired value in **Grid size** edit box to determine the number of grid lines in the X and Y direction of the grid.
- Specify desired maximum distance in **Snapping radius** edit box at which snap grid detects the intersections of grid lines.

- Click on the **Center view** button to use the origin of the current working plane as the center of the 3D view.
- Click on the **Previous** button to reset the working plane to its previous position. Click on the **Next** button to reset the working plane to its next position.
- After specifying desired parameters, on clicking desired plane, the active plane will be displayed in the toolbar along with the grid view; refer to Figure-182.

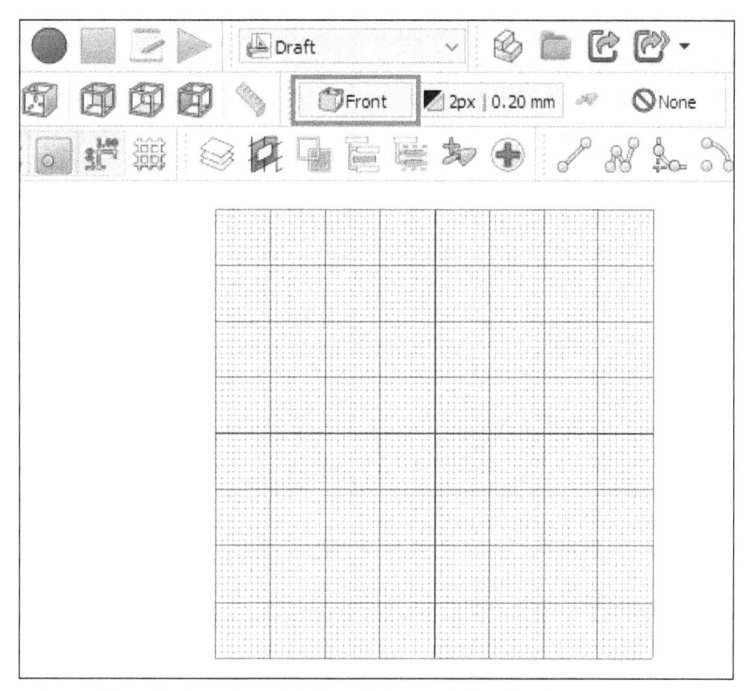

Figure-182. Active plane displayed along with grid view

Applying Style settings

The **Set style** tool allows to set default visual properties like line color or line width to be used by all draft objects created after this tool is used and bulk apply these styles to selected objects. The procedure to use this tool is discussed next.

- Click on the **Set style** tool from **Toolbar** in the **Draft** workbench; refer to Figure-183. The **Style settings** dialog will be displayed in the **Tasks** panel of **Combo View** with the **Shape** tab parameters; refer to Figure-184.

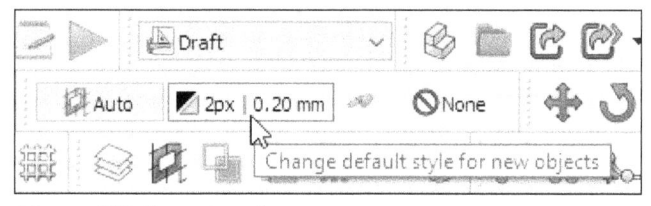

Figure-183. Set style tool

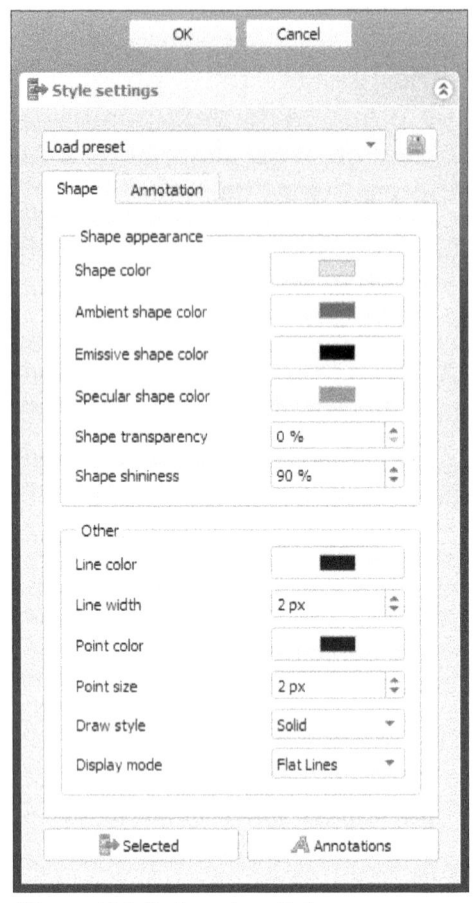

Figure-184. Style settings dialog

Figure-185. Annotation tab of Style settings dialog

- Select desired preset to be loaded from **Load preset** drop-down to use predefined settings.
- Click on the **Save the current style as a preset** button and specify desired name of the new style in the **Save style** dialog box.
- Click on the **Shape color** button and select desired color for the faces from **Select Color** dialog box displayed.
- Other parameters in the **Shape appearance** area have been discussed earlier.
- Click on the **Line color** and **Point color** button from **Other** area and select desired color for the line and point, respectively from the **Select Color** dialog box displayed.
- Specify desired width for the line and size of the point in the **Line width** and **Point size** edit boxes, respectively.
- Select desired line style from **Draw style** drop-down and select desired display mode for the faces from **Display mode** drop-down.
- Click on the **Annotation** tab of the dialog, the parameters will be displayed as shown in Figure-185.
- Select desired font for texts, dimensions, and labels from **Font name** drop-down.
- Specify desired height of texts, dimension texts, and label texts in the **Font size** edit box.
- Specify desired spacing of line for multi-line texts and labels in the **Line spacing factor** edit box.

- Specify desired annotation scale multiplier value in the **Scale multiplier** edit box. If the scale is 1:100, the multiplier is 100.
- Specify desired width of line and size of arrow in the **Line width** and **Arrow size** edit boxes, respectively.
- Select desired type of dimension arrows from **Arrow style** drop-down.
- Click on the **Line and arrow color** button and select desired color for line and arrow from **Select Color** dialog box displayed.
- Select the **Show unit** check box from **Dimensions** area to show the unit suffix on dimension texts.
- Specify desired unit to use for dimensions in the **Unit override** edit box.
- Specify desired length of dimension line below the extension lines in the **Dim line overshoot** edit box.
- Specify desired length of extension lines in the **Ext line length** edit box.
- Specify desired length of extension lines above the dimension line in the **Ext line overshoot** edit box.
- Specify desired distance between the dimension line and the dimension text in the **Text spacing** edit box.
- After setting desired parameters in the dialog, select the object or multiple objects from the **Model Tree** or from the 3D view area on which you want to apply the style; refer to Figure-186.

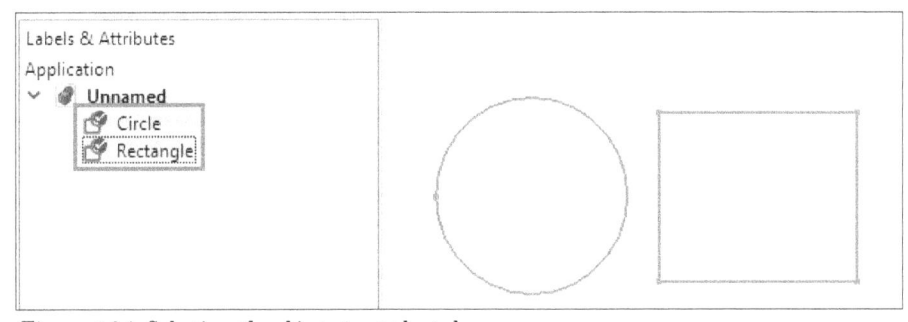

Figure-186. Selecting the objects to apply style

- Click on the **Selected** button from the dialog. The style will be applied to the selected objects; refer to Figure-187.

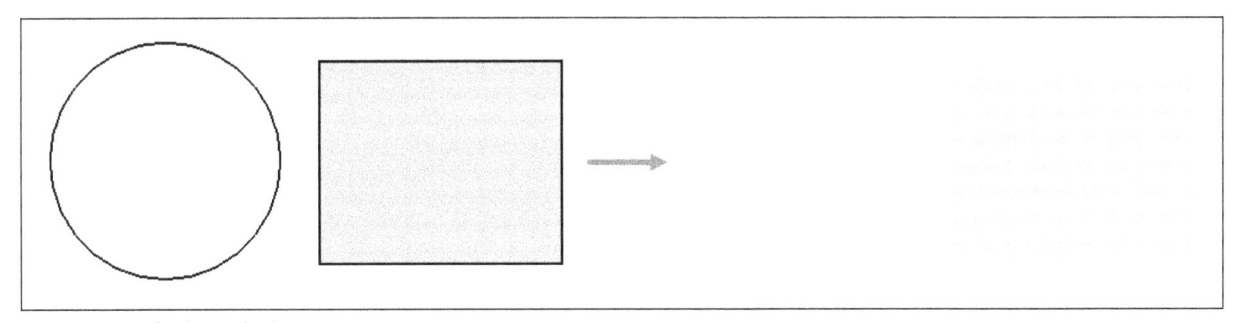

Figure-187. Style applied

- Click on the **Annotations** button from the dialog. The style will be applied to all the texts, dimensions, and labels.

Toggle construction mode

The **Toggle construction mode** toggles the draft construction mode on or off. The construction geometry is comprised of lines, points, and other shapes that serve as references or snapping elements which are helpful in building your main geometry. The procedure to use this tool is discussed next.

- Click on the **Toggle construction mode** button from **Toolbar** in the **Draft** workbench; refer to Figure-188.

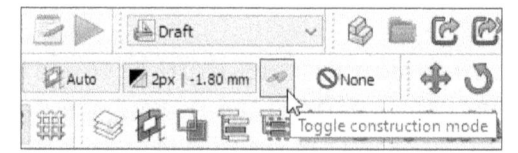

Figure-188. Toggle construction mode button

- Create some objects in the 3D view area as desired. Now, it will be created in construction mode; refer to Figure-189.

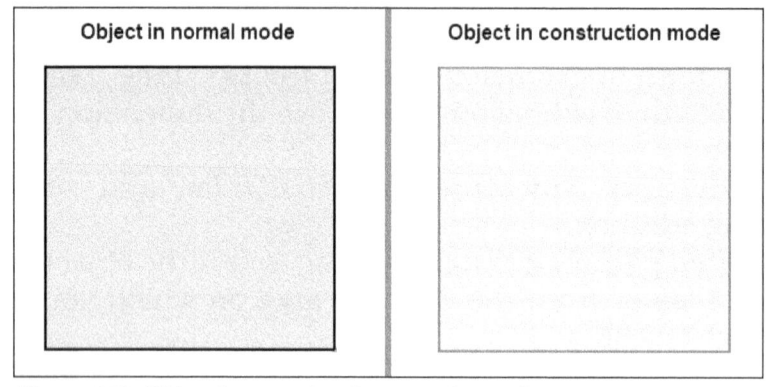

Figure-189. Objects in normal and construction mode

- Click on the **Toggle construction mode** button again to go back to normal mode.

Creating Auto group

The **Auto group** tool changes the active layer, or optionally, the active group or groups like arch object. New draft and arch objects will be automatically placed in this active layer or group. Assigning a group or layer can help in The procedure to use this tool is discussed next.

- First, create the layer in the active document with the help of **Layer** tool as discussed earlier.
- Click on the **Auto group** tool from **Toolbar** in the **Draft** workbench; refer to Figure-190. The active layers created will be displayed in the drop-down; refer to Figure-191.

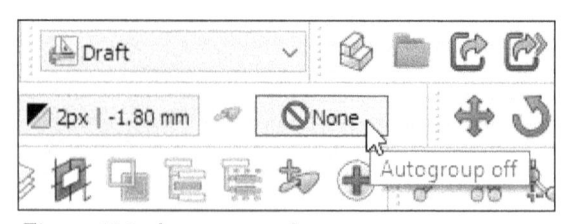

Figure-190. Auto group tool

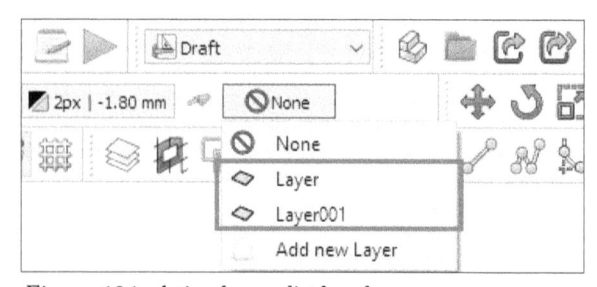

Figure-191. Active layers displayed

- Select **None** option to work without an active layer.
- Select desired layer from the existing layers to make it active.
- Select **Add new Layer** option to create a new layer.

Draft Snap toolbar

The **Draft Snap toolbar** allows to select various snapping mode. All the buttons in this toolbar are toggle buttons. This toolbar is available in the **Draft** workbench; refer to Figure-192. The procedures to use the tools of this toolbar are discussed next.

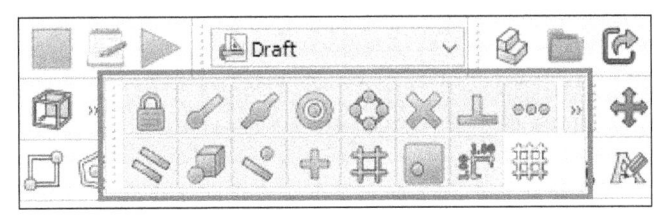

Figure-192. Draft Snap toolbar

Toggle Snap

The **Toggle Snap** tool allows you to activate or deactivate globally the Draft Snap methods. The procedure to use this tool is discussed next.

- If you want to make the snapping methods available then click on the **Toggle Snap** button from **Toolbar** in the **Draft** workbench. The snapping will be displayed while creating the objects; refer to Figure-193.

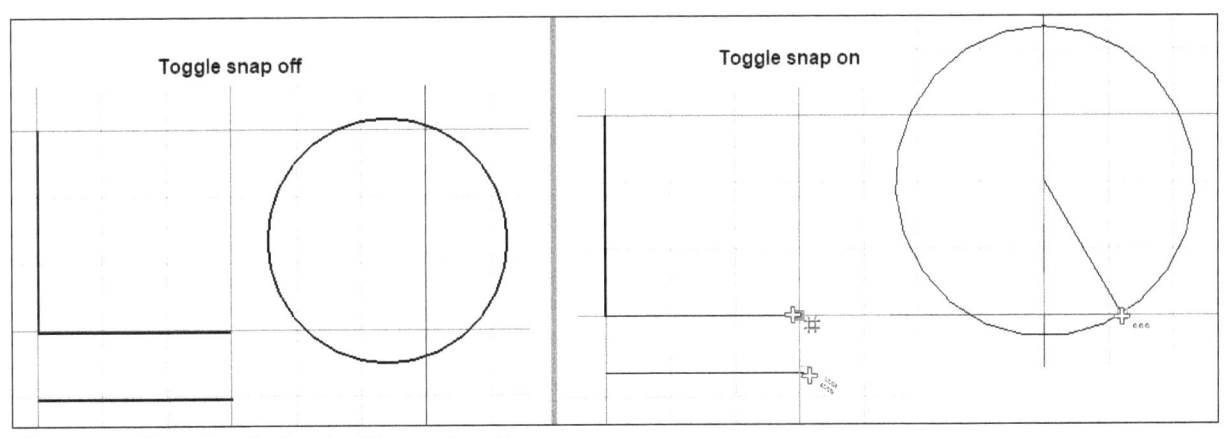

Figure-193. Snapping displayed while creating objects

Endpoint

The **Endpoint** snap button activates snaps to the endpoints of line, arc, and spline segments.

Midpoint

The **Midpoint** snap button activates snaps to the middle point of line and arc segments.

Center

The **Center** snap button activates snaps to the center point of arcs and circles.

Angle

The **Angle** snap button activates snaps to the points of circles and arcs at specific increments of 30° and 45° on the arc; this includes 0°, 60°, 90°, 180°, 210°, 270°, and other integer multiples.

Intersection

The **Intersection** snap button activates snaps to the intersection of two line or arc segments. Hover the mouse over the two desired objects to activate their intersection snaps.

Perpendicular

The **Perpendicular** snap button activates snaps to a line or edge or to an extension of it to produce a line that is perpendicular to that edge.

Extension

The **Extension** snap button activates snaps to a point on an imaginary line that extends beyond the endpoints of line segments. Hover the mouse over desired object to activate its extension snap.

Parallel

The **Parallel** snap button activates snaps on an imaginary line parallel to a line segment. Hover the mouse over desired object to activate its parallel snap.

Special

The **Special** snap button activates snaps to the special location points defined by a particular object such as an Arch Wall.

Near

The **Near** snap button activates snaps to the closest point or edge on the nearest object.

Ortho

The **Ortho** snap button activates snaps to a point on an imaginary line that originates from the previous point and extends infinitely at specific increments of 45°; this includes 0°, 90°, 135°, 180°, 225°, 270°, and other integer multiples.

Grid

The **Grid** snap button activates snaps to the intersection of two grid lines, if the grid is visible.

Working Plane

The **Working Plane** button always places a snapped point on the current working plane even if you also use another snapping method and select a point outside of that working plane. In other words, it projects an external snapping point to the current working plane.

Dimensions

The **Dimensions** snap button shows temporary X and Y dimensions between the current point and the last snapped point on screen while snapping.

Toggle Grid

The **Toggle Grid** tool allows you to show and hide the grid defined in the **Draft Preferences** or with the **Draft Select Plane** tool. The procedure to use this tool is discussed next.

- If you want to set the appearance of the grid then click on the **Preferences** tool from the **Edit** menu; refer to Figure-194. The **Preferences** dialog box will be displayed; refer to Figure-195.

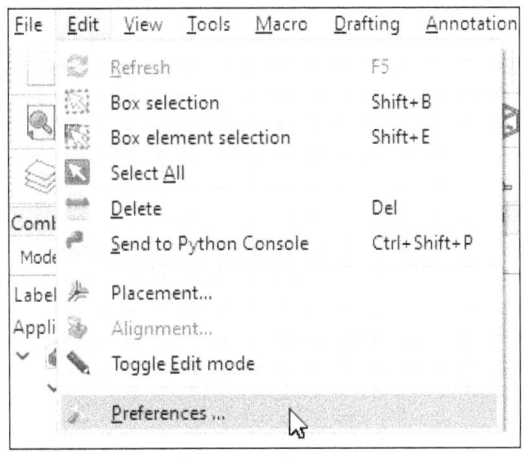

Figure-194. Preferences tool

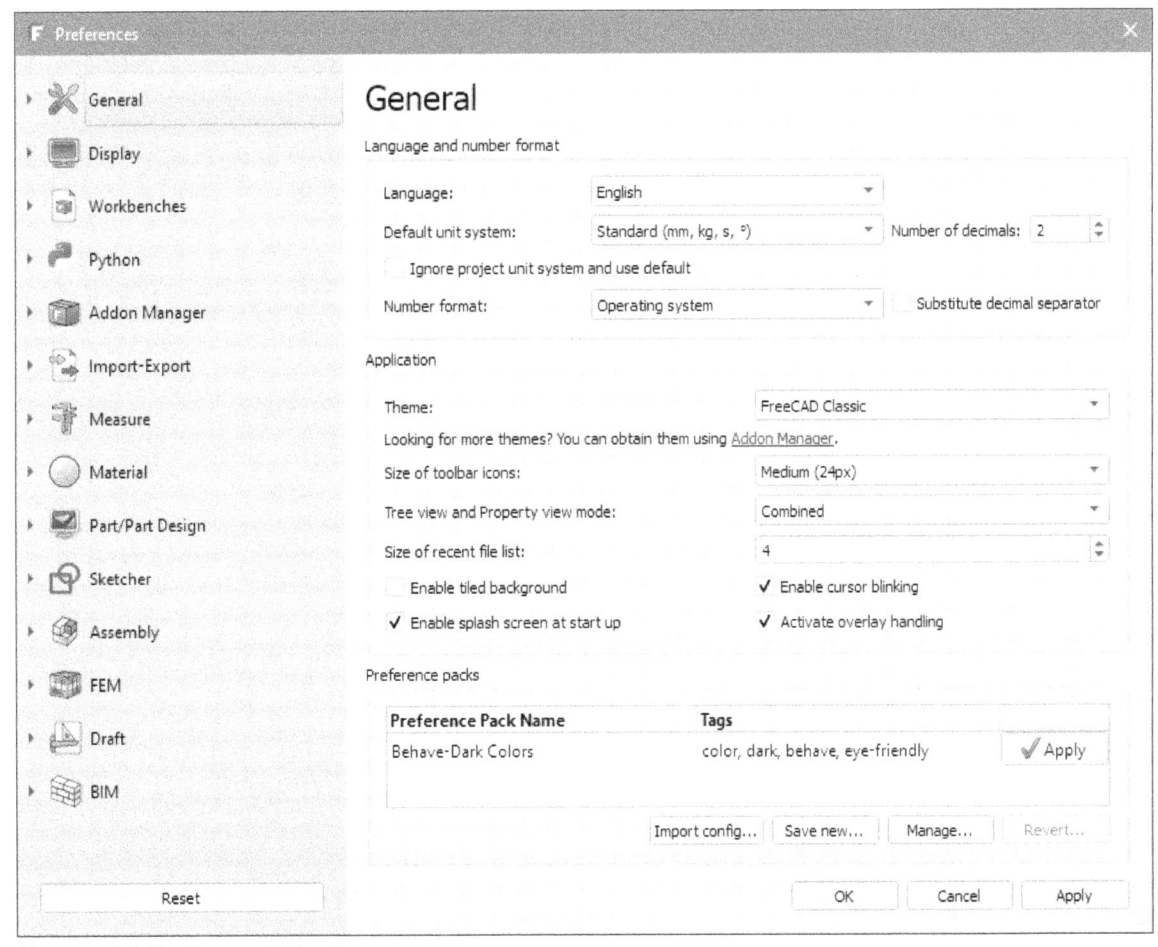

Figure-195. Preferences dialog box

- Click on the **Grid and snapping** option from **Draft** node of the dialog box. The parameters related to **Grid and snapping** will be displayed.
- Specify desired values in the **Main lines every**, **Grid spacing**, **Grid size**, **Grid transparency** and **Grid color** edit boxes from **Grid** area; refer to Figure-196.

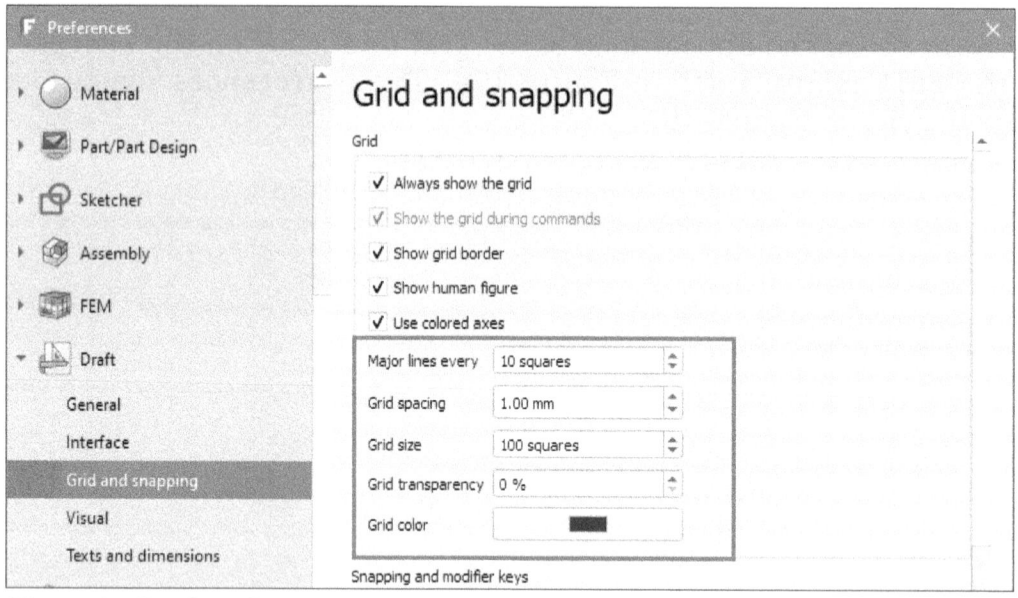

Figure-196. Specifying the grid parameters

- Click on **Apply** and then **OK** button from the dialog box to apply the changes.
- To make the grid visible, click on the **Toggle Grid** button from **Toolbar** in the **Draft** workbench. The grid will be displayed in the 3D view area; refer to Figure-197.

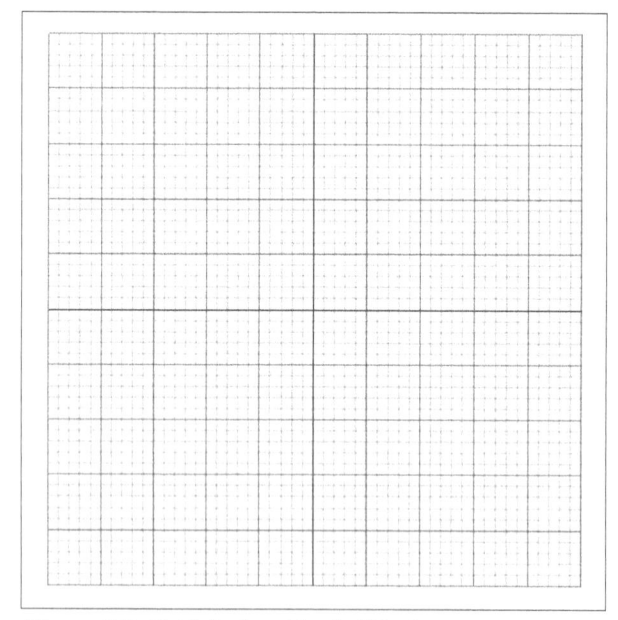

Figure-197. Grid displayed in the 3D view area

- Click on the **Toggle Grid** button again to turn off the draft grid.

Draft utility tools toolbar

The **Draft utility tools toolbar** is available in the **Draft** workbench; refer to Figure-198. The procedures to use the tools of this toolbar are discussed next.

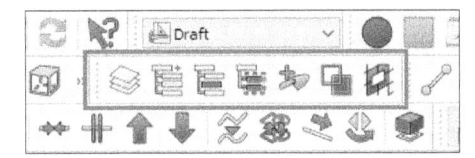

Figure-198. Draft utility tools toolbar

Manage layers

The **Manage layers** tool creates a new layer in the current document to which objects can be added to control object visibility and color. The procedure to use this tool is discussed next.

- Click on the **Manage layers** tool from **Toolbar** in the **Draft** workbench; refer to Figure-199. A layer will be created in the layer container of the **Model Tree**; refer to Figure-200.

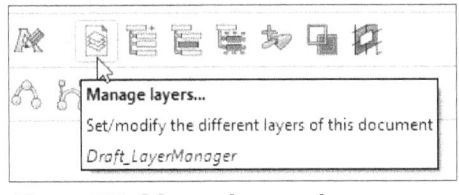

Figure-199. Manage layers tool

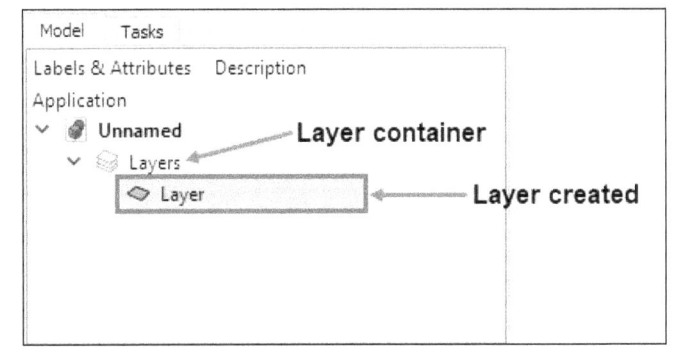

Figure-200. Layer created in the layer container

- Select the layer created, the **Property editor** dialog will be displayed with the properties of layer; refer to Figure-201.

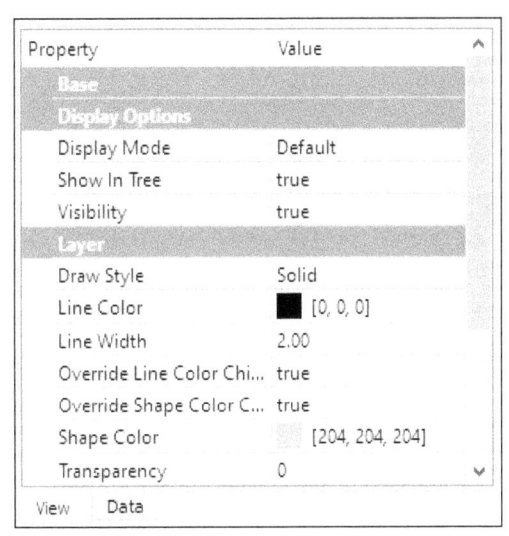

Figure-201. Property editor dialog with layer parameters

- Change the properties of the layer like **Line Color**, **Line Width**, **Shape Color**, and so on as desired; refer to Figure-202.

Figure-202. Properties of the layer changed

- Now, create desired object and drag it into the layer. The properties of the object changed according to the properties of layer; refer to Figure-203.

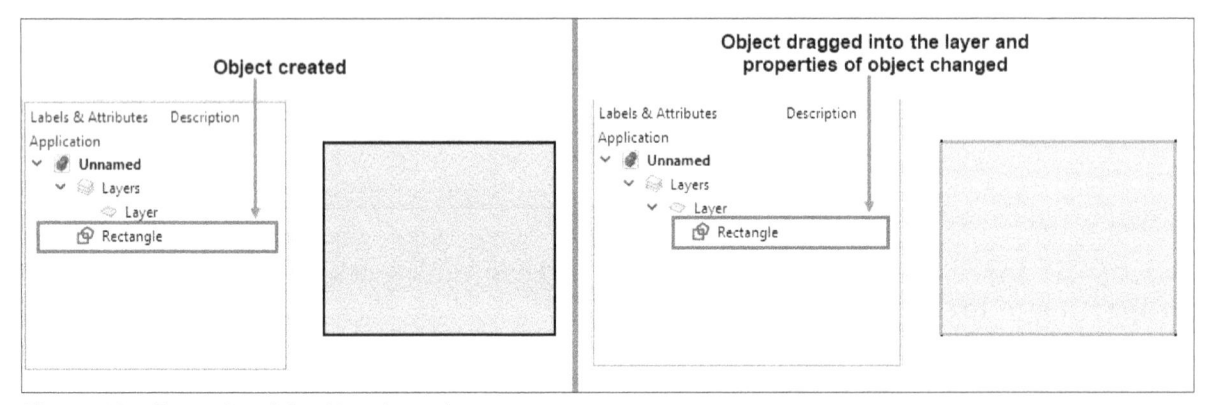

Figure-203. Properties of the object changed

Add a new named group

The **Add a new named group** tool is used to create a named group and moves selected objects to that group. The procedure to use this tool is discussed next.

- Select one or more objects which you want to add to the group; refer to Figure-204.

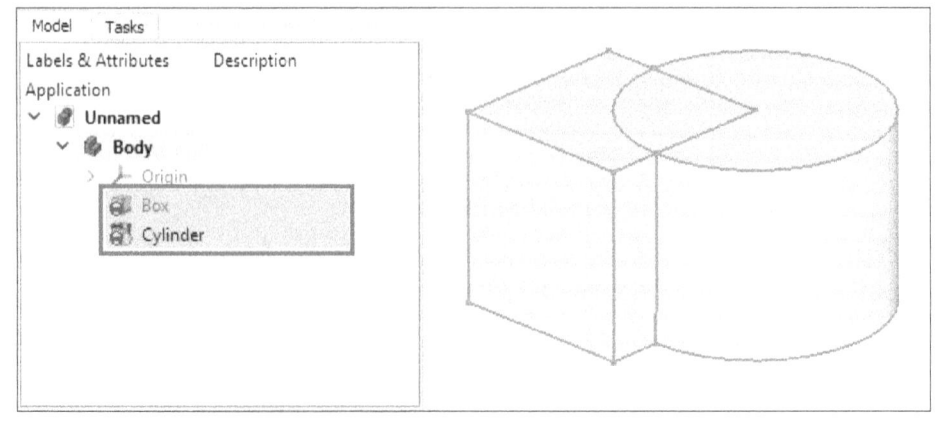

Figure-204. Selecting the objects

- Click on the **Add a new named group** tool from **Toolbar** in the **Draft** workbench; refer to Figure-205. The **Add group** dialog will be displayed in the **Tasks** panel of **Combo View**; refer to Figure-206.

Figure-205. Add a new named group tool

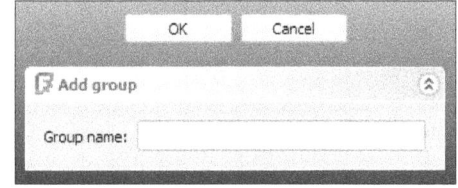

Figure-206. Add group dialog

- Specify desired name of the group in the **Group name** edit box and click on the **OK** button from the dialog. The selected objects will be moved to the created group; refer to Figure-207.

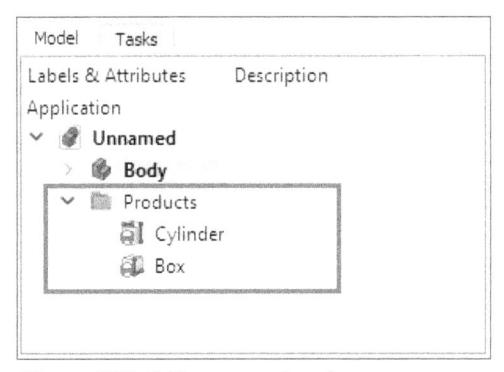

Figure-207. Objects moved to the group

Move to group

The **Move to group** tool quickly adds selected objects to an existing group. The procedure to use this tool is discussed next.

- First, create a group with **Group** tool as discussed earlier in this book and select the object which you want to add to the group; refer to Figure-208.

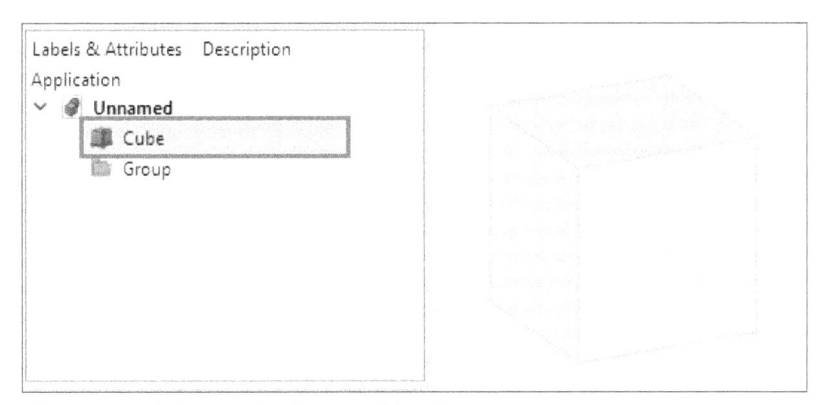

Figure-208. Selecting the object to add to the group

- Click on the **Move to group** tool from **Toolbar** in the **Draft** workbench; refer to Figure-209. A menu will be displayed with the two options near the cursor; refer to Figure-210.

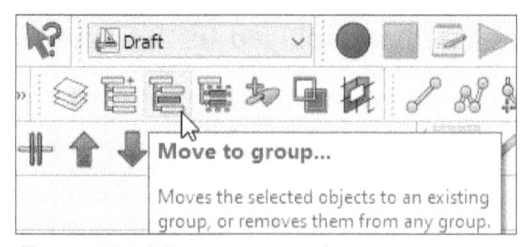

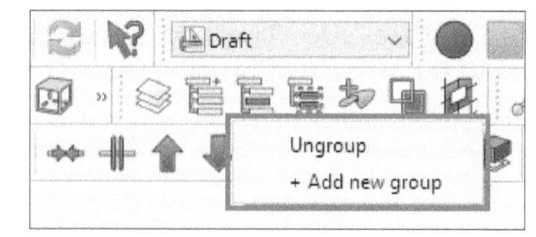

Figure-209. Move to group tool   *Figure-210. Two options displayed*

- Click on the **+ Add new group** option from the menu displayed. The **Add group** dialog will be displayed as discussed earlier.
- Specify desired name of the group in the **Group name** edit box and click on the **OK** button. The object will be added to the group; refer to Figure-211.

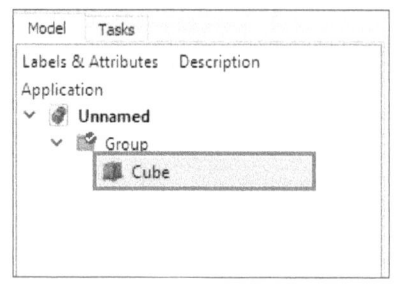

Figure-211. Object added to the group

- If you want to ungroup the object then select the object and click on the **Ungroup** option from the menu displayed. The object will be moved out of the group.

Selecting group contents

The **Select group** tool selects the contents of a selected group or layer. The procedure to use this tool is discussed next.

- Create and select the group or one or more objects in the group from the **Model Tree**; refer to Figure-212.

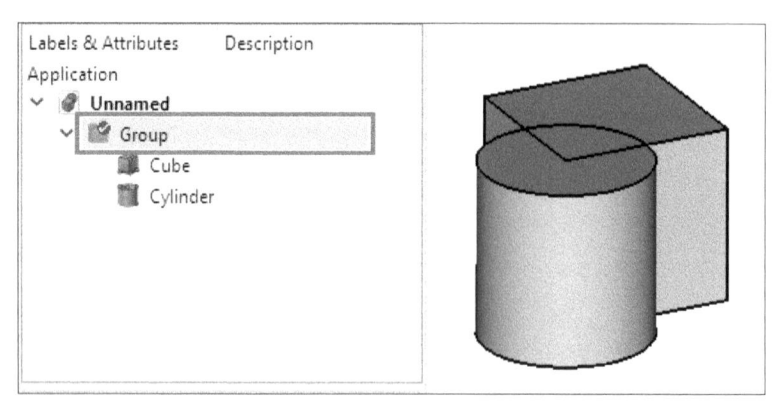

Figure-212. Selecting the group

- Click on the **Select group** tool from **Toolbar** in the **Draft** workbench; refer to Figure-213. All contents of the group will be selected; refer to Figure-214.

Figure-213. Select group tool

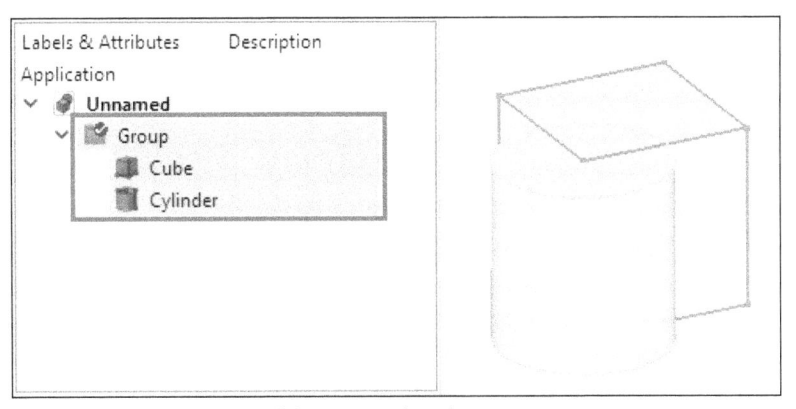

Figure-214. All contents of the group selected

Adding to the construction group

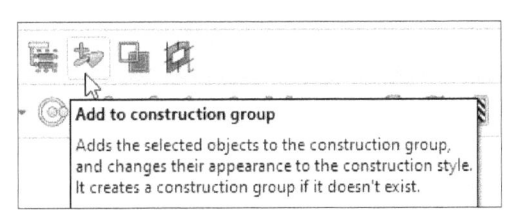

The **Add to construction group** tool add the selected objects to the construction group. The procedure to use this tool is discussed next.

- Select the objects from the **Model Tree** or from the 3D view area which you want to add to the construction group; refer to Figure-215.

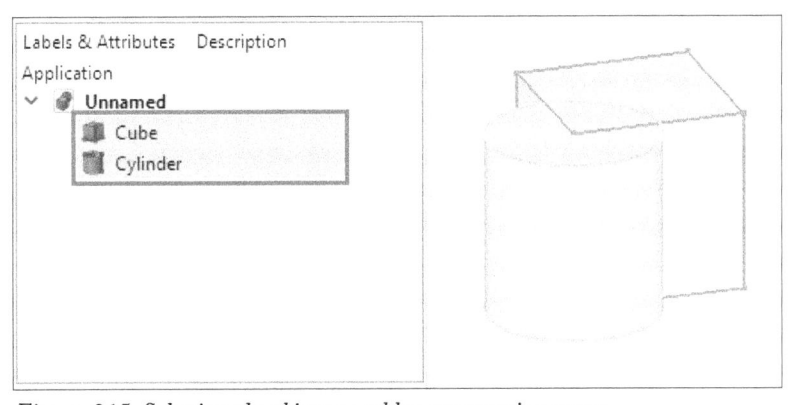

Figure-215. Selecting the objects to add to construction group

- Click on the **Add to construction group** tool from **Toolbar** in the **Draft** workbench; refer to Figure-216. The construction group will be created and the objects will be moved to the construction group; refer to Figure-217.

Figure-216. Add to construction group tool

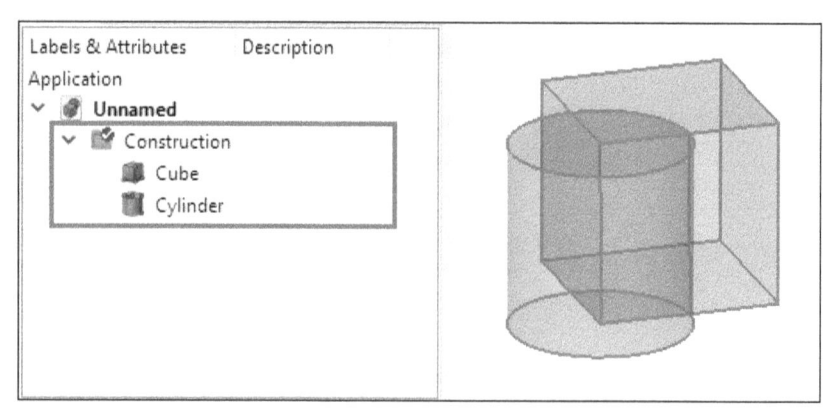

Figure-217. Objects added to the construction group

Toggle normal/wireframe display

The `Toggle normal/wireframe display` tool switches the display mode of selected objects between "`Flat Lines`" and "`Wireframe`". The procedure to use this tool is discussed next.

- Select desired object from the **Model Tree** or from the 3D view area of which you want to toggle the display mode; refer to Figure-218.

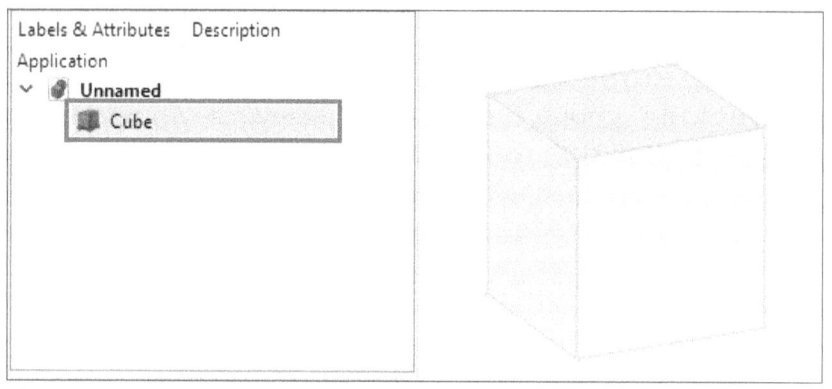

Figure-218. Selecting the object to toggle the display mode

- Click on the `Toggle normal/wireframe display` tool from **Toolbar** in the **Draft** workbench; refer to Figure-219. The display mode of the object will be changed from flat lines to wireframe; refer to Figure-220.

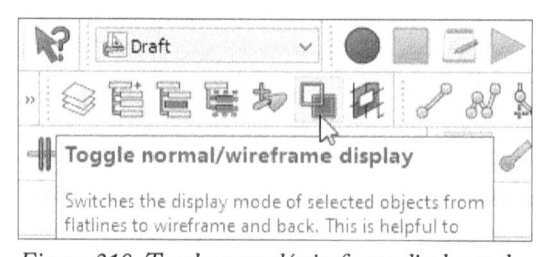

Figure-219. Toggle normal/wireframe display tool

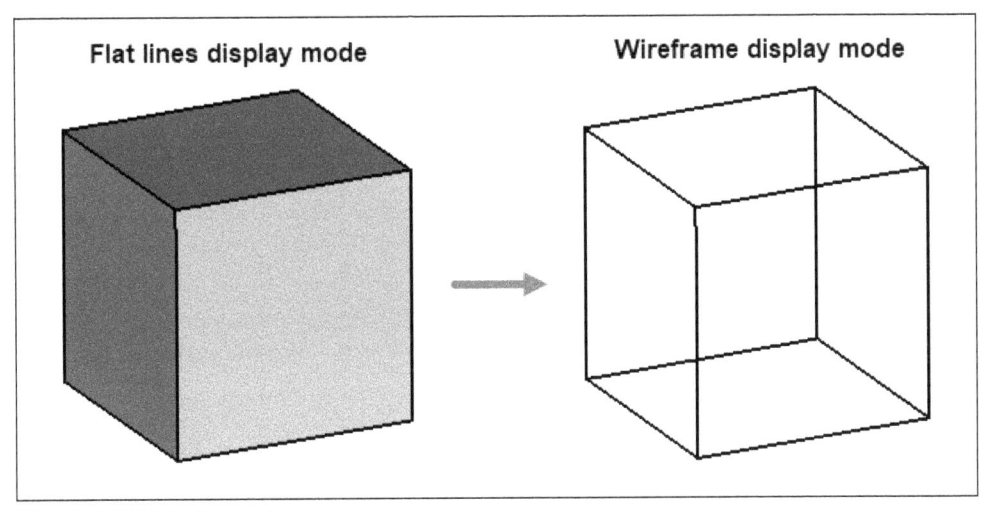

Figure-220. Display mode changed

- Select the object and click on **Toggle normal/wireframe display** tool again to change the display mode from wireframe to flat lines.

Creating working plane proxy

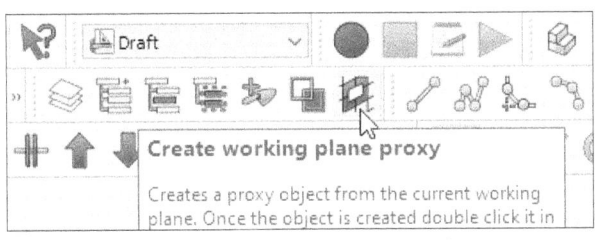

The **Create working plane proxy** tool creates a working plane proxy to save the current working plane position. A working plane proxy can be used to quickly restore a working plane. The procedure to use this tool is discussed next.

- Create and set the position of working plane as desired in the 3D view area; refer to Figure-221.

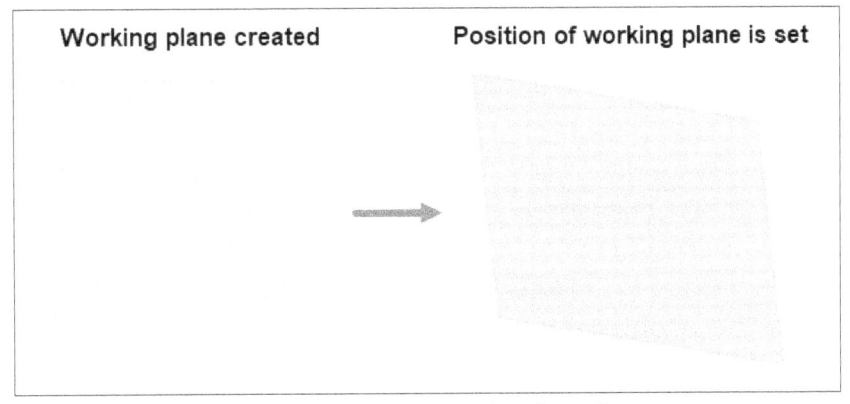

Figure-221. Creating and setting the position of working plane

- Click on the **Create working plane proxy** tool from **Toolbar** in the **Draft** workbench; refer to Figure-222. The working plane proxy will be created; refer to Figure-223.

Figure-222. Create working plane proxy tool

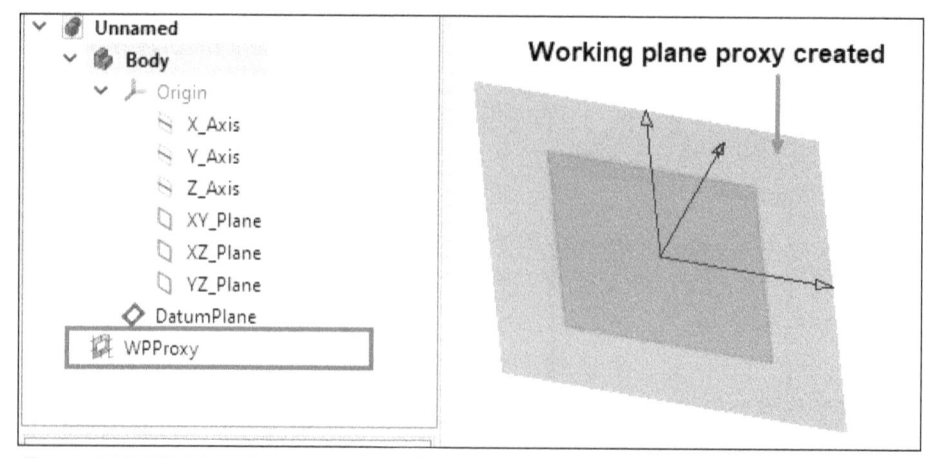

Figure-223. Working plane proxy created

- Now, change the position of working plane as desired and double-click on the working plane proxy created in the **Model Tree**. The position of the working plane will be restored; refer to Figure-224.

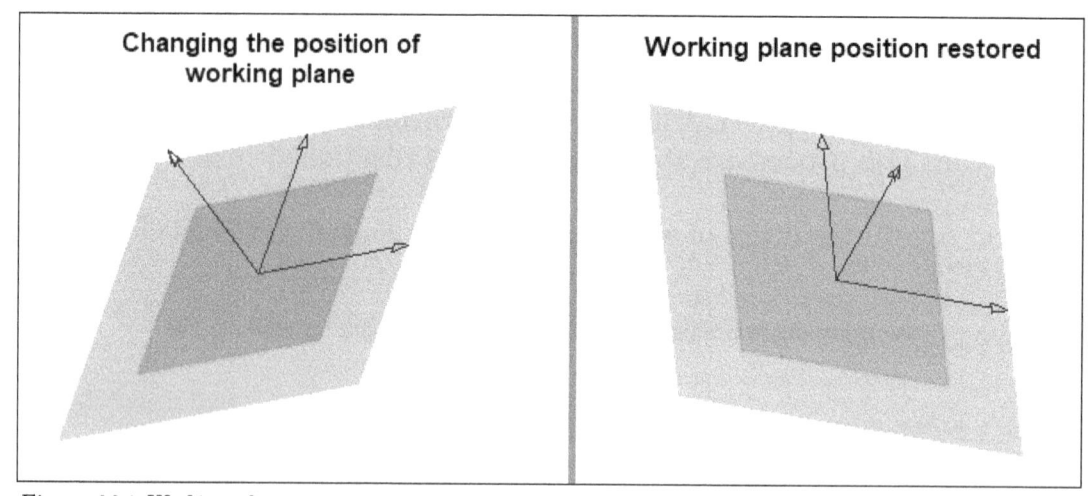

Figure-224. Working plane position restored

Healing the objects

The **Heal** tool heals the problematic objects found in old files of FreeCAD. It tries to recreate the old objects from scratch and transfer their properties to the new objects.

- Select one or more problematic objects from the **Model Tree** or from the 3D view area which you want to heal. If no objects are selected, the entire document will be processed.
- Click on the **Heal** tool from **Utilities** drop-down in the **Menu** bar of **Draft** workbench; refer to Figure-225. The healing process will be started or if no errors are found, the tool will do nothing.

FreeCAD 1.0 Black Book   6-79

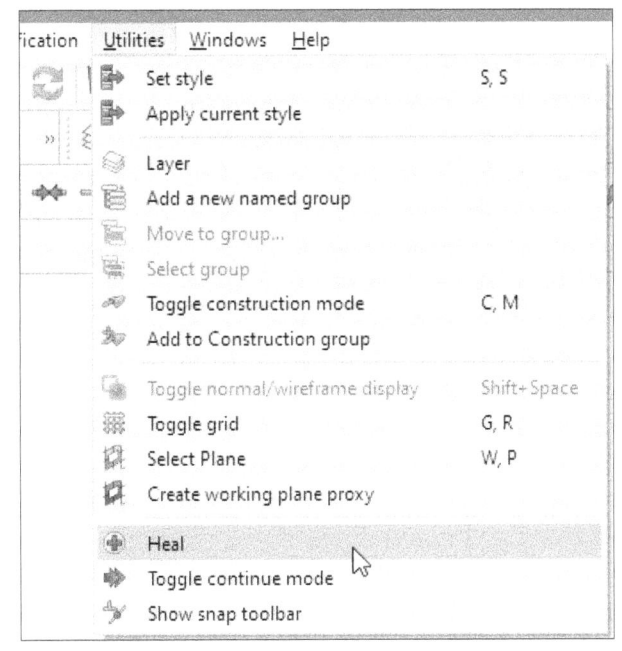

Figure-225. Heal tool

PRACTICAL 1

Create the sketch as shown in Figure-226.

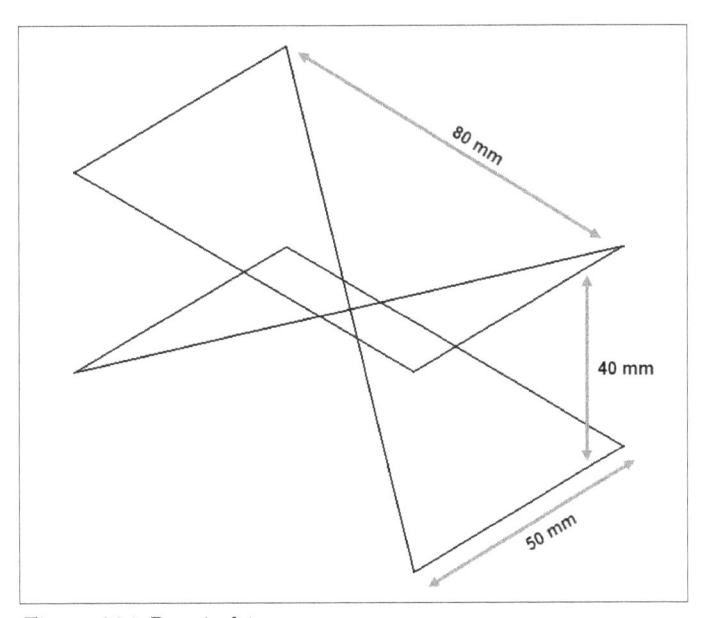

Figure-226. Practical 1

Steps to be performed:
- Start a new part file.
- Select a sketching plane and activate sketching mode.
- Create the sketch using **Rectangle** and **Line** tool.
- Apply the dimensions using dimensioning tools.
- Save the file.

Starting a New Sketch

- Start FreeCAD if not started yet.

- Select the **Draft** option from the **Switch between workbenches** drop-down in the **Toolbar**. The draft environment will be displayed.

Creating Sketch

- Click on the **Rectangle** tool from the **Toolbar** or from the **Drafting** menu to activate rectangle tool. You will be asked to specify start point of line.
- Create a rectangle of width **50** mm and length **80** mm; refer to Figure-227.
- Select the rectangle created, click on the **Clone** tool from the **Toolbar** to create the clone of rectangle and specify the distance of **40** mm between both the rectangles; refer to Figure-228.

Figure-227. Creating rectangle

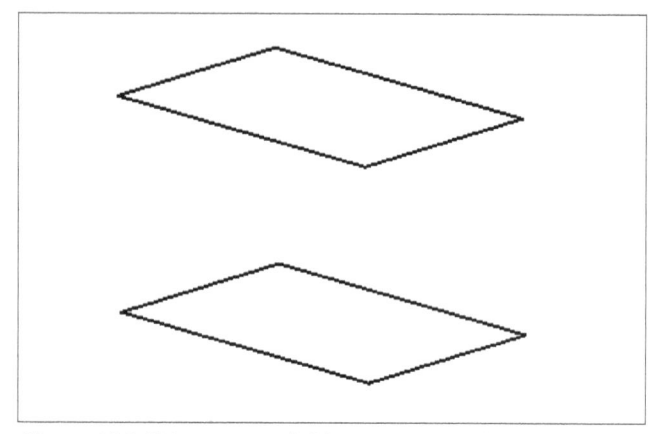

Figure-228. Distance specified between rectangles

- Select the rectangle one by one from the **Model Tree** and downgrade them into edges using **Downgrade** tool; refer to Figure-229.

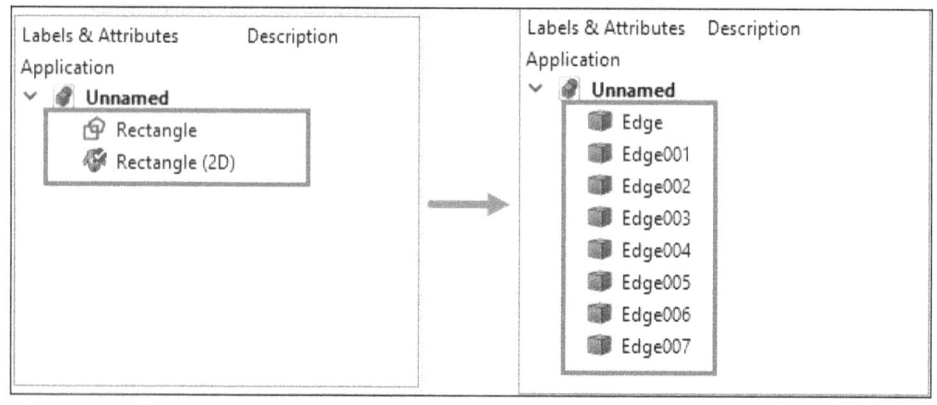

Figure-229. Rectangles downgraded

Deleting the Edges

- Select the edges of the sketch and click on the **Delete** button from the keyboard to delete the selected edges; refer to Figure-230.

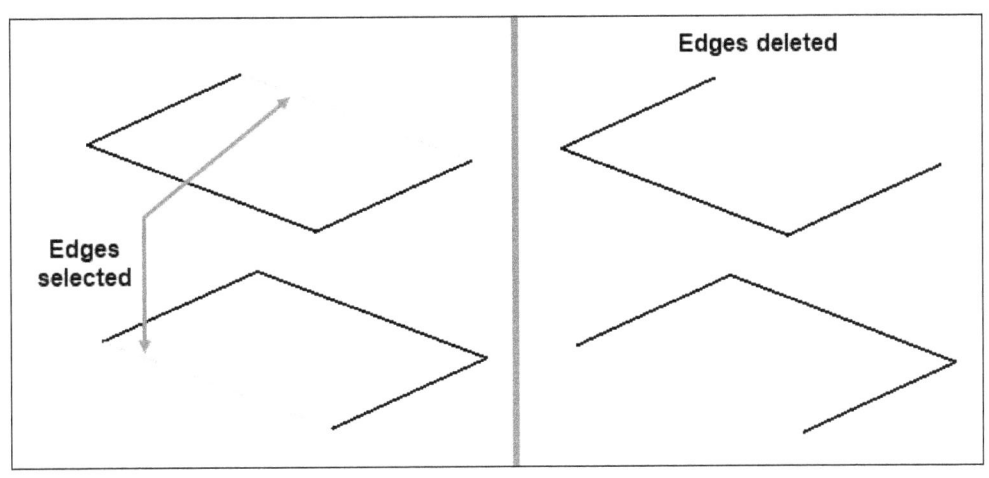

Figure-230. Edges deleted

Creating the lines

- Click on the **Line** tool from the **Toolbar** and create the lines joining open ends of the sketch. The sketch will be created; refer to Figure-231.

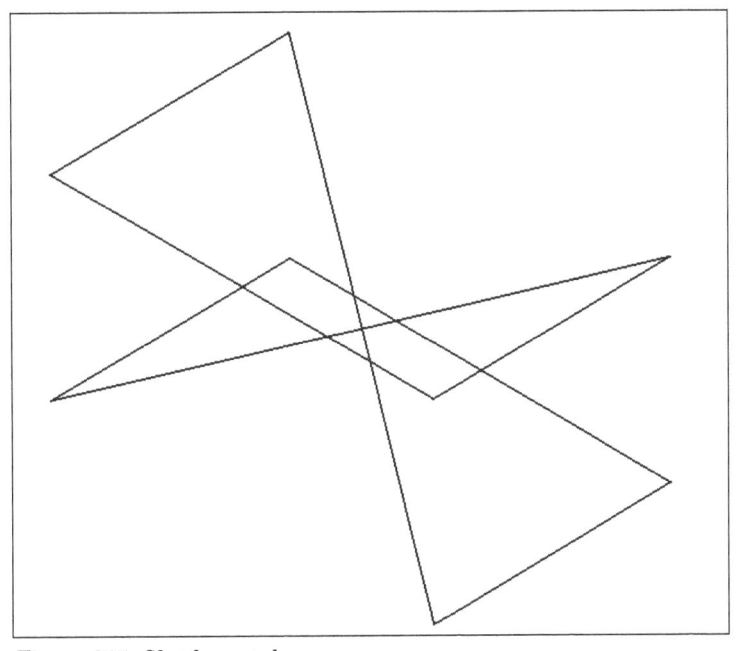

Figure-231. Sketch created

SELF ASSESSMENT

Q1. What does the Line tool create in the Draft workbench?

a) A circle
b) A straight line
c) A polyline
d) A fillet

Q2. Which option must be selected in the Fillet tool to create a chamfer instead of a fillet?

a) Create arc
b) Create circle
c) Create chamfer
d) Delete original objects

Q3. Which tool allows you to create a circular arc using four points in the Draft workbench?

a) Arc by 3 Points tool
b) Arc tool
c) Circle tool
d) Ellipse tool

Q4. Which tool creates a regular polygon by defining the center and radius in the Draft workbench?

a) Rectangle tool
b) Polygon tool
c) B-Spline tool
d) Circle tool

Q5. Which tool creates a Bezier curve using control points in the Draft workbench?

a) Cubic Bezier Curve tool
b) B-Spline tool
c) Bezier Curve tool
d) Ellipse tool

Q6. What is the primary function of the Facebinder tool?

a) To create a new object
b) To create a surface from selected faces of a solid object
c) To modify the object color
d) To create a polyline

Q7. Which tool creates a circle by defining the center and radius in the Draft workbench?

a) Ellipse tool
b) Circle tool
c) Polygon tool
d) Arc tool

Q8. What is the function of the "Filled" checkbox in the Polyline tool?

a) It creates a filled polygon
b) It closes the polyline to form a face
c) It sets the polyline's thickness
d) It defines the angle of the polyline

Q9. Which tool in FreeCAD is used to create a 3D letter from a text string?

A) Hatch Tool
B) Shape String Tool
C) Text Tool
D) Dimension Tool

Q10. Which file type is selected when creating a hatch in FreeCAD?

A) .txt
B) .pdf
C) .PAT
D) .dxf

Q11. What is the primary purpose of the Annotation Style Editor tool in FreeCAD?

A) To create 3D objects
B) To apply visual properties to annotation objects
C) To export annotations
D) To reset object transformations

Q12. What does the Label tool in FreeCAD display?

A) Length and volume of an object
B) A multi-line text box with a leader line and arrow
C) Coordinates of a point in 3D space
D) Object name

Q13. Which option can be selected in the Annotation Style Editor to display the unit next to the dimension value?

A) Show unit
B) Unit override
C) Scale multiplier
D) Line width

Q14. In the Rotate tool, what does the base angle define?

A) The rotation direction
B) The reference point
C) The distance to rotate
D) The initial angle before rotating

Q15. Which option in the Scale tool ensures that the object is scaled uniformly along all axes?

A) Copy
B) Uniform scaling
C) Modify subelements
D) Working plane orientation

Q16. What is the function of the Offset tool in FreeCAD?

A) Moves an object to a new position
B) Resizes an object proportionally
C) Moves an object perpendicular to itself by a specified distance
D) Rotates an object around a base point

Q17. Which tool in FreeCAD is used to trim or extend lines and wires?

A) Stretch Tool
B) Rotate Tool
C) Trimex Tool
D) Scale Tool

Q18. What does the Clone tool do in the Draft workbench?

A) Creates a single copy of an object
B) Creates linked copies of the selected object, so changes in the original object affect all clones
C) Deletes the selected object
D) Creates an array of objects in different directions

Q19. What is the main function of the Path Array tool in Draft workbench?

A) To place copies of an object along a specified path
B) To create an array in the form of concentric circles
C) To distribute elements along predefined layers
D) To generate a polar array with a specified angle

Q20. What is the purpose of the Point Array tool in Draft workbench?

A) To create an array along a path
B) To distribute copies of an object along selected points
C) To fuse overlapping elements in an array
D) To rotate an object around a fixed point

Q21. How does the Path Twisted Array tool differ from the standard Path Array tool?

A) It places copies along a straight line only
B) It creates a path with a defined twist angle
C) It only creates circular arrays
D) It is used for placing copies along concentric circles

Q22. What does the Set Slope tool in Draft workbench do?

A) It increases the Z coordinate of all points
B) It slants or inclines lines by adjusting their Z coordinate
C) It moves objects along a defined path
D) It creates a circular array of objects

Q23. What is the result of using the Downgrade tool in Draft workbench?

A) It merges different elements into a single object
B) It breaks faces and deconstructs wires into their individual elements
C) It creates a B-spline from a wire
D) It upgrades a wire to a closed shape

Q24. What does the Convert Wire to B-Spline tool do?

A) Converts a B-spline into a wire
B) Converts a wire into a B-spline
C) Converts a draft object to a sketch
D) Converts an array to a path

Q25. What action does the Split tool perform in the Draft workbench?

A) It joins multiple wires into a single object
B) It upgrades a wire to a face
C) It splits a wire at a specified location
D) It creates a new wire from a B-spline

Q26. What does the Upgrade tool do in Draft workbench?

A) Breaks faces into individual components
B) Converts a wire into a closed shape or a face
C) Moves objects along a path
D) Creates a polar array

Q27. What does the Flip Dimension tool do in the Draft workbench?

a) It rotates the dimension text by 90 degrees around the dimension line.
b) It rotates the dimension text 180 degrees around the dimension line.
c) It flips the dimension line to the opposite side.
d) It changes the font of the dimension text.

Q28. In the Shape 2D View tool, how is a 2D projection created?

a) By rotating the 3D object to the desired projection and clicking the Shape 2D View tool.
b) By manually drawing the 2D projection in the Draft workbench.
c) By selecting a 3D object and clicking on the View Projection button.
d) By creating a new 3D object from the selected 2D projection.

Q29. What is the purpose of the Working Plane tool in the Draft workbench?

a) It changes the position of the camera in the 3D view.
b) It sets a working plane on a standard view or selected face to build draft shapes.
c) It allows drawing freehand lines on any surface.
d) It hides the working plane in the 3D view.

Q30. Which of the following tools is used to apply default visual properties such as line color and font size in the Draft workbench?

a) Working Plane tool
b) Set Style tool
c) Toggle Construction Mode tool
d) Auto Group tool

Q31. What does the Toggle Construction Mode tool do?

a) It toggles between creating solid and hollow objects.
b) It enables or disables the draft construction mode where geometry can be used as references or snapping elements.
c) It switches the display of the 3D view between shaded and wireframe modes.
d) It automatically groups the objects created in the current session.

Q32. Which tool in the Draft workbench allows you to change the active layer or group for newly created objects?

a) Set Style tool
b) Auto Group tool
c) Manage Layer tool
d) Heal tool

Q33. What does the Toggle Snap tool do in the Draft Snap toolbar?

a) It enables or disables snapping to grid lines.
b) It toggles the snap modes for all snapping methods.
c) It activates a specific snapping method like Midpoint or Endpoint.
d) It toggles between showing and hiding the grid.

Q34. Which of the following snap modes is activated by the Perpendicular snap button?

a) Snaps to the endpoint of a line.
b) Snaps to the center of an arc.
c) Snaps to the intersection of two line segments.
d) Snaps to create a line that is perpendicular to an existing line or edge.

Q35. What does the Manage Layer tool in the Draft workbench allow you to do?

a) It creates a new layer in the current document and allows assigning objects to it.
b) It merges multiple layers into a single group.
c) It adjusts the visibility of a selected layer in the 3D view.
d) It applies color styles to existing layers.

Q36. What is the function of the Add to Construction Group tool?

a) It adds selected objects to a construction layer for easier management.
b) It adds selected objects to the construction group, making them available as reference geometry.
c) It creates a new construction layer and moves objects to it.
d) It adds selected objects to a predefined construction group for visualization.

Q37. Which of the following snap buttons activates snapping to the intersection of two line or arc segments?

a) Endpoint
b) Midpoint
c) Intersection
d) Perpendicular

Q38. What is the function of the Working Plane Proxy tool?

a) It creates a new working plane based on the current 3D view.
b) It saves the current working plane position to restore it later.
c) It changes the position of the working plane dynamically during the session.
d) It deletes any saved working plane position.

Q39. Which tool is used to create a new layer in the current document and manage its properties?

a) Manage Layer tool
b) Heal tool
c) Toggle Construction Mode tool
d) Add to Group tool

Q40. What is the purpose of the Toggle Normal/Wireframe Display tool?

a) It changes the visual style of selected objects between flat lines and wireframe mode.
b) It switches the object's display mode from 2D to 3D.
c) It resets the current object to the default view.
d) It locks the object in a wireframe view for all users.

For Student Notes

Chapter 7

BIM Modeling

Topics Covered

The major topics covered in this chapter are:

- *Introduction to BIM Workbench*
- *BIM Tools*
- *Axis Tools*
- *Panel Tools*
- *Pipe Tools*
- *Material Tools*

INTRODUCTION

The **BIM** workbench provides a modern building information modeling (BIM) workflow to FreeCAD with support for features like fully parametric architectural entities such as walls, beams, roofs, windows, stairs, pipes, and furniture. It supports industry foundation class (IFC) files and production of 2D floor plans in combination with the **TechDraw** workbench; refer to Figure-1.

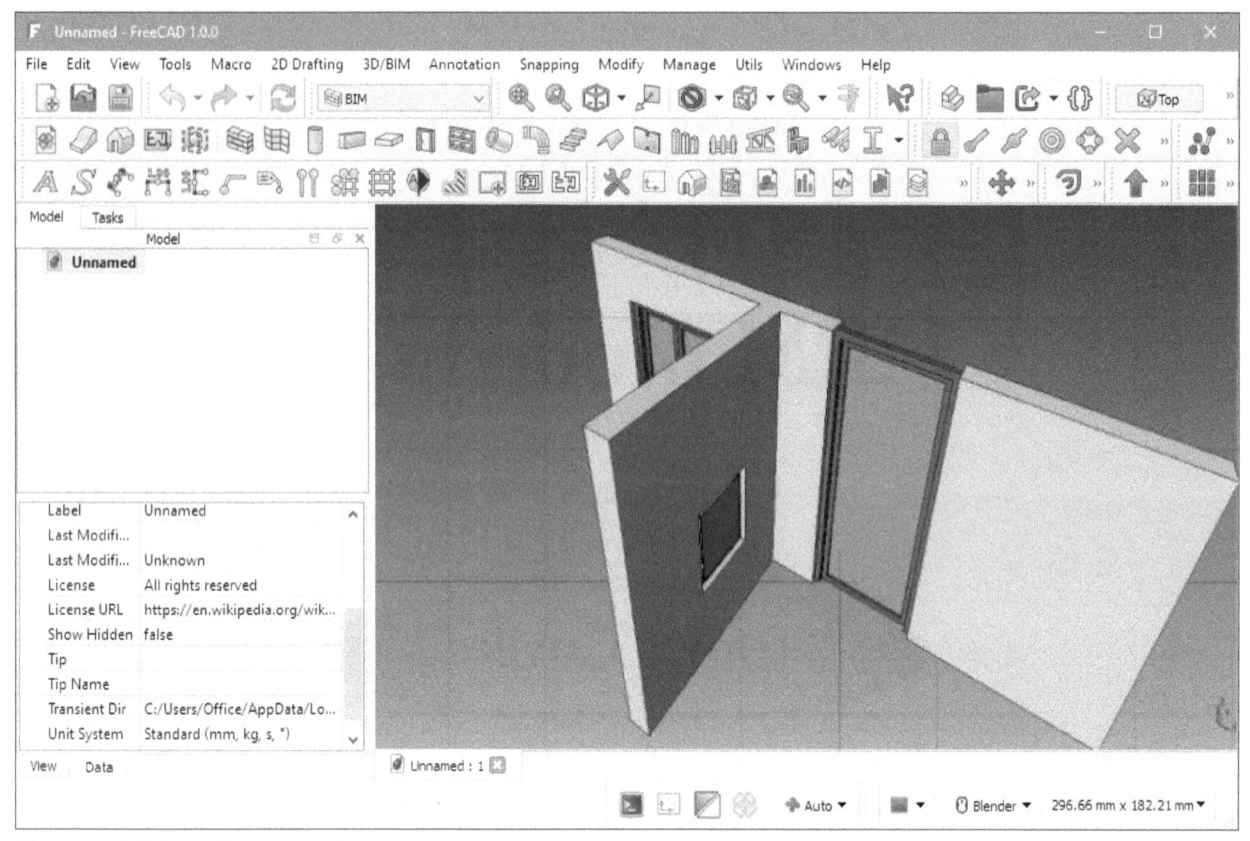

Figure-1. BIM workbench interface

The BIM functionality of FreeCAD is now progressively split into this **BIM** workbench which holds basic architectural tools and the **BIM** workbench which is available from the **Addon Manager**. This **BIM** workbench adds a new interface layer on top of the **BIM** tools with the aim of making the BIM workflow more intuitive and user-friendly.

STARTING BIM WORKBENCH

The BIM workbench imports all tools from the **Draft** workbench as it uses its 2D objects to build 3D parametric architectural objects. Nevertheless, Arch can also use solid shapes created with other workbenches like **Part** and **Part Design**.

- To start a new BIM file, click on the **New** button from **File** menu and select **BIM** workbench from **Switch between workbenches** drop-down in the **Toolbar**; refer to Figure-2. The tools related to **BIM** workbench will be displayed in the **Toolbar**; refer to Figure-3.

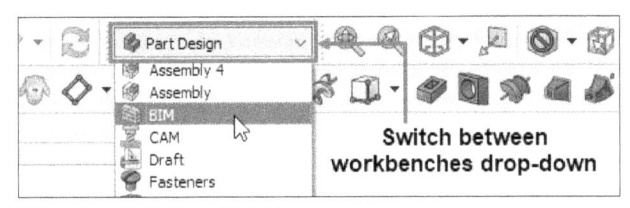

Figure-2. BIM workbench

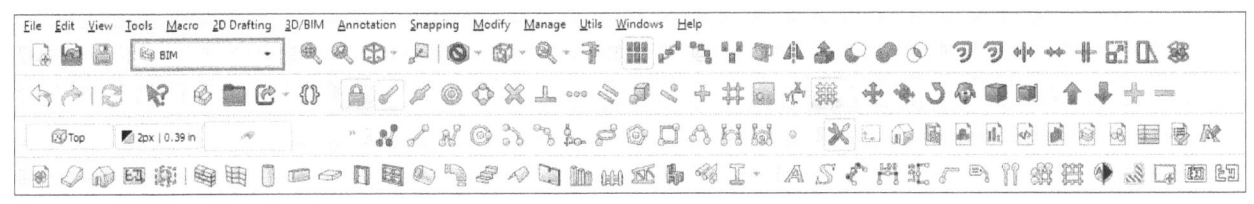

Figure-3. BIM workbench tools

BIM TOOLS

The **BIM** tools are used for creating architectural objects. These tools are available in the **Toolbar** of **BIM** workbench; refer to Figure-4. These tools are discussed next.

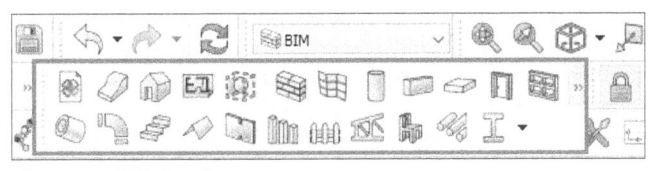

Figure-4. BIM tools

Creating Project

The **Project** tool creates a project including selected objects. It is a special object suitable to add better compatibility with IFC (Industry Foundation Classes) files. Every IFC file is required to contain an IFC project entity. The procedure to use this tool is discussed next.

- Click on the **Project** tool from **Toolbar** in the **BIM** workbench; refer to Figure-5. The **Default structure** dialog box will be displayed; refer to Figure-6.

Figure-5. Project tool

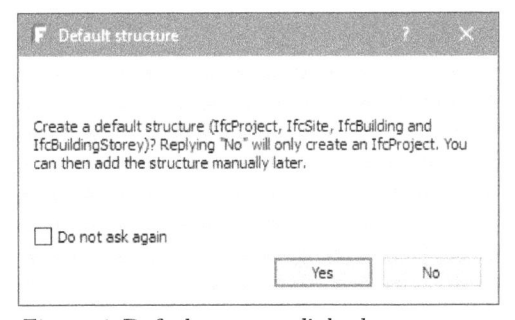

Figure-6. Default structure dialog box

- Click on the **Yes** button to create a default structure, viz. IfcProject, IfcSite, IfcBuilding, and IfcStorey); refer to Figure-7. And click on the **No** button to create only the IfcProject.
- Now, if you want to add object to the project then drag & drop the object to the project which you want to add in the **Model Tree**; refer to Figure-8.

Figure-7. Default structure created

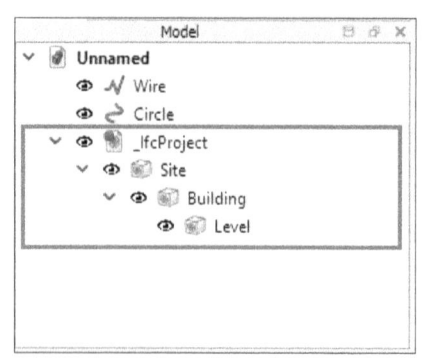

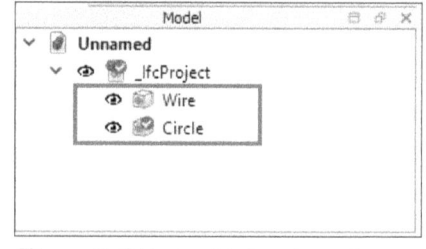

Figure-8. Objects added to the project

Creating Site

The **Site** tool is a special object that combines properties of a standard FreeCAD group object and BIM objects. It is particularly suited for representing a whole project site or terrain. The procedure to use this tool is discussed next.

- Select one or more building objects from the **Model Tree** which you want to include in a site; refer to Figure-9.

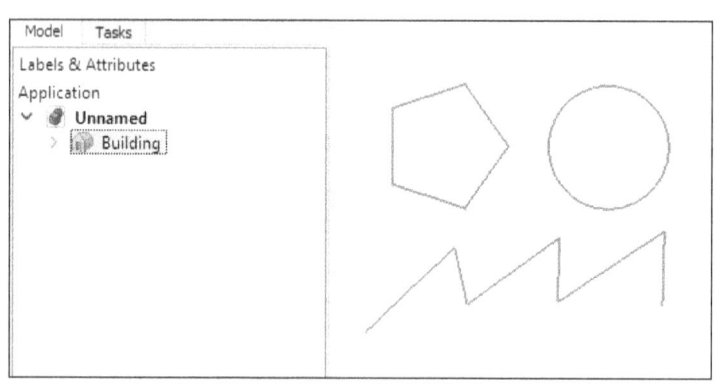

Figure-9. Selecting the building object

- Click on the **Site** tool from **Toolbar** in the **BIM** workbench; refer to Figure-10. The site will be created and the building object will be included in the site; refer to Figure-11.

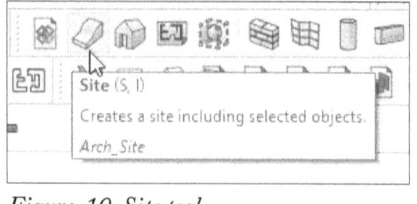

Figure-10. Site tool

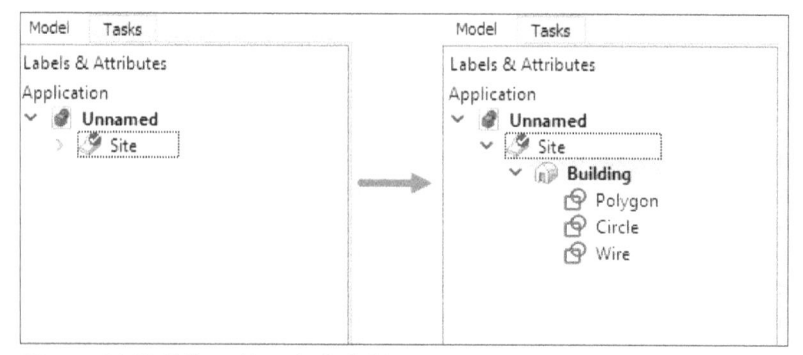

Figure-11. Building objects included in a site

- If you want to add more objects to the site then drag and drop the object to it or use **Add component** tool from the **Toolbar** which will be discussed later in this chapter. If you want to remove the objects then drag and drop the object out of it or use **Remove component** tool from the **Toolbar** which will be discussed later in this chapter.

- If you want to edit the properties of site object then select the site object created from the **Model Tree**. The **Property editor** dialog will be displayed in the **Model** panel of **Combo View** with the parameters related to site; refer to Figure-12.

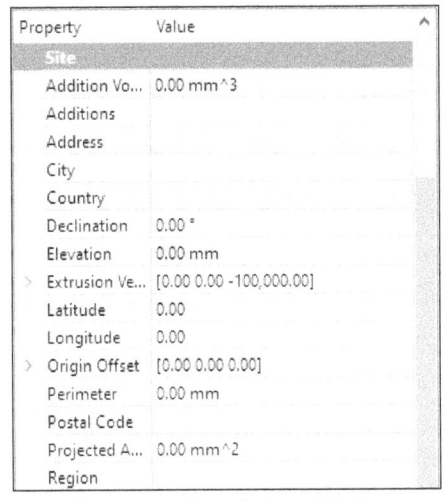

Figure-12. Property editor dialog with site parameters

- Enter desired value in **Addition Volume** and **Subtraction volume** edit boxes from **Site** section of the dialog to specify the volume of earth to be added to this terrain and to be removed from this terrain, respectively.
- Specify desired address, city, country, postal code, and region of this site in the **Address**, **City**, **Country**, **Postal code**, and **Region** edit boxes, respectively.
- Enter desired value in **Latitude** and **Longitude** edit boxes to specify the latitude and longitude of this site, respectively.
- Enter desired value in **x**, **y**, and **z** edit boxes from **Extrusion Vector** cascading menu to use the extrusion vector when performing boolean operations.
- Specify desired perimeter length of this terrain in the **Perimeter** edit box.
- Enter desired value in **Projected Area** edit box to specify area of the projection of this object onto the XY plane.
- Select desired option from **Remove Splitter** drop-down to specify whether to remove the splitters from resulting shape or not.
- Click on the **Terrain** button to specify base terrain of this site.
- Enter desired url in the **Url** edit box to show this site in a mapping website.
- On specifying the parameters in **Property editor** dialog, the properties of site object will be modified.

Creating Building

The **Building** tool is a special type of FreeCAD group object particularly suited for representing a whole building unit. They are mostly used to organize your model by containing floor objects. The procedure to use this tool is discussed next.

- Select one or more objects from the **Model Tree** or from the 3D view area which you want to include in a building; refer to Figure-13.

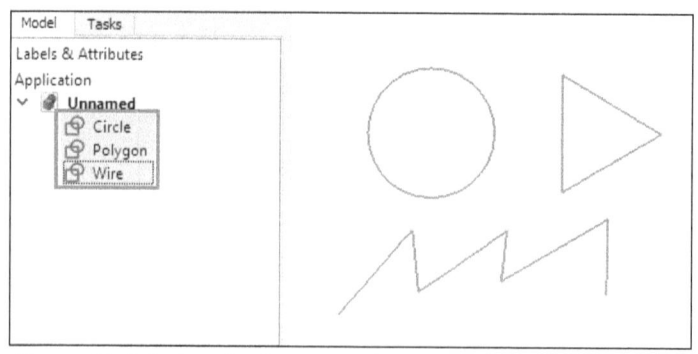

Figure-13. Selecting the objects to include in a building

- Click on the **Building** tool from **Toolbar** in the **BIM** workbench; refer to Figure-14. The building will be created in the **Model Tree** and all the selected objects will be included in the building; refer to Figure-15.

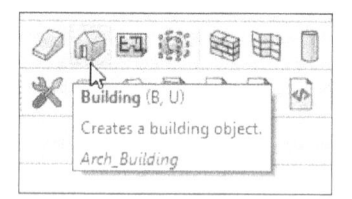

Figure-14. Building tool

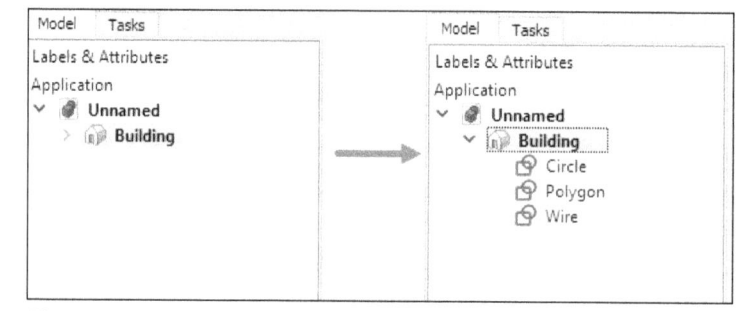

Figure-15. Objects included in building

- If you want to add more objects to the building then drag and drop the object to it or use **Add component** tool from the **Toolbar** which will be discussed later in this chapter.
- If you want to remove the objects then drag and drop the object out of it or use **Remove component** tool from the **Toolbar** which will be discussed later in this chapter.
- The parameters in the **Property editor** dialog to edit the properties of building object are same as discussed for previous tool.

Creating Level

The **Level** tool is used to create Floor/Storey/Levels as well as all kinds of situations where different Arch/BIM objects need to be grouped and that group might need to be handled as one object or replicated. The procedure to use this tool is discussed next.

- Select one or more objects from **Model Tree** or from the 3D view area which you want to include in a new building part; refer to Figure-16.

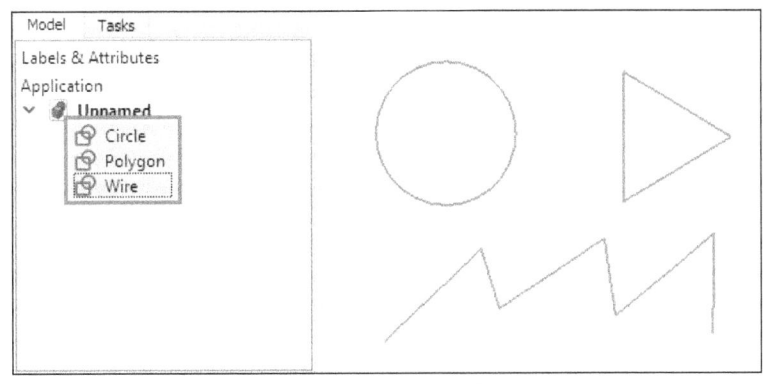

Figure-16. Selecting the objects to include in a floor

- Click on the **Level** tool from **Toolbar** in the **BIM** workbench; refer to Figure-17. The building part will be created in the **Model Tree** and all the selected objects will be included in the building part; refer to Figure-18.

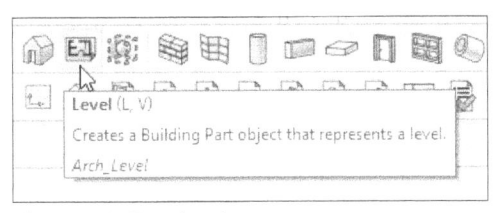

Figure-17. Level tool

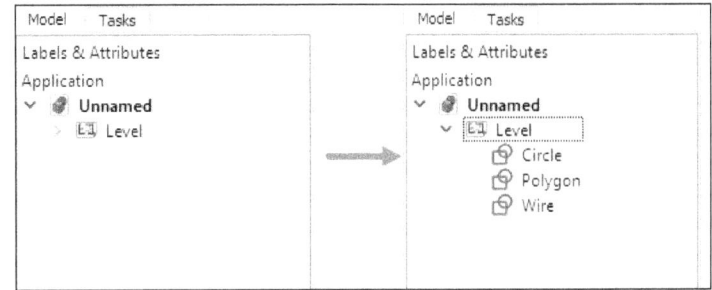

Figure-18. Objects included in a level

- If you want to add more objects to the level then drag and drop the object to it or use **Add component** tool from the **Toolbar** which will be discussed later in this chapter.
- If you want to remove the objects then drag and drop the object out of it or use **Remove component** tool from the **Toolbar** which will be discussed later in this chapter.
- If you want to edit the properties of building part then select the building part created from **Model Tree**. The **Property editor** dialog will be displayed in the **Model** panel of **Combo View** with the parameters related to building part; refer to Figure-19.

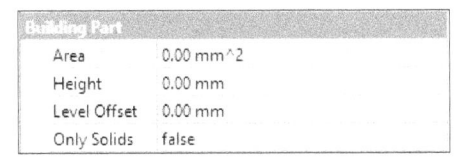

Figure-19. Property editor dialog with building part parameters

- Specify computed floor area of the building part in the **Area** edit box from **Building Part** section of the dialog.
- Specify desired height of the building part and of its children objects in the **Height** edit box.
- Enter desired value in **Level Offset** edit box. This value is added to the **Placement.Base.z** attribute of the building part to indicate a vertical offset without actually moving the object.
- Select desired IFC type of this object from **Ifc Role** drop-down in the **Component** section of the dialog.

- Specify desired description and tag for this object in the **Description** and **Tag** edit boxes of the dialog, respectively.
- On specifying the parameters in **Property editor** dialog, the properties of building part object will be modified.

Creating Space

The **Space** tool allows you to define an empty volume either by basing it on a solid shape or by defining its boundaries or a mix of both. The space object always defines a solid volume. The procedure to use this tool is discussed next.

- Select the existing solid object or faces on the boundary object from **Model Tree** or from the 3D view area which you want to create a space object; refer to Figure-20.

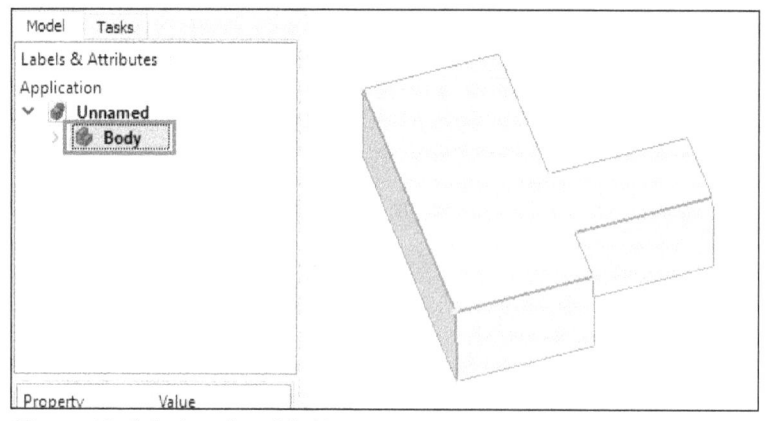

Figure-20. Selecting the solid object

- Click on the **Space** tool from **Toolbar** in the **BIM** workbench; refer to Figure-21. The space object will be created; refer to Figure-22.

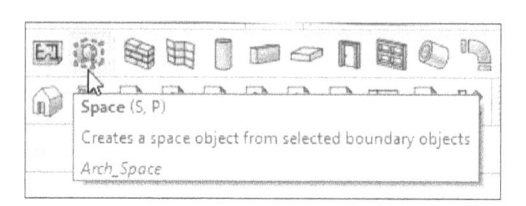

Figure-21. Space tool

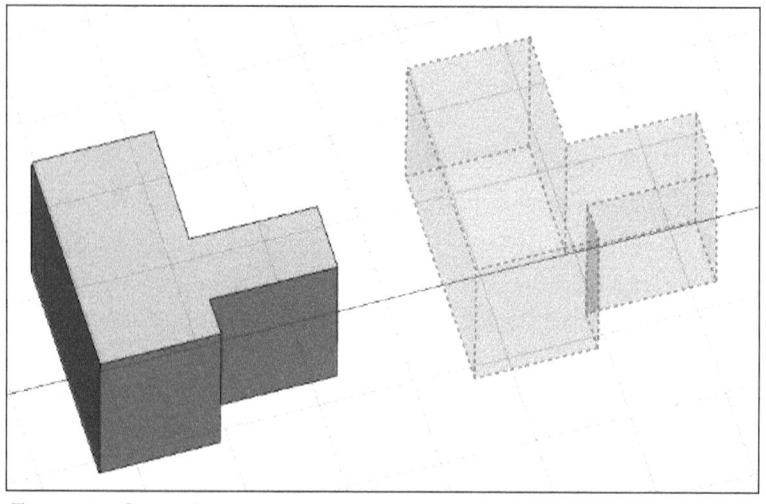

Figure-22. Space object created

- If you want to edit the space object then double-click on the space object created from the **Model Tree**. The **Component** dialog will be displayed in the **Tasks** panel of **Combo View**; refer to Figure-23.
- Specify desired parameters from the **Component** dialog as discussed earlier.
- Click on **OK** button to close the dialog.
- If you want to edit the parameters of space object then select the space object created from **Model Tree**. The **Property editor** dialog will be displayed in the **Model** panel of **Combo View** with parameters related to space object; refer to Figure-24.

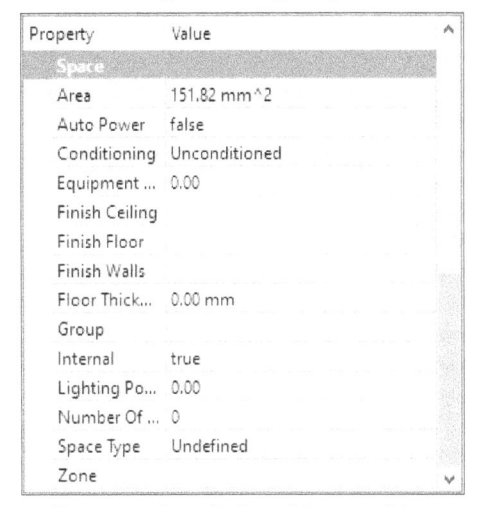

Figure-24. Property editor dialog with space object parameters

Figure-23. Component dialog

- Select desired option from **Auto Power** drop-down in the **Space** section of the dialog to specify whether the equipment power will be automatically filled by the equipment included in this space or not.
- Select desired option from **Conditioning** drop-down to specify the type of air conditioning of this space.
- Enter desired value in **Equipment Power** edit box to specify the electric power needed by the equipment of this space in watts.
- Enter desired values in **Finish Ceiling**, **Finish Floor**, and **Finish Walls** edit boxes to specify finishing of ceiling, finishing of floor, and finishing of walls of this space, respectively.
- Specify desired thickness of the floor finish in the **Floor Thickness** edit box.
- Objects that are included inside this space such as furniture, will be displayed in the list box available by clicking on the **Group** button.
- Select desired option from **Internal** drop-down to specify whether this space is internal or external.
- Enter desired value in **Lighting Power** edit box to specify the electric power needed to light this space in watts.
- Specify desired number of people who typically occupy this space in the **Number Of People** edit box.
- Select desired type of this space from **Space Type** drop-down in the dialog.
- Specify desired number of decimals to use for calculated texts in the **Decimals** edit box.

- Specify desired size of the first line of text in the **First Line** edit box.
- Specify desired name, size, and color of the text in the **Font Name**, **Font Size**, and **Text Color** edit boxes of the dialog, respectively.
- Enter desired value in **Line Spacing** edit box to specify the space between lines of text.
- Select desired option from **Show Unit** drop-down to specify whether to show the unit suffix or not.
- Specify desired text to show in the **Text** edit box.
- Select desired option from **Text Align** drop-down to specify the justification of the text.
- Enter desired value in **x**, **y**, and **z** edit boxes from **Text Position** cascading menu to specify the position of text along x, y, and z directions, respectively.
- On specifying the parameters in **Property editor** dialog, the properties of space object will be modified.

Creating Wall

The **Wall** tool builds a wall object from scratch or on top of any other shape-based or mesh-based object. A wall can be built without any base object in which case it behaves as a cubic volume using length, width, and height properties. When built on top of an existing shape, a wall can be based on a linear 2D object, a flat face, a solid, or a mesh. The procedure to use this tool is discussed next.

- Click on the **Wall** tool from **Toolbar** in the **BIM** workbench; refer to Figure-25. The **First point of wall** and **Wall options** dialogs will be displayed in the **Tasks** panel of **Combo View** along with the plus ⊕ sign in place of original cursor; refer to Figure-26. You will be asked to specify the first point.

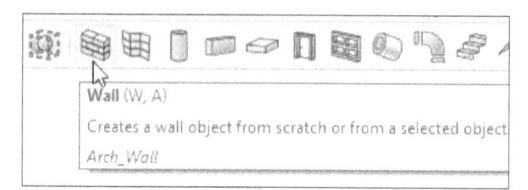

Figure-25. Wall tool

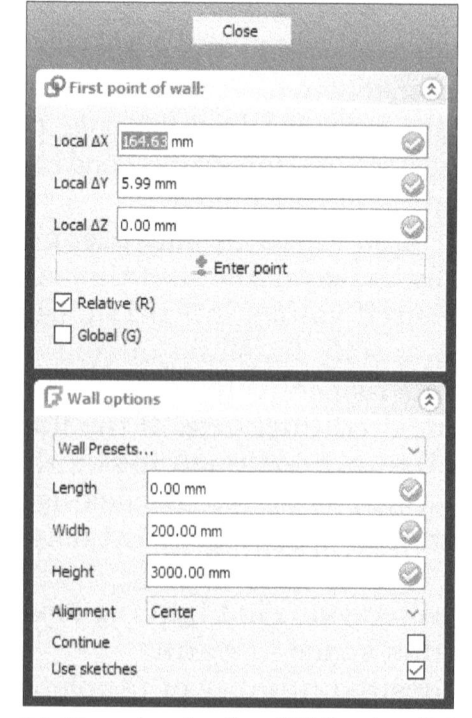

Figure-26. First point of wall and Wall options dialogs

- Specify desired width and height of the wall in the **Width** and **Height** edit boxes from **Wall options** dialog, respectively.

- Specify the alignment of a wall on its baseline by selecting desired option from **Alignment** drop-down in the **Wall options** dialog.
- After specifying parameters in **Wall options** dialog, click in the 3D view area to specify the first point or enter desired values for x, y, and z coordinates in the **Local ΔX**, **Local ΔY**, and **Local ΔZ** edit boxes from **First point of wall** dialog, respectively.
- After specifying coordinates for the first point in dialog, click on the **Enter point** button from **First point of wall** dialog. The **Next point** dialog will be displayed asking you to specify the second point.
- Specify desired angle in the **Angle** edit box at which the wall to be created and select the **Angle** check box to lock the current angle.
- Move the cursor away and click at desired location to specify the second point as well as length of the wall or enter desired values for the coordinates in their respective edit boxes.
- After specifying coordinates for the second point in dialog, click on the **Enter point** button again. The wall will be created; refer to Figure-27.

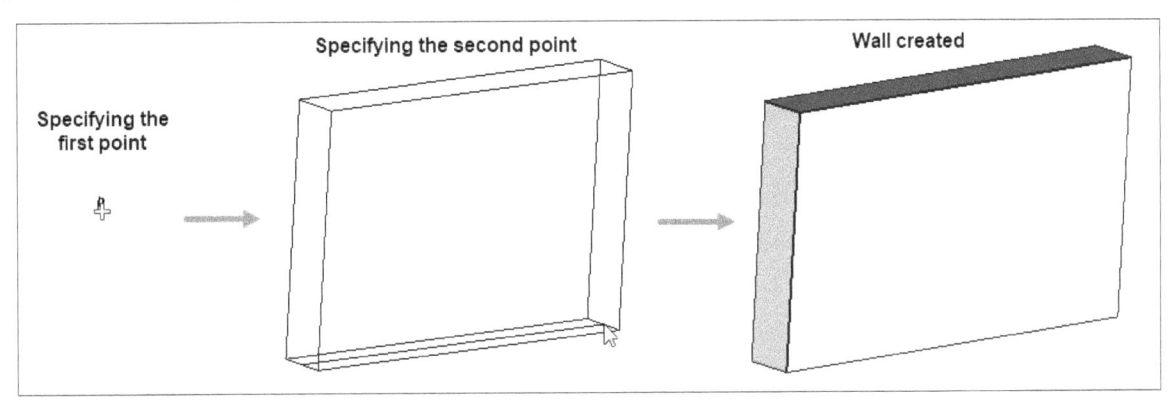

Figure-27. Wall created

- Click on **Close** button to close the dialog.
- If you want to create a wall using base object then select the base geometry objects (Draft object, sketch, etc.) from **Model Tree** or from the 3D view area to create a wall; refer to Figure-28.

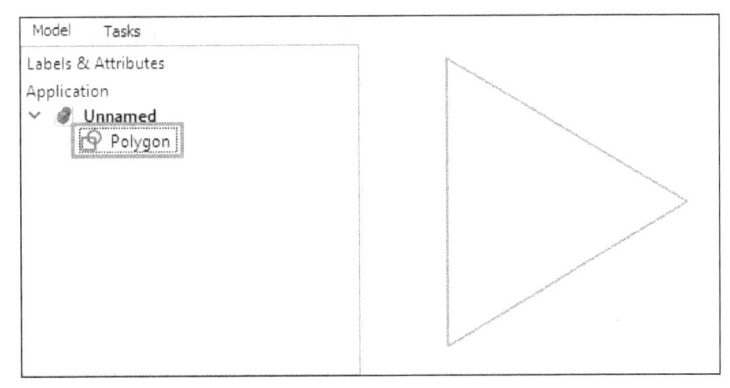

Figure-28. Selecting base object

- Click on the **Wall** tool as discussed earlier. The wall will be created; refer to Figure-29.

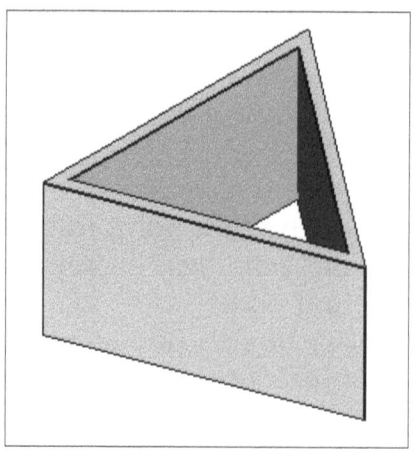

Figure-29. Wall created

- If you want to edit the properties of wall then select the wall created from the **Model Tree**. The **Property editor** dialog will be displayed in the **Model** panel of **Combo View** with the parameters related to wall; refer to Figure-30.

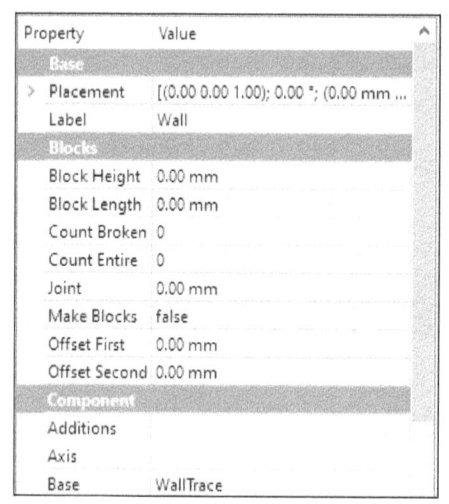

Figure-30. Property editor dialog with wall parameters

- Specify desired height and length of each block in the **Block Height** and **Block Length** edit boxes from **Blocks** section of the dialog, respectively.
- Specify the size of joints between each block in the **Joint** edit box.
- Select desired option from **Make Blocks** drop-down whether to make the wall generate blocks or not.
- Specify the horizontal offset of first line and second line of blocks in the **Offset First** and **Offset Second** edit boxes, respectively.
- Select desired option for the alignment of the wall on its baseline from **Align** drop-down in the **Wall** section of the dialog.
- Enter desired value in the **Face** edit box to specify index of the face from the base object to use. If the value is not set or 0, the whole object is used.
- Specify desired length, width, and height of the wall in the **Length**, **Width**, and **Height** edit boxes, respectively.
- Enter desired value in the **Offset** edit box to specify the distance between the wall and its baseline.
- Enter desired value in the **x**, **y**, and **z** edit boxes from **Normal** cascading menu to specify extrusion for the wall along x, y, and z direction, respectively.

- On specifying parameters in the **Property editor** dialog, the properties of wall will be modified; refer to Figure-31.

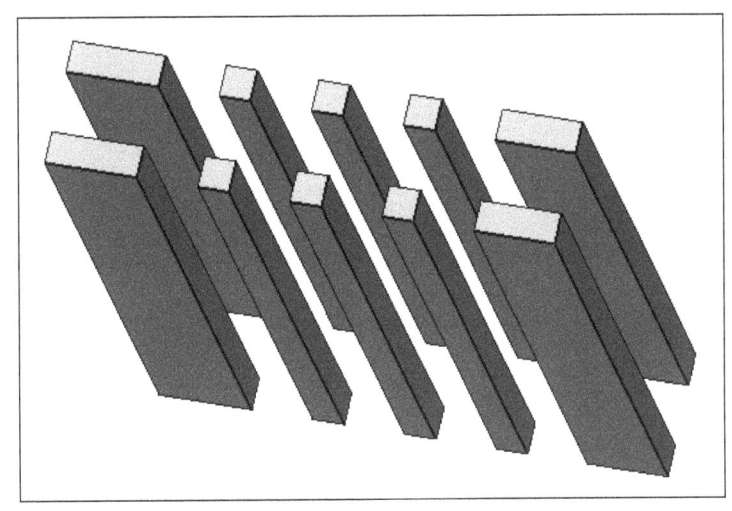

Figure-31. Wall edited

Creating Curtain Wall

The **Curtain Wall** tool creates a curtain wall by subdividing a base face into quadrangular faces, then creating vertical mullions on the vertical edges, horizontal mullions on the horizontal edges, and filling the spaces between mullions with panels. The procedure to use this tool is discussed next.

- Click on the **Curtain Wall** tool from **Toolbar** in the **BIM** workbench; refer to Figure-32. The **Point** dialog will be displayed in the **Tasks** panel of **Combo View** along with the plus sign in place of original cursor; refer to Figure-33. You will be asked to specify the first point.

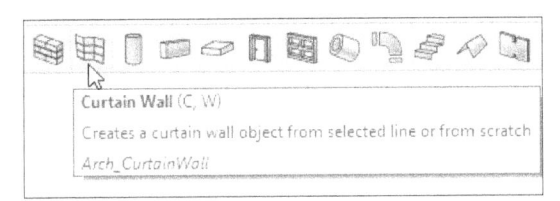

Figure-32. Curtain Wall tool

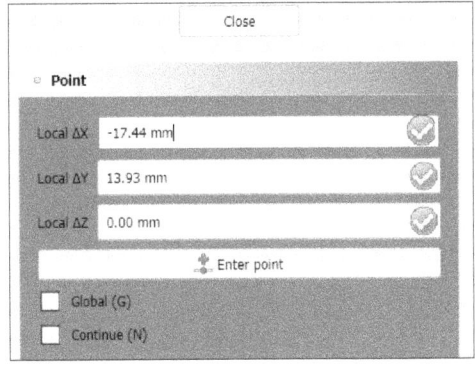

Figure-33. Point dialog

- Click in the 3D view area to specify the first point or enter desired values for the x, y, and z coordinates in the **Local ΔX**, **Local ΔY**, and **Local ΔZ** edit boxes of the **Point** dialog, respectively.
- After specifying coordinates for the first point, click on the **Enter point** button from the dialog. You will be asked to specify the second point.
- Move the cursor away and click at desired location to specify the second point or enter desired values for the coordinates in their respective edit boxes in the dialog.
- After specifying coordinates for the second point, click on the **Enter point** button again from the dialog. The curtain wall will be created; refer to Figure-34.

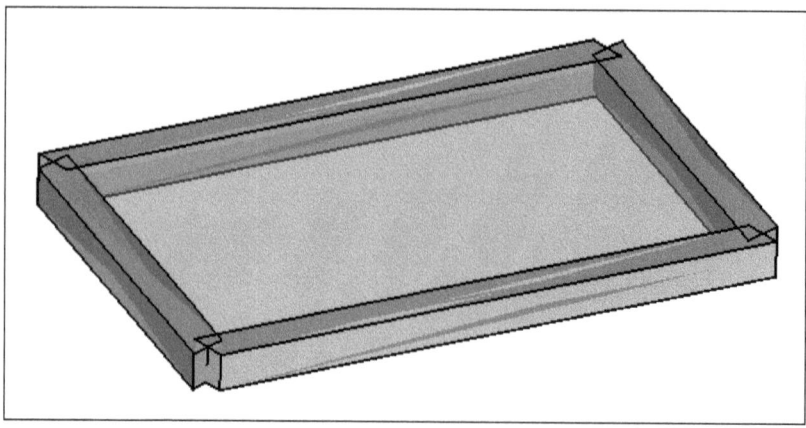

Figure-34. Curtain wall created

- If you want to create the wall from the selected object, sketch, etc. then create and select the object or sketch and click on the **Curtain Wall** tool. The curtain wall will be created.
- If you want to edit the properties of curtain wall then select the curtain wall created from the **Model Tree**. The **Property editor** dialog will be displayed in the **Model** panel of **Combo View** with the parameters related to curtain wall; refer to Figure-35.

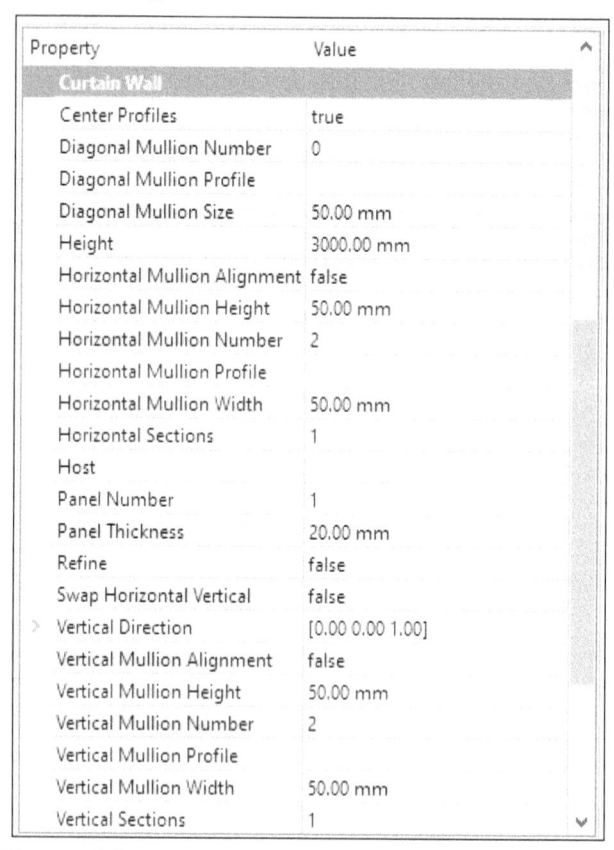

Figure-35. Property editor dialog with curtain wall parameters

- Select desired option from **Center Profiles** drop-down from **Curtain Wall** area of the dialog to specify whether to center the profile over the edges or not.
- Specify desired number of diagonal mullions in the **Diagonal Mullion Number** edit box.
- Click on the **Diagonal Mullion Profile** button to specify the profile for diagonal mullions.

- Specify desired size of the diagonal mullions in the **Diagonal Mullion Size** edit box.
- Specify desired height of the curtain wall in the **Height** edit box.
- Select desired option from **Horizontal Mullion Alignment** drop-down to specify whether the profile of the horizontal mullions gets aligned with the surface or not.
- Specify desired number, height, and width of the horizontal mullions profile in the **Horizontal Mullion Number**, **Horizontal Mullion Height**, and **Horizontal Mullion Width** edit boxes of the dialog, respectively.
- Click on the **Horizontal Mullion Profile** button to specify the profile for horizontal mullions.
- Specify desired number of horizontal sections of this curtain wall in the **Horizontal Sections** edit box.
- Click on the **Host** button to specify the host object for this curtain wall.
- Specify desired number of panels in the **Panel Number** edit box.
- Specify desired thickness of the panels in the **Panel Thickness** edit box.
- Select desired option from **Refine** drop-down to specify whether to perform subtractions between components or not, so none overlap.
- Select desired option from **Swap Horizontal Vertical** drop-down to specify whether to swap horizontal and vertical lines or not.
- Enter desired values in the **x**, **y**, and **z** edit boxes from **Vertical Direction** cascading menu to specify the vertical direction reference to be used by this object to deduce vertical/horizontal directions.
- Select desired option from **Vertical Mullion Alignment** drop-down to specify whether the profile of the vertical mullions get aligned with the surface or not.
- Specify desired number, height, and width of the vertical mullions profile in the **Vertical Mullion Number**, **Vertical Mullion Height**, and **Vertical Mullion Width** edit boxes of the dialog, respectively.
- Click on the **Vertical Mullion Profile** button to specify the profile for vertical mullions.
- Specify desired number of vertical sections of this curtain wall in the **Vertical Sections** edit box.
- On specifying desired parameters in the **Property editor** dialog, the properties of the curtain wall will be modified; refer to Figure-36.

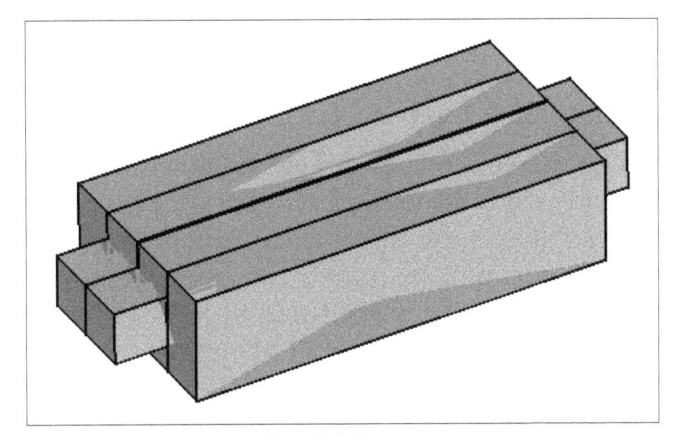

Figure-36. Curtain wall modified

Column

The **Column** tool allows you to build structural elements such as columns or beams, by specifying their width, length, and height, or by basing them on a 2D profile (face, wire, or sketch). The procedure to use this tool is discussed next.

- Click on the **Column** tool from **Toolbar** in the **BIM** workbench; refer to Figure-37. Multiple dialogs will be displayed in the **Tasks** panel of **Combo View** along with the plus ⊕ sign and a structure of column in place of original cursor; refer to Figure-38. You will be asked to specify placement point for structure.

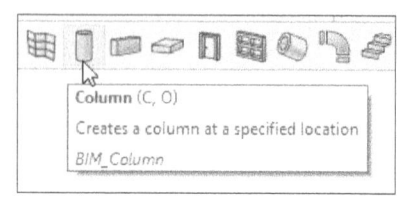

Figure-37. Column tool

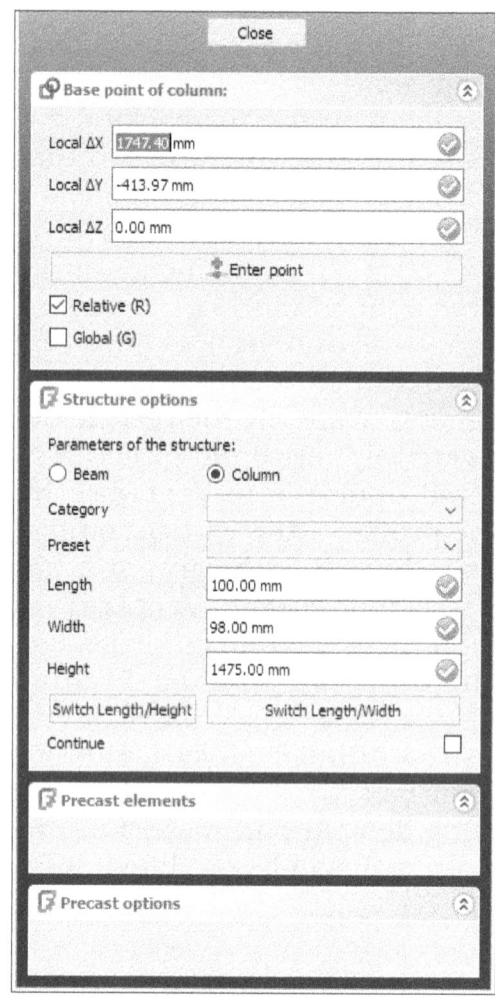

Figure-38. Multiple dialogs displayed

- Select desired drawing mode from **Drawing mode** area of the **Structure options** dialog. In **Beam** mode, you will be asked to specify two points in 3D view area or by entering coordinates. The new structural object will span between these two points. In **Column** mode, you will be asked to specify one point in 3D view area or by entering coordinates. The new structural object will be placed at that point.
- Select desired presets of structure from **Category** drop-down in the **Structure options** dialog.
- Select desired size of the preset from **Preset** drop-down in the **Structure options** dialog.
- On selecting the **Precast concrete** option from **Category** drop-down, the parameters related to the selected structure and preset will be displayed in **Precast elements** dialog and **Precast options** dialog in the **Tasks** panel of **Combo View**; refer to Figure-39 and Figure-40.

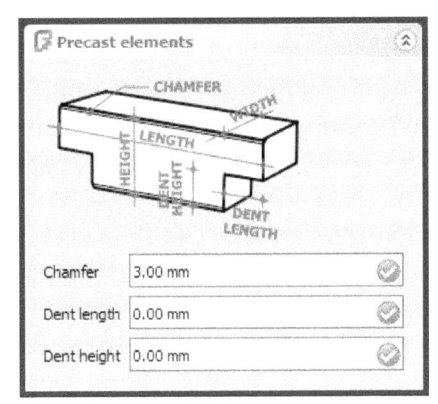

Figure-39. Precast elements dialog

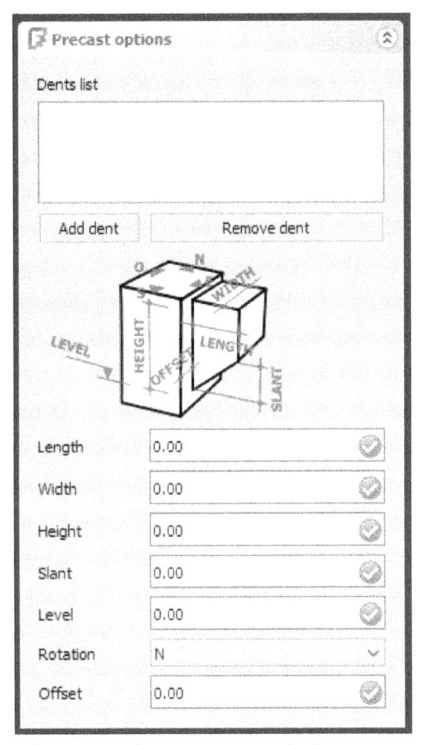

Figure-40. Precast options dialog

- Specify desired parameters in **Precast elements** and **Precast options** dialogs.
- Specify desired length, width, and height of the structure in the **Length**, **Width**, and **Height** edit boxes from **Structure options** dialog, respectively.
- Click on the **Switch Length/Height** or **Switch Length/Width** button to switch between length and height values or length and width values, respectively.
- After specifying parameters in all the dialogs, click in the 3D view area to specify the point for creating the structure or enter desired values for the x, y, and z coordinates in the **Local ΔX**, **Local ΔY**, and **Local ΔZ** edit boxes of the **Base point of column** dialog, respectively.
- After specifying coordinates for the point in dialog, click on the **Enter point** button from **Base point of column** dialog. The structure object will be created; refer to Figure-41.

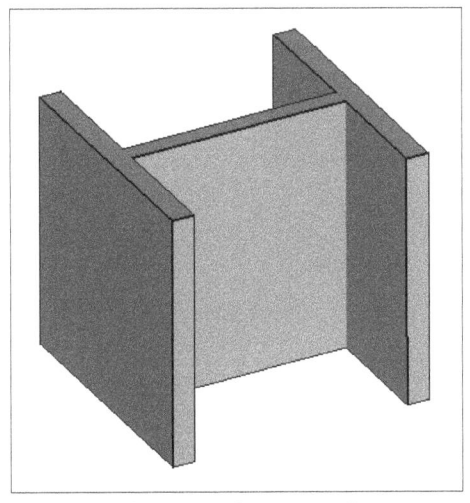

Figure-41. Structure object created

- Click on **Close** button to close the dialog.

- The procedure for creating a structure using base object is same as discussed for previous tool.
- If you want to edit the properties of structure object then select the structure object created from the **Model Tree**. The **Property editor** dialog will be displayed in the **Model** panel of **Combo View** with the parameters related to structure object; refer to Figure-42.

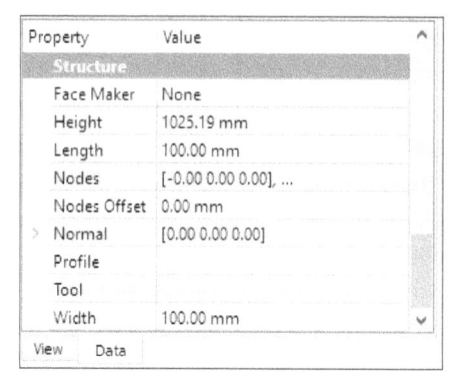

Figure-42. Property editor dialog with structure parameters

- Select desired option from **Face Maker** drop-down in the **Structure** section of the dialog to specify the type of face generation algorithm to be used for building the profile.
- Specify desired height, length, and width of the structure in the **Height**, **Length**, and **Width** edit boxes, respectively.
- Specify desired structural nodes of this element in the **Vectors** dialog box displayed by clicking on the **Nodes** button.
- Enter desired value in the **Nodes Offset** edit box to specify the offset between the centerline and the nodes line.
- Enter desired values in x, y, and z edit boxes of **Normal** cascading menu to specify the extrusion of base face of this structure along x, y, and z directions, respectively.
- Click on the **Tool** button to select an optional extrusion path which can be any type of wire.
- On specifying the parameters in the **Property editor** dialog, the properties of structure object will be modified; refer to Figure-43.

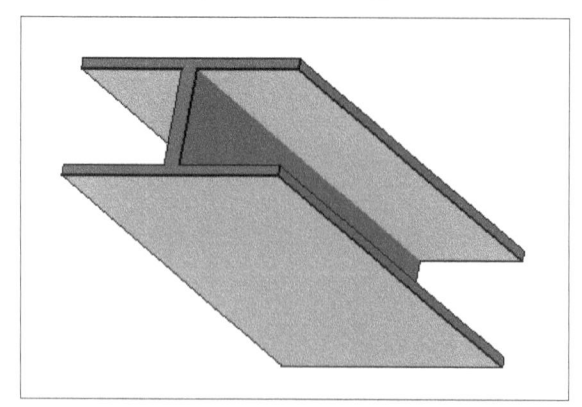

Figure-43. Structure object edited

Similarly, you can use the **Beam** tool from toolbar to create beams.

Creating Structural System

The **Structural System** tool is used to create a structural system object from a selected structure and axis. The procedure to use this tool is discussed next.

- Select a structure and an axis from which you want to create a structural system; refer to Figure-44.

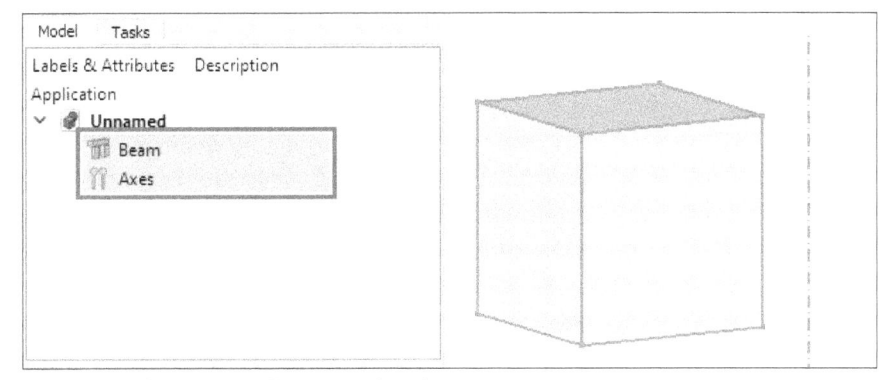

Figure-44. Structure and an axis selected

- Click on the **Structural System** tool from **Structure tools** cascading menu in the **Utils** menu of **BIM** workbench; refer to Figure-45. The structural system will be created; refer to Figure-46.

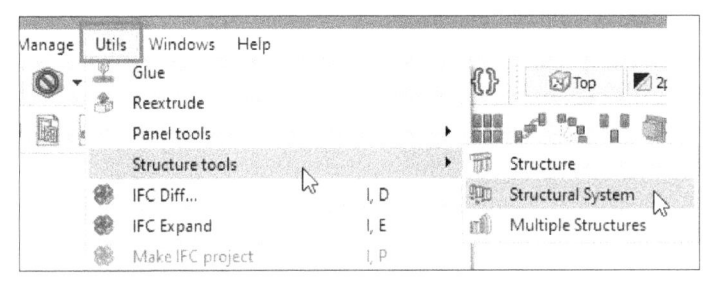

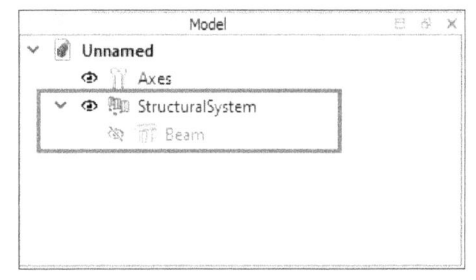

Figure-45. Structural System tool

Figure-46. Structural system created

Creating Multiple Structures

The **Multiple Structures** tool is used to create multiple structure objects from a selected base using each selected edge as an extrusion path. The procedure to use this tool is discussed next.

- First, select the base object and then select an edge as an extrusion path; refer to Figure-47.

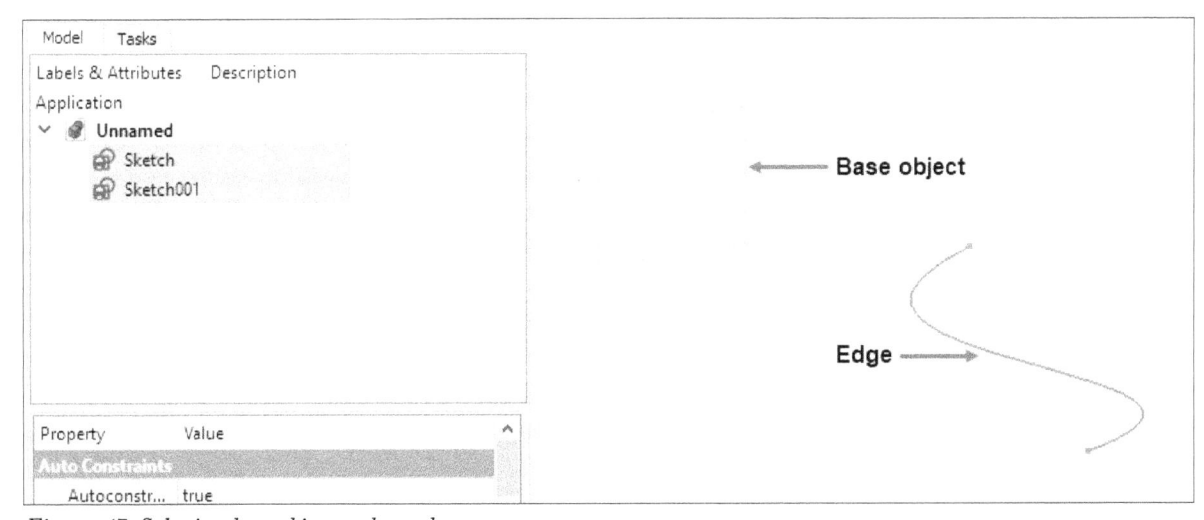

Figure-47. Selecting base object and an edge

- Click on the **Multiple Structures** tool from **Structure tools** cascading menu in the **Utils** menu of **BIM** workbench; refer to Figure-48. The structure object will be created; refer to Figure-49.

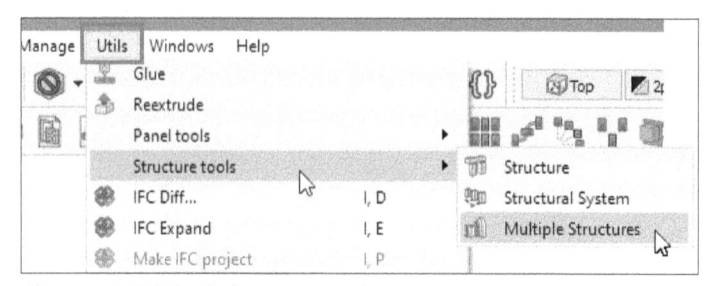

Figure-48. Multiple Structures tool

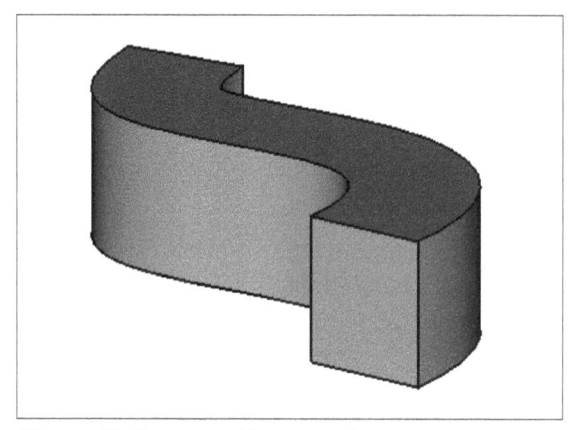

Figure-49. Structure object created

Slab

The **Slab** tool is a create a horizontal structure from a planar shape. The procedure to use this tool is discussed next.

- Select desired planar shape in the 3D view area or from the **Model Tree** from which you want to create a slab; refer to Figure-50.

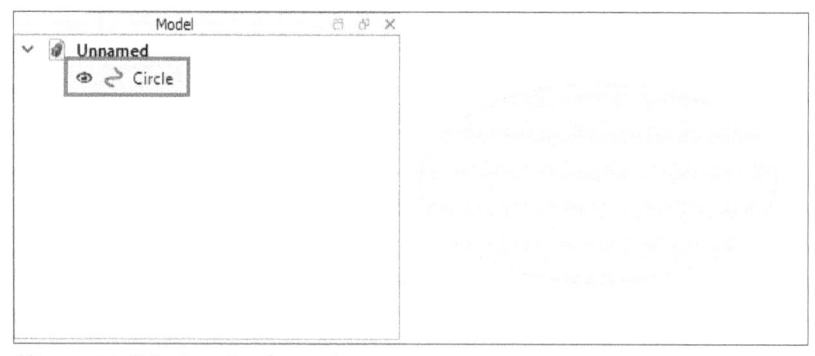

Figure-50. Selecting the planar shape

- Click on the **Slab** tool from **Toolbar** in the **BIM** workbench; refer to Figure-51. The slab will be created; refer to Figure-52.

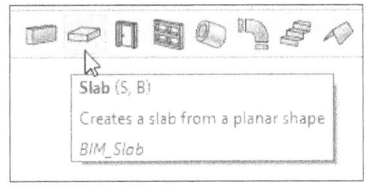

Figure-51. Slab tool

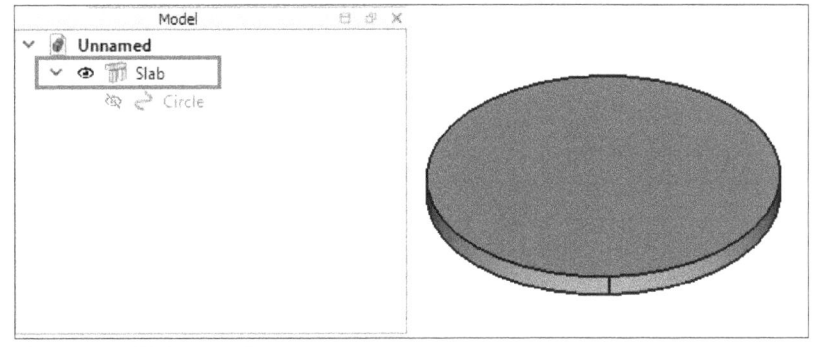

Figure-52. Slab created

Creating Window

A **Window** is a base object for all kinds of embed-able objects such as windows and doors. The **Window** tool features several presets; this allows the user to create common types of windows and doors with certain edit-able parameters without the need for the user to create the base 2D objects and components manually. The procedure to use this tool is discussed next.

- Click on the **Window** tool from **Toolbar** in the **BIM** workbench; refer to Figure-53. The **Point** and **Window options** dialogs will be displayed in the **Tasks** panel of **Combo View** along with the plus ⊞ sign and structure of window in place of original cursor; refer to Figure-54.

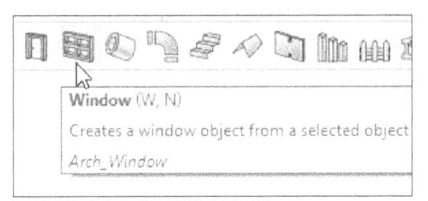

Figure-53. Window tool

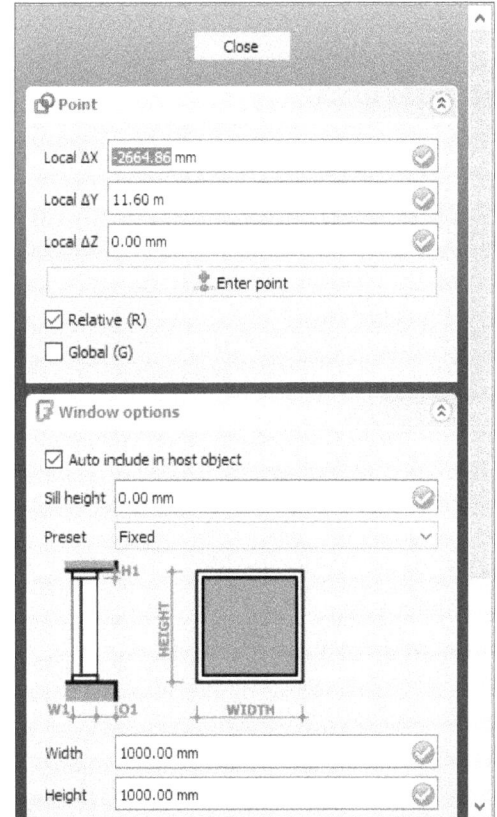

Figure-54. Point and Window options dialogs

- Select **Auto include in host object** check box to insert the window into host object on creation from **Windows options** dialog.
- Specify the height of sill for the window in the **Sill height** edit box of the dialog.

- Select desired preset for the window or door from **Preset** drop-down in the dialog.
- Specify desired width and height of the window in the **Width** and **Height** edit boxes of the dialog.
- After specifying all the parameters in the dialog, you will be asked to specify the location of point.
- Click in the 3D view area to specify the point or enter desired values for the x, y, and z coordinates in the **Local ΔX**, **Local ΔY**, and **Local ΔZ** edit boxes of the **Point** dialog, respectively.
- After specifying coordinates for the point in dialog, click on the **Enter point** button from the **Point** dialog. The window will be created; refer to Figure-55.

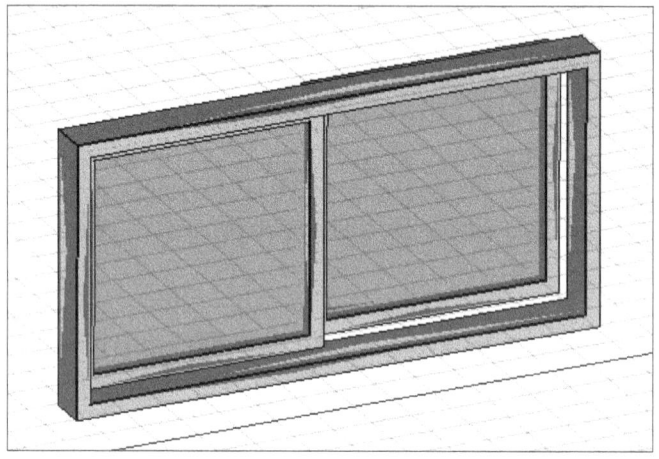

Figure-55. Window created

- If you want to create, modify, or delete the components of a window then double click on the window created from the **Model Tree**. The **Window elements** and **Component** dialog will be displayed in the **Tasks** panel of **Combo View**; refer to Figure-56.

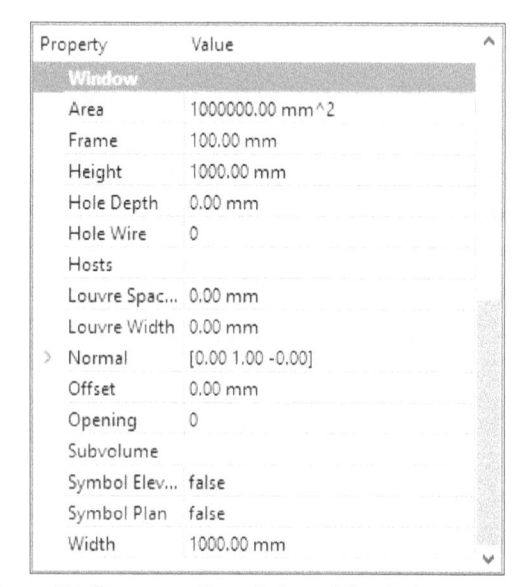

Figure-57. Property editor dialog with window parameters

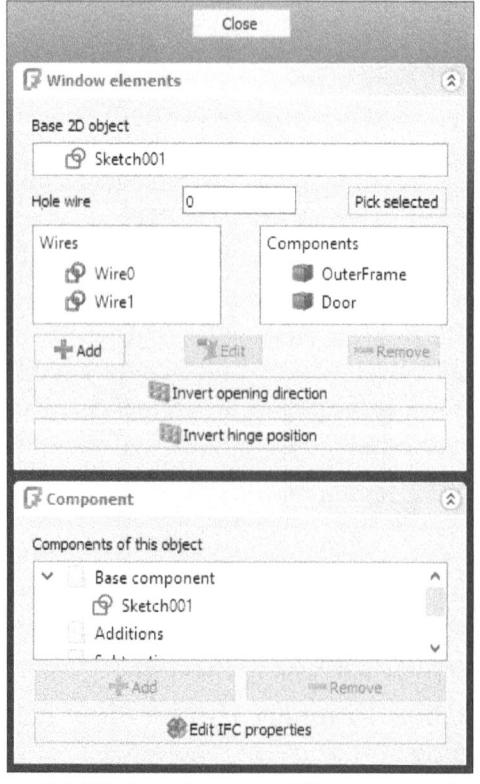

Figure-56. Window elements and Component dialogs

- If you want to edit the properties of window then select the window created from **Model Tree**. The **Property editor** dialog will be displayed in the **Model** panel of **Combo View** with parameters related to window; refer to Figure-57.
- Specify desired height and width of the window in the **Height** and **Width** edit boxes from **Window** section of the dialog, respectively.
- Specify desired depth of the hole created by this window in its host object in the **Hole Depth** edit box.
- Enter desired value in **Host Wire** edit box to specify the number of wire from the base object that is used to create a hole in the host object of this window.
- Enter desired value in **Louvre Spacing** edit box to define the spacing between louvre elements.
- Enter desired value in **Louvre Width** edit box to define the size of the louvre elements.
- Enter desired value in **Opening** edit box to define opening mode of all the components and to define a hinge provided in them or in an earlier component in the list.
- Select desired value from **Symbol Elevation** drop-down to specify whether to show 2D opening symbol in elevation or not.
- Select desired value from **Symbol Plan** drop-down whether to show 2D opening symbol in plan or not.
- On specifying desired value in the **Property editor** dialog, the properties of window will be modified; refer to Figure-58.

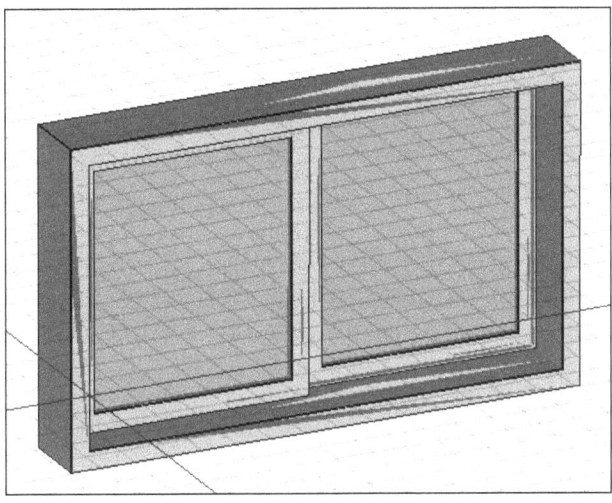

Figure-58. Window edited

Door

The **Door** tool is used to create a door at a given location. The procedure to use this tool is same as discussed for **Window** tool.

Creating Pipe

The **Pipe** tool allows you to create pipes from scratch or from selected objects. You can use geometry tools in Draft workbench to create base wire for pipes. The procedure to use this tool is discussed next.

- Select a linear line, wire, or an open sketch from the **Model Tree** or from the 3D view area with which you want to create a pipe; refer to Figure-59.

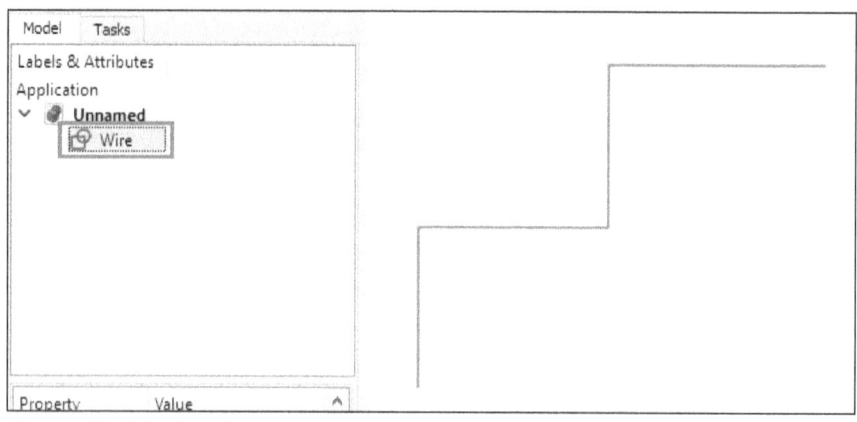

Figure-59. Selecting the wire to create pipe

- Click on the **Pipe** tool from **Toolbar** in the **BIM** workbench; refer to Figure-60. The pipe will be created; refer to Figure-61.

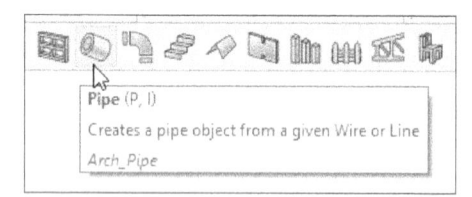

Figure-60. Pipe tool

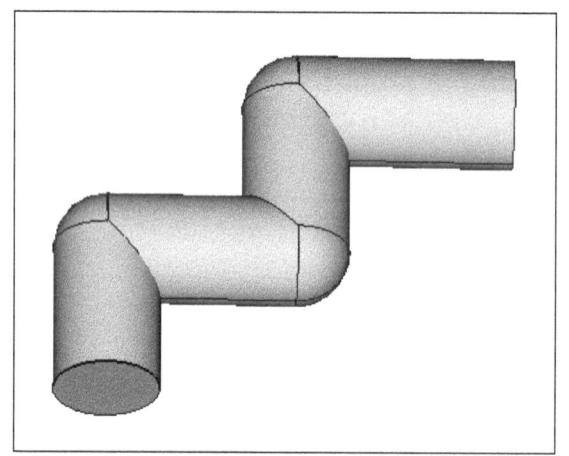

Figure-61. Pipe created

- If you want to edit the properties of pipe then select the pipe created from **Model Tree**. The **Property editor** dialog will be displayed in the **Model** panel of **Combo View** with parameters related to pipe; refer to Figure-62.

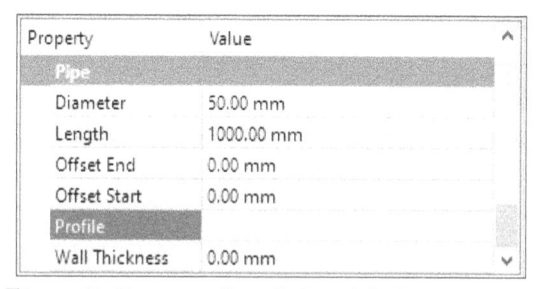

Figure-62. Property editor dialog with pipe parameters

- Specify desired diameter and length of pipe in the **Diameter** and **Length** edit boxes of the dialog, respectively.
- Specify desired offset from the end point and offset from the start point in the **Offset End** and **Offset Start** edit boxes, respectively.
- The **Profile** edit box specifies the base profile of this pipe. If profile is not specified then the pipe will be cylindrical.
- Specify desired wall thickness of this pipe in the **Wall Thickness** edit box.
- On specifying parameters in the **Property editor** dialog, the properties of pipe will be modified; refer to Figure-63.

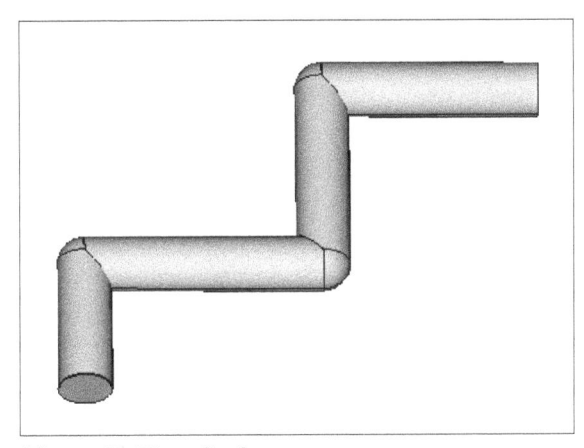

Figure-63. Pipe edited

Creating Connector

The **Connector** tool allows you to create corner or tee connection between two or three selected pipes. The procedure to use this tool is discussed next.

- Select two or three pipes joined at common vertex from the **Model Tree** or from the 3D view area.
- Click on the **Connector** tool from **Toolbar** in the **BIM** workbench; refer to Figure-64. The pipe connector will be created; refer to Figure-65.

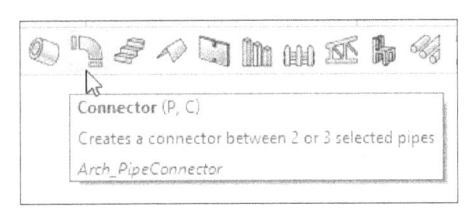

Figure-64. Connector tool

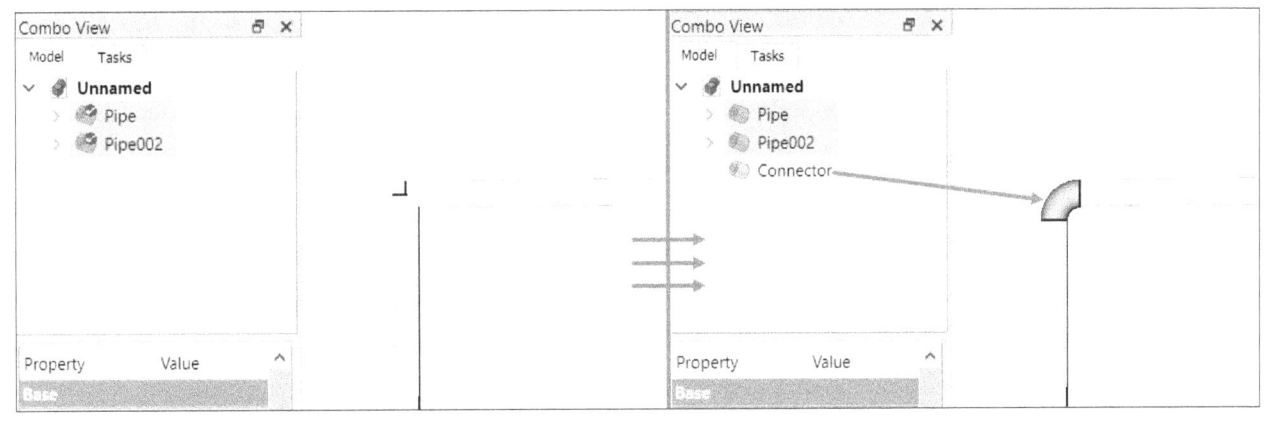

Figure-65. Pipe connector created

- You can edit the curvature radius of connector in the **Radius** edit box from **Pipe Connector** section of **Property editor** dialog.

Creating Stairs

The **Stairs** tool allows you to automatically build several types of stairs. Stairs can be built from scratch or from a straight line in which case, the stairs follow the line. If the line is not horizontal but has a vertical inclination, the stairs will also follow its slope. The procedure to use this tool is discussed next.

- Click on the **Stairs** tool from **Toolbar** in the **BIM** workbench; refer to Figure-66. The stairs will be created; refer to Figure-67.

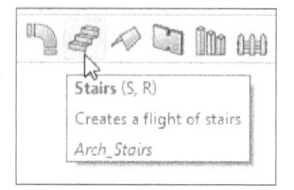

Figure-66. Stairs tool

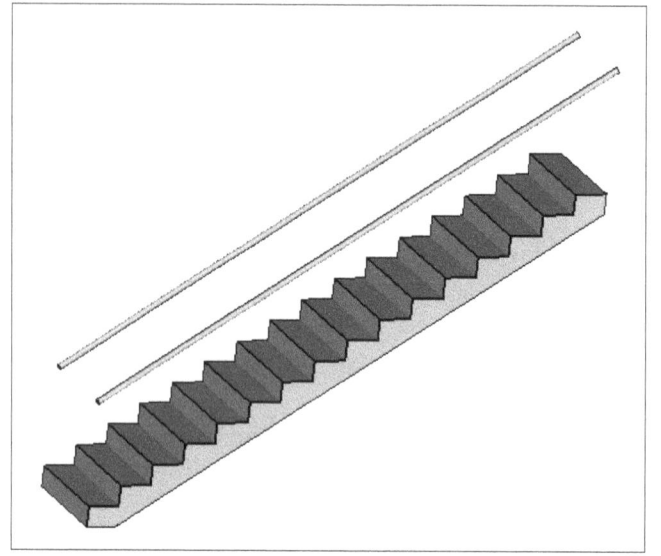

Figure-67. Stairs created

- If you want to edit the properties of stairs then select the stairs created from the **Model Tree**. The **Property editor** dialog will be displayed in the **Model** panel of **Combo View** with the parameters related to stairs; refer to Figure-68.

Figure-68. Property editor dialog with stairs parameters

- Select desired option from **Align** drop-down in the **Stairs** section of dialog to specify the alignment of these stairs on their baseline, if applicable.
- Specify desired height, length, and width of the stairs in the **Height**, **Length**, and **Width** edit boxes of the dialog, respectively.
- Specify desired size of nosing in the **Nosing** edit box from **Steps** section of the dialog.
- Specify number of steps (risers) in these stairs in the **Number of Steps** edit box.
- Specify desired height of the risers in the **Riser Height** edit box of the dialog.
- Specify depth of the treads in the **Tread Depth** edit box of the dialog.
- Specify thickness of the treads in the **Tread Thickness** edit box.

- Select desired type of landings from **Landings** drop-down in the **Structure** section of the dialog.
- Specify desired offset between the border of the stairs and the structure in the **Stringer Offset** edit box.
- Specify width of the stringers in the **Stringer Width** edit box of the dialog.
- Select desired type of structure of these stairs from **Structure** drop-down.
- Specify desired thickness of the structure from **Structure Thickness** edit box.
- Select desired type of winders from **Winders** drop-down in the dialog.
- On specifying the parameters in **Property editor** dialog. The properties of stairs will be modified; refer to Figure-69.

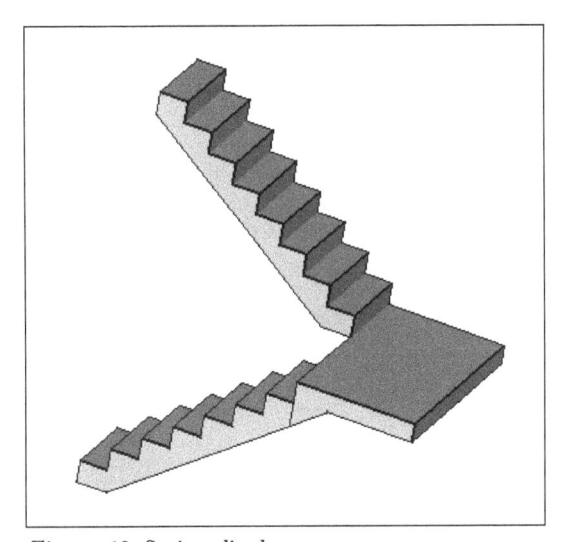

Figure-69. Stairs edited

Creating Roof

The **Roof** tool is used to create a sloped roof using selected wire. The created roof object is parametric keeping its relationship with the base object. The procedure to use this tool is discussed next.

- Select the wire from the **Model Tree** or from the 3D view area on which you want to create the roof; refer to Figure-70.

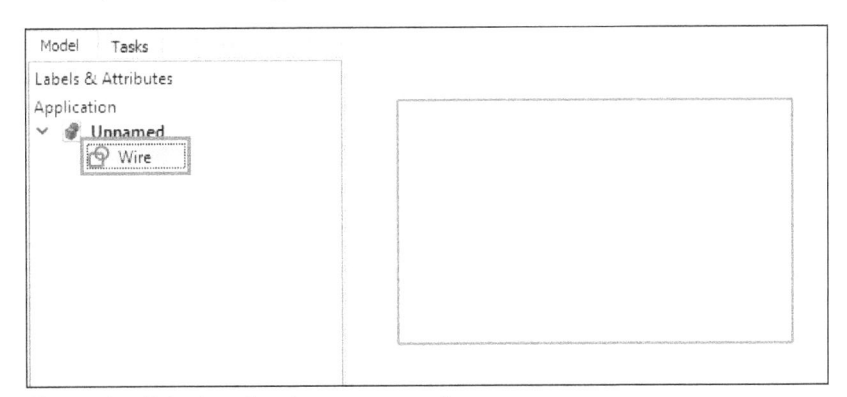

Figure-70. Selecting the wire to create roof

- Click on the **Roof** tool from **Toolbar** in the **BIM** workbench; refer to Figure-71. The default roof will be created; refer to Figure-72.

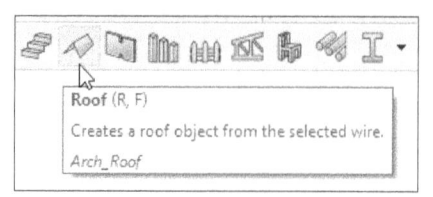

Figure-71. Roof tool

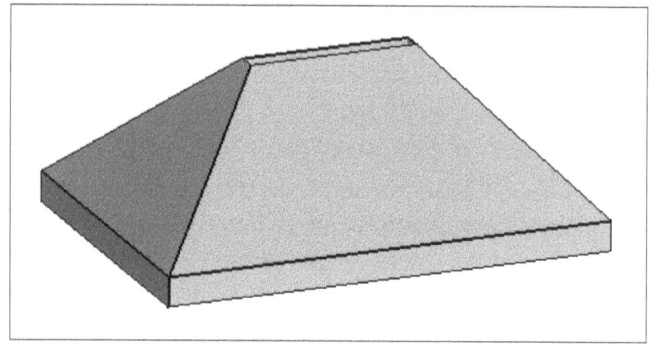

Figure-72. Default roof created

- If you want to edit the roof then select the roof created from the **Model Tree** and double-click on it. The **Roof** dialog will be displayed in the **Tasks** panel of **Combo View**; refer to Figure-73.

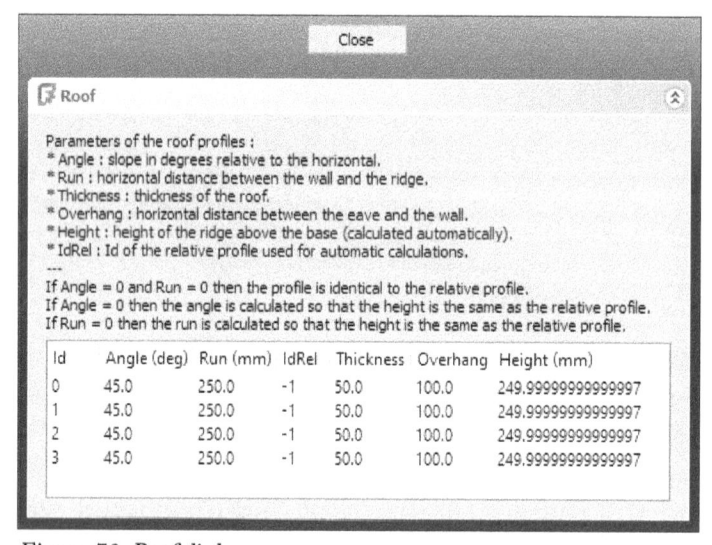

Figure-73. Roof dialog

- Specify desired values in the **Angle**, **Run**, **IdRel**, **Thickness**, **Overhang**, and **Height** edit boxes of the dialog. The roof will be edited.
- If you want to edit the properties of roof then select the roof created from the **Model Tree**. The **Property editor** dialog will be displayed in the **Model** panel of **Combo View** with the parameters related to roof object; refer to Figure-74.

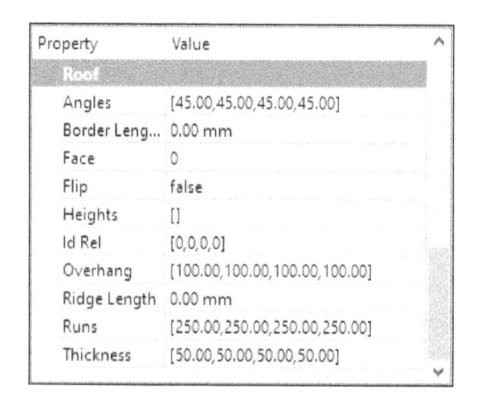

Figure-74. Property editor dialog with roof parameters

- Enter desired values in the **Angles** edit box to specify slope angle of the roof pane (an angle for each edge in the wire).

- Enter desired value in the **Face** edit box to specify face index of the base object to be used.
- Enter desired values in **Id Rel** edit box to specify relation Id of the slope angle of the roof.
- Enter desired values in **Overhang** edit box to specify overhang of the roof pane (an overhang for each edge in the wire).
- Enter desired values in **Runs** edit box to specify width of the roof pane (a run for each edge in the wire).
- Enter desired values in **Thickness** edit box to specify thickness of the roof pane (a thickness for each edge in the wire).
- On specifying parameters in the **Property editor** dialog, the properties of roof object will be modified.

Creating Panel

The **Panel** tool allows you to build all kinds of panel-like elements, typically for panel constructions but also for all kinds of objects that are based on a flat profile. The procedure to use this tool is discussed next.

- Click on the **Panel** tool from **Toolbar** in the **BIM** workbench or from the **Panel tools** cascading menu in the **Utils** menu; refer to Figure-75. The **Point** and **Panel options** dialog will be displayed in the **Tasks** panel of **Combo View** along with the plus ⊕ sign and a structure of panel in place of original cursor; refer to Figure-76. You will be asked to specify the location of panel.

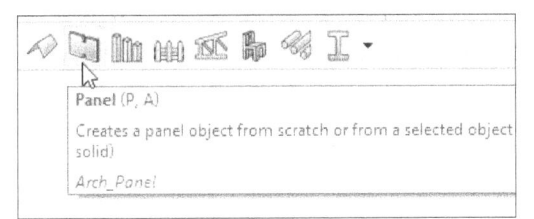

Figure-75. Panel tool

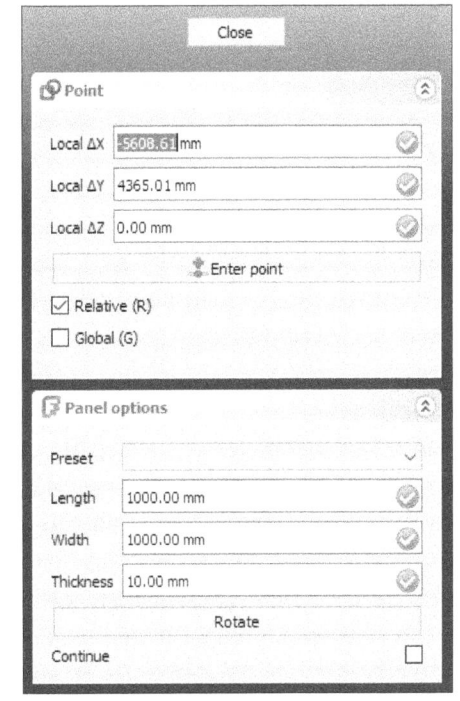

Figure-76. Point and Panel options dialogs

- Select desired preset of panel from **Preset** drop-down in the **Panel options** dialog.
- Specify desired length, width, and thickness of panel in the **Length**, **Width**, and **Thickness** edit boxes of the dialog, respectively.
- Click on **Rotate** button from the dialog to rotate the panel at desired location.
- After specifying parameters in the **Panel options** dialog, click in the 3D view area to specify the location of panel or enter desired values for x, y, and z coordinates in the **Local ΔX**, **Local ΔY**, and **Local ΔZ** edit boxes of the **Point** dialog, respectively.

- After specifying the coordinates in dialog, click on **Enter point** button. The panel will be created; refer to Figure-77.

Figure-77. Panel created

- If you want to edit the properties of panel then select the panel created from the **Model Tree**. The **Property editor** dialog will be displayed in the **Model** panel of **Combo View** with the parameters related to panel; refer to Figure-78.

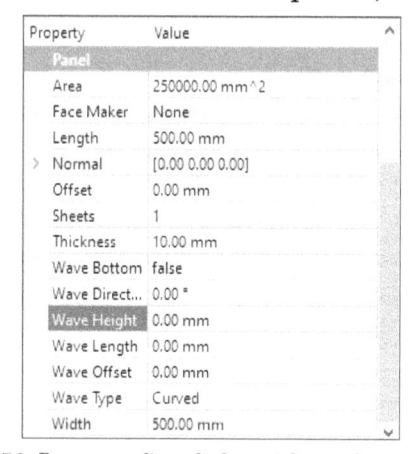

Figure-78. Property editor dialog with panel parameters

- Specify desired length, width, thickness, and area of the panel in the **Length**, **Width**, **Thickness**, and **Area** edit boxes of the dialog, respectively.
- Specify the number of sheets of material of which the panel is made in the **Sheets** edit box.
- Specify desired length, height, and orientation of the waves for corrugated panels in the **Wave Length**, **Wave Height**, and **Wave Direction** edit boxes of the dialog, respectively.
- Select desired type of the wave for corrugated panel from **Wave Type** drop-down in the dialog.
- Select desired option from **Wave Bottom** drop-down to specify whether the bottom wave of the panel is flat or not.
- On specifying the parameters in the **Property editor** dialog, the properties of panel will be modified; refer to Figure-79.

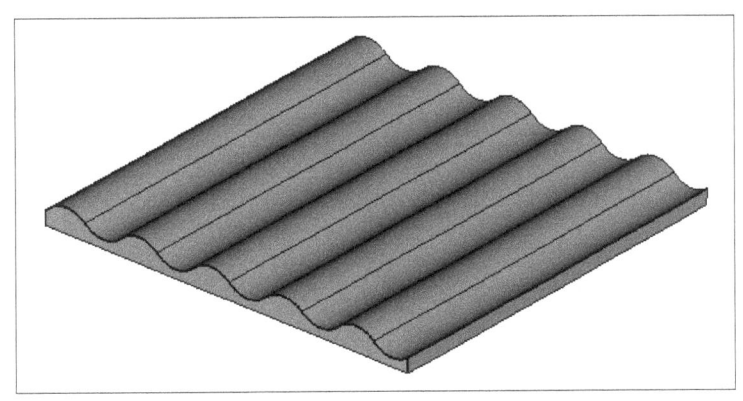

Figure-79. Panel edited

Creating Panel Cut

The **Panel Cut** tool creates a flat 2D view of a panel to be included in the 3D document of a panel sheet or directly exported to DXF. The procedure to use this tool is discussed next.

- Select one or more panel objects from **Model Tree** or from the 3D view area; refer to Figure-80.

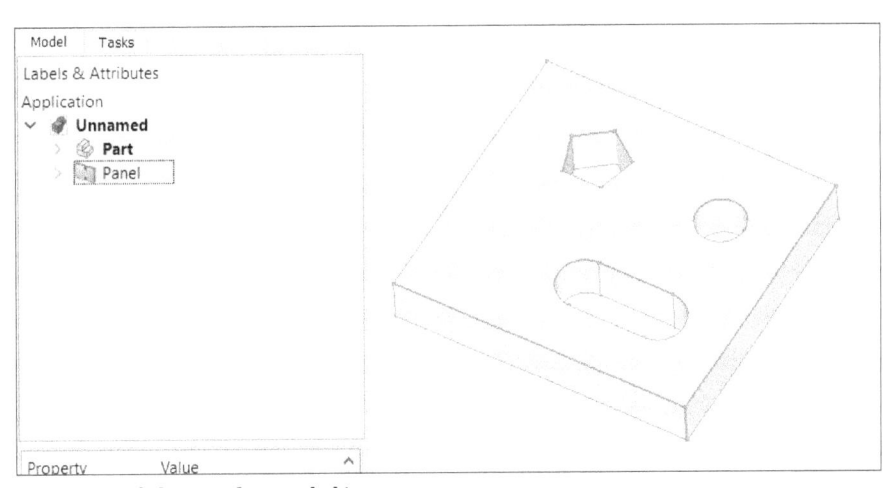

Figure-80. Selecting the panel object

- Click on the **Panel Cut** tool from **Panel tools** cascading menu in the **Utils** menu of **BIM** workbench; refer to Figure-81. The panel cut object will be created; refer to Figure-82.

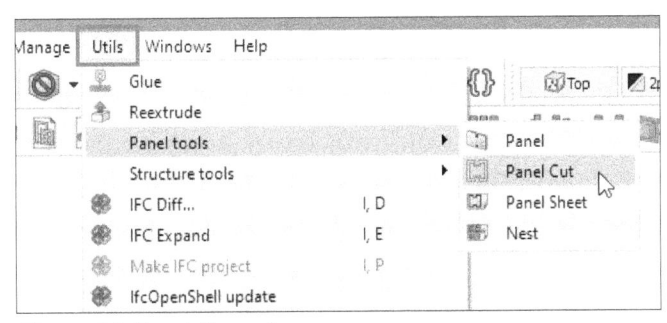

Figure-81. Panel Cut tool

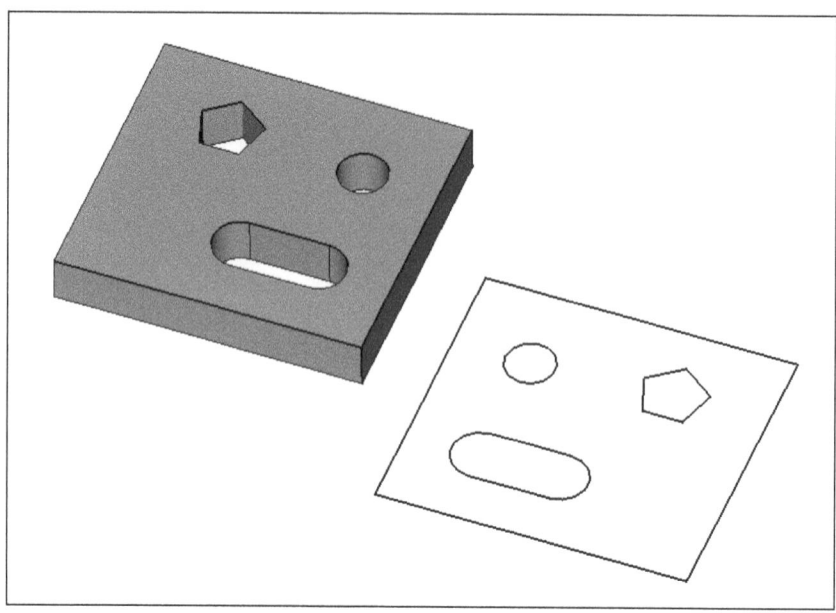

Figure-82. Panel cut object created

- If you want to edit the properties of panel cut then select the panel cut created from the **Model Tree**. The **Property editor** dialog will be displayed in the **Model** panel of **Combo View** with the parameters related to panel cut; refer to Figure-83.

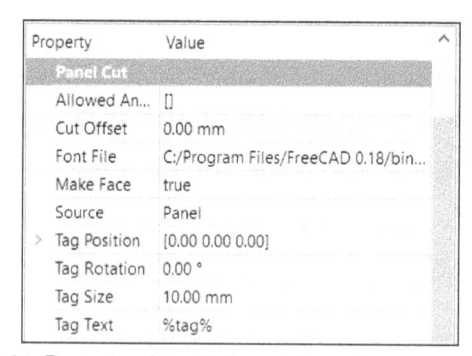

Figure-83. Property editor dialog with panel cut parameters

- Select desired font file to specify font of the tag text from **Font File** edit box in the **Panel Cut** section of the dialog.
- Select desired option from **Make Face** drop-down to specify whether the panel cut will be a face or a wire.
- Enter desired values in **x**, **y**, and **z** edit boxes from **Tag Position** cascading menu to specify the position of tag text along x, y, and z directions, respectively.
- Enter desired value in **Tag Rotation** edit box to specify the rotation of tag text.
- Specify desired size of the tag text in the **Text Size** edit box.
- Specify desired text to display in the **Tag Text** edit box of the dialog.
- Enter desired value in **Margin** edit box from **Arch** section to specify the margin that can be displayed outside the panel cut shape.
- Select desired option from **Show Margin** drop-down to specify whether to turn the display of the margin on/off.
- On specifying parameters in the **Property editor** dialog, the properties of panel cut will be modified.

Creating Panel Sheet

The **Panel Sheet** tool allows you to build a 2D sheet, including any number of panel cut objects or any other 2D object such as those made by the **Draft** workbench and **Sketcher** workbench. The Panel Sheet is typically made to layout cuts to be made by a CNC machine. These sheets can then be exported to a DXF file. The procedure to use this tool is discussed next.

- Select one or more panel cut objects or any other 2D objects from the **Model Tree** or from the 3D view area from which you want to create a panel sheet; refer to Figure-84.

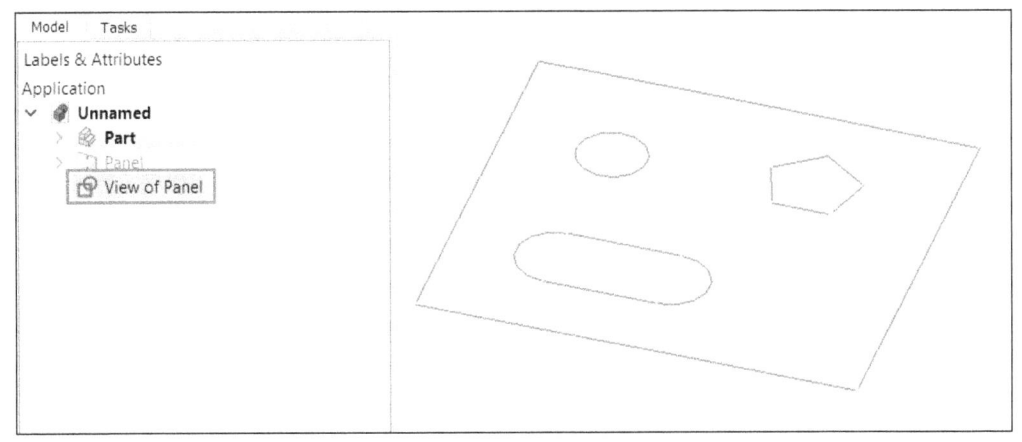

Figure-84. Selecting the panel cut object

- Click on the **Panel Sheet** tool from **Panel tools** cascading menu in the **Utils** menu of **BIM** workbench; refer to Figure-85. A panel sheet of the object will be created; refer to Figure-86.

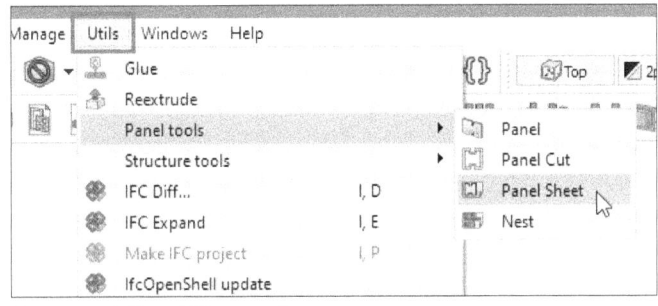

Figure-85. Panel Sheet tool

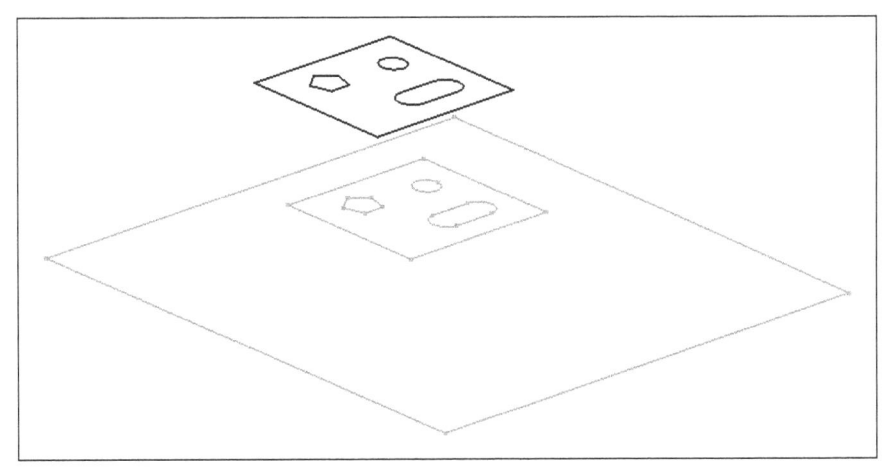

Figure-86. Panel sheet created

- If you want to edit the properties of panel sheet then select the panel sheet created from the **Model Tree**. The **Property editor** dialog will be displayed in the **Model** panel of **Combo View** with parameters related to panel sheet; refer to Figure-87.

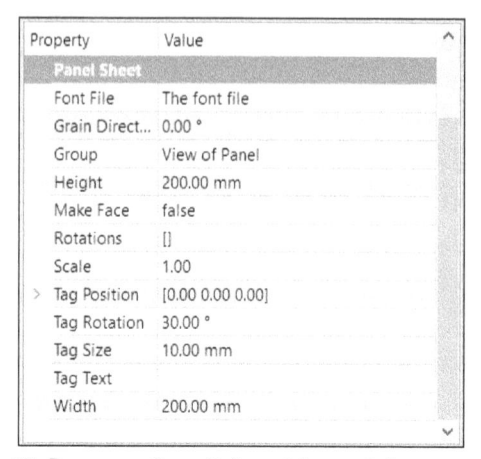

Figure-87. Property editor dialog with panel sheet parameters

- Select desired font file to specify font of the tag text from **Font File** edit box in the **Panel Sheet** section of the dialog.
- Enter desired value in **Grain direction** edit box which allows you to inform the main direction of the panel fiber.
- Specify desired height and width of the sheet in the **Height** and **Width** edit boxes of the dialog, respectively.
- Select desired option from **Make Face** drop-down to specify whether the panel cut will be a face or a wire.
- Enter desired value in **Scale** edit box to specify the scale applied to each panel view.
- Enter desired values in **x**, **y**, and **z** edit boxes from **Tag Position** cascading menu to specify the position of tag text along x, y, and z directions, respectively.
- Enter desired value in **Tag Rotation** edit box to specify the rotation of tag text.
- Specify desired size of the tag text in the **Text Size** edit box.
- Specify desired text to display in the **Tag Text** edit box of the dialog.
- On specifying parameters in the **Property editor** dialog, the properties of panel sheet will be modified.

Creating Nest

The **Nest** tool allows you to select a flat shape to be a container and a series of other flat shapes to be organized inside the space defined by the container shape. The procedure to use this tool is discussed next.

- Click on the **Nest** tool from **Panel tools** cascading menu in the **Utils** menu of **BIM** workbench; refer to Figure-88. The **Nesting** dialog will be displayed in the **Tasks** panel of **Combo View**; refer to Figure-89.

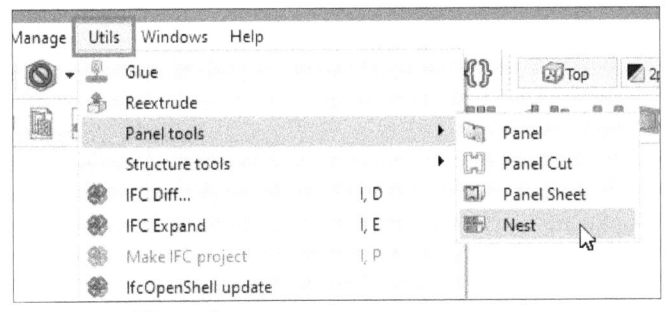

Figure-88. Nest tool

Figure-89. Nesting dialog

- Select a flat, rectangular object from **Model Tree** or from the 3D view area which is to be container; refer to Figure-90.

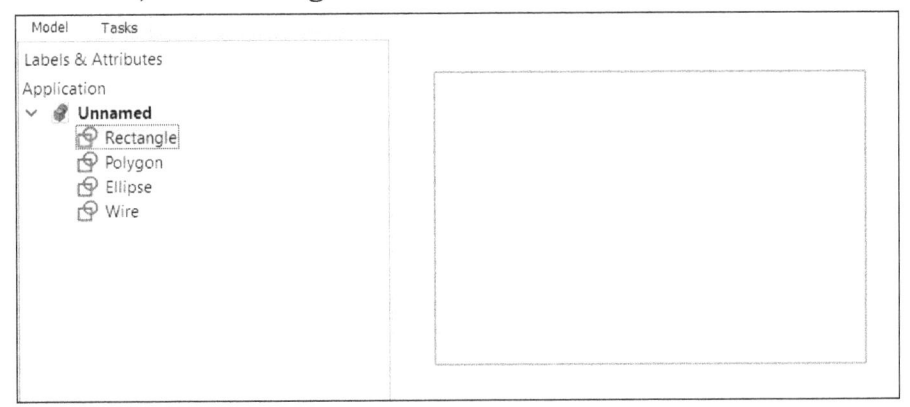

Figure-90. Selecting the rectangular object

- Click on **Pick selected** button from **Container** area of the dialog. The selected object will be displayed in the **Container** selection box.
- Select the other flat objects by holding the **CTRL** key that you want to place inside the container from **Model Tree** or from the 3D view area. Note that these objects must all be flat and in the same plane as the container.
- After selecting the objects, click on **Add selected** button to add it in the **Shapes** list box of the dialog.
- Specify desired tolerance value in the **Tolerance** edit box from **Nesting parameters** area.
- Enter desired value in **Arcs subdivisions** edit box to specify the number of segments to divide non-linear edges for calculations.
- Enter desired values in **Rotations** edit box to specify the angles to rotate the shapes.
- After specifying the parameters, click on **Start** button from **Nesting operation** area to start the calculation process.
- If you want to stop the calculation process then click on the **Stop** button.

- At the end of calculation process, click on the **Preview** button to create a temporary preview of the result.
- Click on **OK** button from the dialog to apply the result. The nest of the objects will be created; refer to Figure-91.

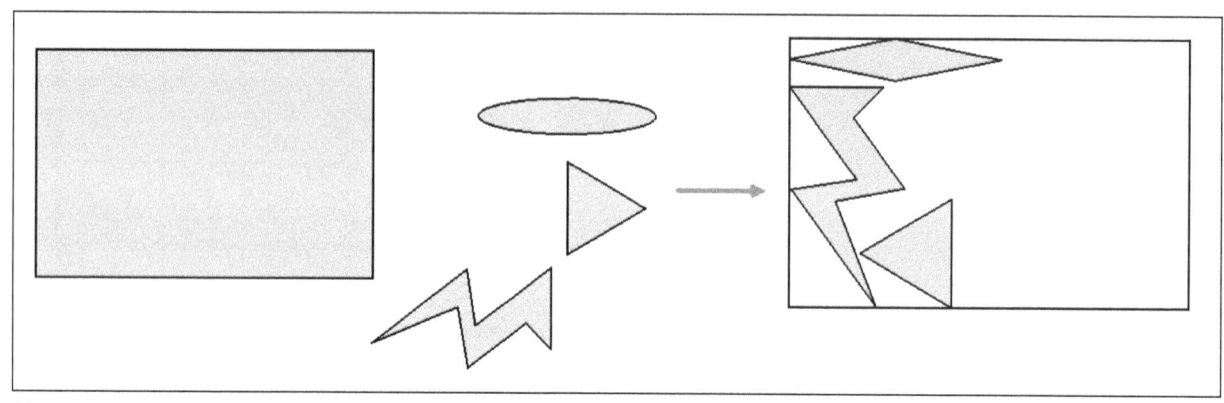

Figure-91. Nest created

Creating Frame

The **Frame** tool is used to build all kinds of frame objects based on a profile and a layout. The profile is extruded along the edges of the layout which can be any 2D object such as a sketch or a draft object. It is especially useful to create railings or frame walls. The procedure to use this tool is discussed next.

- First, create a layout object and a profile object from which you want to create a frame; refer to Figure-92.

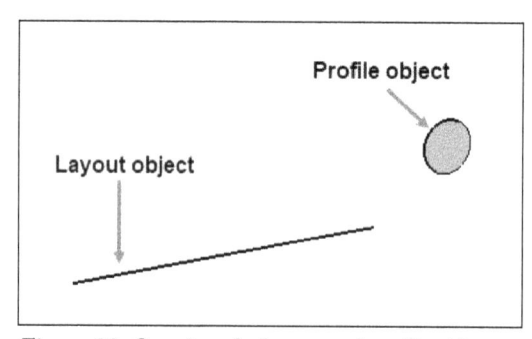

Figure-92. Creating the layout and profile object

- Select the created layout object and then while holding the **CTRL** key, select the profile object from the **Model Tree** or from the 3D view area.
- Click on the **Frame** tool from **Toolbar** in the **BIM** workbench; refer to Figure-93. The frame object will be created; refer to Figure-94.

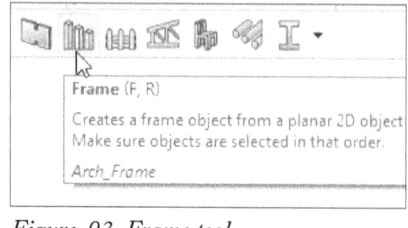

Figure-93. Frame tool

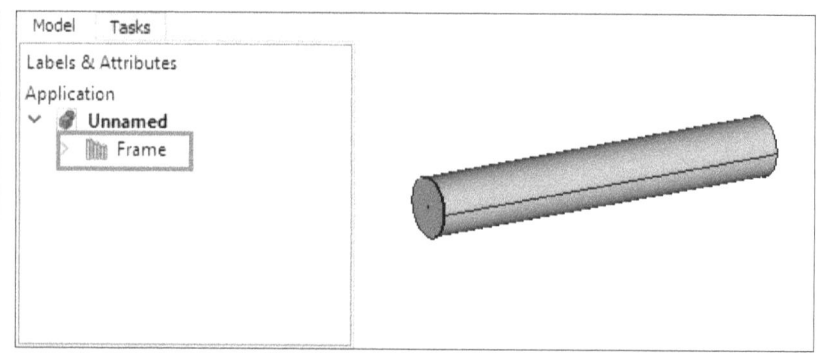

Figure-94. Frame object created

- If you want to edit the properties of frame object then select the frame object created from the **Model Tree**. The **Property editor** dialog will be displayed in the **Model** panel of **Combo View** with the parameters related to frame object; refer to Figure-95.

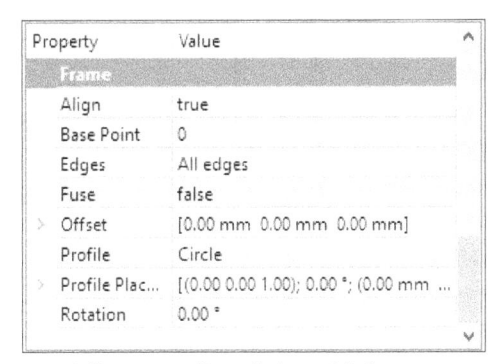

Figure-95. Property editor dialog with frame parameters

- Select desired option from **Align** drop-down in the **Frame** section of the dialog to specify whether the profile must be rotated to have its normal axis aligned with each edge or not.
- Select desired option from the **Edges** drop-down to specify the type of edges to be considered.
- Select desired option from **Fuse** drop-down to specify whether the geometry is fused or not.
- Enter desired values in **x**, **y**, and **z** edit boxes from **Offset** cascading menu to specify the distance between layout object and the frame object.
- The **Profile** edit box specifies the profile used to build this frame.
- Enter desired value in **Rotation** edit box to specify the rotation of the profile around its extrusion axis.
- On specifying parameters in the **Property editor** dialog, the properties of the frame object will be modified.

Creating Fence

The **Fence** tool creates a fence object by repeating single fence post and section along a given path. Note that you need to create separate bodies for section and fence post using Part Design workbench and then convert them into two separate structures by using **Structure** tool in **Arch** workbench. For creating path, we have used Polyline tool of Draft workbench. The procedure to use this tool is discussed next.

- Create a single fence post, a single section, and a path to be followed by the fence using desired workbench; refer to Figure-96.

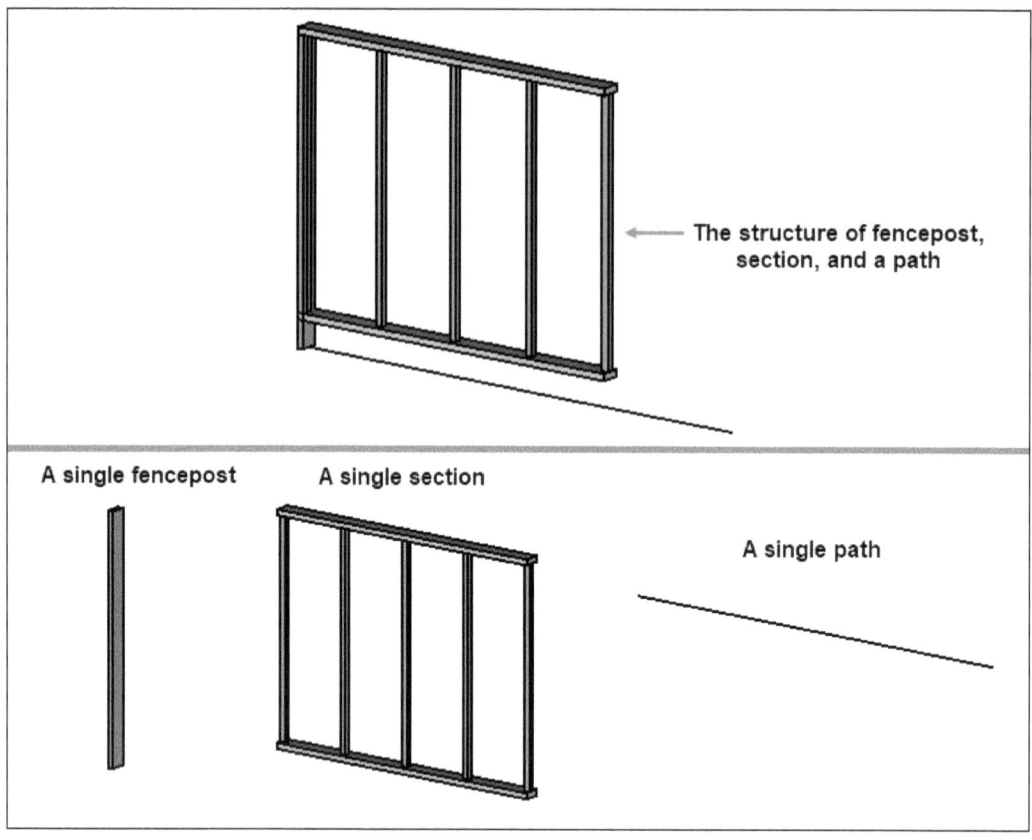

Figure-96. Structure to create fence

- Now, select the section, post, and path in exactly that order and click on the **Fence** tool from **Toolbar** in the **BIM** workbench; refer to Figure-97. The fence object will be created; refer to Figure-98.

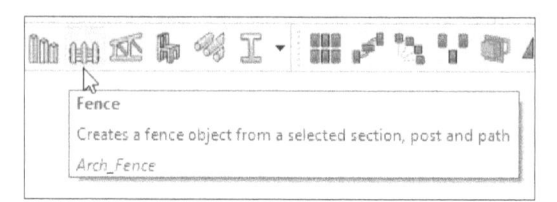

Figure-97. Fence tool

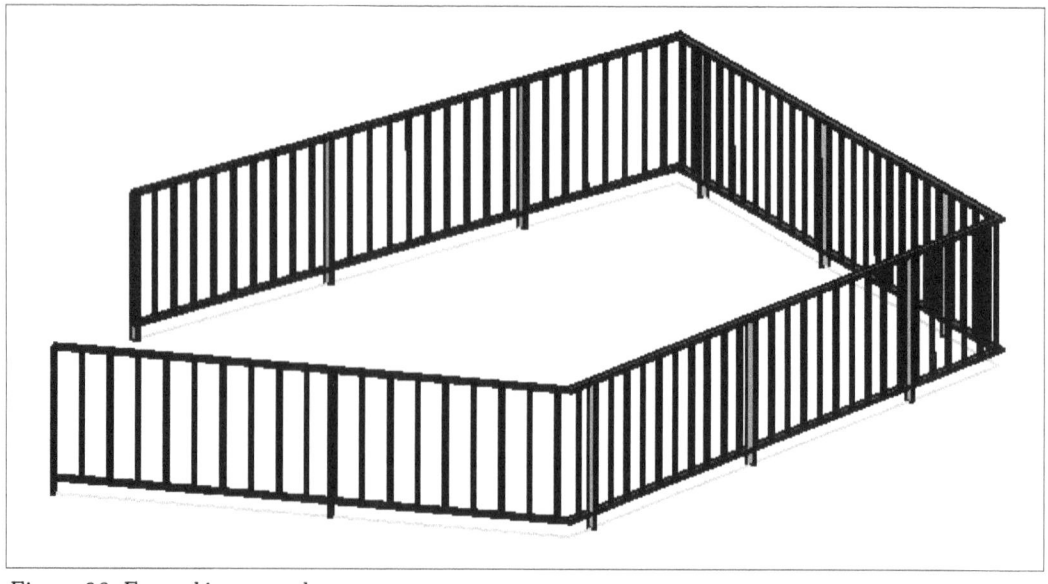

Figure-98. Fence object created

Creating Truss

The **Truss** tool creates a truss object, either from a selected linear object or from scratch, i.e. if no object is selected when launching the command. The procedure to use this tool is discussed next.

- Create and select the line from the **Model Tree** or from the 3D view area from which you want to create a truss; refer to Figure-99.

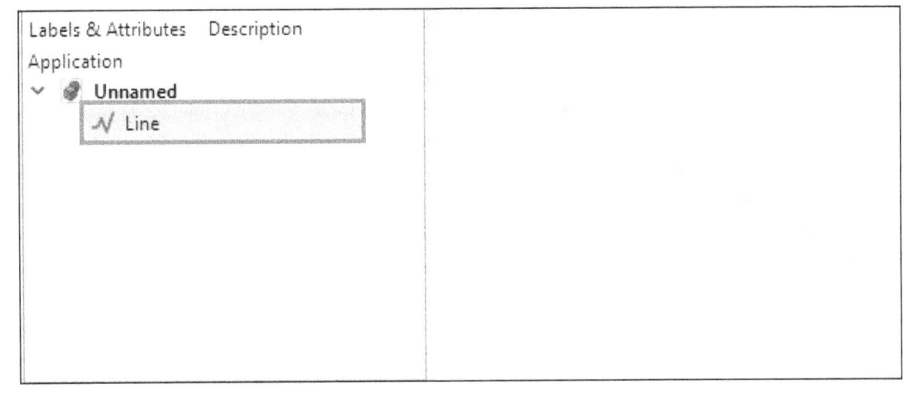

Figure-99. Selecting the line to create truss

- Click on the **Truss** tool from **Toolbar** in the **BIM** workbench; refer to Figure-100. The truss object will be created; refer to Figure-101.

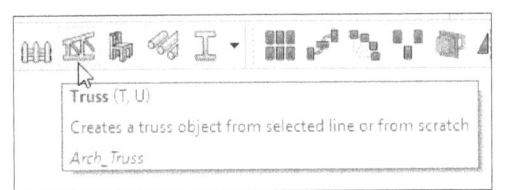

Figure-100. Truss tool

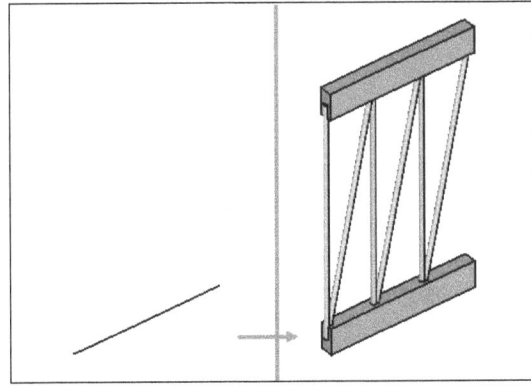

Figure-101. Truss object created

Or,

- Click on the **Truss** tool from **Toolbar** in the **BIM** workbench as discussed earlier. The **Point** dialog will be displayed in the **Tasks** panel of **Combo View** along with the plus sign in place of original cursor; refer to Figure-102. You will be asked to specify the first point.

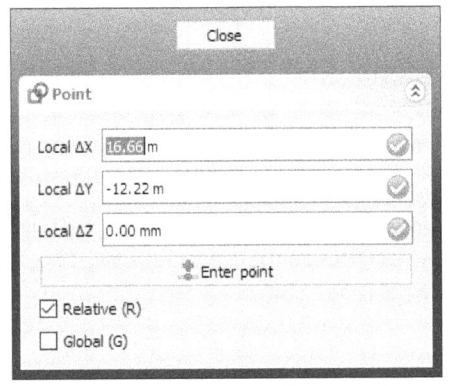

Figure-102. Point dialog of Truss tool

- Click in the 3D view area to specify the first point or enter desired values for the x, y, and z coordinates in the **Local ΔX**, **Local ΔY**, and **Local ΔZ** edit boxes of the dialog, respectively.
- After specifying coordinates for the first point, click on the **Enter point** button in the dialog. You will be asked to specify the second point.
- Move the cursor away and click at desired location to specify the second point or enter desired values for the coordinates in their respective edit boxes.
- After specifying coordinates for the second point, click on the **Enter point** button again from the dialog. The truss object will be created.
- If you want to edit the properties of truss object then select the truss object created from the **Model Tree**. The **Property editor** dialog will be displayed in the **Model** panel of **Combo View** with the parameters related to truss object; refer to Figure-103.

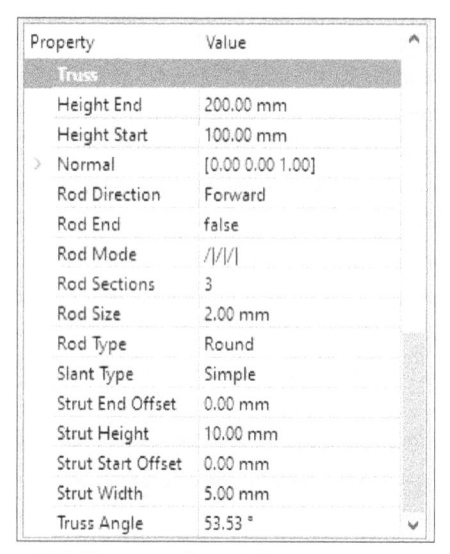

Property	Value
Truss	
Height End	200.00 mm
Height Start	100.00 mm
Normal	[0.00 0.00 1.00]
Rod Direction	Forward
Rod End	false
Rod Mode	/|/|/|
Rod Sections	3
Rod Size	2.00 mm
Rod Type	Round
Slant Type	Simple
Strut End Offset	0.00 mm
Strut Height	10.00 mm
Strut Start Offset	0.00 mm
Strut Width	5.00 mm
Truss Angle	53.53 °

Figure-103. Property editor dialog with truss parameters

- Specify desired height of the truss at the end and start position in the **Height End** and **Height Start** edit boxes from **Truss** section of the dialog, respectively.
- Enter desired values in the **x, y,** and **z** edit boxes from **Normal** cascading menu to specify the normal direction of the truss.
- Select desired direction of the rods from **Rod Direction** drop-down.
- Select desired option from **Rod End** drop-down to specify whether the truss has a rod at its endpoint or not.
- Select desired rod mode to draw the rods from **Rod Mode** drop-down.
- Specify desired number of rod sections in the **Rod Sections** edit box.
- Specify desired diameter or side of the rods in the **Rod Size** edit box.
- Select desired type of the middle element of the truss from **Rod Type** drop-down.
- Select desired slant type of the truss from **Slant Type** drop-down.
- Enter desired values in **Strut End Offset** and **Strut Start Offset** edit boxes to specify end offset and start offset for the top strut, respectively.
- Specify desired height and width of the main top and bottom elements of the truss in the **Strut Height** and **Strut Width** edit boxes of the dialog, respectively.
- Specify desired angle of the truss in the **Truss Angle** edit box.
- After specifying desired parameters in the dialog, the properties of the truss object will be modified; refer to Figure-104.

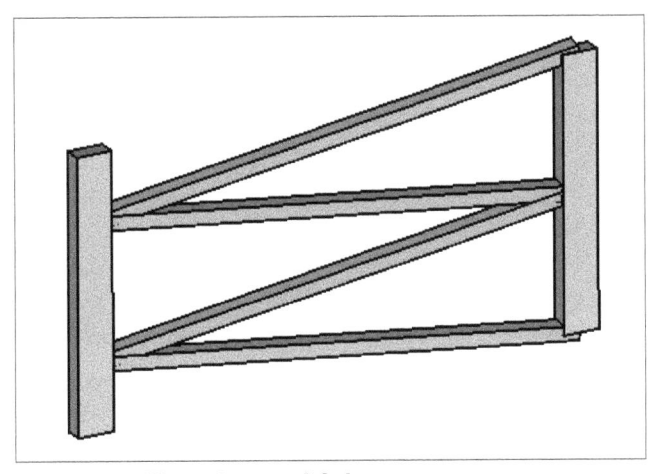

Figure-104. Truss object modified

Creating Equipment

The **Equipment** tool offers you a simple and convenient way to insert non-structural, standalone elements such as pieces of furniture, hydro-sanitary equipments, or electrical appliances to your projects. The procedure to use this tool is discussed next.

- Select the existing solid object or a mesh object from **Model Tree** or from the 3D view area; refer to Figure-105.

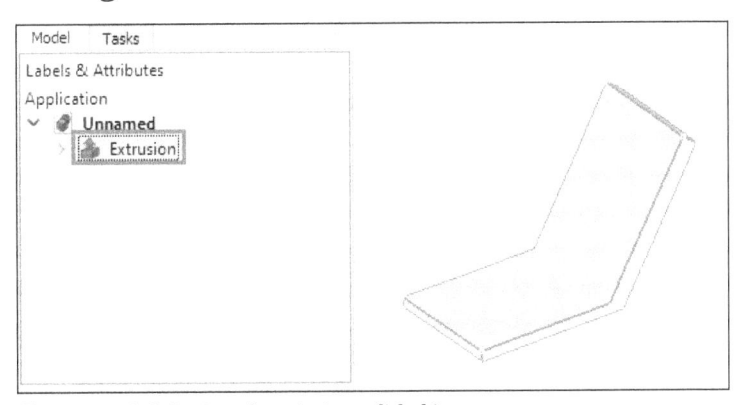

Figure-105. Selecting the existing solid object

- Click on the **Equipment** tool from **Toolbar** in the **BIM** workbench; refer to Figure-106. The non-structural element will be enclosed in an equipment object; refer to Figure-107.

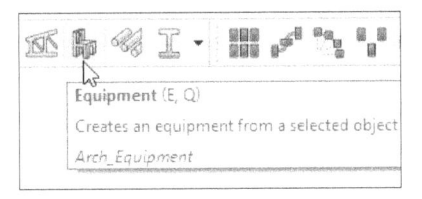

Figure-106. Equipment tool

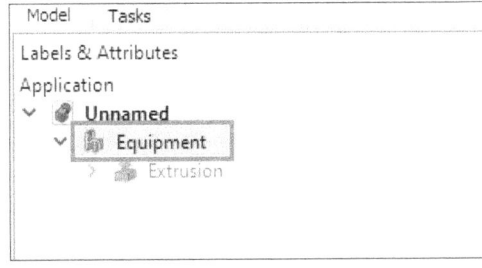

Figure-107. Equipment object created

- If you want to edit the properties of equipment then select the equipment created from **Model Tree**. The **Property editor** dialog will be displayed in the **Model** panel of **Combo View** with parameters related to equipment; refer to Figure-108.

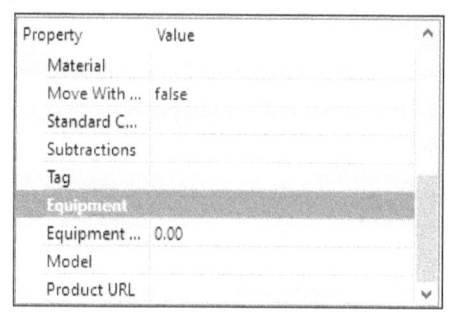

Figure-108. Property editor dialog with equipment parameters

- Enter desired value in **Equipment Power** edit box from **Equipment** section of the dialog to specify the electric power needed by the equipment of this model in watts.
- Specify desired description of the model for this equipment in the **Model** edit box.
- Enter desired URL of the product page where more information about this equipment can be found in the **Product URL** edit box of the dialog.
- On specifying parameters in the **Property editor** dialog, the properties of equipment will be modified.

Creating Custom Rebar

The **Custom Rebar** tool allows you to place reinforcing bars inside structure objects. Rebar objects are based on 2D profiles such as draft objects and sketches that must be drawn on a face of the structural object. The procedure to use this tool is discussed next.

- First, create the structure object using **Structure** tool as discussed earlier; refer to Figure-109.

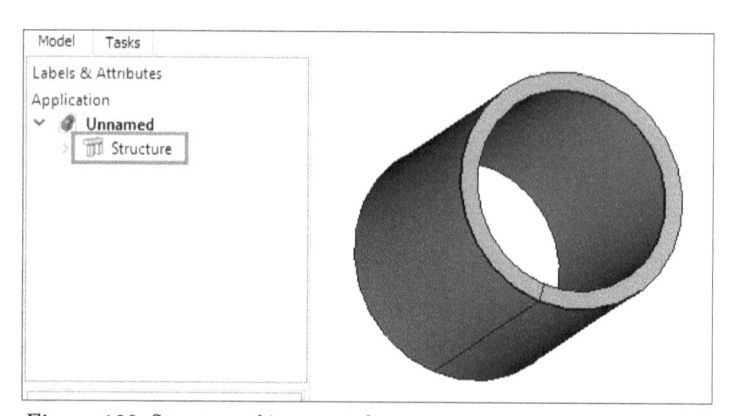

Figure-109. Structure object created

- Switch to **Sketcher** workbench from **Switch between workbenches** drop-down in the **Toolbar**. The tools related to **Sketcher** workbench will be displayed.
- Select desired face of the structural object on which you want to create the sketch; refer to Figure-110.

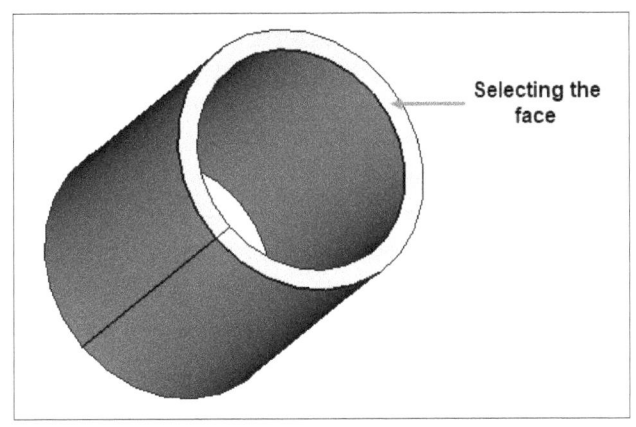

Figure-110. Selecting the face of structural object

- Click on the **New Sketch** tool from **Toolbar** in the **Sketcher** workbench and create desired sketch.
- After creating the sketch, switch to **BIM** workbench from **Switch between workbenches** drop-down.
- Select the newly created sketch from the **Model Tree** or from the 3D view area; refer to Figure-111.

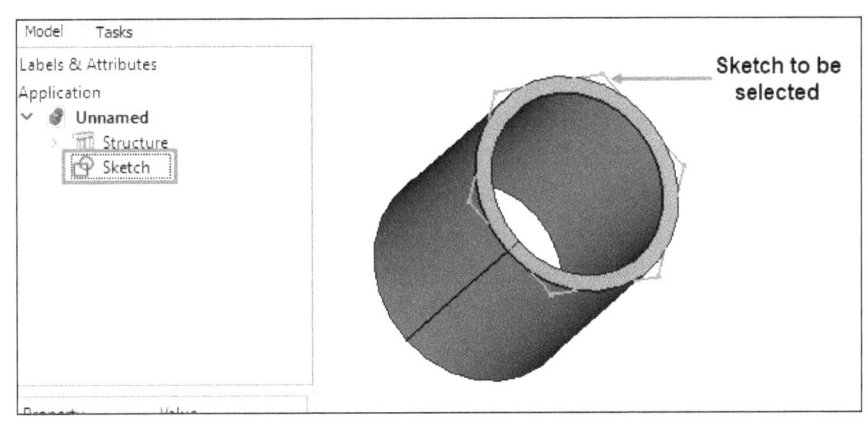

Figure-111. Sketch selected to create rebar

- Click on the **Custom Rebar** tool from **Toolbar** in the **BIM** workbench; refer to Figure-112. The rebar object will be created; refer to Figure-113.

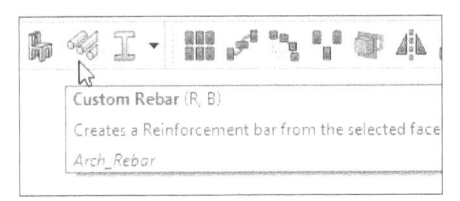

Figure-112. Custom Rebar tool

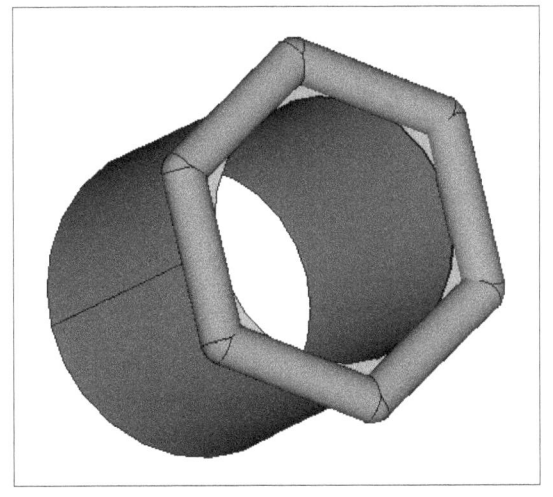

Figure-113. Rebar object created

- If you want to edit the properties of rebar object then select the rebar object created from the **Model Tree**. The **Property editor** dialog will be displayed in the **Model** panel of **Combo View** with the parameters related to rebar object; refer to Figure-114.

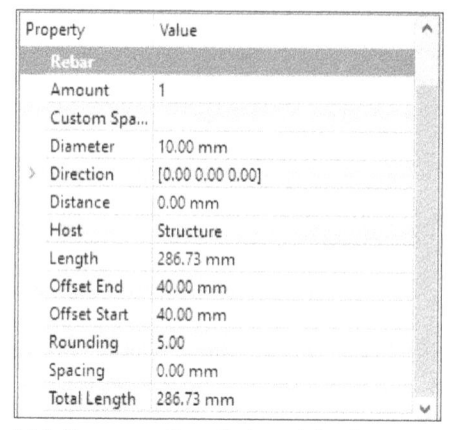

Figure-114. Property editor dialog with rebar parameters

- Specify desired amount of bars in the **Amount** edit box from **Rebar** section of the dialog.
- Specify desired diameter of the bars in the **Diameter** edit box.
- Enter desired value in **x**, **y**, and **z** edit boxes from **Direction** cascading menu to specify the spreading of bars along x, y, and z direction, respectively.
- Specify desired offset distance between the border of the structural object and the first bar in the **Offset Start** edit box and between the border of the structural object and the last bar in the **Offset End** edit box of the dialog.
- Specify desired rounding value to be applied to the corner of the bars, expressed in times the diameter in **Rounding** edit box.
- Enter desired value in **Spacing** edit box to specify the distance between the axis of each bar.
- On specifying the parameters in **Property editor** dialog, the properties of rebar object will be modified; refer to Figure-115.

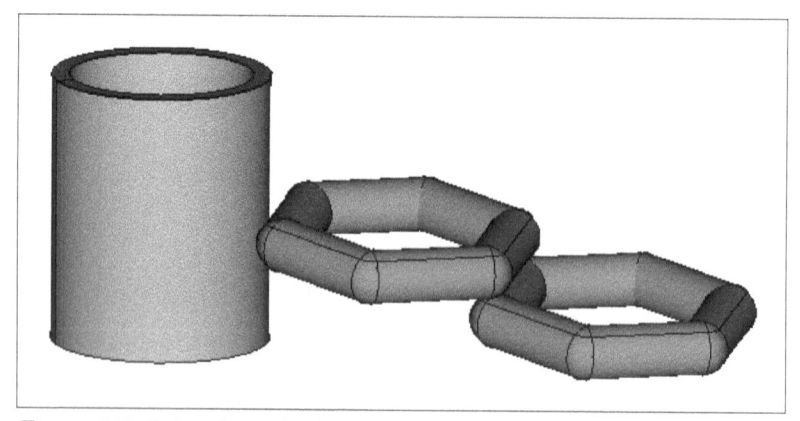

Figure-115. Reber object edited

Creating Profile

The **Profile** tool creates a parametric 2D profile object. This object can then be used as a base in different other tools that perform extrusions. The procedure to use this tool is discussed next.

- Click on the **Profile** tool from the **Generic 3D tools** drop-down in the **Toolbar** of the **BIM** workbench; refer to Figure-116. The **Create profile** and **Profile settings** dialogs will be displayed in the **Tasks** panel of **Combo View**; refer to Figure-117.

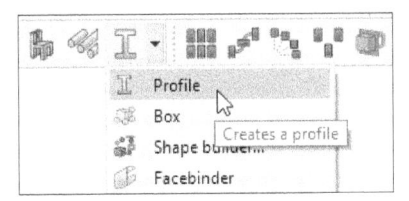

Figure-116. Profile tool

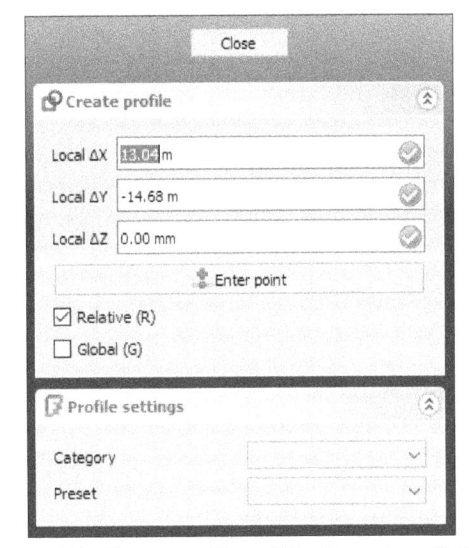

Figure-117. Create profile and Profile settings dialogs

- Select desired category of the profile to be create from **Category** drop-down in the **Profile settings** dialog.
- Select desired preset of the profile from **Preset** drop-down in the **Profile settings** dialog.
- Click in the 3D view area to specify the location of profile or enter desired values for the x, y, and z coordinates in the **Local ΔX**, **Local ΔY**, and **Local ΔZ** edit boxes from the **Create profile** dialog, respectively.
- After specifying coordinates for the location of profile, click on the **Enter point** button from the dialog. The profile object will be created; refer to Figure-118.
- If you want to edit the properties of profile object then select the profile object created from the **Model Tree**. The **Property editor** dialog will be displayed in the **Model** panel of **Combo View** with the parameters related to profile object; refer to Figure-119.

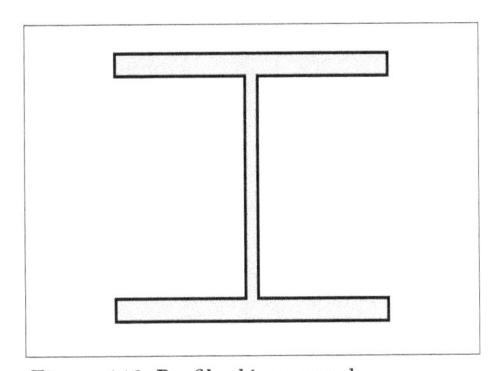

Figure-118. Profile object created

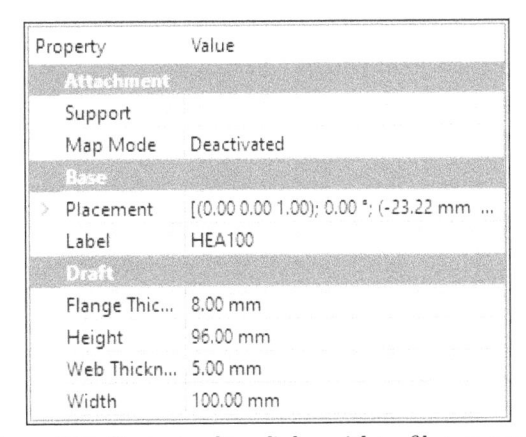

Figure-119. Property editor dialog with profile parameters

- Specify desired thickness of the flanges in the **Flange Thickness** edit box from **Draft** section of the dialog.
- Specify desired height and width of the beam in the **Height** and **Width** edit boxes of the dialog, respectively.
- Specify desired thickness of the web in the **Web Thickness** edit box of the dialog.

- After specifying desired parameters in the dialog, the properties of the profile object will be modified; refer to Figure-120.

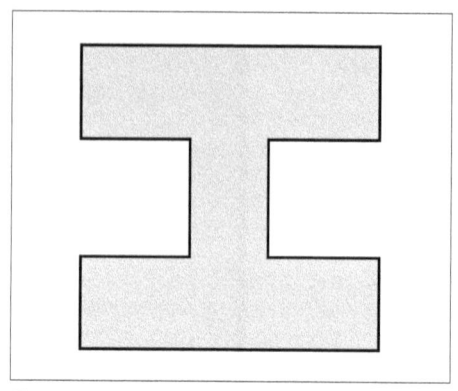

Figure-120. Profile object modified

Box

The **Box** tool is used to create a box by specifying its dimensions graphically when creating the object. The procedure to use this tool is same as discussed for **Panel** tool.

Shape builder

The **Shape builder** tool is used to create more complex shapes from various parametric geometric primitives. The procedure to use this tool has been discussed in **Chapter 4**.

Facebinder

The **Facebinder** tool creates a surface object from the selected faces of a solid object. The procedure to use this tool has been discussed in **Chapter 6**.

Objects library

The **Objects library** tool allows to insert an equipment or furniture object. The procedure to use this tool is discussed next. Note that this tool requires the **Parts Library** workbench which can be downloaded from **Addon Manager**.

- Click on the **Objects library** tool from the **Generic 3D tools** drop-down in the **Toolbar** of **BIM** workbench; refer to Figure-121. The **Library browser** dialog will be displayed in the **Tasks** panel; refer to Figure-122.

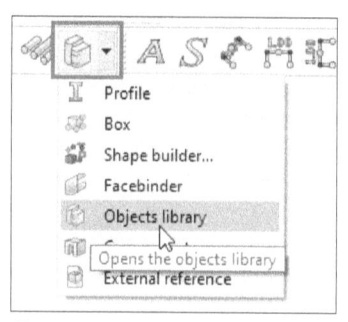

Figure-121. Objects library tool

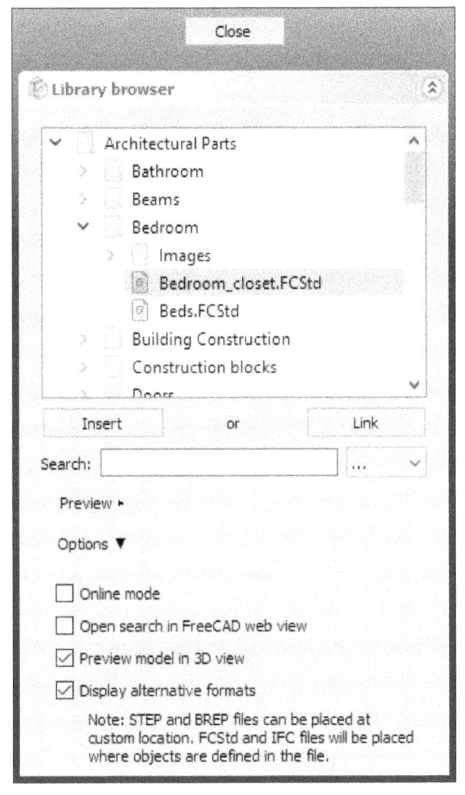

Figure-122. Library browser dialog

- Select desired part in the dialog to be inserted in the 3D view area as shown in Figure-122.
- Click on the **Link** button to link the selected object in the current document.
- Click on the **Search external websites** button from **Search** area and select desired website from the drop-down to insert the parts from that website.
- The **Preview** area of the dialog displays the preview of the selected part in the dialog.
- On selecting the **Online mode** check box from **Options** node, the parts library do not need to be installed. The parts will be inserted online.
- Select the **Open search in FreeCAD web view** check box to open the search result's in the web browser of FreeCAD.
- Select the **Preview model in 3D view** check box to display the 3D preview of the selected part.
- Select **Display alternative formats** check box to display the alternative file formats for library parts.
- After specifying desired parameters, click on the **Insert** button. The part will be inserted and the dialog will be closed; refer to Figure-123.

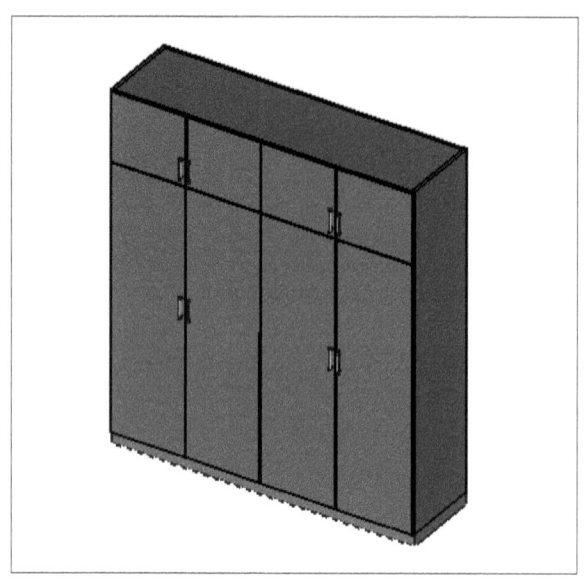

Figure-123. Part inserted

Component

The **Component** tool is used to create an architectural component from any part based object. The procedure to use this tool is same as discussed for **Slab** tool.

Adding an External Reference

The **External reference** tool allows you to place an object in the current document that copies its shape and colors from a part-based object stored in another FreeCAD file. If that FreeCAD file changes, the reference object is marked to be reloaded. The procedure to use this tool is discussed next.

- Click on the **External reference** tool from **Generic 3D tools** drop-down in the **Toolbar** of **BIM** workbench; refer to Figure-124. The **External reference** dialog will be displayed in the **Tasks** panel of **Combo View**; refer to Figure-125.

Figure-124. External reference tool

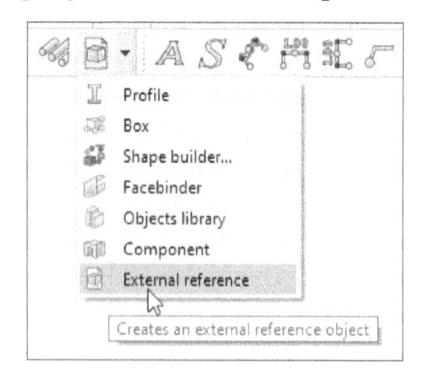

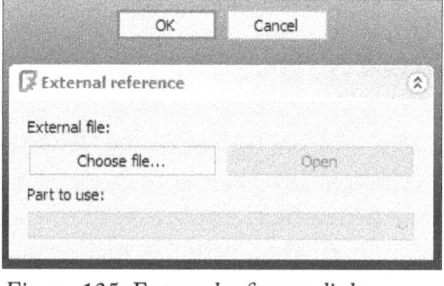

Figure-125. External reference dialog

- Click on the **Choose file** button from **External file** area of the dialog. The **Choose reference file** dialog box will be displayed; refer to Figure-126.

FreeCAD 1.0 Black Book   7-49

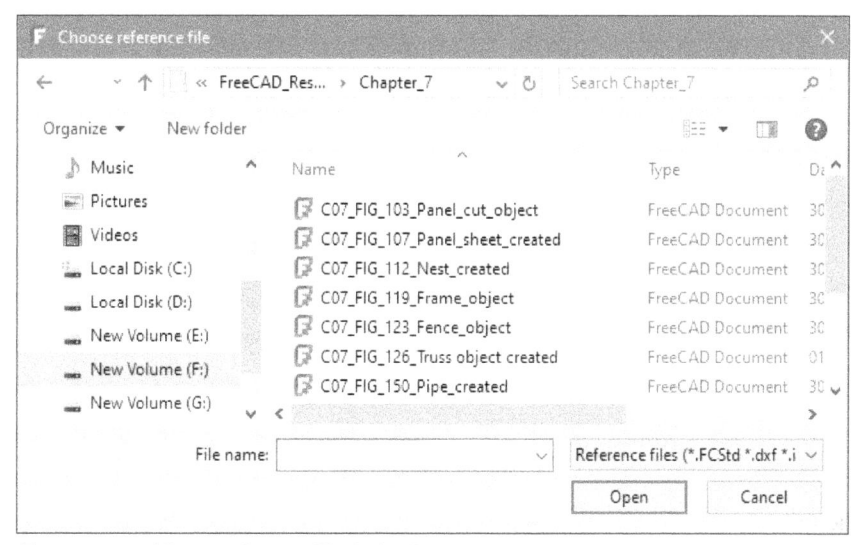

Figure-126. Choose reference file dialog box

- Select the existing FreeCAD file and click on **Open** button from the dialog box. The selected file will display in the **External file** area and the objects of selected reference file will be available in the **Part to use** drop down of **External reference** dialog; refer to Figure-127.

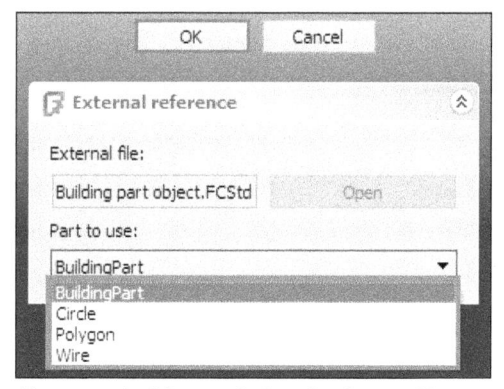

Figure-127. Objects of selected reference file

- Select one of the included part-based objects from the drop-down list and click on **OK** button from the dialog. The selected reference object will be added in the **Model Tree**; refer to Figure-128.

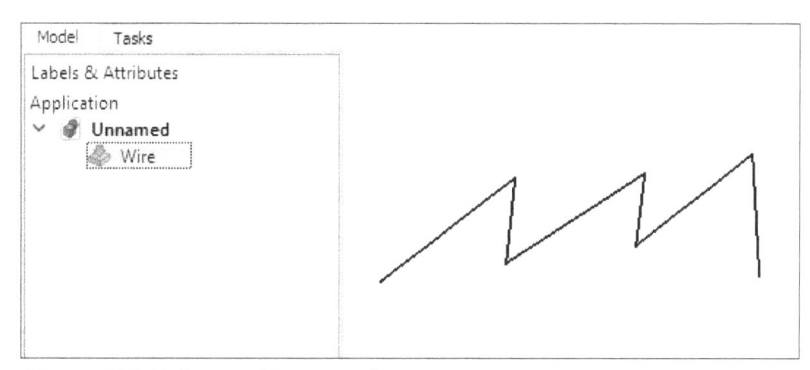

Figure-128. Reference object opened

- The reference object can be moved and rotated, the current position will be retained after reloading the object.
- If the original object gets moved in containing file, this movement will reflect in the reference object.

- If you want to open the containing file of reference object then select reference object and click **RMB** in the **Model Tree**. A shortcut menu will be displayed; refer to Figure-129.

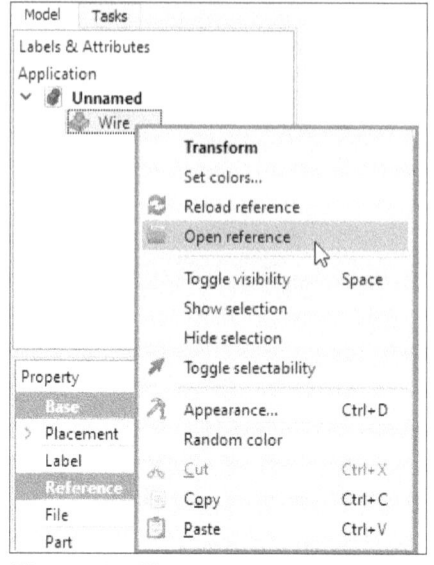

Figure-129. Shortcut menu

- Click on the **Open reference** option from the menu. The file containing the reference object will be opened in the new tab; refer to Figure-130.

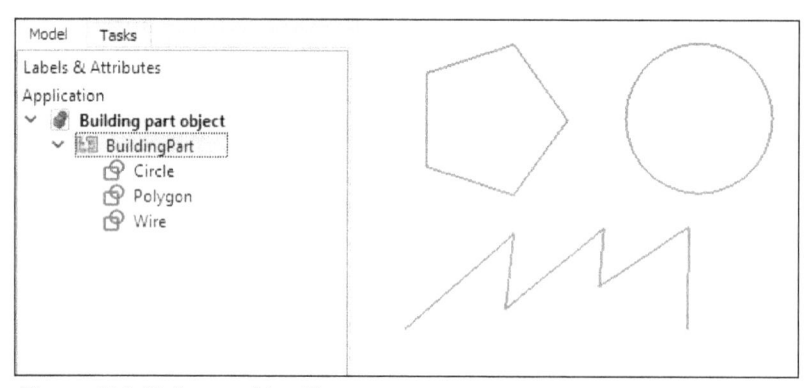

Figure-130. Reference object file

Adding the Component

The **Add Component** tool allows you to add shape-based objects to an Arch component, such as a **Wall** or **Structure**. These objects then make part of the Arch component and allow you to modify its shape but keeping its base properties such as height and width. The procedure to use this tool is discussed next.

- First, select the objects from the **Model Tree** or from the 3D view area which you want to add to the component and then by holding **CTRL** key, select the component, such as a wall or structure; refer to Figure-131.

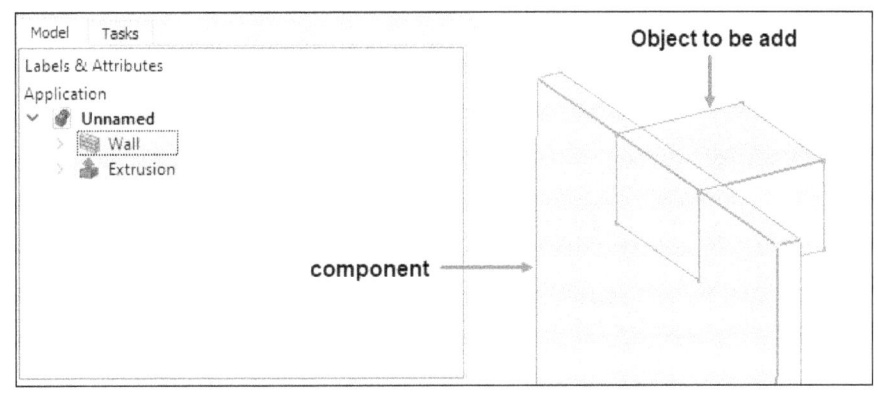

Figure-131. Selecting the object to add

- Click on the **Add component** tool from **Toolbar** in the **BIM** workbench; refer to Figure-132. The object will be added to the Arch component; refer to Figure-133.

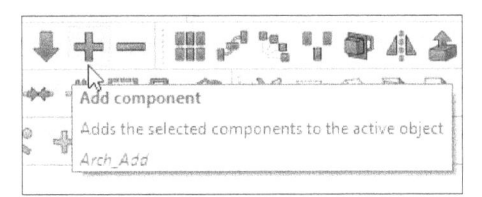

Figure-132. Add component tool

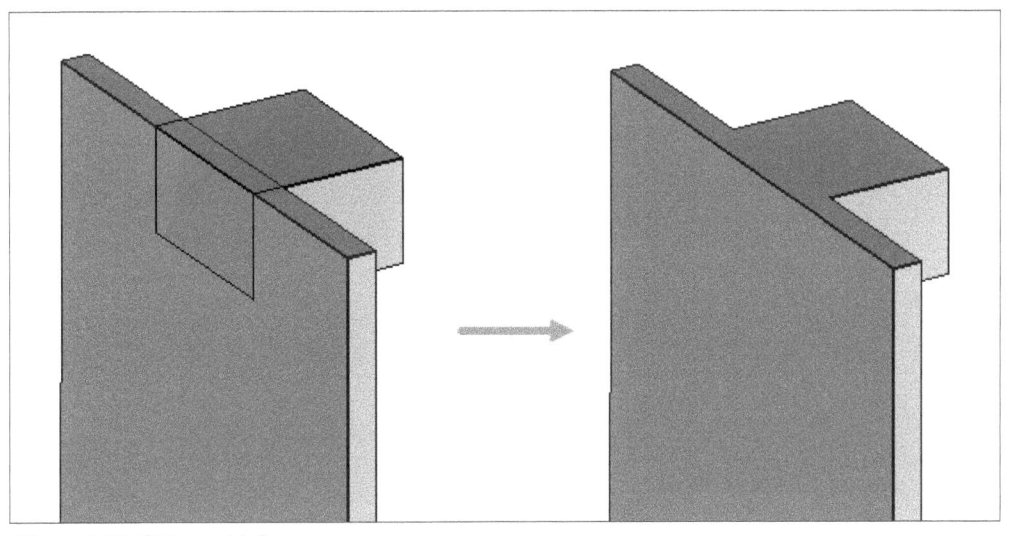

Figure-133. Object added

Removing the Component

The **Remove Component** tool allows you to remove a subcomponent from an Arch component, for example, remove a box that has been added to a wall. The procedure to use this tool is discussed next.

- First, select the object from the **Model Tree** or from the 3D view area which you want to remove from the component and then by holding **CTRL** key, select the component from the **Model Tree** or from the 3D view area; refer to Figure-134.

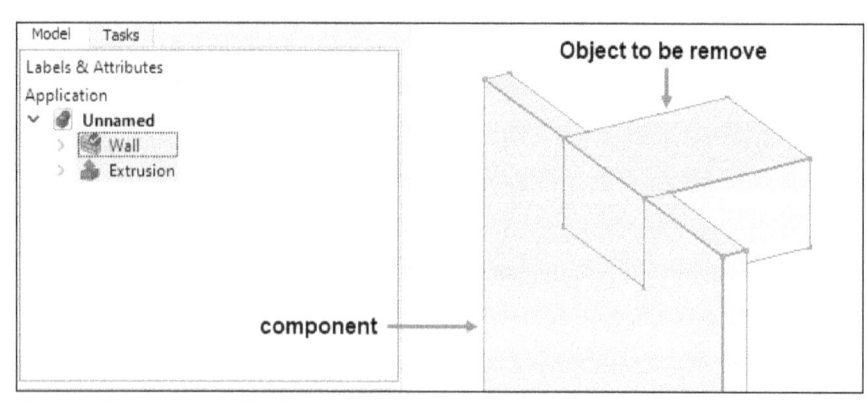

Figure-134. Selecting the object to remove

- Click on the **Remove component** tool from **Toolbar** in the **BIM** workbench; refer to Figure-135. The object will be removed; refer to Figure-136.

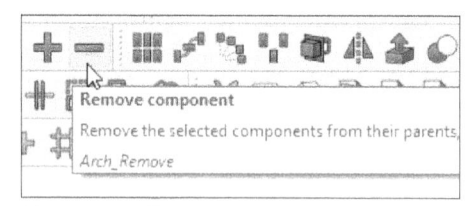

Figure-135. Remove component tool

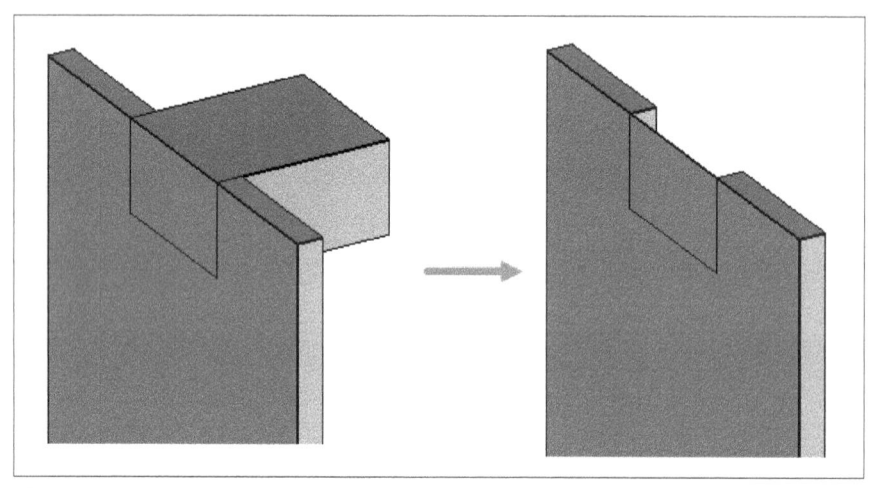

Figure-136. Object removed

Creating Axis

The **Axis** tool allows you to place a series of axes in the current document. The axes serve mainly as references to snap objects but can also be used together with **AxesSystems**, and can also be referenced by other BIM objects to create parametric arrays, for example, of beams or columns. **Grids** can also be used in place of axes. The procedure to use this tool is discussed next.

- Click on the **Axis** tool from **Toolbar** in the **BIM** workbench; refer to Figure-137. The axis will be created; refer to Figure-138.

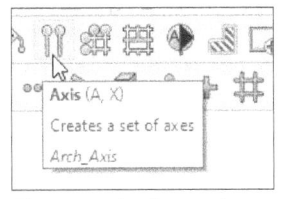

Figure-137. Axis tool

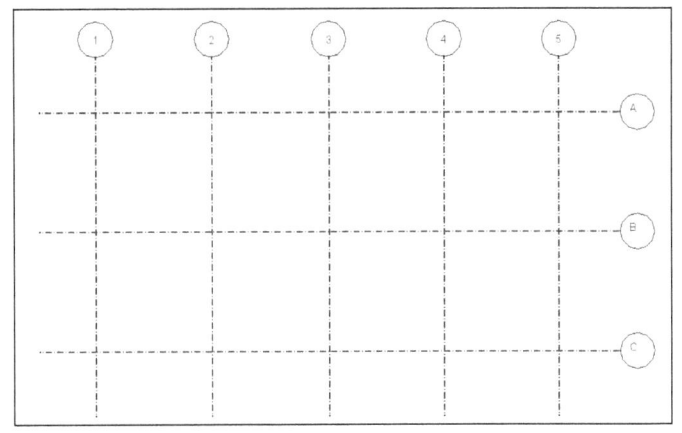

Figure-138. Axis created

- If you want to edit the parameters of axis then double-click on the axis created. The **Axes** dialog will be displayed in the **Tasks** panel of **Combo View**; refer to Figure-139.

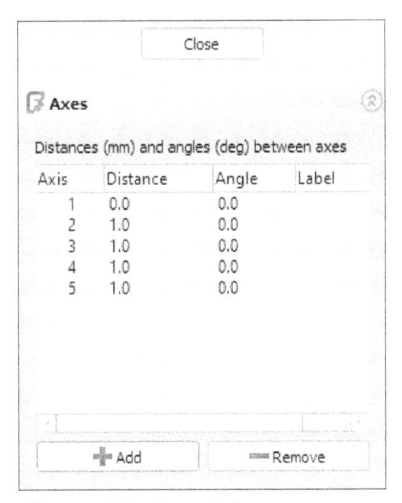

Figure-139. Axes dialog

- Click on **Add** button from the dialog to add the number of axes and if you want to remove the axis then select the axis from the dialog which you want to remove and click on **Remove** button.
- Specify the distance and angle of axis in the **Distance** and **Angle** edit boxes of the dialog, respectively.
- Click on **Close** button to close the dialog.
- If you want to edit the properties of axis then select the axis created from the **Model Tree**. The **Property editor** dialog will be displayed in the **Model** panel of **Combo View** with the parameters related to axis; refer to Figure-140.

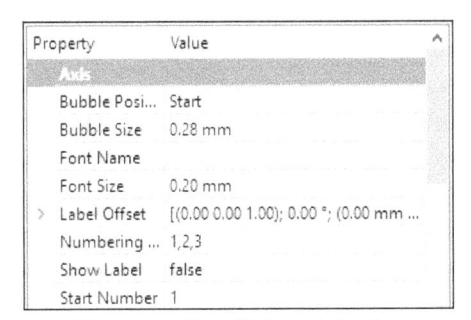

Figure-140. Property editor dialog with axis parameters

- Specify desired length of the axes in the **Length** edit box from **Axis** section of the dialog.
- Select desired option from **Bubble Position** drop-down to specify where the bubble is placed on the axis.
- Specify desired size of the axis bubbles in the **Bubble Size** edit box of the dialog.
- Specify desired font name to draw the bubble number and/or labels in the **Font Name** edit box.
- Specify desired size of the label text in the **Font Size** edit box of the dialog.
- Select desired option from **Numbering Style** drop-down to specify how the axes are numbered; 1, 2, 3, A, B, C, etc.
- Select desired option from **Show Label** drop down to specify whether to turn the display of the label texts on or off.
- On specifying parameters in the **Property editor** dialog, the properties of axis will be modified.

Creating Axis System

The **Axis System** tool allows you to combine several axis to the document. This is useful to define the intersection points between the different axes. Arch objects can then use this system to duplicate their shape on the different intersection points. The procedure to use this tool is discussed next.

- Select the axes from the **Model Tree** or from the 3D view area; refer to Figure-141.

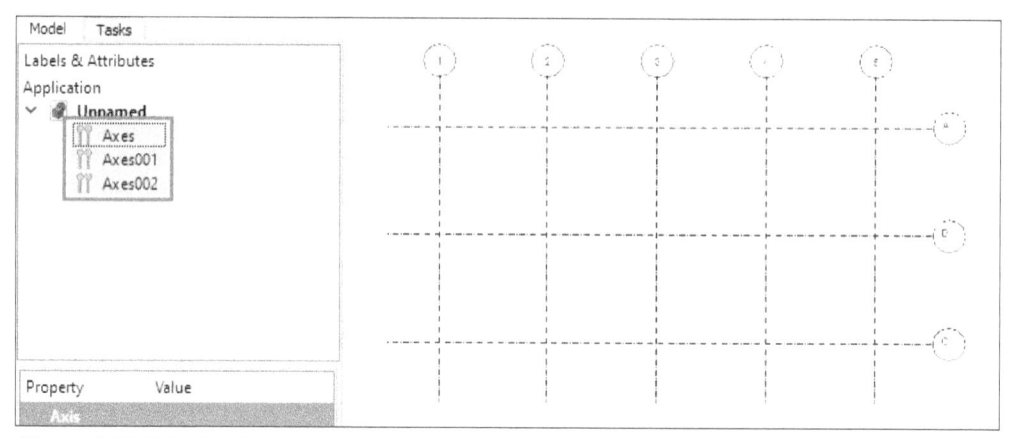

Figure-141. Selecting the axes

- Click on the **Axis System** tool from **Toolbar** in the **BIM** workbench; refer to Figure-142. The Axis system will be created and displayed in the **Model Tree**; refer to Figure-143.

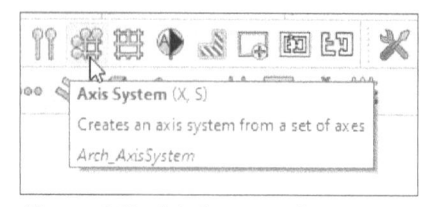

Figure-142. Axis System tool

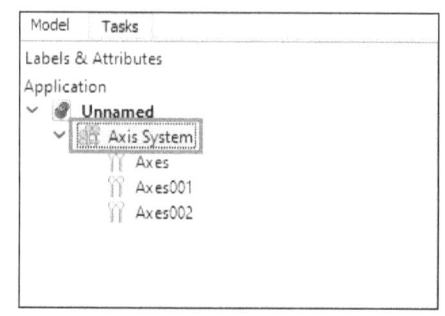

Figure-143. Axis system created

- If you want to edit the parameters of axis system then double-click on the axis system created from the **Model Tree**. The **Axes** dialog will be displayed in the **Tasks** panel of **Combo View**; refer to Figure-144.

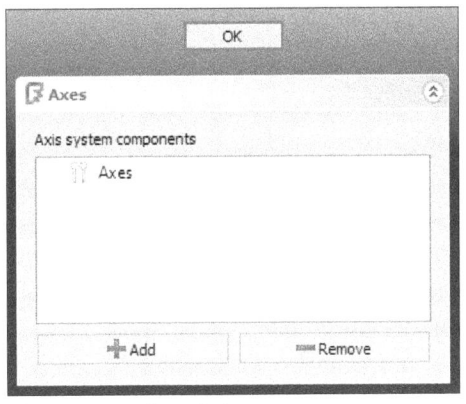

Figure-144. Axes dialog

- If you want to add more axes in the axis system then select the axis from the **Model Tree** and click on **Add** button from the **Axes** dialog and if you want to remove the axis then select the axis from **Axis system components** area of the dialog and click on **Remove** button from the dialog.
- Click on **OK** button to close the dialog.

Creating Grid

The **Grid** tool allows you to place a grid-like object in the document. This object is meant to serve as a base to build Arch objects that need a regular but complex frame, such as windows, curtain walls, column grids, railings, etc. The procedure to use this tool is discussed next.

- Click on the **Grid** tool from **Toolbar** in the **BIM** workbench; refer to Figure-145. The grid will be created and displayed in the **Model Tree**; refer to Figure-146.

Figure-145. Grid tool

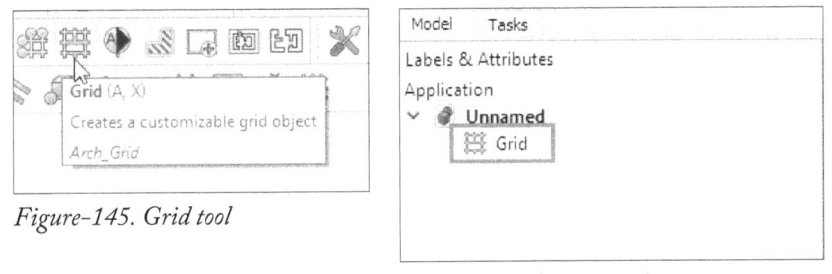

Figure-146. Grid created

- To edit the parameters of grid, double-click on the grid created from the **Model Tree**. The **Grid** dialog will be displayed in the **Tasks** panel of **Combo View**; refer to Figure-147.
- Specify desired width and height of the grid in the **Total width** and **Total height** edit boxes of the dialog, respectively.
- Click on the **Add row** button to add the number of rows in the grid and click on **Add col** button to add the number of columns in the grid. The rows and columns will be displayed in the **Rows/Columns** area of the dialog; refer to Figure-148.

Figure-147. Grid dialog

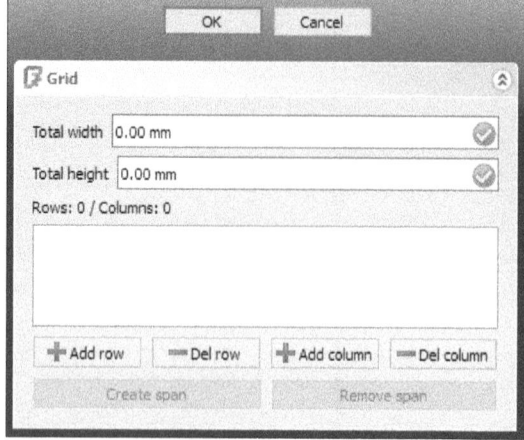

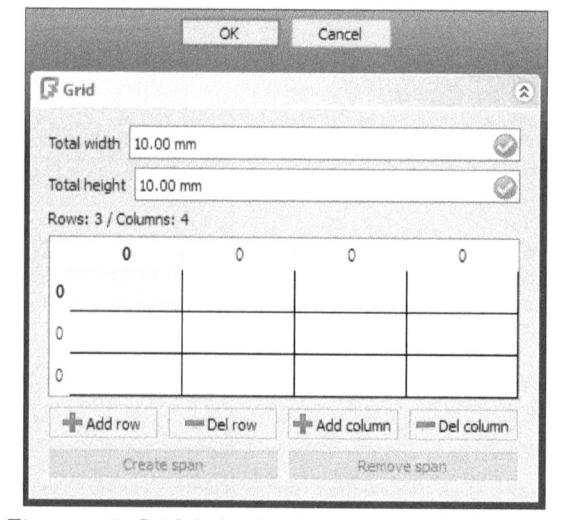

Figure-148. Grid dialog displaying rows and columns

- If you want to remove the row or column then select the row or column which you want to remove from the dialog and click on **Del row** or **Del col** buttons, respectively.
- Click on **OK** button from the dialog. The grid will be created; refer to Figure-149.

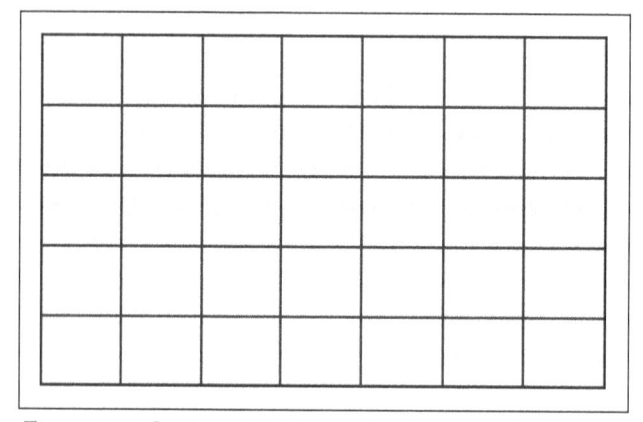

Figure-149. Grid created

- If you want to edit the properties of grid then select the grid created from the **Model Tree**. The **Property editor** dialog will be displayed in the **Model** panel of **Combo View** with the parameters related to grid; refer to Figure-150.

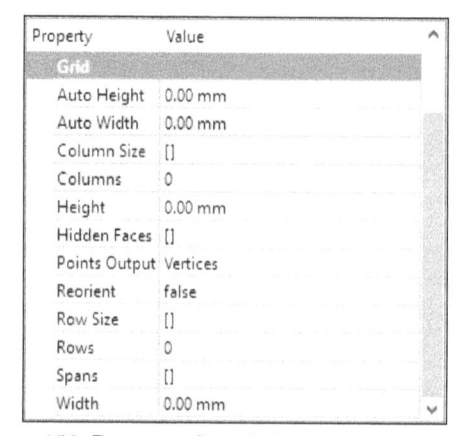

Figure-150. Property editor dialog with grid parameters

- Enter desired values in **Auto Height** and **Auto Width** edit boxes from **Grid** section of the dialog to create automatic row divisions and automatic column divisions, respectively.
- Specify desired size of columns and rows in the **Column Size** and **Row Size** edit boxes of the dialog, respectively.
- Specify desired number of columns and rows in the **Columns** and **Rows** edit boxes, respectively.
- Specify total height and total width of the grid in the **Height** and **Width** edit boxes, respectively.
- Enter desired value in **Hidden Faces** edit box to specify indices of faces to hide.
- Select desired option from **Points Output** drop-down to specify the type of 3D points produced by this grid object.
- Select desired option from **Reorient** drop-down to specify whether the grid must reorient its children along edge normals or not.
- On specifying parameters in the **Property editor** dialog, the properties of grid will be modified; refer to Figure-151.

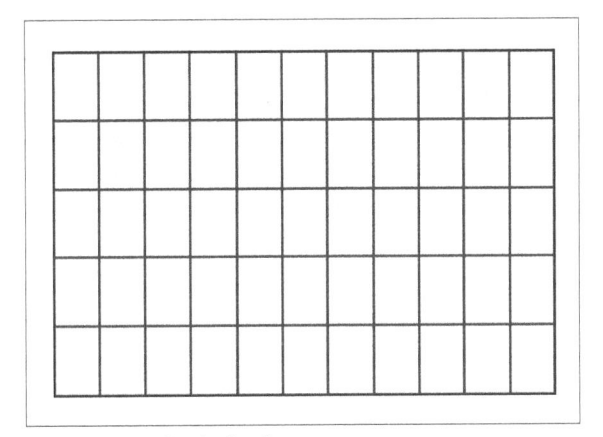

Figure-151. Grid edited

Creating Section Plane

The **Section Plane** tool is used to create section plane for generating section views. You can define the placement of section plane according to the current working plane and it can be relocated and reoriented by moving and rotating it, until it describes the 2D view you want to obtain. The Section plane object will only consider a certain set of objects, not all the objects of the document. The procedure to use this tool is discussed next.

- Set desired working plane to define where you want to place the section plane as discussed earlier.
- Select the object from the **Model Tree** or from the 3D view area which you want to include in the section view; refer to Figure-152.

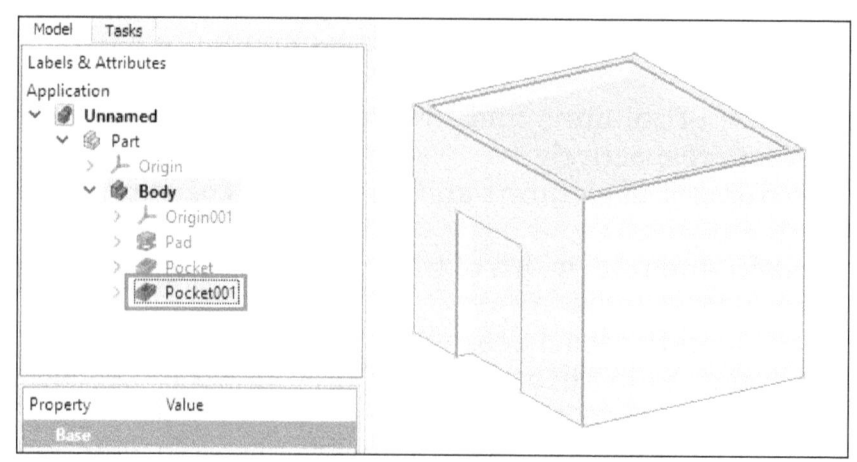

Figure-152. Selecting the object to include in section view

- Click on the **Section Plane** tool from **Toolbar** in the **BIM** workbench; refer to Figure-153. The section plane will be created; refer to Figure-154.

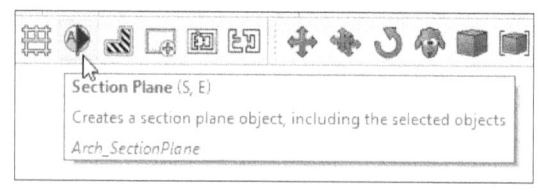

Figure-153. Section Plane tool

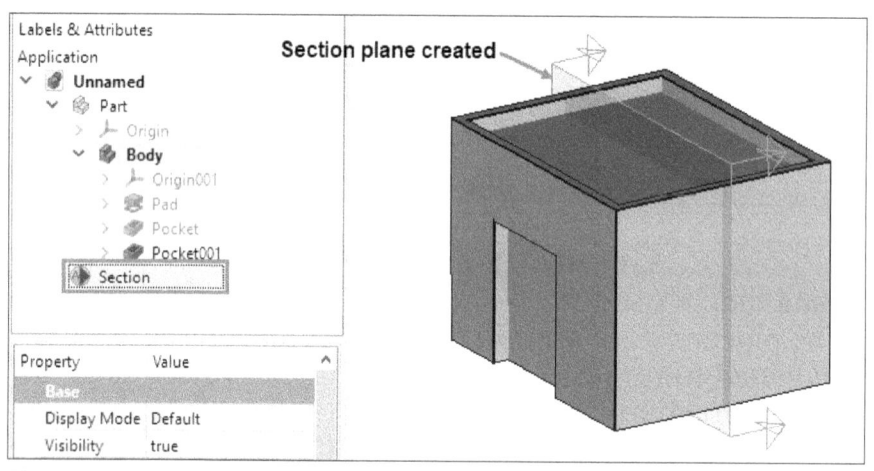

Figure-154. Section plane created

- You can move or rotate the section plane in desired position using **Move** tool as discussed earlier.
- If you want to edit the section plane then double-click on the section plane created from the **Model Tree**. The **Section plane settings** dialog will be displayed in the **Tasks** panel of **Combo View**; refer to Figure-155.

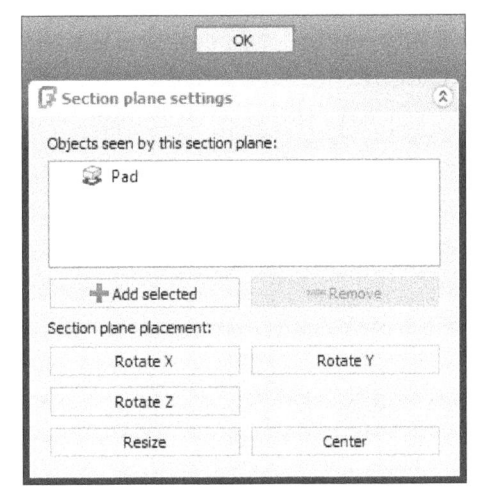

Figure-155. Section plane settings dialog

- The object selected for section plane will be displayed in the **Objects seen by this section plane** list box of the dialog.
- If you want to add more objects to be section then select the object from the **Model Tree** and click on **Add** button from the dialog.
- If you want to remove the objects then select the object from the dialog and click on **Remove** button.
- To rotate the section plane along x, y, or z direction then click on the **Rotate X**, **Rotate Y**, or **Rotate Z** button from **Section plane placement** area of the dialog, respectively.
- Click on the **Resize** button to modify the size of section plane.
- Click on the **Center** button to place the section plane at the center of object.
- Click on **OK** button to close the dialog.
- You can edit the properties of section plane using **Property editor** as discussed earlier.
- To create a section view of the object, select the section plane from the **Model Tree** or from the 3D view area and click on the **Shape 2D View** tool from the **Toolbar**. The 2D section view will be created; refer to Figure-156.

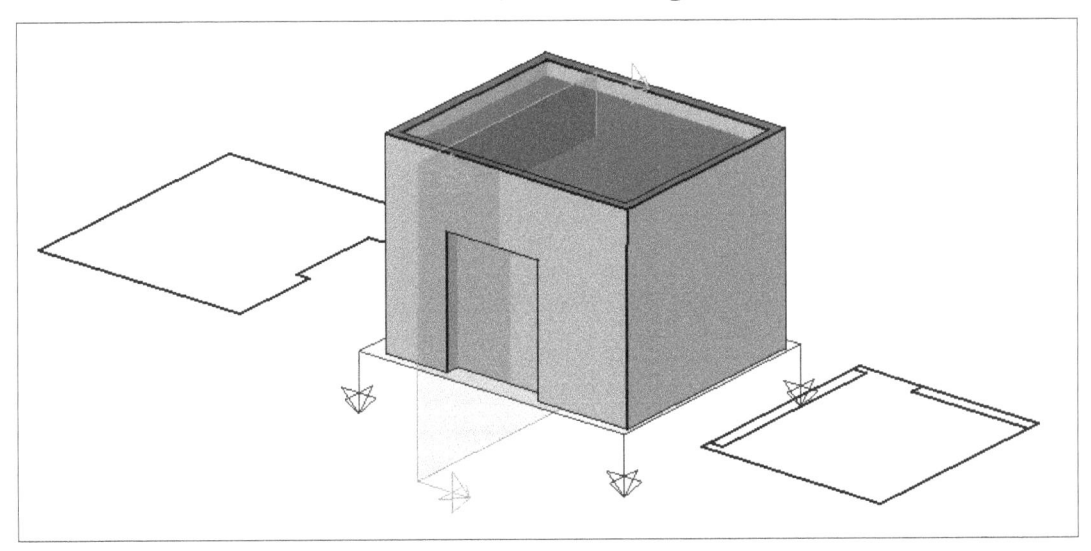

Figure-156. Section view created

- If you want to edit the properties of section then select the section plane created from the **Model Tree**. The **Property editor** dialog will be displayed in the **Model** panel of **Combo View** with parameters related to section plane; refer to Figure-157.

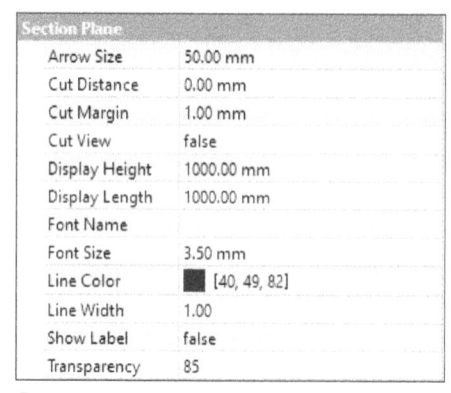

Figure-157. Property editor dialog with section plane parameters

- Select desired option from **Only Solids** drop-down from **Section Plane** section of the dialog to specify whether the non-solid objects in the set will be disregarded or not.
- Specify desired size of the arrows of the section plane gizmo in the 3D view area in the **Arrow Size** edit box of the dialog.
- Specify desired height and length of the section plane gizmo in the 3D view area in the **Display Height** and **Display Length** edit boxes, respectively.
- Select desired option from **Cut View** drop-down to specify whether the whole 3D view will be cut at the location of this section plane or not.
- On specifying parameters in the **Property editor** dialog, the properties of section plane will be modified.

Creating Material

The **Material** tool allows you to add materials to the active document and attribute a material to the **BIM** object. Materials are stored into a **Materials** folder in the active document. The procedure to use this tool is discussed next.

- Select one or more objects from the **Model Tree** or from the 3D view area to which you want to attribute a new material; refer to Figure-158.

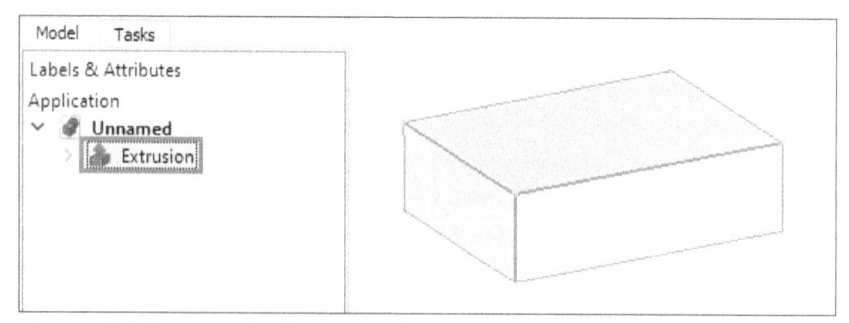

Figure-158. Selecting the object to attribute material

- Click on the **Material** tool from **Toolbar** in the **BIM** workbench; refer to Figure-159. The **BIM material** dialog will be displayed in the **Tasks** panel of **Combo View**; refer to Figure-160.

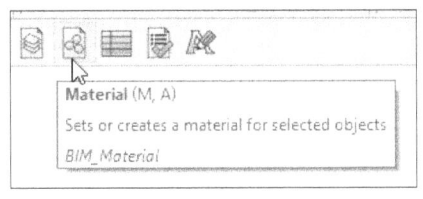

Figure-159. Material tool

Figure-160. BIM material dialog

- Select one of the preset materials to be used from **Choose preset** drop-down in the dialog.
- Select desired option from **Copy existing** drop-down to copy the values from an existing material in the document.
- Specify desired name for the material in the **Name** edit box of the dialog.
- Click on the **Edit button** available next to the **Name** edit box. The **Material Editor** dialog box will be displayed which allows you to edit many additional properties and add your own custom ones; refer to Figure-161.

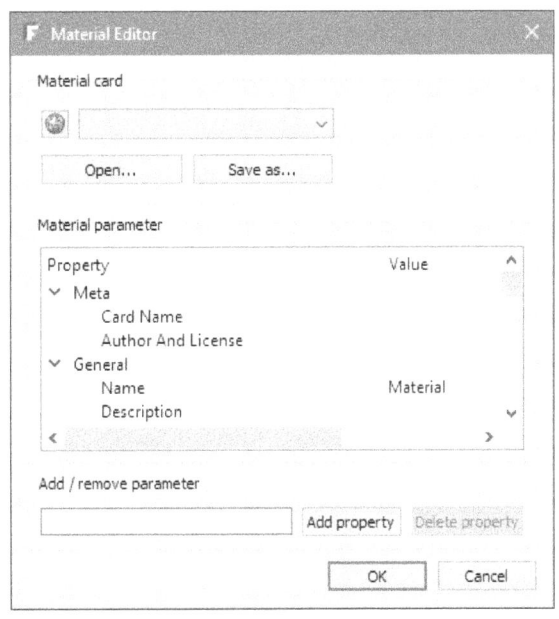

Figure-161. Material Editor dialog box

- Specify more detailed description of the material in the **Description** edit box of the **BIM material** dialog.
- Specify desired display color for the material which will be applied to all objects that use that material by clicking on the **Color** edit button.
- Specify desired display color for the material, which will be applied on **TechDraw** pages, when an object with this material is cut, and the **Display materials** property of the containing section plane is set to **True**.
- Specify desired transparency value for this material in the **Transparency** edit box.

- Specify a name and reference number of a specification system such as Masterformat or Omniclass in the **Standard code** edit box of the dialog.
- Enter desired url in the **URL** edit box where more information about the material can be found.
- Select desired parent material from the **Parent** drop-down in the dialog.
- Click on **OK** button from the dialog. The material will be applied to the object.

Creating Multi-Material

The **Multi-Material** option defines a list of materials with their names and thickness values. This multi-materials list can then be added to a BIM object instead of a single BIM material. The procedure to use this tool is discussed next.

- First, add a material to the active object in the **Model Tree** or in the 3D view area which you want to attribute a new multi-material; refer to Figure-162.

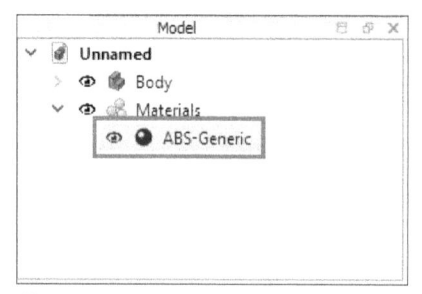

Figure-162. Material applied

- Again, click on the **Material** tool from **Toolbar** in the **BIM** workbench as discussed earlier. The **Select material** dialog box will be displayed in the **Tasks** panel of **Combo View**; refer to Figure-163.

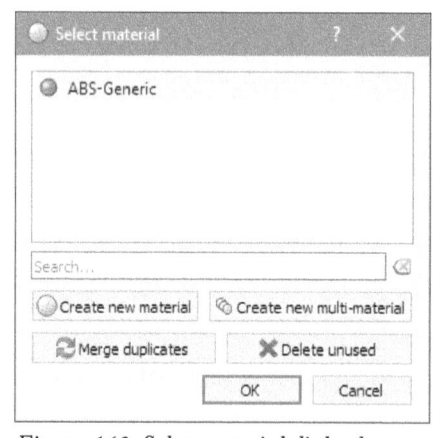

Figure-163. Select material dialog box

- Click on the **Create new material** button to create a new material in the **Model Tree** which can be edited in the **BIM material** dialog displayed on double clicking on that newly material created.
- Click on the **Merge duplicates** button to merge the duplicate materials.
- Click on the **Delete unused** button to delete the unused materials.
- Click on the **Create new multi-material** button. The **Multi-material definition** dialog will be displayed in the **Tasks** panel; refer to Figure-164.

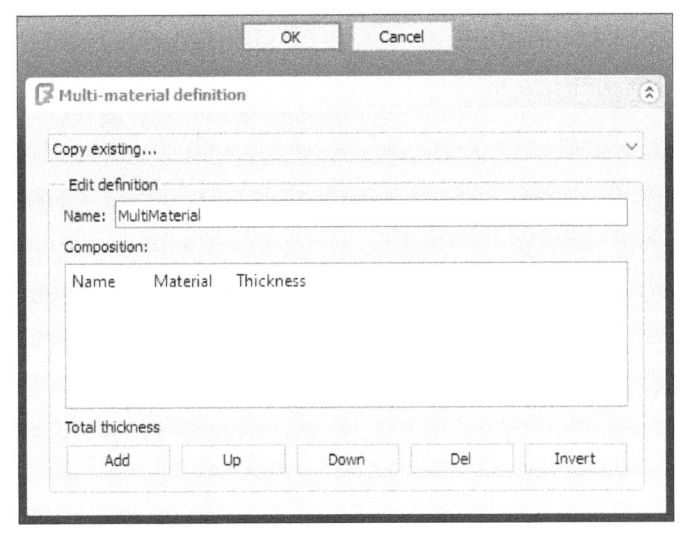

Figure-164. Multimaterial definition dialog

- Select desired option from **Copy existing** drop-down to copy the values from an existing material in the document.
- Specify desired name for the material in the **Name** edit box from **Edit definition** area of the dialog.
- The **Composition** list is the list of the different material layers that compose this multi-material.
- Click on the **Add** button to add a new layer to the list.
- Click on the **Up**, **Down**, **Del**, and **Invert** buttons to move a selected layer up, to move a selected layer down, to delete a selected layer, and to invert the selected layer, respectively.
- Double-click on the layer name to edit the name of layer.
- Select desired material available in the same document from the **Material** drop-down.
- Enter desired thickness value in the **Thickness** edit box of the dialog.
- Click on **OK** button to close the dialog. The Multi-material will be applied to the object.

Creating Schedule

The **Schedule** tool allows you to create and automatically populate a spreadsheet with contents gathered from the model. The procedure to use this tool is discussed next.

- First, you need to create or open a FreeCAD document which contains some objects; refer to Figure-165.

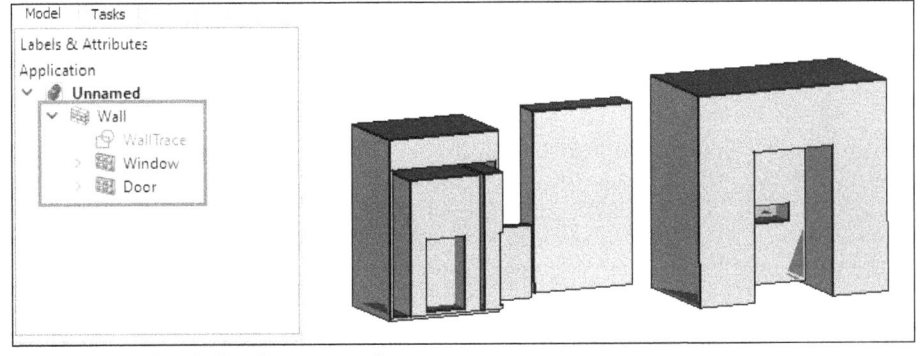

Figure-165. FreeCAD objects created

- Click on the **Schedule** tool from **Toolbar** in the **BIM** workbench; refer to Figure-166. The **Schedule definition** dialog box will be displayed; refer to Figure-167.

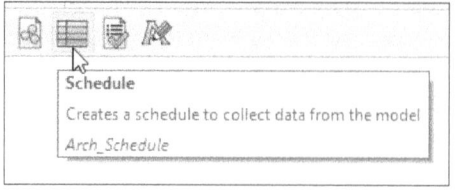

Figure-166. Schedule tool

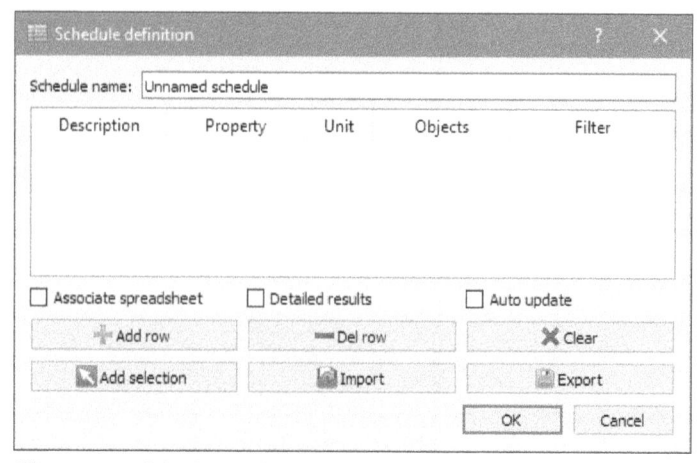

Figure-167. Schedule definition dialog box

- Specify desired name for the schedule in the **Schedule name** edit box.
- On selecting the **Associate spreadsheet** check box, an associated spreadsheet containing the results will be maintained together with this schedule object.
- On selecting the **Detailed results** check box, additional lines will be filled with each object considered.
- Select the **Auto update** check box to update the schedule and the associated spreadsheet on recomputing the document.
- Click on **Add row** button to add a new row, click on **Del row** button to delete the selected row, and click on **Clear** button to delete all the existing rows.
- Click on the **Add selection** button to add objects currently selected in the document.
- Click on **Import** button to build the list created in another spreadsheet application and import that as a csv file here.
- Click on **Export** button to export the contents of Result spreadsheet to a csv file.
- Double-click each cell from that line in the dialog to specify desired values.
- Specify desired description for this operation in the **Description** edit box.
- Enter desired value in the **Property** edit box to retrieve for each object or to count the objects.
- Enter desired unit in the **Unit** edit box to express the resulting value.
- Specify the list of object names to be considered by this operation in the **Objects** edit box.
- Specify desired list of filters in the **Filter** edit box. Each filter is written in the form; filter:value.
- After specifying all the parameters, click on **OK** button from the dialog; refer to Figure-168. A new schedule object will be added to the document which contains a result spreadsheet; refer to Figure-169.

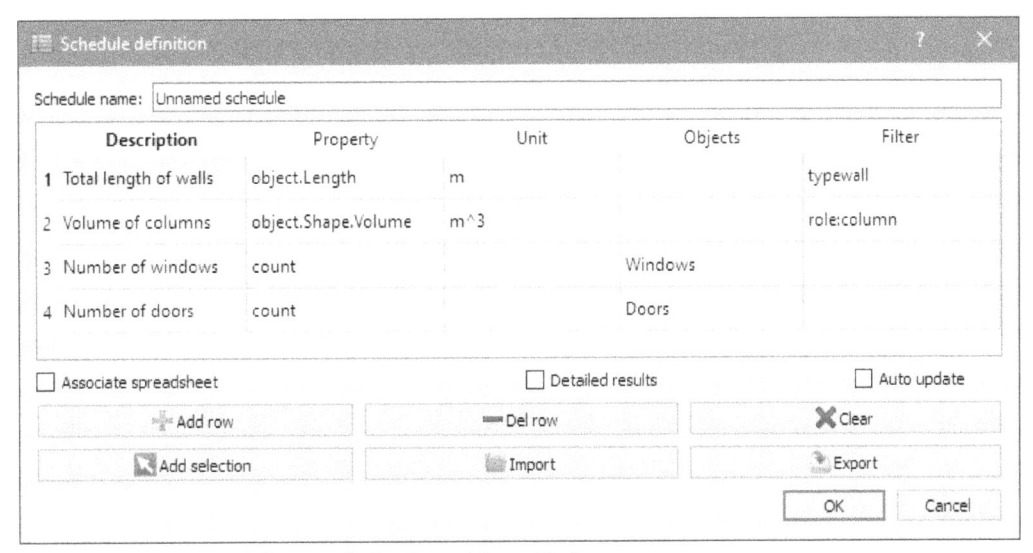

Figure-168. Schedule definition dialog box with specified parameters

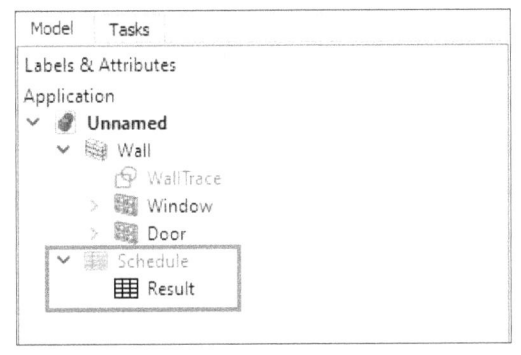

Figure-169. Schedule object added

- If you want to edit the schedule then double-click on the **Schedule** created from the **Model Tree**.
- If you want to get the results in spreadsheet itself then double-click on the **Result** created from the **Model Tree**. A new result file in spreadsheet will be opened; refer to Figure-170.

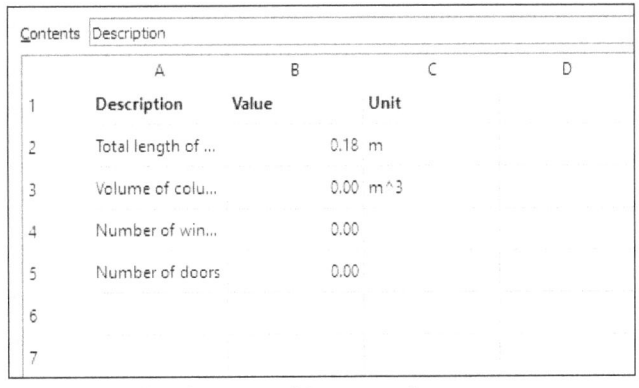

Figure-170. Result in spreadsheet opened

Cutting the object with Plane

The **Cut with plane** tool allows you to cut an object according to a plane. The procedure to use this tool is discussed next.

- First, select the object which you want to cut from the **Model Tree** or from the 3D view area and then by holding **CTRL** key, select face of the second object from 3D view area by which the first object will cut; refer to Figure-171.

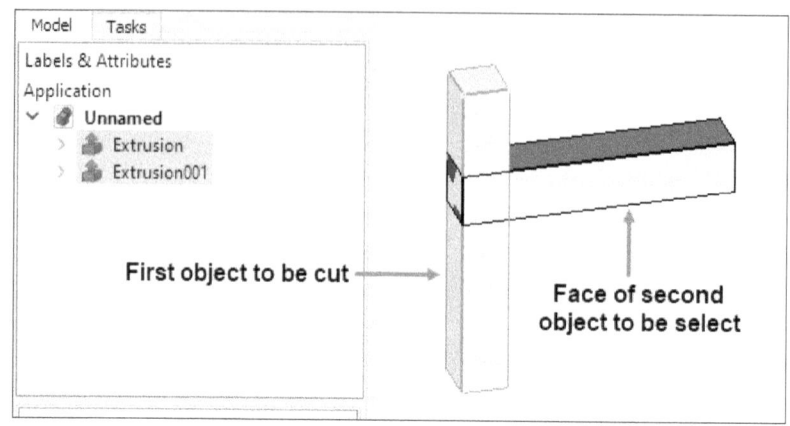

Figure-171. Selecting the objects to be cut

- Click on the **Cut with plane** tool from **Toolbar** in the **BIM** workbench; refer to Figure-172. The **Cut Plane** dialog will be displayed in the **Tasks** panel of **Combo View** along with the cut selection box; refer to Figure-173.

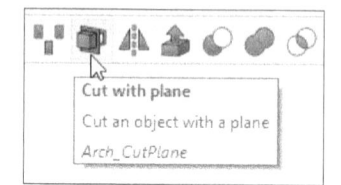

Figure-172. Cut with plane tool

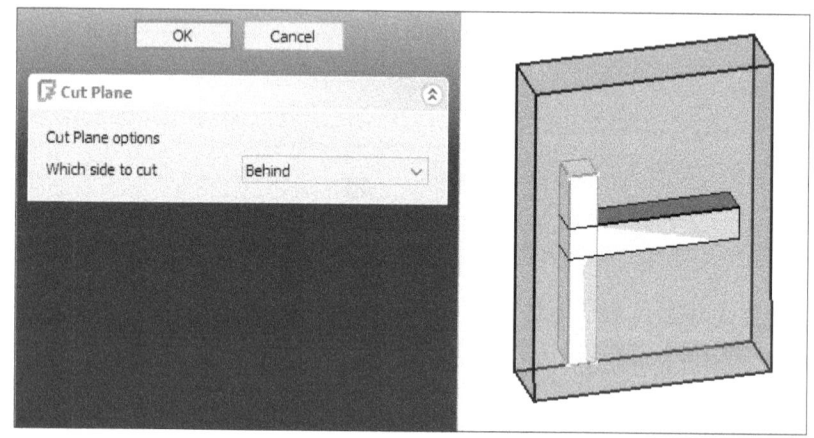

Figure-173. Cut Plane dialog with cut selection box

- Select **Behind** option to cut the object behind the normal face and select **Front** option to cut the object in front of normal face from **Which side to cut** drop-down in **Cut Plane options** section of the dialog.
- Click on **OK** button from the dialog. The object will be cut; refer to Figure-174.

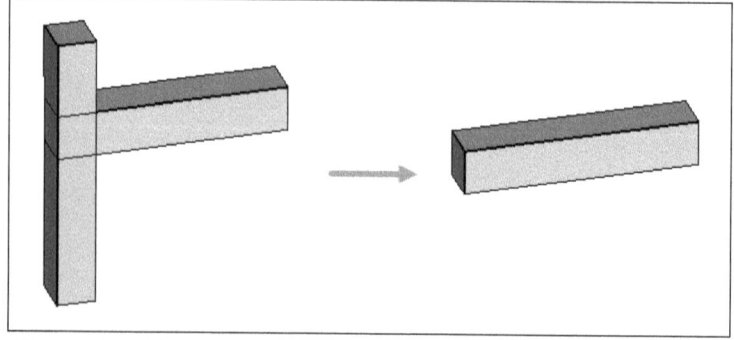

Figure-174. Object cut

Survey

The **Survey** tool enters a special surveying mode which allows you to quickly grab measurements and information from a model and transfer that information to other applications. The procedure to use this tool is discussed next.

- Click on the **Survey** tool from **Utils** menu in the **Ribbon** of **BIM** workbench; refer to Figure-175. The **Survey** dialog will be displayed in the **Tasks** panel of **Combo View**; refer to Figure-176.

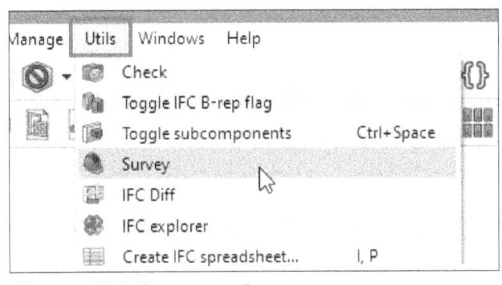

Figure-175. Survey tool

Figure-176. Survey dialog

- Click on the vertices, edges, faces of the object, or double-click on the whole object from the 3D view area to get the measurement. The label will be displayed showing the measurement of selected object in the 3D view area as well as in the **Survey** dialog; refer to Figure-177.

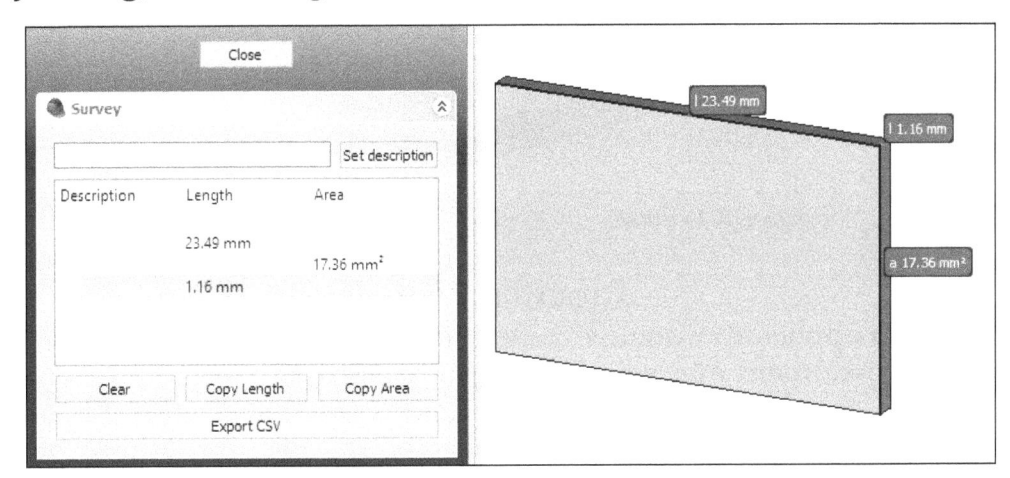

Figure-177. Measurement displayed

- If you want to add description for the measurement then select the measurement created from the dialog and specify desired description in the box available next to the **Set description** button.
- After specifying the description, click on the **Set description** button from the dialog. The description for that measurement will be added; refer to Figure-178.

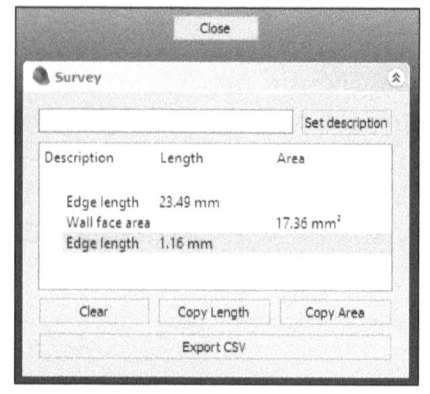

Figure-178. Description added

- Click on the **Export CSV** button from the dialog to export the contents of the dialog in csv file which can be opened in any spreadsheet application.
- Click on **Close** button to close the dialog.

PRACTICAL

Create the model as shown in Figure-179.

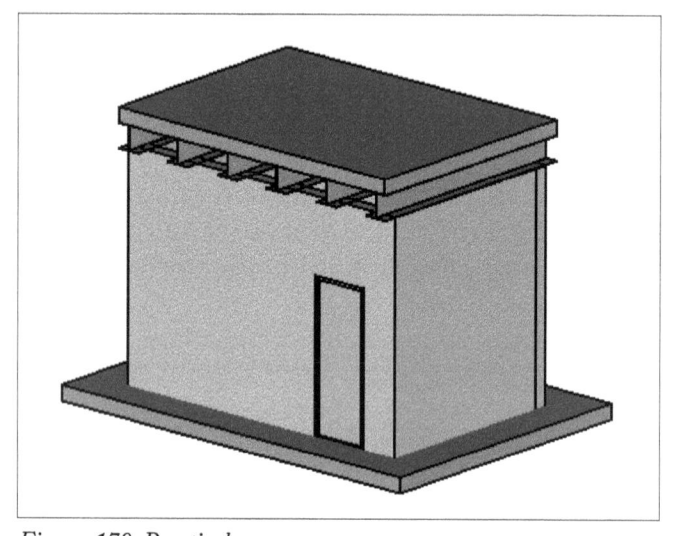

Figure-179. Practical

Creating Wall

- Start FreeCAD application using desktop icon if not started yet.
- Create a new document and switch to **BIM** workbench.
- Select the **Top** plane and create the sketch using **Line** tool as shown in Figure-180.

Figure-180. Creating first sketch

- Select all the lines created from the **Model Tree** and click on the **Wall** tool from the **Toolbar**. The wall will be created; refer to Figure-181.

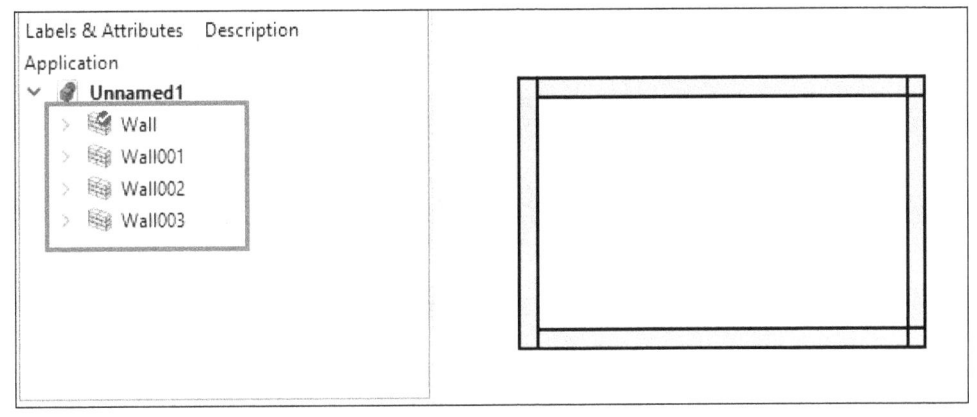

Figure-181. Walls created

- Select all the walls created from the **Model Tree** and click on the **Add component** tool from the **Toolbar**. A single wall will be created; refer to Figure-182.

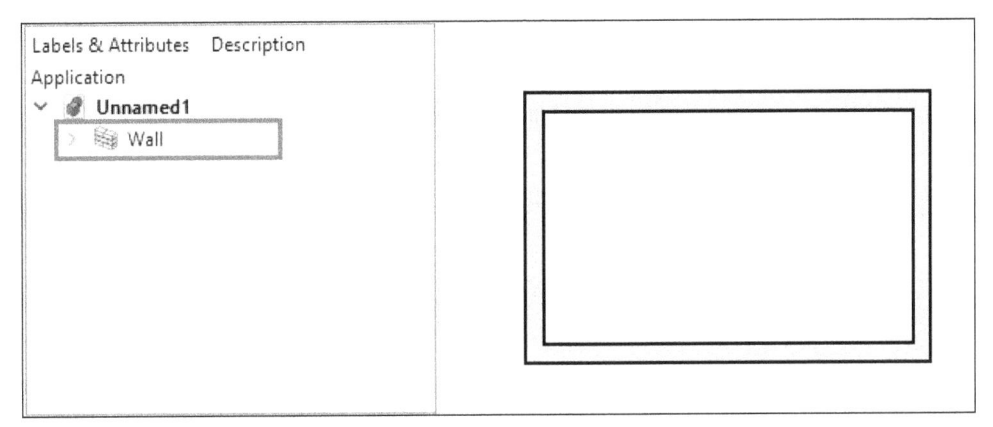

Figure-182. Single wall created

Creating Structure Object

- Click on the **Rectangle** tool from the **Toolbar** and create the rectangle on the **Top** plane as shown in Figure-183.

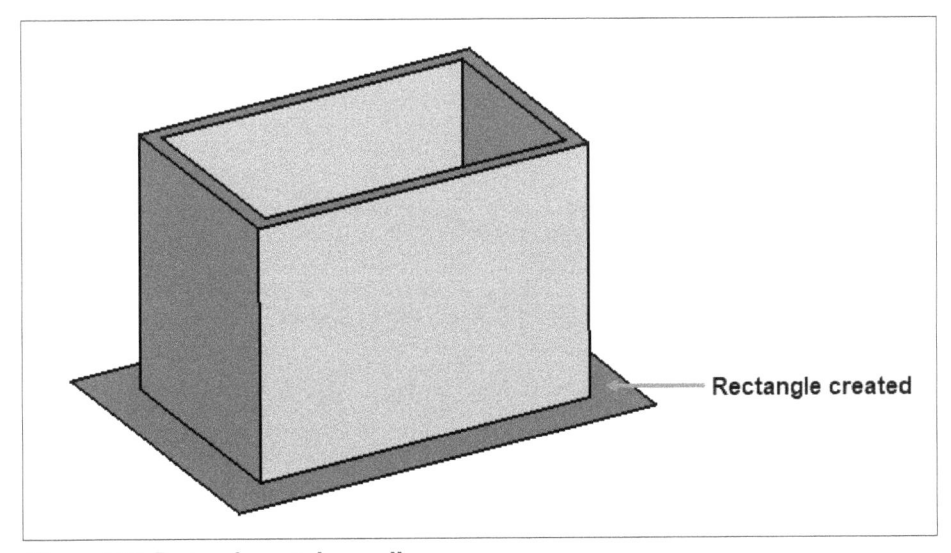

Figure-183. Rectangle created on wall

- Select the rectangle from the **Model Tree** and click on the **Column** tool from the **Toolbar**. The column object will be created; refer to Figure-184.

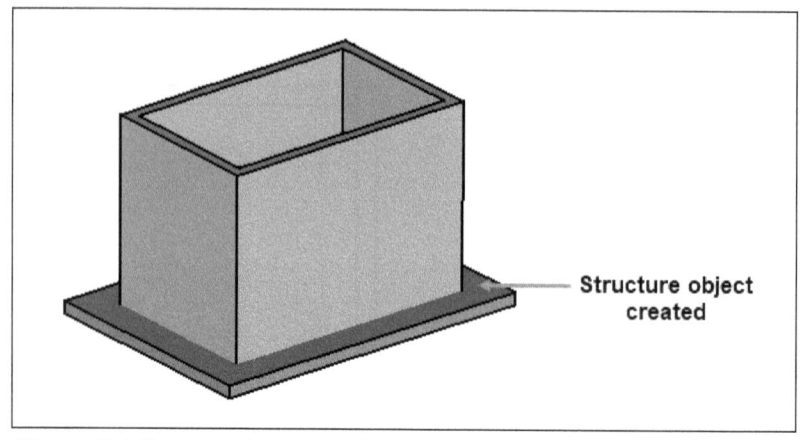

Figure-184. Structure object created

Creating Door

- Click on the **Window** tool from **Toolbar** and create a simple door as shown in Figure-185.

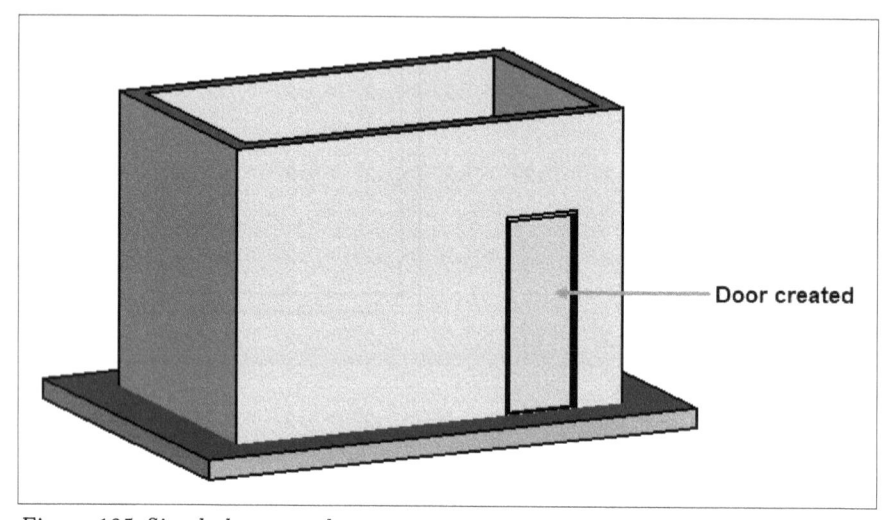

Figure-185. Simple door created

Creating Second Structure

- Click on the **Column** tool from the **Toolbar** and create the structure as shown in Figure-186.

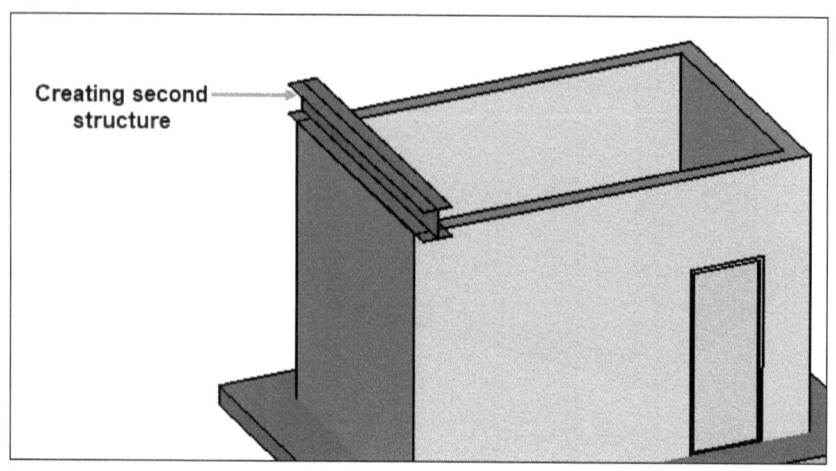

Figure-186. Second structure created

Creating Array

- Select the structure recently created, click on the **Array** tool from the **Toolbar** and create an array of the structure as shown in Figure-187.

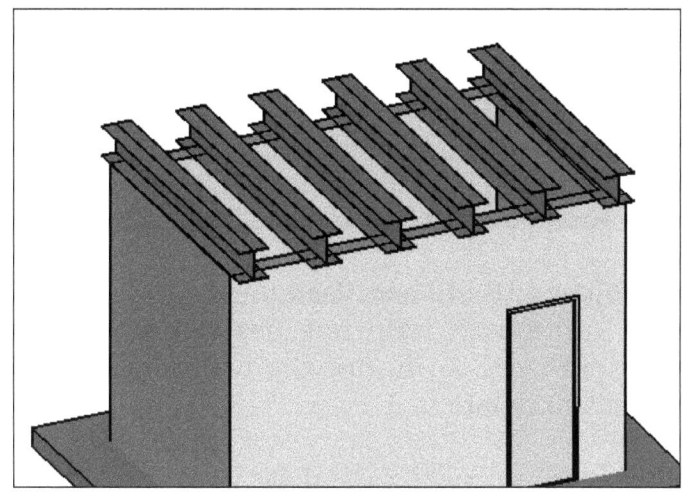

Figure-187. Array of the structure created

Creating Third Structure

- Click on the **Rectangle** tool and create a rectangle on the **Top** plane as shown in Figure-188.

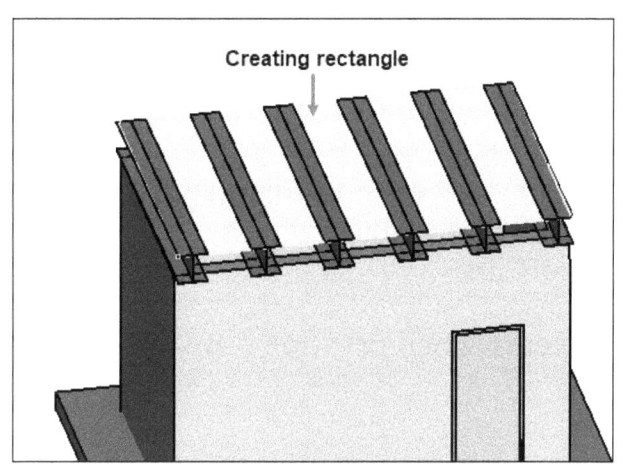

Figure-188. Rectangle created for third structure

- Select the recently created rectangle and create structure object using **Column** tool. The model of house will be created; refer to Figure-189.

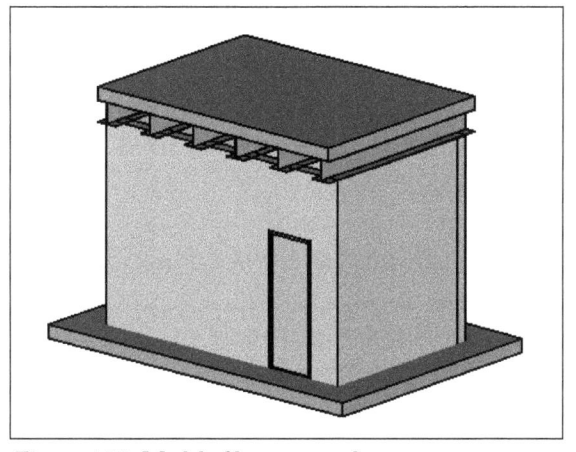

Figure-189. Model of house created

SELF ASSESSMENT

Q1. What is the main purpose of the Project tool in FreeCAD?

A) To create a building model
B) To create a native IFC project in the current document
C) To add objects to a building
D) To edit the properties of a site

Q2. How do you modify the properties of a site object after it is created?

A) Right-click the site object and choose "Delete"
B) Select the site object and modify its properties in the Property editor dialog
C) Use the "Add component" tool to change the properties
D) Click on the site object to delete it

Q3. What is the Building tool primarily used for in FreeCAD?

A) To create a terrain
B) To organize a model by containing floor objects
C) To create a wall object
D) To define the base terrain for a site

Q4. When creating a floor using the Level tool, what is required?

A) Only one object from the Model Tree
B) Select one or more objects you want to include in a floor
C) The object must be a free group object
D) The object must be a terrain

Q5. Which dialog box appears when creating a Space in FreeCAD?

A) Property editor dialog
B) Component dialog
C) Space options dialog
D) Window dialog box

Q6. What does the Wall tool in FreeCAD allow you to do?

A) Create a wall by specifying only its height
B) Build a wall on top of a base object or create one from scratch
C) Create a window or door object
D) Define the properties of a space

Q7. Which of the following is true about the Curtain Wall tool?

A) It is used to create a solid wall from scratch
B) It subdivides a base face into quadrangular faces and adds mullions
C) It only works with the Terrain object
D) It is used to create structural beams

Q8. How do you edit the properties of a pipe after it has been created?

A) Double-click the pipe in the 3D view
B) Right-click the pipe and select "Edit"
C) Select the pipe and modify its properties in the Property editor dialog
D) Use the "Remove component" tool

Q9. What is the purpose of the Slab tool in FreeCAD?

A) To create a floor object
B) To create a column from scratch
C) To create a shape from a planar shape
D) To create a window with preset options

Q10. What is the function of the Pipe Connector tool in BIM workbench?

a) To create pipes
b) To create corner or tee connections between two or three selected pipes
c) To change the diameter of pipes
d) To create junctions for electrical wiring

Q11. What happens when you use the Roof tool in BIM workbench?

a) It creates a flat roof
b) It creates a sloped roof based on selected wire
c) It creates a curved roof
d) It creates a roof without any parameters

Q12. What must be selected before using the Frame tool in BIM workbench?

a) A wall
b) A layout object and a profile object
c) A single fence post and section
d) A 3D model of the object

Q13. What is required to create a Fence using the Fence tool?

a) A pre-created 3D model
b) A fence post, a section, and a path
c) A panel and structure objects
d) A roof and truss combination

Q14. Which of the following can the Truss tool create?

a) A staircase
b) A truss object from a selected linear object or from scratch
c) A frame object
d) A roof structure

Q15. In the Equipment tool, what can be specified in the Property editor dialog?

a) The number of steps
b) The power required by the equipment
c) The slope of the roof
d) The diameter of reinforcing bars

Q16. When using the Rebar tool, what is needed before placing reinforcing bars inside a structure?

a) A sketch created in the Sketcher workbench
b) A fence post
c) A pre-defined truss
d) A frame object

Q17. What does the Profile tool in BIM workbench create?

a) A 3D object
b) A parametric 2D profile object
c) A curved roof structured)
d) A structural frame

Q18. What is the purpose of the External Reference tool in BIM workbench?

a) To add components to an Arch object
b) To link to external files and import objects from them
c) To create new 3D objects
d) To add materials to the project

Q19. Which of the following objects can be added using the Add Component tool?

a) Roof
b) Shape-based objects to an Arch component
c) Rebar
d) Trusses

Q20. What does the Property editor dialog allow you to modify after creating a Roof?

a) The length and width of the roof
b) The overhang, angle, thickness, and height of the roof
c) The number of steps
d) The number of sheets of material in the roof

Q21. What is the primary purpose of the Remove Component tool in the BIM workbench?

A) To add a subcomponent to an Arch component
B) To remove a subcomponent from an Arch component
C) To modify an Arch component's properties
D) To create a new Arch component

Q22. Which of the following is NOT a feature of the Axis tool in the BIM workbench?

A) Allows the placement of axes as references for snapping objects
B) Can be used with Axes Systems to create parametric arrays
C) Enables the modification of axis parameters like distance and angle
D) Creates a grid-like structure for building walls

Q23. What does the Axis System tool in the BIM workbench allow you to do?

A) Create a grid of parallel lines
B) Combine multiple axes to define intersection points
C) Modify the appearance of axis bubbles
D) Add additional rows and columns to a grid

Q24. What is the primary purpose of the Grid tool in the BIM workbench?

A) To generate automatic floor levels in the building
B) To place a grid-like object as a reference for constructing Arch elements
C) To create axes for snapping objects together
D) To visualize structural systems

Q25. Which dialog allows you to modify the properties of a section plane in the BIM workbench?

A) Section Plane Placement Dialog
B) Section Plane Settings Dialog
C) Section Plane Properties Dialog
D) Section Plane Configuration Dialog

Q26. When creating a structure with the Structure tool, what must you specify first?

A) The size of the structure
B) The placement point for the structure
C) The type of material for the structure
D) The extrusion path for the structure

Q27. How do you create multiple structures from a selected base using the Multiple Structures tool?

A) Select the base object, then choose multiple edges for extrusion paths
B) Choose a preset structure, then apply it to all edges
C) Select an axis system and click "Multiple Structures"
D) Drag and drop the base object into the Model Tree

Q28. In the Building Part tool, what happens after selecting objects and clicking the Building Part button?

A) The selected objects are removed from the Model Tree
B) A new Building Part is created, grouping the selected objects together
C) The properties of the selected objects are automatically modified
D) A new floor level is created for the selected objects

Q29. What does the Nest tool allow you to do in the BIM workbench?

A) Group multiple structures into one object
B) Create a 3D model of nested panels
C) Organize flat shapes inside a container shape
D) Generate a cutout from a panel

Q30. When creating a Panel Cut, how do you specify the tag's position?

A) By selecting the tag from the Model Tree
B) By adjusting the Tag Position settings in the Panel Cut dialog
C) By entering coordinates in the 3D view area
D) By rotating the panel cut object

Q31. What action do you take to edit the properties of a panel sheet after it has been created?

A) Right-click the panel sheet and select "Edit Properties"
B) Double-click the panel sheet from the Model Tree
C) Open the Property editor dialog from the Toolbar
D) Select the panel sheet and click the "Edit" button in the Panel Sheet dialog

Q32. How do you modify the size of the section plane gizmo in the 3D view?

A) By adjusting the Arrow Size in the Property editor dialog
B) By resizing the section plane object in the 3D view
C) By entering new values in the Section Plane dialog
D) By rotating the section plane

Q33. What is the function of the Panel Sheet tool in the BIM workbench?

A) To create a 2D sheet that includes panel cuts and 2D objects
B) To place structural elements in the grid system
C) To add texture and material to a panel object
D) To create a 3D structure using panel cuts

Q34. Which of the following options allows you to copy values from an existing material while creating a new material?

a) Choose preset drop-down
b) Copy existing drop-down
c) Parent drop-down
d) Edit button next to the Name edit box

Q35. When creating a multi-material, which button is used to add a new layer to the Composition list?

a) Add row button
b) Add button
c) Import button
d) Up button

Q36. What is the function of the "Composition list" in the Multi-Material tool?

a) It stores the names of materials
b) It defines the material layers and their properties
c) It displays the thickness of each material
d) It defines the parent material

Q37. In the Schedule tool, what happens when you select the "Auto update" checkbox?

a) The spreadsheet will be updated manually
b) The schedule will automatically update when the document is recomputed
c) The schedule will not update automatically
d) The schedule will not be saved

Q38. What type of information can be added to a Schedule using the "Property" edit box?

a) The name of the object
b) The unit of measurement
c) The description of the material
d) The property value to retrieve or count for each object

Q39. Which tool allows you to grab measurements from the model and transfer them to other applications?

a) Material tool
b) Multi-Material tool
c) Survey tool
d) Schedule tool

Q40. What is the purpose of the "Export CSV" button in the Survey tool?

a) To export the model data
b) To export the contents of the survey dialog to a CSV file
c) To export the schedule data
d) To export object properties

For Student Notes

Chapter 8

CAM Workbench

Topics Covered

The major topics covered in this chapter are:

- *Introduction to CAM Workbench*
- *Create Jobs*
- *Tool Manager*
- *G-Code Generation*
- *Basic CAM Operations*
- *CAM Modification Tools*
- *Path Utilities*
- *Post-Processing*

INTRODUCTION

The **CAM** workbench is used to generate machine codes for CNC machines using a FreeCAD 3D model; refer to Figure-1. You can use these codes on CNC machines such as mills, lathes, laser cutters, and so on to perform real machining on workpiece. These machine codes are also called G-codes.

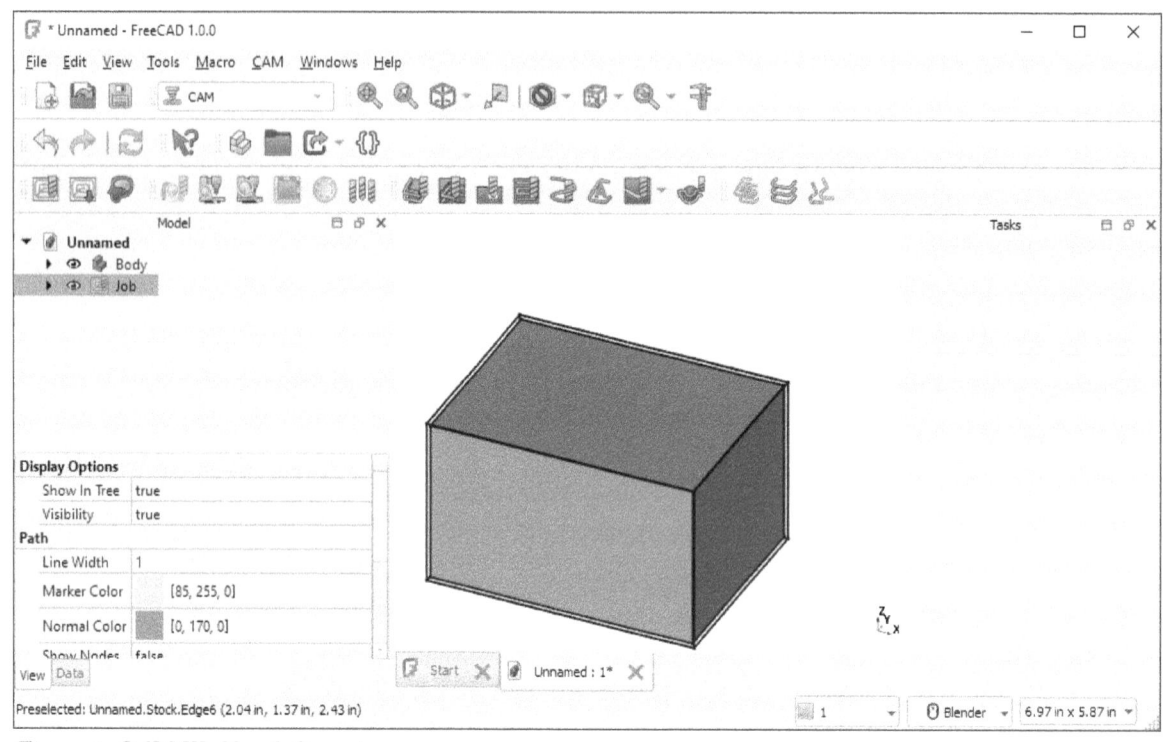

Figure-1. CAM Workbench Overview

The FreeCAD CAM workbench creates these machine instructions as follows:

- A 3D model of the part to be achieved after performing machining is created in modeling workbenches of FreeCAD.
- A Job is created in CAM Workbench. This job contains all the information required to generate necessary G-Code for a CNC machine like Stock material, list of cutting tools to be used for machining with cutting parameters, machine characteristics, and so on.
- Tools are selected as required by the Job Operations.
- Machining paths are created using operations like contour and pocket. These path objects use internal FreeCAD G-Code dialect independent of the CNC machine.
- Machining paths are exported to G-codes matching to your machine. This step is called post processing which uses a set of conversion codes called Post-processor specific to your machine.

STARTING CAM WORKBENCH

The CAM workbench generates G-Code defining the paths required to mill the Project represented by the 3D model on the target mill in the CAM Job Operations FreeCAD G-Code dialect which is later translated to the appropriate dialect for the target CNC controller by selecting the appropriate postprocessor.

- To start a new file in CAM workbench, click on the **New** button from **File** menu and select **CAM** workbench from **Switch between workbenches** drop-down in the **Toolbar**; refer to Figure-2. The tools related to CAM workbench will be displayed in the **Toolbar**; refer to Figure-3.

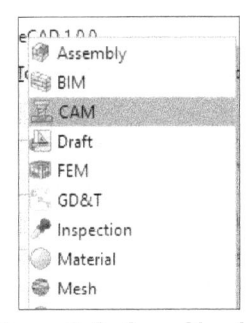

Figure-2. Path workbench

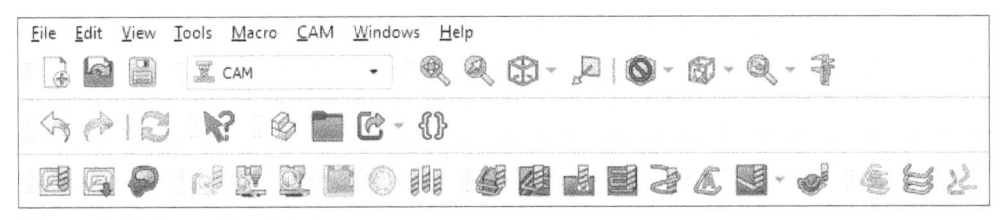

Figure-3. Path workbench tools

JOBS, TOOLS, AND GENERATING G-CODE

After creating model, the first important step for generating CNC program is setting up a job. In FreeCAD, job means defining machine, tool controllers, geometry orientation, and so on. The tools to perform job setup are available in the **Toolbar** of **Path** workbench; refer to Figure-4. These tools are discussed next.

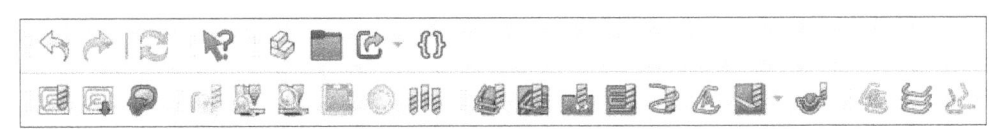

Figure-4. Tools for creating jobs and generating G codes

Creating Job

The **Job** tool is used to create a new machining job in the active document. The Job contains following information:

- A list of Tool-Controller definitions, specifying the geometry, Feeds, and Speeds for the CAM Operations Tools.
- A Workflow sequential list of CAM Operations.
- A Base Body—a clone used for offset.
- A Stock, representing the raw material that will be milled to get final model.
- A SetupSheet containing inputs used by the CAM Operations, including static values and formulas.
- Configuration parameters specifying the output G-Code job's destination CAM, file name and extension, the Postprocessor—used to generate the appropriate dialect for the target CNC Controller, and so on.

The procedure to use this tool is discussed next.

- Open or create the 3D model for which you want to create the job; refer to Figure-5.

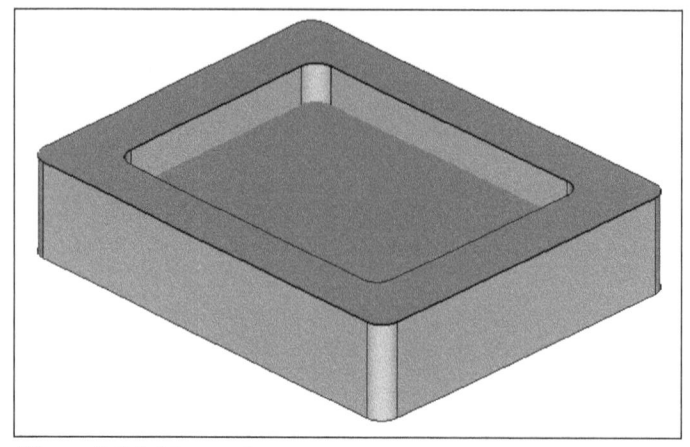

Figure-5. Model created

- Click on the **Job** tool from **Toolbar** in the CAM workbench; refer to Figure-6. If you are running this tool for first time and your unit system does not have minute set as unit for time then a warning box is displayed; refer to Figure-7. Click on the OK button and later set the unit system accordingly. The **Create Job** dialog box will be displayed; refer to Figure-8. You will be asked to select the object.

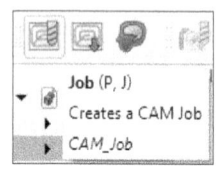

Figure-6. Job tool

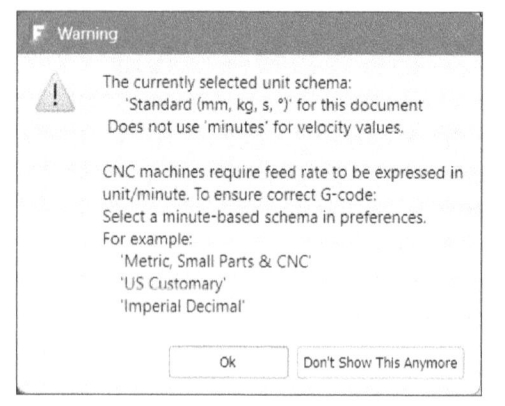

Figure-7. Warning message box

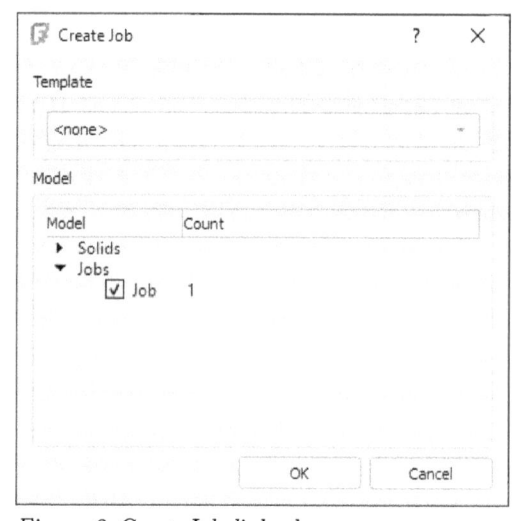

Figure-8. Create Job dialog box

- Select the object from **Model** list box in the dialog box for which you want to create a job and click on **OK** button from the dialog box. The **Job Edit** dialog will be displayed in the **Tasks** panel of **Combo View** with the **Setup** tab opened by default and the model will be enclosed by a bounding box; refer to Figure-9.

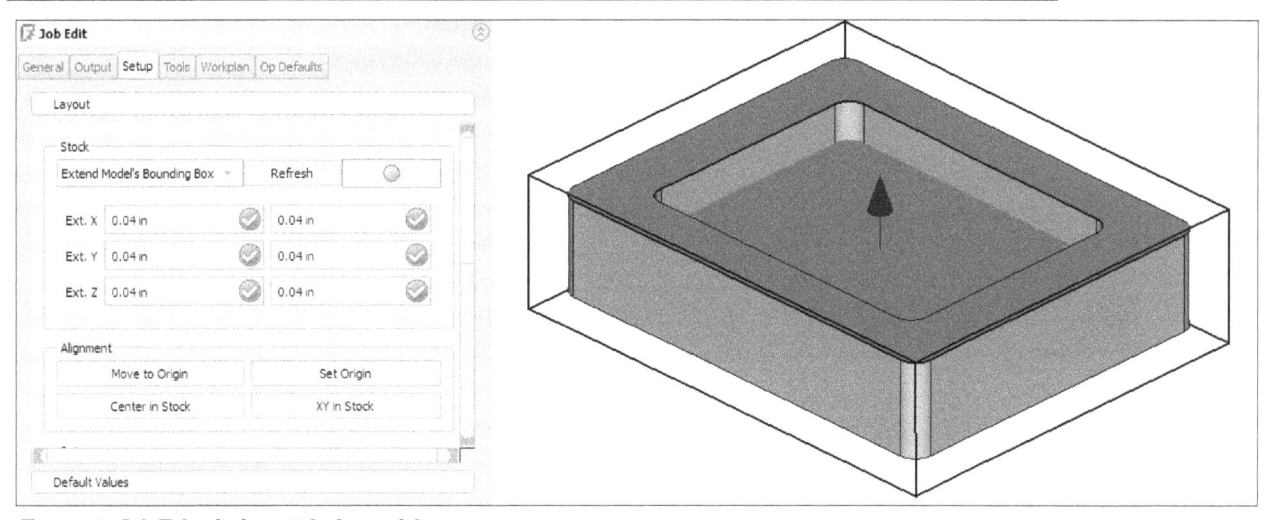

Figure-9. Job Edit dialog with the model

Defining Stock

- Select desired option from the drop-down in the **Stock** area of **Layout** section to define shape and size of the raw material to be machined. Select the **Create Box** option to create stock in box shape using parameters specified in **Length**, **Width**, and **Height** edit boxes; refer to Figure-10.

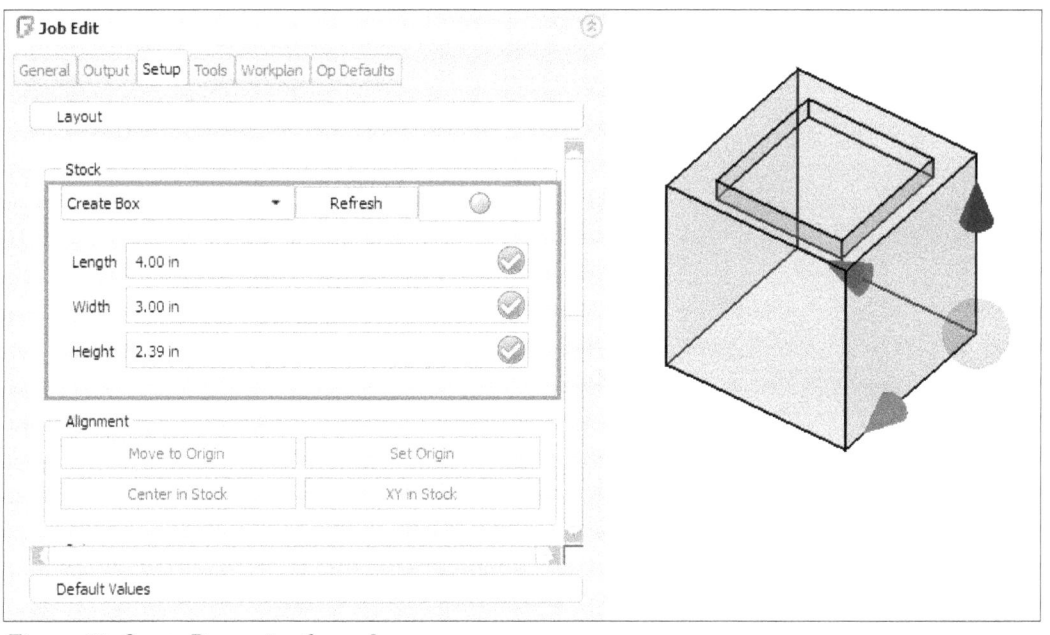

Figure-10. Create Box option for stock

- Select the **Create Cylinder** option from the drop-down if you want to create a cylindrical stock using specified radius and height parameters; refer to Figure-11.

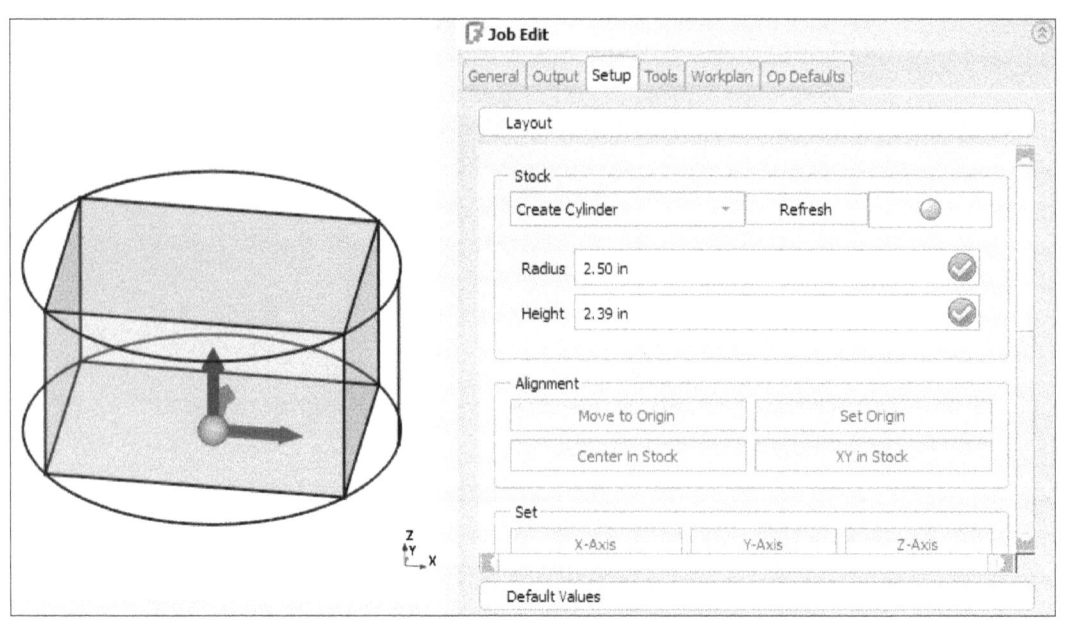

Figure-11. Create Cylinder option for stock

- Select the **Extend Model's Bound Box** option from the drop-down if you want to create stock by extending X, Y, and Z sides of base model by specified values; refer to Figure-12.

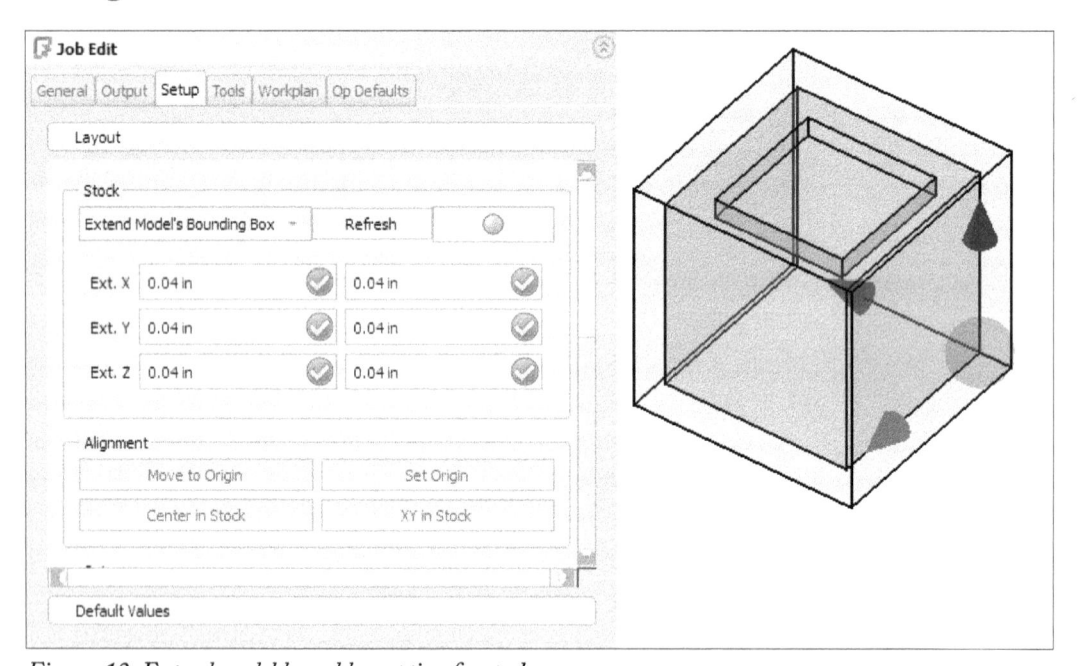

Figure-12. Extend model bound box option for stock

- Select the **Use Existing Solid** option from the drop-down to use existing solid currently available in the drawing area. On selecting this option, you can select any existing solid feature from the second drop-down of **Stock** area; refer to Figure-13.

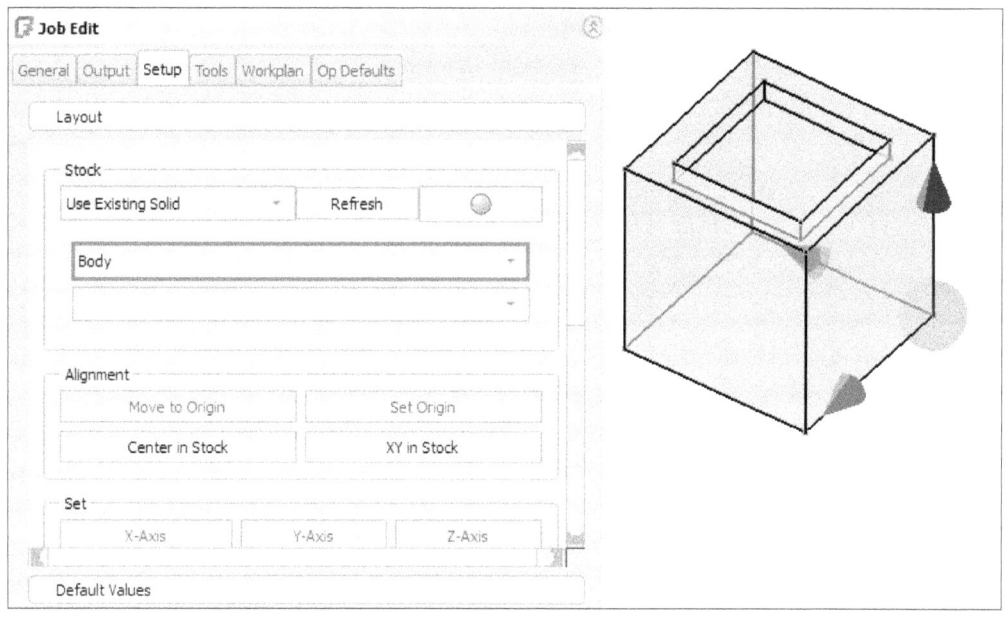

Figure-13. Using existing solid body

- Select a face of the model to specify alignment of model, the options in **Alignment**, **Set**, and other areas of dialog will become active; refer to Figure-14.

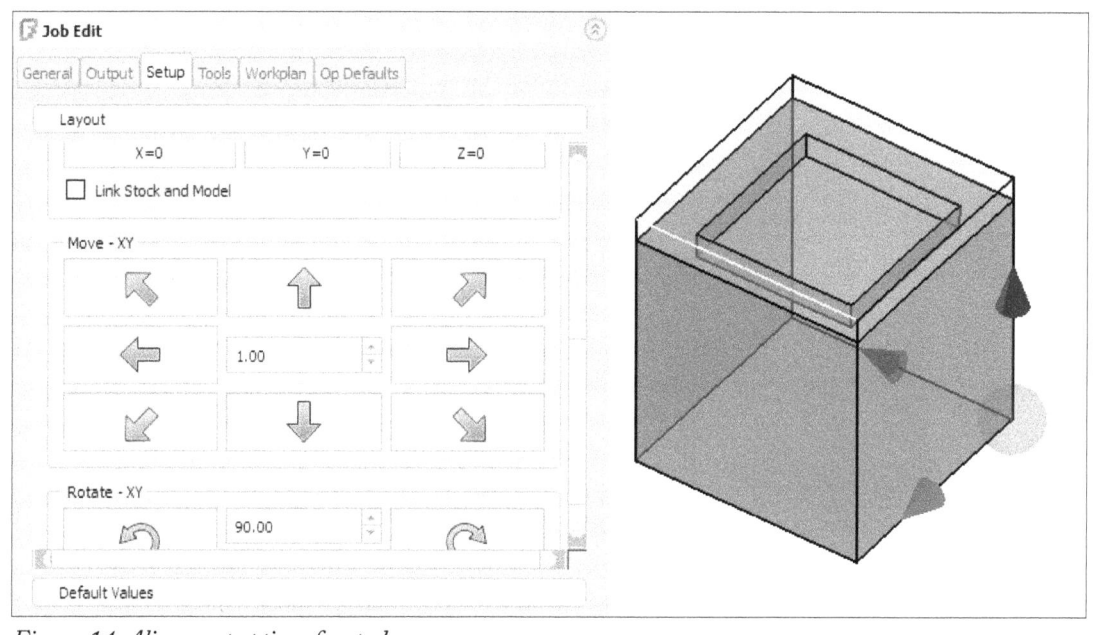

Figure-14. Alignment options for stock

- Select the **Center in Stock** button from the **Alignment** area to place the model in the center of stock. Select the **XY in Stock** button from the **Alignment** area to place the base of model within stock.
- Select desired button from the **Set** area to set X, Y, and Z as 0.
- Using the buttons in the **Move - XY** and **Rotate - XY** areas, you can move or rotate the model with respect to stock.

General Tab

- Click on the **General** tab from the dialog. The options in the **General** tab will be displayed; refer to Figure-15.

Figure-15. General tab of Job Edit dialog

- Specify desired name of job in the **Label** edit box from **Job** section of **General** tab.
- The **Model** edit box displays the base object which defines final shape of model for generating the paths of the job.
- If you want to add more models for creating a job then click on the **Edit** button below **Model** edit box. The **Model Selection** dialog box will be displayed; refer to Figure-16.

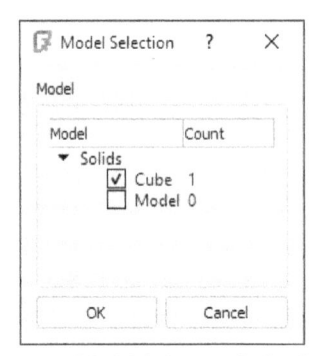

Figure-16. Model Selection dialog box

- Select desired model and click on **OK** button from the **Model Selection** dialog box. The model will be added in the **Job Edit** dialog.
- You can add some notes to the job in the **Description** edit box of the **General** tab of the dialog.
- Click on the **Template Export** button at the bottom in the dialog to export current specified parameters as a template. The options will be displayed as shown in Figure-17.

Figure-17. Template Export options

- Select check boxes from the dialog to specify which parameters are to be exported in the template and click on the **Export** button. The `Path - Job Template` dialog box will be displayed; refer to Figure-18.

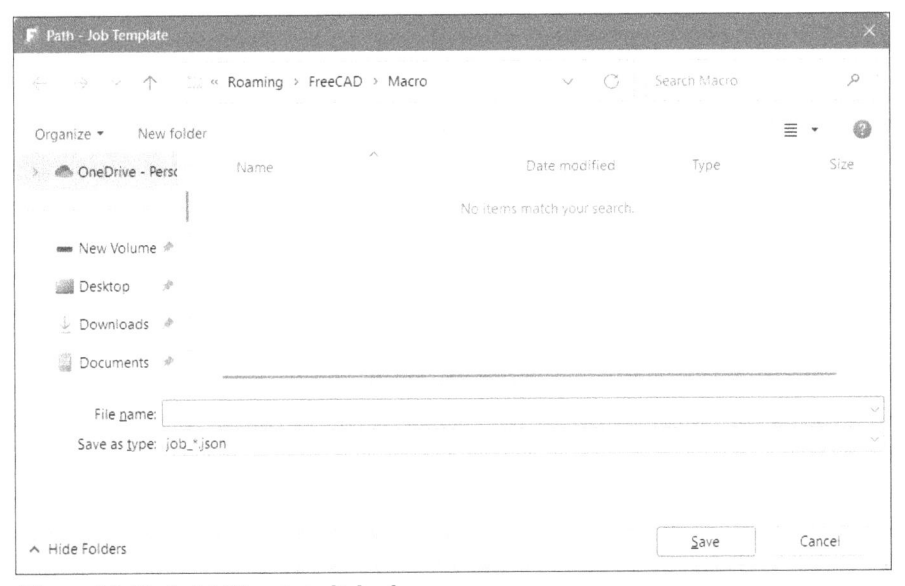

Figure-18. Path-Job Template dialog box

- Specify desired name for template in the `File name` edit box and save the template file at desired location.

Output Tab

- Click on the **Output** tab from `Job Edit` dialog. The options related to output tab will be displayed; refer to Figure-19.

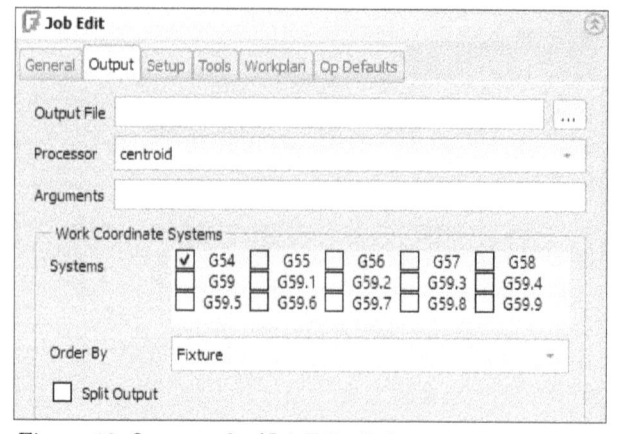

Figure-19. Output tab of Job Edit dialog

- Set desired name, extension, and the file path of the G-Code output in the **Output File** edit box of the **Output** tab.
- Select desired post processor for your machine from **Processor** drop-down.
- You can add desired arguments for the post processor as needed in the **Arguments** edit box of **Output** tab in the dialog.
- Select desired check boxes from **Systems** area of **Work Coordinate Systems** section in the **Output** tab of the dialog to specify the G-Code output for defining coordinate systems.
- Select desired option from **Order By** drop-down of **Work Coordinate Systems** section of the **Output** tab. Select **Fixture** option to perform all the operations in the first coordinate system before switching to the second. Select **Tool** option to minimize the tool changes. A tool change will be done, then all operations will be performed in all coordinate systems before changing tools. Select **Operation** option to perform each operation in all coordinate systems before moving to the next operation.
- Select the **Split Output** check box to write the G-code to multiple output files as controlled by the 'order by' property. For example, if ordering by Fixture, the first output file will be for the first fixture and separate file for the second.

Tools Tab

- Click on the **Tools** tab from **Job Edit** dialog to specify parameters related to cutting tools displayed in the dialog; refer to Figure-20.

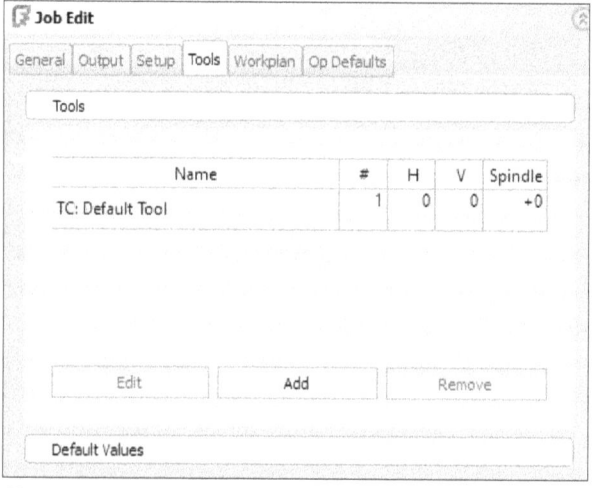

Figure-20. Tools tab of Job Edit dialog

FreeCAD 1.0 Black Book   8-11

- You can add the tool which you need for the operations of this job from **Tool** dialog box displayed by clicking on the **Add** button from **Tools** section of **Tools** tab; refer to Figure-21.

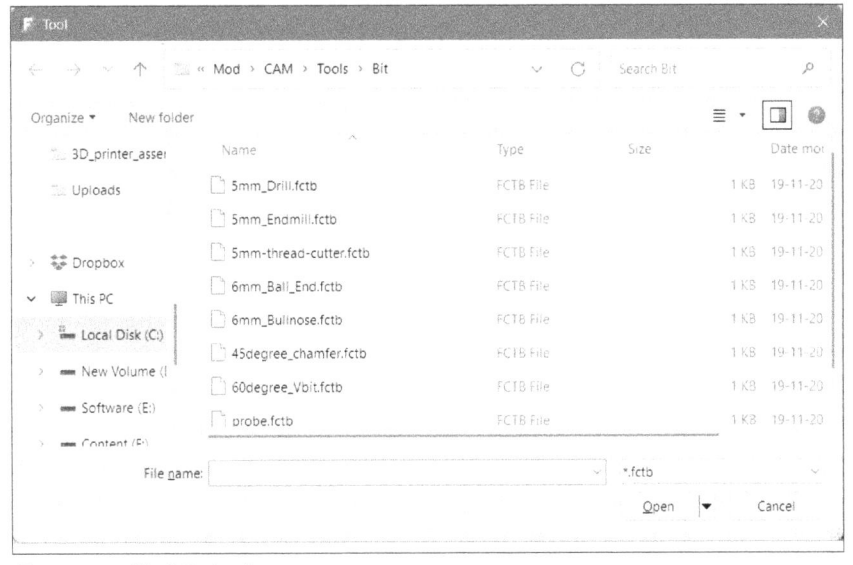

Figure-21. Tool dialog box

- The added tool will be displayed in the tool list box of **Tools** section in **Tools** tab of the **Job Edit** dialog.
- If you want to edit the tool then select the tool from the tool list and click on the **Edit** button from **Tools** section. The **Tool Controller Editor** dialog box will be displayed; refer to Figure-22.
- After adding a tool, you can set/change the feed rate and spindle speed if you need a different feed rate in this job.
- Similarly, specify desired parameters in **Workplan** and **Op Defaults** tab of the **Job Edit** dialog.
- Click on **OK** button from **Job Edit** dialog. The Job will be created and displayed in the **Model Tree**; refer to Figure-23.

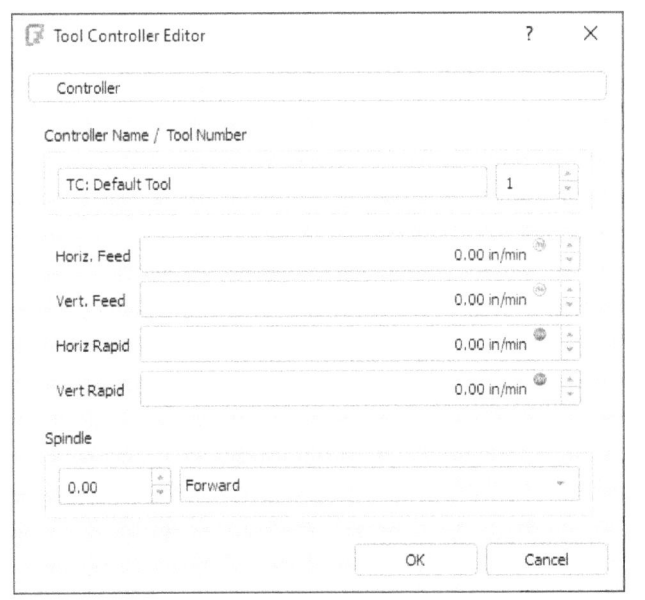

Figure-23. Job created

Figure-22. Tool Controller Editor dialog box

ToolBit Library Editor

The **ToolBit Library editor** tool is used to create and manage cutting tools used in performing machining operations in the Path workbench. The procedure to use or manage cutting tools is discussed next.

- Click on the **ToolBit Library editor** tool from **CAM** menu in the **Toolbar**; refer to Figure-24. If you are using this tool for the first time then you will be asked to specify the location for setting up tool library.

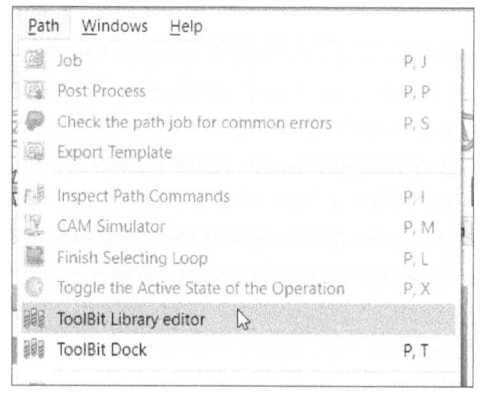

Figure-24. ToolBit Library editor tool

- Click on the **Yes** button from the information box. The **Choose a writeable location for your toolbits** dialog box will be displayed; refer to Figure-25.

Figure-25. Choose a writeable location for your toolbits dialog box

- Select desired location/folder to specify location where toolbit files will be saved for library and click on the **Select Folder** button. A message box will be displayed for creating subdirectories like Library and Shape.
- Click on the **Yes** button from the message box. A message box to copy all example files of toolbits will be displayed.
- Click on the **Yes** button from the message box. The **Library** dialog box will be displayed; refer to Figure-26. Note that the **ToolBit Library editor** tool will be active in **Toolbar** only after you have created a job setup as discussed earlier.

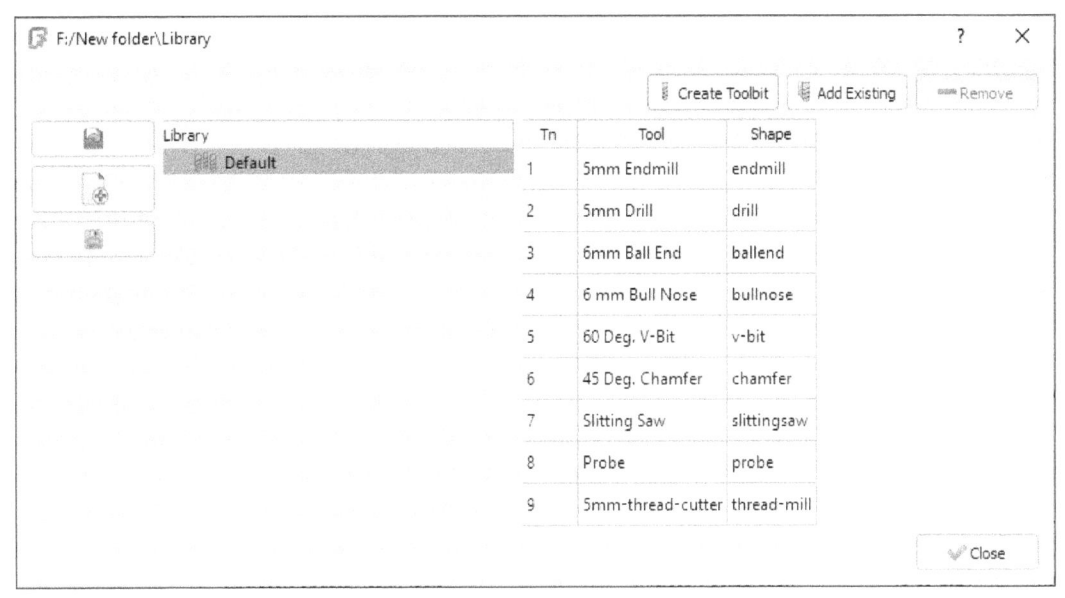

Figure-26. Tool Library dialog box

- Click on the **Add New Tool Table** button from the dialog box to import a cutting tool table from an XML-file. The **Save toolbit library** dialog box will be displayed; refer to Figure-27.

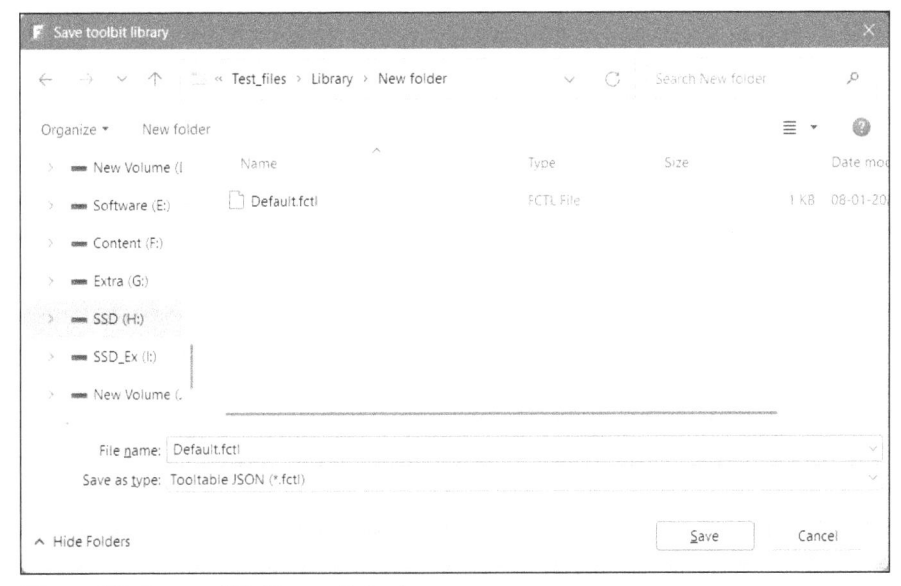

Figure-27. Save toolbit library dialog box

- Specify desired name of library file in the **File name** edit box and click on the **Save** button. The new library will get added in the library list of dialog box.
- You can also import library files from selected folder. To do so, click on **Open** button from the dialog box. The **Tool Library Path** dialog box will be displayed; refer to Figure-28.

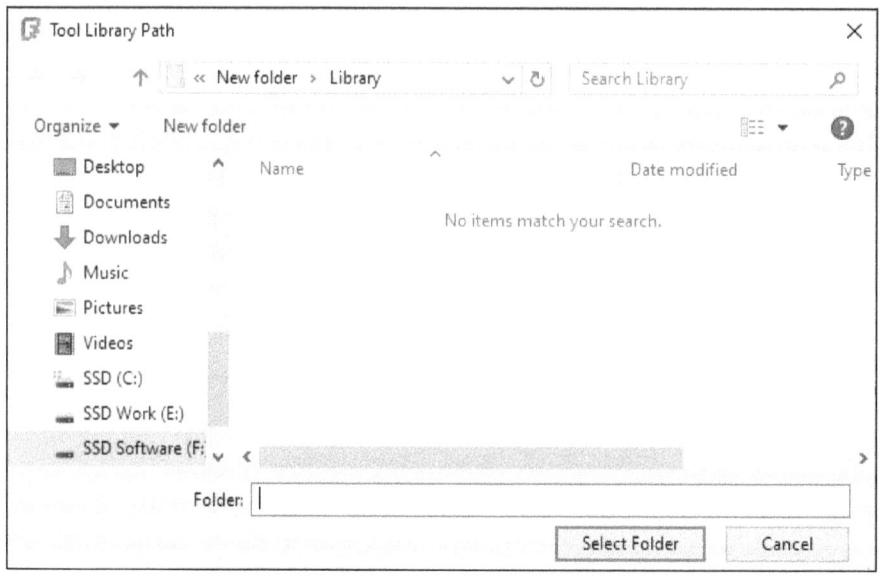

Figure-28. Tool Library Path dialog box

- Select the folder in which library files are available and click on the **Select Folder** button. The libraries of selected folder will be displayed in the dialog box.
- Click on the **Save** button from dialog box to save the toolbit library at desired location. The **Save toolbit library** dialog box will be displayed as discussed earlier.
- If you want to delete a tool from library then select the tool which you want to delete from the tools list box and click on **Remove** button.
- To create a cutting tool, click on **Create Toolbit** button. The **Select Tool Shape** dialog box will be displayed; refer to Figure-29.

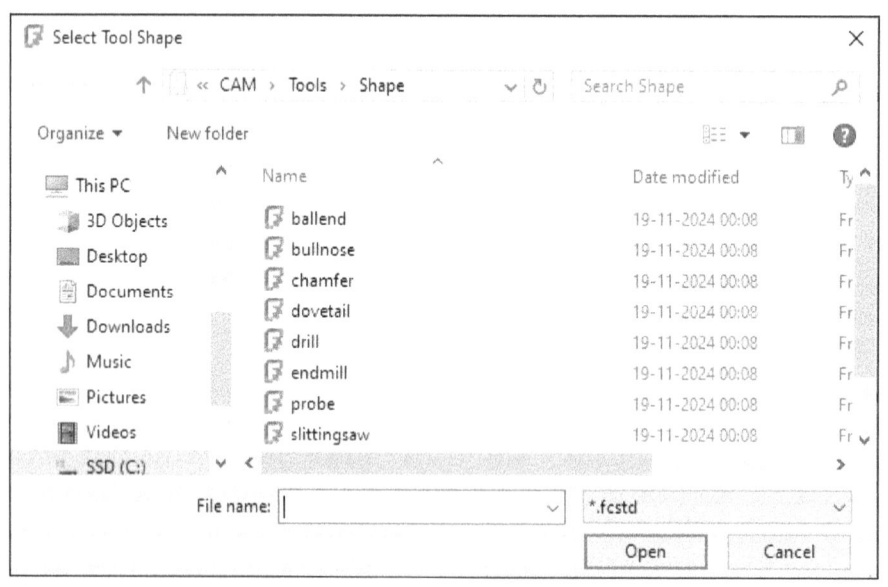

Figure-29. Select Tool Shape dialog box

- Select desired tool shape template from the dialog box and click on the **Open** button. The **Tool** dialog box will be displayed for saving new toolbit; refer to Figure-30.

FreeCAD 1.0 Black Book   8-15

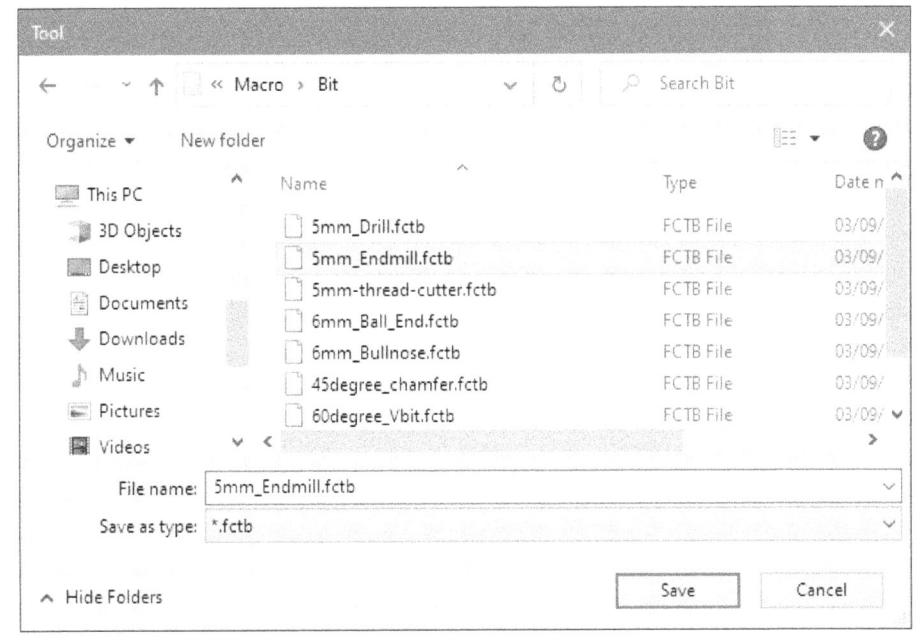

Figure-30. Tool dialog box

- Specify desired name for toolbit in the **File name** edit box and click on the **Save** button. The new tool will be added in the list.
- Select the newly added tool and double-click on it under **Tool** column. The options to modify cutting tool will be displayed; refer to Figure-31.

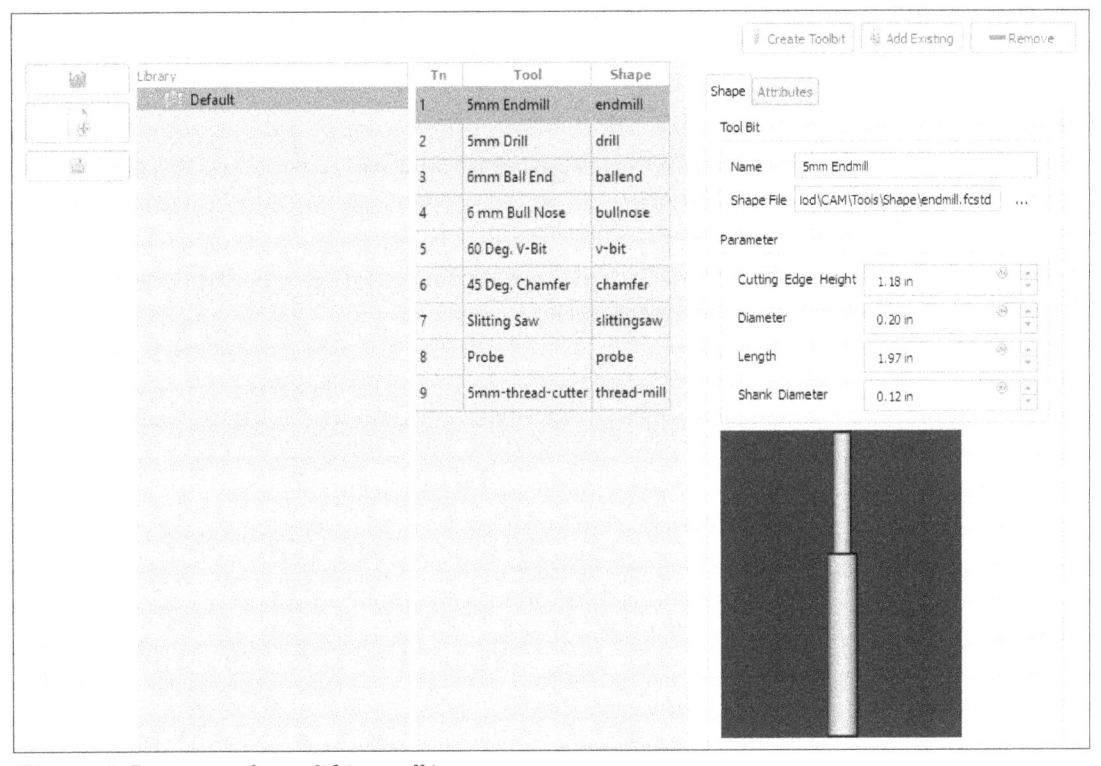

Figure-31. Parameters for modifying toolbit

- Specify desired parameters for cutting tool like diameter, length, cutting edge height, and so on in the edit boxes of **Shape** tab.
- Click on the **Attributes** tab in the dialog box to specify parameters like chipload, flutes, and material; refer to Figure-32.

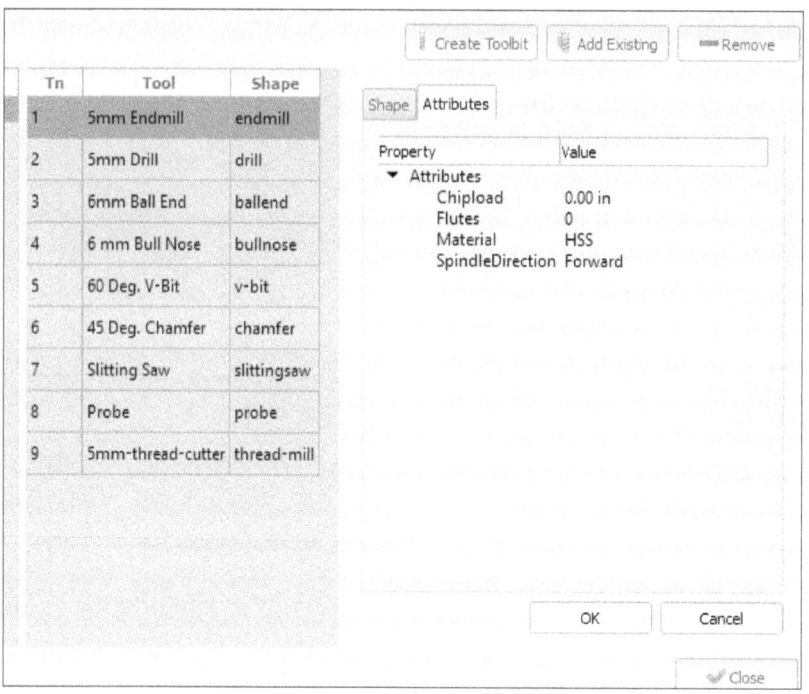

Figure-32. Attributes tab

- Specify desired parameters for the tool in the dialog box and click on **OK** button from the dialog box. The parameters specified for the tool will be applied and displayed in the library.
- Click on the **Add Existing** button to add already existing tools in the library.
- Click on the **Close** button from the **Tool Library** dialog box.

Exporting Template

The **Export Template** provides a convenient mechanism to save commonly used job definitions from within an existing Job. This facilitates the setup of future jobs that are largely similar by allowing job template import during the job creation process. The procedure to use this tool is discussed next.

- Click on **Export Template** tool from **CAM** menu in the **Toolbar**; refer to Figure-33. The **Job Template Export** dialog box will be displayed; refer to Figure-34.

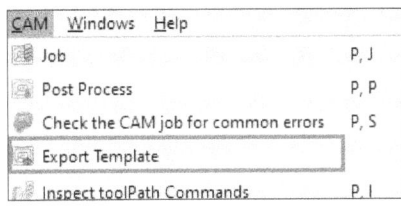

Figure-33. Export Template tool

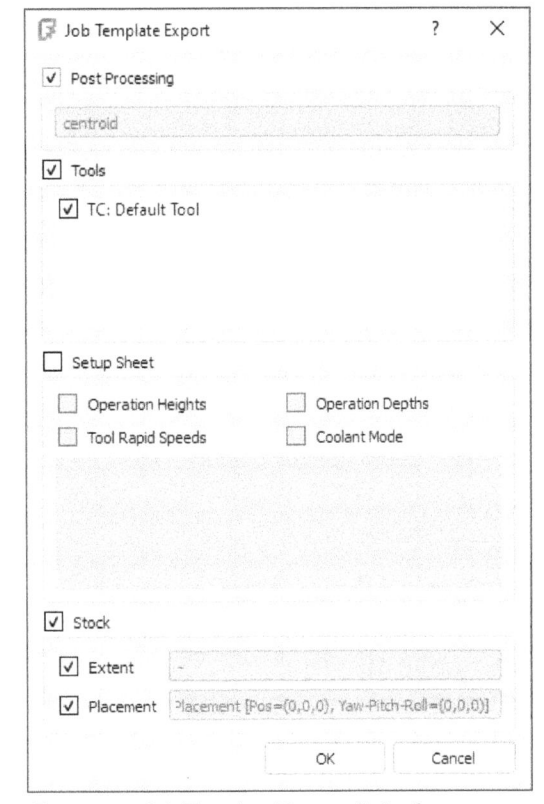

Figure-34. Job Template Export dialog box

- Select **Post Processing** check box from the dialog box to include all post processing settings in the template.
- Select the **Tools** check box to store tool controller definitions in the template.
- Select the check boxes of all the tool controllers which should be included in the template from **Tools** area of the dialog box.
- Select **Setup Sheet** check box to include values of the Setup Sheet in the template.
- Select **Operation Heights** check box from **Setup Sheet** area to include the default heights for operations in the template.
- Select **Operation Depths** check box from **Setup Sheet** area to include values of the setup sheet in the template.
- Select **Tool Rapid Speeds** check box to include the default rapid tool speeds in the template.
- Select **Coolant Mode** check box to include the default coolant mode in the template.
- Select **Stock** check box to include the creation of stock in the template. If a template does not include a stock definition, the default stock creation algorithm will be used.
- Select **Extent** check box to include the current size settings for the stock object in the template.
- Select **Placement** check box to store the current placement of the stock solid in the template.
- Click on **OK** button from the dialog box. The **Path-Job Template** dialog box will be displayed to save the template; refer to Figure-35.

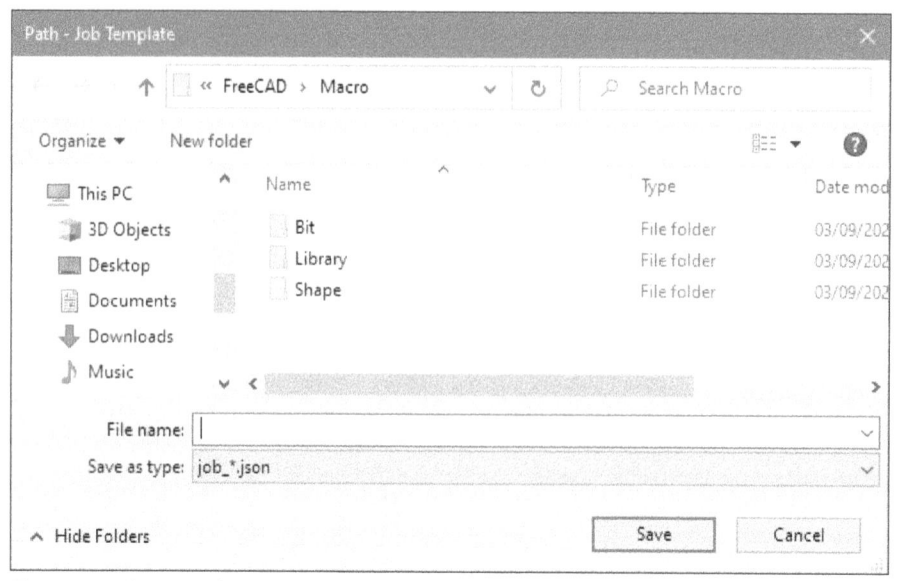

Figure-35. Path-Job Template dialog box

- Specify name for the template file in the `File name` edit box and specify desired location to save the file.
- Click on **Save** button from the dialog box. The template will be saved.

BASIC CAM OPERATIONS

Basic path operations are Contour, Profile Faces, and Profile Edges operations used to remove material from the stock. These tools are available in the **Toolbar** of **CAM** workbench; refer to Figure-36. These tools are discussed next.

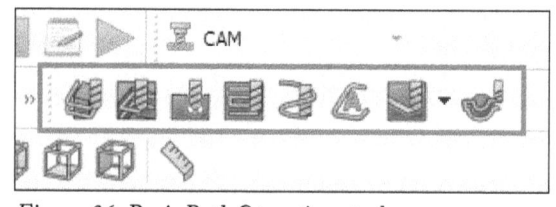

Figure-36. Basic Path Operations tool

Creating Contour Operation

The **Profile** tool creates a contour operation based on selected features of the model. It creates a simple external contour cut for 3D parts. The procedure to use this tool is discussed next.

- Click on the **Profile** tool from **Toolbar** in the **CAM** workbench; refer to Figure-37. The **Choose a Tool Controller** dialog box will be displayed if you have not specified tool controller earlier for this operation while setting up the job; refer to Figure-38.

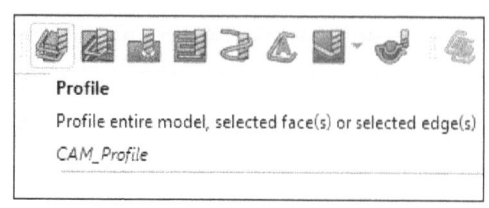

Figure-37. Profile tool

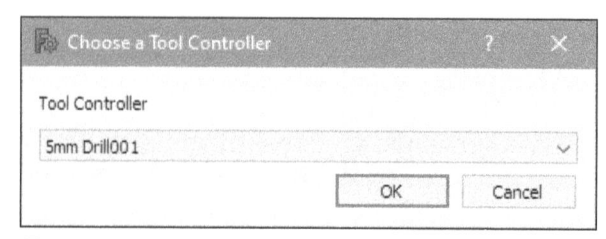

Figure-38. Choose a Tool Controller dialog box

- Select desired tool from the drop-down and click on the **OK** button. The **Profile** dialog will be displayed in the **Tasks** panel of **Combo View** and the model will be enclosed within the bounding box; refer to Figure-39.

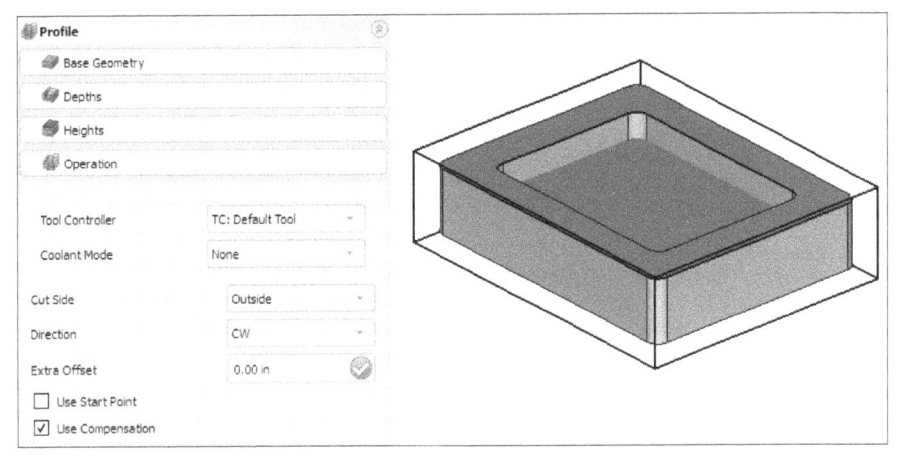

Figure-39. Profile dialog and the model enclosed within bounding box

- The tool controller specified while creating the Job for this operation will be displayed in the **Tool Controller** drop-down in the **Operation** tab of **Profile** dialog.
- Select desired coolant mode to be used for this operation from **Coolant Mode** drop-down.
- Select desired option from **Cut Side** drop-down to specify whether the profile should be performed inside or outside the base geometry features.
- Specify desired direction in which the profile is to be machined from **Direction** drop-down.
- Enter desired value in the **Extra Offset** edit box to specify the amount of extra material left by this operation in relation to the target shape.
- Select desired option from **Enable Rotation** drop-down to gain access to pockets or areas not normal to Z axis. This allows to perform 4-axis and 5-axis machining.
- Select the **Use Start Point** check box if you want to use specified point as starting point for operation. Select **Use Compensation** check box to offset the profile operation by tool radius. The offset direction is determined by the cut side.
- Select one or more features in the 3D view area and click on **Add** button from **Base Geometry** tab of the dialog to add them as the base items for this operation.
- Enter desired values in **Start Depth** and **Final Depth** edit boxes of **Depths** tab in the dialog to specify start and final depth of tool for the operation, respectively.
- Enter desired value in **Step Down** edit box from **Depths** tab of the dialog to specify incremental step down of tool for each cutting pass of operation.
- Enter desired value in **Safe Height** edit box from **Height** tab of the dialog to specify the height above which Rapid motions are allowed.
- Enter desired value in **Clearance Height** edit box from **Height** tab of the dialog to specify the height needed to clear clamps and obstructions.
- Click on **Apply** button and then **OK** button from the dialog. The Contour operation path will be generated; refer to Figure-40.

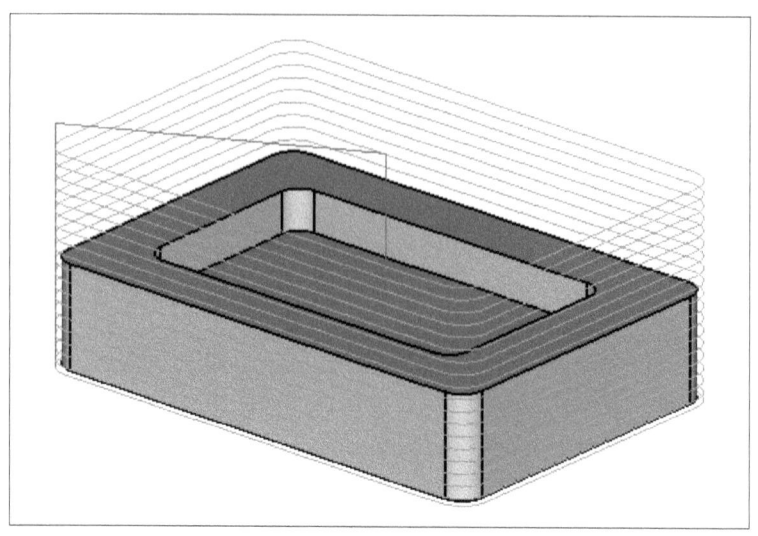

Figure-40. Contour operation path generated

Creating Pocket Object

The **Pocket Shape** tool creates pocket machine operation for selected bottom faces or walls of one or more pockets in the model. The procedure to use this tool is discussed next.

- Click on the **Pocket Shape** tool from **Toolbar** in the **CAM** workbench; refer to Figure-41. The **Pocket Shape** dialog will be displayed in the **Tasks** panel of **Combo View** and the model will be enclosed within the bounding box; refer to Figure-42.

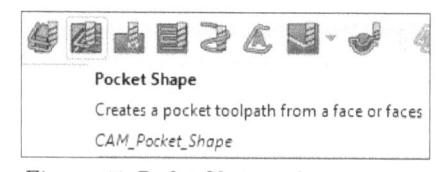

Figure-41. Pocket Shape tool

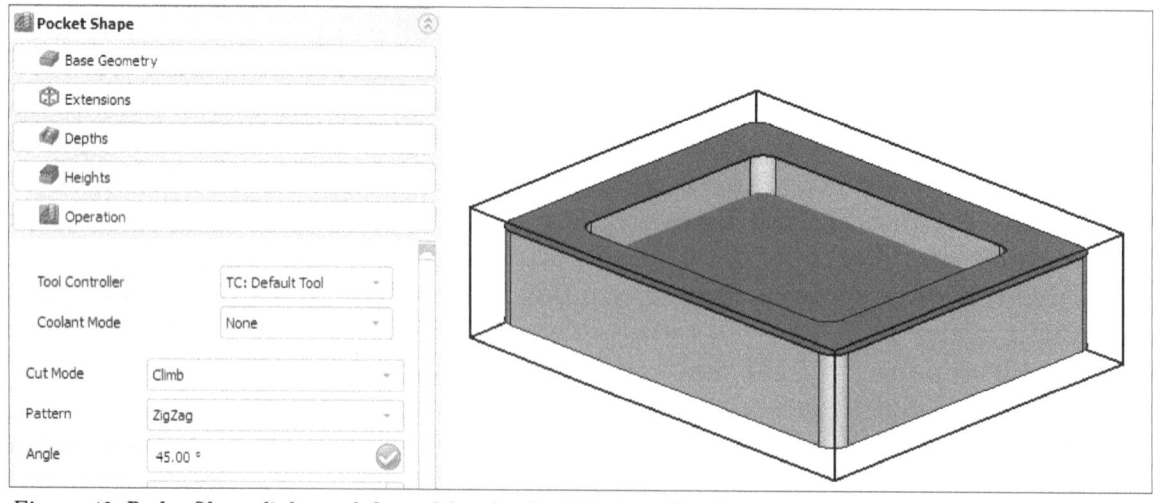

Figure-42. Pocket Shape dialog and the model enclosed within bounding box

- Select the bottom face or the walls of an object in the 3D view area for which you want to create pocket machining operation and click on **Add** button from **Base Geometry** tab of the dialog to add them to the list as the base geometries for this operation; refer to Figure-43.

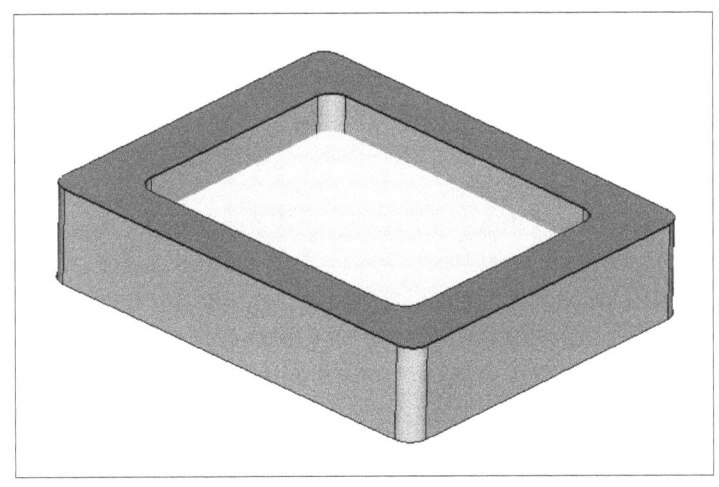

Figure-43. Selecting bottom face of the model

- Specify desired depth of final cut of the operation in the **Finish Depth** edit box from **Depths** tab of the dialog. Specify the value 0 to perform finishing operation.
- Select desired cutting mode from **Cut Mode** drop-down in the **Operation** tab of the dialog. The difference between conventional and climb cutting modes is shown in Figure-44.

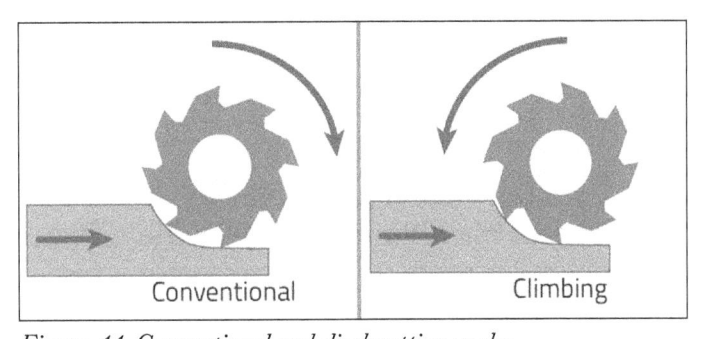

Figure-44. Conventional and climb cutting modes

Note: Conventional Milling V/S Climb Milling
Characteristics of Conventional Milling:
- The width of the chip starts from zero and increases as the cutter finishes slicing.
- The tooth meets the workpiece at the bottom of the cut.
- Upward forces are created that tend to lift the workpiece during face milling.
- More power is required to conventional mill than climb mill.
- Surface finish is worse because chips are carried upward by teeth and dropped in front of cutter. There's a lot of chip recutting. Flood cooling can help!
- Tools wear faster than with climb milling.
- Conventional milling is preferred for rough surfaces.
- Tool deflection during Conventional milling will tend to be parallel to the cut.

Characteristics of climb milling:
- The width of the chip starts at maximum and decreases.
- The tooth meets the workpiece at the top of the cut.
- Chips are dropped behind the cutter--less recutting.
- Less wear with tools lasting up to 50% longer.
- Improved surface finish because of less recutting.
- Less power required.

- Climb milling exerts a down force during face milling which makes work-holding and fixtures simpler. The down force may also help to reduce chatter in thin floors because it helps brace them against the surface beneath.
- Climb milling reduces work hardening.
- It can, however, cause chipping when milling hot rolled materials due to the hardened layer on the surface.
- Tool deflection during Climb milling will tend to be perpendicular to the cut, so it may increase or decrease the width of cut and affect accuracy.
- There is a problem with climb milling which is that it can get into trouble with backlash if cutter forces are great enough. The issue is that the table will tend to be pulled into the cutter when climb milling. If there is any backlash, this allows leeway for the pulling in the amount of the backlash. If there is enough backlash, and the cutter is operating at capacity, this can lead to breakage and potentially injury due to flying shrapnel.

Some worthwhile rules of thumb:

- When cutting half the cutter diameter or less, you should definitely climb mill (assuming your machine has low or no backlash and it is safe to do so!).

- Up to 3/4 of the cutter diameter, it doesn't matter which way you cut.

- When cutting from 3/4 to 1x the cutter diameter, you should prefer conventional milling.

- Select desired option from the **Pattern** drop-down in the **Operation** tab to specify the pattern in which the tool bit will move to clear the material. It is always better to select the pattern which resembles with walls of pocket.
- Enter desired value in the **Angle** edit box from **Operation** tab to specify the angle in which the pattern is applied.
- Enter desired value in **Step Over Percent** edit box to specify the amount by which the cutting tool will move laterally on each cutting pass in the pattern (specified in percent of the tool diameter).
- Specify desired distance value in the **Pass Extension** edit box from **Operation** tab by which, the facing operation will be extended beyond the boundary shape.
- Select **Use Outline** check box from the **Operation** tab if the operation uses the outline of the selected base geometry and ignores all holes and islands.
- Select **Min Travel** check box from the **Operation** tab to use 3D Sorting of Path (when multiple base geometries used).
- Select **Show All** check box from **Extensions** tab to visualize all potential extensions. Enabled extensions are in purple and not enabled extensions are in yellow.
- Select **Extend Corners** check box from **Extensions** tab to extend the corner between two edges of a pocket.
- Specify desired extent of the dimension in the **Default Length** edit box from **Extensions** tab.
- A list box in the **Extensions** tab shows the tree of existing edges and their potential extensions.
- Other parameters in the **Pocket Shape** dialog have been discussed earlier.
- Click on **Apply** button and then **OK** button from the dialog. The Pocket shape operation path will be generated; refer to Figure-45.

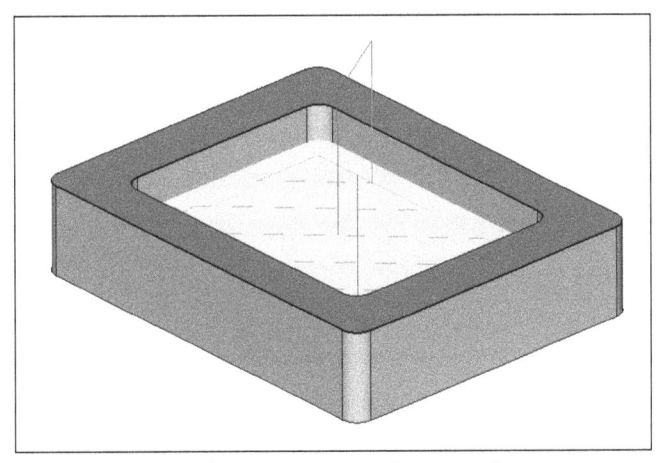

Figure-45. Pocket shape operation path generated

Performing Drill Operation

The **Drilling** tool generates a drilling operation in the job. The procedure to use this tool is discussed next.

- Click on the **Drilling** tool from **Toolbar** in the **CAM** workbench; refer to Figure-46. The **Choose a Tool Controller** dialog box will be displayed if no controller has been selected for this operation during job setup; refer to Figure-47.

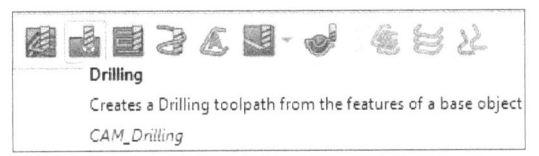

Figure-46. Drilling tool

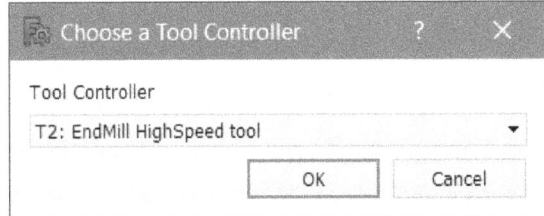

Figure-47. Choose a Tool Controller selection window

- Select desired drilling tool from **Tool Controller** drop-down and click on **OK** button from the selection window. The **Drilling** dialog will be displayed in the **Tasks** panel of **Combo View** and the model will be enclosed within the bounding box; refer to Figure-48.

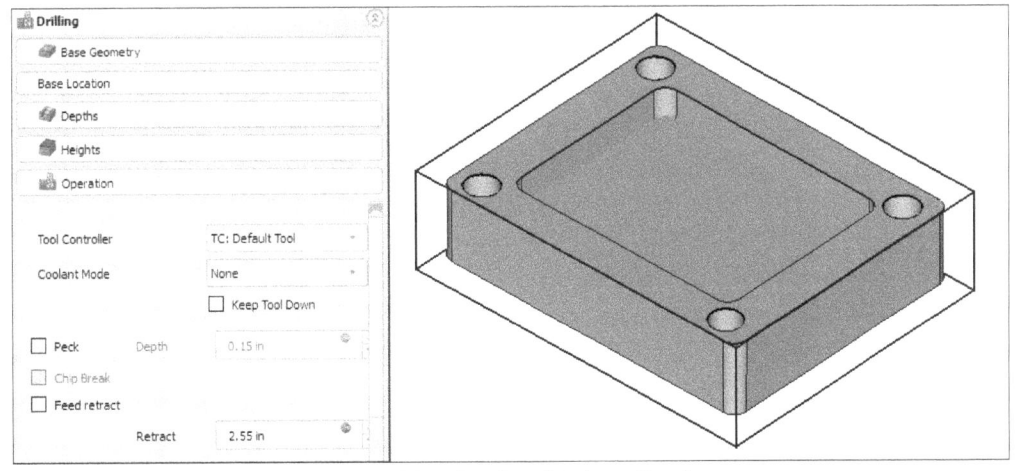

Figure-48. Drilling dialog and the model enclosed within bounding box

- Select the edges or/and faces of the hole feature of the model in the 3D view area on which you want to create the drilling operation; refer to Figure-49.

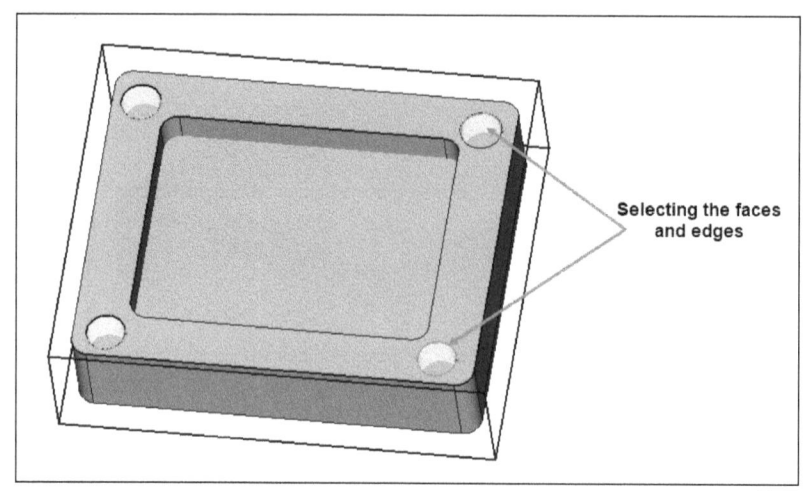

Figure-49. Selecting the faces and edges of hole feature

- After selecting the elements of the hole feature, click on **Add** button from the **Base Geometry** tab of dialog to add them in the selection list box.
- If you want to remove the element then select the element from the list box and click on **Remove** button from the **Base Geometry** tab.
- Click on the **Reset** button from the **Base Geometry** tab to delete all the current elements from the list and fills the list with all circular holes eligible for the drilling operation.
- Click on **Add** button from **Base Location** tab of the dialog. The direction edit boxes will be displayed in the tab and you will be asked to specify the locations.
- Click in the 3D view area at desired locations to specify the locations where you want to drill on the model or enter desired values in **Global X**, **Global Y**, and **Global Z** edit boxes to specify the drilling locations on the model.
- After specifying values in the edit boxes, click on **Save** button to save the drilling location in the list box or click on **Close** button to close the direction edit boxes.
- To remove the specified location, select the location from the list and click on **Remove** button from **Base Location** tab.
- Click on **Edit** button from **Base Location** tab to edit the specified location.
- Select **Peck** check box and enter desired value in the edit box from **Operation** tab to specify incremental drill depth before retracting to clear chips in Peck cycle of drilling.
- Select **Feed retract** check box to retract at the hole at the given inseted of rapid move. Enter desired value in **Retract** edit box from **Operation** tab of the dialog to specify the incremental height where feed starts above the model surface.
- Select **Dwell** check box and enter desired value in the edit box to specify the time of pause between peck cycles.
- Other parameters in the **Drilling** dialog have been discussed earlier. Click on **Apply** button and then **OK** button from the dialog. The drilling operation path will be generated; refer to Figure-50.

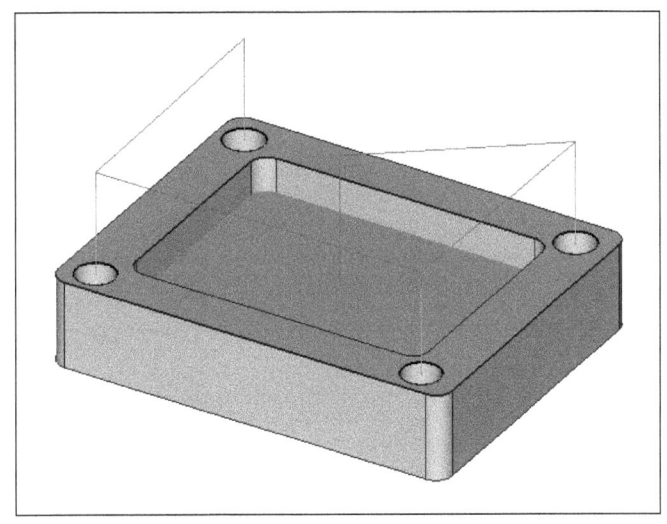

Figure-50. Drilling operation path generated

Creating Face Milling

The **Face** tool creates toolpath to remove material from top flat face of the model. The procedure to use this tool is discussed next.

- Click on the **Face** tool from **Toolbar** in the **CAM** workbench; refer to Figure-51. The **MillFace** dialog will be displayed in the **Tasks** panel of **Combo View** and the model will be enclosed within the bounding box; refer to Figure-52.

Figure-51. Face Mill tool

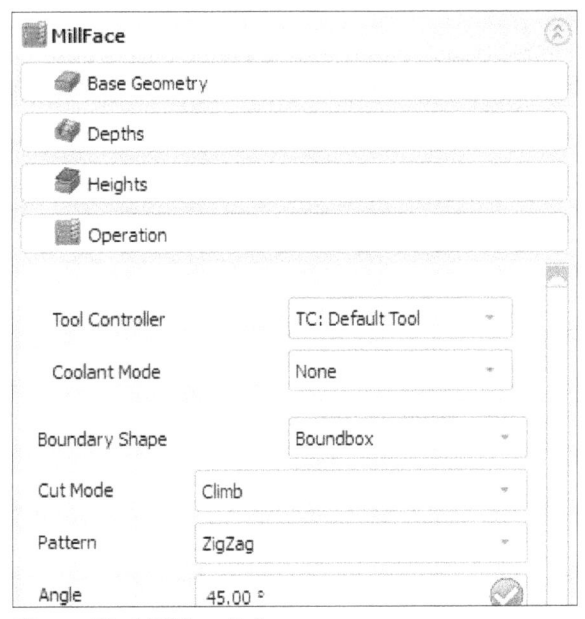

Figure-52. MillFace dialog

- Select one or more features of the model in the 3D view area and click on **Add** button from **Base Geometry** tab of the dialog to add them to the list as the base geometries for this operation; refer to Figure-53.

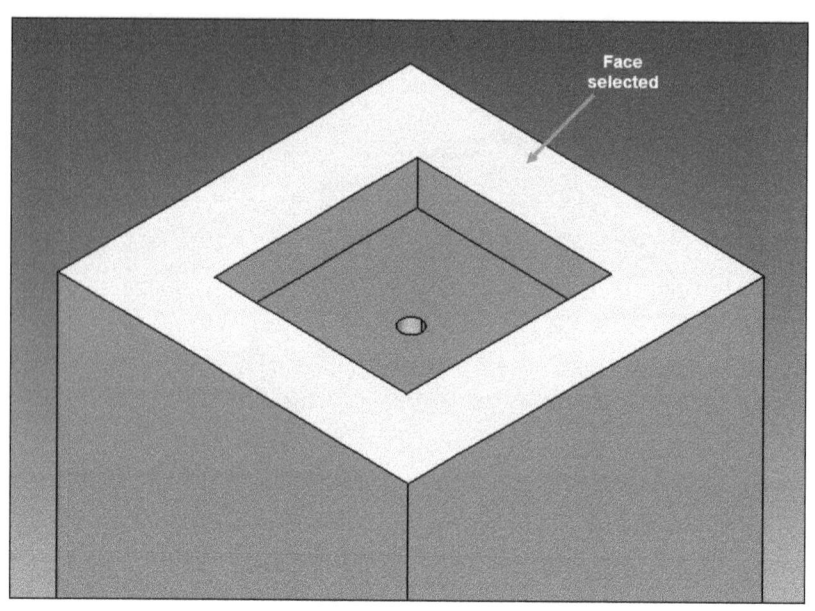

Figure-53. Selecting face of the model

- Select desired option from **Boundary Shape** drop-down in the **Operation** tab whether the facing should be restricted by the actual shape of the selected face or the bounding box should be faced off.
- Enter desired value in **Material Allowance** edit box from **Operation** tab of the dialog to specify the amount of material that should be left by this operation in relation to the target shape.
- Other parameters in the **MillFace** dialog have been discussed earlier. Click on **Apply** button and then **OK** button from the dialog. The Face Mill operation path will be generated; refer to Figure-54.

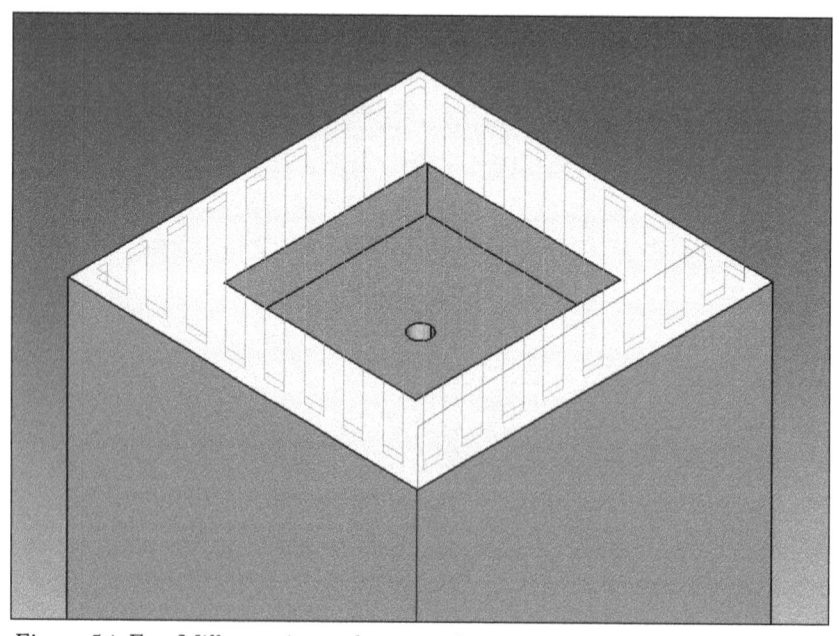

Figure-54. Face Mill operation path generated

Creating Helical Operation

The **Helix** tool is used to create a helical path. This operation is useful to machine circular vertical surfaces of the model. The helix path gets appended automatically to a helical clearing operation in the Job. The procedure to use this tool is discussed next.

- Click on the **Helix** tool from **Toolbar** in the **CAM** workbench; refer to Figure-55. The **Helix** dialog will be displayed in the **Tasks** panel of **Combo View** and the model will be enclosed within the boundary box; refer to Figure-56.

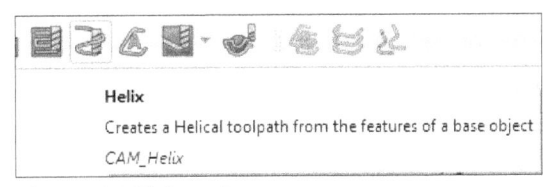

Figure-55. Helix tool

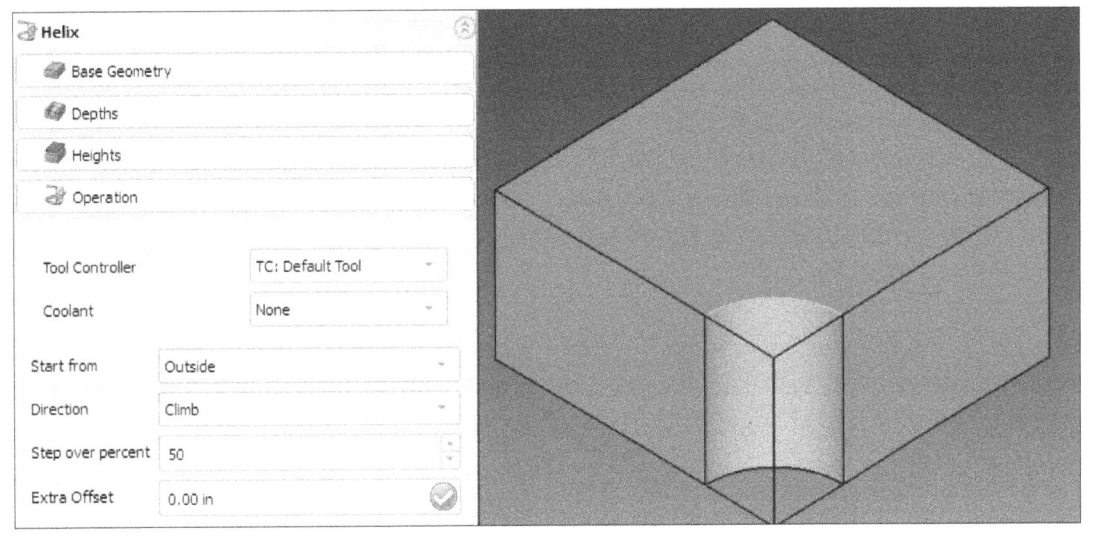

Figure-56. Helix dialog and the model enclosed within boundary box

- Select the edges of features to be machined by the helical operation and click on **Add** button from **Base Geometry** tab of the dialog to add them in the list as the base geometries for this operation.
- Specify appropriate values in the **Depths** section of dialog; refer to Figure-60. Note that you can activate the files by clicking on the arrow buttons next to edit boxes.

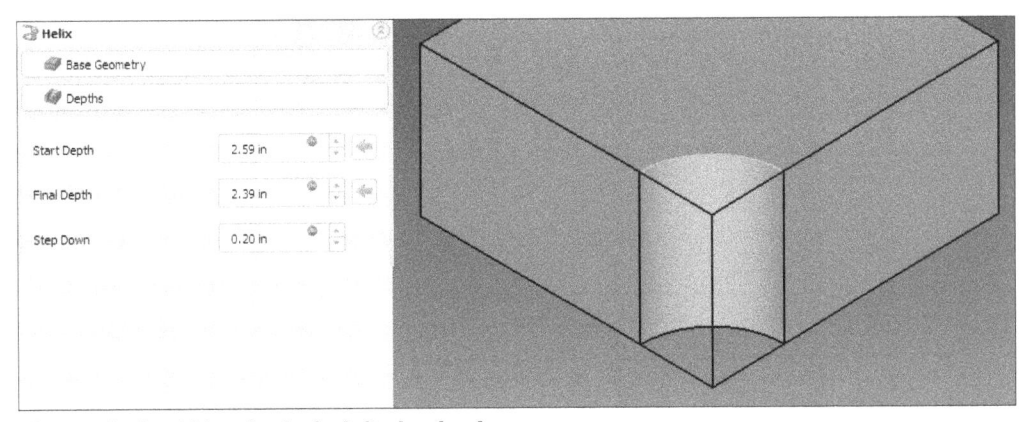

Figure-57. Specifying depths for helical toolpath

- Specify all the other parameters in the **Helix** dialog as discussed earlier.
- Click on **Apply** button and then **OK** button from the dialog. The helical operation path will be generated; refer to Figure-58.

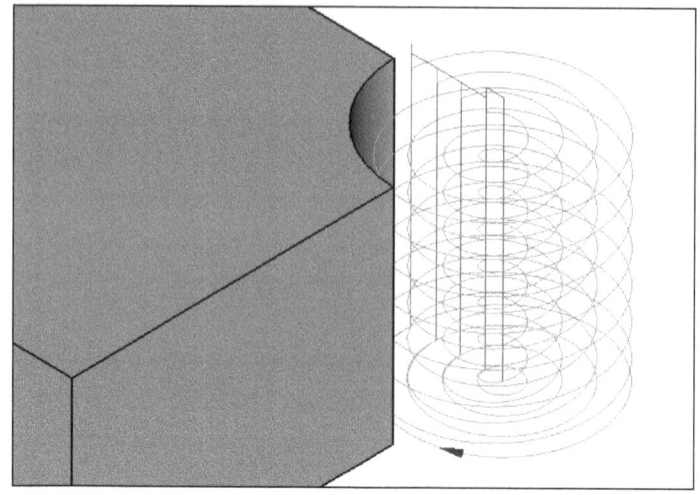

Figure-58. Helical operation path generated

Creating Adaptive Operation

The **Adaptive** tool uses an adaptive algorithm to create clearing and profiling paths that manage cutter engagement so that engagement and material removal never exceed a maximum value. This toolpath uses all the possible cutting strategies available in FreeCAD for removing large stock of material. The procedure to use this tool is discussed next.

- Click on the **Adaptive** tool from **Toolbar** in the **CAM** workbench; refer to Figure-59. The **Adaptive** dialog will be displayed in the **Tasks** panel of **Combo View** and the model will be enclosed within the bounding box; refer to Figure-60.

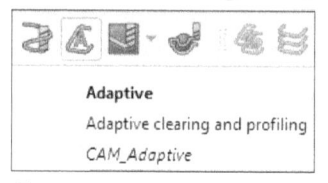

Figure-59. Adaptive tool

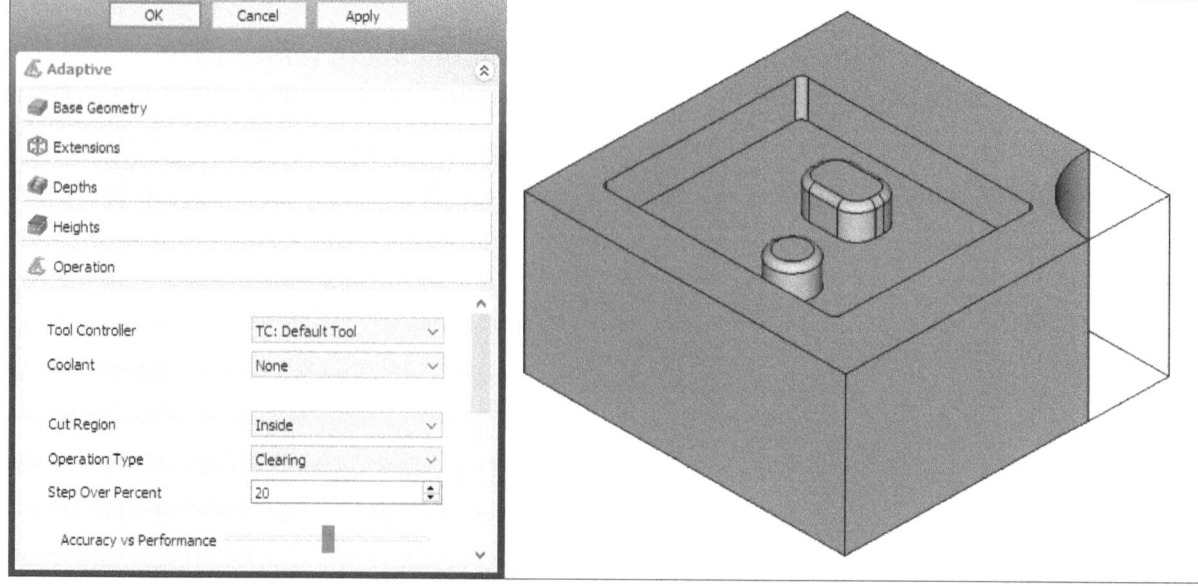

Figure-60. Adaptive dialog and the model enclosed within bounding box

- Select one or more faces of the model in the 3D view area to be machined (generally base face) and click on **Add** button from **Base Geometry** tab of the dialog to add them to the list as the base geometries for this operation; refer to Figure-61.

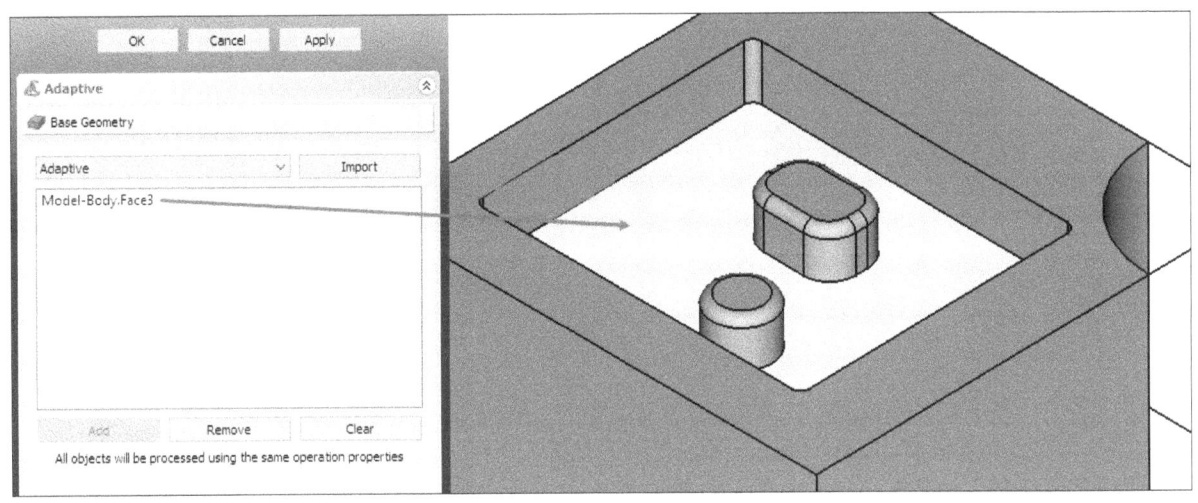

Figure-61. Selecting the face for adaptive operation

- Select desired option from **Cut Region** drop-down in the **Operation** tab to specify the cutting of model inside or outside of the selected shape.
- Select desired type of adaptive operation from **Operation Type** drop-down in the **Operation** tab of the dialog.
- Specify the percent of cutter diameter to step over on each pass in the **Step Over Percent** edit box of **Operation** tab.
- Slide the **Accuracy vs Performance** slider from **Operation** tab at desired position to influence the calculation performance vs stability and accuracy.
- Specify desired angle of the helix ramp entry in the **Helix Ramp Angle** edit box of **Operation** tab.
- Specify desired angle of the helix entry cone in the **Helix Cone Angle** edit box of the **Operation** tab.
- Enter desired value in **Helix Max Diameter** edit box from **Operation** tab to specify the limit helix entry diameter. If limit larger than tool diameter or 0, tool diameter is used.
- Enter desired value in **Lift Distance** edit box from **Operation** tab to specify how much to lift the tool up during the rapid linking moves over cleared regions. If linking path is not clear, tool is raised to clearance height.
- Enter desired value in **Keep Tool Down Ratio** edit box from **Operation** tab to specify maximum length of keep tool down linking path compared to direct distance between point. If exceeded, link will be done by raising the tool to clearance height.
- Enter desired value in **Stock to Leave** edit box to specify how much material to leave (i.e. for finishing operation).
- Select the **Force Clearing Inside-Out** check box from **Operation** tab to specify the force plunging into material inside and clearing towards the edges.
- Other parameters in the **Adaptive** dialog have been discussed earlier.
- Click on **Apply** button and then **OK** button from the dialog. The Adaptive operation path will be generated; refer to Figure-62.

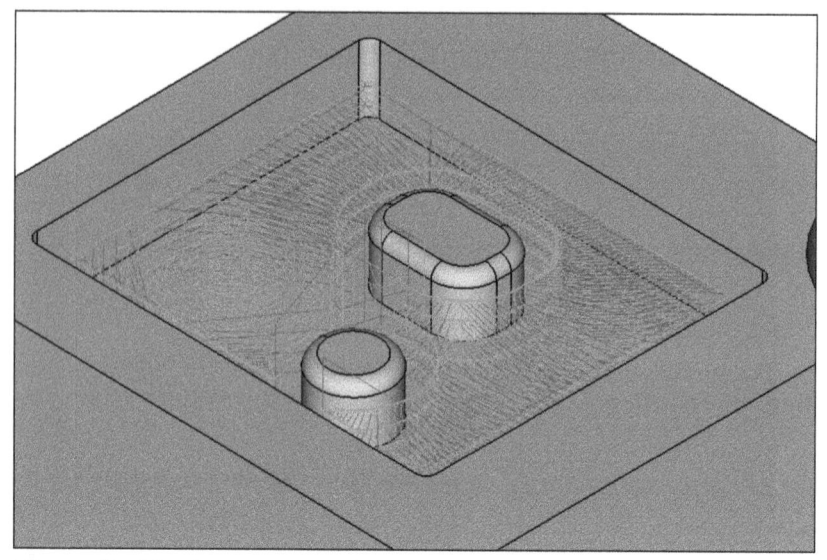

Figure-62. Adaptive operation path generated

Creating an Engraving Path

The **Engrave** tool is primarily used for engraving a **Draft ShapeString** onto a part like writing name on part. However, it may be useful for other kinds of 2D toolpaths. Note that if you want to engrave string on face of model then you will need a new job setup for string. The procedure to perform engraving of a string is discussed next.

- Create the string of text as desired on the face of model in **Draft** workbench.
- Click on the **Job** tool from the **Toolbar** in **CAM** workbench. The **Create Job** dialog box will be displayed.
- Select the check box for **ShapeString** in **2D** node of list in the dialog box; refer to Figure-63 and click on the **OK** button. The **Job Edit** dialog will be displayed in the **Tasks** panel of **Combo View**.

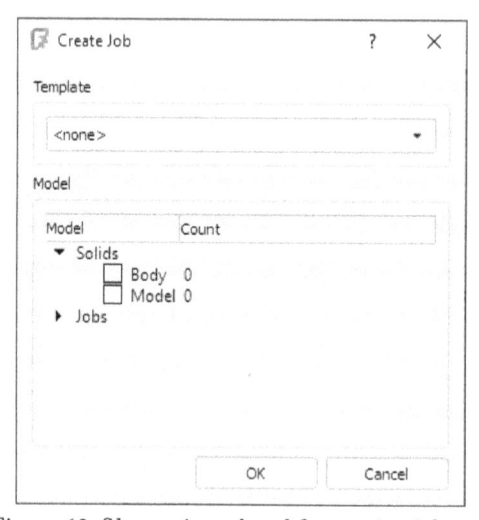

Figure-63. Shapestring selected for creating job

- Set the stock, cutting tool, and other parameters as discussed earlier and click on the **OK** button from the dialog box. A new job with name Job001 will be added in the **Model Tree**.
- Click on the **Engrave** tool from **Engraving Operations** drop-down in the **Toolbar** of **Path** workbench; refer to Figure-64. The **Choose a Path Job** dialog box will be displayed.

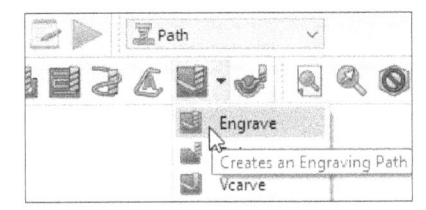

Figure-64. Engrave tool

- Select the newly created job from the drop-down and click on the **OK** button. The **Choose a Tool Controller** dialog box will be displayed if there are multiple tools available. Select desired engraving tool from the drop-down and click on the **OK** button. The **Engrave** dialog will be displayed in the **Tasks** panel of **Combo View** and the model will be enclosed within the bounding box; refer to Figure-65.

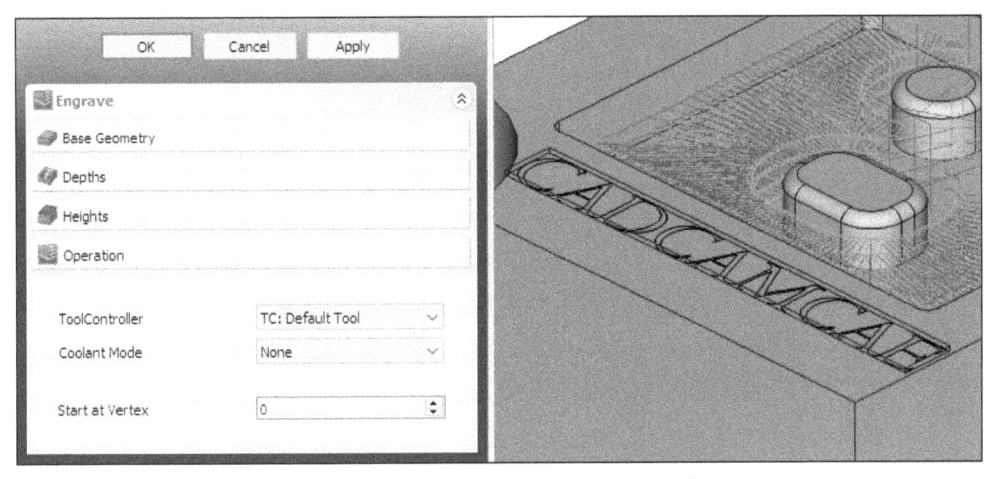

Figure-65. Engrave dialog and the model enclosed within bounding box

- Select faces of strings or every edge of the strings from the model in the 3D view area and click on **Add** button from **Base Geometry** tab of the dialog to add them to the list as the base geometries for this operation; refer to Figure-66.

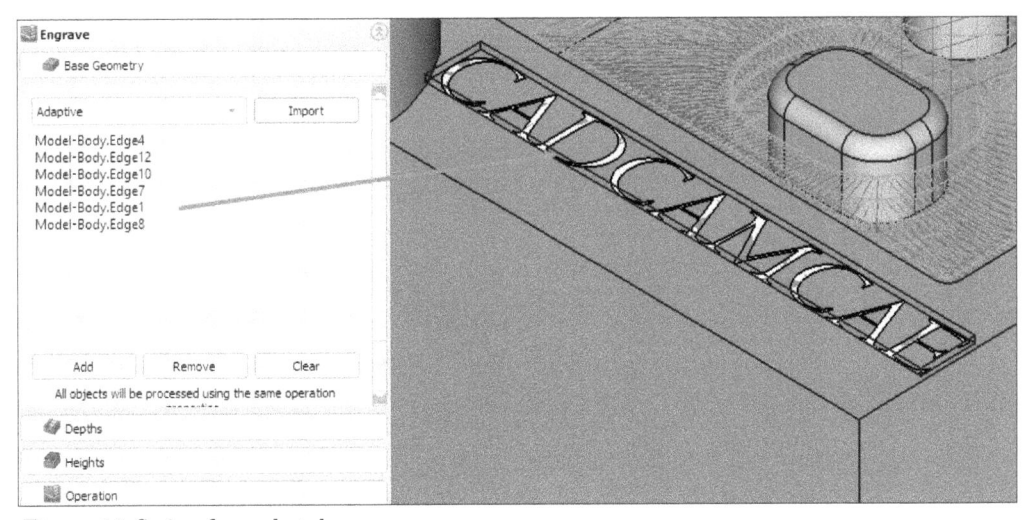

Figure-66. String faces selected

- Specify desired tool to be used for this operation from **ToolController** drop-down in the **Operation** tab of the dialog.
- Specify desired vertex number of the underlying shape string at which engraving should start in the **Start at Vertex** edit box of **Operation** tab in the dialog.
- Other parameters in the **Engrave** dialog have been discussed earlier.

- Click on **Apply** button and then **OK** button from the dialog. The Engraving operation path will be generated at specified depth; refer to Figure-67.

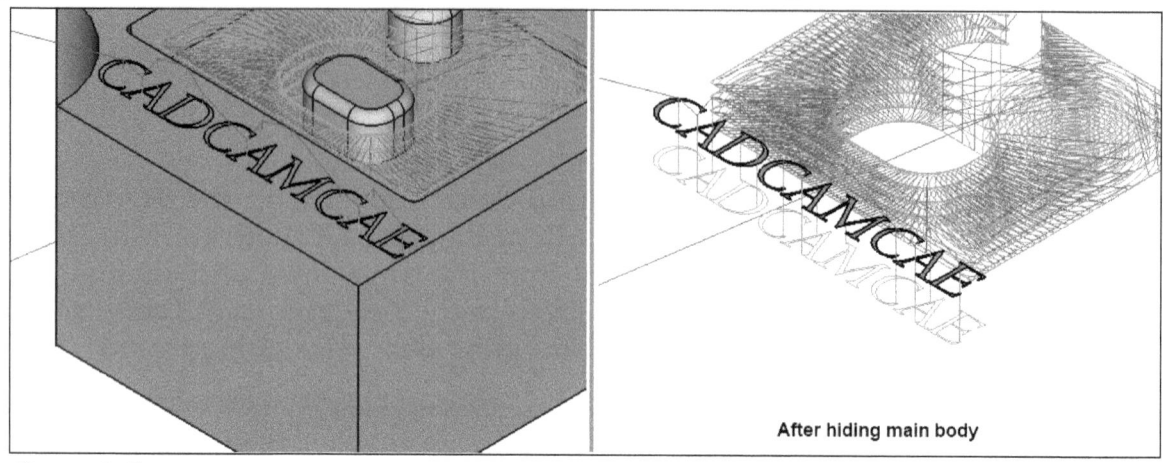

Figure-67. Engraving operation path generated

Creating Deburring Operation

The **Deburr** tool creates a deburr path along edges or around faces. The procedure to use this tool is discussed next.

- Click on the **Deburr** tool from **Engraving Operations** drop-down in the **Toolbar** of **CAM** workbench; refer to Figure-68. The **Deburr** dialog will be displayed in the **Tasks** panel of **Combo View** and the model will be enclosed within the boundary box; refer to Figure-69.

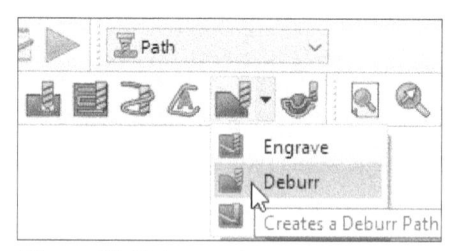

Figure-68. Deburr tool

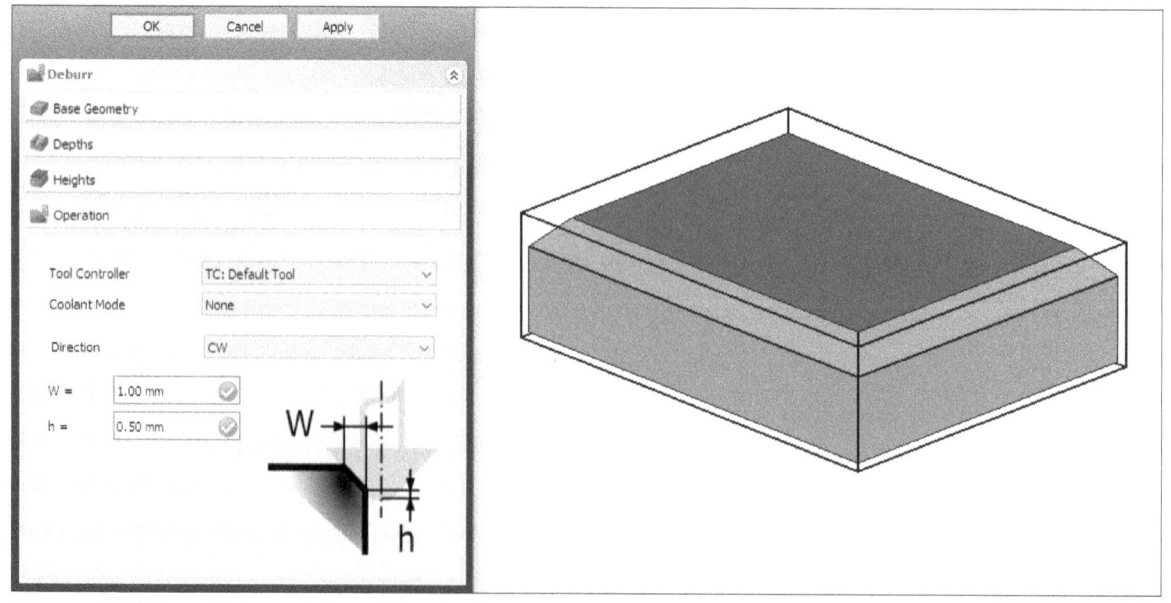

Figure-69. Deburr dialog and the model enclosed within bounding box

- Select the edges and/or faces of the model in the 3D view area on which you want to create the deburring operation and click on **Add** button from **Base Geometry** tab of the dialog to add them to the list as the base geometries for this operation; refer to Figure-70.

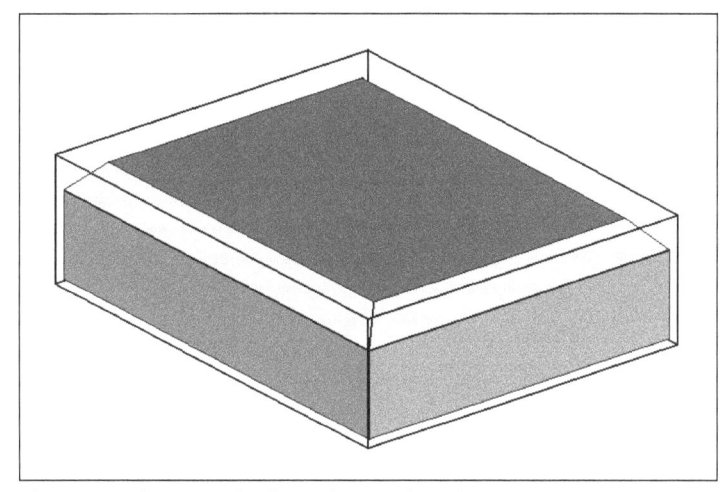

Figure-70. Selecting the faces of model for deburring

- Select desired tool to be used for this operation from **Tool Controller** drop-down in the **Operation** tab of dialog.
- Select desired direction in which the profile is to be performed from **Direction** drop-down in the **Operation** tab.
- Specify desired width of chamfer cut in the **W** edit box from **Operation** tab of the dialog.
- Specify desired extra depth of tool immersion in the **h** edit box from **Operation** tab of the dialog.
- Other parameters in the **Deburr** dialog have been discussed earlier.
- Click on **Apply** button and then **OK** button from the dialog. The Deburring operation path will be generated; refer to Figure-71.

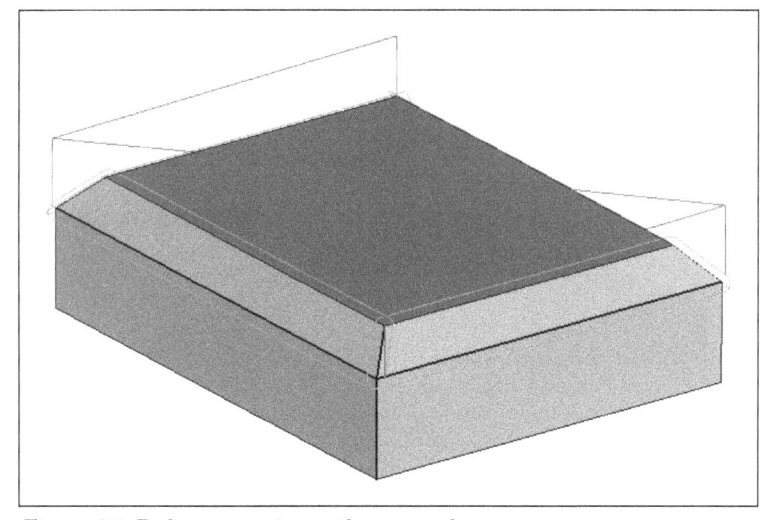

Figure-71. Deburr operation path generated

Creating Vcarve

The **Vcarve** tool is primarily for center-line engraving a Draft ShapeString onto a part. The procedure to use this tool is same as discussed for **Engrave** tool. Use the V-bit cutting tool to perform the operation.

Creating 3D Pocket Operation

The **3D Pocket** tool inserts a path 3D Pocket object into the job. A 3D pocket takes into account the bottom surface of the pocket. The procedure to use this tool is discussed next.

- Click on the **3D Pocket** tool from **Toolbar** in the **Path** workbench; refer to Figure-72. The **Pocket 3D** dialog will be displayed in the **Tasks** panel of **Combo View** and the model will be enclosed within the bounding box; refer to Figure-73.

Figure-72. 3D Pocket tool

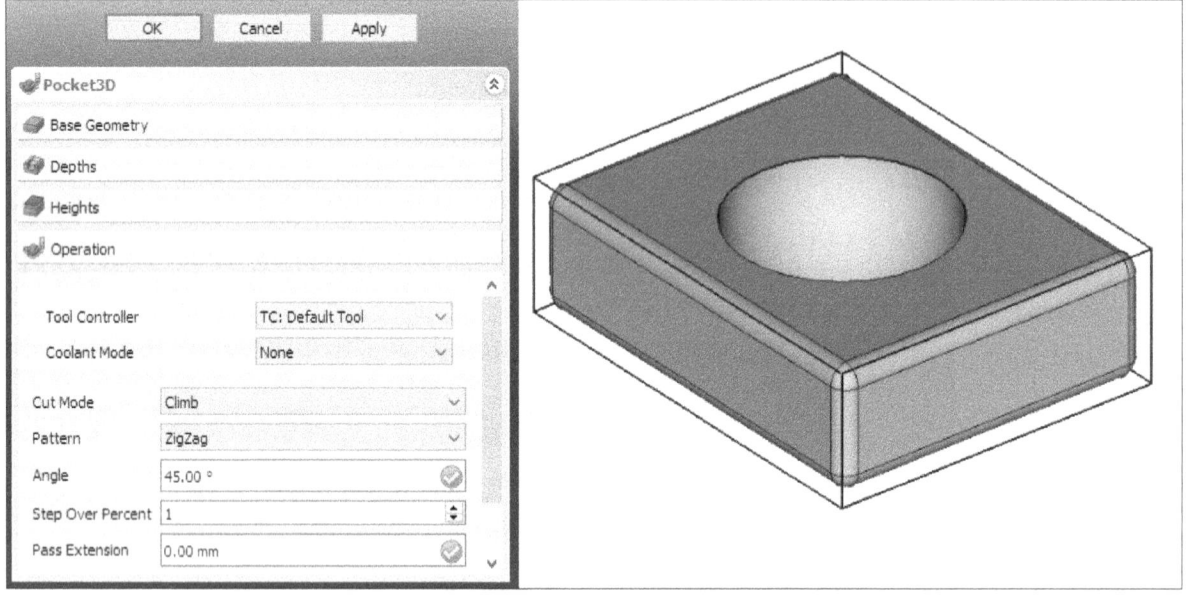

Figure-73. Pocket 3D dialog and the model enclosed within bounding box

- Select one or more faces from the model in the 3D view area and click on **Add** button from **Base Geometry** tab of the dialog to add them to the list as the base geometries for this operation; refer to Figure-74.

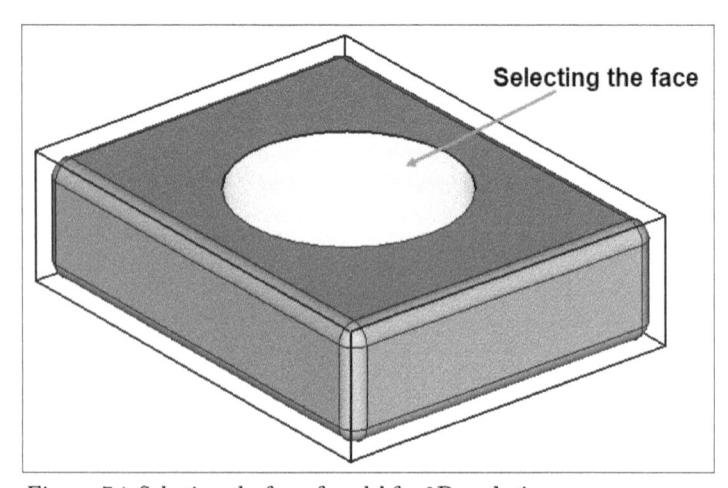

Figure-74. Selecting the face of model for 3D pocketing

- Specify all the parameters in the **Pocket 3D** dialog as discussed earlier.

- Click on **Apply** button and then **OK** button from the dialog. The 3D Pocket operation path will be generated; refer to Figure-75.

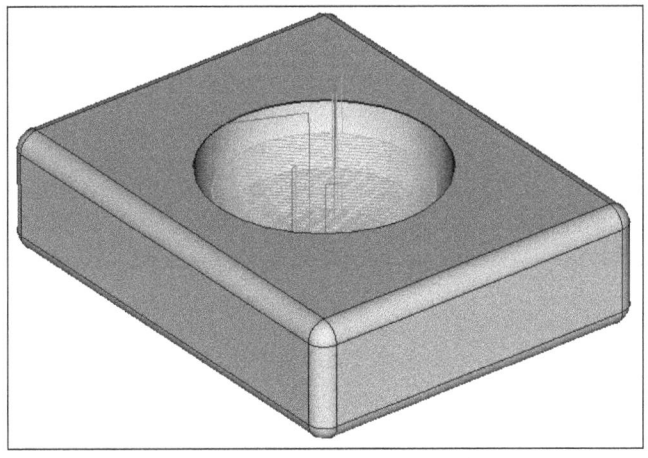

Figure-75. 3D Pocket operation path generated

PATH MODIFICATION TOOLS

There are three tools in **Toolbar** of **Path** workbench used to create multiple copies of the selected toolpaths. These tools are discussed next.

Creating Copy

The **Copy** tool creates a copy of selected path operation. The procedure to use this tool is discussed next.

- Select desired path operation from the **Model Tree** for which you want to create the copy or multiple copies; refer to Figure-76.

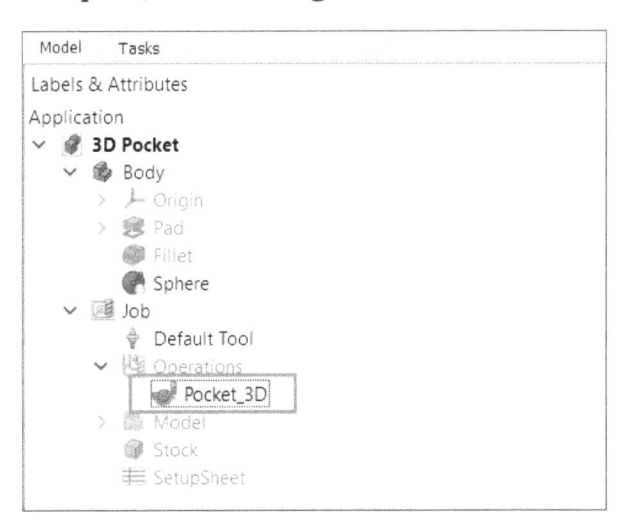

Figure-76. Selecting the path operation

- Click on the **Copy** tool from **Toolbar** in the **Path** workbench; refer to Figure-77. The copy of selected path operation will be created; refer to Figure-78.

Figure-77. Copy tool

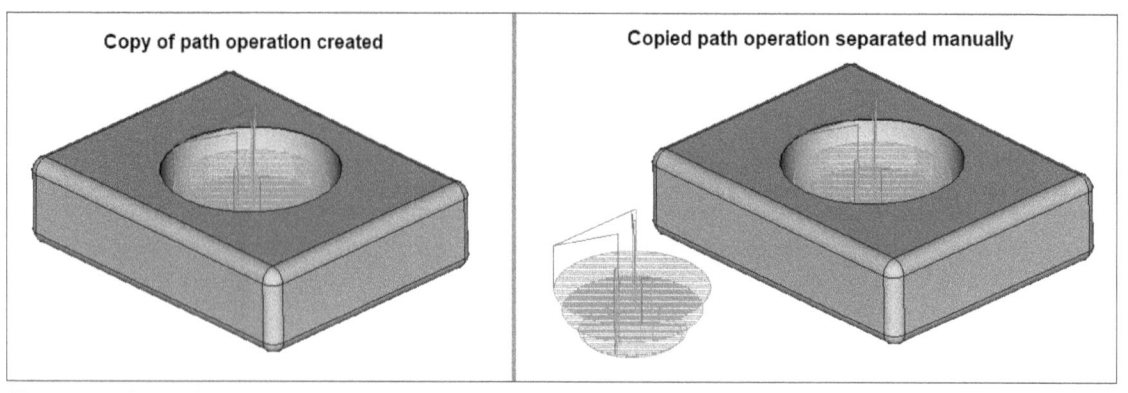

Figure-78. Copy of path operation created

- You can edit the properties of copied path operation using **Property editor** dialog as discussed earlier.

Creating Array

The **Array** tool creates a new path by duplicating another path several times at a certain interval distance. The procedure to use this tool is discussed next.

- Select desired path operation created from the **Model Tree** of which you want to create array; refer to Figure-79.

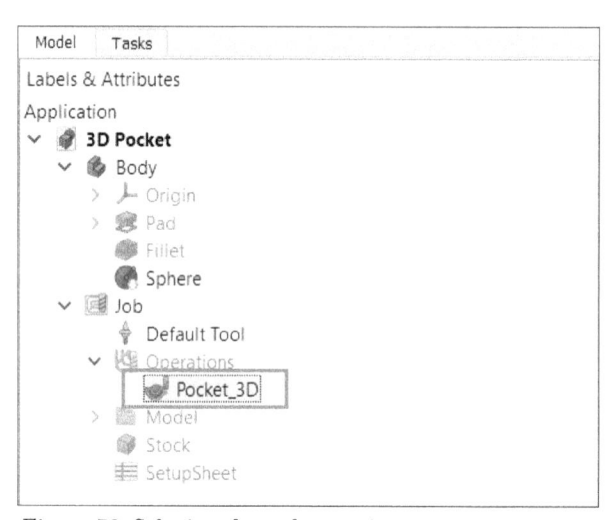

Figure-79. Selecting the path operation

- Click on the **Array** tool from **Toolbar** in the **Path** workbench; refer to Figure-80. The array of path operation will be created; refer to Figure-81.

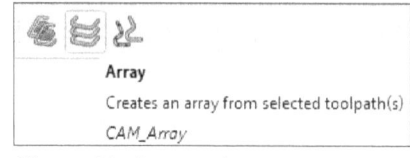

Figure-80. Array tool

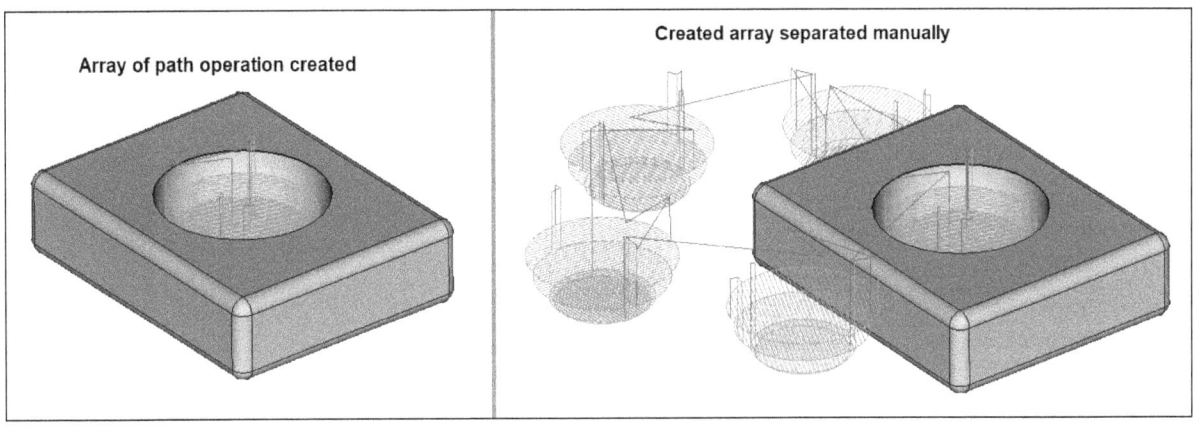

Figure-81. Array of path operation created

- You can edit the properties of array of path operation created from **Property editor** dialog as discussed earlier.

Creating Simple Copy

The **Simple Copy** tool creates a non-parametric copy of a given path. The procedure to use this tool is similar to discussed for **Copy** tool.

PATH UTILITIES

The tools in **Path Utilities** section of **Toolbar** are used to inspect, manage, and simulate toolpaths. These tools are discussed next.

Inspection of G-Code

The **Inspect Path Commands** tool allows inspection of the internal FreeCAD G-code dialect contents of a Path Operation Object. The procedure to use this tool is discussed next.

- Select desired path operation from the **Model Tree** which you want to inspect for G-code; refer to Figure-82.

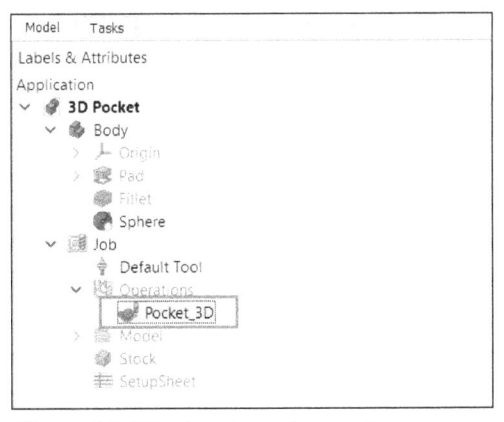

Figure-82. Selecting the path operation

- Click on the **Inspect Path Command** tool from **Toolbar** in the **Path** workbench; refer to Figure-83. The **FreeCAD** dialog box will be displayed showing the G-code of path operation; refer to Figure-84.

Figure-83. G-Code Inspector tool

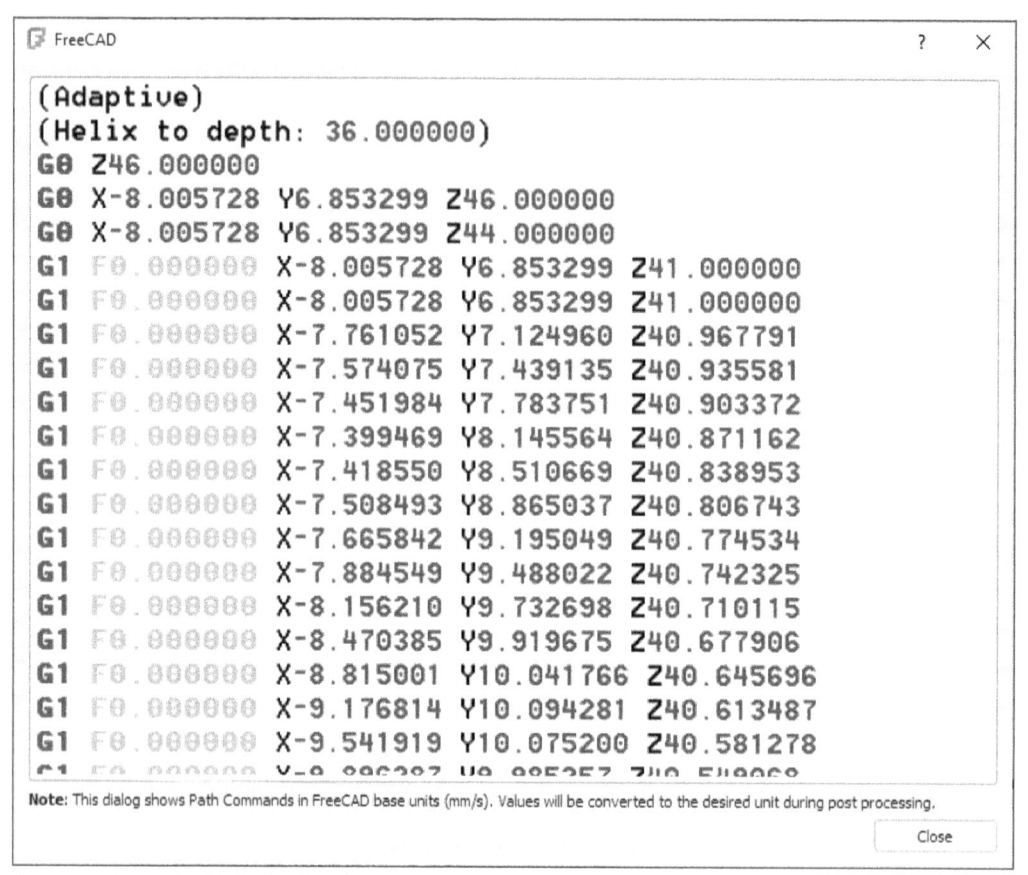

Figure-84. FreeCAD dialog box showing the G code

- Make any changes in the dialog box as desired and click on **OK** button from the dialog box to apply the changes.

CAM Simulator

The **CAM simulator** tool is used to simulate selected operations on 3D model along the G-Code paths, subtracting material from the stock, where the stock and tool overlap, providing visualization of the Job. This allows detection and isolation of errors prior to running the job on a mill. The procedure to use this tool is discussed next.

- Click on the **CAM simulator** tool from **Toolbar** in the **Path** workbench; refer to Figure-85. The **Path Simulator** dialog will be displayed in the **Tasks** panel of **Combo View** and the model will be enclosed within the stock, ready to be simulate; refer to Figure-86.

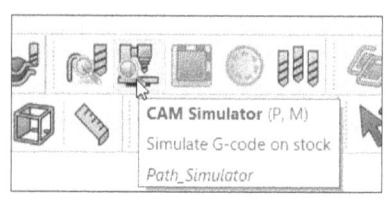

Figure-85. Cam simulator tool

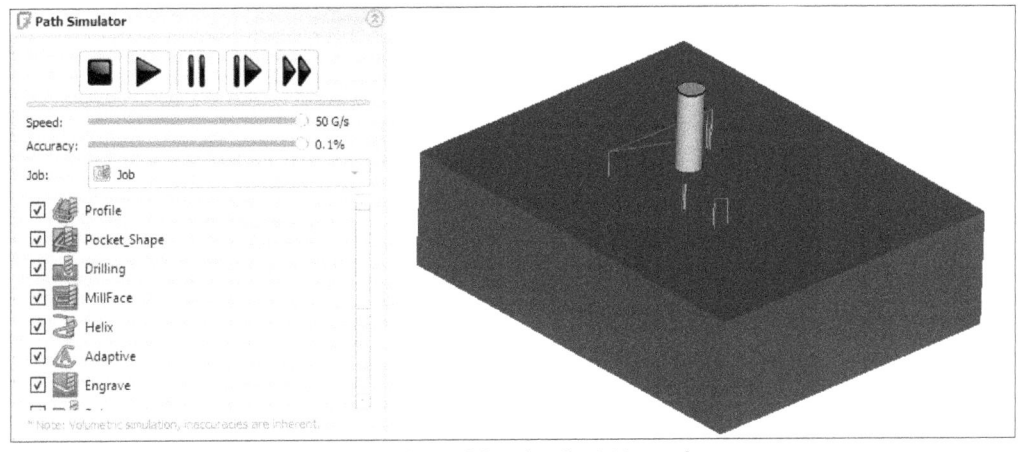

Figure-86. Path Simulator dialog and the model enclosed within stock

- All the operations created that are to be simulated will be selected by default in the **Operation list** area of the dialog.
- Deselect the operations which you do not want to be simulate from the **Operation list**.
- Select desired Job which you want to use as the basis of the simulation from **Job** drop-down in the dialog.
- Press the ▶ **Play** button to play or playback an animation of the operations.
- Press the ■ **Stop** button to stop the animation.
- Press the ‖ **Pause** button to pause animation for troubleshooting purposes.
- Press the ‖▶ **Single-Step** button for slowing down the animation, this functionality helps troubleshooting and resolving specific cuts and/or movements.
- Press the ▶▶ **Fast-Forward** button to increase the speed substantially.
- Tune the speed of simulation by sliding the **Speed** slider from the dialog.
- Tune the accuracy of simulation by sliding the **Accuracy** slider from the dialog.
- After performing the simulation, the object will be displayed as shown in Figure-87.

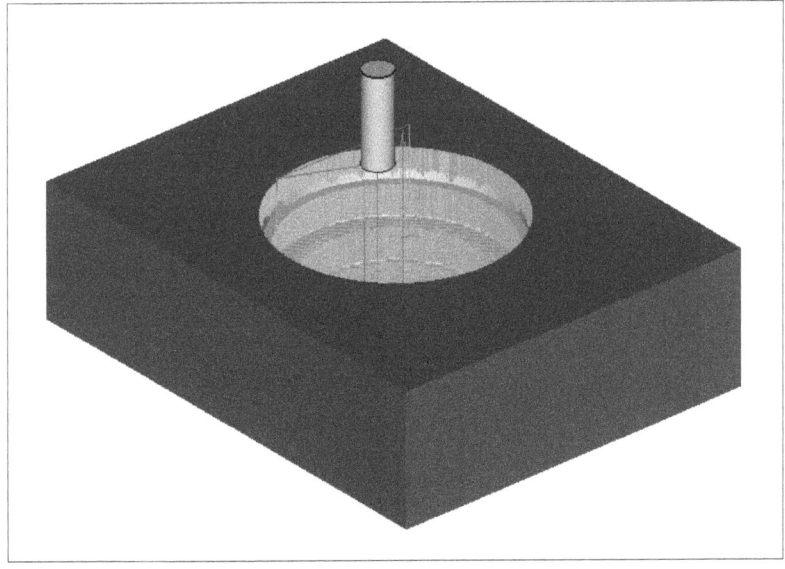

Figure-87. Object simulated

- Click on the **OK** button from the dialog to finish the simulation process.

New CAM Simulator

The Simulator GL tool is a new alternative to CAM Simulator. It's based on low-level Open GL functions. To eliminate interference with the 3D view of FreeCAD, it works in a separate window with a separate Open GL context. It's meant to be faster and more precise than the old simulator. The procedure to use this tool is discussed next.

- Click on the **New CAM simulator** tool from **Toolbar** in the **Path** workbench; refer to Figure-88. The **Path Simulator** dialog will be displayed in the **Tasks** panel of **Combo View**; refer to Figure-89.

Figure-88. New CAM Simulator tool

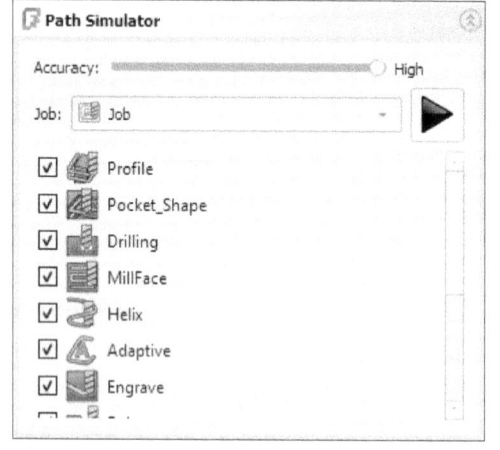

Figure-89. New CAM Simulator dialog box

- Tune the accuracy of simulation by sliding the **Accuracy** slider from the dialog.
- Press the ▶ **Play** button from the Path Simulator dialog box, a separate **FreeCAD** window will be displayed; refer to Figure-90.

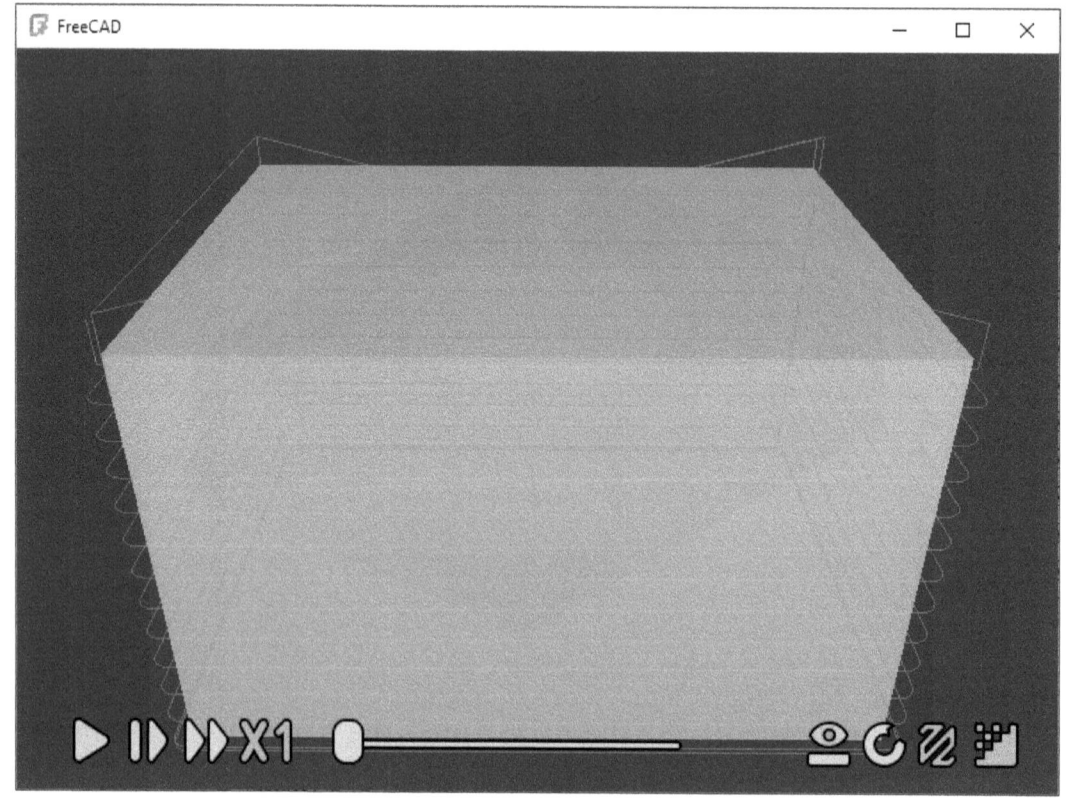

Figure-90. Freecad simulator page

- Press the ▶ **Play** button to play or playback an animation of the operations.
- Press the ▶| **Single-Step** button for slowing down the animation, this functionality helps troubleshooting and resolving specific cuts and/or movements.
- Press the ▶▶ **Fast-Forward** button to increase the speed substantially.
- Click on the **Show/Hide** base model to hide the animation.
- Click on the **Rotate** button to control the 3D view with the current FreeCAD mouse control.
- Click on the button to show/hide toolpaths.
- Click on the button to show the shadow of the rotating pole.

FINISH SELECTING LOOP

The **Finish Selecting Loop** tool is used to complete a loop from two selected edges. The procedure to use this tool is discussed next.

- Select two edges that share a common vertex as shown in the Figure-91.

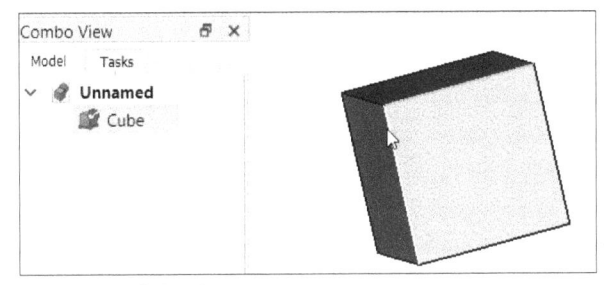

Figure-91. Select loop

- Click on the **Finish Selecting Loop** tool from the **Toolbar** in the path workbench; refer to Figure-92. Full loop of selected edges will be selected; refer to Figure-93.

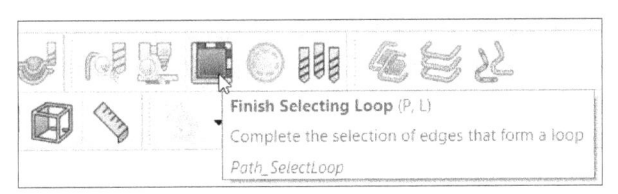

Figure-92. Finish selecting tool

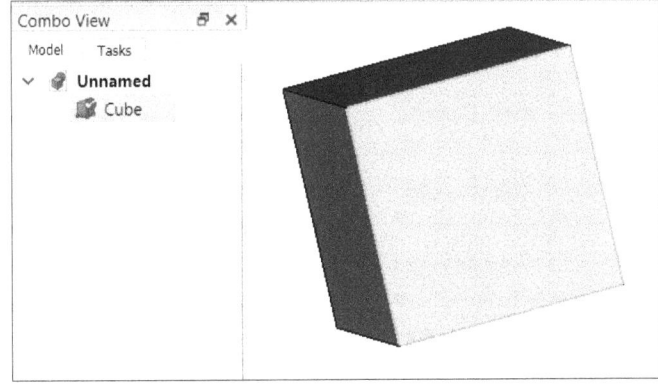

Figure-93. Loop created

POST-PROCESSING

The **Post Process** command exports the selected Path Jobs to a G-code file. Each CNC controller speaks a specific G-Code dialect-correct Postprocessor to translate the final output from the agnostic internal FreeCAD G-Code dialect. The procedure to use this tool is discussed next.

- Click on the **Edit** menu and select the **Preferences** option. The **Preferences** dialog box will be displayed; refer to Figure-94.

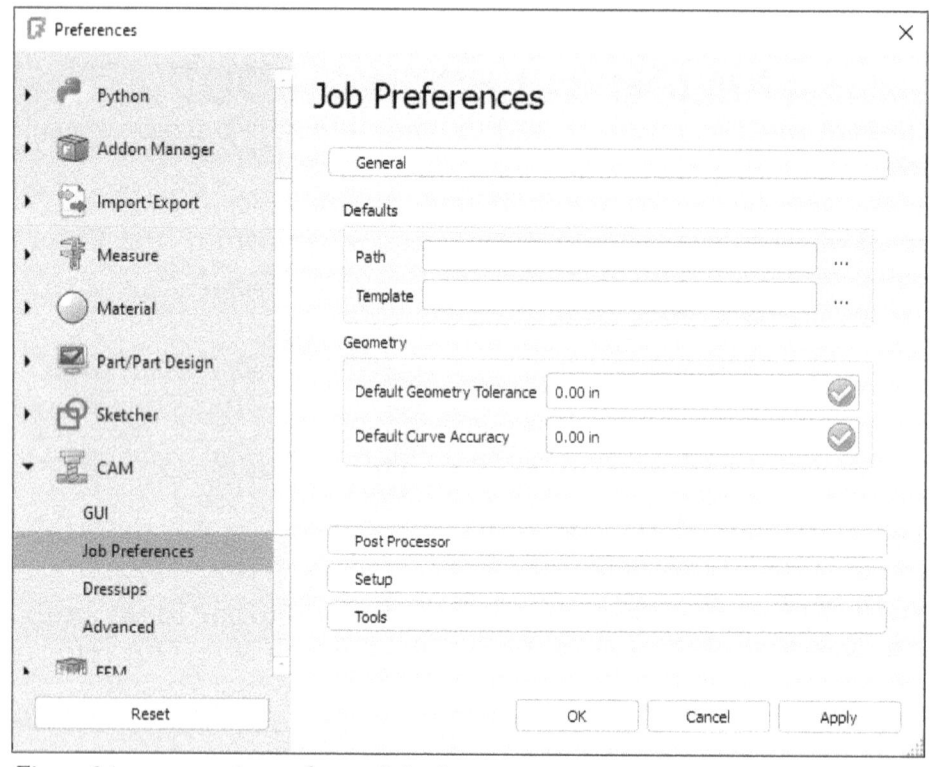

Figure-94. post processing preference dialog box

- Select desired option from **Default Post Process** drop-down in the **Preferences** tab of **Path** option available in the left side of the dialog box. After specifying desired parameters click on the **Ok** button from the dialog box.

- Select the Path Job created from the **Model Tree** which you want to export; refer to Figure-95.

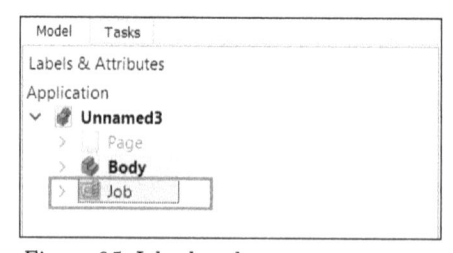

Figure-95. Job selected

Click on the **Post Process** tool from **Toolbar** in the **Path** workbench; refer to Figure-96. The **FreeCAD** dialog box will be displayed showing the postprocessor properties of the job; refer to Figure-97. The **Output File** dialog box will be displayed asking you to specify the file name and directory; refer to Figure-98.

FreeCAD 1.0 Black Book   8-43

Figure-96. Post Process tool

Figure-97. FreeCAD dialog box showing the post processor properties

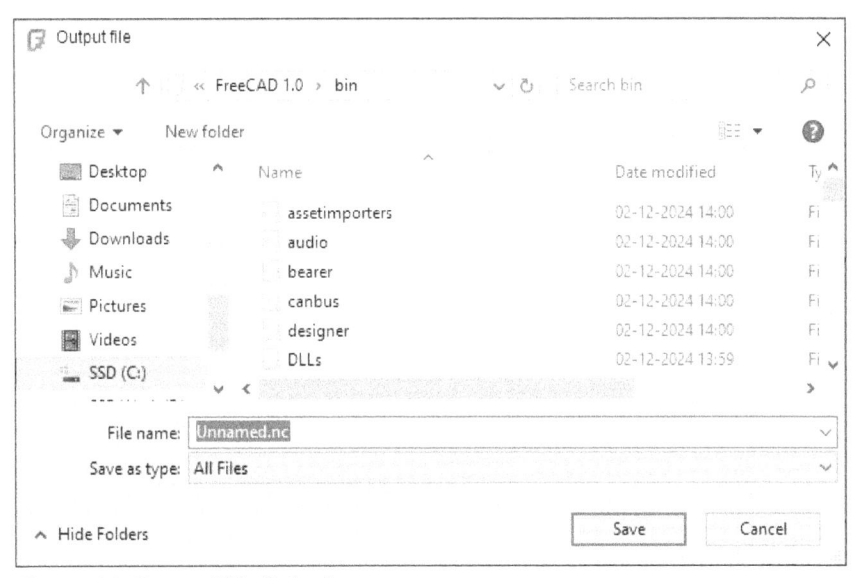

Figure-98. Output File dialog box

- Specify desired name of output file in the **File name** edit box and specify desired directory to save the file.
- Click on **Save** button to save the output file.

Check the CAM Job for common error

The **Check the CAM Job for common error** tool is used to checks the selected jobs for missing values. The procedure to use this tool is discussed next.

- Click on the **Post Process** tool from **Toolbar** in the **CAM** workbench; refer to Figure-99. The **Save Sanity Check Report** dialog box will be displayed; refer to Figure-100.
- Save the file with desired name in your local drive. Once saved, the file will open automatically in your default web browser.

Figure-99. Check CAM for error tool

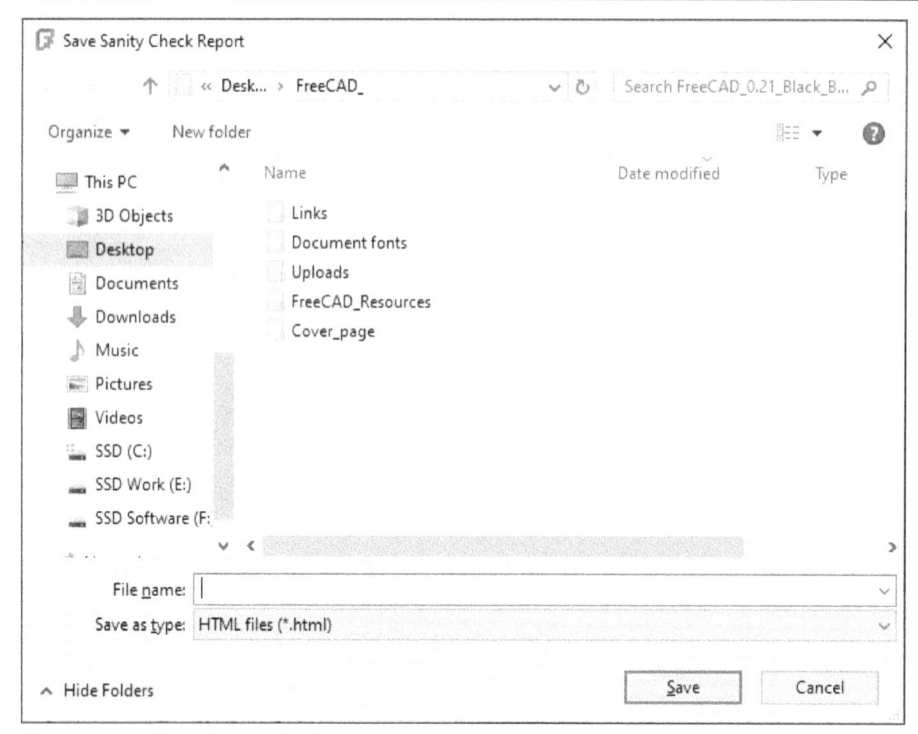

Figure-100. Save Sanity check report dialog box

PATH DRESSUP TOOLS

The tools in **Path Dressup** cascading menu of **CAM** menu are used to modify various aspects of toolpaths like starting axis, boundary, entry/exit of toolpath, and so on; refer to Figure-101. These tools are discussed next.

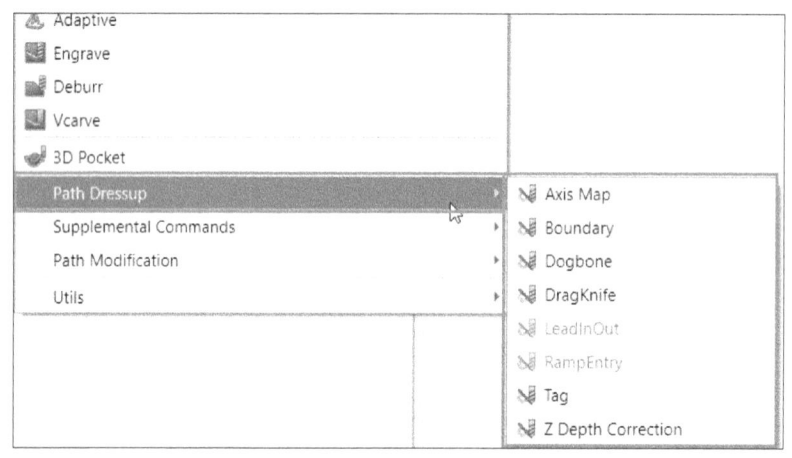

Figure-101. Path Dressup cascading menu

Axis Map

The **Axis Map** tool is used to switch one axis with another rotational axis in the toolpath to accommodate last minute changes in the machining orientation. For example, after creating machining setup if you want to convert X axis to Rotational A axis for multi-axis machine then you can switch them using **Axis Map** tool. The procedure to use this tool is given next.

- Select the toolpath from **Model** panel in the **Combo View** for which you want to change the axis and click on the **Axis Map** tool from the **Path Dressup** cascading menu of **CAM** menu. The **AxisMap Dressup** dialog will be displayed; refer to Figure-102.

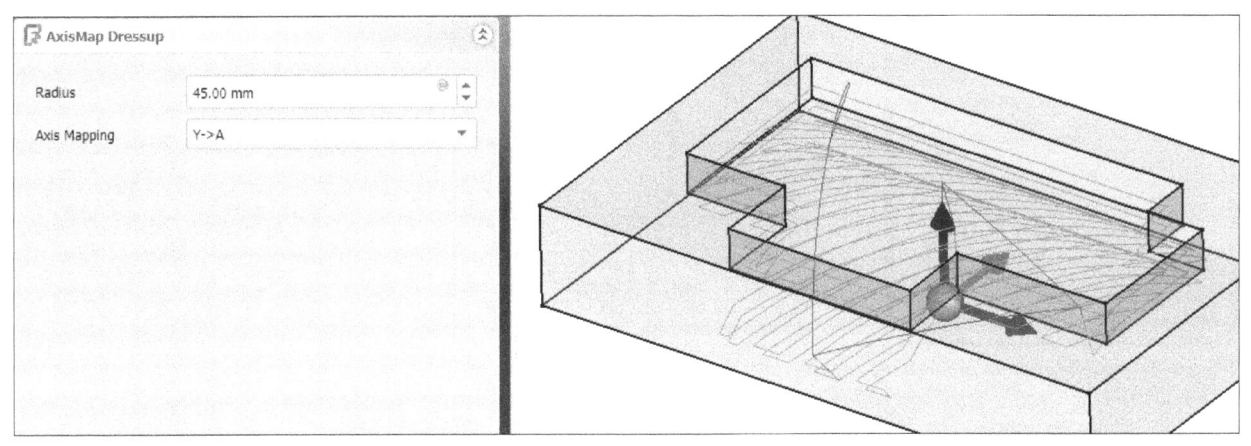

Figure-102. AxisMap Dressup dialog

- Select desired option from the **Axis Mapping** drop-down to define which axis is to be mapped.
- Specify desired value in the **Radius** edit box to define the radius by which toolpath will rotate about selected axis. (Axis Y in our case).
- After setting desired parameters, click on the **OK** button from the dialog to apply changes.

Modifying Boundary for Toolpath

The **Boundary** tool in **Path Dressup** cascading menu is used to change the boundary parameters for the toolpath. If you want to decrease the machining area for a selected toolpath then you can use this tool. The procedure to use this tool is given next.

- Click on the **Boundary** tool from the **Path Dressup** cascading menu of the **CAM** menu. The **Form** dialog will be displayed in the **Combo View**; refer to Figure-103.

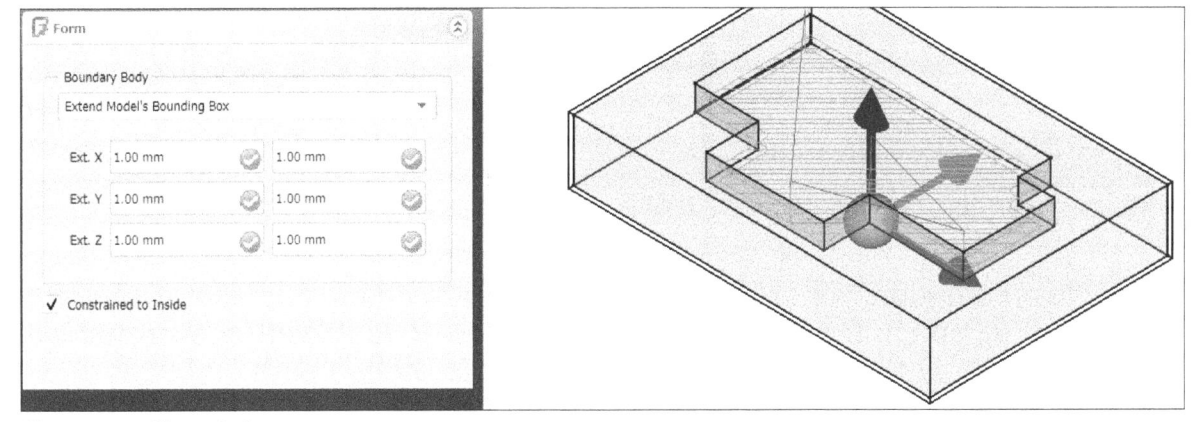

Figure-103. Form dialog

- Set desired parameters in the dialog as discussed earlier to define machining boundaries and click on the **OK** button. The toolpath will be modified accordingly; refer to Figure-104.

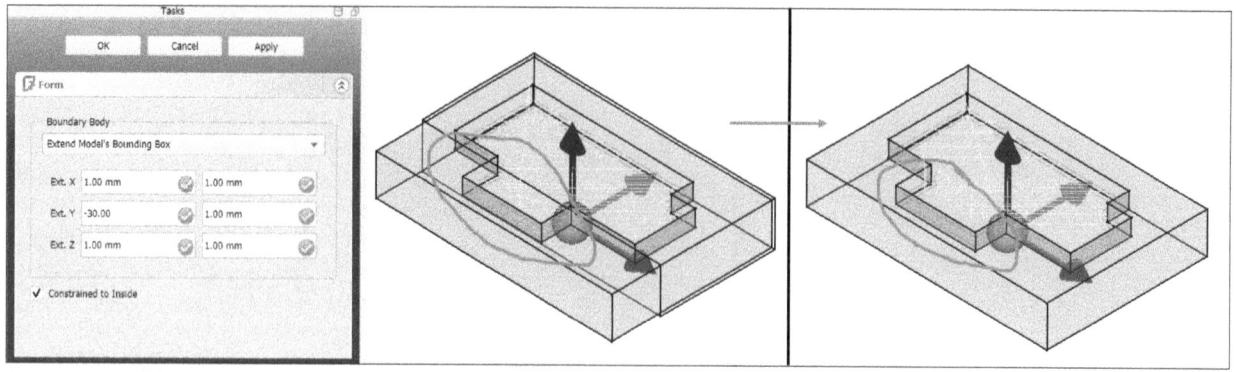

Figure-104. Changing boundary for toolpath

Applying Dogbone Overcuts

The **Dogbone** tool is used to add overcuts to the toolpath so that a cylindrical cutting tool fully machines acute corners. Generally, cutting tool will avoid the walls the model where it can collide with stock but when this dressup is used, it will machine such areas so you should be careful when using this function. The procedure to use this tool is given next.

- Click on the **Dogbone** tool from the **Path Dressup** cascading menu of the **CAM** menu. The **Dogbones** dialog will be displayed with preview of overcuts added to toolpath; refer to Figure-105.

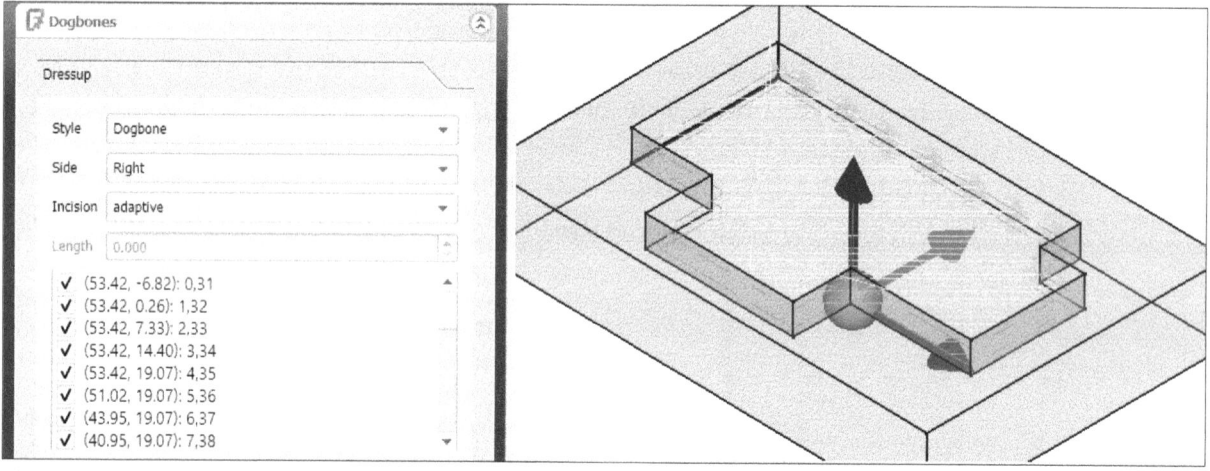

Figure-105. Dogbones dialog

- Select desired option from the **Style** drop-down to define shape of overcuts.
- Select desired option from the **Side** drop-down to define whether overcuts will be created at the right side of toolpath or left side of the toolpath.
- Select desired option from the **Incision** drop-down to define algorithm to be used for making overcuts. Select the **adaptive** option if you want the overcuts to follow curvature of walls. Select the **fixed** option if you want to make overcuts of same size. Select the **custom** option if you want to manually define the extension beyond the walls upto which overcuts will be created and specify the value in Length edit box.
- Clear check boxes for overcuts that you do not want to include in toolpath.
- Click on the **OK** button from the dialog to apply changes.

Applying Dragknife Dressup

The **Dragknife** tool is used to generate offsetted cutting passes for generating cutout from sheetmetal, vinyl, cardboards, and other soft materials. When this dressup is applied, cutting tool edge will follow the toolpath rather than center of cutting tool. You can manually define the offset values for this tool. The procedure to use this tool is given next.

- Click on the **Dragknife** tool from the **Path Dressup** cascading menu of the **CAM** menu. The **Dragknife Dressup** dialog will be displayed in the **ComboView** with preview of toolpath change; refer to Figure-106.

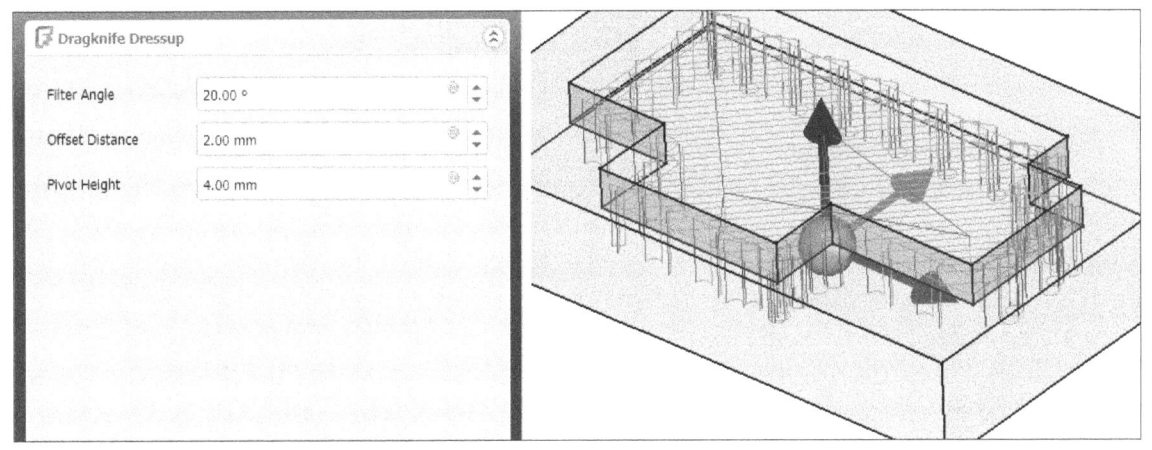

Figure-106. Dragknife Dressup dialog

- Specify desired value in the **Filter Angle** edit box to define the minimum angle limit below which corner action will be inserted in toolpath to compensate for sharp turns in toolpath.
- Specify desired value in the **Offset Distance** edit box to define the distance by which cutting passes will be offset from cutting tool center point in toolpath.
- Specify desired value in the **Pivot Height** edit box to define the height above toolpath plane at which cutting tool will move in stock before starting next corner action.
- After setting desired parameters, click on the **OK** button to apply dressup changes.

Modifying Lead In/Out

The **LeadInOut** tool is used to modify entering and exiting segment of toolpath. The procedure to use this tool is given next.

- Click on the **LeadInOut** tool from the **Path Dressup** cascading menu of **CAM** menu. The **LeadInOut** dialog will be displayed; refer to Figure-107.

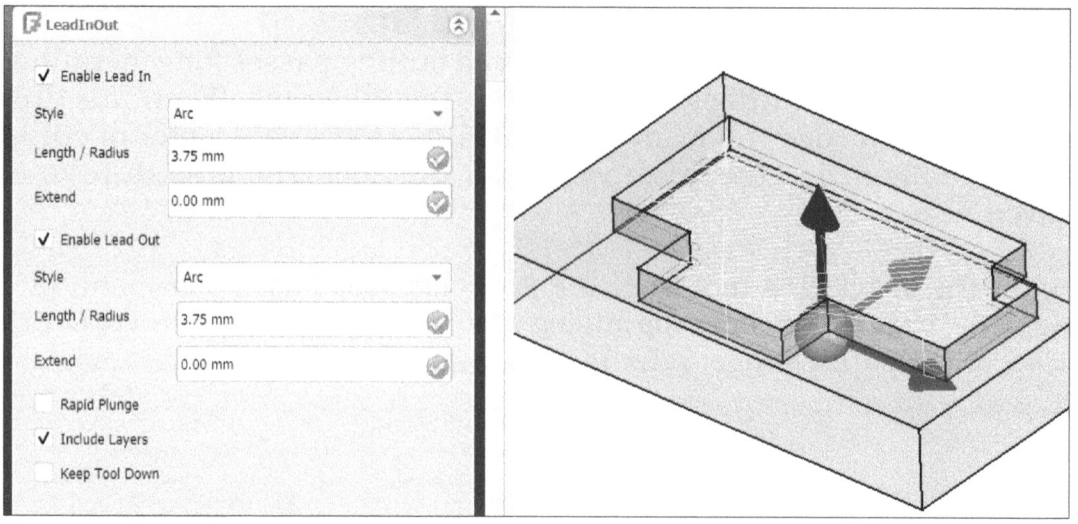

Figure-107. LeadInOut dialog

- Select the **Enable Lead In** check box to enable adjustments for entry segment of toolpath.
- Select desired option from the **Style** drop-down to define shape of lead-in segment. Select the **Arc** option if you want to create an arc segment at the entry point of specified radius. Select the **Tangent** option from the **Style** drop-down to tangentially extend the entry segment of toolpath by specified length. Select the **Perpendicular** option from the **Style** drop-down to perpendicularly extend the toolpath by specified value. After selecting desired option, set the related parameter in the **Length/Radius** edit box.
- Specify desired value in the **Extend** edit box to define extension length for the lead-in.
- Similarly, you can set adjustment of exit segment of toolpath by selecting the **Enable Lead Out** check box.
- Select the **Rapid Plunge** check box to perpendicularly plunge the cutting tool in stock at entry point with maximum feedrate.
- Select the **Include Layers** check box to add new layers for lead in and lead out.
- Select the **Keep Tool Down** check box to minimize extra non-cutting moves in toolpath.
- After setting desired parameters, click on the **OK** button from the dialog.

Defining Ramp Entry

The **RampEntry** tool is used to make entry segment of toolpath formed as ramp. On clicking this tool from **Path Dressup** cascading menu of the **CAM** menu, the ramp segment will be applied to selected toolpath. Select this newly added ramp entry, the properties of toolpath will be displayed as shown in Figure-108. Select desired option from the **Ramp Feedrate** drop-down to define the speed at which cutting operation will be performed for ramp entry. Similarly, set the entry angle and other related parameters in the **Path** section of **Properties** box.

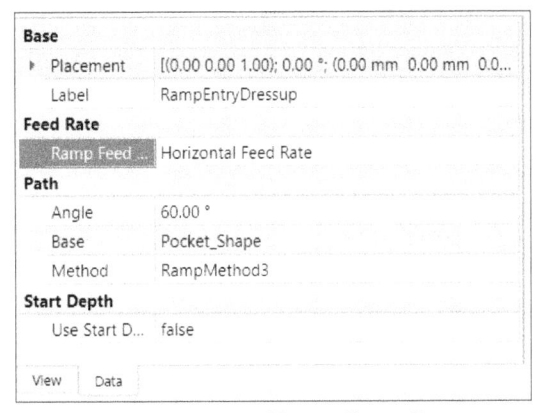

Figure-108. Properties of Ramp Entry Dressup

Defining Holding Tags

The **Tags** tool is used to generate holding tags for workpiece so that the workpiece stays on machine bed during machining. Generating tags makes sure that toolpaths do not cut these holding tags. The procedure to use this tool is given next.

- Click on the **Tags** tool from the **Path Dressup** cascading menu of the **CAM** menu. The **Holding Tags** dialog will be displayed; refer to Figure-109.

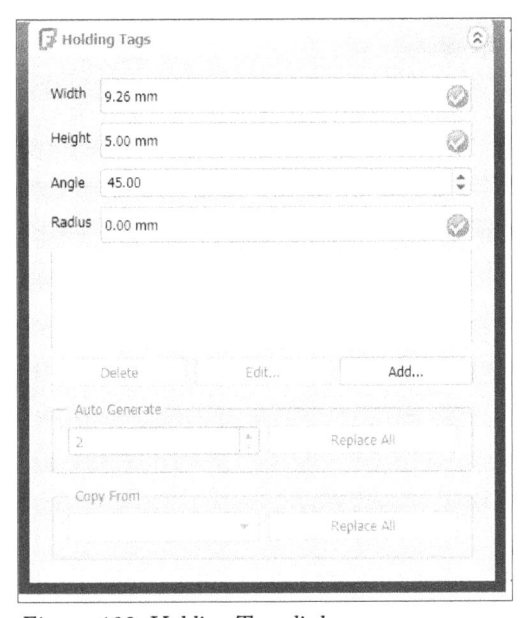

Figure-109. Holding Tags dialog

- Set desired parameters in the **Width**, **Height**, **Angle**, and **Radius** edit boxes to define size of tag and click on the Add button from dialog. You will be asked to specify location of tag.
- Set desired parameters in the **Global X**, **Global Y**, and **Global Z** edit boxes; and click on the **Save** button. The tag will be generated.
- After adding desired tags, click on the **OK** button from the dialog.

Performing Z Depth Correction

The **Z Depth Correction** tool is used to correct the Z depth value of toolpath based on data generating by probing tool. The procedure to use this tool is given next.

- Click on the **Z Depth Correction** tool from the **Path Dressup** cascading menu of the **CAM** menu. The **Z Depth Correction** dialog will be displayed; refer to Figure-110.

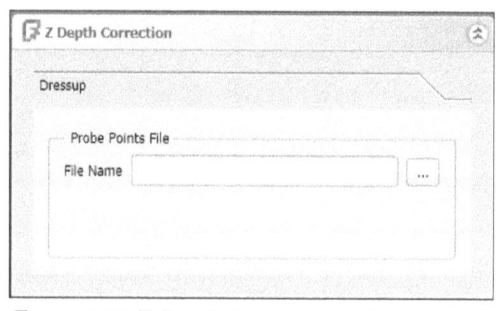

Figure-110. Z Depth Correction dialog

- Click on the **Browse** button for **File Name** edit box and select desired file.
- Click on the **OK** button from the dialog. The Z depth adjustment will be applied.

SUPPLEMENTAL COMMANDS

The tools in **Supplemental Commands** cascading menu are used to perform various auxiliary manufacturing operations like adding fixture to machining setup, adding comments in toolpath output, generating probe data, and so on. Various tools of this cascading menu are discussed next.

Adding Fixture in Setup

The **Fixture** tool is used to add representation of a fixture in the machine setup. Click on the **Fixture** tool from the **Supplemental Commands** cascading menu of the **CAM** menu. The fixture will be added in the **Operations** section of the **Model** panel in **ComboView**. Specify desired parameters in the **Properties Box** to define fixture code, active status of fixture, and so on.

Adding Comments

The **Comment** tool is used to add desired comments in manufacturing setup. Click on the Comment tool from the **Supplemental Commands** cascading menu of the **CAM** menu. The Comment feature will be added in the **Operations** section of machining setup. Select the feature from **ComboView** and type desired comment in the **Comment** field; refer to Figure-111.

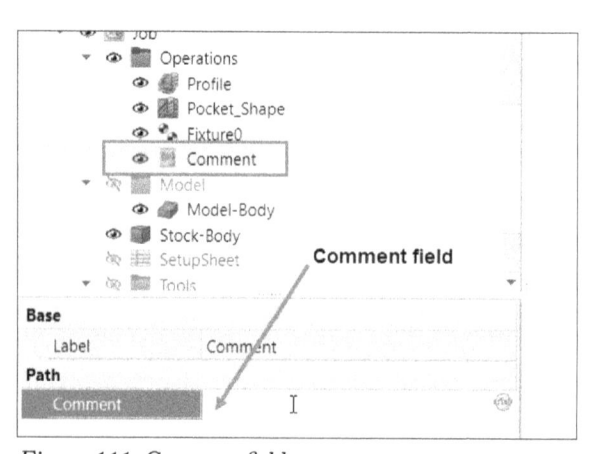

Figure-111. Comment field

Adding Stop Command

The **Stop** tool is used to add optional or mandatory stop to the machining program. Click on the **Stop** tool the **Supplemental Commands** cascading menu of the **CAM** menu. The **Stop** feature will be added in the **Operations** section of **ComboView**. Select this feature and set desired option in the **Stop** field to define whether stop will be optional or mandatory; refer to Figure-112.

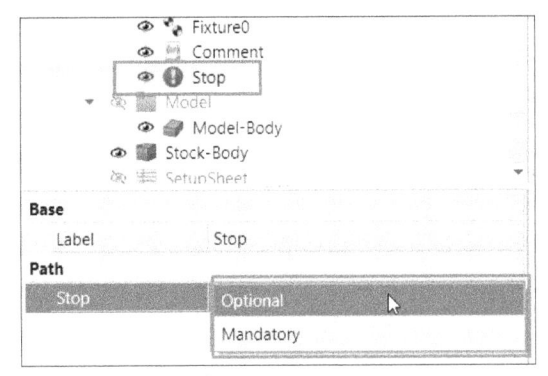

Figure-112. Stop feature options

Adding Custom G-Codes

The **Custom** tool is used to add desired custom G-codes to selected machining operation. The procedure to use this tool is given next.

- Select desired cutting operation to which you want to add custom codes and click on the **Custom** tool from the **Supplemental Commands** cascading menu of the **CAM** menu. The **Custom** dialog will be displayed in the **ComboView**; refer to Figure-113.

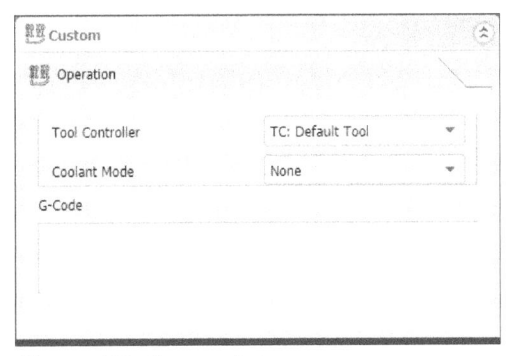

Figure-113. Custom dialog

- Select desired option from the **Tool Controller** drop-down to define cutting tool for custom codes.
- Select desired option from the **Coolant Mode** drop-down to set coolant mode.
- Set desired custom code in the **G-Code** field and click on the **OK** button to apply codes.

Probe Toolpath

The **Probe** tool is used to check surface and generate probe data. The procedure to use this tool is given next.

- Click on the **Probe** tool from the **Supplemental Commands** cascading menu of the **CAM** menu. The **Probe** dialog will be displayed in the **ComboView**; refer to Figure-114.

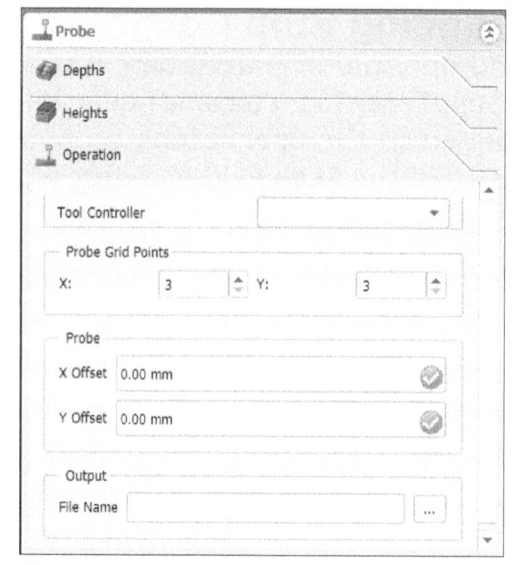

Figure-114. Probe dialog

- Set desired parameters in the **Depths** and **Heights** section of the dialog.
- Select desired probing tool from the **Tool Controller** drop-down.
- Set desired grid points for the probing of surface in the spinners of **Probe Grid Points** section of the dialog.
- Set desired offset values in the **X Offset** and **Y Offset** edit boxes of the **Probe** area if needed.
- Click on the **Browse** button from **Output** section in the dialog and set name & location for output file of probe in the **Select Output File** dialog box; refer to Figure-115. Note that if you are not able to directly save the file then you can create an empty text file at desired location with desired name and then select it from the dialog box to use as probe data file.
- After setting desired parameters, click on the **OK** button to generate the operation.

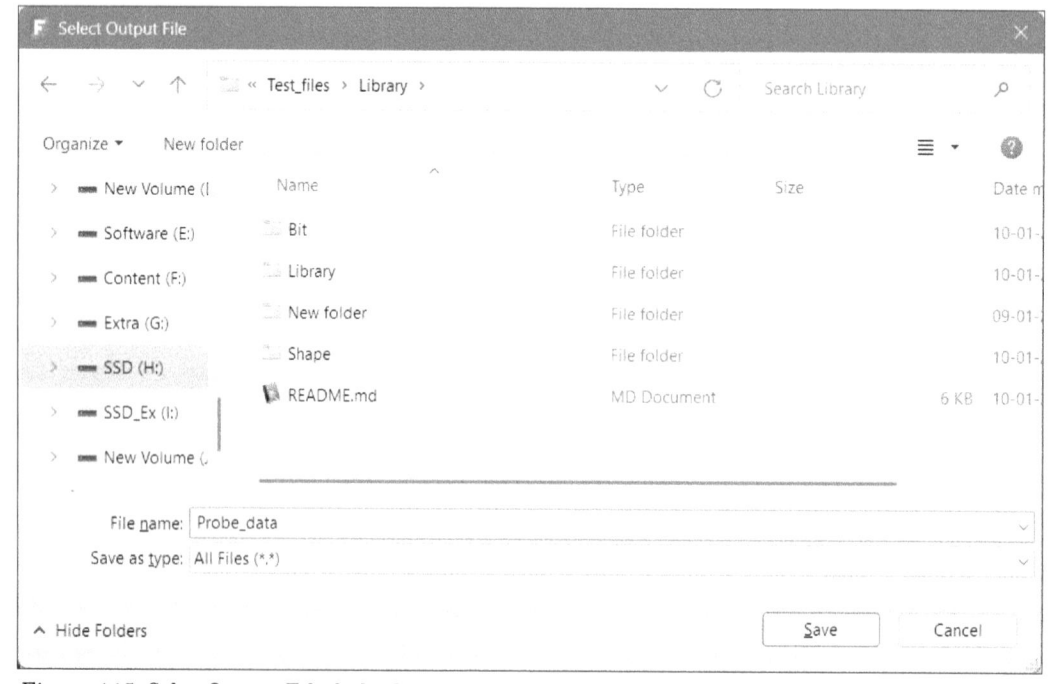

Figure-115. Select Output File dialog box

PRACTICAL

Generate the toolpath of the given model shown in Figure-116.

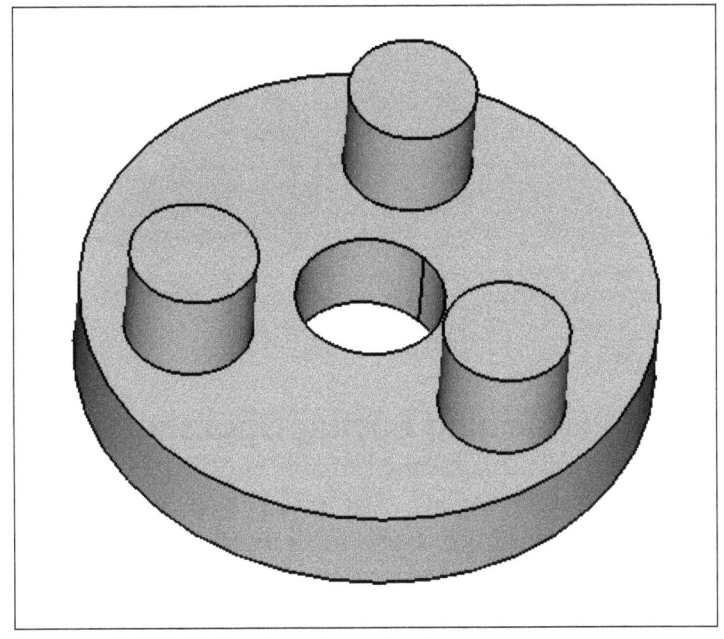

Figure-116. Model for generating toolpaths

Starting a New File

- Start FreeCAD if not started yet and open the model file for this practical from resources.
- Select the **CAM** option from the **Switch between workbenches** drop-down in the **Toolbar**. The path environment will be displayed.

Creating Stock

- Click on the **Job** tool from the **Toolbar**. The **Create Job** dialog box will be displayed.
- Select the object which you want to create a job and click on **OK** button from the dialog box. The **Job Edit** dialog will be displayed along with the model enclosed by a bounding box; refer to Figure-117.

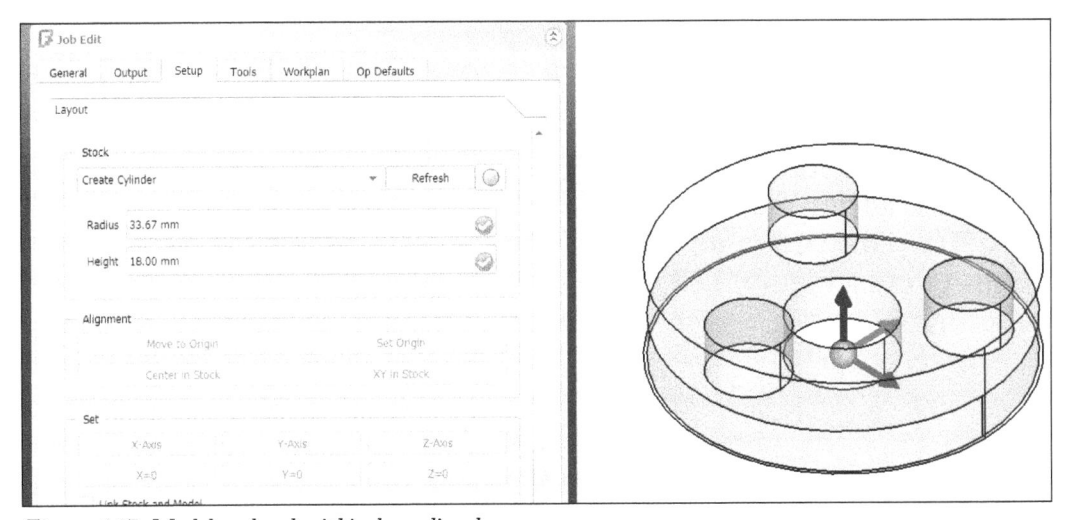

Figure-117. Model enclosed within bounding box

- Specify the parameters in the dialog as shown in Figure-117 and click on **OK** button. The path job will be created; refer to Figure-118.

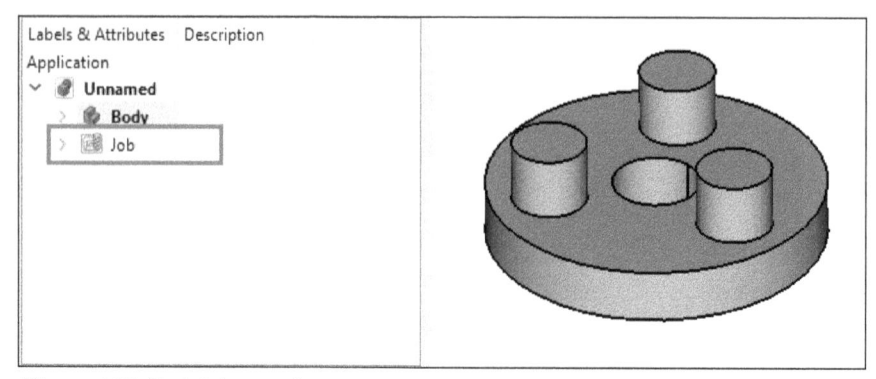

Figure-118. Path job created

Performing Facing Operation

- Removing material from the top face of stock is the first step in machining. Click on the **Face** tool from the **CAM** menu. The **MillFace** dialog will be displayed.
- Select top face of model (any of the boss feature) and click on the **Add** button from **Base Geometry** section in the dialog; refer to Figure-119.

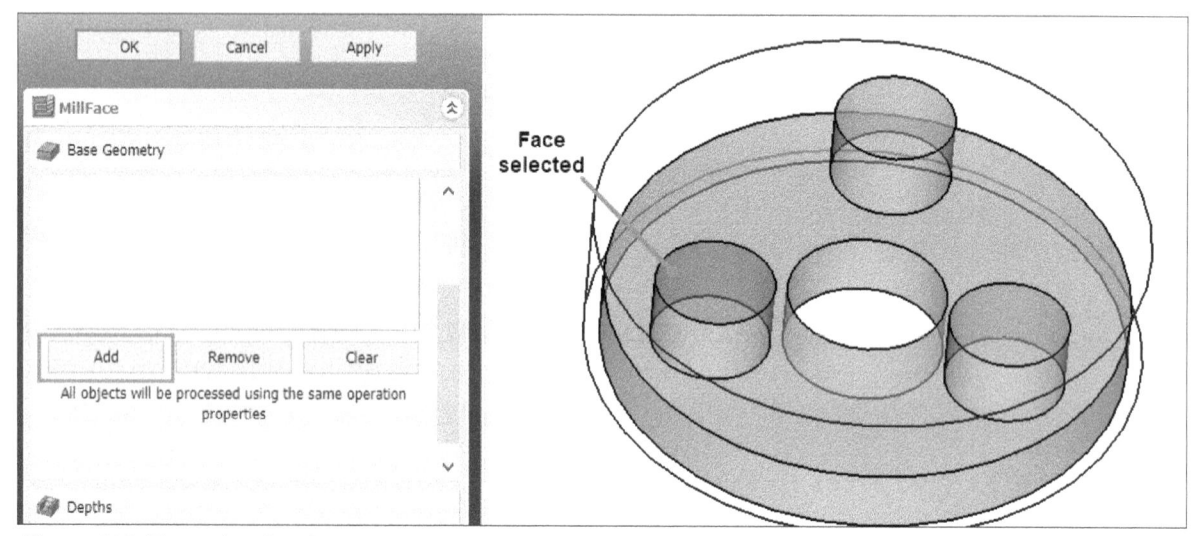

Figure-119. Face to be selected

- Expand the **Operations** section and select desired cutting tool from the **Tool Controller** drop-down.
- Select the **Stock** option from the **Boundary Shape** drop-down to machine up to stock boundaries for facing operation.
- Set the other parameters as desired and click on the **OK** button to create facing operation; refer to Figure-120.

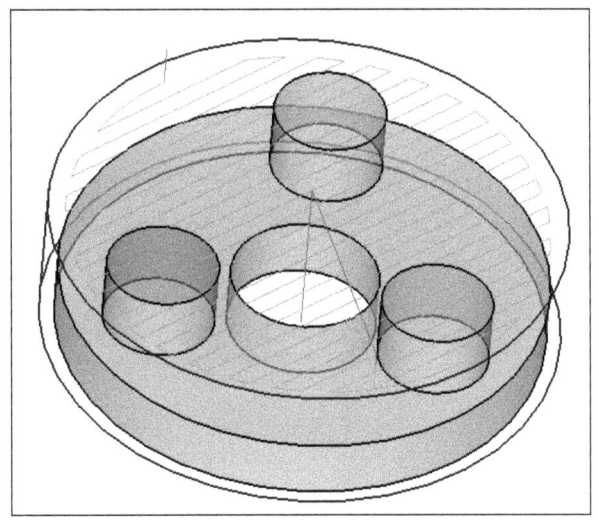

Figure-120. After facing operation

- Select the **MillFace** option from **Job->Operations** category of **Combo View** and click on **Toggle the Active State of the Operation** tool from **Toolbar** or **Path** menu. This will hide the toolpaths from graphics area.

Creating Pocket Operation

- Click on the **Pocket Shape** tool from the **CAM** menu and select face of the model as base geometry for pocket operation as shown in Figure-121. Click on the **Add** button from **Base Geometry** section in the dialog to add it to operation.

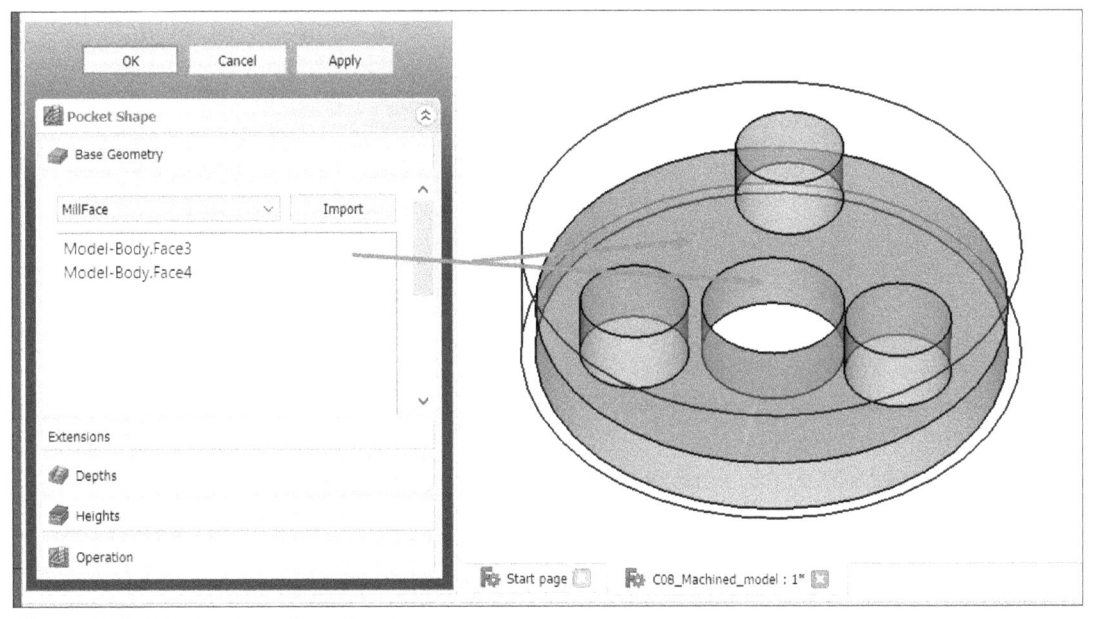

Figure-121. Selecting face of model as base geometry

- Specify other parameters for pocket operation in the dialog as shown in Figure-122 and click on the **OK** button. The pocket operation path of the model will be created; refer to Figure-123.

Figure-122. Parameters for pocket operation

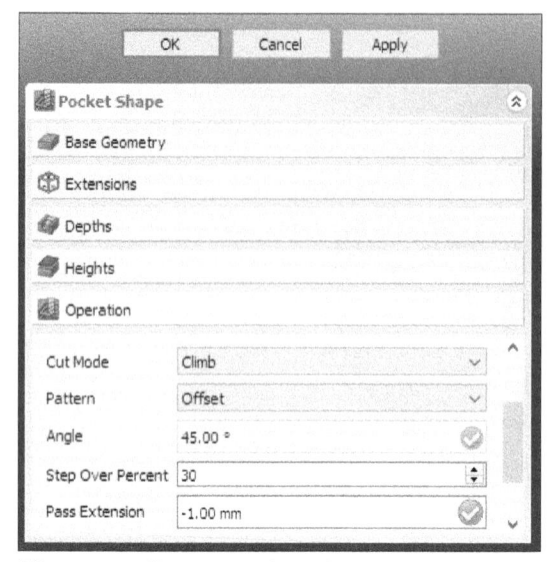

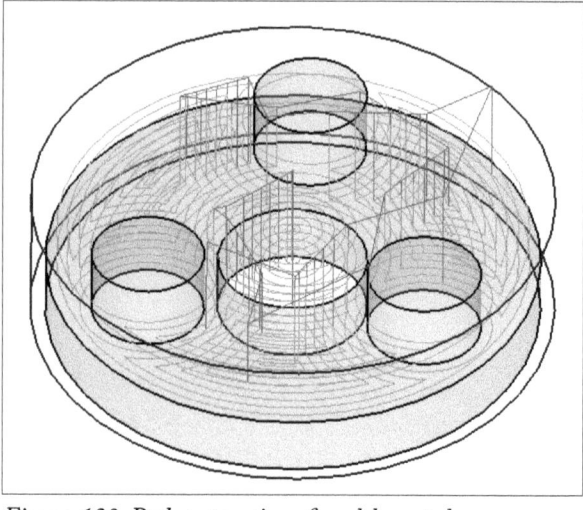

Figure-123. Pocket operation of model created

Simulating the Path

- Click on the **CAM Simulator** tool from the **Toolbar**. The **Path Simulator** dialog will be displayed along with the pocket shape tool; refer to Figure-124.

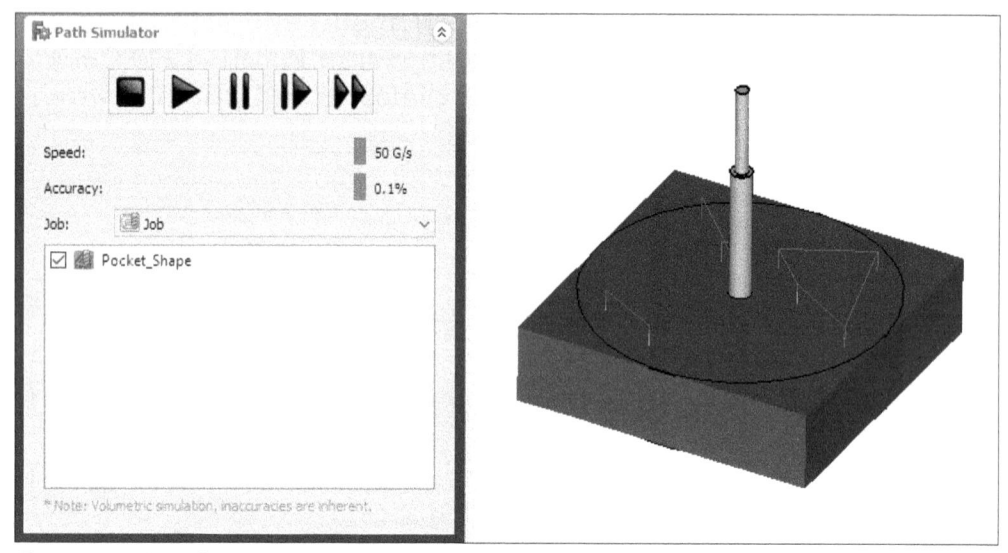

Figure-124. Path Simulator dialog to simulate pocket operation

- Click on the **Activate/resume simulation** button to simulate the pocket shape operation. The pocket operation will be simulated; refer to Figure-125.

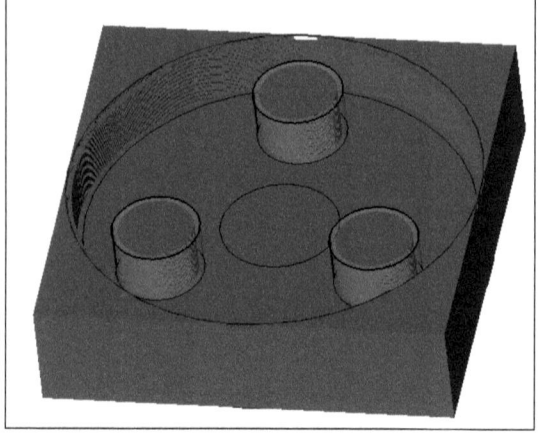

Figure-125. Pocket operation simulated

SELF ASSESSMENT

Q1. What is the primary purpose of the ToolBit Library Editor in FreeCAD?

A) To create and manage cutting tools for the Path workbench
B) To import job templates for machining operations
C) To set up machine configurations
D) To create and manage 3D models

Q2. Which dialog box allows you to import a cutting tool table from an XML file in the ToolBit Library editor?

A) Choose Tool Controller dialog
B) Save Toolbit Library dialog
C) Tool Library Path dialog
D) Library dialog box

Q3. What do you do to delete a tool from the ToolBit Library?

A) Right-click on the tool and select "Delete"
B) Click the Remove button after selecting the tool
C) Double-click on the tool and press delete
D) Click the "Remove Tool" option from the context menu

Q4. Which checkbox in the Exporting Job Template dialog includes the tool controller definitions in the template?

A) Tools
B) Setup Sheet
C) Stock
D) Coolant Mode

Q5. In FreeCAD's Path CAM workbench, what operation does the Profile tool perform?

A) Pocket machining operation
B) Creating a contour operation for 3D parts
C) Drilling holes in a model
D) Face milling on a 3D model

Q6. What is the purpose of the "Extra Offset" in the Profile dialog for a contour operation?

A) To specify the starting point for the tool
B) To leave extra material for the operation
C) To offset the profile operation by tool radius
D) To specify the depth of cut

Q7. In the Pocket Shape operation, which cutting mode reduces tool wear and improves surface finish?

A) Conventional Milling
B) Climb Milling
C) Pecking
D) Helix Milling

Q8. Which option in the Drilling dialog allows for incremental drill depth before retracting?

A) Feed retract
B) Peck
C) Dwell
D) Clearance Height

Q9. What does the Face Mill tool in FreeCAD's Path CAM workbench do?

A) Removes material from the top flat face of the model
B) Creates a contour cut for 3D parts
C) Drills holes in the model
D) Creates a helical path for circular vertical surfaces

Q10. Which operation tool in FreeCAD allows for the creation of a helical path for machining circular vertical surfaces?

A) Adaptive Tool
B) Profile Tool
C) Helix Tool
D) Face Mill Tool

Q11. What does the Adaptive tool use to create clearing and profiling paths?

A) Constant cutter engagement
B) A pre-set fixed path
C) An adaptive algorithm managing cutter engagement
D) A single cutting strategy

Q12. Which operation tool is used to remove material from the top flat face of the model in the Path CAM workbench?

A) Helical Operation
B) Drilling Operation
C) Face Milling
D) Adaptive Operation

Q13. What is the primary function of the Engrave tool in the Path workbench?

A) To create a 3D pocket operation
B) To engrave a Draft Shape String onto a part
C) To simulate tool path operations
D) To inspect G-code in FreeCAD

Q14. Which tool is used to simulate the operations on a 3D model along the G-code paths in the Path workbench?

A) CAM Simulator
B) Inspect Path Command
C) New CAM Simulator
D) Post Process tool

Q15. What is the purpose of the Deburr tool in the Path workbench?

A) To create a deburring path along edges or faces
B) To engrave text on a part
C) To simulate the toolpath
D) To create a 3D pocket operation

Q16. What cutting tool is used when performing the Vcarve operation?

A) Engraving tool
B) V-bit cutting tool
C) Chamfer tool
D) 3D Pocket tool

Q17. Which dialog box is displayed when creating a 3D pocket operation in the Path workbench?

A) Pocket 3D dialog box
B) Deburr dialog box
C) Create Job dialog box
D) Vcarve dialog box

Q18. How can you inspect the G-code of a path operation in FreeCAD?

A) Use the Inspect Path Command tool
B) Use the CAM Simulator tool
C) Use the Post Process tool
D) Use the Array tool

Q19. Which tool in the Path workbench is used to export a Path Job to a G-code file?

A) Post Process tool
B) CAM Simulator tool
C) Inspect Path Command tool
D) Check CAM Job for common errors

Q20. What is the main advantage of using the New CAM Simulator over the old one?

A) It supports more tool types
B) It works in a separate window and is faster and more precise
C) It allows inspection of G-code
D) It can create copies of operations

Q21. What is the function of the Finish Selecting Loop tool in the Path workbench?

A) To create multiple copies of a path
B) To complete a loop from two selected edges
C) To simulate the toolpath
D) To generate a 3D pocket operation

Q22. Which tool in the Path workbench helps in checking jobs for missing values or errors?

A) CAM Simulator
B) Check CAM Job for common errors
C) Array tool
D) Post Process tool

Q23. What must be done before you can engrave a string on the face of a model in FreeCAD?

A) Create a new job setup
B) Select the engraving tool
C) Simulate the toolpath
D) Inspect the G-code

Q24. What does the Array tool in the Path workbench do?

A) Creates a single copy of a path operation
B) Creates a path by duplicating another path multiple times at specific intervals
C) Simulates toolpaths on a 3D model
D) Generates a 3D pocket operation path

Q25. What is the purpose of the "Apply" button in the Engrave dialog?

A) To start the simulation of the path
B) To generate the engraving operation path at the specified depth
C) To save the G-code output
D) To add the base geometries to the operation

Q26. In the CAM Simulator, which button is used to slow down the animation for troubleshooting specific cuts or movements?

A) Pause button
B) Play button
C) Single-Step button
D) Fast-Forward button

Q27. Which of the following tools is used to simulate toolpath operations on a 3D model in the Path workbench?

A) Post Process tool
B) New CAM Simulator tool
C) Inspect Path Command tool
D) Finish Selecting Loop tool

Q28. What type of cutting tool is typically used with the Vcarve operation?

A) Chamfer tool
B) V-bit cutting tool
C) End mill tool
D) Engraving tool

Q29. Which dialog box is displayed after clicking on the Copy tool in the Path workbench?

A) Property Editor dialog
B) New Job dialog box
C) Path Job dialog box
D) Job Edit dialog box

Q30. What is the purpose of the "Base Geometry" tab in the Engrave and Deburr dialogs?

A) To select the tool controller
B) To select the base geometries (edges or faces) for the operation
C) To simulate the path operation
D) To specify the depth of the engraving or deburring operation

Q31. What is the main advantage of the "Simple Copy" tool in the Path workbench?

A) It allows you to create parametric copies
B) It creates a non-parametric copy of the selected path
C) It generates a 3D pocket operation
D) It simulates the toolpath

Q32. When using the Deburr tool, which parameter specifies the width of the chamfer cut?

A) Direction
B) W (Width)
C) h (extra depth)
D) Tool Controller

Q33. What does the Post Process tool in the Path workbench do?

A) It exports the selected Path Job to a G-code file
B) It simulates the toolpath
C) It checks the CAM Job for common errors
D) It allows you to inspect the G-code

Q34. In the Path workbench, which tool is used to complete a loop from two selected edges that share a common vertex?

A) Inspect Path Command
B) Finish Selecting Loop
C) CAM Simulator
D) Post Process tool

Q35. Which dialog box is displayed after clicking on the "Post Process" tool in the Path workbench?

A) Inspect Path Command dialog
B) Path Job dialog box
C) FreeCAD dialog box showing postprocessor properties
D) Property Editor dialog

Q36. What tool should be used to simulate toolpath operations with better speed and accuracy, based on OpenGL functions?

A) CAM Simulator tool
B) New CAM Simulator tool
C) Path Job tool
D) Inspection of G-code tool

Q37. Which of the following tools can be used to check for missing values in a selected job in the Path workbench?

A) CAM Simulator
B) Check CAM Job for common errors
C) Post Process tool
D) Finish Selecting Loop tool

Q38. What is the primary use of the "Inspection of G-Code" tool in FreeCAD?

A) To simulate the toolpath
B) To view the internal FreeCAD G-code dialect contents of a Path Operation Object
C) To generate the final G-code output
D) To create new path operations

Q39. What is the primary function of the "Tool Controller" drop-down menu in the Engrave dialog?

A) To select the depth of the engraving operation
B) To select the tool to be used for the operation
C) To define the geometry for the engraving path
D) To specify the width of the chamfer cut

Q40. In the Path workbench, what happens when you click the "Play" button in the CAM simulator?

A) The toolpath is simulated without any changes
B) The simulation of the job's operations begins with an animation
C) It stops the toolpath simulation
D) It generates the G-code for the operation

For Student Notes

Chapter 9

FEM Workbench

Topics Covered

The major topics covered in this chapter are:

- *Introduction to FEM Workbench*
- *Starting FEM Workbench*
- *Creating Analysis Container*
- *Analysis Types in FreeCAD*
- *Defining Material*
- *Applying Constraints*
- *Creating Geometry Elements*
- *Meshing*
- *Solvers*
- *Post Processing*

INTRODUCTION TO FEM

The **FEM** Workbench provides a modern finite element analysis (FEA) workflow for FreeCAD. This means all the tools to perform an analysis are combined into one graphical user interface (GUI); refer to Figure-1.

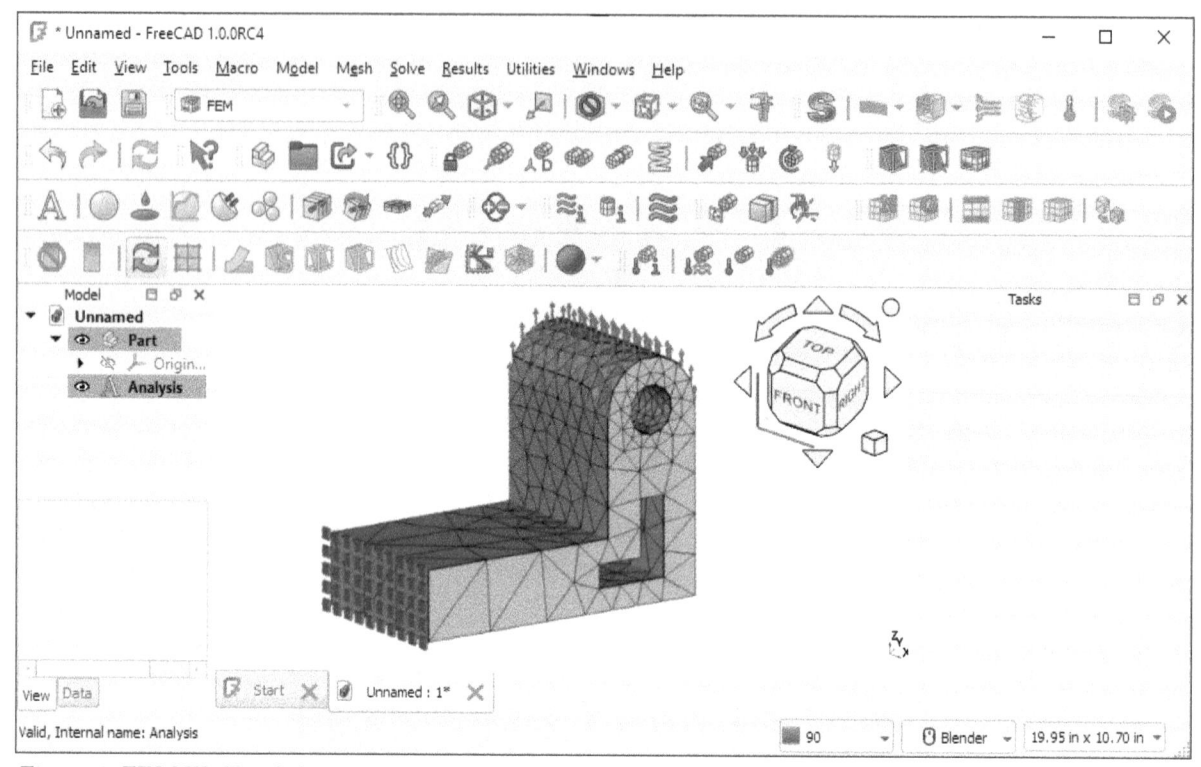

Figure-1. FEM Workbench Overview

The workflow for FEM is shown in Figure-2. The steps to carry out a finite element analysis are given next.

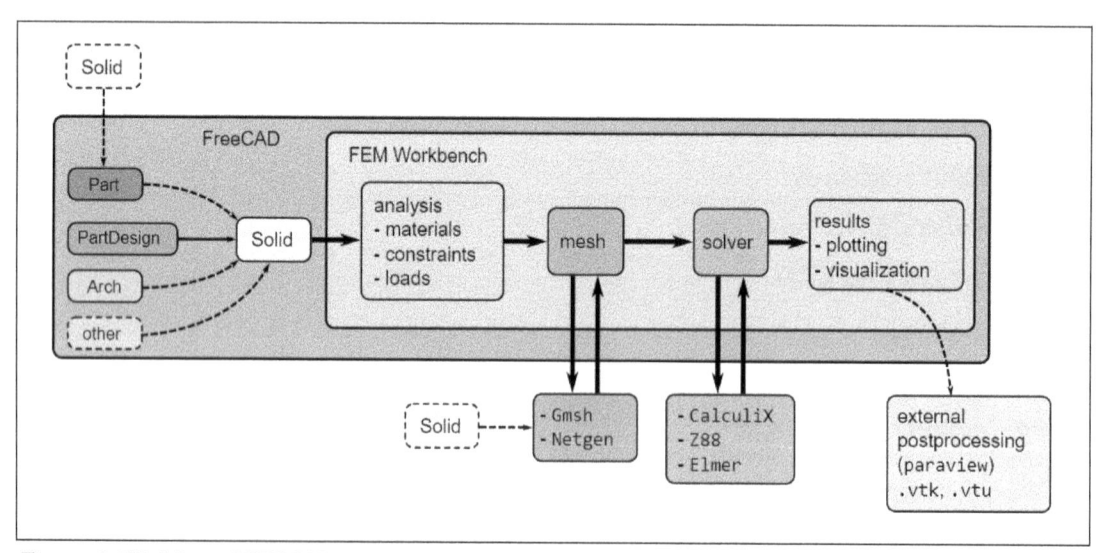

Figure-2. Workflow of FEM Workbench

1. **Preprocessing**: Setting up the analysis problem mainly on plain paper or in your mind.
2. **Modeling the geometry**: Creating the geometry with FreeCAD, or importing it from a different application.
3. **Creating an analysis**: This step is also called Defining Boundary Conditions. It involves adding simulation constraints such as loads and fixed supports to the

geometric model. Adding materials to the parts of the geometric model. Creating a finite element mesh for the geometrical model, or importing it from a different application.
4. **Solving**: Running an external solver from within FreeCAD.
5. **Postprocessing**: Visualizing the analysis results from within FreeCAD, or exporting the results so they can be post processed with another application.

STARTING FEM WORKBENCH

The **FEM** Workbench calls two external programs to perform meshing of a solid object and perform the actual solution of the finite element problem.

- To start a new **FEM** file, click on the **New** button from **File** menu and select **FEM** workbench from **Switch between workbenches** drop-down in the **Toolbar**; refer to Figure-3. The tools related to **FEM** workbench will be displayed in the **Toolbar**; refer to Figure-4.

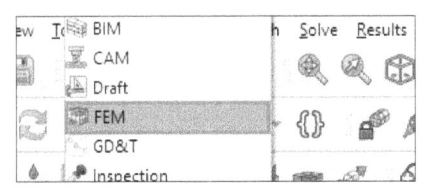

Figure-3. FEM Workbench

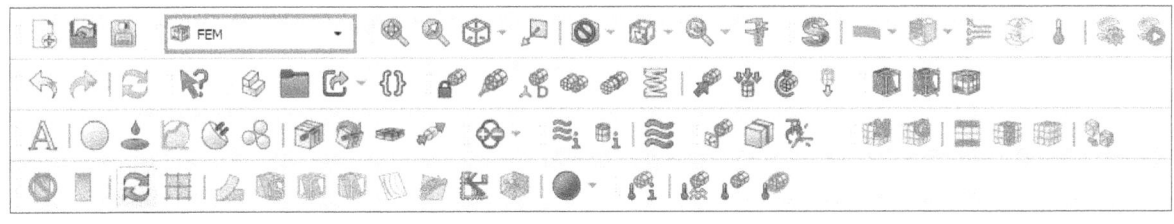

Figure-4. FEM Workbench tools

CREATING ANALYSIS CONTAINER

The **FEM** Analysis in FreeCAD could be seen as a Container that holds all objects of a Finite Element Analysis. It is mandatory to have an analysis container which holds all the needed objects. At least one of the following objects is needed for a mechanical analysis: material, fixed constraint, force constraint, or pressure constraint. The procedure to use this tool is discussed next.

- Click on the **Analysis Container** tool from **Toolbar** in the **FEM** workbench; refer to Figure-5. A new analysis will be created and displayed in the **Model Tree**; refer to Figure-6.

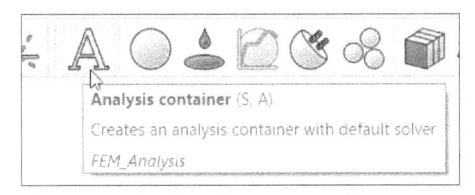

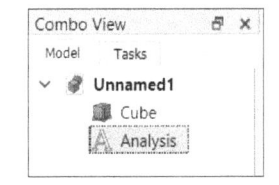

Figure-5. Analysis Container tool *Figure-6. New Analysis created*

Some details about common analyses available in FreeCAD are given next.

Static Analysis

This is the most common type of analysis we perform. In this analysis, loads are applied to a body due to which the body deforms and the effects of the loads are transmitted throughout the body. To absorb the effect of loads, the body generates internal forces and reactions at the supports to balance the applied external loads. These internal forces and reactions cause stress and strain in the body. Static analysis refers to the calculation of displacements, strains, and stresses under the effect of external loads, based on some assumptions. The assumptions are as follows.

1. All loads are applied slowly and gradually until they reach their full magnitudes. After reaching their full magnitudes, load will remain constant (i.e. load will not vary against time).
2. Linearity assumption: The relationship between loads and resulting responses is linear. For example, if you double the magnitude of loads, the response of the model (displacements, strains, and stresses) will also double. You can make linearity assumption if:

- All materials in the model comply with Hooke's Law that is, stress is directly proportional to strain.
- The induced displacements are small enough to ignore the change in stiffness caused by loading.
- Boundary conditions do not vary during the application of loads. Loads must be constant in magnitude, direction, and distribution. They should not change while the model is deforming.

If the above assumptions are valid for your analysis then you can perform **Linear Static Analysis**. For example, a cantilever beam fixed at one end and force applied on other end; refer to Figure-7.

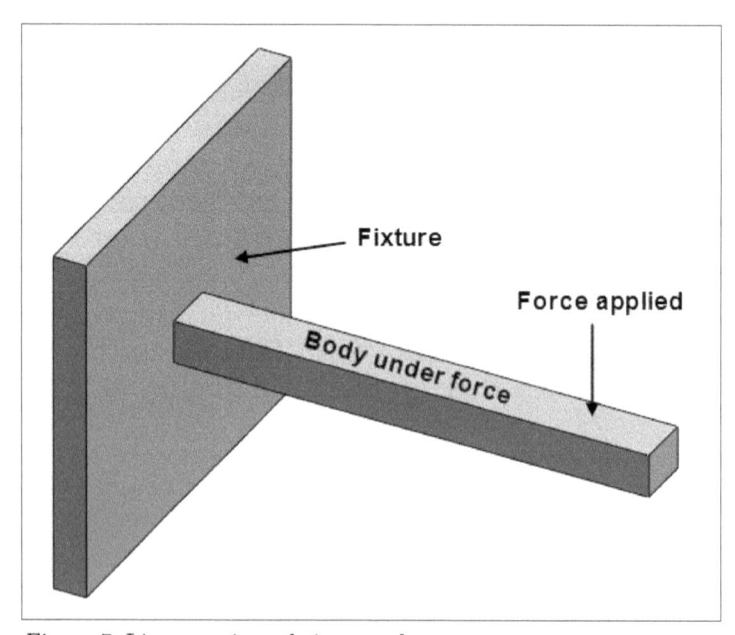

Figure-7. Linear static analysis example

If the above assumptions are not valid then you need to perform the **Non-Linear Static analysis**. For example, an object attached with a spring being applied under forces.

Frequency Analysis (Modal Analysis)

By its very nature, vibration involves repetitive motion. Each occurrence of a complete motion sequence is called a "cycle." Frequency is defined as so many cycles in a given time period. "Cycles per seconds" or "Hertz". Individual parts have what engineers call "natural" frequencies. For example, a violin string at a certain tension will vibrate only at a set number of frequencies, which is why you can produce specific musical tones. There is a base frequency in which the entire string is going back and forth in a simple bow shape.

Harmonics and overtones occur because individual sections of the string can vibrate independently within the larger vibration. These various shapes are called "modes". The base frequency is said to vibrate in the first mode and so on up the ladder. Each mode shape will have an associated frequency. Higher mode shapes have higher frequencies.

The most disastrous kinds of consequences occur when a power-driven device such as a motor for example, produces a frequency at which an attached structure naturally vibrates. This event is called "resonance." If sufficient power is applied, the attached structure will be destroyed. Note that ancient armies, which normally marched "in step," were taken out of step when crossing bridges. Should the beat of the marching feet align with a natural frequency of the bridge, it could fall down. Engineers must design machines so that resonance does not occur during regular operation of machines. This is a major purpose of Modal Analysis. Ideally, the first mode has a frequency higher than any potential driving frequency. Frequently, resonance cannot be avoided, especially for short periods of time. For example, when a motor comes up to speed, it produces a variety of frequencies. So, it may pass through a resonant frequency.

Thermal analysis

There are three mechanisms of heat transfer. These mechanisms are Conduction, Convection, and Radiation. Thermal analysis calculates the temperature distribution in a body due to some or all of these mechanisms. In all three mechanisms, heat flows from a higher-temperature medium to a lower temperature one. Heat transfer by conduction and convection requires the presence of an intervening medium while heat transfer by radiation does not.

There are two modes of heat transfer analysis.

Steady State Thermal Analysis

In this type of analysis, we are only interested in the thermal conditions of the body when it reaches thermal equilibrium, but we are not interested in the time it takes to reach this status. The temperature of each point in the model will remain unchanged until a change occurs in the system. At equilibrium, the thermal energy entering the system is equal to the thermal energy leaving it. Generally, the only material property that is needed for steady state analysis is the thermal conductivity.

Transient Thermal Analysis

In this type of analysis, we are interested in knowing the thermal status of the model at different instances of time. A thermos designer, for example, knows that the temperature of the fluid inside will eventually be equal to the room temperature(steady state), but he is interested in finding out the temperature of the fluid as a function of time. In addition to the thermal conductivity, we also need to specify density, specific heat, initial temperature profile, and the period of time for which solutions are desired.

DEFINING MATERIAL FOR SOLID

The **Material for solid** tool specifies material properties for the model. The procedure to use this tool is discussed next.

- Click on the **Material for solid** tool from **Toolbar** in the **FEM** workbench; refer to Figure-8. The **FEM material** and **Geometry reference selector** dialog will be displayed in the **Tasks** panel of **Combo View**; refer to Figure-9 and Figure-10, respectively.

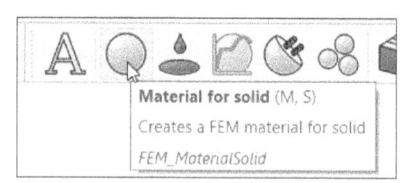

Figure-8. Material for Solid tool

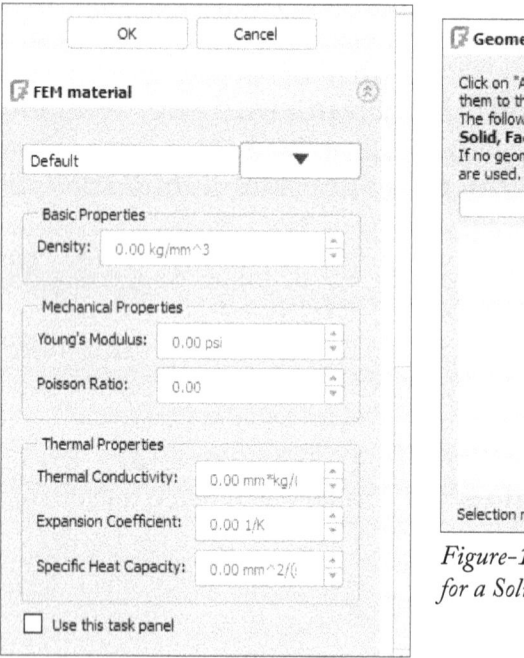

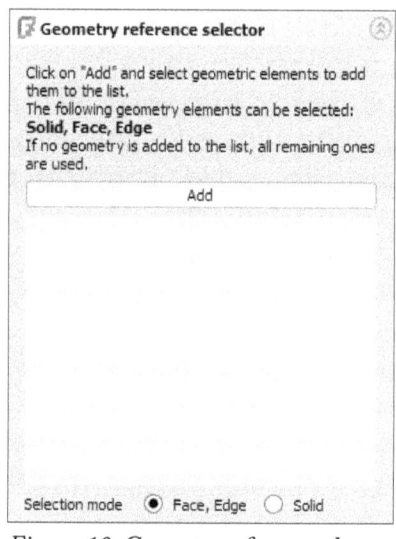

Figure-10. Geometry reference selector for a Solid Face Edge dialog

Figure-9. FEM material dialog

- Select desired material option from the **Material card** drop-down in **FEM material** dialog to be used as material for the model during analysis.
- Select desired radio button from **Selection mode** section of the **Geometry reference selector for a Solid, Face, Edge** dialog to define whether you want to apply material to single face/edge or solid object. Generally, the **Solid** radio button is used from this section. After selecting desired radio button, click on the **Add** button from the dialog and select the object to which you want to apply the material. The object will be added in the list; refer to Figure-11.

Note that you can modify the properties of material directly in **FEM material** dialog if you have selected the **use this task panel** check box in the dialog.

- Click on the **OK** button from the **Tasks** panel in **Combo View** to apply material.

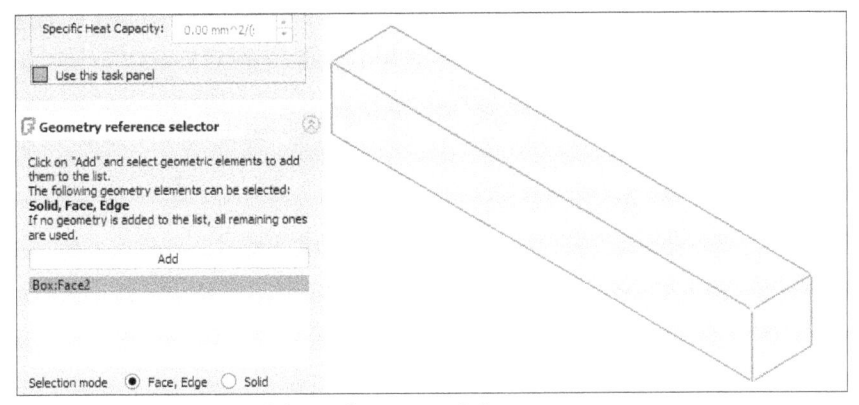

Figure-11. Object selected for applying material

Editing Material

The **Material editor** tool is used to modify the parameters of material like density, Young's Modulus, and so on. The procedure to edit material is given next.

- Click on the **Material editor** tool from the **Materials** cascading menu in the **Model** menu. The **Materials** dialog box will be displayed; refer to Figure-12.

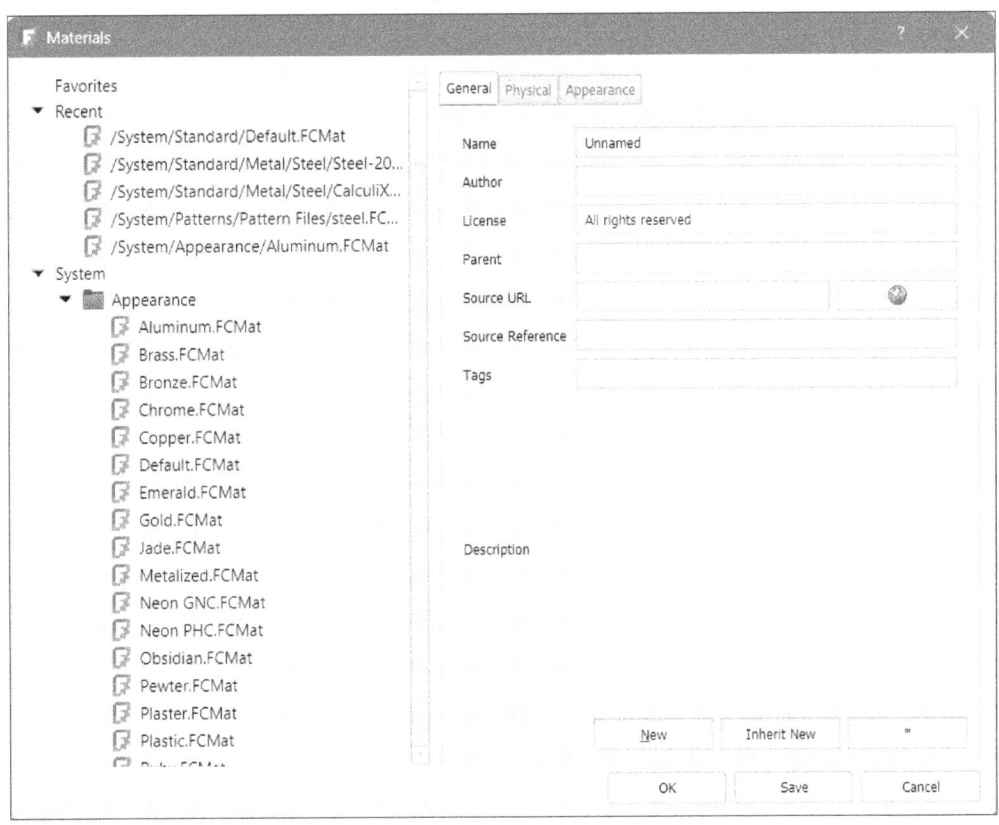

Figure-12. Material Editor dialog box

- Select desired material from the left side to modify its parameter.
- Set desired general parameters in the **General** tab of the dialog box like name of material, description, tags, author, and so on.
- Select the **Physical** tab from right area in the dialog box to add/modify physical properties of the material like density, hardness, linear elasticity, and so on.

- Click on the **Add physical model** (+) button from the **Physical** tab to add a new property. The `Material Models` dialog box will be displayed; refer to Figure-13.
- Select desired property from left area and set the parameter as desired. Click on the **OK** button from the `Material Models` dialog box to apply changes. The `Materials` dialog box will be displayed again. Set desired values for added properties. If you want to delete a property then select desired property from the dialog box and click on the **Delete property** button.
- Select the **Appearance** tab from the dialog box and set desired parameters for appearance of material in the dialog box.

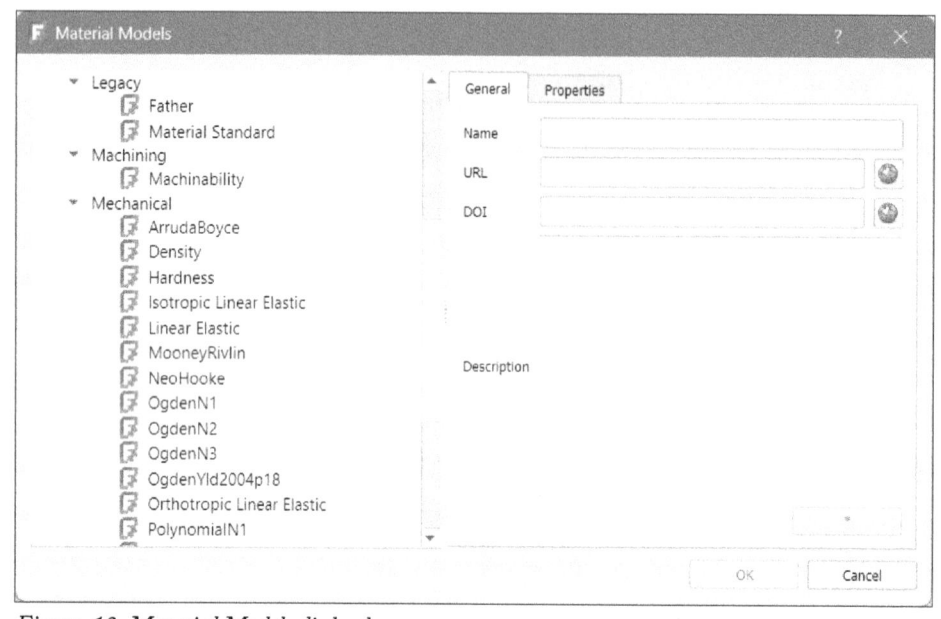

Figure-13. Material Models dialog box

- Click on the **New** button from the **General** tab in the dialog box if you want to generate a new material. Select a base material from left and click on the `Inherit New` button from the dialog box if you want to generate a new material based on properties of another material.
- Click on the **Add to favorites** button from the dialog box to add selected material in the list of favorite materials.
- After setting desired parameters, click on the **OK** button from the dialog box to exit.

After applying material to the model, the next step in FEM is to apply constraints to the model. Applying constraints is a part of defining boundary conditions. First, we will discuss the constraints which restrict the movements of model (Degree of Freedom) like fixing a face, allowing sliding of a face, and so on and then we will discuss the load constraints applied on model to check their effect.

APPLYING CONSTRAINTS

There are various toolbars in FEM workbench to apply constraints to the model; refer to Figure-14. You can also access the tools to apply constraints from **Model** menu in the FEM workbench; refer to Figure-15. The tools in these toolbars are discussed next.

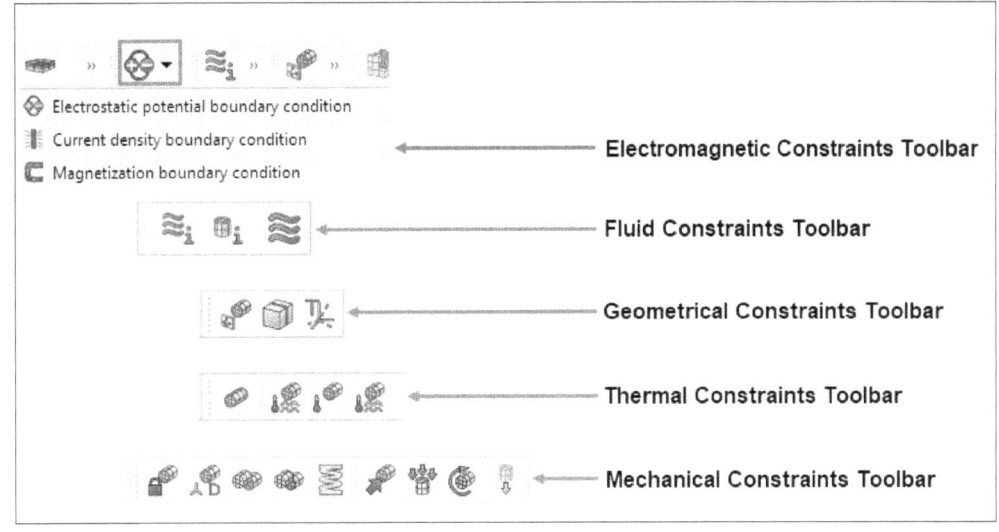

Figure-14. Constraint Toolbars

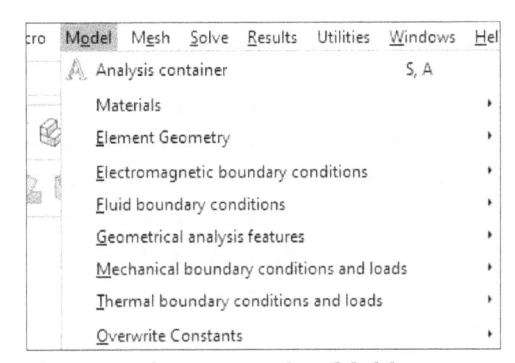

Figure-15. Constraint tools in Model menu

Mechanical Constraints

Mechanical constraints are used to stop physical movements of selected objects in specified directions. The tools to apply mechanical constraints are available in the **Mechanical Constraints Toolbar** and **Mechanical Constraints** cascading menu of **Model** menu; refer to Figure-16. These tools are discussed next.

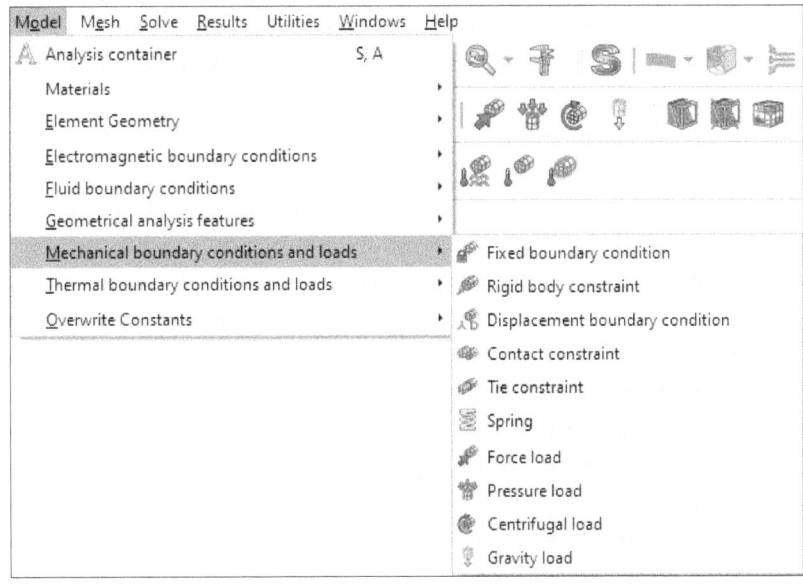

Figure-16. Mechanical Constraints cascading menu

Applying Fix Constraint

The **Fixed Boundary Condition** tool in **Mechanical boundary condition and loads** toolbar/cascading menu is used to stop movement of selected object in specified directions. The procedure to apply this constraint is given next.

- Click on the **Fixed Boundary Condition** tool from the **Mechanical boundary condition and loads** toolbar or **Model** menu. The **Analysis feature parameters** dialog will be displayed in the **Tasks** panel of **Combo View**; refer to Figure-17.

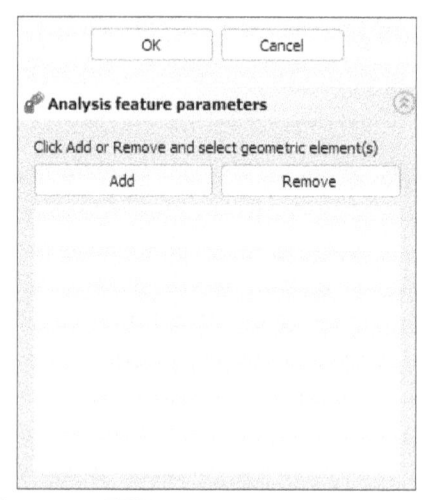

Figure-17. FEM constraint parameters dialog

- Click on the **Add** button from the dialog and select the face(s) whose movements are to be fixed. Red boxes will be displayed attached to the selected faces showing that they are fixed; refer to Figure-18.

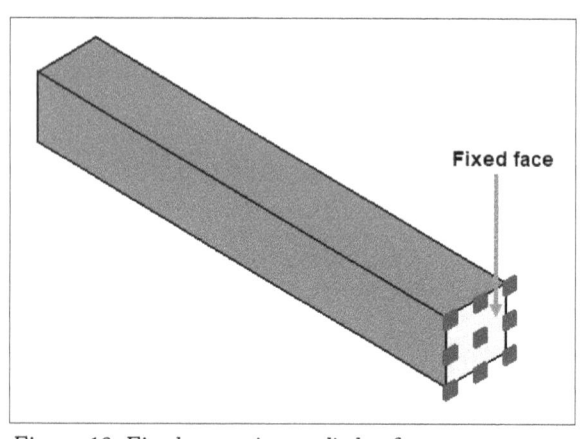

Figure-18. Fixed constraint applied to face

- After selecting desired faces, click on the **OK** button from the dialog to apply constraint.

Applying Rigid Body Constraint

The **Rigid Body Constraints** tool is used to constrain the motion of the nodes of a selected geometrical entity to a reference node. The procedure to use this tool is discussed next.

- Click on the **Rigid Body Constraints** tool from the **Mechanical boundary conditions and loads Toolbar** or **Model** menu. The **Analysis Feature parameters** dialog box will be displayed as shown in Figure-19.

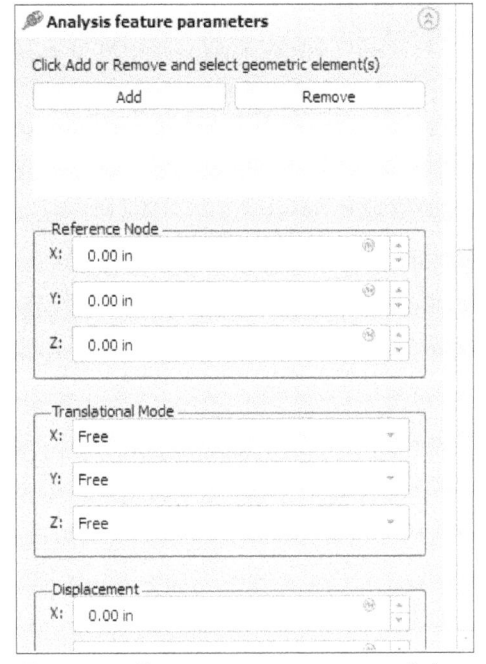

Figure-19. Fem constraint parameters dialog for rigid body

- Click on the **Add** button from the dialog and select the face(s)/edge(s)/vertices on which you want to apply constraint.
- Below the selection list box, there are seven sections in the dialog box: **Reference node, Translational mode, Displacement, Force, Rotational Mode, Rotation,** and **Moment**. Specify desired values for X,Y, and Z components of these parameters in the edit boxes.
- After specifying desired parameters, click on the **OK** button from the dialog to apply the constraints. The annotation for rigid body will be displayed; refer to Figure-20.

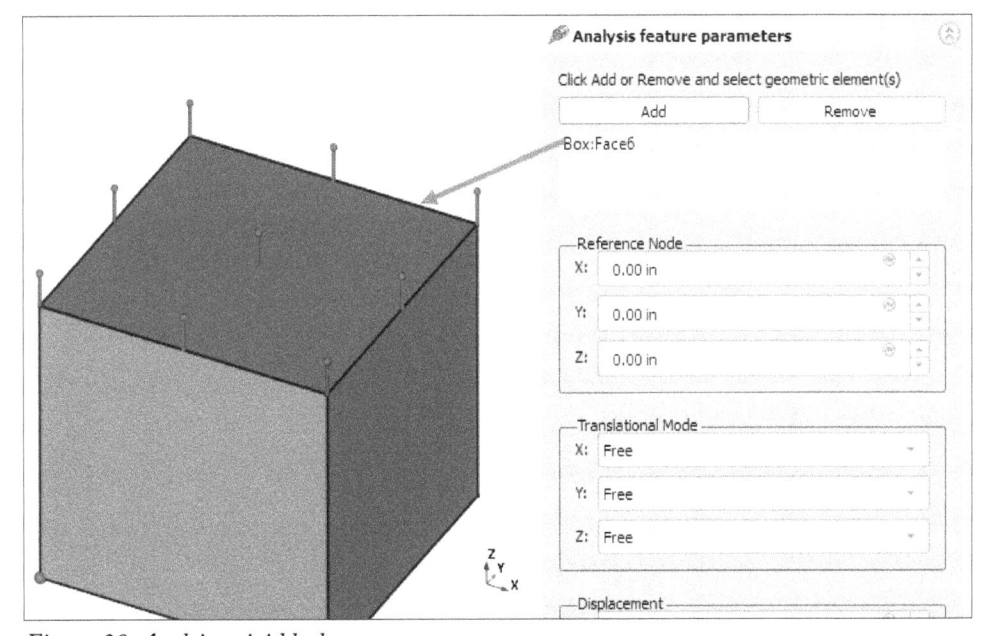

Figure-20. Applying rigid body

Applying Displacement Constraint

The **Displacement boundary condition** tool is used to specify the degree of freedom for selected objects in 3 linear and 3 rotation directions. The procedure to apply displacement constraint is given next.

- Click on the **Displacement boundary condition** tool from the **Mechanical boundary conditions and loads Toolbar** or **Model** menu. The **Analysis Feature parameters** dialog will be displayed as shown in Figure-21.

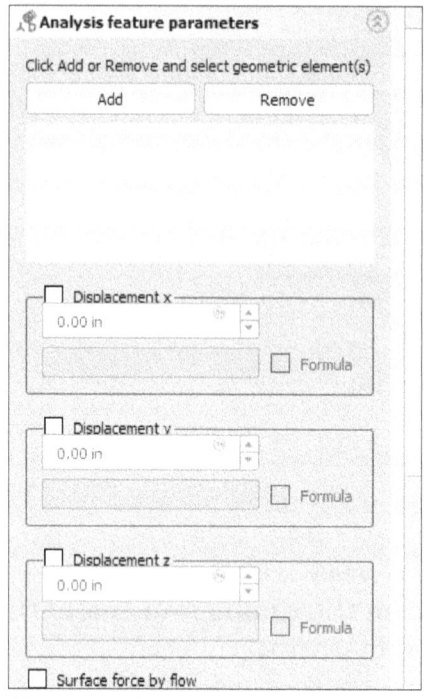

Figure-21. FEM constraint parameters dialog for displacement constraint

- Click on the **Add** button from the dialog and select the face(s) on which you want to apply constraint.
- Below the selection list box, there are six sections in the dialog box : **Displacement x**, **Displacement y**, **Displacement z**, **Rotation x**, **Rotation y**, and **Rotation z**. Select the check boxes for respective sections to activate them and input desired load values in the edit boxes. Note that rotation constraint can be applied for beam and shell elements only.
- If you are using Elmer Solver for solving analysis then you can select the **Formula** check box for the parameter and specify the value in the form of a formula dependent on other parameters of analysis setup. For example if you want to set displacement along x dependent on time in seconds as D(t) = 2xt then formula should be specified in the field as :

    Variable "time"; Real MATC "2*tx"
    This code has the following syntax:

    The prefix Variable specifies that the displacement is not a constant but a variable
    The variable is the current time
    The displacement values are returned as Real (floating point) values
    The MATC is a prefix for the Elmer solver indicating that the following code is a formula
    The tx is always the name of the variable in MATC formulas, no matter that tx in our case is actually t.

- Select the **Surface force by flow** check box if you are using Elmer solver to automatically generate displacement values based on surface force applied to selected geometry.
- After setting desired parameters, click on the **OK** button from the dialog to apply the constraint. The annotation for displacement constraint will be shown on the model; refer to Figure-22.

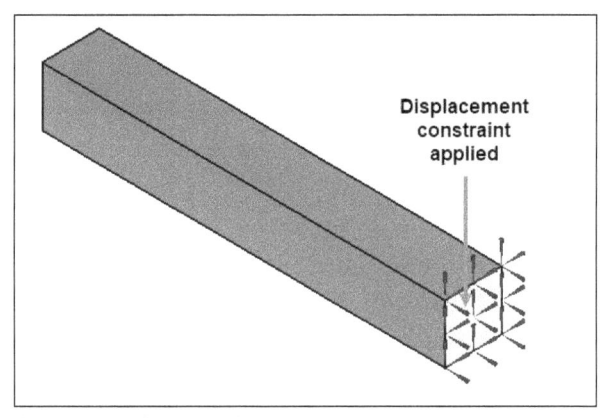

Figure-22. Applying displacement constraint

Applying Contact Constraint

The **Contact constraint** is applied to the model for making two selected faces in contact during the analysis. You can also specify the contact stiffness and friction coefficient between the faces when applying this constraint. This constraint is useful when you are working on an assembly of components. In assembly, you can define the parameters for contact between faces of two components. The procedure to apply this constraint is given next.

- Click on the **Contact constraint** tool from the **Mechanical boundary conditions and loads Toolbar** or **Model** menu in **FEM** workbench. The **Analysis feature parameters** dialog will be displayed as shown in Figure-23.

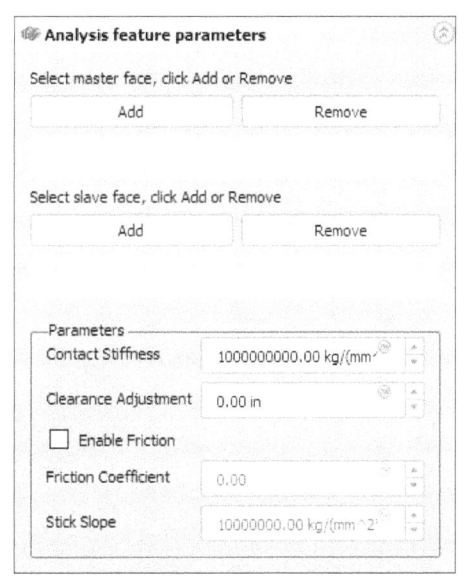

Figure-23. FEM constraint parameters dialog for contact constraint

- Select the face of first body to be used as master face and click on the **Add** button for **Select master face** section in the dialog; refer to Figure-24.

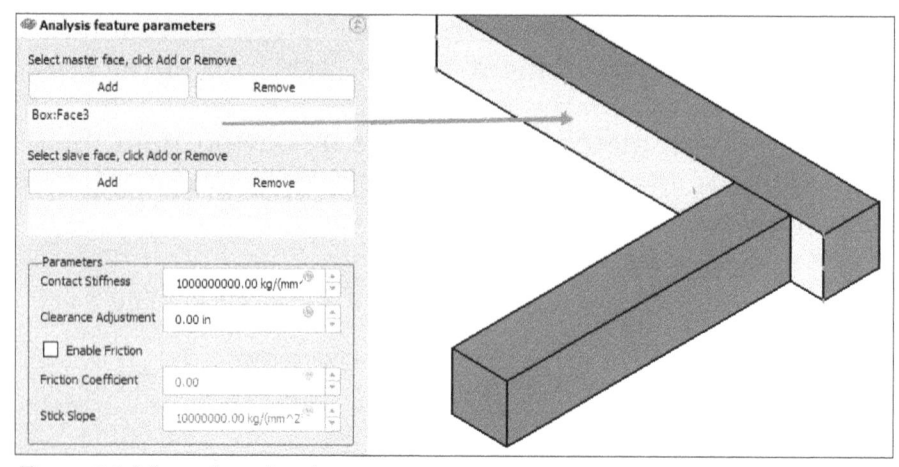

Figure-24. Master face selected

- Now, hide the master body by selecting any of its face and then pressing **SPACEBAR** from keyboard, so that you can select other face touching the master body.
- Select the slave face from other body and click on the **Add** button from **Select slave face** section of the dialog; refer to Figure-25.

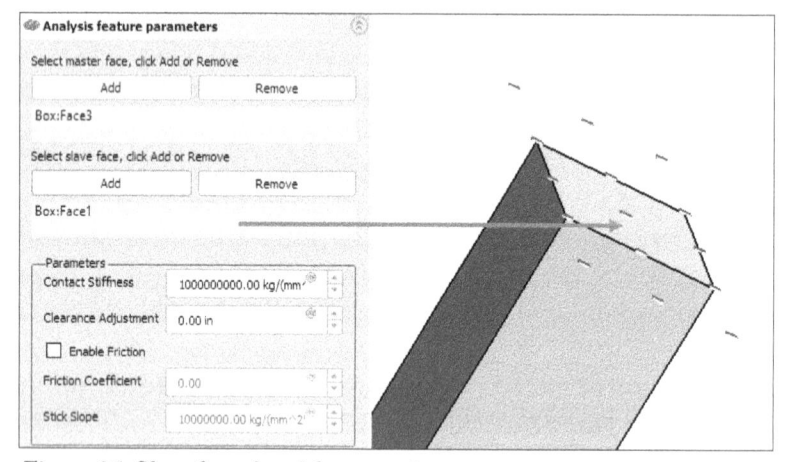

Figure-25. Slave face selected for constraint

- Open the **Model** panel in **Combo View** and select the hidden body to be displayed again and press **SPACEBAR**. The hidden body will be displayed again; refer to Figure-26.

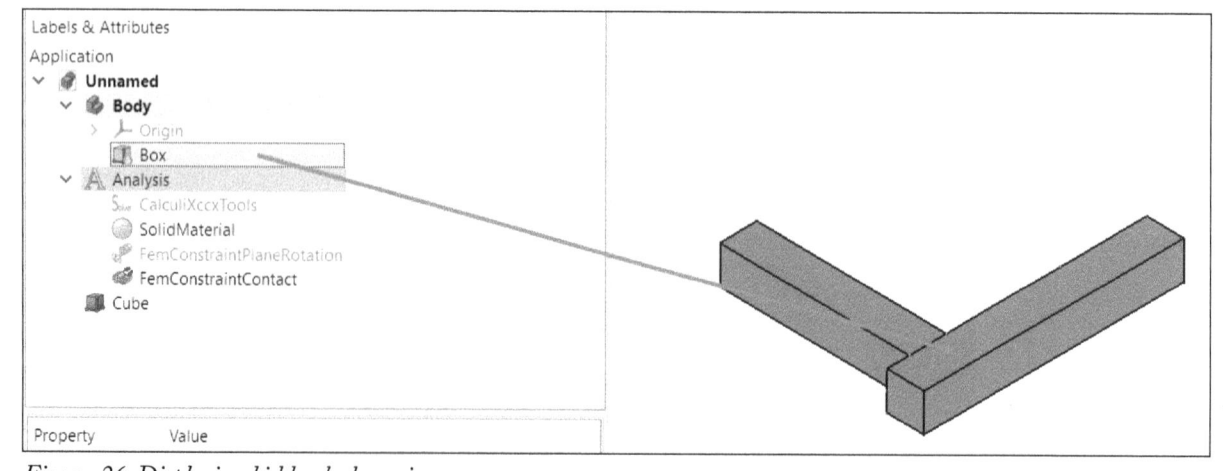

Figure-26. Displaying hidden body again

- Switch back to **Tasks** panel in the **Combo View** and specify desired value in **Contact Stiffness** edit box to define stiffness of contact point up to which the two faces

will remain in contact under load. Specify desired value of clearance adjustment in the **Clearance Adjustment** edit box to define the distance from master face upto which all nodes will be considered as master face nodes.
- Select **Enable Friction** check box to specify the value of friction coefficient and stick slope.
- Click in the **Friction coefficient** edit box and specify desired value of friction coefficient between two faces.
- Specify desired value in the **Stick Slope** edit box to define contact stiffness.
- After setting desired parameters, click on the **OK** button from the dialog.

Applying Tie Constraint

The **Tie Constraint** tool defines a tie constraint that connects two selected surfaces in such a way that they cannot separate or slide on each other throughout the analysis. It is similar two bonded contact between two faces. The procedure to use this tool is discussed next.

- Click on the **Tie Constraint** tool from **Toolbar** or **Mechanical boundary conditions and loads** cascading menu of **Model** menu in the **FEM** workbench. The **Tie parameter** and **Geometry reference selector** dialog will be displayed in the **Tasks** panel of **Combo View**; refer to Figure-27.

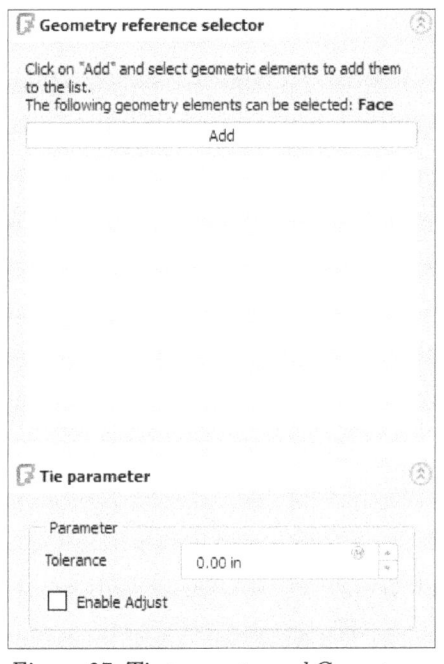

Figure-27. Tie parameter and Geometry reference selector for a Face dialogs

- Click on the **Add** button from **Geometry reference selector** dialog and select the surface of the model in the 3D view area which you want to add to define tie constraint. The selected surface will be displayed in the list area of the dialog.
- Click on the **Add** button again and select the another surface between which you want to define tie constraint.
- Specify desired tolerance value in the **Tolerance** edit box from **Parameter** area in the **Tie parameter** dialog. Select **Enable Adjust** check box to allow the slave surface nodes to be automatically moved so that they lie on the master surface.
- Click on the **OK** button from the dialog. The tie constraint will be applied to the model; refer to Figure-28.

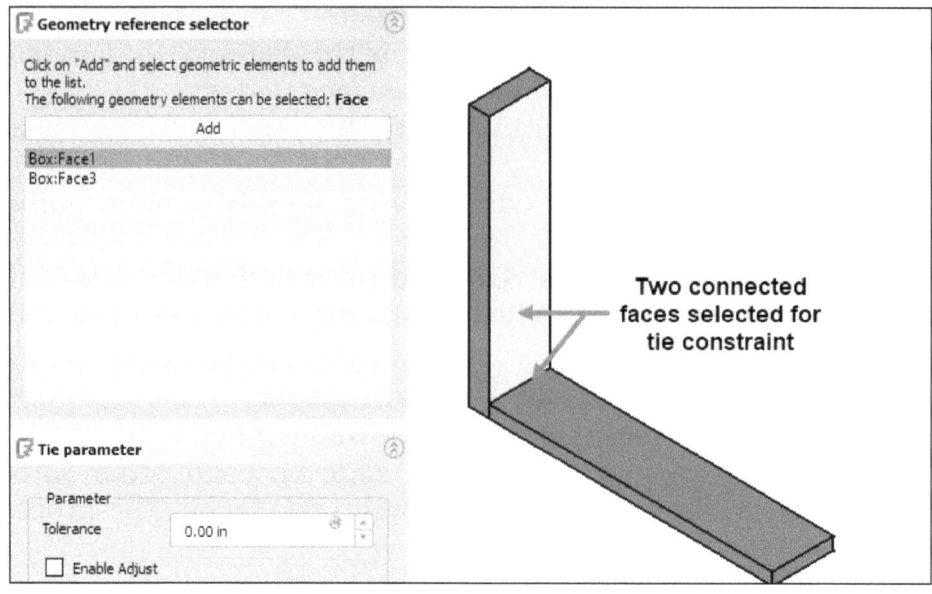

Figure-28. Tie constraint applied to the model

Applying Spring Constraint

The **Spring** tool is used to apply constraint for a spring acting on a face. You can use this constraint to represent a soft base of the part that can absorb energy like soil or rubber mat. The procedure to use this constraint is given next.

- Click on the **Spring** tool from the **Toolbar** or **Mechanical boundary conditions and loads** cascading menu of **Model** menu in **FEM** workbench. The **Analysis feature parameters** dialog will be displayed; refer to Figure-29.

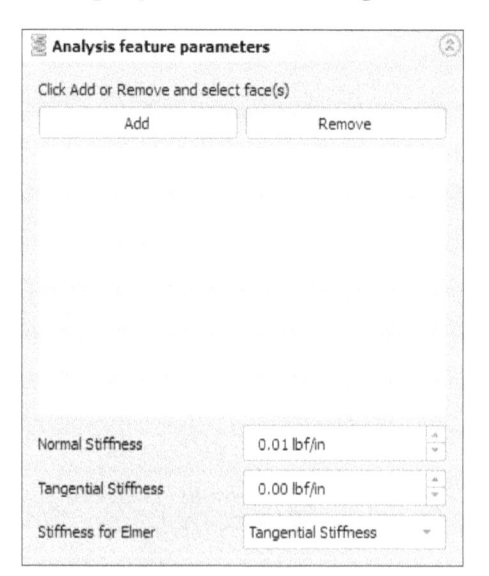

Figure-29. FEM constraint parameters dialog for spring constraint

- Click on the **Add** button from the dialog and select the face model you want to be connected by a spring; refer to Figure-30.

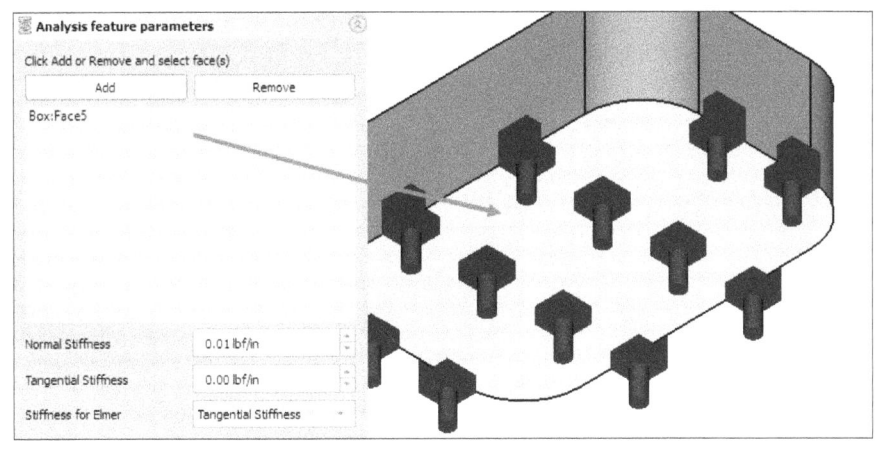

Figure-30. Selecting the face for spring constraint

- Specify desired value in the **Normal Stiffness** edit box.
- Specify desired value in the **Tangential Stiffness** edit box.
- Specify desired option from the **Stiffness For Elmer** drop-down in dialog box.
- After specifying desired parameters, click on the **OK** button from the dialog.

Applying Force Constraint

The **Force load** is used to apply force load on selected face of specified value. The procedure to use this constraint is given next.

- Click on the **Force load** tool from the **Toolbar** or **Mechanical boundary conditions and loads** cascading menu of **Model** menu in **FEM** workbench. The **Analysis feature parameters** dialog will be displayed; refer to Figure-31.

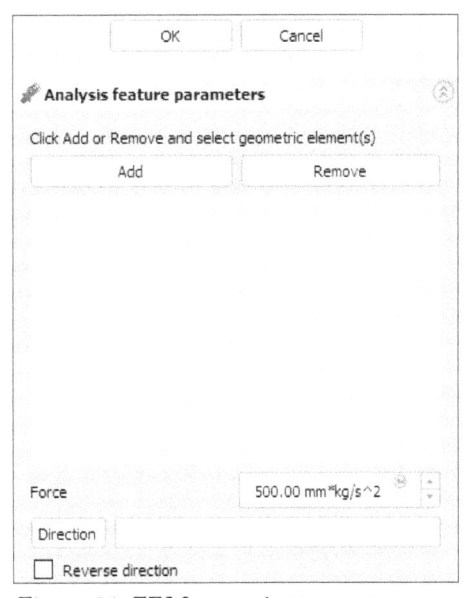

Figure-31. FEM constraint parameters dialog for force constraint

- Click on the **Add** button to add geometry for applying force.
- Select desired face from the model on which you want to apply the force and specify desired force value in the **Force** edit box.
- Select the **Reverse direction** check box to reverse direction of load.
- If you want to define desired direction for load then click on the **Direction** button and select desired reference; refer to Figure-32.

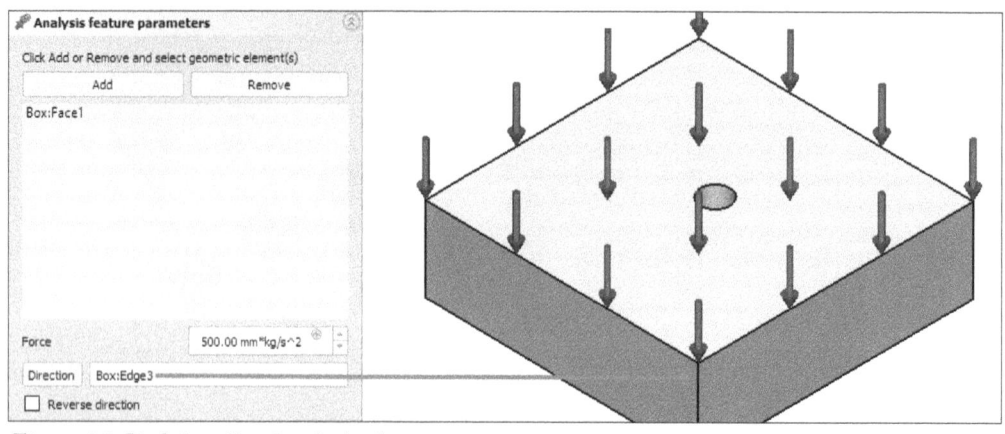

Figure-32. Defining direction for load

- After setting desired parameters, click on the **OK** button from the dialog.

Applying Pressure Load

The **Pressure Load** tool is used to define pressure load on selected face/surface. The procedure to use this tool is given next.

- Click on the **Pressure Load** tool from the **Toolbar** or **Model** menu in **FEM** workbench. The `Analysis feature parameters` dialog will be displayed; refer to Figure-31.

Figure-33. FEM constraint parameters dialog for pressure constraint

- Select the face(s) on which you want to specify pressure load and specify desired value in the **Pressure** edit box of the dialog.
- Specify the other parameters as discussed earlier and click on the **OK** button to apply load.

Applying Centrif Constraint

The `Centrifugal load` tool is used to create constraint of centrifugal force. The procedure to use this tool is given next.

- Click on the `Centrifugal load` tool from the **Model** menu or from the **Toolbar**. The `Centrif parameter`, `Geometry reference selector`, and `Geometry reference selector` dialogs will be displayed; refer to Figure-32, Figure-33, and Figure-34.

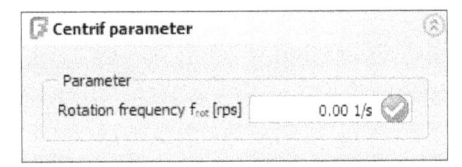

Figure-34. Centrif parameter dialog

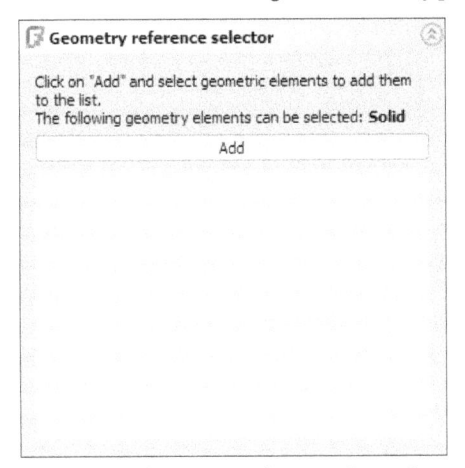

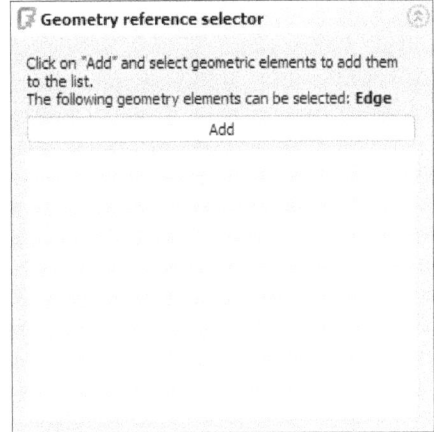

Figure-35. Geometry reference selector for a Solid dialog

Figure-36. Geometry reference selector for a Edge dialog

- Specify desired revolutions per second in the **Rotation frequency f_rot [rps]** edit box from **Centrif parameter** dialog.
- Click on the **Add** button from **Geometry reference selector** dialog and select desired solid object on which you want to apply the constraint; refer to Figure-37.

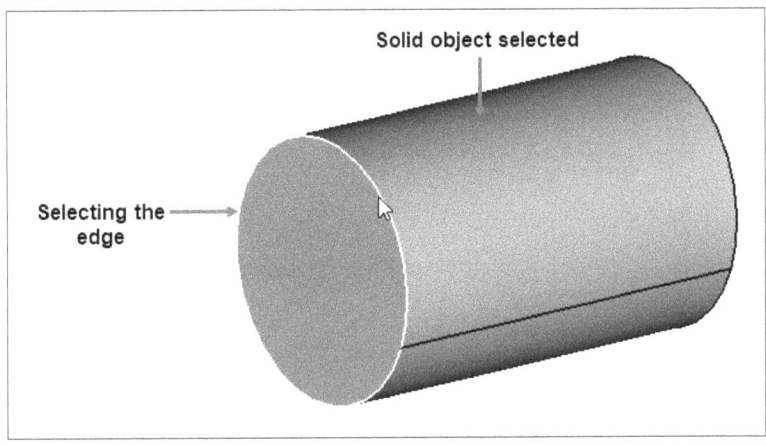

Figure-37. Selecting the elements for centrif constraint

- Click on the **Add** button from **Geometry reference selector** dialog and select desired edge of the solid object; refer to Figure-37.
- After specifying desired parameters, click on the **OK** button from the dialog. The centrif constraint will be applied; refer to Figure-38.

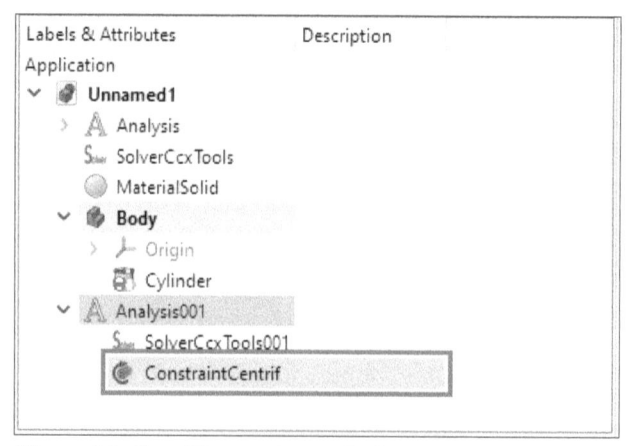

Figure-38. Centrif constraint applied

Applying Gravity Load Constraint

The **Gravity load** is applied on the model to check the effect of weight of the model in various situations. The procedure to use this tool is given next.

- Click on the **Gravity Load** tool from the **Model** menu or from the **Toolbar**. The self weight constraint will be created and displayed in the **Model Tree**.
- If you want to modify the parameters of self weight then select it from the **Model Tree** and set desired parameters in the **Properties** box; refer to Figure-37.

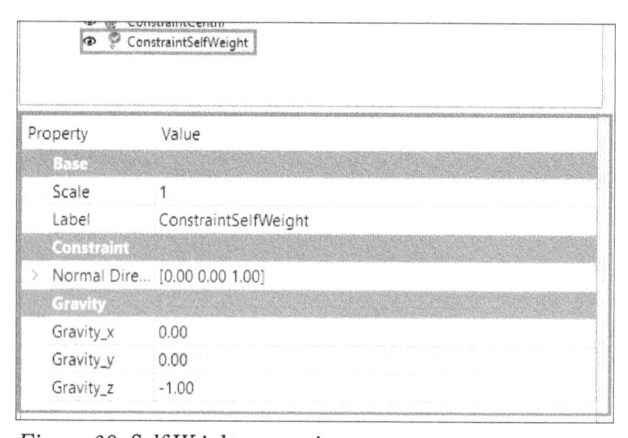

Figure-39. Self Weight properties

Geometrical Constraints

The tools to apply geometrical constraints are available in the **Geometrical Constraints Toolbar** and **Geometrical analysis features** cascading menu of **Model** menu; refer to Figure-40. These tools are discussed next.

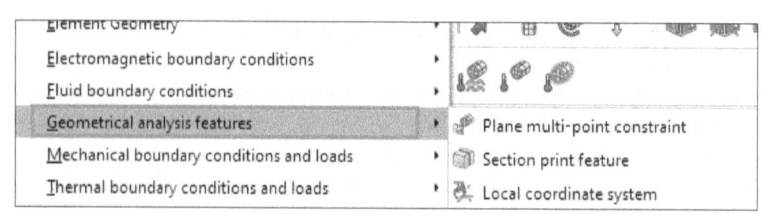

Figure-40. Geometrical Constraints cascading menu

Applying Plane Rotation Constraint

The **Plane multi-point constraint** tool is applied on selected faces to allow rotation of model in plane parallel to selected face but restrict translation in any direction.

This type of constraint is useful for fixing gears in an analysis so that they are free to rotate. The procedure to apply this constraint is given next.

- Click on the **Plane multi-point constraint** tool from the **Toolbar** or **Geometrical analysis features** cascading menu in the **Model** menu. The **Analysis feature parameters** dialog will be displayed in the **Tasks** panel; refer to Figure-41.
- Select the face on which you want to apply plane rotation constraint and click on the **Add** button from the dialog. The annotation of constraint will be displayed on the model face; refer to Figure-42.

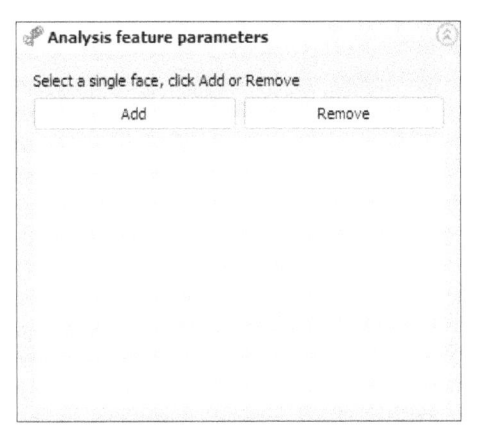

Figure-41. FEM constraint parameters dialog for plane rotation constraint

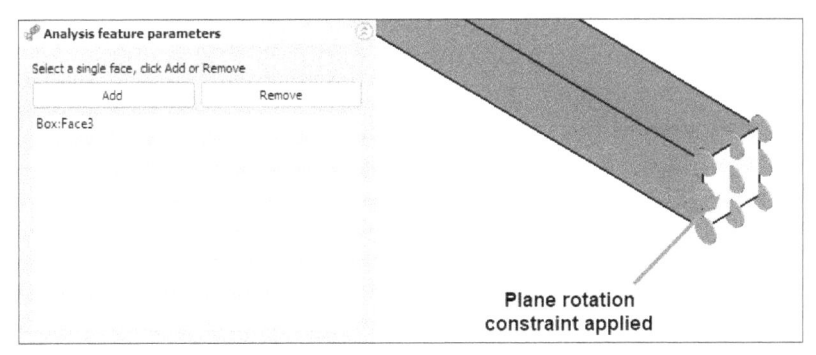

Figure-42. Applying plane rotation constraint

- After adding faces, click on the **OK** button from the dialog to apply constraints and exit the dialog.

Applying Section Print Constraint

The **Section print feature** tool is used to print internal section forces at nodes of section surface. The procedure to use this tool is discussed next.

- Click on the **Section print feature** tool from the **Toolbar** or **Geometrical analysis feature** cascading menu in the **Model** menu. The **Section Print parameter** and **Geometry reference selector** dialogs will be displayed in the **Tasks** panel of **Combo View**; refer to Figure-43.

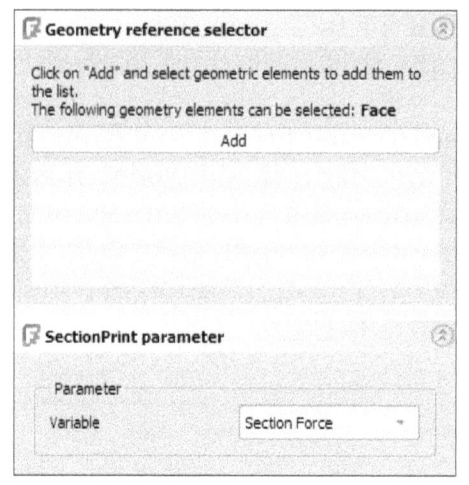

Figure-43. SectionPrint parameter and Geometry reference selector for a Face dialogs

- Click on the **Add** button from dialog and select the face to be used for checking internal forces.
- Specify desired option from the **Variable** drop-down in the **Parameter** section.
- After specifying desired parameters, click on the **OK** button to apply constraint.

Applying Local Coordinate System

The **Local coordinate system** tool is used to transform coordinate system for faces on which displacement constraint is applied earlier. Using this constraint, you can modify the direction of displacement to desired orientation with respect to original coordinate system. The procedure to use this tool is given next.

- Click on the **Local coordinate system** tool from the **Toolbar** or **Model** menu in **FEM** workbench. The **Analysis feature parameters** dialog will be displayed in the **Tasks** panel of **Combo View**; refer to Figure-44.

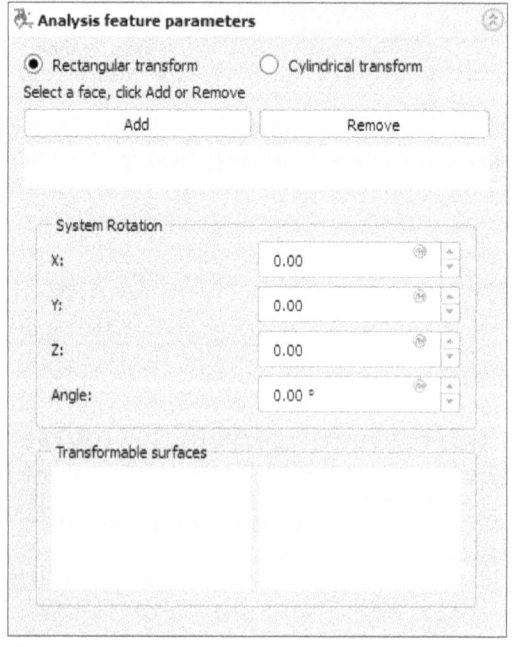

Figure-44. FEM constraint parameters dialog for transform constraint

- The surfaces/faces that can be used for applying transform constraint are available in the **Transformable surfaces** list box at the bottom in the dialog. Note that these are the faces on which displacement constraint has been applied.
- Select the face from model on which transform constraint is to be applied and click on the **Add** button from the dialog. The current orientation of coordinate system will be displayed on the face; refer to Figure-45.

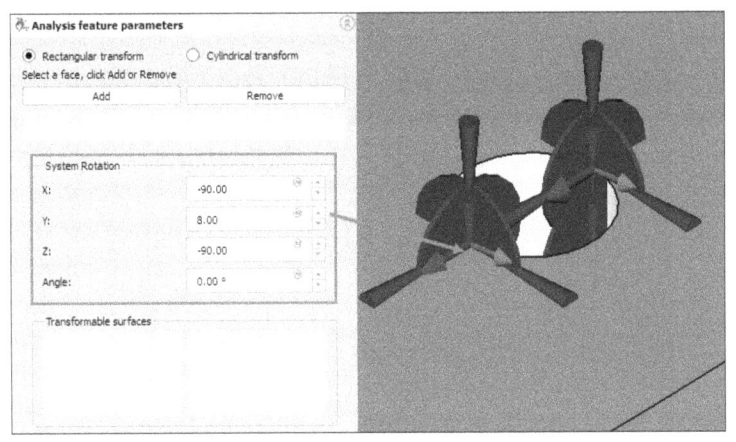

Figure-45. Orientation of coordinate system

- By default, the **Rectangular transform** radio button is selected, so you can apply rotation to each axis in Cartesian coordinate system. You can select the **Cylindrical** radio button to use Polar coordinate system for transformation.
- Set desired parameters in the **X**, **Y**, **Z,** and **Angle** edit boxes in the **System Rotation** section. The coordinate system will orient according to the values specified in these edit boxes.
- Click on the **OK** button from the dialog to apply constraint.

Thermal boundary condition and loads

The **Thermal** constraints are used to define various thermal conditions like temperature, heat flux, body heat source, and so on. The tools to apply thermal constraints are available in the **Thermal Constraints Toolbar** and **Thermal Constraints** cascading menu of the **Model** menu; refer to Figure-46. These tools are discussed next.

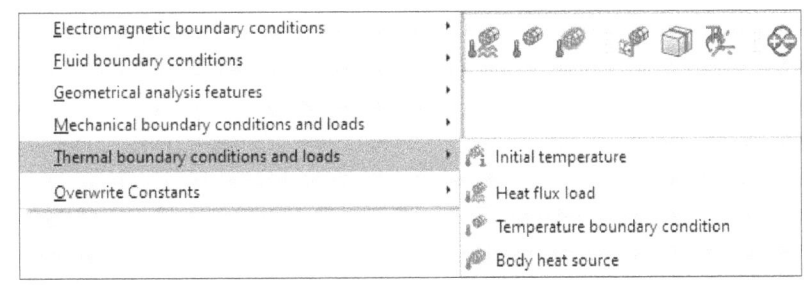

Figure-46. Thermal Constraints cascading menu

Defining Initial Temperature Constraint

The **Initial temperature** tool is used to apply initial temperature to the model for thermal analysis. The procedure to use this tool is given next.

- Click on the **Initial temperature** tool from the **Thermal boundary conditions and loads** cascading menu. The **Analysis feature parameters** dialog will be displayed; refer to Figure-47.

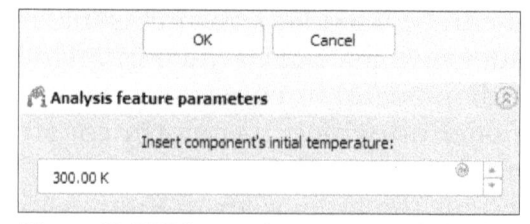

Figure-47. FEM constraint parameters dialog for initial temperature constraint

- Specify desired value of initial temperature in the edit box of dialog. Note that you need to specify temperature in Kelvin (K) unit.
- After specifying desired value, click on the **OK** button to apply constraint.

Applying Heat Flux Constraint

Heat flux load is the rate of heat energy transfer from selected face(s). The **Heat flux load** tool in **Thermal Constraints** cascading menu of **Model** menu is used to define the heat flux for selected faces. The procedure to apply this constraint is given next.

- Click on the **Heat flux load** tool from **Thermal boundary condition and load** cascading menu of **Model** menu. The **Analysis feature parameters** dialog will be displayed as shown in Figure-48.

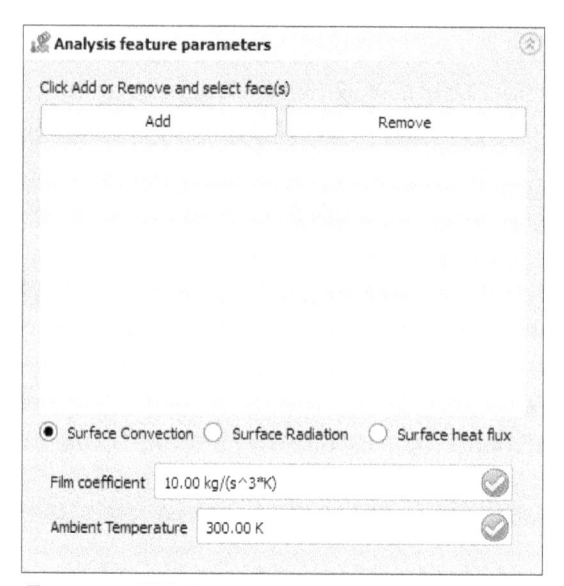

Figure-48. FEM constraint parameters dialog for heat flux constraint

- Select the **Surface Convection** radio button if you want to define heat convection rate for selected face/surface at specified temperature. Heat convection is the amount of heat passing through selected face/surface. Select **Surface Radiation** radio button to define the specification of radiation heat transfer of a surface at absolute temperature. Select the **Surface heat flux** radio button if you want to specify heat energy transfer rate.
- Specify the parameters as desired in the edit boxes of the dialog.
- Select the face on which you want to apply heat flux constraint and click on the **Add** button from the dialog. If you want to apply heat flux to multiple faces then you can select multiple faces while holding the **CTRL** key.
- After setting desired parameters, click on the **OK** button from the dialog. The constraint will be applied; refer to Figure-49.

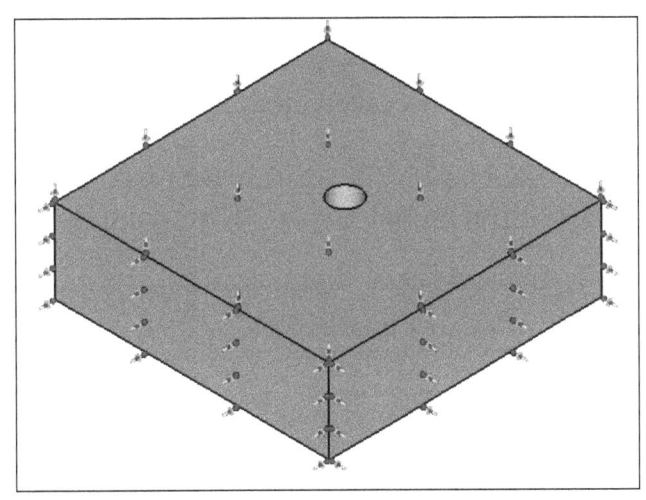

Figure-49. Heat flux constraint applied on faces

Applying Temperature Constraint

The **Temperature boundary condition** tool is used to apply fix temperature to selected faces. Applying fix temperature to a face makes that face a heat source for analysis. The procedure to apply this constraint is given next.

- Click on the **Temperature boundary condition** tool from the **Thermal boundary condition and load** cascading menu in the **Model** menu. The **Analysis feature parameters** dialog will be displayed; refer to Figure-50.

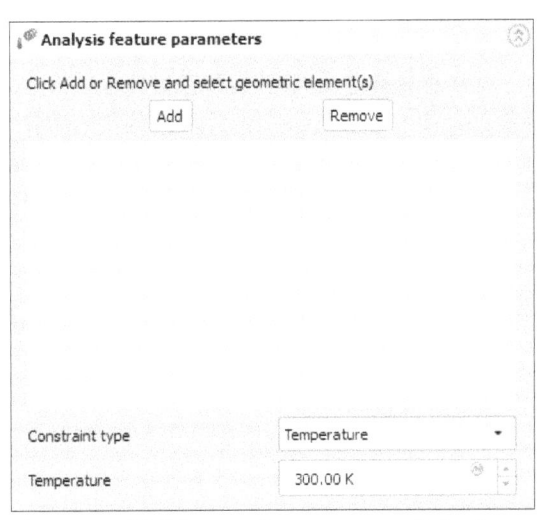

Figure-50. FEM constraint parameters dialog for thermal constraint

- Select the face(s) to which you want to apply thermal constraint and click on the **Add** button.
- Select **Temperature** option from **Constraint type** drop-down to specify temperature of selected face(s).
- Select **Cflux** option from the drop-down to specify the total heat energy of the selected face.
- Specify desired value of **Temperature/Cflux** in the edit box below the drop-down.
- After setting desired parameters, click on the **OK** button from the dialog to apply the constraint.

Applying Body Heat Source Constraint

The **Body Heat Source** constraint is used to define the heat generated by model. The procedure to apply this constraint is given next.

- Click on the **Body heat source** tool from the **Thermal boundary conditions and loads** cascading menu of the **Model** menu. The constraint will be applied to the model.
- Select the constraint from the **Model Tree** and modify the parameters as desired in the **Property editor**; refer to Figure-51.

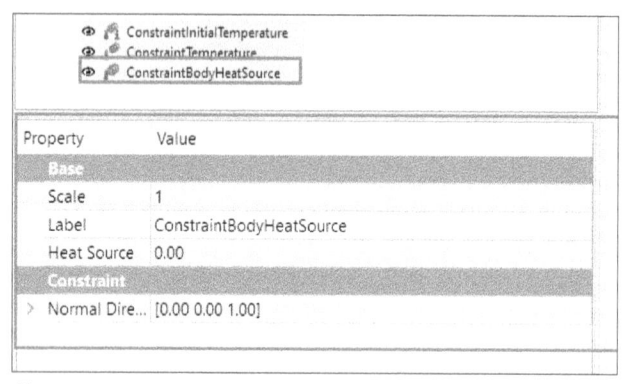

Figure-51. Property editor for body heat source constraint

- Specify desired value of heat in the **Heat Source** edit box. Click anywhere in the model area to exit the editing mode.

Fluid Constraints

The **Fluid** constraints are used to define boundary conditions for fluid flow in the model. These tools are available in the **Fluid Constraints** toolbar and **Fluid boundary conditions** cascading menu of the **Model** menu; refer to Figure-52. These tools are discussed next.

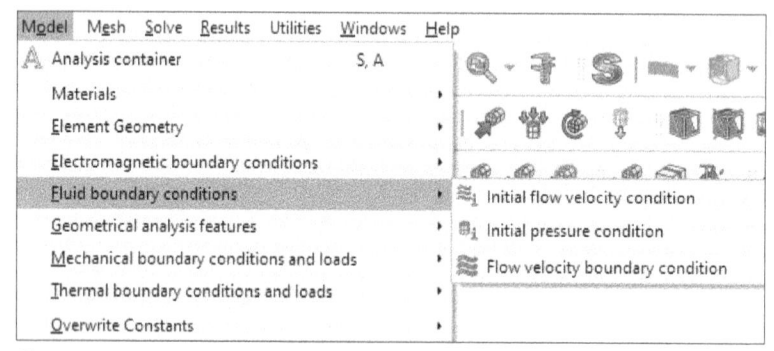

Figure-52. Fluid Constraints cascading menu

Applying Initial Flow Velocity Condition

The **Initial flow velocity condition** tool is used to specify initial velocity for flow of fluid in the model. This constraint is useful for fluid flow analysis. The procedure to use this tool is given next.

- Click on the **Initial flow velocity condition** tool from the **Fluid boundary conditions** cascading menu of the **Model** menu. The **Constraint Properties** dialog will be displayed; refer to Figure-53.

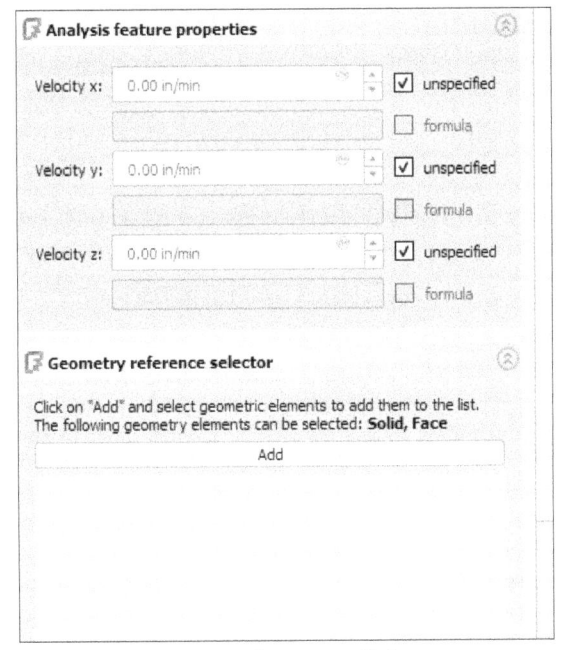

Figure-53. Constraint Properties dialog

- Clear the **unspecified** check box(es) for desired velocity direction(s) and specify the value(s) in respective edit box(es).
- Click on the **Add** button from `Geometry reference selector` dialog to add the selected geometric elements to the list. Select **Face** radio button from `Selection mode` area to select the face as a geometric element. Select **Solid** radio button to select the solid body as a geometric element.
- After setting desired parameters, click on the **OK** button from the dialog to apply constraint.

Applying Initial Pressure Condition

The `Initial pressure condition` tool is used to create an initial pressure constraints for a fluid flow analysis. The procedure to use this tool is given next.

- Click on the **Initial pressure condition** tool from the **Fluid boundary condition** cascading menu of the **Model** menu. The `Analysis feature properties` dialog will be displayed; refer to Figure-54.

Figure-54. Dialog for Initial Pressure Condition

- Specify desired value of initial pressure in the **Pressure** edit box.
- Specify desired parameters in the **Geometry reference** selector dialog as discussed earlier.
- Click on the **OK** button from the dialog to apply constraints.

Applying Flow Velocity Boundary Condition

The **Flow velocity boundary condition** tool is used to specify velocity of fluid through selected face. The procedure to use this tool is given next.

- Click on the **Flow velocity boundary condition** tool from the **Fluid boundary condition** cascading menu of the **Model** menu. The **Analysis feature properties** and **Geometry reference selector** dialogs will be displayed in the **Tasks** panel of **Combo View**; refer to Figure-57.

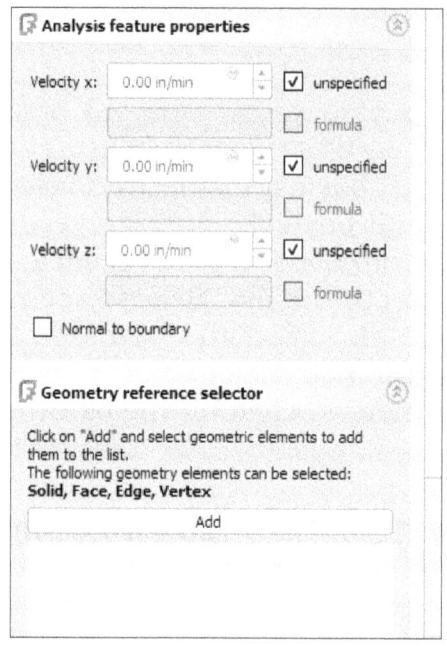

Figure-55. Analysis feature properties and Geometry reference selector dialogs

- Clear the unspecified check box(es) for desired velocity direction(s) and specify the value(s) in respective edit box(es).
- Select the faces, edges, or vertices for which you want to define flow velocity and click on the **Add** button.
- Set the other parameters as discussed earlier and click on the **OK** button.

Electromagnetic Constraints

The tools to apply electromagnetic conditions are available in the **Electromagnetic boundary conditions Toolbar** and **Electromagnetic boundary conditions** cascading menu of the **Model** menu; refer to Figure-56.

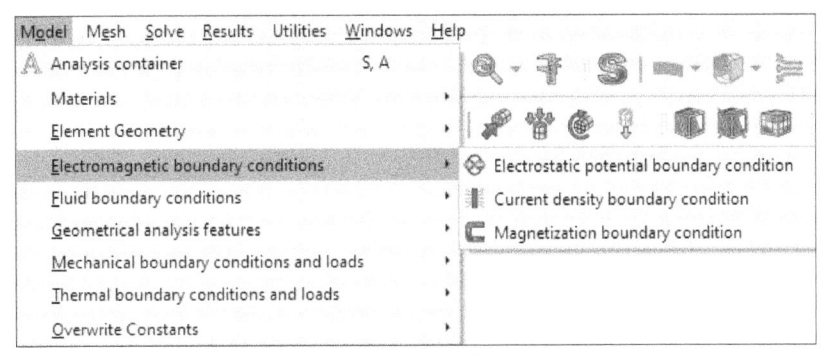

Figure-56. Electromagnetic Constraints cascading menu

Applying Electrostatic Potential Boundary Condition

The **Electrostatic potential boundary condition** tool is used to specify electrical potential (voltage) at selected faces/edges/vertices. The procedure to use this tool is given next.

- Click on the **Electrostatic potential boundary condition** tool from **Electromagnetic boundary conditions** cascading menu of the **Model** menu. The **Analysis feature properties** dialog will be displayed; refer to Figure-57.

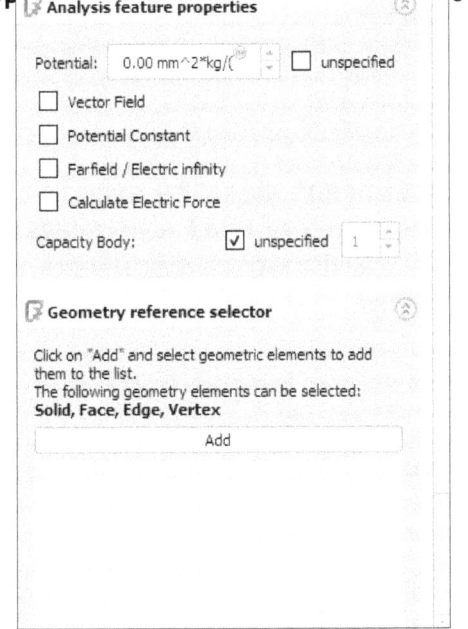

Figure-57. Dialog for Electrostatic Potential Condition

- Select the faces/edges/vertices on which you want to apply electrostatic potential constraint and click on the **Add** button from the dialog.
- Clear the unspecified check box for **Potential** edit box and specify desired value of voltage.
- Select the **Potential Constant** check box from the dialog to keep voltage constant in analysis.
- After setting desired parameters, click on the **OK** button from the dialog to apply the constraint.

Current Density Boundary Condition

The **Current density boundary condition** tool is used to create an FEM constraint for defining the current density. The procedure to use this tool is given next.

- Click on the **Current density boundary condition** tool from **Electromagnetic boundary conditions** cascading menu of the **Model** menu. The **Analysis feature properties** dialog will be displayed; refer to Figure-58.

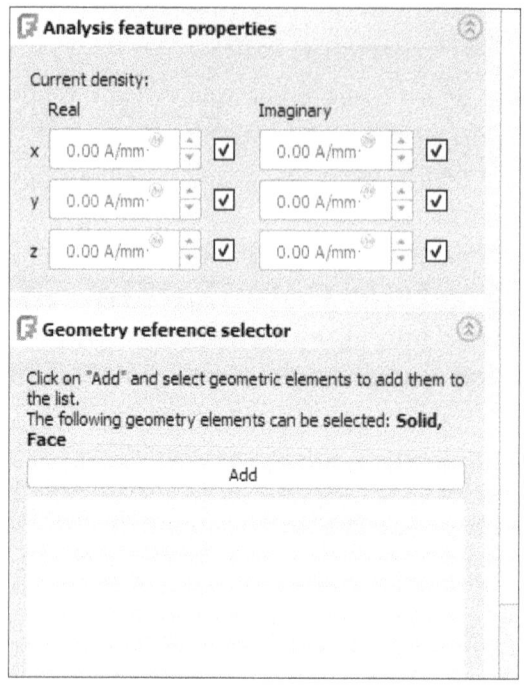

Figure-58. Dialog for Current Density Boundary Condition

- Clear the check boxes from **Current density** area of the dialog to specify desired values in the **x**, **y**, and **z** edit boxes of **Real** and **Imaginary** sections.
- Specify other parameters in the dialog as discussed earlier.
- Click on the **OK** button from the dialog to apply the constraint.

Magnetization boundary condition

The **Magnetization boundary condition** tool is used to restrict the magnetism value of selected solid/face at specified value. The procedure to use this constraint is given next.

- Click on the **Magnetization boundary condition** tool from **Electromagnetic boundary conditions** cascading menu of the **Model** menu. The **Analysis feature properties** dialog will be displayed; refer to Figure-59.

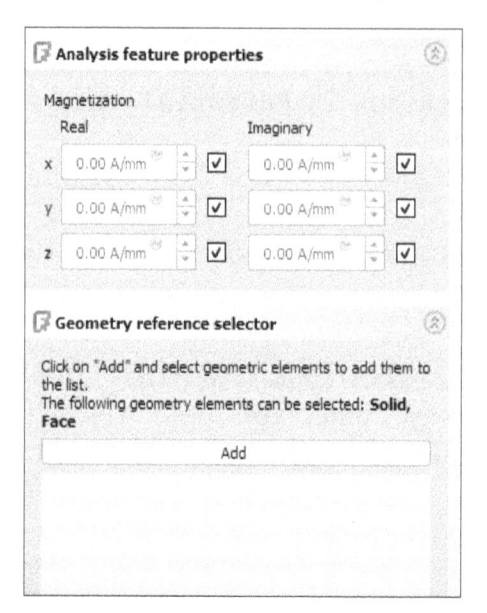

- Clear the check boxes from **Magnetization** area of the dialog to specify desired values in the **x**, **y**, and **z** edit boxes of **Real** and **Imaginary** sections.
- Specify other parameters in the dialog as discussed earlier.
- Click on the **OK** button from the dialog to apply the constraint.

Applying Constant Vacuum Permittivity

Vacuum permittivity is used to define ability of medium to pass current in dielectric environment. Click on the **Constant Vacuum Permittivity** tool from the **Overwrite Constants** cascading menu of **Model** menu to add the parameter in current analysis. You can modify the value of parameter using **Property** dialog after selecting this constant from **Model** panel in **Combo View**; refer to Figure-60.

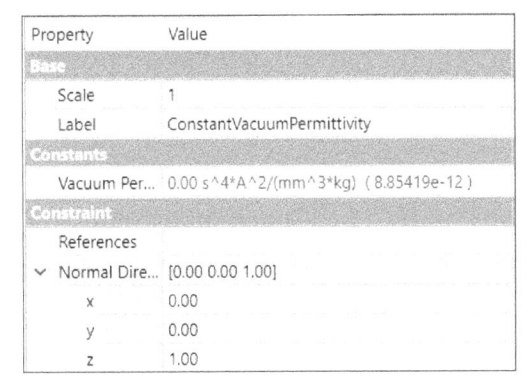

Figure-60. Applying constant vacuum permittivity

CREATING GEOMETRY ELEMENTS

In FreeCAD FEM, you can create 2D elements of specified thickness by using the tools in **Element Geometry** toolbar or **Element Geometry** cascading menu of the **Model** menu; refer to Figure-61. These tools are discussed next.

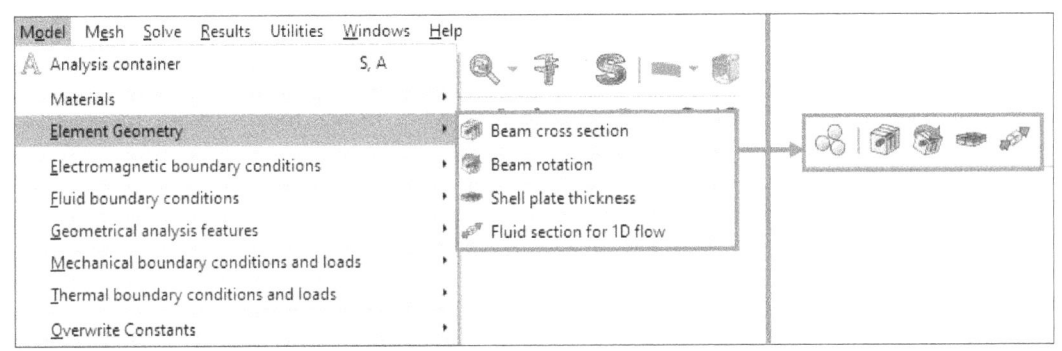

Figure-61. Element Geometry tools

Creating Beam Cross-section

The **Beam cross section** tool is used to create beam elements with different type of cross sections like circular beam, rectangular beam, and pipe beam. To create beam elements, you need to select reference edges from the model. These edges can be of a surface or a solid. The procedure to use this tool is given next.

- Click on the **Beam cross section** tool from the **Element Geometry** cascading menu of **Model** menu. The **Beam section parameter** and **Geometry reference selector** dialogs will be displayed; refer to Figure-62.

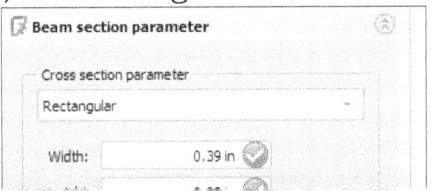

- Select desired option from the **Cross section parameter** drop-down in the dialog. Select the **Rectangular** option from the drop-down to create rectangular beam element. Select the **Circular** option to create circular rod elements. Select the **Pipe** option from the drop-down to create tube like elements.
- After selecting desired option from the drop-down, specify the related parameters in edit box(es) below the drop-down.
- After setting desired parameters in the dialog, click on the **Add** button from the **Geometry reference selector** dialog and select the edge on which you want to create the element. Note that you need to click on **Add** button each time before you select an edge for creating element.
- Click on the **OK** button from the dialog to create the elements.

Applying Rotation to Beam Elements

You can change the default orientation of beam elements by specified rotation value using the **Beam rotation** tool. The procedure to use this tool is given next.

- Click on the **Beam rotation** tool from the **Element Geometry** cascading menu of the **Model** menu. The **Element Rotation 1D** feature will be added in the **Model Tree**.
- Select the **Element Rotation 1D** feature from **Model Tree**. The **Property Editor** dialog will be displayed; refer to Figure-63.

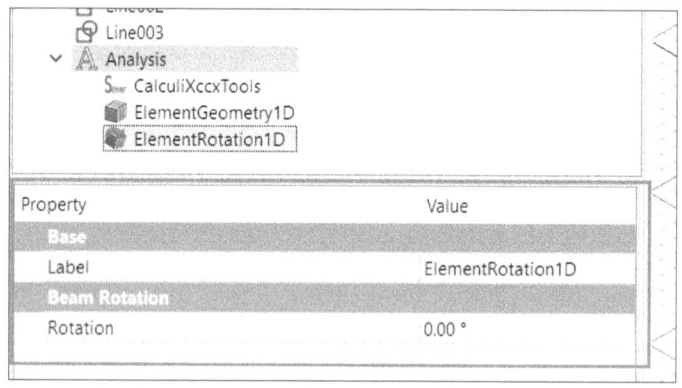

Figure-63. Beam Rotation Element properties

- Specify desired parameters in the **Property Editor** and click in the empty area of drawing to exit editing model.

Defining Shell Plate Thickness

The **Shell plate thickness** tool is used to define thickness of shell element based on selected face. The procedure to use this tool is given next.

- Click on the **Shell plate thickness** tool from the **Element Geometry** cascading menu of the **Model** menu. The **Shell thickness parameter** and **Geometry reference selector** dialogs will be displayed in the **Tasks** panel of **Combo View**; refer to Figure-64.

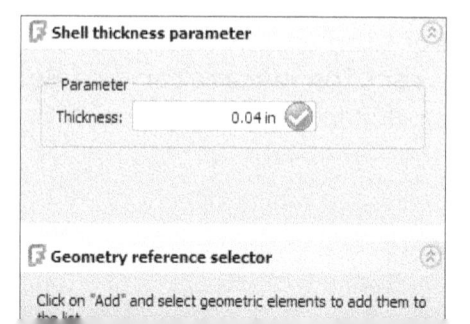

- Specify desired value of thickness for shell element in the **Thickness** edit box of the dialog.
- Click on the **Add** button from the dialog and select the face on which you want to create the shell element of specified thickness.
- After setting desired parameters, click on the **OK** button from the dialog to create the geometry.

Creating Fluid Section for 1D Flow

The **Fluid section for 1D flow** tool is used to create section in the model for fluid flow. The procedure to use this tool is given next.

- Click on the **Fluid section for 1D flow** tool from the **Element Geometry** cascading menu of the **Model** menu. The **Form** and **Geometry reference selector** dialogs will be displayed; refer to Figure-65.

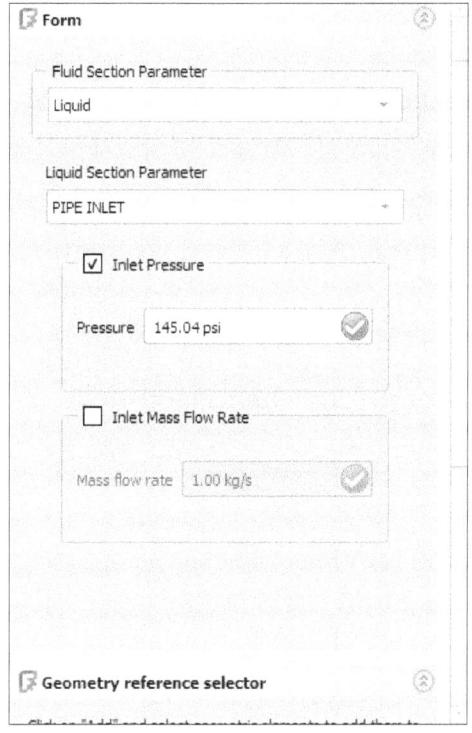

Figure-65. Dialogs for 1D fluid section

- Select desired option from the **Fluid Section Parameter** drop-down to define whether fluid section is for liquid, gas, or it is open for both liquid and gas. The other parameters will be displayed based on selected option.
- If you have selected **Liquid** option for fluid section then select desired option from the **Liquid Section Parameter** drop-down to specify whether the section is at inlet, outlet, diaphragm, or any other section of fluid domain.
- Specify the other parameter as needed and click on the **Add** button from the dialog. You will be asked to select an edge to define location of section.
- Select desired edge and click on the **OK** button from the dialog to create the section.

Note that the functionality to work with Beam elements has not been added yet up to version 1.0, so you will not be able to work with them in software.

MESHING

Meshing is the base of FEM (Finite Element Method) which is one of the methods used for FEA. Meshing divides the solid/shell models into elements of finite size and shape. These elements are joined at some common points called nodes. These nodes define the load transfer from one element to other element. Meshing is a very crucial step in design analysis because it directly affects the accuracy of analysis. The automatic mesher in the software generates a mesh based on a global element size, tolerance, and local mesh control specifications. Mesh control lets you specify different sizes of elements for components, faces, edges, and vertices.

The software estimates a global element size for the model taking into consideration its volume, surface area, and other geometric details. The size of the generated mesh (number of nodes and elements) depends on the geometry and dimensions of the model, element size, mesh tolerance, mesh control, and contact specifications. In the early stages of design analysis where approximate results may suffice, you can specify a larger element size for a faster solution. For a more accurate solution, a smaller element size may be required.

Meshing generates 3D tetrahedral solid elements, 2D triangular shell elements, and 1D beam elements. A mesh consists of one type of elements unless the mixed mesh type is specified. Solid elements are naturally suitable for bulky models. Shell elements are naturally suitable for modeling thin parts (sheet metals), and beams and trusses are suitable for modeling structural members.

There are two plug-ins in FreeCAD to perform meshing: Netgen and Gmsh. The tools to perform meshing using these plug-ins are available in the **Mesh** menu; refer to Figure-66. The tools in this menu are discussed next.

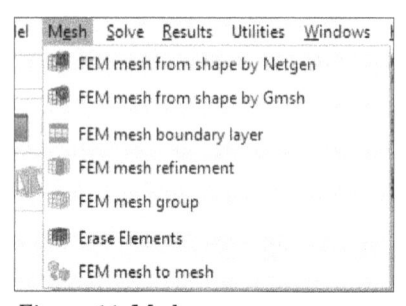

Figure-66. Mesh menu

Creating FEM mesh using Netgen Mesher

The **FEM mesh from shape by Netgen** tool is used to create mesh from selected shape(body) using Netgen plug-in. The procedure to use this tool is given next.

- Click on the **FEM mesh from shape by Netgen** tool from the **Mesh** menu after selecting the body. The **Tet Parameter** dialog will be displayed; refer to Figure-67.

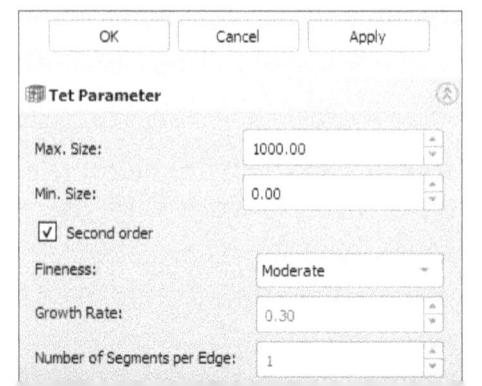

- Specify desired value in **Max. Size** and **Min. Size** edit box to define maximum size of element that can be generated in current mesh. Note that this size can be length, width, or height of the element.
- Select the **Second order** check box if you want to generate elements in mesh with additional nodes that define curvature of segments; refer to Figure-68.

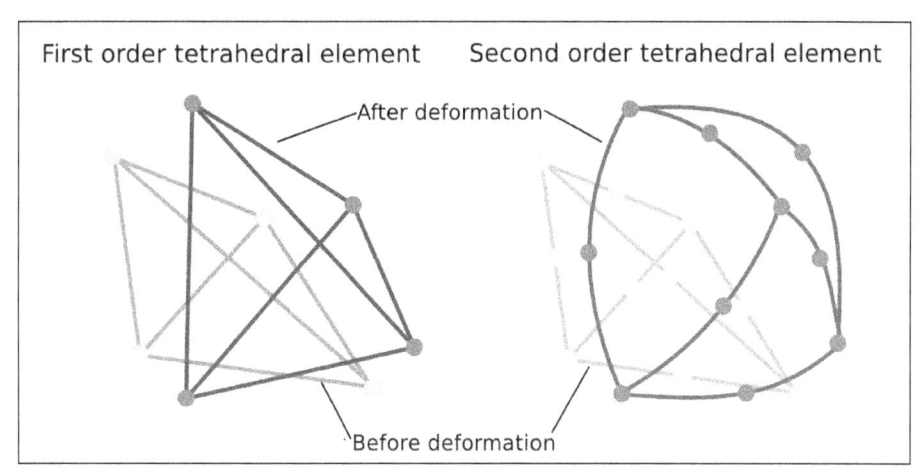

Figure-68. Difference between first order and second order mesh

- Select desired option from the **Fineness** drop-down to define smoothness in mesh elements at edges. A finer mesh gives better results of analysis but it also increases the computational time for analysis. Figure-69 shows same object mesh with both very coarse and very find mesh types.

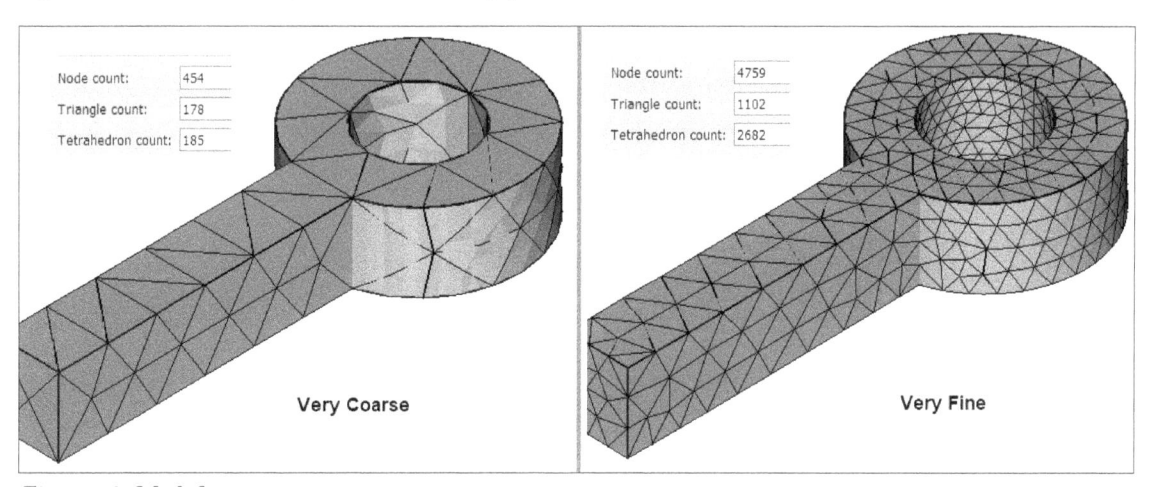

Figure-69. Mesh fineness

Note that a simple move from very coarse to very fine mesh can exponentially increase the number of elements as in our case. There is a certain threshold in meshing fineness after which you get a marginal accuracy at the loss of a lot of computing power and time. So, you should start with a moderate fineness before moving to finer side when performing analysis.

- If you want to manually specify the smoothness of your mesh then select the **UserDefined** option from the **Fineness** drop-down. The edit boxes below the drop-down will become active.
- Specify desired value in **Growth Rate** edit box to define increase in edge length of element when moving away from edges or complex regions of the model.
- Specify desired value in **Number of Segments per Edge** edit box to define number of segments in which the edges of model will be divided in mesh. Note that the mesh

will be a compromise between maximum element size and number of segments per edge values. So, you may or may not get the definite number of segments per edge depending on the model.
- Specify desired value in **Number of Segments per Radius** edit box to define number of segments in circular/arc type edges.
- Select the **Optimize** check box to automatically increase number of elements near edges and complex regions.
- After setting desired parameters, click on the **OK** button from the dialog. The mesh will be generated; refer to Figure-70.

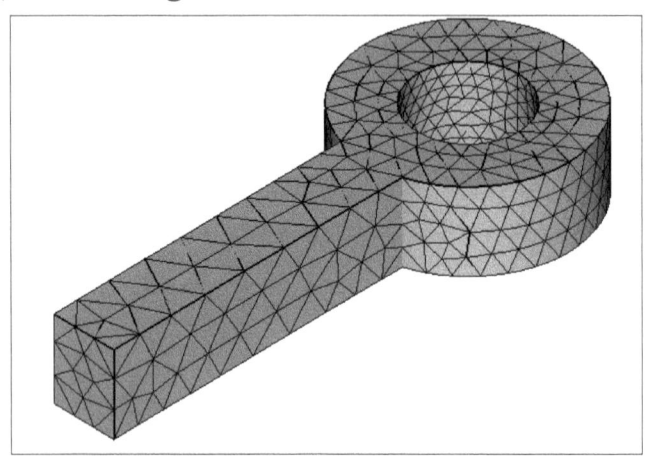

Figure-70. Mesh generated

Creating FEM mesh using Gmsh Mesher

The **FEM mesh from shape by Gmsh** tool is used to create mesh using the Gmsh plug-in. The procedure to use this tool is given next.

- Click on the **FEM mesh from shape by Gmsh** tool from **Mesh** menu after selecting the body. The **FEM Mesh by Gmsh** dialog will be displayed; refer to Figure-71.

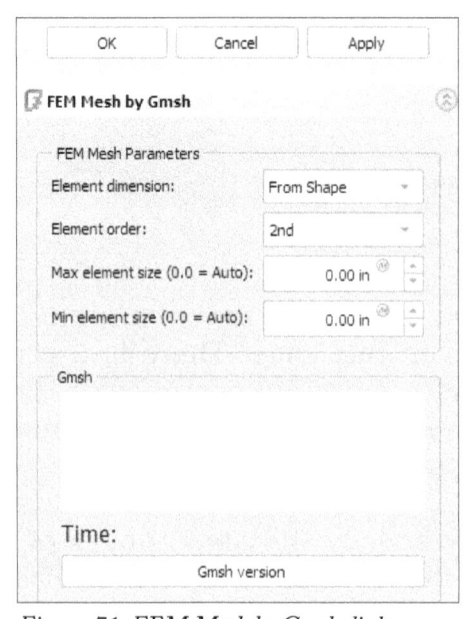

Figure-71. FEM Mesh by Gmsh dialog

- Select desired option from the **Element dimension** drop-down to define what type of mesh elements are to be created. By default, **From Shape** option is selected in this drop-down, so dimension of element is automatically decided based on selected object. For example, for a 3D solid body, 3D elements will be generated in the mesh. You can manually specify the dimensions of element by selecting 1D, 2D, or 3D option from the drop-down.

The element types available in FreeCAD FEM are as follows:

1D Elements:

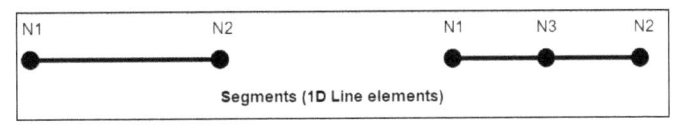

Figure-72. 1D elements

2D Elements:

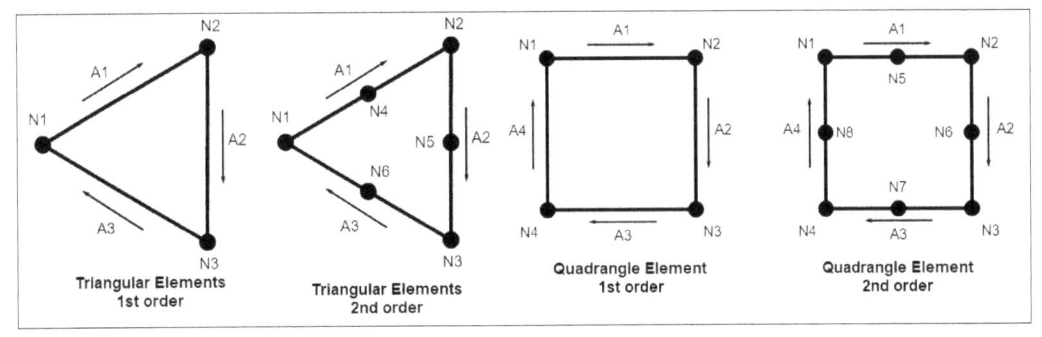

Figure-73. 2D elements

3D Elements:

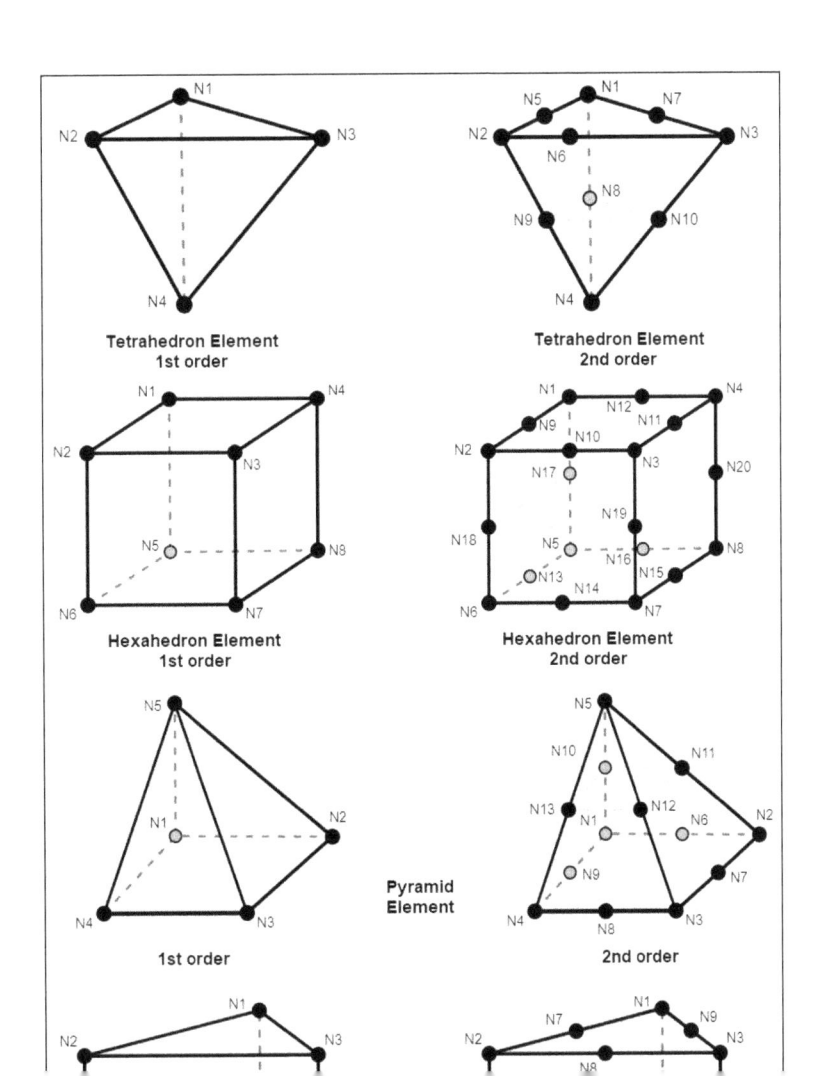

- Specify desired values in **Max element size** and **Min element size** edit boxes to define the maximum and minimum element size in mesh.
- After setting desired parameters, click on the **Apply** button to generate mesh.

Creating Boundary Mesh Layer

The **FEM mesh boundary layer** tool is used to create a layer of mesh with different element size from main mesh at the boundary of model. The procedure to use this tool is given next.

- Click on the **FEM mesh boundary layer** tool from the **Mesh** menu after selecting the **FEMMeshGmsh** feature (mesh created by Gmsh) from **Model Tree**. The **Mesh boundary layer settings** and **Geometry reference selector** dialogs will be displayed; refer to Figure-75.

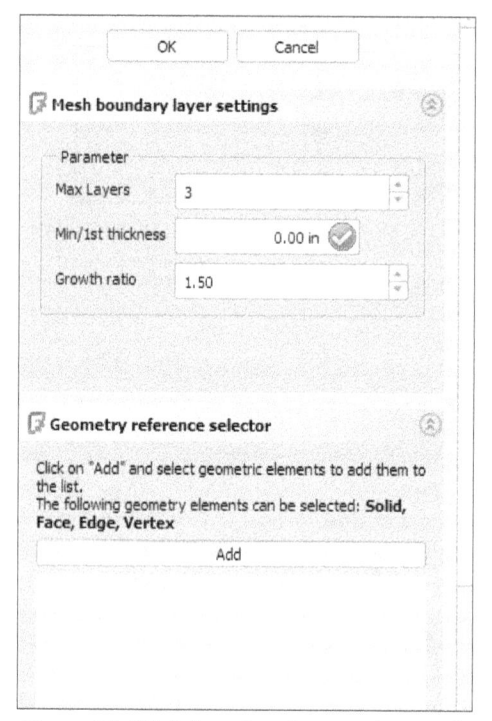

Figure-75. Mesh boundary layer settings and Geometry reference selector dialogs

- Specify desired value in the **Max Layers** edit box to define number of layers to be meshed by boundary layer settings.
- Specify desired value in the **Min/1st thickness** edit box to define the thickness of 1st layer in the mesh boundary.
- Specify desired value in **Growth ratio** edit box to specify the rate of increase/decrease in element size while moving away from boundary regions.
- Click on the **Add** button and select the solid/face/edge/vertex at which you want to apply boundary layer mesh.
- After setting desired parameters, click on the **OK** button from the dialog. The mesh will be created.

Create FEM mesh refinement

The **Mesh refinement** tool is used to create meshing for specific region of model with different element size from main mesh. This tool is useful for creating finer mesh in complex regions of model. The procedure to use this tool is given next.

- Click on the **Mesh refinement** tool from the **Mesh** menu. The **Mesh refinement** and **Geometry reference selector** dialogs will be displayed; refer to Figure-76.

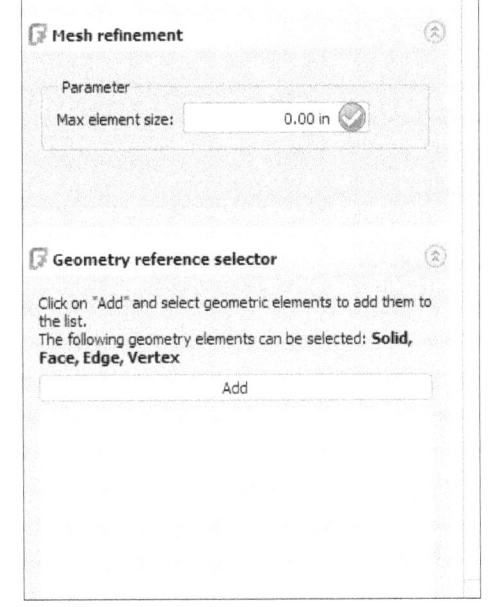

Figure-76. Mesh refinement and Geometry reference selector dialogs

- Specify desired value of maximum element size for refined mesh in the **Max element size** edit box.
- Select desired radio button for **Selection mode** and add the geometry region to be refined.
- After setting desired parameters, click on the **OK** button from the dialog.

Creating FEM mesh group

The **FEM mesh group** tool is used to create a group of mesh for exporting to other applications. The procedure to use this tool is given next.

- Click on the **FEM mesh group** tool from the **Mesh** menu. The **Mesh group** and **Geometry reference selector** dialogs will be displayed; refer to Figure-77.

- Select desired radio button from **Mesh group** dialog to define whether name is used for exporting mesh or label.
- Select desired face, edge, vertex, or solid object to be added in mesh group. If no object is selected then all the unused objects will be selected for mesh group.
- After setting desired parameters, click on the **OK** button to create mesh group.

Creating Erase Elements

The **Erase Elements** tool is used to create a set of selected nodes for checking results of analysis later on those nodes. The procedure to use this tool is given next.

- Click on the **Erase Elements** tool from the **Mesh** menu. The **TaskObjectName** and **Nodes set** dialogs will be displayed; refer to Figure-78.

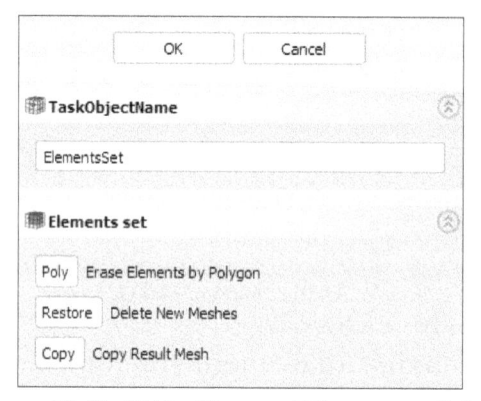

Figure-78. TaskObjectName and Elements set dialogs

- Specify desired name of node set in the **TaskObjectName** edit box.
- Click on the **Poly** button from the **Elements set** dialog to create a polygonal boundary around mesh elements to be deleted and then draw polygon by consecutive clicks; refer to Figure-79.

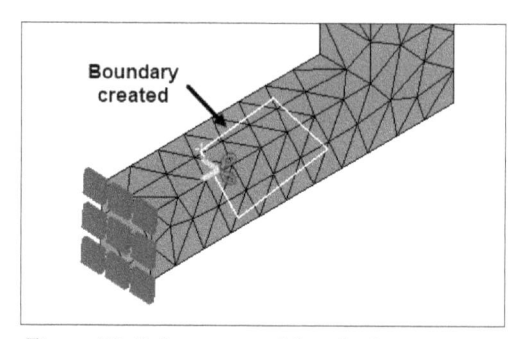

Figure-79. Polygon created for selection

- After drawing polygon, right-click in the drawing area. A shortcut menu will be displayed as shown in Figure-80. Select the **Inner** option to select all the nodes falling inside a polygon and select the **Outer** option if you want to select the nodes outside the polygon. Selected nodes will be displayed in green color; refer to Figure-81.

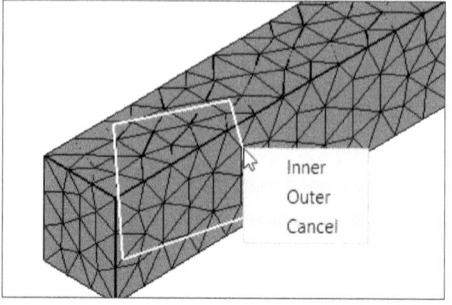

Figure-80. Shortcut menu for polygon selection

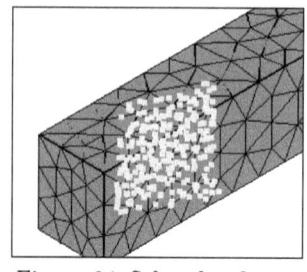

Figure-81. Selected nodes

- Click on the **Restore** button if you want to return to original mesh.
- Click on the **Copy** button to create copy of the result mesh generating after performing the analysis.
- After setting desired parameters, click on the **OK** button from the dialog to apply operation.

FEM Mesh to Polygonal Mesh Model Conversion

The Mesh created in FEM workbench is not a model mesh. It is representation of original model in the form of integration of small elements. You can convert the FEM mesh to a model mesh by using the **FEM mesh to mesh** tool in **Model** menu. The procedure to use this tool is given next.

- Click on the **FEM mesh to mesh** tool from the **Model** menu after selecting **FEMMeshGmsh** feature from the **Model Tree**. The FEM mesh will be converted to mesh model; refer to Figure-82.

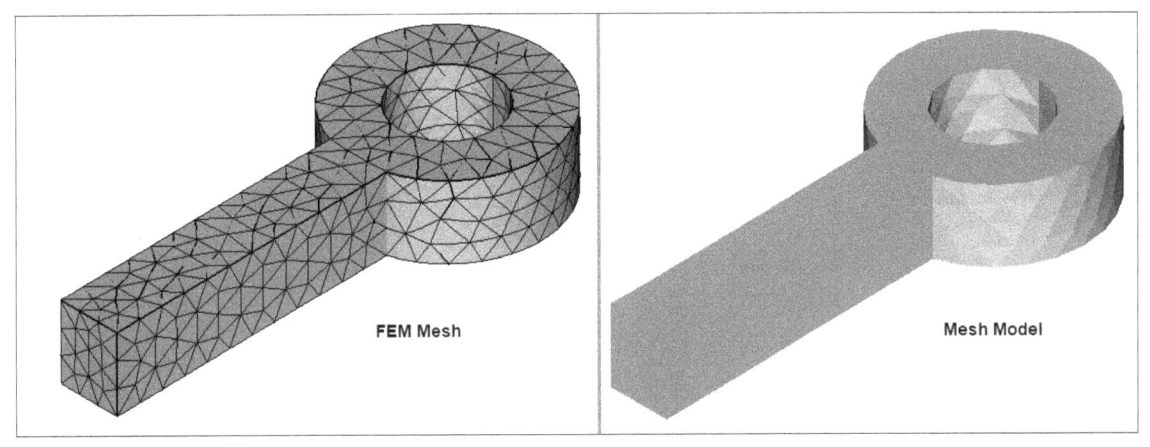

Figure-82. Mesh model conversion

SOLVERS

FreeCAD FEM has many analysis solvers available in **Solve** menu; refer to Figure-83. Some of these solvers are in experimental stage and some are specific to analysis type. Various solvers available in FreeCAD are discussed next.

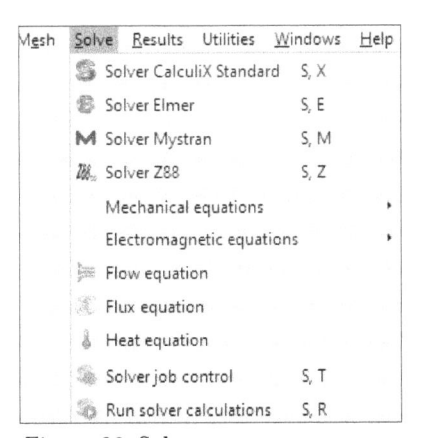

Figure-83. Solve menu

Calculix Standard Solver

CalculiX is a package designed to solve field problems. The method used by this solver is the finite element method. With CalculiX, Finite Element Models can be built, calculated, and post-processed. The solver is able to do linear and non-linear calculations. Static, dynamic, and thermal solutions can be obtained by this solver. Because the solver makes use of the abaqus input format, it is possible to use commercial pre-processors as well with this solver. The pre-processor of this solver is able to write mesh related data for nastran, abaqus, ansys, code-aster and for the free-cfd codes like duns, ISAAC and OpenFOAM. The procedure to use this solver is given next.

- Click on the **Solver CalculiX Standard** option from the **Solve** menu. The **CalculiXccxTools** feature will be added in the **Model Tree**. Note that when you start a mechanical analysis then **CalculiXccxTools** feature is added in the **Model Tree** by default.
- Double-click on the **CalculiXccxTools** feature. The **Mechanical analysis** dialog will be displayed in **Tasks** panel of **Combo View**; refer to Figure-84.

Figure-84. Mechanical analysis dialog

- Click on the **Write .inp file** button to create an input file for solver if you have specified all the other parameters of analysis like constraints, material, and so on in the model. This input file will tell the solver, what is to be calculated and what parameters are known.
- On clicking the **Write .inp file** button, the **Run CalculiX** button will become active and overview of input file writing process will be displayed; refer to Figure-85. Note that if there are parameters like material of body, constraints applied on the model, and loads which have not been defined yet then you will get an error message. Also, you can perform analysis on single body with one mesh feature. Multiple meshes are not allowed till writing this book in FreeCAD. Most of the time, mesh is major cause of error in solving the analysis, so make sure the mesh is created without error. Easiest way in FreeCAD to create a fine mesh is by using the **FEM mesh from shape by Netgen** tool if you are not familiar with advanced mesh terms.

FreeCAD 1.0 Black Book   9-43

- Click on the **Run CalculiX** button from the dialog. The resulting mesh will be generated if solver successfully runs. You may not get results if model is inconsistent, parameters specified are not feasible, or all the parameters are not defined.
- After running the analysis, click on the **Close** button to exit the dialog. The **Result mesh** feature will be added in the **Model Tree**; refer to Figure-86.

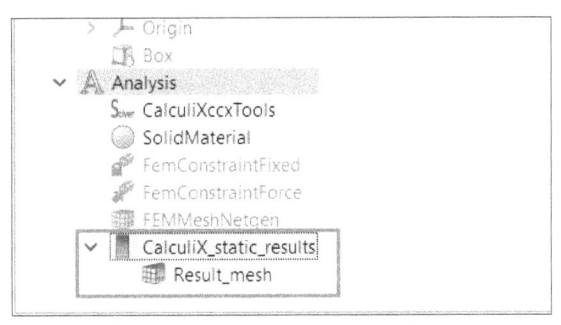

Figure-86. Result mesh feature generated

- To view results of analysis, select the CalculiX static results (or CCX results in some cases) feature from the **Model Tree** and click on the **Show Results** option from the **Results** menu; refer to Figure-87. The **Show result** dialog will be displayed result mesh preview; refer to Figure-88.

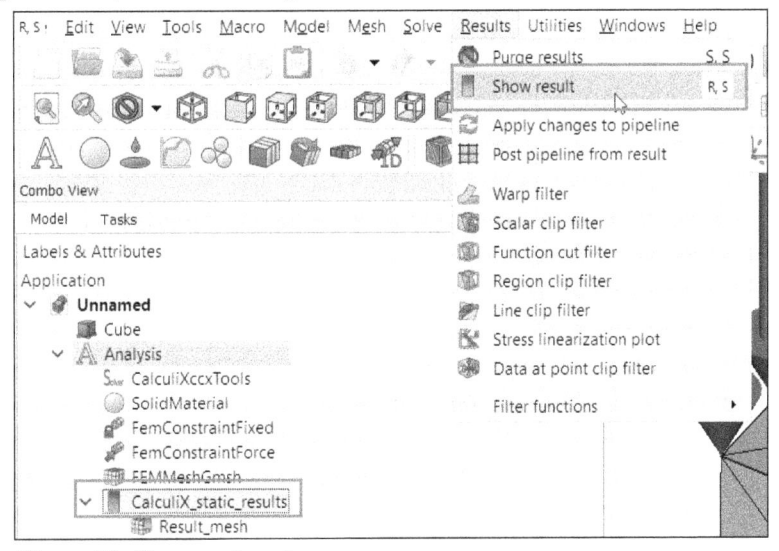

Figure-87. Show result option

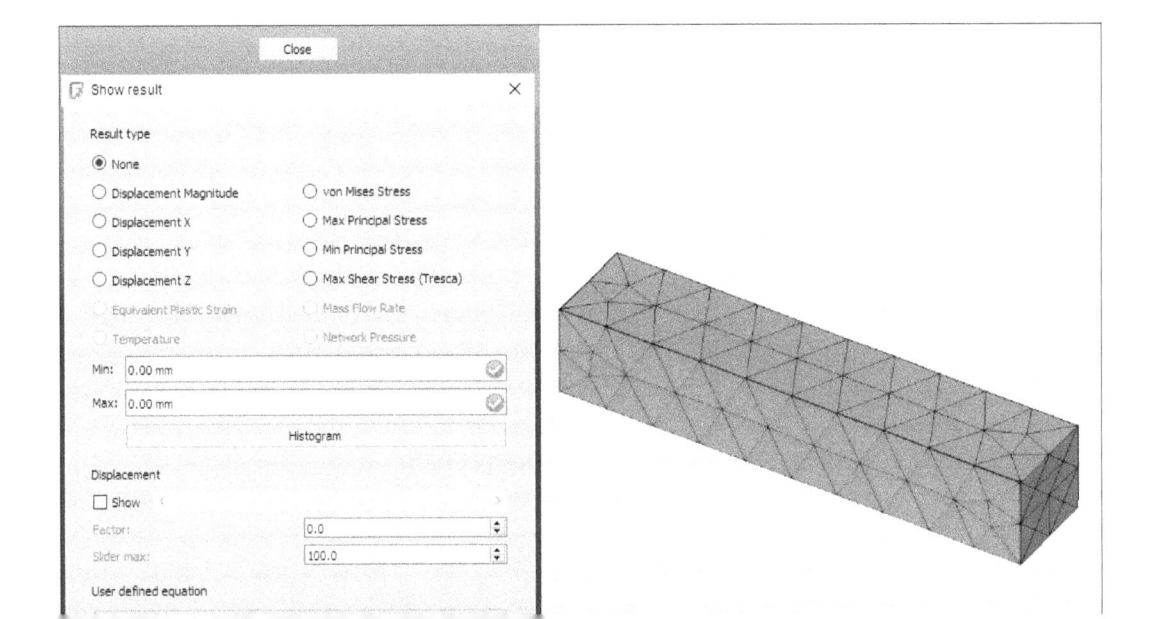

- Select desired radio button from the **Result type** area of the dialog to check respective result.
- Select the **Show** check box from **Displacement** area of dialog and set desired magnification factor using slider to check displacement cause due to load.
- You can check the Minimum and Maximum value of selected result type from the **Min** and **Max** edit boxes in the dialog.
- Click on the **Close** button from the top in the dialog to exit.

Elmer Solver

Elmer is an open source multi-physics simulation software mainly developed by CSC - IT Center for Science (CSC). Elmer development was started 1995 in collaboration with Finnish Universities, research institutes, and industry. After it's open source publication in 2005, the use and development of Elmer has become international.

Elmer includes physical models of fluid dynamics, structural mechanics, electromagnetism, heat transfer, and acoustics. These are described by partial differential equations which Elmer solves by the Finite Element Method (FEM). **Note that in FreeCAD, you can use the Elmer Solver with mesh generated by using Gmsh Mesher only.** Although this solver now comes pre-installed but if you somehow do not get it with software, you need to install the Elmer solver with its binary files which will be linked to FreeCAD GUI. Elmer requires two components to be interfaced with FreeCAD:

- ElmerGrid is the interface handling meshes.
- ElmerSolver is handling the computation.

The procedure to install this solver is given next.

- Go to the CSC binaries resources for Elmer: **https://www.nic.funet.fi/pub/sci/physics/elmer/bin/**
- Download and install the version best suited to your Operating System (Windows 64 bits, Linux) refer to Figure-89.

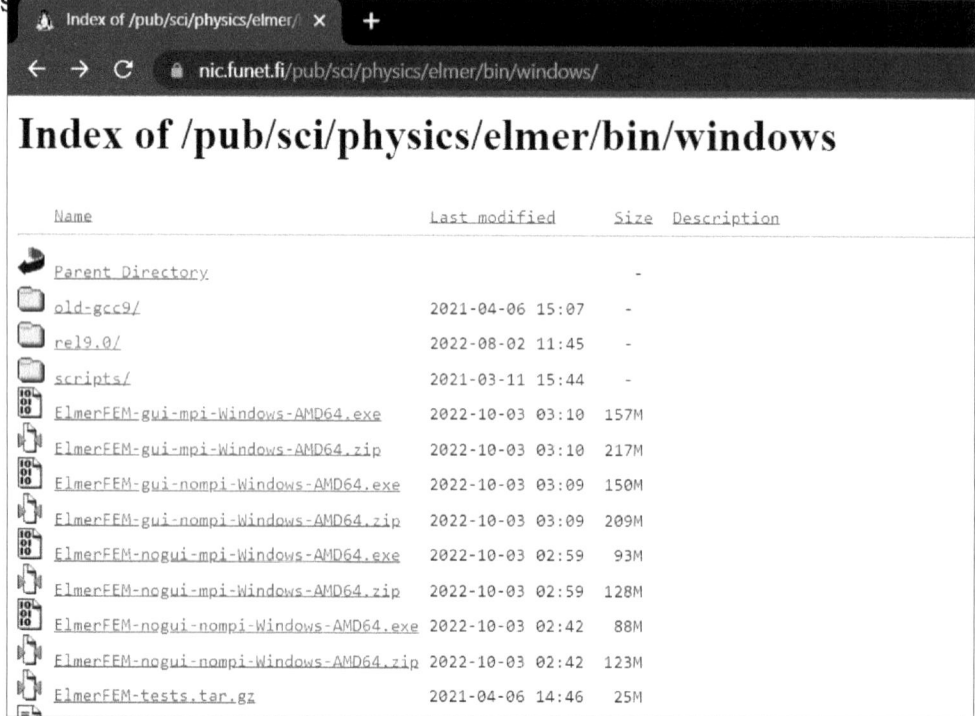

Figure-89. Elmer binaries for Windows

- After installing the Elmer solver, go to **Edit → Preferences → FEM → Elmer** in menu and link the correct path for both ElmerGrid and ElmerSolver; refer to Figure-90.

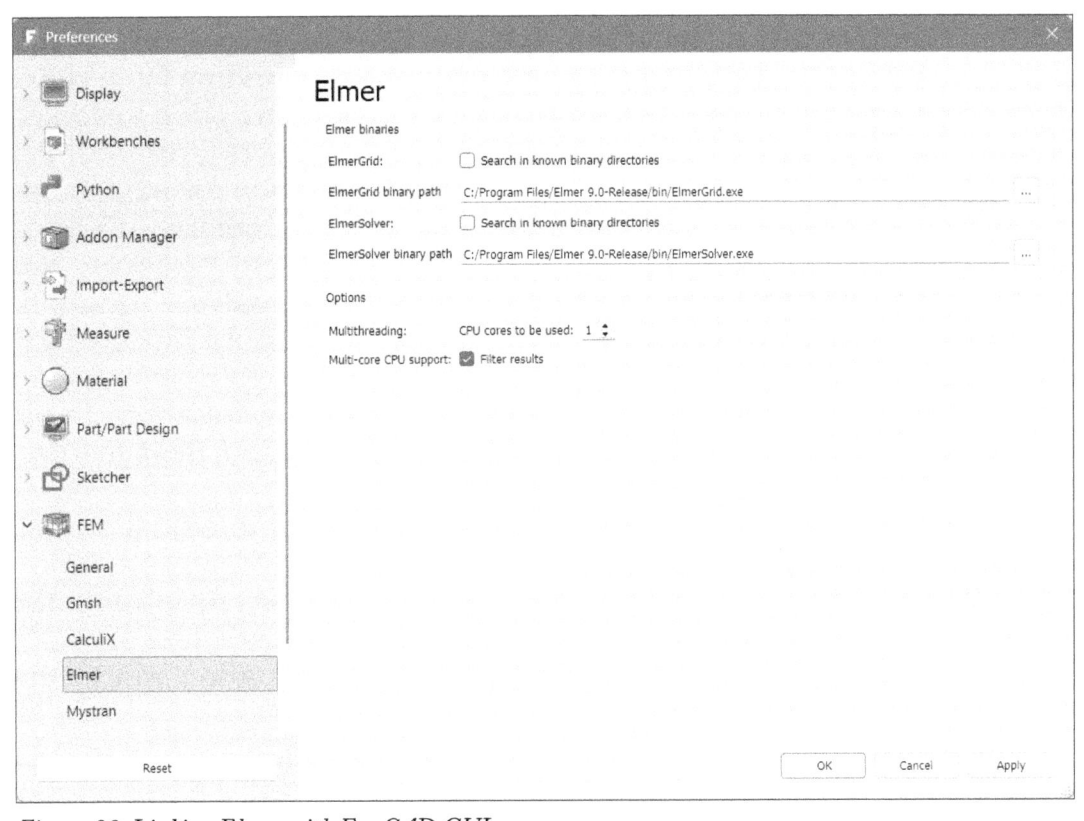

Figure-90. Linking Elmer with FreeCAD GUI

Now, we are ready to use the solver for performing analysis. The procedure to use this solver is given next.

- Create the analysis setup including mesh created by Gmsh, thermal/mechanical/electrostatic/fluid loads applied on model, and material applied to model which has properties related to analysis; refer to Figure-91. After that, click on the **Solver Elmer** tool from the **Solve** menu. The **SolverElmer** feature will be added in the **Model Tree**.

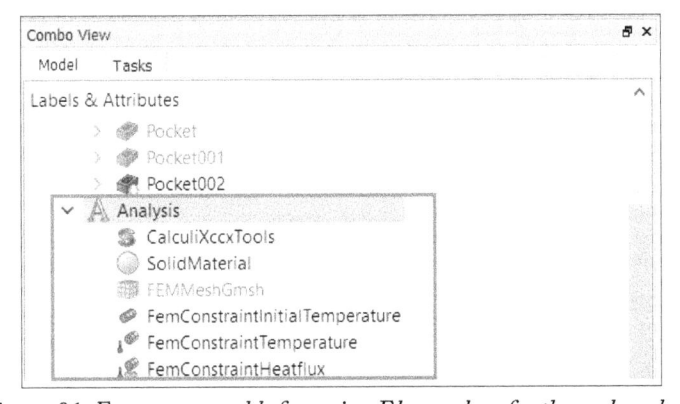

Figure-91. Features created before using Elmer solver for thermal analysis

For using Elmer solver, you need to create equations related to the analysis. There are six types of equations available in FreeCAD for different type of analyses; refer to Figure-92.

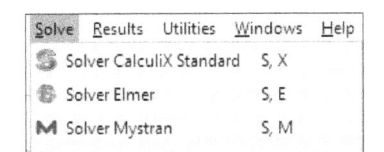

- Select the **Elasticity equation** option from **Mechanical equations** cascading menu of the **Solve** menu to perform linear mechanical analyses. Select the **Deformation equation** option to perform nonlinear mechanical analyses (deformations).
- Select the **Electrostatic equation** option from **Electromagnetic equations** cascading menu to add electrostatic analysis equation for finding force between two charged static objects. Select the **Electricforce equation** option to add equation for finding electric force between two moving charged objects/particles. Select the **Magnetodynamic equation** option to create a FEM Equation for magnetodynamic forces. Select the **Magnetodynamic2D equation** option to create a FEM equation for 2D magnetodynamic forces.
- Select the **Flow equation** option to add fluid dynamics analysis equation.
- Select the **Flux equation** option to add equation for finding out flux of any selected scalar field variable.
- Select the **Heat equation** option to add thermal analysis equation. In our case, we are using **Heat equation** option. On selecting this option, **Heat** feature will be added below **Elmer** solver in **Model Tree**.
- Select the **Solver job control** option to open the menu to adjust and start the selected solver.
- Select the **Run Solver Calculations** option to run the selected solver of the active analysis.
- Select the feature from **Model Tree**. Properties of the equation will be displayed in the **Property Ed**

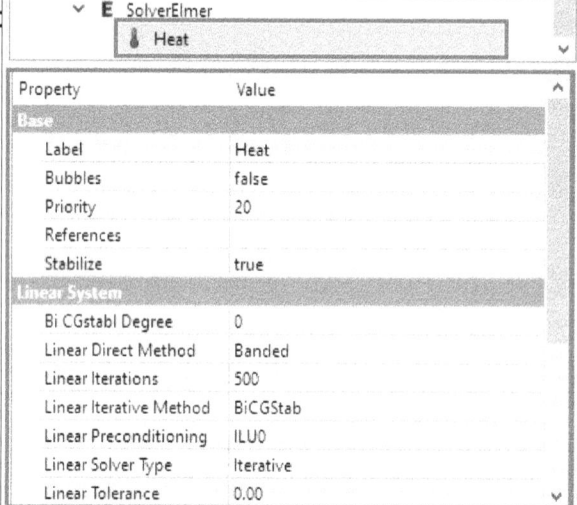

Figure-93. Properties for heat equation

- Specify desired parameters in the **Property Editor** dialog. Various parameters for equation are given next:

--Base Section--
Label : Used to specify name of the equation.
Bubbles : Used to stabilize solution of equation depending on compressibility of model. If you are working on compressible fluids then you should set this value to True.
Priority : Used to define priority level of current equation with respect to other equations to be solved by Elmer solver. A higher value of priority means respective equation will be solved first.
References : Used to change the linked object.
Stabilize : Used to specify whether Stabilized finite element method will be used or RFB (Residual Free Bubble) stabilization will be used.

--Linear System--

BiCGstabl Degree : In numerical linear algebra, the biconjugate gradient stabilized method, often abbreviated as BiCGSTAB, is an iterative method developed by H. A. van der Vorst for the numerical solution of nonsymmetric linear systems. Used to solve iterative linear equations for 3D analysis cases in FreeCAD. Set desired value to define stabilization order for the solver.

Linear Direct Method : Select desired type of linear direct method for solving linear FEM equations. For stress analysis, a direct solver is used instead of an iterative solver. It is often difficult for the iterative solver to find a solution for a structure that contains parts with varying stiffness properties. Use the **Banded** option when matrix to be formed for equation is dense (In simple words, a small region is to be analyzed with minimum void areas). Use the **umfpack** option when problem is complex with load free matrices in equation. If you do not know what to select then go with **umfpack**, although it will take more resources but it will do the trick.

Linear Iterative Method : Select desired option from the drop-down to define which iterative method is to be used for solving linear equations if the problem becomes complex or the matrix of equation is large. Each iterative method has its pros and cons which you can browse on Internet.

Linear Preconditioning : Linear Preconditioning is used to form a matrix easily solvable by selected iteration method. Preconditioning is typically related to reducing a condition number of the problem. The options available in FreeCAD try to form Incomplete Lower unitriangular and Upper triangular (ILU) matrices for preconditioning.

Linear Solver Type : Used to define which type of solver will be used for solving current analysis problem. Select the **Direct** option if problem is small scale linear and select the **Iterative** option if problem is complex or large.

Linear Tolerance : Used to define the accuracy up to which solution is needed. The value specified here also acts as goal for convergence if true solution is not available.

--Nonlinear System--

Nonlinear Iterations : Used to specify the number of iterations to be performed for finding solution of problem.

Nonlinear Newton After Iterations : Used to specify when Newton method will be used for solving equations in terms of iterations. If a certain number of specified iterations have been solved then their solutions will be used as input for Newton method to further solve the problem.

Nonlinear Newton After Tolerance : Used to specify when Newton method will be introduced for solving equation in terms of tolerance. If specified tolerance in solution has been achieved then Newton method will further solve the equation for convergence.

Nonlinear Tolerance : Used to specify the tolerance allowed in final solution of problem.

Relaxation Factor : The relaxation factor is used to speed up convergence of iteration. You can assume relaxation factor as a value which removes the amount of error in solution of each iteration due to numerous unknown real conditions.

--Steady State--

Steady State Tolerance : Used to define the accuracy in results to be achieved after performing the analysis.

- After setting desired parameters in the **Property Editor** dialog, select the

SolverElmer feature from the **Model Tree**. The options in **Property Editor** will be displayed as shown in Figure-94.

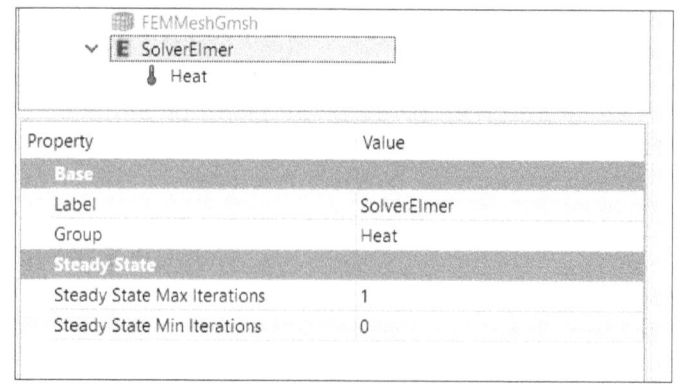

Figure-94. SolverElmer properties

- Specify the name and group for feature in the **Label** and **Group** fields, respectively.
- Specify desired values in the **Steady State Max Iterations** and **Steady State Min Iterations** fields to define maximum and minimum number of iterations to be solved when you are running coupled analysis (Coupled analysis involve multiple variables which are interdependent like temperature of body is function of current in analysis and current value changes with temperature of body as well). Generally, default value is **1** for max and **0** for min when you are solving for one variable only. If your analysis is coupled then specify desired value for iterations up to which you will be able to achieve solution of coupled equations.
- After setting desired solver parameters, double-click on the **SolverElmer** feature from **Model Tree**. The **Solver Control** dialog will be displayed in **Tasks** panel; refer to Figure-95.

Figure-95. Solver Control dialog

- Click on the **Write** button from dialog to generate input file. Since, we are using external solver, so input file parameters will not be displayed in the dialog. If you want to check the input file then browse to the location of **Working Directory** edit box in the dialog and open the **case.sif** file in any Word processor program.
- Click on the **Run** button from the dialog after writing input file. If all the parameters specified are feasible then results will be generated with Output file; refer to Figure-96. You can also check the results of analysis by double-clicking on the **SolverElmerResult** node in **Model** tab of **Combo View** dialog.

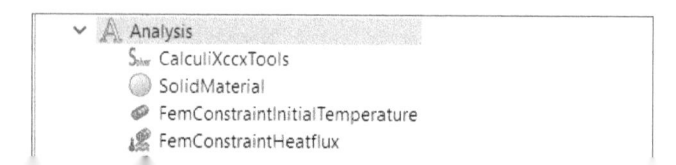

Note that sometimes, you may not get the results displayed in FreeCAD GUI. In that case, reach to the result file (*.vtu) of analysis in temporary folder of Elmer solver analysis and open it in ElmerGUI application which we have installed earlier. You will file the address of temporary result file in **Report View** panel of application window; refer to Figure-97.

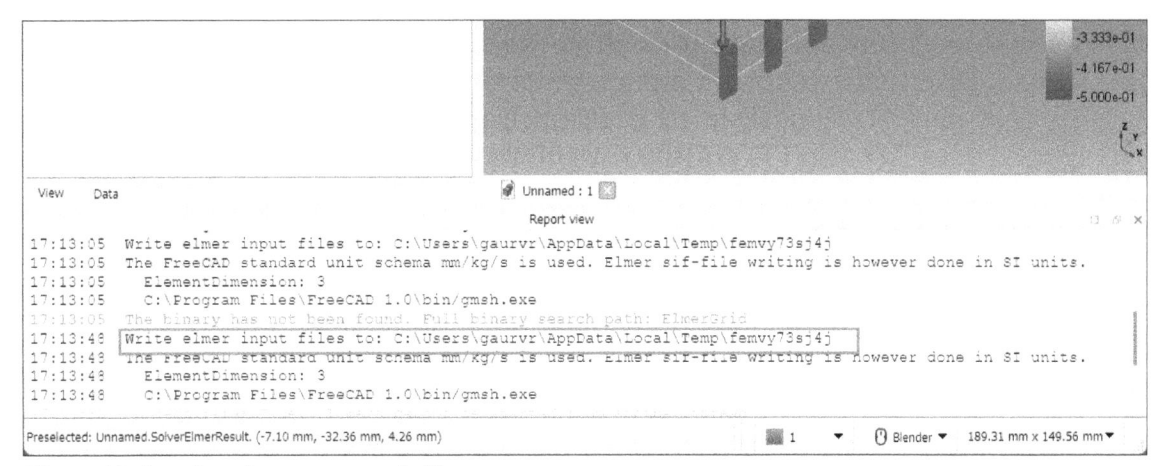

Figure-97. Location of temporary result file

ParaView is also a good post processor for analyzing results of analysis. You can download ParaView from https://www.paraview.org/download web page. Figure-98 shows the result of our current analysis in ParaView.

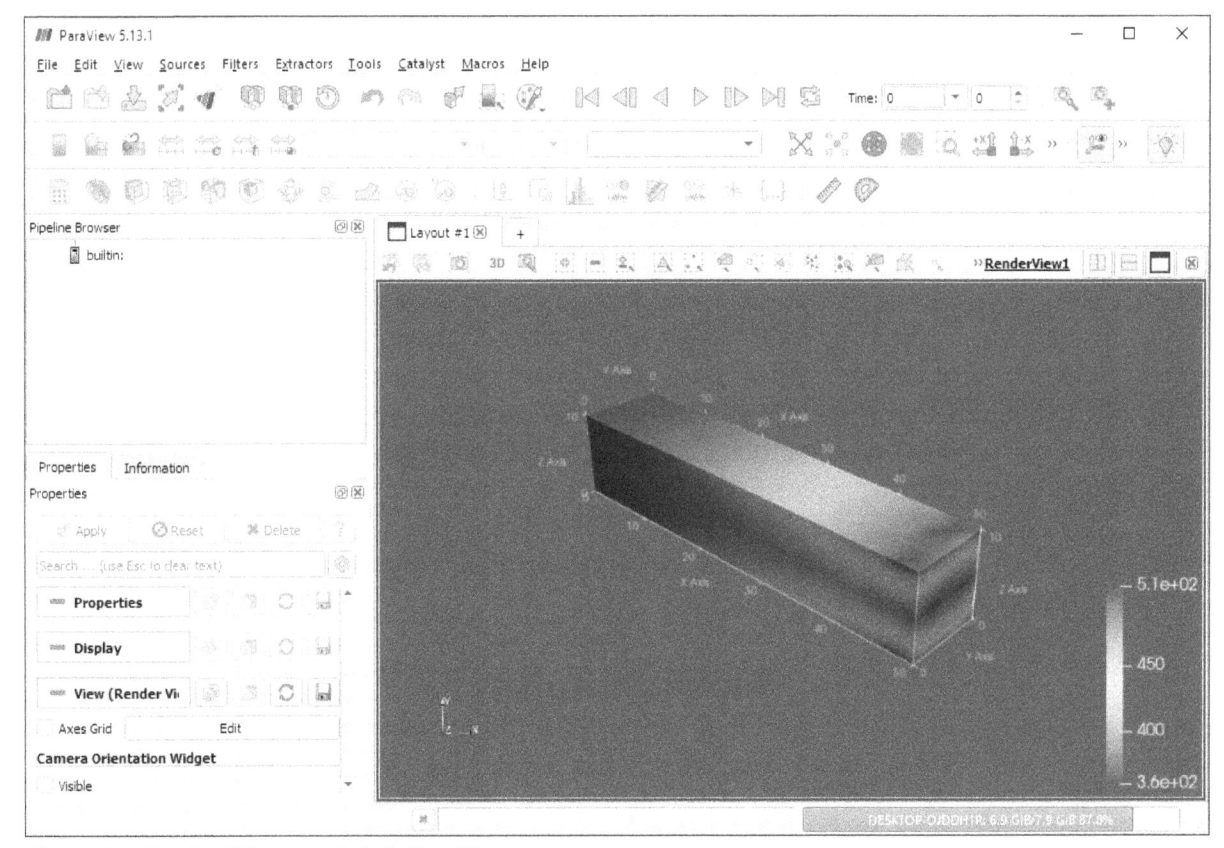

Figure-98. Result of Elmer analysis in ParaView

Similarly, you can use the other equations of Elmer Solver for performing related analyses.

Z88 Solver

The software was developed by Frank Rieg, a professor for engineering design and CAD at the University of Bayreuth. Originally written in FORTRAN 77, the program was ported to the programming language C in the early 1990s.

There are two programs for finite element analysis:

Z88OS is available as free software including the source code under the GNU General Public License. Due to the modular structure of the program and the open availability of the source code, it is possible to develop customized extensions, add-ons and several special case 2D and 3D continuum elements (e.g. anisotropic shell element).

Z88Aurora originally described the user interface of the Z88 finite element analysis program. After several additions and further development, it now comprises a significantly larger range of functionality than Z88OS. Z88Aurora is freeware, however the source code is not publicly available. Z88Aurora's current version contains several computation modules:

In the case of linear static analyses, it is assumed that the result is proportional to the applied forces.
Nonlinear analyses are used for nonlinear geometries and nonlinear materials.
Using thermal and thermomechanical analyses, it is possible to not only compute results about temperature or heat currents, but also thermomechanical displacements and stresses.

By utilizing natural frequency simulation, natural frequencies and the resulting oscillations can be determined.

A contact module makes it possible to simulate interacting parts and assemblies. An integrated part management tool enables an effective handling of assemblies. There are options to simulate a glued connection or a friction-free connection and the contact discretization (type of contact: node-surface-contact, or surface-surface-contact), the mathematical imposition method (lagrange method, perturbed lagrange method, or penalty method), and the direction of contact stiffness (normal or tangential direction) can be changed via the contact settings. This module only supports tetrahedrons and hexahedrons with linear or quadratic shape functions. Additionally, the module is only available for linear mechanical strength analyses.

The procedure to use Z88 solver in FreeCAD is similar to using CalculiX solver which has been discussed earlier.

Mystran Solver

Mystran is an open source general purpose solver used to perform linear finite element analysis. The primary author of this solver Dr. Bill Case has been working on development of this solver for approximately 30 years till he died in 2021. In a linear analysis, displacement, forces, and stresses are directly proportional to applied load. Mystran is acronym of **My Structural Analysis** which signifies it can solve wide variety of FEA problems. Note that most of the structural analyses modeled for Nastran can be solved in Mystran. There is a large community of developers working on development of this solver. You can know more about the solver from website https://www.mystran.com/.

POST PROCESSING IN FREECAD FEM

The tools in **Results** menu of **FEM** workbench are used to perform post processing, so that you can analyze the results. The tools of this menu are active only when you performed the analysis and result of solver is selected in the **Model Tree**; refer to Figure-99. Various tools of this menu are discussed next.

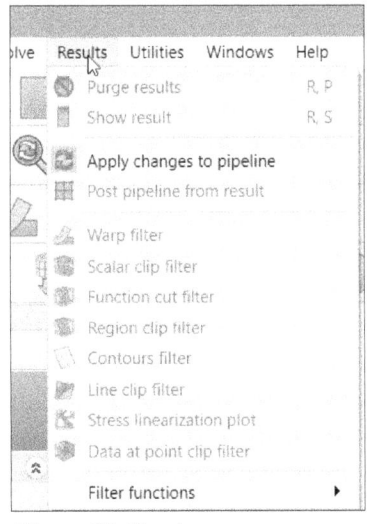

Figure-99. Results menu

Purge Results

The **Purge results** tool is used to delete the results generated by using **Show result** tool. Note that this tool is active only after you have used the **Show result** tool.

Show Result

The **Show result** tool from the **Results** menu to generate analysis results. The procedure to use this tool has been discussed earlier.

Creating Post Processing Pipeline for Results

The **Post pipeline from result** tool is used to create post processing pipelines from the results. The procedure to use this tool is given next.

- Click on the **Post pipeline from result** tool from the **Results** menu after selecting solver's result feature from the **Model Tree**. The **Pipeline** feature will be added in the **Model Tree**.
- Double-click on this feature to modify its parameters. The **Result display options** dialog will be displayed; refer to Figure-100.

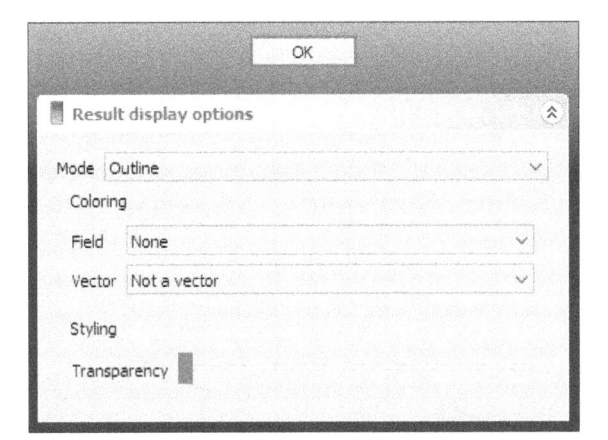

- Select desired option from the **Mode** drop-down to define what type of pipeline plot you want to generate using the result parameters. Select the **Outline** option to generate only border lines of the model. Select the **Nodes** option to generate colored nodes for analyzing results. Select the **Surface** option to generate result surface with multicolored annotations. Select the **Surface with Edges** option to generate result surface with wireframe edges. Select the **Wireframe** option to generate wireframe model of results. Select the **Wireframe (surface only)** option to generate wireframe model over the surface of model for analyzing results.
- Select desired option from the **Field** drop-down to define result parameter to be displayed over the pipeline plot. Preview of result will be displayed in the modeling

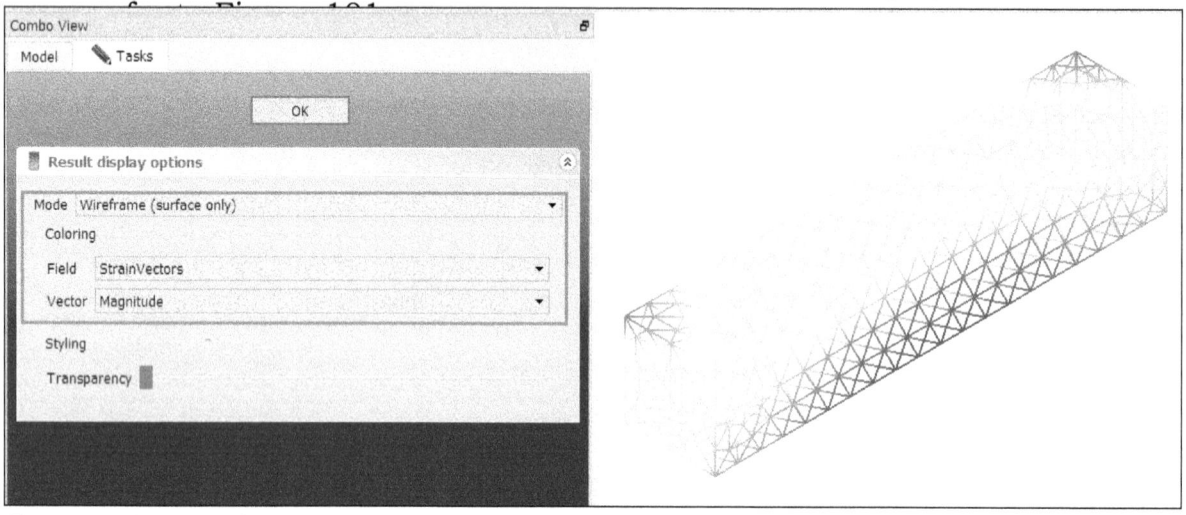

Figure-101. Preview of pipeline result

- Select desired option from the **Vector** drop-down to define which component of vector is to be used for generating result model.
- Click on the **OK** button from the dialog to generate pipeline result.

Applying Warp Filter

The **Warp filter** tool is used to distort the result model, so that higher values are further away from base model as compared to lower result values. The procedure to use this tool is given next.

- Click on the **Warp filter** tool from the **Results** menu after selecting **Pipeline** feature from the **Model Tree**. The **Warp options** and **Result display options** dialogs will be displayed; refer to Figure-102.

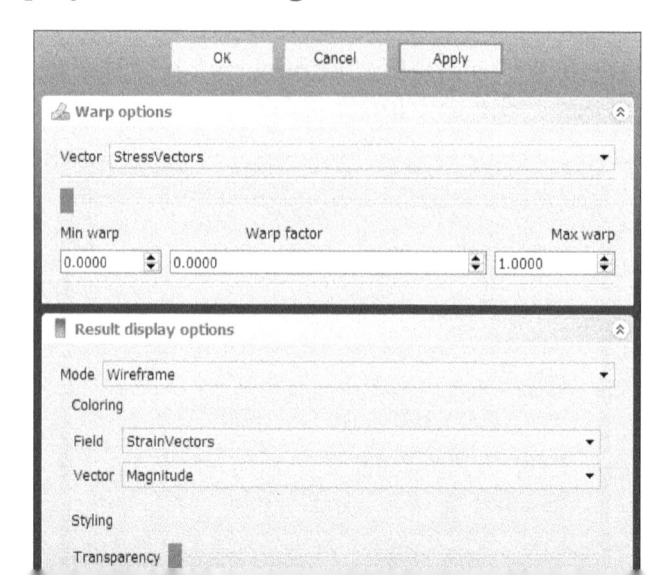

- Select desired option from the **Vector** drop-down to specify which result vector is to be displayed in warped model.
- Set the other parameters as discussed earlier in the **Result display options** dialog.
- Use the warp slider and warp related parameters in the **Warp options** dialog to check warped model of results; refer to Figure-103.

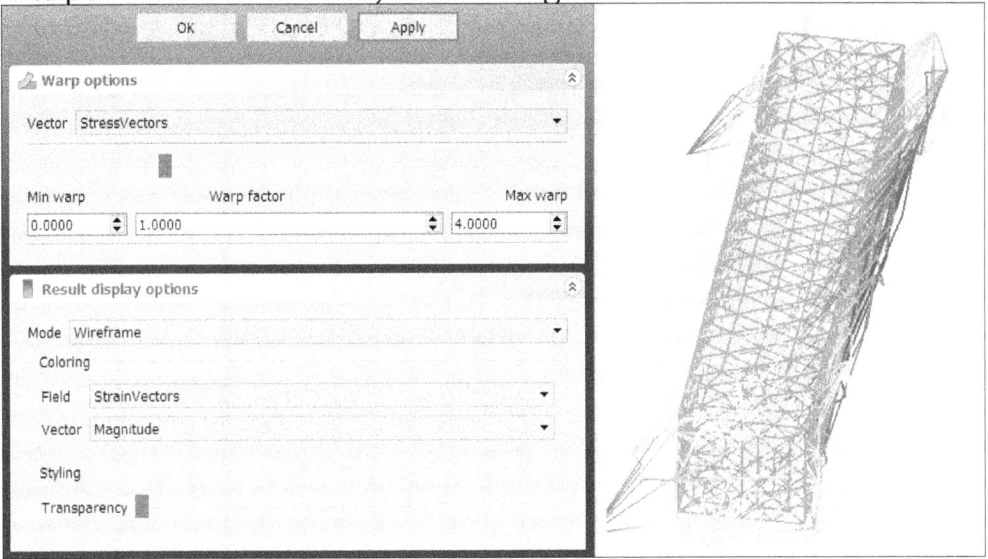

Figure-103. Warped model

- After setting desired parameters, click on the **OK** button from the dialog.

Applying Scalar Clip Filter

The **Scalar clip filter** tool is used to check internal section of model for result parameters. The procedure to use this tool is given next.

- Click on the **Scalar clip filter** tool from the **Results** menu after selecting **Pipeline** feature from **Model Tree**. The **Scalar clip options** and **Result display options** dialogs will be displayed.
- Set the parameters as discussed earlier in the dialogs and use the clipping slider for checking internal regions of model for analysis results; refer to Figure-104.

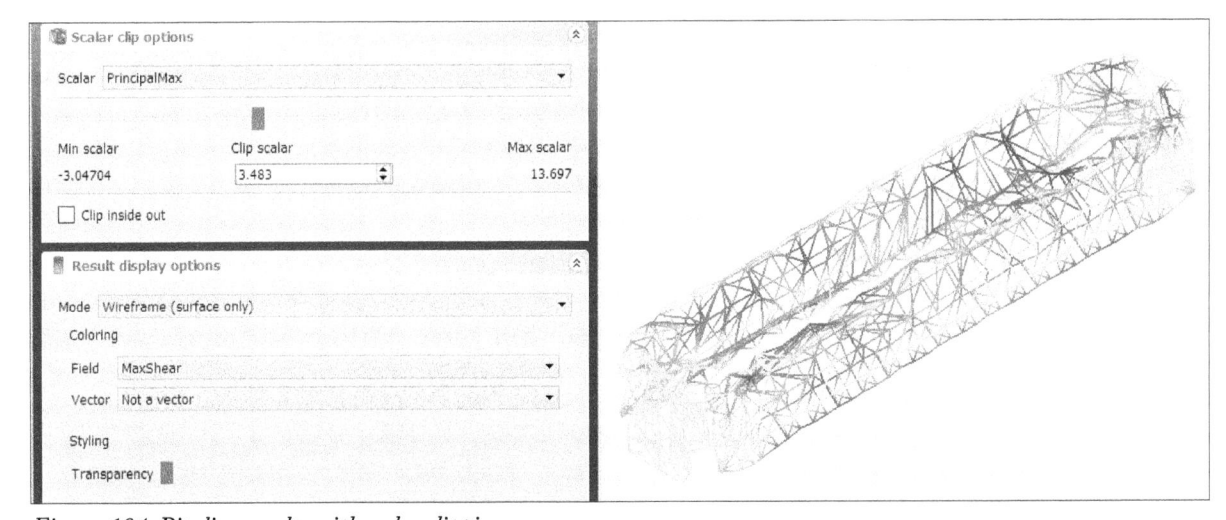

Figure-104. Pipeline results with scalar clipping

- Click on the **OK** button from the dialog to create the result model and exit dialog.

Applying Function Cut Filter

The **Function cut filter** tool is used to check the pipeline results of analysis along a plane or over the surface of a sphere. The procedure to use this tool is given next.

- Click on the **Function cut filter** tool from the **Results** menu after selecting the **Pipeline** feature from **Model Tree**. The **Function cut, choose implicit function** dialog will be displayed along with **Result display options** dialog; refer to Figure-105.

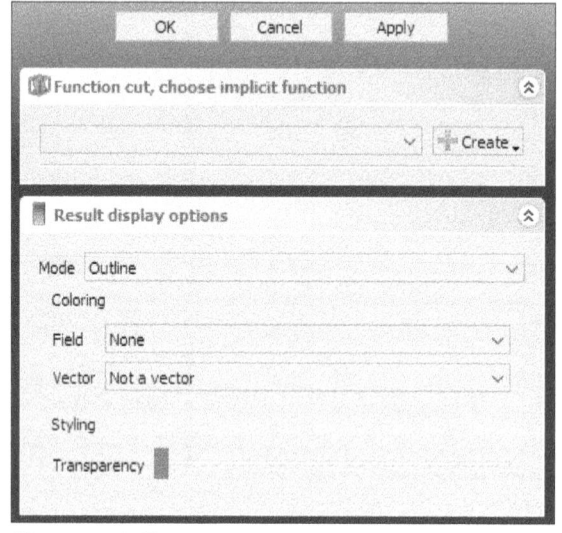

Figure-105. Function cut dialog

- Click on the **Create** button from the dialog and select desired option to define scope of result output geometrically. Select the **Plane** option if you want to display result on a plane intersecting with the model. Select the **Sphere** option if you want to check results on surface of a sphere intersecting with the model. We have selected **Plane** option in our case for tutorial.
- On selecting the **Plane** option, parameters related to location of plane will be displayed in the dialog; refer to Figure-106.

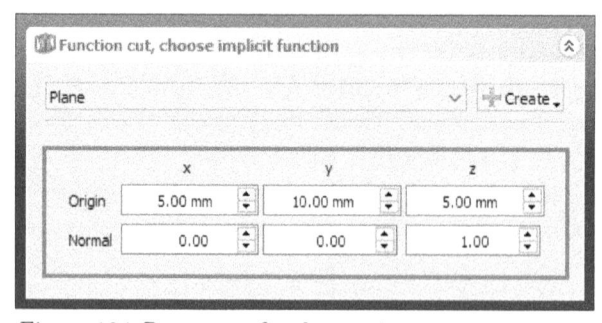

Figure-106. Parameters for plane option

- Set desired location and orientation for plane using the **Origin** and **Normal** edit boxes.
- Set desired options in the **Result display options** dialog as discussed earlier. Preview of the plot will be displayed; refer to Figure-107.

- Click on the **OK** button from the dialog to create the plot and exit the dialog.

Region Clip Filter

The **Region Clip Filter** tool is used to clip a field using a sphere or a plane cutting through the model. The procedure to use this tool is given next.

- Click on the **Region clip filter** tool from the **Results** menu after selecting the **Pipeline** feature from **Model Tree**. The **Clip region, choose implicit function** dialog will be displayed along with **Result display options** dialog; refer to Figure-108.

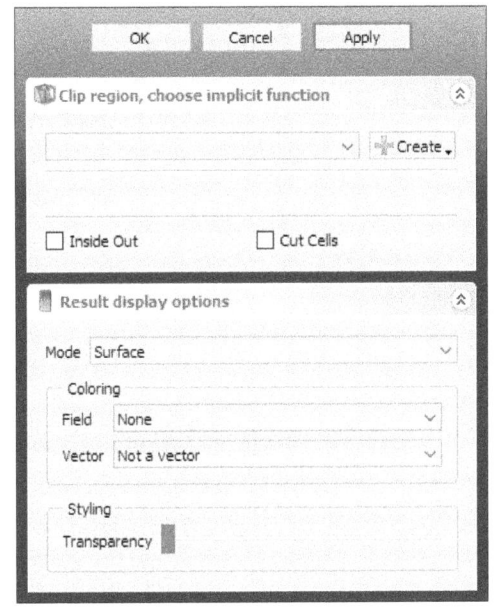

Figure-108. Clip region dialog

- Click on the **Create** button from the dialog and select desired option to define scope of result output geometrically. Select the **Plane** option if you want to display result on a plane intersecting with the model. Select the **Sphere** option if you want to check results on surface of a sphere intersecting with the model. We have selected **Plane** option in our case for tutorial.
- On selecting the **Plane** option, parameters related to location of plane will be displayed in the dialog; refer to Figure-109.

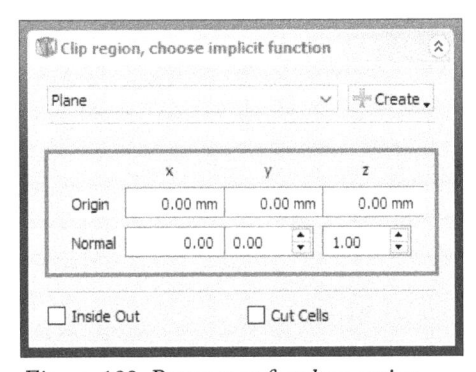

Figure-109. Parameters for plane option

- Set desired location and orientation for plane using the **Origin** and **Normal** edit boxes.
- Select **Inside Out** check box to invert the cut.
- Select **Cut Cells** check box to smoothen the clipped region by removing parts of

finite elements that are sticking out.
- Set desired options in the **Result display options** dialog as discussed earlier.

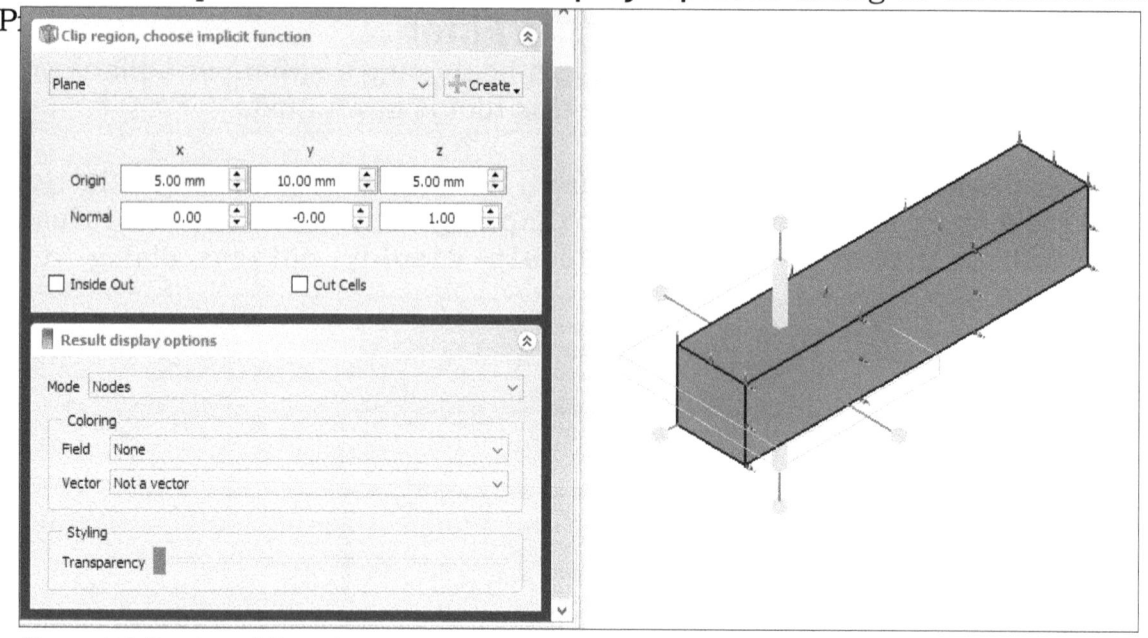

Figure-110. Preview of clip region

- Click on the **OK** button from the dialog to create the plot and exit the dialog.

Contours filter

The **Contours filter** tool is used to create iso-contours and iso-lines in the results mesh. The procedure to use this tool is discussed next.

- Click on the **Contours filter** tool from the **Results** menu after selecting the Pipeline feature. The **Contours filter options** dialog will be displayed; refer to Figure-111.

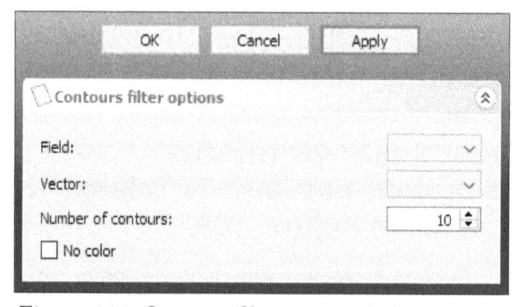

Figure-111. Contours filter options dialog

- Select desired results field to be drawn from **Field** drop-down.
- Select desired vector component to be used from **Vector** drop-down.
- Specify desired number of contours to be created in the **Number of contours** edit box.
- Select the **No color** check box to not color the contour lines.
- After specifying desired parameters, click on the **OK** button from the dialog.

Line Clip Filter

The **Line clip filter** tool is used to generate line plot of desired result parameter. The procedure to use this tool is given next.

- Click on the **Line clip filter** tool from the **Results** menu after selecting the **Pipeline** feature. The **Data along a line options** dialog will be displayed; refer to Figure-112.

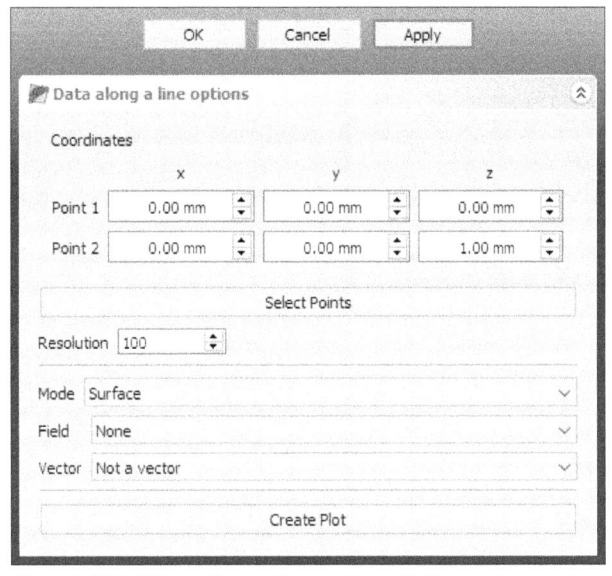

Figure-112. Data along a line options dialog

- Click on the **Select Points** button from the dialog and one by one select two points from model to define result output limits.
- Set the parameters as discussed earlier in **Resolution**, **Mode**, **Field**, and **Vector** drop-downs.
- Click on the **Create Plot** button to generate the plot. The plot will be created and preview of plot will be displayed in the model; refer to Figure-113.

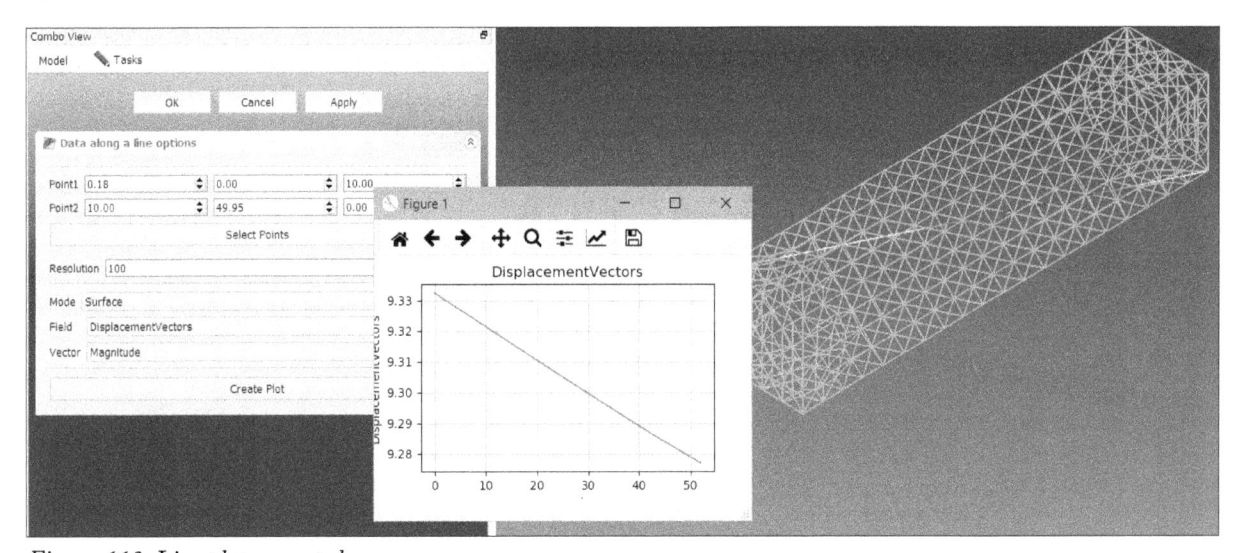

Figure-113. Line plot generated

- Using the **Zoom** button in the **Figure 1** dialog box, you can zoom into a specific region of plot by drawing a box.
- Using the **Pan** button you can move left/right/up/down in the plot to check values.
- The **Back** and **Forward** buttons are used to switch between previous and next views of plot.
- Click on the **Configure subplots** button to modify various spacings in the plot. On clicking this button, the **FreeCAD** dialog box will be displayed as shown in Figure-114. Set desired values and click on the **Close** button.

- Click on the **Edit axes, curve, and image parameters** button from the dialog box. The **Figure options** dialog will be displayed; refer to Figure-115.

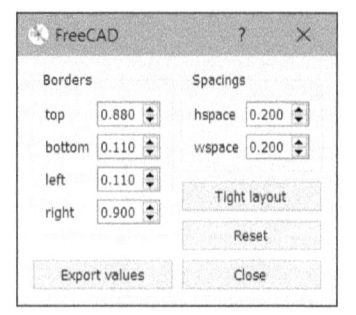

Figure-114. FreeCAD dialog box

Figure-115. Figure options dialog box

- Set desired parameters in **Axes** and **Curves** tab of the dialog box and click on the **OK** button to exit the dialog box.
- Click on the **Save** button from the **Figure 1** dialog to save current plot as a png file.
- Close the dialog after checking plot and click on the **OK** button from the **Data along a line** dialog to exit the tool.

Similarly, you can use other post processing tools in **Results** menu. Sometimes, you may get errors in postprocessing due to some programming needs of software. In such cases, your other post processor ParaView is available to help.

PRACTICAL

Perform finite element analysis of the given model; refer to Figure-116.

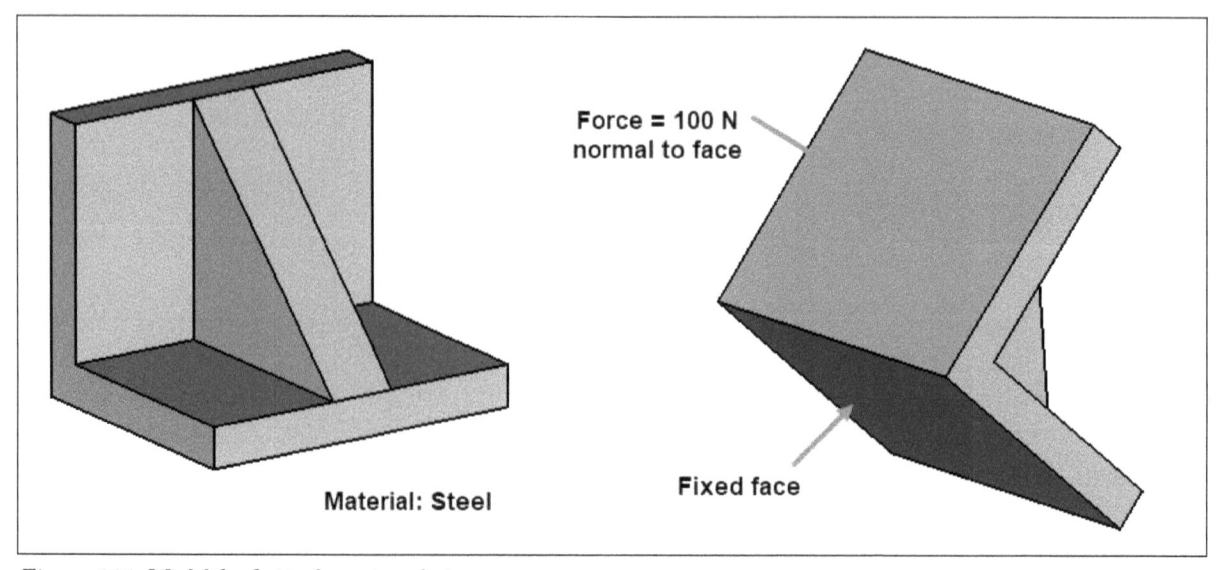

Figure-116. Model for finite element analysis

Starting a New File

- Start FreeCAD if not started yet.
- Select the **FEM** option from the **Switch between workbenches** drop-down in the **Toolbar**. The path environment will be displayed.

Creating Analysis Container

- Click on the **Analysis Container** tool from the **Toolbar**. The Analysis container will be created in the **Model Tree**; refer to Figure-117.

Defining the Material

- Click on the **Material For Solid** tool from the **Toolbar**. The **FEM Material** dialog will be displayed.
- Select the **Calculix - Steel** material from the **Material card** drop-down and specify the parameters in the dialog as shown in Figure-118.

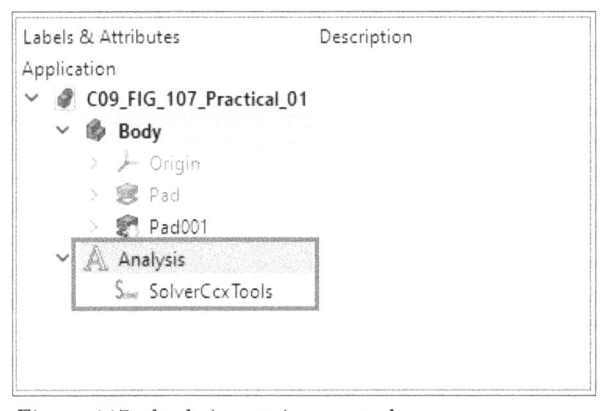

Figure-117. Analysis container created

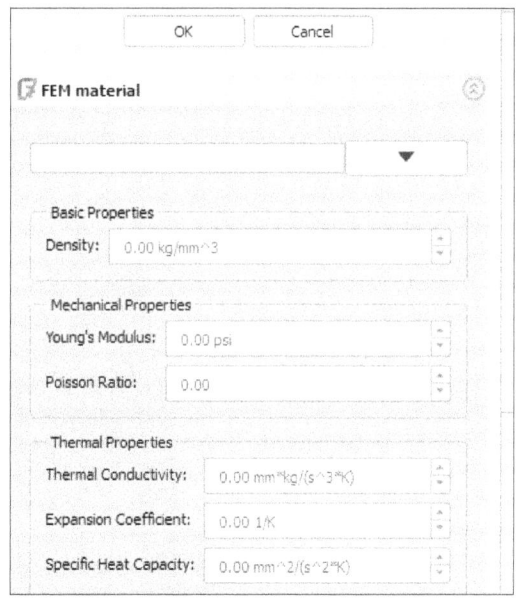

Figure-118. Material defined

- Scroll down in dialog, select the **Solid** radio button and click on the **Add** button. You will be asked to select face of the body.
- Select the model to apply material and click on the **OK** button. The material will be defined.

Creating FEM Mesh from shape

- Select the body and click on the **FEM mesh from shape by Gmsh** tool from the **Toolbar**. The **FEM Mesh by Gmsh** dialog will be displayed.
- Click on the **Apply** button from the dialog. The FEM mesh will be created; refer to Figure-119.
- Click on the **OK** button from the dialog. The FEM mesh will be displayed in the Model Tree.

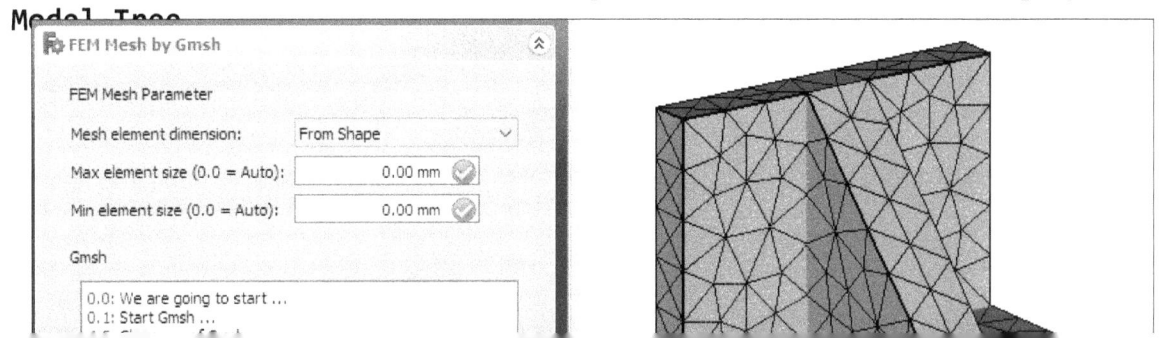

Defining Fixed Constraint

- Click on the **Constraint fixed** tool from **Toolbar**. The **FEM constraint parameters** dialog will be displayed.
- Select the face of model to apply the fixed constraint as shown in Figure-120, and click on **Add** button from the dialog. The fixed constraint will be applied; refer to Figure-121.

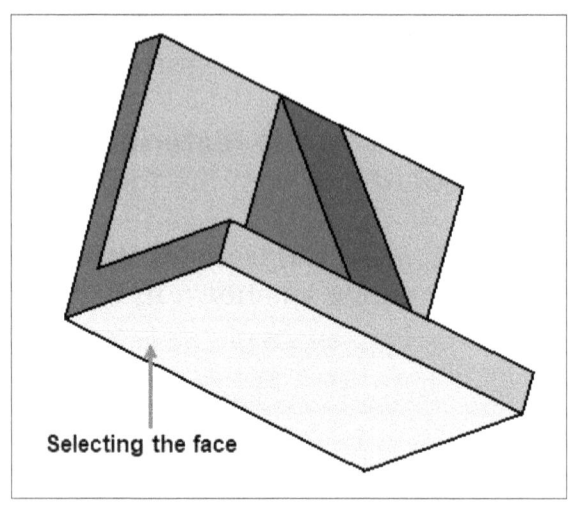

Figure-120. Selecting the face to apply fixed constraint

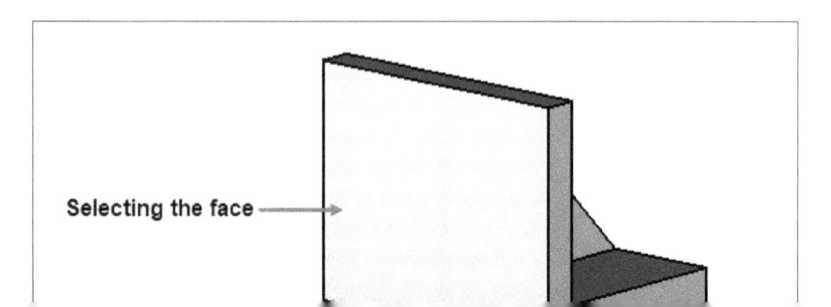

Figure-121. Fixed constraint applied

- Click on the **OK** button to close the dialog.

Defining Force Constraint

- Click on the **Constraint force** tool from the **Toolbar**. The **FEM constraint parameters** dialog will be displayed.
- Select the face of model to apply the force constraint as shown in Figure-122, and click on the **Add** button from the dialog. The force constraint will be applied; refer to Figure-123.

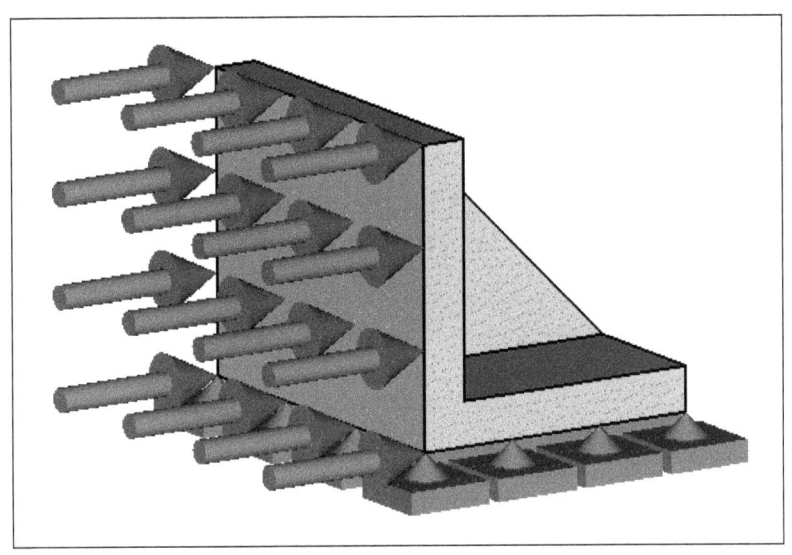

Figure-123. Force constraint applied

- Specify the load value as **100** in **Load** edit box and click on the **OK** button to close the dialog.

Running the Solver

- Select the solver from the **Model Tree** to run the calculations; refer to Figure-124, and click on the **Run solver calculations** tool from the **Toolbar** or **Solve** menu. The result will be displayed; refer to Figure-125.

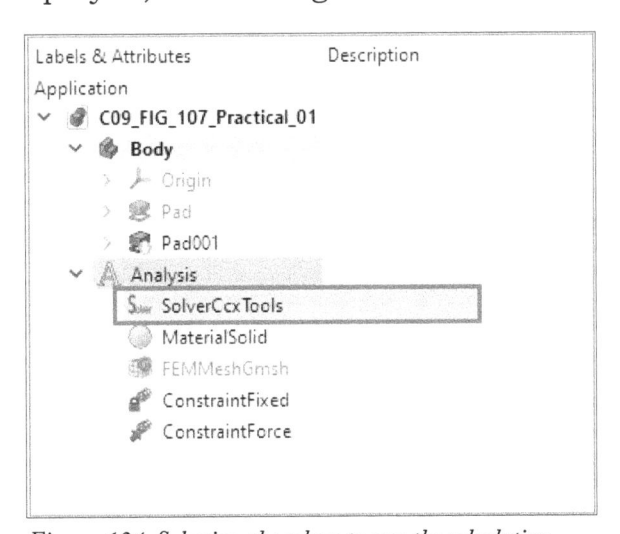

Figure-124. Selecting the solver to run the calculation

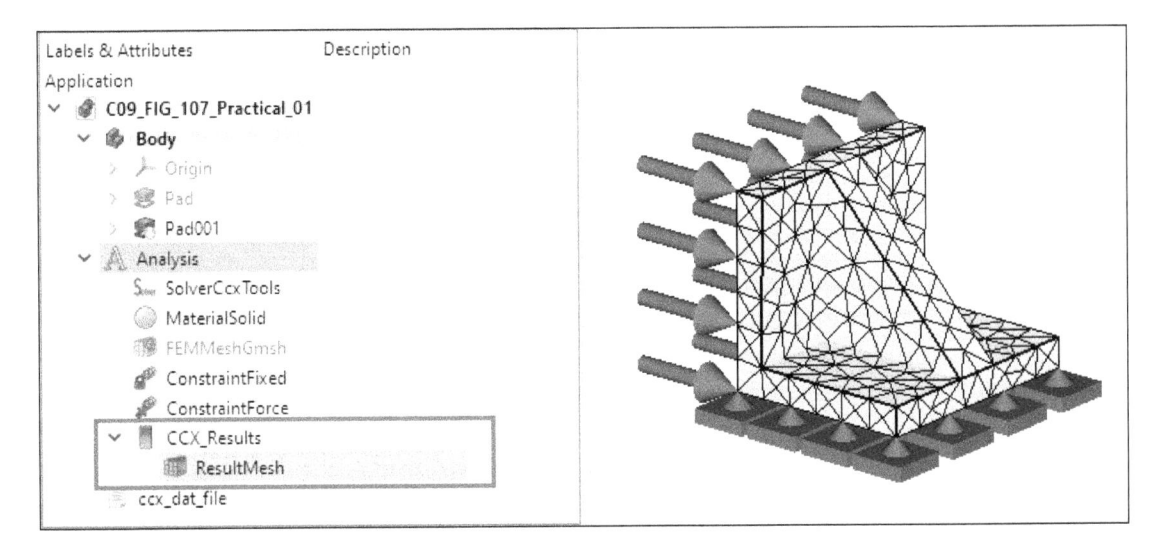

Tip: While applying constraints, you may get error like Face already selected when selecting a face to apply load/constraint. In such cases, save the file, close it and then reopen it to apply load/constraint. This can be software programming fault for retaining objects in cache memory.

SELF ASSESSMENT

Q1. What is the primary material property needed for Steady State Thermal Analysis?

A) Thermal conductivity
B) Specific heat
C) Density
D) Thermal expansion

Q2. In a Static Analysis, which assumption is made regarding the applied loads?

A) The loads are applied rapidly
B) The loads vary over time
C) The loads are applied slowly and remain constant after reaching their full magnitude
D) The loads are always dynamic

Q3. Which type of analysis is used when the relationship between loads and responses is linear?

A) Non-Linear Static Analysis
B) Frequency Analysis
C) Thermal Analysis
D) Linear Static Analysis

Q4. What is the main goal of Modal (Frequency) Analysis?

A) To calculate stresses and strains
B) To determine how an object vibrates at different frequencies
C) To calculate the temperature distribution in a body
D) To model heat transfer through conduction

Q5. Which constraint is used to stop the movement of a selected object in specified directions in FEM?

A) Rigid Body Constraint
B) Displacement Boundary Condition
C) Fixed Boundary Condition
D) Tie Constraint

Q6. What is the purpose of the Rigid Body Constraint in FEM?

A) To apply a fixed force on the object
B) To restrict the movement of the object
C) To connect two surfaces in an assembly
D) To constrain the motion of the nodes of a selected geometrical entity

Q7. Which constraint allows for the free rotation of a model while restricting translation in any direction?

A) Contact Constraint
B) Displacement Boundary Condition
C) Plane Rotation Constraint
D) Fixed Boundary Condition

Q8. What does the Contact Constraint tool in FEM workbench do?

A) It restricts displacement of an object in all directions
B) It connects two faces and defines contact stiffness and friction coefficient
C) It applies a force to the object
D) It modifies material properties

Q9. Which of the following is used to apply a force load on a selected face of the model?

A) Pressure Load
B) Force Constraint
C) Tie Constraint
D) Displacement Boundary Condition

Q10. In the FEM workbench, how is the material applied to the model?

A) By selecting a geometry reference and assigning a material from the material dialog
B) By clicking on the "Apply Material" button
C) By entering material properties directly in the geometry dialog
D) By manually adjusting the temperature of the model

Q11. In the FEM workbench, which tool is used to modify material properties such as Young's Modulus and density?

A) Material for Solid Tool
B) Use FreeCAD Material Editor
C) Apply Constraints Tool
D) Model Parameters Tool

Q12. What is the primary purpose of performing a Frequency Analysis?

A) To calculate thermal conductivity
B) To determine the deformation of a body under load
C) To calculate stress and strain in a structure
D) To determine the natural frequencies of vibration and avoid resonance

Q13. Which analysis type is concerned with the heat transfer in a body due to conduction, convection, and radiation?

A) Thermal Analysis
B) Static Analysis

C) Frequency Analysis
D) Modal Analysis

Q14. What is the main purpose of the Local coordinate system tool in FEM workbench?

A) To rotate the entire model
B) To transform the coordinate system for faces with displacement constraint applied
C) To change the material properties of the model
D) To set the boundary conditions for thermal analysis

Q15. In the Transform Constraint Local coordinate system tool, which radio button is selected by default for transformation?

A) Cylindrical
B) Rectangular
C) Spherical
D) Polar

Q16. Which tool is used to define boundary conditions for fluid flow analysis in FreeCAD FEM workbench?

A) Fluid boundary condition tool
B) Bearing constraint tool
C) Gear constraint tool
D) Thermal constraint tool

Q17. What option should you select from the Boundary drop-down list to specify physical conditions for a wall in fluid analysis?

A) Inlet
B) Outlet
C) Wall
D) Freestream

Q18. Which tool is used to apply a pulley load equivalent to a specified pulley load in FreeCAD FEM?

A) Bearing constraint tool
B) Gear constraint tool
C) Pulley constraint tool
D) Fluid flow condition tool

Q19. What unit is required when specifying the initial temperature in the Constraint Initial temperature tool?

A) Fahrenheit
B) Celsius
C) Kelvin
D) Rankine

Q20. The Heat flux load defines the rate of:

A) Fluid flow
B) Electrical potential
C) Heat energy transfer from selected face(s)
D) Magnetic field

Q21. The Body Heat Source constraint in FreeCAD is used to:

A) Specify initial flow velocity for fluid
B) Define heat generated by the model during analysis
C) Apply a pressure boundary condition
D) Set electrostatic potential

Q22. The Electrostatic Potential Constraint tool is used to:

A) Define the electrical potential (voltage) at selected faces/edges/vertices
B) Specify the current density for the model
C) Set the magnetization value
D) Define fluid boundary conditions

Q23. What is the function of the Beam cross-section tool in FreeCAD?

A) To define boundary conditions for fluid analysis
B) To create beam elements with various cross-sections
C) To apply thermal constraints to the model
D) To set electromagnetic boundary conditions

Q24. What is the first step in creating a Beam cross-section element in FreeCAD?

A) Specify the material type
B) Select the desired cross-section type
C) Define the boundary conditions for the beam
D) Select the faces to apply the constraint

Q25. What is the purpose of the Beam Rotation tool in the software?

A) To define the thickness of shell elements
B) To change the default orientation of beam elements
C) To create mesh for fluid flow
D) To generate 3D solid elements

Q26. Which tool is used to define the thickness of shell elements based on a selected face?

A) Beam Rotation tool
B) Shell Plate Thickness tool
C) Fluid Section for 1D Flow tool
D) Mesh Boundary Layer tool

Q27. What does meshing do in Finite Element Analysis (FEA)?

A) It simulates fluid flow
B) It divides the model into smaller elements for analysis
C) It rotates beam elements
D) It creates mesh for the shell thickness

Q28. When using the FEM mesh from shape by Netgen tool, which parameter controls the smoothness of mesh elements at edges?

A) Max. Size
B) Fineness
C) Tet Parameter
D) Growth Rate

Q29. Which tool allows you to create a mesh with different element sizes at the boundary of the model?

A) FEM Mesh Group
B) FEM Mesh Region Refinement
C) FEM Mesh Boundary Layer
D) FEM Mesh from Shape by Gmsh

Q30. What is the purpose of the FEM Mesh Region Refinement tool?

A) To create boundary mesh layers
B) To refine mesh for a specific region of the model
C) To create mesh groups for export
D) To define fluid sections in a model

Q31. What is the role of the FEM mesh group tool?

A) To refine mesh elements in complex regions
B) To create a group of mesh elements for export to other applications
C) To apply boundary layers to the mesh
D) To define shell plate thickness

Q32. How can you convert the FEM mesh to a mesh model in FreeCAD?

A) By using the FEM mesh to model tool
B) By clicking the FEM mesh to mesh tool in the Model menu
C) By selecting the mesh and pressing the delete key
D) By using the Create Mesh from Model tool

Q33. Which solver in FreeCAD uses the Abaqus input format and is capable of solving linear and nonlinear calculations?

A) Elmer Solver
B) Z88 Solver
C) Mystran Solver

D) CalculiX Standard Solver

Q34. In the CalculiX Standard Solver, what does clicking the "Write .inp file" button do?

A) Creates an input file for the solver with specified parameters
B) Runs the solver immediately
C) Deletes the input file
D) Converts the mesh into a model

Q35. Which solver requires two components, ElmerGrid and ElmerSolver, to be installed and linked in FreeCAD?

A) Z88 Solver
B) CalculiX Solver
C) Elmer Solver
D) Mystran Solver

Q36. What is the purpose of the "Purge Results" tool in FreeCAD FEM workbench?

A) To delete the results generated by the solver
B) To run the solver calculations
C) To preview the results of the analysis
D) To create a new analysis

Q37. In the Elmer Solver, which equation type would you select to perform a linear mechanical analysis?

A) Elasticity equation
B) Deformation equation
C) Electrostatic equation
D) Magnetodynamic equation

Q38. Which solver in FreeCAD is based on a modular structure and allows custom extensions and special case elements?

A) Elmer Solver
B) Mystran Solver
C) Z88 Solver
D) CalculiX Solver

Q39. What should you do before using the Elmer solver in FreeCAD?

A) Install the ElmerSolver component
B) Install the Gmsh Mesher tool
C) Install the Elmer solver and link both ElmerGrid and ElmerSolver in preferences
D) Create the mesh from the Shape tool

Q40. When using the Post Pipeline tool for results, which display option can be selected to generate a wireframe model of results?

A) Surface with Edges
B) Wireframe
C) Surface
D) Nodes

For Student Notes

Chapter 10

Tech Drawing

Topics Covered

The major topics covered in this chapter are:

- *Introduction to TechDraw Workbench*
- *Starting New Drawing Page*
- *Inserting Views*
- *Creating Section Views and Detail Views*
- *Applying Annotations*
- *Clip Groups*
- *Linking Dimensions of 3D Geometry*
- *Exporting Drawing Pages*

INTRODUCTION TO TECHDRAW WORKBENCH

The **TechDraw** workbench of FreeCAD is used to create technical drawings from the model earlier created in the software using other workbenches. To start creating technical drawings, select the **TechDraw** option from the **Switch between workbenches** drop-down in the **Toolbar**; refer to Figure-1. On selecting the TechDraw workbench, tools will be displayed similar to Figure-2.

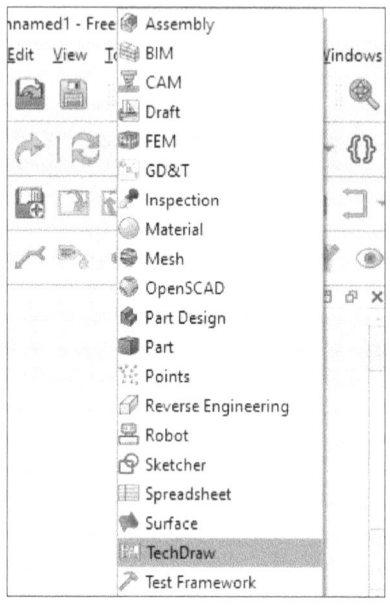

Figure-1. TechDraw workbench option

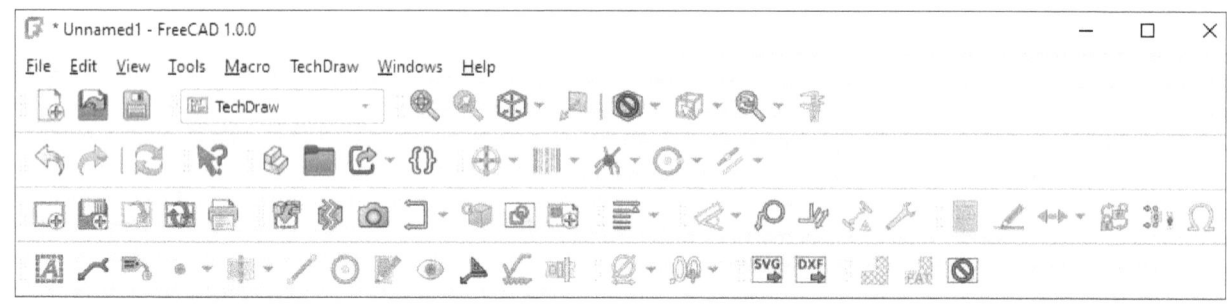

Figure-2. TechDraw toolbar

Various tools of this workbench are discussed next.

STARTING A NEW DRAWING PAGE

The **Insert Default Page** tool is used to generate a drawing page using default drawing template. The procedure to use this tool is given next.

- Click on the **Insert Default Page** tool from the **Toolbar** in **TechDraw** workbench; refer to Figure-3. A new drawing page will be created and displayed in the interface; refer to Figure-4.

Figure-3. Insert default page tool

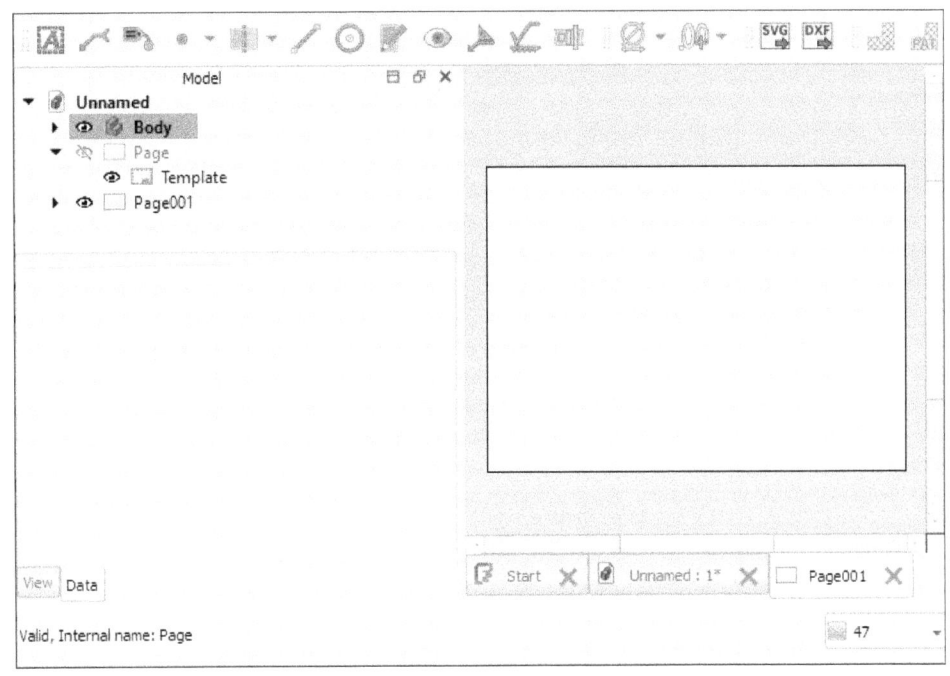

Figure-4. Drawing page generated

- Select the **Page** feature from the **Model Tree**. The properties of page will be displayed as shown in Figure-5.

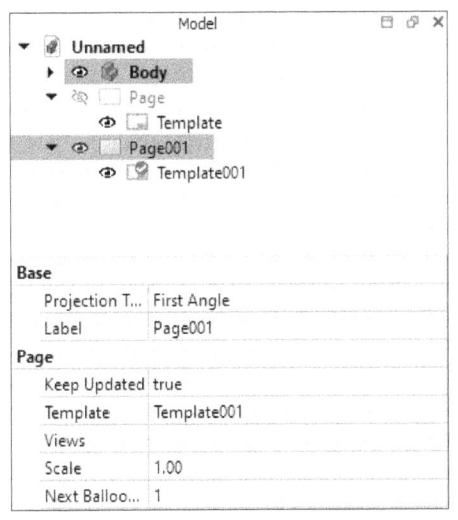

Figure-5. Properties for page feature

- Select desired option from the **Projection Type** field in **Property Editor** dialog. Select the **First Angle** option from the field if you want to use First Angle projection for generating views. Select the **Third Angle** option from the field if you want to use Third Angle projection for generating views.

First Angle Projection and Third Angle Projection
Figure-6 shows an object with different view directions say, a, b, c, d, e, and f.

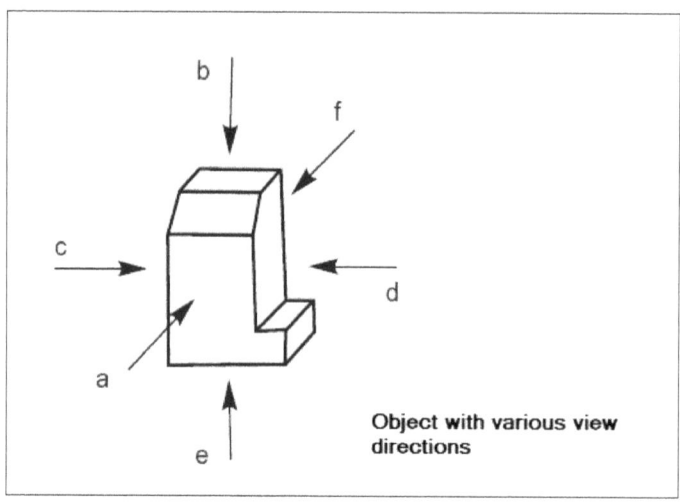

Figure-6. Object with view directions

Here,
1. View in the direction a = view from the front
2. View in the direction b = view from top
3. View in the direction c = view from the left
4. View in the direction d = view from the right
5. View in the direction e = view from bottom
6. View in the direction f = view from the back

In First Angle projection, these views are arranged as shown in Figure-7. In Third Angle projection, these views are arranged as shown in Figure-8.

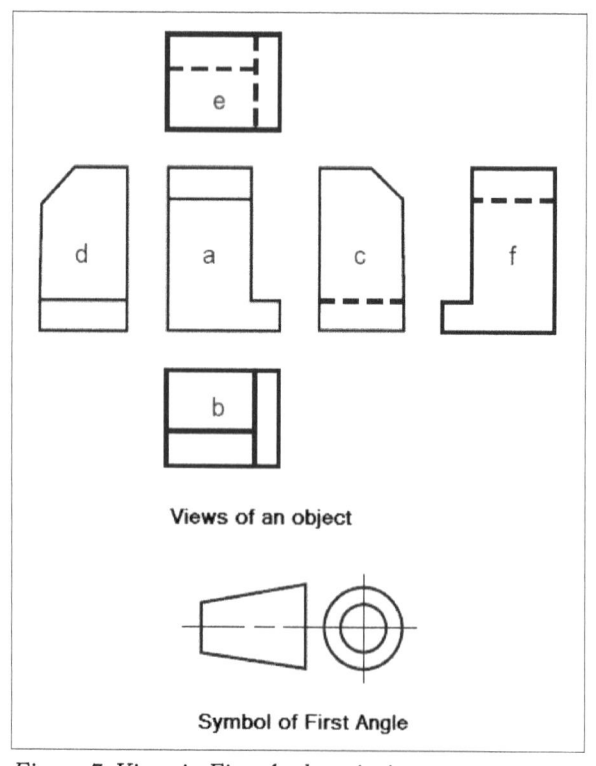

Figure-7. Views in First Angle projection

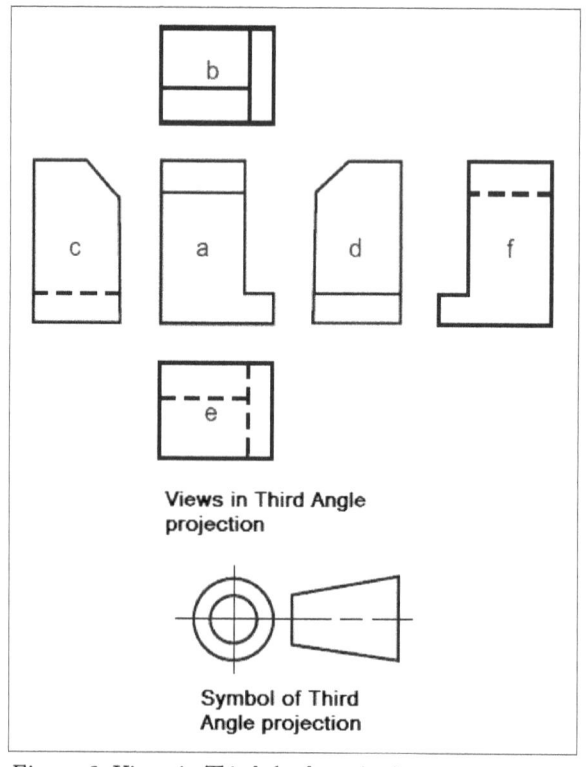

Figure-8. Views in Third Angle projection

- Specify desired name of page in the **Label** field of **Property Editor**.

- If you want to change template of drawing page and click on the **Template** link button from **Template** field of **Property Editor**. The **Template** feature will be selected in **Property Editor** and related options will be displayed; refer to Figure-9.

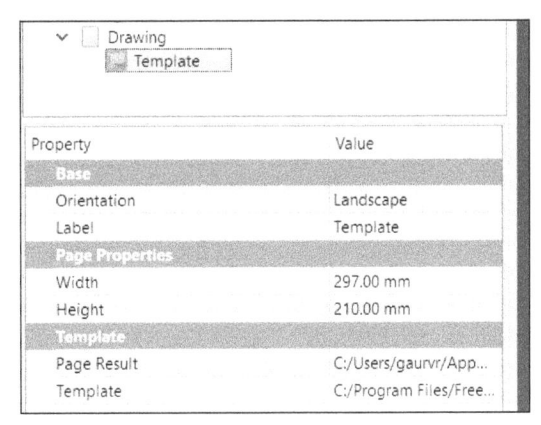

Figure-9. Properties of template

- Click on the **Browse** button in **Template** field of **Property Editor** and select desired template file.
- You can specify the scale of drawing views in the **Scale** field of **Property Editor** for **Page**.

INSERTING NEW PAGE OF SELECTED TEMPLATE

The **Insert Page using Template** tool is used to create new page using selected template. A template includes various features like size of page, orientation of page, scale of views, and so on. The procedure to use this tool is given next.

- Click on the **Insert Page using Template** tool from the **Toolbar** or **TechDraw** menu of **TechDraw** workbench; refer to Figure-10. The **Select a Template File** dialog box will be displayed; refer to Figure-11.

Figure-10. Insert page using template tool

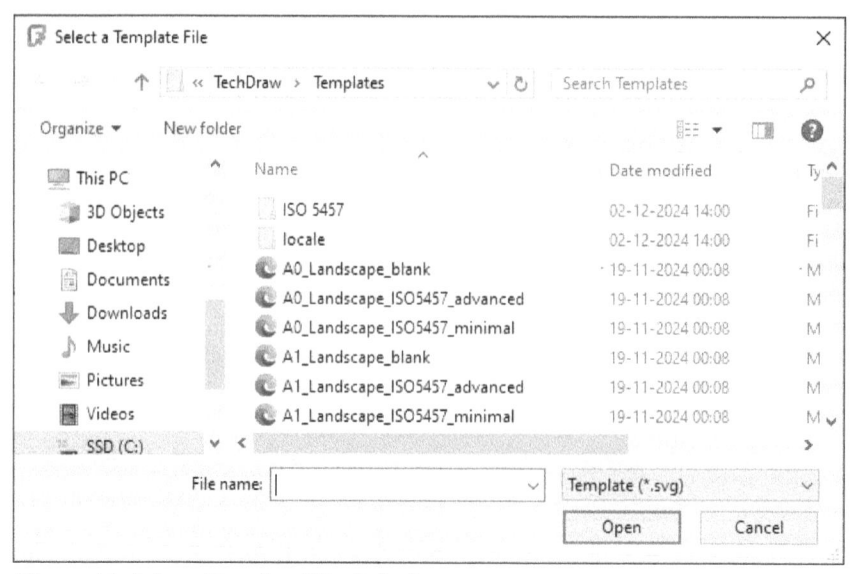

Figure-11. Select a Template File dialog box

- Select desired template from the dialog box and click on the **Open** button. The page will be created and displayed in the application; refer to Figure-12.

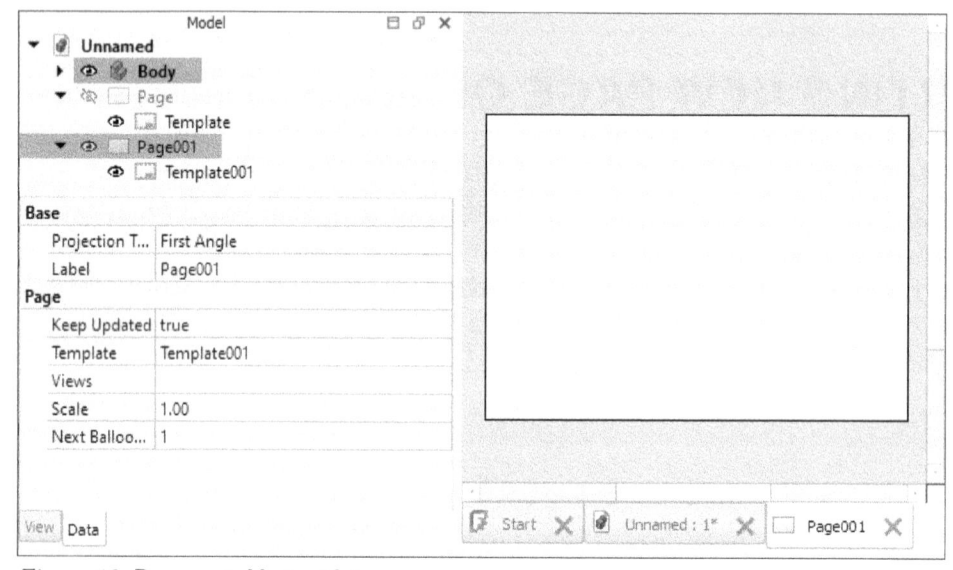

Figure-12. Page created by template

UPDATE TEMPLATES FIELDS

The **Update templates fields** tool fills the editable text fields in the title block of the current page. The procedure to use this tool is discussed next.

- Select desired page from the **Model Tree** which you want to update and click on the **Update templates fields** tool from the **Toolbar** or **TechDraw** menu in the **TechDraw** workbench; refer to Figure-13.

Figure-13. Update templates fields tool

FreeCAD 1.0 Black Book   10-7

REDRAW PAGE

The **Redraw Page** tool forces an update of the selected page. The procedure to use this tool is discussed next.

- Select desired page from the **Model Tree** which you want to update and click on the **Redraw Page** tool from the **Toolbar** or **TechDraw** menu in the **TechDraw** workbench; refer to Figure-13. The selected page will be updated.

Figure-14. Redraw page tool

Print All Pages

The **Print All Pages** tool is used to send all pages in a document to the printer. The procedure to use this tool is discussed next.

- Click on the **Print All Pages** tool from the **Toolbar** or **TechDraw** menu in the **TechDraw** workbench. The **Print** dialog box will be displayed; refer to Figure-15.

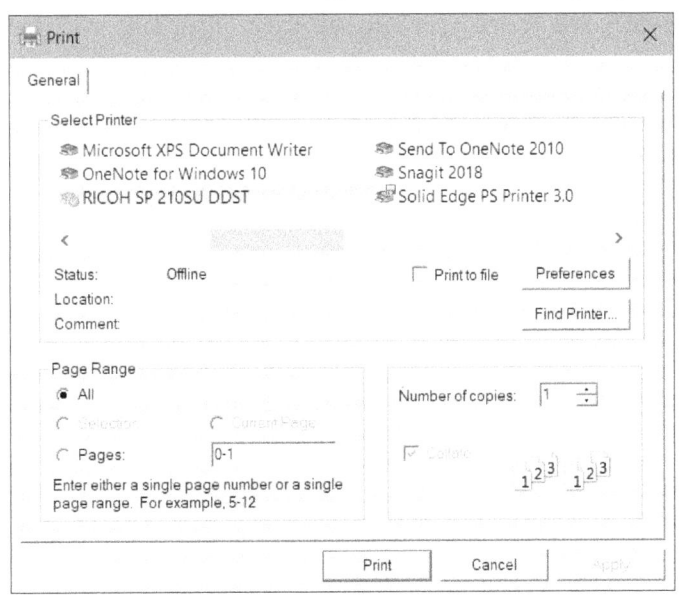

Figure-15. Print dialog box

- Specify desired parameters in the dialog box to print the page and click on the **Print** button. The **Save Print Output As** dialog box will be displayed; refer to Figure-16.

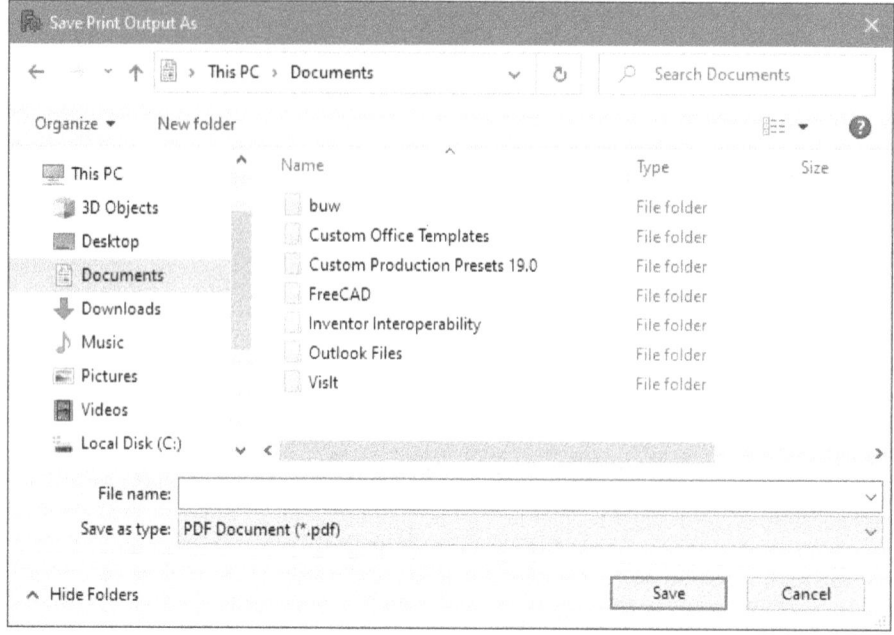

Figure-16. Save Print Output As dialog box

- Specify desired directory to save the file and specify desired name of file in the **File name** edit box.
- Click on the **Save** button from the dialog box. The file will be saved.

INSERTING VIEW IN PAGE

The **Insert View** tool is used to insert current view of model in the page. The procedure to use this tool is given next.

- Select the **Model** tab from the application menu and orient the model to desired view to be placed in the drawing; refer to Figure-17.

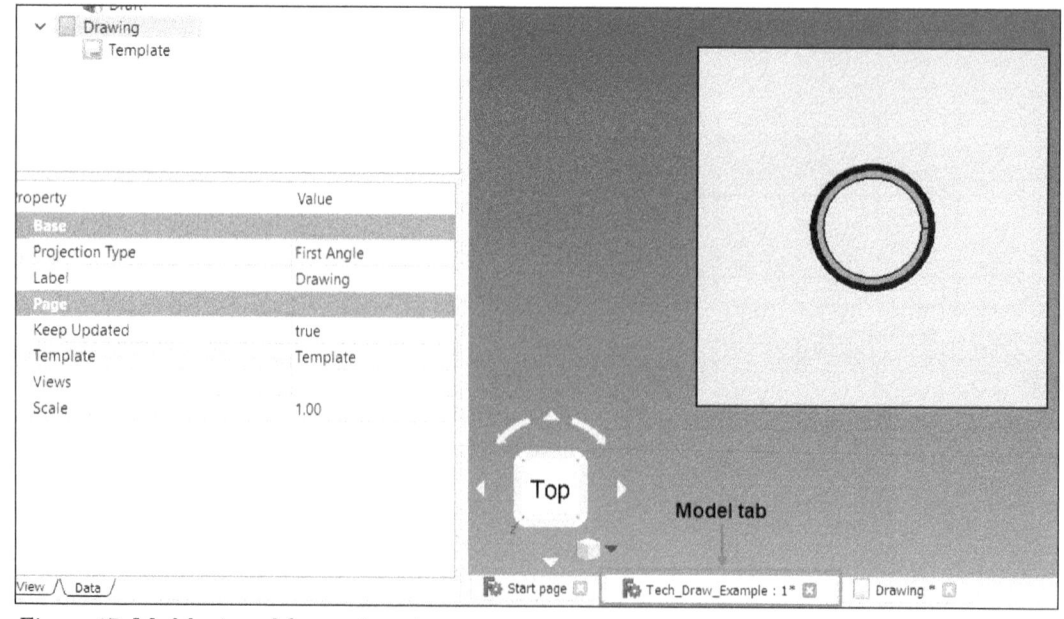

Figure-17. Model oriented for creating view

- Now, select the **Page** tab from the application window, select the model whose views are to be placed from **Model Tree**, and click on the **Insert View** tool from the **Toolbar** or **TechDraw** menu; refer to Figure-18. The current view of model will be placed in the page; refer to Figure-19. Note that the view generated has same orientation as defined in Figure-17.

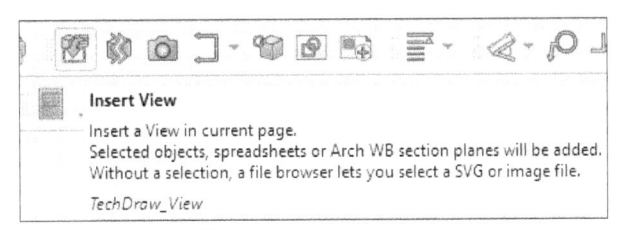

Figure-18. Insert View tool

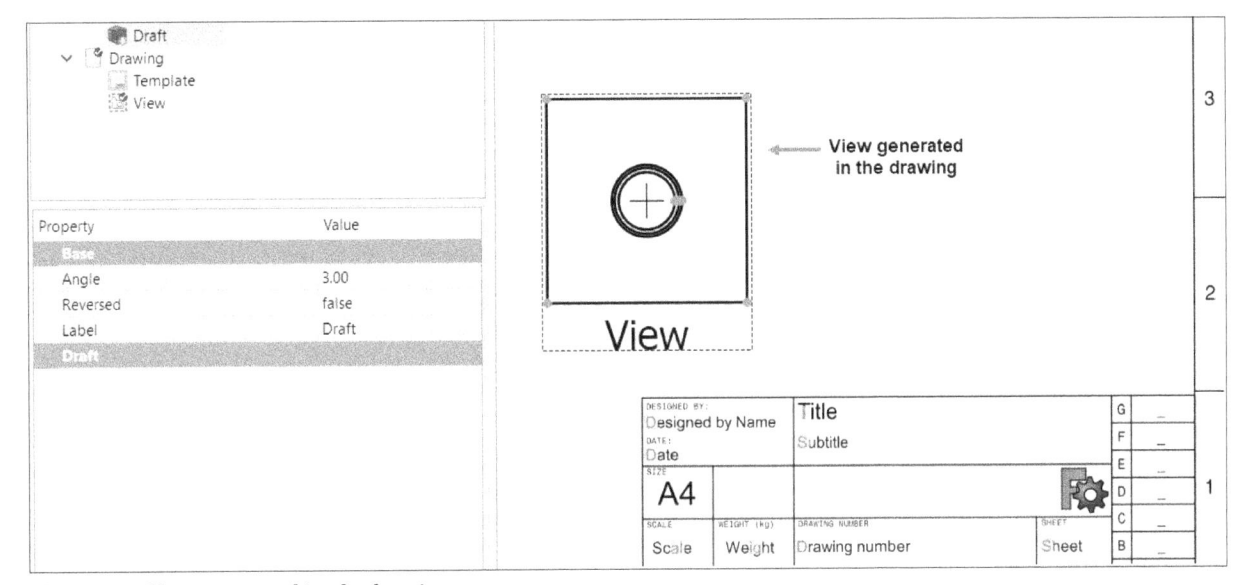

Figure-19. View generated in the drawing

- You can move this view by selecting the dotted boundary and dragging it to desired location.
- On selecting the view from **Model Tree**, various properties of the view are displayed in **Property Editor**; refer to Figure-20.

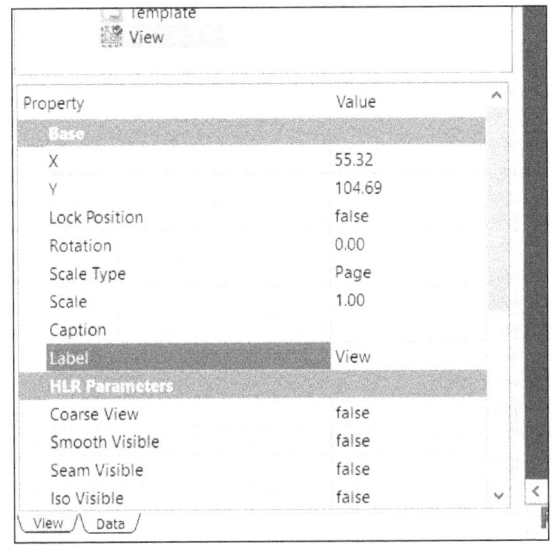

Figure-20. Properties of view

10-10   FreeCAD 1.0 Black Book

- Specify desired values in **X** and **Y** edit boxes to define location of view on page.
- Select **True** option for **Lock Position** field if you want to lock the position of view on page and do not want it to move.
- Specify desired value in **Rotation** field to rotate selected view and click on the **Recompute** button. The rotated view will be generated.
- Set desired (**True** or **False**) value for various display objects in **HLR Parameters** section of **Property Editor**.
- By default, orthographic views are generated from the model but if you want to generate perspective view of model then select the true option from **Perspective** field in the **Projection** section. You can also specify the Focus distance for perspective view in the **Focus** field.

INSERT BROKEN VIEW

The **Insert Broken View** tool inserts a broken view for very long parts that do not fit in drawing view area. The procedure to use this tool is discussed next.

- Select the model from the application menu and orient the model to desired view to be placed in the drawing; refer to Figure-21.

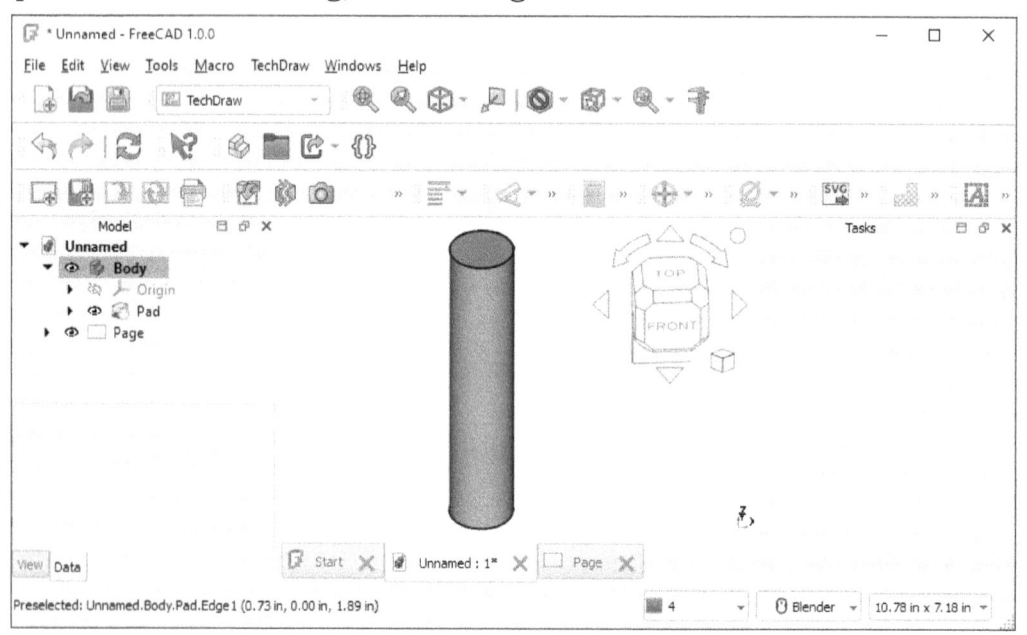

Figure-21. Creating the model in broken view

- Select the **Sketcher** workbenches from drop-down and click on the **Create Sketch** button from the toolbar.
- Select Front plane of the drawing area, click on the **Line** tool from the toolbar and create horizontal lines to the model; refer to Figure-22.

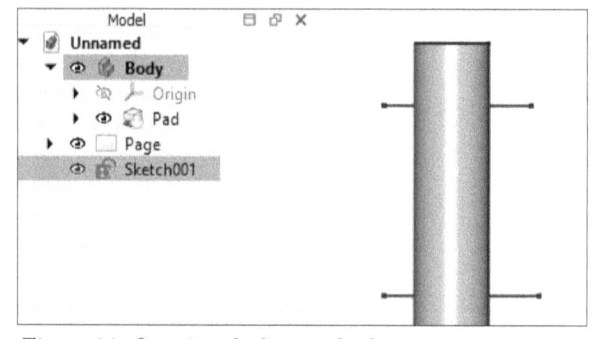

Figure-22. Creating the lines in broken view

- Now, select the **TechDraw** workbench and click on the **Insert Default Page** from the toolbar. Select the body from the **Model** panel in **Combo View** and then click on the **Insert view** tool. The front view will be displayed; refer to Figure-23.

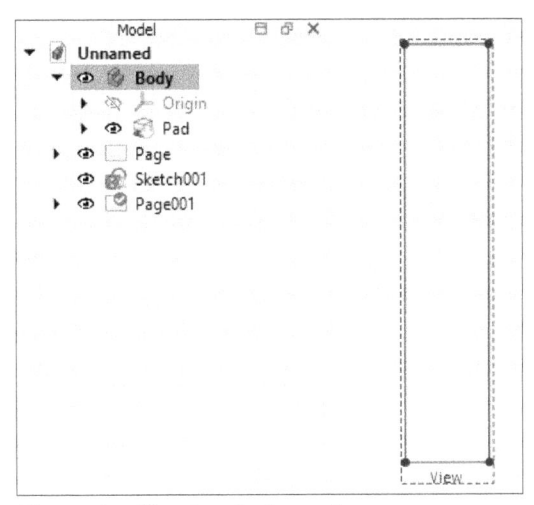

Figure-23. Showing the front view

- After inserting the view, select the body and sketch of the **Model** panel as shown in Figure-24, and then click on the **Broken view** tool from the toolbar. The broken view will be displayed; refer to Figure-24.
- You can edit the broken view by double-clicking on the **BrokenView** feature from the **Model** panel in **Combo View**. The **Part View** dialog will be displayed; refer to Figure-25.

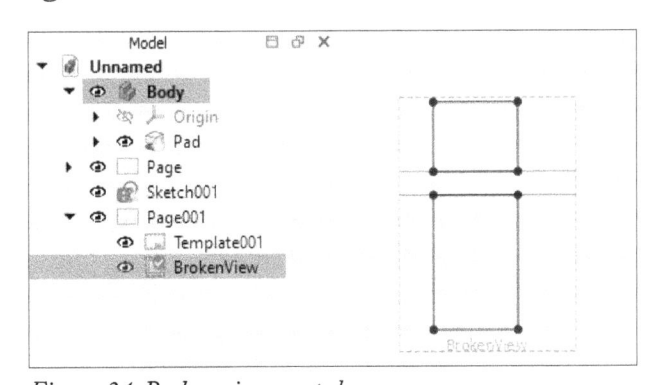

Figure-24. Broken view created

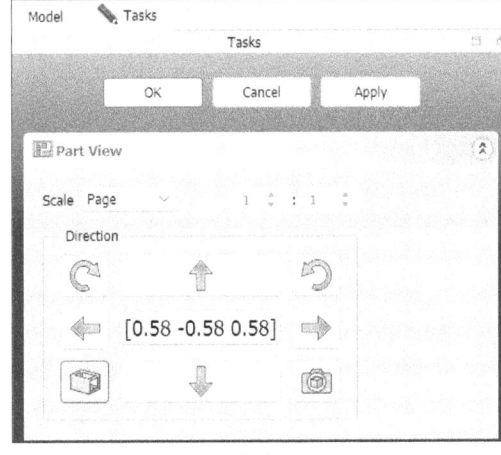

Figure-25. Part View dialog

- Click on the **Rotate up**, **Rotate down**, **Rotate right**, and **Rotate left** buttons to rotate the model in view by 90 degrees in respective orientation.
- Click on the **Spin clock wise** and **Spin counter clock wise** buttons to rotate the view in current view plane by 90 degrees.
- Select desired face of model from view and click on the button to orient the view along selected faces of model.
- Click on the button to orient the model view along default front plane of the model.
- Click on the **Current primary view direction** button to modify the view direction for the broken view. An input box will be displayed as shown in Figure-26. Set desired parameters in the input box to define the camera direction for current view and click on the **OK** button from the input box to apply the changes.

Figure-26. Direction input box

- Select desired option from the **Scale** drop-down to define scale value for the view. Select the **Page** option to decide scale value based on page parameters. Select the **Automatic** option to let software decide the scale value based on size of objects and views. Select the **Custom** option from the drop-down if you want to manually define the scale value. In this case, **Scale Numerator** and **Scale Denominator** spinners will be displayed. Set higher value in **Scale Numerator** as compared to **Scale Denominator** to increase model view size and set lower value in **Scale Numerator** to decrease the model view size.
- After specifying desired parameters click on the **OK** button from the dialog box.

INSERTING ACTIVE VIEW IN PAGE

The **Insert Active View (3D View)** tool inserts a copy of a 3D window into a drawing page. The procedure to use this tool is discussed next.

- Select the tab of model from the application menu and orient the model to desired view to be placed in the drawing; refer to Figure-27.

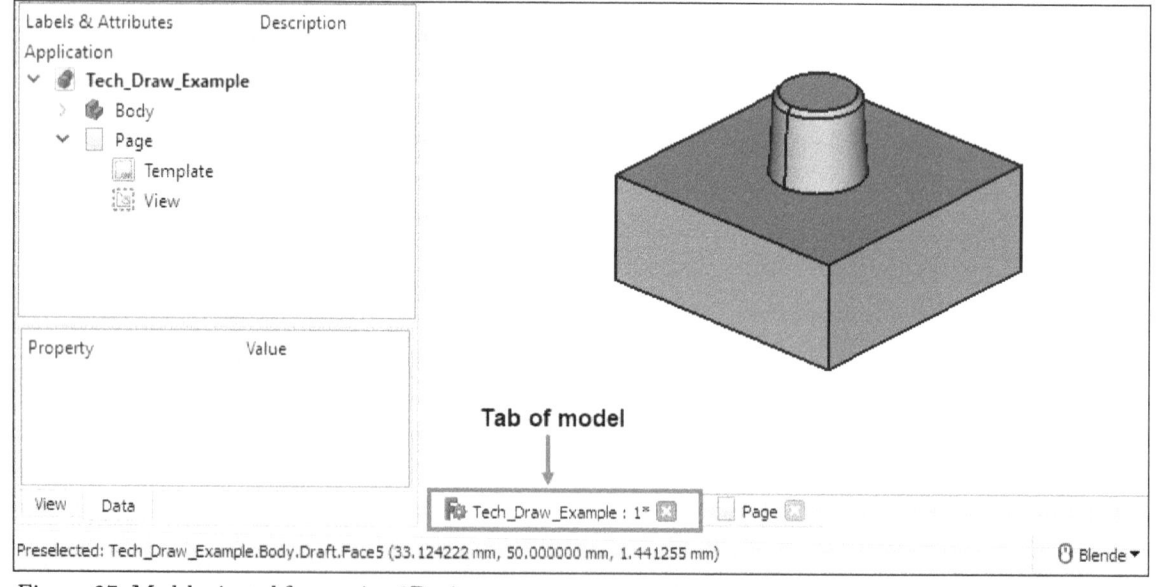

Figure-27. Model oriented for creating 3D view

- Click on the **Insert Active View (3D View)** tool from **Toolbar** or **TechDraw** menu in the **TechDraw** workbench; refer to Figure-28. The **ActiveView to TD View** dialog will be displayed in the **Tasks** panel of **Combo View**; refer to Figure-29.

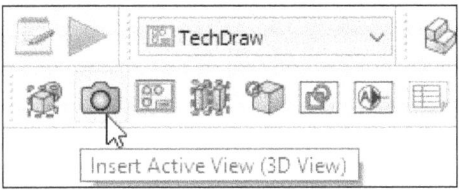

Figure-28. Insert active view 3D view tool

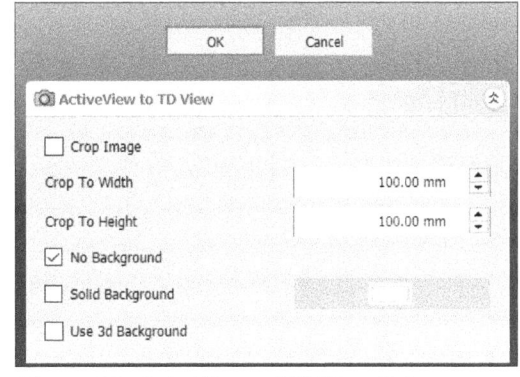

Figure-29. ActiveView to TD View dialog

- Specify desired width and height of generated view in the **Width** and **Height** edit boxes of the dialog.
- Enter desired value in the **Border** edit box to specify minimal distance of the object from the top and left view border.
- Select the **Background** check box and click on the **Background color** button to specify the color for the background.
- Enter desired value in **Line Width** edit box to specify desired thickness of individual lines in the generated view.
- Select desired drawing style from **Render Mode** drop-down.
- After specifying desired parameters in the dialog, click on the **OK** button. The active 3D view will be created and displayed in the **Page** tab; refer to Figure-30.

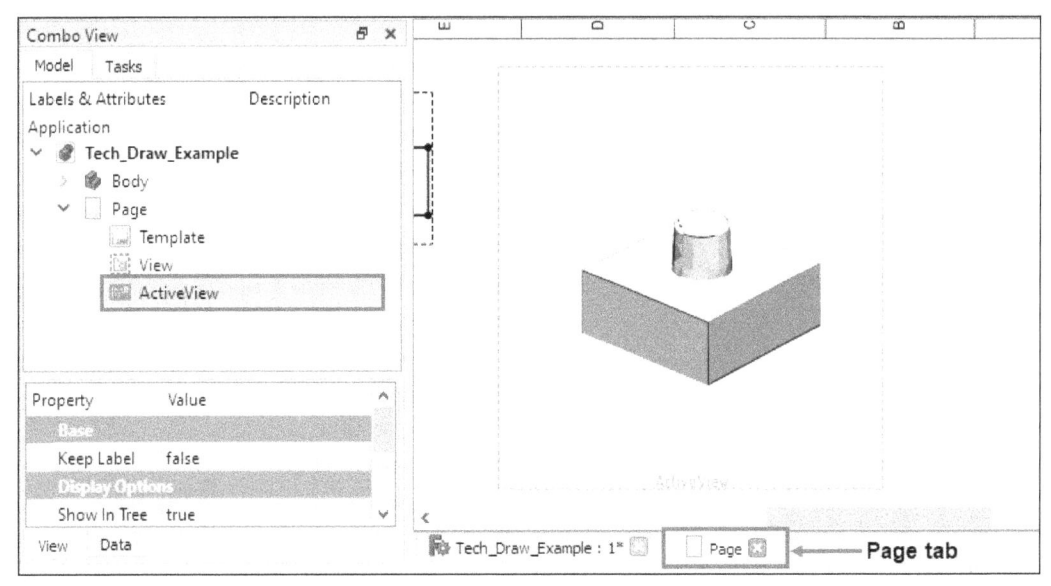

Figure-30. Active 3D view created

INSERTING PROJECTION GROUP

The **Insert projection group** tool is used to create multiple views (like front, left, and top views) of selected model object. The procedure to use this tool is given next.

- Click on the **Insert Projection Group** tool from the **Toolbar** or from the **TechDraw** menu after selecting the model from **Model Tree**; refer to Figure-31. The **Projection Group** dialog will be displayed with preview of base view; refer to Figure-32.

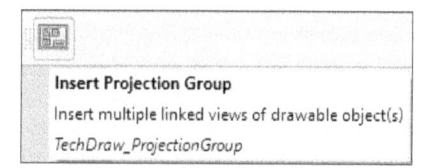

Figure-31. Insert multiple linked views tool

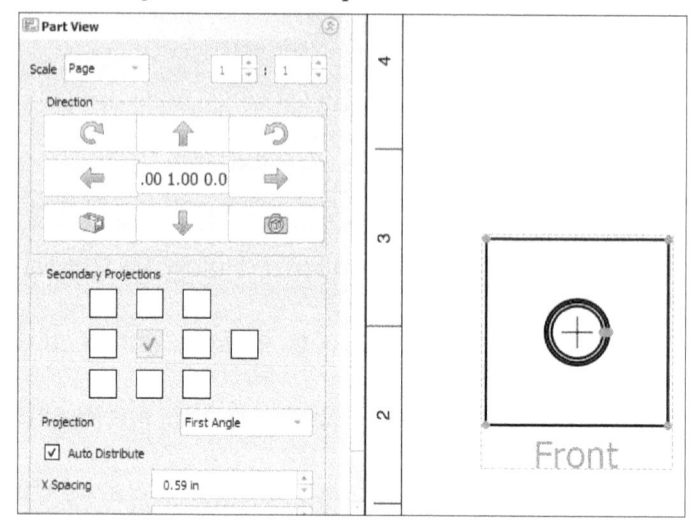

Figure-32. Projection Group dialog

- Select check boxes for views to be generated. Preview of views will be displayed accordingly in the drawing area; refer to Figure-33.

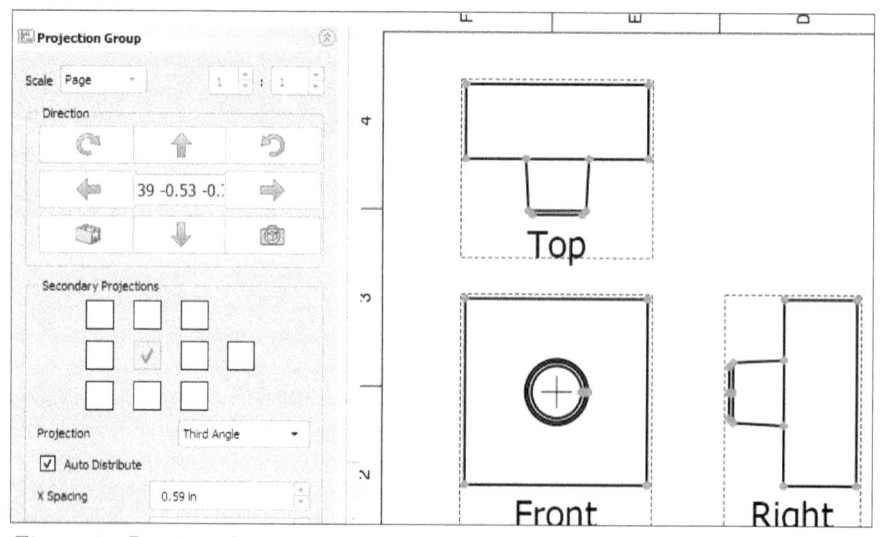

Figure-33. Preview of views

- Select desired option from the **Scale** drop-down to define whether you want to use Automatic scaling, Scale parameter set for page, or you want to specify custom scale values in the **Custom Scale** edit boxes. The views will be generated as per the scale option selected.
- Select desired option from the **Projection** drop-down to specify whether you want to use First Angle projection or Third Angle projection.
- Select the **Auto Distribute** check box to distribute the projections automatically using the given X/Y Spacing.
- Enter desired value in **X Spacing** and **Y Spacing** edit box to specify horizontal distance and vertical distance between border of projections, respectively.
- After setting desired parameters, click on the **OK** button from the dialog to create the views.

INSERT SECTION VIEWS

The **Section View** tool is used to insert section view using selected view in the drawing area. The procedure to use this tool is given next.

- Click on the **Section View** tool from the **Toolbar** or **TechDraw** menu after selecting a view from the drawing area; refer to Figure-34. The **Create Section View** dialog will be displayed; refer to Figure-35.

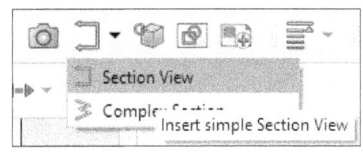

Figure-34. Section View tool

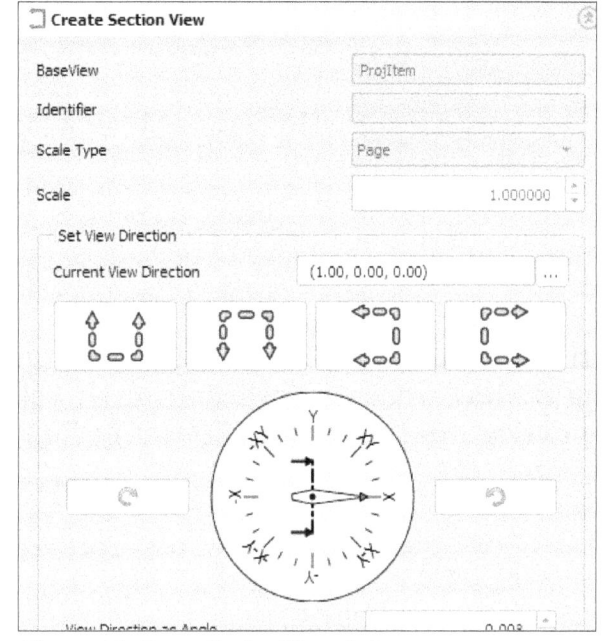

Figure-35. Create Section View dialog

- Select desired button from the dialog to specify the direction in which the section view will be generated. Preview of the section view will be displayed if **Live Update** check box is selected in the **Preview** area of the dialog; refer to Figure-36.
- You can change the section line angle by specifying desired value in the **View Direction as Angle** edit box of the dialog.
- Use the edit boxes in **Section Plane Location** area of dialog to move section line in desired direction.
- Click on the **OK** button from the dialog to create the section view.

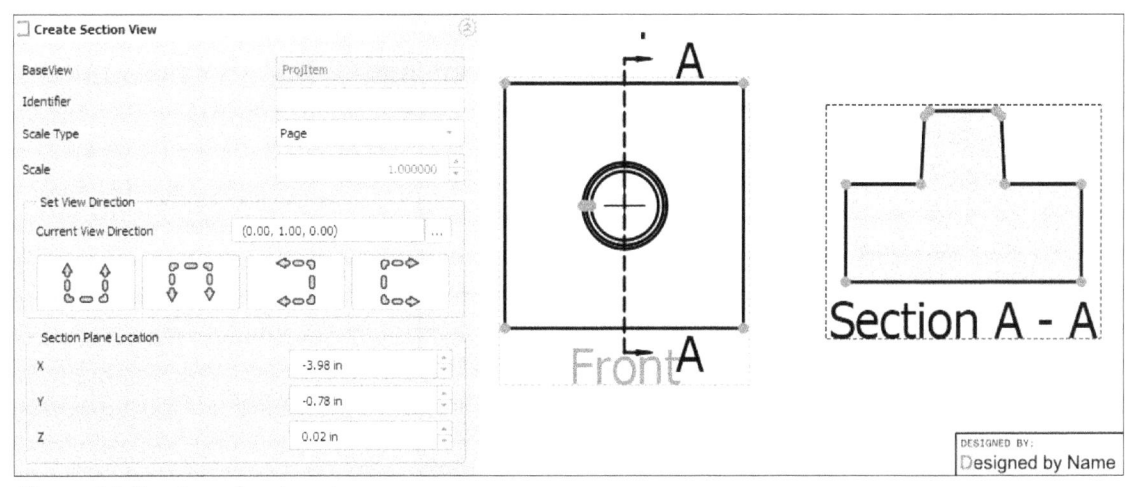

Figure-36. Preview of section view

Insert Complex Section View

The **Complex Section** tool is used to insert a complex cross section view based on selected open profile object (line/arc/multiline created in Draft workbench) and view. The procedure to use this tool is given next.

- Select a profile object passing through model created in draft workbench on the model and view from which cross-section view is to be generated; refer to Figure-37.

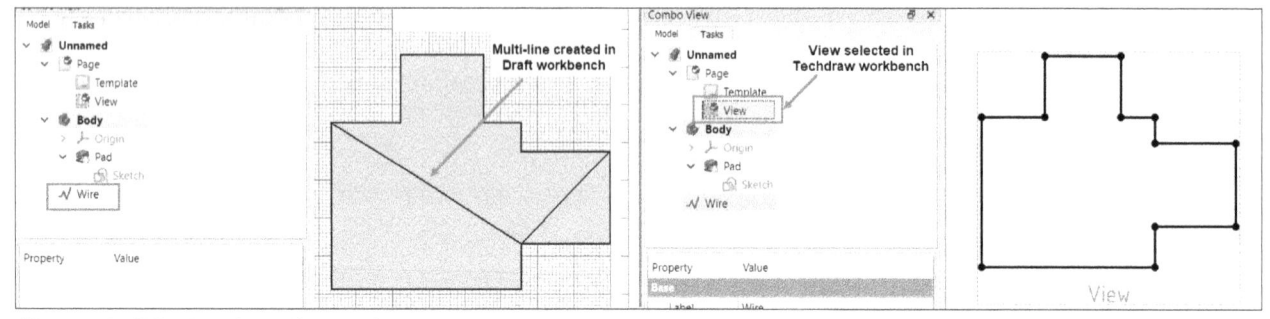

Figure-37. Selecting objects for complex cross-section

- Click on the **Complex Section** tool from the **Toolbar** or **TechDraw** menu after selecting a view from the drawing area; refer to Figure-38. The **New Complex Section** dialog will be displayed; refer to Figure-39

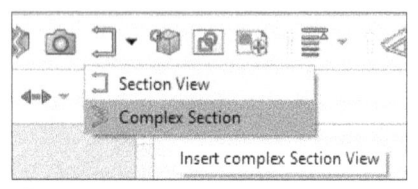

Figure-38. complex section view

Figure-39. New complex section dialog box

- Click on the **OK** button from the dialog without modifying any parameters. The complex cross-section view will be generated; refer to Figure-40.

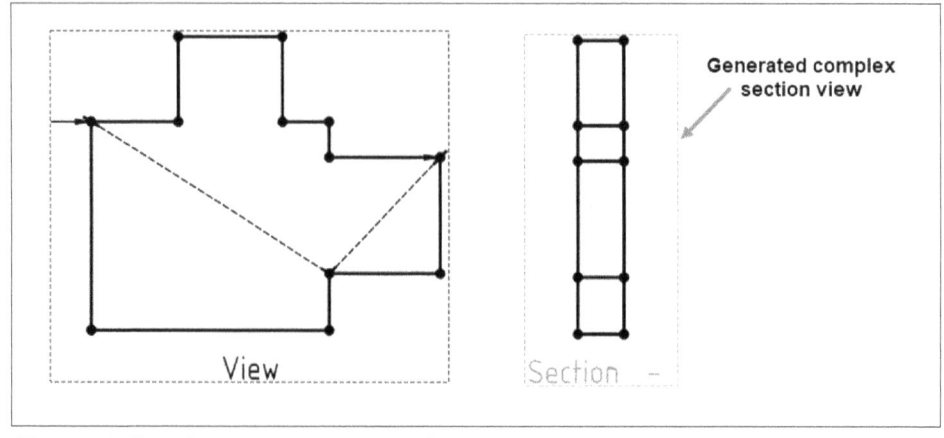

Figure-40. Complex section view generated

INSERTING DETAIL VIEW

The **Insert Detail View** tool is used to create a magnified view of the model for defined region. The procedure to use this tool is given next.

- Click on the **Insert Detail View** tool from the **Toolbar** or **TechDraw** menu after selecting a view; refer to Figure-41. The **New Detail View** dialog will be displayed in the **Tasks** panel of **Combo View** along with the preview of detail view; refer to Figure-42.

Figure-41. Insert Detail View tool

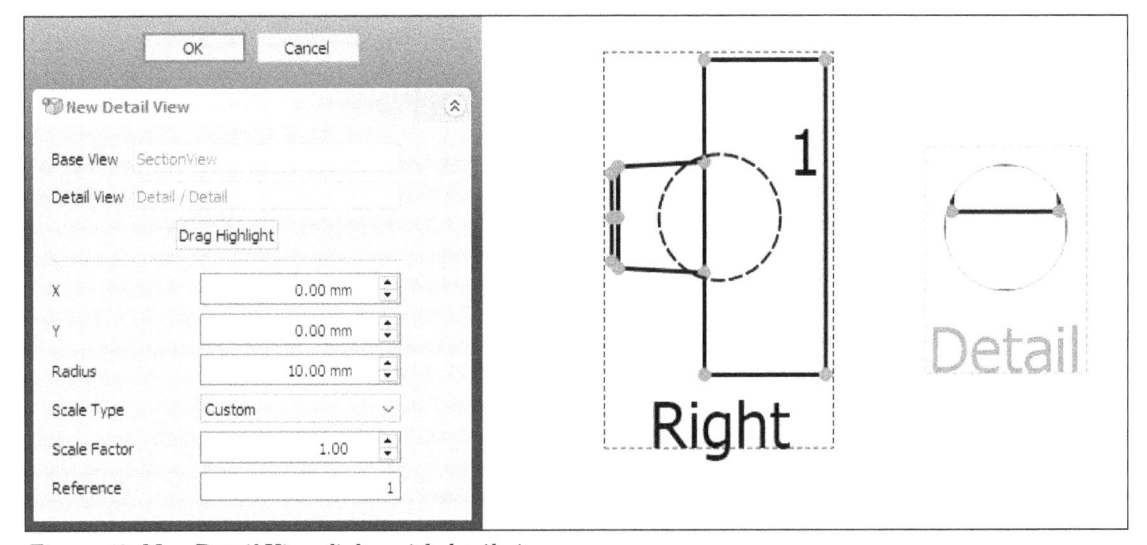

Figure-42. New Detail View dialog with detail view

- The **Base View** box displays the view on which the detail view is based.
- The **Detail View** box displays the view which is being created.
- Click on the **Drag Highlight** button to highlight the detail origin border bold and with the label drag.
- Enter desired value in **X** and **Y** edit boxes to specify the x and y position of detail highlight within view, respectively.
- Specify desired size of detail view in the **Radius** edit box.
- Select desired type of scale from **Scale Type** drop-down. Select **Page** option from the drop-down to use the scale factor of drawing page. On selecting the **Automatic** option, in case, the detail view would be larger than the page, it will be scaled down to fit into the page. Select **Custom** option from the drop-down to set the custom scale factor value in the **Scale Factor** edit box.
- Enter desired value in **Reference** edit box to specify an identifier to indicate the area of the base view that is displayed.
- After specifying desired parameters, click on the **OK** button from the dialog to create the detail view.
- If you want to edit the properties of detail view then select the detail view created from the **Model Tree**, the **Property Editor** dialog will be displayed in the **Model** panel of **Combo View**; refer to Figure-43.

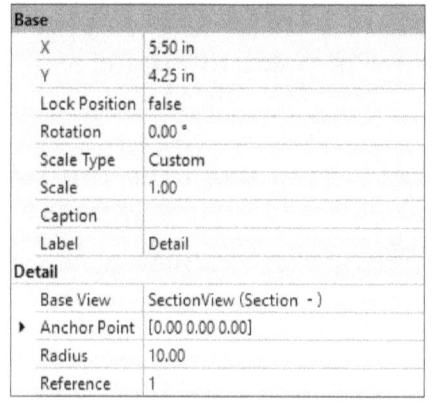

Figure-43. Property editor dialog with detail view parameters

- Specify desired value of scale factor in the **Scale** field of **Property Editor** to increase or decrease the size of detail view.
- Expand the **Anchor Point** field and specify desired location parameters in **x, y**, and **z** fields of **Property Editor** to specify location whose detail view will be generated with respect to center of selected main view; refer to Figure-44.

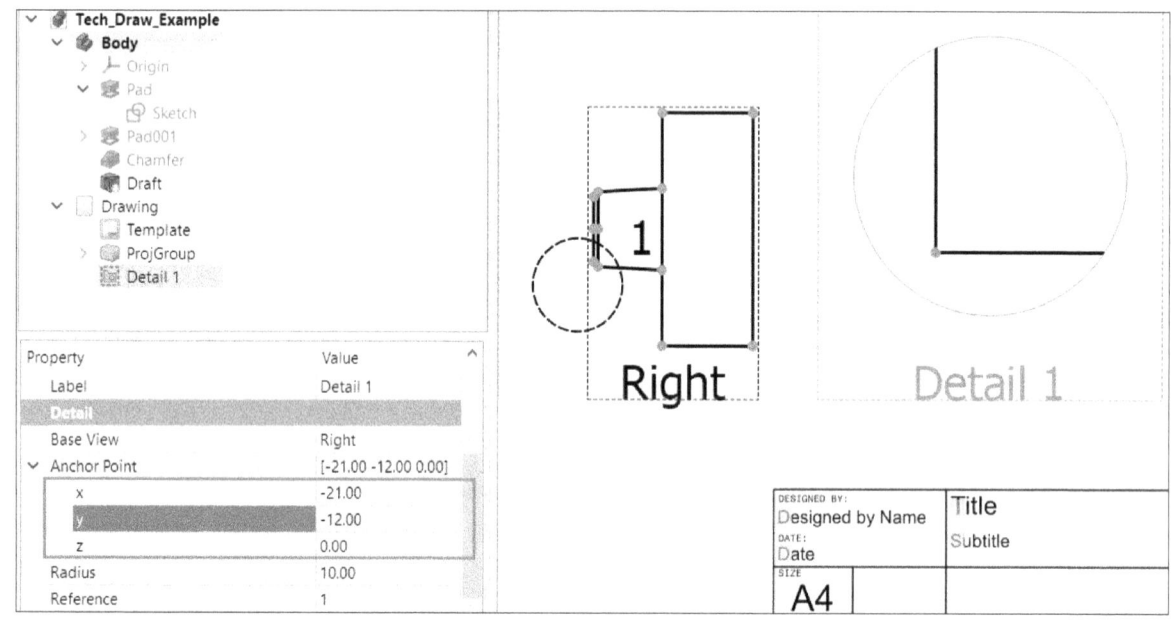

Figure-44. Setting location of detail view circle

- Set the other parameters of **Property Editor** as discussed earlier and click in the empty area of drawing to exit.

INSERTING DRAFT WORKBENCH OBJECT

The **Insert Draft Workbench Object** tool is used to create a view of objects in Draft workbench. The procedure to use this tool is given next.

- Click on the **Insert Draft Workbench Object** tool from the **Toolbar** or **TechDraw** menu after selecting the **Draft** workbench objects to be inserted in the view; refer to Figure-45. The draft views will be generated for selected objects; refer to Figure-46.

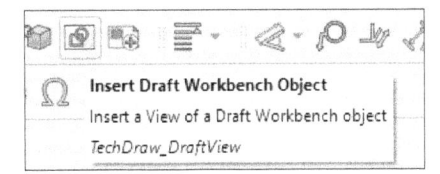

Figure-45. Insert Draft Workbench Object tool

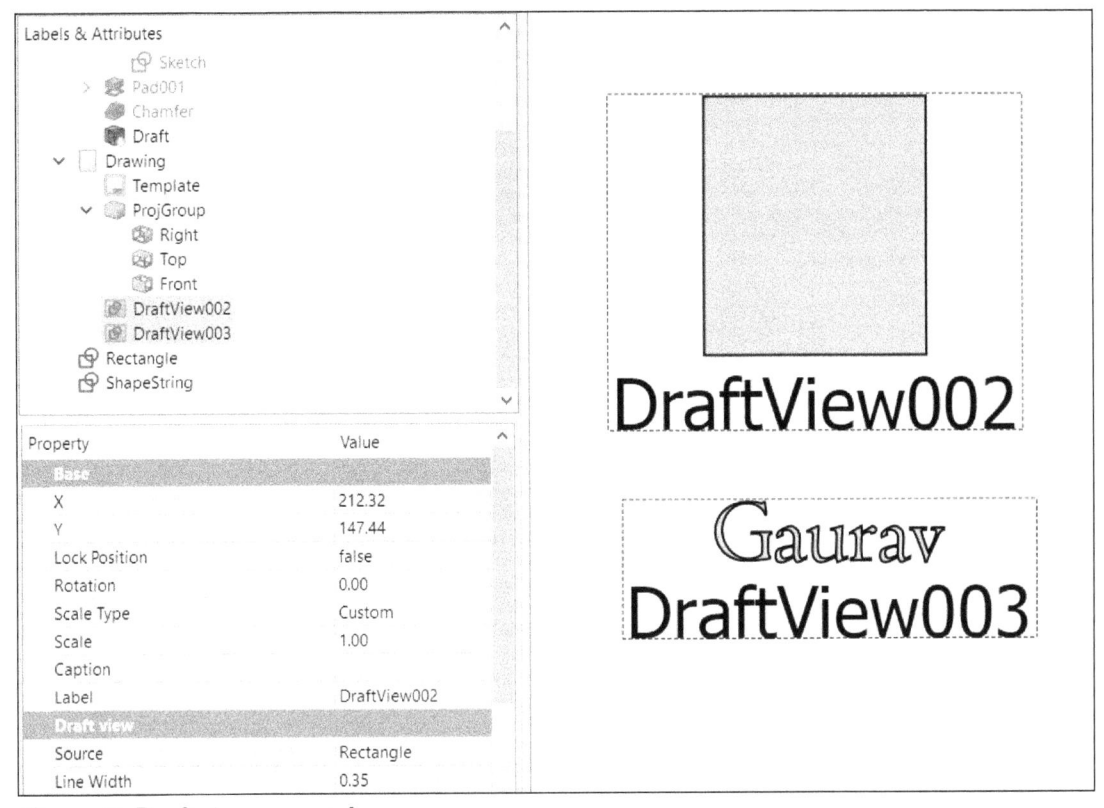

Figure-46. Draft views generated

- Set the parameters as discussed earlier in the **Property Editor** section of the **Combo View**.

Insert Arch Workbench Object

The **Insert Arch Workbench Object** tool is used to insert a section view of model created in Arch workbench.

- Select the section view from the architectural model; refer to Figure-47 and click on the **Insert Arch Workbench Object** tool from the **Toolbar** or **TechDraw** menu; refer to Figure-48. The architectural section view will be generated in the drawing; refer to Figure-49.

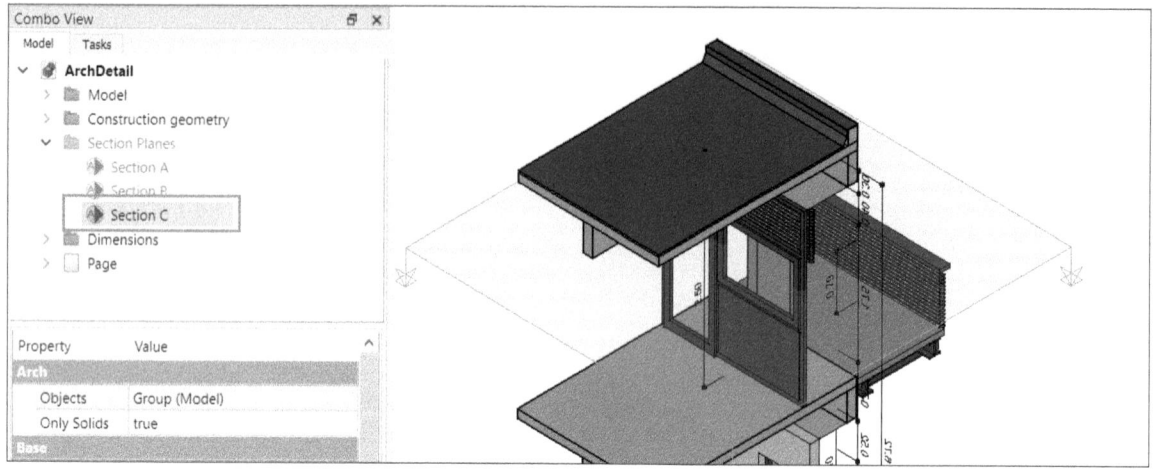

Figure-47. Section view selected

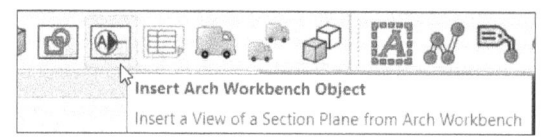

Figure-48. Insert arch workbench object

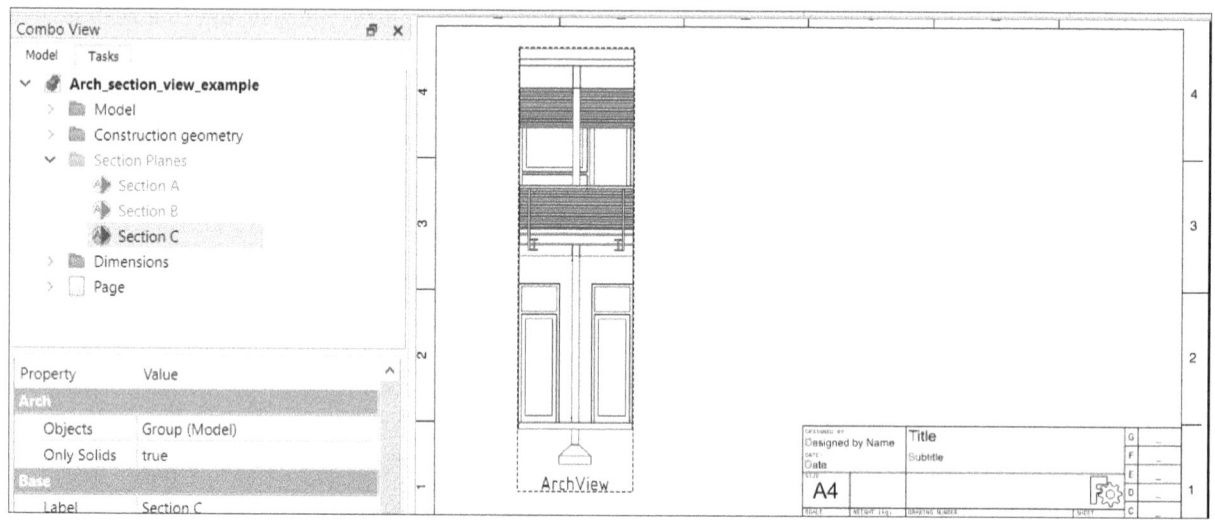

Figure-49. Arch view generated

Insert Spreadsheet View

The **Insert Spreadsheet view** tool is used to place view of selected spreadsheet in the drawing area. The procedure to use this tool is given next.

- Make sure you have already created a spreadsheet to be inserted in the drawing view using the Spreadsheet workbench; refer to Figure-50. Select the spreadsheet to be inserted in the drawing and click on the **Insert Spreadsheet View** tool from the **Toolbar** or **TechDraw** menu in the **TechDraw** workbench; refer to Figure-51. The sheet view will be placed in the drawing; refer to Figure-52.

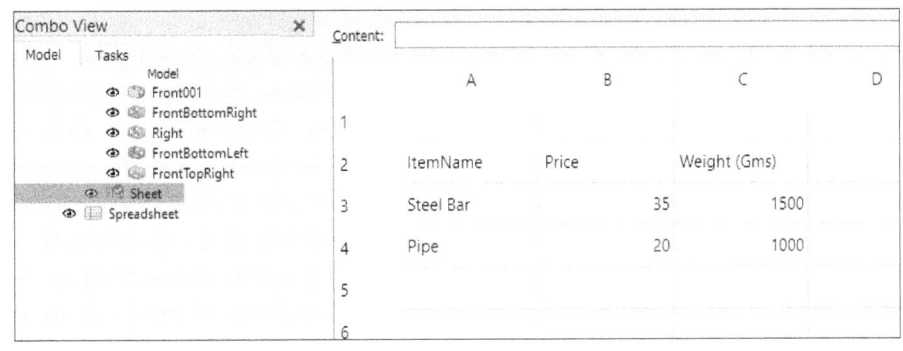

Figure-50. Spreadsheet created

Figure-51. Insert spreadsheet view

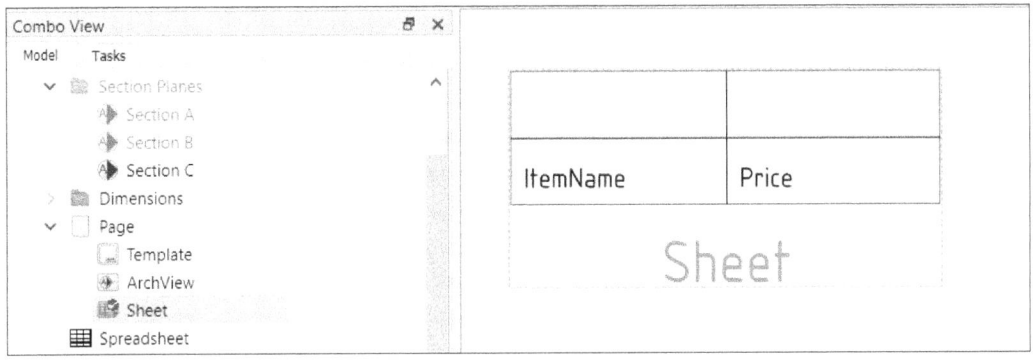

Figure-52. Sheet view placed in drawing

- Select the **Sheet** view from **Combo View** and modify the **Cell Start** & **Cell End** parameters in **Property Editor** to include desired cells in the sheet view; refer to Figure-53.

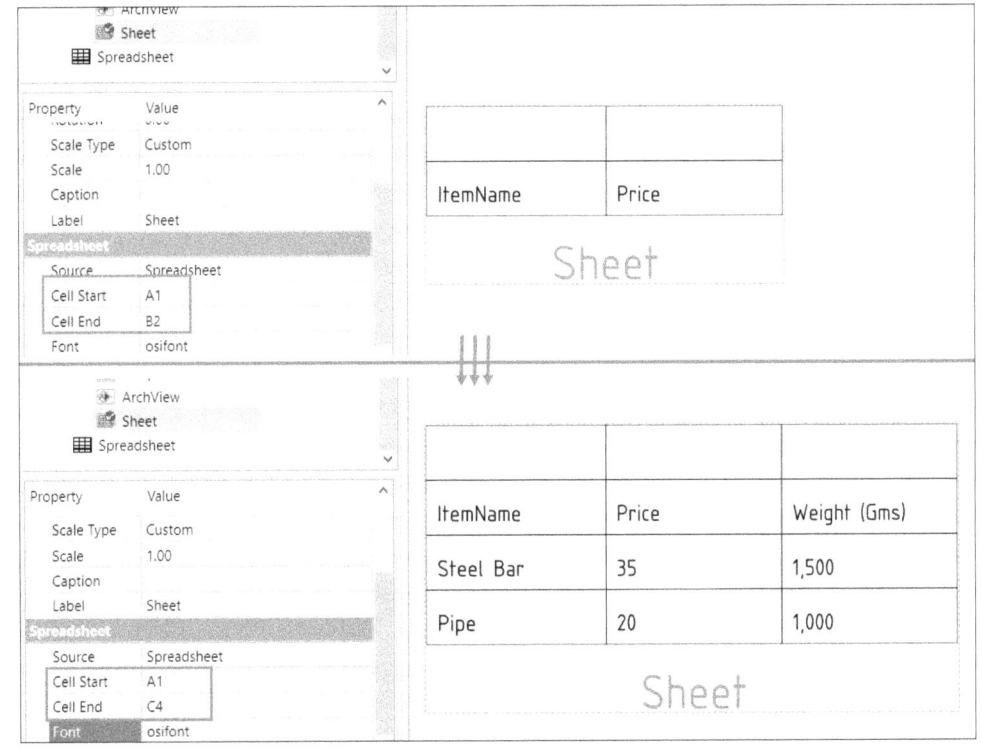

Figure-53. Sheet view after modifying Cell End parameter

MOVING THE VIEW

The **Move View** tool moves a view and all its dependents (Balloons, Dimension, etc.) to a different page. The procedure to use tool is discussed next.

- Click on the **Move View** tool from the **Toolbar** or **Techdraw** menu; refer to Figure-54. The **Move View to a different Page** dialog will be displayed; refer to Figure-55.

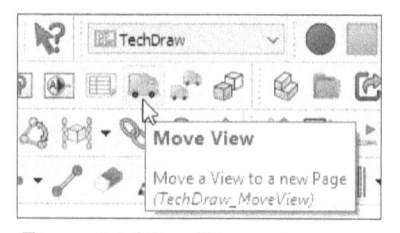

Figure-54. Move View tool

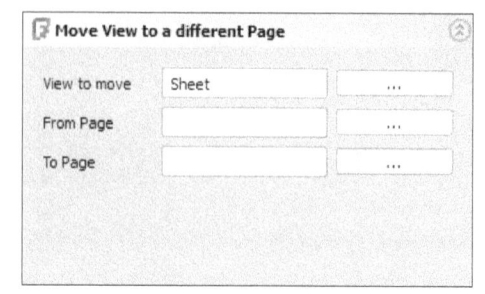

Figure-55. Move View to a different Page dialog

- Click on the button [...] next to **View to move** selection box. The **Select View** dialog box will be displayed; refer to Figure-56.

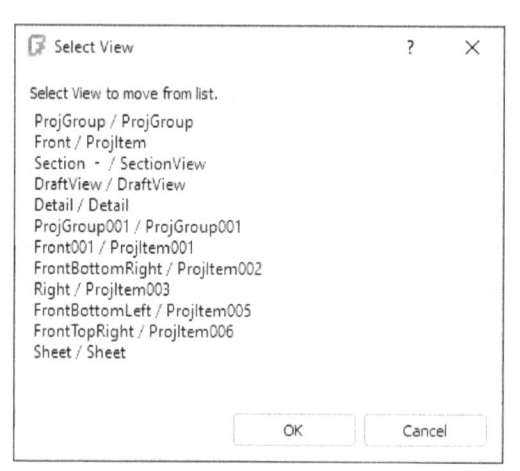

Figure-56. Select View dialog box

- Select desired view from the list box which you want to move to another page; refer to Figure-56 and click on the **OK** button. The selected view will be displayed in the **View to move** selection box of the dialog.
- Click on the [...] button next to **From Page** selection box. The **Select Page** dialog box will be displayed; refer to Figure-57.

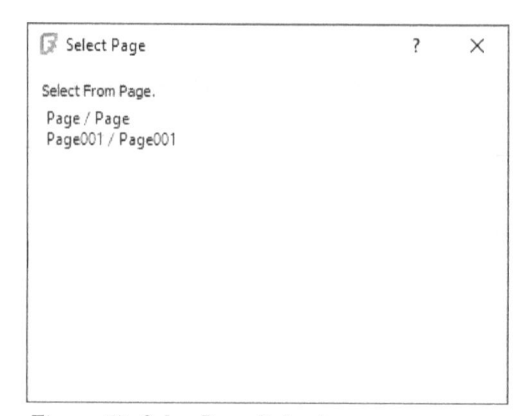

Figure-57. Select Page dialog box

- Select desired page from the list box from which you want to move the view; refer to Figure-57 and click on the **OK** button. The selected page will be displayed in the **From Page** selection box of the dialog.
- Click on the button [...] next to **To Page** selection box. The **Select Page** dialog box will be displayed; refer to Figure-58.

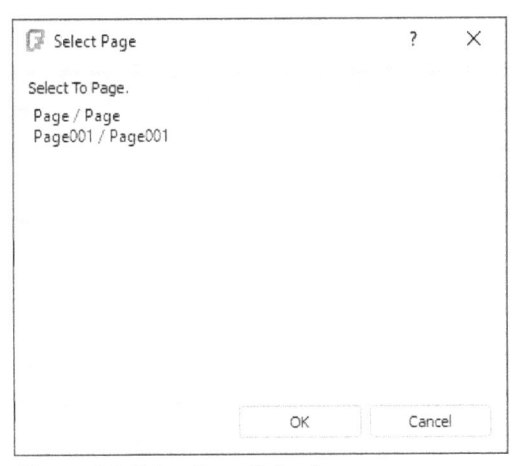

Figure-58. Select Page dialog box

- Select desired page from the list box to which you want to move the view and click on the **OK** button. The selected page will be displayed in the **To Page** selection box of the dialog.
- After specifying the parameters, click on the **OK** button from the dialog. The selected view will be move to the another page; refer to Figure-59.

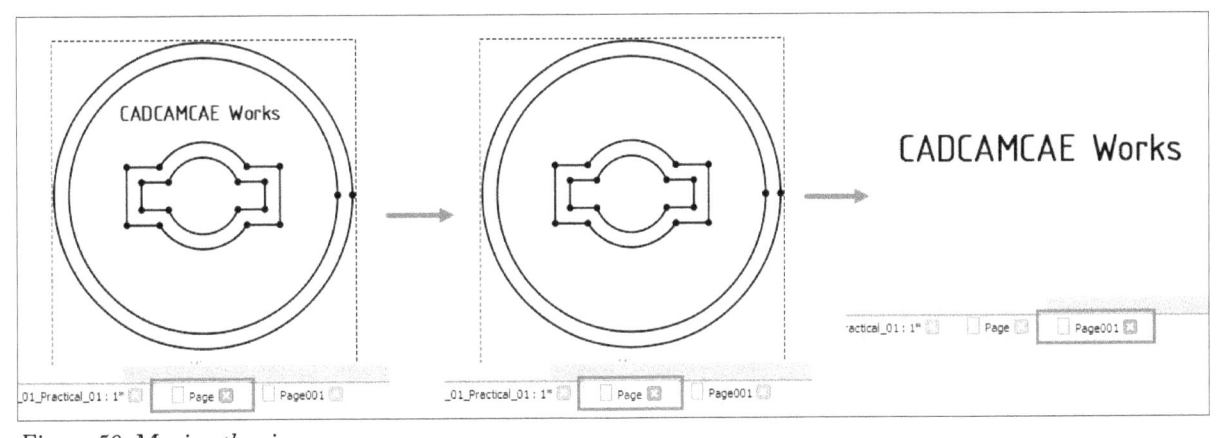

Figure-59. Moving the view

SHARING THE VIEW

The **Share View** tool makes a view and all its dependents (Balloons, Dimensions, etc.) visible on a second page. The procedure to use this tool is same as discussed for **Move View** tool.

PROJECTING THE SHAPE

The **Project Shape** tool creates projections of shapes. The projections are created in the 3D view, and not on a page. The procedure to use this tool is discussed next.

- Select desired feature from **Model Tree** or from the 3D view area which you want to project; refer to Figure-60.

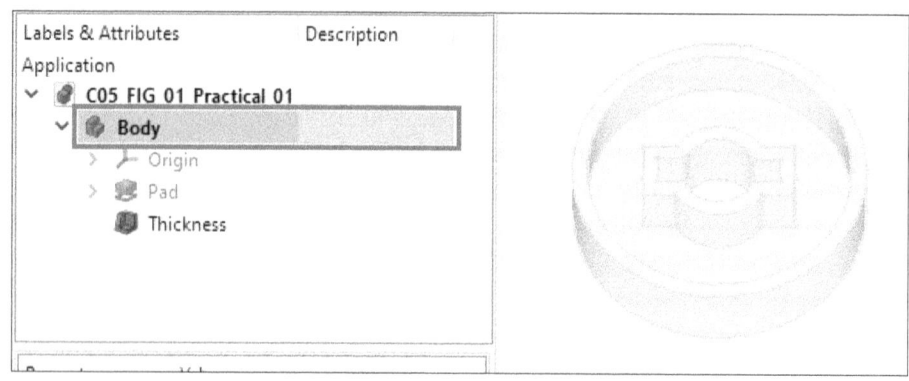

Figure-60. Selecting the feature

- Click on the **Project Shape** tool from the **Toolbar** or **TechDraw** menu; refer to Figure-61. The **Project shapes** dialog will be displayed; refer to Figure-62.

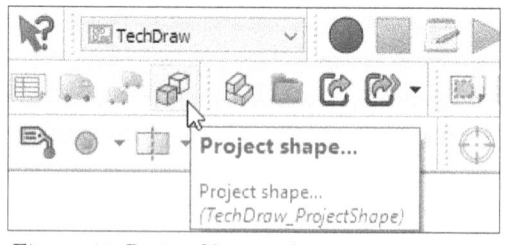

Figure-61. Project Shape tool

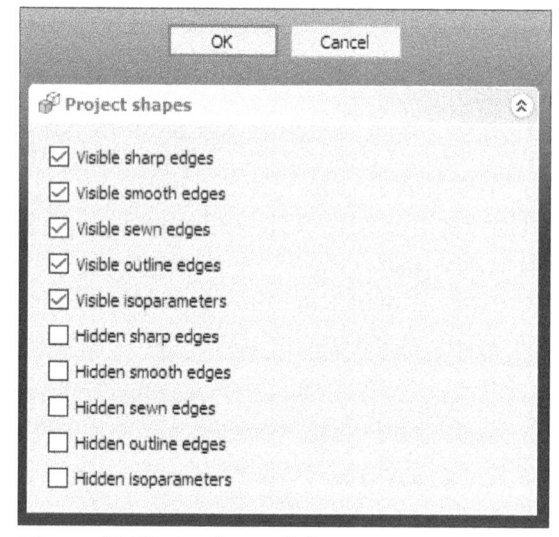

Figure-62. Project shapes dialog

- Select desired check boxes in the dialog and click on the **OK** button. The selected feature will be projected; refer to Figure-63.

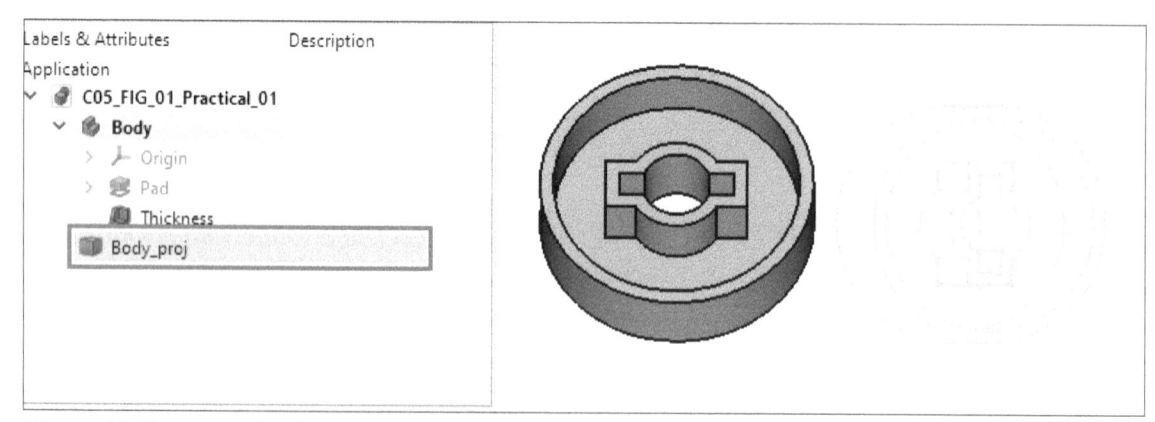

Figure-63. Feature projected

ADJUST STACKING ORDER OF VIEW

Stack Top View

The **Stack Top** tool is used to move view to the top of the stacking order. The stacking order controls the apparent depth of views on a page. For example if two views are overlapping then you can move selected view to top of another view.

- Select the view from model which you want to stack at top of other views and click on the **Stack Top** tool from the **Toolbar** or **TechDraw** menu; refer to Figure-64. The view will come at top; refer to Figure-65.

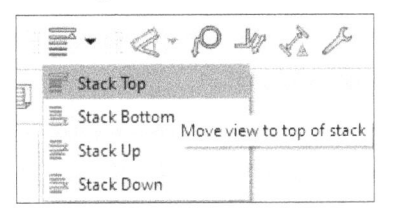

Figure-64. Stack top view

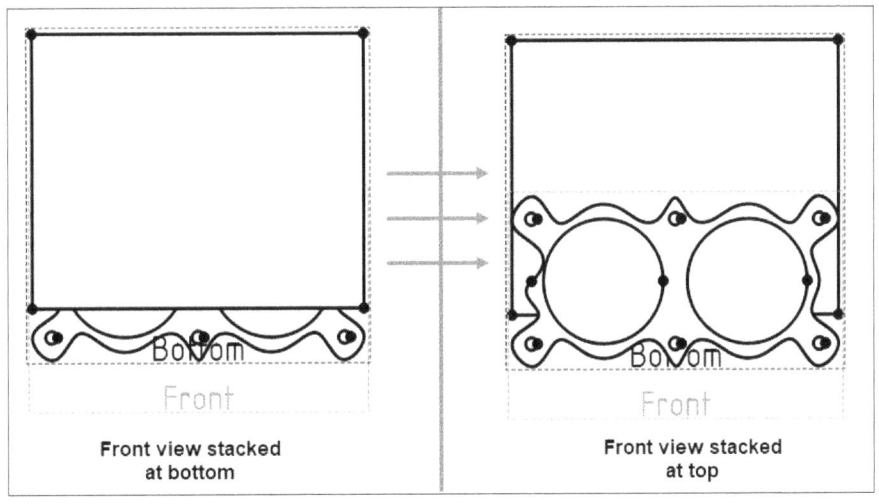

Figure-65. Stacking view at top

Stack Bottom View

The **Stack Bottom** tool is used to moves views to the bottom in the stacking order.

Stack UP View

The **Stack Up** tool is used to move view up one level in the stacking order.

Stack Down View

The **Stack Down** tool is used to move views down one level in the stacking order.

INSERTING ANNOTATIONS

The **Insert Annotation** tool is used to insert desired text in the drawing view. The procedure to use this tool is given next.

- Click on the **Insert Annotation** tool from the **Toolbar** or **TechDraw** menu; refer to Figure-66. The annotation will be placed in drawing area and properties of annotation will be displayed in the **Property Editor**; refer to Figure-67.

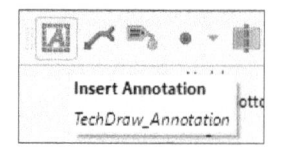

Figure-66. Insert Annotation tool

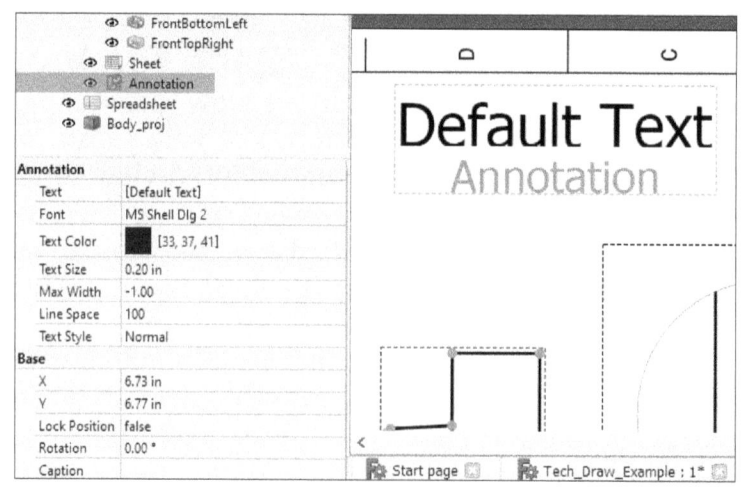

Figure-67. Annotation created

- Click in the **Text** field and then click on the **Browse** button. The **List** dialog box will be displayed; refer to Figure-68.

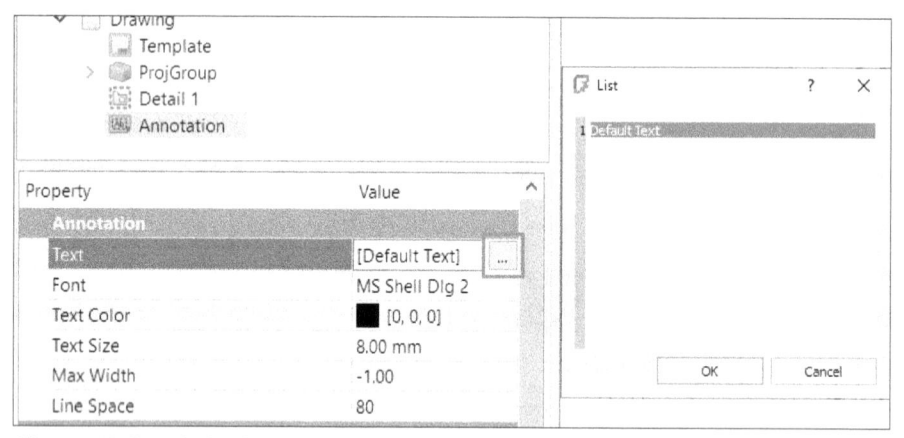

Figure-68. List dialog box

- Specify desired text in the edit box and click on the **OK** button from the dialog box. The annotation will be modified accordingly.
- Specify other parameters like font, text color, text size, and so on in the **Property Editor**.

ADDING LEADER LINE TO VIEW

The **Add Leaderline to View** tool adds a line to a view. Other annotation objects can be connected to the leaderline to form complex annotations. The procedure to use this tool is discussed next.

- Select desired view from the **Model Tree** or from the 3D view area to which you want to add leaderline; refer to Figure-69.

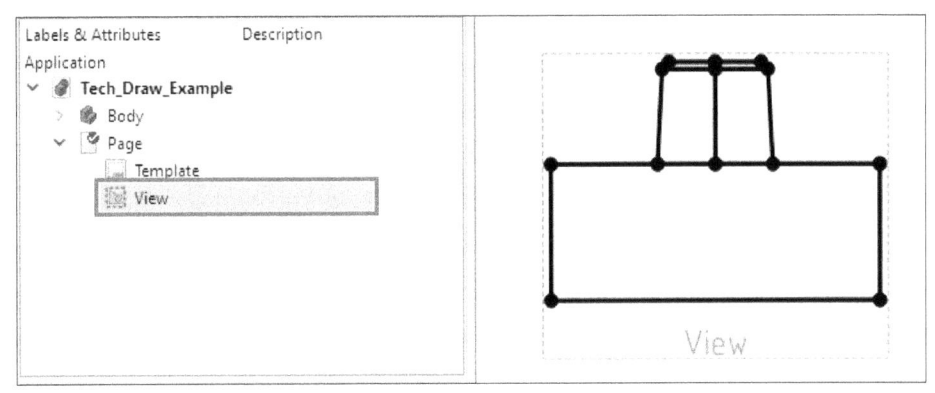

Figure-69. Selecting the view to add leaderline

- Click on the **Add Leaderline to View** tool from the **Toolbar** or **TechDraw** menu; refer to Figure-70. The **New Leader Line** dialog will be displayed in the **Tasks** panel of **Combo View**; refer to Figure-71.

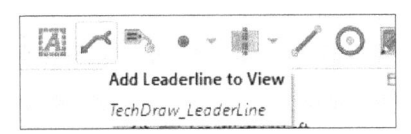

Figure-70. Add Leaderline to View tool

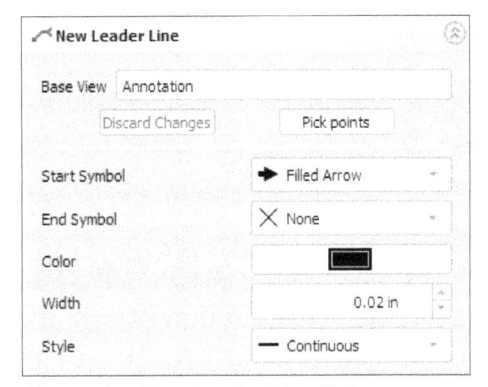

Figure-71. New Leader Line dialog

- Click on the **Pick points** button from the dialog and click in the 3D view area at desired location to specify the starting point of leaderline. The line will be attached to the cursor; refer to Figure-72.

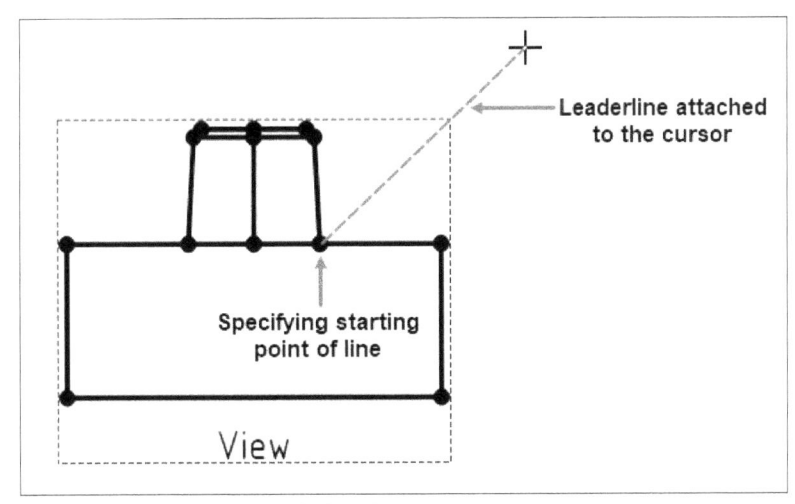

Figure-72. Leaderline attached to the cursor

- Move the cursor away and click at desired location to specify the another point of leaderline.
- Now, click on the **Save Points** button from the dialog to finish the line creation or move the cursor away and click at desired locations to create more line segments.

- Select desired symbol from **Start Symbol** and **End Symbol** drop-downs which you want at the start and end of line, respectively.
- Click on the **Color** button in the dialog to specify desired color for the leaderline.
- Specify desired width of the leaderline in the **Width** edit box.
- Select desired style for the leaderline from **Style** drop-down.
- After specifying desired parameters, click on the **OK** button from the dialog to create the leaderline; refer to Figure-73.

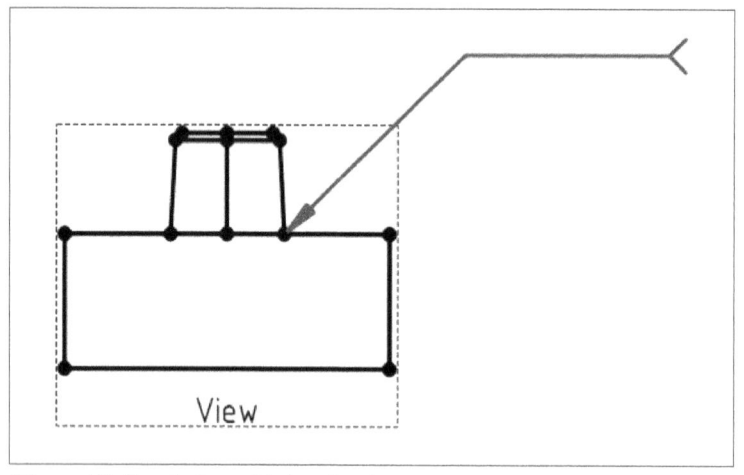

Figure-73. Leaderline added to the view

INSERTING RICH TEXT ANNOTATION

The **Insert Rich Text Annotation** tool adds a formatted annotation block to a leaderline or a view. The procedure to use this tool is discussed next.

- Click on the **Insert Rich Text Annotation** tool from **Toolbar** or **TechDraw** menu; refer to Figure-74. The **Rich text creator** dialog will be displayed in the **Tasks** panel of **Combo View**; refer to Figure-75.

Figure-74. Insert Rich Text Annotation tool

Figure-75. Rich text creator dialog

- Specify desired maximal width of the text in the **Max. Width** edit box of the dialog.
- Enter desired annotation text which you want to create in the **Input** box of the dialog.
- Click on the **Start Rich Text Editor** button from the dialog. The **Rich text editor** dialog will be displayed with the text written in the **Input** box; refer to Figure-76.

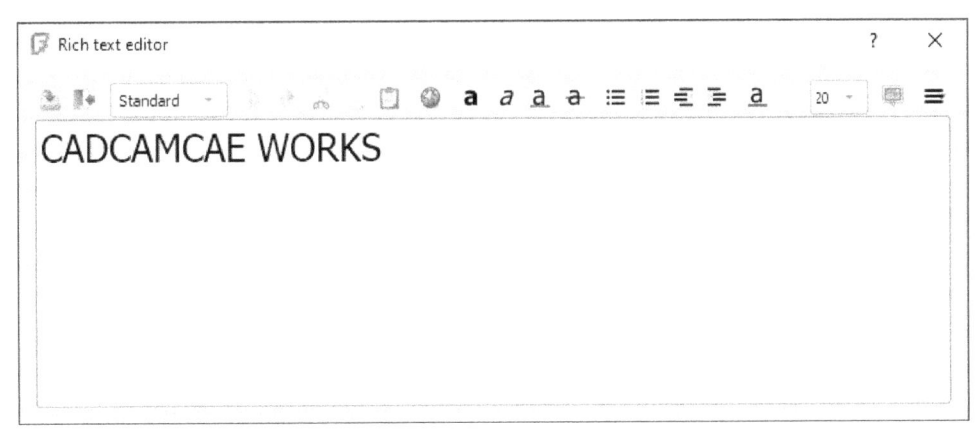

Figure-76. Rich text editor dialog box

- Specify desired parameters in the **Rich text editor** dialog and click on the **Save changes** button to save all the changes created in the dialog. The dialog will be closed.
- Select the **Show Frame** check box from **Rich text creator** dialog. The options below the check box will be activated.
- Click on the **Color** button to select desired color for the text.
- Specify desired width for the text in the **Width** edit box of the dialog.
- Select desired style for the text from the **Style** drop-down.
- After specifying desired parameters, click on the **OK** button from the dialog. The rich text annotation will be created; refer to Figure-77.

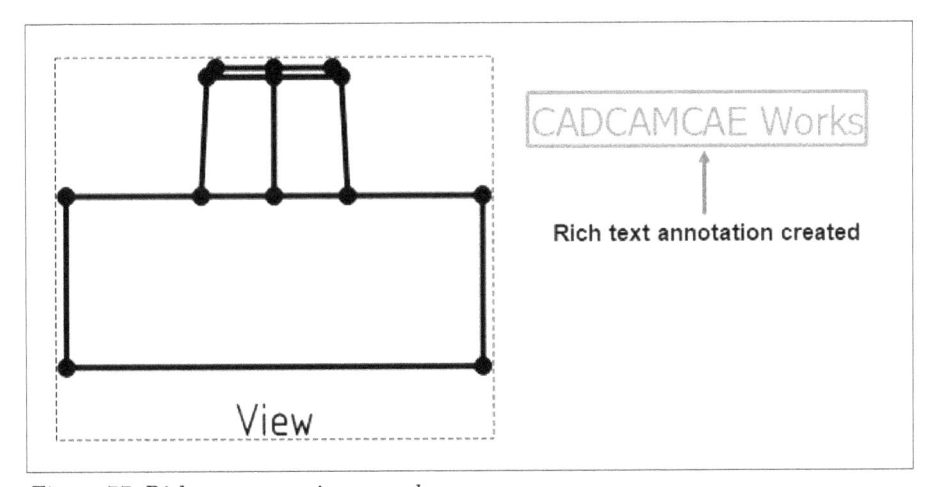

Figure-77. Rich text annotation created

ADDING COSMETIC VERTEX

The **Add Cosmetic Vertex** tool adds a vertex which is not part of the source geometry to a view. This vertex behaves like any other vertex and can be used for dimensioning. The procedure to use this tool is discussed next.

- Select desired view from the **Model Tree** or from the 3D view area to which you want to add the vertex; refer to Figure-78.

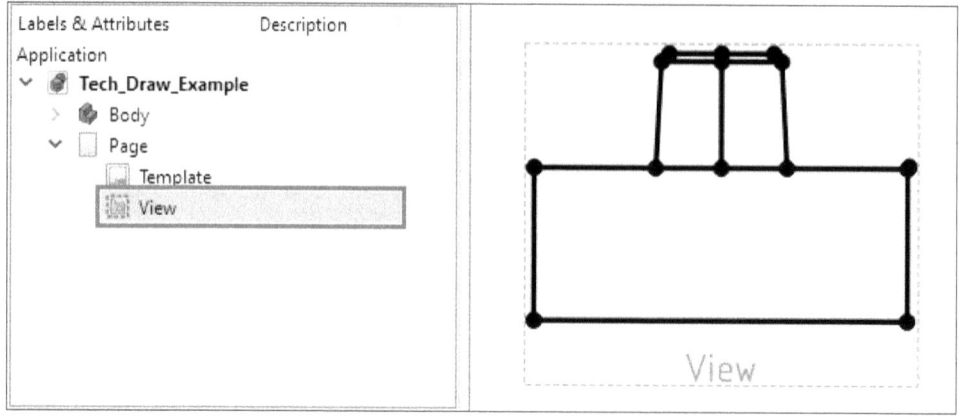

Figure-78. Selecting the view to add vertex

- Click on the **Add Cosmetic Vertex** tool from the drop-down in the **Toolbar** or **TechDraw** menu; refer to Figure-79. The **New Cosmetic Vertex** dialog will be displayed in the **Tasks** panel of **Combo View**; refer to Figure-80.

Figure-79. Add Cosmetic Vertex tool

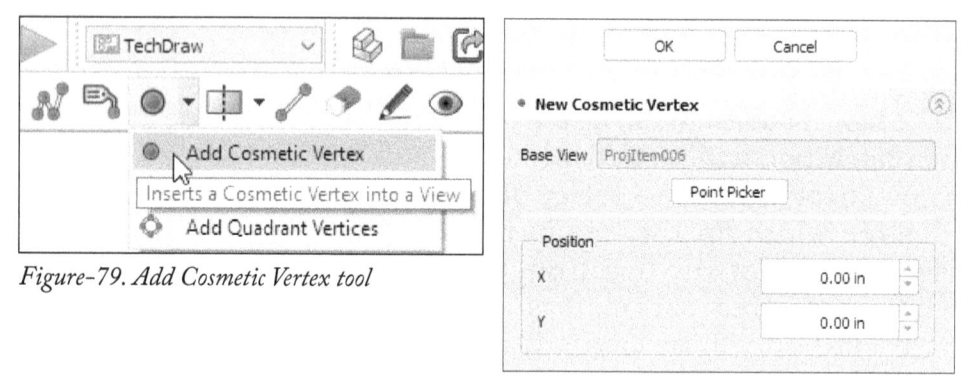

Figure-80. New Cosmetic Vertex dialog

- Click on the **Point Picker** button from the dialog and click at desired location in the 3D view area to place the point or enter desired values in the **X** and **Y** edit boxes from **Position** area of the dialog.
- After clicking at desired location for the point, click on the **OK** button from the dialog. The cosmetic vertex will be added to the view; refer to Figure-81.

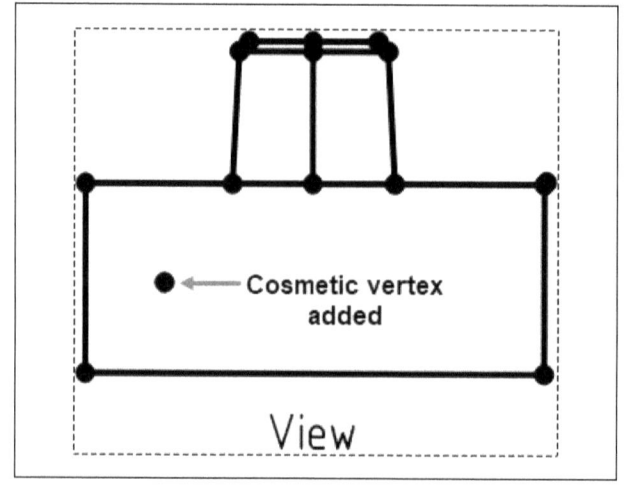

Figure-81. Cosmetic vertex added to the view

- If you want to remove the added cosmetic vertex then select the cosmetic vertex added to the view from the 3D view area and click on the **Remove Cosmetic Object** tool from the **Toolbar** or **TechDraw** menu; refer to Figure-82. The cosmetic vertex will be removed.

Figure-82. Remove Cosmetic Object tool

ADDING MIDPOINT VERTICES

The **Add Midpoint Vertices** tool adds cosmetic vertices at the midpoints of one or more edges. The procedure to use this tool is discussed next.

- Select one or more edges of desired view to which you want to add the midpoint vertex or vertices; refer to Figure-83.

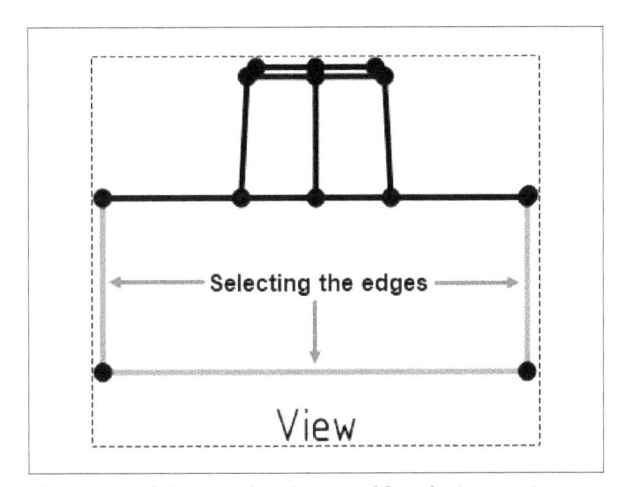

Figure-83. Selecting the edges to add midpoint vertices

- Click on the **Add Midpoint Vertices** tool from the drop-down in the **Toolbar** or **TechDraw** menu; refer to Figure-84. The midpoint vertices will be added to the view; refer to Figure-85.

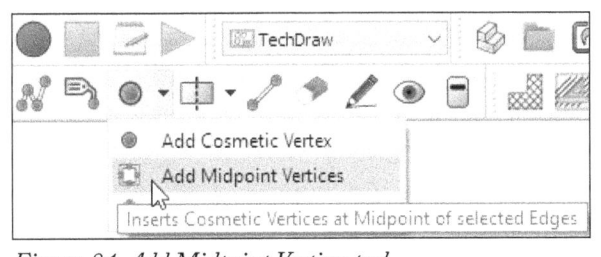

Figure-84. Add Midpoint Vertices tool

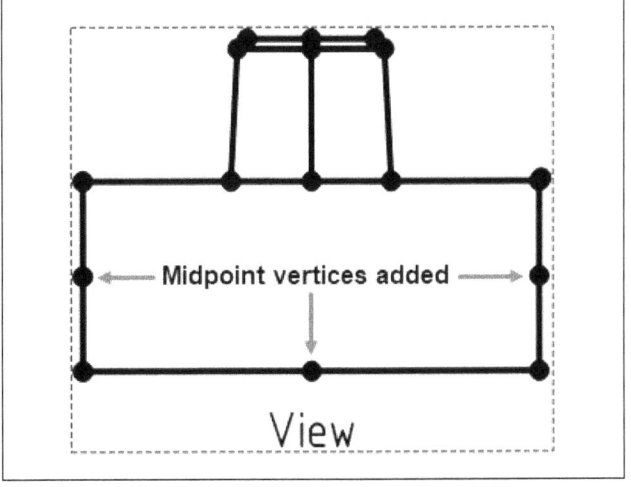

Figure-85. Midpoint vertices added to the view

- You can remove the midpoint vertices using **Remove Cosmetic Object** tool as discussed earlier.

ADDING QUADRANT VERTICES

The **Add Quadrant Vertices** tool adds cosmetic vertices at the 90/180/270° points of a circular edge. The 0° vertex should already be there as a geometric vertex. The procedure to use this tool is discussed next.

- Select one or more circular edges of desired view to which you want to add the quadrant vertices; refer to Figure-86.

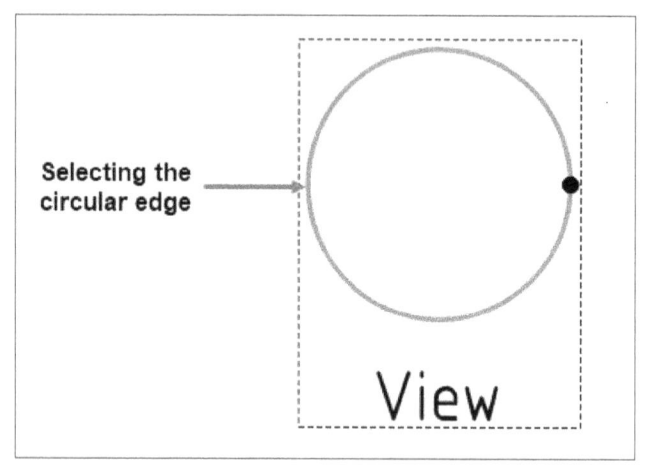

Figure-86. Selecting the circular edge to add quadrant vertices

- Click on the **Add Quadrant Vertices** tool from the drop-down in the **Toolbar** or **TechDraw** menu; refer to Figure-87. The quadrant vertices will be added to the view; refer to Figure-88.

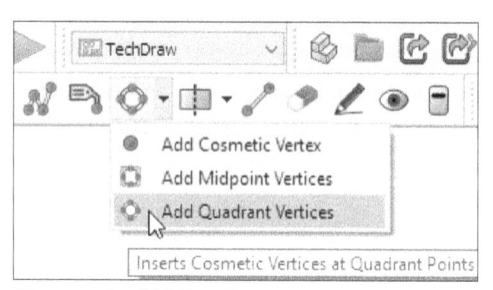

Figure-87. Add Quadrant Vertices tool

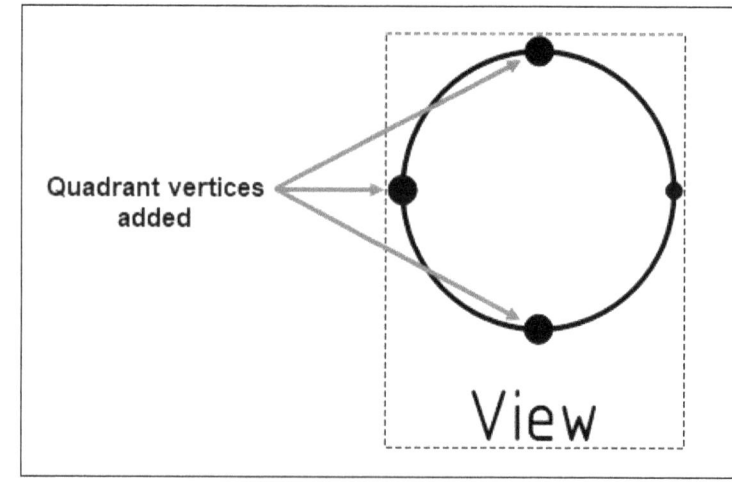

Figure-88. Quadrant vertices added to the view

- You can remove the quadrant vertices using **Remove Cosmetic Object** tool as discussed earlier.

ADDING CENTERLINE TO FACES

The **Add Centerline to Faces** tool adds a centerline to selected faces. The procedure to use this tool is discussed next.

- Select one or more faces of desired view to which you want to add centerline; refer to Figure-89.

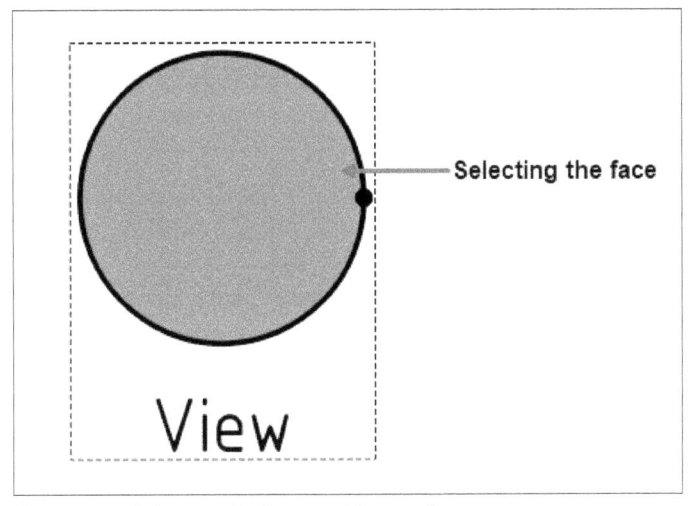

Figure-89. Selecting the face to add centerline

- Click on the **Add Centerline to Faces** tool from the drop-down in the **Toolbar** or **TechDraw** menu; refer to Figure-90. The **Create Center Line** dialog will be displayed in the **Tasks** panel of **Combo View**; refer to Figure-91.

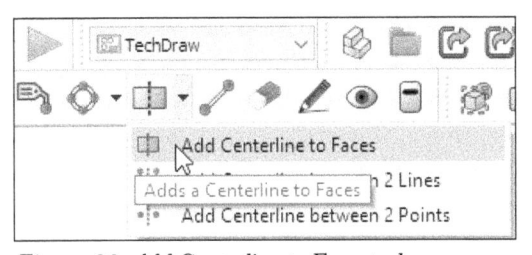

Figure-90. Add Centerline to Faces tool

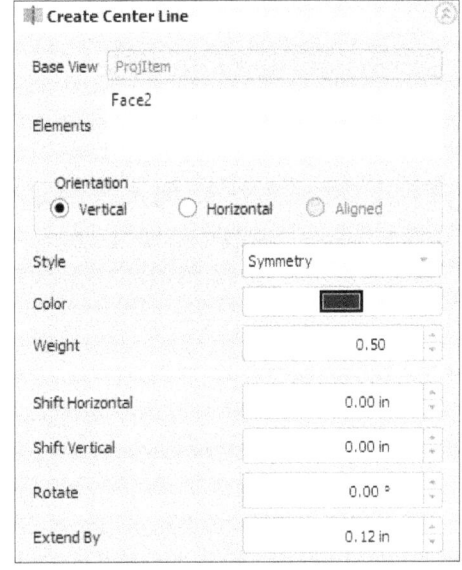

Figure-91. Create Center Line dialog

- The selected face will be displayed in the **Elements** area of the dialog.
- Select desired radio button from **Orientation** area of the dialog. Select **Vertical** radio button to orient the centerline from Top to Bottom line. Select **Horizontal** radio button to orient the centerline from left to right line. Select **Aligned** radio button to orient the centerline between lines or points.
- Select desired style for the centerline from **Style** drop-down.
- Click on the **Color** button to specify desired color for the centerline.
- Enter desired value in the **Weight** edit box to specify thickness of the centerline.
- Specify desired value in **Shift Horizontal** edit box to move the centerline left or right of its normal position.
- Specify desired value in **Shift Vertical** edit box to move the centerline up or down from its normal position.
- Specify desired value in **Rotate** edit box to rotate the centerline around its center.

- Specify desired value in **Extend By** edit box to make the centerline longer by this amount.
- After specifying desired parameters, click on the **OK** button from the dialog. The centerline will be added to the face; refer to Figure-92.

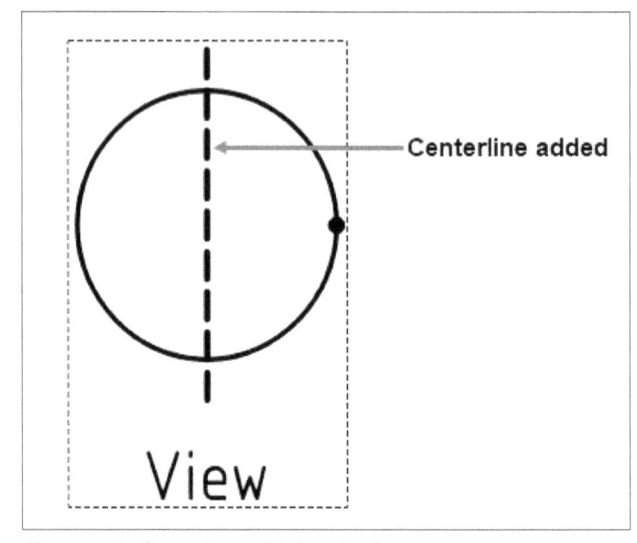

Figure-92. Centerline added to the face

- You can remove the centerline using **Remove Cosmetic Object** tool as discussed earlier.

ADDING CENTERLINE BETWEEN 2 LINES

The **Add Centerline between 2 Lines** tool adds a centerline between two edges. The procedure to use this tool is discussed next.

- Select the two edges of desired view between which you want to add the centerline; refer to Figure-93.

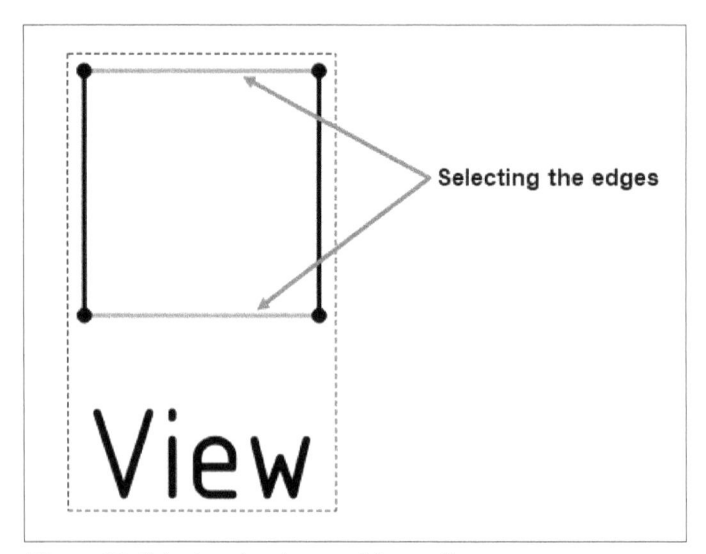

Figure-93. Selecting the edges to add centerline

- Click on the **Add Centerline between 2 Lines** tool from the drop-down in the **Toolbar** or **TechDraw** menu; refer to Figure-94. The **Create Center Line** dialog will be displayed in the **Tasks** panel of **Combo View** as discussed earlier.

FreeCAD 1.0 Black Book 10-35

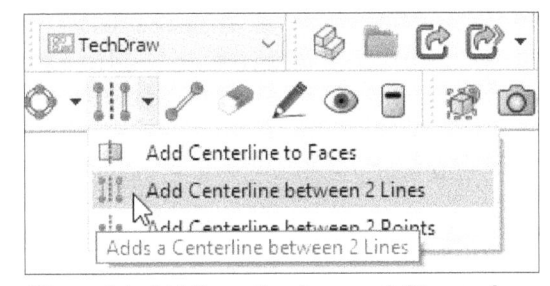

Figure-94. Add Centerline between 2 Lines tool

- Specify desired parameters in the dialog as discussed earlier and click on **OK** button. The centerline between the two edges will be added; refer to Figure-95.

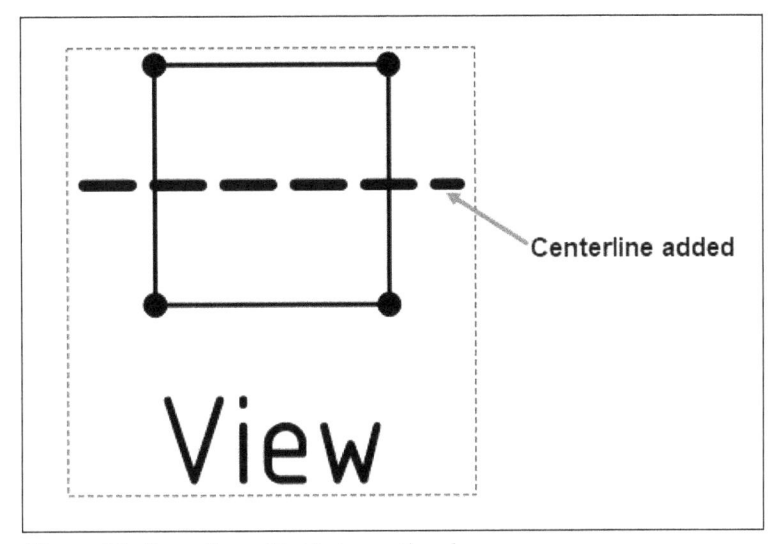

Figure-95. Centerline added between the edges

- You can remove the centerline using **Remove Cosmetic Object** tool as discussed earlier.

ADDING CENTERLINE BETWEEN 2 POINTS

The **Add Centerline between 2 Points** tool adds a centerline between two vertices (Points). The procedure to use this tool is discussed next.

- Select the two vertices of desired view between which you want to add the centerline; refer to Figure-96.

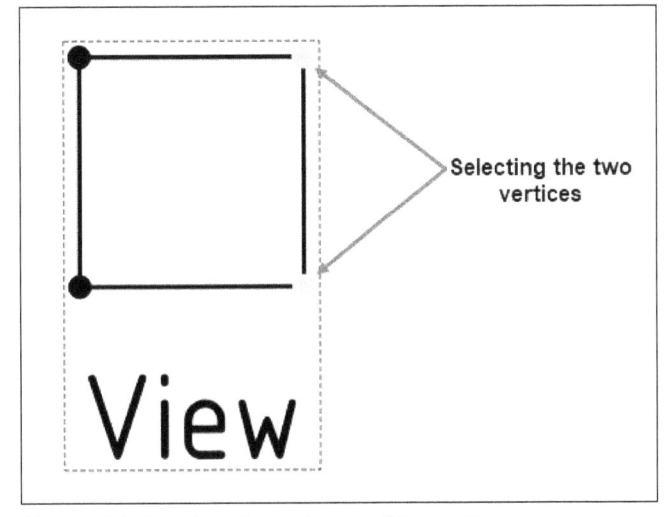

Figure-96. Selecting the vertices to add centerline

- Click on the **Add Centerline between 2 Points** tool from the drop-down in the **Toolbar** or **TechDraw** menu; refer to Figure-97. The **Create Center Line** dialog will be displayed in the **Tasks** panel of **Combo View** as discussed earlier.

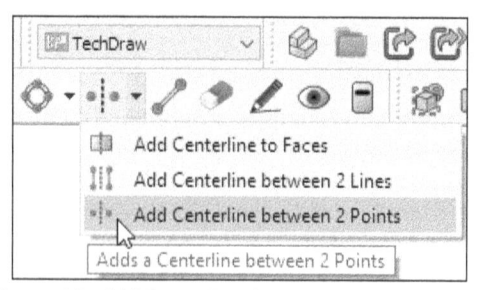

Figure-97. Add Centerline between 2 Points tool

- Specify desired parameters in the dialog as discussed earlier and click on the **OK** button. The centerline between the two points will be added; refer to Figure-98.

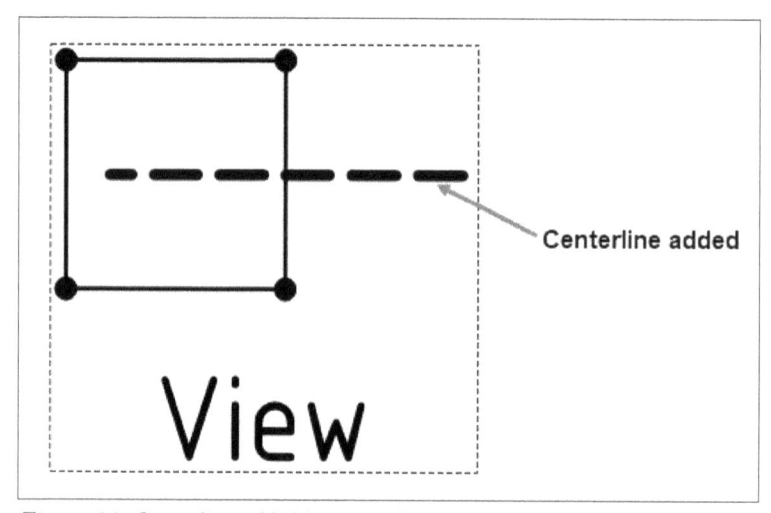

Figure-98. Centerline added between the points

- You can remove the centerline using **Remove Cosmetic Object** tool as discussed earlier.

ADDING COSMETIC LINE THROUGH 2 POINTS

The **Add Cosmetic Line Through 2 Points** tool adds a cosmetic line between two vertices (points). The vertices can be 2D or 3D. The resulting line can be used for dimensioning. The procedure to use this tool is discussed next.

- Select the two vertices of desired view or in the 3D view area between which you want to add the cosmetic line; refer to Figure-99.

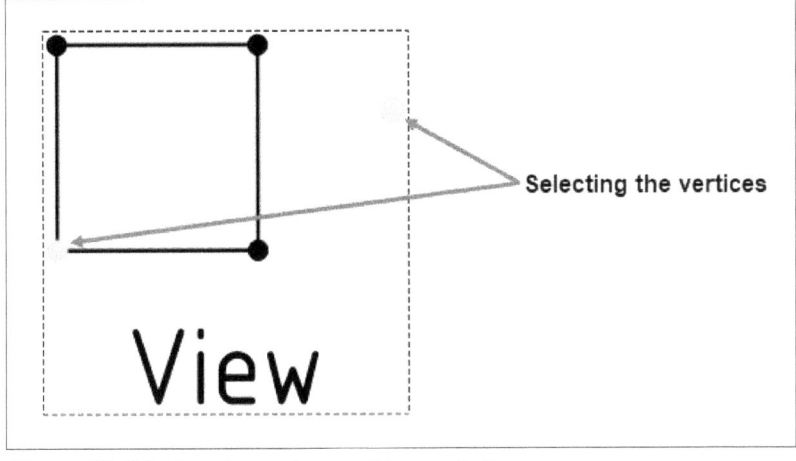

Figure-99. Selecting the vertices to add cosmetic line

- Click on the **Add Cosmetic Line Through 2 Points** tool from **Toolbar** or **TechDraw** menu; refer to Figure-100. The **Create Cosmetic Line** dialog will be displayed in the **Tasks** panel of **Combo View**; refer to Figure-101.

Figure-100. Add Cosmetic Line Through 2 Points tool

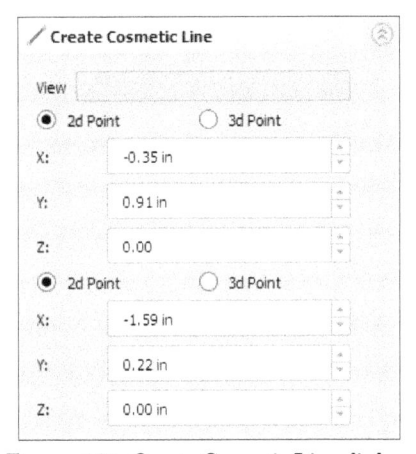

Figure-101. Create Cosmetic Line dialog.

- Select desired radio button from **2D Point** and **3D Point** radio buttons in the dialog.
- Enter desired values in the **X**, **Y**, and **Z** edit boxes to specify the coordinates of line in the x, y, and z directions, respectively.
- After specifying desired parameters, click on the **OK** button from the dialog. The cosmetic line will be created between the points; refer to Figure-102.

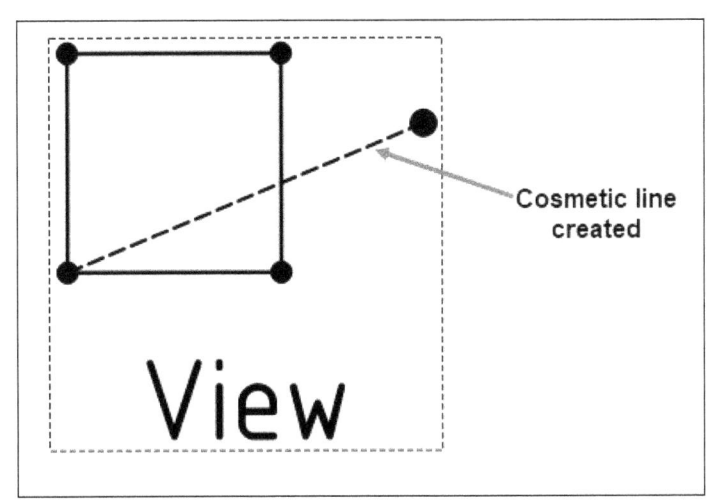

Figure-102. Cosmetic line between the points created

- You can remove the cosmetic line using **Remove Cosmetic Object** tool as discussed earlier.

ADD COSMETIC CIRCLE

The **Add Cosmetic Circle** tool is used to add a cosmetic circle at selected center point. The point can be 2D or 3D. The procedure to use this tool is discussed next.

- Select a vertex/point to define center point of the circle from the view; refer to Figure-103 and click on the **Add Cosmetic Circle** tool from the **TechDraw Annotation** toolbar or **TechDraw>Add Lines** menu. The **Create Cosmetic Circle** dialog box will be displayed; refer to Figure-104.

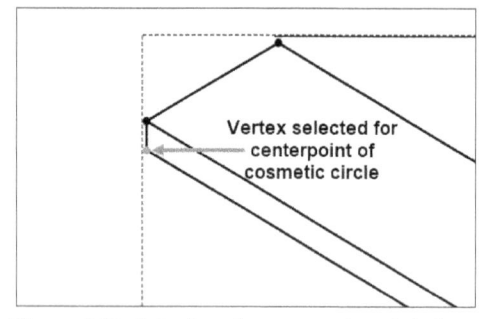

Figure-103. Selecting the center point of circle

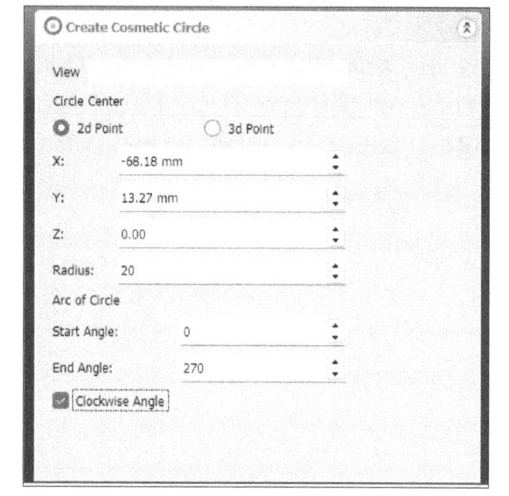

Figure-104. Create Cosmetic Circle dialog

- Select **2d Point** radio button to treat the selected center point as a 2d point within the parent view. Select **3d Point** radio button to treat selected center point as a 3d point and project onto the parent view.
- Specify desired values in **X**, **Y**, and **Z** edit boxes to define the position of center point if you want to modify it. Similarly, specify the radius, start angle and end angle of cosmetic circular arc in **Radius**, **Start Angle**, and **End Angle** edit boxes respectively.
- Select **Clockwise Angle** check box to make an arc from an start angle to end angle in clock wise direction.
- After specifying desired parameters, click on the **OK** button from the dialog. The cosmetic arc/circle will be created; refer to Figure-105.

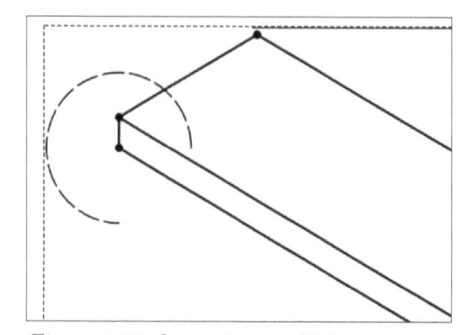

Figure-105. Cosmetic arc added

CHANGING APPEARANCE OF LINE

The **Change Appearance of selected Lines** tool changes the appearance of edges. The procedure to use this tool is discussed next.

- Select one or more edges of desired view of which you want to change the appearance; refer to Figure-106.

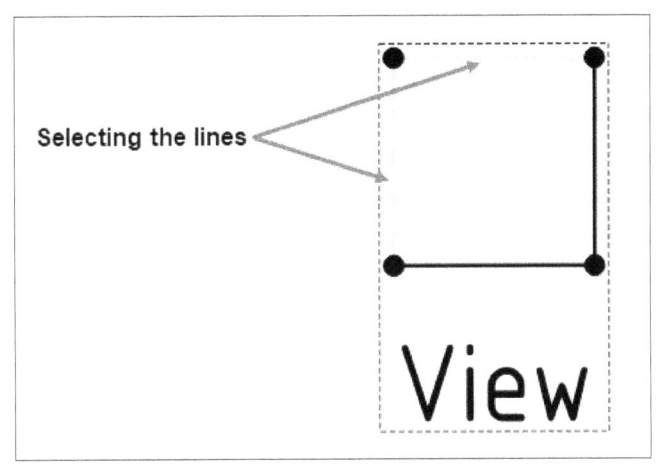

Figure-106. Selecting the lines to change the appearance

- Click on the **Change Appearance of Lines** tool from **Toolbar** or **TechDraw** menu; refer to Figure-107. The **Line Decoration and Restore Invisible Lines** dialogs will be displayed in the **Tasks** panel of **Combo View**; refer to Figure-108.

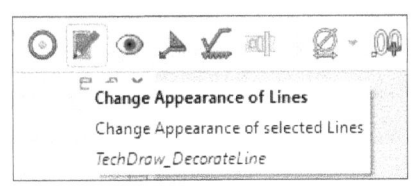

Figure-107. Change Appearance of selected Lines tool

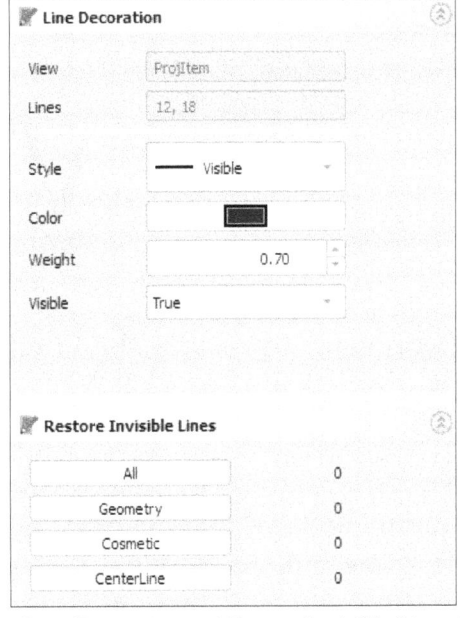

Figure-108. Line Decoration and Restore Invisible Lines dialogs

- Select desired style for the lines from **Style** drop-down.
- Click on the **Color** button to specify desired color for the lines.
- Enter desired value in the **Weight** edit box to specify desired thickness for the lines.
- Select desired option from **Visible** drop-down to specify whether the line should be visible or not.
- After specifying desired parameters, click on the **OK** button from the dialog. The Appearance of the lines will be changed; refer to Figure-109.

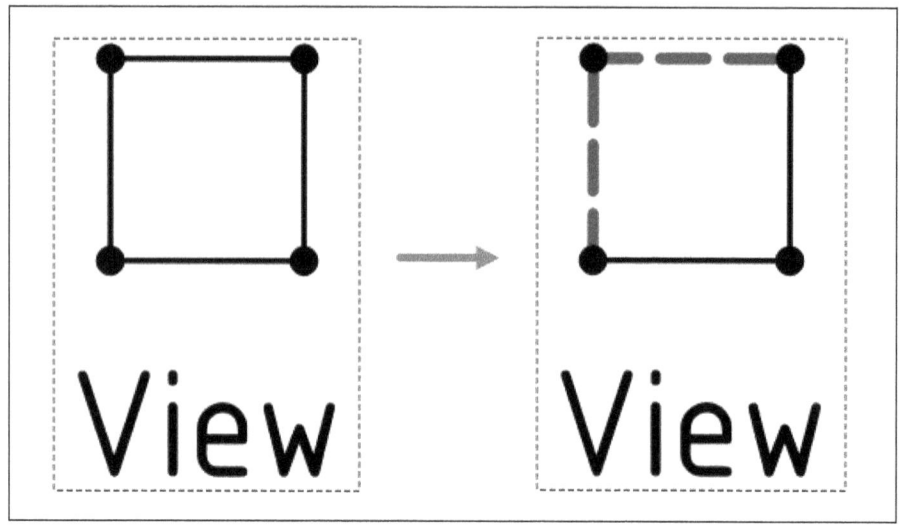

Figure-109. Appearance of the lines changed

- If you select the **False** option from **Visible** drop-down and click on the **OK** button then the lines will become invisible; refer to Figure-110.

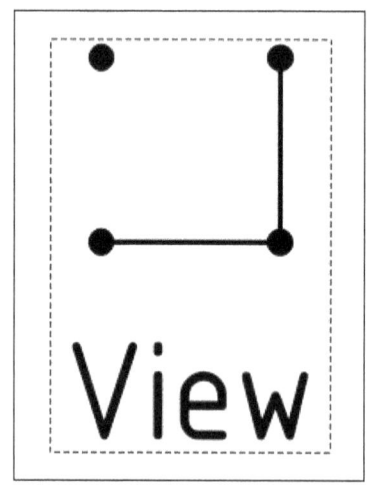

Figure-110. Lines become invisible

- Now, if you want to restore the lines then select the view from the **Model Tree** or from the 3D view area and click on the tool again. The **Restore Invisible Lines** dialog will be displayed.
- Click on desired button from the dialog to restore the lines. Click on the **All** button to restore all the invisible lines of the view. Click on the **Geometry** button to restore the invisible geometry lines. Click on the **Cosmetic** button to restore the invisible cosmetic lines. Click on the **CenterLine** button to restore the invisible centerlines.
- After specifying desired parameters, click on the **OK** button from the dialog. The invisible lines will be restored.

You can also restore the invisible lines using **Show/Hide Invisible Edges** tool.

SHOW/HIDE INVISIBLE EDGES

The **Show/Hide Invisible Edges** tool is used to temporarily show and hide invisible lines in a View. Lines can be made invisible with the **TechDraw DecorateLine** tool. Note that "invisible" is a cosmetic state, not to be confused with hidden lines which are geometric constructs.

- Click on the **Show/Hide Invisible Edges** tool from the **Toolbar** or **TechDraw** menu; refer to Figure-111. All invisible lines in the view are shown/hidden based on tool selection.

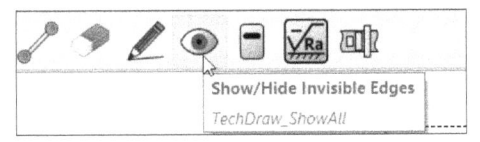

Figure-111. Show hide invisible edges

ADDING WELDING INFORMATION TO LEADERLINE

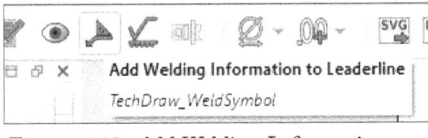

The **Add Welding Information to Leaderline** tool adds welding specifications to an existing leader line. The procedure to use this tool is discussed next.

- Select desired leaderline from the 3D view area to which you want to add welding information; refer to Figure-112.

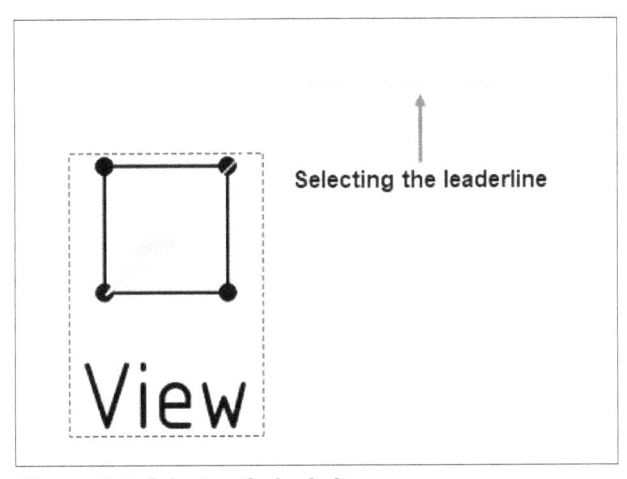

Figure-112. Selecting the leaderline

- Click on the **Add Welding Information to Leaderline** tool from the **Toolbar** or **TechDraw** menu; refer to Figure-113. The **Create Welding Symbol** dialog will be displayed in the **Tasks** panel of **Combo View**; refer to Figure-114.

Figure-113. Add Welding Information to Leaderline tool

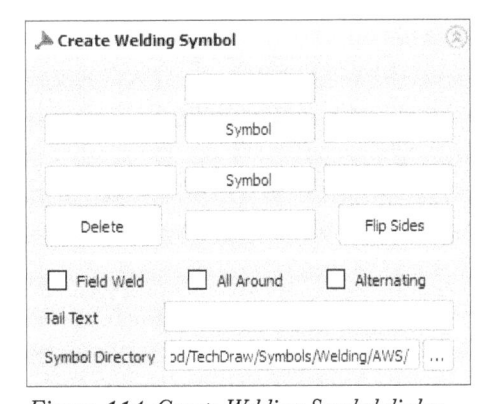

Figure-114. Create Welding Symbol dialog

- Click on the **Pick arrow side symbol** button. The **Select a symbol** dialog box will be displayed; refer to Figure-115.

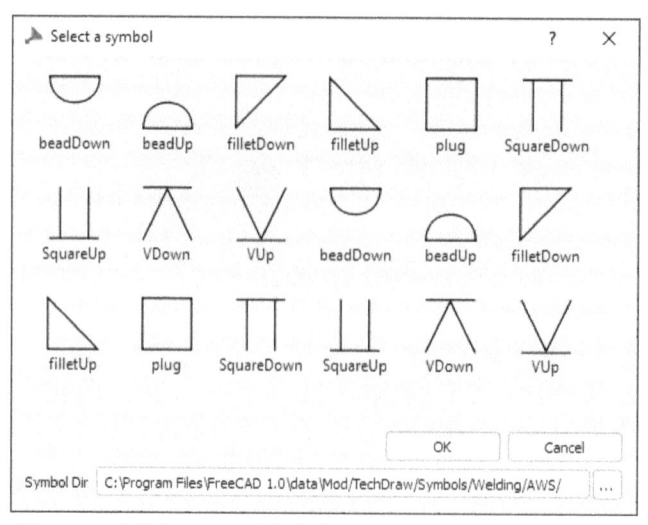

Figure-115. Select a symbol dialog

- Select desired symbol from the dialog box or from the symbol directory in your system and click on the **OK** button. The `Pick arrow side symbol` button will be replaced with the selected symbol.
- Specify desired text in the `Text above arrow side symbol`, `Text before arrow side symbol`, and `Text after arrow side symbol` edit boxes to place the text above, before, and after the arrow side symbol, respectively.
- In the same way, click on the `Pick other side symbol` button and select desired symbol from the `Select a symbol` dialog box.
- Specify desired text in the `Text below other side symbol`, `Text before other side symbol`, and `Text after other side symbol` edit boxes to place the text below, before, and after the other side symbol, respectively.
- Click on the `Delete` button to remove the other side symbol.
- Click on the `Flip Sides` button to flip the sides of the symbol.
- Select the `Field Weld` check box to add the Field Weld symbol (flag) at the kink in the leader line.
- Select the `All Around` check box to add the All Around symbol (circle) at the kink in the leader line.
- Select the `Alternating` check box to offset the lower symbol to indicate alternating welds.
- Specify desired text at the end of symbol in the `Tail Text` edit box.
- Click on the `Symbol Directory` button to specify directory to welding symbols to be used for the symbol selection.
- After specifying desired parameters, click on the **OK** button from the dialog. The weld information will be added to the leader line; refer to Figure-116.

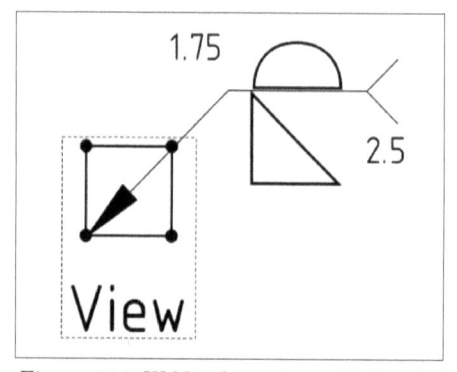

Figure-116. Weld information added to the leaderline

Create Surface Finish Symbol

The **Create a surface finish symbol** tool is used to add a surface finish symbol in drawing page.

- Click on the **Create a Surface finish symbol** tool from the **Toolbar** or **TechDraw** menu; refer to Figure-117. The **Surface Finish Symbols** dialog box will be displayed; refer to Figure-118.
- Select desired symbol from the dialog to create surface finish symbol.
- Specify desired rotation angle for the symbol in the **Symbol angle** edit box.

Figure-117. create a surface finish symbol

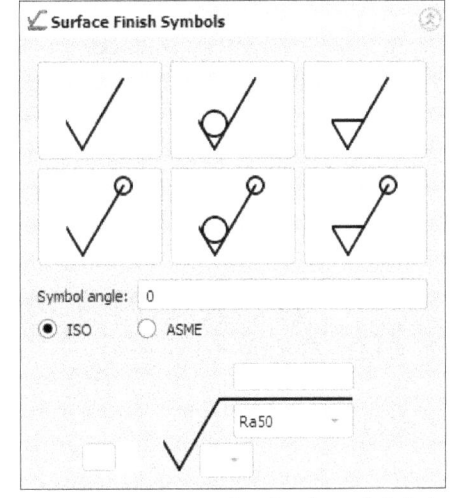

Figure-118. create a surface finish symbol dialog box

- Select desired radio button to select either **ISO** or **ASME** standard.
- Specify the other parameter as discussed earlier and click on the **OK** button from the dialog.

Add Hole or Shaft Fit

The **Add Hole or Shaft Fit** tool is used to add hole or shaft tolerance using ISO 286 to a linear dimension or a diameter dimension.

- Select the linear/diameter dimension from drawing and click on the **Add Hole or Shaft Fit** tool from the **Toolbar** or **TechDraw** menu; refer to Figure-119. The **Hole/Shaft Fit ISO 286** dialog will be displayed; refer to Figure-120.

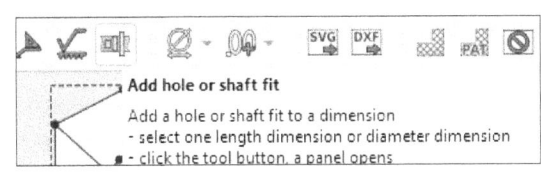

Figure-119. Add hole or shaft tool

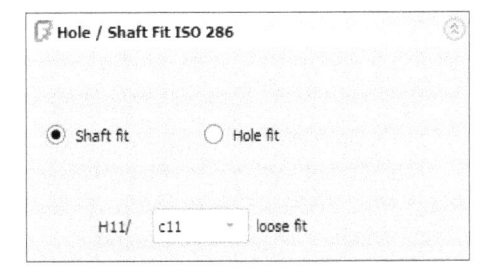

Figure-120. Add hole or shaft fit dialog

- Select the **shaft fit** or **hole fit** radio button as applicable in the dialog and select fit option from the drop-down in the dialog.

- Click on the **OK** button from the dialog to apply fit tolerance; refer to Figure-121.

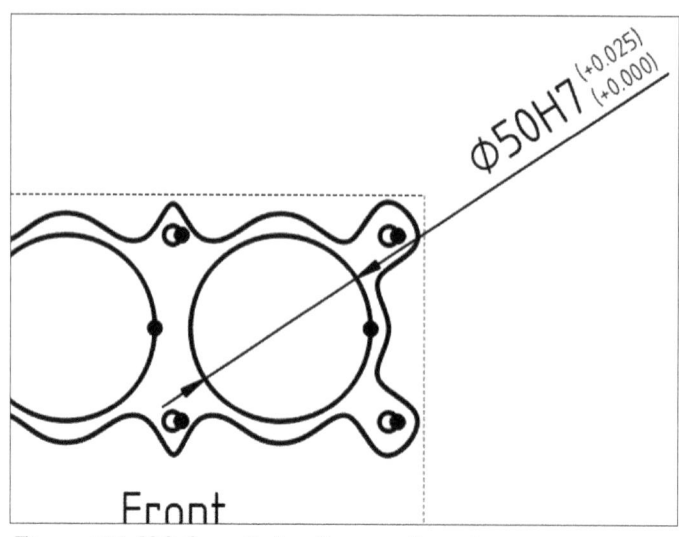

Figure-121. ISO fit applied to diameter dimension

WORKING WITH CLIP GROUP

The **Insert Clip Group** tool is used to insert a clipping group in the drawing. You can add or remove objects from the clip group by dragging them inside or out of the group. The procedure to use this tool is given next.

- Click on the **Insert Clip Group** tool from the **TechDraw Views** toolbar or **TechDraw > TechDraw Views** menu; refer to Figure-122. A clip group will be inserted in the drawing; refer to Figure-123.

Figure-122. Insert Clip Group tool

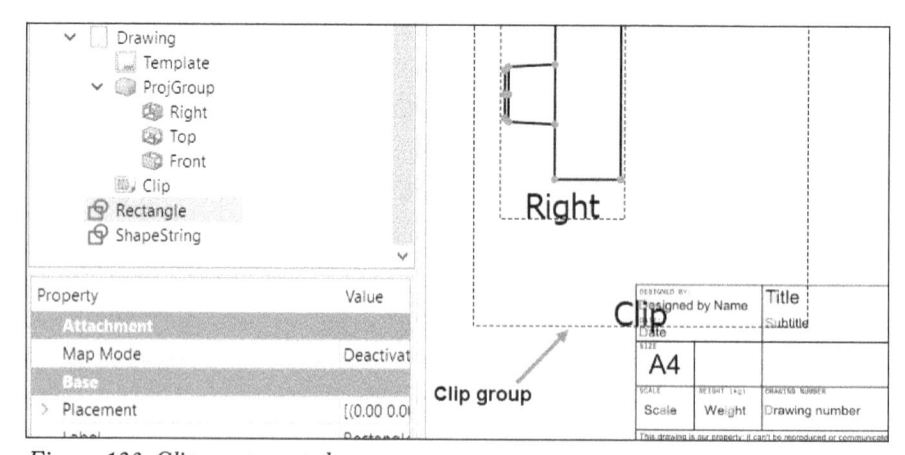

Figure-123. Clip group created

- You can drag the clip group to desired location as discussed for drawing views.
- Select the clip group recently created, select the object/view to be added in the clip group while holding the **CTRL** key and then click on the **Add a View to Clip group** tool from the **Toolbar** or **TechDraw** menu; refer to Figure-124. Selected view will be clipped and added in the clip group.

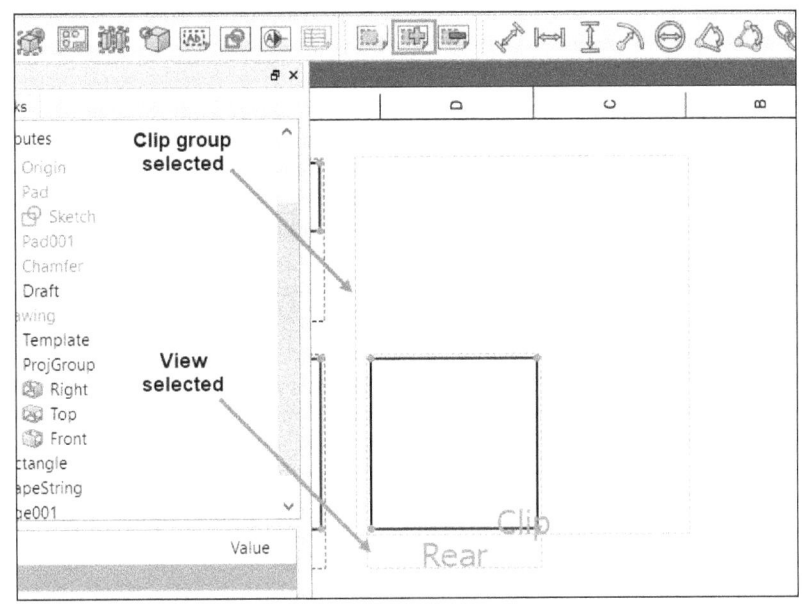

Figure-124. Adding view to clip group

- If you want to remove a view from clip group then select it from the clip group and click on the **Remove a View from Clip group** tool from the **Toolbar** or **TechDraw** menu. The view will be removed from clip group; refer to Figure-125.

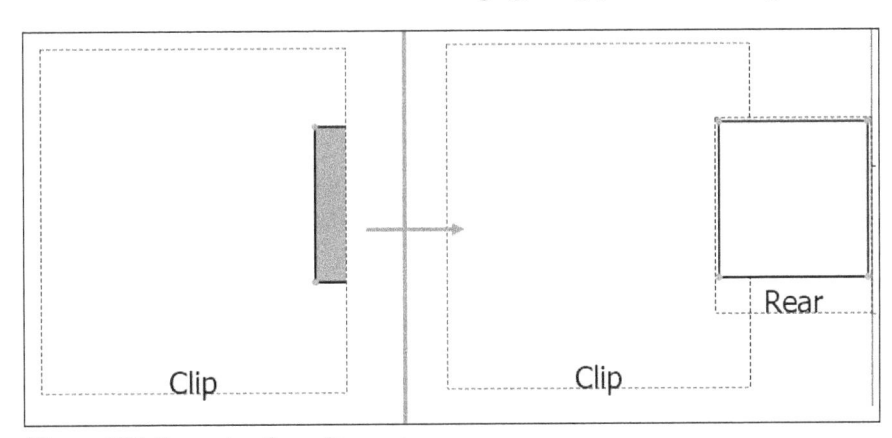

Figure-125. Removing from clip group

Note that you can use **CTRL+Z** and **CTRL+Y** shortcut keys to undo or redo previous operations.

INSERTING DIMENSIONS

The **Insert Dimension** tool is used to apply different types of dimensions based on selected entities. For example, if you have selected an arc then radius dimension will be applied, if you have selected a line then length dimension will be applied, if you have selected two vertices then distance dimension will be applied, and so on. The procedure to use this tool is discussed next.

- After selecting desired entity(ies) to be dimensioned, click on the **Insert Dimension** tool from the **Insert Dimension** drop-down of the **TechDraw Dimensions** toolbar or **TechDraw > Dimensions** menu; refer to Figure-126. The dimension will be generated; refer to Figure-127.

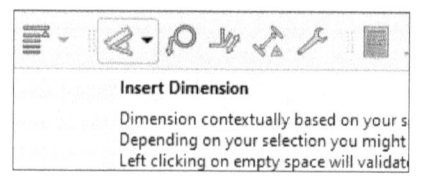

Figure-126. Insert dimension tool

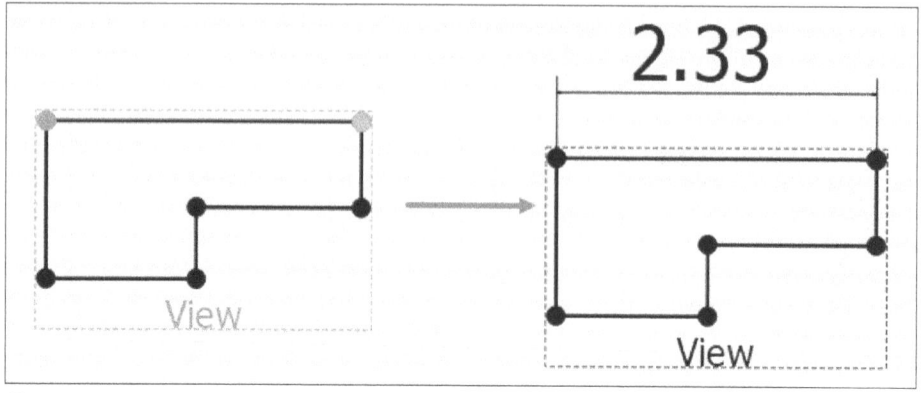

Figure-127. Insert dimension created

INSERTING LENGTH DIMENSION

The `Insert Length Dimension` tool is used to apply a length dimension based on selected two vertices, an edge, or two edges. The procedure to use this tool is given next.

- After selecting two vertices, two edges (between which length dimension is to be created), or an edge to be dimensioned, click on the `Insert Length Dimension` tool from the `Insert Dimension` drop-down of the `TechDraw Dimensions` toolbar or `TechDraw > Dimensions` menu; refer to Figure-128. The dimension will be generated; refer to Figure-129 and Figure-130.

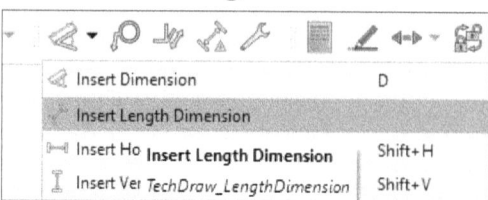

Figure-128. Insert Length Dimension tool

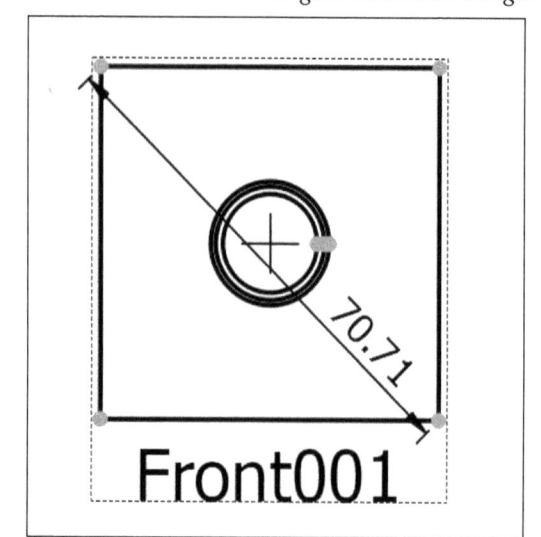

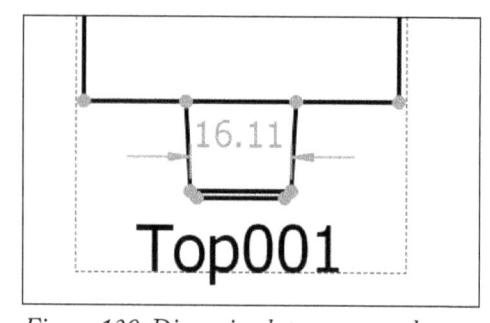

Figure-130. Dimension between two edges

Figure-129. Dimension generated between two vertices

Similarly, you can use other dimensioning tools to create horizontal distance dimension, vertical distance dimension, radius dimension, diameter dimension, angle dimension, 3 point angle dimensions, horizontal extent dimension, vertical extent dimension, balloon dimension, and landmark dimension.

LINKING DIMENSION TO 3D GEOMETRY

The **Link dimension to 3D geometry** tool is used to link selected dimension of **TechDraw** workbench to an edge or vertices of the 3D model. On doing so, the dimension in TechDraw changes as per the actual dimension in 3D model for same edge/vertices. The procedure to use this tool is given next.

- Create the dimension of edge/vertices that you want to be linked in **TechDraw** workbench page; refer to Figure-131.

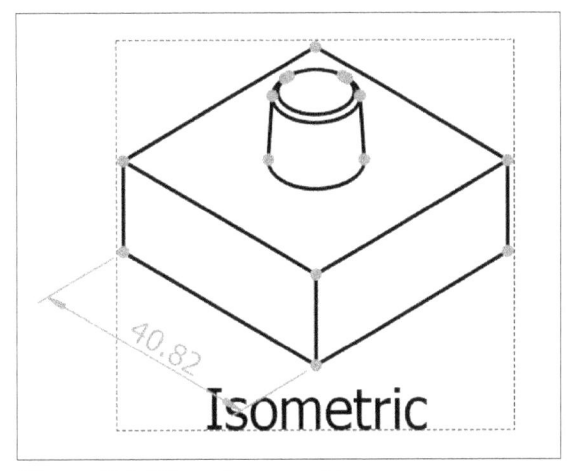

Figure-131. Dimension created in page

- Select the model tile at the bottom of drawing area and select the edge that you want to be linked to available dimensions along with page from **Model Tree**; refer to Figure-132.

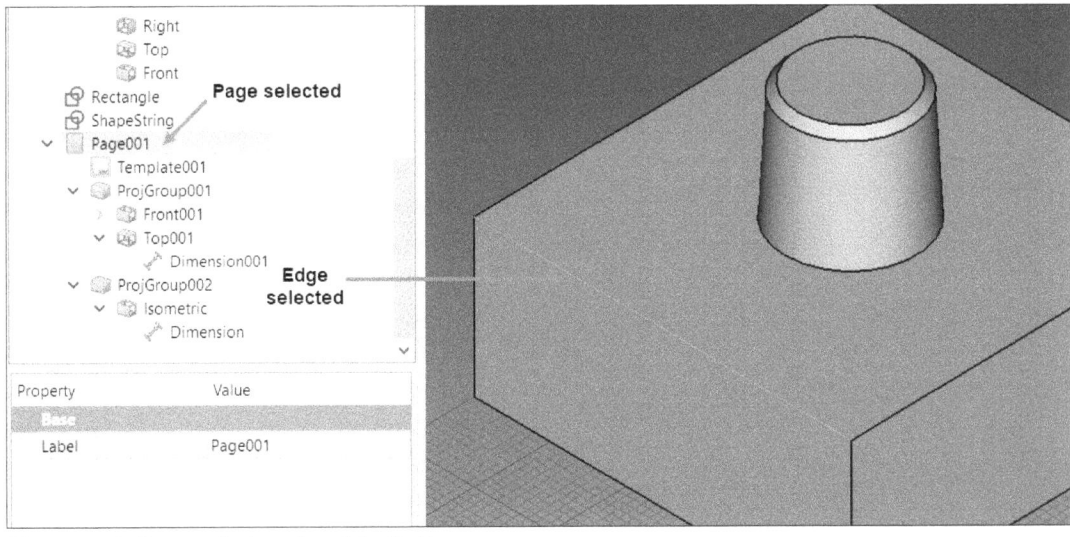

Figure-132. Page and edge selected for linking

- After selecting the edge and page, click on the **Link dimension to 3D geometry** tool from the **Toolbar** or **TechDraw > Dimensions** menu; refer to Figure-133. The **Link Dimension** dialog will be displayed; refer to Figure-134.

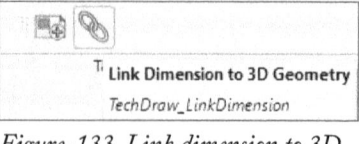

Figure-133. Link dimension to 3D geometry tool

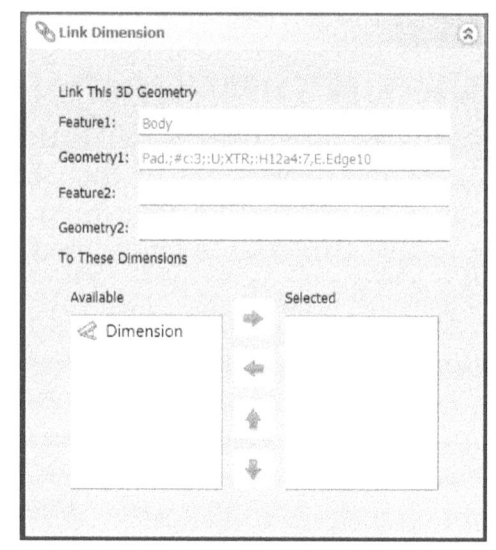

Figure-134. Link Dimension dialog

- Select the dimension to be linked from the **Available** area in the dialog and click on the **Add** button. The dimension will be linked to selected edge.
- After adding desired dimension for linking, click on the **OK** button from the dialog. The linked dimension will reflect actual dimension of the edge in 3D model.

Insert Balloon Annotation

The **Insert balloon annotation** tool is used to add balloons with leader line in a drawing.

- Select desired object from the drawing to which balloon will be attached and click on the **Insert balloon annotation** tool from the **Toolbar** or **TechDraw** menu; refer to Figure-135. The balloon will be applied to the object; refer to Figure-136. Press **ESC** to exit the tool.

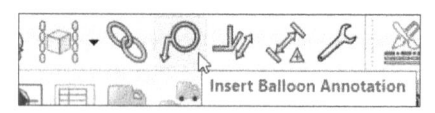

Figure-135. Insert Balloon Annotation

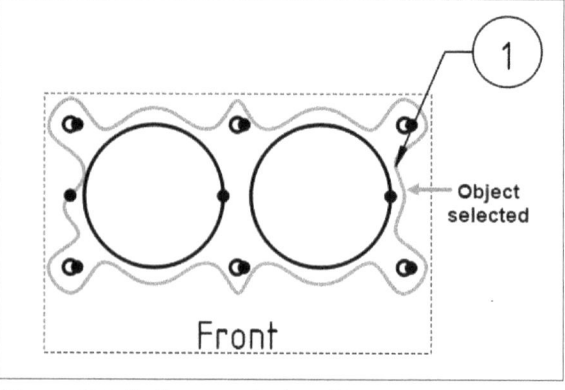

Figure-136. Balloon attached to object

- To move the bubble of a balloon, press and hold the left mouse button on its center and drag the mouse.
- To change the properties of balloon, double-click it in the drawing. The **Balloon** dialog will be displayed; refer to Figure-137. Set the parameters as desired and click on the **OK** button from the dialog.

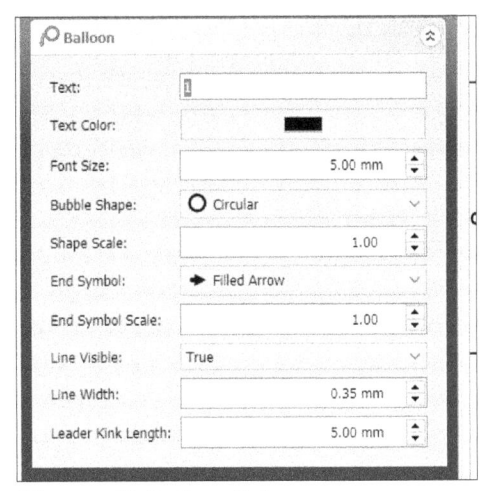

Figure-137. Balloon dialog

Note that position of a balloon is relative to its Source View and uses the same scale factor.

Axonometric Length Dimension

The **Axonometric Length Dimension** tool is used to add a length dimension to selected entity in axonometric view. The dimension may be the length of an edge or the distance between two points. The procedure to use this tool is given next.

- Select the line/edge to be dimensioned and another line/edge to define direction of extension lines. Click on the **Axonometric Length Dimension** tool from the **Toolbar** or **TechDraw** menu; refer to Figure-138. The dimension will be generated; refer to Figure-139.

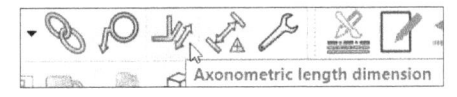

Figure-138. Axonometric length dimension

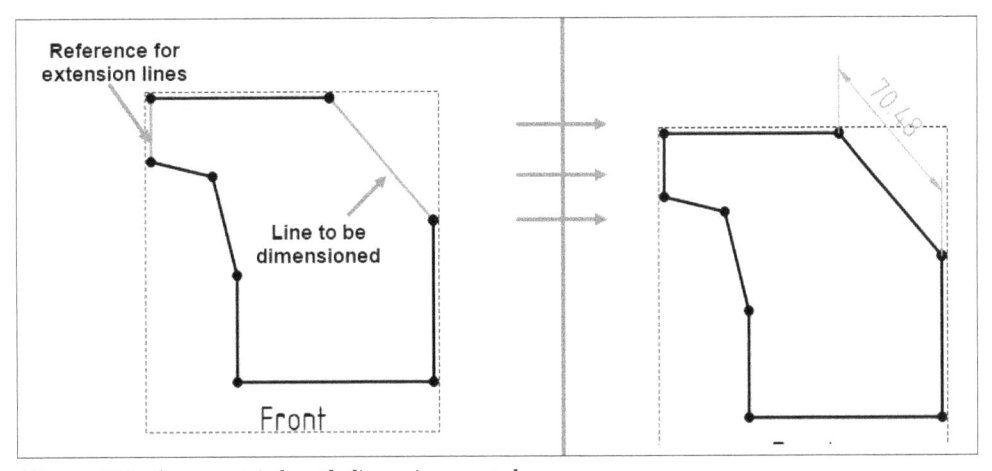

Figure-139. Axonometric length dimension created

Inserting Landmark Dimension

The **Insert Landmark Dimension** tool adds a linear dimension to the view. The dimension is based on two point objects from the 3D model. The procedure to use this tool is discussed next.

- Select desired model from the **Model** panel in **Combo View** and orient the model to desired orientation for placement in the drawing; refer to Figure-140.

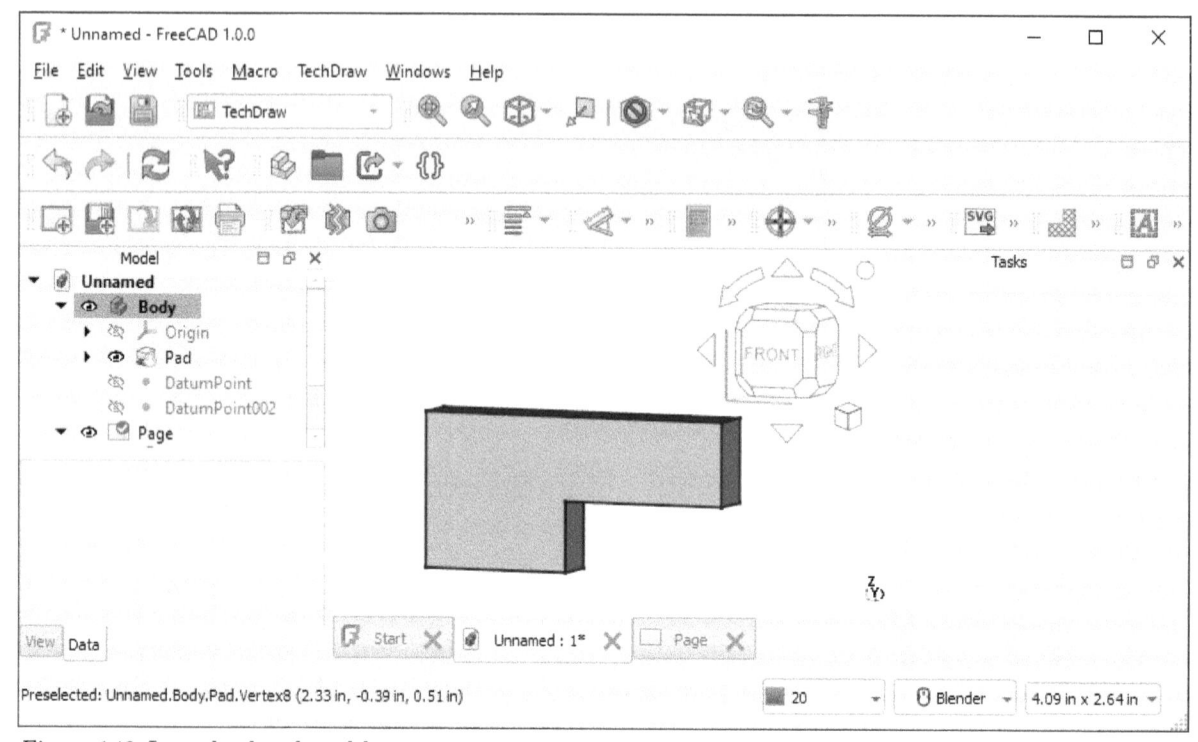

Figure-140. Insert landmark model

- Click on the **Create a datum point** tool from the toolbar in **Part Design** workbench and create two datum points on the model; refer to Figure-141.

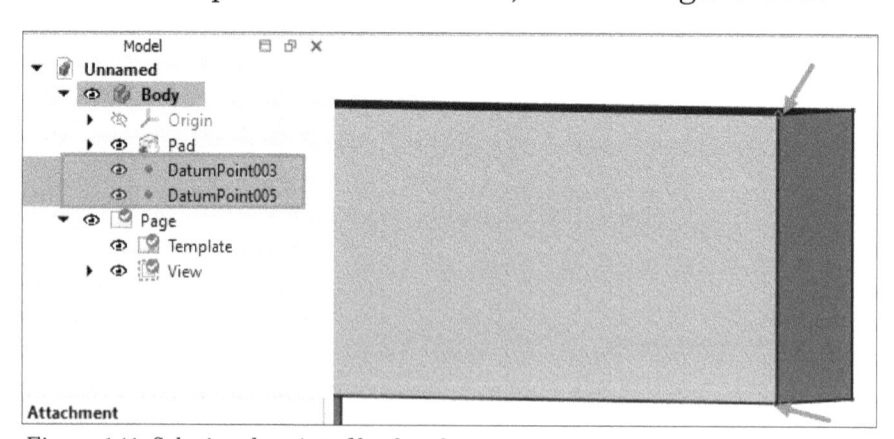

Figure-141. Selecting the point of landmark

- Now, switch to the **TechDraw** workbench.
- Select two datum points recently created and view in which you want to generate the dimension from the **Model** panel in **Combo View**; refer to Figure-142 and click on the **Insert Landmark Dimension** tool. The landmark dimension will be created; refer to Figure-143.

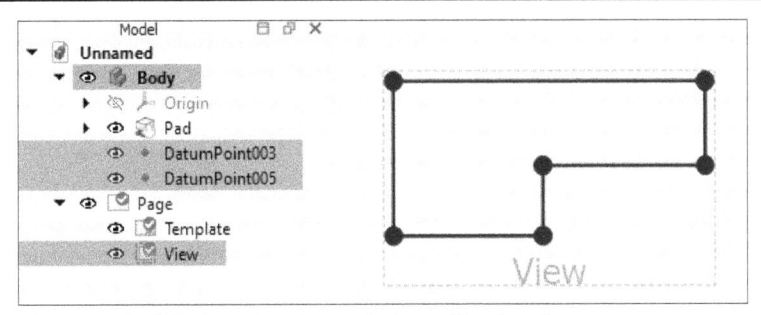

Figure-142. Selecting the point and view of landmark

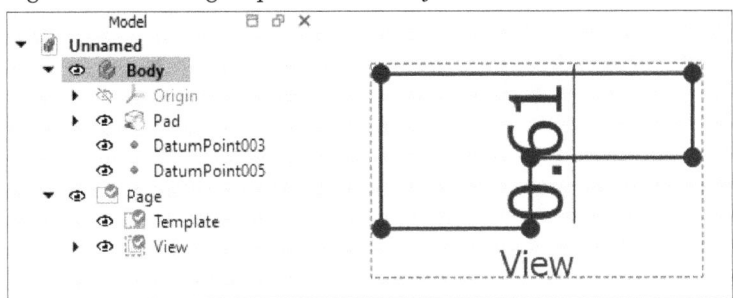

Figure-143. Insert landmark dimension created

Repair Dimension Reference

The **Repair Dimension Reference** tool is used to adjust the 2D or 3D geometry references of a dimension after the model has been modified.

- Select the dimension whose reference is to be repaired and click on the **Repair Dimension Reference** tool from the **Toolbar** or **TechDraw** menu; refer to Figure-144. The **Dimension Repair** dialog will be displayed; refer to Figure-145.

Figure-144. Repair Dimension Reference tool

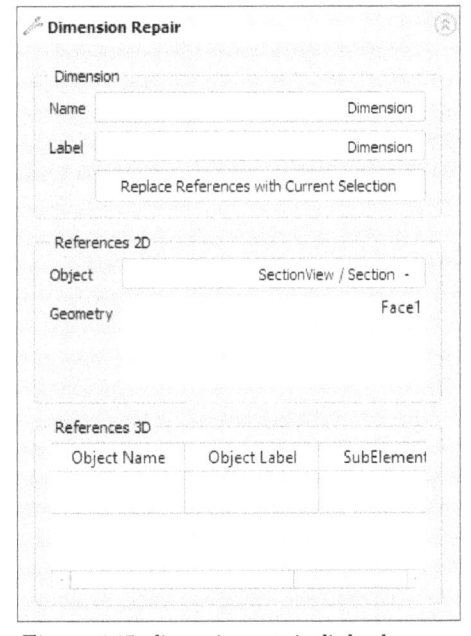

Figure-145. dimension repair dialog box

- Select new geometry references to be dimensioned and click on the **Replace References with Current Selection** button.
- Press the **OK** button to update the dimension; refer to Figure-146.

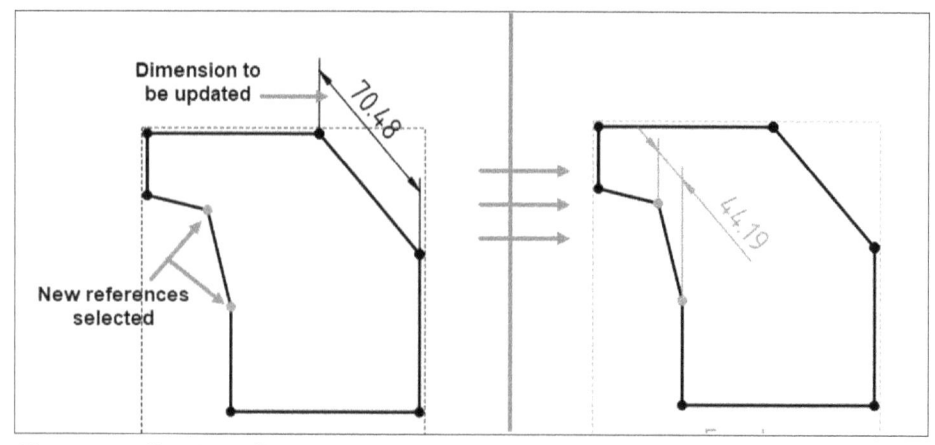

Figure-146. Repairing dimension

EXPORTING PAGE TO SVG FILE

The **Export Page as SVG** tool is used to export selected technical drawing page as an SVG image file. The procedure to use this tool is given next.

- Click on the **Export Page as SVG** tool from the **Toolbar** or **TechDraw** menu; refer to Figure-147. The **Export page as SVG** dialog box will be displayed; refer to Figure-148.

Figure-147. Export page as SVG file tool

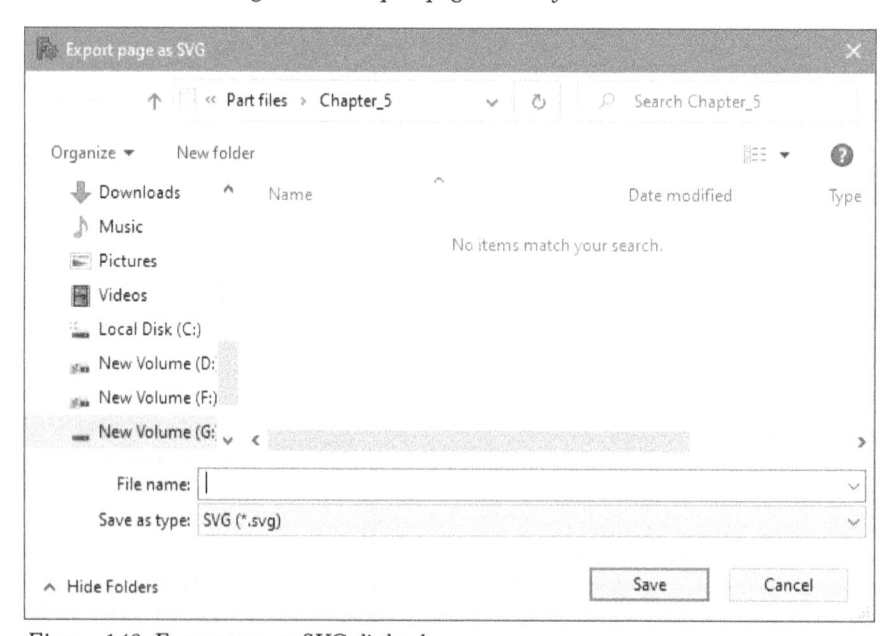

Figure-148. Export page as SVG dialog box

- Specify desired name of file to be exported in the **File name** edit box and click on the **Save** button to save the file. You can open this svg file in your default web browser.

EXPORTING PAGE AS DXF FILE

The **Export Page as DXF** tool is used to export current drawing page as DXF file generally used by CAD programs for generating engineering drawings. The procedure to use this tool is given next.

- Click on the **Export Page as DXF** tool from the **Toolbar** or **TechDraw** menu; refer to Figure-149. The **Save DXF File** dialog box will be displayed; refer to Figure-150.

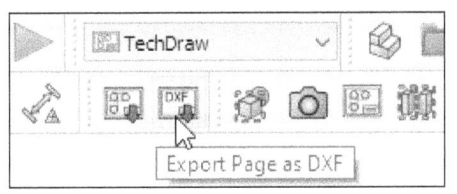

Figure-149. Export page as DXF tool

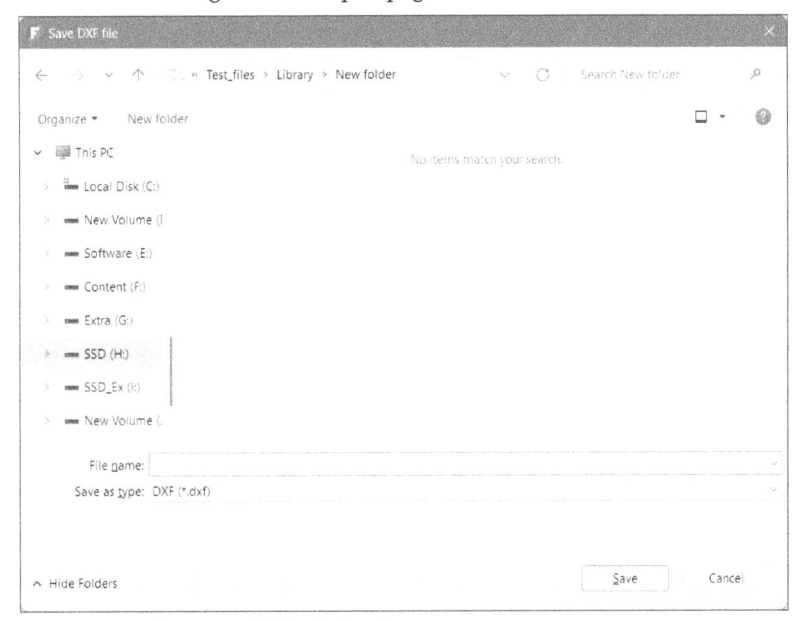

Figure-150. Save DXF File dialog box

- Specify desired name of file in the **File name** edit box and click on the **Save** button from the dialog box. The file will be created. You can open this file using any Engineering drawing CAD software.

APPLYING HATCH PATTERN

The **Hatch a Face using Image file** tool is used to apply hatch pattern to selected faces. The procedure to use this tool is given next.

- Select the face on which you want to apply hatch pattern and click on the **Hatch a Face using Image file** tool from the **TechDraw Decoration** toolbar or **TechDraw > Hatching** menu; refer to Figure-151. The **Create Face Hatch** dialog will be displayed in the **Tasks** panel of **Combo View** along with the preview of hatch pattern; refer to Figure-152.

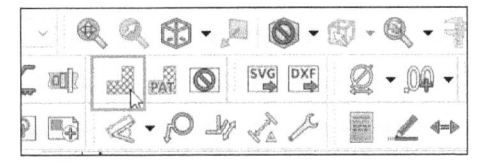

Figure-151. Hatch face using image file tool

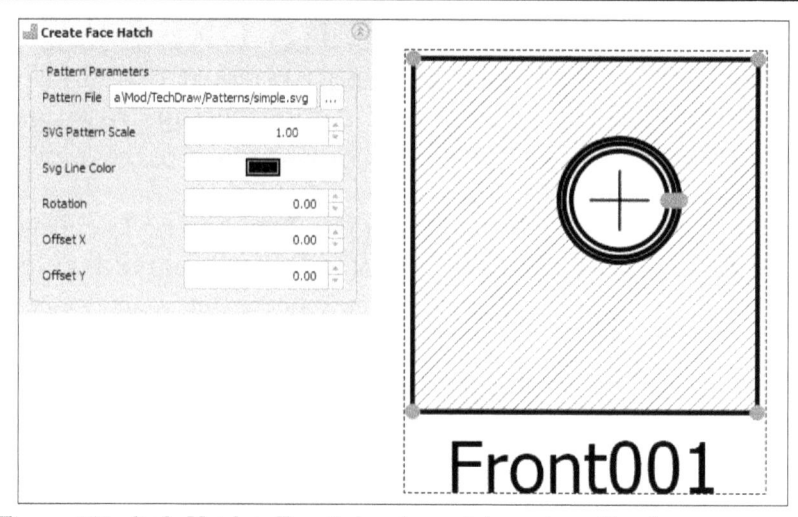

Figure-152. Apply Hatch to Face dialog along with preview of hatch pattern

- Click on the **Browse** button for **Pattern File** box from **Define your pattern** area of the dialog. The **Select a file** dialog box will be displayed; refer to Figure-153.

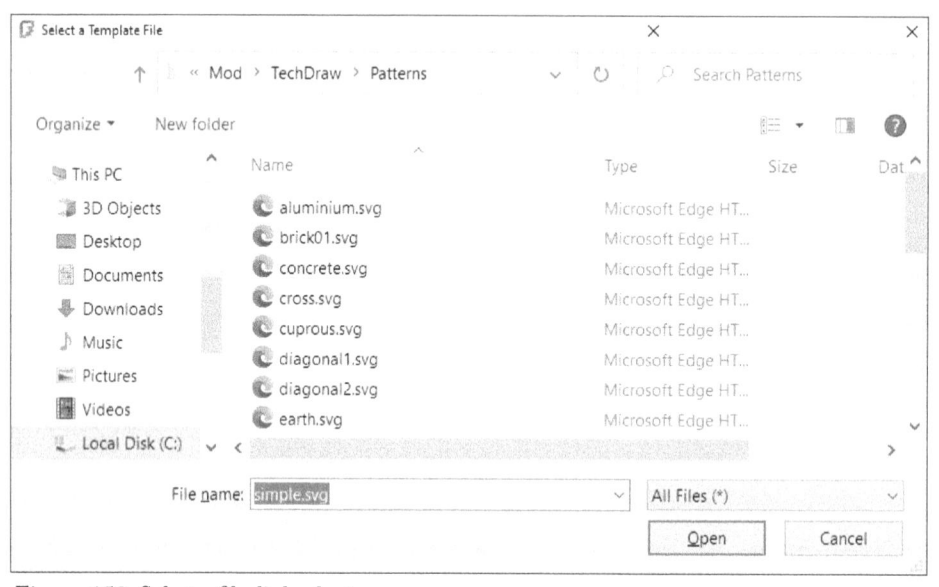

Figure-153. Select a file dialog box

- Select desired image file of hatch pattern from the dialog box and click on the **Open** button. The hatch pattern in drawing will be modified, accordingly.
- Enter desired value in **SVG Pattern Scale** edit box to enlarges/shrinks the pattern.
- Click on the **SVG Line Color** button to specify the color of pattern lines.
- Specify desired values in **Rotation**, **Offset X**, and **Offset Y** edit boxes to reorient the pattern.
- After specifying desired parameters, click on the **OK** button from the dialog to create the pattern.

APPLYING GEOMETRIC HATCH TO A FACE

The **Apply Geometric Hatch to Face** tool is used to apply hatch based on specified geometric parameters. The procedure to use this tool is given next.

- Select a face and click on the **Apply Geometric Hatch to Face** tool from the **TechDraw Decoration toolbar** or **TechDraw > Hatching** menu; refer to Figure-154. The **Apply Geometric Hatch to Face** dialog will be displayed; refer to Figure-155.

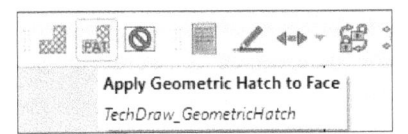

Figure-154. Apply geometric hatch to face tool

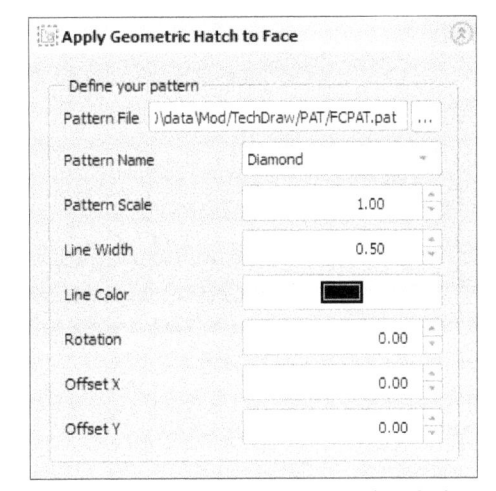

Figure-155. Apply Geometric Hatch to Face dialog

- Select desired option from the **Pattern Name** drop-down to define pattern to be created for hatching.
- Specify the scale, line weight, and line color for pattern in respective fields of the dialog.
- The other options of the dialog box as discussed earlier.
- After setting desired parameters, click in the face on which you want to apply the pattern and click on the **OK** button from the dialog to create the pattern.

INSERTING SYMBOL FROM AN SVG FILE

The **Insert SVG Symbol** tool is used to insert a symbol using a local svg file available. The procedure to use this tool is given next.

- Click on the **Insert SVG Symbol** tool from the **Toolbar** or **TechDraw > TechDraw Views** menu; refer to Figure-156. The **Choose an SVG file to open** dialog box will be displayed; refer to Figure-157.

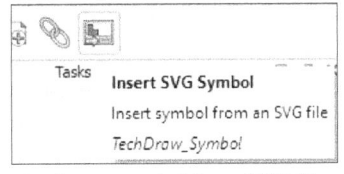

Figure-156. Insert symbol from SVG file tool

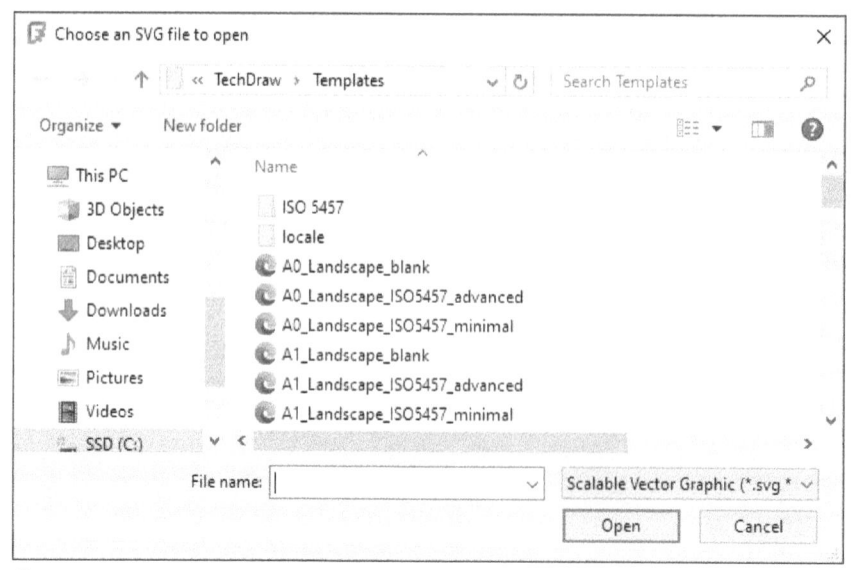

Figure-157. Choose an SVG file to open dialog box

- Select desired symbol file and click on the **Open** button. The symbol will be inserted at the center of the page; refer to Figure-158.

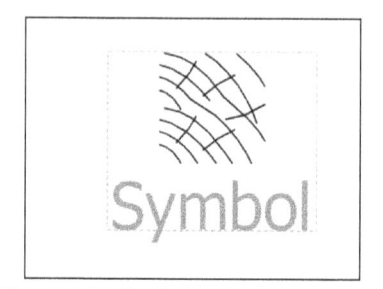

Figure-158. Symbol inserted in drawing

INSERTS A BITMAP FROM A FILE INTO A PAGE

The **Inserts Bitmap Image** tool is used to insert an image file in the page. The procedure to use this tool is given next.

- Click on the **Inserts Bitmap Image** tool from the **Toolbar** or **TechDraw > TechDraw Views** menu; refer to Figure-159. The **Select an Image File** dialog box will be displayed; refer to Figure-160.

Figure-159. Insert bitmap image tool

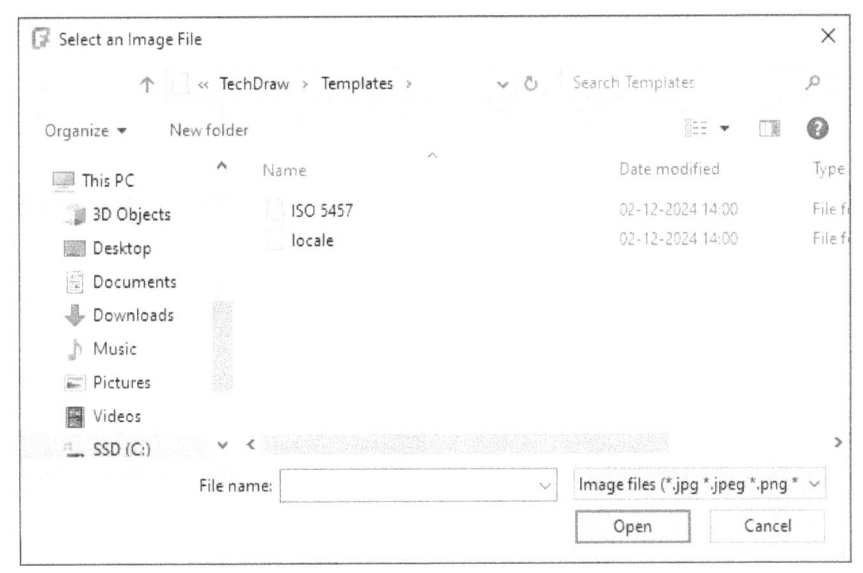

Figure-160. Select an Image File dialog box

- Select desired image file to be inserted in the page and click on the **Open** button. The selected image file will be inserted in the drawing. Note that only PNG, BMP, and JPG/JPEG formats are supported for image insertion.
- Select the **Image** feature from the **Model Tree** and specify desired parameters in the **Property Editor**.
- You can specify width and height of image in the **Width** and **Height** edit boxes of the **Image** section in **Property Editor**.

TURNING VIEW FRAMES ON/OFF

The **Turn View Frames On/Off** tool is used to turn ON or turn OFF frames around the views in drawing page. Click on the **Turn View Frames On/Off** tool from **Toolbar** or **TechDraw > TechDraw Views** menu to toggle display of frames around the views; refer to Figure-161.

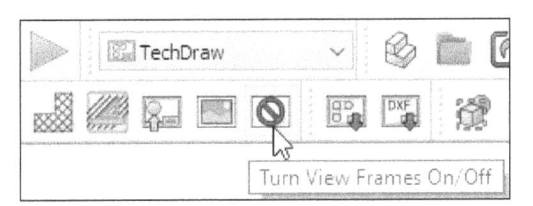

Figure-161. Turn view frames on off tool

SELECTING LINE ATTRIBUTES, CASCADE SPACING, AND DELTA DISTANCE

This tool is used to select the attributes (style, width, and color) for new cosmetic lines and centerlines, and specifies the cascade spacing and delta distance. The procedure to use this tool is discussed next.

- Click on the **Select Line Attributes, Cascade Spacing, and Delta Distance** tool from the **TechDraw Attribute Toolbar** or **Techdraw > Extensions: Attributes/Modifications** menu; refer to Figure-162. The **Select line attributes** dialog will be displayed; refer to Figure-163.

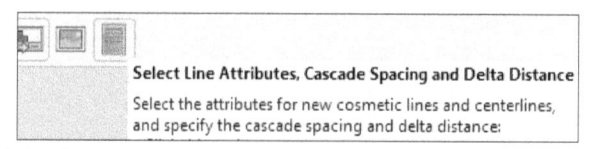

Figure-162. Select Line Attributes Cascade Spacing and Delta Distance tool

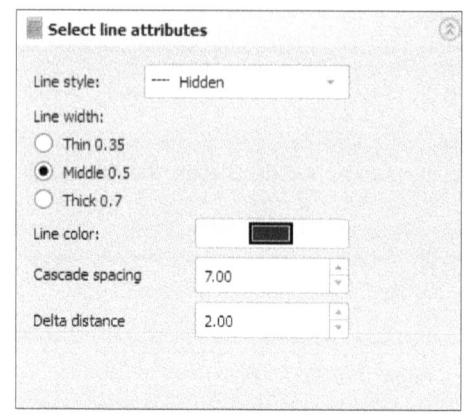

Figure-163. Select line attributes dialog

- Specify desired line style, line width, and line color by selecting desired radio button from **Line style**, **Line width**, and **Line color** sections of the dialog, respectively.
- Specify desired values in the **Cascade spacing** and **Delta distance** edit boxes of the dialog.
- After specifying desired parameters, click on the **OK** button from the dialog.

CHANGING LINE ATTRIBUTES

The **Change Line Attributes** tool is used to change the attributes (style, width, and color) of cosmetic lines and centerlines. The procedure to use this tool is discussed next.

- Select desired lines or centerlines for which you want to change the attributes; refer to Figure-164.

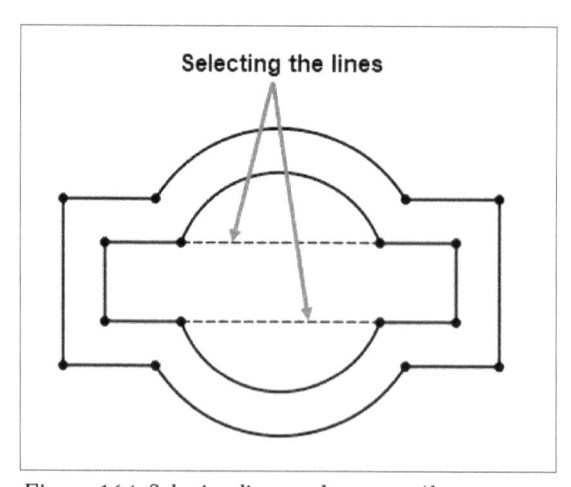

Figure-164. Selecting lines to change attributes

- Click on the **Change Line Attributes** tool from **Toolbar** or **Techdraw > Extensions: Attributes/Modifications** menu; refer to Figure-165. The line attributes will be changed to specified attributes created by using the **Select Line Attributes, Cascade Spacing, and Delta Distance** tool; refer to Figure-166.

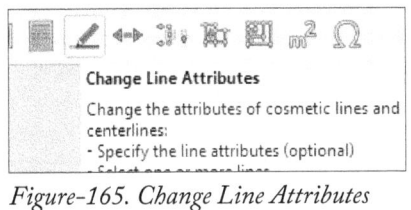

Figure-165. Change Line Attributes tool

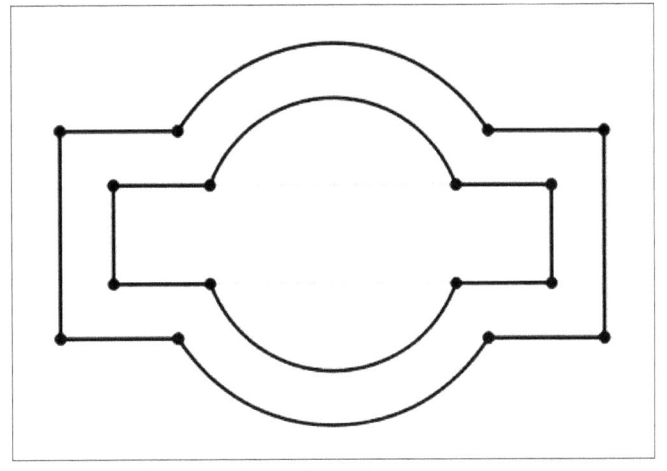

Figure-166. Line attributes changed

EXTENDING LINE

The **Extend Line** tool extends a cosmetic line or centerline at both ends. The procedure to use this tool is discussed next.

- Select desired line which you want to extend; refer to Figure-167.

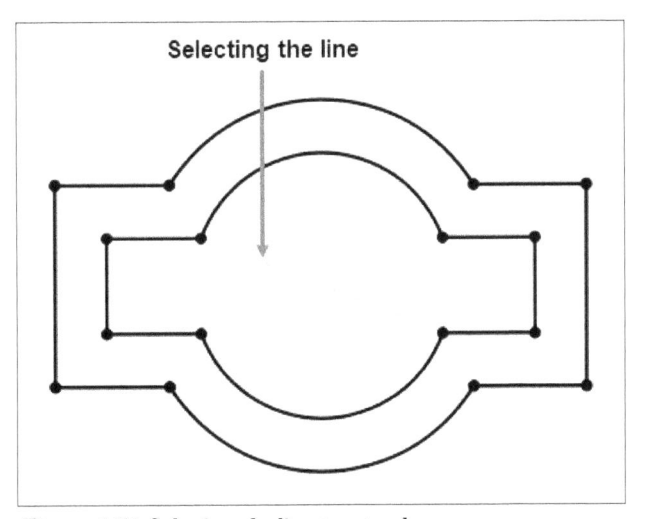

Figure-167. Selecting the line to extend

- Click on the **Extend Line** tool from drop-down in the **Toolbar** or **Techdraw > Extensions: Attributes/Modifications** menu; refer to Figure-168. The line will be extended from both the ends; refer to Figure-169.

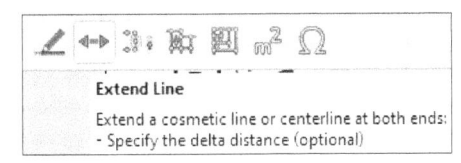

Figure-168. Extend Line tool

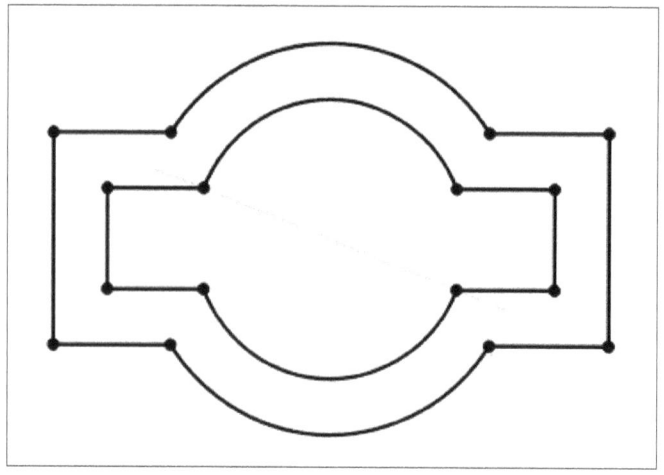

Figure-169. Line extended from both the ends

SHORTENING THE LINE

The **Shorten Line** tool is used to shorten a cosmetic line or centerline at both ends. The procedure to use this tool is same as discussed for **Extend Line** tool; refer to Figure-170.

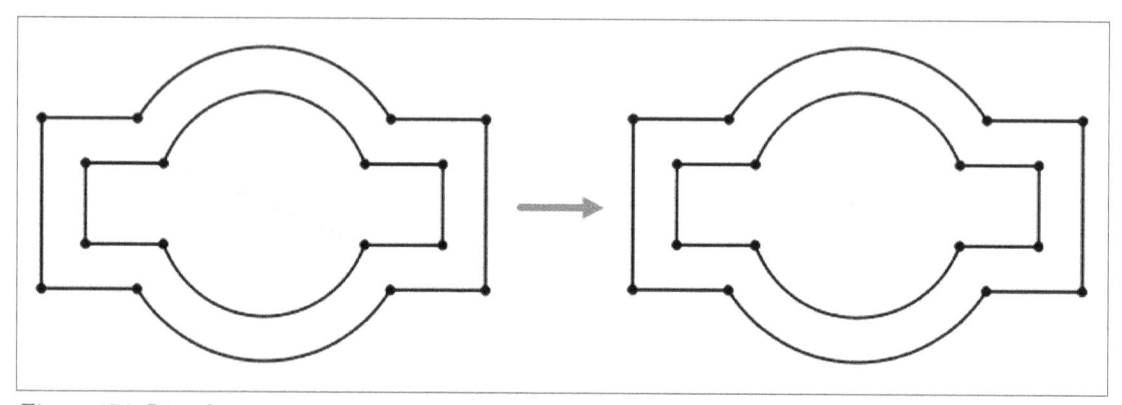

Figure-170. Line shorten

LOCKING/UNLOCKING THE VIEW

The **Lock/Unlock View** tool is used to lock or unlock the position of a view. A locked view has a fixed position relative to the page or the group it belongs to.

POSITIONING SECTION VIEW

The **Position Section View** tool is used to align a section view orthogonally with its source view. The procedure to use this tool is discussed next.

- Select the section view which you want to align orthogonally; refer to Figure-171.

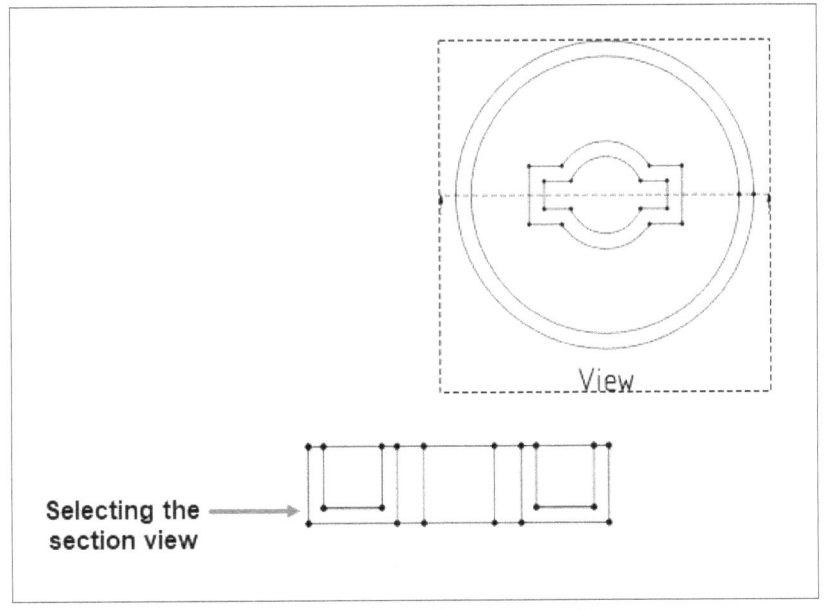

Figure-171. Selecting the section view to align orthogonally

- Click on the **Position Section View** tool from **Toolbar** or **Techdraw > Extensions: Attributes/Modifications** menu; refer to Figure-172. The section view will be aligned orthogonally with the source view; refer to Figure-173.

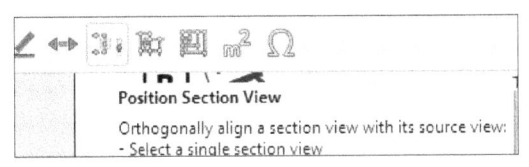

Figure-172. Position Section View tool

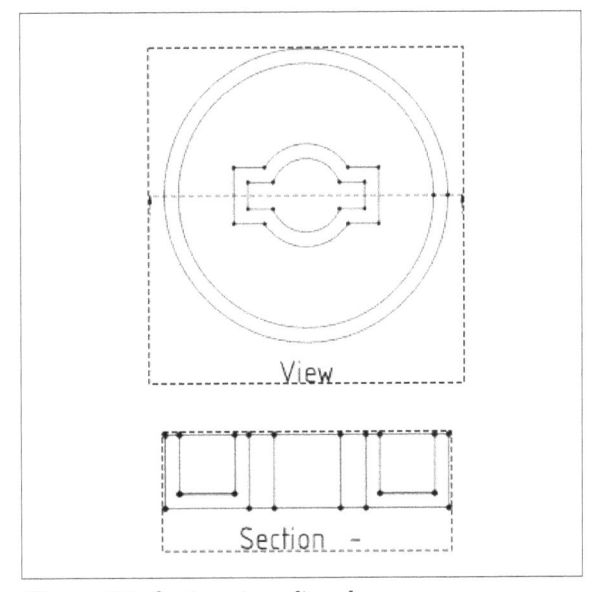

Figure-173. Section view aligned

POSITIONING HORIZONTAL CHAIN DIMENSIONS

The **Position Horizontal Chain Dimensions** tool is used to align horizontal dimensions to create a chain dimension. The procedure to use this tool is discussed next.

- Select two or more horizontal dimensions which you want to align to create a chain dimensions; refer to Figure-174.

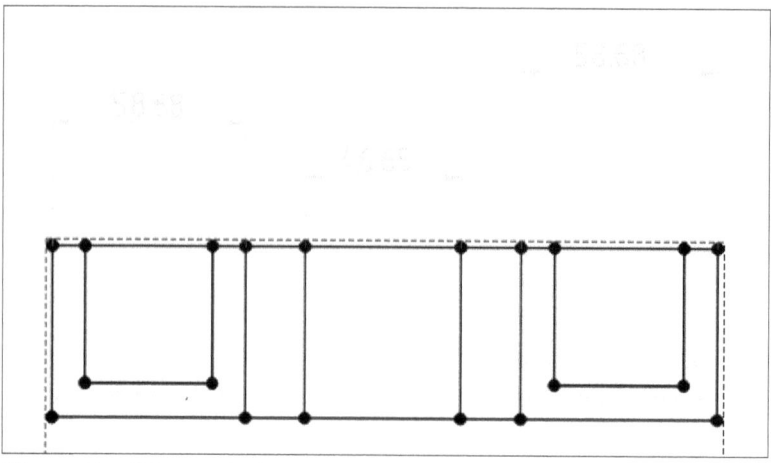

Figure-174. Selecting the horizontal dimensions to align

- Click on the **Position Horizontal Chain Dimensions** tool from drop-down in the **Toolbar** or **Techdraw > Extensions: Attributes/Modifications** menu; refer to Figure-175. The dimensions will be aligned to create chain dimensions; refer to Figure-176.

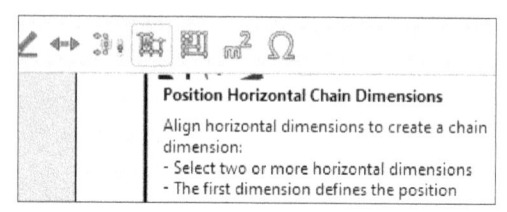

Figure-175. Position Horizontal Chain Dimensions tool

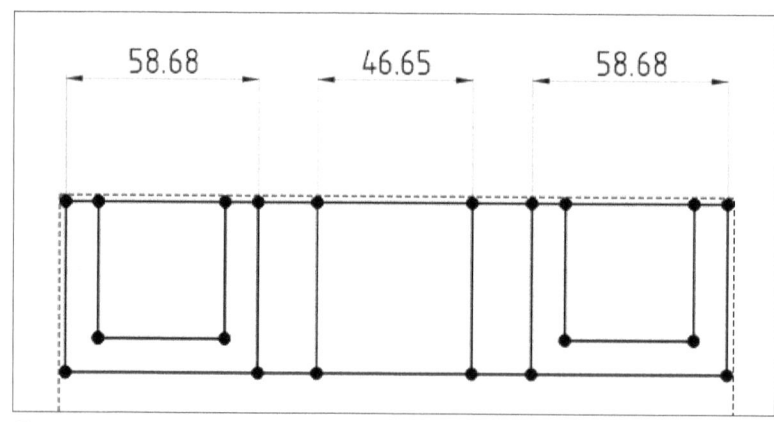

Figure-176. Dimensions aligned

POSITIONING VERTICAL CHAIN DIMENSIONS

The **Position Vertical Chain Dimensions** tool is used to align vertical dimensions to create a chain dimension. The procedure to use this tool is same as discussed for previous tool; refer to Figure-177.

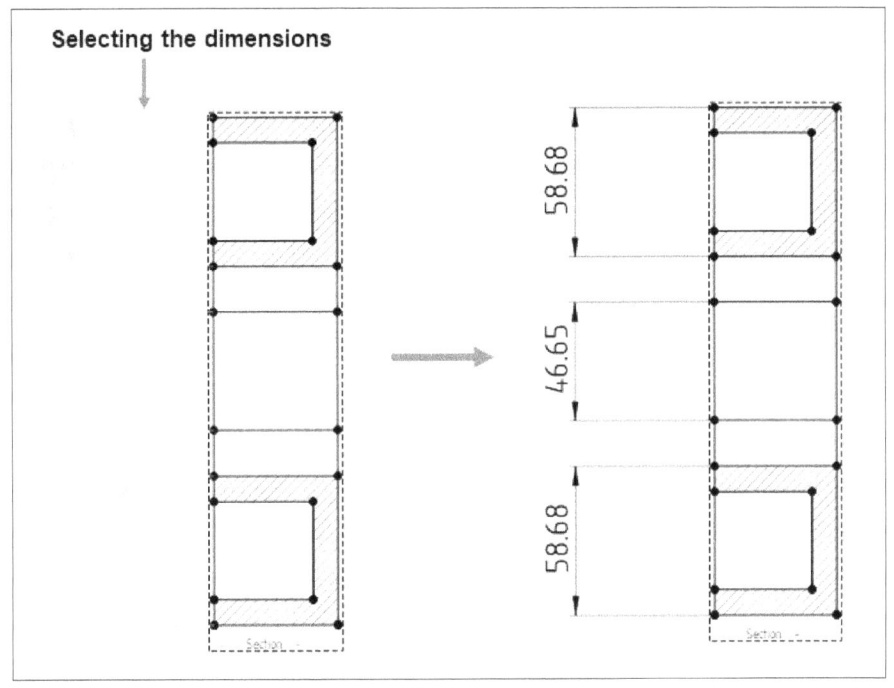

Figure-177. Vertical dimensions aligned

POSITIONING OBLIQUE CHAIN DIMENSIONS

The **Position Oblique Chain Dimensions** tool is used to align oblique dimensions to create a chain dimension. The procedure to use this tool is same as discussed for previous tool; refer to Figure-178.

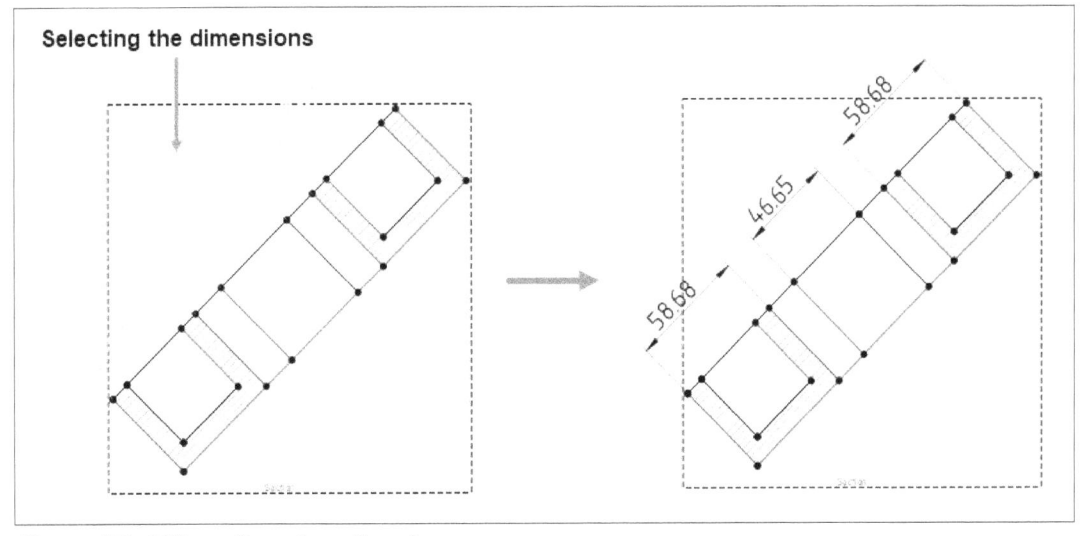

Figure-178. Oblique dimensions aligned

CASCADING HORIZONTAL DIMENSIONS

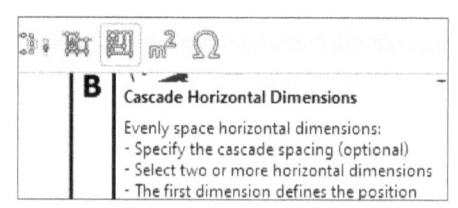

The **Cascade Horizontal Dimensions** tool is used to evenly space the horizontal dimensions. The procedure to use this tool is discussed next.

- Select two or more horizontal dimensions which you want to space evenly; refer to Figure-179.

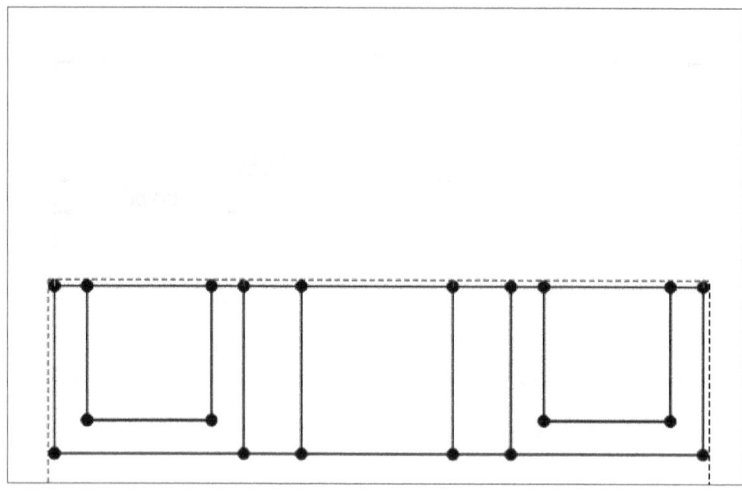

Figure-179. Selecting horizontal dimensions to space evenly

- Click on the **Cascade Horizontal Dimensions** tool from drop-down in the **Toolbar** or **Techdraw > Extensions: Attributes/Modifications** menu; refer to Figure-180. The dimensions will be evenly spaced; refer to Figure-181.

Figure-180. Cascade Horizontal Dimensions tool

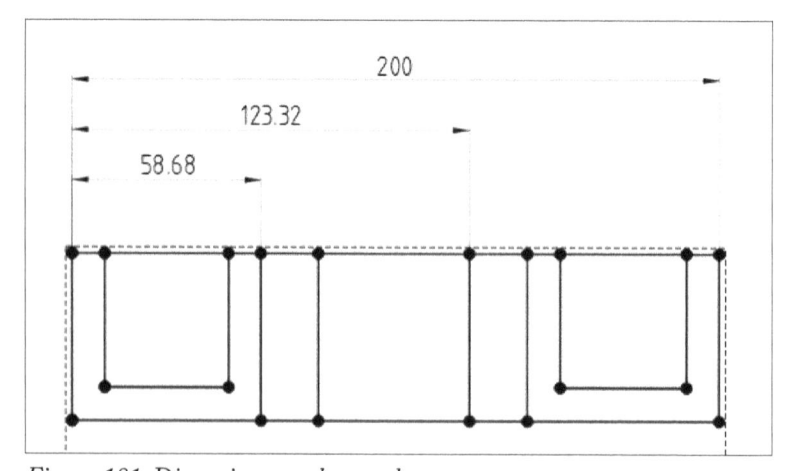

Figure-181. Dimensions evenly spaced

CASCADING VERTICAL DIMENSIONS

The **Cascade Vertical Dimensions** tool is used to evenly space the vertical dimensions. The procedure to use this tool is same as discussed for previous tool; refer to Figure-182.

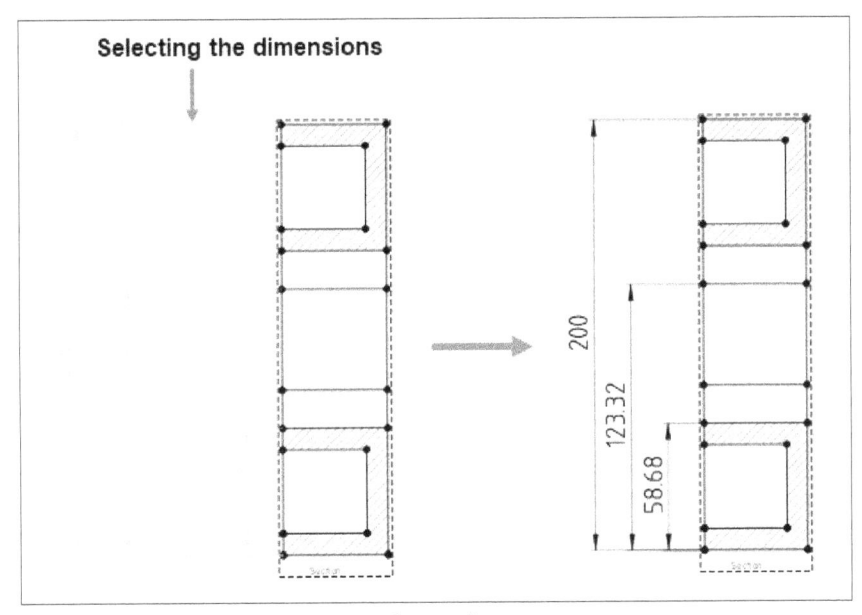

Figure-182. Vertical dimensions evenly spaced

CASCADING OBLIQUE DIMENSIONS

The **Cascade Oblique Dimensions** tool is used to evenly space the oblique dimensions. The procedure to use this tool is same as discussed for previous tool; refer to Figure-183.

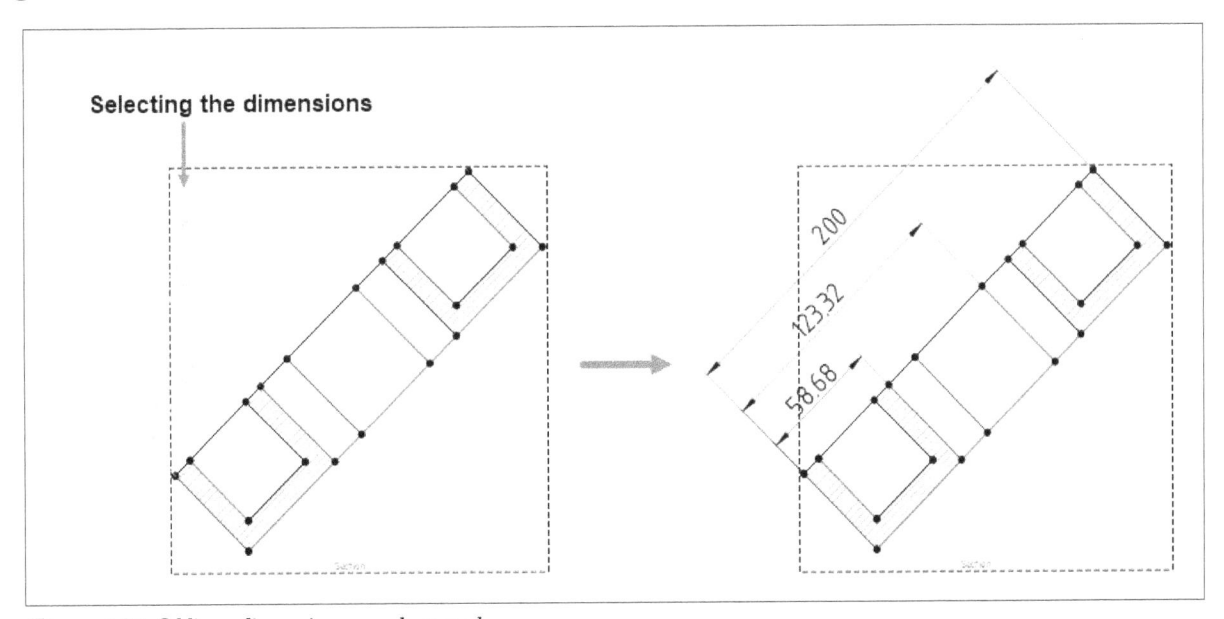

Figure-183. Oblique dimensions evenly spaced

CALCULATING THE AREA OF SELECTED FACES

The **Calculate the area of selected faces** tool is used to calculate the area of selected faces and inserts an area annotation. The tool can calculate the area of faces with only straight edges. The procedure to use this tool is discussed next.

- Select one or more faces with only straight edges; refer to Figure-184.

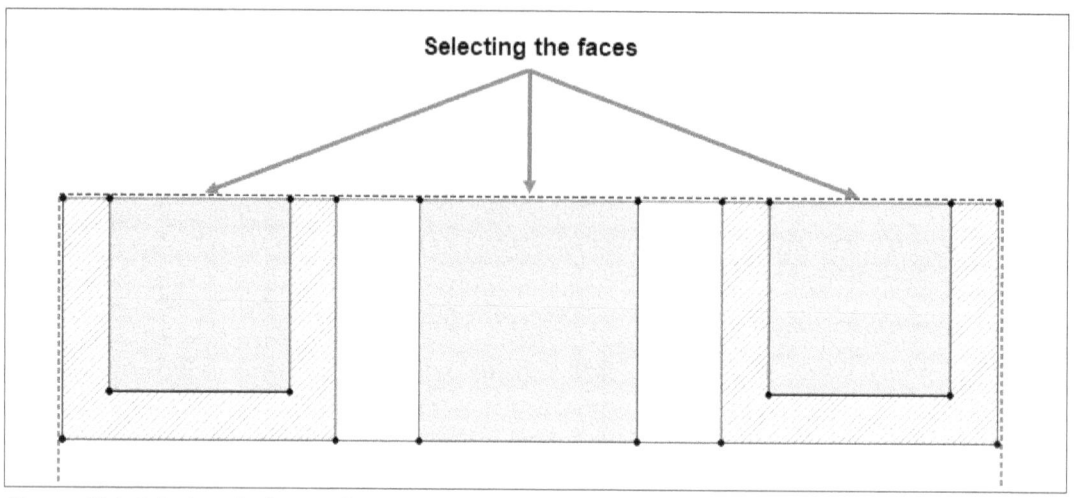

Figure-184. Selecting the faces with straight edges

- Click on the **Calculate the area of selected faces** tool from **Toolbar** or **Techdraw > Extensions: Attributes/Modifications** menu; refer to Figure-185. The area of selected faces will be calculated; refer to Figure-186.

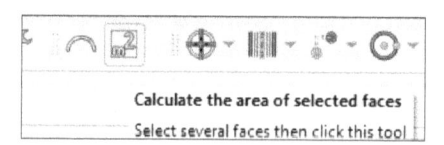

Figure-185. Area Annotation tool

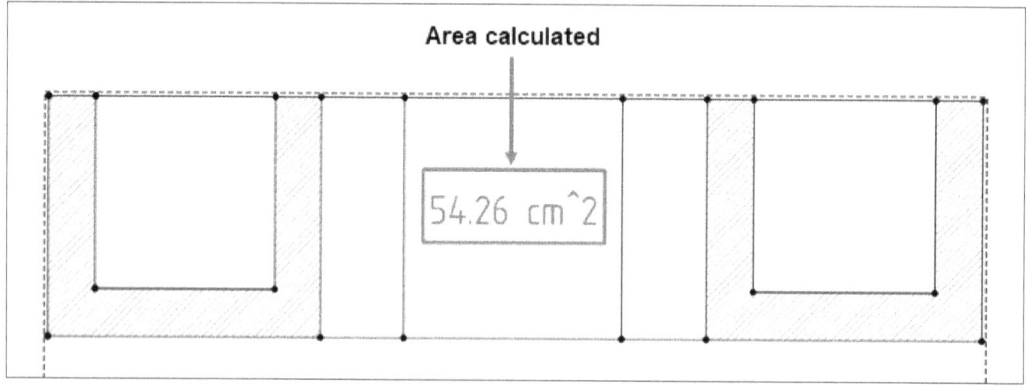

Figure-186. Area calculated of selected faces

CALCULATE THE ARC LENGTH OF SELECTED EDGES

The **Calculate the arc length of selected edges** tool is used to calculate the arc length of selected edges and inserts an arc length annotation. The procedure to use this tool is discussed next.

- Select the edge from the drawing; refer to Figure-187.

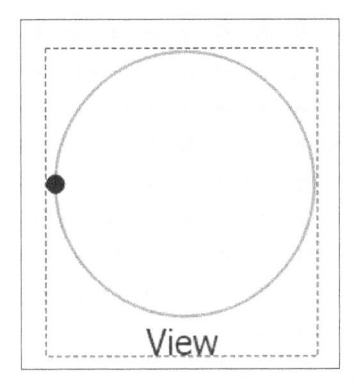

Figure-187. Selecting the edge for arc length

- After selecting the edge, click on the **Calculate the arc length of selected edges** tool from the toolbar or **Techdraw > Extensions: Attributes/Modifications** menu; refer to Figure-188. The feature will be created; refer to Figure-189.

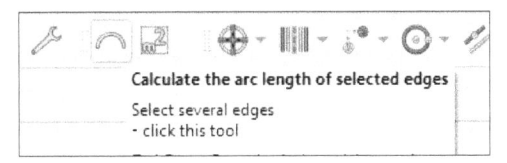

Figure-188. Calculate the arc length of selected edge tool

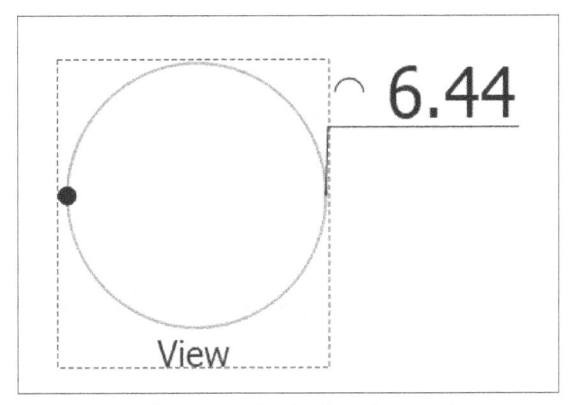

Figure-189. Calculate the Arc length created

CUSTOMIZING FORMAT LABEL

The **Customize Format Label** tool is used to customize the formatting of a balloon text or dimension text. The procedure to use this tool is discussed next.

- Select a balloon object or a dimension object from **Model Tree** or from drawing view area; refer to Figure-190.

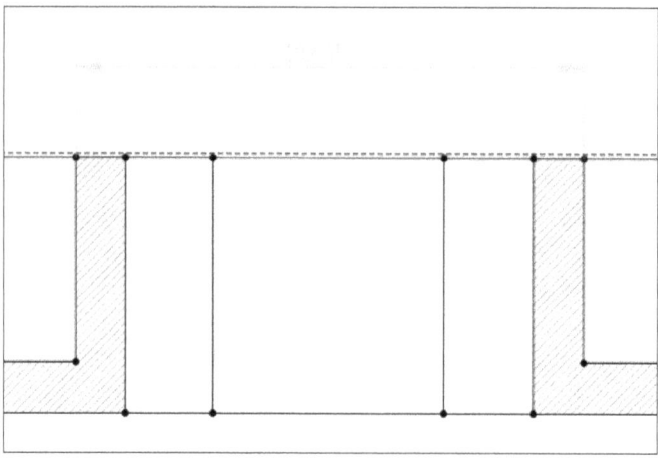

Figure-190. Selecting the dimension to customize

- Click on the **Customize Format Label** tool from **Toolbar** or **Techdraw > Extensions: Attributes/Modifications** menu; refer to Figure-191. The **Customize Format** dialog will be displayed; refer to Figure-192.

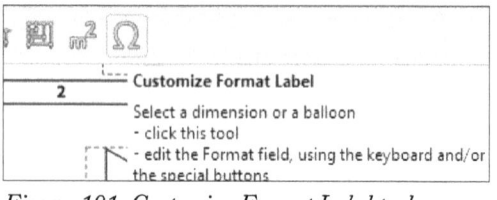

Figure-191. Customize Format Label tool

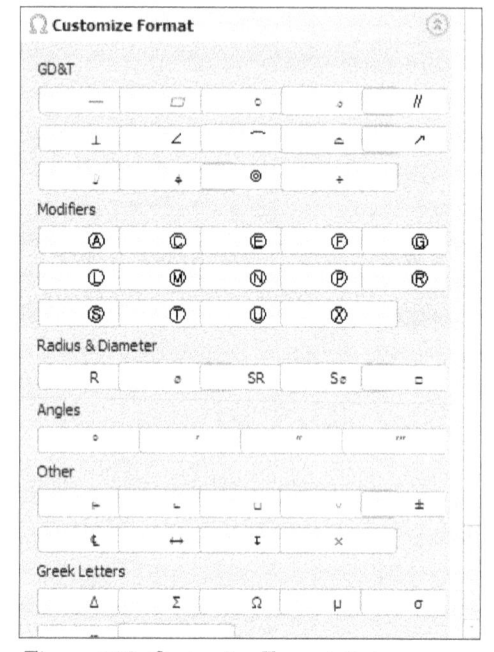

Figure-192. Customize Format dialog

- Select desired options from **GD&T**, **Modifiers**, **Radius & Diameter**, **Angles**, **Other** and **Greek Letters** sections of the dialog to customize the formatting of selected dimension. The selected options will be displayed in the **Format** edit box of the dialog.
- The **Preview** section of the dialog displays the result of the current formatting.
- After specifying desired parameters, click on the **OK** button from the dialog. The formatting of dimension will be customized; refer to Figure-193.

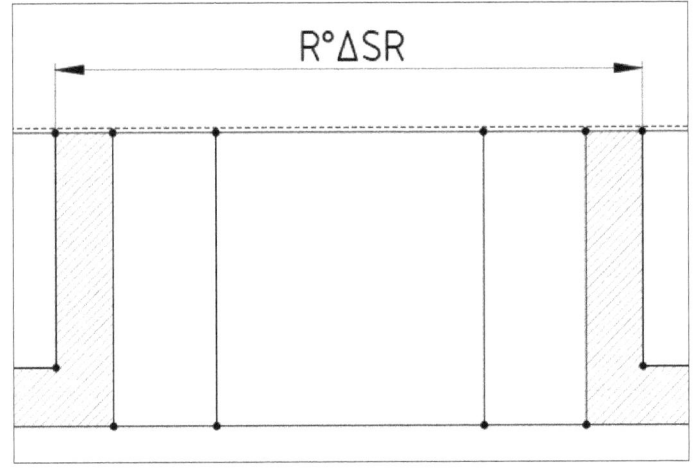

Figure-193. Formatting of dimension customized

ADDING CIRCLE CENTERLINES

The **Add Circle Centerlines** tool is used to add centerlines to circles and arcs. The procedure to use this tool is discussed next.

- Select one or more circles or arcs to which you want to add centerlines; refer to Figure-194.

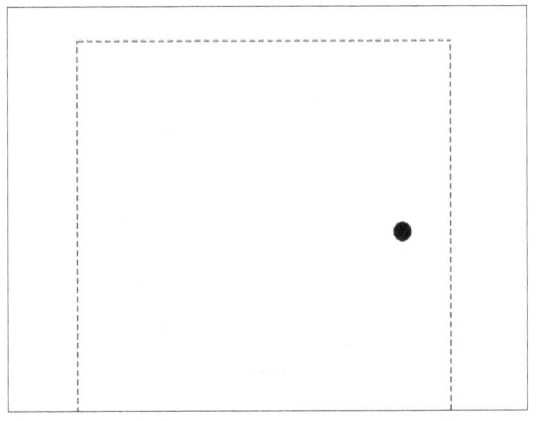

Figure-194. Selecting the circle to add centerlines

- Click on the **Add Circle Centerlines** tool from drop-down in the **Toolbar** or **Techdraw > Extensions: Centerlines/Threading** menu; refer to Figure-195. The centerlines will be added to the circle; refer to Figure-196.

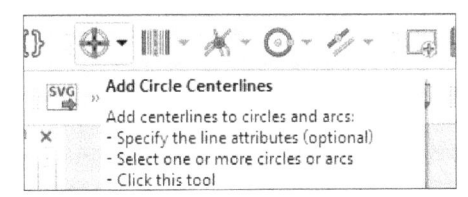

Figure-195. Add Circle Centerlines tool

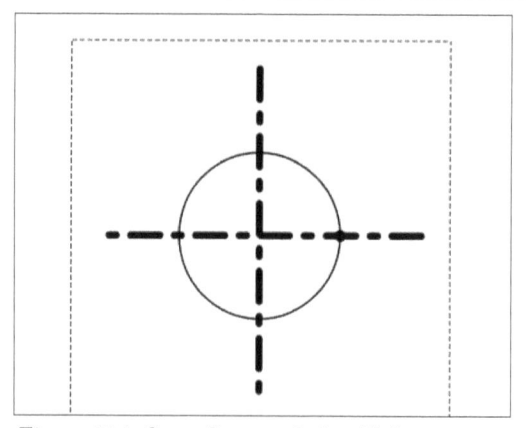

Figure-196. Centerlines on circle added

ADDING BOLT CIRCLE CENTERLINES

The **Add Bolt Circle Centerlines** tool is used to add centerlines to a circular pattern of circles. The procedure to use this tool is same as discussed for **Add Circle Centerlines** tool; refer to Figure-197.

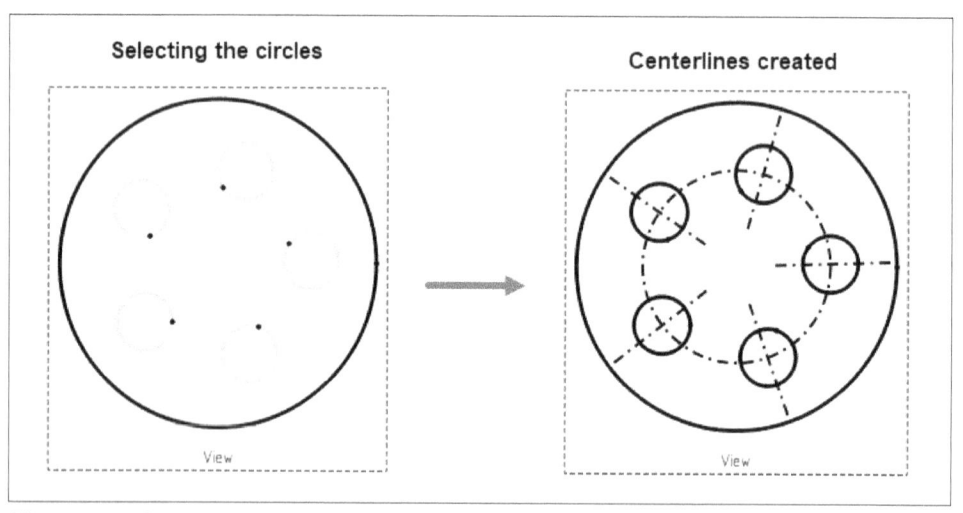

Figure-197. Centerlines on circular pattern created

ADDING COSMETIC THREAD HOLE SIDE VIEW

The **Add Cosmetic Thread Hole Side View** tool is used to add a cosmetic thread to the side view of a hole. The procedure to use this tool is discussed next.

- Select two parallel lines of the side view of a hole; refer to Figure-198.

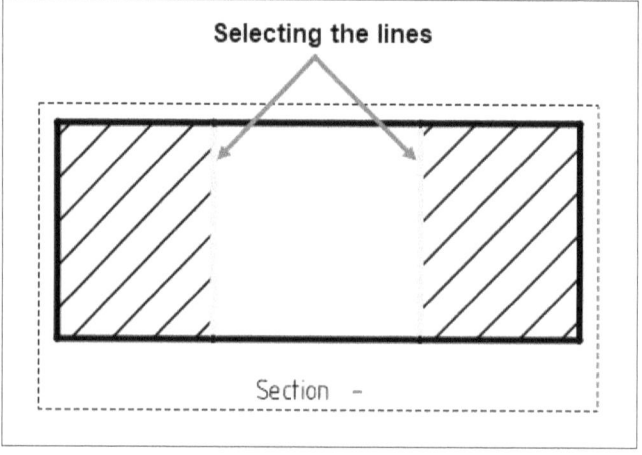

Figure-198. Selecting two parallel lines to add cosmetic thread

- Click on the **Add Cosmetic Thread Hole Side View** tool from drop-down in the **Toolbar** or **Techdraw > Extensions: Centerlines/Threading** menu; refer to Figure-199. The cosmetic threads will be added to the side view of a hole; refer to Figure-200.

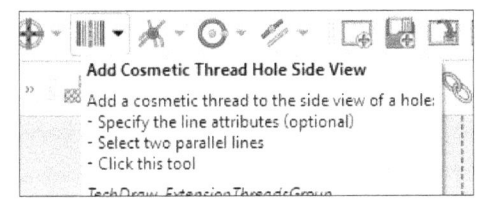

Figure-199. Add Cosmetic Thread Hole Side View tool

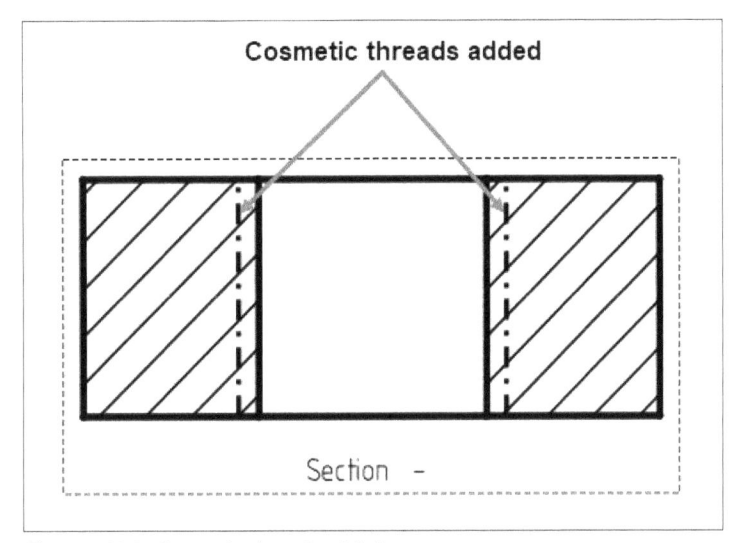

Figure-200. Cosmetic threads added

ADDING COSMETIC THREAD HOLE BOTTOM VIEW

The **Add Cosmetic Thread Hole Bottom View** tool is used to add a cosmetic thread to the top or bottom view of holes. The procedure to use this tool is same as discussed for previous tool; refer to Figure-201.

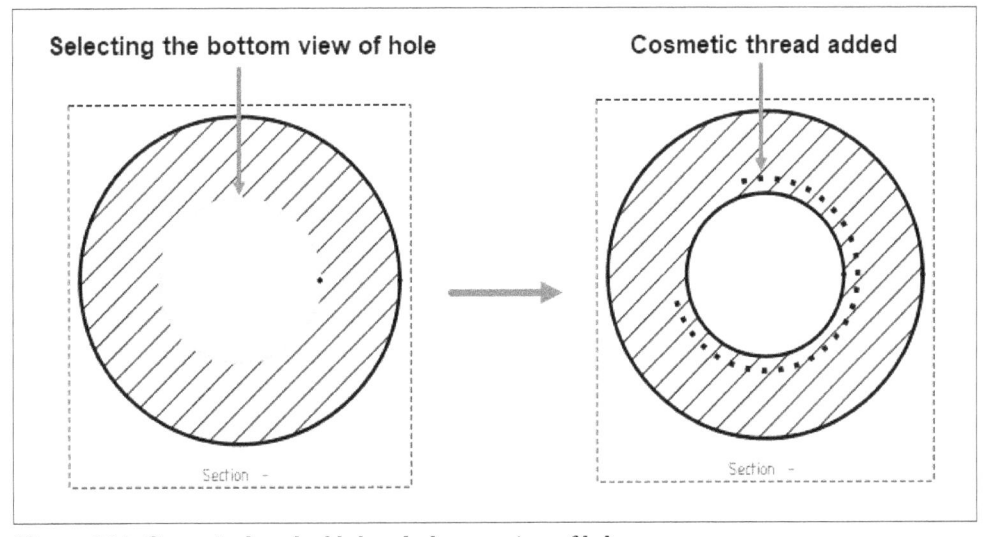

Figure-201. Cosmetic thread added to the bottom view of hole

ADDING COSMETIC THREAD BOLT SIDE VIEW

The **Add Cosmetic Thread Bolt Side View** tool is used to add cosmetic thread to the side view of a bolt/screw/rod. The procedure to use this tool is same as discussed for previous tool; refer to Figure-202.

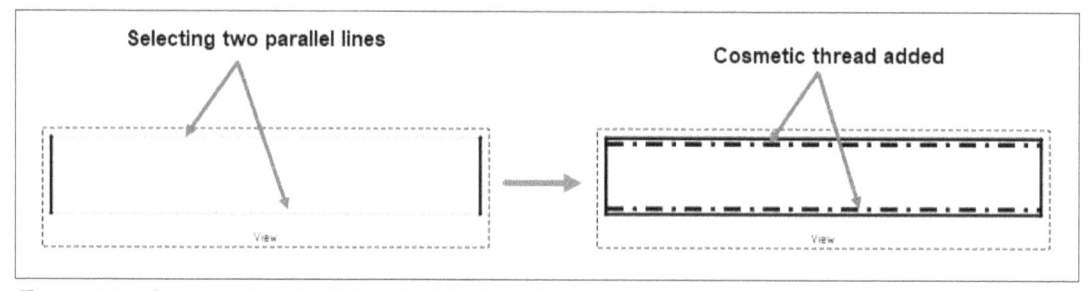

Figure-202. Cosmetic thread added to the side view of rod

ADDING COSMETIC THREAD BOLT BOTTOM VIEW

The **Add Cosmetic Thread Bolt Bottom View** tool is used to add a cosmetic thread to the top or bottom view of bolts/screws/rods. The procedure to use this tool is same as discussed for previous tool; refer to Figure-151.

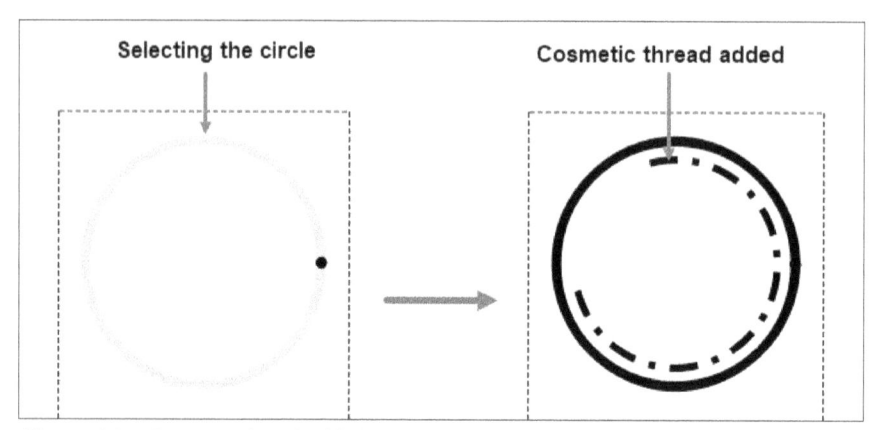

Figure-203. Cosmetic thread added to the bottom view of rod

ADDING COSMETIC INTERSECTION VERTEX(ES)

The **Add Cosmetic Intersection Vertex(es)** tool is used to add cosmetic vertex(es) at the intersection of selected edges. The procedure to use this tool is discussed next.

- Select two edges and a circle to which you want to add vertices; refer to Figure-204.

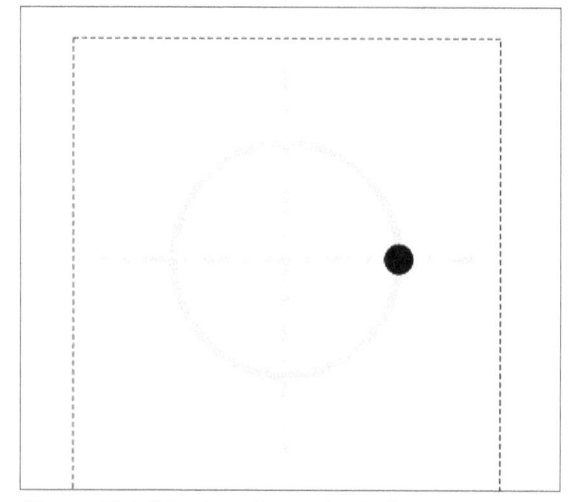

Figure-204. Selecting edges and a circle

- Click on the **Add Cosmetic Intersection Vertex(es)** tool from **Toolbar** or **Techdraw > Extensions: Centerlines/Threading** menu; refer to Figure-205. The vertices will be added at the intersection of the edges; refer to Figure-206.

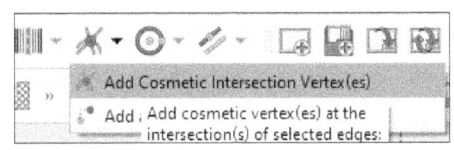

Figure-205. Add Cosmetic Intersection Vertex(es) tool

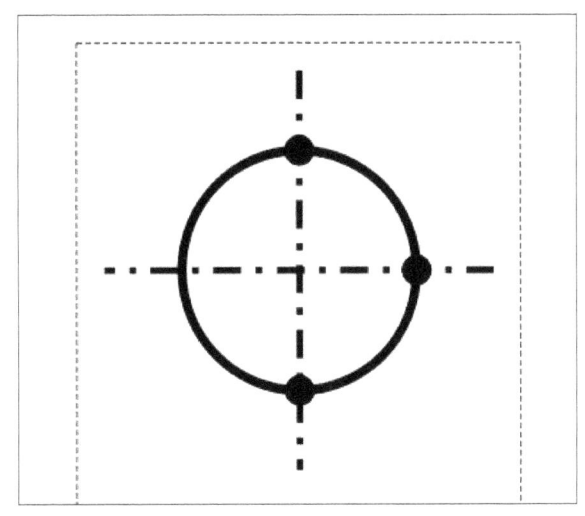

Figure-206. Vertices added at the intersection of edges

ADDING AN OFFSET VERTEX

The **Add an offset vertex** tool is used to add a vertex at a specified offset from selected vertex. The procedure to use this tool is discussed next.

- Select the vertex using which you want to create cosmetic offset vertex; refer to Figure-207 and click on the **Add an offset vertex** tool from the toolbar or **Techdraw > Extensions: Centerlines/Threading** menu; refer to Figure-208. The **Add offset vertex** dialog box will be displayed.

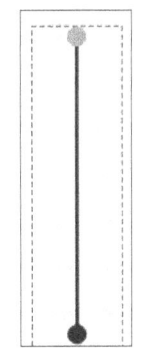

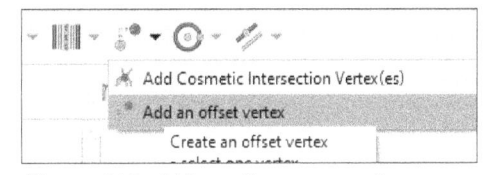

Figure-208. Add an offset vertex tool

Figure-207. Selecting the vertex for offset

- Specify desired values in **X-Offset** and **Y-Offset** edit boxes to define offset distances in along X and Y directions, respectively and click on the **OK** button from the dialog box. The vertex will be generated; refer to Figure-209.

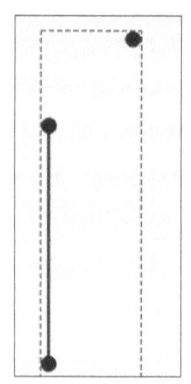

Figure-209. Add an offset vertex created

ADDING COSMETIC CIRCLE

The **Add Cosmetic Circle** tool is used to add a cosmetic circle based on two vertexes. The procedure to use this tool is discussed next.

- Select the first vertex to specify the center point of the circle and select the second vertex to specify the radius of the circle; refer to Figure-210.

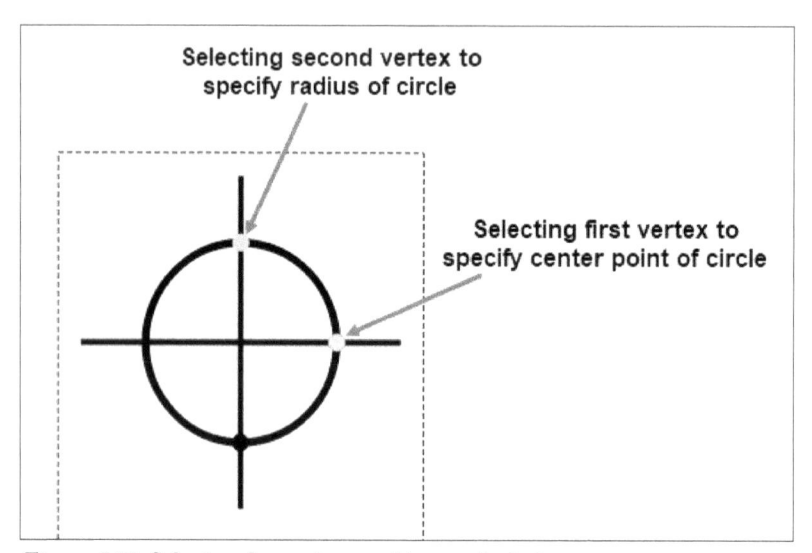

Figure-210. Selecting the vertices to add cosmetic circle

- Click on the **Add Cosmetic Circle** tool from drop-down in the **Toolbar** or **Techdraw > Extensions: Centerlines/Threading** menu; refer to Figure-211. The cosmetic circle will be added; refer to Figure-212.

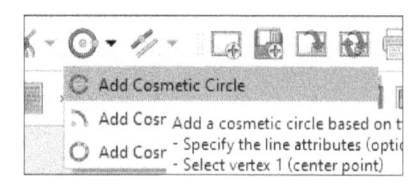

Figure-211. Add Cosmetic Circle tool

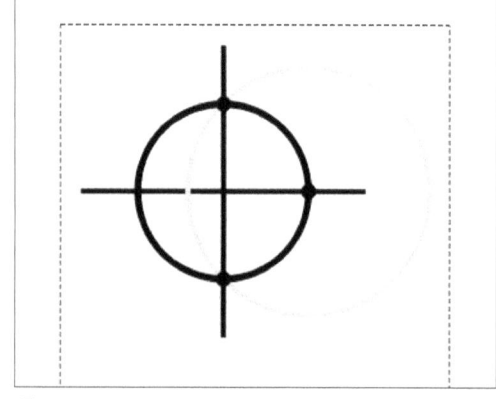

Figure-212. Cosmetic circle added

ADDING COSMETIC ARC

The **Add Cosmetic Arc** tool is used to add a cosmetic counter clockwise arc based on three vertexes. The procedure to use this tool is same as discussed for previous tool; refer to Figure-213.

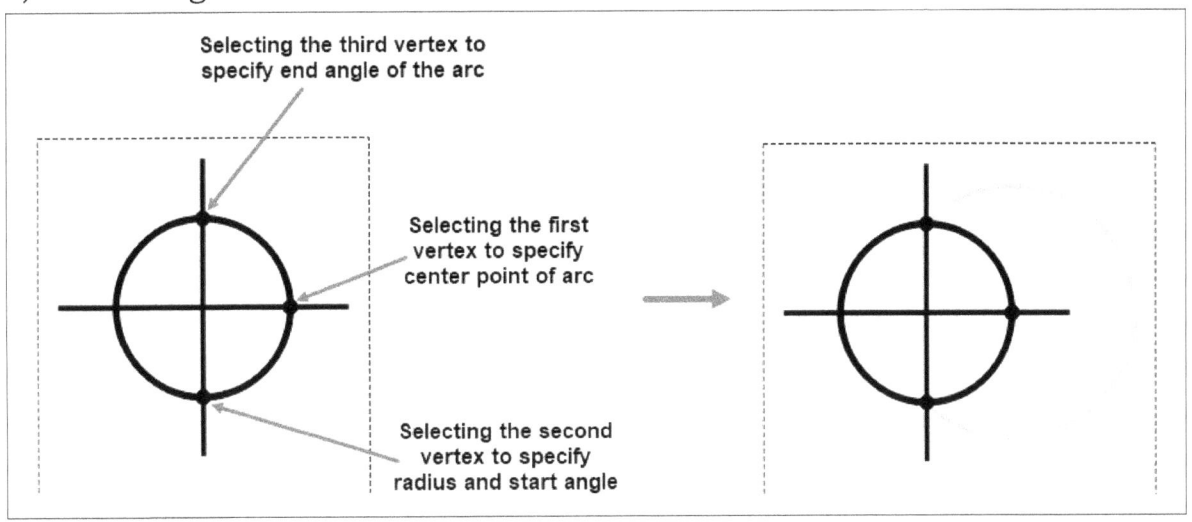

Figure-213. Cosmetic arc added

ADDING COSMETIC CIRCLE 3 POINTS

The **Add Cosmetic Circle 3 Points** tool is used to add a cosmetic circle based on three vertexes. The procedure to use this tool is same as discussed for previous tool; refer to Figure-214.

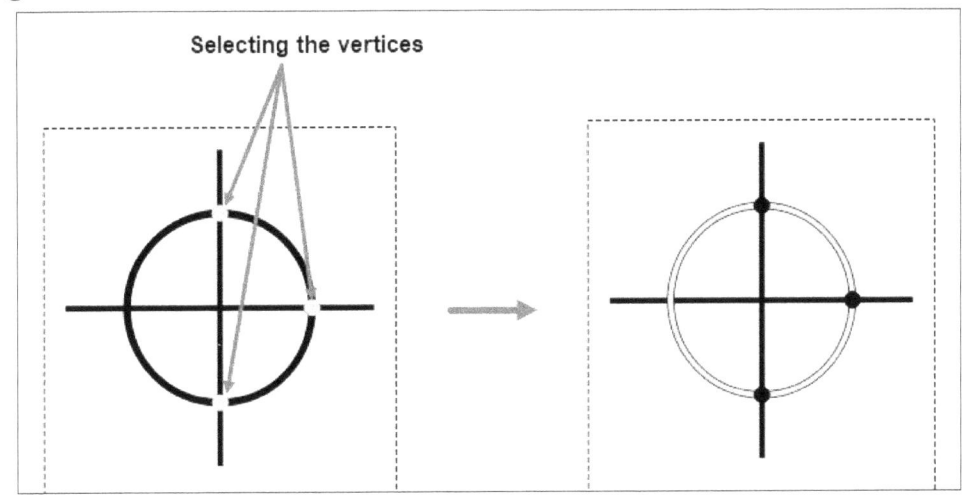

Figure-214. Cosmetic circle from 3 points added

ADDING COSMETIC PARALLEL LINE

The **Add Cosmetic Parallel Line** tool is used to add a cosmetic line parallel to another line through a vertex. The procedure to use this tool is discussed next.

- Select a straight edge to specify the direction and length of line and then select a vertex to specify the position of line; refer to Figure-215.

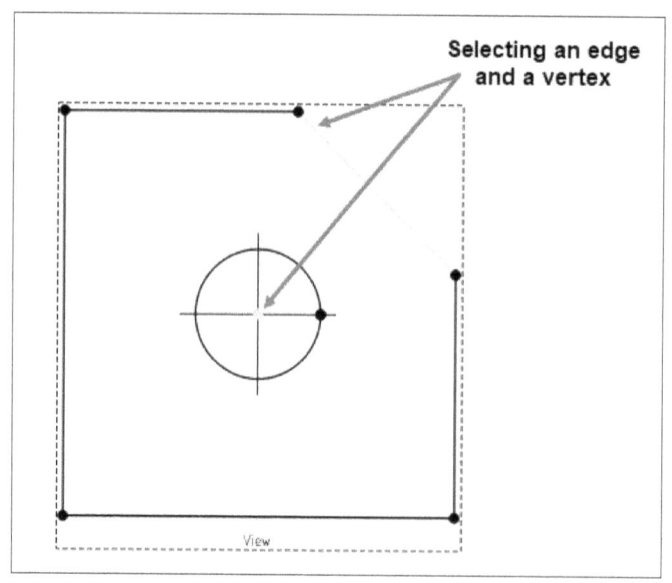

Figure-215. Selecting an edge and a vertex to add a cosmetic parallel line

- Click on the **Add Cosmetic Parallel Line** tool from drop-down in the **Toolbar** and **Techdraw** menu; refer to Figure-216. The cosmetic parallel line will be added; refer to Figure-217.

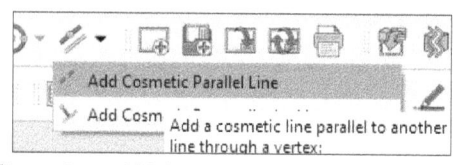

Figure-216. Add Cosmetic Parallel Line tool

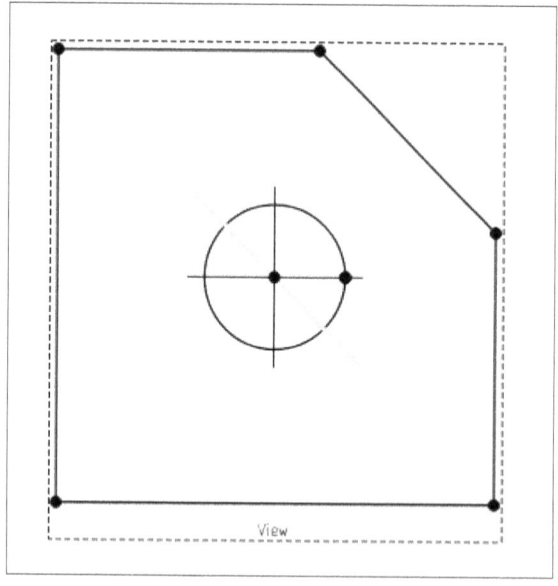

Figure-217. Cosmetic parallel line added

ADDING COSMETIC PERPENDICULAR LINE

The **Add Cosmetic Perpendicular Line** tool is used to add a cosmetic line perpendicular to another line through a vertex. The procedure to use this tool is same as discussed for previous tool; refer to Figure-218.

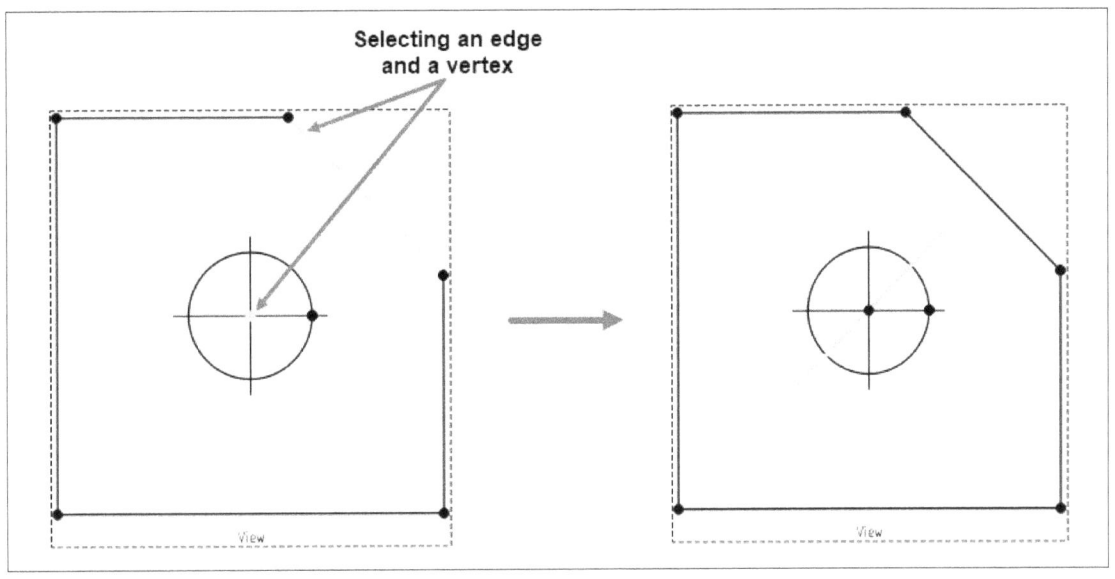

Figure-218. Cosmetic perpendicular line added

CREATING HORIZONTAL CHAIN DIMENSIONS

The **Create Horizontal Chain Dimensions** tool is used to create horizontal chain dimensions: a sequence of aligned dimensions. The procedure to use this tool is discussed next.

- Select three or more vertices to create horizontal chain dimensions; refer to Figure-219. The two left-most vertices specify the position of chain dimension.

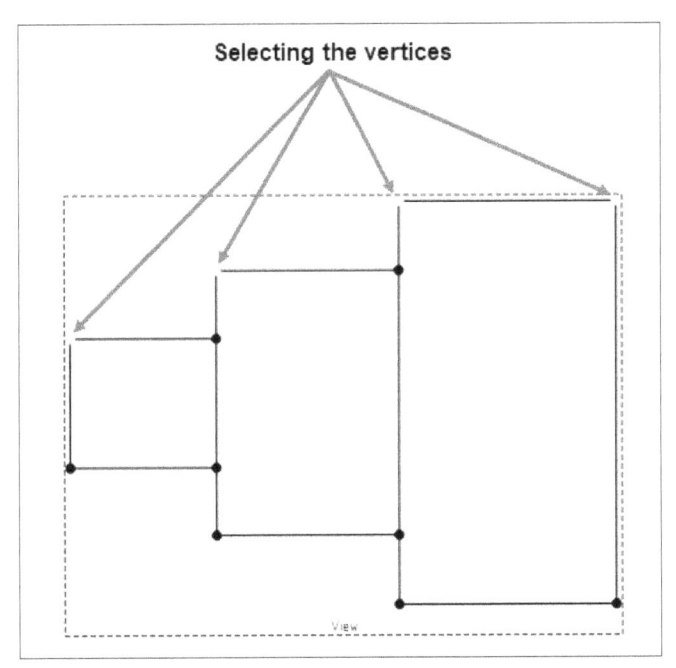

Figure-219. Selecting the vertices to create horizontal chain dimensions

- Click on the **Create Horizontal Chain Dimensions** tool from drop-down in the **Toolbar** or **Techdraw > Extensions: Dimensions** menu; refer to Figure-220. The horizontal chain dimensions will be created; refer to Figure-221.

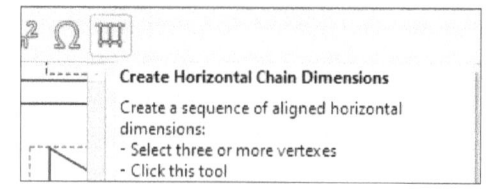

Figure-220. Create Horizontal Chain Dimensions tool

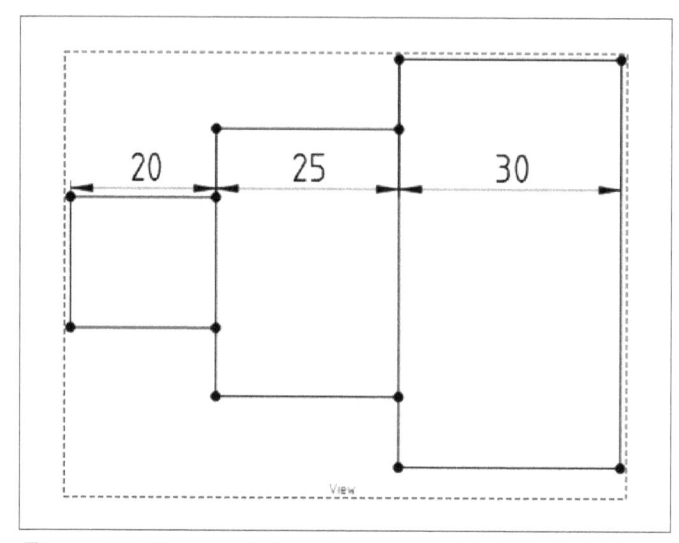

Figure-221. Horizontal chain dimensions created

CREATING VERTICAL CHAIN DIMENSIONS

The **Create Vertical Chain Dimensions** tool is used to create vertical chain dimensions: a sequence of aligned dimensions. The procedure to use this tool is same as discussed for previous tool; refer to Figure-222.

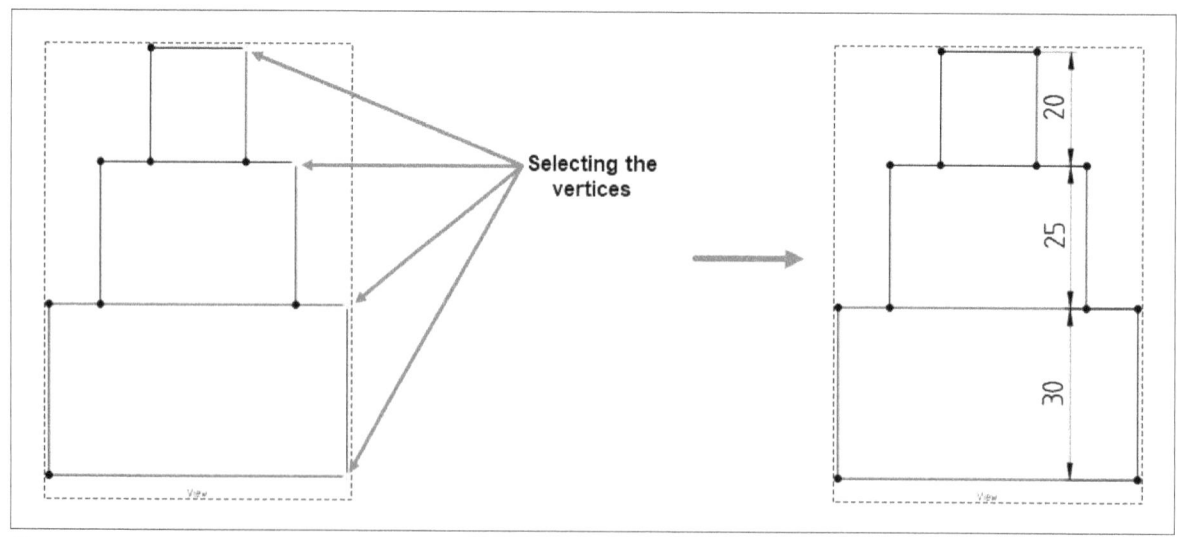

Figure-222. Vertical chain dimensions created

CREATING OBLIQUE CHAIN DIMENSIONS

The **Create Oblique Chain Dimensions** tool is used to create oblique chain dimensions; a sequence of aligned dimensions. The procedure to use this tool is same as discussed for previous tool; refer to Figure-223.

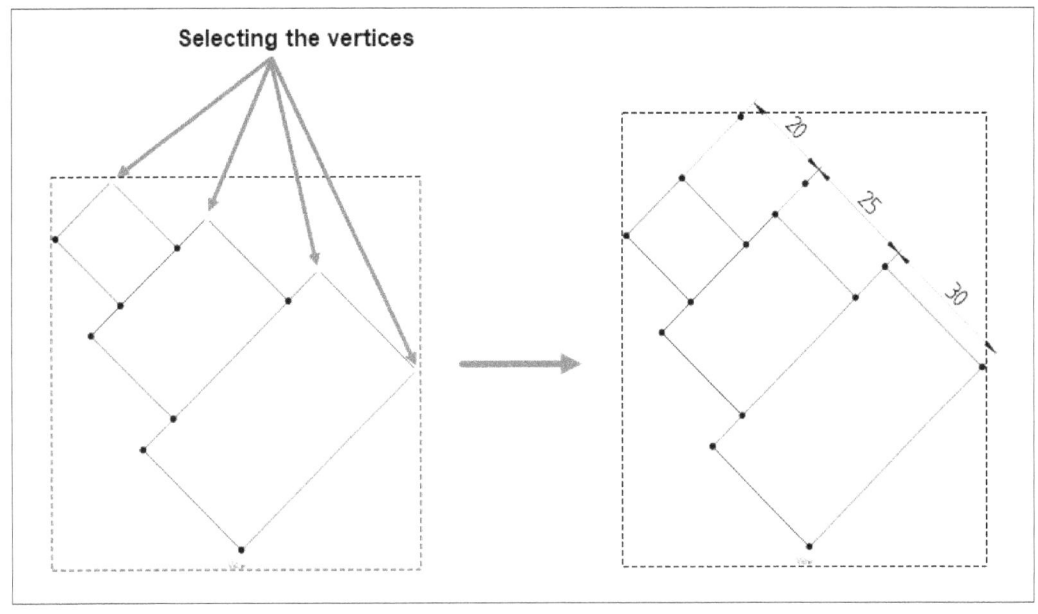

Figure-223. Oblique chain dimensions created

CREATING HORIZONTAL COORDINATE DIMENSIONS

The **Create Horizontal Coordinate Dimensions** tool is used to create horizontal coordinate dimensions: multiple evenly spaced dimensions starting from the same baseline. The procedure to use this tool is discussed next.

- Select three or more vertices to create horizontal coordinate dimensions; refer to Figure-224. The selection order of first two vertices specifies the position of the baseline.

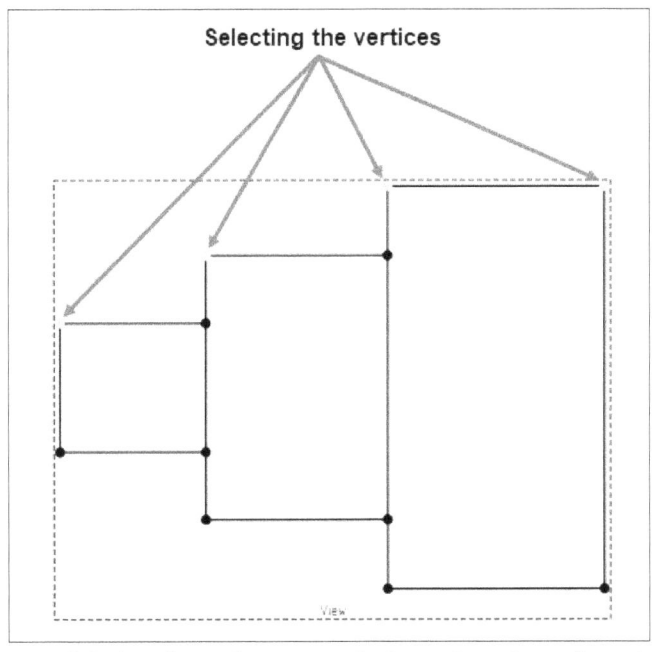

Figure-224. Selecting the vertices to create horizontal coordinate dimensions

- Click on the **Create Horizontal Coordinate Dimensions** tool from drop-down in the **Toolbar** or **Techdraw > Extensions: Dimensions** menu; refer to Figure-225. The horizontal coordinate dimensions will be created; refer to Figure-226.

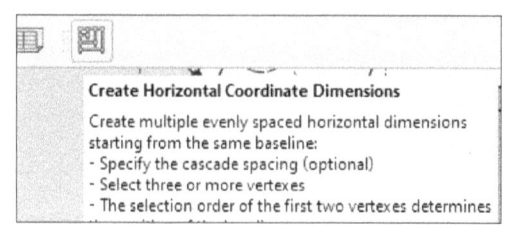

Figure-225. Create Horizontal Coordinate Dimensions tool

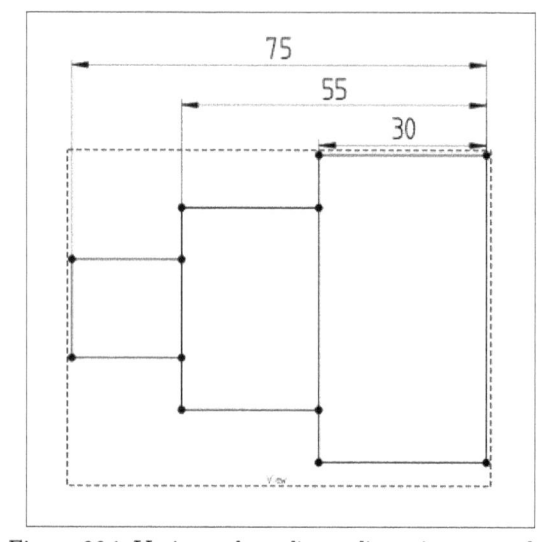

Figure-226. Horizontal coordinate dimensions created

CREATING VERTICAL COORDINATE DIMENSIONS

The **Create Vertical Coordinate Dimensions** tool is used to create vertical coordinate dimensions: multiple evenly spaced dimensions starting from the same baseline. The procedure to use this tool is same as discussed for previous tool; refer to Figure-227.

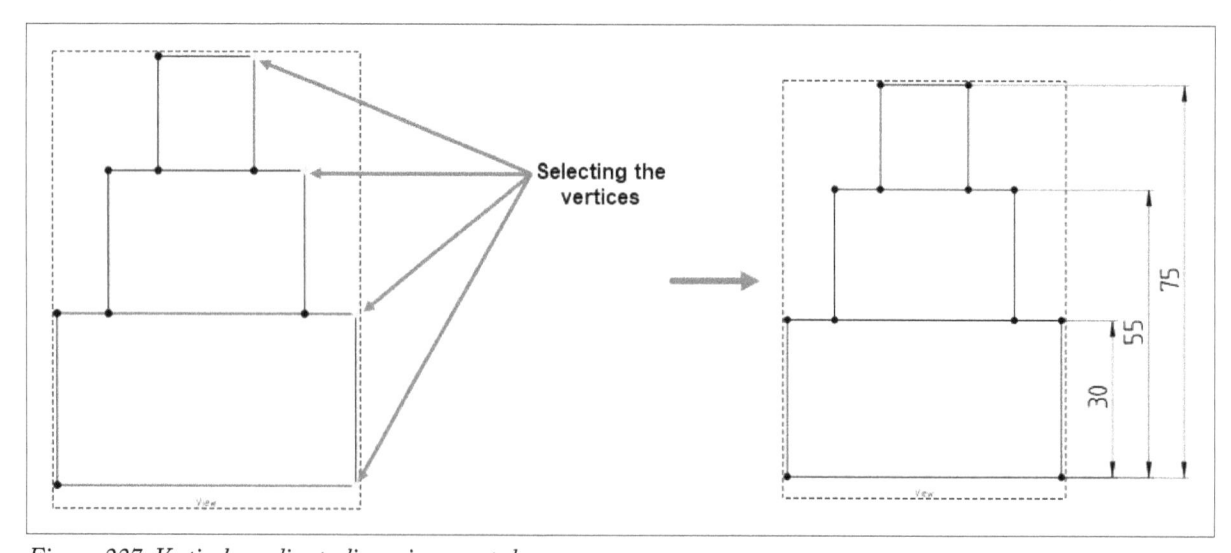

Figure-227. Vertical coordinate dimensions created

CREATING OBLIQUE COORDINATE DIMENSIONS

The **Create Oblique Coordinate Dimensions** tool is used to create oblique coordinate dimensions: multiple evenly spaced dimensions starting from the same baseline. The procedure to use this tool is same as discussed for previous tool; refer to Figure-228.

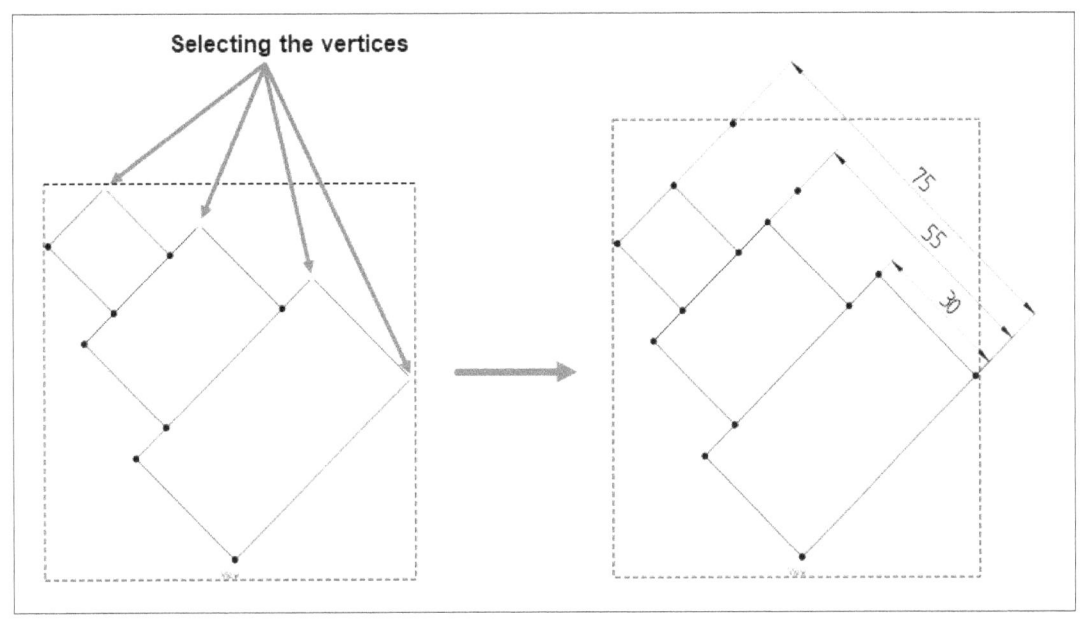

Figure-228. Oblique coordinate dimensions created

CREATING HORIZONTAL CHAMFER DIMENSION

The **Create Horizontal Chamfer Dimension** tool is used to create a horizontal size and angle dimension for a chamfer. The procedure to use this tool is discussed next.

- Select two vertices to create a horizontal dimension for a chamfer; refer to Figure-229.

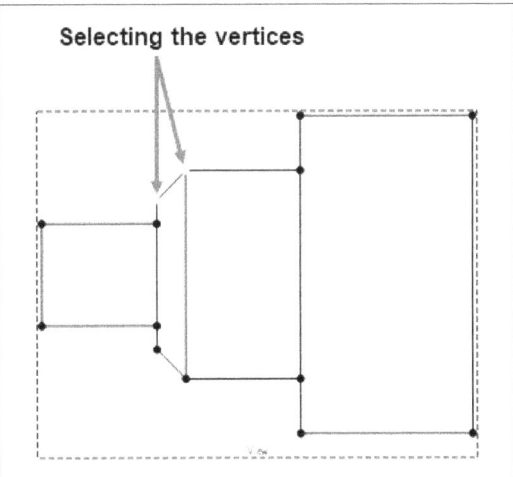

Figure-229. Selecting the vertices to create horizontal chamfer dimension

- Click on the **Create Horizontal Chamfer Dimension** tool from drop-down in the **Toolbar** or **Techdraw > Extensions: Dimensions** menu; refer to Figure-230. The horizontal chamfer dimension will be created; refer to Figure-231.

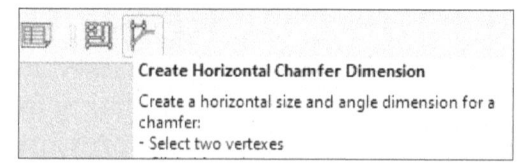

Figure-230. Create Horizontal Chamfer Dimension tool

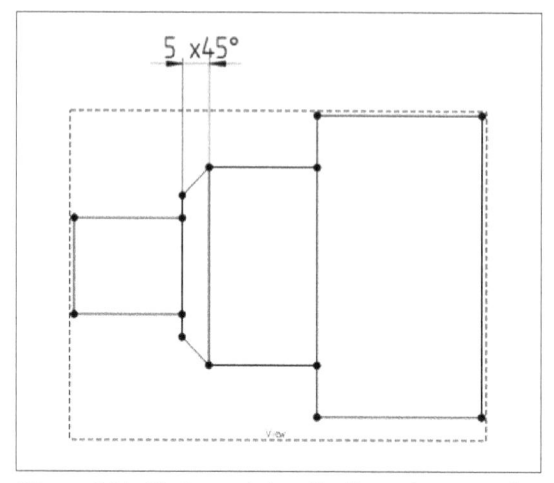

Figure-231. Horizontal chamfer dimension created

CREATING VERTICAL CHAMFER DIMENSION

The **Create Vertical Chamfer Dimension** tool is used to create a vertical size and angle dimension for a chamfer. The procedure to use this tool is same as discussed for previous tool; refer to Figure-232.

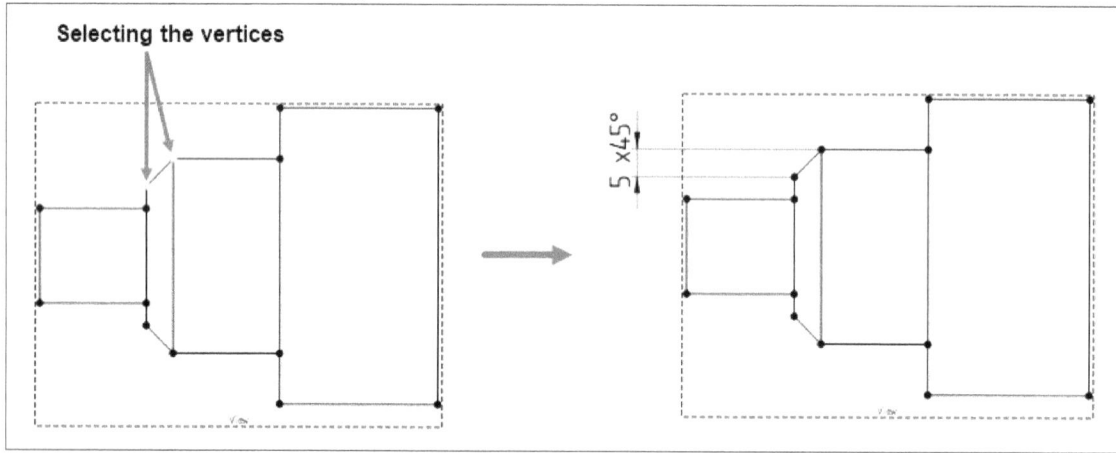

Figure-232. Vertical chamfer dimension created

INSERTING AREA ANNOTATION

The **Insert Area Annotation** tool is used to add an area dimension to a face in a drawing area. The procedure to use this tool is discussed next.

- Select the face of model in drawing view to which you want to apply area dimension; refer to Figure-233.

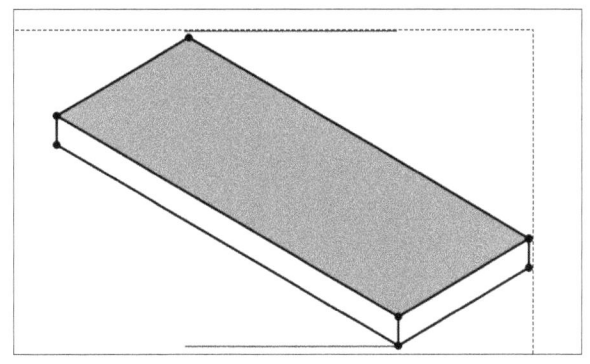

Figure-233. Selecting the face to create area

- Click on the **Insert Area Annotation** tool from the **Toolbar** or **Techdraw > Dimensions** menu; refer to Figure-234. The area will be created; refer to Figure-235.

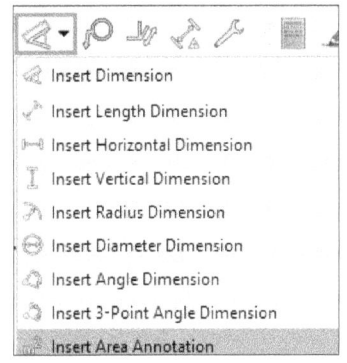

Figure-234. Insert Area annotation tool

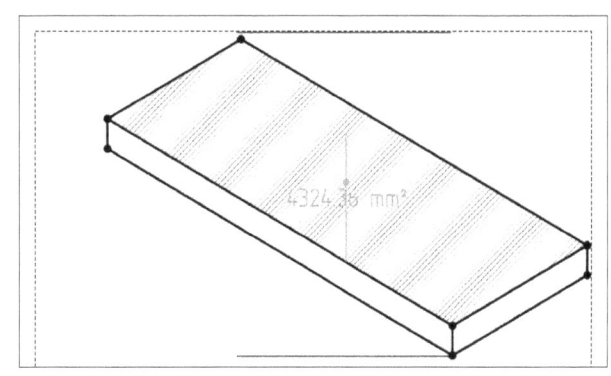

Figure-235. Insert area annotation created

CREATING ARC LENGTH DIMENSION

The **Create Arc Length Dimension** tool is used to create an arc length dimension. The procedure to use this tool is discussed next.

- Select an arc to create an arc length dimension; refer to Figure-236.

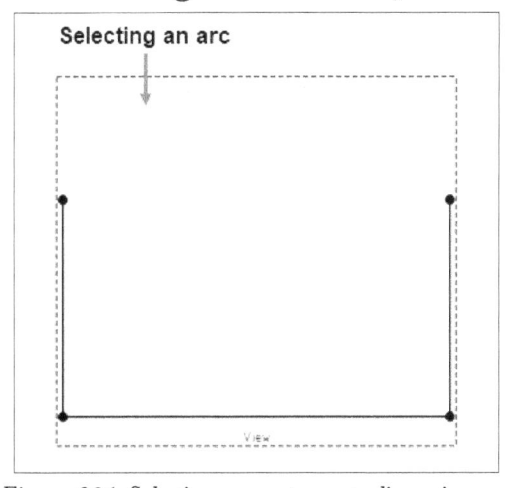

Figure-236. Selecting an arc to create dimension

- Click on the **Create Arc Length Dimension** tool from **Toolbar** or **Techdraw > Extensions: Dimensions** menu; refer to Figure-237. The arc length dimension will be created; refer to Figure-238.

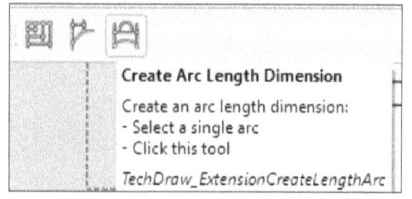

Figure-237. Create Arc Length Dimension tool

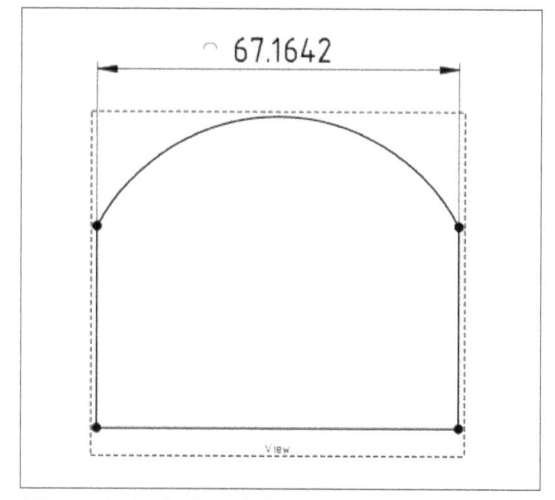

Figure-238. Arc length dimension created

INSERTING 'ø' PREFIX

The **Insert 'ø' Prefix** tool is used to insert a 'ø' symbol at the beginning of the dimension text. The procedure to use this tool is discussed next.

- Select one or more dimension to insert symbol at the beginning of that dimension; refer to Figure-239.

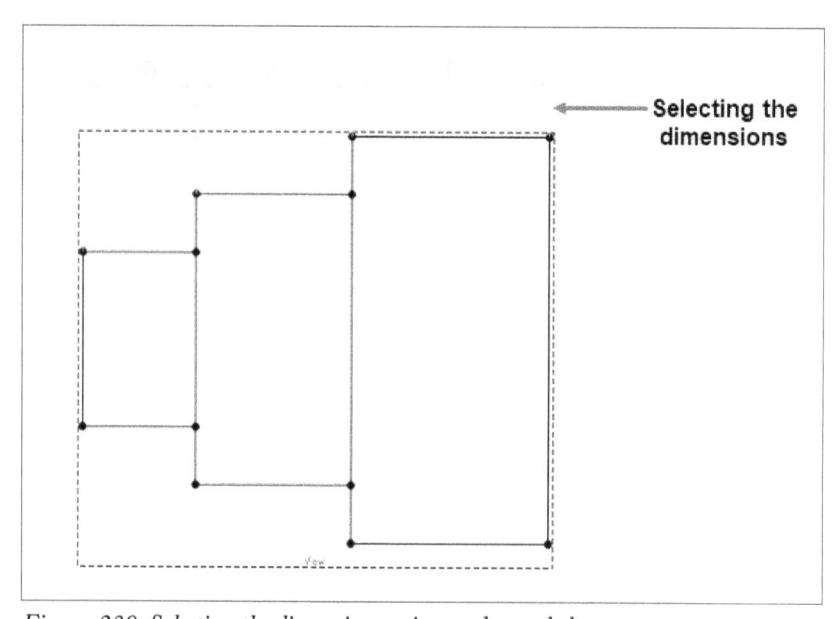

Figure-239. Selecting the dimensions to insert the symbol

- Click on the **Insert 'ø' Prefix** tool from drop-down in the **Toolbar** or **Techdraw > Extensions: Dimensions** menu; refer to Figure-240. The symbol will be inserted at the beginning of the dimension; refer to Figure-241.

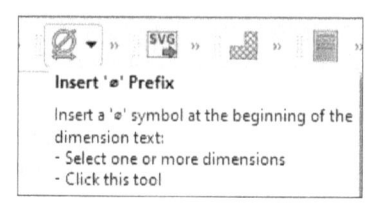

Figure-240. Insert ø' Prefix tool

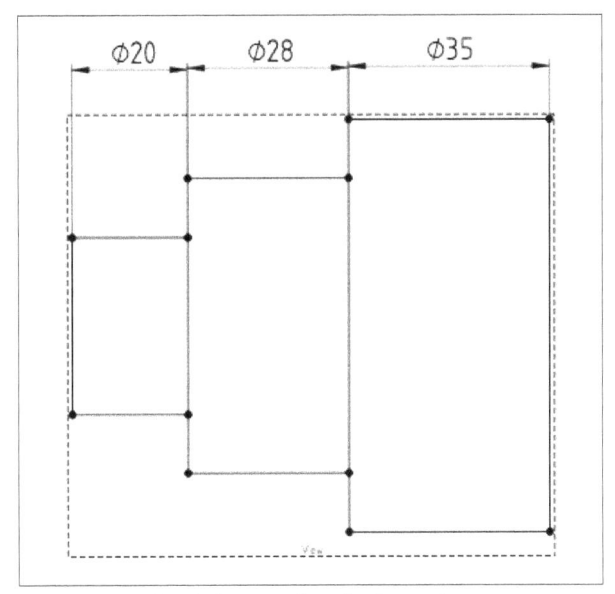

Figure-241. Symbol inserted to the dimension

INSERTING '□' PREFIX

The **Insert '□' Prefix** tool is used to insert a '□' symbol at the beginning of the dimension text. The procedure to use this tool is same as discussed for previous tool; refer to Figure-242.

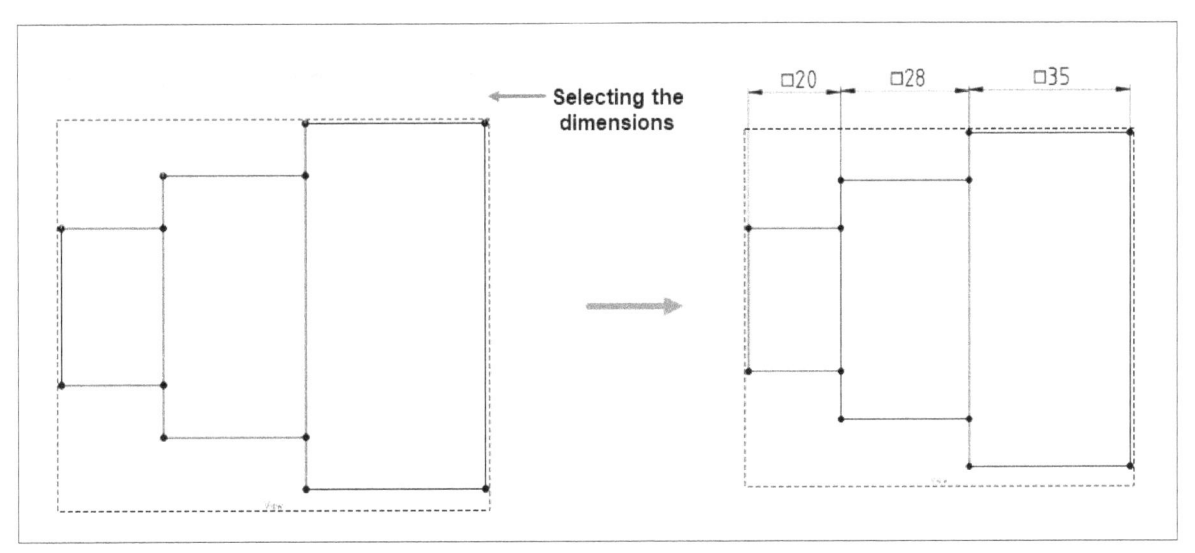

Figure-242. Symbol inserted to the dimension

INSERTING NX PREFIX

The **Insert nx Prefix** tool is used to inserts a repeated feature count at the beginning of the dimension text. The procedure to use this tool is discussed next.

- Select one or more dimension to insert symbol at the beginning of that dimension; refer to Figure-243.

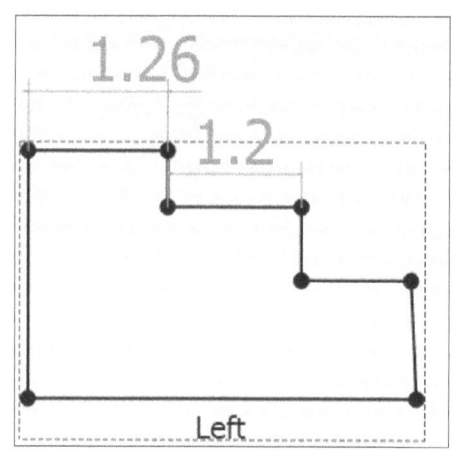

Figure-243. Selecting the dimension of insert nx

- Select the dimension and click on the **Insert 'nx' Prefix** tool from the toolbar. The **Change Editable Field** dialog box will be displayed; refer to Figure-244.

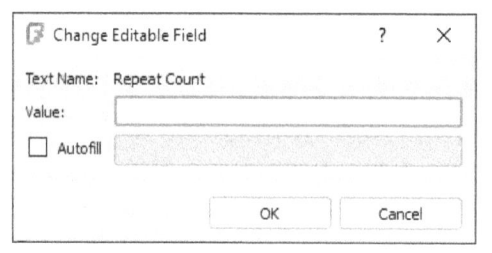

Figure-244. Change etiable field dialog box

- Set desired value in the **Value** edit box.
- Select the **Autofill** check box to automatically fill the value of this field.
- After specifying desired parameters, click on the **OK** button from the dialog box. The feature will be displayed; refer to Figure-245.

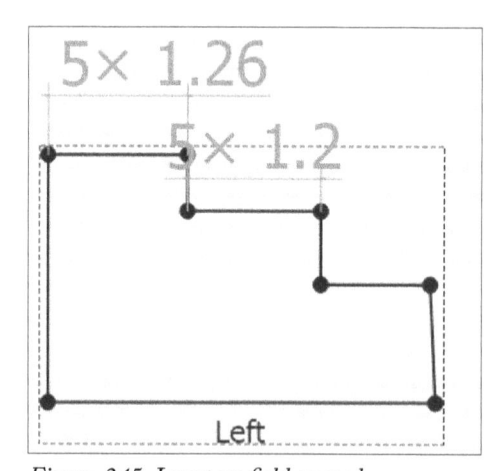

Figure-245. Insert nx field created

REMOVING PREFIX

The **Remove Prefix** tool is used to remove all prefix symbols from the dimension text. The procedure to use this tool is same as discussed for previous tool; refer to Figure-246.

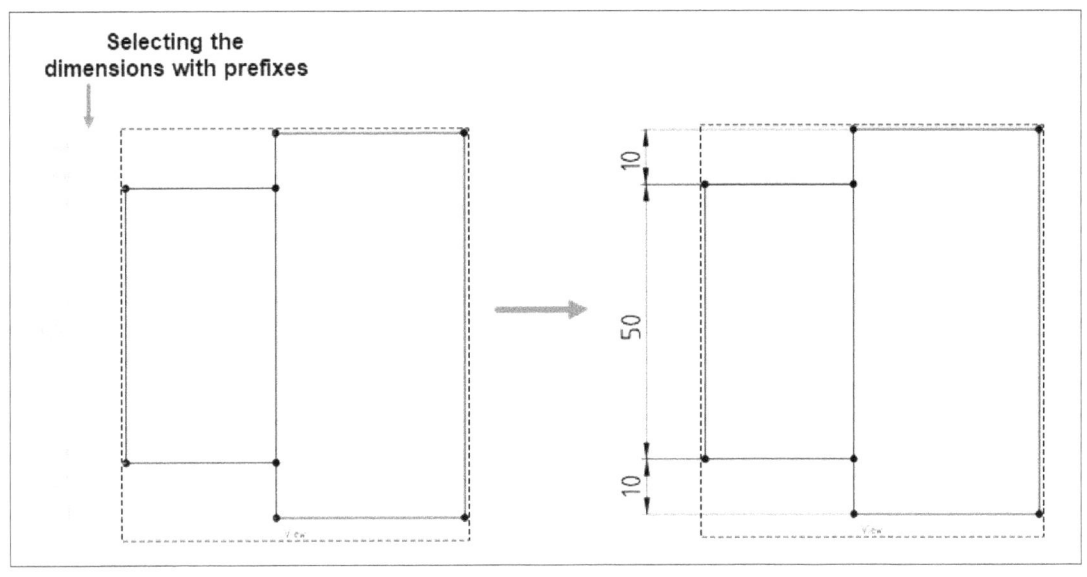

Figure-246. Prefixes removed

INCREASING DECIMAL PLACES

The **Increase Decimal Places** tool is used to increase the number of decimal places of the dimension text. The procedure to use this tool is discussed next.

- Select one or more dimensions to increase the number of decimal places; refer to Figure-247.

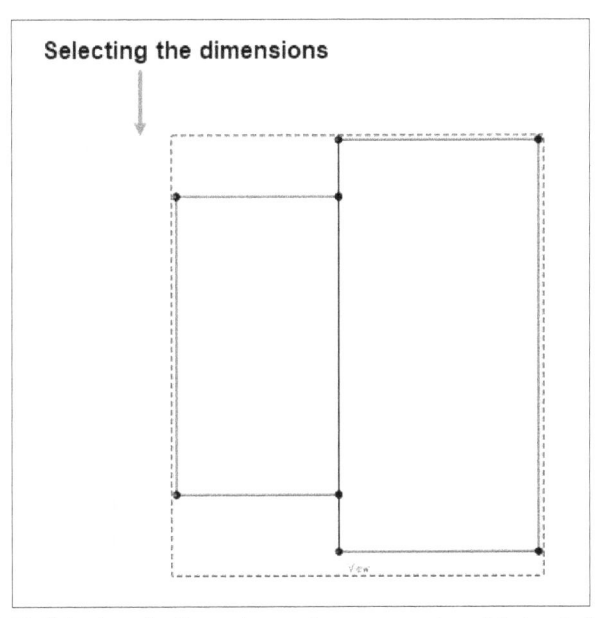

Figure-247. Selecting the dimensions to increase number of decimal places

- Click on the **Increase Decimal Places** tool from drop-down in the **Toolbar** or **Techdraw > Extensions: Dimensions** menu; refer to Figure-248. The number of decimal places of the dimensions will be increased by one; refer to Figure-249.

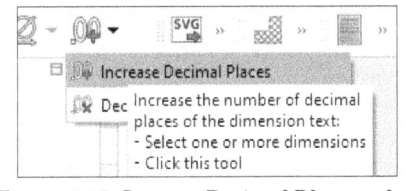

Figure-248. Increase Decimal Places tool

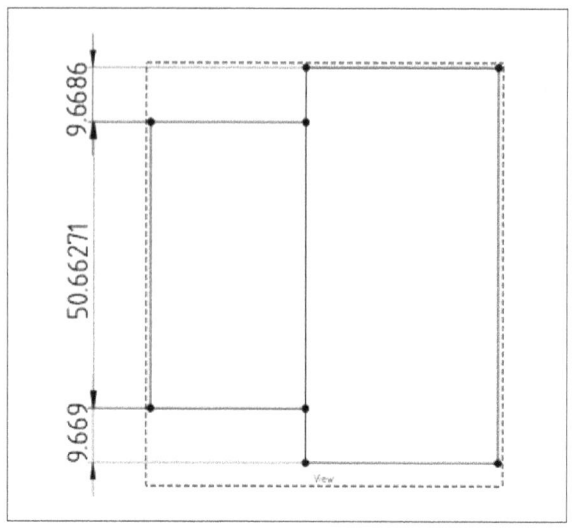

Figure-249. The number of decimal places of dimensions increased

DECREASING DECIMAL PLACES

The **Decrease Decimal Places** tool is used to decrease the number of decimal places of the dimension text. The procedure to use this tool is same as discussed for previous tool; refer to Figure-250.

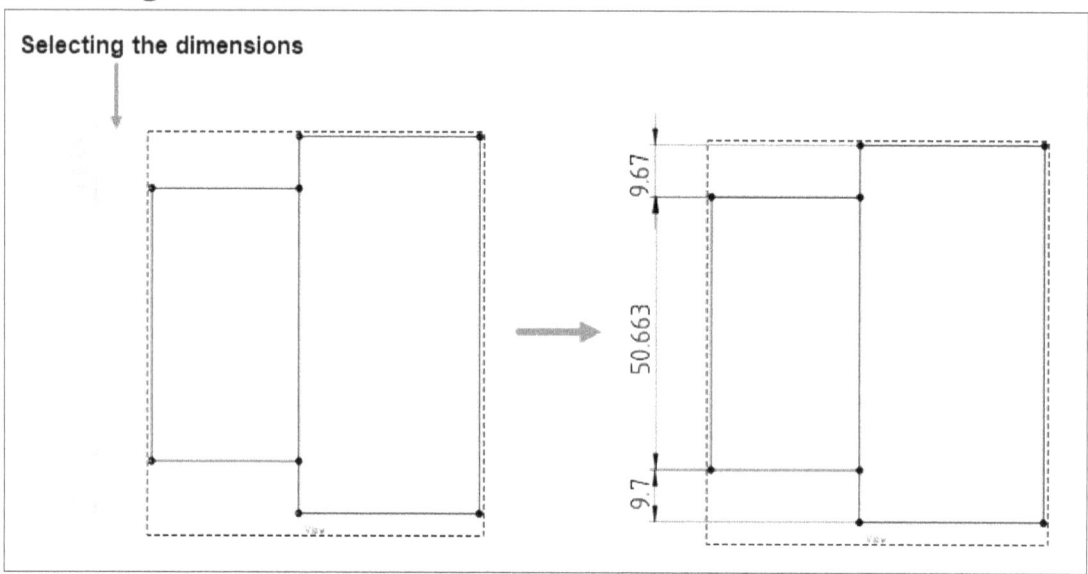

Figure-250. The number of decimal places of dimensions decreased

PRACTICAL

Create the drawing views of the given model; refer to Figure-251.

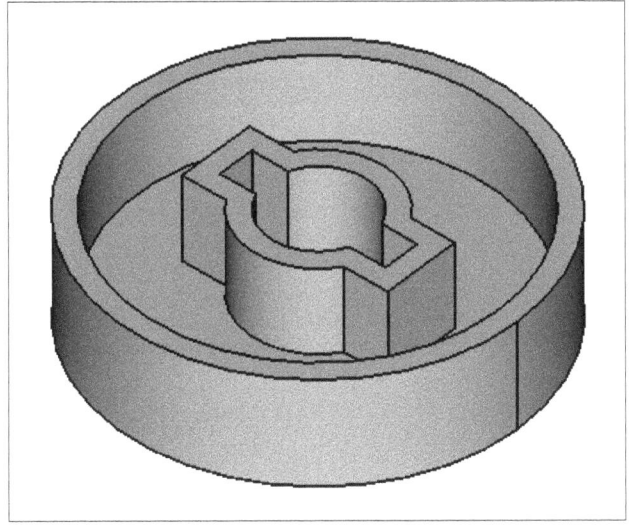

Figure-251. Practical

Starting a New File

- Start FreeCAD if not started yet and open the model file to be used for creating drawing views
- Select the **TechDraw** option from the **Switch between workbenches** drop-down in the **Toolbar**. The drawing environment will be displayed.

Inserting a View

- Click on the **Insert Default Page** tool from the **Toolbar**. A new drawing page will be created.
- Select the **Model** tile from the application window and orient the model to desired view to be placed in the drawing; refer to Figure-252.
- Now, select the **Page** tile from the application window and select the model whose views are to be placed from **Model Tree**; refer to Figure-253.

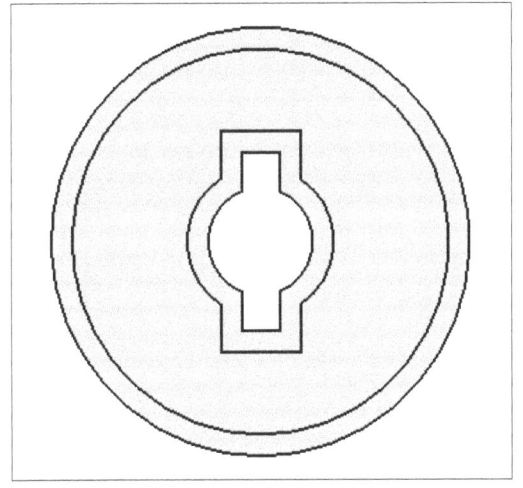

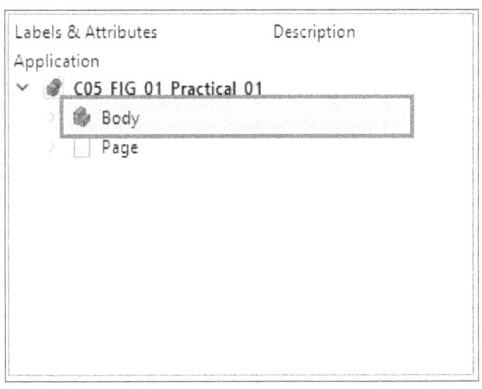

Figure-253. Selecting the model

Figure-252. Orientation of model

- After selecting model, click on the **Insert View** tool from the **Toolbar**. The **Part View** dialog will be displayed.
- Specify the parameters in the dialog as shown in Figure-254, and click on the **OK** button. The views of the model will be created; refer to Figure-255.

Figure-254. Specifying parameters

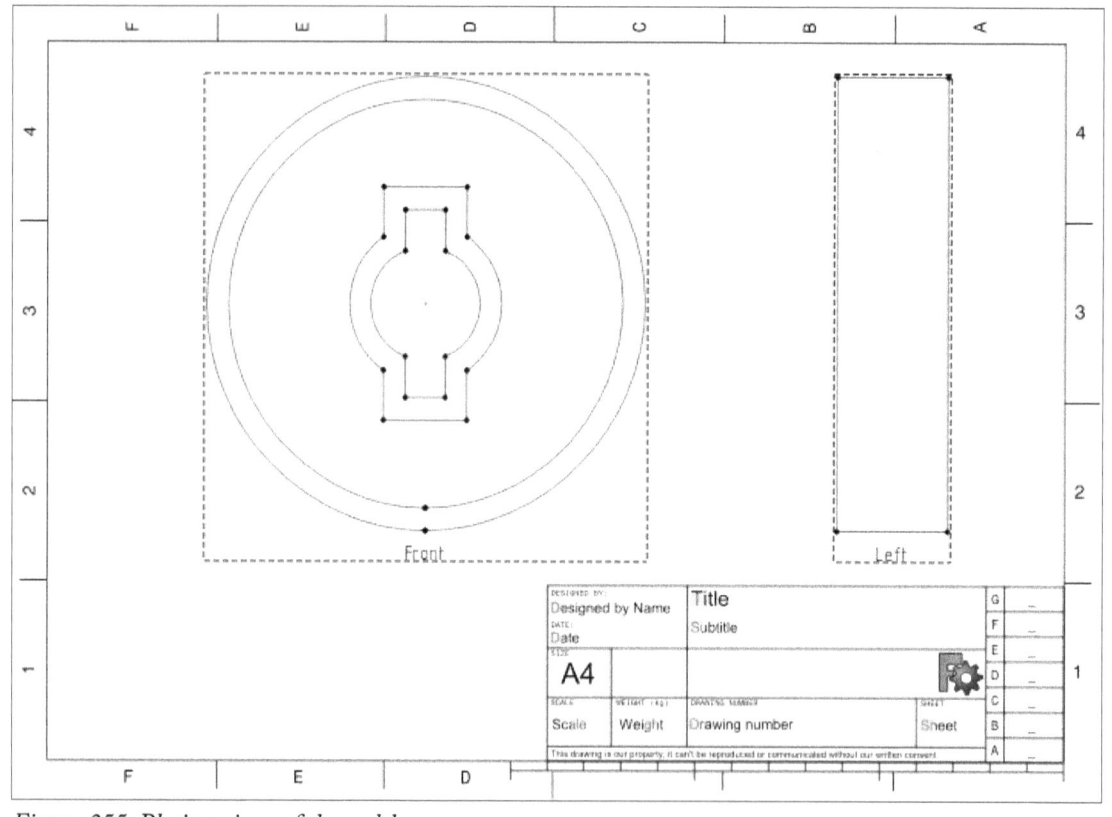

Figure-255. Placing views of the model

Inserting the Dimensions

- Select the entities of the model one by one and place respective dimensions using the tools of **TechDraw Dimensions** toolbar. The drawing view of the model with dimensions will be created; refer to Figure-256.

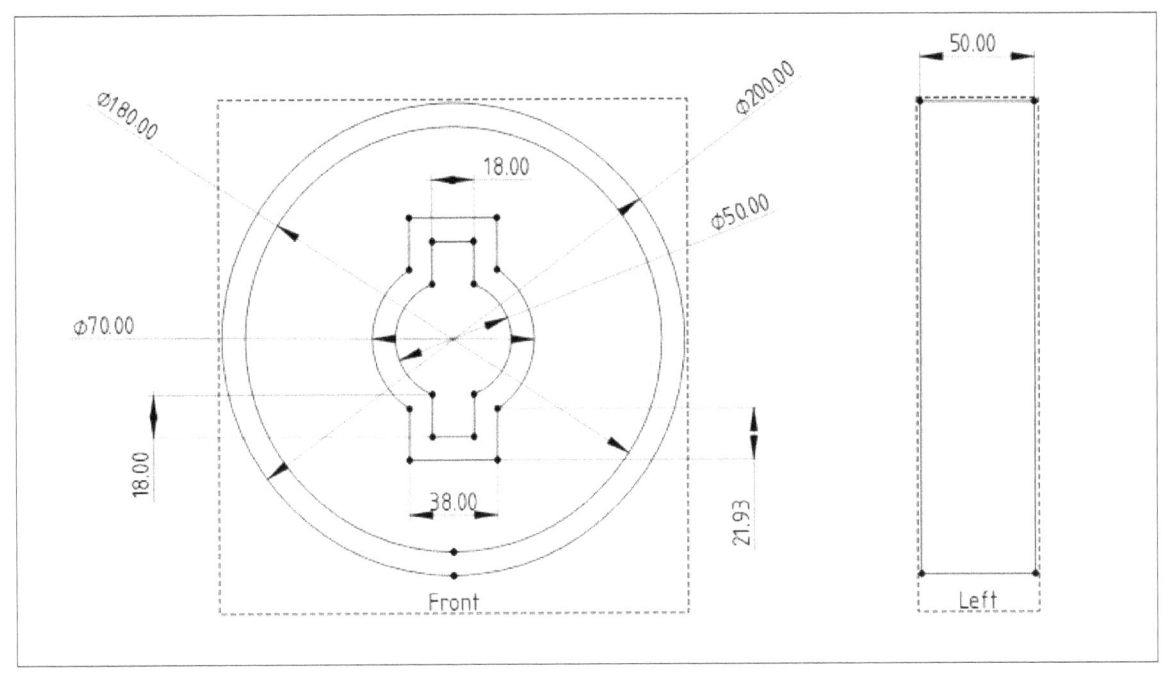

Figure-256. Drawing view of the model created

Practice

Create the model and drawing as shown in Figure-257.

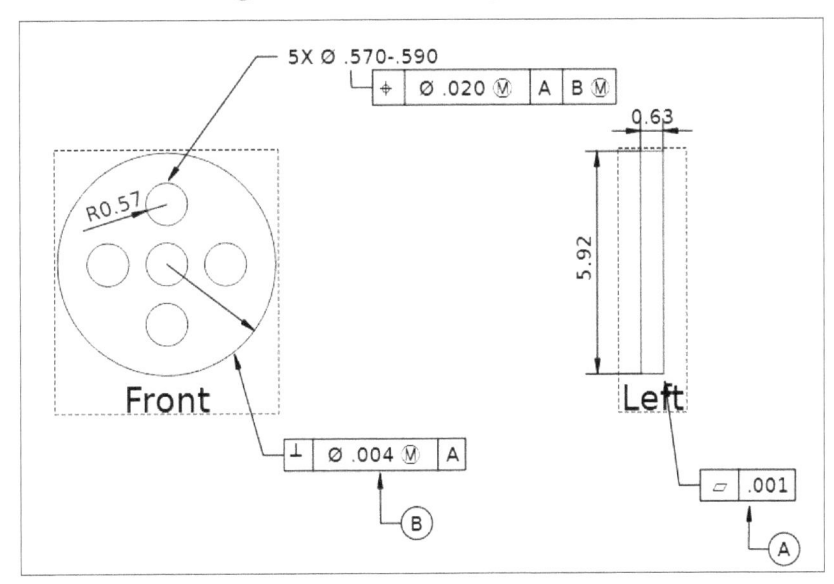

Figure-257. GD&T

SELF-ASSESSMENT

Q1. What is the purpose of the "Insert Default Page" tool in TechDraw workbench?

a) To generate a page with custom templates
b) To create a drawing page using the default drawing template
c) To update the properties of the page
d) To insert a view in the page

Q2. How can you specify the scale of drawing views for a page in TechDraw?

a) In the Print dialog box
b) In the Scale field of the Property Editor
c) By clicking on the Insert View tool
d) By selecting the Template button

Q3. Which tool is used to insert a view based on a selected template in TechDraw?

a) Insert Page Using Template
b) Insert Default Page
c) Insert View
d) Insert Projection Group

Q4. What does the "Redraw Page" tool do in TechDraw?

a) Inserts a new page
b) Forces an update of the selected page
c) Saves the current drawing
d) Prints all pages

Q5. What is the purpose of the "Insert Broken View" tool in TechDraw?

a) To create a magnified section of a view
b) To insert a broken view based on a part or object
c) To generate a complex section view
d) To insert a 3D active view

Q6. Which tool is used to insert a 3D view into a drawing page in TechDraw?

a) Insert Active View (3D View)
b) Insert Projection Group
c) Insert Section Views
d) Insert Detail View

Q7. What is the function of the "Insert Complex Section View" tool in TechDraw?

a) To insert a simple section view
b) To create a section view based on a selected profile object and view
c) To insert a broken view
d) To insert a 3D active view

Q8. What does the "Insert Detail View" tool do in TechDraw?

a) It inserts a detailed view of the entire model
b) It magnifies a specific region of a view for more details
c) It inserts a broken section of the view
d) It creates a complex section view

Q9. What is the purpose of the "Insert Draft Workbench Object" tool in TechDraw?

a) To create views for Draft workbench objects
b) To insert a perspective view
c) To generate a complex section view
d) To update page properties

Q10. Which workbench is used to create section views for architectural models in TechDraw?

a) Draft Workbench
b) Arch Workbench
c) Sketcher Workbench
d) Model Workbench

Q11. What is the purpose of the "Insert Spreadsheet View" tool in TechDraw?

A) To add annotations to the drawing
B) To place a view of the selected spreadsheet in the drawing area
C) To create a new spreadsheet in the drawing
D) To move a view to another page

Q12. What should be modified in the Property Editor after selecting the Sheet view in TechDraw?

A) The drawing style
B) The Cell Start & Cell End parameters
C) The color of the view
D) The font of the annotations

Q13. Which tool is used to move a view and all its dependents (such as balloons and dimensions) to a different page?

A) Share View tool
B) Move View tool
C) Stack Top tool
D) Add Leaderline to View tool

Q14. Which tool moves a view to the top of the stacking order?

A) Stack Bottom tool
B) Stack Up tool
C) Stack Top tool
D) Move View tool

Q15. What can be modified using the "Insert Annotation" tool in TechDraw?

A) The font, text color, and size of annotations
B) The position of the view on the page
C) The stacking order of the view
D) The page layout

Q16. Which tool is used to add a leaderline to a view in TechDraw?

A) Insert Annotation tool
B) Add Leaderline to View tool
C) Add Cosmetic Vertex tool
D) Stack Down View tool

Q17. The "Insert Rich Text Annotation" tool in TechDraw allows you to:

A) Add a basic text annotation to a view
B) Add a formatted annotation block to a leaderline or view
C) Project shapes onto the drawing
D) Move views between pages

Q18. Which of the following tools adds a centerline between two vertices in TechDraw?

A) Add Centerline to Faces tool
B) Add Centerline between 2 Lines tool
C) Add Centerline between 2 Points tool
D) Add Cosmetic Line through 2 Points tool

Q19. In the "Change Appearance of Selected Lines" tool, which dialog box allows you to restore invisible lines?

a) Line Decoration dialog
b) Restore Invisible Lines dialog
c) Surface Finish Symbols dialog
d) Link Dimension dialog

Q20. What does the "Show/Hide Invisible Edges" tool do?

a) Changes the color of lines
b) Temporarily hides or shows invisible lines
c) Adds welding symbols to leaderlines
d) Restores previously invisible lines

Q21. What symbol is added to a leaderline by selecting the "Field Weld" checkbox in the "Add Welding Information to Leaderline" tool?

a) All Around symbol
b) Alternating symbol
c) Field Weld symbol (flag)
d) Cosmetic symbol

Q22. In the "Create Surface Finish Symbol" tool, which standard can be selected for surface finish symbols?

a) ISO or ASME
b) DIN or ANSI
c) ISO or JIS
d) ANSI or JIS

Q23. The "Add Hole or Shaft Fit" tool in TechDraw applies tolerance based on which standard?

a) ISO 286
b) ASME
c) ANSI
d) DIN

Q24. Which tool in TechDraw is used to insert a clipping group into the drawing?

a) Insert Clip group
b) Show/Hide Invisible Edges
c) Insert Dimension
d) Insert Landmark Dimension

Q25. What is the primary function of the "Insert Dimension" tool in TechDraw?

a) To apply tolerance to a dimension
b) To add annotations to the drawing
c) To apply the dimension to two vertices, edges, or a line
d) To link a dimension to a 3D model

Q26. What does the "Insert Balloon Annotation" tool in TechDraw do?

a) Inserts a leaderline to a dimension
b) Adds a balloon with a leader line to a drawing object
c) Applies a surface finish symbol to an object
d) Adds a geometric hatch pattern

Q27. The "Axonometric Length Dimension" tool is used to add a length dimension to which type of view?

a) 2D view
b) Isometric view
c) Axonometric view
d) Perspective view

Q28. Which tool is used to add a linear dimension based on two point objects from the 3D model in TechDraw?

a) Insert Length Dimension
b) Insert Landmark Dimension
c) Insert Balloon Annotation
d) Repair Dimension Reference

Q29. What does the "Repair Dimension Reference" tool do?

a) Changes the style of the dimension
b) Links dimensions to a 3D model
c) Adjusts geometry references of a dimension after model modification
d) Restores invisible lines

Q30. Which tool is used to export a selected TechDraw page as an SVG file?

a) Export page as DXF file
b) Export page to SVG file
c) Insert Bitmap Image
d) Insert Symbol from SVG file

Q31. Which file formats are supported for inserting a bitmap image in TechDraw?

a) PNG, JPG/JPEG, BMP
b) TIFF, GIF, PNG
c) JPG/JPEG, SVG, BMP
d) PNG, BMP, PDF

Q32. What action does the "Hatch a Face using image file" tool perform?

a) Adds a surface finish symbol to a face
b) Applies a geometric hatch pattern
c) Applies a hatch pattern using an image file
d) Adds a leaderline to a dimension

Q33. Which dialog box is used when applying a geometric hatch pattern to a face in TechDraw?

a) Apply Geometric Hatch to Face dialog
b) Apply Hatch to Face dialog
c) Surface Finish Symbols dialog
d) Create Welding Symbol dialog

Q34. In the "Insert Symbol from an SVG file" tool, what happens after selecting a symbol file?

a) The symbol is inserted at the center of the page
b) The symbol is inserted into the drawing as a new layer
c) The symbol is converted to a bitmap format
d) The symbol is applied as a hatch pattern

Q35. What does the "Change Line Attributes" tool do?

A) It calculates the area of selected faces.
B) It changes the attributes (style, width, and color) of selected cosmetic lines and centerlines.
C) It adds cosmetic circles to the design.
D) It extends or shortens lines.

Q36. Which tool is used to extend a cosmetic line or centerline at both ends?

A) Change Line Attributes tool
B) Extend Line tool
C) Cascade Spacing tool
D) Position Section View tool

Q37. What is the function of the "Position Section View" tool?

A) It changes the attributes of a line.
B) It extends or shortens a line.
C) It aligns a section view orthogonally with its source view.
D) It locks or unlocks a view.

Q38. Which tool is used to create a chain of aligned dimensions for horizontal dimensions?

A) Cascade Horizontal Dimensions tool
B) Position Horizontal Chain Dimensions tool
C) Calculate Arc Length tool
D) Add Circle Centerlines tool

Q39. Which tool is used to evenly space two or more horizontal dimensions?

A) Cascade Horizontal Dimensions tool
B) Add Bolt Circle Centerlines tool
C) Position Vertical Chain Dimensions tool
D) Add Cosmetic Parallel Line tool

Q40. What is the purpose of the "Calculate the Area of Selected Faces" tool?

A) To calculate the area of curved edges
B) To calculate the area of selected faces with only straight edges
C) To calculate the volume of a 3D object
D) To calculate the circumference of a circle

FOR STUDENT NOTES

Chapter 11

Miscellaneous Workbench

Topics Covered

The major topics covered in this chapter are:

- *Introduction to Surface Workbench*
- *Creating Surfaces using Boundary Curves*
- *Introduction to Mesh Workbench*
- *Introduction to OpenSCAD Workbench*
- *Introduction to Plot Workbench*
- *Introduction to Points Workbench*
- *Introduction to Sheet Metal Workbench*

INTRODUCTION TO SURFACE WORKBENCH

The tools in **Surface** workbench are used to create surfaces using closed loop curve sections and boundary curves. Select the **Surface** option from **Switch between workbenches** drop-down in the **Toolbar**. The tools to create surfaces will be displayed in the **Toolbar**; refer to Figure-1.

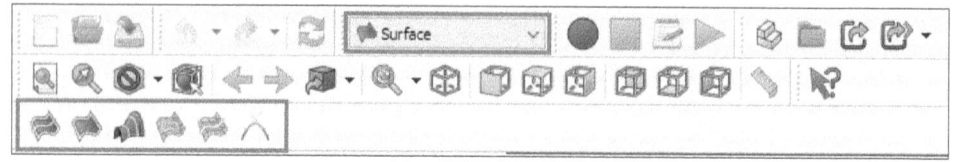

Figure-1. Tools in Surface workbench

Creating Fill Surface using Curves and Vertices

The **Filling** tool is used to create surface using selected curves and vertices. The procedure to use this tool is given next.

- Click on the **Filling** tool from the **Surface** menu or **Toolbar** in **Surface** workbench; refer to Figure-2. The **Boundaries** dialog will be displayed; refer to Figure-3.

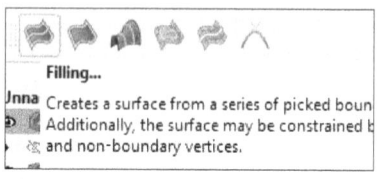

Figure-2. Filling tool

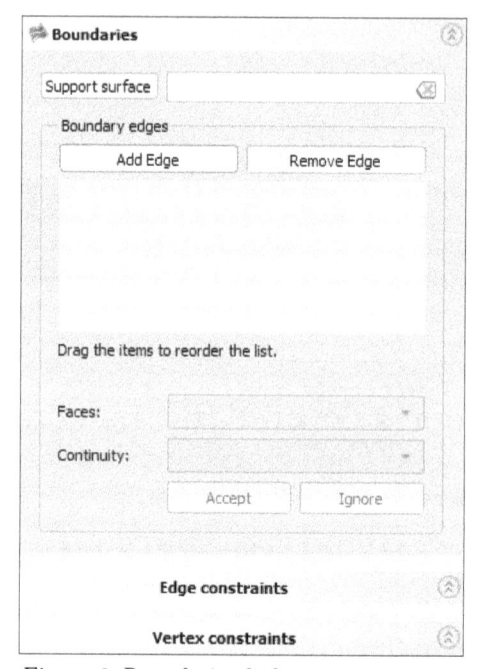

Figure-3. Boundaries dialog

- Click on the **Add Edge** button from dialog and select all the edges/curves forming a closed region for fill surface. Preview of the surface will be displayed; refer to Figure-4.

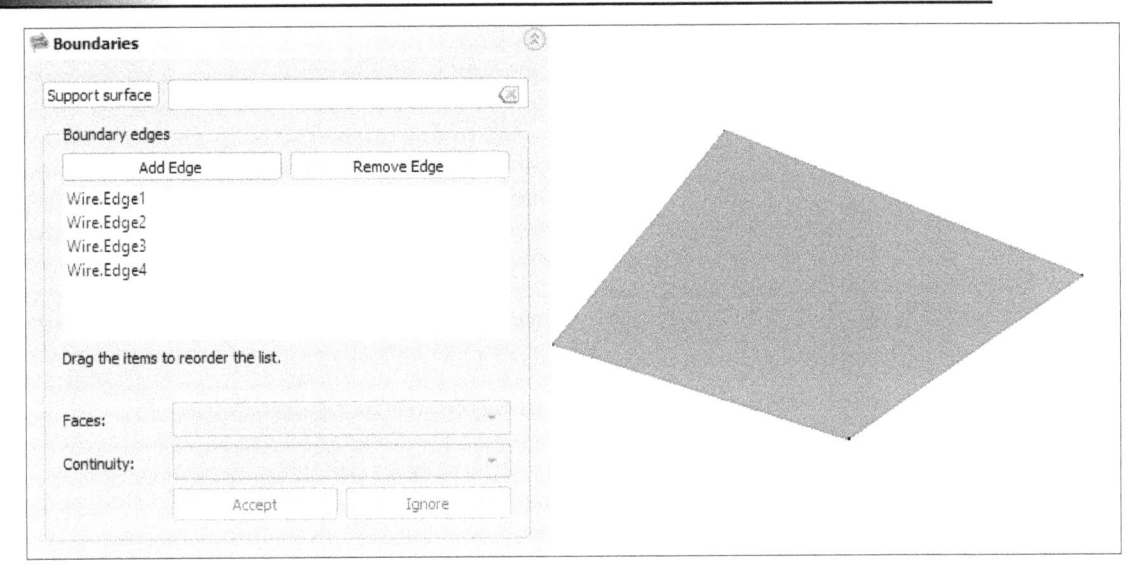

Figure-4. Preview of fill surface

- Expand the **Edge constraints** dialog in the **Tasks** panel and click on the **Add Edge** button from this dialog to define the curves to be used as guide for creating surface.
- Select the curves from the drawing area. Note that the curves can be from sketch or draft workbenches. The curves may or may not be intersecting with the boundary curves. The preview of surface will be displayed; refer to Figure-5.

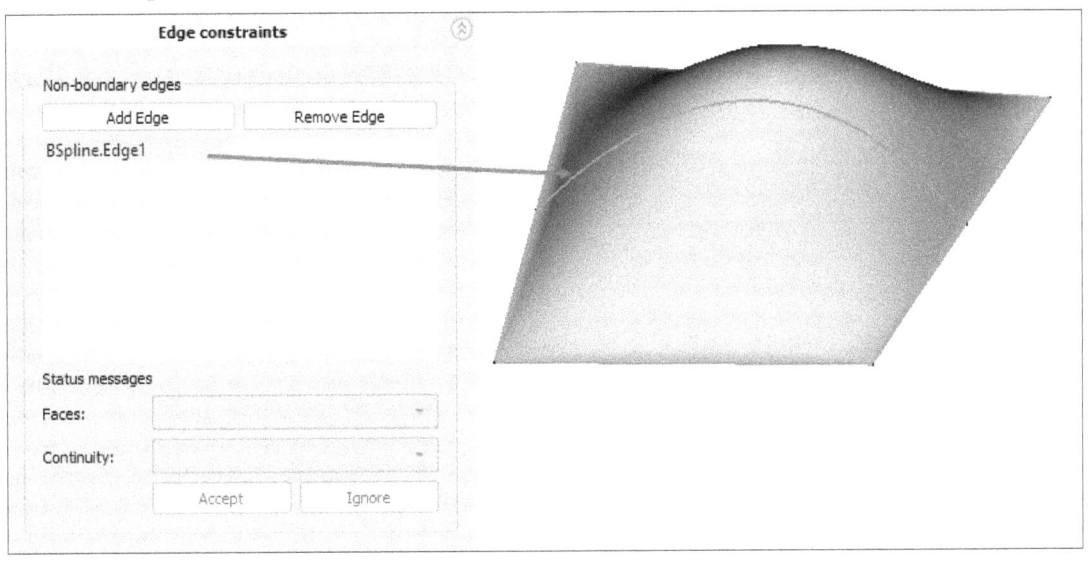

Figure-5. Preview of surface displayed

- Similarly, you can use vertices to refine shape of surface.
- After setting desired parameters, click on the **OK** button to create the surface.

Creating Surface using Boundary Curves

The **Fill boundary curves** tool is used to create surface using 2, 3, or 4 boundary edges. The procedure to use this tool is given next.

- Click on the **Fill boundary curves** tool from the **Surface** menu or **Toolbar** in **Surface** workbench; refer to Figure-6. The **Surface** dialog will be displayed; refer to Figure-7.

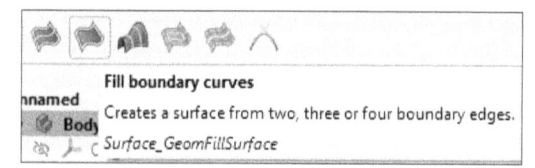

Figure-6. Fill boundary curves tool

Figure-7. Surface dialog

- Select the **Stretch** radio button if you want to create surface as combination of flat patches.
- Select the **Coons** radio button if you want to surface with better finish compared to stretch but lesser finish as compared to curved.
- Select the **Curved** radio button if you want to create surfaces having finish following curvature of curves.
- After selecting desired radio button, click on the **Add Edge** button from the dialog and select two or more curve in succession. Preview of surface will be displayed; refer to Figure-8.

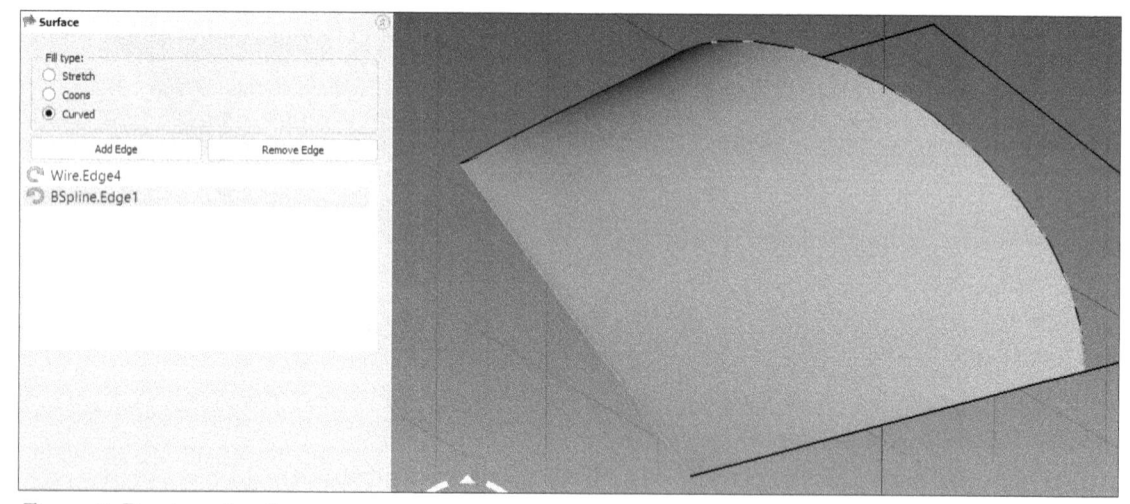

Figure-8. Preview of surface

- Click on the **OK** button from the dialog to create the surface.

Creating Surface using Edges

The **Sections** tool is used to create a surface from edges that represent transversal sections of a surface. The procedure to use this tool is discussed next.

- Click on the **Sections** tool from **Surface** menu or **Toolbar** in **Surface** workbench; refer to Figure-9. The **Sectional edges** dialog will be displayed in the **Tasks** panel of **Combo View**; refer to Figure-10.

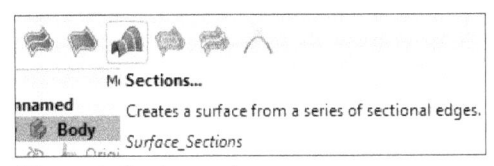

Figure-9. Sections tool

Figure-10. Sectional edges dialog

- Click on the **Add Edge** button from the dialog and select two or more desired edges or curves in the 3D view area. The preview of surface will be displayed; refer to Figure-11.

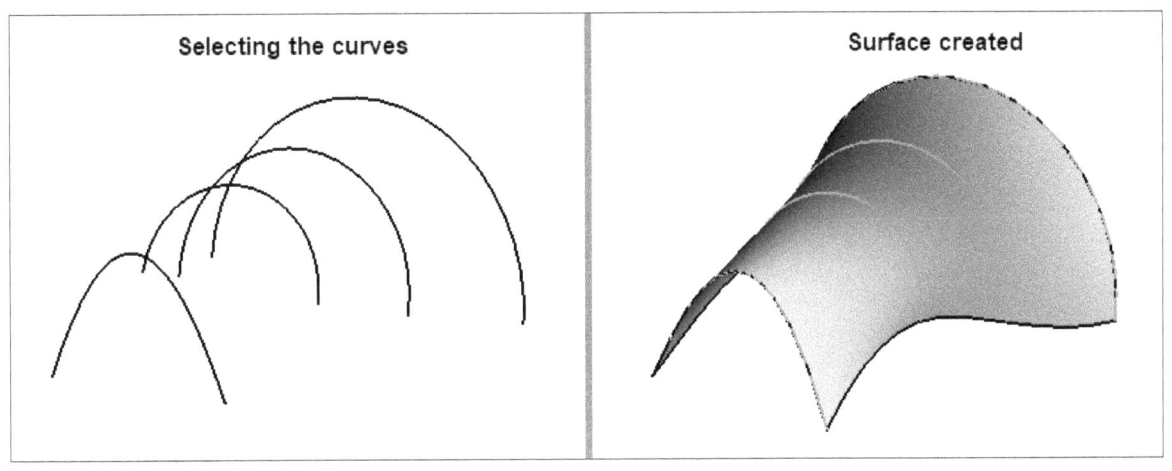

Figure-11. Preview of surface using curves

- After getting desired shape, click on the **OK** button from the dialog to create the surface.

Creating Surface by Extending a Face

The **Extend face** tool is used to create a surface by extending selected model face. The procedure to use this tool is given next.

- Select the face/surface you want to use for created extended surface and click on the **Extend face** tool from the **Surface** menu; refer to Figure-12. Preview of extended surface will be displayed; refer to Figure-13.

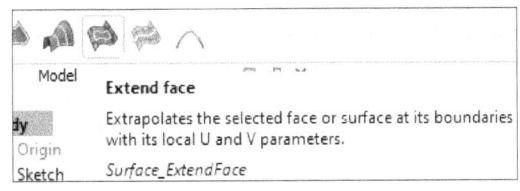

Figure-12. Extend face tool

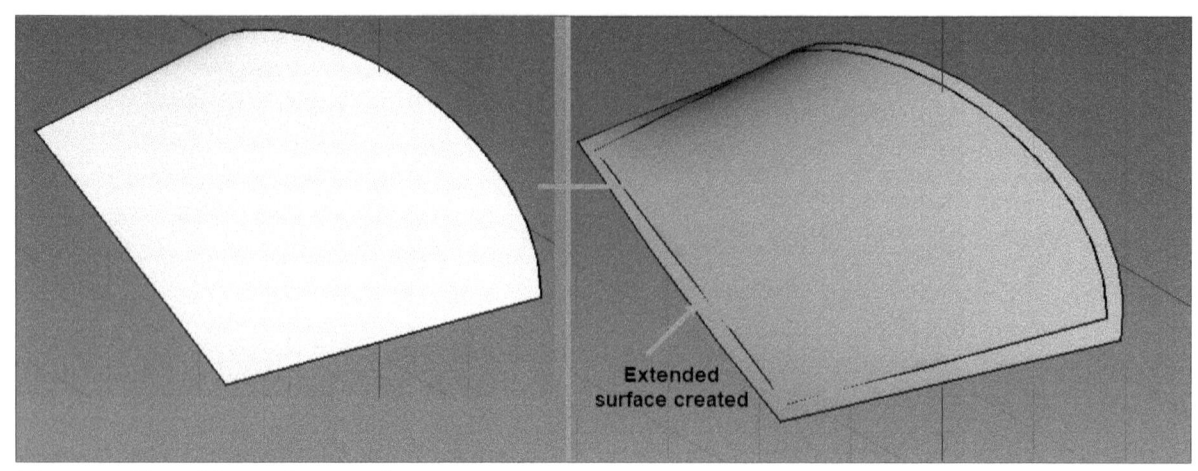

Figure-13. Extend face surface created

- Select the newly created surface from drawing area or **Model Tree** and specify desired parameters in the **Property Editor** to specify extension length in various directions; refer to Figure-14.

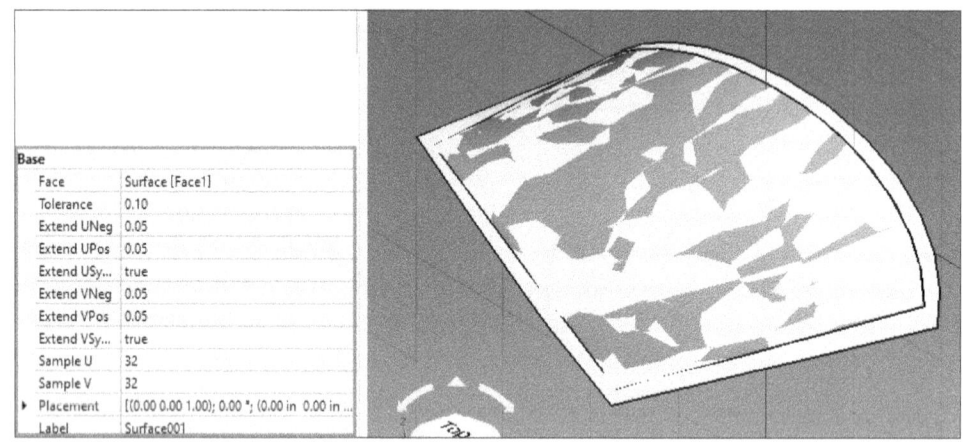

Figure-14. Properties of extended face surface

Curve on mesh

The **Curve on mesh** tool is used to create approximated spline segments on top of a selected mesh. Note that Curve on mesh tool is active only if mesh object is available. The procedure to use this tool is discussed next.

- Click on the **Curve on mesh** tool from the **Surface** menu or **Toolbar** in **Surface** workbench; refer to Figure-15. The **Curve on mesh** dialog will be displayed; refer to Figure-16.

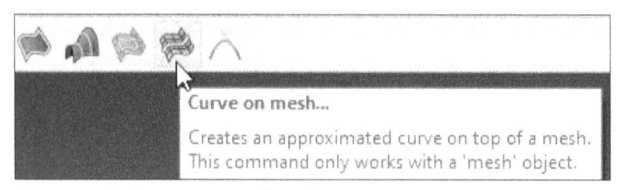

Figure-15. Curve on mesh tool

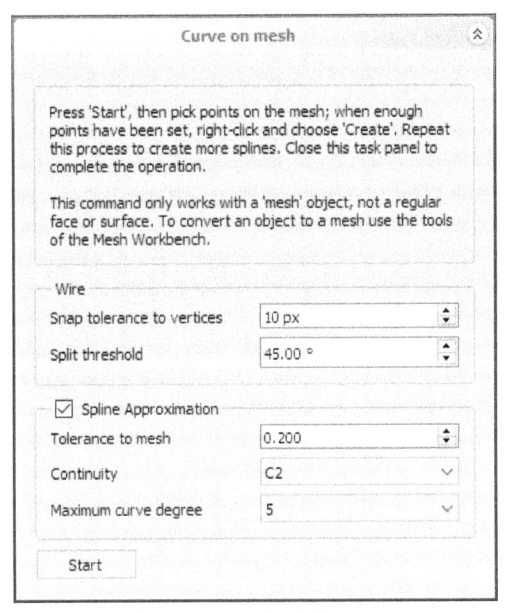

Figure-16. Curve on mesh dialog

- Click on the **Start** button from the dialog. The curve symbol will be attached to the cursor and you will be asked to create the curve on the mesh object.
- Click on the mesh object at desired location to specify first point of curve. Move the cursor away and click at desired location(s) to specify second point or multiple points of curve.
- After specifying desired number of points, right-click. A list of options will be displayed; refer to Figure-17.

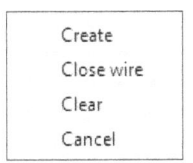

Figure-17. Options on clicking RMB

- Click on the **Create** option to create the curve. The curve on the mesh object will be created; refer to Figure-18. Click on the **Close wire** option to join the last point to the first point if at least three points have been specified. Click on the **Clear** option to erase the specified points and allow to specify new points. Click on the **Cancel** option to erase the specified points and stop specifying new points.

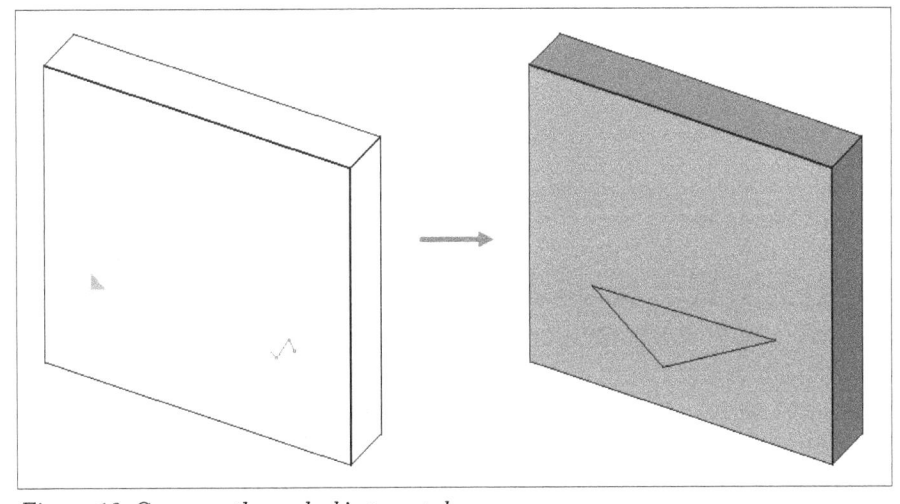

Figure-18. Curve on the mesh object created

- Specify desired minimum distance value between one point and another point in the **Snap tolerances to vertices** edit box from **Wire** area of the dialog.
- Specify desired angular deviation from one point in the mesh to another point in the **Split threshold** edit box from **Wire** area of the dialog.
- Select **Spline Approximation** check box to create spline objects.
- The **Tolerance to mesh** edit box display the imperfections of the mesh. The smaller the number, the mesh will be very fine.
- Select desired option from **Continuity** drop-down to determine the continuity of the spline.
- Select desired option from **Maximum curve degree** drop-down to determine the maximum degree of the spline to approximate the surface.
- Click on the **Close** button to close the dialog.

Blend Curve

The **Blend Curve** tool is used to create a bezier curve between two edges, with desired continuity. The procedure to use this tool is given next.

- Select two edges from the 3D view area and click on the **Blend Curve** tool from the **Surface** menu or **Toolbar** in **Surface** workbench; refer to Figure-19. The blend curve will be created; refer to Figure-20.

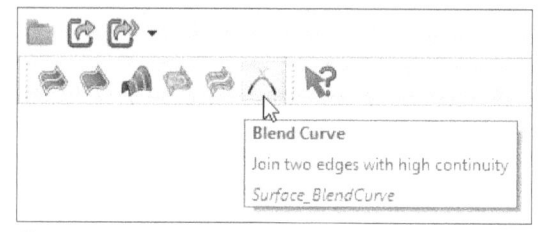

Figure-19. Blend Curve tool

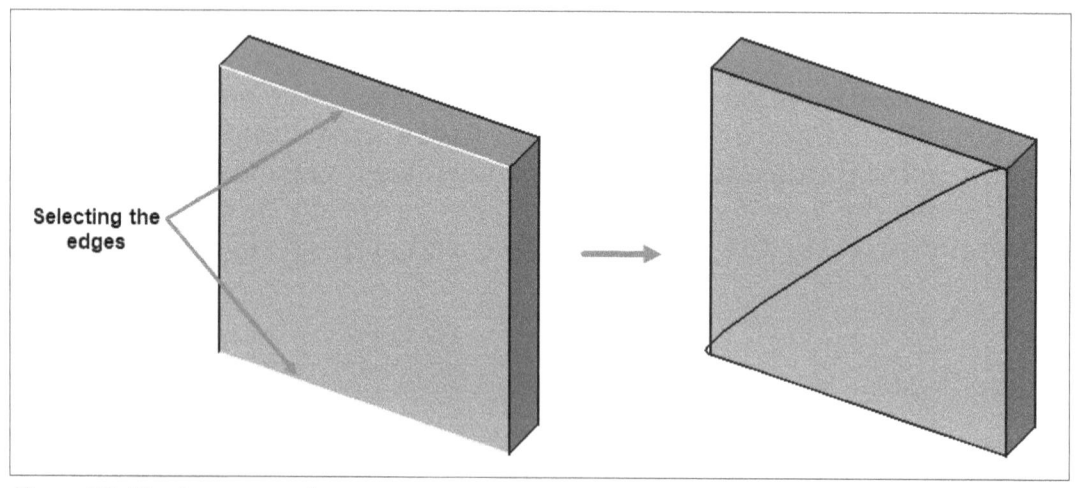

Figure-20. Blend curve created

INTRODUCTION TO MESH WORKBENCH

The **Mesh** workbench of FreeCAD is used to import, create, and modify mesh objects generated in FreeCAD as well as other CAD software. Select the **Mesh Design** option from the **Switch between workbenches** drop-down to activate the **Mesh** workbench. The tools will be displayed as shown in Figure-21. Various tools in this workbench are discussed next.

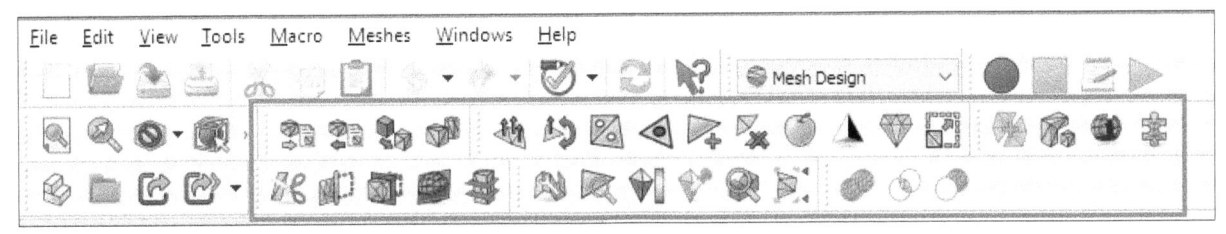

Figure-21. Mesh workbench tools

Importing Mesh Objects

The **Import mesh** tool is used to import mesh objects created in other CAD software like Alias, Inventor, and so on. The procedure to use this tool is given next.

- Click on the **Import mesh** tool from the **Meshes** menu or **Toolbar**. The **Import mesh** dialog box will be displayed; refer to Figure-22.

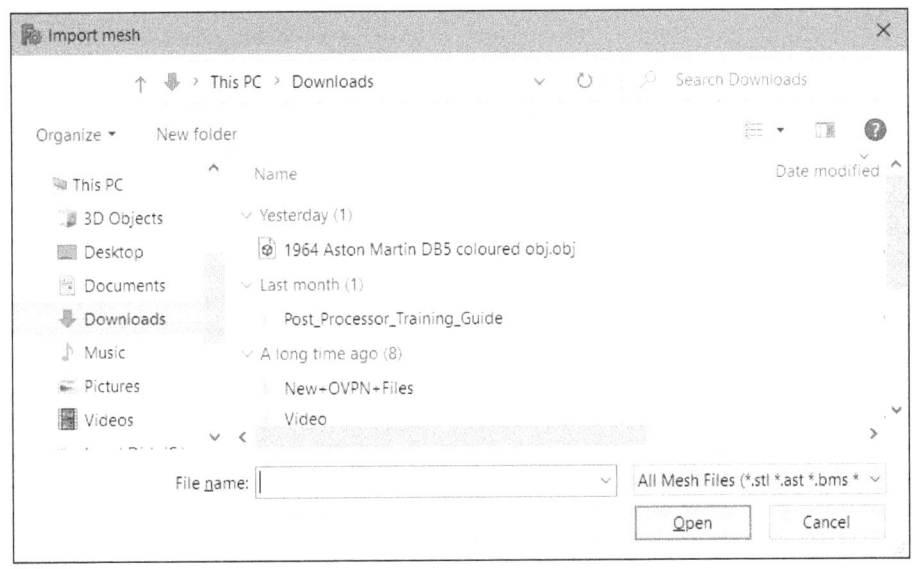

Figure-22. Import mesh dialog box

- Select desired mesh model file and click on the **Open** button. The mesh model will be displayed in the drawing area; refer to Figure-23.

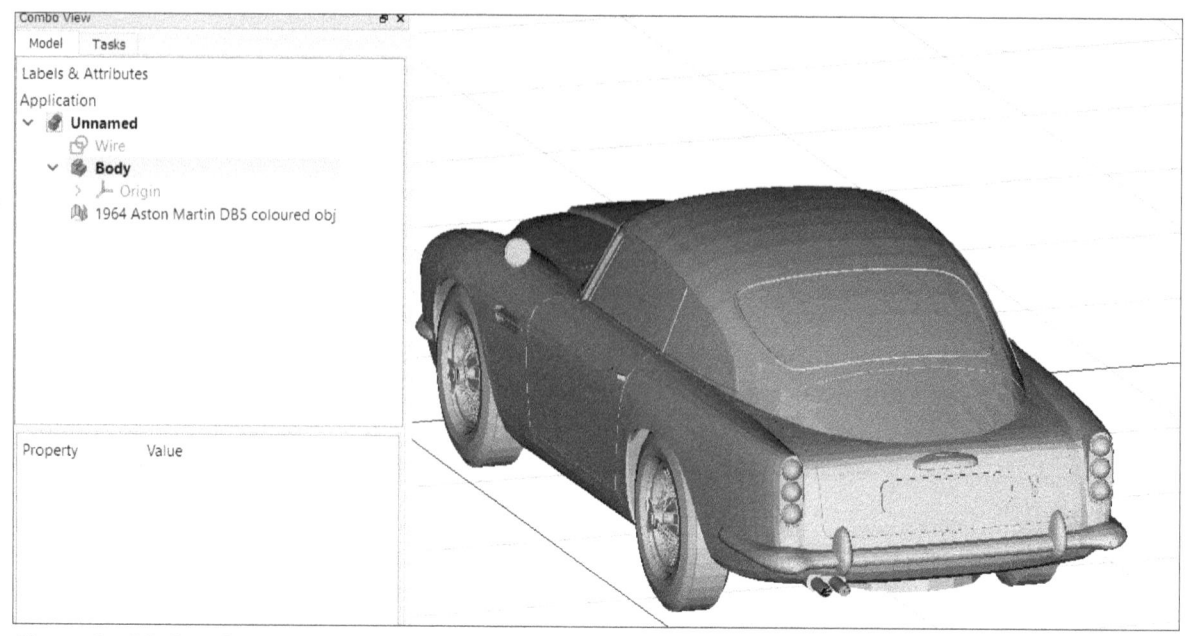

Figure-23. Mesh model imported

Exporting Mesh Model

The **Export mesh** tool is used to export mesh models created in FreeCAD for other CAD software. The procedure to use this tool is given next.

- Select the mesh model created in FreeCAD from **Model Tree** and click on the **Export mesh** tool from **Meshes** menu. The **Export mesh** dialog box will be displayed; refer to Figure-24.

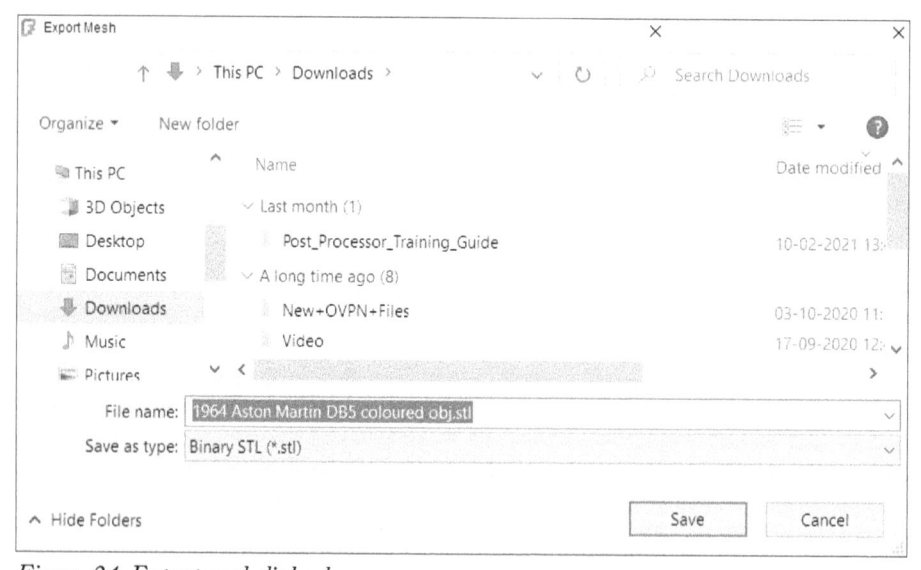

Figure-24. Export mesh dialog box

- Select desired format in which you want to export the file in **Save as type** drop-down.
- Specify desired name of file in the **File name** edit box and click on the **Save** button to export the file.

Creating Mesh from Shape

The **Create mesh from shape** tool is used to create mesh from selected solid body created using **Part** or **Part Design** workbench. The procedure to use this tool is given next.

- Click on the **Create mesh from shape** tool from **Meshes** menu after selecting solid body to be converted. The **Tessellation** dialog will be displayed; refer to Figure-25.

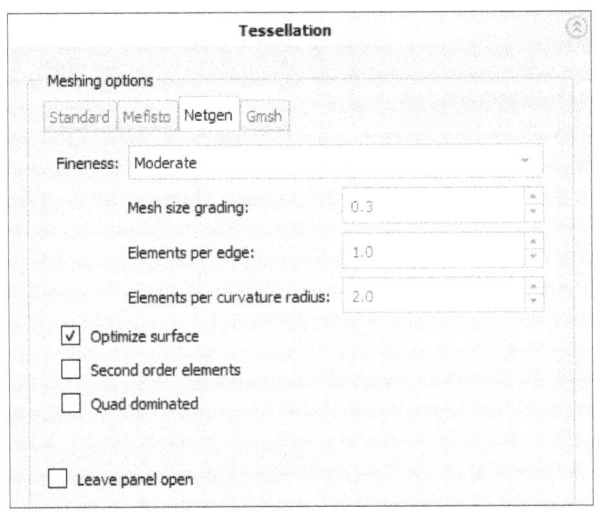

Figure-25. Tessellation dialog

- Select desired tab from the **Meshing options** area to specify method to be used for generating mesh.
- Select the **Netgen** tab to choose fineness of mesh from predefined mesh qualities; refer to Figure-25. You can select **Very coarse**, **Coarse**, **Moderate**, **Fine**, and **Very fine** to use presets of mesh quality. If you want to manually define mesh quality then select the **User defined** option and specify related parameters in the edit boxes below the drop-down.
- Select the **Standard** tab if you want to specify surface deviation in the mesh to define quality of mesh; refer to Figure-26. Note that the smaller value of surface deviation gives finer mesh. The minimum value that can be specified for surface deviation is 0.001.

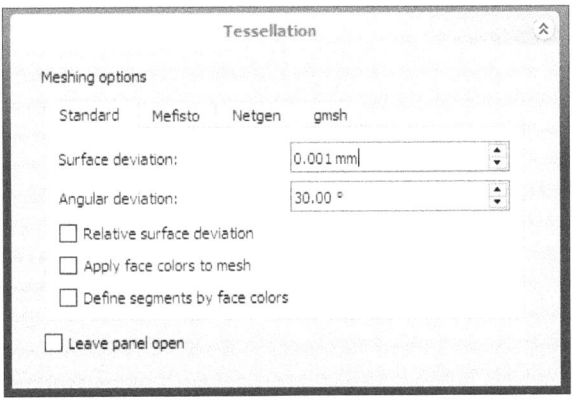

Figure-26. Standard mesh refinement

- Select the **Mefisto** tab if you want to specify maximum edge length in mesh to define quality of mesh; refer to Figure-27. Note that the smaller value of edge length gives finer mesh.

- Select the **gmsh** tab if you want to use gmsh mesher to define the quality of mesh; refer to Figure-28. Note that for Linux user, the external gmsh module is required.

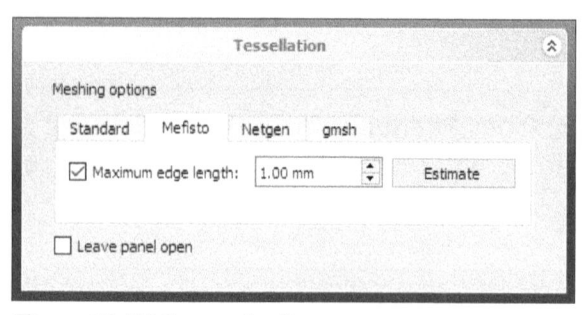

Figure-27. Mefisto mesh refinement

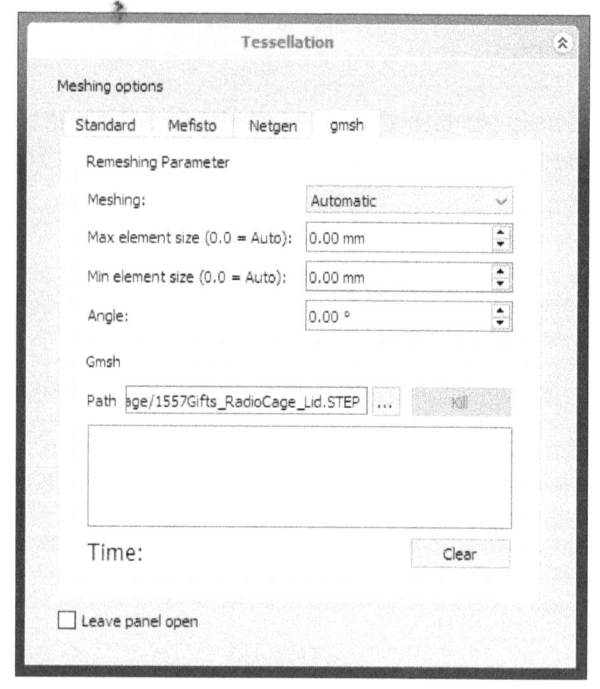

Figure-28. Gmsh mesh refinement

- Specify the other parameters as desired in the dialog and click on the **OK** button. The mesh model will be generated.

Refinement

The **Refinement** tool remeshes a mesh object using the Gmsh mesher. The new mesh can be finer or coarser. The procedure to use this tool is discussed next.

- Select desired mesh object from the **Model Tree** or from the 3D view area which you want to remesh; refer to Figure-29.

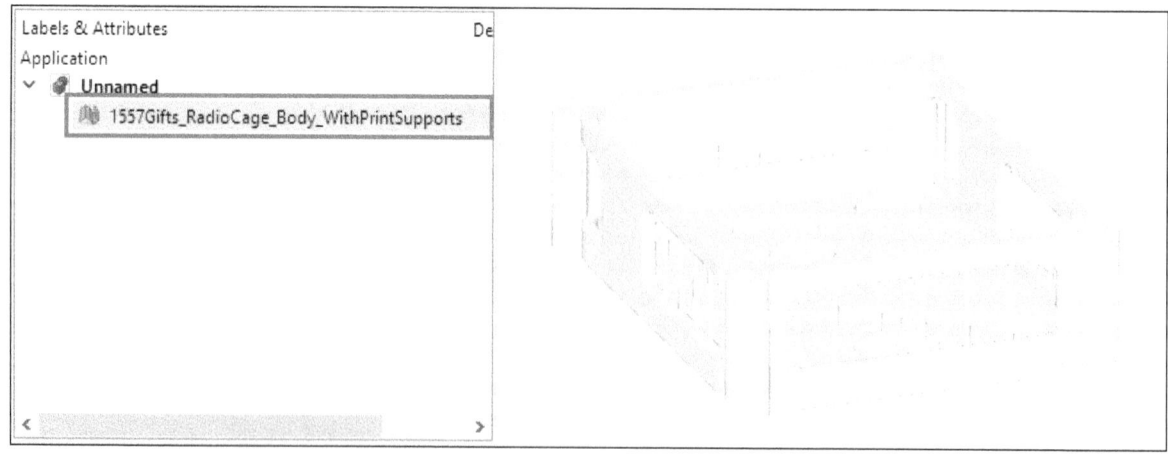

Figure-29. Object selected to remesh

- Click on the **Refinement** tool from **Meshes** menu or **Toolbar** in the **Mesh Design** workbench. The **Remesh by gmsh** dialog will be displayed in the **Tasks** panel of **Combo View**; refer to Figure-30.

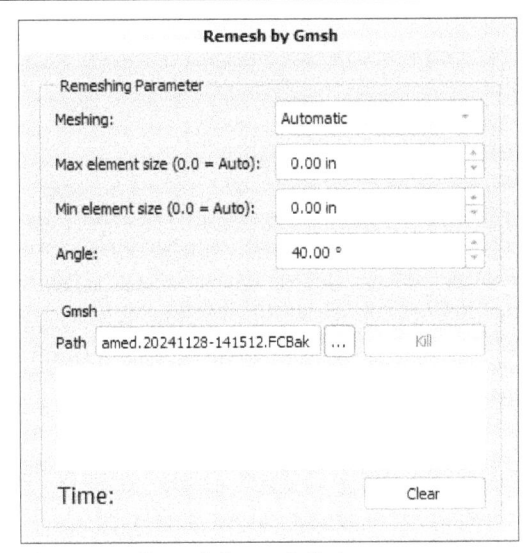

Figure-30. Remesh by gmsh dialog

- Select desired type of mesh from **Meshing** drop-down in the dialog.
- Enter desired value in the **Max element size** and **Min element size** edit boxes to specify the maximum and minimum size of the element, respectively.
- Specify desired angle of the element in the **Angle** edit box.
- Click on the **Path** button and specify the path for gmsh application.
- After specifying desired parameters, click on the **Apply** button. The gmsh operation will be running and the details of remeshing will be displaying in the box of the dialog; refer to Figure-31.

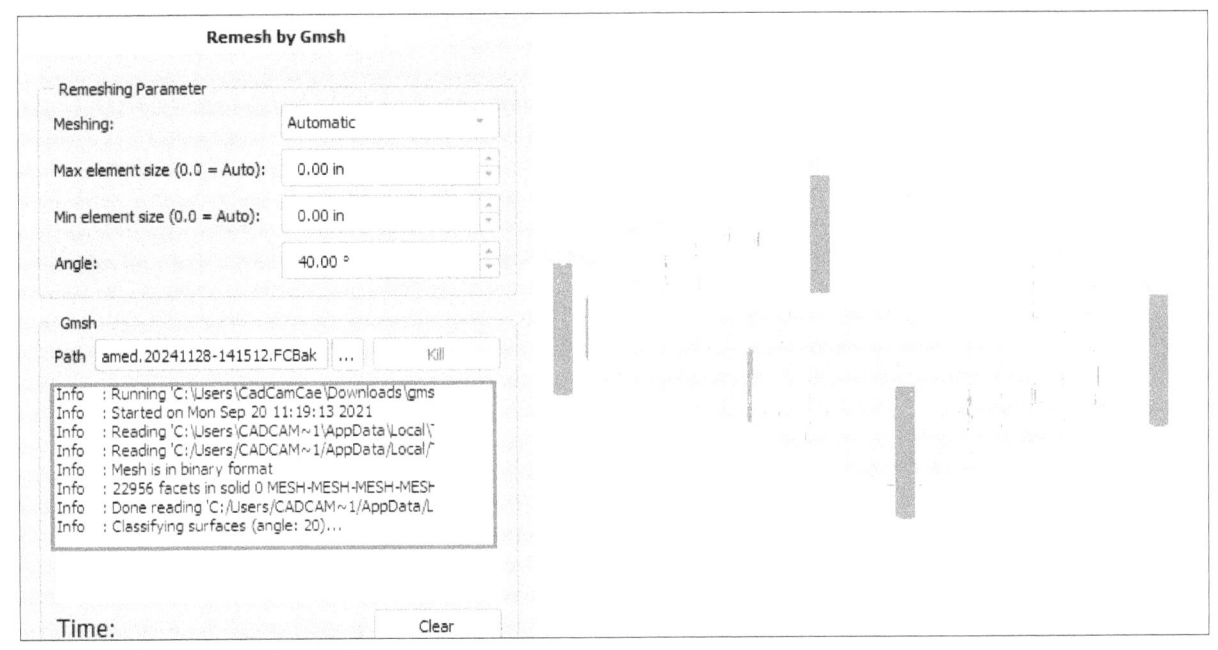

Figure-31. Running gmsh operation

- After the procedure is finished, click on the **Close** button to exit the tool.

Mesh Analysis Tools

The tools in **Analyze** cascading menu of **Meshes** menu are used to analyze and repair various aspects of mesh model imported/created in FreeCAD; refer to Figure-32. Various tools in this menu are discussed next.

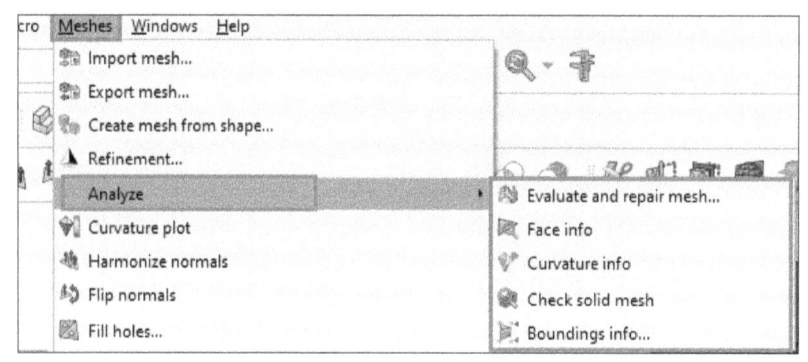

Figure-32. Analyze cascading menu

Evaluating and Repairing Mesh Model

The **Evaluate and repair mesh** tool is used to evaluate various aspects of mesh and repair them. The procedure to use this tool is given next.

- Click on the **Evaluate and repair mesh** tool from **Analyze** cascading menu of the **Meshes** menu. The **Evaluate & Repair Mesh** panel will be displayed; refer to Figure-33.

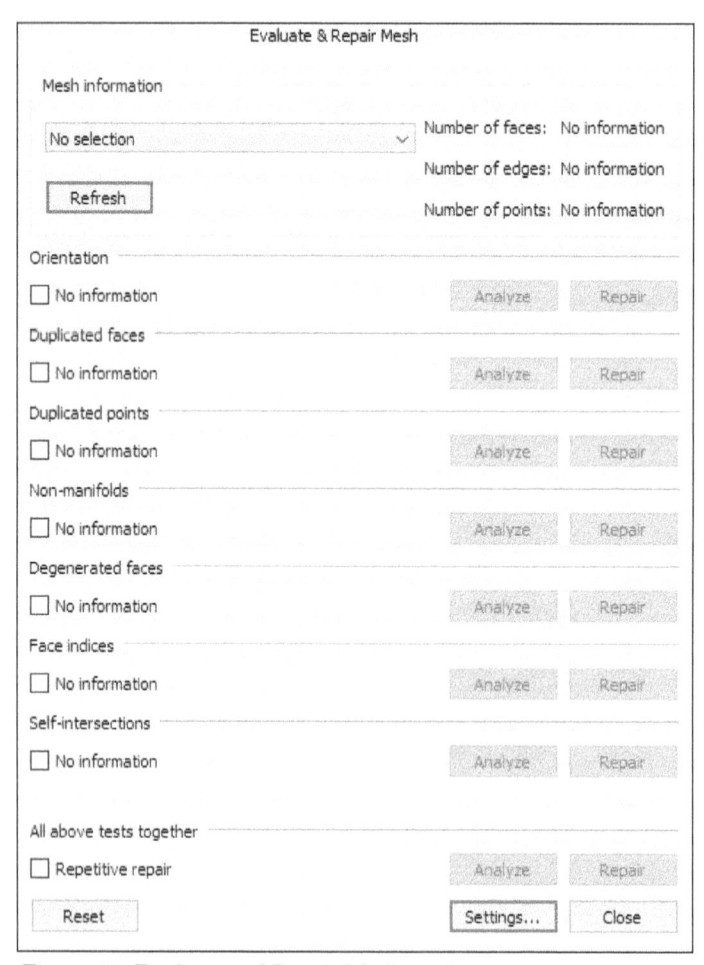

Figure-33. Evaluate and Repair Mesh panel

- There are various sections of panel to analyze various aspects of mesh model like orientation, duplicate faces, degenerated faces, and so on. Click on the **Analyze** button from the panel to check related results. If you want to analyze model based on all the analyses then click on the **Analyze** button from the **All above tests together** area of panel. The result of analyses will be displayed in respective sections of panel.
- Click on the **Repair** button from the panel to perform repairs in the model.
- Click on the **Settings** button from the panel to modify evaluation settings. The **Evaluation settings** dialog box will be displayed; refer to Figure-34.

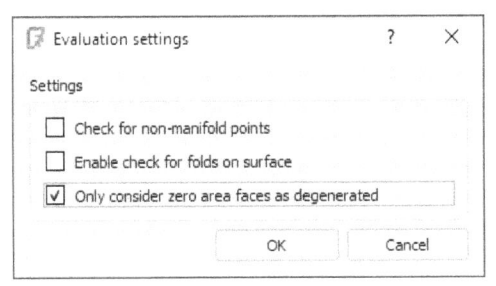

Figure-34. Evaluation settings dialog box

- Select the **Check for non-manifold points** check box to check mesh model for non-manifold points. Non-manifold points are locations of model where you cannot define precision of vertices.
- Select the **Enable check for folds on surface** check box to check for folds in the surface of mesh.
- Select the **Only consider zero area faces as degenerated** check box to define which faces are to be considered as degenerated.
- After setting desired parameters, click on the **OK** button.
- Click on the **Close** button from the panel to exit the panel.

Face Info

The **Face info** tool is used to check information of a face in mesh. Using this tool, you can index of selected face. The procedure to use this tool is given next.

- Click on the **Face info** tool from the **Analyze** cascading menu of **Meshes** menu. The cursor will change to pick face cursor.
- Click on desired faces of mesh model. Indexes of the faces will be displayed; refer to Figure-35.

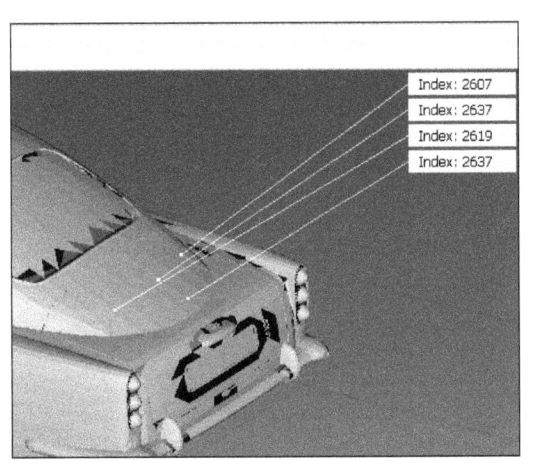

Figure-35. Indexes of faces displayed

- Right-click in the empty area of the drawing and select **Leave info mode** option from the shortcut menu.

Similarly, you can use the **Curvature info** tool of **Analyze** cascading menu in **Meshes** menu to check curvature information of mesh model. Note that this tool is active only when curvature plot has been generated.

Checking Solid Mesh

The **Check solid mesh** tool is used to check whether mesh is a solid mesh body or not. On clicking this tool, the **Solid Mesh** dialog box will be displayed showing information of mesh; refer to Figure-36.

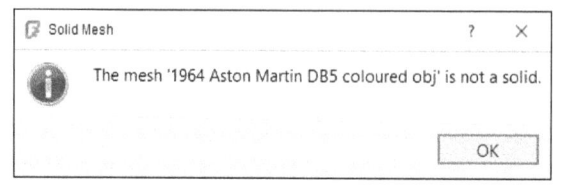

Figure-36. Solid Mesh dialog box

Checking Boundings Info

The **Boundings info** tool is used to check bounding parameters of the mesh body to define minimum and maximum point of the mesh; refer to Figure-37.

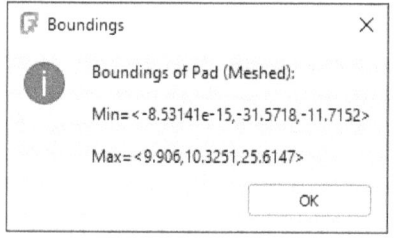

Figure-37. Boundings dialog box

Generating Curvature Plot

The **Curvature plot** tool is used to generate curvature plot for selected mesh model. Click on this tool after selecting mesh body, the curvature plot will be generated and displayed in the drawing area.

Harmonizing Normals

The **Harmonize normals** tool is used to make normals of faces in mesh share the same pattern. Click on the **Harmonize normals** tool from the **Meshes** menu after selecting the mesh body. All the normals of faces in mesh will point in same direction.

Flipping Normals of Mesh Faces

The **Flip normals** tool in **Meshes** menu is used to flip normal direction of mesh faces. After selecting mesh model, click on the **Flip normals** tool from the menu.

Filling Holes in Mesh

The **Fill holes** tool in **Meshes** menu is used to fill holes in the mesh based on specified edges. On clicking this tool, the **Fill holes** dialog box will be displayed; refer to Figure-38. Specify minimum number of edges to be considered for filling holes and click on the **OK** button. All the holes in mesh will be filled.

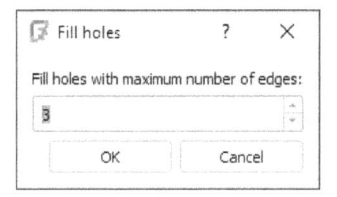

Figure-38. Fill holes dialog box

Closing Holes

The **Close hole** tool is used to close holes in selected faces of mesh. The procedure to use this tool is given next.

- Click on the **Close hole** tool from the **Meshes** menu. The cursor will change to face selection cursor.
- Click on the faces with empty holes/regions. The holes in selected faces will be filled.
- Right-click in the drawing area and select **Leave hole-filling mode** option to exit the hole closing mode.

Adding Triangles to Mesh

The **Add triangle** tool is used to add more triangles to the mesh body. On clicking this tool, you will be asked to select the points where more triangles for tessellation are to be added. One by one specify three points on the faces of mesh body to create a triangle. On selecting the points, a shortcut menu will be displayed; refer to Figure-39. Select the **Add triangle** option from shortcut menu. A triangle will be created in mesh.

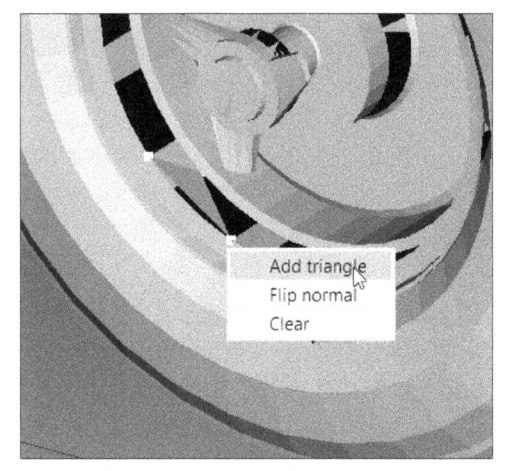

Figure-39. Shortcut menu for adding triangles

- Right-click in the drawing area and select **Finish** button to exit triangle creation mode.

Removing Components of Mesh

The **Remove components** tool is used to remove desired sections of the mesh model. The procedure to use this tool is given next.

- Click on the **Remove components** tool from the **Meshes** menu. The **Remove components** dialog will be displayed; refer to Figure-40.

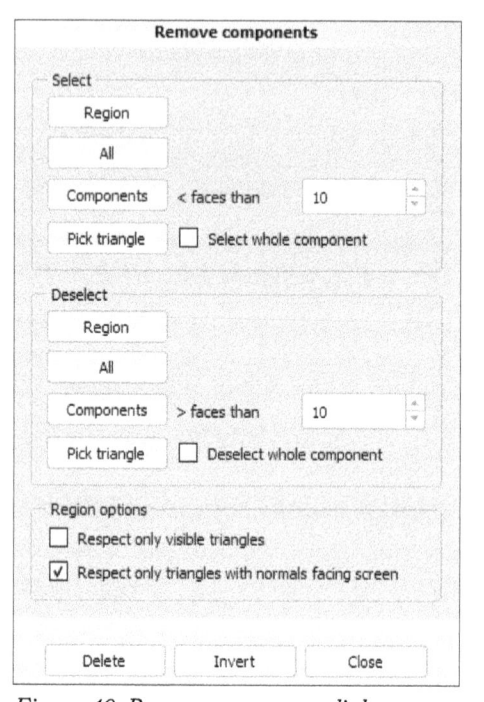

Figure-40. Remove components dialog

- Select the **Region** button from **Select** section and click & drag the cursor over faces of the mesh body to select them; refer to Figure-41. The faces in created region will get selected; refer to Figure-42.

Figure-41. Selecting region on mesh body

Figure-42. Region selected

- If you want to select full body of mesh then select the **All** button. If you want to select specified number of faces from center of body then specify number of faces to be selected in the **< faces than** edit box and click on the **Components** button. If you want to individually select triangles from the mesh body then click on the **Pick triangle** button and select desired faces.
- If you want to deselect faces then you can deselect the faces in the same way from the **Deselect** section of dialog.
- Select the **Respect only visible triangles** check box to select or deselect visible triangles from the model.

- Select the **Respect only triangles with normals facing screen** check box to only select/deselect the triangles which have normals pointing towards the screen.
- After selecting desired faces, click on the **Delete** button to delete faces.

Removing Components by Hand

The **Remove components by hand** tool is used to remove selected faces of the mesh model. The procedure to use this tool is given next.

- Click on the **Remove components by hand** tool from the **Meshes** menu. The cursor will change to a hand sign for selection.
- Click on desired faces of mesh model to select them and right-click in the drawing area. A shortcut menu will be displayed; refer to Figure-43.

Figure-43. Shortcut menu for deleting faces

- Select the **Delete selected faces** option from the shortcut menu to delete selected faces. Select the **Clear selected faces** option to remove all faces from selection and start a new selection set. Select the **Leave removal mode** option to exit the face removal mode.

Creating Mesh Segments

The **Create mesh segments** tool is used to generate segments from the mesh model. The procedure to use this tool is given next.

- After selecting the mesh model, click on the **Create mesh segments** tool from the **Meshes** menu. The **Mesh segmentation** dialog will be displayed; refer to Figure-44.

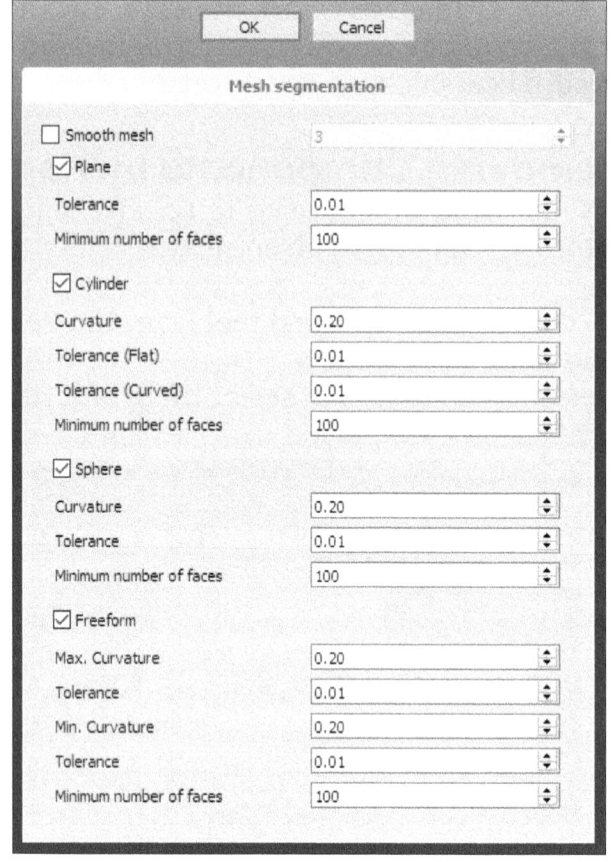

Figure-44. Mesh segmentation dialog

- Select check boxes for object types to be created in segmenting the mesh and specify related tolerances.
- After setting desired parameters, click on the **OK** button from the dialog. The segments will be created; refer to Figure-45.

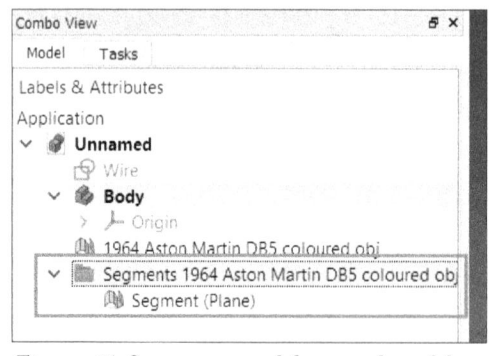

Figure-45. Segment created from mesh model

Similarly, you can use the **Create mesh segments from best-fit surfaces** tool to generate segments of mesh body.

Smoothening Mesh Model

The **Smooth** tool is used to smoothen selected mesh surface by specified degrees. The procedure to use this tool is given next.

- Select the mesh model from **Model Tree** or drawing area and then select the **Smooth** tool from the **Meshes** menu. The **Smoothing** dialog will be displayed; refer to Figure-46.

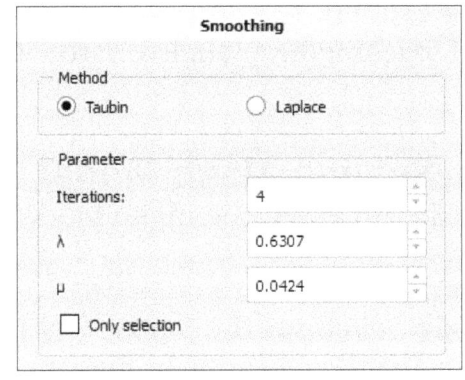

Figure-46. Smoothing dialog

- Select the **Taubin** radio button to use taubin smoothing and select the **Laplace** radio button to use Laplacian smoothing. Figure-47 shows difference between two smoothing methods.

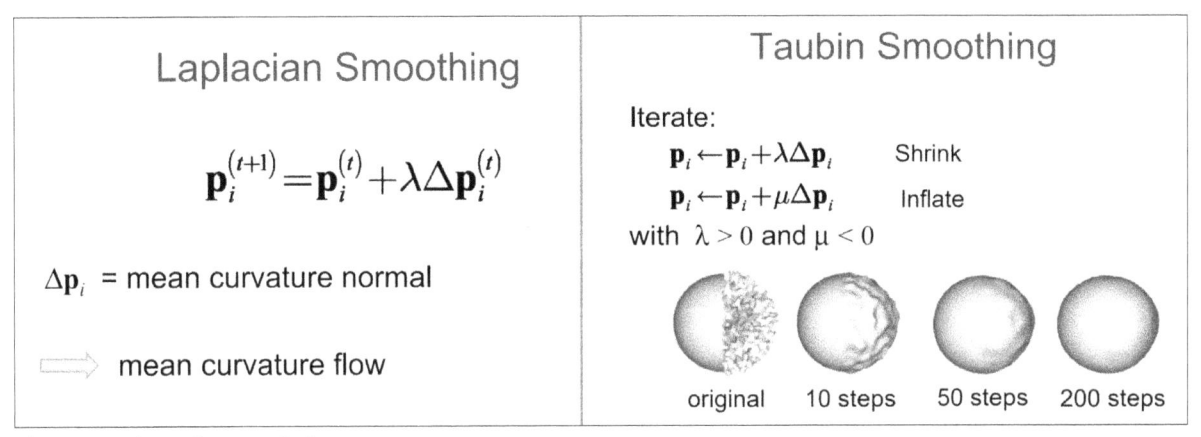

Figure-47. Smoothing methods

- Specify desired parameters in the dialog and click on the **OK** button to smoothen mesh model.

Decimating Mesh Model

The **Decimation** tool reduces the number of faces in mesh objects. The procedure to use this tool is discussed next.

- After selecting mesh model, click on the **Decimation** tool from the **Meshes** menu. The **Decimating** dialog will be displayed; refer to Figure-48.

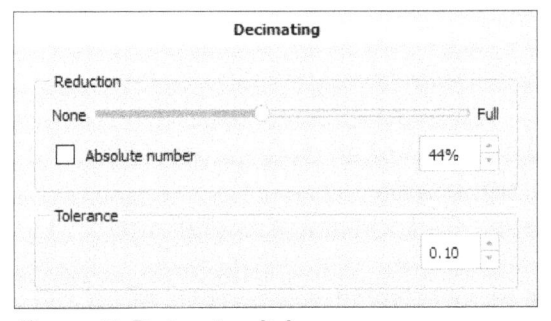

Figure-48. Decimating dialog

- Move the slider towards **None** or **Full** as desired to specify the reduction of faces.
- Select the **Absolute number** check box to specify the reduction of absolute number of faces by entering desired value in the edit box of **Reduction** area in the dialog.

- Specify desired tolerance value in the edit box of **Tolerance** area in the dialog.
- After specifying desired parameters, click on the **OK** button to complete the procedure.

Scaling Mesh Model

The **Scale** tool is used to scale up/down the mesh model. The procedure to use this tool is given next.

- After selecting mesh model, click on the **Scale** tool from the **Meshes** menu. The **Scaling** dialog box will be displayed; refer to Figure-49.

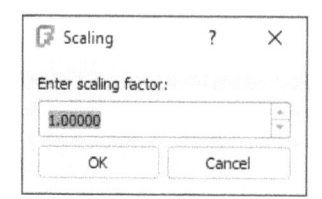

Figure-49. Scaling dialog box

- Specify desired value of scale factor in the edit box and click on the **OK** button. A value more than **1** will increase the size of mesh model and a value less than **1** but greater than **0** will decrease the size of mesh model.

Creating Regular Mesh Solids

The **Regular solid** tool is used to create regular mesh solids. The procedure to use this tool is given next.

- After selecting the mesh body, click on the **Regular solid** tool from **Meshes** menu. The **Regular Solid** dialog box will be displayed; refer to Figure-50.

Figure-50. Regular Solid dialog box

- Select desired option from the **Solid** drop-down to define shape of solid and specify related parameters in the dialog box.
- After setting parameters, click on the **Create** button to create solid object at the origin.
- After creating solid, click on the **Close** button from the dialog box to exit.

Performing Boolean Operations on Meshes

The tools in **Boolean** cascading menu of **Meshes** menu are used to perform boolean operations like performing union, intersection, and difference of two mesh objects.

Note that you need to install OpenSCAD using web link https://openscad.org and specify path for exe file of OpenSCAD in **Preferences** dialog box. To do so, first switch to OpenSCAD workbench and then use the **Preferences** dialog box to set the openscad.exe path. The procedure to perform boolean operations is given next.

- Select two meshes from the **Model Tree** on which you want to perform boolean operations. The tools in **Boolean** cascading menu of **Meshes** menu will become active.
- Click on the **Union** tool from the cascading menu if you want to join two mesh bodies; refer to Figure-51.

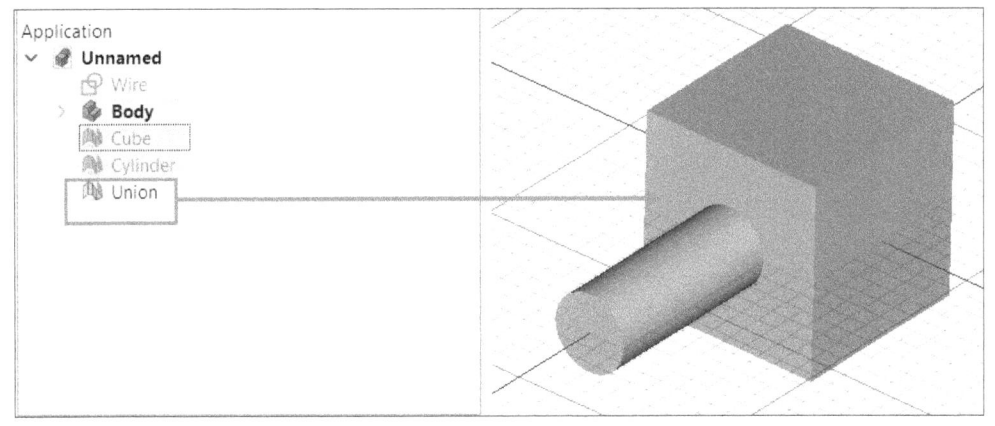

Figure-51. Union body created

- Click on the **Difference** tool to subtract 2nd selected object from 1st selected object. The **Difference** feature will be displayed; refer to Figure-52.

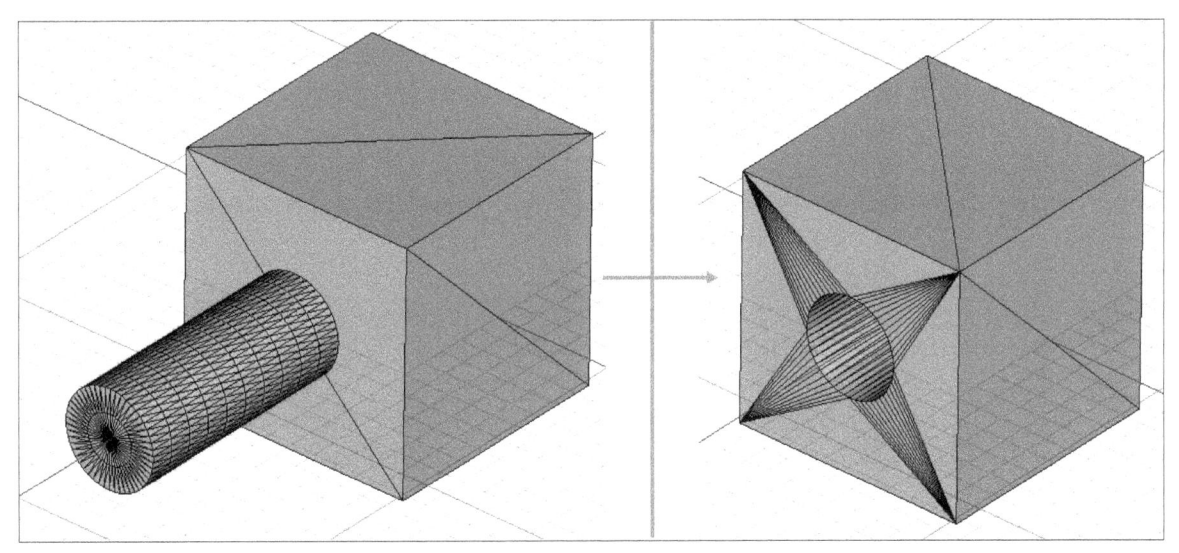

Figure-52. Difference feature

- Click on the **Intersection** tool from the **Boolean** cascading menu after selecting two mesh objects, the region common in both objects will be created as intersection object; refer to Figure-53.

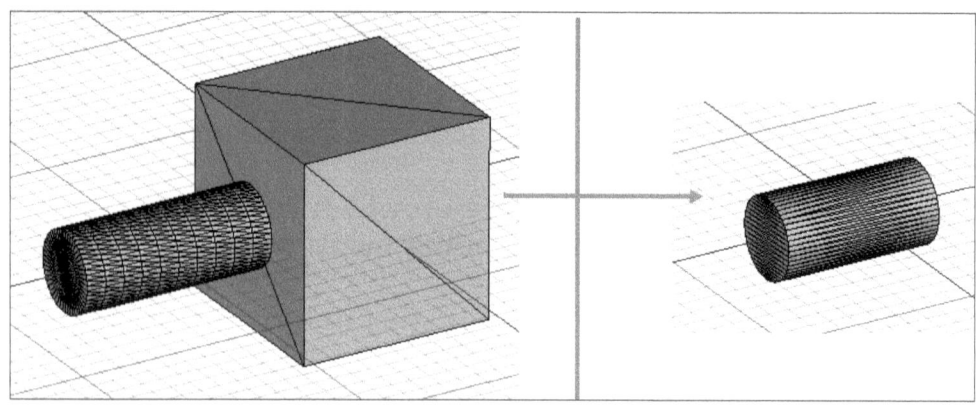

Figure-53. Intersection feature created

Mesh Cutting Operations

The tools in **Cutting** cascading menu of **Meshes** menu are used to cut mesh object using plane, section, or polygon selection; refer to Figure-54. Various tools in this cascading menu are discussed next.

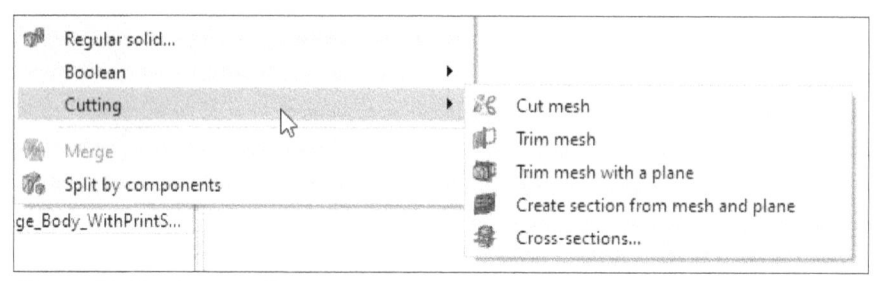

Figure-54. Cutting cascading menu

Cut Mesh Tool

The **Cut Mesh** tool is used to cut a mesh by drawing a polygon selection box. The procedure to use this tool is given next.

- Click on the **Cut Mesh** tool from the **Cutting** cascading menu of the **Meshes** menu after selecting the mesh body to be cut. The cursor will change to a scissor icon and you will be asked to draw a polygonal selection box.
- Draw a polygon on the body by specifying end points and right-click in the drawing area. A shortcut menu will be displayed; refer to Figure-55.

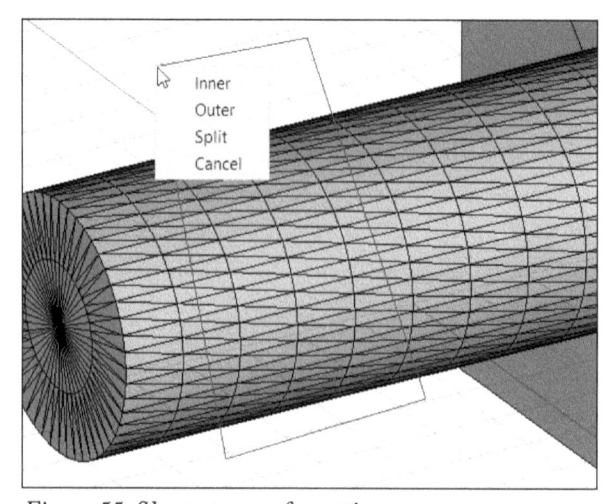

Figure-55. Shortcut menu for cutting

- Select the **Inner** option from shortcut menu to delete the mesh elements falling inside drawn polygon; refer to Figure-56. Select the **Outer** option from shortcut menu to delete mesh elements falling outside the selection polygon. Select the **Split** option from shortcut menu to delete the area falling inside selection polygon and create a new body. Select the **Cancel** option to abort the operation.

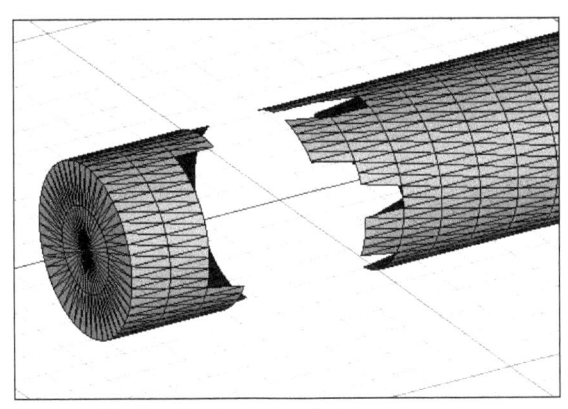

Figure-56. Cutting inner mesh elements

Trim Mesh Tool

The **Trim mesh** tool is used to trim a portion of mesh body using drawn selection polygon. Note that when you use **Cut mesh** tool then full elements falling inside selection polygon will be removed but when you use **Trim mesh** tool then these elements will also be trimmed at the boundary of selection polygon. The procedure to use **Trim mesh** tool is given next.

- Click on the **Trim mesh** tool from **Cutting** cascading menu of the **Meshes** menu after selecting the mesh body to be trimmed. You will be asked to draw a selection polygon.
- Create the polygon by specifying end points and right-click. A shortcut menu will be displayed.
- Select desired option from the shortcut menu to trim the mesh body; refer to Figure-57. The options in shortcut menu are same as discussed earlier.

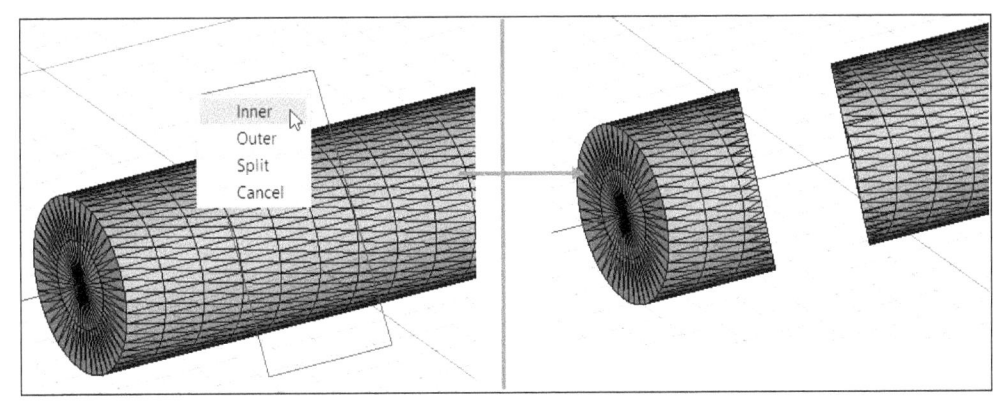

Figure-57. Trimming mesh

Trim mesh with a plane

The **Trim mesh with a plane** tool is used to trim mesh body using selected plane. Note that the plane must be created using the **Create primitives** tool of **Part** workbench; refer to Figure-58. The procedure to use this tool is given next.

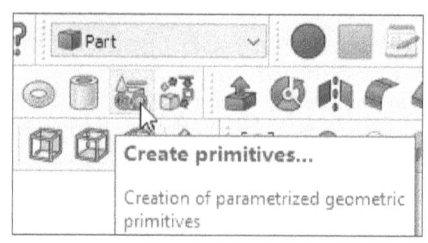

Figure-58. Create primitives tool

- Click on the **Trim mesh with a plane** tool from the **Cutting** cascading menu of the **Meshes** menu after selecting mesh body and an intersecting plane. The **Trim by plane** dialog box will be displayed; refer to Figure-59.

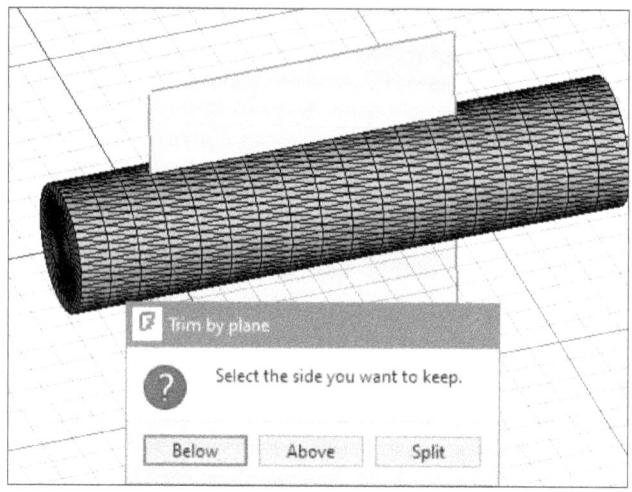

Figure-59. Trim by plane dialog box

- Select the **Below** button from the dialog box to remove mesh body on the front side of plane and keep back side of mesh body.
- Select the **Above** button from the dialog box to remove mesh body on the back side of plane and keep front side of mesh body.
- Select the **Split** button from the dialog box to split the mesh body into two pieces.

Create section from mesh and plane

The **Create section from mesh and plane** tool is used to generate section curves at the intersection of mesh body and plane. The procedure to use this tool is given next.

- After selecting plane and mesh body, click on the **Create section from mesh and plane** tool from the **Cutting** cascading menu of the **Meshes** menu. The section will be created; refer to Figure-60.

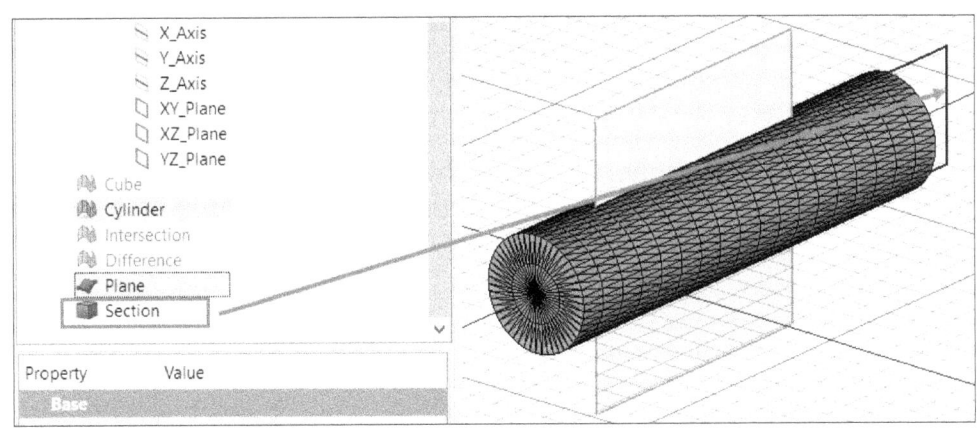

Figure-60. Section created

Creating Cross Sections

The **Cross section** tool creates multiple cross-sections across mesh objects. The cross-sections are taken parallel to one of the main global planes (XY, XZ, or YZ). The procedure to use this tool is discussed next.

- After selecting one or more mesh objects, click on the **Cross-sections** tool from the **Cutting** cascading menu of the **Meshes** menu. The **Cross sections** dialog will be displayed; refer to Figure-61.

Figure-61. Cross sections dialog

- Select desired guiding plane from **XY, XZ,** or **YZ** radio button to be used to create the cross-sections.
- Enter desired value in the **Position** edit box to specify the position of guiding plane from the origin. The default position is based on the center of the bounding box of the selected mesh objects.
- Select the **Sections** check box to create multiple cross-sections.
- Select **On both sides** check box to create the cross sections on both sides of the guiding plane.
- Specify desired number of cross sections in the **Count** edit box.
- Specify desired distance between the cross sections in the **Distance** edit box.
- Select **Connect edges if distance less than** check box and specify desired value in the activated edit box.
- After specifying desired parameters, click on the **OK** button from the dialog to complete the procedure.

Merging Mesh

The **Merge** tool is used to combine two or more mesh bodies to form a single mesh. The procedure to use this tool is given next.

- Select the mesh bodies to be merged from the **Model Tree** and click on the **Merge** tool from the **Meshes** menu. The merged mesh body will be created; refer to Figure-62.

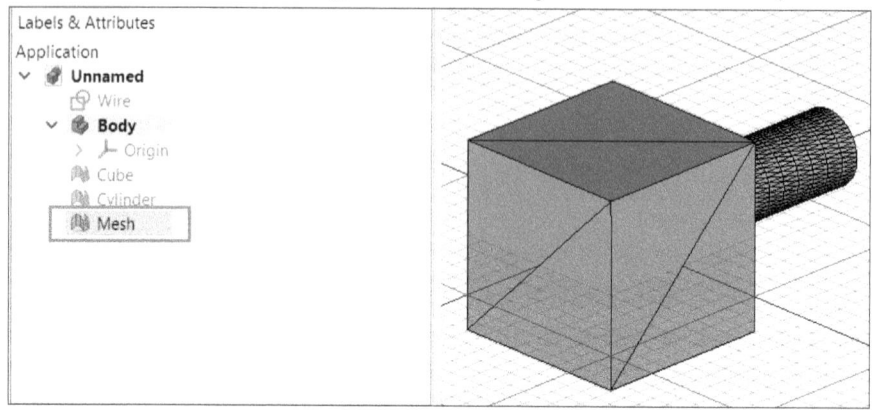

Figure-62. Mesh body created

Splitting Mesh Object

The **Split by components** tool splits a mesh object into its components. The procedure to use this tool is discussed next.

- Select the mesh bodies to be split from the **Model Tree** and click on the **Split by components** tool from the **Meshes** menu. The mesh body will be split to its components; refer to Figure-63.

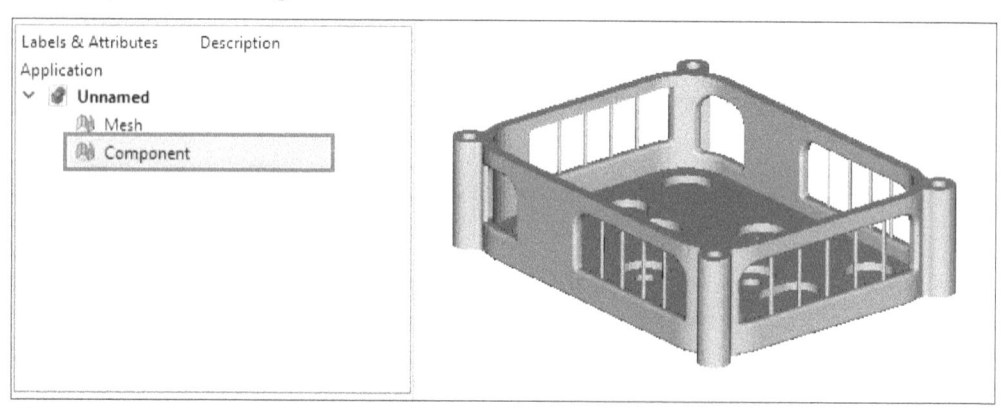

Figure-63. Mesh body split into components

Unwrap Mesh

The **Unwrap Mesh** tool creates a flat representation of a mesh object by unwrapping or unfolding it. The procedure to use this tool is discussed next.

- Select desired mesh body from the **Model Tree** to unwrap and click on the **Unwrap Mesh** tool from the **Meshes** menu. The meshed body will be unwrapped as a Part Feature; refer to Figure-64.

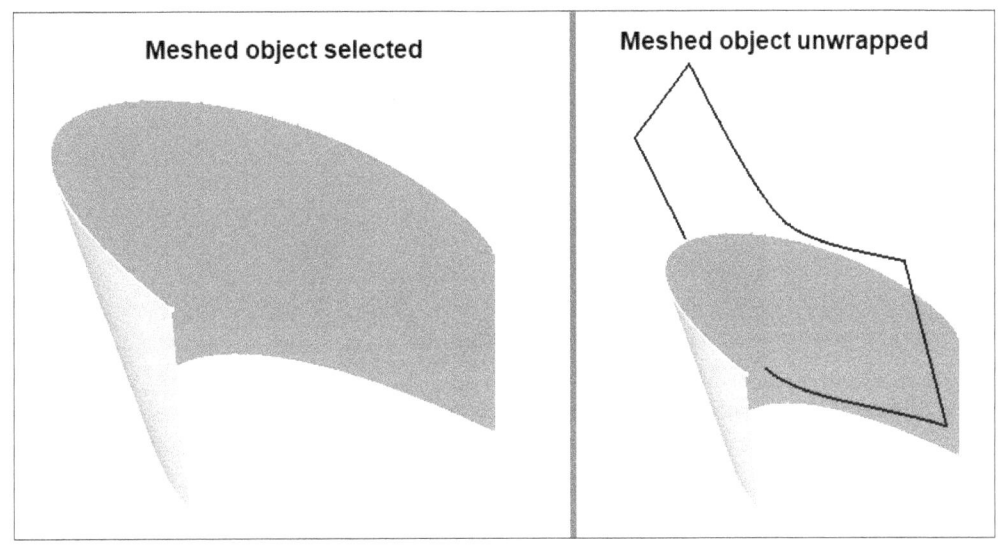

Figure-64. Meshed object unwrapped

Unwrap Face

The **Unwrap Face** tool is used to create a flat representation of a face of a shape object. The procedure to use this tool is same as discussed for **Unwrap Mesh** tool.

ROBOT WORKBENCH

The tools in **Robot** workbench are used to simulate path of robot movements. A robot is generally used for performing repeating tasks like welding an edge, tightening some screws, and so on which have predefined paths. The tools in **Robot** workbench are shown in Figure-65. Various tools of this workbench are discussed next.

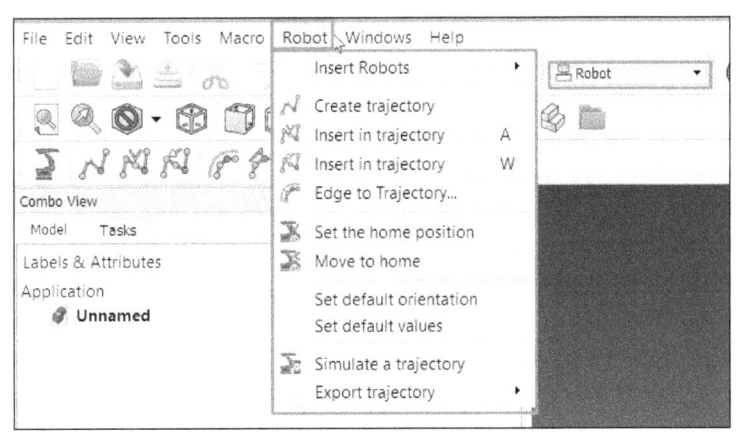

Figure-65. Robot workbench tools

Inserting Robot

The tools in **Insert Robots** cascading menu of **Robot** menu are used to insert different type of robots for path simulation; refer to Figure-66. Select desired robot from the menu for which you want to simulate the toolpath. The robot will be displayed in drawing area; refer to Figure-67.

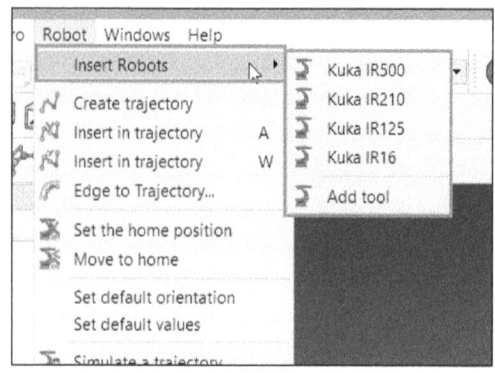

Figure-66. Insert Robots cascading menu

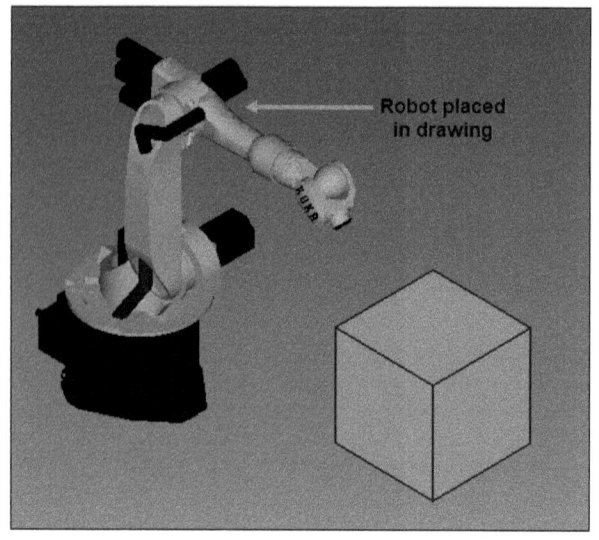

Figure-67. Robot placed in drawing

Creating Robot Trajectory

The **Create trajectory** tool is used to create an empty trajectory path for robot. Click on the **Create trajectory** tool from the **Robot** menu. An empty trajectory feature will be added in the **Model Tree**; refer to Figure-68.

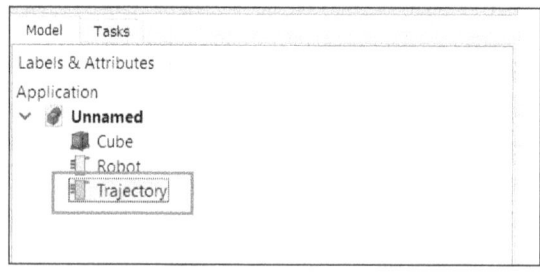

Figure-68. Empty trajectory added

Adding Edges to Trajectory

The **Edge to Trajectory** tool is used to add selected edges to the trajectory. The procedure to use this tool is given next.

- Click on the **Edge to Trajectory** tool from the **Robot** menu. The dialogs to add edges to trajectory will be displayed; refer to Figure-69.

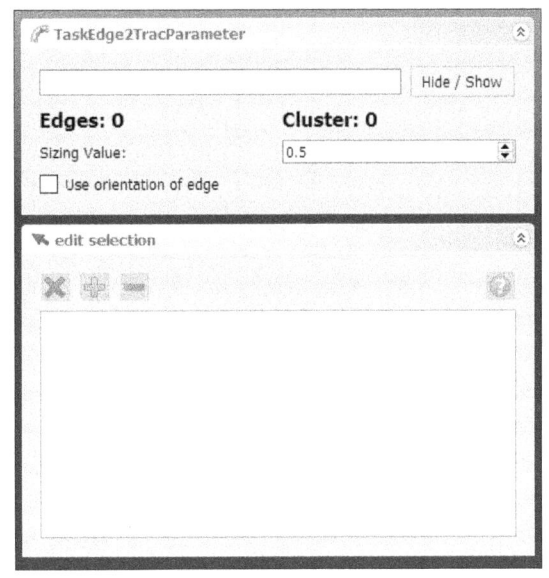

Figure-69. Dialogs for adding edges to trajectory

- Select the edges to be added in the trajectory while holding the **CTRL** key; refer to Figure-70.

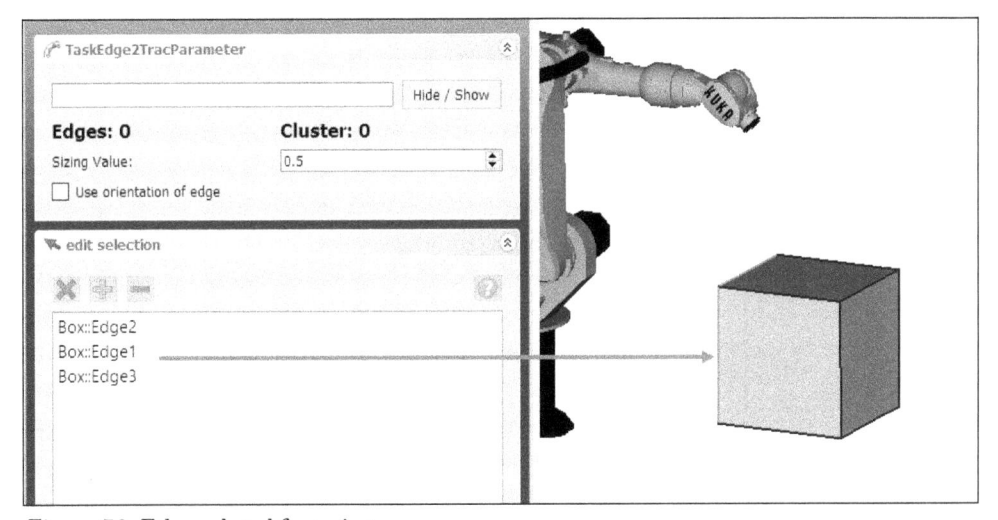

Figure-70. Edges selected for trajectory

- Click on the **Apply** button from the dialog after selecting edges and click on the **OK** button to add edges to trajectory.

Setting Default Orientation of Robot

The `Set default orientation` tool is used to set initial placement position and orientation of the selected robot. The procedure to use this tool is given next.

- Click on the **Set default orientation** tool from the **Robot** menu. The **Placement** dialog box will be displayed; refer to Figure-71.

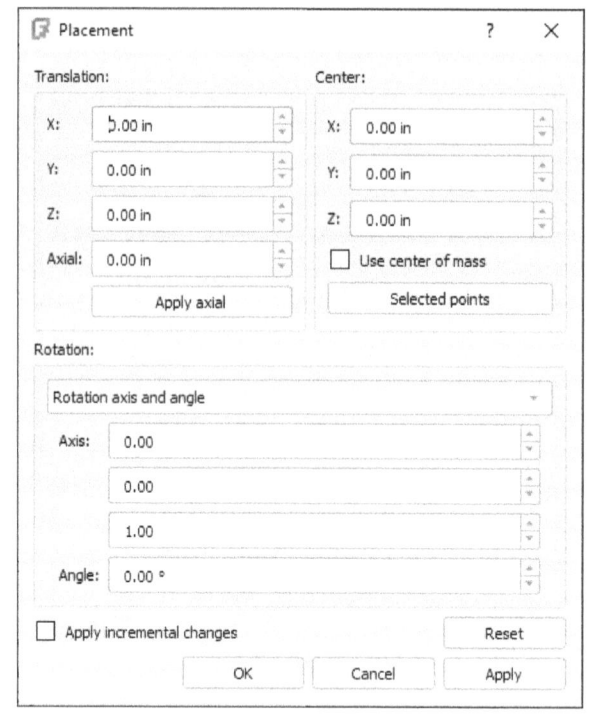

Figure-71. Placement dialog box

- Set desired parameters in the **Translation** area to move robot to desired location.
- Set desired values in **X**, **Y**, and **Z** edit boxes of **Center** area to define location of center of Robot for translation.
- In the **Rotation** area, you can rotate robot about selected axis by specifying desired values in the **Axis** edit boxes and define rotation angle in the **Angle** edit box.
- After setting desired changes, click on the **Apply** button to apply orientation and click on the **OK** button to exit the dialog box.

Setting Default Values for Robot

The **Set default values** tool is used to define default movement parameters for robot like velocity of movement, acceleration, and so on. Click on the **Set default values** tool from the **Robot** menu after selecting robot from **Model Tree**. The **Set default speed** dialog box will be displayed; refer to Figure-72. Specify desired value of speed and click on the **OK** button. The **Set default continuity** dialog box will be displayed; refer to Figure-73. Select the **False** option if you want robot movements to be in fragments. Select the **True** option if you want robot movements to be continuous. After selecting desired option, click on the **OK** button. The **Set default acceleration** dialog box will be displayed; refer to Figure-74. Specify desired value in the edit box for acceleration and click on the **OK** button.

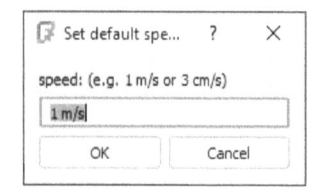

Figure-72. Set default speed dialog box

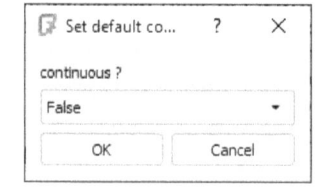

Figure-73. Set default continuity dialog box

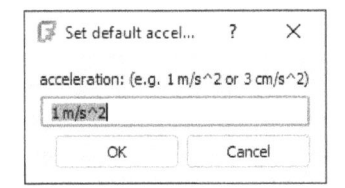

Figure-74. Set default acceleration dialog box

Adding Tool to Robot

To add a tool to Robot arm, first create the model of tool using **Part Design** workbench. Now, select the **Robot** and part model created as tool from **Model Tree**, switch to **Tasks** panel in the **Combo View** and click on the **Add tool** option from **Robot tools** dialog; refer to Figure-75. The selected body will be added as tool to robot arm; refer to Figure-76.

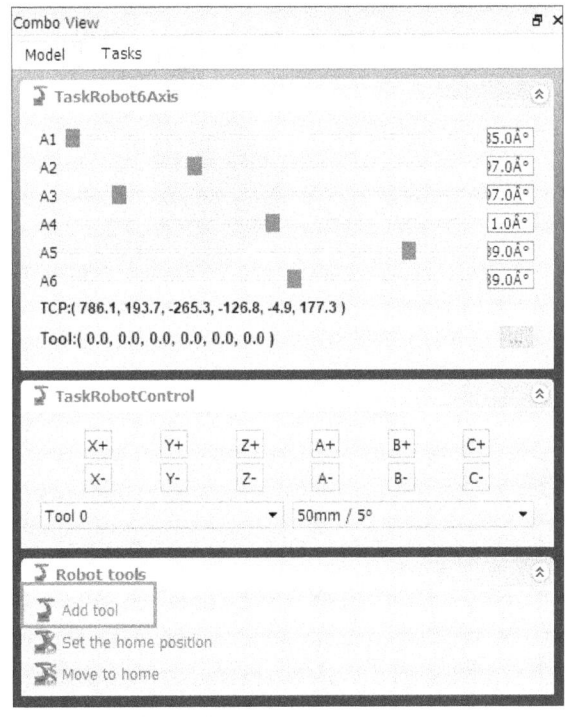

Figure-75. Add tool option

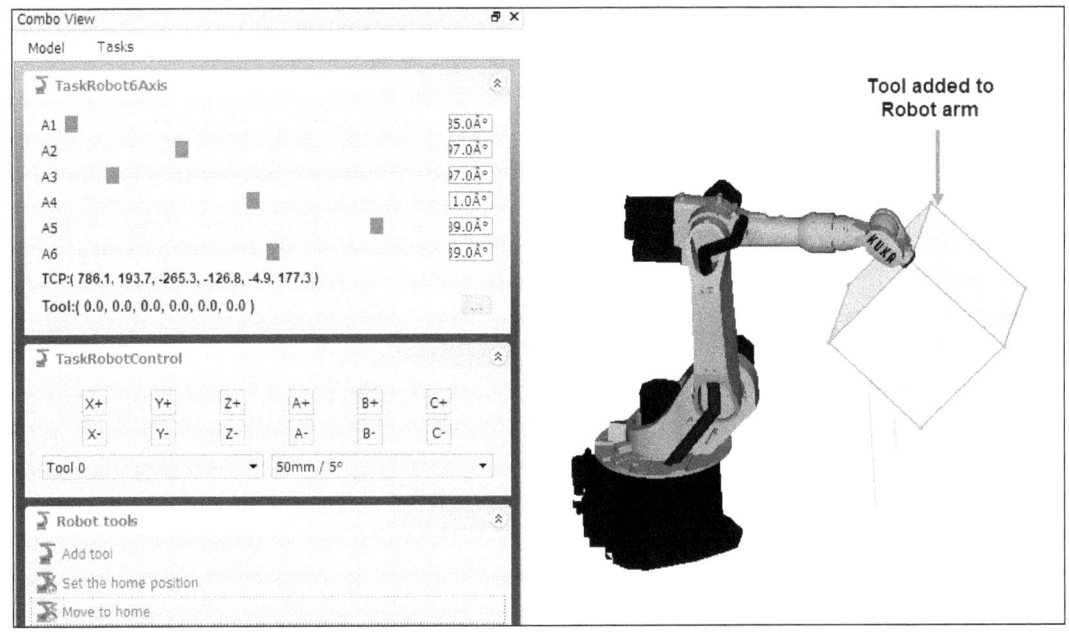

Figure-76. Tool added to robot arm

Simulating Robot Trajectory

The **Simulate a trajectory** tool is used to simulate movements of robot on the trajectory earlier created. Select the Robot and trajectory features from the **Model Tree** and click on the **Simulate a trajectory** tool from the **Robot** menu. Dialogs will be displayed in the **Tasks** panel to simulate robot trajectory; refer to Figure-77. You can move the slider in **Trajectory** dialog to check movement of robot arm or use the buttons to run simulation.

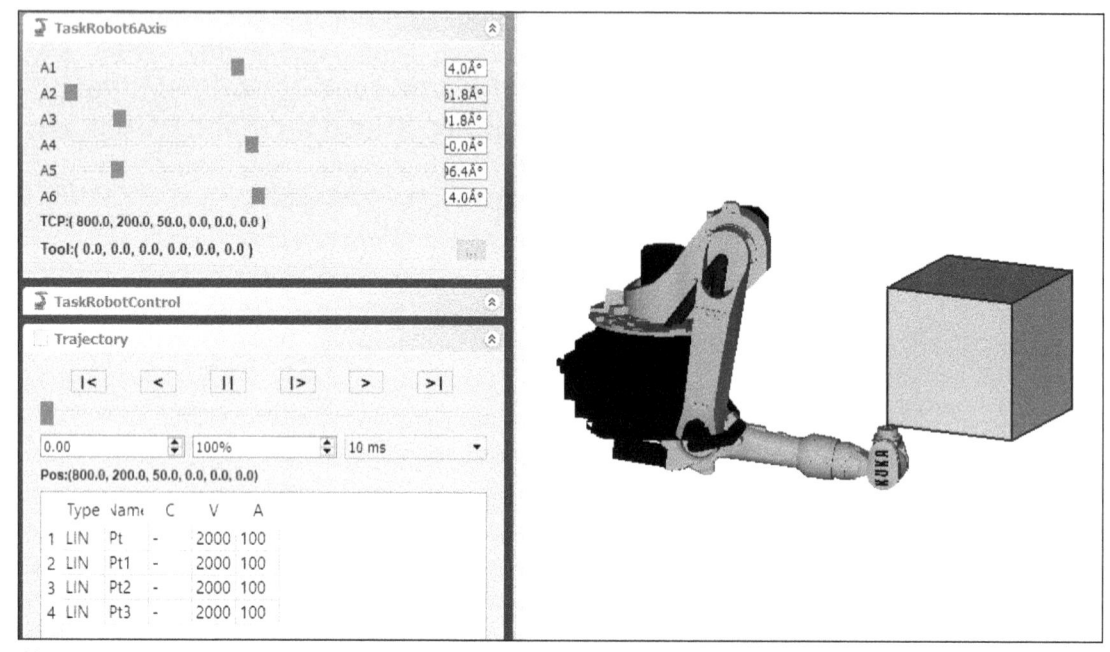

Figure-77. Dialogs for simulating robot

3D PRINTING TOOLS WORKBENCH

The tools in **3D Printing Tools** workbench are used to scale and modify mesh models. To start **3D Printing Tools** workbench, select the **3D Printing Tools** option from the **Switch between workbenches** drop-down in the **Toolbar**; refer to Figure-78. Make sure to active the **3D Printing Tools** addon from the **Addon Manager**. Various tools in this workbench are discussed next.

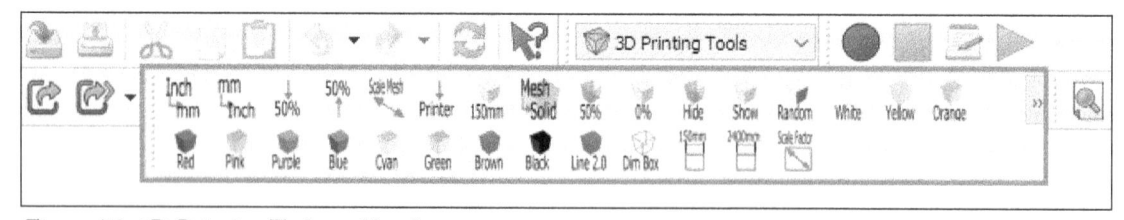

Figure-78. 3D Printing Tools workbench

Converting Imperial Mesh to Metric and vice-versa

Select the mesh model and click on the **Converts Imperial Mesh to Metric** button from the **Toolbar**. The imperial mesh model will be converted to metric mesh model.

Similarly, click on the **Converts Metric Mesh to Imperial** button from the **Toolbar**. The metric mesh model will be converted to imperial mesh model.

Reducing and scaling mesh by 50%

Select the mesh model and click on the **Reduces Mesh by 50%** button from the **Toolbar**. The mesh model will be scaled down by 50%.

Similarly, click on the **Scale Mesh by 50%** button from the **Toolbar**. The mesh model will be scaled up by 50%.

Scaling mesh by variable scale factor

Select the mesh model and click on the **Scale Geometry by variable scale factor** button from the **Toolbar**. The mesh model will be scaled by variable scale factor.

Reducing mesh by 95%

Select the mesh model and click on the **Reduces Mesh by 95% to fit 150mm Printer Bed** button from the **Toolbar**. The mesh model will be scaled down by 95% to fit 150mm Printer Bed.

Creating mesh box

Select the mesh model and click on the **Create mesh box size of Printer** button from the **Toolbar**. The mesh box by the size of the Printer Bed will be generated.

Converting mesh to refined solids

Select the mesh model and click on the **Module to convert Mesh to Refined Solids** button from the **Toolbar**. The mesh model will be converted to refined solid model.

Changing objects transparency

Select the mesh model and click on the **Changes objects 50% Transparency** button from the **Toolbar**. The transparency of mesh model will be changed to 50%.

Similarly, click on the **Changes objects 0% Transparency** button from the **Toolbar**. The transparency of mesh model will be changed to 0%.

Hiding or showing the selected objects

Select the mesh model and click on the **Hides Selected Objects** button from the **Toolbar**. The selected mesh model will be hidden.

Similarly, click on the **Unhides Selected Objects** button from the **Toolbar**. The selected mesh model will be shown.

Modifying objects to random color

Select the mesh model and click on the **Modifies objects to Random Colour** button from the **Toolbar**. The color of model will be changed to 11 different colors.

Changing color of objects

Select the mesh model and click on desired color buttons, viz. **White**, **Yellow**, **Orange**, **Red**, **Pink**, **Purple**, **Blue**, **Cyan**, **Green**, **Brown**, and **Black** from the **Toolbar** to change the respective colors of the model.

Changing the line width

Select the mesh model and click on the **Changes Line Width to 2.0** button from the **Toolbar**. The line width of the selected model will be changed to 2.00.

Turning on or off the Bounding box size

Create and select the bounding box of the mesh model and then click on the **Turn on and off the Bounding box size** button from the **Toolbar**. The bounding box size of the model will be turned on or off.

Defining Printer bed size

Click on the **Change default Bed Length Size** button from the **Toolbar**. The **Printer Bed Length** dialog box will be displayed; refer to Figure-79. Specify desired Printer Bed Size in the edit box of the dialog box and click on the **OK** button. The value of size will be changed.

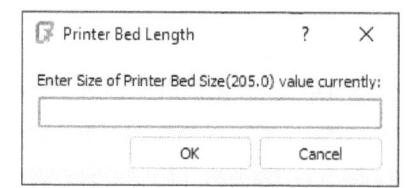

Figure-79. Printer Bed Length dialog box

Defining Die length size

Click on the **Change default Die Length Size** button from the **Toolbar**. The **Die Size** dialog box will be displayed; refer to Figure-80. Specify desired size of die to scale down in the edit box of the dialog box and click on the **OK** button. The value of size will be changed.

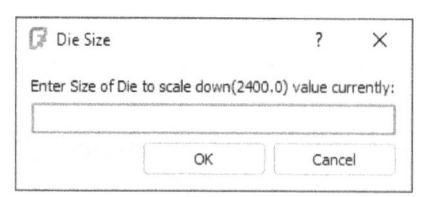

Figure-80. Die Size dialog box

Defining the Scale factor

Click on the **Change default Variable scale factor** button from the **Toolbar**. The **Scale Factor** dialog box will be displayed; refer to Figure-81. Specify desired value of the scale factor of die to scale in the edit box of the dialog box. The value of scale factor of die will be changed.

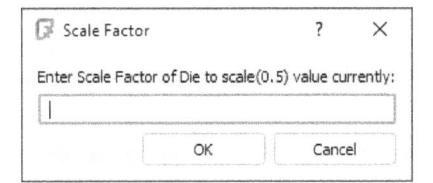

Figure-81. Scale Factor dialog box

SHEET METAL WORKBENCH

The Sheetmetal workbench is external workbench to be added in FreeCAD using the Addon manager. After adding workbench, select the **Sheet Metal** option from the **Switch between workbenches** drop-down of the toolbar to activate sheet metal design tools; refer to Figure-82. Various tools of this workbench are discussed next.

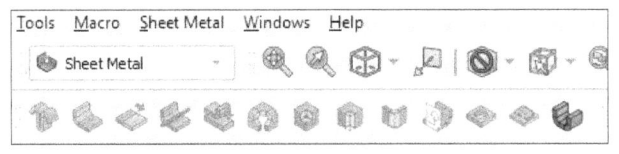

Figure-82. Sheet Metal toolbar

Making Base Wall

The **Make Base Wall** tool is used to create base flange for sheetmetal design. Note that you must have a closed sketch for using this tool. The procedure to use this tool is given next.

- Make sure there is a closed loop sketch in the graphics area created using Sketcher workbench. Select the sketch from **Model** tab in **Combo View** and click on the **Make Base Wall** tool from **Toolbar** or **Sheet Metal** menu. The base wall/flange will be created; refer to Figure-83.

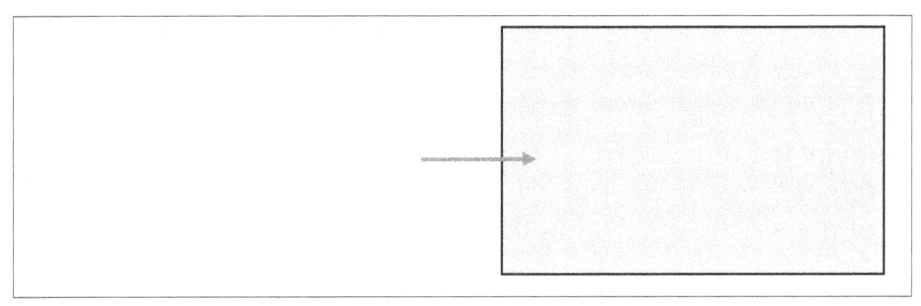

Figure-83. Base wall created

Creating Bend Wall

The **Make Wall** tool is used to create a bend wall using selected edge(s). The procedure to use this tool is given next.

- Select the edges of base flange to be used for creating bend walls; refer to Figure-84 and click on the **Make Wall** tool from the **Toolbar** or **Sheet Metal** menu. The preview of bend walls will be displayed along with **Flange Parameters** dialog; refer to Figure-85.

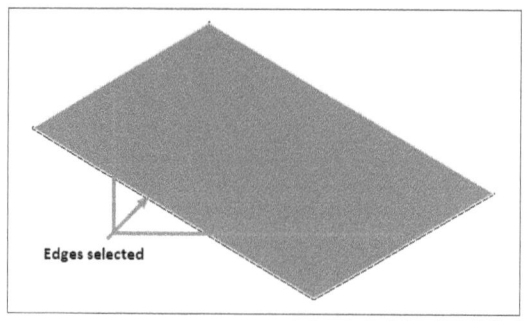

Figure-84. Creating bend wall

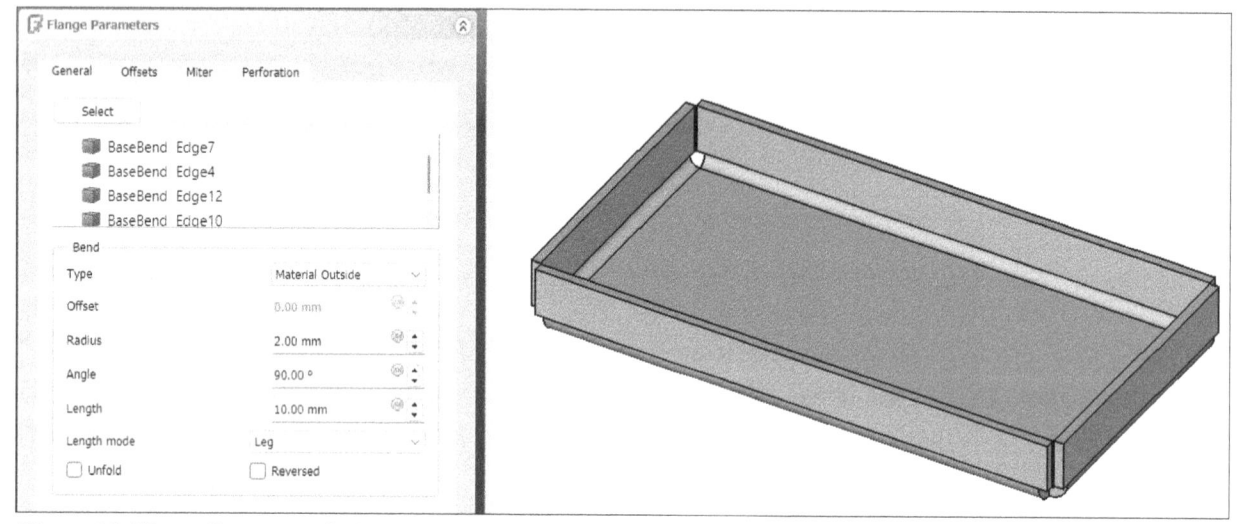

Figure-85. Flange Parameters dialog

- Set desired option in the **Type** drop-down to define the location for bends whether they are outside the selected edges, inside the edges, or offset by specified distance.
- Set desired value in the **Radius** edit box to define the radius of bends of flange walls.
- Specify desired value in the **Angle** and **Length** edit boxes to define bend angle and length of flange walls.
- Select the **Unfold** check box to display flange walls in unfolded state.
- Select the **Reversed** check box to flip the direction of the flange walls.
- Select the **Offsets** tab from the **Flange Parameters** dialog to define gaps and extensions of the side walls. Using the options in **Relief Cuts** area, you can define the shape and size of relief cuts provided at the corners of sheetmetal folded part.
- Select the **Miter** tab to define parameters related to miter joints in flanges; refer to Figure-86.

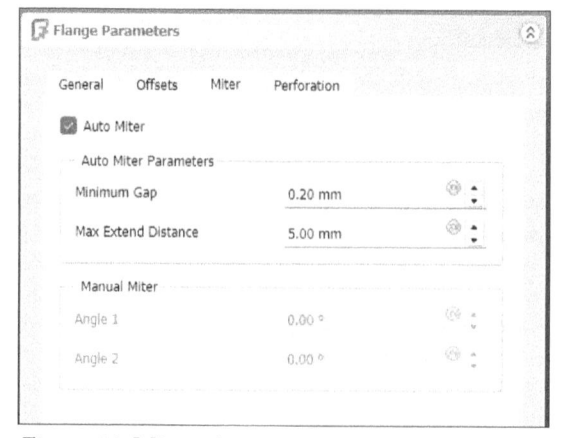

Figure-86. Miter tab

- Select the **Auto Miter** check box to automatically apply miter joints at the intersections of flange walls. Clear this check box if you want to manually define the miter angles in **Manual Miter** area of the dialog.
- If the **Auto Miter** check box is selected then you need to specify the gap between flange walls at corner and extension distance in respective edit boxes so that flange walls are in contact.
- Select the **Perforation** tab from the dialog to define parameters for splitting bend in multiple segments; refer to Figure-87. Select the **Perforate** check box to split bend in multiple segments. The options in the **Perforation Parameters** area of the dialog will become active.
- Specify desired value in the **Angle** edit box to define the angle at which cuts will be created using centerline of bend as reference; refer to Figure-88.
- Similarly, specify the length of first cut, maximum cut length and maximum tab length in respective edit boxes.

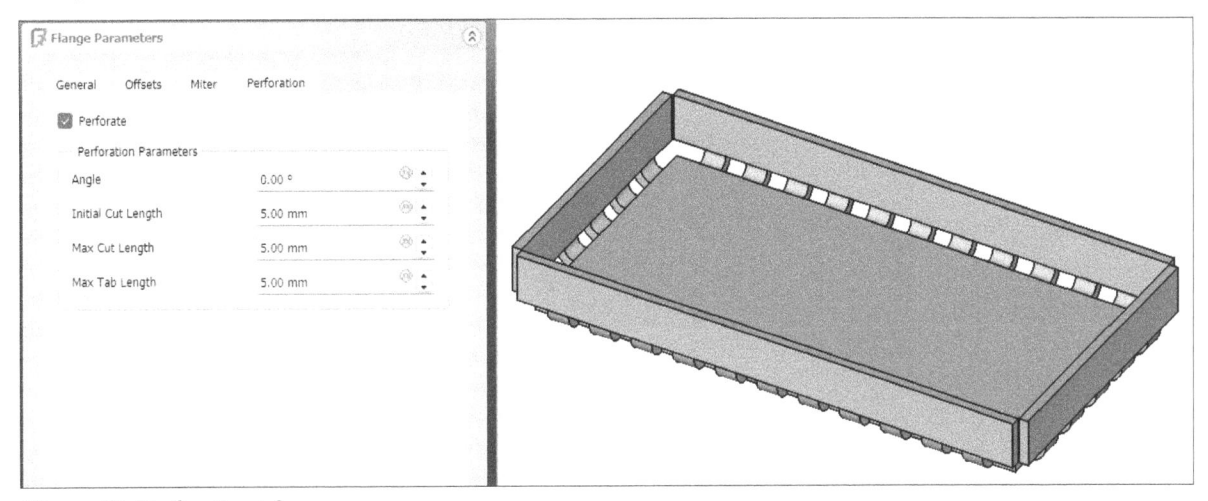

Figure-87. Perforation tab

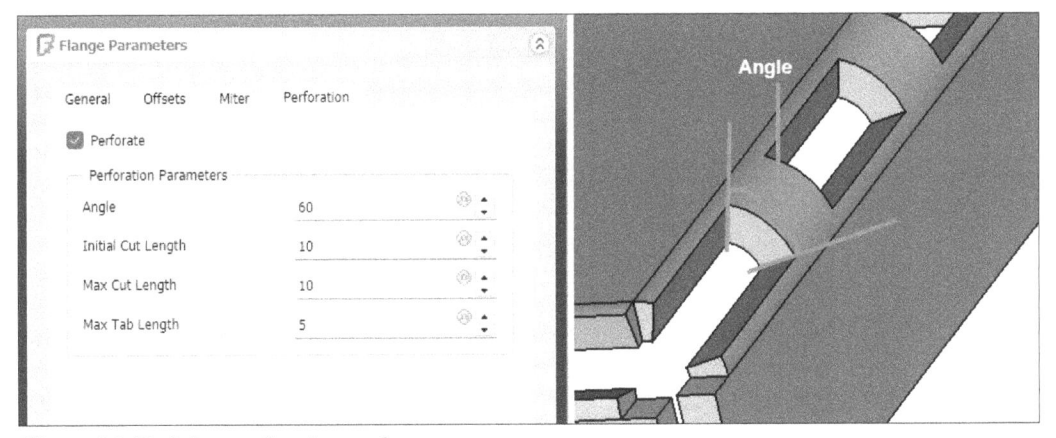

Figure-88. Defining perforation angle

- After setting desired parameters, click on the **OK** button from the dialog to create the flange walls.

Extending Sheet Metal Flange Face

The **Extend Face** tool is used to extend selected edge of flange by specified value. The procedure to use this tool is given next.

- Select the edge of flange wall to be extended and click on the **Extend Face** tool from the **Toolbar** or **Sheet Metal** menu. The edge will be extended; refer to Figure-89 and the **Extend Parameters** dialog will be displayed; refer to Figure-90.
- Set desired value in the **Length** edit box to define extension length of flange. Similarly, specify desired values in the **Offset A** and **Offset B** edit boxes to offset top and bottom faces of extended flange, respectively.
- Select the **Refine** check box to combine extended flange with main body.
- After setting desired parameters, click on the **OK** button to apply extension.

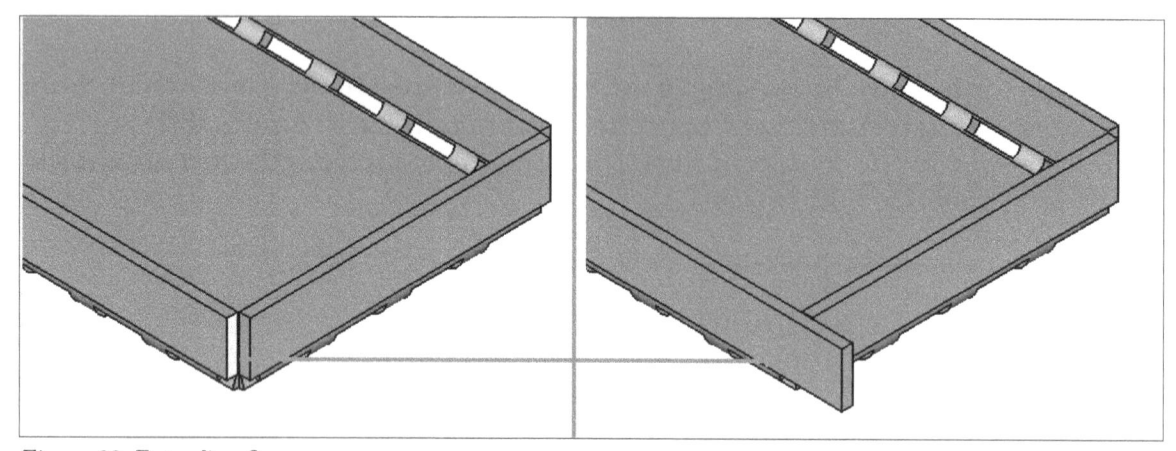

Figure-89. Extending flange

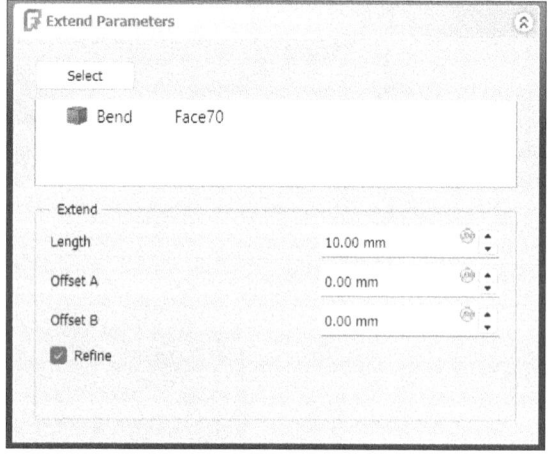

Figure-90. Extend Parameters dialog

Applying Sketch Bend

The **Fold a Wall** tool is used to bend selected flange using sketch line. The procedure to use this tool is given next.

- Select face of flange and sketch line from graphics area. The **Fold a Wall** tool will become active. Click on the tool from the **Toolbar** or **Sheet Metal** menu. The sketch bend will be created; refer to Figure-91 and **Fold on sketch line parameters** dialog will be displayed.

- Select the **Flip Direction** check box to change the side to be bend.
- Select the **Unbend** check box to keep the bend in unbent state.
- Set desired parameters in the **Bend Radius** and **Bend Angle** edit boxes to define radius and angle of bends.

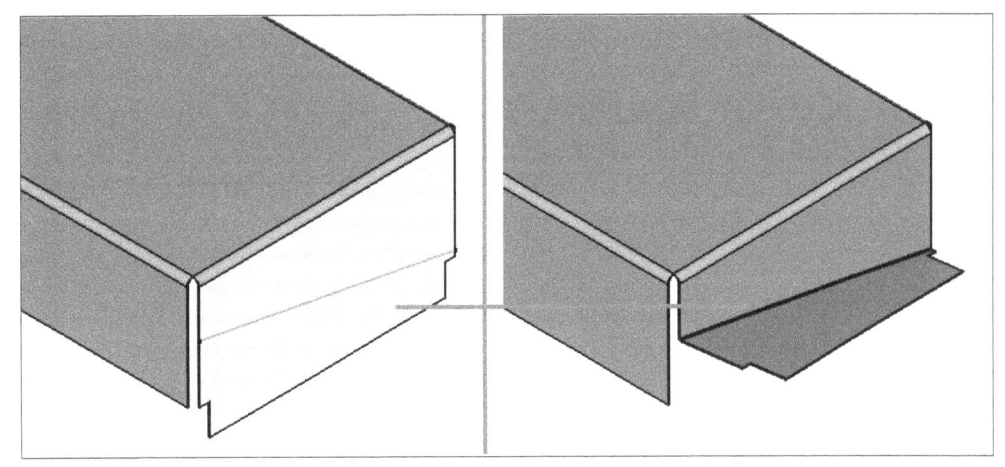

Figure-91. Applying sketch bend

Unfolding Sheet Metal Bends

The **Unfold** tool is used to unfold all the bends of the model while keeping a selected face fixed. The procedure to use this tool is given next.

- Select the face of sheet metal part that you want to be fixed and click on the **Unfold** tool from the **Toolbar** or **Sheet Metal** menu. The **Unfold sheet metal object** dialog will be displayed in the **Tasks** panel of **Combo View**; refer to Figure-92.

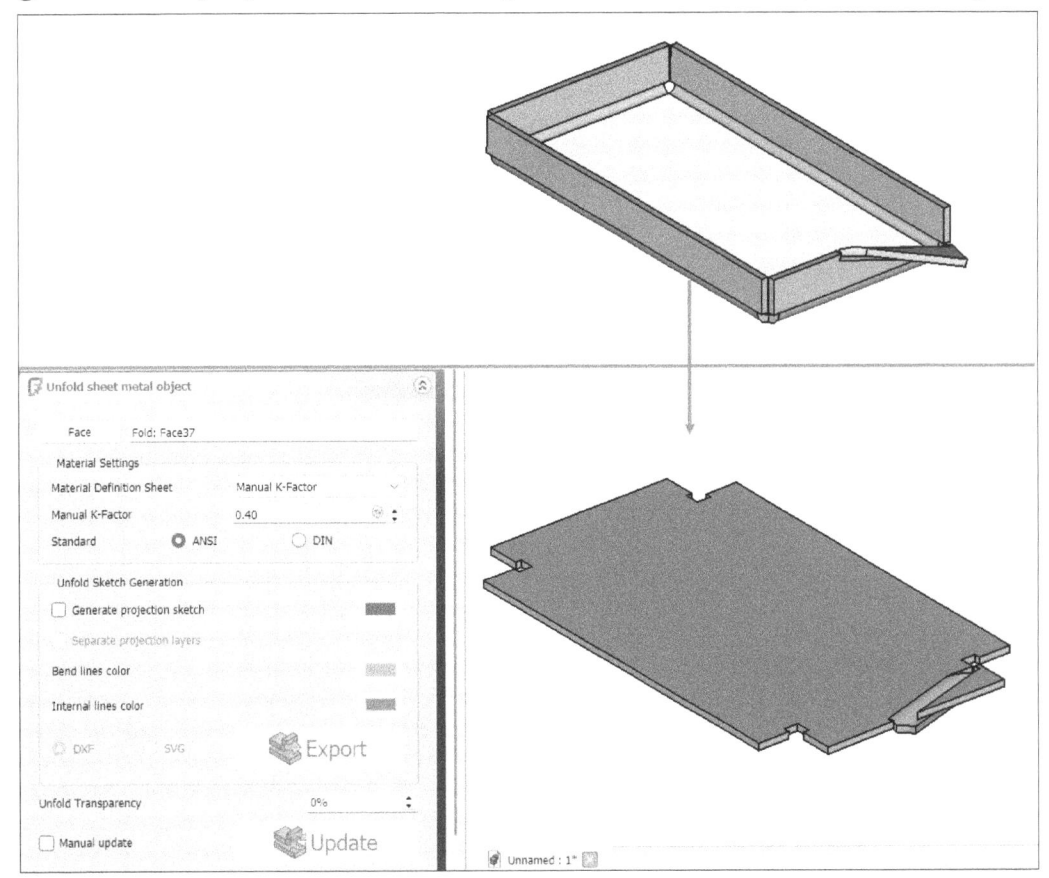

Figure-92. Face selected

- Select the **Generate projection sketch** check box to create projection sketch of unfolded sheet metal part.
- Select the **Separate projection layers** check box to create a new layer for projection.
- Click on color button from the dialog to change it.
- Select the **Manual K-factor** check box to manually specify K-factor for sheet metal design.
- Select the **Use Material Definition Sheet** check box if you want to use sheet metal data specified in material properties.
- After setting desired parameters, click on the **OK** button from the dialog. The unfolded model will be generated.

Adding Corner Relief

The **Add Corner Relief** tool is used to apply cut at the intersection for easy folding/unfolding of bends at intersection. The procedure to use this tool is given next.

- Select two edges of bends intersecting at a corner point; refer to Figure-93. The **Add Corner Relief** tool will become active.
- Click on the **Add Corner Relief** tool from **Toolbar** or **Sheet Metal** menu. The **Generate Sheet Metal base shape** dialog will be displayed; refer to Figure-94.
- Select desired radio button from the **Relief Type** drop-down to define whether you want to create a circular shaped, square shaped, or custom sketch shaped relief. After defining shape, set desired parameters in the **Relief Size** area to define size of relief.
- Click on the **OK** button from the dialog to create the feature.

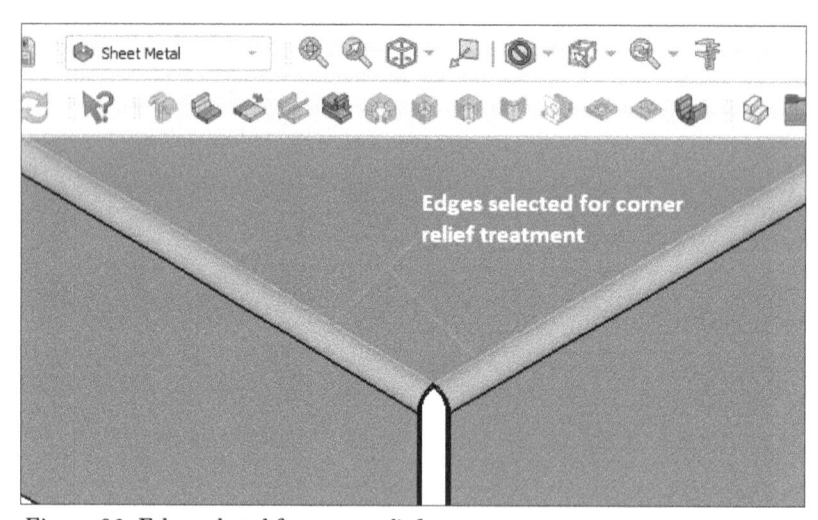

Figure-93. Edges selected for corner relief

Figure-94. Generate Sheet Metal base shape dialog

Applying Solid Corner Relief

The **Make Relief** tool is used to apply corner relief at solid corner of model. The procedure to use this tool is given next.

- Select corner point of a flange/wall. The **Make Relief** tool will become active. Click on the tool from the **Toolbar** or **Sheet Metal** menu. The relief will be created; refer to Figure-95 and **Add Corner Relief on Solid** dialog will be displayed.

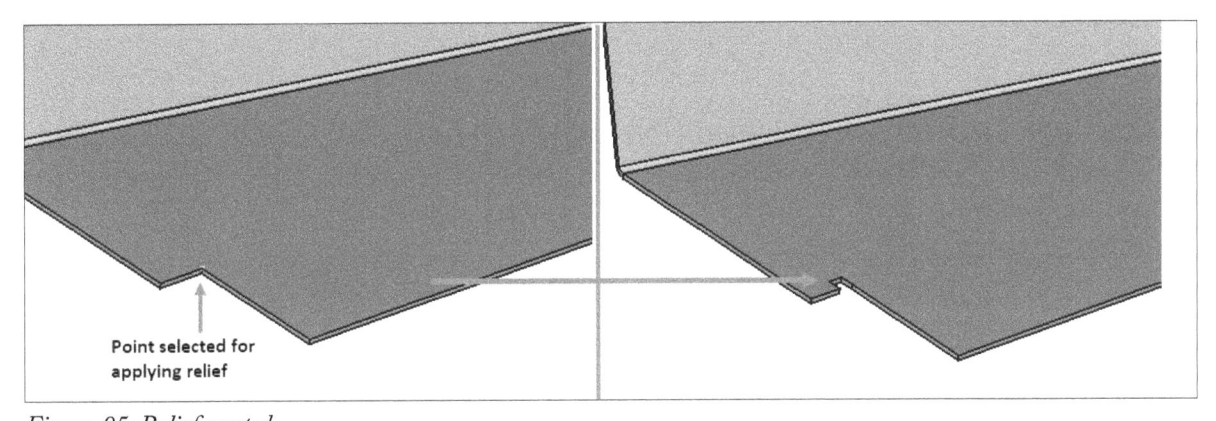

Figure-95. Relief created

- Set desired value of relief size in the **Size** edit box of the **Extend** area and click on the **OK** button from the dialog.

Applying Rip Feature

The rip feature is applied to cut gap at the edges of a solid body when you are converting a solid into sheet metal part. The **Make Junction** tool is used to create rip feature. The procedure to use this tool is given next.

- Select the edge of solid body which you want to convert into a sheetmetal part. The **Make Junction** tool will become active; refer to Figure-96. Select the tool from **Toolbar** or **Sheet Metal** menu. The rip feature will be created; refer to Figure-97 and **Add Junction Parameters** dialog will be displayed.

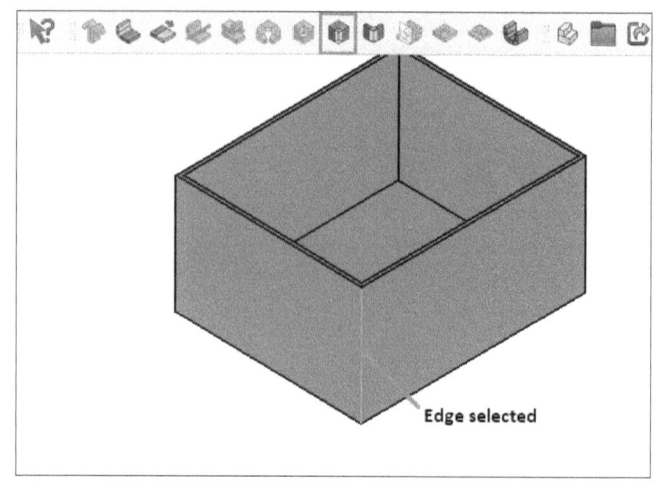

Figure-96. Edge selected for rip

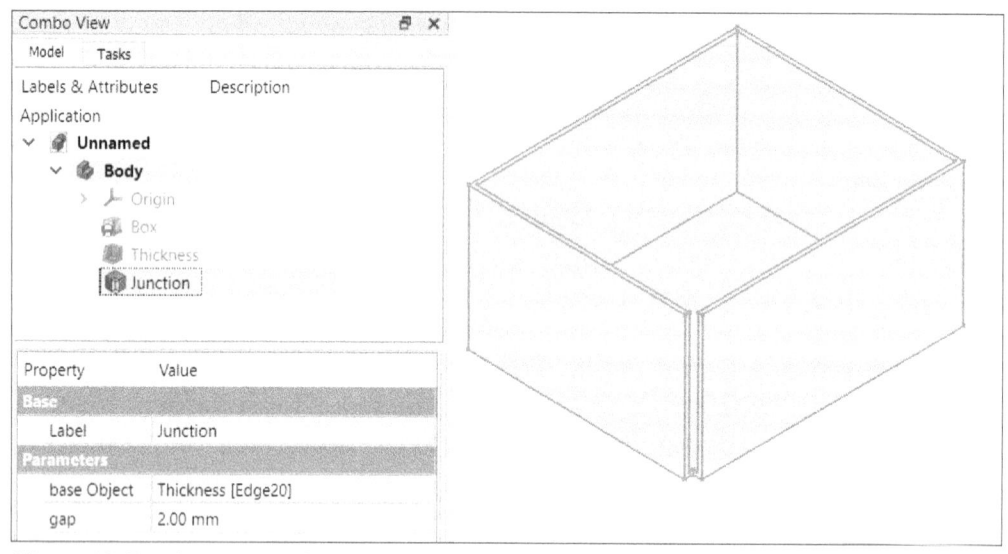

Figure-97. Rip feature created

- Set desired value in the **Width** edit box of the **Extend** area in the dialog to define the width of rip feature.
- Click on the **OK** button from the dialog to create feature.

Creating Bend

The **Make Bend** tool is used to create bend at the intersection of two faces of a solid body. This tool is also useful in converting a solid into sheet metal part. The procedure to use this tool is given next.

- Select an edge of a solid body to be converted into sheet metal part. The **Make Bend** tool will become active; refer to Figure-98.

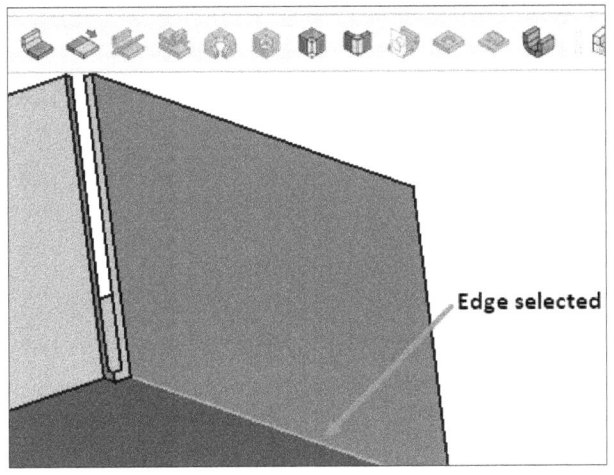

Figure-98. Make Bend tool

- Select the tool from **Toolbar** or **Sheet Metal** menu. The selected edge will be converted into a bend and **Bend sharp corner Parameters** dialog will be displayed.
- Specify desired value in the **Radius** edit box to define bend radius; refer to Figure-99. Click on the **OK** button from dialog to create the feature.

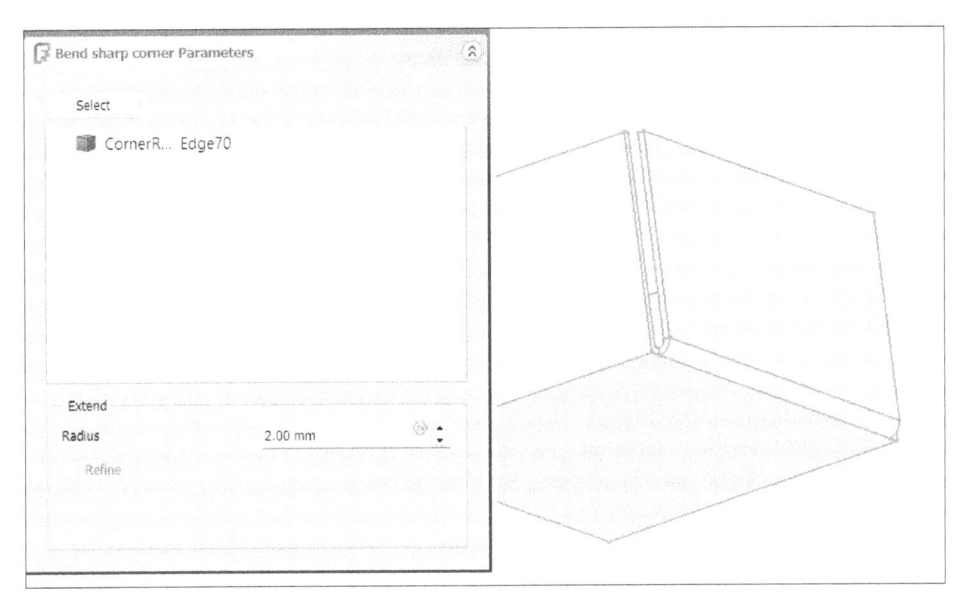

Figure-99. Bend parameters

Making Extruded Cut

The **Sketch On Sheet metal** tool is used to create an extruded cut from the sheet metal part. The procedure to use this tool is given next.

- Select a closed loop sketch created on sheet metal part face and sheet metal part from the graphics area. The **Sketch On Sheet metal** tool will become active; refer to Figure-100.
- Click on the tool from **Toolbar** or **Sheet Metal** menu. The extruded cut feature will be displayed; refer to Figure-101.

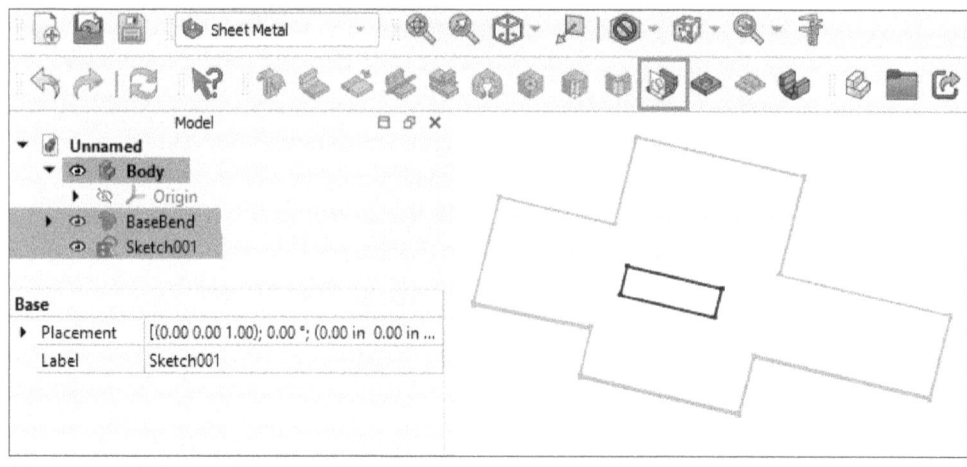

Figure-100. Selection for extruded cut

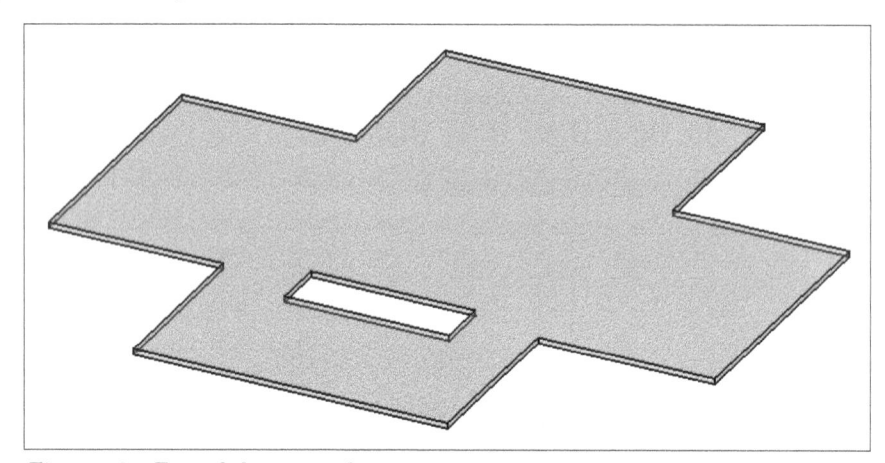

Figure-101. Extruded cut created

Applying Forming Tool

A forming tool in sheet metal design is used to indent its shape of sheet metal part surface. The procedure to use this tool is given next.

- Select the face of sheet metal part and face of solid body to be used as forming tool. The **Make Forming in Wall** tool will become active; refer to Figure-102.

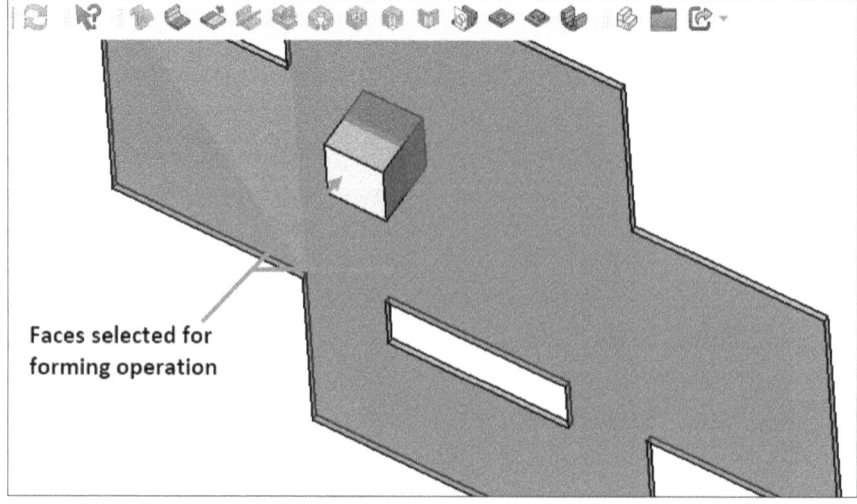

Figure-102. Make Forming in Wall tool

- Click on the tool from the **Toolbar** or **Sheet Metal** menu. The form of solid body face will be applied; refer to Figure-103.

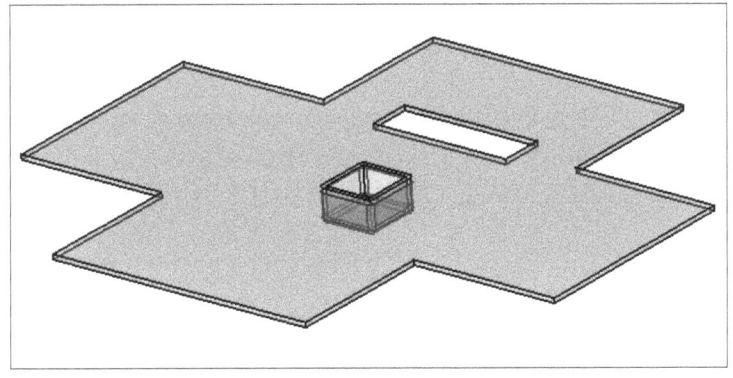

Figure-103. Forming

Extruded Cutout

The **Extruded Cutout** tool is used to cutout the sketch from sketch extrusion. The procedure to use this tool is discussed next.

- Select the face from the model and then select a closed loop sketch to be used as cutout; refer to Figure-104. The **Extruded Cutout** tool will become active in toolbar.

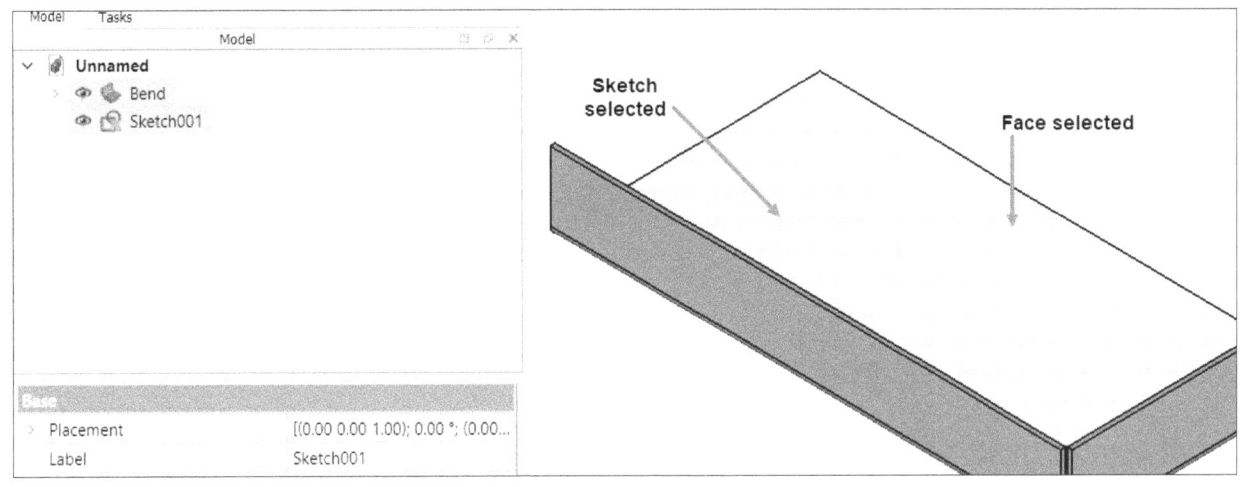

Figure-104. Selecting the face and part of extruded cutout

- Click on the tool from toolbar. The preview of cutout will be displayed with **Extruded Cutout Parameters** dialog; refer to Figure-105.

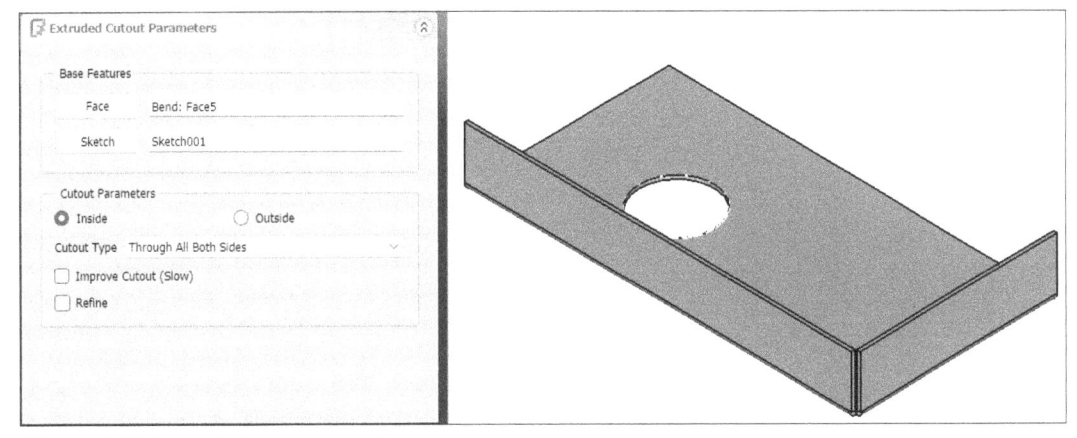

Figure-105. Extruded cutout created

- Select the **Inside** radio button if you want to remove the region inside the sketch. Select the **Outside** radio button if you want to remove the region outside the sketch.
- Select desired option from the **Cutout Type** drop-down to define extrude limits of the cutout. Set the value of extrude height/depth depending on option selected in the drop-down.
- Select the **Improve Cutout** check box to generate improved cutout graphic. Similarly, select the **Refine** check box to refine edges of cutout.
- After setting desired parameters, click on the **OK** button from the dialog.

Add Base Shape

The **Add base shape** tool is used to add the basic sheet metal object. The procedure to use this tool is discussed next.

- Click on the **Add base shape** tool from the toolbar; refer to Figure-106. The **Generate Sheet Metal base shape** dialog box will be displayed; refer to Figure-107.

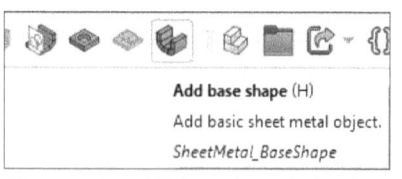

Figure-106. Add base shape dialog box

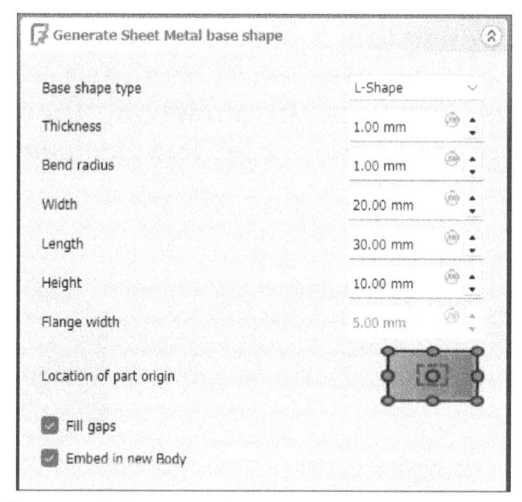

Figure-107. Generate sheet metal base shape dialog box

- Select desired shape from the **Base shape type** drop-down in the dialog box to create respective type of base sheetmetal shape.
- Set desired values of thickness, bend radius, width, length, height, and flange width in respective edit boxes of the dialog.
- Select **Fill gaps** check box to extend sides and flange to close all gaps.
- Select the **Embed in new Body** check box to create the base shape as a new body in **Model** panel of **Combo View**.
- After setting desired parameters, click on the **OK** button from the dialog box. The shape will be created.

There are many other workbenches that can be used with FreeCAD to perform various tasks. You can download them and add them to the software using FreeCAD website links.

Self Assessment

Q1. What is the first step to create a surface using the Filling tool?

A) Click on the Sections tool
B) Click on the Filling tool from the Surface menu or Toolbar
C) Select the mesh object
D) Click on the Add Edge button from the dialog

Q2. What can be used to refine the shape of the surface after defining the curves?

A) Boundaries
B) Vertexes
C) Mesh
D) Tessellation

Q3. Which radio button should you select if you want a surface with a better finish, compared to a stretch, but lesser than a curved finish?

A) Stretch
B) Coons
C) Curved
D) Flat

Q4. Which tool is used to create a surface from edges that represent transversal sections of a surface?

A) Curve on Mesh
B) Sections tool
C) Extend face tool
D) Filling tool

Q5. Which tool is used to create a surface by extending a selected model face?

A) Curve on Mesh tool
B) Blend Curve tool
C) Extend face tool
D) Sections tool

Q6. What is required for the Curve on Mesh tool to be active?

A) A mesh object must be available
B) The surface must be flat
C) The mesh object should be from the sketch workbench
D) The object must be a solid body

Q7. Which workbench is used to import, create, and modify mesh objects in FreeCAD?

A) Part Design Workbench
B) Surface Workbench
C) Mesh Workbench
D) Draft Workbench

Q8. Which dialog is used to create a mesh from a solid body in FreeCAD?

A) Export Mesh dialog
B) Tessellation dialog
C) Surface dialog
D) Remesh dialog

Q9. What does the Refinement tool in the Mesh workbench allow you to do?

A) Remesh a mesh object using the Gmsh mesher
B) Import mesh models from other CAD software
C) Create a mesh from a shape
D) Analyze and repair mesh models

Q10. What is the purpose of the Evaluate and Repair Mesh tool in the Mesh workbench?

A) To create a mesh from shape
B) To analyze and repair aspects of a mesh model
C) To remesh a mesh object
D) To import mesh objects from other CAD software

Q11. What is the primary function of the Face Info tool in the Meshes menu?

a) To check the curvature of the mesh
b) To check the index of selected faces in the mesh
c) To harmonize normals of the mesh faces
d) To create new faces in the mesh

Q12. What does the Curvature Info tool require to be active?

a) Mesh faces must be selected.
b) A curvature plot must have been generated.
c) The mesh must be decimated.
d) A robot must be inserted.

Q13. The Flip Normals tool is used for:

a) Changing the shape of the mesh.
b) Flipping the normal direction of mesh faces.
c) Creating a curvature plot.
d) Adding new triangles to the mesh.

Q14. What is the purpose of the Remove Components by Hand tool?

a) To remove selected sections of the mesh.
b) To delete specific faces of the mesh.
c) To manually add new triangles to the mesh.
d) To smoothen the mesh.

Q15. The 'Create Mesh Segments' tool is used for:

a) Merging mesh objects
b) Creating mesh body segments with specific tolerances
c) Creating mesh curvature plots
d) Filling holes in the mesh

Q16. What does the Decimation tool do in Meshes menu?

a) It scales the mesh model up or down.
b) It reduces the number of faces in the mesh model.
c) It smoothens the mesh model.
d) It merges two or more mesh bodies.

Q17. Which tool would you use to combine multiple mesh bodies into a single mesh?

a) Merge tool
b) Split Mesh tool
c) Remove Components tool
d) Unwrap Mesh tool

Q18. What does the Unwrap Mesh tool do?

a) It removes selected faces of a mesh model.
b) It unravels a mesh object into a flat representation.
c) It scales the mesh model.
d) It creates mesh segments.

Q19. Which of the following tools is used to create multiple cross-sections of a mesh object?

a) Trim Mesh tool
b) Merge tool
c) Cross Section tool
d) Create Mesh Segments tool

Q20. What is the purpose of the "Set Default Orientation" tool in FreeCAD?

A) To set the initial placement position and orientation of the selected robot
B) To simulate robot movements
C) To add a tool to the robot arm
D) To define default movement parameters like velocity and acceleration

Q21. When setting the default orientation of a robot, which of the following actions can be done in the "Rotation area"?

A) Define location of the robot center for translation
B) Move the robot to a desired location
C) Rotate the robot about selected axis by specifying values in the Axis and Angle edit boxes
D) Adjust the speed of robot movement

Q22. What does the "Set Default Values" tool allow you to define for a robot?

A) The color and shape of the robot
B) Default movement parameters such as velocity and acceleration
C) The size of the robot's workspace
D) The type of tool to attach to the robot

Q23. What is required to add a tool to a robot in FreeCAD?

A) The tool must be created using the Part Design workbench
B) The tool must be pre-assembled in the Robot menu
C) The tool must be selected from the Model Tree
D) The tool must be created using the Sketcher workbench

Q24. Which tool in FreeCAD simulates the movements of a robot along a trajectory?

A) Set Default Orientation
B) Simulate a trajectory
C) Add Tool to Robot
D) Set Default Values

Q25. In the 3D Printing Tools workbench, which tool is used to convert a metric mesh model to an imperial mesh model?

A) Scale Mesh by 50%
B) Converts Metric Mesh to Imperial
C) Reduces Mesh by 95%
D) Changes objects 50% Transparency

Q26. What does the "Reduces Mesh by 50%" tool do in the 3D Printing Tools workbench?

A) Scales up the mesh model by 50%
B) Reduces the mesh model size by 50%
C) Converts the mesh from imperial to metric units
D) Changes the transparency of the mesh

Q27. In the Sheet Metal Workbench, what is required before using the "Make Base Wall" tool?

A) The model must be selected from the Model tab
B) There must be a closed loop sketch created in the Sketcher workbench
C) The bend radius must be specified
D) The tool must be converted to sheet metal

Q28. Which tool in the Sheet Metal Workbench is used to create a bend at the intersection of two faces of a solid body?

A) Make Bend
B) Make Relief
C) Make Wall
D) Make Base Wall

Q29. What does the "Extend Face" tool do in the Sheet Metal Workbench?

A) Extends the length of the sheet metal
B) Extends the edge of a flange by a specified value
C) Applies a forming tool to the surface
D) Unfolds the sheet metal part

Q30. What happens when the "Unfold" tool is used in the Sheet Metal Workbench?

A) It bends the sheet metal
B) It applies a corner relief
C) It unfolds all the bends of the model while keeping a selected face fixed
D) It creates a base wall for the sheet metal

Q31. How do you add a corner relief to sheet metal bends?

A) By selecting two edges of bends intersecting at a corner
B) By using the "Make Bend" tool
C) By folding a wall with a sketch line
D) By selecting the sheet metal part face

Q32. In the Sheet Metal Workbench, which tool is used to apply corner relief at a solid corner of a model?

A) Make Junction
B) Make Relief
C) Add Corner Relief
D) Make Forming Tool

Q33. What does the "Make Junction" tool do in the Sheet Metal Workbench?

A) Creates a bend in the sheet metal
B) Applies a solid corner relief
C) Creates a gap at the edges of a solid body when converting to sheet metal
D) Adds a base shape to the sheet metal

Q34. What is the purpose of the "Forming Tool" in the Sheet Metal Workbench?

A) To indent the shape of a sheet metal part surface
B) To create a base shape
C) To bend sheet metal
D) To add a corner relief

Q35. In the Sheet Metal Workbench, what is necessary to use the "Make Bend" tool?

A) The solid body must be converted into sheet metal
B) A closed loop sketch must be present
C) The bend radius must be defined
D) The edge of a solid body must be selected

Q36. What does the "Make Wall" tool in the Sheet Metal Workbench create?

A) A bend wall from a selected edge of a base flange
B) A solid corner relief
C) A base flange for the sheet metal
D) A projection sketch

Q37. What is the purpose of the "Sketch Bend" tool in the Sheet Metal Workbench?

A) To create a bend at the intersection of two faces
B) To unfold all bends of a sheet metal part
C) To apply a forming tool to the sheet metal surface
D) To create a bend in a sheet metal part using a sketch line

Q38. What happens when the "Make Junction" tool is used in the Sheet Metal Workbench?

A) It creates a bend at the intersection of two faces
B) It adds a corner relief to sheet metal bends
C) It creates a rip feature in the sheet metal
D) It converts a solid body to a sheet metal part

Q39. What tool in the Sheet Metal Workbench is used to make an extruded cut from a sheet metal part?

A) Make Bend
B) Sketch on Sheet Metal
C) Make Junction
D) Make Relief

Q40. What does the "Make Forming in Wall" tool do in the Sheet Metal Workbench?

A) It applies a forming tool to the wall of a sheet metal part
B) It creates a bend at the intersection of two faces
C) It adds a corner relief to the sheet metal
D) It unfolds the sheet metal part

Chapter 12

Miscellaneous Tool

Topics Covered

The major topics covered in this chapter are:

- *Edit menu*
- *View menu*
- *Tools menu*
- *Macro menu*

INTRODUCTION

In previous chapters, you have learned about various workbenches of FreeCAD. There are some tools in FreeCAD which are common for all the workbenches and are used to modify basic parameters of software like changing orientation of model, moving object in 3D space, and so on. These tools are available in various menus of FreeCAD which are discussed next.

EDIT MENU

The tools in **Edit** menu are used to perform common tasks like Undo, Redo, Copy, Paste, and so on; refer to Figure-1. These tools are discussed next.

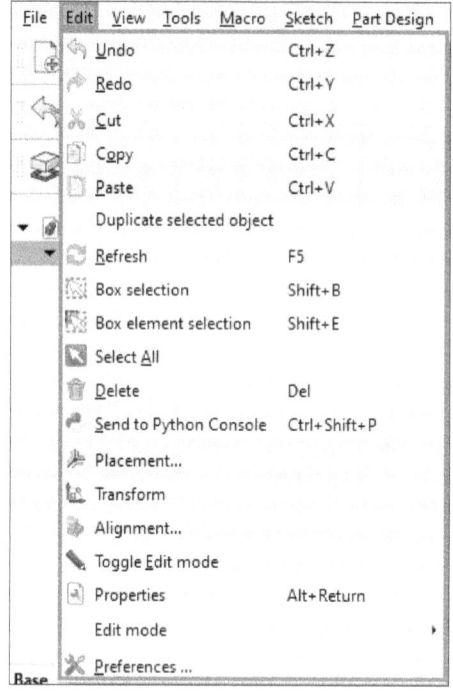

Figure-1. Edit menu

Undo Operation

The **Undo** tool in **Edit** menu is used to reverse the effect of any operation performed recently in the software. For example, you have drawn a line in FreeCAD sketcher and dragged it to a different location but now you want it to be back at its original location then you can use the **Undo** tool to do so. You can also use the shortcut key **CTRL+Z** to perform undo operation.

Redo Operation

The **Redo** tool in **Edit** menu is used to reverse the Undo operation. You can also use the shortcut key **CTRL+Y** to perform redo operation.

Cut, Copy, and Paste Operations

The **Cut** tool in **Edit** menu is used to copy selected objects in temporary memory of software and delete them from original location. You can also use **CTRL+X** shortcut keys for cut operation. The **Copy** tool in **Edit** menu is used to copy selected objects in temporary memory of software while keeping the original objects intact. You can also use **CTRL+C** shortcut keys for copy operation. After selecting **Cut** or **Copy** tool, click on the **Paste** tool from **Edit** menu to place copy of the objects in temporary memory. You can also use the **CTRL+V** shortcut keys for paste operation.

Duplicating Selected Objects

The **Duplicate selection** tool is used to generate a duplicate copy of selected objects. This tool is useful when creating assemblies using same components.

Refresh Operation

The **Refresh** tool is used to recompute any changes made in the model. Sometimes, changes made in the model do not reflect automatically in drawing area. For refreshing operation, you need to first mark an object for refreshing by selecting the object, right-clicking on it, and then selecting the **Mark to recompute** option from shortcut menu; refer to Figure-2. After marking an object to recompute, select the **Refresh** tool from **Edit** menu or press **F5** key from keyboard to update changes in the model.

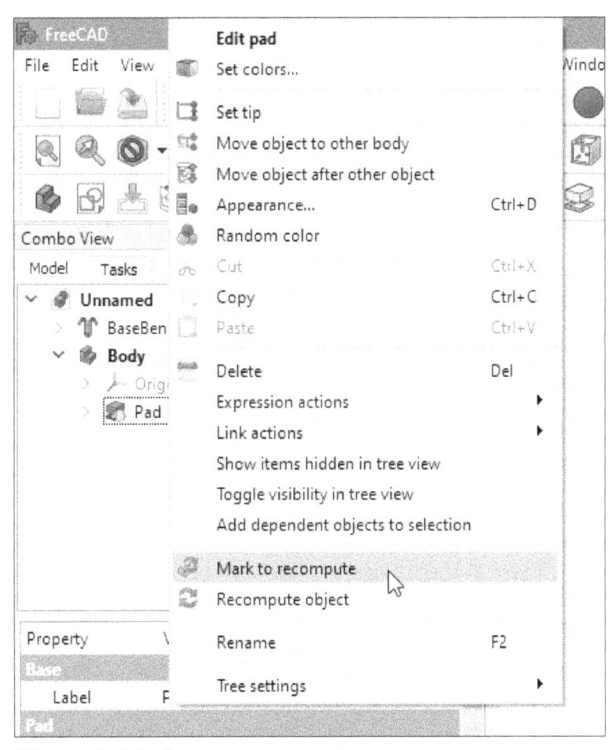

Figure-2. Mark to recompute option

Box Selection

The **Box selection** tool is used to select bodies which fall inside the boundaries of box selected created. To use box selection, select the **Box selection** tool from **Edit** menu and draw a box by dragging cursor; refer to Figure-3. Similarly, use the **Box element selection** tool to select the geometry/feature falling inside the box.

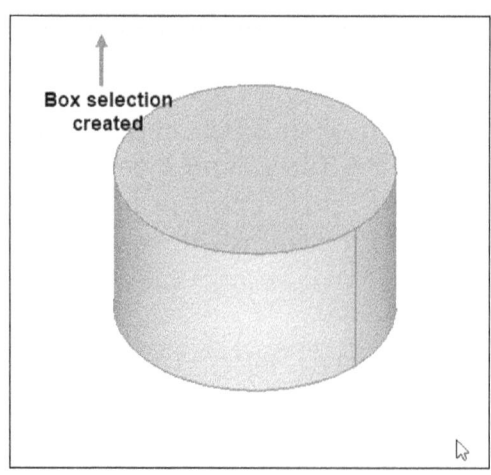

Figure-3. Box selection created

Select All

The **Select All** tool is used to select all the objects in drawing area.

Deleting Objects

The **Delete** tool in **Edit** menu is used to delete selected objects. You can select the objects from drawing area as well as **Model Tree**. You can also use the shortcut key **DELETE** from keyboard to delete selected objects.

Sending to Python Console

The **Send to Python Console** tool creates a variable in the Python console referencing a selected object. If a subshape of the object is selected, two additional variables are created, one referencing the shape of the object and the other referencing the subshape itself. The variables and the code involved can be used to develop python code. You can also use the shortcut key **CTRL+SHIFT+P** from the keyboard.

Modifying Placement of Objects

The **Placement** tool in **Edit** menu is used to modify placement location and orientation of selected object. The procedure to do so is given next.

- Select the object whose placement is to be modified and click on the **Placement** tool from **Edit** menu. The **Placement** dialog will be displayed in **Tasks** panel of **Combo View**; refer to Figure-4.

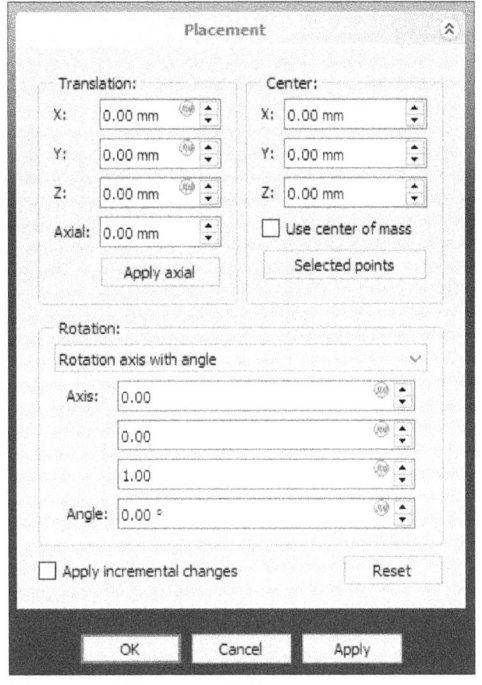

Figure-4. Placement dialog

The options in this dialog have been discussed earlier. Note that you may need to refresh the model after performing placement operation.

Transform

The **Transform** tool is used to apply rotation and translation by specified increments to an object. The procedure to use this tool is discussed next.

- Select the feature that you want to be modified and click on the **Transform** tool from the **Edit** menu. The related dialog for modifying feature will be displayed. If you have selected an object/body which do not have any dialog associated with it then **Translation Increment** and **Rotation Increment** edit boxes will be displayed in the dialog; refer to Figure-5.

Figure-5. Transform dialog

- Set the parameters as desired and use the drag handles displayed on the object to transform.
- Click on the **OK** button from the dialog.

Alignment Operation

The **Alignment** tool is used to align two selected objects about selected points. The procedure to use this tool is given next.

- Select the two objects that you want to align and click on the **Alignment** tool from the **Edit** menu. The two selected objects will be displayed in different viewports of drawing area; refer to Figure-6.

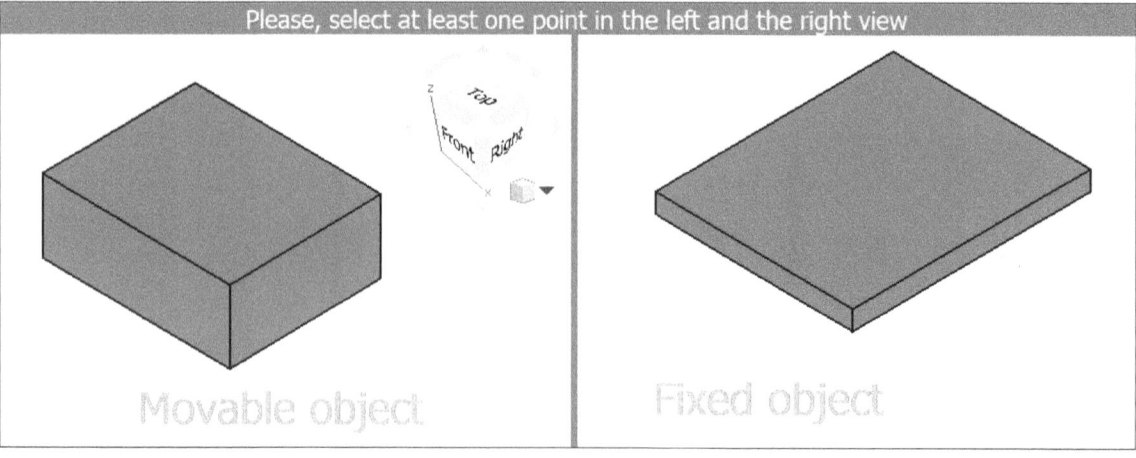

Figure-6. Objects displayed for alignment

- Select equal number of points on both the objects (movable as well as fixed object) and right-click in the drawing area. A shortcut menu will be displayed; refer to Figure-7.

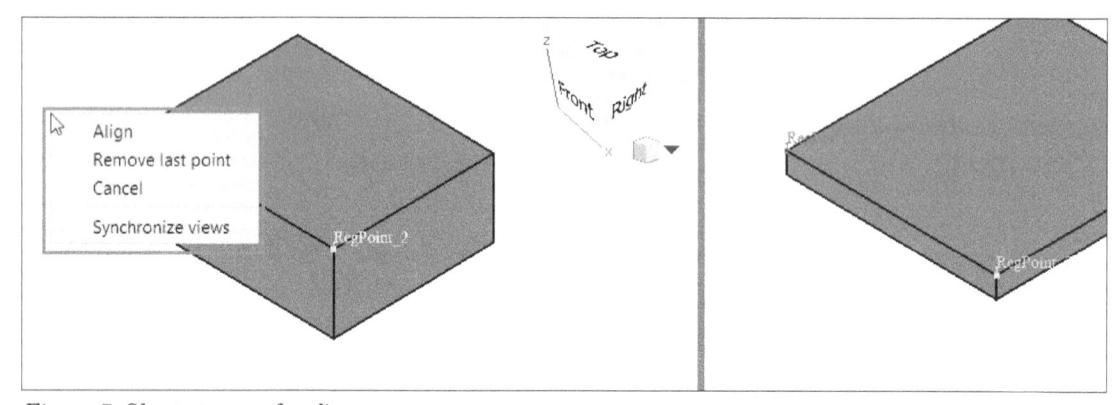

Figure-7. Shortcut menu for alignment

- Select the **Align** option from the shortcut menu to perform alignment operation. The two objects will be aligned as per the points selected on the bodies.

Activating Edit Mode

The **Toggle Edit mode** tool is used to modify a feature created in FreeCAD using its related dialog. The procedure to use this tool is given next.

- Select the feature that you want to be modified and click on the **Toggle Edit mode** tool from the **Edit** menu. The related dialog for modifying feature will be displayed; refer to Figure-8.

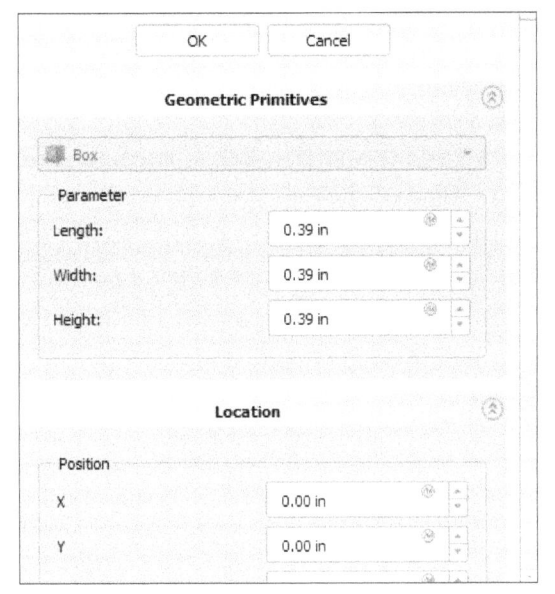

Figure-8. Increments dialog

- Set the parameters as desired and click on the **OK** button from the dialog.

Properties
The **Properties** tool displays the **Properties** box in the **Combo View**.

Edit Mode options
The options in the **Edit Mode** cascading menu define the edit mode to be used when an object is double-clicked in the **Model Tree**; refer to Figure-9.

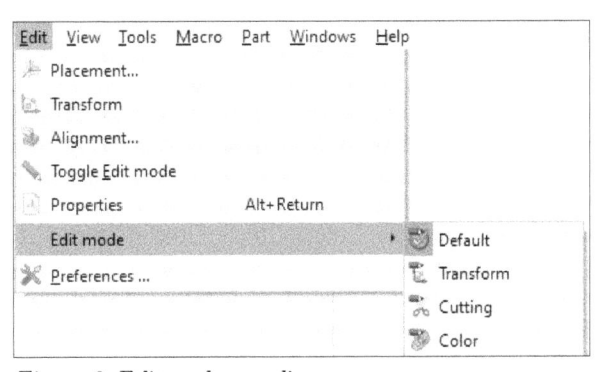

Figure-9. Edit mode cascading menu

- If you select the **Default** option then **Primitive parameters** dialog will be displayed on double-clicking the object in the **Model Tree**.
- If you select the **Transform** option then **Increments** dialog will be displayed on double-clicking the object in the **Model Tree**.
- If you select the **Color** option then **Set color per face** dialog will be displayed on double-clicking the object in the **Model Tree**.

The **Cutting** option in **Edit mode** cascading menu is still in experiment mode. All the dialogs have been discussed earlier in this book.

Preferences

The **Preferences** tool is used to modify underlying parameters of FreeCAD application. The procedure to use this tool is given next.

- Click on the **Preferences** tool from the **Edit** menu. The **Preferences** dialog box will be displayed; refer to Figure-10.

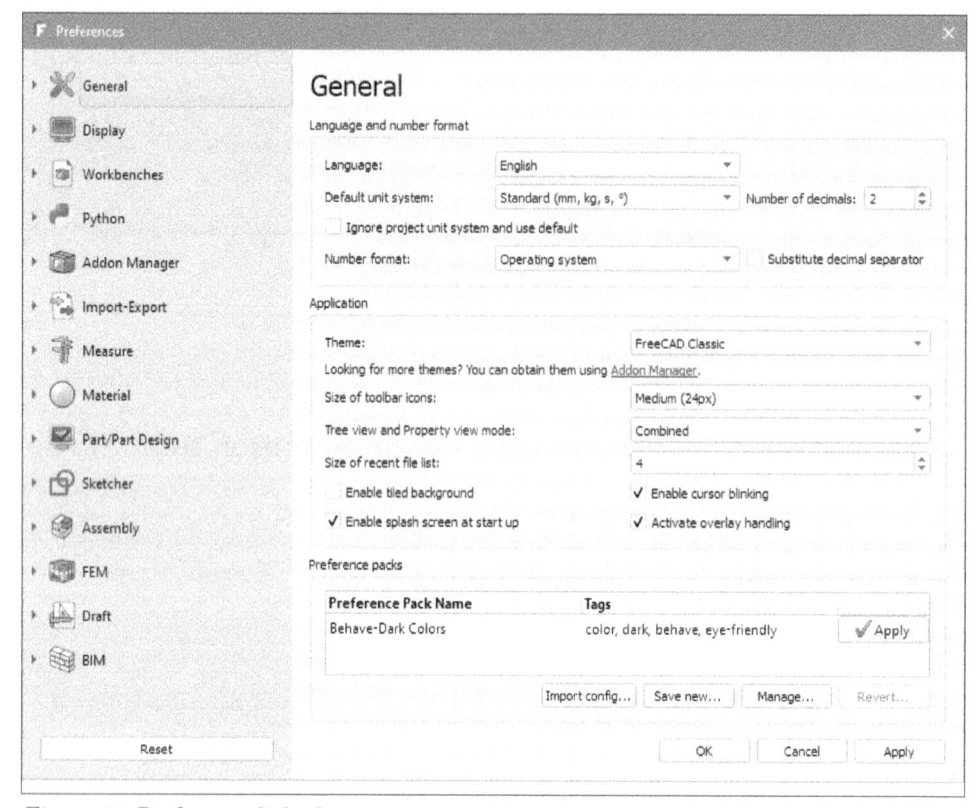

Figure-10. Preferences dialog box

The options in the dialog box are categorized in various pages like **General** page, **Display** page, **Python** page, and so on. To modify options related to a category, select the respective page option from the left area of the dialog box. Some of the most common options of this dialog box are discussed next.

General Page

The options in **General** page are used to define general parameters for interface of application, documents, code editor, and so on. These options are discussed next.

General tab

- Select desired option from the **Language** drop-down to specify language used by application interface.
- Select desired option from the **Unit system** drop-down to define unit system used for creating model.
- Specify desired value in the **Number of decimals** edit box to define precision up to which you want to specify values in FreeCAD.
- Select **Ignore project unit system and use default** check box to ignore project unit systems and use the default unit system set in **Preferences** dialog box.
- If you are using **Building US (ft-in/sqft/cft)** unit system then you need to specify minimum fractional value allowed in the **Minimum fractional inch** drop-down.

- Select desired option from **Number format** drop-down to specify the format of number. Select **Operating system** option to define the decimal separator by the selected operating system. Select **Selected language** option to define the decimal separator by the selected FreeCAD interface language. Select **C/POSIX** option to use the point as the decimal separator.
- Select **Substitute decimal separator** check box to substitute the numerical keypad decimal separator with the local separator.
- Select desired option from the **Style sheet** drop-down to define color and style of interface of application. For example, select the **Dark-blue** option to make background of application dark and selected tools/options highlighted in blue color.
- Select desired option from the **Size of toolbar icons** drop-down to define size of buttons in the toolbar.
- Select desired option from **Tree view mode** drop-down to customize how tree view is shown in the panel.
- Set desired value in the **Size of recent file list** edit box to specify number of recently worked files being displayed in **Recent files** cascading menu of **File** menu.
- Select the **Enable tiled background** check box to display background as tiled image.
- Select **Enable cursor blinking** check box to enable the text cursor to be blink.
- Select the **Enable splash screen at start up** check box to display splash screen showing modules being loaded in software when start the FreeCAD software.
- Select **Activate overlay handling** check box to handle docked windows as transparent overlays.

Document tab

The options in **Document** tab are used to specify settings related to documents in FreeCAD; refer to Figure-11.

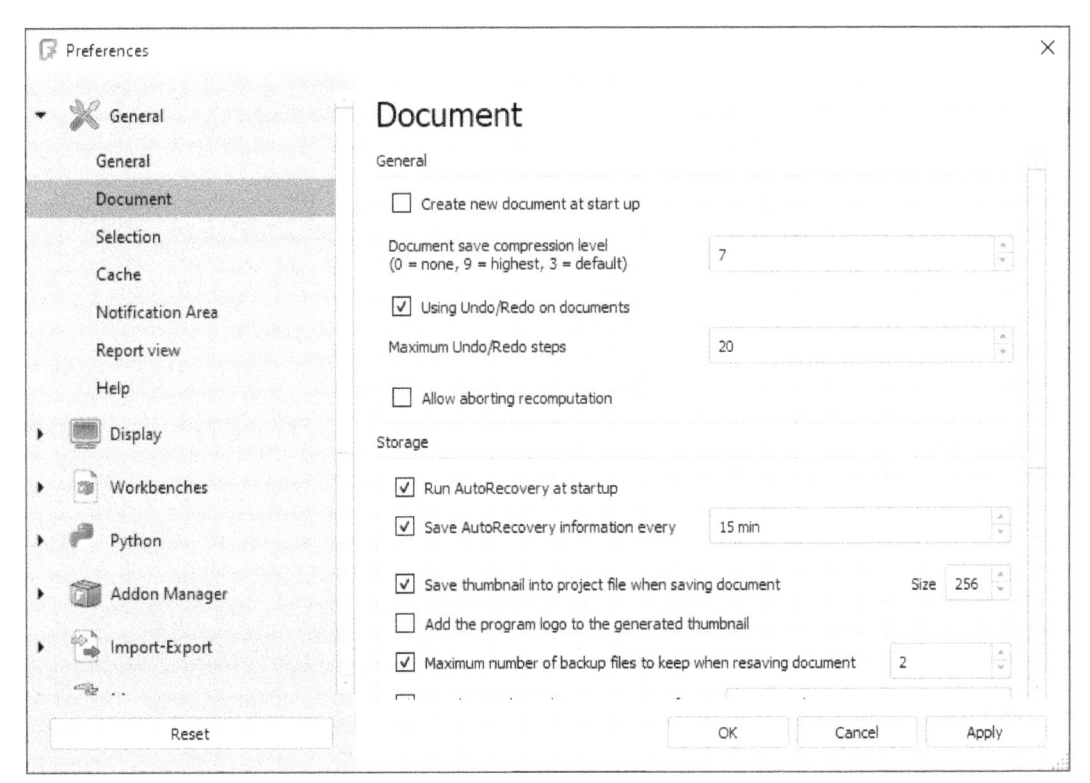

Figure-11. Document tab

- Select the **Create new document at start up** check box to automatically create a new document when you start the software.
- Set desired value in the **Document save compression level** edit box to define the compression level being used when saving the file. A higher value for compression will make saved files occupy less memory but it will also take longer to open the files in application. The minimum compression value is **0** and maximum compression value is **9**.
- Select the **Using Undo/Redo on documents** check box to allow undo/redo operations on the model. You can specify maximum number of undo/redo steps allowed in the **Maximum Undo/Redo steps** edit box.
- Select **Allow aborting recomputation** check box to allow the user to abort document recomputation by pressing **ESC**. This feature may slightly increase recomputation time.
- Select the **Run AutoRecovery at startup** check box to check for auto recovered documents at the startup.
- Select the **Save AutoRecovery information every** check box to automatically save file at regular interval specified in the edit box next to check box.
- Select the **Save thumbnail into project file when saving document** check box to add a thumbnail of model with project file while saving it.
- Select the **Add the program logo to the generated thumbnail** check box to also add logo of application with the thumbnail of model.
- Select the **Maximum number of backup files to keep when resaving document** check box to define the number of documents to be created as backup for a model.
- On selecting the **Use date and FCBak extension** check box, the backup files will get extension '**.FCBak**' and file names get date suffix according to the specified format in the **Date format** edit box.
- Select the **Allow duplicate object labels in one document** check box to allow duplicate labels of features in the **Model Tree** of a document.
- On selecting the **Disable partial loading of external linked objects** check box, the only referenced objects and their dependencies will be loaded when a linked document is auto-opened together with the main document.
- Specify desired values in **Author name** and **Company** edit boxes to define name of model author and designing company.
- If you select the **Set on save** check box then you will be asked to specify author name and company information while saving model.
- Similarly, you can specify license information in the **Default license** and **License URL** edit boxes.

Selection tab

The options in the **Selection** tab are used to specify the settings related to selection of documents in FreeCAD; refer to Figure-12.

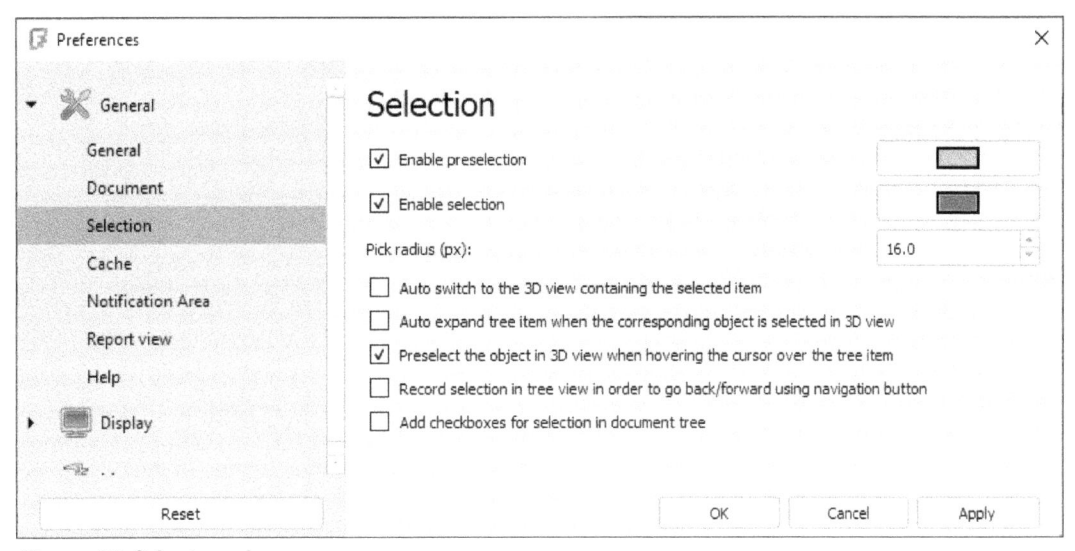

Figure-12. Selection tab

- Specify desired value in the **Pick radius** edit box to define the circle radius within which objects will be selected when clicked at the cursor location.
- Select the **Auto switch to the 3D view containing the selected item** check box to enable the **Tree view SyncView** mode. If this mode is on, selecting an object from a different document in the Tree view automatically activates that document's last used docked 3D view.
- Select the **Auto expand tree item when the corresponding object is selected in 3D view** check box to enable the **Tree view SyncSelection** mode. If this mode is on, selecting an object in a 3D view will automatically expand the Tree view to show that object.
- Select the **Preselect the object in 3D view when mouse over the tree item** check box to enable the **Tree view PreSelection** mode. If this mode is on, an object that the mouse is over in the Tree view will be preselected in all 3D views belonging to that object's document.
- Select the **Record selection in tree view in order to go back/forward using navigation button** check box to enable the **Tree view RecordSelection** mode. If this mode is on, each Tree view selection is stored for future use.
- On selecting the **Add checkboxes for selection in document tree** check box, each document tree item will get a check box. This is useful for selecting multiple items on a touchscreen.

Cache tab

The options in **Cache** tab are related to the cache directory where FreeCAD stores temporary files; refer to Figure-13. Specify desired location for the **Location** edit box to store the temporary files. Select desired option from **Check periodically at program start** drop-down. Select desired option from **Cache size limit** drop-down so that the user will be notified if the cache size exceeds the specified limit. Click on the **Check now** button to check the current cache size stored.

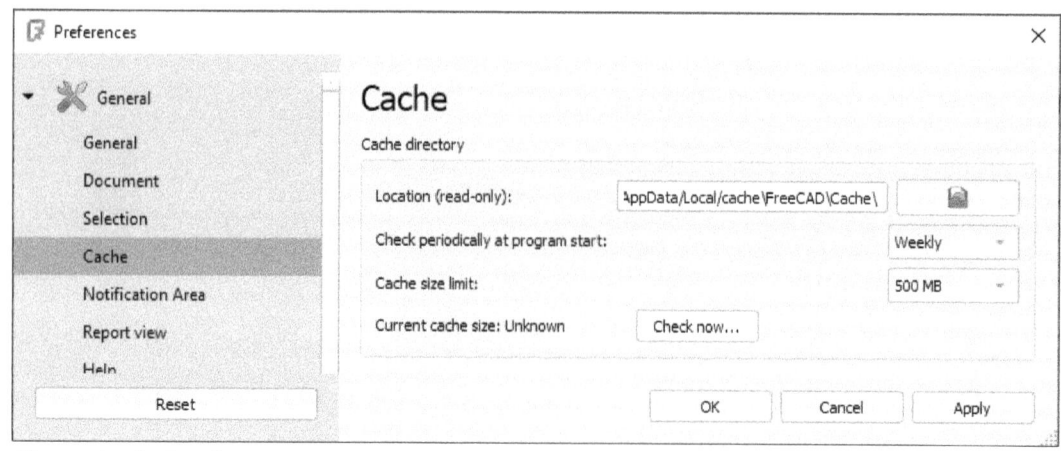

Figure-13. Cache tab

Notification Area tab

The options in **Notification Area** tab are used to control the notifications in FreeCAD; refer to Figure-14.

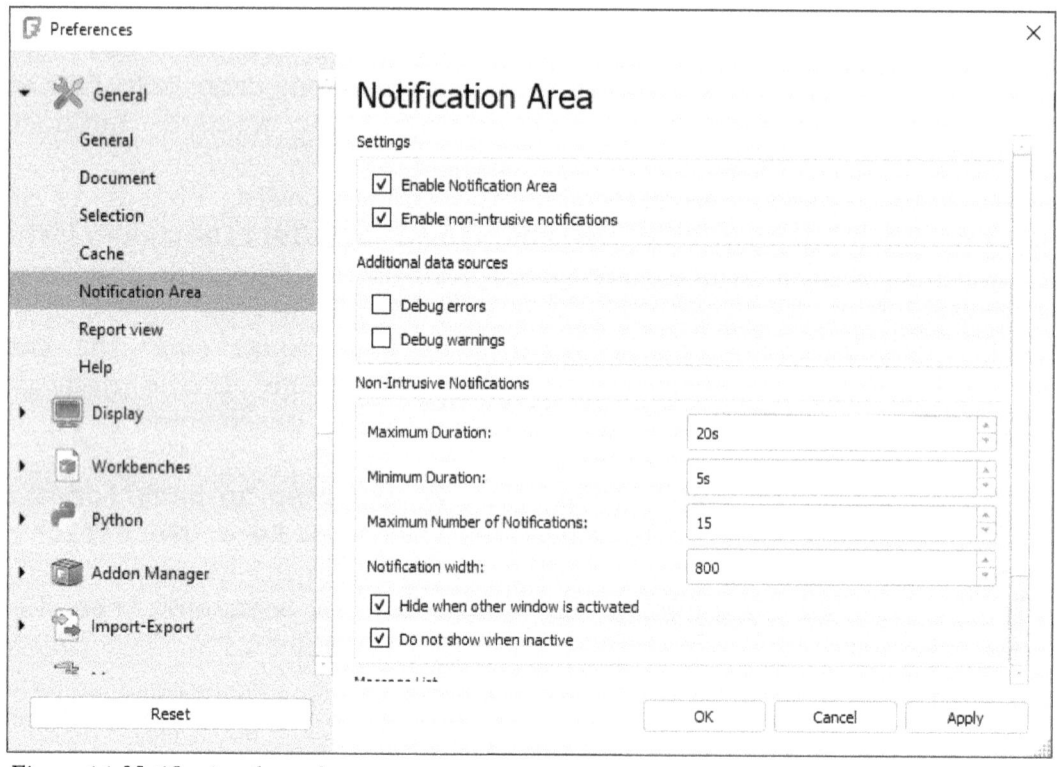

Figure-14. Notification Area tab

- Select **Enable Notification Area** check box from **Settings** area to display Notification Area in the status bar.
- Select **Enable non-intrusive notifications** check box from **Settings** area to display non-intrusive notifications next to the Notification Area in the status bar.
- Select **Debug errors** check box from **Additional data sources** area to display errors in the Notification Area.
- Select **Debug warnings** check box from **Additional data sources** area to display warnings in the Notification Area.
- Specify desired maximum or minimum duration during which notifications are shown in the **Maximum Duration** or **Minimum Duration** edit boxes from the **Non Intrusive Notifications** area, respectively.

- Specify desired maximum number of notifications to be shown simultaneously in the **Maximum Number of Notifications** edit box.
- Specify desired width of the notification area in pixels in the **Notification width** edit box.
- Select **Hide when other window is activated** check box to disappear open notifications when another window is activated.
- Select **Do not show when inactive** check box to disappear the notifications if the FreeCAD window is not the active window.
- Specify desired value in the **Maximum Messages (0 = no limit)** edit box from **Message List** area to keep maximum number of messages in the list.
- Select **Auto-remove User Notifications** check box to remove the notifications from the message list.

Report view tab

The options in **Report view** tab are used to define how information will be displayed in the **Report view** panel; refer to Figure-15.

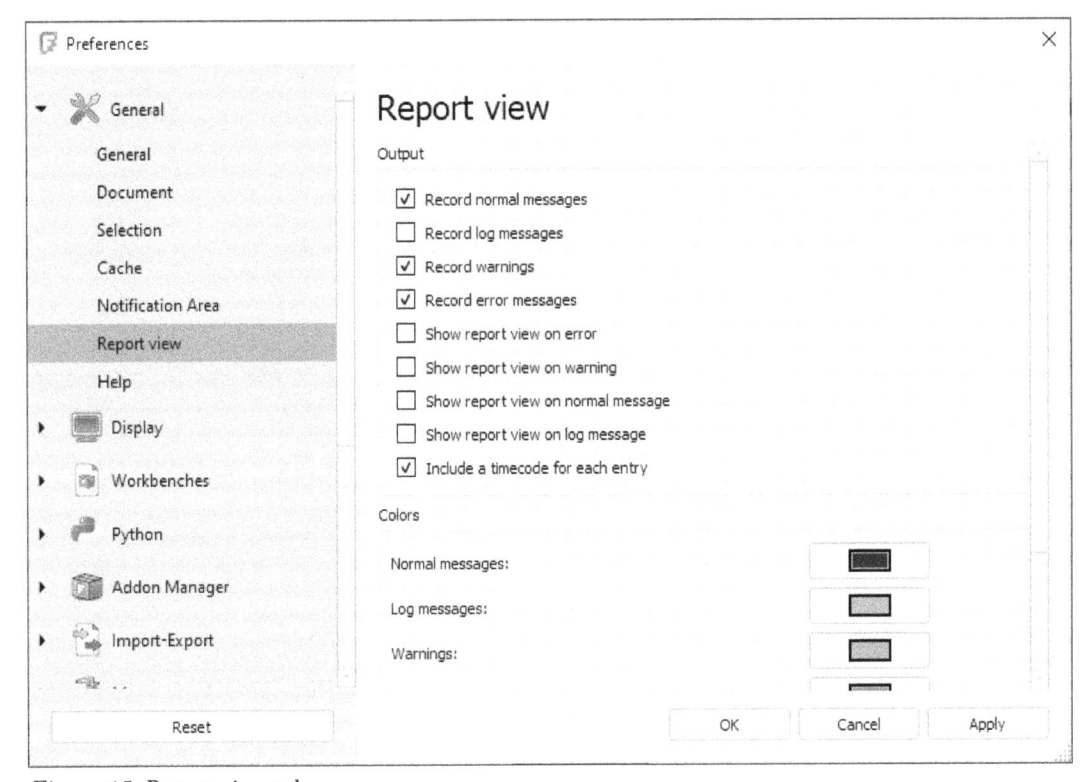

Figure-15. Report view tab

- Select the **Record normal messages** check box to record the normal messages in a file.
- Select the **Record log messages** check box to record the log messages displayed in **Report view** panel as a log file.
- Select the **Record warnings** check box to record warning messages in a file.
- Select the **Record error messages** check box to record error messages in a file.
- Select the **Show report view on error** check box to display the **Report View** panel on-screen showing the error, when an error has occurred.
- Select the **Show report view on warning** check box to display the **Report View** panel on-screen showing the warning, when a warning has occurred.
- Select the **Show report view on normal message** check box to display the **Report View** panel on-screen showing the message, when a normal message has occurred.

- Select the **Show report view on log message** check box to display the **Report View** panel on-screen showing the log message, when the log message has occurred.
- Select **Include a timecode for each entry** check box to include a timecode for each report.
- Set desired colors for Normal messages, Log messages, Warnings, and Errors in the **Colors** area of the dialog box.
- Select the check boxes from **Python interpreter** area of the tab to redirect outputs and errors of Python in **Report view** panel.

Help

The options in **Help** tab are used to define how help information will be displayed; refer to Figure-16.

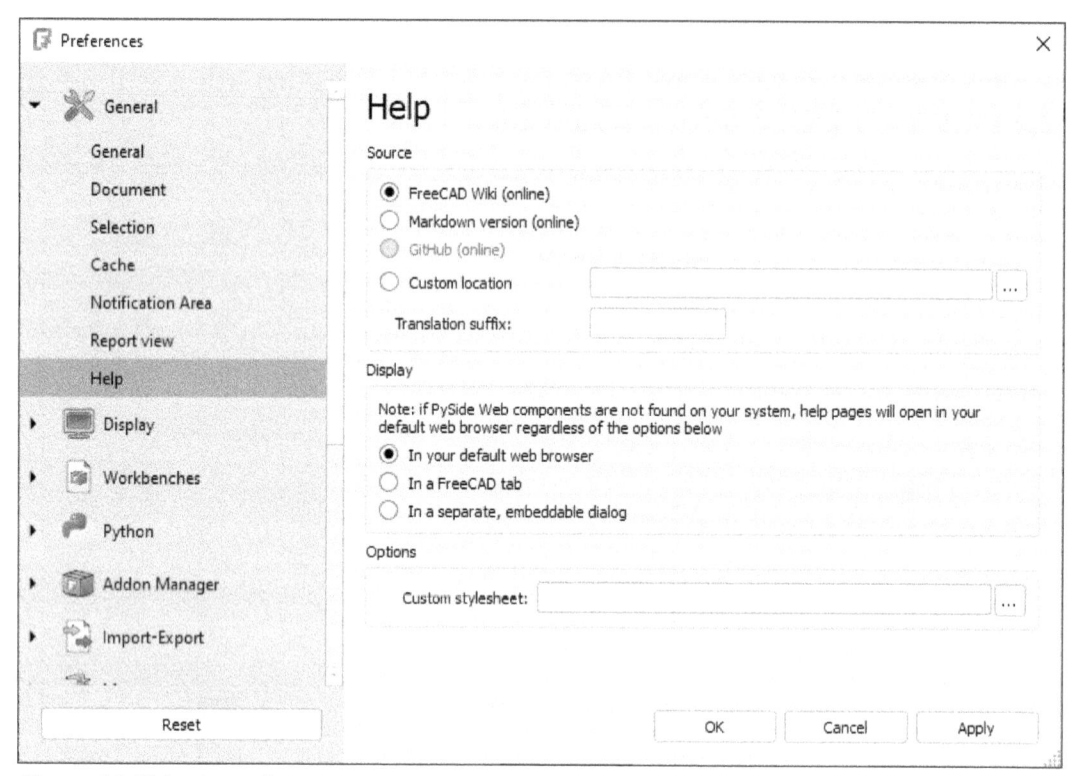

Figure-16. Help view tab

- Select the **Custom location** radio button to enter the path where the downloaded FreeCAD documentation can be found.
- Specify desired value in **Translation suffix** edit box if you want to add a language suffix to the end of help page name like Freecad/HN where HN represents Hindi.
- Select the **In your default web browser** radio button to the display help documentation in your default web browser.
- Select the **In a FreeCAD tab** to display the documentation in a new tab in FreeCAD.
- Select the **In a separate, embeddable** dialog radio button to display the documentation in the separate dialog area.
- Select **Custom stylesheet** to specify an optional custom style sheet.

Display Page

The options in the **Display** page of the dialog box are used to set parameters for display of various objects in the interface; refer to Figure-17.

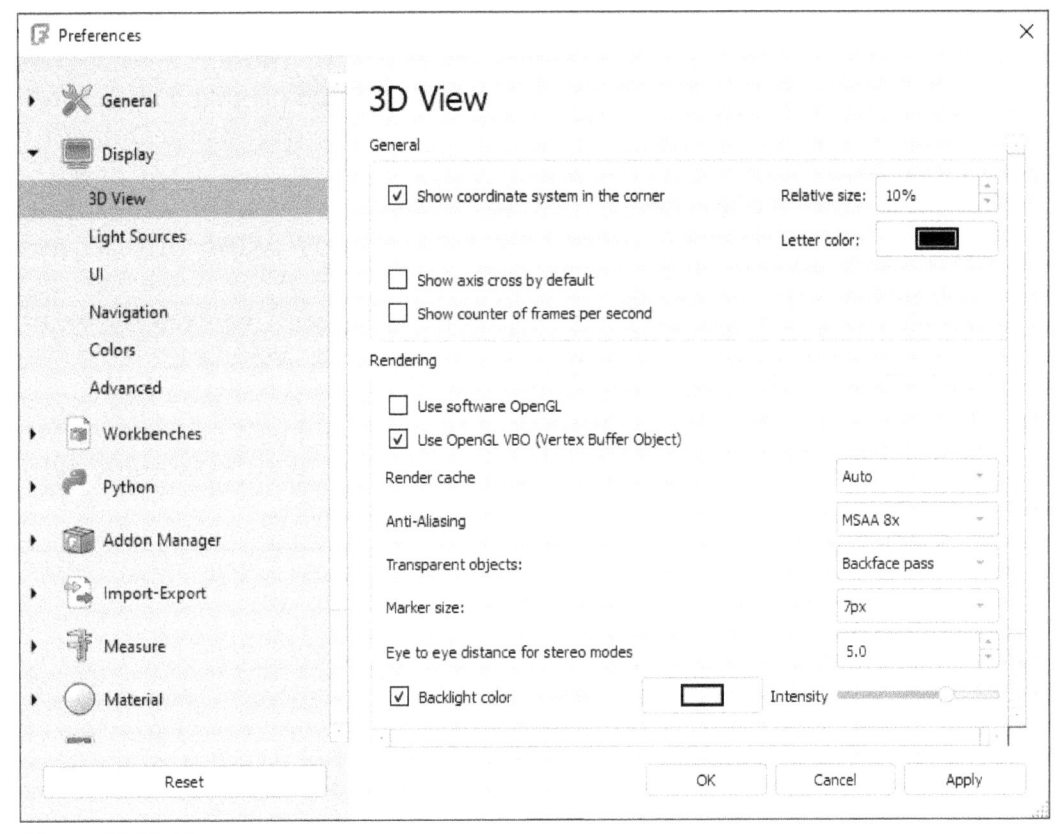

Figure-17. Display page

3D View tab
- Select the `Show coordinate system in the corner` check box to display Coordinate system at the bottom right corner of the drawing area. Specify desired value in the `Relative size` edit box from `General` area in the tab.
- Select the `Show axis cross by default` check box to show the axis cross by default at file opening or creation.
- Select the `Show counter of frames per second` check box to display a counter showing number of frames running in the software per second.
- Select `Use software OpenGL` check box for troubleshooting graphics card and driver problems.
- Select `Use OpenGL VBO (Vertex Buffer Object)` check box to use OpenGL buffer of your system for rendering objects in the drawing area.
- Select desired option from `Render cache` drop-down. Select `Auto` option from the drop-down to let the Coin3D decide where to cache. Select `Distributed` option to manually turn on cache for all view provider root nodes. Select `Centralized` option to manually turn off cache in all nodes of all view providers, and only cache at the scene graph root node.
- Select desired option from the `Anti-Aliasing` drop-down to smoothen jagged edges of the model.
- Select desired option from `Transparent objects` drop-down to specify render type of transparent objects.
- Select desired option from the `Marker size` drop-down to define size of point/constraint marks.

- Specify desired value in **Eye to eye distance for stereo modes** edit box to define distance between two eye for enhancing 3D view of model. For a mechanical engineer, it hardly makes any difference but if you are creating different views of model for each eye then you need to specify the value carefully. Default value for this parameter is 5.
- Select the **Backlight color** check box to define the color in which backside of model will be highlighted and select desired color by using button next to the check box.
- Select desired radio button from the **Camera type** area to specify whether you want to render model in perspective view or orthographic view.

Light source

The options in the **Light source** page are used to modify the direction for light source; refer to Figure-18.

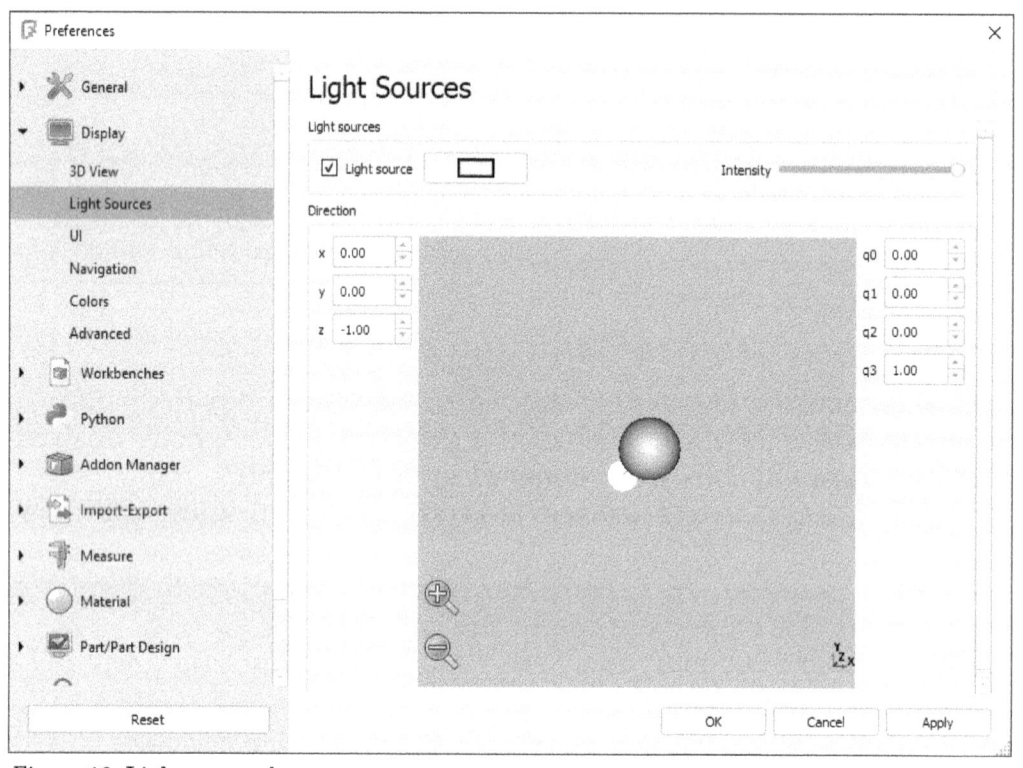

Figure-18. Light source tab

- Click on ▭ button to specify color of directional light source.
- Move the **Intensity** slider to change the intensity of the light.
- Specify desired value in the **X, Y,** and **Z** edit boxes from the **Direction** section to define source location.
- Specify desired values in the **q0, q1, q2,** and **q3** edit boxes to define focus location for light source.

UI

The option in the **UI** page are used to perform theme customization; refer to Figure-19.

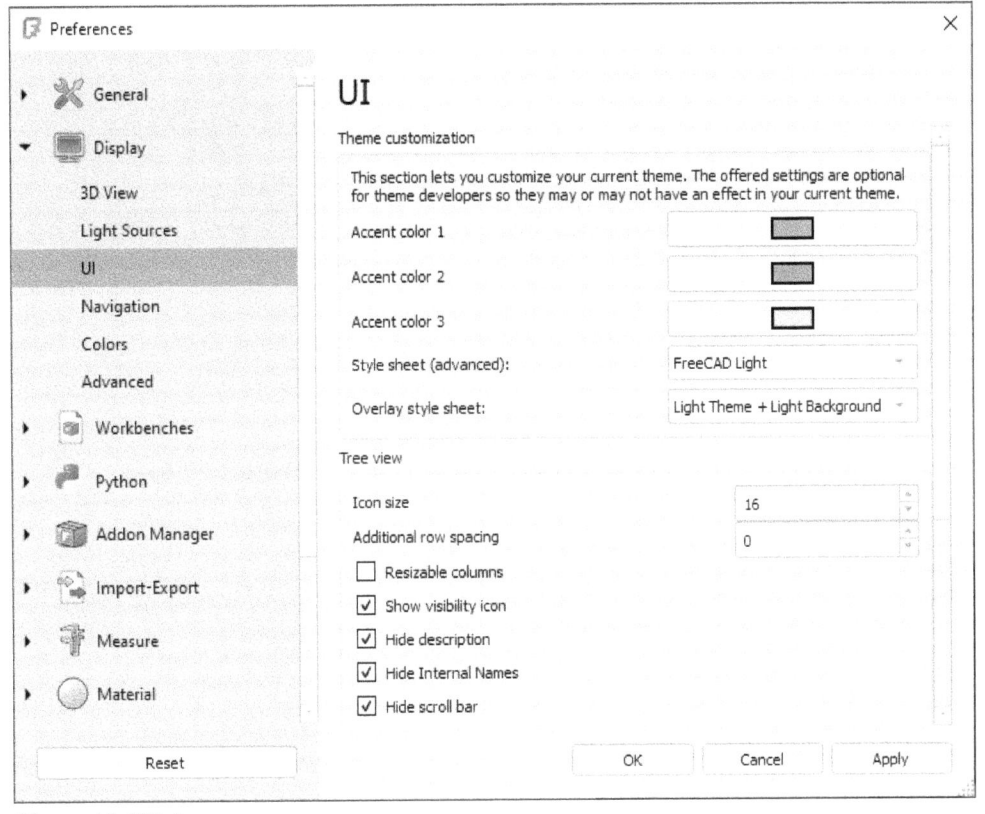

Figure-19. UI theme

- Specify desired theme color from the **Accent color 1, Accent color 2,** and **Accent color 3**, respectively.
- Select desired option from the **Style sheet (advanced)** and **Overlay style sheet** drop-down from the dialog to use predefine style settings for geometric objects and user interface.
- Specify desired values in the **Icon size** and **Additional row spacing** edit boxes.
- Select the **Resizable columns** check box to allow manual resizing of columns in the **Combo View**.
- Select the **Show visibility icon** check box to show the visibility status of objects in **Model** panel of **Combo View**.
- Select the **Hide description** check box to hide column with object description in **Model** panel of **Combo View**.
- Select the **Hide Internal Names** check box to hide internal names of objects in **Model** panel of **Combo View**.
- Select the **Hide scroll bar** check box to hide scroll bar in the **Combo View**.

Navigation tab

The options in the **Navigation** tab are used for the settings related to navigation of elements; refer to Figure-20.

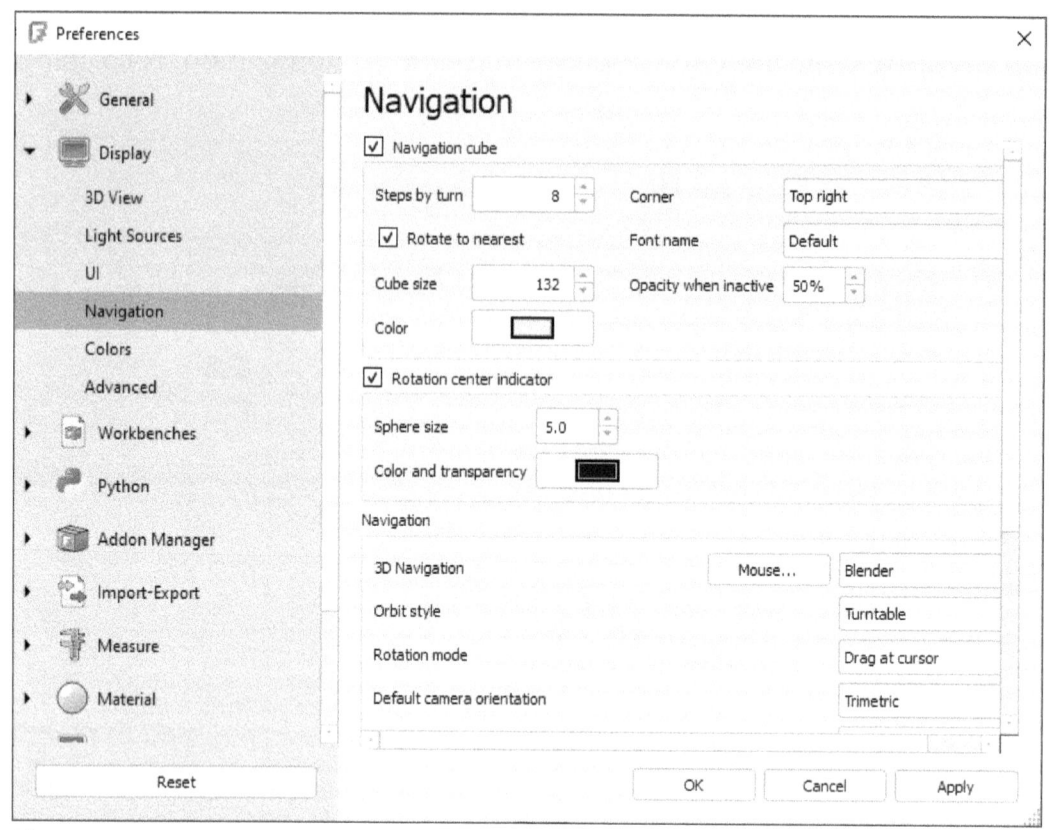

Figure-20. Navigation tab

- Select the **Navigation cube** check box to place 3D navigation cube for changing view. Enter desired value in **Steps by turn** edit box to specify number of steps by turn when using arrows. Select the **Rotate to nearest** check box to rotate to nearest possible state when clicking a cube face. Enter desired value in **Cube size** edit box to specify the size of navigation cube. Select desired option from the **Corner and Font name** drop-down to define location of Cube in drawing area.
- Select desired option from the **3D Navigation** drop-down in the **Navigation** area of the tab to specify what type of mouse gestures will be used in FreeCAD for rotating view, pan view, zoom view, or other orientation functions. You can find some common CAD application styles in this drop-down. Click on the **Mouse** button next to drop-down to check functions of different mouse buttons for selected style.
- Select desired option from the **Orbit style** drop-down to define how object will rotate in 3D orbit. Select the **Turntable** option from the drop-down to allow rotation of object as if it is placed on a turn table. Select the **Trackball** option to rotate the object freely in 3D orbit.
- Select desired option from **Rotation mode** drop-down to define the rotation center.
- Select desired option from the **Default Camera Orientation** drop-down to define which orientation will be applied to the model by default when you start a new document.
- Enter desired value in the **Camera zoom** edit box to set the camera zoom for new documents. The value is the diameter of the sphere to fit on the screen.
- Select the **Enable animation** check box to animate the change of views in drawing area.
- Select the **Zoom at cursor** check box to zoom the object at the location of cursor.
- Enter desired value in the **Zoom step** edit box to specify how much the object will be zoomed.

- Select the **Invert zoom** check box to reverse the style of zoom when you scroll up/down in drawing.
- Select the **Disable touchscreen tilt gesture** check box to disable tilting of screen when you perform pinch zooming.

Colors tab

The options in the **Colors** tab are used to modify colors for selection and background of drawing area; refer to Figure-21.

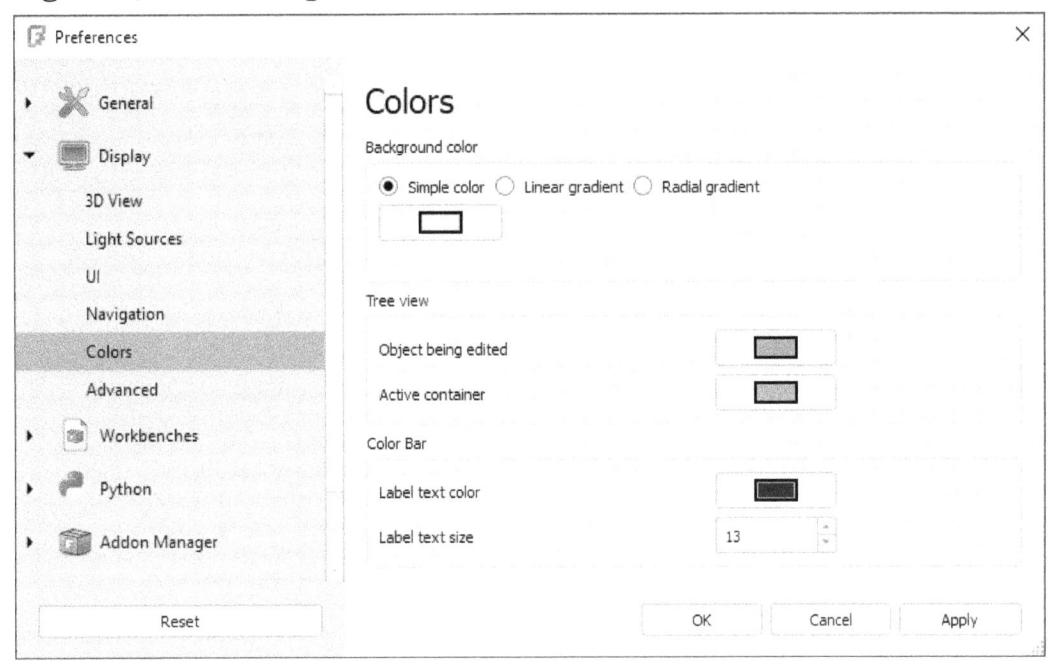

Figure-21. Colors tab

- Select the **Enable preselection highlighting** check box to highlight the object under cursor in drawing area. You can specify the color of highlighting by selecting the button next to the check box.
- Select the **Enable selection highlighting** check box to highlight the object selected from drawing area in color set next to the check box.
- Select the **Simple color** radio button from the **Background color** area to apply simple color in the background of drawing area. You can select desired color by using button below the radio button.
- Select the **Linear gradient** radio button to apply vertical color gradient in the background of drawing area by specifying colors in the **Top** and **Bottom** section.
- Select the **Radial gradient** radio button to apply radial color gradient in the background of drawing area by specifying colors in the **Central** and **End** section.
- If you want to create a three color gradient background then select the **Middle color** check box.
- Click on the **Object being edited** color button from **Tree view** area of the tab to define the background color for objects in the tree view that are currently edited.
- Click on the **Active container** color button to define the background color for active containers in the tree view.
- Similarly, you can set colors for objects in **Model Tree** by using options in the **Tree View** area of the dialog box.

The options of **Mesh view** tab can be modified in the same way. After setting desired parameters, click on the **OK** button from the dialog box to apply changes.

Python Page

The options in the **Python** page of the dialog box are discussed next; refer to Figure-22.

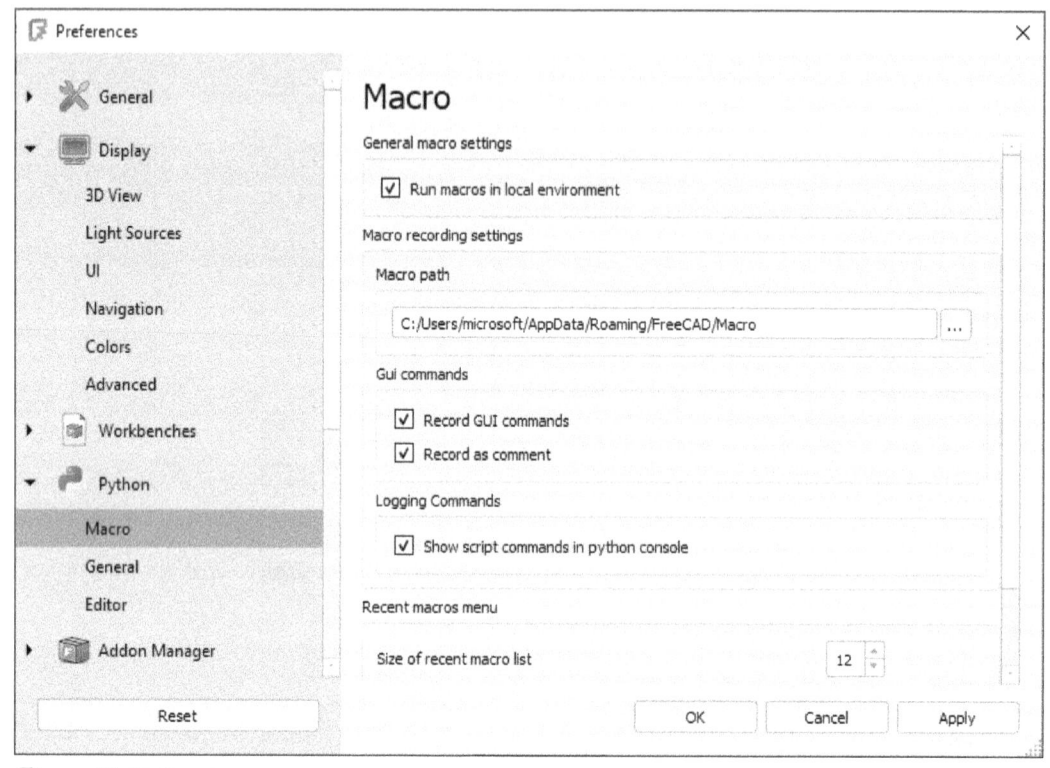

Figure-22. Python page

Macro tab

The options in the **Macro** tab are used to define path and parameters for running macros in FreeCAD; refer to Figure-22.

- Select the **Run macros in local environment** check box to run the macros internally in FreeCAD environment.
- Specify desired location in the **Macro path** area to define location where macro file will be saved and searched.
- Select check boxes from the **Gui commands** area to define whether graphic user interface commands and comments will be recorded in the macro.
- Select the **Show script commands in python console** check box to show scripts in python console when you activate a tool.
- Specify desired number of macros which should be listed in recent macros list in the **Size of recent macro list** edit box from **Recent macros menu** area of the dialog box.
- Specify desired number of recent macros which should have shortcuts in the **Keyboard shortcut count** edit box.
- Specify the default keys for the macro in the **Keyboard Modifiers** edit box.

General tab

The options in **General** tab are used to control FreeCAD processes and create & modify objects; refer to Figure-23. Select `Enable word wrap` check box from `Settings` area of the tab to wrap the words when they exceed available horizontal space in Python console. If you select the `Enable block cursor` check box then the cursor shape will be a block. Select the `Save history` check box to save Python history across sessions.

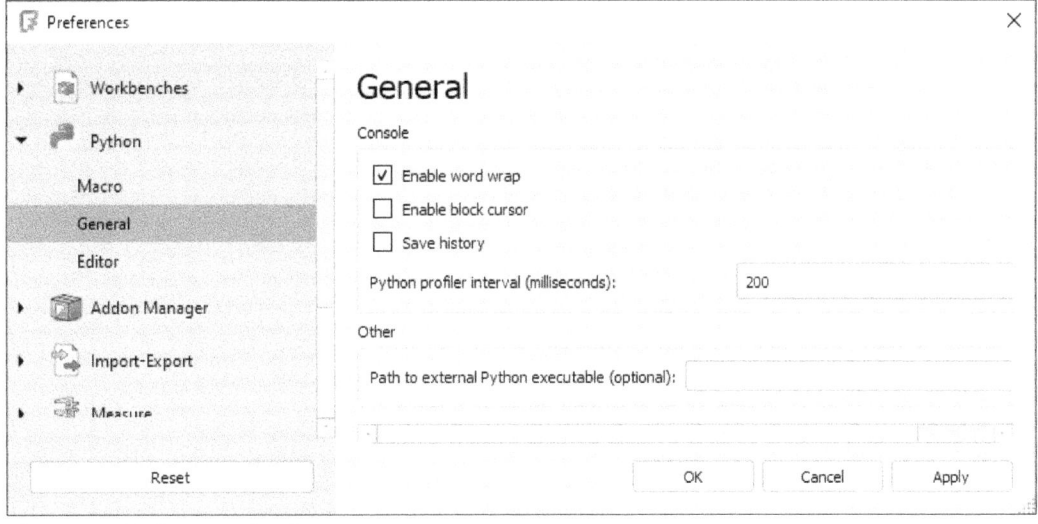

Figure-23. Python console tab

Editor tab

The options in **Editor** tab are used to specify font type and size for different types of texts in Python Console; refer to Figure-24. Select desired text type from `Display Items` list box and specify related parameters in the right area of dialog box. Select the `Enable line numbers` check box from the `Options` area of this tab to display line numbers in Python Console. Select `Enable block cursor` check box to allow the cursor to be a block shape. Similarly, you can specify the indent sizes in the `Indentation` area of the tab.

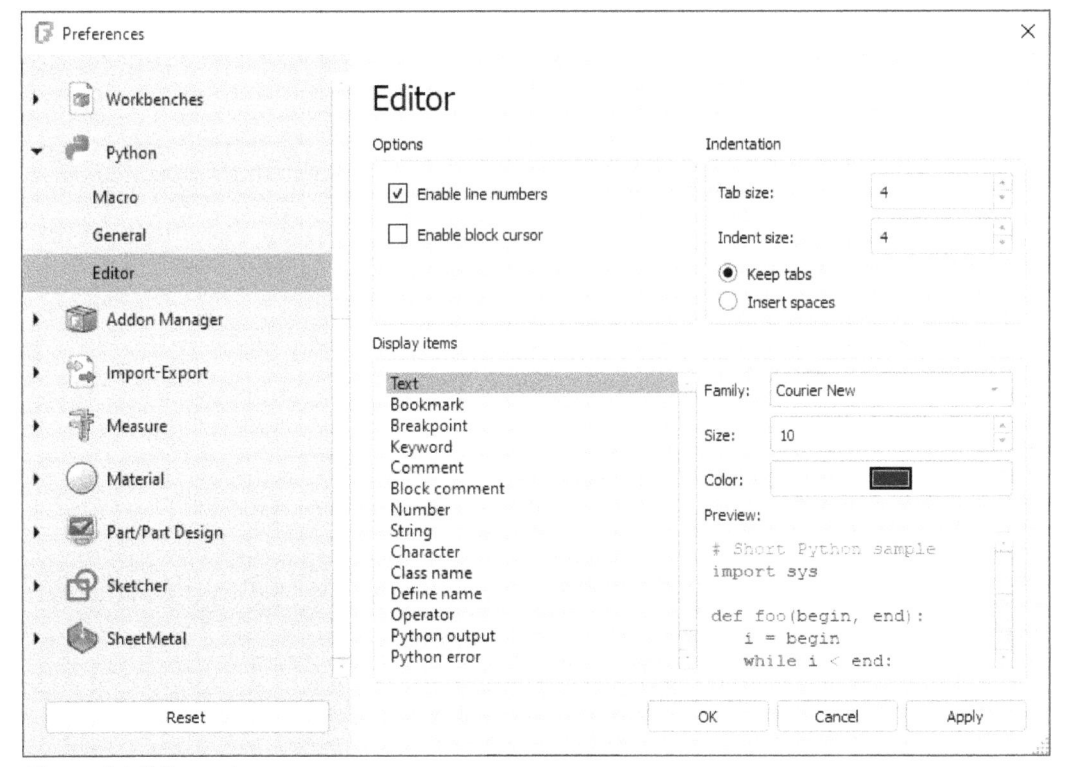

Figure-24. Editor tab

VIEW MENU

The tools in the **View** menu are used to create and manage views of drawing area. You can also display/hide various elements of user interface of software by using the tools in this menu; refer to Figure-25. Some of the important tools in this menu have been discussed earlier in Chapter 1. The other tools in this menu are discussed next.

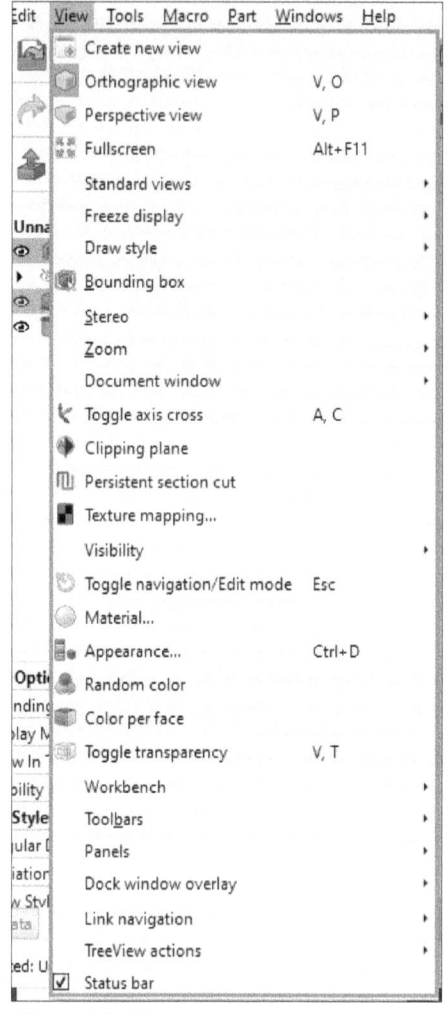

Figure-25. View menu

Stereoscopic Views

The tools in the **Stereo** cascading menu are used to display the model stereoscopic color style; refer to Figure-26.

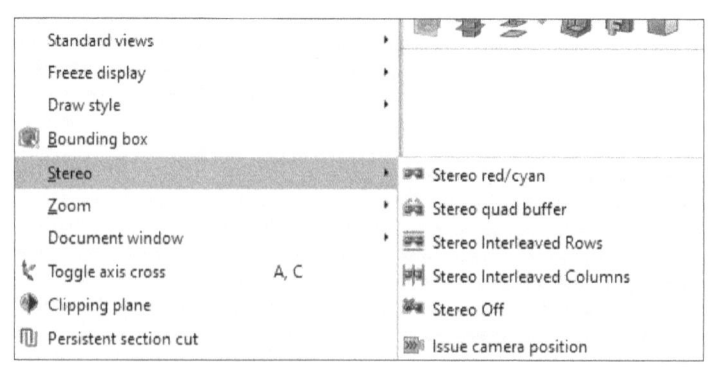

Figure-26. Stereo cascading menu

- Select the **Stereo red/cyan** option from the cascading menu to apply red and cyan color filters on the model, so that stereoscopic 3D effect can be achieved when viewing model using 3D glasses.
- Select the **Stereo quad buffer** option from the cascading menu to switch to Quad buffer stereo display mode if supported by your hardware. Quad-buffered stereo mode provides separate left and right eye frame buffers. It allows the window display in stereo while all other windows functions appear as normal.
- Select the **Stereo Interleaved Rows** or **Stereo Interleaved Columns** option to apply stereo color filters alternatively to every pixel row or column in the display, respectively.
- Select the **Stereo Off** option to stop stereoscopic display of model.
- Select the **Issue camera position** tool from the cascading menu to display current position of camera in 3D space. Note that the parameters will be displayed in **Report view** panel.

Zoom Options

The options in the **Zoom** cascading menu are used to zoom in/out and zoom specific objects; refer to Figure-27. Select the **Zoom In** option to enlarge the view of model and select the **Zoom Out** option to diminish the view of model. If you want to zoom on a specific object then select the **Box zoom** option and create a selection box around the object to be zoomed at.

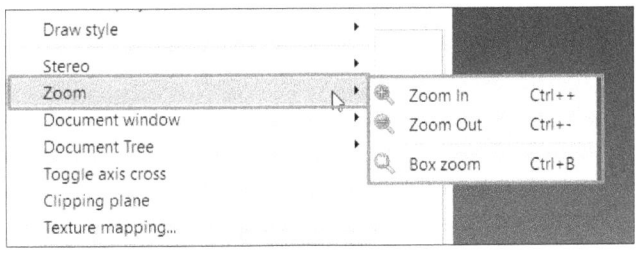

Figure-27. Zoom cascading menu

Document Window Options

The options in the **Document window** cascading menu are used to dock, undock, and full screen the document windows where models are displayed; refer to Figure-28. Select the **Undocked** option from menu to make document window free floating in interface. Select the **Docked** option to make document window attached as tile in interface. Select the **Fullscreen** option to display document window on full screen. You can also use **F11** shortcut key to enter and exit full screen mode.

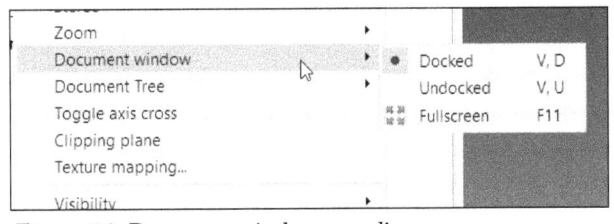

Figure-28. Document window cascading menu

TreeView actions Options

The options in the **TreeView actions** cascading menu are used to modify display parameters for document tree; refer to Figure-29. Select the **Single Document** option from cascading menu to display features of current document in **Model Tree**. Select the **Multi Document** option from the cascading menu to display features of all the documents currently open in **Model Tree**. Select the **Collapse/Expand** option from the cascading menu to expand or collapse all the feature nodes displayed in **Model Tree**. Select the **Initiate dragging** option to initiate a drag operation for selected objects in the Tree view. Select the **Go to selection** option to scroll the tree view to the first created object in a 3D view selection. The other options in the cascading menu have been discussed earlier in this chapter.

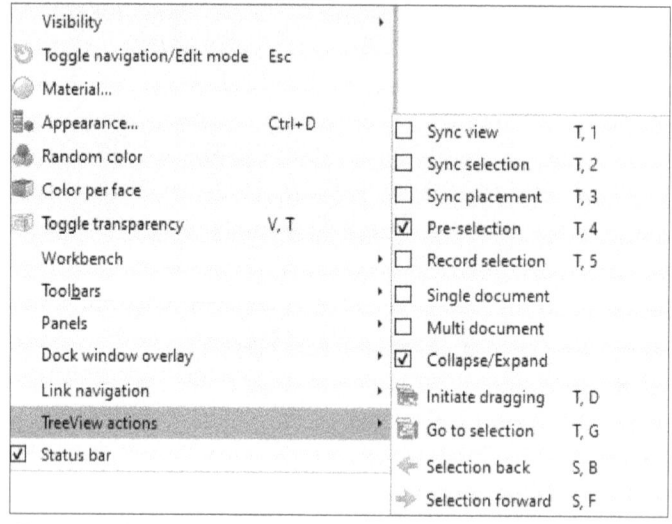

Figure-29. TreeView actions cascading menu

Toggle Axis Cross

The **Toggle Axis Cross** option in **View** menu is used to toggle global coordinate system On/Off. When the Axis Cross is On, it will be displayed at the center of drawing area as shown in Figure-30.

Figure-30. Axis cross

Bounding Box

The **Bounding box** tool in the **View** menu toggles the global bounding box highlighting mode. If this mode is switched on, selected objects are marked in a 3D view with a highlighted bounding box even if their selection style is set to 'Shape'.

Clipping Plane

The **Clipping plane** tool in **View** menu is used to clip the model for checking interior of model. The procedure to use this tool is given next.

- Click on the **Clipping plane** tool from the **View** menu. The **Clipping** dialog box will be displayed; refer to Figure-31.

Figure-31. Clipping dialog box

- Select desired check box from the **Clipping** dialog to define in which direction you want to perform clipping of model. For example, select the **Clipping X** check box to clip model along X direction.
- After selecting check box, specify desired offset value to define offset distance of clipping plane from origin parallel to selected direction; refer to Figure-32.

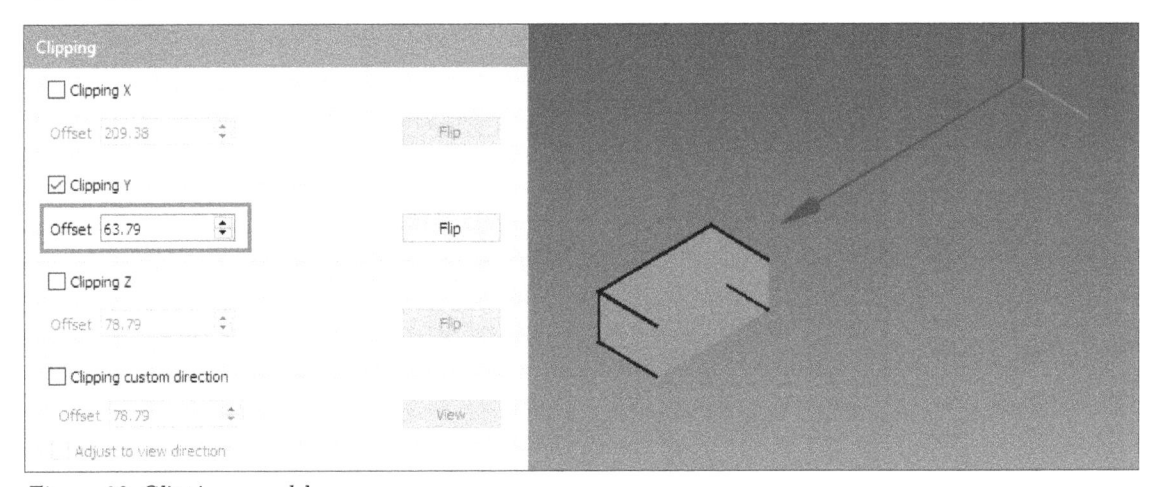

Figure-32. Clipping a model

- Using the **Flip** button, you can reverse the direction of clipping.
- Click on the **Close** button from the dialog to exit the tool.

Persistent section cut

The **Persistent section cut** tool is used to create a persistent cut of objects and assemblies. This tool only works for Part and PartDesign objects and assemblies of those. The procedure to use this tool is discussed next.

- Click on the **Persistent section cut** tool from the **View** menu. The **Section Cutting** dialog will be displayed; refer to Figure-33.

Figure-33. Section Cutting dialog

- Select desired check box from the dialog to define in which direction you want to perform section cutting of model. For example, select the **Cutting X** check box to cut model along **X** direction.
- After selecting check box, specify desired offset value to define offset distance of cutting plane from origin parallel to selected direction; refer to Figure-34.

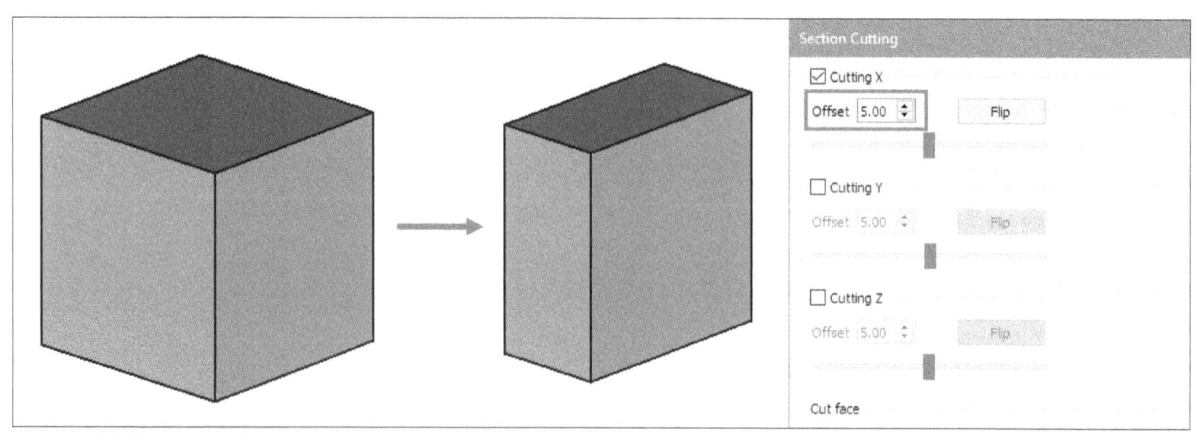

Figure-34. Section cutting a model

- Using the **Flip** button, you can reverse the direction of cutting.
- Click on the **Color** button from **Cut face** section of the dialog to select the color for cut objects.

- Using the **Transparency** slider, adjust the transparency of cut face.
- If **Auto** check box is selected, the color and transparency of the cut face will be taken from the cut objects.
- The **Cut intersecting objects** check box allows to cut objects that intersects each other.
- Click on the **Refresh view** button to refresh the list of visible objects.
- Select the **Keep only cuts visible when closing** check box to visible only created cuts, when dialog is closed.
- Click on the **Close** button from the dialog to exit the tool.

Texture Mapping

The **Texture mapping** tool is used to apply texture on the model. The procedure to use this tool is given next.

- Click on the **Texture mapping** tool from the **View** menu. The **Texture** dialog will be displayed; refer to Figure-35.

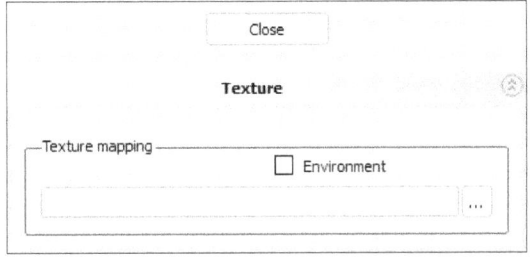

Figure-35. Texture dialog

- Select the **Environment** check box to apply selected image file on all the faces of model as environment.
- Click on the browse button next to edit box in the dialog to selected desired texture image file.
- Click on the **Close** button to exit the tool and texture display mode.

Visibility Options

The options in **Visibility** cascading menu are used to display or hide various objects in the drawing area; refer to Figure-36. Note that some of the options in this cascading menu are available when you have selected an object from drawing area.

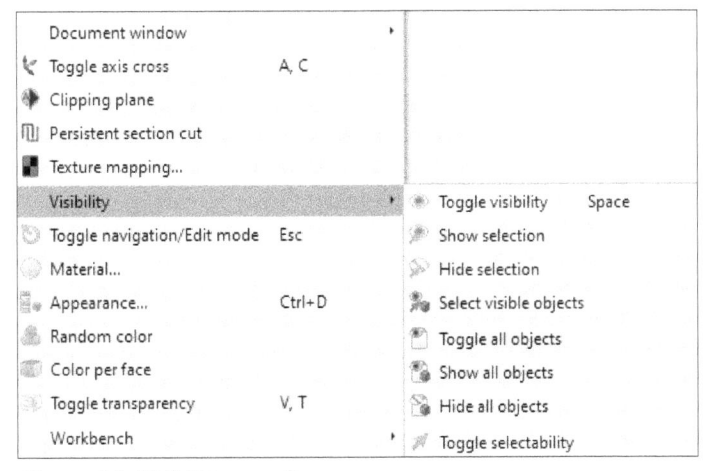

Figure-36. Visibility cascading menu

- Select the **Toggle visibility** option from the **Visibility** cascading menu to hide selected component if it is visible and display the component if it is hidden in drawing area. You can also use the **SPACEBAR** key to toggle visibility. Note that although an object is hidden in drawing area, you can still select it from **Model Tree**. Hidden objects are displayed as washed out Black & White icons in the **Model Tree**.
- Select the **Show selection** option from the cascading menu to display selected object in drawing area.
- Select the **Hide selection** option from the cascading menu to hide selected object in drawing area.
- Select the **Select visible objects** option to select all the objects currently visible in drawing area.
- Select the **Toggle all objects** option from the **Visibility** cascading menu to hide currently visible objects and display currently hidden objects.
- Select the **Show all objects** option from the cascading menu to display all the objects of **Model Tree** in drawing area.
- Select the **Hide all objects** option from the cascading menu to hide all the objects of **Model Tree** from drawing area.
- Select the **Toggle selectability** option to make currently selected objects unselectable in drawing area. Note that you can still select those objects from **Model Tree**. If you have made an object unselectable then you can use this tool again to make it selectable.

The **Toggle navigation/Edit mode** tool is used to switch the editing mode of a tool temporarily for rotation/reorientation of model.

The **Appearance** tool is used to apply different colors and appearances to the model. This tool has been discussed earlier.

Apply Random Colors to Model

The **Random color** tool is used to apply random colors to the faces of selected model. After selecting model, click on the **Random color** tool from the **View** menu.

Workbench Options

The options in **Workbench** cascading menu are used to switch between various workbenches. You can also use the **Switch between workbenches** drop-down of toolbar to select desired workbench.

Showing/Hiding Toolbars

The options in **Toolbars** cascading menu are used to display/hide various toolbars from the interface of software; refer to Figure-37.

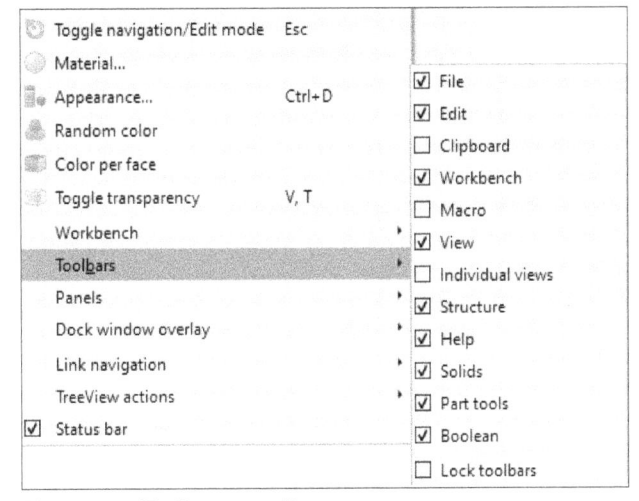

Figure-37. Toolbars cascading menu

Select desired option from the cascading menu to display respective toolbar. Note that toolbars for options which have tick mark before them are showing in interface.

Showing/Hiding Panels

The options in the **Panels** cascading menu are used to display/hide various panels of the interface like report view, panel view, selection view, and so on; refer to Figure-38.

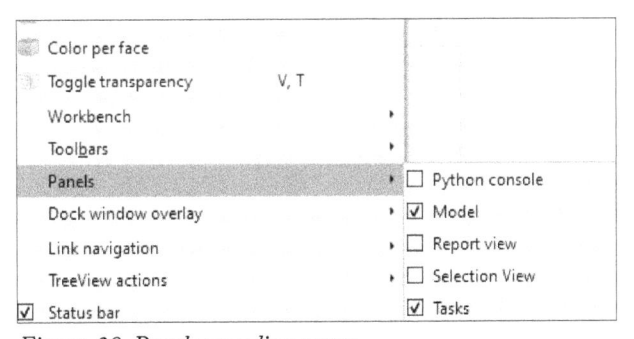

Figure-38. Panels cascading menu

Select desired option from **Panels** cascading menu to display respective panel in interface.

Status Bar

The **Status Bar** is used to display information like coordinates, mouse style, and so on. Select the **Status Bar** option from **View** menu to display **Status Bar** in interface.

MACRO MENU

The options in **Macro** menu are used to create and manage macros of FreeCAD. Macros are set of instructions for software to automate repetitive tasks; refer to Figure-39. For example, if you want to perform scale up of 10 sketches one by one then you can create a macro to do so after recording for just one sketch. Various tools of this menu are discussed next.

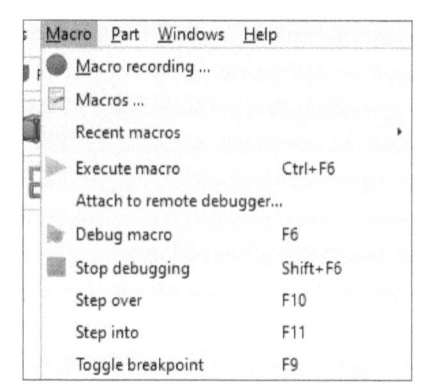

Figure-39. Macro menu

Recording Macro

The **Macro recording** tool in **Macro** menu is used to record sequence of operations performed in FreeCAD. The procedure to use this tool is given next.

- Click on the **Macro recording** tool from the **Model** menu. The **Macro recording** dialog box will be displayed; refer to Figure-40.

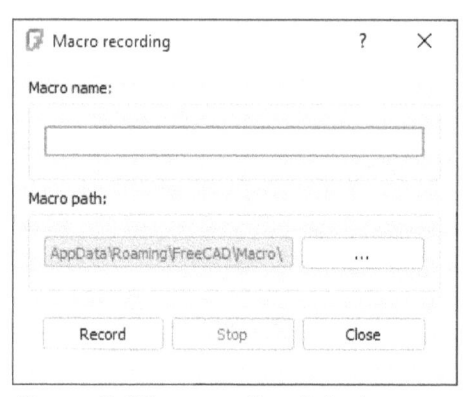

Figure-40. Macro recording dialog box

- Specify desired name for macro in the **Macro name** edit box.
- Click on the **Browse** button for **Macro path** edit box to specify location for saving macro and click on the **Record** button. The macro recording mode will start.
- Perform desired operations in the drawing area to be recorded and then click on the **Stop macro recording** tool from **Macro** menu. The macro will be created and saved at earlier specified location.

Executing and Managing Macros

The **Macros** tool in **Macro** menu is used to execute, edit, delete, or perform other tasks for macros. The procedure to use this tool is given next.

- Click on the **Macros** tool from the **Macro** menu. The **Execute macro** dialog box will be displayed; refer to Figure-41, with the list of macros available in location specified in **User macros location** edit box.

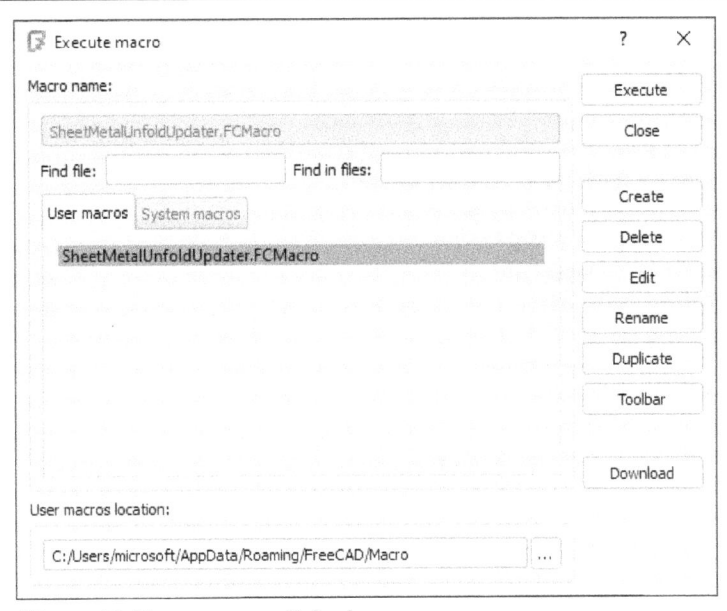

Figure-41. Execute macro dialog box

- Select desired macro from the list and click on the **Execute** button to run the macro.
- If you want to edit a macro then select it from list and click on the **Edit** button in the **Execute macro** dialog box. The `Macro Editor document` window will be displayed; refer to Figure-42. Use the Python coding to modify macro and save the file.

Figure-42. Macro Editor dialog box

- You can use the **Delete**, **Rename**, and **Duplicate** buttons in dialog box to delete, rename, or create duplicate copy of selected macro, respectively.
- Click on the **Toolbar** button to set up a macro in custom global toolbar.
- Click on the **Download** button to import macros available online. The `Addon Manager` dialog box will be displayed. Select `Macros` option from `Show addons containing` drop-down to check available macros; refer to Figure-43. Select desired macro from list and execute or Install/update as desired.

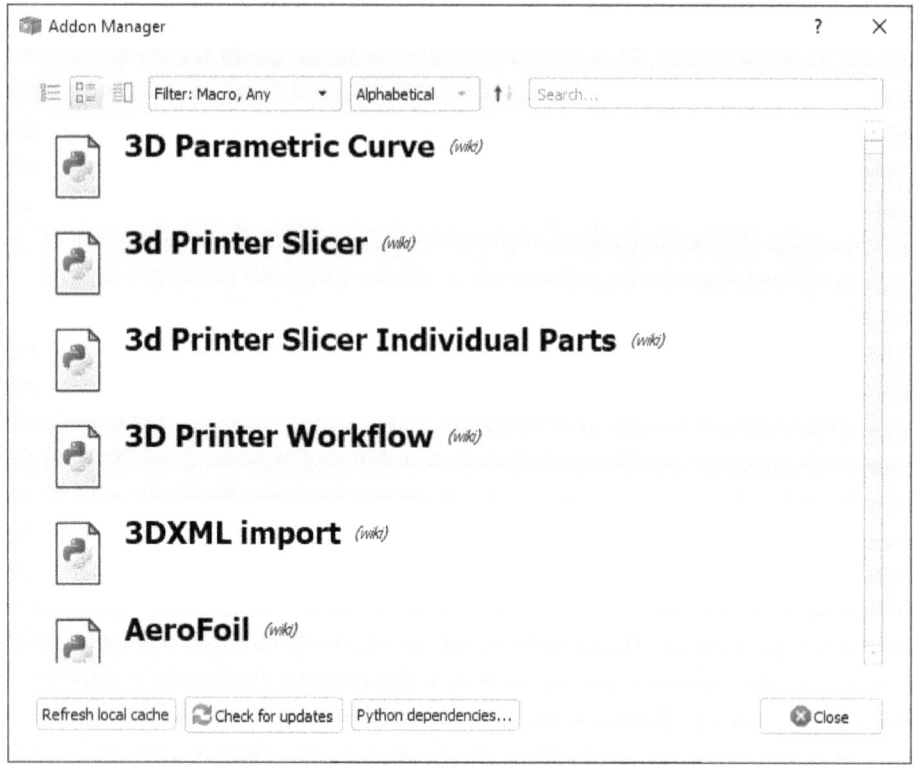

Figure-43. Macros option in Addon manager

- Click on the **Close** button to exit dialog boxes.

SELF ASSESSMENT

Q1. What does the "Undo" tool in the Edit menu do in FreeCAD?

A) Redoes the last undone operation
B) Reverses the effect of the most recent operation
C) Cuts the selected objects
D) Duplicates the selected objects

Q2. What is the shortcut key for the Undo operation in FreeCAD?

A) CTRL+Y
B) CTRL+Z
C) CTRL+C
D) CTRL+X

Q3. What does the "Redo" tool in the Edit menu do in FreeCAD?

A) Reverses the Undo operation
B) Duplicates the selected object
C) Deletes the selected object
D) Refreshes the model

Q4. Which of the following shortcut keys is used for the Redo operation in FreeCAD?

A) CTRL+Z
B) CTRL+X
C) CTRL+Y
D) CTRL+C

Q5. What is the purpose of the "Cut" tool in the Edit menu in FreeCAD?

A) Copies selected objects without removing them
B) Deletes selected objects
C) Copies selected objects and removes them from the original location
D) Duplicates selected objects

Q6. What does the "Paste" tool in the Edit menu do in FreeCAD?

A) Deletes selected objects
B) Places a copy of the objects from temporary memory into the drawing area
C) Creates a duplicate of the selected object
D) Recomputes the changes in the model

Q7. What does the "Duplicate selected objects" tool do in FreeCAD?

A) Creates a duplicate copy of selected objects
B) Deletes the selected objects
C) Moves the selected objects to a new location
D) Changes the properties of the selected objects

Q8. How can the "Refresh" tool be used in FreeCAD?

A) To update changes in the model after marking an object for recomputation
B) To create a duplicate copy of an object
C) To undo the last operation
D) To delete the selected object

Q9. What does the "Box Selection" tool do in FreeCAD?

A) Selects all objects in the drawing area
B) Selects objects that fall inside the boundaries of a drawn box
C) Deletes selected objects
D) Refreshes the model

Q10. Which tool in the Edit menu is used to select all objects in the drawing area in FreeCAD?

A) Select All
B) Box Selection
C) Delete
D) Duplicate

Q11. What does the "Delete" tool in the Edit menu do in FreeCAD?

A) Removes the selected object from the drawing area
B) Duplicates the selected object
C) Refreshes the model
D) Aligns the selected objects

Q12. What is the shortcut key for the "Delete" operation in FreeCAD?

A) CTRL+Z
B) DELETE key
C) CTRL+X
D) CTRL+Y

Q13. What is the function of the "Send to Python Console" tool in FreeCAD?

A) Displays a report in the Python console
B) Creates a variable in the Python console referencing a selected object
C) Aligns selected objects
D) Modifies placement of selected objects

Q14. What does the "Placement" tool in the Edit menu do in FreeCAD?

A) Changes the color of the selected object
B) Modifies the location and orientation of a selected object
C) Creates a duplicate of the selected object
D) Deselects the selected object

Q15. What operation is performed using the "Transform" tool in FreeCAD?

A) Changes the scale of the selected object
B) Applies rotation and translation increments to an object
C) Refreshes the model
D) Modifies the properties of the selected object

Q16. What does the "Alignment" tool do in FreeCAD?

A) Changes the color of the selected object
B) Aligns two selected objects about selected points
C) Refreshes the model
D) Duplicates the selected object

Q17. What does the "Toggle Edit Mode" tool in FreeCAD allow you to do?

A) Modify the properties of a selected object
B) Modify a feature created in FreeCAD using its related dialog
C) Undo the last operation
D) Align two selected objects

Q18. What does the "Preferences" tool in the Edit menu allow you to do in FreeCAD?

A) Modify the appearance and settings of FreeCAD application
B) Delete selected objects
C) Undo the last operation
D) Modify the placement of selected objects

Q19. What does the "General Page" in the Preferences dialog allow you to modify in FreeCAD?

A) The number of decimals for displaying values
B) The color of the toolbar icons
C) The size of the recent file list
D) The number of undo steps

Q20. What does the "Selection" tab in the Preferences dialog help configure in FreeCAD?

A) The appearance of the document
B) The settings related to selecting objects in FreeCAD
C) The location for saving temporary files
D) The maximum number of notifications

Q21. What is the purpose of the "Cache" tab in the Preferences dialog in FreeCAD?

A) To modify the appearance of the FreeCAD interface
B) To manage the location and size of temporary files stored by FreeCAD
C) To adjust the document save compression level
D) To configure undo/redo operations

Q22. What does the "Notification Area" tab in Preferences control in FreeCAD?

A) How error messages are displayed in FreeCAD
B) The location where temporary files are saved
C) The duration and behavior of notifications in the status bar
D) The style of the toolbar icons

Q23. Which option in the "3D View" tab displays the coordinate system at the bottom right corner of the drawing area?

a) Show axis cross by default
b) Show counter of frames per second
c) Show coordinate system in the corner
d) Use OpenGL VBO

Q24. In the "UI" tab, which checkbox allows you to manually resize columns in the tree view?

a) Show visibility icon
b) Resizable columns
c) Hide scroll bar
d) Hide description

Q25. Which option in the "Navigation" tab defines the rotation behavior of the object in 3D orbit?

a) Orbit style
b) Camera zoom
c) Enable animation
d) Rotation mode

Q26. Which option in the "Zoom" cascading menu allows you to zoom on a specific object by creating a selection box?

a) Zoom In
b) Zoom Out
c) Box zoom
d) Select visible objects

Q27. Which option in the "View" menu is used to toggle the global coordinate system On/Off?

a) Bounding Box
b) Clipping Plane
c) Toggle Axis Cross
d) Persistent section cut

Q28. Which checkbox in the "Colors" tab defines the background color for active containers in the tree view?

a) Active container color
b) Object being edited color
c) Background color
d) Highlighting color

Q29. What is the default value for "Eye to eye distance" in the "3D View" tab for stereo modes?

a) 2
b) 5
c) 10
d) 15

Q30. In the "Macro" tab, what does the "Run macros in local environment" checkbox do?

a) Runs macros externally
b) Runs macros within the FreeCAD environment
c) Defines location for storing macros
d) Enables macro execution history

Q31. Which of the following is used to record a series of operations in FreeCAD?

a) Execute macro
b) Macro recording tool
c) Macro path
d) Python script editor

Q32. Which option in the "View" menu is used to apply random colors to the faces of a selected model?

a) Apply Random Colors to Model
b) Appearance
c) Texture Mapping
d) Bounding Box

Q33. What does the "Persistent section cut" tool do in the "View" menu?

a) Clips the model for interior inspection
b) Applies a texture on the model
c) Creates a persistent cut of objects and assemblies
d) Highlights bounding boxes of selected objects

Q34. Which option in the "TreeView actions" menu expands or collapses all the feature nodes displayed in Model Tree?

a) Collapse/Expand
b) Initiate dragging
c) Go to selection
d) Toggle visibility

Q35. Which option in the "View" menu is used to hide or display the component currently selected in the drawing area?

a) Toggle visibility
b) Show selection
c) Hide selection
d) Toggle navigation

Q36. What is the purpose of the "Clipping plane" tool in the View menu?

a) To zoom into a specific object
b) To clip the model along a selected direction
c) To toggle axis cross
d) To apply random colors to the model

Q37. Which option in the "Display Page" dialog box allows you to display the coordinate system at the bottom right corner of the drawing area?
A) Show axis cross by default
B) Show coordinate system in the corner
C) Show counter of frames per second
D) Show transparency of objects

Q38. In the "Navigation" tab, which of the following options allows you to zoom the object at the location of the cursor?

A) Enable animation
B) Zoom at cursor
C) Invert zoom
D) Rotate to nearest

Q39. Which option in the "View Menu" allows you to apply random colors to the faces of a selected model?

A) Appearance tool
B) Apply Random Colors to Model
C) Texture Mapping
D) Bounding Box

Q40. In the "Python Page," which checkbox allows the user to wrap words that exceed the available horizontal space in the Python console?

A) Enable block cursor
B) Enable word wrap
C) Save history
D) Enable line numbers

Chapter 13

Assembly Workbench

Topics Covered

The major topics covered in this chapter are:

- *Introduction to Assembly Workbench*
- *Starting New Assembly*
- *Adding New Parts, Bodies, and Groups*
- *Editing Part Information*
- *Inserting Parts and Fasteners*
- *Creating and Importing Datum Features*
- *Creating and Modifying Configurations*
- *Animating Assembly*

INTRODUCTION TO ASSEMBLY WORKBENCH

The **Assembly** workbench of FreeCAD is used to perform assembly of different bodies contained in a single file or in multiple documents. To start creating assembly, select the **Assembly** option from the **Switch between workbenches** drop-down in the **Toolbar**; refer to Figure-1. On selecting the Assembly workbench, tools will be displayed as shown in Figure-2.

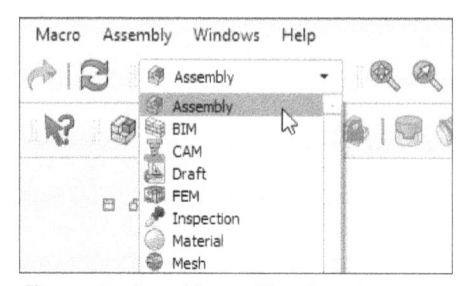

Figure-1. Assembly workbench option

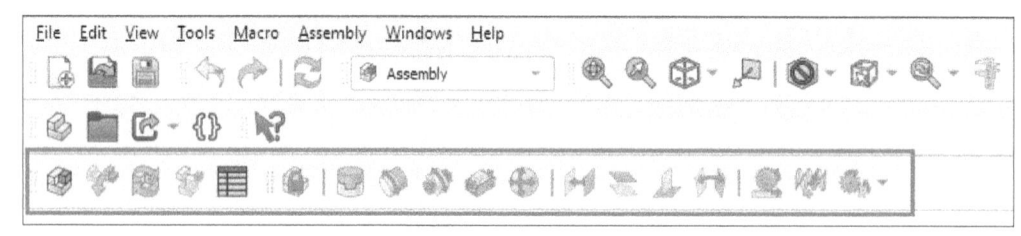

Figure-2. Assembly toolbar

Various tools of this workbench are discussed next.

CREATE ASSEMBLY

The **Create Assembly** tool is used to creates a root assembly in the current document, or a sub-assembly in a pre-existing active assembly. The procedure to use this tool is discussed next.

- Click on the **Create Assembly** tool from the **Toolbar** or **Assembly** menu; refer to Figure-3. A new model will be created; refer to Figure-4.

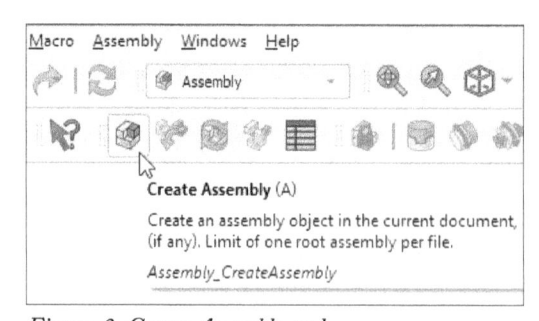

Figure-3. Create Assembly tool

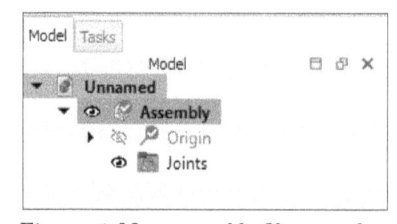

Figure-4. New assembly file created

INSERT COMPONENT

The **Insert Component** tool is used to insert a component into the active assembly. The procedure to use this tool is discussed next.

- Click on the **Insert Component** tool from **Toolbar** or **Assembly** menu; refer to Figure-5. The **Insert Component** dialog box will be displayed; refer to Figure-6.

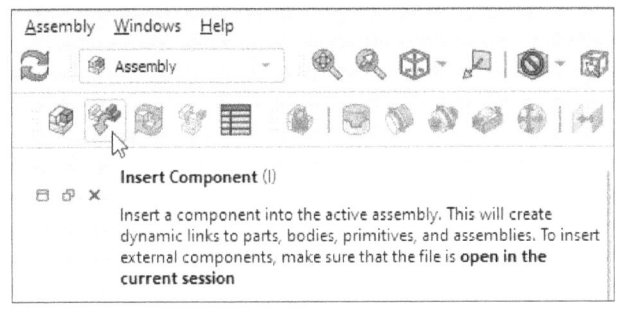

Figure-5. Insert Component tool

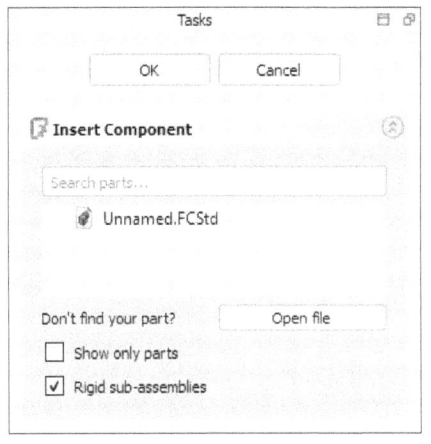

Figure-6. Insert Component dialog box

- Click on the **Open file** button. The **Select FreeCAD documents to import parts from** dialog box will be displayed; refer to Figure-7.

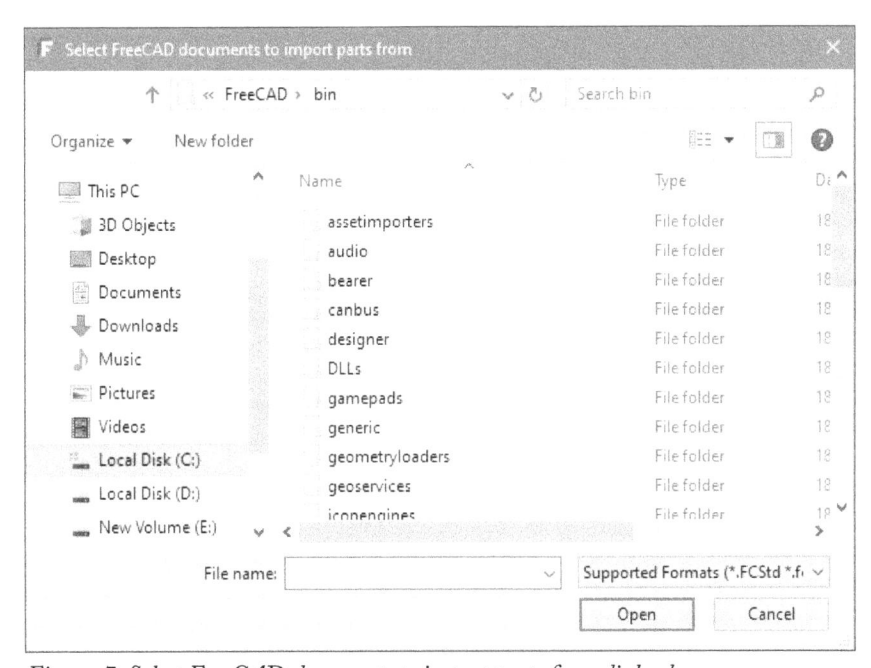

Figure-7. Select FreeCAD documents to import parts from dialog box

- Select desired component from the dialog box to be inserted into the active assembly and click on the **Open** button. The components will be inserted; refer to Figure-8.

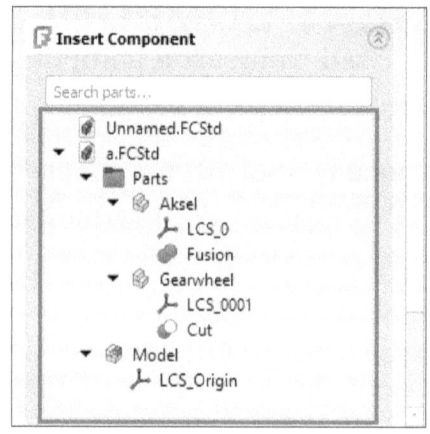

Figure-8. Components inserted

- Click on desired component to be inserted from the dialog box. The **Save Document** dialog box will be displayed asking you to save the document; refer to Figure-9.

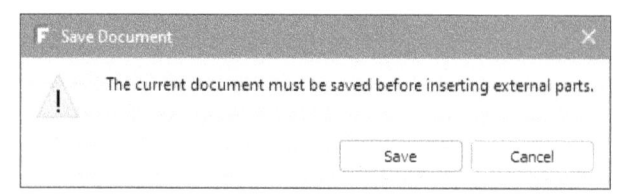

Figure-9. Save Document dialog box

- Click on **Save** button. The **Save FreeCAD Document** dialog box will be displayed; refer to Figure-10.

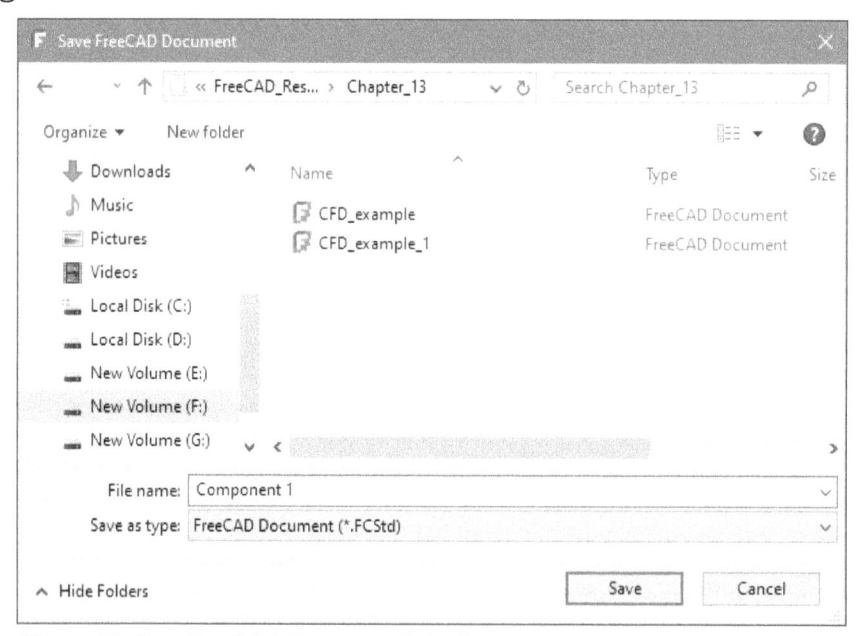

Figure-10. Save FreeCAD Document dialog box

- Specify desired name of file in the **File name** edit box and click on the **Save** button. The file will be saved and the **Ground Part?** dialog box will be displayed asking you to ground the first inserted part automatically; refer to Figure-11.

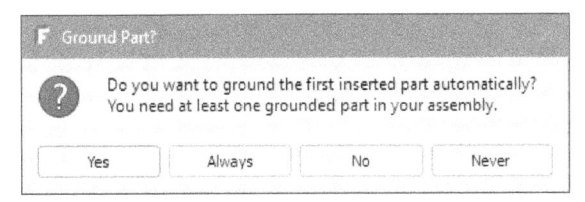

Figure-11. Ground Part dialog box

- Click on **Yes** button, the component will be grounded.
- Click on desired component(s) from the **Insert Component** dialog box to be inserted. The component(s) will be inserted. Click on the component multiple times in Insert Component dialog until you get desired number of component(s).
- Select **Show only parts** check box to display only parts in the list box.
- Select **Rigid sub-assemblies** check box to consider the inserted sub-assemblies as a solid.
- After specifying desired parameters, click on the **OK** button. The components will be inserted; refer to Figure-12.

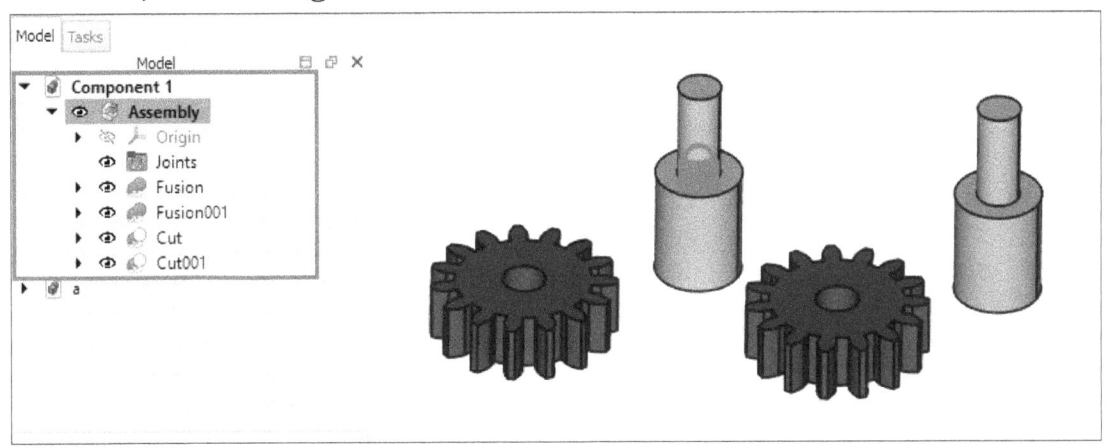

Figure-12. Components inserted

SOLVE ASSEMBLY

The **Solve Assembly** tool is used to solve the current active assembly based on applied constraints. For example, you may have applied fixed joint to a body but it will still be free to move until you click on the **Solve Assembly** tool to enforce the constraints.

TOGGLE GROUNDED

The **Toggle grounded** tool is used to ground/fix selected body/part permanently which lock its position in the assembly, preventing any movement or rotation.

CREATE A FIXED JOINT

The **Create a Fixed Joint** tool is used to create a joint locking two assembly parts together, preventing any movement or rotation but can be also used to define other types of joints. The procedure to use this tool is discussed next.

- Insert the parts in the graphics window as discussed earlier and click on the **Create a Fixed Joint** tool from **Toolbar** or **Assembly** menu; refer to Figure-13. The **Create Joint** dialog box will be displayed; refer to Figure-14.

Figure-13. Create a Fixed Joint tool

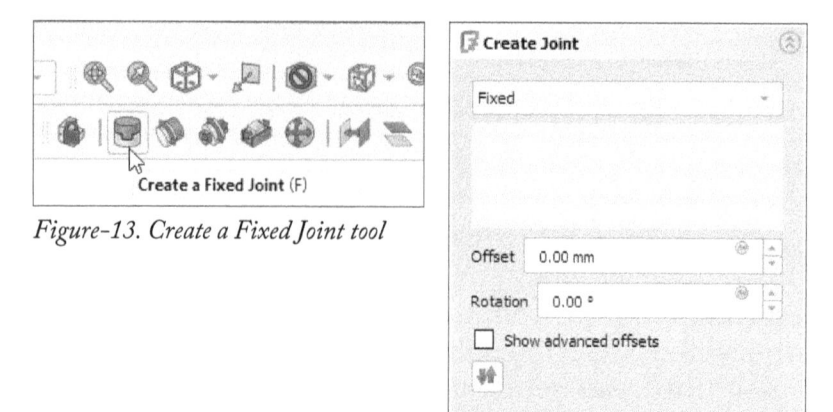

Figure-14. Create Joint dialog box

- Select two geometric entities of two different parts to be fixed from the graphics window. As you select the geometry entity of second part, the parts will be joint at their selected entities; refer to Figure-15.

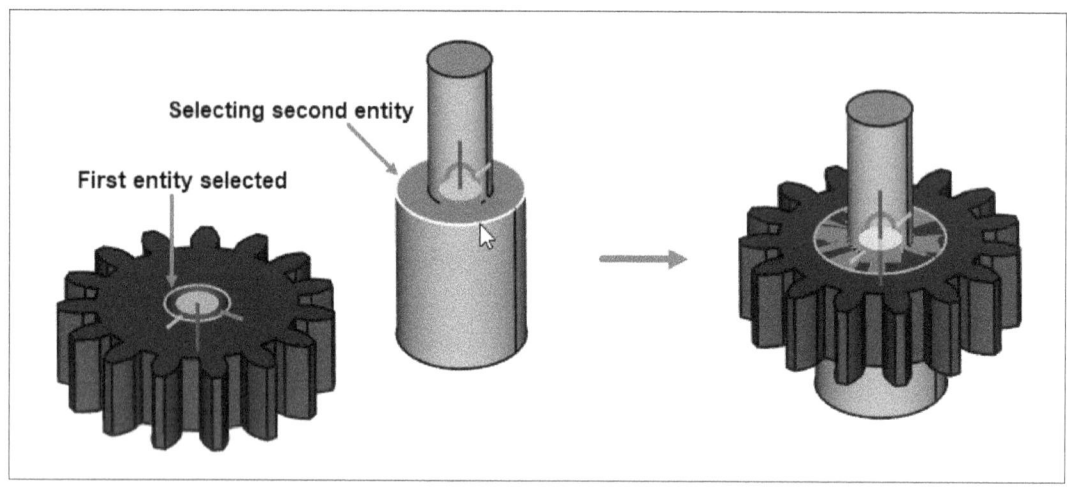

Figure-15. Creating the fixed joint

- Specify desired offset and rotation values in the **Offset** and **Rotation** edit boxes, respectively.
- Select **Show advanced offsets** check box, the **Offset1** and **Offset2** edit boxes will be displayed to specify the advanced offset parameters.
- Click on the **Offset1** edit box, the **Placement** dialog box will be displayed; refer to Figure-16.

Figure-16. Placement dialog box

- Specify desired parameters in the dialog box to place the component and click on the **OK** button from **Placement** dialog box.
- Click on the [icon] button to reverse the direction of the joint.
- After specifying desired parameters, click on **OK** button from the **Create Joint** dialog box. The component will be placed at the specified location.

CREATE REVOLUTE JOINT

The **Create Revolute Joint** tool allow rotation around a single axis between two selected parts. The procedure to use this tool is discussed next.

- Insert the parts in the graphics window as discussed earlier and click on the **Create Revolute Joint** tool from **Toolbar** or **Assembly** menu or from **Joint Type** drop-down in the **Create Joint** dialog box as discussed earlier. The **Revolute** parameters will be displayed in the **Create Joint** dialog box; refer to Figure-17.

Figure-17. Revolute parameters in Create Joint dialog box

- Select two geometric entities of two different parts to be fixed from the graphics window. As you select the geometry entity of second part, the parts will be joint at their selected entities; refer to Figure-18.

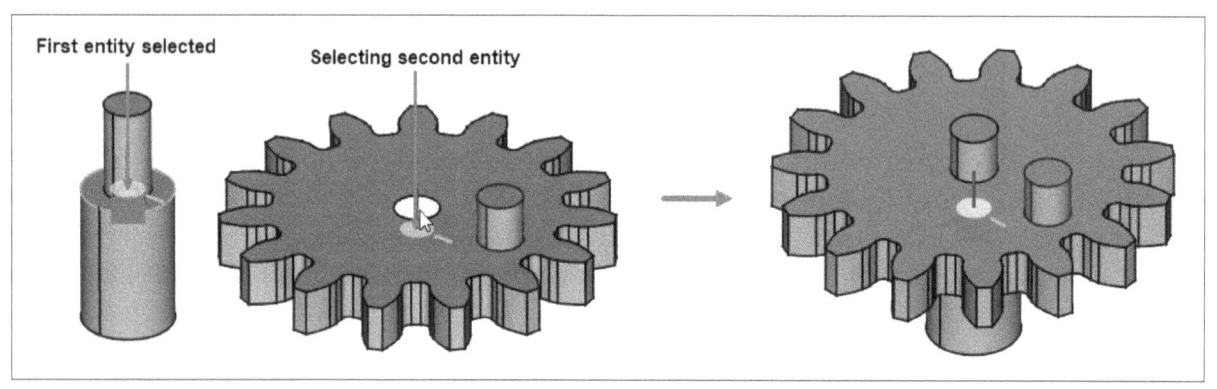

Figure-18. Applying Revolute Joint

- Select **Min angle** and **Max angle** check boxes from **Limits** area of the dialog box and specify limit for minimum and maximum angle of revolution in their respective edit boxes.
- Other parameters in the dialog box are same as discussed earlier.
- After specifying desired parameters, click on **OK** button from the **Tasks Manager**.

CREATE CYLINDRICAL JOINT

The **Create Cylindrical Joint** tool is used to create cylindrical joint between two selected parts, allowing rotation around a single axis and also a movement along the same axis. The procedure to use this tool is same as discussed in previous tool.

CREATE SLIDER JOINT

The **Create Slider Joint** tool is used to create a slider joint between two selected parts, allowing a linear movement along a single axis while restricting rotation. The procedure to use this tool is discussed next.

- Insert the parts in the graphics window as discussed earlier and click on the **Create Slider Joint** tool from **Toolbar** or **Assembly** menu or from **Joint Type** drop-down in the **Create Joint** dialog box as discussed earlier. The **Slider** parameters will be displayed in the **Create Joint** dialog box; refer to Figure-19.

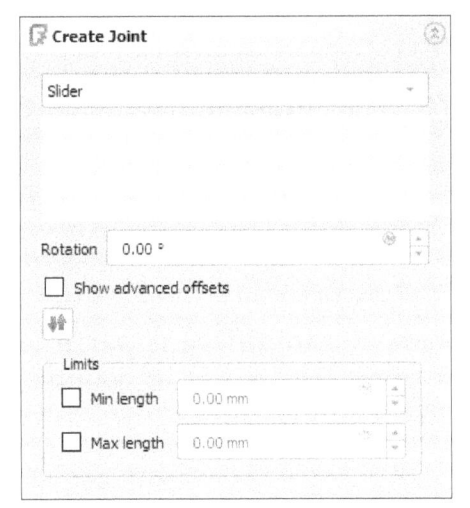

Figure-19. Slider parameters in Create Joint dialog box

- Select two geometric entities of two different parts to be fixed from the graphics window. As you select the geometry entity of second part, the parts will be joint at their selected entities; refer to Figure-20.

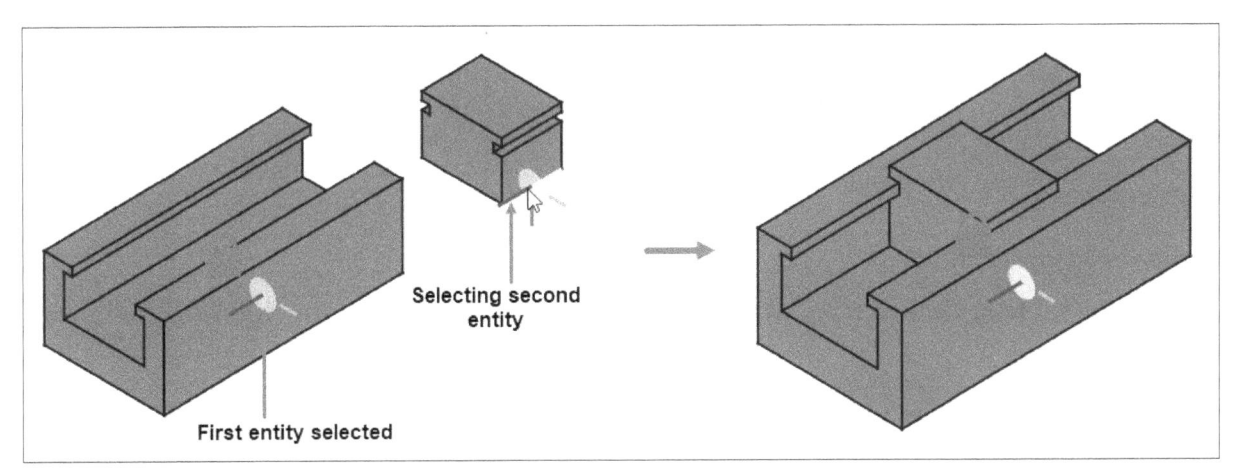

Figure-20. Applying slider joint

- Specify desired angle of rotation to rotate the component while applying joint in the **Rotation** edit box.
- Select **Min length** and **Max length** check boxes from **Limits** area of the dialog box and specify limit for minimum and maximum length of sliding in their respective edit boxes.
- Other parameters in the dialog box is same as discussed earlier.
- After specifying desired parameters, click on **OK** button from the **Tasks Manager**.

CREATE BALL JOINT

The **Create Ball Joint** tool is used to create a ball joint between two selected parts at a single point, allowing free rotation around the point while keeping both parts connected at this point. The procedure to use this tool is discussed next.

- Insert the parts in the graphics window as discussed earlier and click on the **Create Ball Joint** tool from **Toolbar** or **Assembly** menu or from **Joint Type** drop-down in the **Create Joint** dialog box as discussed earlier. The **Ball** joint parameters will be displayed in the **Create Joint** dialog box; refer to Figure-21.

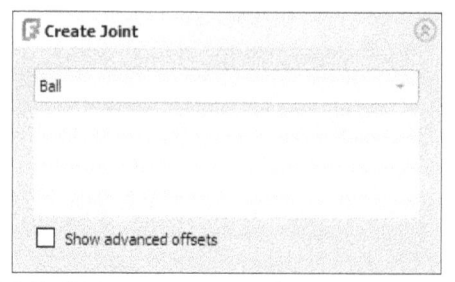

Figure-21. Ball joint parameters in Create Joint dialog box

- Select two geometric entities of two different parts to be fixed from the graphics window. As you select the geometry entity of second part, the parts will be joint at their selected entities; refer to Figure-22.

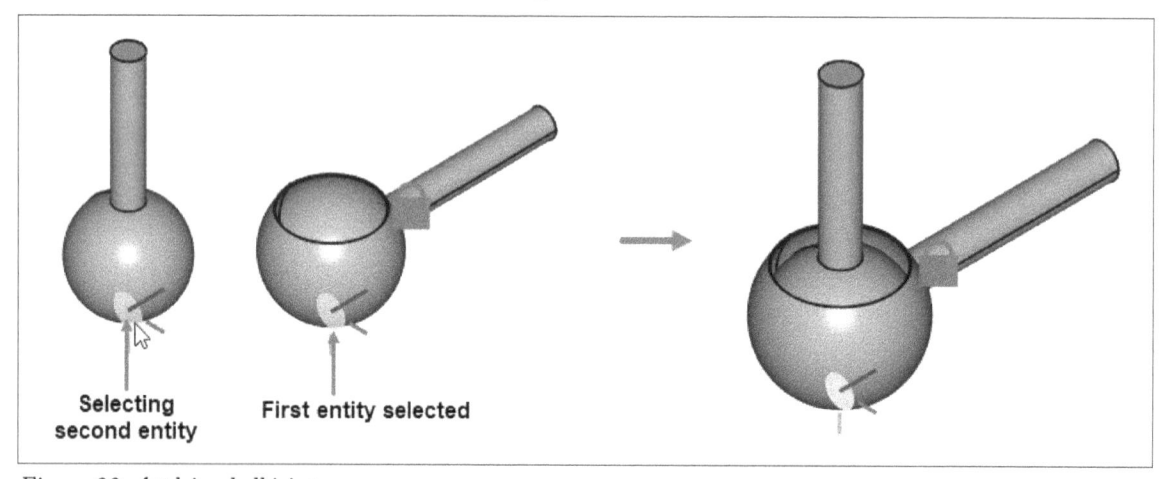

Figure-22. Applying ball joint

- After specifying desired parameters, click on **OK** button from the **Tasks Manager**.

CREATE DISTANCE JOINT

The **Create Distance Joint** tool is used to create a distance joint between two selected parts, fixing the distance between both parts. The procedure to use this tool is discussed next.

- Insert the parts in the graphics window as discussed earlier and click on the **Create Distance Joint** tool from **Toolbar** or **Assembly** menu or from **Joint Type** drop-down in the **Create Joint** dialog box as discussed earlier. The **Distance** joint parameters will be displayed in the **Create Joint** dialog box; refer to Figure-23.

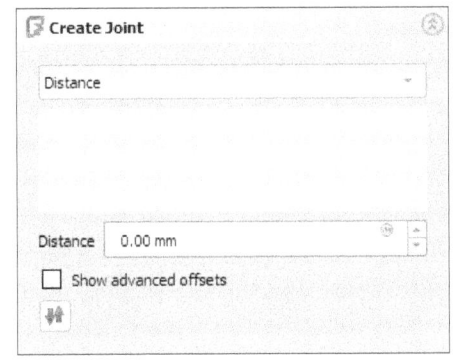

Figure-23. Distance joint parameters in Create Joint dialog box

- Select two geometric entities of two different parts to be fixed from the graphics window. As you select the geometry entity of second part, the parts will be joint at their selected entities; refer to Figure-24.

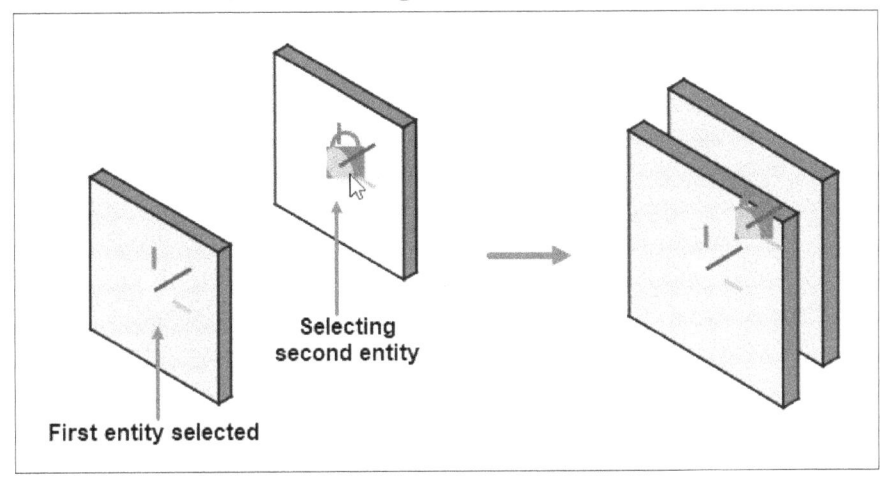

Figure-24. Applying distance joint

- Enter desired value of distance in the **Distance** edit box to specify the distance between the parts.
- Other parameters in the dialog box is same as discussed earlier.
- After specifying desired parameters, click on **OK** button from the **Tasks Manager**.

Similarly, you can create **Parallel**, **Perpendicular**, and **Angle** joints.

CREATE RACK AND PINION JOINT

The **Create Rack and Pinion Joint** tool is used to create a rack and pinion joint that couples the translation of a part of a slider joint and the rotation of a part of a revolute joint. The procedure to use this tool is discussed next.

- Insert the parts in the graphics window as discussed earlier; refer to Figure-25.

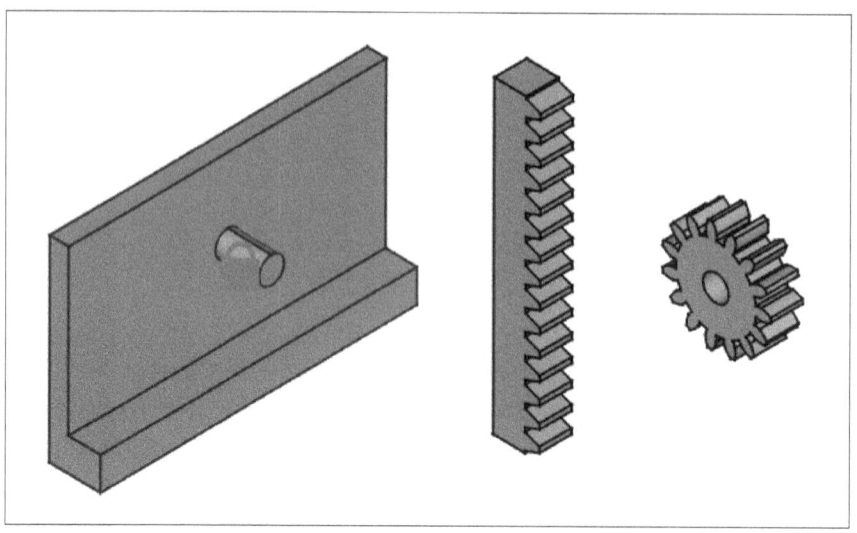

Figure-25. Parts inserted for rack and pinion joint

- First, apply the slider joint between base and rack and specify the parameters in the dialog box as shown in Figure-26.

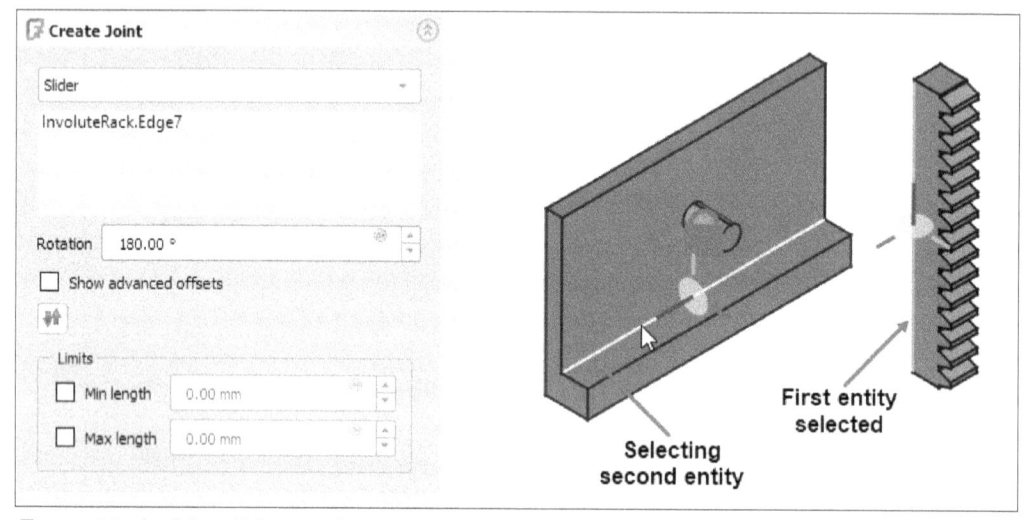

Figure-26. Applying slider joint between rack and base

- Now, apply revolute joint between base and pinion as shown in Figure-27.

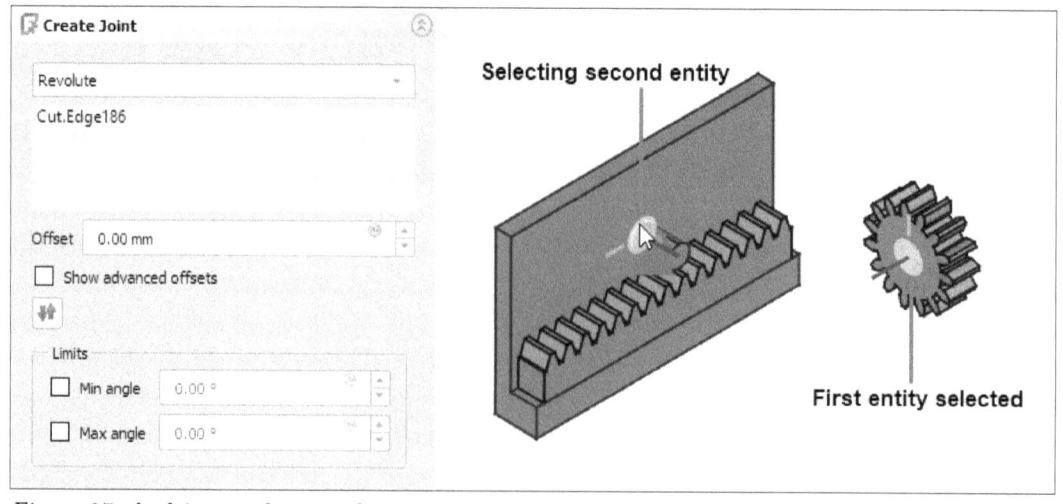

Figure-27. Applying revolute joint between base and pinion

- Now, click on the **Create Rack and Pinion Joint** tool from **Toolbar** or **Assembly** menu or from **Joint Type** drop-down in the **Create Joint** dialog box as discussed earlier. The **RackPinion** joint parameters will be displayed in the **Create Joint** dialog box; refer to Figure-28.

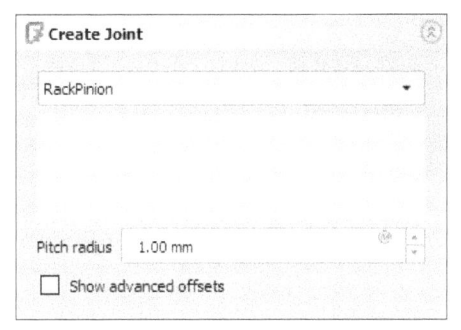

Figure-28. RackPinion parameters in Create Joint dialog box

- Apply the rackpinion joint between rack and pinion and specify the parameters in the dialog box as shown in Figure-29. As you select the geometry entity of the rack, the parts will be joint at their selected entities.

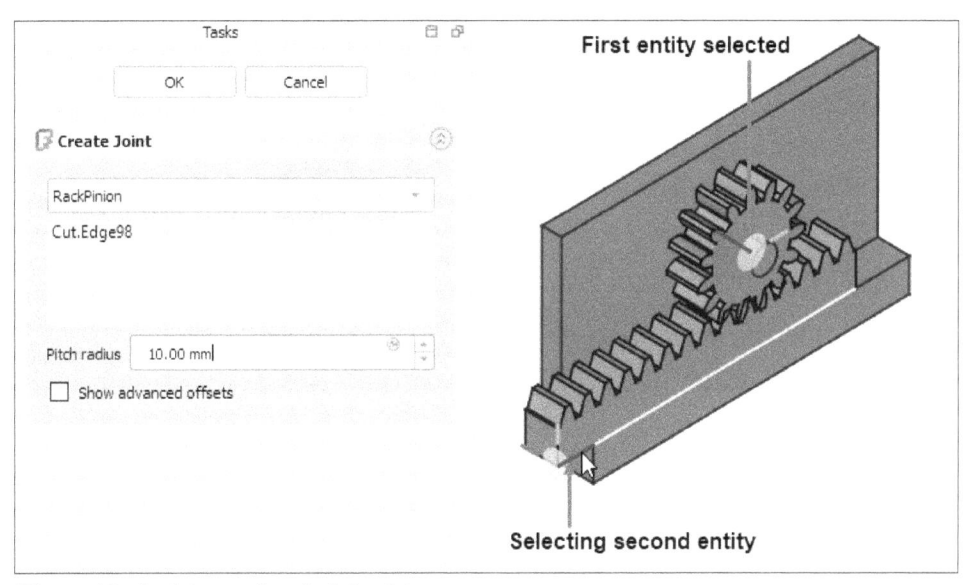

Figure-29. Applying rack and pinion joint

- Enter desired pitch radius in the **Pitch radius** edit box to specify the pitch circle of the pinion.
- Other parameters in the dialog box is same as discussed earlier.
- After specifying desired parameters, click on **OK** button from the **Tasks Manager**.

CREATE SCREW JOINT

The **Create Screw Joint** tool is used to create a screw joint that couples the translation of a part of a slider joint and the rotation of a part of a revolute joint. The procedure to use this tool is discussed next.

- Insert the parts in the graphics window as discussed earlier and apply fixed joint between parts and sketch; refer to Figure-30.

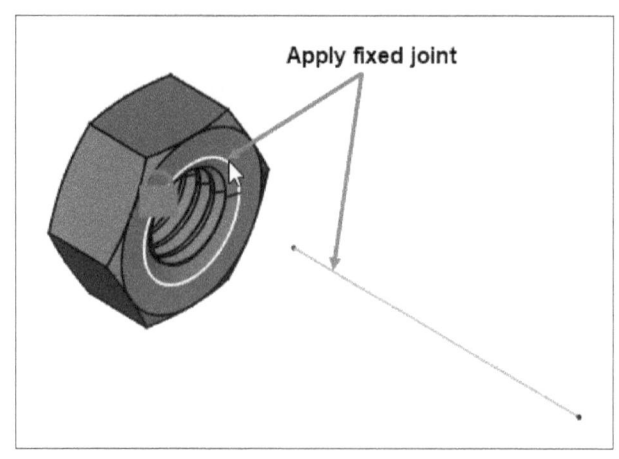

Figure-30. Applying fixed joint

- Then, apply slider joint between the sketches as shown in Figure-31.

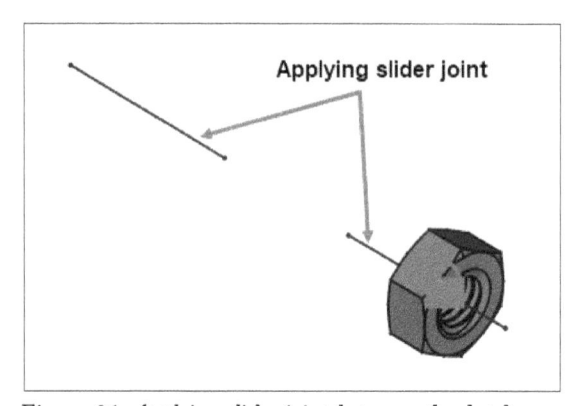

Figure-31. Applying slider joint between the sketches

- Then, apply revolute joint between screw and a sketch as shown in Figure-32.

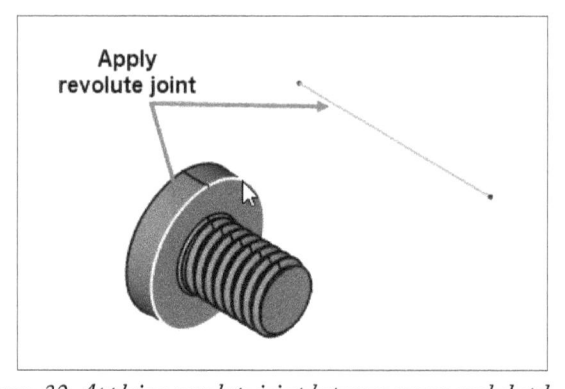

Figure-32. Applying revolute joint between screw and sketch

- Now, click on the **Create Screw Joint** tool from **Toolbar** or **Assembly** menu or from **Joint Type** drop-down in the **Create Joint** dialog box as discussed earlier. The **Screw** joint parameters will be displayed in the **Create Joint** dialog box.
- Apply the screw joint between the parts and specify the parameters in the dialog box as shown in Figure-33. As you select the geometry entity of second part, the parts will be joint at their selected entities.

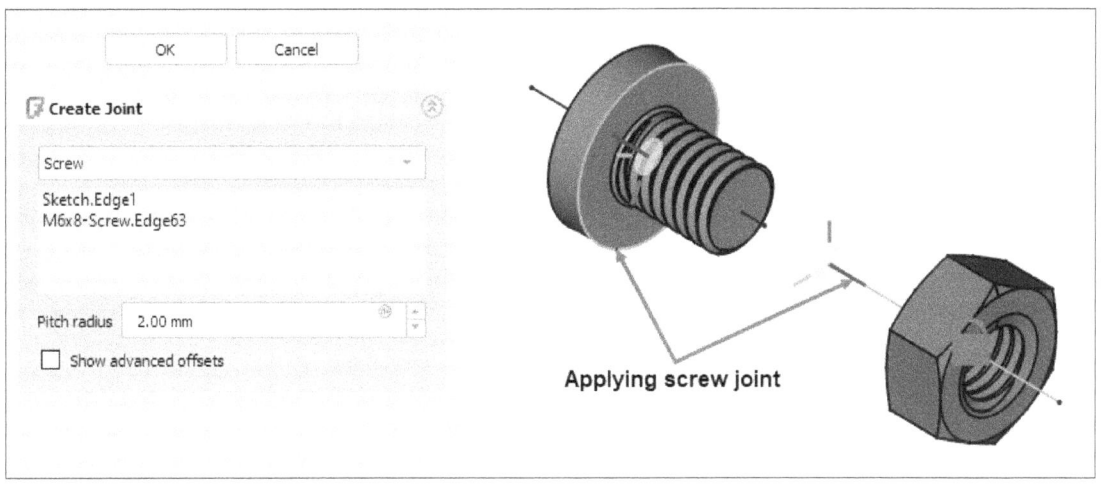

Figure-33. Applying screw joint

- Enter desired pitch radius in the **Pitch radius** edit box to specify the pitch of a screw.
- Other parameters in the dialog box is same as discussed earlier.
- After specifying desired parameters, click on **OK** button from the **Tasks Manager**.

CREATE GEARS JOINT

The **Create Gears Joint** tool is used to create a gears joint that couples the rotation of two parts of two different revolute joints. The procedure to use this tool is discussed next.

- Insert the parts in the graphics window as discussed earlier and apply revolute joint between parts and sketch; refer to Figure-34. The parts will be joint.

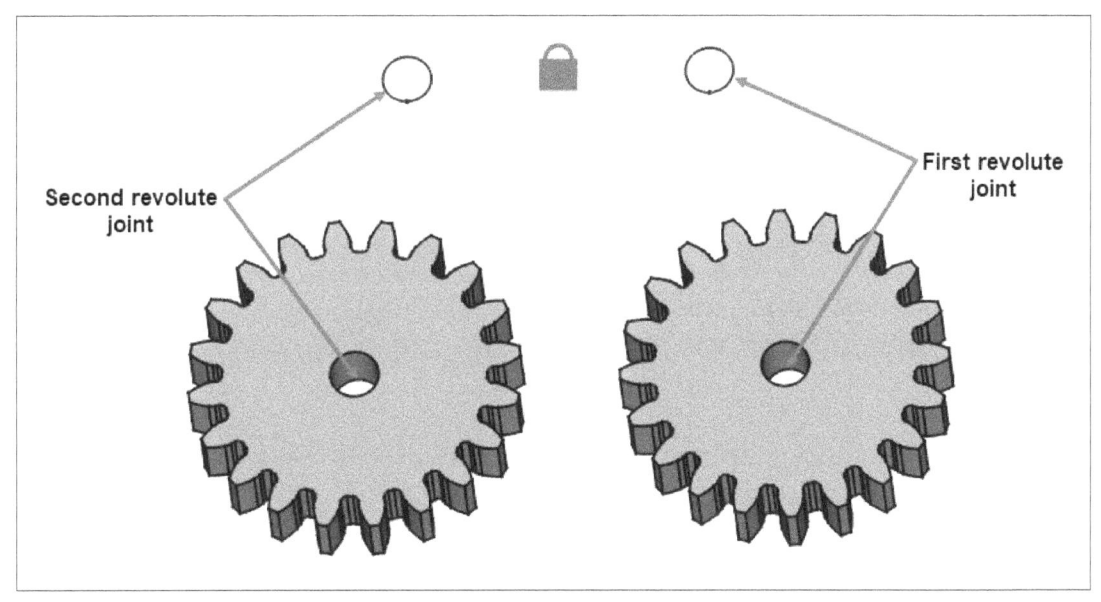

Figure-34. Creating revolute joints

- Now, click on the **Create Gears Joint** tool from **Toolbar** or **Assembly** menu or from **Joint Type** drop-down in the **Create Joint** dialog box as discussed earlier. The **Gears** joint parameters will be displayed in the **Create Joint** dialog box; refer to Figure-35.

Figure-35. Gears joint parameters in Create Joint dialog box

- Apply the gears joint between the parts and specify the parameters in the dialog box as shown in Figure-36. As you select the geometry entity of second part, the parts will be joint at their selected entities.

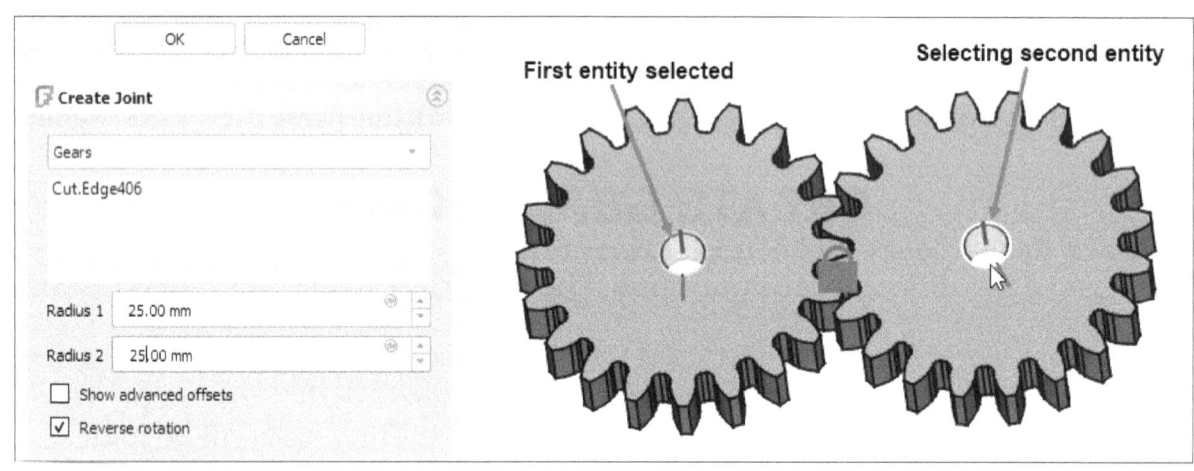

Figure-36. Applying gears joint

- Enter desired radius values in the **Radius 1** and **Radius 2** edit boxes to specify the pitch circle of gears.
- Other parameters in the dialog box is same as discussed earlier.
- After specifying desired parameters, click on **OK** button from the **Tasks Manager**.

CREATE BELT JOINT

The **Create Belt Joint** tool is used to create a belt joint that couples the rotation of two parts of two different revolute joints. The procedure to use this tool is discussed next.

- Insert the parts in the graphics window as discussed earlier and apply revolute joint between parts and sketch; refer to Figure-37. The parts will be joint.

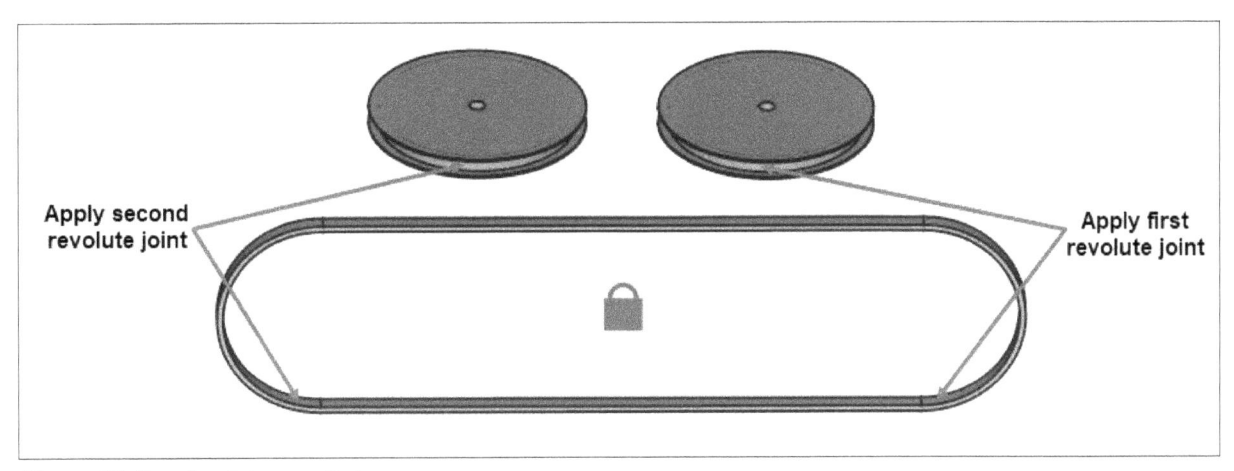

Figure-37. Revolute joints applied

- Now, click on the **Create Belt Joint** tool from **Toolbar** or **Assembly** menu or from **Joint Type** drop-down in the **Create Joint** dialog box as discussed earlier. The **Belt** joint parameters will be displayed in the **Create Joint** dialog box.
- Apply the belt joint between the parts and specify the parameters in the dialog box as shown in Figure-38. As you select the geometry entity of second part, the parts will be joint at their selected entities.

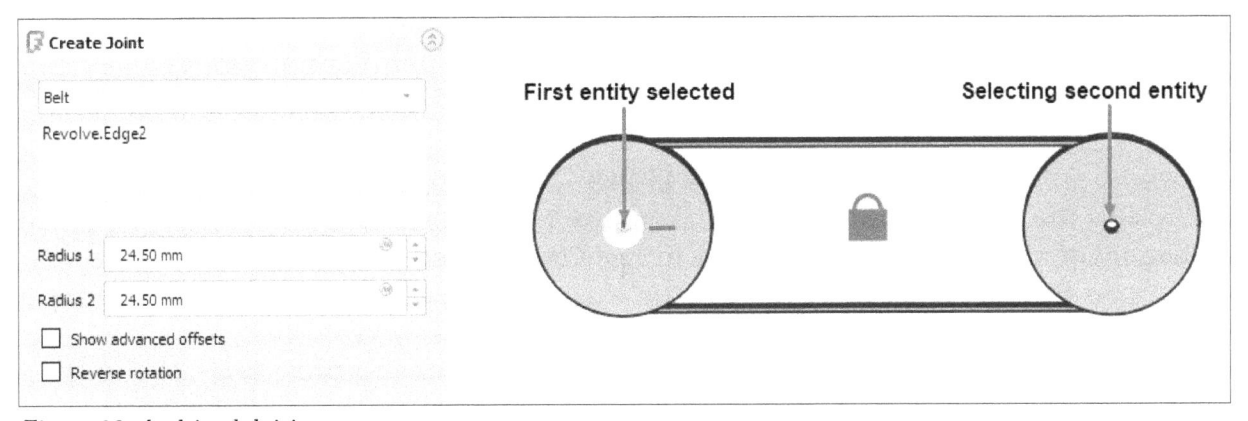

Figure-38. Applying belt joint

- Enter desired radius values in the **Radius 1** and **Radius 2** edit boxes to specify the pitch circle of gears.
- Other parameters in the dialog box is same as discussed earlier.
- After specifying desired parameters, click on **OK** button from the **Tasks Manager**.

PRACTICAL

Create the assembly of double-slider mechanism as shown in Figure-39.

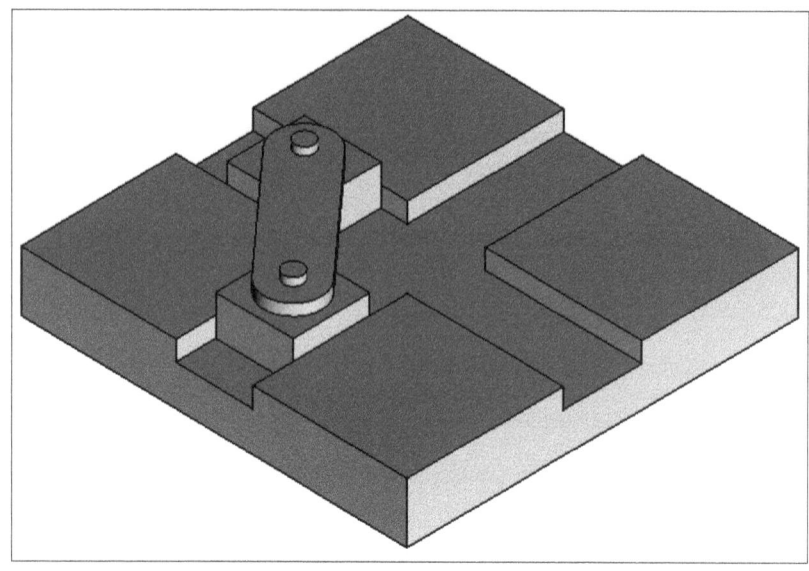

Figure-39. Assembly Practical

Starting A New Assembly

- Start FreeCAD, if not started yet. Click on the **New** tool from the **File** menu. A new document will open.
- Select the `Assembly` option from the `Switch between workbenches` drop-down to change the environment to Assembly workbench.
- Click on the **Create Assembly** tool from the **Assembly** menu or toolbar. The assembly container will be added in the **Model** panel of **Combo View**.

Placing the Base Component

- Click on the **Insert Component** tool from the **Assembly** menu or toolbar. The **Insert Component** dialog will be displayed in the **Combo View**.
- Click on the **Open file** button from the dialog. The **Select FreeCAD documents to import parts from** dialog will be displayed.
- Select the **Base** part from the resource folder for this chapter; refer to Figure-40 and click on the **Open** button. The part will be added in the dialog.

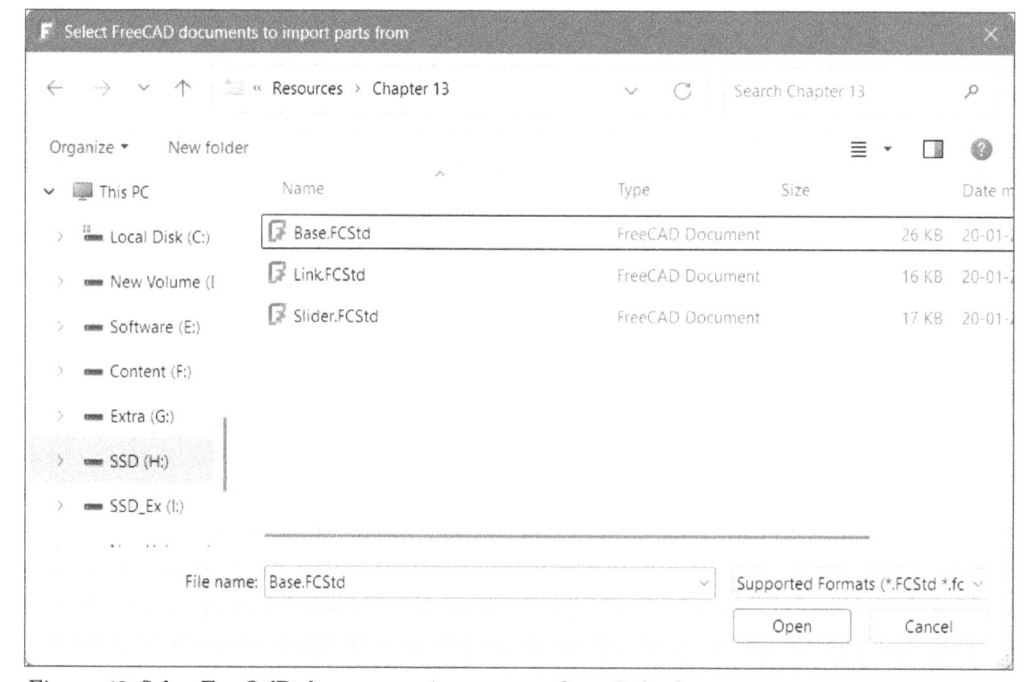

Figure-40. Select FreeCAD documents to import parts from dialog box

- Click on the Body feature of Base.FCStd part in the **Insert Component** dialog. If you have not saved the assembly file then a message box will be displayed asking you to save the file. Save the file by desired name. Preview of Base part will be displayed with the **Ground Part?** dialog box; refer to Figure-41.

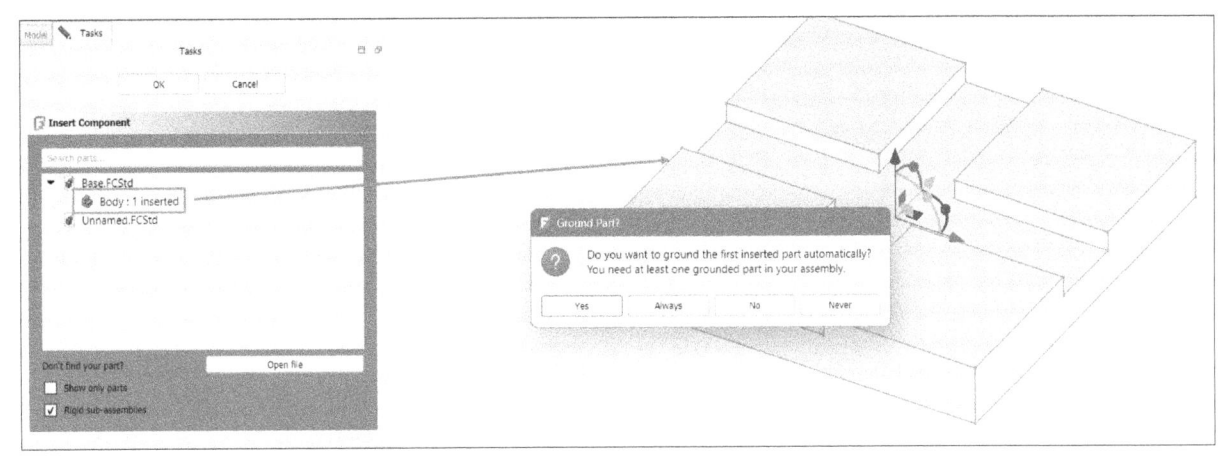

Figure-41. Base part with Ground Part dialog box

- Select the **Yes** button from the dialog to fix the base component at its current location. A lock sign will be displayed on the part in graphics area.

Inserting Sliders

- Click on the **Open file** button from the dialog again and double-click on Slider.FCStd part file from the dialog box. The component will be added in the **Insert Component** dialog.
- Select the Body feature for Slider.FCStd part from the dialog. The part will be added in the graphics area.
- Drag the Slider part out of Base part using drag handles; refer to Figure-42.

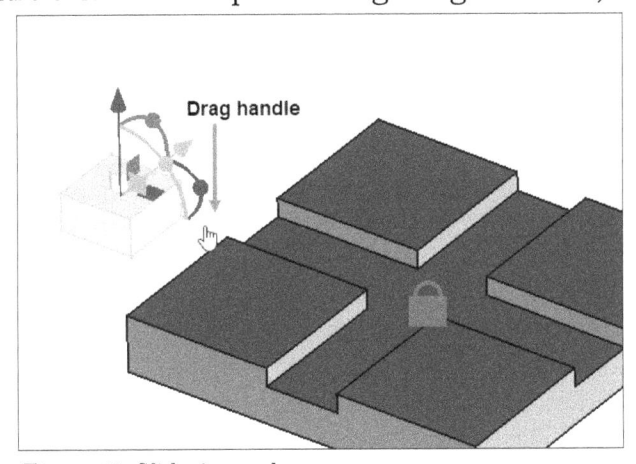

Figure-42. Slider inserted

- Click again on the Body feature of Slider.FCStd part to insert another Slider component in the graphics area and move it to open area. Click on the **OK** button from the dialog.
- Click on the **Create Slider Joint** tool from the toolbar. The Slider options will be displayed in **Create Joint** dialog.
- Select the edge of Base component and Slider component as shown in Figure-43. The component will be constrained; refer to Figure-44. Click on the **OK** button to create the joint.

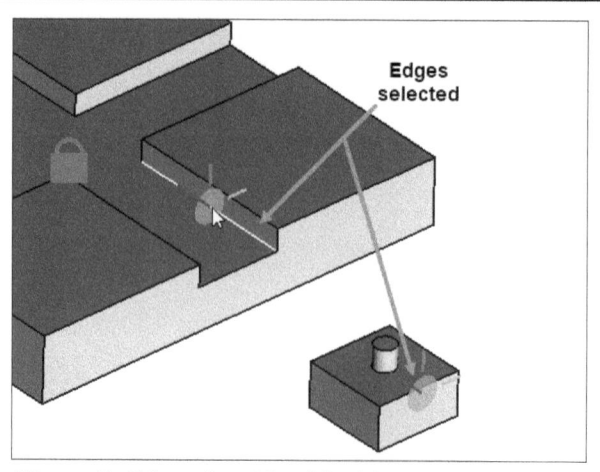

Figure-43. Edges selected for slider joint

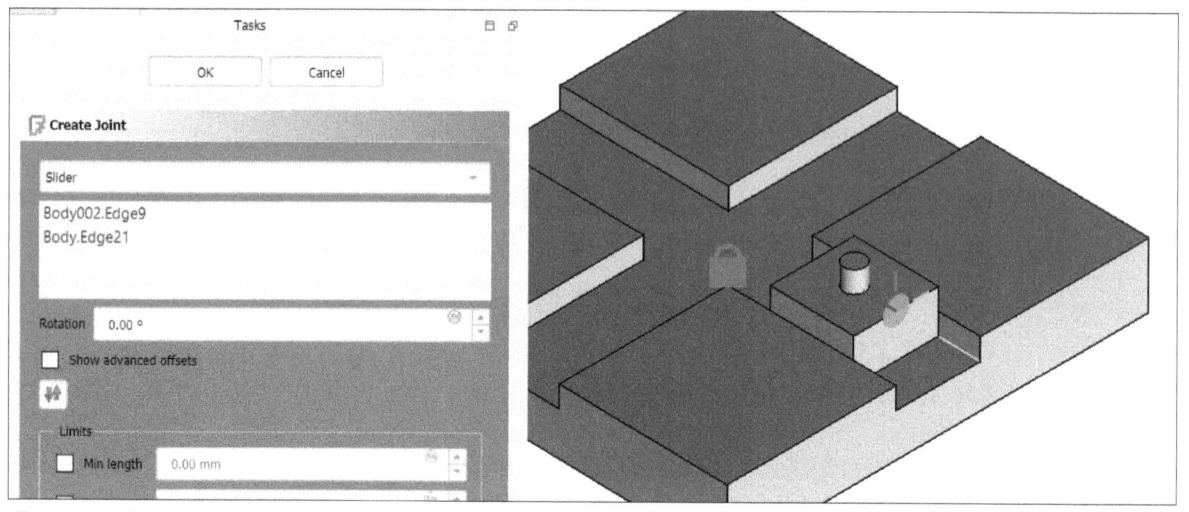

Figure-44. Preview of applied slider joint

- Similarly, apply the slider joint to another Slider component.

Inserting Link Part

- Click on the **Insert Component** tool from the toolbar and click on the **Open file** button from the dialog. The **Select FreeCAD documents to import parts from** dialog box will be displayed.
- Double-click on the **Link.FCStd** file from the resources folder in the dialog box. The part will be added in the **Insert Component** dialog.
- Select the Body feature from the Link.FCStd node in the **Insert Component** dialog. The Link part will be added in the graphics area.
- Using the drag handles move the component in empty area and click on the **OK** button from the dialog.
- Click on the **Create Cylindrical Joint** tool from the **Assembly** menu or toolbar. The **Create Joint** dialog will be displayed with options related to cylindrical joint.
- Select the round edges of Link part and Slider part as shown in Figure-45. The joint will be created.
- Click on the **OK** button from the dialog to apply the joint.

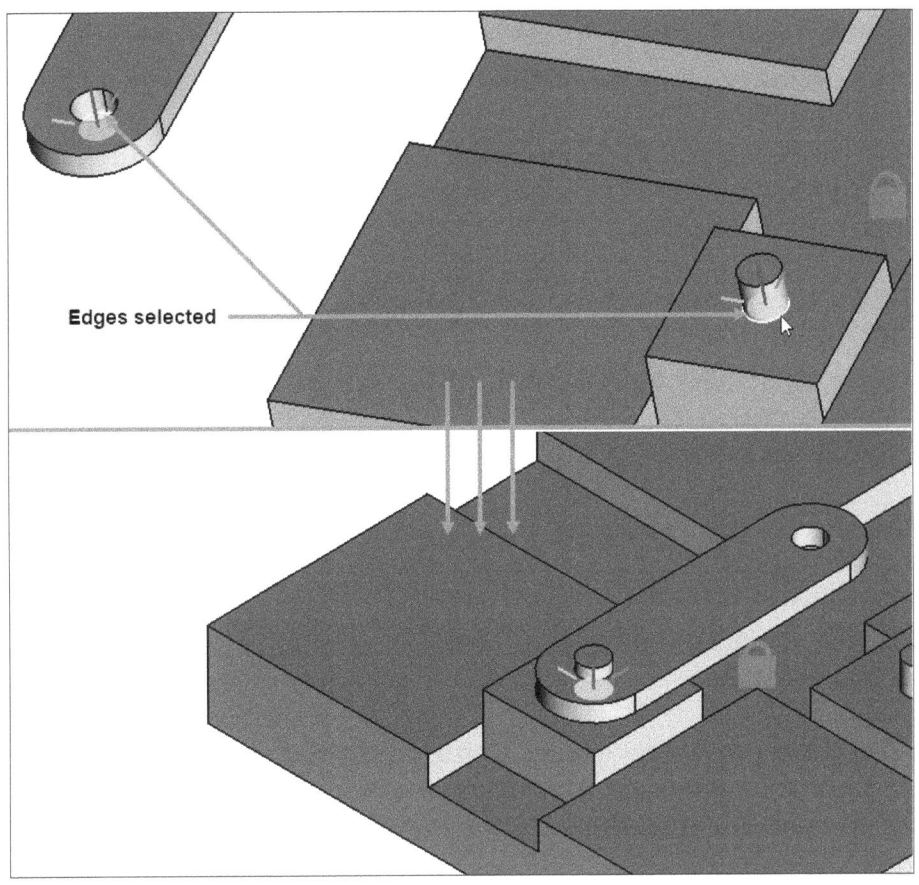

Figure-45. Edges selected for cylindrical joint

- Similarly, apply the cylindrical joint to the other end of Link part; refer to Figure-46.

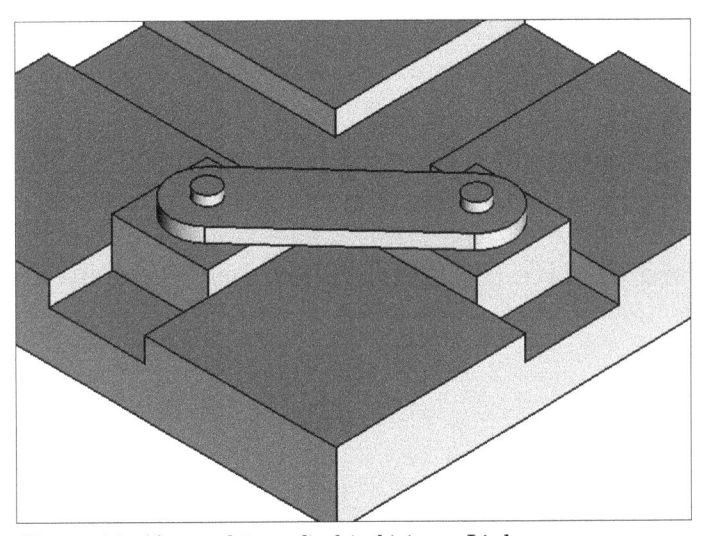

Figure-46. After applying cylindrical joint to Link part

You can animate the slider mechanism by dragging one of the Slider.

SELF ASSESSMENT

Q1. What is the primary purpose of the Assembly workbench in FreeCAD?
A. To create 3D models
B. To assemble different bodies from single or multiple documents
C. To render graphics
D. To apply textures to models

Q2. How can you access the Assembly workbench in FreeCAD?
A. From the File menu
B. By selecting the Assembly option from the Switch between workbenches drop-down
C. From the View menu
D. By using the Tools menu

Q3. What does the Create Assembly tool do in FreeCAD?
A. It creates a fixed joint
B. It solves the assembly
C. It creates a root assembly or sub-assembly
D. It imports components

Q4. What is the function of the Insert Component tool?
A. To create a joint
B. To insert a component into the active assembly
C. To solve the assembly
D. To ground a component

Q5. What dialog box is displayed when selecting a component to insert into an assembly?
A. Placement dialog box
B. Ground Part? dialog box
C. Save Document dialog box
D. Select FreeCAD documents to import parts from dialog box

Q6. Which tool is used to solve the active assembly based on applied constraints?
A. Toggle Grounded
B. Create Fixed Joint
C. Solve Assembly
D. Insert Component

Q7. What does the Toggle Grounded tool do?
A. It grounds or fixes the selected body/part permanently
B. It solves the assembly constraints
C. It creates a fixed joint
D. It moves components

Q8. What is the function of the Create a Fixed Joint tool?
A. To define rotation limits for a component
B. To lock two assembly parts together, preventing movement or rotation
C. To insert a new component
D. To create sub-assemblies

Q9. What does the Create Revolute Joint tool allow?
A. Linear movement along a single axis
B. Rotation around a single axis between two parts
C. Rotation around a single point
D. Fixed distance between two parts

Q10. Which joint allows linear movement along a single axis while restricting rotation?
A. Revolute Joint
B. Slider Joint
C. Ball Joint
D. Cylindrical Joint

Q11. What does the Create Ball Joint tool enable?
A. Rotation around a single axis
B. Movement along a single axis
C. Free rotation around a single point
D. Fixed distance between two parts

Q12. What is the primary purpose of the Create Distance Joint tool?
A. To fix the distance between two parts
B. To create a rack and pinion mechanism
C. To create cylindrical movement
D. To allow free rotation

Q13. What type of joint couples the translation of a slider joint with the rotation of a revolute joint?
A. Cylindrical Joint
B. Distance Joint
C. Screw Joint
D. Ball Joint

Q14. Which joint is used to synchronize the rotation of two parts with different radii?
A. Ball Joint
B. Gears Joint
C. Slider Joint
D. Belt Joint

Q15. What parameter is entered in the Pitch radius edit box when creating a Screw Joint?
A. Distance between parts
B. Radius of the gear
C. Pitch of the screw
D. Offset values

For Student Notes

Chapter 14

Introduction to Computational Fluid Dynamics

Topics Covered

The major topics covered in this chapter are:

- *Introduction to CFD*
- *Basic Properties of Fluids, Types of Fluids, and Thermodynamic Properties of Fluids*
- *Variations Of Navier-Strokes Equation*
- *Steps Of Computational Fluid Dynamics*
- *Introduction To Cfdof*
- *Activating Cfdof Workbench*
- *Installing Cfdof Related Software*
- *Performing Cfd On A Model*

INTRODUCTION

In this chapter, you will know various aspects of CFD for various practical problems. But, keep in mind that all computer software work on same concept of GIGO which means Garbage In - Garbage Out. So, if you have specified any wrong parameter while defining properties of analysis then you will not get the correct results. This problem demands a good knowledge of Fluid Mechanics so that you are well conversant with the terms of classical fluid mechanics and can relate the results to the theoretical concepts. In this chapter, we will discuss the basics of Fluid Mechanics and we will try to relate them with analysis wherever possible.

BASIC PROPERTIES OF FLUIDS

There are various basic properties required while performing analysis on fluid. These properties are collected by performing experiments in labs. Most of these properties are available in the form of tables in Steam Tables or Design Data books. These properties are explained next.

Mass Density, Weight Density, and Specific Gravity

- Density or Mass Density is the mass of fluid per unit volume. In SI units, mass is measured in kg and volume is measured in m³. So, mathematically we can say,

$$\text{Density (or Mass Density) } \rho = \frac{Mass\ of\ Fluid}{Volume\ Occupied\ by\ Fluid} \text{ kg/m}^3$$

- If you are asked for weight density then multiply mass by gravity coefficient. Mathematically it can be expressed as:

$$\text{Weight Density } w = \frac{Mass\ of\ Fluid \times Gravity\ Coefficient}{Volume\ Occupied\ by\ Fluid} \text{ N/m}^3$$

- Most of the time, fluid density is available as **Specific Gravity**. Specific gravity is the ratio of weight density of fluid to weight density of water in case of liquid. In case of gases, it is the ratio of weight density of fluid to weight density of air. Note that weight density of water is 1000 kg/m³ at 4 °C and weight density of air is 1.225 kg/m³ at 15 °C.

Note that as the temperature of liquid rises, its density is reduced and vice-versa. But as the temperature of gas rises, its density is increased and vice-versa.

Viscosity

Viscosity is the coefficient of friction between different layers of fluid. In other terms, it is the shear stress required to produce unit rate of shear strain in one layer of fluid. Mathematically it can be expressed as:

$$\mu = \frac{\tau}{\left(\frac{dx}{dy}\right)} \text{ N.s/m}^2 \text{ or Pa.s}$$

Where, μ is viscosity, τ is shear stress (or force applied tangentially to the layer of fluid) and (dx/dy) is the shear strain.

As the density of fluid changes with temperature so does the viscosity. The formula for viscosity of fluid at different temperature is given next.

For Liquids, $$\mu = \mu_0 \frac{1}{1 + \alpha t + \beta t^2}$$

For Gases, $$\mu = \mu_0 + \alpha t - \beta t^2$$

here, μ_0 is viscosity at 0 °C
α and β are constants for fluid (for water α is 0.03368 and β is 0.000221)
(for air α is 5.6x10⁻⁸ and β is 1.189x10⁻¹⁰)
t is the temperature

PROBLEM ON VISCOSITY

Dynamic viscosity of lubricant oil used between shaft and sleeve is 8 poise. The shaft has a diameter of 0.4 m and rotates at 250 r.p.m. Find out the power lost due to viscosity of fluid if length of sleeve is 100 mm and thickness of oil film is 1.5 mm; refer to below figure.

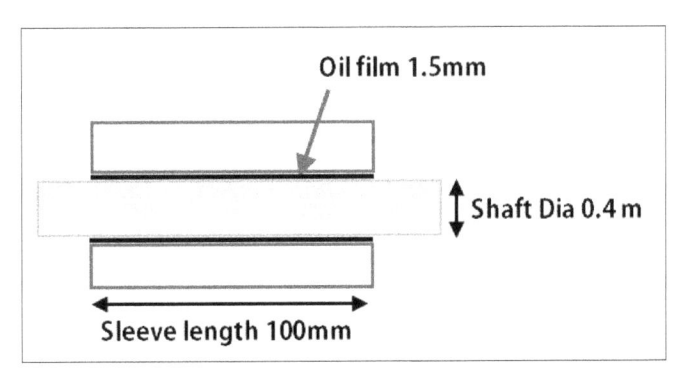

Solution:

Viscosity μ = 8 poise = 8/10 N.s/m² = 0.8 N.s/m²

Tangential velocity of shaft $u = \frac{\pi \times D \times N}{60} = \frac{\pi \times 0.4 \times 250}{60}$ = 5.236 m/s

Using the relation, $\tau = \mu \frac{dx}{dy}$

where dx is 5.236
and dy is 1.5x10⁻³

$$\tau = 0.8 \times \frac{5.236}{1.5 \times 10^{-3}} = 2792.53 \text{ N/m}^2$$

Shear force F = τ x Area

$$F = \tau \times \pi D \times L = 2792.53 \times \pi \times 0.4 \times 100 \times 10^{-3} = 350.92 \text{ N}$$

Torque (T) = Force x Radius = 350.92 x 0.2 = 70.184 N.m

Power = 2 π.N.T/60 = (2 π x 250 x 70.184)/60 = 1837.41 W Ans.

Now, you may ask how this problem relates with CFD. As discussed earlier, the viscosity changes with temperature and as fluid flows through pipe or comes in contact with rolling shaft, its temperature rises. In such cases, CFD gives the approximate viscosity and temperature of fluids in the system at different locations. This data later can be used to find solution for other engineering problems.

TYPES OF FLUIDS

There are mainly 5 types of fluids:

Ideal Fluids: These fluids are incompressible and have no viscosity which means they flow freely without any resistance. This category of fluid is imaginary and used in some cases of calculations.

Real Fluids: These are the fluids found in real world. These fluids have viscosity values as per their nature and can be compressible in some cases.

Newtonian Fluids: Newtonian fluids are those in which shear stress is directly proportional to shear strain. In a specific temperature range, water, gasoline, alcohol etc. can be Newtonian fluids.

Non-Newtonian Fluids: Those fluids in which shear stress is not directly proportional to shear strain. Most of the time Real Fluids fall in this category.

Ideal Plastic Fluids: Those fluids in which shear stress is more than yield value and so fluid deforms plastically. The shear stress in these fluids is directly proportional to shear strain.

THERMODYNAMIC PROPERTIES OF FLUID

Most of the liquids are not considered as compressible in general applications as their molecules are already bound closely to each other. But, Gas have large gap between their molecules and can be compressed easily relative to liquids. As we pick pressure to compress the gas, other thermodynamic properties also come into play. The relationship between Pressure, Temperature, and specific Volume is given by;

$$P.\forall = RT$$

P = Absolute pressure of a gas in N/m²
$\forall$ = Specific Volume = 1/ρ
R = Gas Constant (for Air is 287 J/Kg-K
T = Absolute Temperature
ρ = Density of gas

If the density of gas changes with constant temperature then the process is called Isothermal process and if density changes with no heat transfer then the process is called Adiabatic process.

For Isothermal process, $p/ρ$ = Constant
For Adiabatic process, $p/ρ^k$ = Constant

Here, k is Ratio of specific heat of a gas at constant pressure and constant volume (1.4 for air).

Universal Gas Constant

By Pressure, Temperature, volume equation,

$$p.\forall = nMRT$$

Here,

p = Absolute pressure of a gas in N/m²
$\forall$ = Specific Volume = 1/ρ
n = Number of moles in a Volume of gas
M = Mass of gas molecules/ Mass of Hydrogen atom = n x m (m is mass gas in kg)
R = Gas Constant (for Air is 287 J/Kg-K
T = Absolute Temperature

MxR is called Universal Gas constant and is equal to 8314 J/kg-mole K for water.

Compressibility of Gases

Compressibility is reciprocal of bulk modulus of elasticity K, which is defined as ratio of Compressive stress to volumetric strain.

Bulk Modulus = Increase in pressure/ Volumetric strain

$$K = -(dp/d\forall)x\forall$$

Vapour Pressure and Cavitation

When a liquid converts into vapour due to high temperature in a vessel then vapours exert pressure on the walls of vessel. This pressure is called **Vapour pressure**.

When a liquid flows through pipe, sometimes bubbles are formed in the flow. When these bubbles collapse at the adjoining boundaries then they erode the surface of tube due to high pressure burst of bubble. This erosion is in the form of cavities at the surface of tube and the phenomena is called **Cavitation**.

PASCAL'S LAW

Pascal's Law states that pressure at a point in static fluid is same in all directions. In mathematical form $p_x=p_y=p_z$ in case of static fluids.

FLUID DYNAMICS

Up to this point, the rules stated in this chapter were for static fluid that is fluid at rest. Now, we will discuss the rules for flowing fluid.

Bernoulli's Incompressible Fluid Equation

Bernoulli's equation states that the total energy stored in fluid is always same in a closed system. In the language of mathematics,

$$\boxed{p + \frac{1}{2}\rho V^2 + \rho g h = constant}$$

Here, p is pressure
    ρ is density
    V is velocity of fluid
    h is height of fluid
    g is gravitational acceleration

Eulerian and Lagrangian Method of Analysis

There are two different points of view in analyzing problems in fluid mechanics. The first view, appropriate to fluid mechanics, is concerned with the field of flow and is called the eulerian method of description. In the eulerian method, we compute the pressure field p(x, y, z, t) of the flow pattern, not the pressure changes p(t) that a particle experiences as it moves through the field.

The second method, which follows an individual particle moving through the flow, is called the lagrangian description. The lagrangian approach, which is more appropriate to solid mechanics, will not be treated in this book. However, certain numerical analyses of sharply bounded fluid flows, such as the motion of isolated fluid droplets, are very conveniently computed in lagrangian coordinates.

Fluid dynamic measurements are also suited to the eulerian system. For example, when a pressure probe is introduced into a laboratory flow, it is fixed at a specific position (x, y, z). Its output thus contributes to the description of the eulerian pressure field p(x, y, z, t). To simulate a lagrangian measurement, the probe would have to move downstream at the fluid particle speeds; this is sometimes done in oceanographic measurements, where flow meters drift along with the prevailing currents.

Now, we know the two methods of analyzing fluid mechanics problem. But, there are further three approaches for these two methods by which problems are derived to solution. These approaches are:

1. Control Volume
2. Differential
3. Experimental

Control volume analysis, is accurate for any flow distribution but is often based on average or "one dimensional" property values at the boundaries. It always gives useful "engineering" estimates. In principle, the differential equation approach can be applied to any problem. Only a few problems, such as straight pipe flow, yield to exact analytical solutions. But the differential equations can be modeled numerically, and the flourishing field of computational fluid dynamics (CFD) can now be used to give good estimates for almost any geometry. Finally, the dimensional analysis applies to any problem, whether analytical, numerical, or experimental. It is particularly useful to reduce the cost of experimentation.

Since the Differential Equation approach is more concerned to CFD so we will discuss this approach a little deeper.

DIFFERENTIAL APPROACH OF FLUID FLOW ANALYSIS

As discussed earlier, in this approach, the fluid is divided in to very small finite number of elements via a computer process called meshing. Various equations for different properties of a fluid are given next.

Acceleration

Acceleration **a** can be given as:

$$\mathbf{a} = \frac{d\mathbf{V}}{dt} = \mathbf{i}\frac{du}{dt} + \mathbf{j}\frac{dv}{dt} + \mathbf{k}\frac{dw}{dt}$$

Various components of acceleration **a** are:

$$a_x = \frac{du}{dt} = \frac{\partial u}{\partial t} + u\frac{\partial u}{\partial x} + v\frac{\partial u}{\partial y} + w\frac{\partial u}{\partial z} = \frac{\partial u}{\partial t} + (\mathbf{V} \cdot \nabla)u$$

$$a_y = \frac{dv}{dt} = \frac{\partial v}{\partial t} + u\frac{\partial v}{\partial x} + v\frac{\partial v}{\partial y} + w\frac{\partial v}{\partial z} = \frac{\partial v}{\partial t} + (\mathbf{V} \cdot \nabla)v$$

$$a_z = \frac{dw}{dt} = \frac{\partial w}{\partial t} + u\frac{\partial w}{\partial x} + v\frac{\partial w}{\partial y} + w\frac{\partial w}{\partial z} = \frac{\partial w}{\partial t} + (\mathbf{V} \cdot \nabla)w$$

Summing these into a vector, we obtain the total acceleration as

$$a = \frac{dV}{dt} = \frac{\partial V}{\partial t} + \left(u\frac{\partial V}{\partial x} + v\frac{\partial V}{\partial y} + w\frac{\partial V}{\partial z}\right) = \frac{\partial V}{\partial t} + (V.\nabla)V$$

Similarly, you can divide other parameters as vector like Force, Pressure, Temperature, and so on.

INTRODUCTION TO CFD

The CFD stands for Computational Fluid Dynamics. Computational Fluid Dynamics constitutes a new "third approach" in the philosophical study and development of the whole discipline of fluid dynamics. In the seventeenth century, the foundations for experimental fluid dynamics were laid in France and England. The eighteenth and nineteenth centuries saw the gradual development of theoretical fluid dynamics, again primarily in Europe. As a result, throughout most of the twentieth century the study and practice of fluid dynamics involved the use of theory on the one hand and pure experiment on the other hand.

However, to keep things in context, CFD provides a new third approach-but nothing more than that. It nicely and synergistically complements the other two approaches of pure theory and pure experiment, but it will never replace either of these approaches. There will always be a need for theory and experiment. The future advancement of fluid dynamics will rest upon a balance of all three approaches, with computational fluid dynamics helping to interpret and understand the results of theory, experiment and vice-versa; refer to Figure-1.

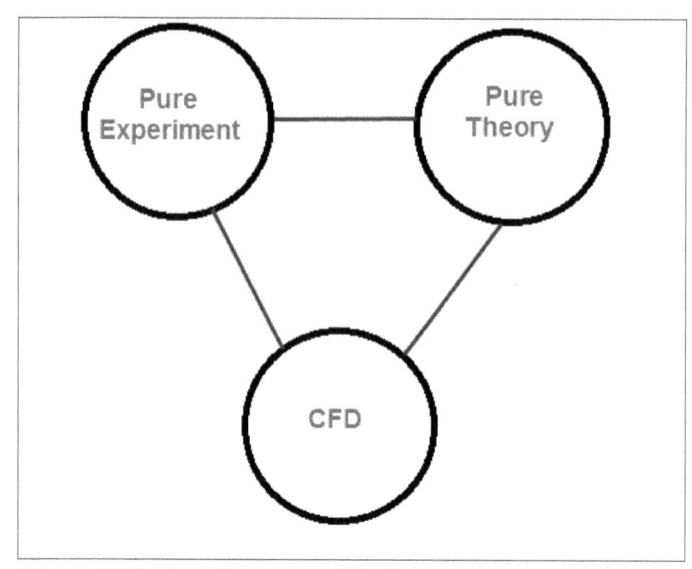

Figure-1. The three dimensions of fluid dynamics

Finally, we note that computational fluid dynamics is commonplace enough today that the composition CFD is universally accepted for the phrase "Computational Fluid Dynamics". We will use this composition throughout the chapter.

Computational fluid dynamics (CFD) is the use of applied math, physics, and computational software package to examine how a gas or liquid flows, also as how the gas or liquid affects objects as it flows past. Computational fluid dynamics is based on the Navier-Stokes equations. These equations describe how the velocity, pressure, temperature, and density of a moving fluid are related.

Navier-Stokes equation

The Navier-Stokes equations are the elemental partial differentials equations that describe the flow of incompressible fluids. Using the rate of stress and rate of strain, it can be shown that the parts of a viscous force F during a non rotating frame are given by-

$$\begin{aligned}\frac{F_i}{V} &= \frac{\partial}{\partial x_j}\left[\eta\left(\frac{\partial u_i}{\partial x_j}+\frac{\partial u_j}{\partial x_i}\right)+\lambda\delta_{ij}\nabla\cdot\mathbf{u}\right] \\ &= \frac{\partial}{\partial x_j}\left[\eta\left(\frac{\partial u_i}{\partial x_j}+\frac{\partial u_j}{\partial x_i}-\tfrac{2}{3}\delta_{ij}\nabla\cdot\mathbf{u}\right)+\mu_B\delta_{ij}\nabla\cdot\mathbf{u}\right],\end{aligned}$$

Where η is the dynamic viscosity, λ is the second viscosity coefficient, δ_{ij} is the Kronecker delta, $\nabla\cdot\mathbf{u}$ is the divergence, μ_B is the bulk viscosity and Einstein summation has been used to sum over j=1,2,and 3.

- **Dynamic Viscosity**

Dynamic viscosity is the force required by a fluid to overcome its own internal molecular friction so the fluid can flow. In other words, dynamic viscosity is defined as the tangential force per unit area required to move the fluid in one horizontal plane with reference to other plane with a unit velocity whereas the fluid's molecules maintain a unit distance apart.

A parameter η is defined as

$$[\text{Shear Stress}] = \eta [\text{Strain Rate}]$$

Written explicitly.

$$\sigma = \eta \dot{e} = \eta \frac{1}{l} \frac{dl}{dt} = \eta \frac{u}{l},$$

Where l is the length scale and u is the velocity scale. In cgs η has units of g cm^{-1} s^{-1}. Dynamic viscocity is related to kinematic viscocity ν by

$$\eta = \rho \nu$$

Where ρ is the density.

- **Second Viscosity Coefficient**

For a compressible fluid, i.e. one for which $\nabla \cdot \mathbf{u} = 0$, where $\nabla \cdot \mathbf{u}$ is the divergence Σ of the velocity field, the stress tensor of the fluid can be written

$$S_{ij} = \eta \left(\frac{\partial u_i}{\partial x_j} + \frac{\partial u_j}{\partial x_i} \right) + \lambda \delta_{ij} \nabla \cdot \mathbf{u},$$

Where δ_{ij} is the Kronecker delta, η is the dynamic viscosity, and λ is the second coefficient of viscosity. λ is analogous to the first Lame constant. For an incompressible fluid, the term involving λ drops out from the equation, so λ can be ignored.

- **Kronecker Delta**

The simplest interpretation of the Kronecker delta is as the discrete version of the delta function defined by

$$\delta_{ij} = \begin{cases} 1 & i = j \\ 0 & i \neq j \end{cases}$$

The Kronecker delta is implemented in the Wolfram language as KroneckerDelta[i, j], as well as in a generalized form KroneckerDelta[i, j, ...] that returns 1 if all arguments are equal and 0 otherwise.
It has the contour integral representation

$$\delta_{mn} = \frac{1}{2 \pi i} \oint_\gamma z^{m-n-1} \, dz,$$

Where γ is a contour corresponding to the unit circle and m and n are integers.

In three space, the Kronecker delta satisfies the identities

$$\delta_{ii} = 3$$
$$\delta_{ij}\epsilon_{ijk} = 0$$
$$\epsilon_{ipq}\epsilon_{jpq} = 2\delta_{ij}$$
$$\epsilon_{ijk}\epsilon_{pqk} = \delta_{ip}\delta_{jq} - \delta_{iq}\delta_{jp},$$

where Einstein summation is implicitly assumed. i,j=1,2,3, and ϵ_{ijk} is the permutation symbol.

Technically, the Kronecker delta is a tensor defined by the relationship

$$\delta_i^k \frac{\partial x_i'}{\partial x_k}\frac{\partial x_l}{\partial x_j'} = \frac{\partial x_i'}{\partial x_k}\frac{\partial x_k}{\partial x_j'} = \frac{\partial x_i'}{\partial x_j'}.$$

Since, by definition, the coordinates x_i and x_j are independent for i=j,

$$\frac{\partial x_i'}{\partial x_j'} = \delta''_j,$$

So

$$\delta''_j = \frac{\partial x_i'}{\partial x_k}\frac{\partial x_l}{\partial x_j'}\delta_l^k,$$

and δ''_j is really a mixed second-rank tensor. It satisfies

$$\delta_{ab}{}^{jk} = \epsilon_{abi}\epsilon^{jki}$$
$$= \delta_a^j\delta_b^k - \delta_a^k\delta_b^j$$
$$\delta_{abjk} = g_{aj}g_{bk} - g_{ak}g_{bj}$$
$$\epsilon_{aij}\epsilon^{bij} = \delta_{ai}{}^{bi}$$
$$= 2\delta_a^b.$$

- **Divergence**

The divergence of a vector field **F**, denoted div(**F**) or $\nabla \cdot \mathbf{F}$ (the notation used in this work), is defined by a limit of the surface integral

$$\nabla \cdot \mathbf{F} \equiv \lim_{V \to 0} \frac{\oint \mathbf{F} \cdot d\mathbf{a}}{V}$$

Where the surface integral provides the value of **F** integrated over a closed small boundary surface $S = \partial V$ surrounding a volume component V that is taken to size zero using a limiting method. The divergence of vector field is thus a scalar field. If $\nabla \cdot \mathbf{F} = 0$, then the filed is alleged to be a divergence less field. The symbol ∇ is referred to as nabla or del.
- **Bulk Viscosity**

The Bulk viscosity μ_B of a fluid is defined as

$$\mu_B = \lambda + \tfrac{2}{3}\mu,$$

Where λ is the second viscosity coefficient and μ is the shear viscosity.

- **Reynolds Number**

The Reynolds number for a flow through a pipe is defined as

$$\mathrm{Re} \equiv \frac{\rho \bar{u} d}{h} = \frac{\bar{u} d}{\nu},$$

Where ρ is the density of the fluid, **u** is the velocity scale, d is the pipe diameter and v is the kinematic viscosity of the fluid. Poiseuille (laminar) flow is experimentally found to occur for **Re<30**. At larger
Reynolds numbers flow become turbulent.

- **Density**

A measure of a substance's mass per unit of volume. For a substance with mass **m** and volume **V**,

$$\rho \equiv \frac{m}{V}.$$

For a body weight w_a placed in a fluid of weight w_w,

$$\rho = G_s \rho_w = \frac{w_a}{w_a - w_w} \rho_w,$$

Where G_s is the specific gravity for an ideal gas,

$$\rho = \frac{mP}{kT}.$$

- **Kinematic Viscosity**

A coefficient which describes the diffusion of momentum. Let η be the dynamic viscosity, then

$$\nu \equiv \frac{\eta}{\rho}.$$

The unit of kinematic viscosity is Stoke, equal to 1 cm² s⁻¹
V_{water} = 1.0 X 10⁻⁶ m² s⁻¹ = 0.010 cm² s⁻¹
V_{air} = 1.5 X 10⁻⁵ m² s⁻¹ = 0.15 cm² s⁻¹

The story of CFD starts with Navier-Strokes equation. This equation is based on conservation laws of mass, momentum, and energy. These law's can be defined as:

CONSERVATION OF MASS

The conservation of mass law states that the mass in the control volume can be neither created nor destroyed in accordance with physical laws. The conservation of mass, also expressed as Continuity Equation, states that the mass flow difference throughout system between inlet- and outlet-section is zero. In equation terms, we can write as:

$$\frac{D\rho}{D_t} + \rho\left(\nabla \cdot \vec{V}\right) = 0$$

Where, ρ is density, V is velocity and gradient operator ∇

$$\vec{\nabla} = \vec{i}\frac{\partial}{\partial x} + \vec{j}\frac{\partial}{\partial y} + \vec{k}\frac{\partial}{\partial z}$$

When density of fluid is constant, the flow is assumed as incompressible and then this equation represents a steady state process:

$$\frac{D\rho}{D_t} = 0 \quad \text{so,} \quad \nabla \cdot \vec{V} = \frac{\partial u}{\partial x} + \frac{\partial v}{\partial y} + \frac{\partial w}{\partial z} = 0$$

here, u, v, and w are components of velocity at point (x,y,z) at time t.

Note that incompressible fluid is also called Newtonian fluid when stress/strain curve is linear.

CONSERVATION OF MOMENTUM

The momentum in a control volume is kept constant, which implies conservation of momentum that we call 'The Navier-Stokes Equations'. The description is set up in accordance with the expression of Newton's Second Law of Motion:

F=m.a

where, F is the net force applied to any particle, a is the acceleration, and m is the mass. In case the particle is a fluid, it is convenient to divide the equation to volume of particle to generate a derivation in terms of density as follows:

$$\rho\frac{DV}{D_t} = f = f_{body} + f_{surface}$$

in which f is the force exerted on the fluid particle per unit volume, and f_{body} is the applied force on the whole mass of fluid particles as below:

$f_{body} = \rho \cdot g$

Where, ρ is density, g is gravitational acceleration. External forces which are deployed through the surface of fluid particles, $f_{surface}$ is expressed by pressure and viscous forces as shown below:

$$f_{surface} = \nabla \cdot \tau_{ij} = \frac{\partial \tau_{ij}}{\partial x_i} = f_{pressure} + f_{viscous}$$

where τ_{ij} is expressed as stress tensor. According to the general deformation law of Newtonian viscous fluid given by Stokes, τ_{ij} is expressed as

$$\tau_{ij} = -p\delta_{ij} + \mu \left(\frac{\partial u_i}{\partial x_j} + \frac{\partial u_j}{\partial x_i} \right) + \delta_{ij} \lambda \nabla \cdot V$$

Hence, Newton's equation of motion can be specified in the form as follows:

$$\rho \frac{DV}{D_t} = \rho \cdot g + \nabla \cdot \tau_{ij}$$

Navier-Stokes equations of Newtonian viscous fluid in one equation gives:

$$\underbrace{\rho \frac{DV}{D_t}}_{I} = \underbrace{\rho \cdot g}_{II} - \underbrace{\nabla p}_{III} + \underbrace{\frac{\partial}{\partial x_i} \left[\mu \left(\frac{\partial v_i}{\partial x_j} + \frac{\partial v_j}{\partial x_i} \right) + \delta_{ij} \lambda \nabla \cdot V \right]}_{IV}$$

I : Momentum convection

II: Mass force

III: Surface force

IV: Viscous force

where, static pressure ρ and gravitational force ρ.g. The equation is convenient for fluid and flow fields both transient and compressible. D/D_t indicates the substantial derivative as follows:

$$\frac{D(\,)}{D_t} = \frac{\partial(\,)}{\partial t} + u \frac{\partial(\,)}{\partial x} + v \frac{\partial(\,)}{\partial y} + w \frac{\partial(\,)}{\partial z} = \frac{\partial(\,)}{\partial t} + V \cdot \nabla(\,)$$

If the density of fluid is accepted to be constant, the equations are greatly simplified in which the viscosity coefficient μ is assumed constant and $\nabla \cdot V = 0$ in equation. Thus, the Navier-Stokes equations for an incompressible three-dimensional flow can be expressed as follows:

$$\rho \frac{DV}{Dt} = \rho g - \nabla p + \mu \nabla^2 V$$

For each dimension when the velocity is V(u,v,w):

$$\rho\left(\frac{\partial u}{\partial t} + u\frac{\partial u}{\partial x} + v\frac{\partial u}{\partial y} + w\frac{\partial u}{\partial z}\right) = \rho g_x - \frac{\partial p}{\partial x} + \mu\left(\frac{\partial^2 u}{\partial x^2} + \frac{\partial^2 u}{\partial y^2} + \frac{\partial^2 u}{\partial z^2}\right)$$

$$\rho\left(\frac{\partial v}{\partial t} + u\frac{\partial v}{\partial x} + v\frac{\partial v}{\partial y} + w\frac{\partial v}{\partial z}\right) = \rho g_y - \frac{\partial p}{\partial y} + \mu\left(\frac{\partial^2 v}{\partial x^2} + \frac{\partial^2 v}{\partial y^2} + \frac{\partial^2 v}{\partial z^2}\right)$$

$$\rho\left(\frac{\partial w}{\partial t} + u\frac{\partial w}{\partial x} + v\frac{\partial w}{\partial y} + w\frac{\partial w}{\partial z}\right) = \rho g_z - \frac{\partial p}{\partial z} + \mu\left(\frac{\partial^2 w}{\partial x^2} + \frac{\partial^2 w}{\partial y^2} + \frac{\partial^2 w}{\partial z^2}\right)$$

p, u, v, and w are unknowns where a solution is sought by application of both continuity equation and boundary conditions. Besides, the energy equation has to be considered if any thermal interaction is available in the problem.

CONSERVATION OF ENERGY

The Conservation of Energy is the first law of thermodynamics which states that the sum of the work and heat added to the system will result in the increase of energy of the system:

$$dE_t = dQ + dW$$

where dQ is the heat added to the system, dW is the work done on the system, and dE_t is the increment in the total energy of the system. One of the common types of energy equation is:

$$\rho\left[\underbrace{\frac{\partial h}{\partial t}}_{I} + \underbrace{\nabla \cdot (hV)}_{II}\right] = \underbrace{-\frac{\partial p}{\partial t}}_{III} + \underbrace{\nabla \cdot (k\nabla T)}_{IV} + \underbrace{\phi}_{V}$$

I : Local change with time

II: Convective term

III: Pressure work

IV: Heat flux

V: Heat dissipation term

The Navier-Stokes equations have a non-linear structure and various complexities so it is hardly possible to conduct an exact solution of those equations. Thus, with regard to the physical domain, both approaches and assumptions are partially applied to simplify the equations. Some assumptions also need to be applied to provide a reliable model in which the equation is carried out to further step in terms of complexity such as turbulence.

VARIATIONS OF NAVIER-STROKES EQUATION

The solution of the Navier-Stokes equations can be realized with either analytical or numerical methods. The analytical method is the process that only compensates solutions in which non-linear and complex structures of the Navier-Stokes equations are ignored within several assumptions. It is only valid for simple / fundamental cases such as Couette flow, Poisellie flow, etc. Almost every case in fluid dynamics comprises non-linear and complex structures in the mathematical model which cannot be ignored to sustain reliability. Hence, the solution of the Navier-Stokes equations are carried out with several numerical methods. Various parameters based on which Navier-Strokes equation can vary are given next.

Time Domain

The analysis of fluid flow can be conducted in either steady (time-independent) or unsteady (time-dependent) condition depending on the physical incident. In case the fluid flow is steady, it means the motion of fluid and parameters do not rely on change in time, the term $\partial()/\partial\tau=0$ where the continuity and momentum equations are re-derived as follows:

Continuity equation:

$$\frac{\partial(\rho u)}{\partial x} + \frac{\partial(\rho v)}{\partial y} + \frac{\partial(\rho w)}{\partial z} = 0$$

The Navier-Stokes equation in x direction:

$$\rho\left(u\frac{\partial u}{\partial x} + v\frac{\partial u}{\partial y} + w\frac{\partial u}{\partial z}\right) = \rho g_x - \frac{\partial p}{\partial x} + \mu\left(\frac{\partial^2 u}{\partial x^2} + \frac{\partial^2 u}{\partial y^2} + \frac{\partial^2 u}{\partial z^2}\right)$$

While the steady flow assumption negates the effect of some non-linear terms and provides a convenient solution, variation of density is a hurdle that keeps the equation in a complex formation.

Compressibility

Due to the flexible structure of fluids, the compressibility of particles is a significant issue. Despite the fact that all types of fluid flow are compressible in a various range regarding molecular structure, most of them can be assumed to be incompressible in which the density changes are negligible. Thus, the term $\partial\rho/\partial t = 0$ is thrown away regardless of whether the flow is steady or not, as below:

Continuity equation:

$$\frac{\partial u}{\partial x} + \frac{\partial v}{\partial y} + \frac{\partial w}{\partial z} = 0$$

The Navier-Stokes equation in x direction:

$$\rho\left(\frac{\partial u}{\partial t} + u\frac{\partial u}{\partial x} + v\frac{\partial u}{\partial y} + w\frac{\partial u}{\partial z}\right) = \rho g_x - \frac{\partial p}{\partial x} + \mu\left(\frac{\partial^2 u}{\partial x^2} + \frac{\partial^2 u}{\partial y^2} + \frac{\partial^2 u}{\partial z^2}\right)$$

As incompressible flow assumption provides reasonable equations, the application of steady flow assumption concurrently enables us to ignore non-linear terms where ∂()/∂t=0. Moreover, the density of fluid in high speed cannot be accepted as incompressible in which the density changes are important. "The Mach Number" is a dimensionless number that is convenient to investigate fluid flow, whether incompressible or compressible.

$$\boxed{Ma = \frac{V}{a} \leq 0.3}$$

where, Ma is the Mach number, V is the velocity of flow, and a is the speed of sound at 340.29 m/s at sea level.

As in above equation, when the Mach number is lower than 0.3, the assumption of incompressibility is acceptable. On the contrary, the change in density cannot be negligible in which density should be considered as a significant parameter. For instance, if the velocity of a car is higher than 100 m/s, the suitable approach to conduct credible numerical analysis is the compressible flow. Apart from velocity, the effect of thermal properties on the density changes has to be considered in geophysical flows.

Low and High Reynolds Numbers

The Reynolds number, the ratio of inertial and viscous effects, is also effective on Navier-Stokes equations to truncate the mathematical model. While **Re** -> ∞, the viscous effects are presumed negligible and viscous terms in Navier-Stokes equations are removed. The simplified form of Navier-Stokes equation, described as Euler equation, can be specified as follows.

The Navier-Stokes equation in x direction:

$$\boxed{\rho \left(\frac{\partial u}{\partial t} + u\frac{\partial u}{\partial x} + v\frac{\partial u}{\partial y} + w\frac{\partial u}{\partial z} \right) = \rho g_x - \frac{\partial p}{\partial x}}$$

Even though viscous effects are relatively important for fluids, the inviscid flow model partially provides a reliable mathematical model as to predict real process for some specific cases. For instance, high-speed external flow over bodies is a broadly used approximation where inviscid approach reasonably fits. While Re<<1, the inertial effects are assumed negligible where related terms in Navier-Stokes equations drop out. The simplified form of Navier-Stokes equations is called either creeping flow or Stokes flow.

The Navier-Stokes equation in x direction:

$$\boxed{\rho g_x - \frac{\partial p}{\partial x} + \mu \left(\frac{\partial^2 u}{\partial x^2} + \frac{\partial^2 u}{\partial y^2} + \frac{\partial^2 u}{\partial z^2} + \right) = 0}$$

Having tangible viscous effects, creeping flow is a suitable approach to investigate the flow of lava, swimming of microorganisms, flow of polymers, lubrication, etc.

Turbulence

The behavior of the fluid under dynamic conditions is a challenging issue that is compartmentalized as laminar and turbulent. The laminar flow is orderly at which motion of fluid can be predicted precisely. Except that, the turbulent flow has various hindrances, therefore it is hard to predict the fluid flow which shows a chaotic behavior. The Reynolds number, the ratio of inertial forces to viscous forces, predicts the behavior of fluid flow whether laminar or turbulent regarding several properties such as velocity, length, viscosity, and also type of flow. Whilst the flow is turbulent, a proper mathematical model is selected to carry out numerical solutions. Various turbulent models are available in literature and each of them has a slightly different structure to examine chaotic fluid flow.

Turbulent flow can be applied to the Navier-Stokes equations in order to conduct solutions to chaotic behavior of fluid flow. Apart from the laminar, transport quantities of the turbulent flow, it is driven by instantaneous values. Direct numerical simulation (DNS) is the approach to solving the Navier-Stokes equation with instantaneous values. Having district fluctuations varies in a broad range, DNS needs enormous effort and expensive computational facilities. To avoid those hurdles, the instantaneous quantities are reinstated by the sum of their mean and fluctuating parts as follows:

$$instantaneous\ value = \overline{mean\ value + fluctuating\ value'}$$

$$u = \bar{u} + u'$$

$$v = \bar{v} + v'$$

$$w = \bar{w} + w'$$

$$T = \bar{T} + T'$$

where u, v, and w are velocity components and T is temperature.

Instead of instantaneous values which cause non-linearity, carrying out a numerical solution with mean values provides an appropriate mathematical model which is named "The Reynolds-averaged Navier-Stokes (RANS) equation". The fluctuations can be negligible for most engineering cases which cause a complex mathematical model. Thus, RANS turbulence model is a procedure to close the system of mean flow equations. The general form of The Reynolds-averaged Navier-Stokes (RANS) equation can be specified as follows:

Continuity equation:

$$\frac{\partial \bar{u}}{\partial x} + \frac{\partial \bar{v}}{\partial y} + \frac{\partial \bar{w}}{\partial z} = 0$$

The Navier-Stokes equation in x direction:

$$\rho \left(\frac{\partial \bar{u}}{\partial t} + \bar{u}\frac{\partial \bar{u}}{\partial x} + \bar{v}\frac{\partial \bar{u}}{\partial y} + \bar{w}\frac{\partial \bar{u}}{\partial z} \right) = \rho g_x - \frac{\partial \bar{p}}{\partial x} + \mu \left(\frac{\partial^2 u}{\partial x^2} + \frac{\partial^2 u}{\partial y^2} + \frac{\partial^2 u}{\partial z^2} \right)$$

The turbulence model of RANS can also vary regarding methods such as k-omega, k-epsilon, k-omega-SST, and Spalart-Allmaras which have been used to seek a solution for different types of turbulent flow.

Likewise, large eddy simulation (LES) is another mathematical method for turbulent flow which is also comprehensively applied for several cases. Tough LES ensures more accurate results than RANS, it requires much more time and computer memory. As in DNS, LES considers to solve the instantaneous Navier-Stokes equations in time and three-dimensional space.

Favre Time Averaging

In Favre-averaging, the time averaged equations can be simplified significantly by using the density weighted averaging procedure suggested by Favre. If Reynolds time averaging is applied to the compressible form of the Navier-Stokes equations, some difficulties arise. In particular, the original form of the equations is significantly altered. To see this, consider Reynolds averaging applied to the continuity equation for compressible flow.

Favre time averaging can be defined as follows. The instantaneous solution variable, ϕ, is decomposed into a mean quantity, $\tilde{\phi}$, and fluctuating component, ϕ'', as follows:

$$\phi = \tilde{\phi} + \phi''$$

The Favre time-averaging is then

$$\overline{\rho\phi(x_i, t)} = \frac{1}{T}\int_{t-T/2}^{t+T/2} \rho(x_i, t')\phi(x_i, t')\,dt' = \overline{\rho\tilde{\phi}} + \overline{\rho\phi''} = \bar{\rho}\tilde{\phi}$$

Where,

$$\tilde{\phi}(x_i, t) \equiv \frac{1}{\bar{\rho}T}\int_{t-T/2}^{t+T/2} \rho(x_i, t')\phi(x_i, t')\,dt', \quad \overline{\rho\phi''} \equiv 0$$

Favre-Averaged Navier-Stokes (FANS) Equations can be given as:

Continuity Equation:

$$\frac{\partial}{\partial t}(\bar{\rho}) + \frac{\partial}{\partial x_i}(\bar{\rho}\tilde{u}_i) = 0$$

Momentum Equation:

$$\frac{\partial}{\partial t}(\bar{\rho}\tilde{u}_i) + \frac{\partial}{\partial x_j}(\bar{\rho}\tilde{u}_i\tilde{u}_j + \bar{p}\delta_{ij}) = \frac{\partial}{\partial x_j}\left(\bar{\tau}_{ij} - \overline{\rho u_i'' u_j''}\right)$$

Favre-Averaged Reynolds Stress Tensor:

$$\lambda = -\overline{\rho u_i'' u_j''}$$

Turbulent Kinetic Energy:

$$\frac{1}{2}\overline{\rho u_i'' u_i''} = -\frac{1}{2}\lambda_{ii} = \bar{\rho}\tilde{k}$$

STEPS OF COMPUTATIONAL FLUID DYNAMICS

The process of solving Computational Fluid Dynamics can be defined in 5 stages which are: Creating Mathematical Model, Discretization of model, Analyzing with Numerical Scheme, Getting solution, and Post Processing (Visualization). These steps are discussed next.

Creating Mathematical Model

In this stage, various equations are defined based on physical properties, boundary conditions, and other real-world parameters of the problem. The mathematical model generally consists of partial differential equations, integral equations, and combinations of both. In this case, it will be Navier-Stokes equation with other mathematical equations to define boundary conditions and thermodynamic properties. Note that at this step there will be some assumptions which converting real problem into mathematical model like material might be assumed isotropic, heat conditions might be considered adiabatic and so on. These assumptions should generate permissible level of error in the solution and as a designer you should be aware of the quantum of those errors.

Discretization of Model

While studying the behavior of fluid, you will find that there are infinite number of particles with different physical properties in the fluid stream. These infinite particles will generate infinite equations to solve if mathematical model is to be solved directly which is not possible. So, we convert the mathematical model into finite number of elements and then we can solve equations for each of the element. In case of CFD, elements are called cells. The process of converting a mathematical model into finite element equations is called discretization. In simple words, the partial differential and integral equations are converted into algebraic equations of **A.x = B**.

There are many methods to perform discretization of mathematical model like Finite Difference Method, Finite Volume Method, and Finite Element Method which use mesh type structuring of the model. There are also methods which do not use mesh like Smooth Particle Hydrodynamics (SPH) and Finite Pointset Method (FPM). Note that there is again loss of information at the stage of discretization.

Analyzing with Numerical Schemes

Numerical scheme is the complete setup of equations with all the parameters and boundary conditions defined as needed. The scheme also includes numerical methods by which these equations will be solved and limiting points upto which the equations will be solved. When you start analyzing the problem with specified numerical scheme, there is no input required from your side in system; everything is automatic by computer. Every numerical scheme need to satisfy some basic requirements which are consistency, stability, convergence, and accuracy.

Solution

At this stage, we get the solution of equations as different basic variables like pressure, speed, volume, and so on. The resulting flow variables are obtained at each grid point/mesh point whether the scheme is time dependent (transient) or steady.

Visualization (Post-processing)

At this stage, everything is based on interpretation of designer. Various desired parameters are derived from basic calculated parameters at this stage. You also need to check whether results of CFD analysis are realistic or not.

Now, one question left here to understand in more detail is discretization. Discretization can be performed by various methods like FDM, FVM, FEM, and so on as discussed earlier. Out of these methods, Finite Difference Method and Finite Volume Methods are the most used methods for CFD. Here, we will discuss FDM in detail.

FINITE DIFFERENCE METHOD

At discretization stage, the model need to be converted into numerical grid with different cells defined by nodes; refer to Figure-2. For a 2D problem, i and j will be used for direction in horizontal and vertical directions.

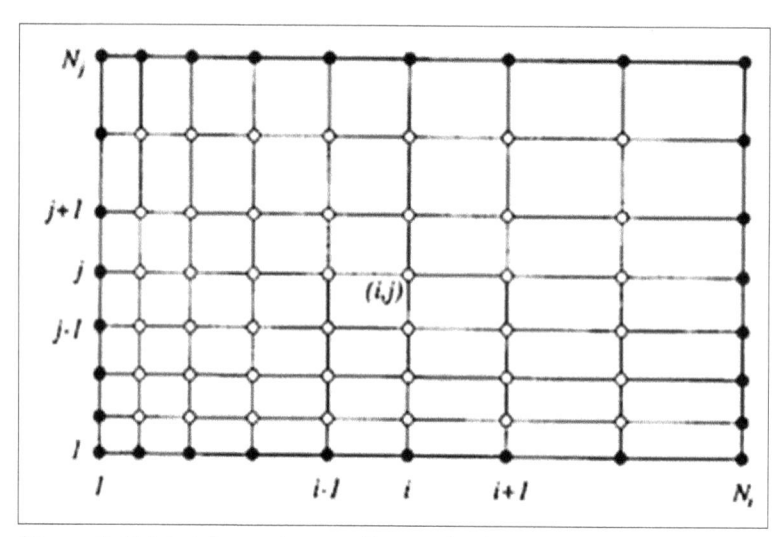

Figure-2. Grid with cartesian coordinates of nodes

If you recall the definition of derivative: $\left(\dfrac{\partial u}{\partial x}\right)_{x_i} = \lim_{\Delta x \to 0} \dfrac{u(x_i + \Delta x) - u(x_i)}{\Delta x}$

The equation represents slope of tangent to a curve u(x) which can be geometrically represented as shown in Figure-3.

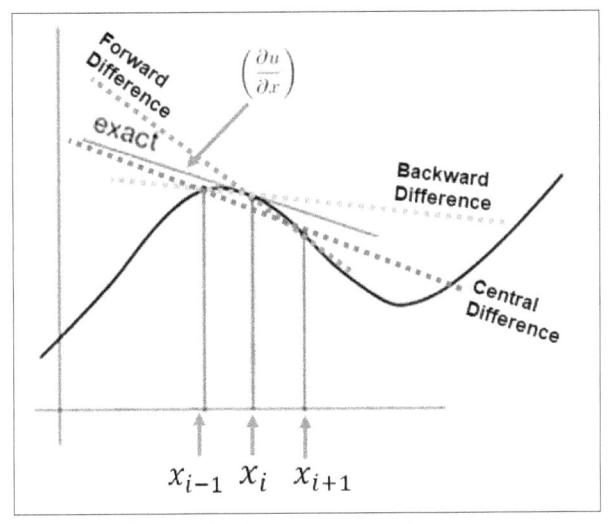

Figure-3. Geometrical interpretation of derivative

Now, assume that we have value of x_i, x_{i+1}, and x_{i-1} then there are three ways in which we can find the slope of curve.

Backward Difference which uses x_i and x_{i-1}.
Forward Difference which uses x_i and x_{i+1}.
Central Difference which uses x_{i+1} and x_{i-1}.

If we assume that Δx is difference between two consecutive nodes then lower the value of Δx more accurate we will get the value of slope.

For a uniform grid where Δx is same for each node then based on

Backward Difference : $\dfrac{du}{dx} \cong \dfrac{u_i - u_{i-1}}{\Delta x}$

Forward Difference : $\dfrac{du}{dx} \cong \dfrac{u_{i+1} - u_i}{\Delta x}$

Central Difference : $\dfrac{du}{dx} \cong \dfrac{u_{i+1} - u_{i-1}}{2\Delta x}$

Now, if you expand the backward difference derivative using Taylor's series then

Consider a function u(x) and its derivative at point x,

$$\left(\dfrac{\partial u}{\partial x}\right)_{x_i} = \lim_{\Delta x \to 0} \dfrac{u(x_i + \Delta x) - u(x_i)}{\Delta x}$$

If u(x + Δx) is expanded in Taylor series about u(x), we obtain

$$u(x + \Delta x) = u(x) + \Delta x \dfrac{\partial u(x)}{\partial x} + \dfrac{(\Delta x)^2}{2} \dfrac{\partial^2 u(x)}{\partial x^2} + \dfrac{(\Delta x)^3}{3!} \dfrac{\partial^3 u(x)}{\partial x^3} + \cdots$$

If we substitute this value in derivative equation then

$$\frac{\partial u(x)}{\partial x} = \lim_{\Delta x \to 0} \left(\frac{\partial u(x)}{\partial x} + \frac{\Delta x}{2} \frac{\partial^2 u(x)}{\partial x^2} + \cdots \right)$$

Based on this equation, we can write u in Taylor Series at i+1 and i-1 as:

$$u_{i+1} = u_i + \Delta x \left(\frac{\partial u}{\partial x}\right)_i + \frac{\Delta x^2}{2}\left(\frac{\partial^2 u}{\partial x^2}\right)_i + \frac{\Delta x^3}{3!}\left(\frac{\partial^3 u}{\partial x^3}\right)_i + \frac{\Delta x^4}{4!}\left(\frac{\partial^4 u}{\partial x^4}\right)_i + \cdots$$

$$u_{i-1} = u_i - \Delta x \left(\frac{\partial u}{\partial x}\right)_i + \frac{\Delta x^2}{2}\left(\frac{\partial^2 u}{\partial x^2}\right)_i - \frac{\Delta x^3}{3!}\left(\frac{\partial^3 u}{\partial x^3}\right)_i + \frac{\Delta x^4}{4!}\left(\frac{\partial^4 u}{\partial x^4}\right)_i + \cdots$$

Re-writing the above two equations, we get:

Forward difference : $\left(\frac{\partial u}{\partial x}\right)_i = \frac{u_{i+1} - u_i}{\Delta x} + O(\Delta x)$

Backward Difference : $\left(\frac{\partial u}{\partial x}\right)_i = \frac{u_i - u_{i-1}}{\Delta x} + O(\Delta x)$

Central Difference : $\left(\frac{\partial u}{\partial x}\right)_i = \frac{u_{i+1} - u_{i-1}}{2\Delta x} + O(\Delta x^2)$

Note that central difference takes second derivative of delta x so the value of error is lesser than compared to forward or backward difference. For example if forward or backward difference gives an error of 0.001 then second derivative will be 0.001 x 0.001 which gives 0.000001. This error value is a lot lesser.

You can learn more about these methods in books dedicated to Finite Difference Method and solution techniques.

INTRODUCTION TO CFDOF

In this chapter, you will learn about the basic use of CfdOF workbench. This workbench is not the part of native FreeCAD but you can use Addon Manager to use this workbench. The CfdOF workbench acts as front-end of OpenFOAM CFD toolkit. OpenFoam is the open source free CFD software developed by OpenCFD Ltd. Various features supported by CfdOF workbench in FreeCAD are given next.

Flow physics
- Incompressible, laminar flow (simpleFoam, pimpleFoam)
- Support for various RANS, LES and DES turbulent flow models
- Incompressible free-surface flow (interFoam, multiphaseInterFoam)
- Compressible buoyant flow (buoyantSimpleFoam, buoyantPimpleFoam)
- High-speed compressible flow (HiSA)
- Porous regions and porous baffles
- Basic material database
- Flow initialization with a potential solver
- Solution of arbitrary passive scalar transport functions

Meshing
- Cut-cell Cartesian meshing with boundary layers (cfMesh)
- Cut-cell Cartesian meshing with baffles (snappyHexMesh)
- Tetrahedral meshing using Gmsh, including conversion to polyhedral dual mesh
- Post-meshing check mesh
- Support for dynamic mesh adaptation for supported solvers

Post-processing and monitoring
- Postprocessing using Paraview
- Basic support for force-based function objects (Forces, Force Coefficients)
- Basic support for probes

Other features
- Runs on Windows 7-11 and Linux
- Unit/regression testing
- Case builder using an extensible template structure
- Macro scripting

ACTIVATING CFDOF WORKBENCH

The option to activate CfdOF workbench is available in the Addon manager. The procedure to activate this workbench is given next.

- Click on the **Addon manager** tool from the **Tools** menu. The **Addon Manager** window will be displayed; refer to Figure-4.

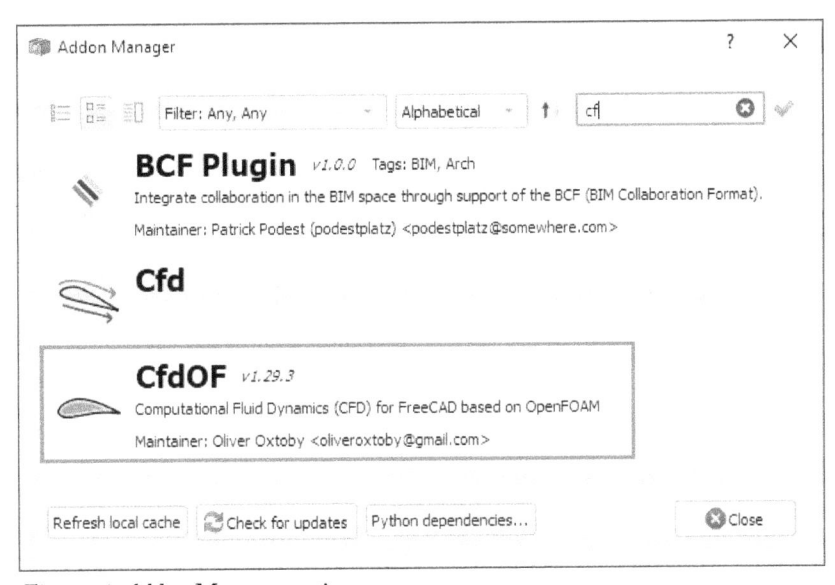

Figure-4. Addon Manager options

- Select the **CfdOF** option from the **Addon Manager** and click on the **Install** button. A message box will be displayed after installation is complete; refer to Figure-5.

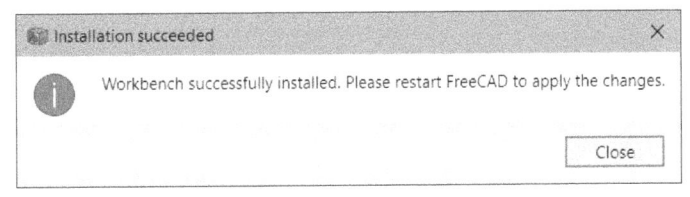

Figure-5. Installation message

- Click on the **Close** button to exit the message box. Close the **Addon Manager** and restart the software.

INSTALLING CFDOF RELATED SOFTWARE

For running CfdOF workbench, you need to install some necessary software to efficiently perform the analysis. After installing the CfdOF addon, you will be asked to restart the software. The procedure to install related software is given next.

- Restart the FreeCAD software in administrator mode. To do so, right-click on FreeCAD from Start menu and select **Run as administrator** option from right-click shortcut menu; refer to Figure-6.

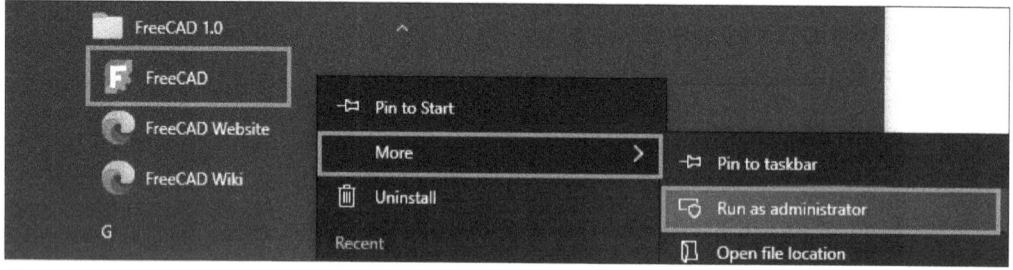

Figure-6. Right-click menu

- Select the **Preferences** option from the **Edit** menu. The **Preferences** dialog box will be displayed.
- Select the **CfdOF** option from the left area. The options will be displayed as shown in Figure-7.

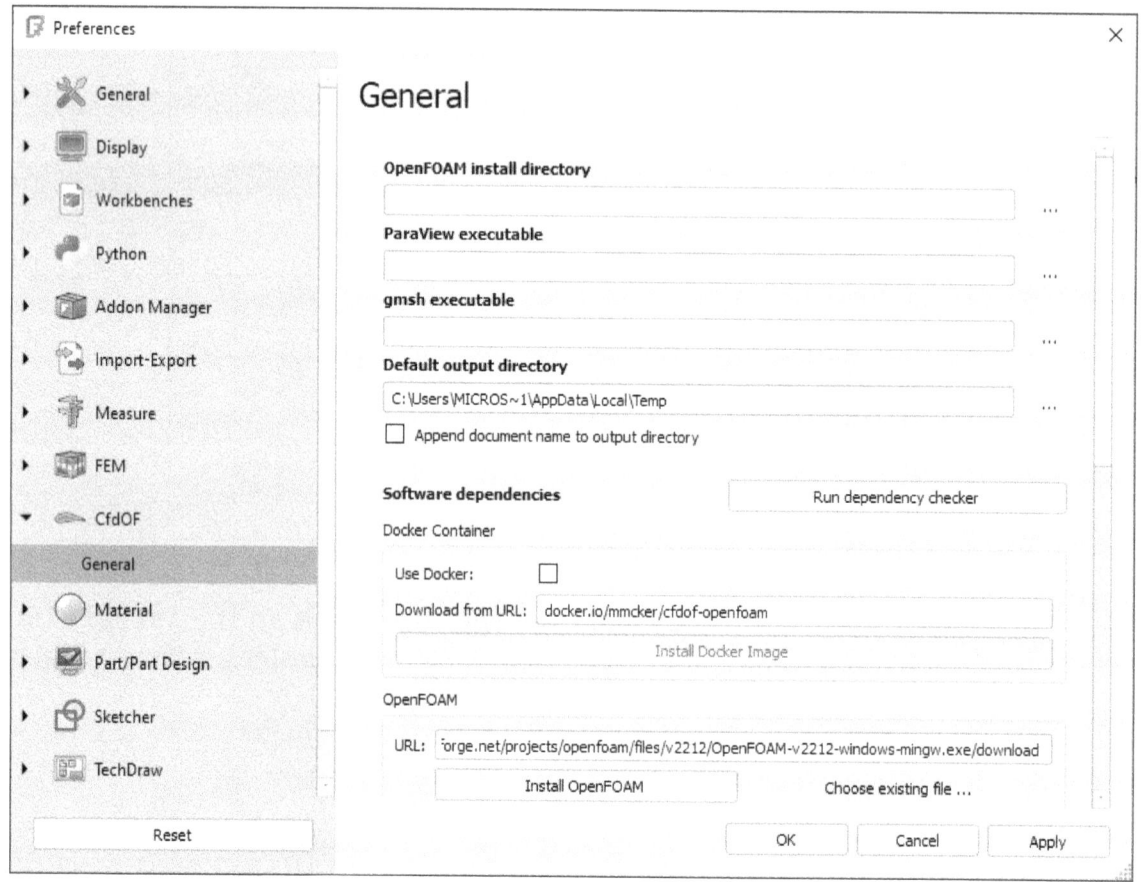

Figure-7. CfdOF options

- Click on the **Browse** button for `OpenFOAM install directory` field and select/create a folder location for installing the OpenFOAM software; refer to Figure-8.
- After selecting desired folder, click on the **Select Folder** button from the dialog box. The path will be added in the `OpenFOAM install directory` field.

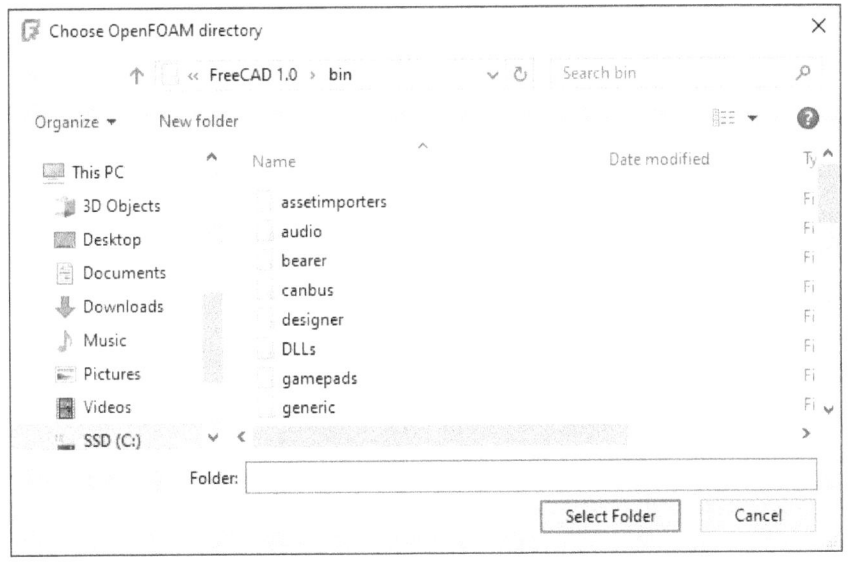

Figure-8. CFD software install folder

- Click on the **Install OpenFOAM** button from the **OpenFOAM** area at bottom in the **Preferences** dialog box. The dialog box for installing OpenFOAM will be displayed; refer to Figure-9.

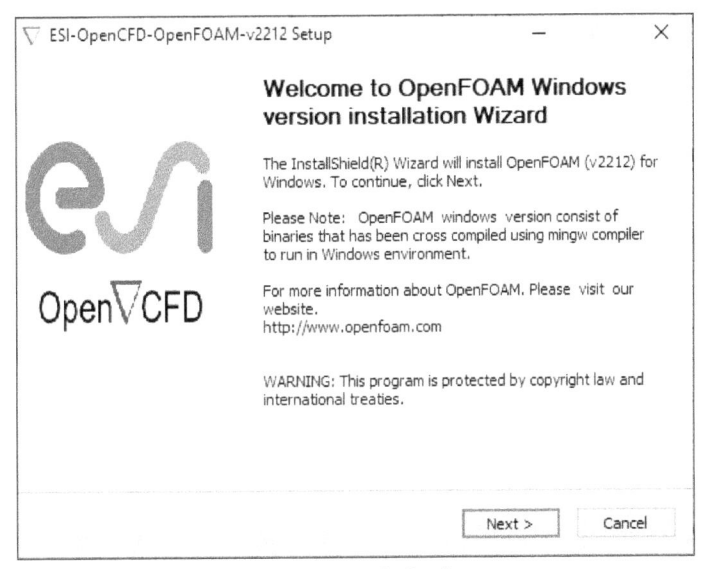

Figure-9. OpenFOAM installation dialog box

- Click on the **Next** button and follow next instructions to install the software. Make sure to set same location for installation in the dialog box; refer to Figure-10. After setting location, click on the **Install** button and install the software.

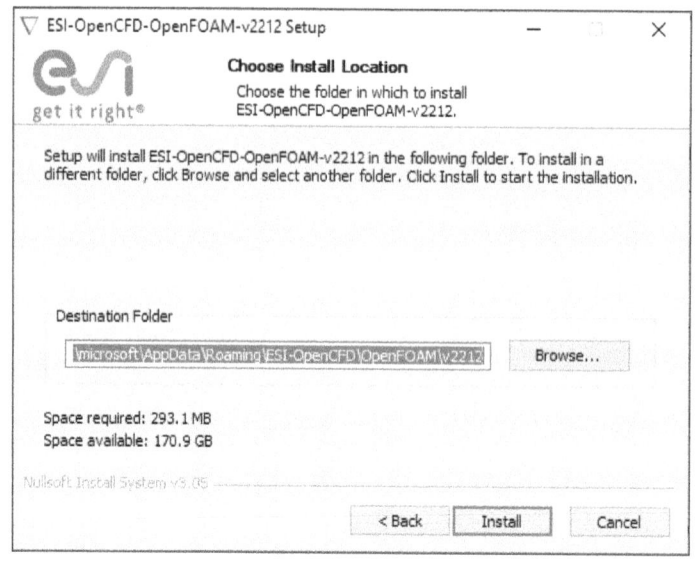

Figure-10. Installation location for OpenFOAM

- Similarly, install ParaView, cfMesh, and HISA software.
- After installing all requisite software, set path for executable of gmsh and ParaView.
- Click on the **OK** button from the **Preferences** dialog box to apply settings.

PREPARING MODEL FOR CFD

In terms of CAD, there are two ways to prepare model for CFD. The model in which internal CFD is to be performed is generally a hollow tube with lid faces at both the ends (outlet/inlet) making it airtight. For external CFD, you can use any solid model to check aerodynamics and other CFD aspects. In this step, you need to close any openings in the model and make sure there is continuity in the model for fluid flow.

An example on performing CFD is given next.

PERFORMING CFD ON A MODEL

The common steps involved in performing CFD analysis on a model are given next.

- Create/open the model on which you want to perform CFD analysis; refer to Figure-11.

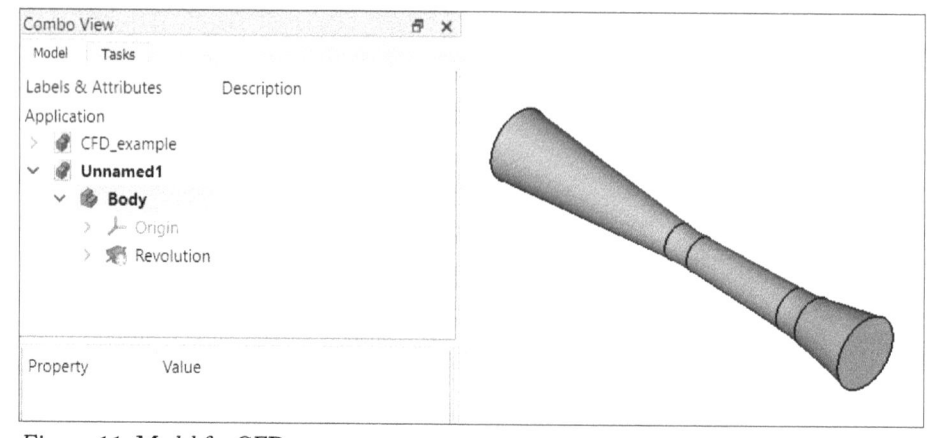

Figure-11. Model for CFD

- Select the **CfdOF** option from the **Workbench** drop-down in the **Toolbar**. The tools to perform CFD analysis will be displayed.
- Click on the **Analysis container** tool from the **CfdOF** toolbar. The options related to CFD will be added in the **Model Tree**.

Applying Boundary Conditions

- Select the face of model as shown in Figure-12 and click on the **Fluid boundary** tool from the **CfdOF** toolbar. The **CFD boundary condition** dialog will be displayed in the **Tasks** pane.

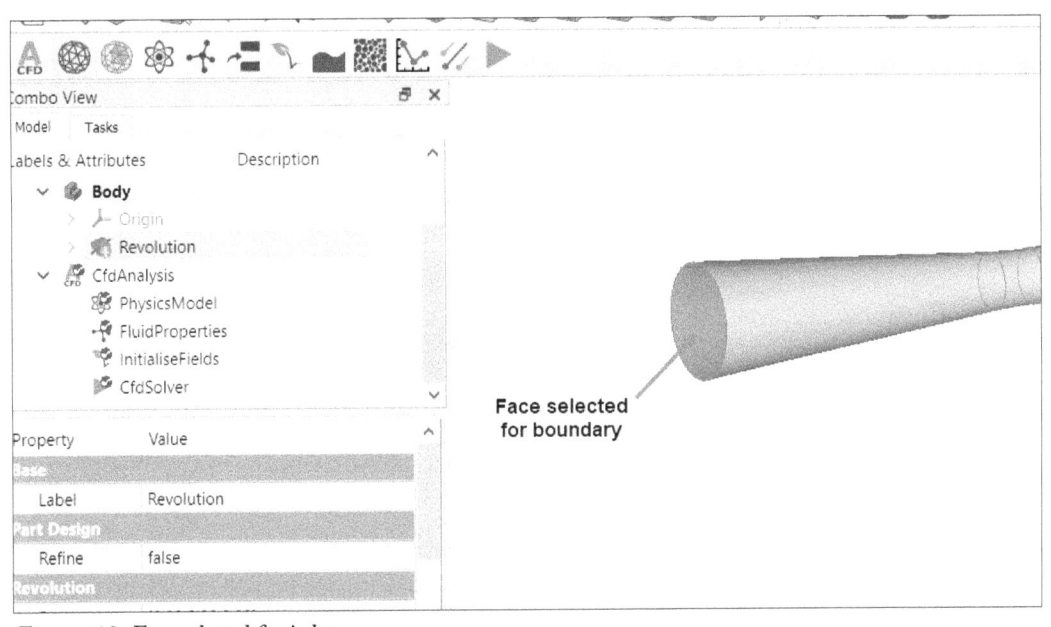

Figure-12. Face selected for inlet

- Select the **Inlet** option from the **Boundary** drop-down to define fluid inlet condition. Select desired option from the **Sub-type** drop-down to define how fluid flow will be defined. Select the **Uniform velocity** option to define velocity of fluid flowing uniformly through inlet. Select the **Volumetric flow rate** option to define flow of fluid by volume of fluid passing through inlet per second. Generally, unit for volumetric flow rate is m^3/s. Select the **Mass flow rate** option from the drop-down to define fluid flow rate in kg/s or lbs/s. Select the **Total pressure** option from the **Sub-type** drop-down to define pressure at which fluid will enter the domain. Select the **Static pressure** option from the drop-down to specify pressure of the fluid when it is not moving.
- Set desired parameters in the edit boxes of **CFD boundary condition** dialog based on sub-type selected. After setting desired parameters, click on the **OK** button to apply the boundary condition.
- Similarly, apply the outlet CFD boundary condition for other face; refer to Figure-13.

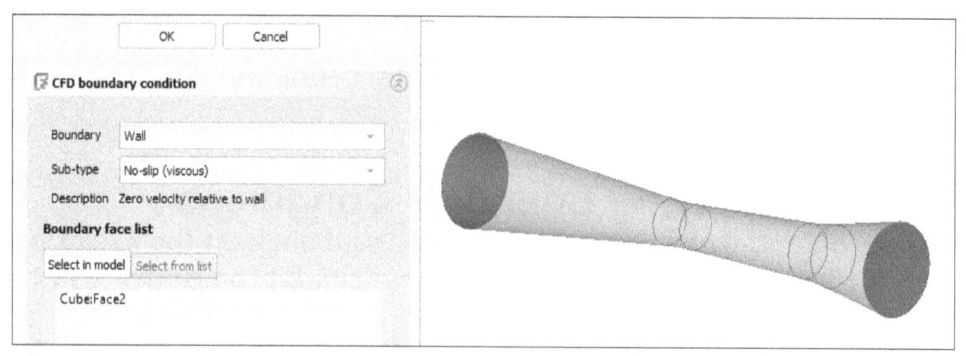

Figure-13. Outlet cfd boundary condition

- Select the faces to be used as wall for analysis and specify parameters as shown in Figure-14.

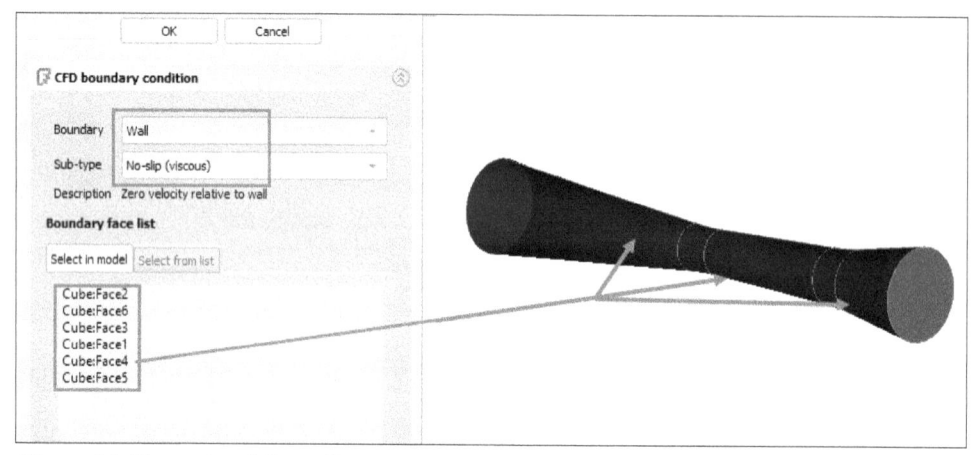

Figure-14. Faces selected for wall

- After setting parameters, click on the **OK** button to apply wall boundary conditions.

Creating CFD Mesh

- Select the body from the **Model Tree** in **Combo View** and click on the **CFD Mesh** tool from the **Toolbar**; refer to Figure-15. The **CFD Mesh** dialog will be displayed in the **Tasks** panel of **Combo View**; refer to Figure-16.

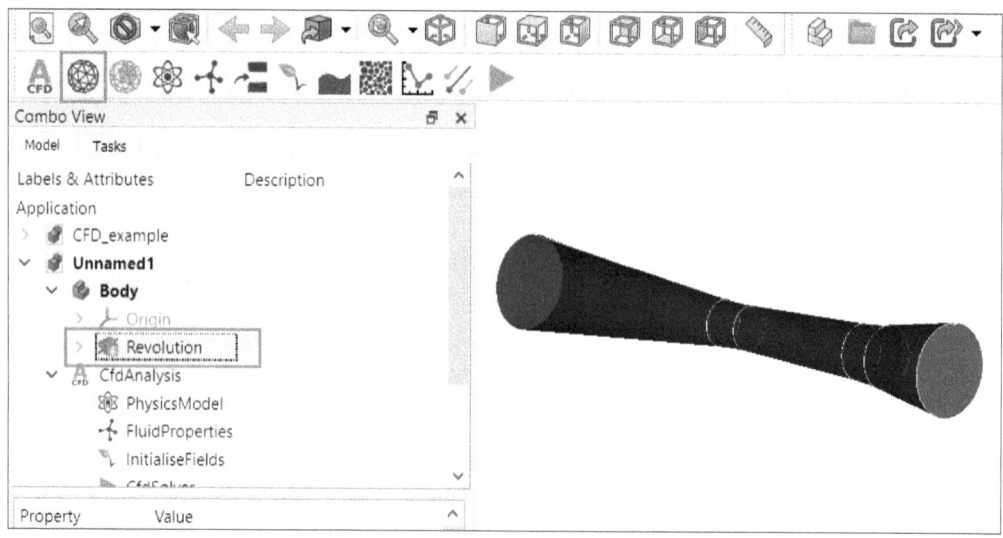

Figure-15. CFD mesh tool

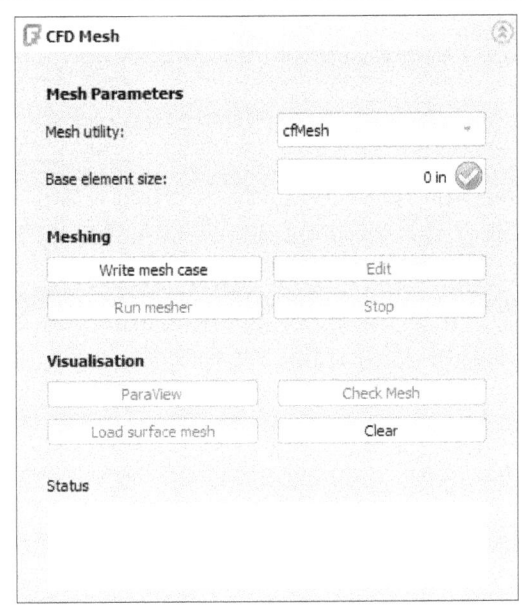

Figure-16. CFD Mesh dialog

- Set desired value of mesh element size in the **Base element size** edit box and click on the **Write mesh case** button. The status of mesh case writing will be displayed in the **Status** section of **CFD Mesh** dialog.
- Once you get the Mesh case written successfully message in the **Status** section of dialog, click on the **Run mesher** tool from **Meshing** section of dialog. Note that it may take some time for mesher to generate the cfd mesh. After the meshing is complete, the status "**Meshing completed**" will be displayed in the **Status** section of dialog.
- Click on the **Load surface mesh** button from the **Visualisation** section in **CFD Mesh** dialog to check preview of CFD mesh. You can also use the **Paraview** tool to check mesh.

Mesh refinement

The **Mesh refinement** tool is used to control mesh element size/type in specific region. The procedure to use this tool is discussed next.

- Select meshed object from the **Model Tree** and click on the **Mesh refinement** tool from the **Toolbar**; refer to Figure-17. The **Mesh refinement** dialog will be displayed; refer to Figure-18.

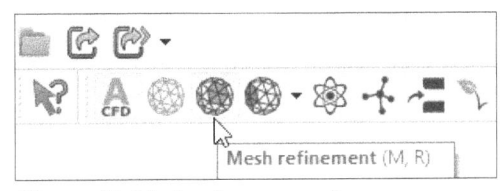

Figure-17. Mesh refinement tool

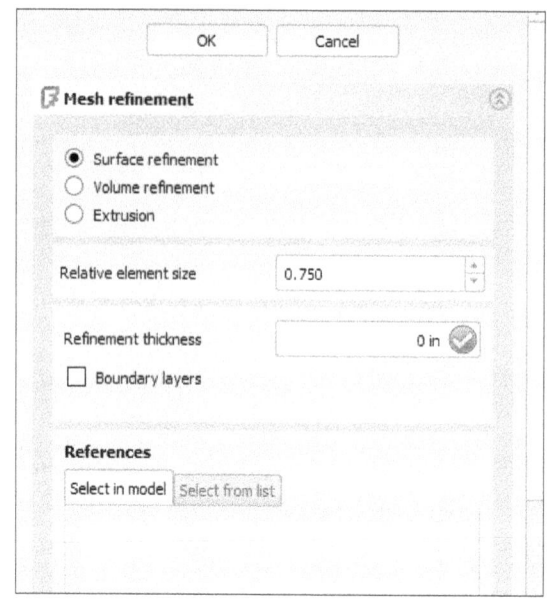

Figure-18. Mesh refinement dialog

- Specify desired parameters in the dialog to refine the meshed object and click on the **OK** button. The meshed object will be refined; refer to Figure-19.

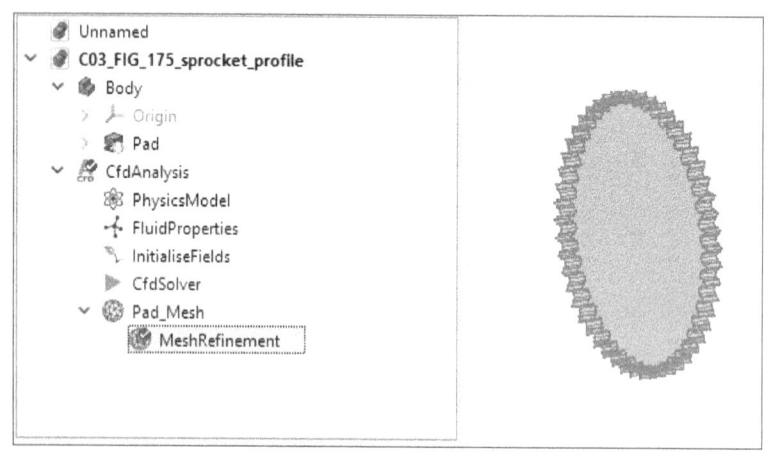

Figure-19. Meshed object refinement

Setting Physics Model

- Select the CfdAnalysis container from the **Model Tree** in **Combo View** and click on the **Select models** tool from the **Toolbar**. The **Select physics model** dialog will be displayed in the **Tasks** pane; refer to Figure-20.

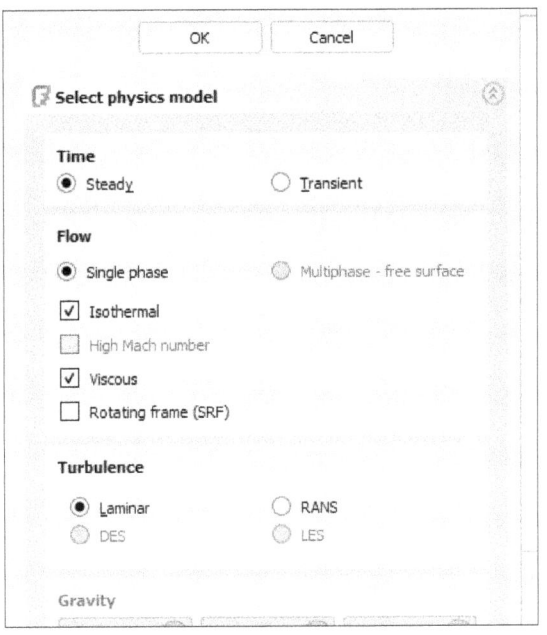

Figure-20. Select physics model dialog

- Set desired parameters in the dialog to define type of the study. Here, we will be using steady, single phase, incompressible, and viscous fluid analysis with RANS turbulence. After setting parameters, click on the **OK** button.

Adding Fluid Properties

- Click on the **Add fluid properties** tool from the **Toolbar**. The **Fluid properties** dialog will be displayed; refer to Figure-21.

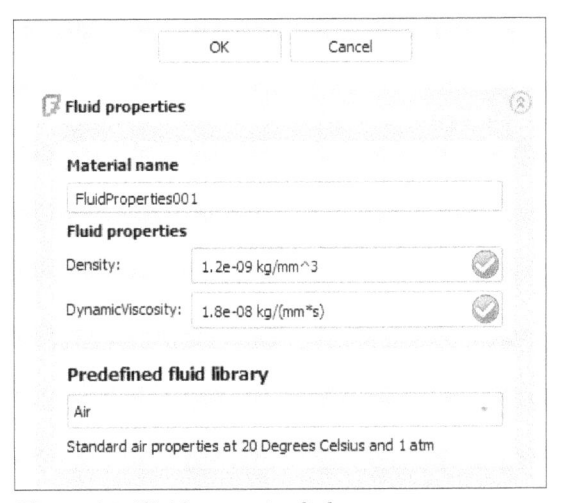

Figure-21. Fluid properties dialog

- Select desired option from the **Predefined fluid library** drop-down to define the fluid to be used in analysis. In our case, we are using Water as fluid. After setting parameters, click on the **OK** button.

Setting Initial Internal Flow Conditions

- Click on the **Initialise** tool from the **Toolbar**. The **Initialise flow field** dialog will be displayed; refer to Figure-22.

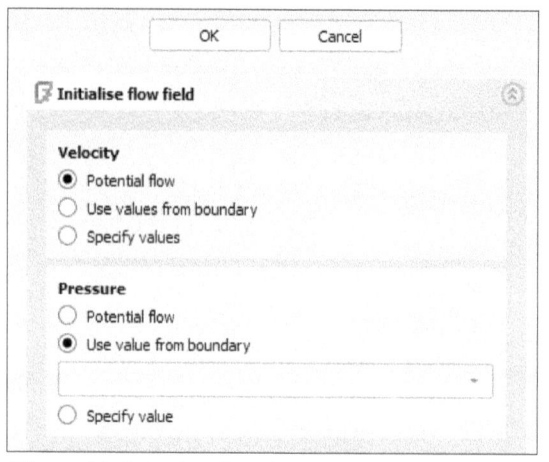

Figure-22. Initialise flow field dialog

- Set desired parameters in the dialog and click on the **OK** button to apply flow conditions.

Porous Zone

The **Porous Zone** tool is used to make selected body porous. A porous material allows restricted flow of fluid. The procedure to use this tool is given next.

- Click on the **Porous Zone** tool from the **Toolbar**. The **CFD Zone** dialog will be displayed; refer to Figure-23.

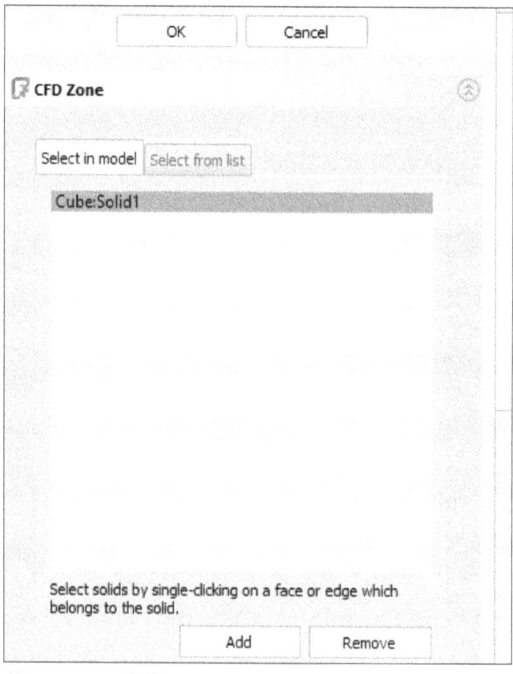

Figure-23. CFD zone dialog box

- Select desired model from the **Select in model** tab in the 3D drawing area and then click on the **Add** button from the dialog to apply porous properties to selected model.
- Select the **Darcy-Forchhemier coefcients** option from the **Porous correlation** drop-down to specify viscous and inertial drag tensors by giving their principal component and direction.

- Select the **Staggered tube bundle (Jakob)** option from the drop-down to specify geometry of parallel tube with struggled layer for defining porous parameters.
- Set desired parameters in the dialog and click on the **OK** button to apply flow conditions.

Reporting function

The **Reporting function** tool is used to define the report parameters and style for showing results of analysis.

- Click on the **Reporting function** tool from the Toolbar. The **CFD reporting function** will be displayed; refer to Figure-24.

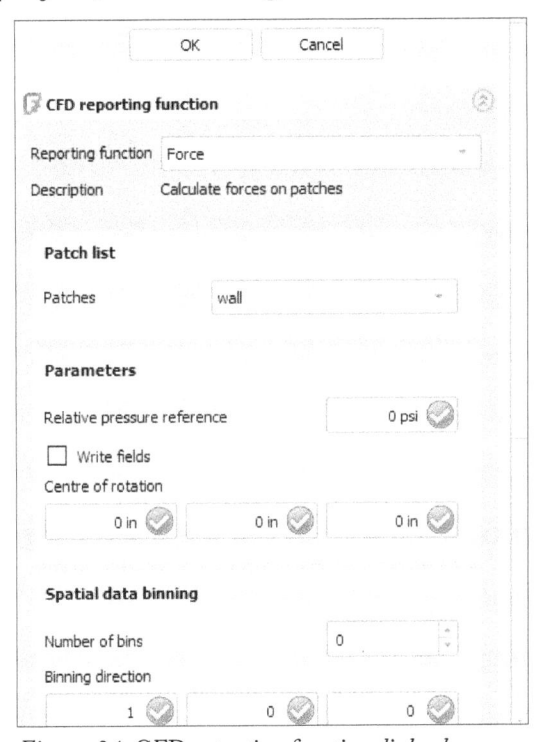

Figure-24. CFD reporting function dialog box

- Select the **Force** option from the **Reporting function** drop-down to calculate force on patches. Select the **Force Coefficients** option from the drop-down to calculate force coefficients from patches. Select the **Probes** option from the drop-down to calculate sample fields at specified locations.
- Select desired option from the **Patches** drop-down in the **Patch list** section to define whether inlet patches will be displayed on monitor or outlet.
- Set desired value in the **Relative pressure reference** edit box of the **Parameters** section to define reference pressure value related to which result parameter will be reported.
- Specify desired value in the **Center of rotation** edit box to define center for moments parameter.
- Specify desired values in the **Number of bins** and **Binning direction** edit boxes from the **Spatial data binning** edit box to define number of data sets to be generated for the result parameter.
- Select the **Cumulative** check box to combine the spatial data in single set.
- Set the other parameters as discussed earlier and click on the **OK** button to apply flow conditions.

Cfd scalar transport function

The `Cfd scalar transport function` tool is used to define scalar function parameters for the fluid particles. The procedure to use this tool is given next.

- Click on the `Cfd transport function` tool from the toolbar. The `Scalar transport function` dialog box will be displayed; refer to Figure-25.

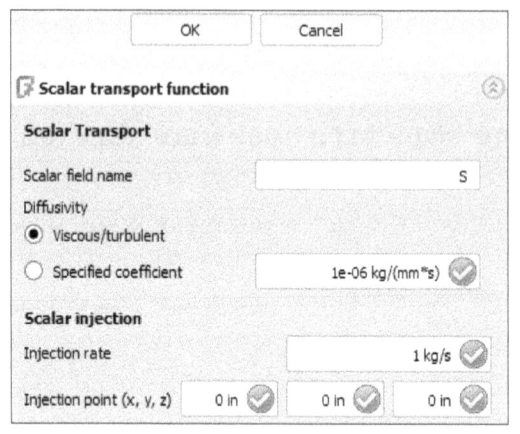

Figure-25. Scaler transport function dialog box

- Specify desired value in the `Scalar field name` edit box to define name of scalar parameter.
- Select desired radio button from the `Diffusivity` section to define the rate of particles passing through material.
- Set desired rate and direction of injection in the edit boxes of the `Scalar injection` section.
- After setting desired parameters, click on the `OK` button from the dialog.

Solving the Analysis

- Click on the `Solver job control` tool from the `Toolbar`. The `Analysis control` dialog will be displayed in the `Tasks` pane; refer to Figure-23.

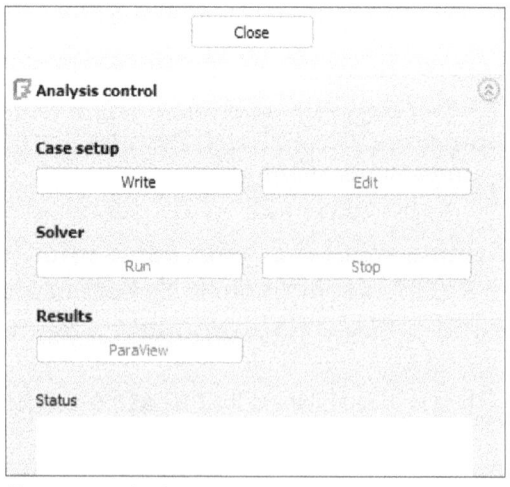

Figure-26. Analysis control dialog

- Click on the `Write` tool from the `Case setup` section of the dialog. The case file will be written and status will be displayed in the `Status` section of the dialog.
- Click on the `Run` tool from the `Solver` section of the dialog to perform solution of analysis. If your mesh is not updated before analysis then a message box will be displayed asking you to remesh the case. Select `Yes` button from the message box.

- Once the analysis is solved, the result will be displayed; refer to Figure-26.

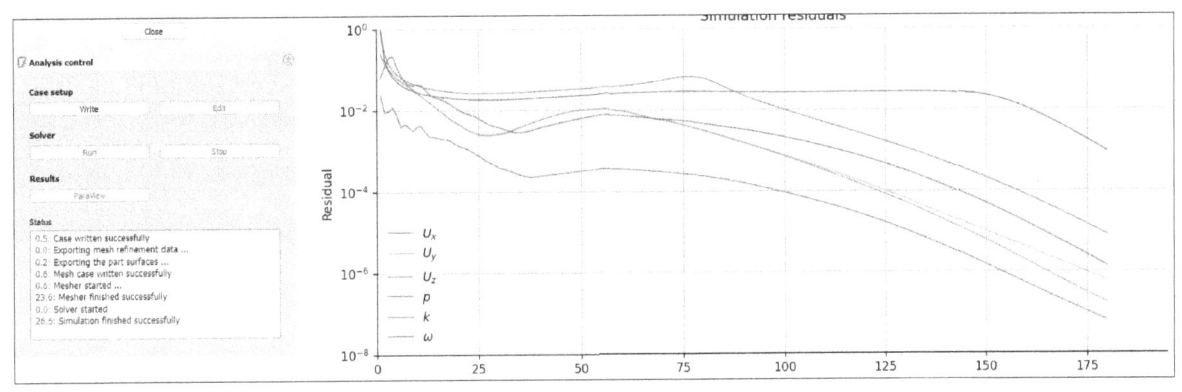

Figure-27. CFD analysis result

- Click on the **Paraview** button from the **Results** section of **Analysis control** dialog. Results of the analysis will be displayed; refer to Figure-27.

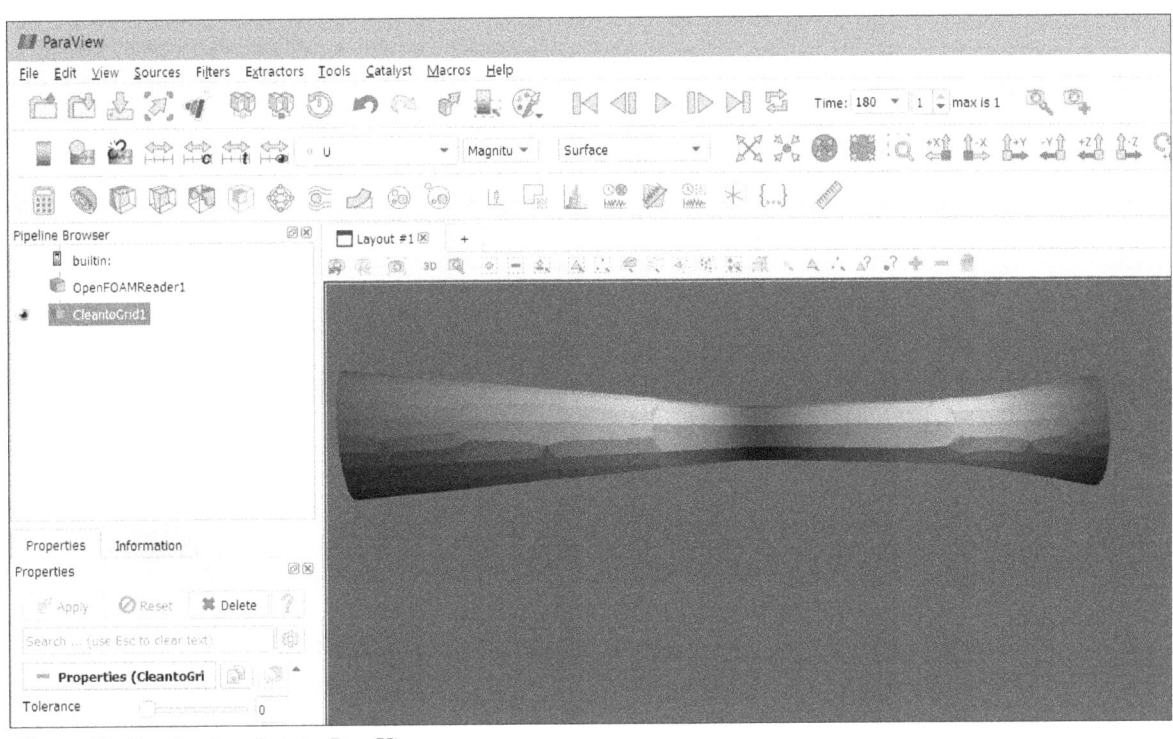

Figure-28. Result of analysis in ParaView

SELF ASSESSMENT

Q1. What does GIGO stand for in the context of CFD?

A. Good Input, Good Output
B. Garbage In, Garbage Out
C. General Input, General Output
D. Gravitational Input, Gravitational Output

Q2. Which property of fluid is represented by the ratio of the weight density of fluid to the weight density of water for liquids?

A. Mass Density
B. Specific Gravity
C. Viscosity
D. Weight Density

Q3. How is weight density calculated for a fluid?

A. By multiplying the mass by gravity coefficient
B. By dividing the mass by gravity coefficient
C. By multiplying mass density by volume
D. By multiplying mass by velocity

Q4. What happens to the viscosity of a fluid as its temperature increases?

A. It remains constant
B. It decreases for liquids and increases for gases
C. It increases for liquids and decreases for gases
D. It increases for both liquids and gases

Q5. Which of the following is a type of fluid where shear stress is directly proportional to shear strain?

A. Newtonian Fluids
B. Non-Newtonian Fluids
C. Ideal Fluids
D. Ideal Plastic Fluids

Q6. What is the formula for calculating the pressure in an isothermal process?

A. P/ρ = Constant
B. $P/\rho k$ = Constant
C. $P = \rho RT$
D. $P = V/T$

Q7. In an adiabatic process, what is the relationship between pressure and density?

A. p/ρ = Constant
B. $p/\rho k$ = Constant
C. pV = Constant
D. pT = Constant

Q8. What is the Universal Gas Constant for air?

A. 8314 J/kg-mole K
B. 287 J/kg-K
C. 1.225 J/kg-K
D. 273 J/kg-mole K

Q9. In the context of Pascal's Law, what does it state about the pressure in a static fluid?

A. Pressure is uniform at all depths
B. Pressure is the same in all directions at a point
C. Pressure increases with height
D. Pressure is greater at the surface

Q10. Bernoulli's equation is used to describe which of the following?

A. Energy conservation in a moving fluid
B. Temperature distribution in a fluid
C. The motion of individual fluid particles
D. The viscosity of a fluid

Q11. Which method is more suitable for analyzing the pressure field in fluid mechanics, according to the Eulerian method?

A. Describing the flow through individual particles
B. Computing pressure changes experienced by a moving particle
C. Analyzing the field of flow at specific points
D. Focusing on the velocity of individual fluid elements

Q12. What does the term "Control Volume" refer to in fluid mechanics analysis?

A. An exact analytical solution for all flow cases
B. A system where properties are averaged over boundaries
C. A method based on theoretical assumptions
D. A numerical simulation of fluid behavior

Q13. Which of the following is a key approach in CFD?

A. Control Volume
B. Experimental Method
C. Differential Method
D. All of the above

Q14. In CFD, what is the purpose of meshing in the differential approach?

A. To simplify the equations
B. To divide the fluid into small finite elements for analysis
C. To measure fluid pressure
D. To calculate the temperature changes

Q5. Which of the following is true about the Navier-Stokes equations?

A. They describe the velocity of fluid at different points
B. They apply only to incompressible fluids
C. They are a set of partial differential equations that describe fluid flow
D. They can only be solved analytically

Q16. What is dynamic viscosity in a fluid?

A. The resistance to flow caused by fluid molecules
B. The force exerted by the fluid on surrounding objects
C. The rate at which fluid particles move
D. The pressure difference across a fluid layer

Q17. How is Reynolds number calculated?

A. By dividing dynamic viscosity by kinematic viscosity
B. By multiplying velocity with pipe diameter
C. By dividing density, velocity, and pipe diameter by kinematic viscosity
D. By multiplying pressure and temperature

Q18. What does the Kronecker delta represent in the context of fluid dynamics?

A. The relation between velocity and pressure
B. A tensor that represents the direction of flow
C. A discrete version of the delta function
D. The rate of change of pressure

Q19. Which of the following is true for compressible fluids?

A. They are incompressible at all conditions
B. Their density can change under pressure
C. They have no viscosity
D. They always follow ideal gas laws

Q20. What is the primary use of CFD in engineering?

A. To determine theoretical fluid behavior
B. To analyze fluid flow and interaction with objects
C. To measure fluid properties experimentally
D. To calculate the ideal temperature of a fluid

Q21. Which of the following is the primary assumption made when assuming fluid flow is incompressible?

A) The density of the fluid is constant
B) The fluid does not change temperature
C) The velocity of the fluid is constant
D) The flow is always steady

Q22. What is the Mach number used for in fluid dynamics?

A) To calculate the Reynolds number
B) To determine if the flow is compressible or incompressible
C) To measure the temperature of the fluid
D) To evaluate the viscosity of the fluid

Q23. At what Mach number does the assumption of incompressibility for fluid flow become unacceptable?

A) 0.1
B) 0.3
C) 1.0
D) 2.5

Q24. In which scenario is the compressible flow model most appropriate?

A) High-speed flow over a body
B) Low-speed steady flow
C) Flow through a laminar pipe
D) Flow through an incompressible fluid

Q25. What does the Reynolds number indicate in fluid dynamics?

A) The ratio of inertial forces to viscous forces
B) The temperature of the fluid
C) The density of the fluid
D) The speed of sound in the fluid

Q26. When the Reynolds number (Re) is much greater than 1, what is assumed in fluid flow?

A) The flow is laminar
B) The flow is turbulent
C) The flow is incompressible
D) The flow is steady

Q27. What is the term used for the simplified form of the Navier-Stokes equations when viscous effects are negligible?

A) Euler equations
B) Reynolds equations
C) Stokes equations
D) Mach number equations

Q28. In the case of low Reynolds numbers (Re << 1), what type of flow is most applicable?

A) Inviscid flow
B) Stokes flow (creeping flow)
C) Turbulent flow
D) Steady laminar flow

Q29. What does the RANS equation represent in fluid dynamics?

A) The instantaneous Navier-Stokes equations
B) The time-averaged Navier-Stokes equations
C) The simplified equation for steady flow
D) The equation for turbulent flow without averaging

Q30. Which of the following is an example of a commonly used turbulence model for fluid dynamics simulations?

A) Direct Numerical Simulation (DNS)
B) Reynolds-averaged Navier-Stokes (RANS)
C) Inviscid model
D) Euler model

Q31. Which method provides a more accurate solution for turbulent flow but requires more computational resources?

A) RANS
B) Direct Numerical Simulation (DNS)
C) Large Eddy Simulation (LES)
D) Stokes flow

Q32. What does Favre-averaging use in its averaging procedure?

A) Time-averaging
B) Density-weighted averaging
C) Reynolds averaging
D) Instantaneous value averaging

Q33. What is the first stage of the Computational Fluid Dynamics (CFD) process?

A) Discretization of the model
B) Post-processing
C) Creating a mathematical model
D) Solution of the equations

Q34. In the discretization stage of CFD, what is the term for the process of converting continuous equations into a finite number of elements?

A) Numerical solution
B) Mesh generation
C) Grid refinement
D) Discretization

Q35. Which of the following numerical methods does NOT require a mesh?

A) Finite Element Method (FEM)
B) Finite Difference Method (FDM)
C) Smooth Particle Hydrodynamics (SPH)
D) Finite Volume Method (FVM)

Q36. What is the main goal of a numerical scheme in CFD?

A) To solve the algebraic equations
B) To discretize the model
C) To define the boundary conditions
D) To specify the parameters for solving the equations

Q37. At which stage of CFD are basic variables like pressure, velocity, and volume solved?

A) Discretization
B) Solution
C) Visualization
D) Analyzing with numerical schemes

Q38. Which method is commonly used to convert a 3D model into a numerical grid for CFD analysis?

A) Finite Element Method (FEM)
B) Finite Volume Method (FVM)
C) Finite Difference Method (FDM)
D) Mesh generation

Q39. What does mesh refinement control in CFD?

A) The accuracy of the solution
B) The type of boundary conditions
C) The size/type of mesh elements in a specific region
D) The computational cost

Q40. Which physics model is typically used in a steady, single-phase, incompressible fluid analysis with RANS turbulence?

A) Euler model
B) Laminar model
C) Incompressible viscous model
D) Compressible turbulent model

Q41. In CFD, what is the purpose of applying boundary conditions?

A) To refine the mesh
B) To specify the fluid properties
C) To define how the fluid interacts with the boundaries of the model
D) To set the flow conditions

Q42. What software is used for post-processing in the CfdOF workbench?

A) MATLAB
B) ParaView
C) OpenFOAM
D) Gmsh

Q43. What is the purpose of using the Porous Zone tool in CFD analysis?

A) To define a model for high-speed flow
B) To model the effect of porous materials or regions within the flow
C) To create mesh refinement in specific regions
D) To define the physical properties of fluids

Q44. What is the next step after setting the fluid properties in CFD?

A) Solving the analysis
B) Setting initial internal flow conditions
C) Reporting results
D) Mesh refinement

Q45. When solving a CFD analysis, what is the first step in the Solver job control tool?

A) Apply boundary conditions
B) Write the case file
C) Select the physics model
D) Initialize the flow conditions

Index

Symbols

2D Offset tool 4-26
2D triangular shell elements 9-34
3D Offset tool 4-24
3D Pocket tool 8-34
3D tetrahedral solid elements 9-34
3D view 1-4

A

Acceleration 14-7
Adaptive tool 8-28
Add a new named group tool 6-72
Add an offset vertex tool 10-73
Add base shape tool 11-48
Add Bolt Circle Centerlines tool 10-70
Add Centerline between 2 Lines tool 10-34
Add Centerline between 2 Points tool 10-35
Add Centerline to Faces tool 10-32
Add Circle Centerlines tool 10-69
Add component tool 7-7
Add Component tool 7-50
Add construction tool 6-75
Add Corner Relief tool 11-42
Add Cosmetic Arc tool 10-75
Add Cosmetic Circle 3 Points tool 10-75
Add Cosmetic Circle tool 10-71, 10-74
Add Cosmetic Intersection Vertex(es) tool 10-72
Add Cosmetic Line Through 2 Points tool 10-36
Add Cosmetic Parallel Line tool 10-75
Add Cosmetic Perpendicular Line tool 10-76
Add Cosmetic Thread Bolt Bottom View tool 10-72
Add Cosmetic Thread Bolt Side View tool 10-72
Add Cosmetic Thread Hole Bottom View tool 10-71
Add Cosmetic Thread Hole Side View tool 10-70
Add Cosmetic Vertex tool 10-29
Add Edge button 3-25
Add fluid properties tool 14-31
Add Hole or Shaft Fit tool 10-43
Additive Box tool 3-26
Additive Cone tool 3-30
Additive Cylinder tool 3-27
Additive Ellipsoid tool 3-31
Additive Helix tool 3-25
Additive loft tool 3-21
Additive pipe tool 3-23
Additive Prism tool 3-33
Additive Sphere tool 3-29
Additive tools 3-18
Additive Torus tool 3-32
Additive Wedge tool 3-35
Add Leaderline to View tool 10-26
Add Midpoint Vertices tool 10-31
Addon Manager tool 1-42
Addons button 12-31
Add Quadrant Vertices tool 10-32
Add text document tool 1-37
Add to group tool 6-73
Add tool option 11-33
Add triangle tool 11-17
Add Welding Information to Leaderline tool 10-41
Alignment tool 12-5
Align to selection tool 1-23
Analysis container tool 14-27
Analysis Container tool 9-3
Analyze cascading menu 11-13
Angle constraint tool 2-27
Angle option 4-51
Angle snap button 6-67
Annotation style editor tool 6-27
Appearance tool 12-28
Apply geometric hatch to a Face tool 10-54
Apply style tool 6-63
Arc by 3 points tool 6-8
Arch tools 7-3
Arch workbench 7-2
Arc of hyperbola tool 2-8
Arc of parabola tool 2-9
Arc tool 6-7
Area Annotation tool 10-66
Area option 4-52
Array tool , 6-41
Aspect ratio button 1-31
Attachment tool 4-63
Auto Group tool 6-66
Auto option 4-50
Auto Power drop-down 7-9
Axis Map tool 8-44
Axis System tool 7-54
Axis tool 7-52
Axonometric cascading menu 1-23
Axonometric Length Dimension tool 10-49

B

Beam cross section tool 9-31
Beam rotation tool 9-32
Bernoulli's equation 14-5
Bezier Curve tool 6-15
Bi CGstabl Degree 9-47
Blend Curve tool 11-8
Blender navigation style 1-45
Block constraint tool 2-36
Body heat source tool 9-25
Boolean cascading menu 11-22
Boolean fragments tool 4-42
Boolean tool 3-70, 4-37
Boolean XOR tool 4-46
Bottom tool 1-24
Boundary tool 8-45
Bounding box tool 12-24
Boundings info tool 11-16
Box Selection tool 4-54, 12-3

Box tool 4-3, 7-46
B-spline by control points tool 2-20
B-Spline tool 6-13
Building tool 7-5
Bulk Modulus 14-5
Bulk Viscosity 14-10

C

Calculate the arc length of selected edges tool 10-66
Calculix Experimental Solver 9-44
Calculix Standard Solver 9-41
Cascade Horizontal Dimensions tool 10-64
Cascade Oblique Dimensions tool 10-65
Cascade Spacing 10-57
Cascade Vertical Dimensions tool 10-65
Cavitation 14-5
Centered rectangle tool 2-13
Center in Stock button 8-7
Center of Mass option 4-52
Center snap button 6-67
Centrif Constraint tool 9-18
CFD 14-7
CFD boundary condition dialog 14-27
CFD mesh tool 14-28
CfdOF option 14-24
CfdOF workbench 14-22
Cfd scalar transport function tool 14-34
Chamfer tool 3-66, 4-16
Change Appearance of selected Lines tool 10-39
Change default Bed Length Size button 11-36
Change default Die Length Size 11-36
Change default Variable scale factor 11-36
Change Line Attributes tool 10-58
Changes objects 50% Transparency 11-35
Check geometry tool 3-7
Check Geometry tool 4-47
Check solid mesh tool 11-16
Choose Icon dialog box 1-42
Circle tool 6-9
Circular array tool 6-45
Clear views tool 1-27
Climb Milling 8-21
Clipping plane tool 12-25
Clone tool 6-41
Close All tool 1-11
Close hole tool 11-17
Close tool 1-10
Coincident constraint tool 2-28
Colors tab 12-19
Column tool 7-16
Comment tool 8-50
Complex Section tool 10-16
Component tool 7-48
Compound Filter tool 4-36
Compressibility 14-5, 14-15
Conduction 9-5
Cone tool 4-5
Connect tool 4-39

Conservation of Energy 14-14
conservation of mass law 14-12
Constant Vacuum Permittivity tool 9-30
Constrain auto radius/diameter tool 2-25
Constrain diameter tool 2-26
Constrain horizontal distance 2-23
Constrain horizontal/vertical tool 2-29
Constrain parallel tool 2-31
Constrain radius tool 2-25
Constrain tangent tool 2-33
Constrain vertical tool 2-31
Construction Mode tool 2-21
Contact Stiffness edit box 9-14
Continuity equation 14-17
Contours filter tool 9-57
Contour tool 8-18
Convection 9-5
Conventional Milling 8-21
Convert Geometry to B-spline tool 2-56
Converts Imperial Mesh to Metric button 11-34
Convert to solid tool 4-57
Copy tool 8-35, 12-3
Create a B-spline drop-down 2-12, 2-19, 2-29, 2-39, 2-41, 2-43
Create a circle drop-down 2-10
Create a clone tool 3-11
Create a datum line tool 3-14
Create a datum plane tool , 3-12
Create a datum point tool , 3-15
Create a Fixed Joint tool 13-5
Create a Line tool 2-5
Create a local coordinate system tool 3-17
Create an arc drop-down 2-6
Create a new sketch button 2-2
Create a polyline tool 2-3
Create arc by 3 points tool 2-7
Create arc by center tool 2-6
Create Arc Length Dimension tool 10-83
Create arc of ellipse tool 2-7
Create arc slot tool 2-18
Create a regular polygon drop-down 2-15, 2-17, 2-57
Create a shape binder tool 3-7
Create a sub-object shape binder tool 3-9
Create a surface finish symbol tool 10-43
Create Ball Joint tool 13-9
Create Belt Joint tool 13-16
Create body tool 3-3
Create carbon copy tool 2-46
Create circle by 3 points tool 2-11
Create circle by center tool 2-10
Create Cylindrical Joint tool 13-8
Create Distance Joint tool 13-10
Create ellipse by 3 points tool 2-12
Create Gears Joint tool 13-15
Create Horizontal Chain Dimensions tool 10-77
Create Horizontal Chamfer Dimension tool 10-81
Create Horizontal Coordinate Dimensions tool 10-79
Create mesh box size of Printer 11-35

Create mesh from shape tool 11-11
Create mesh segments tool 11-19
Create new view tool 1-21
Create Oblique Chain Dimensions tool 10-79
Create Oblique Coordinate Dimensions tool 10-81
Create points object from mesh tool 4-56
Create point tool 2-3
Create Rack and Pinion Joint tool 13-11
Create Revolute Joint tool 13-7
Create ruled surface tool 4-18
Create Screw Joint tool 13-13
Create section from mesh and plane tool 11-26
Create shape element copy tool 4-61
Create shape from mesh tool 4-55
Create simple copy tool 4-60
Create sketch tool 3-3
Create Slider Joint tool 13-8
Create trajectory tool 11-30
Create transformed copy tool 4-61
Create Vertical Chain Dimensions tool 10-78
Create Vertical Chamfer Dimension tool 10-82
Create Vertical Coordinate Dimensions tool 10-80
Cross-sections tool 4-23
Cross section tool 11-27
Cubic bezier curve tool 6-14
Current density boundary condition tool 9-29
Curtain Wall tool 7-13
Curvature plot tool 11-16
Curve on mesh tool 11-6
Customize Format Label tool 10-67
Customize tool 1-39
Custom tool 8-51
Cut Mesh tool 11-24
Cutout tool 4-41
Cut Plane tool 7-65, 7-66
Cut tool 12-3
cycle 9-5
Cylinder tool 4-4

D

Datum tools 3-12
Deburr tool 8-32
Decimation tool 11-21
Decrease Decimal Places tool 10-88
Decrease knot multiplicity tool 2-57
Defeaturing tool 4-48
Delete All Constraints tool 2-54
Delete All Geometry tool 2-53
Delete tool 12-4
Density 14-2, 14-11
Dependency graph tool 1-33
Dependency Graph tool 1-33
Difference tool 11-23
Dimensional constraints 2-22
Dimensions snap button 6-68
Dimension tool 2-22, 6-24
Dimetric tool 1-24
Displacement Constraint tool 9-12

Distance constraint tool 2-24
Distance Free option 4-50
Distance option 4-50
Divergence 14-10
Document window cascading menu 12-23
Dogbone tool 8-46
Door tool 7-23
Downgrade tool 6-55
Download now button 1-3
Draft Snap toolbar 6-67
Draft tool 3-67
Draft to Sketch tool 6-57
Draft tray toolbar 6-61, 6-70
Draft Workbench 6-2
Dragknife tool 8-47
Drawing objects tools 6-3, 6-23
Draw style cascading menu 1-27
Dress-up tools 3-64
Drilling tool 8-23
Duplicate selected objects tool 12-3
Dynamic viscosity 14-8
Dynamic Viscosity 14-8

E

Edge to Trajectory tool 11-30
Edit menu 12-2
Edit parameters tool 1-28
Edit sketch tool 3-5
Edit tool 6-50
Elasticity equation option 9-46
Electrostatic equation option 9-46
Electrostatic potential boundary condition tool 9-28
Element Geometry cascading menu 9-31
Ellipse by center tool 2-11
Ellipse tool 6-10
ElmerGrid 9-44
Elmer Solver 9-44
ElmerSolver 9-44
Embed tool 4-40
Endpoint snap button 6-67
Engrave tool 8-30
Environment check box 12-27
Equal constraint tool 2-34
Equipment tool 7-41
Erase Elements tool 9-40
Evaluate and repair mesh tool 11-14
Explode compound tool 4-35
Export a page to a DXF file tool 10-53
Export a page to an SVG file tool 10-52
Export button 7-64
Export CAD tool 4-54
Export CSV button 7-68
Export dependency graph tool 1-35
Export mesh tool 11-10
Export PDF tool 1-20
Export Template tool 8-16
Export tool 1-15
Extend face tool 11-5

Extend Face tool 11-39
Extend Line tool 10-59
Extend tool 2-44
Extension snap button 6-68
External Geometry tool 2-45
External reference tool 7-48
Extruded Cutout tool 11-47
Extrude tool 4-9

F

Facebinder tool , 6-17
Face Colors tool 4-30
Face info tool 11-15
Face Mill tool 8-25
Favre Time Averaging 14-18
FEM mesh boundary layer tool 9-38
FEM mesh from shape by Gmsh tool 9-36
FEM mesh from shape by Netgen tool 9-34
FEM mesh group tool 9-39
FEM mesh to mesh tool 9-41
FEM Workbench 9-2
Fence tool 7-37
File menu 1-8
Fill Boundary Curves tool 11-3
Fillet tool 4-15, 3-64, 3-56, 2-41, 2-42
Fill holes tool 11-17
Filling tool 11-2
Finish Selecting Loop tool 8-41
Finite Difference Method 14-20
First Angle option 10-3
Fit all tool 1-23
Fit selection tool 1-23
Fix Constraint tool 9-10
Fixture tool 8-50
Flip Dimension tool 6-59
Flip normals tool 11-16
Floor tool 7-6
Flow equation option 9-46
Flow velocity boundary condition tool 9-27
Fluid Constraints 9-26
Fluid section for 1D flow tool 9-33
Flux equation option 9-46
Fold a Wall tool 11-40
Force load tool 9-17
Frame tool 7-36
Freeze display cascading menu 1-24, 1-25
Freeze view tool 1-26
Frequency Analysis 9-5
Front tool 1-24
Fullscreen tool 1-22
Function cut filter tool 9-54

G

G-Code Inspector tool 8-37
Gesture Navigation 1-46
Grain direction edit box 7-34
Graphviz 1-33
Gravity Load tool 9-20

Grid snap button 6-68
Grid tool 7-55
Groove tool 3-40

H

Harmonize normals tool 11-16
Hatch a Face using image file tool 10-53
Hatch tool 6-21
Heal tool 6-78
Heat equation option 9-46
Heat flux load tool 9-23
Helix tool 8-27
Hides Selected Objects 11-35
Hole tool 3-38
Hooke's Law 9-4
Horizontal constraint tool 2-30
Host Wire edit box 7-23

I

Ideal Fluids 14-4
Ideal Plastic Fluids 14-4
Image properties area 1-31
Import button 7-64
Import CAD tool 4-53
Import mesh tool 11-9
Import tool 1-14
Increase Decimal Places tool 10-87
Increase degree tool 2-56
Increase knot multiplicity tool 2-57
Initial flow velocity condition tool 9-26
Initialise tool 14-31
Initial pressure condition tool 9-27
Initial temperature tool 9-23
Inlet option 14-27
Insert Active View (3D View) tool 10-12
Insert a new length dimension tool 10-46
Insert Annotation tool 10-25
Insert Arch Workbench Object tool 10-19
Insert Area Annotation tool 10-82
Insert a View of a Draft Workbench Objects tool 10-18
Insert balloon annotation tool 10-48
Insert Broken View tool 10-10
Insert Clip group tool 10-44
Insert Component tool 13-3
Insert Detail View tool 10-17
Insert Dimension tool 10-45
Insert knot tool 2-58
Insert Landmark Dimension tool 10-50
Insert MIBA radio button 1-31
Insert multiple linked views tool 10-13
Insert new default page tool 10-2, 10-89
Insert new Page using Template tool 10-5
Insert nx Prefix tool 10-85
Insert '□' Prefix tool 10-84
Insert Rich Text Annotation tool 10-28
Insert Robots cascading menu 11-29
Inserts a bitmap from a file into a Page tool 10-56
Insert Spreadsheet view tool 10-20

Insert symbol from a svg file tool 10-55
Insert View in Page tool 10-8
Install OpenFOAM button 14-25
interface 1-4
Intersection snap button 6-68
Intersection tool 11-23
Involute gear tool 3-74
Isometric tool 1-23

J

Job tool 8-3, 8-4
Join Curves tool 2-58
Join tool 6-53

K

Keyboard tab 1-39
Kinematic Viscosity 14-11
Kronecker delta 14-9
Kronecker Delta 14-9

L

Label tool 6-26
Laplace radio button 11-21
Launch editor button 4-32
LeadInOut tool 8-47
Left tool 1-24
Length option 4-51
Linear Direct Method 9-47
Linear Iterative Method 9-47
Linear Pattern tool 3-58
Linear Preconditioning 9-47
Linear Solver Type 9-48
Linear Static Analysis 9-4
Line clip filter tool 9-57
Line tool 2-4, 6-3
Link a dimension to 3D geometry tool 10-47
Load image tool 1-31
Load views tool 1-25
Lock constraint tool 2-28
Lock/Unlock View tool 10-60
Loft tool 4-19
Louvre Spacing edit box 7-23
Louvre Width edit box 7-23

M

Macro menu 12-29
Macro recording tool 12-30
Macros tab 1-41
Macros tool 12-30
Macro tab 12-20
Magnetization boundary condition tool 9-30
Make Base Wall tool 11-37
Make Bend tool 11-44
Make Compound tool 4-34
Make face from wires tool 4-17
Make Forming in Wall tool 11-46
Make Junction tool 11-43

Make Relief tool 11-43
Make Wall tool 11-37
Manage Layer tool 6-71
Map sketch to face tool 3-4
Mass Density 14-2
Mass flow rate option 14-27
Material Solid tool 9-6
Material tool 7-60
Maya-Gesture navigation 1-45
Measure tool 1-43
Measure tools 4-49
Mechanical Constraints 9-9, 9-20
Merge project tool 1-17
Merge tool 11-28
Meshing 9-33
Mesh refinement tool 9-38, 14-29
Mesh workbench 11-8
Midpoint snap button 6-67
Mirrored tool 3-57
Mirror tool 4-13, 6-34
Model Selection dialog box 8-8
Model tab 1-5
Modifies objects to Random Colour button 11-35
Modifying objects tool 6-29
Modifying object tools 4-9
Module to convert Mesh to Refined Solids 11-35
momentum in a control volume 14-12
Move/Array transform tool 2-47
Move tool 6-29
Move View tool 10-22
Multi Document option 12-23
Multi-Material tool 7-62
Multiple Structures tool 7-19
Multi Transform tool 3-61
Mystran Solver 9-51

N

Navier-Stokes equations 14-8
Navigation 1-44
Navigation cube check box 12-18
Near snap button 6-68
Nesting dialog 7-34
Nest tool 7-34
New CAM simulator tool 8-40
Newtonian Fluids 14-4
New tool 1-8
Nonlinear Iterations 9-48
Nonlinear Newton After Iterations 9-48
Nonlinear Newton After Tolerance 9-48
Non-Linear Static analysis 9-4
Non-Newtonian Fluids 14-4

O

Objects library tool 7-46
Offset geometry tool 2-50
Offset tool 6-35
Offset tools 4-24
OpenCascade navigation 1-47

OpenCASCADE Technology 4-2
OpenFOAM install directory 14-25
OpenInventor navigation 1-45
Open Recent cascading menu 1-10
Open tool 1-9
Orthographic view tool 1-21
Ortho snap button 6-68
Output tab 8-9

P

Pad tool 3-18
Panel Cut tool 7-31
Panels cascading menu 12-29
Panel Sheet tool 7-33
Panel tool 7-29
Parallel snap button 6-68
Parameter Editor dialog box 1-28
ParaView 9-50
Paraview button 14-35
Part Design body 3-3
Part Design helper tools 3-2
Part Design Workbench 3-2
Part Workbench 4-2
Pascal's Law 14-5
Paste tool 12-3
Path Array tool 6-46
Path Link array tool 6-47
Path twisted array tool 6-48
Path twisted link array tool 6-49
Periodic B-spline by control points tool 2-20
Perpendicular constraint tool 2-32
Perpendicular snap button 6-68
Persistent section cut tool 12-26
Perspective view tool 1-22
Pipe Connector tool 7-25
Pipe tool 7-23
Placement tool 12-4
Plane multi-point constraint tool 9-20
Pocket Shape tool 8-20
Pocket tool 3-37
Point Array tool 6-47
Point Link array tool 6-48
Point Picker button 10-30
Point tool 2-3, 6-16
Polar array tool 6-43
Polar Pattern tool 3-60
Polygon tool 6-12
Polyline tool 6-4
Porous Zone tool 14-32
Position Horizontal Chain Dimensions tool 10-61
Position Oblique Chain Dimensions tool 10-63
Position option 4-51
Position Section View tool 10-60
Position Vertical Chain Dimensions tool 10-63
Post pipeline from result tool 9-52
Post Process command 8-42

Post Process tool 8-42, 8-43
Preferences tool 12-8
Preprocessing 9-2
Pressure Load tool 9-18
Primitives tool 4-7
Primitive type drop-down 4-7
Print All Pages tool 10-7
Print Preview tool 1-19
Print tool 1-19
Probe tool 8-51
Profile tool 7-44
Project information tool 1-18
Projection On Surface tool 4-28
Project Shape tool 10-23
Project tool 7-3
Project utility tool 1-36
Properties tool 12-7
Property editor 1-5
Purge results tool 9-52
Python console 1-5

R

Radiation 9-5
RampEntry tool 8-48
Random color tool 12-28
Raytracing workbench 11-34
Real Fluids 14-4
Rear tool 1-24
Rebar tool 7-42
Recall working view tool 1-24
Record log messages check box 12-13
Record warnings check box 12-13
Rectangle tool 2-12, 6-11
Redo tool 12-2
Redraw Page tool 10-7
Reduces Mesh by 50% button 11-35
Refinement tool 11-12
Refine shape tool 4-61
Refresh tool 12-3
Region button 11-18
Region Clip Filter tool 9-56
Regular Polygon tool 2-16
Regular solid tool 11-22
Relaxation Factor 9-48
Remove axes alignment tool 2-52
Remove components by hand tool 11-19
Remove components tool 11-18
Remove component tool 7-7
Remove Component tool 7-51
Remove Prefix tool 10-86
Repair Dimension Reference tool 10-51
Reporting function tool 14-33
Report view 1-5
Reset point button 6-44
Resonance 9-5
Restore frozen views dialog box 1-26

Restore view option 1-27
Restore views dialog box 1-25
Reverse shapes tool 4-58
Revert tool 1-13
Revit navigation style 1-45
Revolution tool 3-20
Revolve tool 4-11
Reynolds number 14-16
Reynolds Number 14-11
Right tool 1-24
Rigid Body Constraints tool 9-10
Robot workbench 11-29
Roof tool 7-27
Rotate Left tool 1-24
Rotate/Polar transform tool 2-49
Rotate Right tool 1-24
Rotate tool 6-31
Rounded rectangle tool 2-14

S

Save a Copy tool 1-12
Save All tool 1-13
Save As tool 1-12
Save picture tool 1-30
Save tool 1-11
Save views tool 1-25
Scalar clip filter tool 9-54
Scale Mesh by 50% 11-35
Scale tool 11-22, 4-14, 6-32
Scale transform tool 2-50
Scene inspector tool 1-32
Schedule tool 7-63
Second Viscosity Coefficient 14-9
Section Plane tool 7-57
Section print feature tool 9-21
Sections tool 11-4
Section tool 4-22
Section View tool 10-15
Select All tool 12-4
Select associated geometry tool 2-60
Select Conflicting Constraints tool 2-61
Select Constraints tool 2-60
Select group tool 6-74
Selection view 1-5
Select Line Attributes 10-57
Select Plane tool 6-61
Select Redundant Constraints tool 2-61
Select solver DOFs tool 2-59
Select visible objects option 12-28
Send to Python Console tool 12-4
Set default orientation tool 11-31
Set default values tool 11-32
Set slope tool 6-58
Shaft design wizard tool 3-74
Shape 2D View tool 6-60, 7-59
Shape builder tool 4-8, 7-46

Shape String tool 6-19
Share View tool 10-23
Sheetmetal workbench 11-37
Shell plate thickness tool 9-32
Shorten Line tool 10-60
Show/hide B-spline control point weight tool 2-64
Show/Hide B-spline control polygon tool 2-63
Show/Hide B-spline curvature comb tool 2-64
Show/Hide B-spline degree tool 2-63
Show/Hide B-spline knot multiplicity tool 2-64
Show/hide circular helper for arcs tool 2-61
Show/Hide internal geometry tool 2-61
Show/Hide Invisible Edges tool 10-40
Show result tool 9-52
Show selection option 12-28
Simple Copy tool 8-37
Simulate a trajectory tool 11-34
Simulator tool 8-38
Site tool 7-4
Sketcher B-spline tools 2-62
Sketcher workbench 2-2
Sketch On Sheet metal tool 11-45
Slab tool 7-20
Slice apart tool 4-44
Slice to compound tool 4-45
Slice tool 4-45
Slot tool 2-17
Smooth tool 11-20
Snell's Law constraint tool 2-37
Solid Modeling 3-2
Solve Assembly tool 13-5
Solver job control tool 14-34
Space tool 7-8
Special snap button 6-68
Specific Gravity 14-2
Sphere tool 4-5
Split by components tool 11-28
Split edge tool 2-43
Split objects tools 4-42
Splitting tools drop-down 4-45
Split tool 6-54
Spring tool 9-16
Sprocket tool 3-72
Stack Bottom tool 10-25
Stack Down tool 10-25
Stack Top tool 10-25
Stack Up tool 10-25
Stairs tool 7-26
Standard sizes drop-down list 1-31
Standard views 1-23
Static Analysis 9-4
Static pressure option 14-27
Status bar 1-5
Status Bar 12-29
Stereo cascading menu 12-22
Stereo Off option 12-23
Stereo quad buffer option 12-23

Stereo red/cyan option 12-23
Stock area 8-5
Stop tool 8-51
Store working view tool 1-24
Stretch tool 6-39
Structural System tool 7-18
Subelement highlight tool 6-52
Subtracting tools 3-36
Subtractive Box tool 3-46
Subtractive Cone tool 3-50
Subtractive Cylinder tool 3-47
Subtractive Ellipsoid tool 3-51
Subtractive Helix tool 3-45
Subtractive Loft tool 3-42
Subtractive Pipe tool 3-43
Subtractive Prism tool 3-54
Subtractive Sphere tool 3-49
Subtractive Torus tool 3-52
Subtractive Wedge tool 3-55
Surface option 11-2
Surface workbench 11-2
Survey tool 7-67
Sweep tool 4-20
Switch virtual space tool 2-62
Symmetric constraint tool 2-35
Symmetry tool 2-51

T

Tags tool 8-49
Tangent to surface option 3-14
Tasks tab 1-5
Taubin radio button 11-21
TechDraw option 10-2, 13-2
TechDraw workbench 10-2, 13-2
Temperature boundary condition tool 9-24
Terrain button 7-5
Text tool 6-23
Texture mapping tool 12-27
Thermal analysis 9-5
Thermal Constraints 9-23
Thickness tool 3-68, 4-27
Third Angle option 10-3
third angle projection 5-2
Tie Constraint tool 9-15
Toggle Active constraint tool 2-40
Toggle Axis Cross option 12-24
Toggle construction geometry option 2-5
Toggle Construction Mode 6-65
Toggle display mode tool 6-76
Toggle Driving constraint tool 2-39
Toggle Edit mode tool 12-6
Toggle Grid tool 6-69
Toggle grounded tool 13-5
Toggle navigation/Edit mode tool 12-28
Toggle selectability option 12-28
Toggle Snap tool 6-67
Toggle visibility option 12-28
Toolbars 1-4

Toolbars cascading menu 12-28
Toolbars tab 1-40
Tool Manager tool 8-12
Tools menu 1-28
Top tool 1-24
Torus tool 4-5
Total pressure option 14-27
Touchpad Navigation 1-46
Transformation tools 3-56
Transform Constraint tool 9-22
Transform tool 12-5
Triangle tool 2-16
Trimetric tool 1-24
Trimex tool 6-37
Trim mesh tool 11-25
Trim mesh with a plane tool 11-25
Trim tool 2-43
Truss tool 7-39
Tube tool 4-6
Turbulence 14-17
Turn View Frames On/Off tool 10-57

U

Undo tool 12-2
Unfold tool 11-41
Uniform velocity option 14-27
Union tool 11-23
Units calculator tool 1-38
Universal Gas constant 14-5
Unwrap Face tool 11-29
Unwrap Mesh tool 11-28
Update templates fields tool 10-6
Upgrade tool 6-55
use FreeCAD material editor button 9-7
user.cfg 1-28
User system drop-down 12-8

V

Validate Sketch tool 3-6
Vapour pressure 14-5
Vcarve tool 8-33
Vertical constraint tool 2-40
Vertical distance constraint tool 2-24
View menu 1-21, 12-22
View turntable tool 1-37
Viscosity 14-2
Visibility cascading menu 12-27
Volumetric flow rate option 14-27

W

Wall tool 7-10
Warp filter tool 9-53
Weight Density 14-2
Window tool 7-21
Wire 6-4
Wire to B-Spline tool 6-56
Workbench cascading menu 12-28

Working Plane 6-61
Working Plane button 6-68
Working plane proxy tool 6-77
Write .inp file button 9-42

Z

Z88Aurora 9-50
Z88 Solver 9-50
Z Depth Correction tool 8-49
Zoom cascading menu 12-23

OTHER BOOKS BY CADCAMCAE WORKS

Autodesk Revit 2025 Black Book
Autodesk Revit 2024 Black Book
Autodesk Revit 2023 Black Book
Autodesk Revit 2022 Black Book

Autodesk Inventor 2025 Black Book
Autodesk Inventor 2024 Black Book
Autodesk Inventor 2023 Black Book
Autodesk Inventor 2022 Black Book

Autodesk Fusion 360 Black Book (V 2.0.18477)
Autodesk Fusion 360 PCB Black Book (V 2.0.18719)

AutoCAD Electrical 2025 Black Book
AutoCAD Electrical 2024 Black Book
AutoCAD Electrical 2023 Black Book
AutoCAD Electrical 2022 Black Book

SolidWorks 2025 Black Book
SolidWorks 2024 Black Book
SolidWorks 2023 Black Book
SolidWorks 2022 Black Book

SolidWorks Simulation 2025 Black Book
SolidWorks Simulation 2024 Black Book
SolidWorks Simulation 2023 Black Book
SolidWorks Simulation 2022 Black Book

SolidWorks Flow Simulation 2025 Black Book
SolidWorks Flow Simulation 2024 Black Book
SolidWorks Flow Simulation 2023 Black Book

SolidWorks CAM 2025 Black Book
SolidWorks CAM 2024 Black Book
SolidWorks CAM 2023 Black Book
SolidWorks CAM 2022 Black Book

SolidWorks Electrical 2025 Black Book
SolidWorks Electrical 2024 Black Book

SolidWorks Electrical 2022 Black Book
SolidWorks Electrical 2021 Black Book

SolidWorks Workbook 2022

Mastercam 2023 for SolidWorks Black Book
Mastercam 2022 for SolidWorks Black Book
Mastercam 2017 for SolidWorks Black Book

Mastercam 2025 Black Book
Mastercam 2024 Black Book
Mastercam 2023 Black Book
Mastercam 2022 Black Book

Creo Parametric 11.0 Black Book
Creo Parametric 10.0 Black Book
Creo Parametric 9.0 Black Book
Creo Parametric 8.0 Black Book
Creo Parametric 7.0 Black Book

Creo Manufacturing 11.0 Black Book
Creo Manufacturing 10.0 Black Book
Creo Manufacturing 9.0 Black Book
Creo Manufacturing 4.0 Black Book

ETABS V22 Black Book
ETABS V21 Black Book
ETABS V20 Black Book
ETABS V19 Black Book

Basics of Autodesk Inventor Nastran 2025
Basics of Autodesk Inventor Nastran 2024
Basics of Autodesk Inventor Nastran 2022
Basics of Autodesk Inventor Nastran 2020

Autodesk CFD 2024 Black Book
Autodesk CFD 2023 Black Book
Autodesk CFD 2021 Black Book
Autodesk CFD 2018 Black Book

FreeCAD 1.0 Black Book
FreeCAD 0.21 Black Book
FreeCAD 0.20 Black Book
FreeCAD 0.19 Black Book

LibreCAD 2.2 Black Book